陈运泰院士

学术活动剪影

1962～1966年于北京中国科学院地球物理研究所读研究生，师从曾融生院士（1980年当选中国科学院学部委员）。1965年随曾融生先生迁往中国科学院昆明地球物理研究所，1966年研究生毕业，适逢"文革"，毕业分配延至翌年。1967年分配回北京中国科学院地球物理研究所。此为1994年与曾融生院士在上海举行的"国际地球动力学中的力学问题学术讨论会"期间合影

1970年1月自天津南郊军垦农场回北京中国科学院地球物理研究所工作，1975年任震源物理研究室副主任，在室主任、国际著名地球物理学家傅承义院士（1957年当选中国科学院学部委员）指导下从事震源物理研究工作（此为摄于1975年的影片剪影）

1977年7月随国际著名地球物理学家顾功叙院士（1955年当选中国科学院学部委员）参加在英国达勒姆（Durham）举行的"国际地震学与地球内部物理学协会（IASPEI）和国际火山学与地球内部化学协会（IAVCAI）联合学术大会"，途经伦敦与顾功叙院士合影

1981年1月14日，应国际著名地球物理学家诺波夫（Knopoff, L.）教授（右）之邀，赴美国洛杉矶加州大学（UCLA）地球与行星物理研究所（IGPP）从事访问研究，1983年10月3日回国。2007年9月26日摄于洛杉矶

1981年1月至1983年10月在洛杉矶加州大学（UCLA）地球与行星物理研究所（IGPP）从事访问研究，此为工作中留影

1964年10月3日国际著名地球物理学家、剑桥大学教授、前燕京大学教授赖朴吾（Lapwood, E. R.）访问中国科学院地球物理研究所并做学术报告，报告会后在研究所门前与参会人员合影．前排左起：李善邦教授，傅承义院士，顾功叙院士，赖朴吾教授，王子昌教授；二排左起：林邦慧，王碧泉，李钦祖；二排右1：马恩泽；三排左起：朱传镇，徐果明，林庭煌，陈培善，姚振兴，陈运泰，陈大元

1977年7月参加在英国达勒姆（Durham）举行的"国际地震学与地球内部物理学协会（IASPEI）暨国际火山学与地球内部化学协会（IAVCAI）联合学术大会"，与昆明地球物理研究所董颂声教授（右）在伦敦合影

1982年7月与到访洛杉矶加州大学（UCLA）的钱家栋教授（右）合影

1985年7月当选第二届中国地震学会理事长．中国科学技术协会主席周培源院士（右）和第一届中国地震学会理事长顾功叙院士（中）莅临换届大会指导

1970年2月24日四川大邑 M_S6.2 地震后翌日即赴震区进行考察与流动观测．6月，野外工作结束前在灌县向阳坡地震台前留影（四川省地震局杨晓源研究员摄）

1970年2月24日四川大邑 $M_S6.2$ 地震后翌日即赴震区进行考察与流动观测. 6月, 野外工作结束前在灌县向阳坡地震台前留影. 后排右1: 杨晓源研究员; 后排左2: 陈运泰; 第二排左3: 张端军代表; 第一排左2: 高世玉研究员, 左4: 罗灼礼研究员

1984年10月2日在新疆核基地从事地下核爆炸近场强地面运动观测时留影. 左起: 陈运泰, 王璋博士(时为陈运泰的硕士研究生)

2007年3月29日重访新疆核基地, 在1984年10月2日地下核爆炸爆心处留影

1985年与到访的联邦德国著名地球科学家约翰斯·古登堡·缅因兹大学(Johannes Gutenberg Mainz University)雅可比(Jacoby, W.)教授(右)在研究所门口合影

1985年6月与到访讲学的诺波夫(Knopoff, L.)教授(左3)摄于北大未名湖畔. 左2: 陈运泰; 右1: 臧绍先教授

1985年8月参加在日本东京举行的"国际地震学与地球内部物理学协会(IASPEI)学术大会". 会议期间, 与国际著名地球物理学家京都大学三云健(Mikumo, T.)教授(左)和东京大学笠原庆一(Kasahara, K.)教授(右)讨论问题

1985年8月参加在日本东京举行的"国际地震学与地球内部物理学协会（IASPEI）学术大会"后访问东京大学地震研究所．左起：国家地震局副局长高文学教授，陈运泰，张祖胜教授，郑斯华教授，马瑾院士

1985年8月参加在日本东京举行的"国际地震学与地球内部物理学协会（IASPEI）学术大会"．会议期间，与日本著名地震学家宫村摄三（Miyamura, S.）教授合影

1988年12月赴美国旧金山参加美国地球物理学联合会（AGU）年会，与国际著名地震工程学家刘恢先院士（左）合影

1993年10月29日至31日参加在鸟取举行的"纪念日本鸟取地震50周年暨东亚地震学联合大会"

1993年10月29日至31日出席"纪念日本鸟取地震50周年暨东亚地震学联合大会"．会议期间与冯德益教授（左）和日本北海道大学小山顺二（Koyama, Junji）教授（右）合影

1992年访问东京大学地震研究所与国际著名地球物理学家丸山卓男（Maruyama, T.）教授合影

1994年参加在新西兰惠灵顿举行的"国际地震学与地球内部物理学协会（IASPEI）学术大会"时受到当地土著居民按传统礼节的热烈欢迎. 左起：印度古普塔（Gupta, H.K.）教授，意大利吉亚迪尼（Giardini, D.）教授，俄罗斯尼可拉耶夫（Nikolaev, A. V.）教授；左5起：陈运泰，日本深尾良夫（Fukao, Y.）教授，英国亚当斯（Adams, R. D.）教授

1997年与导师曾融生院士（中），国际著名地球物理学家、美国圣克鲁斯加州大学（UCSC）吴如山教授（左）参加在希腊塞萨洛尼基（Thessaloniki）举行的"国际地震学与地球内部物理学协会（IASPEI）学术大会"时留影

1998年1月8日参加在印度海德拉巴（Hyderabar）举行的"国际地震学与地球内部物理学协会（IASPEI）亚洲地震委员会（ASC）学术大会"时留影

1997年参加在希腊塞萨洛尼基（Thessaloniki）举行的"国际地震学与地球内部物理学协会（IASPEI）学术大会"时留影. 右起：高原教授，金昭九教授（韩国汉阳大学），陈运泰，朱传镇教授

1998年1月8日参加在印度海德拉巴（Hyderabar）举行的"亚洲地震委员会（ASC）学术大会"时与国际著名地球物理学家、国际地震学与地球内部物理学协会（IASPEI）主席、法国弗罗伊德沃（Froidevaux, C.）教授（右）交谈

2001年11月14日昆仑山口西地震现场考察

与尤惠川研究员（右）一道考察2001年11月14日昆仑山口西地震现场（海拔4767米）

2010年12月16日参加旧金山美国地球物理学联合会（AGU）年会期间与美国孟菲斯大学邱哲明教授（左）及夫人（右）合影

2005年与美国南卫理大学（Southern Methodist University）斯图姆（Stump, B. W.）教授参观考察美国地震仪器公司

2004年10月18日至21日参加在亚美尼亚埃里温举行的"第5届亚洲地震委员会（ASC）地震危险性评估与减轻地震风险学术研讨会". 会议期间，考察1988年亚美尼亚斯皮塔克（Spitak）M_S6.8地震途中，与国际著名地球物理学家、日本上田诚也（Uyeda, S.）教授（左）和印度地震学家沙赫（Shah, H.）教授（右）休息时合影

2005年与美国南卫理大学（Southern Methodist University）斯图姆（Stump, B. W.）教授（中）、辽宁省地震局顾浩鼎教授（右）参观沈阳东陵

在2005年中国地震局地球物理研究所授予著名地球物理学家、意大利的里亚斯特大学潘扎（Panza, G.）教授荣誉教授称号的授予仪式上

2005年9月21日至24日参加在昆明举行的"板缘地震学术研讨会暨云南地震工作90周年（1965～2005）"纪念活动，与老同事、老领导马瑾院士（左2）、原局长姜葵教授（左4）、原局长晏凤桐教授（右1）合影

2005年8月26日应邀只身访问考察进行过全球第一次核试验的美国新墨西哥州洛斯·阿拉莫斯沙漠中的阿尔伯克基（Albuquerque, New Mexico）地震实验室山洞内部

2003年6月参加"美国地震学联合研究会（IRIS）"年会期间考察夏威夷火山，坐在刚凝固的熔岩上休息

2005年10月2日至9日参加在智利圣地亚哥（Santiago）举行的"国际地震学与地球内部物理学协会（IASPEI）学术大会"与日本地球物理学家松浦充雄（Matsu'ura, M.）教授（右）在会议上的留影

2006年1月15至16日随中国地震局代表团访问伊朗

2006年1月15日随中国地震代表团访问伊朗国际地震工程学与地震学研究所（IIEES）与所长加弗利－阿什蒂亚尼（Ghafory-Ashtiany, Mohsen）教授（右）合影

2006年1月16日随中国地震代表团访问伊朗德黑兰大学地球物理研究所，和与高孟潭研究员（左3）共同培养的博士、副所长米尔扎伊（Mirzaei, Noorbakhsh）教授（左2）合影. 左1：张志中研究员

2006年1月16日随中国地震代表团访问伊朗德黑兰大学地球物理研究所与所长格坦齐（Gheitanchi, M.）教授（左）合影

参加2006年5月15日至18日在杭州举行的"固体中的机械波"国际学术研讨会的部分中外科学家合影

会见参加2006年5月15日至18日在杭州举行的"固体中的机械波"国际学术研讨会的浙江大学校长潘云鹤院士（右）

参加 2006 年 5 月 15 至 18 日在杭州举行的 "固体中的机械波" 国际学术研讨会的各国科学家合影

2006 年 5 月 15 日在杭州举行的 "固体中的机械波" 国际学术研讨会开幕式上致词

2006 年 8 月 8 日在北京举行的 "西太平洋地球物理会议（WPGM）" 期间会见日本地球物理学家．坐者右起：郦永刚教授，陈运泰，岛崎邦彦教授，山下辉夫教授；立者右起：陈晓非院士

与参加 2006 年 7 月 1 日至 7 日在夏威夷举行的 "西太平洋地球物理会议（WPGM）" 的学生、美国南卫理大学（SMU）周荣茂博士（右）和中国地震局地震预测研究所高原研究员（左）合影

在 2006 年 11 月 7 日至 10 日 "亚洲地震委员会（ASC）第 6 届大会暨预防和减轻地震与海啸灾害学术研讨会" 闭幕式上讲话．右起：印度古普塔（Gupta, H.K.）教授，伊朗格坦齐（Gheitanchi, M.）教授

2007年9月20日参加在美国加州门罗帕克（Menlo Park）举行的"旋转地震学国际研讨会"，与著名波兰地球物理学家泰塞伊尔（Teisseyre, R.）（右2）等国科学家休息时留影

主持2008年7月25日在成都举行的"科学技术与抗震救灾"技术科学论坛. 左起：陈祖煜院士，陈运泰院士，刘盛纲院士

2008年8月6日参加在德国卡斯鲁厄举行的"国际大地测量学与地球物理学联合会（IUGG）执委会会议". 与IUGG秘书长伊斯梅尔-扎德赫（Ismail-Zadeh, Alik）教授（右）合影

2015年6月30日在布拉格举行的IUGG大会期间与国际著名地球物理学家理查兹（Richards, P.）教授（左）合影

"国际大地测量学与地球物理学联合会（IUGG）第20届（2003～2007）执行委员会"全体成员合影（2005年）. 前排坐者左起：日本河野正郎（Kono, Masaru, 前任主席），丹麦汉森（Hansen, A. K., 司库），澳洲比尔（Beer, T., 副主席），以色列沙米尔（Shamir, U., 主席），美国约瑟琳（Joselyn, JoAnn, 秘书长），埃及梯勒伯（Tealeb, Ali, A. A., 执委），中国陈运泰（执委），德国奥斯瓦尔德（Oswald, S., 助理秘书）；后面站立者：IUGG下属各分会主席

2005年"国际大地测量学与地球物理学家联合会（IUGG）第20届（2003～2007）执行委员会"在美国科罗拉多大学举行会议期间与部分成员合影．左起：中国陈运泰（执委），埃及梯勒伯（Tealeb, Ali A. A., 执委），以色列沙米尔（Shamir, U., 主席），德国奥斯瓦尔德（Oswald, S., 助理秘书）

2008年8月3日，"国际大地测量学与地球物理学联合会（IUGG）第21届（2007～2011）执行委员会"在德国卡斯鲁厄（Karlsruhe）举行．此为全体成员合影．左起：陈运泰（执委），印度古普塔（Gupta, H., 副主席），汉森（Hansen, A. K., 司库），比尔（Beer, T., 主席），伊斯梅尔-扎德赫（Ismail-Zadeh, Alik, 秘书长），美国杰克逊（Jakson, D., 执委），埃及梯勒伯（Tealeb, Ali, A. A., 执委）

2007年3月30日中科院院士考察浙江常山国家地质公园

2009年4月6日意大利拉奎拉（L'Aquila）矩震级 M_W6.3 地震后，应意大利政府之邀，作为"国际民防地震预报专家委员会（ICEF）"的成员赴意大利工作，2009年5月12日至13日在拉奎拉举行第一次会议

2010年9月21日中国科学院考察团访问日本学士院．前排坐者左起：盛海涛研究员，陈运泰、朱作言、白以龙、顾逸东等院士；后排立者右1：张恒研究员

2010年9月28日中国科学院考察团访问澳大利亚科学院时留影

2010年12月15日获2010年度AGU"国际奖",在授奖仪式上与AGU主席麦克华登(McPhaden,Mike)(左)合影

2010年12月15日在AGU"国际奖"授奖仪式上发表获奖感言

2010年12月16日AGU授奖宴会上与前来祝贺的IUGG秘书长伊斯梅尔–扎德赫(Ismail-Zadeh,Alik)教授亲切握手

2013年6月28日在澳大利亚布里斯班举行的"AOGS学术大会"上AOGS主席佐竹健治(Satake,Kenji)教授向陈运泰颁发艾克斯福特奖章(Axford Medal Award)

2013年6月28日在澳大利亚布里斯班举行的"AOGS学术大会"上获2012～2013年度艾克斯福特奖(Axford Medal Award)后发表获奖感言

与访问北京大学地球与空间科学学院的国际著名地球物理学家、美国哥伦比亚大学理查兹(Richards,P.)教授(左)合影

2001年10月27日参加三峡地震监测台网验收会时在工地上留影纪念.左起：马宗晋院士，陈运泰，滕吉文院士

中国柴达木盆地资源环境科学钻探工程开工典礼.左起：邹才能院士，戴金星院士，陈运泰，滕吉文院士，李明教授

2009年7月10日与辽宁省地震局顾浩鼎教授（右）访问云南下关中国地震局地震预报试验场

2011年8月8日至12日"亚洲与大洋洲地球科学学会（AOGS）第8届年会"在中国台北举行.会议期间与日本石川有三（Ishikawa, Yuzo）教授（左）、王锦华教授（右）合影

1995年6月15日在辽宁省沈阳市举行"发展中的地震科学研究（海城地震成功预报20周年纪念）会议"，与会人员合影

2015年8月"亚洲与大洋洲地球科学学会（AOGS）第12届学术大会"在新加坡举行.会议期间与AOGS助理秘书长吴俊杰（Wu, Chun-Chieh）教授（左）和北京大学黄清华教授（右）合影

2015年8月7日在新加坡举行的"亚洲与大洋洲地球科学学会（AOGS）第12届年会"期间与王赤院士（左）合影

在2015年8月3日新加坡"AOGS第12届年会"开幕式上致词

2015年8月2日～7日"亚洲与大洋洲地球科学学会（AOGS）第12届年会"在新加坡举行.此为执委会（2014～2016）全体成员合影.前排左起：Higgitt, D.（秘书长），佐竹健治（Satake, Kenji）教授（副主席），陈运泰（主席），Terry, J.（助理司库），Khoo Cheng-Hoon（Meeting Matters）；后排右2：吴俊杰（Wu, Chun-Chieh）教授（助理秘书长）

2015年8月2日～7日"亚洲与大洋洲地球科学学会（AOGS）第12届年会"在新加坡举行.此为执委会（2014～2016）部分执委合影.前排右1：比尔（Beer, T., IUGG主席），前排左3起：赵丰（Chao, Benjamin Fang，候任主席），陈运泰（主席），Higgitt, D.（秘书长），叶有瑄（Ip, Wing-Huen，创会主席）

2005年"国际大地测量学与地球物理学联合会（IUGG）第20届（2003～2007）执行委员会"在美国科罗拉多大学举行.会议期间与著名地球物理学家基斯林格（Kisslinger, Karl）教授（左）及夫人（右）合影

1998年1月20日参观巴黎地球物理学研究所与法国地球物理学家蒙塔尼尔（Montagner, J.-P.）教授（右）等合影

1998年1月19日访问德国艾尔兰根（Erlangen）地震台与塞德尔（Seidl, D.）博士（右）和克林奇（Klinge, K.）博士（左）合影

2014年8月"亚洲与大洋洲地球科学学会（AOGS）第11届学术大会"在新加坡举行. 会议期间与李建平教授（左1）、胡敦欣院士（右2）、日本佐竹健治（Satake, Kenji）教授（右1）合影

2014年8月"亚洲与大洋洲地球科学学会（AOGS）第11届学术大会"在新加坡举行. 会议期间与戴民汉院士（左1）、香港理工大学甘剑平教授（右1）合影

2014年8月"亚洲与大洋洲地球科学学会（AOGS）第11届学术大会"在新加坡举行. 会议期间和与会的印度等国的科学家合影

"国际大地测量学与地球物理学联合会中国国家委员会（IUGG/CNC）"成员与中国科协国际部领导合影. 左起：李建平教授（秘书长），中国科协国际部梁英南副部长，陈运泰（主席），吴国雄院士（副主席）

"国际大地测量学与地球物理学联合会中国国家委员会（IUGG/CNC）"部分成员于西安考察学术大会会址. 左起：李建平教授（IUGG/CNC 秘书长），杨元喜院士 IUGG/CNC 主席），陈运泰（IUGG/CNC 原主席），党亚民教授（IUGG/CNC 候任秘书长）

2018 年 5 月 13 日于成都会见国际著名地球物理学家、瑞士魏斯（Wyss，Max）教授

2018 年 7 月 1 日至 7 日在意大利科莫湖（Lake Como）瓦伦纳（Varenna）恩里科·费米（Enrico Fermi）国际物理学院（International School of Physics）举办的"地震断层力学（Mechanics of earthquake faulting）"讲习班师生合影

与参加 2003 年 9 月 22 日至 25 日在捷克布尔诺（Brno）举行的国际数字地球学会（ISDE）学术大会的部分中国专家合影. 左 2 起：郭华东院士，董宝青教授，陈运泰，陈闽峰博士

参加 2003 年 9 月 23 日在捷克布尔诺（Brno）举行的国际数字地球学会（ISDE）执委会会议

2002年参加在印度举行的第13届TWAS院士大会期间参观泰姬陵

参加2005年11月29日至12月3日在埃及亚历山大港（Alexandria）举行的第16届发展中国家科学院（TWAS）院士大会（2005年12月1日摄）

参加2013年10月1日至4日在阿根廷布宜诺斯艾利斯举行的第24届发展中国家科学院（TWAS）院士大会

参加2013年10月1至4日在阿根廷布宜诺斯艾利斯举行的第24届发展中国家科学院（TWAS）院士大会的部分中国院士．左起：吕永龙院士，傅伯杰院士，杨晓明院士，白春礼院长，陈运泰院士，郭华东院士

2014年11月19日在菲律宾马尼拉举行的"第10届亚洲地震委员会（ASC）学术大会"上致词

2015年5月26日应邀在日本地球科学联合会（JpGU）大会上报告

1987年赴英国爱丁堡参加"S波分裂与地震预测研究10周年纪念会"与布什（Booth, D. C.）博士合影

2014年11月19日在菲律宾马尼拉举行的"第10届亚洲地震委员会(ASC)学术大会上与ASC主席意大利吉亚迪尼（Giardini, D.）教授合影

在"第一届中意双边学术论坛——地球物理、地球动力学暨地震危险性评估研究进展（2010年3月29日至30日，北京）"会议上

2009年10月20日出席在南非德班举行的TWAS大会休息时留影

纪念王子昌先生诞辰100周年暨学术研讨会（2012年5月20日，北大临湖轩）

1994年12月14日中国地震学会第4届理事会成立大会暨第5次学术大会在北京召开，陈运泰当选理事长

1998年10月20日中国地震学会第5届理事会成立大会在江西省井冈山市召开，陈运泰连任理事长

中国石油发展战略高层论坛暨北京大学石油与天然气研究中心理事会学术委员会成立大会（2006年11月19日）

中国科学技术大学地球物理创新团队 2014 年学术交流会（2014 年 2—6 日，合肥）

中国科学院地学部部分常委参观王国维故居合影留念．从左至右：汪品先，陈运泰，赵鹏大，肖序常，戴金星，欧阳自远等院士

院士大会休息时合影（2018 年 6 月 1 日）．左起：高钧院士，陈运泰，戴金星院士，符淙斌院士

中国科协科学家在贵州考察时留影（2010 年 7 月 9 日）

北京科学技术协会与北京理工大学出版社科普工作讨论会与会人员合影．前排左起：陈运泰，胡亚东教授，王绶琯院士

两岸学术交流片断

1988年9月17日与来访的中国台湾科学家合影. 前排右起：中国科学院院长周光召院士，原院长卢嘉锡院士. 后排右起：周昌泓院士，陈运泰，叶永田教授，吴甘美女士

1988年9月17日陪同来北京参加中国科协学术大会的中国台湾地震学家叶永田教授（中）等游览天坛

向到访的中国台湾地震学家介绍中国数字地震台网（1993）

1999年9月21日中国台湾集集M_W7.6地震后，应邀于2000年1月18日赴台考察，在桃源机场受到台湾朋友的热忱欢迎. 左起：陈运泰，中国台湾地球物理学家王乾盈教授，国家地震局何永年副局长，中国台湾朋友李庆华，冯沪生

1998年与到访厦门地震局的中国台湾地球物理学家合影. 左起：福建省地震局副局长林继华研究员，陈运泰，中国台湾地球物理学家张建兴教授和王乾盈教授，厦门科技局王碧惠副局长，厦门地震局叶振民局长

1999年8月16至25日 第三届海峡两岸地震科技研讨会野外考察期间和中国台湾黄柏寿教授（左）与陈朝辉教授（右）合影（1999年8月21日摄于敦煌鸣沙山）

2000年6月1日海峡两岸城市防震减灾研讨会上与中国台湾叶义雄教授合影

2006年3月16日中国台湾梅山地震百周年纪念研讨会期间考察1999年9月21日中国台湾集集$M_W7.6$地震现场

2009年2月15日至23日第六届海峡两岸地震科技研讨会于中国台湾日月潭举行，此为研讨会与会专家全体合影

2009年4月访问中国台湾"中研院"地球科学研究所. 左起：黄柏寿教授，赵里教授，王锦华教授，邱宏智教授，陈运泰，江博明教授（院士、所长）；右2：郑琪琴

2009年访问中国台湾"中大"地球科学学院. 左起：中国地震局郑州物探中心段永红教授，美国圣克鲁斯加州大学（UCSC）吴如山教授，陈浩维教授，陈运泰

教学与科学知识传播活动剪影

"科学与中国"院士专家巡讲团系列报告会（1998）

在中国地震学会地震学术研讨班上发言（2003年10月27日）

2005年4月19日赴福建永泰做"科学与中国"科普报告后和与会师生合影

2006年6月18日"院士专家八闽行"活动留影

2006年6月18日"院士专家八闽行"活动期间与浙江大学教授徐世浙院士（左）留影

2006年6月18日"院士专家八闽行"活动期间与谢毓元院士（左）、福建省地震局局长林思诚教授（右）合影

2007年6月21日与郑琪琴老师参观访问闽江学院并植树纪念

2007年6月21日与郑琪琴老师参观访问闽江学院并植树纪念

应邀在厦门市地震局做科普报告

应邀在杭州浙江省地震局做科学普及报告后与局长苏晓梅研究员（左2）、原局长陈修民研究员（右1）合影

1981～1983年陈运泰于美国洛杉矶加州大学（UCLA）地球与行星物理研究所（IGPP）从事访问研究。此为2010年12月19日与UCLA/IGPP所长、导师、国际著名地球物理学家诺波夫（Knopoff, L.）教授在杭州讲学时合影

2017年7月6日与同在武汉大学"珞珈讲坛"讲演的著名空间物理学家李罗权院士（左）、著名物理学家李家明院士（中）合影

2017年7月访问西北大学大陆动力学国家重点实验室，应邀做大会报告

2003年8月20日赴西藏拉萨做"科学与中国"报告.左起：陈运泰，中科院副院长孙鸿烈院士，西藏自治区党委郭金龙书记

2003年8月20日赴西藏拉萨做"科学与中国"报告.左起：郑度院士，中科院副院长孙鸿烈院士，西藏自治区党委郭金龙书记，陈运泰

2007年4月17日北京大学－中国地震局现代地震科学技术研究中心成立

2015年4月28日美国国家工程院院士、国际著名地球物理学家、国际地震预测先驱者之一、斯坦福大学努尔（Nur, A.）教授访问北京大学地球与空间科学学院，与院长张立飞教授及地球物理系教授合影.左起：胡天跃，周仕勇，陈永顺，张立飞，努尔，陈运泰，蔡永恩，黄清华，宁杰远

2015年3月19日与到北京大学地球与空间科学学院访问的新西兰奥克兰大学专家合影.左起：美国南加州大学南加州地震研究中心郦永刚资深研究员，地球科学工程研究中心马林（Malin, Reed）先生，陈运泰，地球科学工程研究中心主任马林（Malin, Peter）教授，校企服务部门主管普特（Putt, Gary）博士

2008年5月4日北京大学110周年校庆时与原地球物理学系"系友"、北大物理学院大气科学系卢咸池教授（右）合影

与中国科协副主席李静海院士（中）合影

与中国科协副主席冯长根教授合影

2007年6月25日应聘任中国科学院武汉测量与地球物理研究所兼职教授．左：所长孙和平院士

2007年1月18日在院士新春团拜会上．左起：陈运泰，黄荣辉院士，叶大年院士

2009年3月17日"10000个科学难题"天文学和地球科学领域编辑委员会第一次会议合影

在2011年中国科协学术建设新闻发布会上

2008年3月12日第11届全国政协科技组政协委员合影

2008年3月13日与第11届全国政协委员、中国科学院李家春院士合影

2008年3月13日与第11届全国政协委员、中国工程院黄其励院士（左）合影

陈运泰（左）与戎嘉余院士（右）在全国政协会上交谈（2012年3月12日）

2008年3月13日与国务院参事沈梦培学长在全国政协会上合影

2017年12月10日与李德仁院士在两院院士大会上合影

师生谊剪影

1970年代末至1980年初为中国科学院研究生院研究生开课

2018年中国科学院大学地球与行星科学学院建院40周年纪念会上发言.背后立者：学院院长孙文科教授

1995年与加州理工学院（CIT）张家骏博士（1978年至1980年协助傅承义院士培养的硕士）合影

1997年和朱传镇教授（右1）与1980年在国家地震局地球物理研究所从事博士后研究的罗马尼亚马尔扎（Marza, V.）教授（中）合影

与1999年毕业的硕士倪晓希博士（右）在UCLA合影

和北京大学地球与空间科学学院蔡永恩教授等师生合影.左起：蔡永恩教授，胡才博博士，谢周敏博士，吴晶博士，陈运泰

2007年9月23日访问赖斯大学（Rice University）时与学生赵明（左）博士、李旭（右）博士合影

2007年9月23日访问赖斯大学（Rice University）时与学生李旭（左）博士、高伟（右）博士合影

2008年与学生朱新运教授（右）合影

2005年6月23日于新加坡参加亚洲与大洋洲地球科学学会（AOGS）学术大会期间与学生许力生教授（左）合影

2015年6月8日参观考察成都理工大学时与成都理工大学特聘教授余嘉顺教授（右）合影

与学生陈晓非院士（中）和北京大学地球与空间科学学院地球物理系宁杰远教授（右）合影

中国地震局地球物理研究所研究生毕业典礼暨学位授予仪式合影

1950年7月25日厦门粤侨小学（今文安小学）毕业照．第2排右起：严能融，刘以光；第3排右3起：陈镜湖，林永清，黄幼卿，赵碧卿，许宝珍，周荫治，陈运贤，陈运泰．前排坐者左4：卓杰华校长

2006年6月16日厦门大同中学初中1953届（毕业）、高中1956届（毕业）校友聚会留影．前排左7起：黄元祥老师，戴光华老师，刘怡复老师，洪秋和校长，汤维强老师，翁健老师，陈镛老师，房波老师，刘与权副校长，李琪老师，陈祝荣书记

2015年11月27日回母校厦门文安小学（原厦门粤侨小学）．左起：黄阿娜，周荫治，黄幼卿，刘以光，陈运泰，陈运贤，邱永年，许宝珍

2006年6月16日厦门大同中学（初中1953届、高中1956届）校友聚会留影

2010年6月16日在厦门大同中学建校85周年庆典上发言

2010年6月16日在厦门大同中学建校85周年庆典上与学生在一起

2010年6月16日在厦门大同中学建校85周年庆典时与学生合影

1985年大同中学同学会北京分会成立．前排左1：许璇玑；前排左3：陈运泰；后排左起：李正心，许敬行，秦麟征，高亚场，刘永来

厦门大同中学北京同学会部分校友合影．前排左起：陈耀辉夫人，陈耀辉，许敬行，陈运泰，许敬行夫人；后排左起：李正心夫人，林壬子夫人，林壬子，陈震，秦麟征，刘永来

海峡两岸防震减灾学术研讨会在北京香山举行，与会全体专家合影(2008年6月27日)

厦门大同中学建校80周年与原副校长汤维强老师(左)合影(2005年11月26日)

厦门大同中学建校80周年与詹龙标老师(中)和李正心学友(右)合影(2005年11月26日)

厦门大同中学建校 80 周年与许嘉义学友（右）合影 (2005 年 11 月 26 日)

厦门大同中学建校 85 周年与郭子炎学友（左）合影（2010 年 6 月 16 日）

2015 年 9 月 23 日青海西宁青海大学"地震与防震减灾"报告会后和与会部分师生合影

与参加在布拉格举行的第 26 届 IUGG 大会的李建成院士（左 3）等中国专家合影（2015 年 6 月 24 日）

谢礼立院士（左）和陈运泰在衡水市城市发展研讨会上（2018 年 9 月 9 日）

北京大学物理系物理专业校友毕业 50 周年（2012 年 5 月 20 日）同学合影：陈运鸿（左）和陈运泰（右）

北京大学物理系物理专业 56 级 4 班同学 1957 年 10 月摄于北大未名湖畔．前排左起：张顺钧，张南海，陈运泰；后排左起：赵佐政，黄兴让，王正行

北京大学物理系物理专业 56 级 4 班同学 1967 年 7 月摄于北大未名湖畔．左起：张南海，陈运泰，王正行

2012年5月北京. 左起: 王正行, 张南海, 陈运泰

2018年5月4日北京大学120周年校庆大会开会前留影

北京大学地球物理系地球物理专业561班于1961年6月在白家疃地震台实习时部分同学与地震台长姜葵学长（中排左3）合影

中国科学院大学地球与行星科学学院为获"杰出贡献奖"教师颁奖

1998年5月4日庆祝北京大学建校一百周年

北京大学地球物理系地球物理专业56.1班入学50周年联谊会（2006年）

2013年10月19日北京大学物理学科建立100周年庆祝大会时地球物理系地球物理专业56.1班部分校友合影

北京大学地球物理系地球物理专业56.1班入学60周年部分校友合影（2016年5月21日）

2013年10月19日北京大学物理学科建立100周年庆祝大会时地球物理系地球物理专业56.1班部分校友合影

北京大学物理系物理专业56级4班合影（1957年）

厦门大同中学北京同学会部分校友2014年1月27日新春团聚合影

1998年春节厦门大同中学北京同学会部分校友看望原全国人大副委员长、老学长卢嘉锡院士时合影

1998年春节厦门大同中学北京同学会部分校友看望原全国人大副委员长、老学长卢嘉锡院士时亲切交谈

北京大学地球物理系地球物理专业56.1班毕业留念（1962年9月29日）．前排左4起：地球物理教研室主任王子昌教授，副校长王竹溪院士，地球物理系苏士文主任，刘宝诚教授

北京大学物理系物理专业 56 级 4 班入学 60 周年部分校友合影（2016 年 5 月 21 日）

北京大学物理系物理专业 56 级 4 班入学 60 周年部分校友合影（2016 年 5 月 21 日）

北京大学物理系物理专业 56 级（1956 年入学）校友毕业 50 年合影（2012 年 5 月 19 日）

北京大学物理系物理专业 56 级校友入学 60 周年纪念（2016 年 5 月 21 日）

2000年7月重访洛杉矶加州大学

2010年12月24日看望老师、美国洛杉矶加州大学（UCLA）地球与行星物理研究所（IGPP）所长诺波夫（Knopoff, L., 1925.7.1～2011.1.20）教授夫妇

中国科学院院士考察福建纪念

1970年代末至1980年代初在洛杉矶加州大学（UCLA）访问的部分UCLA校友合影. 前排左起：许卓群夫人，陈俊生夫人，曹采芳教授，濮祖荫夫人，陈琦教授；后排左起：陈运泰，陈俊生院士，北京邮电大学原校长朱祥华教授，濮祖荫教授，许卓群教授

中国科协常委参观大理白族民俗馆. 前排左起：陈运泰，民俗馆馆长，铁道部原副部长国林；后排：北京航空航天大学原校长李未院士

中国科学院地学部院士考察深圳纪念

2015年9月3日抗日战争胜利70周年大阅兵纪念

2015年9月22日兰州观象台建立61周年纪念活动留影.左3起：李小军研究员，郝纪川研究员，王兰民研究员，陈运泰

2015年9月22日会见参加兰州观象台建立61周年纪念活动的法国朋友、国际著名地球物理学家、新加坡南洋理工大学（Nanyang Technological University）塔波尼尔（Tapponnier, Paul）教授（左2）等

2010年11月21日厦门大同中学建校85周年庆典留影

2006年6月15日回母校厦门大同中学时与校领导在卢嘉锡老学长塑像前合影.左起：洪秋和校长，陈运泰，陈祝荣书记，刘与权副校长，物理教研组组长范毅老师

2018年4月2日厦门、广州、北京三地的厦门大同中学56级（1956年高中毕业）部分校友聚会

老照片及日常生活剪影

父亲陈云梁先生
（1909—1988）

母亲郭淑惠女士
（1910—1991）

1941 年

1943 年

1944 年

1945 年 12 月

1946 年 1 月

1946 年 7 月

1947 年 5 月

1953 年 7 月

1955 年 8 月

1960 年 9 月

1961 年

1961 年 3 月

1961 年 12 月

1971 年 8 月

1976 年 12 月

工作照（1997 年 1 月 15 日）

参观卡文迪什（Cavendish）实验室（1977 年 7 月）

参观意大利比萨斜塔（2001年1月9日）

智利圣地亚哥留影（2005年10月9日）

参观洛杉矶盖蒂（Getty）博物馆留影

埃及狮身人首像前留影（2005年12月5日）

胡夫金字塔内部留影（2005年12月5日，朱清时院士摄）

陈运泰（左3）和郑琪琴（右1）于云南下关蝴蝶泉留影

郑琪琴于中国台湾台北街头留影（2009年5月）

陈运泰（右）和郑琪琴（左）于象踞石台前留影

陈运泰在塞上草原

陈运泰（右）和郑琪琴（左）于塞上草原留影

郑琪琴（右）漓江漂流时留影

陈运泰（右）和郑琪琴（左）于桂林象鼻山前留影

陈运泰（左）和郑琪琴（右）在云南昆明西山聂耳雕像前合影（2009年7月29日）

陈运泰（左）和郑琪琴（右）游览桂林漓江时合影

1997年6月21日中科院地学部常委会期间与周秀骥(右2)，苏纪兰（右3），汪品先（左2）等院士合影

2001年6月IUGG大会于日本北海道扎幌举行．途经日本洞爷火山口留影

中国-东盟地震海啸预警研讨会（2005年1月25日，北京）

南非好望角留影（2009年1月13日）

陈运泰院士论文选

陈运泰 著

科学出版社
北京

内 容 简 介

陈运泰院士是国际著名的地球物理学家,为地震学的发展做出了杰出的贡献. 从 20 世纪 70 年代起,他开创并发展了中国震源物理过程的研究工作,积极倡导和从事数字地震学研究,增进了对地震破裂过程时空复杂性的认识. 半个多世纪以来,陈运泰院士笔耕不辍,发表了大量的学术论文. 本书选录了陈运泰院士自 1971 年至今所发表的 60 篇论文,内容涉及震源理论、地震预报、地球内部结构等,是陈运泰院士地震工作生涯的学术成果集锦. 另外,还收录了陈运泰院士的工作剪影、培养的博士及硕士名录、论著目录及学生、同事撰写的 8 篇回忆性文章.

本书可作为地球物理学,尤其是地震学相关领域的科研人员、研究生以及其他地学工作者的参考书.

审图号:GS(2020)2311 号

图书在版编目(CIP)数据

陈运泰院士论文选/陈运泰著. —北京:科学出版社,2020.12
ISBN 978-7-03-065655-1

Ⅰ. ①陈… Ⅱ. ①陈… Ⅲ. ①地震学-文集 Ⅳ. ①P315-53

中国版本图书馆 CIP 数据核字(2020)第 118755 号

责任编辑:张井飞 韩 鹏/责任校对:张小霞
责任印制:肖 兴/封面设计:图阅盛世

科学出版社 出版
北京东黄城根北街 16 号
邮政编码:100717
http://www.sciencep.com

北京九天鸿程印刷有限责任公司 印刷
科学出版社发行 各地新华书店经销

*

2020 年 12 月第 一 版 开本:889×1194 1/16
2020 年 12 月第一次印刷 印张:54 1/4 插页:22
字数:1 440 000
定价:598.00 元
(如有印装质量问题,我社负责调换)

出 版 说 明

中国科学院院士陈运泰先生是国际著名的地震学家和地球物理学家. 本书选录了陈运泰院士自 1971 年至 2020 年间在国内外学术期刊、图书中发表的 60 篇学术论文，内容涉及震源理论、地震预报、地球内部结构等. 论文来自多种不同期刊、图书，时间跨度大，原始体例不尽统一. 为尊重原文风貌，收录时对论文各级标题体例、物理量外文符号格式等进行了统一，而对参考文献引用格式等未作统一处理.

序　　言

在中国地震科学不断取得辉煌成就的同时，有这么一个人，他隐身于喧嚣的都市背后，几十年如一日，默默洞察着大地的脉搏，不断探索地球内部的奥秘. 这个人就是国际著名的地球物理学家、中国科学院院士陈运泰.

陈运泰，1940 年 8 月出生于福建厦门，原籍广东潮阳. 陈运泰的父亲经商，母亲是家庭妇女，父母于 20 世纪 20 年代自广东潮阳移居福建厦门. 他们十分重视子女的培养教育. 1950 年，陈运泰考入厦门大同中学（1953 年改名为厦门第四中学，1985 年恢复原名）. 厦门大同中学在当时并非名校，至少不是当地最著名的学校，但在她的爱国、进步、民主、求实的优良校风的熏陶下，在以卢嘉锡先生为代表的杰出校友的榜样影响下，陈运泰如沐春风，学业进步很快.

1956 年，陈运泰以全校最优异的成绩高中毕业，考入中国一流学府——北京大学. 巍巍学府里深厚的文化底蕴、满腹经纶的科学巨匠和涌动的学术思潮充分滋润着青年陈运泰，他如饥似渴地汲取知识的营养. 1958 年分配专业（那时叫"专门化"）时，陈运泰原志愿学习理论物理，但被分配学习地球物理. 他随遇而安，高高兴兴开始了与生气勃勃的大自然打交道的地球物理生涯. 1959 年元月，北大地球物理系正式建立，但学生继续留在物理系学完全部基础课. 北大物理系的基础课学习为他后来的科研生涯打下了坚实的基础. 1962 年毕业于北大地球物理系（六年制）后，随即顺利考入中国科学院地球物理研究所攻读研究生，师从曾融生先生（1980 年当选中国科学院学部委员，1993 年改称院士），研究方向是与地壳上地幔结构有关的非均匀介质中地震波传播理论. 研究生期间，他自强自立，踏实为学，研究工作进展很快.

1965 年 4 月，陈运泰跟随导师曾融生先生来到位于云南昆明的中国科学院昆明地球物理研究所. 云南是一个地震频发的地区，在短短的一年多时间里，陈运泰亲历了包括 1966 年 2 月 5 日东川 M_S6.5 地震在内的多次破坏性地震，并立即赴地震现场考察，逐渐积累了许多有关地震的实际经验. 随后，以发生在人口稠密地区的 1966 年 3 月河北邢台地震为开端的中国大陆一系列强烈地震活动，造成人民生命财产的重大损失. 这些强烈地震所造成的灾害深深地触动了陈运泰的心灵，为此，他立志要从事与地震震源相关的地震学研究，并主动将主要研究方向从地震波传播理论转向震源物理研究.

"如果说我比别人看得更远一些，那是因为我站在了巨人们的肩膀上". 英国著名科学家牛顿一席话揭示了科学传承的重要性. 陈运泰不仅得到曾融生教授的关怀和指导，还得到了傅承义教授（1957 年当选学部委员）、顾功叙教授（1955 年当选学部委员）等的欣赏和指点. 他从这些先辈的经历中提炼精华，用于丰富自己，并引领自己走向成功.

我国是一个多地震的国家. 地震是一种会给人类带来严重灾难的自然现象，是在地球内部特定条件下由构造运动所导致的物理过程. 地震科学的基本任务之一是认识地震发生的原因及其规律，预测地震及其灾害，为人类预防和减轻地震灾害服务.

在以后的近六十年的时间里，陈运泰融入中国地震研究事业的拼搏过程中，以探求地震发生的原因和规律为主要目标，将地震学与其他学科研究相结合，发展地震科学，在实际的观测资料中寻求地震科学前沿问题的新证据，埋首耕耘、奉献智慧，亲眼见证和推动了中国地震科学的建设和发展.

从 20 世纪 70 年代起，陈运泰用地震波、大地形变测量和重力等观测资料反演与综合研究邢台、昭通、海城、唐山等大地震震源过程、地震破裂动力学，并进行天然与人为地震（地下核爆炸、大型工业爆破等）近震源观测以及震源过程反演，增进了对地震破裂过程时空复杂性的了解，是我国震源研究领域的先驱性工作，并因此荣获1978年全国科学大会奖. 此后的十年间，陈运泰综合研究了中国大陆地震活跃期（1966-1976年）许多大地震的震源过程，取得了反映这些大地震震源特性的参量. 这些研究成果对于增进对地震孕育和发生过程的认识，在理论和实践方面都有着重要意义.

从1981年开始，陈运泰主持我国第一支用宽频带数字化地震仪装备起来的近震源强地面运动观测队伍的工作. 在十余年的时间里，这支观测队伍的足迹遍布河北卢龙、云南剑川和禄劝、山西大同、新疆马兰等地，获得了大量高质量的强余震或地下核爆炸的近震源强地面运动记录.

1981-1983年，陈运泰在美国洛杉矶加州大学（University of California, Los Angeles，缩写为UCLA）地球与行星物理研究所（Institute of Geophysics and Planetary Physics，缩写为IGPP）进行访问研究. 其间，他在国际著名地球物理学家李昂·诺波夫（Leon Knopoff）的指导下，在地震震源的静态、准静态和动态裂纹模型、地震序列的模拟等前沿性的理论研究中做出了重要贡献，从理论上指出应力降的非均匀性和摩擦的非均匀性是决定裂纹扩展、停止和愈合的主要控制因素，地震波激发与辐射性质的复杂性则主要来自裂纹愈合过程和传播过程的复杂性，获得国际同行的重视和好评. 他在《英国皇家天文学会地球物理学刊》（Geophysical Journal of the Royal Astronomical Society）和《构造物理学》（Tectonophysics）上发表的一系列论文至今仍不时被引用.

从20世纪90年代以来，陈运泰和他领导的研究组利用全球的数字地震记录对青藏高原发生的一系列重要地震的震源过程进行研究. 他们发展了用地震波形资料反演位错动态分布的方法，并应用于1990年青海共和M_S6.9地震、1996年云南丽江M_S7.0地震、1997年西藏玛尼M_S7.9地震和2001年青海昆仑山口西M_S8.0地震等地震的研究中.

进入21世纪，他再次与国际著名地球物理学家诺波夫合作，从理论上重新审视地震学研究中普遍采用的两条基本原理（即远场位移与动态断层面上的滑移速度成正比，以及断层面上的动态滑移与双力偶等效这两条与地震能量、地震矩的测定密切相关的原理）. 通过严格的分析论证，指出以往的理论忽视了断层厚度的效应，以及在地震破裂过程中动态扩展的断层端部强度弱化区的存在，对国际地震学界沿用达半个世纪之久的这两条基本原理做出了重要的修订和补充.

陈运泰就这样一步一个脚印，实现了人生的积累. 他把人生的每一次机遇和挑战当成进步的阶梯，在不断攀登的过程中，收获了丰硕的"果实"：他的一系列研究成果先后荣获卢森堡大公勋章（1988）、国家自然科学奖三等奖（1987）、国家科技进步奖三等奖两项（1997, 2005）、顾功叙奖（2019）等多项奖励. 2010年，美国地球物理联合会（American Geophysical Union）"鉴于陈运泰数十年来杰出的科学研究成就，鉴于他为中国地震学与大地构造学研究的现代化所做的长期不懈的努力，以及他对国际地球物理界的贡献"（颁奖词），特授予陈运泰美国地球物理联合会（AGU）国际奖（International Award）. 2012年，亚洲与大洋洲地球科学学会（Asia Oceania Geosciences Society，AOGS）为表彰"他在地震震源研究、地震学理论和实践方面取得的巨大学术成就，为国际和亚洲科学界做出的无私服务和为推动AOGS发展付出的不懈努力"（颁奖词），授予他AOGS的最高奖艾克斯福特奖（Axford Medal Award）.

他个人还曾先后荣获"国家有突出贡献中青年专家（1986）"、"全国地震系统先进个人（1996）"、"何梁何利基金科学与技术进步奖（2000）"、"全国地震科技工作先进个人（2007）"等荣誉称号，并当选第十一届全国政协委员，这些都是党和国家对他从事地震科学研究的最高褒奖.

"勇于创新,追求卓越",是陈运泰对待事物的一贯态度,是他能够坦然面对现实,勇敢地憧憬未来的原因. 凭借深厚的数理功底和多年固体地球物理学的研究造诣,他的研究工作始终位于地球物理领域的前沿.

在世纪之交(1997年),根据国际地球科学研究的动向和趋势,陈运泰适时地提出了中国固体地球物理,特别是地震学和地球内部构造的学科发展方向. 他指出:"在科学上更深刻地认识地震的本质,是未来世纪地震科学的发展方向. 在这方面,新的观测技术和新理论是发展的重点."

根据我国地震科学的特点,陈运泰强调地震学研究应与大地测量学和地质学等学科紧密结合,为我国地球科学多学科之间交叉融合构筑了沟通的渠道,为后人开辟了地球科学深入研究的新领域. 他提出:"作为研究地震这一自然现象的科学,地震科学已从传统的地震学发展成为包括地震学、地震大地测量学、地震地质学、岩石力学、复杂系统科学,以及与地震研究有关的信息科学在内的,以认识地震本质、预防和减轻地震灾害、造福人类为目标的多学科交叉融合的现代地震科学." 他的这一思想还直接与减灾和改善环境等密切相关,大大促进了我国固体地球物理学,特别是现代地震科学的发展.

20世纪80年代起,随着数字技术在地震观测中的大量应用,陈运泰在国内积极倡导并身体力行从事数字地震学的研究,这一富有远见的战略思想极大地促进了高技术时代中国地震学的发展. 在他的指导下,中美合作的中国数字地震台网(China Digital Seismograph Network,CDSN)以其高质量的资料和管理水平被国际同行誉为"有记录以来最值得信赖的地震台网".

20世纪80年代,他与美国洛杉矶加州大学诺波夫教授合作,就震源动力学发表一系列高水平的论文,其中提出的"旋转地震学"等基本概念至今仍有深远的学术影响力. 1998年美国洛杉矶南加州大学邓大量教授来北京访问,在学术报告中特地提到陈运泰关于地震波辐射中的旋转和旋转张量问题的最新理论成果,说他来北京就是为了得到这一理论推导的最新内容. 陈运泰的地球物理造诣在国际上的影响可见一斑.

另外,他领导的中国-欧共体合作的"京西北怀来数字地震台网"被国家科委(科技部)确定为中外合作的"窗口项目". 在此基础上,他主持编著的《数字地震学》一书,对以数字地震资料为基础的中国地震学的发展起到很大的促进作用.

20世纪90年代以来,陈运泰开创性地将"地震矩张量"引入中国地震研究. 他投入大量精力研究地震破裂过程反演,并对中国及其周边地区的一些重大地震的震源过程进行了详尽的研究.

同时,他是"率先将数字地震学和强地面运动地震学引入中国这样一个大地震多发、地震观测与地震研究历史悠久的国家的科学家之一". 1998年,在印度海德拉巴召开的亚洲地震委员会(Asian Seismological Commission,ASC)学术大会上,时任国际地震学与地球内部物理学协会(International Association of Seismology and Physics of the Earth′s Interior,IASPEI)主席、著名的地球物理学家法国伏罗依德瓦(C. Froidvaux)向国际同行特别介绍了陈运泰"把数字地震学引入一个地震大国"的贡献.

进入21世纪以来,他一直致力于为地震应急反应服务的快速确定震源破裂过程的工作,力求将理论研究成果与我国地震学研究和防震减灾工作的实际相结合,并取得了实效.

2008年5月12日,汶川$M_S 8.0$地震发生后,他领导的研究组随即下载全球地震台网(Global Seismographic Network,GSN)记录资料,进行了汶川地震破裂过程的反演. 在地震发生后4小时就提交了测定结果,指出汶川地震有4个滑动量集中的区域,这一结果"预测"(遥测)了汶川地震宏观破坏的分布特征并为随后的现场调查所证实,为部署抗震救灾工作提供了宝贵的信息.

步入古稀之年的陈运泰，一刻也没有停止过工作，"活到老，学到老，做到老"．陈运泰至今仍活跃在我国地震科学研究领域的第一线，担任着中国科学院大学地球与行星科学学院资深讲席教授、中国地震局地球物理研究所研究员、名誉所长，北京大学地球与空间科学学院教授、名誉院长，中国地震学会名誉理事长，中国科学院学部主席团成员，中国科协荣誉委员，国际大地测量学与地球物理学联合会（International Union of Geodesy and Geophysics，IUGG）执行局成员，亚洲与大洋洲地球科学学会主席，亚洲地震委员会副主席，国际《地震学刊》（Journal of Seismology）、《国际地球物理学刊》（International Journal of Geophysics）、《科技导报》、《科学》编委，《地震学报》、英文《地震科学》（Earthquake Science）、《世界地震译丛》主编，《地球物理学报》副主编等职．

陈运泰历来十分关注科技人才的培养工作，认为培养一流的科技人才与取得高水平的科研成果同样重要．1978 年我国恢复研究生制度，他在傅承义先生、曾融生先生等前辈的领导下，在中国科学院研究生院一起开设"地球物理学基础"、"地球物理学进展"等课程．

当时百废待兴，没有教材，他们一边上课，一边编写讲义．在这些讲义基础上撰写而成的《地球物理学基础》（傅承义，陈运泰，祁贵仲著）一书于 1985 年由科学出版社出版，1990 年再版，成为我国地球物理学领域最畅销的专业读物之一．陈运泰为研究生院开设震源理论课程，并撰写了"震源理论基础"教材，将现代地震震源理论及时地、系统地介绍给中国地震科学界．

为培养掌握现代地震科学理论的科技人才，他还为地震系统的科技骨干培训班编写了"地震矩张量及其反演"等多种教材，并在这些教材的基础上逐步完善成后来在地震科学界具有广泛影响的著作《数字地震学》．

在研究生培养过程中，陈运泰把高尚的科学道德、严谨的治学态度和无私的奉献精神贯彻始终．身教重于言传，他一向严于律己，凡是要求学生做的，自己一定先行．在学生们的心目中，他既是一位德高望重的科学家，也是一位可敬、可亲的好老师．

1978 年以来，除了 1981–1983 年出国访问期间外，每年都为中国科学院研究生院（今中国科学院大学）和北京大学研究生上课，从不间断，教过的研究生有数百人之多．1981 年以来，亲自培养的硕士与博士研究生 40 余名，其中包括两名伊朗博士生，还有德国、罗马尼亚等国的博士后．他培养的学生不少已经成为国内外地球物理学各个研究领域的中坚力量．2018 年，中国科学院大学授予他地球与行星科学学院"杰出贡献教师"荣誉称号．

本书收录的是陈运泰具有代表性的论文，仅占其论著的一部分．从中可以看出陈运泰勤奋严谨的治学态度，不断开拓进取的精神．

陈运泰的信念是："揭示地球的奥秘，是一代又一代地学工作者所追求的目标，任重而道远．然而，只要我们以求真务实的精神，严谨的科学态度，百折不挠的毅力，团结协作，开展探测、探索，始终遵循认识、实践、再认识的过程，相信一定能为进一步认识地球，揭示其地质过程的发生发展规律，从必然王国向自由王国的迈进，做出应有的贡献．"

"老骥伏枥，志在千里．"工作还在继续，这位老科学家背后承载的是中国地震科学未来的发展……

李世愚

目 录

出版说明

序言

第一部分 学术论文

强震发生的规律性探讨	3
多层弹性半空间中的地震波（一）	10
多层弹性半空间中的地震波（二）	33
根据地面形变的观测研究 1966 年邢台地震的震源过程	46
巧家、石棉的小震震源参数的测定及其地震危险性的估计	63
1976 年 7 月 28 日河北省唐山 7.8 级地震的发震背景及其活动性	87
1975 年 2 月 4 日辽宁省海城地震的震源机制	97
中、小地震体波的频谱和纵、横波拐角频率比	112
由瑞雷波方向性函数研究 1974 年 5 月 11 日云南省昭通地震的震源过程	117
用大地测量资料反演的 1976 年唐山地震的位错模式	129
Variations of gravity before and after the Haicheng earthquake, 1975, and the Tangshan earthquake, 1976	144
唐山地震引起的剩余倾斜场的空间分布和倾斜阶跃	154
地球自由振荡	160
Static shear crack with a zone of slip-weakening	163
The quasistatic extension of a shear crack in a viscoelastic medium	181
Simulation of earthquake sequences	194
Spontaneous growth and autonomous contraction of a two-dimensional earthquake fault	210
Lancang-Gengma earthquake—a preliminary report on the November 6, 1988, event and its aftershocks	224
Inversion of near-source-broadband accelerograms for the earthquake source-time function	232
Seismological studies of earthquake prediction in China: a review	243
卢龙地区 S 波偏振与上地壳裂隙各向异性	270
平面内剪切断层的超 S 波速破裂	280
The China Digital Seismograph Network	286
Source process of the 1990 Gonghe, China, earthquake and tectonic stress field in the northeastern Qinghai-Xizang (Tibetan) Plateau	290
Delineation of potential seismic sources for seismic zoning of Iran	306
A time-domain inversion technique for the tempo-spatial distribution of slip on a finite fault plane with applications to recent large earthquakes in the Tibetan Plateau	323

Source parameter determination of regional earthquakes in the Far East using moment tensor inversion of single-station data ········· 337

Source process of the 4 June 2000 southern Sumatra, Indonesia, earthquake ········· 349

Temporal-spatial rupture process of the 1999 Chi-Chi earthquake from IRIS and GEOSCOPE long-period waveform data using aftershocks as empirical Green's functions ········· 360

Realistic modeling of seismic wave ground motion in Beijing City ········· 386

Estimation of site effects in Beijing City ········· 398

Decade-scale correlation between crustal deformation and length of day: implication to earthquake hazard estimation ········· 411

Deterministic seismic hazard map in North China ········· 423

从全球长周期波形资料反演2001年11月14日昆仑山口地震时空破裂过程 ········· 429

Double-difference relocation of earthquakes in central-western China, 1992—1999 ········· 438

2007年云南宁洱M_S6.4地震震源过程 ········· 464

2008年汶川大地震的时空破裂过程 ········· 475

地震预测：回顾与展望 ········· 485

提取视震源时间函数的PLD方法及其对2005年克什米尔M_W7.6地震的应用 ········· 514

利用阿拉斯加台阵资料分析2008年汶川大地震的破裂过程 ········· 525

2008年汶川大地震震源机制的时空变化 ········· 534

Single-couple component of far-field radiation from dynamical fractures ········· 547

Teleseismic receiver function and surface-wave study of velocity structure beneath the Yanqing-Huailai Basin Northwest of Beijing ········· 565

Shear-wave splitting in the crust beneath the southeast Capital area of North China ········· 585

2010年青海玉树地震震源过程 ········· 595

京津唐地区地壳三维P波速度结构与地震活动性分析 ········· 599

The 2009 L'Aquila M_W6.3 earthquake: a new technique to locate the hypocentre in the joint inversion of earthquake rupture process ········· 612

An inversion of Lg-wave attenuation and site response in the North China region ········· 628

2008年5月12日汶川M_W7.9地震的震源位置与发震时刻 ········· 645

芦山4.20地震破裂过程及其致灾特征初步分析 ········· 655

从汶川地震到芦山地震 ········· 660

Kinematic rupture model and hypocenter relocation of the 2013 M_W6.6 Lushan earthquake constrained by strong-motion and teleseismic data ········· 670

Comment on the paper "Normal and shear stress acting on arbitrarily oriented faults, earthquake energy, crustal GPE change, and the coefficient of friction" by P. P. Zhu ········· 680

断层厚度的地震效应和非对称地震矩张量 ········· 684

Automatic imaging of earthquake rupture processes by iterative deconvolution and stacking of high-rate GPS and strong motion seismograms ········· 695

2014 年云南鲁甸 M_W6.1 地震：一次共轭破裂地震 …………………………………………… 718

2015 年尼泊尔 M_W7.9 地震破裂过程：快速反演与初步联合反演 ………………………… 729

Further studies on the focal mechanism and source rupture process of the 2012 Haida Gwaii, Canada, 7.8 moment magnitude earthquake ………………………………………………………… 737

非对称地震矩张量时间域反演：理论与方法 ……………………………………………………… 755

Inversion of earthquake rupture process: theory and applications ………………………………… 777

第二部分　回忆与祝贺

贺陈运泰院士八十华诞 ……………………………………………………………………………… 813

回忆和陈运泰老师在一起的日子 …………………………………………………………………… 815

陈运泰老师带领我们做野外观测 …………………………………………………………………… 818

我的导师陈运泰院士——永远年轻的科学家 ……………………………………………………… 821

名师指路，终生受益——回忆陈运泰老师对我的教诲 …………………………………………… 823

我的老师陈运泰 ……………………………………………………………………………………… 827

地震科学研究的引路人——陈运泰老师 …………………………………………………………… 830

随笔——为陈老师八十岁生日作 …………………………………………………………………… 833

陈运泰院士培养硕士研究生名录 …………………………………………………………………… 835

陈运泰院士培养博士研究生名录 …………………………………………………………………… 836

陈运泰论著目录（1971–2020） …………………………………………………………………… 837

第一部分　学术论文

强震发生的规律性探讨

陈运泰　林邦慧　顾浩鼎

中国科学院地球物理研究所

为探讨强烈地震发生的规律性，我们从地震活动性的角度分析了强烈地震发生之前地震带上的地震活动与强震之间的关系，并从统计的角度研究了地震带上的极大地震发生的规律．我们对强震发生的规律性的认识是否正确，有待于实践的检验，这里只是一个初步的总结．

1 强震发生之前地震带上地震活动性的变化

在长期的实践中，人们找到地震频度和地震能量这两个量作为地震带或地震区的地震活动性高低的标志．由于小地震的数目通常远远多于大地震，所以地震频度这个量实际上主要是由小地震决定的，忽略了大地震在地震活动过程中所起的作用．相反，由于大小地震所释放的能量差异很大，震级相差一级时所释放的能量相差30倍左右，地震能量这个量实际上主要是由大地震决定的，从而忽略了小地震在地震活动过程中所起的作用．纯用地震频度与地震能量描述地震活动性时存在的上述矛盾，可以用一个介于地震频度和地震能量之间的量加以解决．这个量可以表示为

$$\Sigma(t) = \sum_k N(k) L^k, \quad (1 \leq L \leq 10) \tag{1}$$

式中，$N(k)$表示在时间间隔$t-\Delta t$到t之内能级为k的地震数目（即地震波能量在$10^{k-1/2}$J和$10^{k+1/2}$J之间的地震数目）．显然，若$L=1$，则$\Sigma(t)$等于地震频度；若$L=10$，则$\Sigma(t)$等于地震波总能量．若$L=4.5$，则$l=\lg L=2/3$，此时

$$\Sigma(t) = \sum_k N(k) 10^{2k/3} \tag{2}$$

如果我们认为地震波总能量E正比于地震总能量E_0，即$E \propto E_0$，按照一种估计，$E_0 \propto S^{3/2}$，这里S是地震的断层面面积，因此，$S \propto E^{2/3} \propto 10^{2k/3}$，也就是$L=4.5$时的$\Sigma(t)$相当于地震的断层面总面积．我们称$L=4.5$时的$\Sigma(t)$为"断层面总面积"．

由于$L=4.5$时的$\Sigma(t)$代表一定的物理意义，所以我们在分析地震活动性时用它来描述地震活动性，而暂不用其他别的L值所对应的$\Sigma(t)$．至于L取什么值时所对应的$\Sigma(t)$能最恰当地处理大、小地震在地震过程中所起作用的关系，解决频度和能量之间的矛盾，需要进行更多的实践才能得到正确的认识．这里只是一个初步的尝试．

基于上述考虑，我们分析了我国境内15条地震带上23个强震之前地震活动性随时间的变化：

（1）燕山地震带：1679年9月2日三河、平谷地震，$M=8$.
（2）郯城–庐江地震带：1597年10月6日渤海地震，$M=7\frac{1}{2}$. 1668年7月25日郯城地震，$M=8\frac{1}{2}$.
（3）山西地震带：1695年5月18日临汾地震，$M=7\frac{3}{4}$.
（4）银川–中卫地震带：1739年1月3日平罗地震，$M=8$.

---

* 本文发表于《地震战线》，1971年，第8期，18–24.

(5) 滇西地震带：1948 年 6 月 27 日剑川地震，$M=6\frac{1}{4}$. 1951 年 12 月 21 日剑川地震，$M=6\frac{1}{4}$. 1966 年 9 月 28 日中甸地震，$M=6\frac{1}{2}$.

(6) 滇东地震带：1970 年 1 月 5 日通海地震，$M=7\frac{3}{4}$.

(7) 西藏中部地震带：1951 年 11 月 8 日当雄地震，$M=8$.

(8) 雅鲁藏布江地震带：1915 年 12 月 3 日拉萨东地震，$M=7$.

(9) 察隅地震带：1950 年 8 月 15 日察隅地震，$M=8\frac{1}{2}$.

(10) 康定-甘孜地震带：1923 年 3 月 24 日炉霍地震，$M=7\frac{1}{4}$. 1967 年 8 月 30 日甘孜地震，$M=6.8$.

(11) 华北地震带：1830 年 6 月 2 日磁县地震，$M=7\frac{1}{2}$. 1966 年 3 月 22 日邢台地震，$M=7.2$.

(12) 西海固地震带：1920 年 12 月 16 日海源地震，$M=8\frac{1}{2}$.

(13) 西秦岭地震带：1654 年 7 月 21 日天水地震，$M=8$.

(14) 横断山地震带：1950 年 2 月 3 日西双版纳区中缅边界两个连续发生的地震，$M=7$ 和 $M=6\frac{3}{4}$. 1971 年 4 月 28 日普洱地震，$M=6.7$.

(15) 塔里木南缘地震带：1924 年 7 月 3 日民丰地震，$M=7\frac{1}{4}$ 和 1924 年 7 月 12 日民丰地震，$M=7\frac{1}{4}$.

图 1–图 3 是上述 15 条地震带中的 3 条地震带（滇东地震带、西藏中部地震带和雅鲁藏布江地震带）强震来临之前"断层面总面积"$\Sigma(t)$ 随时间的变化. 为了方便起见，我们不用（1）式，而用

$$\Sigma(t) = \sum_{k=k_0}^{K-1} N(k) L^{k-k_0}, \quad (L=4.5) \tag{3}$$

式中，k_0 取为 11，是进行统计的地震能级的下限，K 是所研究的那个强震所对应的能级，Δt 取为 5 年. 显而易见，（3）式与（1）式并无本质不同，仅仅是单位的差别而已.

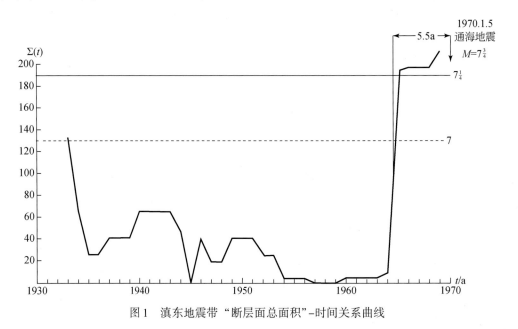

图 1　滇东地震带"断层面总面积"-时间关系曲线

根据对上述 15 条地震带上强震来临之前"断层面总面积"$\Sigma(t)$ 和 t 时间的关系曲线的分析，我们从中总结出以下的初步认识：

(1) 这些地震带上历史上发生的强震，除横断山地震带 1950 年 2 月 3 日的两个地震和塔里木南缘地震带 1924 年 7 月的两个地震外，在每一个强震来临之前若干年，描述地震活动性高低的量——"断层面总面积"$\Sigma(t)$ 曲线有一个显著的"峰"出现. 它表明，在强震来临之前，地震带上的地震活动性显著增高."峰"的出现意味着地震带上孕育着一个强烈地震. 对于横断山地震带 1950 年 2 月 3 日的两个地震

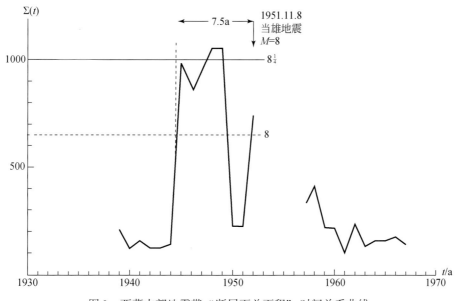

图 2 西藏中部地震带 "断层面总面积"-时间关系曲线

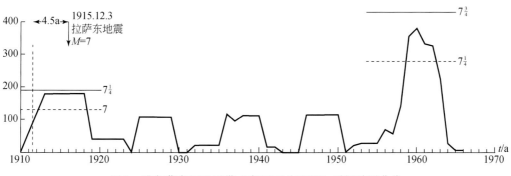

图 3 雅鲁藏布江地震带 "断层面总面积"-时间关系曲线

和塔里木南缘地震带 1924 年 7 月的两个地震，$\Sigma(t)$ 曲线也各有一个显著的"峰"出现，所不同的仅是这个"峰"与连续出现的两个大小相近的地震相对应.

$\Sigma(t)$ 曲线"峰"的出现和发生强震之间的对应关系是明显的，在 15 条地震带中只有两个例外. 一个是西秦岭地震带的 1718 年 6 月 19 日通渭南地震 ($M=7\frac{1}{2}$)，在它之前没有明显的"峰"出现. 一个是燕山地震带 1719 年至 1725 年间出现一个明显的"峰"，但以后并没有与之相应的强震发生. 有"峰"无震与有震无"峰"的例外情况虽占少数，但仍有进一步研究的必要.

(2) 预示着在地震带上即将发生强震的 $\Sigma(t)$ 曲线的"峰"的高度 $\Sigma(t)_\text{峰}$ 和正在孕育的强震的震级 M 有一定的关系. 图 4 在单对数纸上表示这一关系，纵坐标代表 $\Sigma(t)_\text{峰}$ 的对数，横坐标代表相应的强震的震级 M. $\Sigma(t)_\text{峰}$ 的对数和震级 M 的关系大致上是线性的，即：

$$\lg\Sigma(t)_\text{峰} = aM + b \tag{4}$$

a, b 可以根据观测数据用最小二乘法定出，它们分别是：

$$a = 0.724, \quad b = -2.97.$$

(3) 进一步分析 $\Sigma(t)$ 曲线的"峰"的出现与强震的发震时间在数量上的关系. 这个关系可以用下述方法确定. 设 $\Sigma(t)$ 在"峰"的左侧等于 $\Sigma(t)_\text{峰}$ 的一半时到强震发生时的时间间隔为 $\Delta\tau$，则 $\Delta\tau$（我们称它为孕育地震的时间）与正在孕育的强震的震级 M 的关系如图 5 所示．$\Delta\tau$ 与 M 粗略地是对数关系，即：

$$\lg\Delta\tau = cM + d \qquad (5)$$

c, d 也可以根据观测数据用最小二乘法定出，它们分别为

$$c = 0.489, \quad d = -2.58.$$

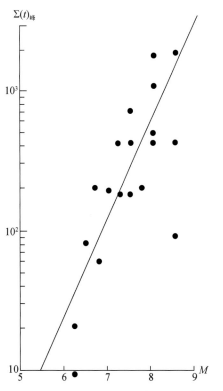

图 4　"断层面总面积"的峰值 $\Sigma(t)_{峰}$ 与震级 M 的关系：$\lg\Sigma(t)_{峰} = 0.724M - 2.97$

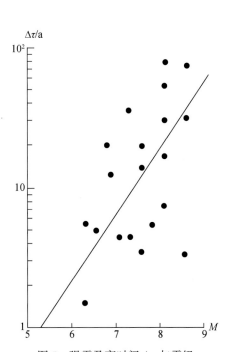

图 5　强震孕育时间 $\Delta\tau$ 与震级 M 的关系：$\lg\Delta\tau = 0.489M - 2.58$

公式（5）所表示的孕育地震的时间 $\Delta\tau$ 与震级 M 的关系是很粗略的．由于引起各地震带地震的构造力之间的差异以及各地震带介质构造之间的差异，孕育同样强度的地震所需要的时间可能差别很大．因此，公式（5）所表示的关系必然比较粗糙，只是在定性方面指出 $\Delta\tau$ 随 M 的增大而增大，而在定量方面，各地震带势必有其独特的特点，与总的变化趋势有一定的差异或偏离，这一差异或偏离应当通过对各地震带地震活动的具体分析才能弄清楚．

根据上述认识，我们分析了这些地震带上近期的地震活动性，认为在 1970 年 1 月 5 日通海地震后，滇东地震带的地震活动性不高，目前没有发生 7 级以上大震的预兆（图 1）．至于西藏中部地震带（图 2），最近地震活动性也不高，也没有发生 7 级以上大震的预兆．与滇东地震带、西藏中部地震带不同，雅鲁藏布江地震带的 $\Sigma(t)$ 曲线在 1951 年 –1968 年间出现一个"峰"，按"峰"的高度和出现时间来推算，1973 年 –1983 年间雅鲁藏布江地震带将有 $7\frac{3}{4}$ 级左右的地震发生．

前面分析的 15 条地震带上的 23 个强烈地震所遵从的规律对这些地震带而言，可能具有一定的普遍性．但是，由于地震现象极其错综复杂，上述认识仍有待于通过再实践加以检验．对其他地震带而言，这些结论是否适用，更必须通过再实践才能获得正确的认识．

2 极大地震发生的概率性规律

上面从地震活动性的角度分析了强震发生之前地震带上地震活动性的变化规律，下面从统计的角度，探讨极大地震发生的概率性规律.

所谓极大地震，是指某一地震带在一定时间间隔（比如说，一年）之内发生的强度最大的地震. 分析表明，极大地震的出现遵从一定的规律，这个规律可以这样表示：设 $\varphi(x)$ 是震级小于或等于 x 的极大地震发生的概率（分布函数），则

$$\varphi(x) = \exp[-\exp(-y)], \quad (y = \alpha(x-u)) \tag{6}$$

$\alpha(>0)$ 和 u 是描述这个分布的两个参数.

求极大地震的分布规律，方法如下：

从历史地震资料中找出逐年的极大地震，设这一系列极大地震共有 N 个（即研究连续 N 年历史地震的极大值分布），按震级由小到大的次序排列成：

$$x_1 \leqslant x_2 \leqslant \cdots \leqslant x_m \leqslant \cdots \leqslant x_N,$$

对应于 x_m 的经验频率 $\varphi(x_m)$ 可由下式计算：

$$\varphi(x_m) = \frac{m}{N+1}, \quad m = 1, 2, \cdots, N \tag{7}$$

由式（6）可以求得相应于 x_m 的 y_m：

$$y_m = -\ln\left(-\ln\frac{m}{N+1}\right), \quad m = 1, 2, \cdots, N \tag{8}$$

由一系列的 (x_m, y_m)，用最小二乘法，可以确定直线方程 $x = y/\alpha + u$ 的两个参数 $1/\alpha$ 和 u，从而也就确定了极大地震的分布函数. 这样就可以估算今后 D 年内最大的极大地震的大小. 首先，由（6）式计算 $N+D$ 年一遇的极大地震的震级，拿它与历史上已经发生过的 N 年的实际资料比较，如果计算出来的 $N+D$ 年一遇的极大地震震级大于历年的极大地震，这就说明这个地震在 $N+D$ 年中的前 N 年一直没有发生，因而在今后的 D 年中很可能发生. 反之，如果计算出来的 $N+D$ 年一遇的极大地震震级小于、等于历年的极大地震，这就说明这个地震在 $N+D$ 年中的前 N 年里已经发生，因而在今后 D 年中再发生大于、等于这么大的地震的可能性不大.

根据上述方法，我们研究了我国 17 条地震带上极大地震的分布规律. 它们是：天山北缘地震带，塔里木南缘地震带，西藏中部地震带，雅鲁藏布江地震带，银川-中卫地震带，西海固地震带，西秦岭地震带，武都-马边地震带，安宁河地震带，滇东地震带，滇西地震带，山西地震带，燕山地震带，信阳-南京地震带，东南沿海地震带，台湾西部地震带以及台湾东部地震带. 为了兼顾 $4\frac{1}{2}$—6 级这个震级范围内也有比较可信、足够的资料，我们所使用的资料起始年份绝大多数是从 19 世纪末、20 世纪初开始，一般所统计的总年数 N 是 60 年左右.

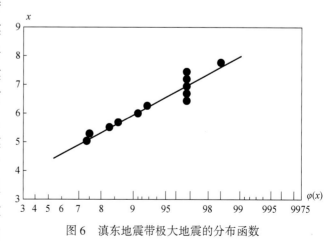

图 6　滇东地震带极大地震的分布函数

图 6-图 8 表示这 17 条地震带中的三条地震带（滇东地震带，西藏中部地震带和雅鲁藏布江地震带）极大地震的分布规律. 在图 6-图 8 中，纵坐标表示极大地震的震级 x，横坐标是 y，即极大地震分布函数

$\varphi(x)$ 的重自然对数：

$$y = -\ln[-\ln\varphi(x)],$$

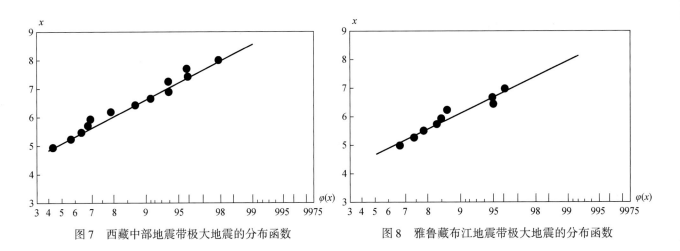

图 7　西藏中部地震带极大地震的分布函数　　　图 8　雅鲁藏布江地震带极大地震的分布函数

横坐标上标明的数字是 $\varphi(x)$ 的数值．根据实际资料得到的结果在图 6–图 8 中用小黑点表示．结果表明 x 和 y 的线性关系很明显，即公式（6）所表示的分布函数可以相当好地反映极大地震的分布规律．我们用最小二乘法根据观测数据确定了这些地震带的极大地震分布函数的两个参数，即 $x=y/\alpha+u$ 中的 $1/\alpha$ 和 u．上述三条地震带的结果列于表 1．

表 1

地震带名称	资料起止年份	N	$1/\alpha$	u	$M_{计算值}$	$M_{最大}$	δM
滇东	1909–1970	62	0.862	4.05	7.7	7.8	−0.1
西藏中部	1921–1969	49	0.800	5.00	8.1	8.0	0.1
雅鲁藏布江	1913–1969	57	0.676	4.57	7.4	7.0	0.4

确定了各地震带的极大地震分布函数之后，就可以估算今后 D 年内最大的极大地震的大小．在表 1 中我们列出上述三条地震带 $N+D$ 年一遇的极大地震震级 $M_{计算值}$ 和（D 取为 5 年）该地震带前 N 年已经发生的最大地震震级 $M_{最大}$，以及 $M_{计算值}$ 与 $M_{最大}$ 的差值 δM．

从表 1 的结果可以看到，对于滇东地震带和西藏中部地震带，$M_{计算值}$ 与 $M_{最大}$ 几乎相等（考虑到震级间隔是 $\frac{1}{4}$ 级，如果震级相差在 $\frac{1}{4}$ 级范围之内，我们便可以认为 $M_{计算值}$ 与 $M_{最大}$ 在误差范围内，是相等的）．这说明，滇东地震带和西藏中部地震带，在 5 年之内发生比表 1 列出的 $M_{最大}$ 还要大的地震的可能性很小．

对于雅鲁藏布江地震带，$M_{计算值}=7.4$，$M_{最大}=7.0$，$M_{计算值}>M_{最大}$，意味着在 1970 年–1974 年间可能发生 7.4 级左右的地震（误差范围是 $\frac{1}{4}$ 级，所以我们可以说：可能发生 $7\frac{1}{4}$–$7\frac{3}{4}$ 级的地震）．

我们的工作还存在许多问题．比如，当某一条地震带极大地震震级的计算值大于实际上已经发生的最大地震震级时，我们能够估算这条地震带上未来若干年里最大的极大地震震级．反过来时，我们只能肯定该地震带发生大于、等于已经发生过的最大地震的可能性不大，而无法预测其大小．这是这个方法的局限性，它有待于进一步在实践中寻求克服的办法．此外，我们还可以看到，在多数情况下，我们用的分布函数可以相当好地反映极大地震的分布规律，但在有些情况下就比较差．这说明，描述极大地震分布规律的分布函数仍有深入研究的必要，以求更正确、更全面地反映客观存在的极大地震的分布规律．最后，我们知道，各地震带极大地震的分布规律是由有限观测时间的观测结果导出的，在我们的工作中，这个观测时间 N 一般是 60 年左右．显然，观测时间不同将会影响结果．这个影响是怎么样的？有多大？如何克服等等问题都有待于通过反复实践才能解决．

多层弹性半空间中的地震波（一）

陈运泰

中国科学院地球物理研究所

1 引言

为了了解地震震源和地球介质的性质，很有必要对地震波的辐射、传播和衰减问题作仔细的分析．作为一种近似，可以暂且忽略地球的曲率，把传播地震波的地球介质视为多层半空间．为简便起见，地震波的衰减问题另作考虑．这样，便需要研究多层、均匀、各向同性和完全弹性半空间中地震震源辐射的地震波传播问题．

用哈斯克尔（Haskell）[1-2]矩阵法解多层介质中弹性波的传播问题是很方便的．如果将哈斯克尔层矩阵略加变动，用元素是无量纲实数的矩阵定义，会给理论分析和数值计算带来一些方便．

在处理一般类型地震震源辐射的地震波传播问题时，妹泽克惟（Sezawa）[3]的表示式及其另一种形式[2]并不太方便．这时，用汉森（Hansen）[4]向量表示运动方程的一般解比较方便[5]．把汉森展开和哈斯克尔矩阵法结合起来，可以求得多层、均匀、各向同性和完全弹性半空间中一般类型的震源引起的位移场表示式（形式解），特别是各种基本类型的偶极源引起的位移场表示式的具体形式．

2 多层弹性半空间中的平面波

2.1 运动方程及其一般解

先考虑多层、均匀、各向同性和完全弹性半空间中平面波的传播问题．所研究的系统如图 1 所示，$z_0 = 0$ 是自由表面，$z = z_i$，$i = 1, 2, 3, \cdots, (n-1)$ 是第 i 层底面（第 i 个分界面）的深度，$d_i = z_i - z_{i-1}$，$i = 1, 2, 3, \cdots, (n-1)$ 是第 i 层的厚度．第 n 层，即弹性半空间．现在要分析平面波在这样的多层弹性半空间中的传播问题．质点的位移 $\boldsymbol{u}$ 满足运动方程：

$$(\lambda+2\mu)\nabla(\nabla \cdot \boldsymbol{u}) - \mu\nabla\times\nabla\times\boldsymbol{u} = \rho\frac{\partial^2 \boldsymbol{u}}{\partial t^2}, \quad (1)$$

而位移 $\boldsymbol{u}$ 的谱 $\boldsymbol{U}$ 满足方程：

$$(\lambda+2\mu)\nabla(\nabla \cdot \boldsymbol{U}) - \mu\nabla\times\nabla\times\boldsymbol{U} = -\rho\omega^2 \boldsymbol{U}, \quad (2)$$

（1）式和（2）式中，λ, μ 是拉梅常数，ρ 是介质的密度，ω 是角频率．

引进由下式定义的位函数 φ, ψ, χ：

$$\boldsymbol{U} = \nabla\varphi + \nabla\times\nabla\times(\psi \boldsymbol{e}_z) + \nabla\times(\chi \boldsymbol{e}_z), \quad (3)$$

式中，$\boldsymbol{e}_z$ 是 z 方向的单位向量．则如果 φ, ψ, χ 满足方程

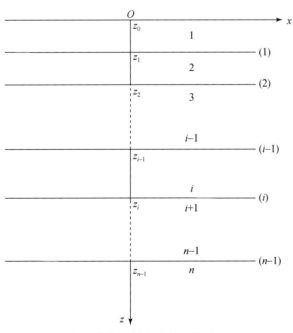

图 1　直角坐标中的多层弹性半空间

---

* 本文发表于《地球物理学报》，1974 年，第 **17** 卷，第 1 期，20-43.

$$\begin{cases} \nabla^2\varphi + k_\alpha^3\varphi = 0, \\ \nabla^2\psi + k_\beta^2\psi = 0, \\ \nabla^2\chi + k_\beta^2\chi = 0, \end{cases} \tag{4}$$

$$\begin{cases} k_\alpha = \dfrac{\omega}{\alpha}, \\ k_\beta = \dfrac{\omega}{\beta}, \end{cases} \tag{5}$$

$$\begin{cases} \alpha = \sqrt{\dfrac{\lambda+2\mu}{\rho}}, \\ \beta = \sqrt{\dfrac{\mu}{\rho}}, \end{cases} \tag{6}$$

那么 **U** 就满足方程（2）.

在直角坐标（x，y，z）中，如果讨论的是仅与坐标 x，z 有关而与 y 无关的二维问题，那么

$$\begin{cases} \nabla\varphi = \dfrac{\partial\varphi}{\partial x}\boldsymbol{e}_x + \dfrac{\partial\varphi}{\partial z}\boldsymbol{e}_z, \\ \nabla\times\nabla\times(\psi\boldsymbol{e}_z) = \dfrac{\partial^2\psi}{\partial x\partial z}\boldsymbol{e}_x + \left(k_\beta^2\psi + \dfrac{\partial^2\psi}{\partial z^2}\right)\boldsymbol{e}_z, \\ \nabla\times(\chi\boldsymbol{e}_z) = -\dfrac{\partial\chi}{\partial x}\boldsymbol{e}_y, \end{cases} \tag{7}$$

$\boldsymbol{e}_x$ 和 $\boldsymbol{e}_y$ 分别是 x 和 y 方向的单位向量. 在直角坐标中，方程（4）的基本解是

$$\begin{cases} \varphi = e^{-ikx \pm az}, \\ \psi = e^{-ikx \pm bz}, \\ \chi = e^{-ikx \pm bz}, \end{cases} \tag{8}$$

式中，k 是波数，

$$\begin{cases} a = \sqrt{k^2 - k_\alpha^2}, & k > k_\alpha, \\ i\sqrt{k_\alpha^2 - k^2}, & k < k_\alpha, \\ b = \sqrt{k^2 - k_\beta^2}, & k > k_\beta, \\ i\sqrt{k_\beta^2 - k^2}, & k < k_\beta. \end{cases} \tag{9}$$

把（8）式代入（7）式，可以得到

$$\begin{cases} \nabla\varphi = (-ik\,\boldsymbol{e}_x \pm a\,\boldsymbol{e}_z)\,e^{-ikx \pm az}, \\ \nabla\times\nabla\times(\psi\boldsymbol{e}_z) = (\mp ikb\,\boldsymbol{e}_x + k^2\,\boldsymbol{e}_z)\,e^{-ikx \pm bz}, \\ \nabla\times(\chi\boldsymbol{e}_z) = (ik\,\boldsymbol{e}_y)\,e^{-ikx \pm bz}. \end{cases} \tag{10}$$

引进三个向量 **B**，**P**，**C**：

$$\begin{cases} \boldsymbol{B} = -i\,e^{-ikx}\boldsymbol{e}_x, \\ \boldsymbol{P} = e^{-ikx}\boldsymbol{e}_z, \\ \boldsymbol{C} = -e^{-ikx}\boldsymbol{e}_y, \end{cases} \tag{11}$$

用这三个向量表示（10）式，可得：

$$\begin{cases} \nabla\varphi = (k\boldsymbol{B}\pm a\boldsymbol{P})e^{\pm az}, \\ \nabla\times\nabla\times(\psi\,\boldsymbol{e}_z) = k(\pm b\boldsymbol{B}+k\boldsymbol{P})e^{\pm bz}, \\ \nabla\times(\chi\,\boldsymbol{e}_z) = (-ik\boldsymbol{C})e^{\pm bz}. \end{cases} \quad (12)$$

如果定义三个向量 $\boldsymbol{L}$，$\boldsymbol{N}$，$\boldsymbol{M}$：

$$\begin{cases} k_\alpha \boldsymbol{L}^\pm = (k\boldsymbol{B}\pm a\boldsymbol{P})e^{\pm az}, \\ k_\beta \boldsymbol{N}^\pm = (\pm b\boldsymbol{B}+k\boldsymbol{P})e^{\pm bz}, \\ k_\beta \boldsymbol{M}^\pm = (k_\beta \boldsymbol{C})e^{\pm bz}, \end{cases} \quad (13)$$

那么可以把方程（2）的解表示成

$$\boldsymbol{U} = (a^+ k_\alpha \boldsymbol{L}^+ + a^- k_\alpha \boldsymbol{L}^-) + (b^+ k_\beta \boldsymbol{N}^+ + b^- k_\beta \boldsymbol{N}^-) + (c^+ k_\beta \boldsymbol{M}^+ + c^- k_\beta \boldsymbol{M}^-), \quad (14)$$

若用 $\boldsymbol{B}$，$\boldsymbol{P}$，$\boldsymbol{C}$ 表示，则为

$$\boldsymbol{U} = (kf_1 + f_2')\boldsymbol{B} + (f_1' + kf_2)\boldsymbol{P} + k_\beta f_3 \boldsymbol{C}. \quad (15)$$

其中，

$$\begin{cases} f_1 = a^+ e^{az} + a^- e^{-az}, \\ f_2 = b^+ e^{bz} + b^- e^{-bz}, \\ f_3 = c^+ e^{bz} + c^- e^{-bz}, \end{cases} \quad (16)$$

$$\begin{cases} f_1' = \dfrac{df_1}{dz} = a(a^+ e^{az} - a^- e^{-az}), \\ f_2' = \dfrac{df_2}{dz} = b(b^+ e^{bz} - b^- e^{-bz}), \\ f_3' = \dfrac{df_3}{dz} = b(c^+ e^{bz} - c^- e^{-bz}), \end{cases} \quad (17)$$

a^+，b^+，c^+ 是上行波的振幅系数，a^-，b^-，c^- 是下行波的振幅系数，它们是待定系数.

应力张量的谱 $S(\boldsymbol{U})$ 由下式决定：

$$S(\boldsymbol{U}) = \lambda \boldsymbol{I}(\nabla\cdot\boldsymbol{U}) + \mu(\nabla\boldsymbol{U}\cdot\boldsymbol{U}\nabla), \quad (18)$$

其中，$\boldsymbol{I}$ 是单位张量. 因此，在 z 平面上应力的谱 $S(\boldsymbol{U})\cdot\boldsymbol{e}_z$ 为

$$S(\boldsymbol{U})\cdot\boldsymbol{e}_z = 2\mu(kf_1' + \Omega f_2)\boldsymbol{B} + 2\mu(\Omega f_1 + kf_2')\boldsymbol{P} + \mu k_\beta f_3' \boldsymbol{C}, \quad (19)$$

$$\Omega = k^2 - \dfrac{k_\beta^2}{2}. \quad (20)$$

(15) 式就是用基向量 $\boldsymbol{B}$，$\boldsymbol{P}$，$\boldsymbol{C}$ 表示的运动方程（2）的一般解，(19) 式就是用 $\boldsymbol{B}$，$\boldsymbol{P}$，$\boldsymbol{C}$ 表示的 z 平面上应力的谱.

2.2 多层弹性半空间的自由表面上的位移谱

根据上面的结果（(15) 式和 (19) 式），可以把第 i 层中位移的谱 $\boldsymbol{U}_i$ 和在 z 平面上应力的谱 $S(\boldsymbol{U}_i)\cdot\boldsymbol{e}_z$ 表示为

$$k\boldsymbol{U} = k(kf_1 + f_2')\boldsymbol{B} + k(f_1' + kf_2)\boldsymbol{P} + kk_\beta f_3 \boldsymbol{C}, \quad (21)$$

$$\dfrac{1}{\mu_1}S(\boldsymbol{U})\cdot\boldsymbol{e}_z = 2\dfrac{\mu}{\mu_1}(kf_1' + \Omega f_2)\boldsymbol{B} + 2\dfrac{\mu}{\mu_1}(\Omega f_1 + kf_2')\boldsymbol{P} + \dfrac{\mu}{\mu_1}k_\beta f_3' \boldsymbol{C}, \quad (22)$$

其中，(21) 式是由 (15) 式两边乘上 k 得到的，(22) 式是由 (19) 式两边除以第一层的刚性系数 μ_1 得

到的. 这样做, 为的是使量纲整齐. 此外, 在 (21), (22) 式以及下面的分析中, 为书写简便起见, 一般都略去了层的指标 i, 只是在必要时才添上. (21) 式和 (22) 式还可以表示成:

$$k\boldsymbol{U} = k\,U_B\boldsymbol{B} + kU_p\boldsymbol{P} + kU_c\boldsymbol{C}, \tag{23}$$

$$\frac{1}{\mu_1}\boldsymbol{S}(\boldsymbol{U})\cdot\boldsymbol{e}_z = \frac{1}{\mu_1}\tau_B\boldsymbol{B} + \frac{1}{\mu_1}\tau_p\boldsymbol{P} + \frac{1}{\mu_1}\tau_c\boldsymbol{C}, \tag{24}$$

其中,

$$\begin{cases} kU_B = k^2\mathrm{ch}\,az(a^+ + a^-) + k^2\mathrm{sh}\,az(a^+ - a^-) + kb\mathrm{ch}\,bz(b^+ - b^-) \\ \qquad + kb\mathrm{sh}\,bz(b^+ + b^-), \\ kU_p = ka\mathrm{sh}\,az(a^+ + a^-) + ka\mathrm{ch}\,az(a^+ - a^-) + k^2\mathrm{sh}\,bz(b^+ - b^-) \\ \qquad + k^2\mathrm{ch}\,bz(b^+ + b^-), \\ \dfrac{1}{\mu_1}\tau_p = 2\dfrac{\mu}{\mu_1}\Omega\mathrm{ch}\,az(a^+ + a^-) + 2\dfrac{\mu}{\mu_1}\Omega\mathrm{sh}\,az(a^+ - a^-) \\ \qquad + 2\dfrac{\mu}{\mu_1}kb\mathrm{ch}\,bz(b^+ - b^-) + 2\dfrac{\mu}{\mu_1}kb\mathrm{sh}\,bz(b^+ + b^-), \\ \dfrac{1}{\mu_1}\tau_B = 2\dfrac{\mu}{\mu_1}ka\mathrm{sh}\,az(a^+ + a^-) + 2\dfrac{\mu}{\mu_1}ka\mathrm{ch}\,az(a^+ - a^-) \\ \qquad + 2\dfrac{\mu}{\mu_1}\Omega\mathrm{sh}\,bz(b^+ - b^-) + 2\dfrac{\mu}{\mu_1}\Omega\mathrm{ch}\,bz(b^+ + b^-), \\ kU_c = kk_\beta\mathrm{ch}\,bz(c^+ + c^-) + kk_\beta\mathrm{sh}\,bz(c^+ - c^-), \\ \dfrac{1}{\mu_1}\tau_c = \dfrac{\mu}{\mu_1}k_\beta b\mathrm{sh}\,bz(c^+ + c^-) + \dfrac{\mu}{\mu_1}k_\beta b\mathrm{ch}\,bz(c^+ - c^-). \end{cases} \tag{25}$$

上式给出了第 i 层中位移的谱和 z 平面上应力的谱跟层中波的振幅系数之间的线性变换. 为简明起见, 以 $\mathbf{H}$ 代表位移的谱和 z 平面上应力的谱构成的列向量. $\mathbf{H}$ 的转置 $\mathbf{H}^\mathrm{T}$ 是:

$$\mathbf{H}^\mathrm{T} = \left[kU_B,\ kU_p,\ \frac{1}{\mu_1}\tau_p,\ \frac{1}{\mu_1}\tau_B,\ kU_c,\ \frac{1}{\mu_1}\tau_c\right]. \tag{26}$$

以 $\mathbf{K}$ 代表波的振幅系数构成的列向量:

$$\mathbf{K}^\mathrm{T} = [a^+ + a^-,\ a^+ - a^-,\ b^+ - b^-,\ b^+ + b^-,\ c^+ + c^-,\ c^+ - c^-]. \tag{27}$$

那么,

$$\mathbf{H} = \mathbf{D}(z)\mathbf{K}. \tag{28}$$

其中,

$$\mathbf{D}(z) = \begin{bmatrix} k^2\mathrm{ch}\,az & k^2\mathrm{sh}\,az & kb\mathrm{ch}\,bz & kb\mathrm{sh}\,bz & 0 & 0 \\ ka\mathrm{sh}\,az & ka\mathrm{ch}\,az & k^2\mathrm{sh}\,bz & k^2\mathrm{ch}\,bz & 0 & 0 \\ 2\dfrac{\mu}{\mu_1}\Omega\mathrm{ch}\,az & 2\dfrac{\mu}{\mu_1}\Omega\mathrm{sh}\,az & 2\dfrac{\mu}{\mu_1}kb\mathrm{ch}\,bz & 2\dfrac{\mu}{\mu_1}kb\mathrm{sh}\,bz & 0 & 0 \\ 2\dfrac{\mu}{\mu_1}ka\mathrm{sh}\,az & 2\dfrac{\mu}{\mu_1}ka\mathrm{ch}\,az & 2\dfrac{\mu}{\mu_1}\Omega\mathrm{sh}\,bz & 2\dfrac{\mu}{\mu_1}\Omega\mathrm{ch}\,bz & 0 & 0 \\ 0 & 0 & 0 & 0 & kk_\beta\mathrm{ch}\,bz & kk_\beta\mathrm{sh}\,bz \\ 0 & 0 & 0 & 0 & \dfrac{\mu}{\mu_1}k_\beta b\mathrm{sh}\,bz & \dfrac{\mu}{\mu_1}k_\beta b\mathrm{ch}\,bz \end{bmatrix}. \tag{29}$$

从上面的结果中可以看到，与向量 $\boldsymbol{B}$，$\boldsymbol{P}$ 相联系的振动和与向量 $\boldsymbol{C}$ 相联系的振动是不耦合的，可以分开讨论．事实上，对于体波来说，与 $\boldsymbol{B}$，$\boldsymbol{P}$ 相联系的波即 P 波和 SV 波，与 $\boldsymbol{C}$ 相联系的波即 SH 波．对于面波来说，前者即瑞雷（Rayleigh）面波，后者即勒夫（Love）面波．

如果把坐标原点置于第 i 层的顶面上，那么令（28）式中的 $z=0$，便得到：

$$\mathbf{H}_{i-1} = \mathbf{E}_i \mathbf{K}_i, \tag{30}$$

其中，$\mathbf{H}_{i-1}$ 表示在第 i 层顶面上（即第 $i-1$ 个分界面上）位移和应力的谱构成的列向量，$\mathbf{K}_i$ 代表第 i 层中的波的振幅系数，而

$$\mathbf{E} = \mathbf{D}(0) = \begin{bmatrix} k^2 & 0 & kb & 0 & 0 & 0 \\ 0 & ka & 0 & k^2 & 0 & 0 \\ 2\dfrac{\mu}{\mu_1}\Omega & 0 & 2\dfrac{\mu}{\mu_1}kb & 0 & 0 & 0 \\ 0 & 2\dfrac{\mu}{\mu_1}ka & 0 & 2\dfrac{\mu}{\mu_1}\Omega & 0 & 0 \\ 0 & 0 & 0 & 0 & kk_\beta & 0 \\ 0 & 0 & 0 & 0 & 0 & \dfrac{\mu}{\mu_1}k_\beta b \end{bmatrix}. \tag{31}$$

令 $z = d_i$，可以得到

$$\mathbf{H}_i = \mathbf{D}_i \mathbf{K}_i, \tag{32}$$

其中，

$$\mathbf{D}_i = \mathbf{D}_i(d_i). \tag{33}$$

由（30），（32）式得：

$$\mathbf{H}_i = \mathbf{a}_i \mathbf{H}_{i-1}, \tag{34}$$

其中，

$$\mathbf{a}_i = \mathbf{D}_i \mathbf{E}_i^{-1}, \tag{35}$$

$\mathbf{E}^{-1}$ 是 $\mathbf{E}$ 的逆矩阵，

$$\mathbf{E}^{-1} = \begin{bmatrix} \dfrac{2}{k_\beta^2} & 0 & -\dfrac{\mu_1}{\mu k_\beta^2} & 0 & 0 & 0 \\ 0 & -\dfrac{2\Omega}{kak_\beta^2} & 0 & \dfrac{\mu_1 k}{\mu a k_\beta^2} & 0 & 0 \\ -\dfrac{2\Omega}{kbk_\beta^2} & 0 & \dfrac{\mu_1 k}{\mu b k_\beta^2} & 0 & 0 & 0 \\ 0 & \dfrac{2}{k_\beta^2} & 0 & -\dfrac{\mu_1}{\mu k_\beta^2} & 0 & 0 \\ 0 & 0 & 0 & 0 & \dfrac{1}{kk_\beta} & 0 \\ 0 & 0 & 0 & 0 & 0 & \dfrac{\mu_1}{\mu b k_\beta} \end{bmatrix}, \tag{36}$$

$$\boldsymbol{a} = \begin{bmatrix} \frac{2\,k^2}{k_\beta^2}C_\alpha - \frac{2\Omega}{k_\beta^2}C_\beta & -\frac{2\Omega k}{ak_\beta^2}S_\alpha + \frac{2kb}{k_\beta^2}S_\beta & -\frac{\mu_1}{\mu}\frac{k^2}{k_\beta^2}C_\alpha + \frac{\mu_1}{\mu}\frac{k^2}{k_\beta^2}C_\beta & \frac{\mu_1}{\mu}\frac{k^3}{ak_\beta^2}S_\alpha - \frac{\mu_1}{\mu}\frac{kb}{k_\beta^2}S_\beta & 0 & 0 \\ \frac{2ka}{k_\beta^2}S_\alpha - \frac{2\Omega k}{b\,k_\beta^2}S_\beta & -\frac{2\Omega}{k_\beta^2}C_\alpha + \frac{2\,k^2}{k_\beta^2}C_\beta & -\frac{\mu_1}{\mu}\frac{ka}{k_\beta^2}S_\alpha + \frac{\mu_1}{\mu}\frac{k^3}{bk_\beta^2}S_\beta & \frac{\mu_1}{\mu}\frac{k^2}{k_\beta^2}C_\alpha - \frac{\mu_1}{\mu}\frac{k^2}{k_\beta^2}C_\beta & 0 & 0 \\ \frac{4\mu\Omega}{\mu_1}\frac{}{k_\beta^2}C_\alpha - \frac{4\mu\Omega}{\mu_1\,k_\beta^2}C_\beta & -\frac{4\mu}{\mu_1}\frac{\Omega^2}{kak_\beta^2}S_\alpha + \frac{4\mu kb}{\mu_1 k_\beta^2}S_\beta & -\frac{2\Omega}{k_\beta^2}C_\alpha + \frac{2\,k^2}{k_\beta^2}C_\beta & \frac{2\Omega k}{ak_\beta^2}S_\alpha - \frac{2kb}{k_\beta^2}S_\beta & 0 & 0 \\ \frac{4\mu ka}{\mu_1 k_\beta^2}S_\alpha - \frac{4\,\mu\Omega^2}{\mu_1 bk_\beta^2}S_\beta & -\frac{4\mu\Omega}{\mu_1\,k_\beta^2}C_\alpha + \frac{4\mu\Omega}{\mu_1 k_\beta^2}C_\beta & -\frac{2ka}{k_\beta^2}S_\alpha + \frac{2\Omega k}{b\,k_\beta^2}S_\beta & \frac{2\,k^2}{k_\beta^2}C_\alpha - \frac{2\Omega}{k_\beta^2}C_\beta & 0 & 0 \\ 0 & 0 & 0 & 0 & C_\beta & \frac{\mu_1 k}{\mu b}S_\beta \\ 0 & 0 & 0 & 0 & \frac{\mu b}{\mu_1 k}S_\beta & C_\beta \end{bmatrix},$$

(37)

$$\begin{cases} C_\alpha = \text{ch}\ ad, \\ C_\beta = \text{ch}\ bd, \\ S_\alpha = \text{sh}\ ad, \\ S_\beta = \text{sh}\ bd. \end{cases} \tag{38}$$

(34) 式表示第 i 层底面位移和应力的谱构成的列向量跟第 i 层顶面位移和应力的谱构成的列向量之间的线性变换. 因为 $\mathbf{H}_i$ 和 $\mathbf{H}_{i-1}$ 的分量的量纲一样, 所以矩阵的各个元素是无量纲量. 因为在矩阵 $\boldsymbol{a}_i$ 的各个元素中, S_α 总是以 $a^{\pm 1}S_\alpha$ 的形式出现, S_β 总是以 $b^{\pm 1}S_\beta$ 的形式出现, 所以不拘 a, b 是实数还是虚数, $\boldsymbol{a}_i$ 的各个元素总是实数. (37) 式定义的层矩阵和哈斯克尔[1-2]定义的层矩阵比较, 具有元素均为实数及无量纲的优点. 这样定义的层矩阵便于进一步理论分析和数值计算, 但并无本质的差别.

根据 (30) 式,

$$\mathbf{K}_n = \mathbf{E}_n^{-1}\mathbf{H}_{n-1}, \tag{39}$$

运用递推关系 (34) 式, 可得

$$\mathbf{K}_n = \mathbf{E}_n^{-1}\boldsymbol{a}_{n-1}\boldsymbol{a}_{n-2}\cdots \boldsymbol{a}_2\boldsymbol{a}_1\mathbf{H}_0, \tag{40}$$

式中, $\mathbf{K}_n$ 是第 n 层 (半无限空间) 中的波的振幅系数列向量, 其转置 $\mathbf{K}_n^\text{T}$ 是:

$$\mathbf{K}_n^\text{T} = [a_n^+ + a_n^-,\ a_n^+ - a_n^-,\ b_n^+ - b_n^-,\ b_n^+ + b_n^-,\ c_n^+ + c_n^-,\ c_n^+ - c_n^-], \tag{41}$$

$\mathbf{H}_0$ 是在自由表面上的位移和应力的谱构成的列向量, 其转置 $\mathbf{H}_0^\text{T}$ 是:

$$\mathbf{H}_0^\text{T} = \left[kU_B(0),\ kU_p(0),\ \frac{1}{\mu_1}\tau_p(0),\ \frac{1}{\mu_1}\tau_B(0),\ kU_c(0),\ \frac{1}{\mu_1}\tau_c(0)\right]. \tag{42}$$

令

$$\mathbf{J} = \mathbf{E}_n^{-1}\mathbf{A}, \tag{43}$$

$$\mathbf{A} = \boldsymbol{a}_{n-1}\boldsymbol{a}_{n-2}\cdots \boldsymbol{a}_2\boldsymbol{a}_1, \tag{44}$$

并考虑到在自由表面上应力为零, 就可得到

$$\begin{cases} a_n^+ + a_n^- = J_{11}kU_B(0) + J_{12}kU_p(0), \\ a_n^+ - a_n^- = J_{21}kU_B(0) + J_{22}kU_p(0), \\ b_n^+ - b_n^- = J_{31}kU_B(0) + J_{32}kU_p(0), \\ b_n^+ + b_n^- = J_{41}kU_B(0) + J_{42}kU_p(0), \\ c_n^+ + c_n^- = J_{55}kU_c(0), \\ c_n^+ - c_n^- = J_{65}kU_c(0), \end{cases} \tag{45}$$

式中，J_{jk}表示矩阵 **J** 的元素．消去a_n^-，b_n^-，c_n^-，可得：

$$\begin{cases} kU_B(0) = \dfrac{2}{F_R(k)} \left[(J_{32}+J_{42}) a_n^+ - (J_{12}+J_{22}) b_n^+ \right], \\ kU_P(0) = \dfrac{2}{F_R(k)} \left[(J_{11}+J_{21}) b_n^+ - (J_{31}+J_{41}) a_n^+ \right], \\ kU_c(0) = \dfrac{2}{F_L(k)} c_n^+, \end{cases} \quad (46)$$

式中，

$$F_R(k) = (J_{11}+J_{21})(J_{32}+J_{42}) - (J_{12}+J_{22})(J_{31}+J_{41}), \quad (47)$$

$$F_L(k) = J_{55}+J_{65}. \quad (48)$$

公式（46）表示在自由表面位移的谱和第 n 层（半空间）中波的振幅系数a_n^+，b_n^+，c_n^+之间的关系．对于体波的传播问题而言，a_n^+代表入射 P 波，b_n^+代表入射 SV 波，c_n^+代表入射 SH 波．

2.3 多层弹性半空间对平面体波的响应谱

以上结果对波数 k 并无限制，不拘 k 是大于k_α，k_β还是小于它们，都是成立的．所以，这个结果无论对于体波还是面波都成立，视 k 的值而定．

若$k<k_{\alpha_n}<k_{\beta_n}$，公式（46）的意义是入射 P 波、SV 波和 SH 波在自由表面产生的位移谱．若$k_{\alpha_n}<k<k_{\beta_n}$，那么$a_n^+=0$，公式（46）的意义是入射 SV 波和 SH 波在自由表面产生的位移谱．若$k>k_{\beta_n}>k_{\alpha_n}$，那么$a_n^+ = b_n^+ = c_n^+ = 0$，若$kU_B(0)$，$kU_P(0)$，$kU_c(0)$ 不等于零，那就要求

$$F_R(k) = 0, \quad (49)$$

$$F_L(k) = 0, \quad (50)$$

这就是面波的频散方程．公式（49）是瑞雷波频散方程，它的根用k_R表示；而公式（50）是勒夫波频散方程，它的根用k_L表示．

由公式（36）可见，当$k>k_{\beta_n}>k_{\alpha_n}$时，$\mathbf{E}_n^{-1}$的各个元素总是实数，因此，由（43）式所决定的 **J** 的各个元素也总是实数．这样，$F_R(k)$ 和 $F_L(k)$ 当$k>k_{\beta_n}>k_{\alpha_n}$时是 k 的实函数．换言之，在讨论多层弹性半空间中面波的传播问题时，矩阵 **J** 的各个元素总是实数．这里，又一次见到用（36）和（37）式定义矩阵$\mathbf{E}^{-1}$和 **a** 的优点．

当只有 P 波入射时，$a_n^+ \neq 0$，$b_n^+ = c_n^+ = 0$，此时，在第 n 层顶面上的位移谱

$$\begin{cases} [kU_B(z_{n-1})]_P = k^2 a_n^+, \\ [kU_P(z_{n-1})]_P = k\, a_n a_n^+, \\ [kU_c(z_{n-1})]_P = 0, \end{cases} \quad (51)$$

所以，

$$\begin{cases} \left[\dfrac{kU_B(0)}{kU_B(z_{n-1})}\right]_P = \dfrac{2}{k^2 F_R(k)}(J_{32}+J_{42}), \\ \left[\dfrac{kU_P(0)}{kU_P(z_{n-1})}\right]_P = -\dfrac{2}{k\, a_n F_R(k)}(J_{31}+J_{41}), \\ [kU_c(0)]_P = 0. \end{cases} \quad (52)$$

当只有 SV 波入射时，$b_n^+ \neq 0$，$a_n^+ = c_n^+ = 0$，此时，在第 n 层顶面上的位移谱

$$\begin{cases} [kU_B(z_{n-1})]_{SV} = k\, b_n b_n^+, \\ [kU_P(z_{n-1})]_{SV} = k^2 b_n^+, \\ [kU_c(z_{n-1})]_{SV} = 0, \end{cases} \quad (53)$$

所以，

$$\begin{cases} \left[\dfrac{kU_B(0)}{kU_B(z_{n-1})}\right]_{\mathrm{SV}} = -\dfrac{2}{k\,b_n F_R(k)}(J_{12}+J_{22}), \\ \left[\dfrac{kU_p(0)}{kU_p(z_{n-1})}\right]_{\mathrm{SV}} = \dfrac{2}{k^2 F_R(k)}(J_{11}+J_{21}), \\ [kU_c(0)]_{\mathrm{SV}} = 0. \end{cases} \quad (54)$$

当只有 SH 波入射时，$c_n^+ \neq 0$，$a_n^+ = b_n^+ = 0$，此时，在第 n 层顶面上的位移谱

$$\begin{cases} [kU_B(z_{n-1})]_{\mathrm{SH}} = 0, \\ [kU_p(z_{n-1})]_{\mathrm{SH}} = 0, \\ [kU_c(z_{n-1})]_{\mathrm{SH}} = kk_{\beta_n} c_n^+, \end{cases} \quad (55)$$

所以，

$$\begin{cases} [kU_B(0)]_{\mathrm{SH}} = 0, \\ [kU_p(0)]_{\mathrm{SH}} = 0, \\ \left[\dfrac{kU_c(0)}{kU_c(z_{n-1})}\right]_{\mathrm{SH}} = \dfrac{2}{kk_{\beta_n} F_L(k)}. \end{cases} \quad (56)$$

由 (11) 式可知 (52)，(54) 和 (56) 式的意义. $\left[\dfrac{kU_B(0)}{kU_B(z_{n-1})}\right]_P$ 是只有 P 波入射时，自由表面的水平位移谱和入射 P 波在第 n 层顶面的水平位移谱之比，$\left[\dfrac{kU_p(0)}{kU_p(z_{n-1})}\right]_P$ 是垂直位移谱之比. $\left[\dfrac{kU_B(0)}{kU_B(z_{n-1})}\right]_{\mathrm{SV}}$ 是只有 SV 波入射时，自由表面的水平位移谱和入射 SV 波在第 n 层顶面的水平位移谱之比，$\left[\dfrac{kU_p(0)}{kU_p(z_{n-1})}\right]_{\mathrm{SV}}$ 是垂直位移谱之比. $\left[\dfrac{kU_c(0)}{kU_c(z_{n-1})}\right]_{\mathrm{SH}}$ 是只有 SH 波入射时，自由表面的 SH 波位移谱和入射 SH 波在第 n 层顶面的位移谱之比.

若以 $U_x(0)$，$U_y(0)$，$U_z(0)$ 表示在自由表面上沿 x，y，z 三个方向的位移谱，$U_x(z_{n-1})$，$U_y(z_{n-1})$，$U_z(z_{n-1})$ 表示入射波在第 n 层顶面上沿 x，y，z 三个方向的位移谱，那么：

$$\begin{cases} \left[\dfrac{U_x(0)}{U_x(z_{n-1})}\right]_P = \dfrac{2}{k^2 F_R(k)}(J_{32}+J_{42}), \\ \left[\dfrac{U_z(0)}{U_z(z_{n-1})}\right]_P = -\dfrac{2}{k\,a_n F_R(k)}(J_{31}+J_{41}), \\ [U_y(0)]_P = 0; \end{cases} \quad (57)$$

$$\begin{cases} \left[\dfrac{U_x(0)}{U_x(z_{n-1})}\right]_{\mathrm{SV}} = -\dfrac{2}{k\,b_n F_R(k)}(J_{12}+J_{22}), \\ \left[\dfrac{U_z(0)}{U_z(z_{n-1})}\right]_{\mathrm{SV}} = \dfrac{2}{k^2 F_R(k)}(J_{11}+J_{21}), \\ [U_y(0)]_{\mathrm{SV}} = 0; \end{cases} \quad (58)$$

$$\begin{cases} [U_x(0)]_{\mathrm{SH}} = 0, \\ [U_z(0)]_{\mathrm{SH}} = 0, \\ \left[\dfrac{U_y(0)}{U_y(z_{n-1})}\right]_{\mathrm{SH}} = \dfrac{2}{kk_{\beta_n} F_L(k)}. \end{cases} \quad (59)$$

有时，把自由表面上的位移谱和入射波的总位移的谱作比较更为方便．不难求得，入射波 P 波的总位移的谱 $[U_t(z_{n-1})]_p$ 等于 $\dfrac{k_{a_n}}{k}[U_x(z_{n-1})]_p$ 或 $\dfrac{k_{a_n}}{ia_n}[U_z(z_{n-1})]_p$，所以

$$\begin{cases} \left[\dfrac{U_x(0)}{U_t(z_{n-1})}\right]_p = \dfrac{2}{k\,k_{a_n}F_R(k)}(J_{32}+J_{42}), \\ \left[\dfrac{U_z(0)}{U_t(z_{n-1})}\right]_p = -\dfrac{2i}{k\,k_{a_n}F_R(k)}(J_{31}+J_{41}), \\ [U_y(0)]_p = 0. \end{cases} \tag{60}$$

入射 SV 波的总位移的谱 $[U_t(z_{n-1})]_{SV}$ 等于 $\dfrac{k_{\beta_n}}{ib_n}[U_x(z_{n-1})]_{SV}$ 或 $-\dfrac{k_{\beta_n}}{k}[U_z(z_{n-1})]_{SV}$，所以

$$\begin{cases} \left[\dfrac{U_x(0)}{U_t(z_{n-1})}\right]_{SV} = -\dfrac{2i}{k\,k_{\beta_n}F_R(k)}(J_{12}+J_{22}), \\ \left[\dfrac{U_z(0)}{U_t(z_{n-1})}\right]_{SV} = -\dfrac{2}{k\,k_{\beta_n}F_R(k)}(J_{11}+J_{21}), \\ [U_y(0)]_{SV} = 0. \end{cases} \tag{61}$$

入射 SH 波的总位移的谱 $[U_t(z_{n-1})]_{SH}$ 就是 $[U_y(z_{n-1})]_{SH}$，所以，对于 SH 波入射的情形，结果仍如 (59) 式．

2.4　多层弹性半空间中任意深度处的位移和应力的谱

知道了 $\mathbf{H}_0$，便可求得在半空间中任意深度处的一点的位置和应力的谱．若以 $\mathbf{H}_i(D_i)$ 表示在第 i 层中距离其顶面 h_i 的平面上位移和应力的谱构成的列向量，$D_i = z_{i-1}+h_i$，那么

$$\mathbf{H}_i(D_i) = \mathbf{a}_i(h_i)\mathbf{a}_{i-1}\mathbf{a}_{i-2}\cdots\mathbf{a}_2\mathbf{a}_1\mathbf{H}_0, \tag{62}$$

令

$$\mathbf{M}' = \mathbf{a}_i(h_i)\mathbf{a}_{i-1}\mathbf{a}_{i-2}\cdots\mathbf{a}_2\mathbf{a}_1, \tag{63}$$

则

$$\mathbf{H}_i(D_i) = \mathbf{M}'\mathbf{H}_0. \tag{64}$$

把上式写成分量形式：

$$\begin{cases} kU_B(D_i) = M'_{11}kU_B(0) + M'_{12}kU_p(0), \\ kU_p(D_i) = M'_{21}kU_B(0) + M'_{22}kU_p(0), \\ \dfrac{1}{\mu_1}\tau_p(D_i) = M'_{31}kU_B(0) + M'_{32}kU_p(0), \\ \dfrac{1}{\mu_1}\tau_B(D_i) = M'_{41}kU_B(0) + M'_{42}kU_p(0), \\ kU_c(D_i) = M'_{55}kU_c(0), \\ \dfrac{1}{\mu_1}\tau_c(D_i) = M'_{65}kU_c(0), \end{cases} \tag{65}$$

将 (46) 式中的 $kU_B(0)$，$kU_p(0)$，$kU_c(0)$ 代入上式后得：

$$\begin{cases} kU_B(D_i) = \dfrac{2}{F_R(k)} \{ [M'_{11}(J_{32}+J_{42}) - M'_{12}(J_{31}+J_{41})] a_n^+ \\ \qquad\qquad - [M'_{11}(J_{12}+J_{22}) - M'_{12}(J_{11}+J_{21})] b_n^+ \}, \\ kU_p(D_i) = \dfrac{2}{F_R(k)} \{ [M'_{21}(J_{32}+J_{42}) - M'_{22}(J_{31}+J_{41})] a_n^+ \\ \qquad\qquad - [M'_{21}(J_{12}+J_{22}) - M'_{22}(J_{11}+J_{21})] b_n^+ \}, \\ kU_c(D_i) = \dfrac{2}{F_L(k)} M'_{55} c_n^+. \end{cases} \quad (66)$$

分别 P 波、SV 波和 SH 波入射三种情形，可得：

P 波入射时，

$$\begin{cases} \left[\dfrac{kU_B(D_i)}{kU_B(0)}\right]_P = M'_{11} - M'_{12}\dfrac{J_{31}+J_{41}}{J_{32}+J_{42}}, \\ \left[\dfrac{kU_p(D_i)}{kU_p(0)}\right]_P = -M'_{21}\dfrac{J_{32}+J_{42}}{J_{31}+J_{41}} + M'_{22}, \\ \left[\dfrac{\frac{1}{\mu_1}\tau_p(D_i)}{kU_B(0)}\right]_P = M'_{31} - M'_{32}\dfrac{J_{31}+J_{41}}{J_{32}+J_{42}}, \\ \left[\dfrac{\frac{1}{\mu_1}\tau_B(D_i)}{kU_p(0)}\right]_P = -M'_{41}\dfrac{J_{32}+J_{42}}{J_{31}+J_{41}} + M'_{42}, \\ [kU_c(D_i)]_P = 0, \\ \left[\dfrac{1}{\mu_1}\tau_c(D_i)\right]_P = 0. \end{cases} \quad (67)$$

SV 波入射时，

$$\begin{cases} \left[\dfrac{kU_B(D_i)}{kU_B(0)}\right]_{SV} = M'_{11} - M'_{12}\dfrac{J_{11}+J_{21}}{J_{12}+J_{22}}, \\ \left[\dfrac{kU_p(D_i)}{kU_p(0)}\right]_{SV} = -M'_{21}\dfrac{J_{12}+J_{22}}{J_{11}+J_{21}} + M'_{22}, \\ \left[\dfrac{\frac{1}{\mu_1}\tau_p(D_i)}{kU_B(0)}\right]_{SV} = M'_{31} - M'_{32}\dfrac{J_{11}+J_{21}}{J_{12}+J_{22}}, \\ \left[\dfrac{\frac{1}{\mu_1}\tau_B(D_i)}{kU_p(0)}\right]_{SV} = -M'_{41}\dfrac{J_{12}+J_{22}}{J_{11}+J_{21}} + M'_{42}, \\ [kU_c(D_i)]_{SV} = 0, \\ \left[\dfrac{1}{\mu_1}\tau_c(D_i)\right]_{SV} = 0. \end{cases} \quad (68)$$

SH 波入射时，

$$\begin{cases} [kU_B(D_i)]_{SH} = 0, \\ [kU_p(D_i)]_{SH} = 0, \\ \left[\dfrac{1}{\mu_1}\tau_p(D_i)\right]_{SH} = 0, \\ \left[\dfrac{1}{\mu_1}\tau_B(D_i)\right]_{SH} = 0, \\ \left[\dfrac{kU_c(D_i)}{kU_c(0)}\right]_{SH} = M'_{55}, \\ \left[\dfrac{\dfrac{1}{\mu_1}\tau_c(D_i)}{kU_c(0)}\right]_{SH} = M'_{65}. \end{cases} \qquad (69)$$

2.5 多层弹性半空间不同深度处的面波的相对激发程度

对于面波的情形，根据（47）-（50）式，可以把瑞雷波频散方程写成：

$$\frac{J_{31}+J_{41}}{J_{32}+J_{42}} = \frac{J_{11}+J_{21}}{J_{12}+J_{22}}, \qquad (70)$$

把勒夫波频散方程写成：

$$J_{55} = -J_{65}, \qquad (71)$$

从而，由（46）式得

$$\frac{k_R U_p(0)}{k_R U_B(0)} = -K, \qquad (72)$$

其中，

$$K = \frac{J_{11}+J_{21}}{J_{12}+J_{22}}. \qquad (73)$$

考虑到（11）式，可得

$$\frac{U_z(0)}{U_x(0)} = -iK. \qquad (74)$$

上式表明，在多层弹性半空间的自由表面，瑞雷波的质点运动不一定是逆进椭圆. 只有当 $K<0$ 时才是逆进椭圆. 如果 $K>0$，则为前进椭圆.

以 $U_x(D_i)$，$U_y(D_i)$，$U_z(D_i)$ 表示在 $z=D_i$ 平面上面波位移谱的三个分量，以 $\tau_x(D_i)$，$\tau_y(D_i)$，$\tau_z(D_i)$ 表示在 $z=D_i$ 平面上面波引起的应力的谱的三个分量，那么由（65）和（72）式可以求得：

$$\begin{cases} \chi'_{R_1} = \dfrac{U_x(D_i)}{U_x(0)} = \dfrac{k_R U_B(D_i)}{k_R U_B(0)} = M'_{11} - K M'_{12}, \\ \chi'_{R_2} = \dfrac{U_z(D_i)}{U_z(0)} = \dfrac{k_R U_p(D_i)}{k_R U_p(0)} = -\dfrac{M'_{21}}{K} + M'_{22}, \\ \chi'_{R_3} = \dfrac{\dfrac{1}{\mu_1}\tau_z(D_i)}{ik_R U_x(0)} = \dfrac{\dfrac{1}{\mu_1}\tau_p(D_i)}{k_R U_B(0)} = M'_{31} - kM'_{32}, \\ \chi'_{R_4} = \dfrac{\dfrac{i'}{\mu_1}\tau_x(D_i)}{k_R U_z(0)} = \dfrac{\dfrac{1}{\mu_1}\tau_B(D_i)}{k_R U_p(0)} = -\dfrac{M'_{41}}{K} + M'_{42}, \\ \chi'_{L_1} = \dfrac{U_y(D_i)}{U_y(0)} = \dfrac{k_L U_c(D_i)}{k_L U_c(0)} = M'_{55}, \\ \chi'_{L_2} = \dfrac{\dfrac{1}{\mu_1}\tau_y(D_i)}{k_L U_y(0)} = \dfrac{\dfrac{1}{\mu_1}\tau_c(D_i)}{k_L U_c(0)} = M'_{65}. \end{cases} \qquad (75)$$

上式表示在第 i 层中距离该层顶面为 h_i 的任一点其面波的位移和应力谱跟自由表面上的位移谱之间的简单关系. χ'_{R_1} 和 χ'_{R_2} 即不同深度处的瑞雷波的相对激发程度, χ'_{L_1} 即不同深度处的勒夫波的相对激发程度. 由 (75) 式还可求得:

$$\frac{U_z(D_i)}{U_x(D_i)} = -i\,K', \tag{76}$$

$$K' = K\frac{\chi'_{R_2}}{\chi'_{R_1}}. \tag{77}$$

和 (74) 式相似, 在任意深度处, 瑞雷波的质点运动是逆进椭圆偏振还是前进椭圆偏振由 $K'<0$ 还是 $K'>0$ 决定.

3 多层弹性半空间中一般类型的震源辐射的地震波 (形式解)

3.1 运动方程在圆柱坐标下的一般解

上面讨论了多层弹性半空间中平面波的传播问题. 引进基向量 **B**, **P**, **C** 代替直角坐标的基向量 e_x, e_y, e_z, 把各层中的位移谱和应力谱通过基向量 **B**, **P**, **C** 表示出来. 通过位移谱和应力谱构成的列向量跟层中波的振幅系数构成的列向量之间的线性变换引进矩阵 **D**(z), **E**, **E**$^{-1}$, **a** 和 **J** 等, 这样引进的矩阵和哈斯克尔 1964 年引进的矩阵相比, 固然没有本质的不同, 但具有量纲整齐、便于理论分析和数值计算的优点. 下面将在这个基础上, 进一步研究多层弹性半空间中一般类型的震源辐射的地震波的传播问题. 通过下面的分析可以看到, 前面分析平面波传播问题时导出的许多结果, 如多层弹性半空间对平面体波的响应谱, 面波的频散方程, 面波在不同深度处的相对激发程度, 瑞雷波的质点运动情况, 等等结果可以方便地移用过来.

先处理点状震源辐射的地震波的传播问题, 然后再推广到非点源情形. 由于问题的性质, 取圆柱坐标 (r, φ, z) 比较方便, z 轴通过点源、垂直于地面, 朝下为正 (图 2).

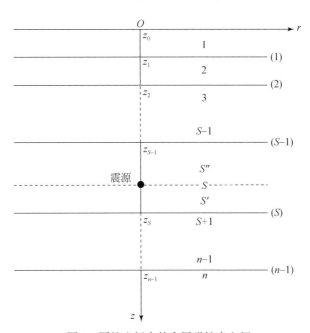

图 2 圆柱坐标中的多层弹性半空间

在研究一般类型震源辐射的地震波的传播问题时,用妹泽克惟的表示式及其另一种形式并不太方便.这时,用汉森向量[4]表示运动方程的一般解比较方便[5].因此,这里将采用汉森向量表示运动方程的一般解.

在圆柱坐标中,方程(4)的基本解是

$$\begin{cases} \varphi = J_m(kr) \begin{array}{c} e^{im\varphi} \\ \frac{1}{i} e^{im\varphi} \end{array} e^{\pm az}, \\ \psi = J_m(kr) \begin{array}{c} e^{im\varphi} \\ \frac{1}{i} e^{im\varphi} \end{array} e^{\pm bz}, \\ \chi = J_m(kr) \begin{array}{c} e^{im\varphi} \\ \frac{1}{i} e^{im\varphi} \end{array} e^{\pm bz}. \end{cases} \quad (78)$$

其中,J_m 是整数阶贝塞尔函数,m 是阶数.因此,

$$\begin{cases} \nabla\varphi = \left[\frac{\partial}{\partial r}J_m(kr)\boldsymbol{e}_r + \frac{im}{r}J_m(kr)\boldsymbol{e}_\varphi \pm a J_m(kr)\boldsymbol{e}_z\right] \begin{array}{c} e^{im\varphi} \\ \frac{1}{i} e^{im\varphi} \end{array} e^{\pm az}, \\ \nabla\times\nabla\times(\psi\boldsymbol{e}_z) = \left[\pm b\frac{\partial}{\partial r}J_m(kr)\boldsymbol{e}_r \pm \frac{imb}{r}J_m(kr)\boldsymbol{e}_\varphi \pm k^2 J_m(kr)\boldsymbol{e}_z\right] \begin{array}{c} e^{im\varphi} \\ \frac{1}{i} e^{im\varphi} \end{array} e^{\pm bz}, \\ \nabla\times(\chi\boldsymbol{e}_z) = \left[\frac{im}{r}J_m(kr)\boldsymbol{e}_r - \frac{\partial}{\partial r}J_m(kr)\boldsymbol{e}_\varphi\right] \begin{array}{c} e^{im\varphi} \\ \frac{1}{i} e^{im\varphi} \end{array} e^{\pm bz}. \end{cases} \quad (79)$$

如果定义三个向量 $\boldsymbol{B}_m$,$\boldsymbol{P}_m$,$\boldsymbol{C}_m$:

$$\begin{cases} \boldsymbol{B}_m = \frac{1}{k}\frac{\partial}{\partial r}J_m(kr) e^{im\varphi} \boldsymbol{e}_r + \frac{im}{kr}J_m(kr) e^{im\varphi} \boldsymbol{e}_\varphi, \\ \boldsymbol{P}_m = J_m(kr) e^{im\varphi} \boldsymbol{e}_z, \\ \boldsymbol{C}_m = -\frac{m}{kr}J_m(kr) e^{im\varphi} \boldsymbol{e}_r - \frac{i}{k}\frac{\partial}{\partial r}J_m(kr) e^{im\varphi} \boldsymbol{e}_\varphi, \end{cases} \quad (80)$$

那么

$$\begin{cases} \nabla\varphi = (k\boldsymbol{B}_m \pm a\boldsymbol{P}_m) e^{\pm az}, \\ \nabla\times\nabla\times(\psi\boldsymbol{e}_z) = k(\pm b\boldsymbol{B}_m \pm k\boldsymbol{P}_m) e^{\pm bz}, \\ \nabla\times(\chi\boldsymbol{e}_z) = (-ik\boldsymbol{C}_m) e^{\pm bz}. \end{cases} \quad (81)$$

本·梅纳赫姆和辛格(Ben-Menahem and Singh)[5]曾经运用汉森展开导出过两层半空间中偶极源引起的位移场(形式解).这里的做法和他们的类似,$\boldsymbol{B}_m$,$\boldsymbol{P}_m$ 的定义和他们的定义一样,只是 $\boldsymbol{C}_m$ 的定义和他们的定义不一样,是他们定义的 $\boldsymbol{C}_m$ 乘 i.这虽然没有本质的不同,但是将使得以后导出的结果形式更为整齐,便于理论分析,并且所得的若干结果将与哈斯克尔用不同方法导出的结果形式一致,便于比较.这点,在下面会再提到.

倘若定义三个向量 $\boldsymbol{L}_m$,$\boldsymbol{N}_m$,$\boldsymbol{M}_m$(叫汉森向量):

$$\begin{cases} k_\alpha \boldsymbol{L}_m = \nabla \varphi, \\ k k_\beta \boldsymbol{N}_m = \nabla \times \nabla \times (\psi \boldsymbol{e}_z), \\ k \boldsymbol{M}_m = i \nabla \times (\chi \boldsymbol{e}_z), \end{cases} \tag{82}$$

那么由（81）和（82）式得：

$$\begin{cases} k_\alpha \boldsymbol{L}_m^\pm = (k \boldsymbol{B}_m \pm a \boldsymbol{P}_m) e^{\pm az}, \\ k_\beta \boldsymbol{N}_m^\pm = (\pm b \boldsymbol{B}_m + k \boldsymbol{P}_m) e^{\pm bz}, \\ k_\beta \boldsymbol{M}_m^\pm = (k_\beta \boldsymbol{C}_m) e^{\pm bz}, \end{cases} \tag{83}$$

因此，运动方程的一般解可以表示为

$$\boldsymbol{U} = \sum_m \int_0^\infty \boldsymbol{U}^m k \mathrm{d}k, \tag{84}$$

$$\boldsymbol{U}^m = (a_m^+ k_\alpha \boldsymbol{L}_m^+ + a_m^- k_\alpha \boldsymbol{L}_m^-) + (b_m^+ k_\beta \boldsymbol{N}_m^+ + b_m^- k_\beta \boldsymbol{N}_m^-) + (c_m^+ k_\beta \boldsymbol{M}_m^+ + c_m^- k_\beta \boldsymbol{M}_m^-), \tag{85}$$

或

$$\boldsymbol{U}^m = (k f_{m_1} + f'_{m_2}) \boldsymbol{B}_m + (f'_{m_1} + k f_{m_2}) \boldsymbol{P}_m + k_\beta f_{m_3} \boldsymbol{C}_m, \tag{86}$$

其中，

$$\begin{cases} f_{m_1} = a_m^+ e^{az} + a_m^- e^{-az}, \\ f_{m_2} = b_m^+ e^{bz} + b_m^- e^{-bz}, \\ f_{m_3} = c_m^+ e^{bz} + c_m^- e^{-bz}, \end{cases} \tag{87}$$

$$\begin{cases} f'_{m_1} = \dfrac{d f_{m1}}{dz} = a(a_m^+ e^{az} - a_m^- e^{-az}), \\ f'_{m_2} = \dfrac{d f_{m2}}{dz} = b(b_m^+ e^{bz} - b_m^- e^{-bz}), \\ f'_{m_3} = \dfrac{d f_{m3}}{dz} = b(c_m^+ e^{bz} - c_m^- e^{-bz}). \end{cases} \tag{88}$$

待定系数 $a_m^\pm$，$b_m^\pm$，$c_m^\pm$ 是 k 的函数，正号代表上行波，负号代表下行波．

比较（86）-（88）式和（15）-（17）式可以看到，在圆柱坐标中，如果用基向量 $\boldsymbol{B}_m$，$\boldsymbol{P}_m$，$\boldsymbol{C}_m$ 表示运动方程的解答，则可将解答表示成和二维情况下直角坐标中运动方程的解答相似的形式．

应力张量的谱 $S(\boldsymbol{U})$ 由（18）式决定，它在 z 平面上的应力的谱 $S(\boldsymbol{U}) \cdot \boldsymbol{e}_z$ 可以表示成

$$S(\boldsymbol{U}) \cdot \boldsymbol{e}_z = \sum_m \int_0^\infty S(\boldsymbol{U}^m) \cdot \boldsymbol{e}_z k \mathrm{d}k, \tag{89}$$

$$S(\boldsymbol{U}^m) \cdot \boldsymbol{e}_z = 2\mu(k f'_{m_1} + \Omega f_{m_2}) \boldsymbol{B}_m + 2\mu(\Omega f_{m_1} + k f'_{m_2}) \boldsymbol{P}_m + \mu k_\beta f'_{m_3} \boldsymbol{C}_m, \tag{90}$$

（90）式和（19）式是相似的．由于（86），（90）式和（15），（19）式的相似性，所以讨论平面波传播问题得到的结果可以非常方便地移用于下面的分析中．

3.2 震源的表示

为了表示一般类型的点状震源，我们把点源看作是位移和应力不连续的奇点．具体地说，设 $(0, 0, h)$ 是点源，我们将其位移谱和应力谱表示为

$$\boldsymbol{U}_0 = \sum_m \int_0^\infty \boldsymbol{U}_0^m k \mathrm{d}k, \tag{91}$$

$$S(\boldsymbol{U}_0) \cdot \boldsymbol{e}_z = \sum_m \int_0^\infty S(\boldsymbol{U}_0^m) \cdot \boldsymbol{e}_z k \mathrm{d}k, \tag{92}$$

$$U_0^m = (kX_{m_1} + X'_{m_2})B_m + (X'_{m_1} + kX_{m_2})P_m + k_\beta X_{m_3} C_m, \tag{93}$$

$$S(U_0^m) \cdot e_z = 2\mu(kX'_{m_1} + \Omega X_{m_2})B_m + 2\mu(\Omega X_{m_1} + kX'_{m_2})P_m + \mu k_\beta X'_{m_3} C_m. \tag{94}$$

式中，

$$\begin{cases} X_{m_1} = i_m e^{-a|z-h|}\epsilon^m, \\ X_{m_2} = j_m e^{-b|z-h|}\epsilon^{m+1}, \\ X_{m_3} = k_m e^{-b|z-h|}\epsilon^m, \end{cases} \tag{95}$$

$$\epsilon = 1, \quad z > h, \\ -1, \quad z < h. \tag{96}$$

对于给定的点源，i_m，j_m，k_m 是确定的．

地震酝酿的过程可以看作是震源区应力集中的过程．当应力逐渐集中达到介质的极限强度时，便发生破裂，破裂面两边的介质突然错动（位错），释放积蓄在断层附近介质中的应变能．根据这一分析，可以把地震震源（至少是浅源地震震源）当作位错源．设断层面（位错面）为 Σ，断层面两边 Σ^+ 和 Σ^- 上某一点 $P(\zeta_1, \zeta_2, \zeta_3)$ 的位移谱分别为 U^+ 和 U^-，那么位错（相对位移）的谱为 ΔU：

$$\Delta U = U^+ - U^- = \Delta U_m e_m, \tag{97}$$

这里，采用哑指标下的求和约定．如果位错前和位错后断层面两边的应力相等，那么在均匀、各向同性和完全弹性介质中由于断层面两边相对错动所引起的空间某一点 $Q(x_1, x_2, x_3)$ 的位移场的谱是[6]：

$$U(Q) = U_m(Q) e_m, \tag{98}$$

$$U_m(Q) = \iint_\Sigma \Delta U_k(P) T_{kl}^m(P, Q) n_l(P) d\Sigma, \tag{99}$$

$n_l(P)$ 是包含 P 点的面积元 $d\Sigma$ 的法线的方向余弦，以 Σ^- 指向 Σ^+ 为正方向，而

$$T_{kl}^m = \lambda \delta_{kl} \frac{\partial}{\partial \zeta_n} G_n^m + \mu \left(\frac{\partial}{\partial \zeta_l} G_k^m + \frac{\partial}{\partial \zeta_k} G_l^m \right), \tag{100}$$

G_k^m 是索米亚那（Somigliana）张量 G 的分量：

$$G = G_k^m e_m e_k, \tag{101}$$

$$G = \frac{1}{4\pi\mu k_\beta^2} \left[\nabla \times \nabla \times \left(\frac{\mathbf{I} e^{-ik_\beta R}}{R} \right) - \nabla \nabla \cdot \left(\frac{\mathbf{I} e^{-ik_\alpha R}}{R} \right) \right], \tag{102}$$

$$R = |\overline{PQ}|. \tag{103}$$

它的意义是在 P 点沿 k 方向的单位简谐集中力在 Q 点引起的 m 方向的位移谱．由（99），（100）式以及 G_k^m 的意义可知，$T_{kl}^m(P, Q)$ 的意义是在 P 点作用的、由 (kl) 定义的力系在 Q 点引起的 m 方向的位移谱（差一个量纲因子）．$T_{kl}^m(P, Q)$ 对 $U_m(Q)$ 的贡献是，若 $k = l$，这个力系由膨胀中心加上在 k 方向的无矩偶极构成，膨胀中心的强度是 $\lambda \Delta U_k n_l d\Sigma$，无矩偶极的强度 m_0 是 $2\mu \Delta U_k n_l d\Sigma$．若 $k \neq l$，这个力系由两对有矩偶极构成，它们在 $x_k x_l$ 平面共面，彼此垂直且力偶矩方向相反，每对有矩偶极的偶极矩 m_0 的大小是 $\mu(\Delta U_k n_l + \Delta U_l n_k) d\Sigma$．通常称 $k = l$ 时的 $T_{kl}^m(P, Q)$ 为 A 型核，$k \neq l$ 时的 $T_{kl}^m(P, Q)$ 为 B 型核．由此可见，剪切位错可以用 B 型核表示，张裂位错可以用 A 型核表示．

既然地震震源可以用断层错动描述，而一般类型的位错在均匀、各向同性和完全弹性的介质中所引起的位移谱（差一个量纲因子）由 $T_{kl}^m(P, Q)$ 表示．所以要表示一般类型的位错源，只需把 T_{kl}^m 通过基向量 B_m，P_m，C_m 表示出来．

通过下述步骤用基向量 B_m，P_m，C_m 表示 T_{kl}^m[6]．首先，将 T_{kl}^m 用球贝塞尔函数表示．然后，用球极坐标中的波函数表示 T_{kl}^m．再把球极坐标中的波函数按柱坐标中的波函数展开，把 T_{kl}^m 用柱面波函数表示出来．最

后，根据 T_{kl} 在直角坐标中的分量和在圆柱坐标中的分量的关系，利用贝塞尔函数的递推关系以及基向量 $\boldsymbol{B}_m$, $\boldsymbol{P}_m$, $\boldsymbol{C}_m$ 的定义，就可以得到 T_{kl} 按基向量 $\boldsymbol{B}_m$, $\boldsymbol{P}_m$, $\boldsymbol{C}_m$ 的展开式. 为节省篇幅，这里略去计算的细节，只列出结果. 结果是：

$$\frac{1}{2}T_{11} = \int_0^\infty \left\{ \left[-\frac{1}{ak}((1-2\gamma)k_\beta^2 - k^2)X - \frac{b}{k}Y \right] \boldsymbol{B}_0 + \left[-\frac{k}{a}X + \frac{b}{k}Y \right] \boldsymbol{B}_2 \right.$$
$$+ \epsilon \left[\frac{1}{k^2}((1-2\gamma)k_\beta^2 - k^2)X + Y \right] \boldsymbol{P}_0 + \epsilon(X - Y)\boldsymbol{P}_2$$
$$\left. + \frac{k_\beta^2}{kb}Y \boldsymbol{C}_2 \right\} k\mathrm{d}k, \tag{104}$$

$$\frac{1}{2}T_{22} = \int_0^\infty \left\{ \left[-\frac{1}{ak}((1-2\gamma)k_\beta^2 - k^2)X - \frac{b}{k}Y \right] \boldsymbol{B}_0 - \left[-\frac{k}{a}X + \frac{b}{k}Y \right] \boldsymbol{B}_2 \right.$$
$$+ \epsilon \left[\frac{1}{k^2}((1-2\gamma)k_\beta^2 - k^2)X + Y \right] \boldsymbol{P}_0 - \epsilon(X - Y)\boldsymbol{P}_2$$
$$\left. + \frac{k_\beta^2}{kb}Y \boldsymbol{C}_2 \right\} k\mathrm{d}k, \tag{105}$$

$$\frac{1}{2}T_{33} = \int_0^\infty \left\{ -\frac{2}{ak}(\Omega X - abY)\boldsymbol{B}_0 + \frac{2\epsilon}{k^2}(\Omega X - k^2 Y)\boldsymbol{P}_0 \right\} k\mathrm{d}k, \tag{106}$$

$$\frac{1}{2}T_{23} = \int_0^\infty \left\{ \frac{2i\epsilon}{k^2}(k^2 X - \Omega Y)\boldsymbol{B}_1 - \frac{2i}{kb}(abX - \Omega Y)\boldsymbol{P}_1 - \frac{i\epsilon k_\beta^2}{k^2}Y \boldsymbol{C}_1 \right\} k\mathrm{d}k, \tag{107}$$

$$\frac{1}{2}T_{31} = \int_0^\infty \left\{ -\frac{2\epsilon}{k^2}(k^2 X - \Omega Y)\boldsymbol{B}_1 + \frac{2}{kb}(abX - \Omega Y)\boldsymbol{P}_1 + \frac{\epsilon k_\beta^2}{k^2}Y \boldsymbol{C}_1 \right\} k\mathrm{d}k, \tag{108}$$

$$\frac{1}{2}T_{12} = \int_0^\infty \left\{ \frac{i}{ka}(k^2 X - abY)\boldsymbol{B}_2 + i\epsilon(-X + Y)\boldsymbol{P}_2 - \frac{i k_\beta^2}{kb}Y \boldsymbol{C}_2 \right\} k\mathrm{d}k, \tag{109}$$

式中，

$$X = -\frac{k^2}{8\pi k_\beta^2} e^{-|z-h|a}, \tag{110}$$

$$Y = -\frac{k^2}{8\pi k_\beta^2} e^{-|z-h|b}, \tag{111}$$

$$\gamma = -\frac{\lambda + \mu}{\lambda + 2\mu}. \tag{112}$$

这个结果跟本·梅纳赫姆和辛格[5]的结果形式上略有不同. 由于这里定义的基向量 $\boldsymbol{C}_m$ 为他们的 $i\boldsymbol{C}_m$，所以这里的 $\boldsymbol{C}_m$ 前的系数等于他们的 $\boldsymbol{C}_m$ 前的系数乘 $-i$. 这个差别不是本质的差别，但观察一下 (104)-(109) 诸式可见，基向量 $\boldsymbol{B}_m$, $\boldsymbol{P}_m$, $\boldsymbol{C}_m$ 前面的 i 是整齐的. 下面将看到，这样表示很便于结果的表达和分析.

比较 (93) 式和 (104)-(109) 诸式，就可确定出各种偶极源的系数 i_m, j_m, k_m. 结果见表 1. 表 1 的系数 k_m 是本·梅纳赫姆和辛格的 k_m 乘 $-i$，而 i_m, j_m 相同. 最后一栏是爆炸源，即 (11)+(22)+(33) 的情形.

3.3 多层弹性半空间中一般类型的震源辐射的地震波（形式解）

前面将圆柱坐标中运动方程的一般解通过基向量 $\boldsymbol{B}_m$, $\boldsymbol{P}_m$, $\boldsymbol{C}_m$ 表示出来，并且将一般类型的位错源的位移谱和应力谱通过 $\boldsymbol{B}_m$, $\boldsymbol{P}_m$, $\boldsymbol{C}_m$ 表示出来. 在这个基础上，可以用哈斯克尔矩阵法求多层弹性半空间中一般类型的震源辐射的地震波.

表 1 系数 i_m, j_m, k_m 的数值

	kl						
	(11)	(22)	(33)	(23)	(31)	(12)	(11) + (22) + (33)
i_0	$\dfrac{(1-2\gamma)k_\beta^2 - k^2}{4\pi a k_\beta^2}$	$\dfrac{(1-2\gamma)k_\beta^2 - k^2}{4\pi a k_\beta^2}$	$\dfrac{\Omega}{2\pi a k_\beta^2}$	0	0	0	$-\dfrac{4\gamma-1}{4\pi a}$
j_0	$-\dfrac{k}{4\pi k_\beta^2}$	$-\dfrac{k}{4\pi k_\beta^2}$	$\dfrac{k}{2\pi k_\beta^2}$	0	0	0	0
k_0	0	0	0	0	0	0	0
i_1	0	0	0	$-\dfrac{ik}{2\pi k_\beta^2}$	$\dfrac{k}{2\pi k_\beta^2}$	0	0
j_1	0	0	0	$-\dfrac{i\Omega}{2\pi b k_\beta^2}$	$\dfrac{\Omega}{2\pi b k_\beta^2}$	0	0
k_1	0	0	0	$\dfrac{i}{4\pi k_\beta}$	$-\dfrac{1}{4\pi k_\beta}$	0	0
i_2	$\dfrac{k^2}{4\pi a k_\beta^2}$	$-\dfrac{k^2}{4\pi a k_\beta^2}$	0	0	0	$-\dfrac{ik^2}{4\pi a k_\beta^2}$	0
j_2	$\dfrac{k}{4\pi k_\beta^2}$	$-\dfrac{k}{4\pi k_\beta^2}$	0	0	0	$-\dfrac{ik}{4\pi k_\beta^2}$	0
k_2	$-\dfrac{k}{4\pi b k_\beta}$	$\dfrac{k}{4\pi b k_\beta}$	0	0	0	$\dfrac{ik}{4\pi b k_\beta}$	0

根据公式（84），（86），（89），（90），把第 i 层中的位移谱 U_i 和在 z 平面上的应力谱 $S(U_i) \cdot e_z$ 表示为

$$U = \sum_m \int_0^\infty U^m k \mathrm{d}k, \tag{113}$$

$$S(U) \cdot e_z = \sum_m \int_0^\infty S(U^m) \cdot e_z k \mathrm{d}k, \tag{114}$$

$$kU^m = k(kf_{m_1} + f'_{m_2})B_m + k(f'_{m_1} + kf_{m_2})P_m + kk_\beta f'_{m_3}C_m, \tag{115}$$

$$\frac{1}{\mu_1}S(U^m) \cdot e_z = 2\frac{\mu}{\mu_1}(kf'_{m_1} + \Omega f_{m_2})B_m + 2\frac{\mu}{\mu_1}(\Omega f_{m_1} + kf'_{m_2})P_m + \frac{\mu}{\mu_1}k_\beta f'_{m_3}C_m. \tag{116}$$

(113)–(116) 式中，均略去了层的指标 i。f_{m_1}, f_{m_2}, f_{m_3} 及 f'_{m_1}, f'_{m_2}, f'_{m_3} 如下所示：

$$\begin{cases} f_{m_1} = a_m^+ e^{az} + a_m^- e^{-az}, \\ f_{m_2} = b_m^+ e^{bz} + b_m^- e^{-bz}, \\ f_{m_3} = c_m^+ e^{bz} + c_m^- e^{-bz}, \end{cases} \tag{117}$$

$$\begin{cases} f'_{m_1} = \dfrac{\mathrm{d}f_{m_1}}{\mathrm{d}z} = a(a_m^+ e^{az} - a_m^- e^{-az}), \\ f'_{m_2} = \dfrac{\mathrm{d}f_{m_2}}{\mathrm{d}z} = b(b_m^+ e^{bz} - b_m^- e^{-bz}), \\ f'_{m_3} = \dfrac{\mathrm{d}f_{m_3}}{\mathrm{d}z} = b(c_m^+ e^{bz} - c_m^- e^{-bz}). \end{cases} \quad (118)$$

也略去了层的指标 i. a_m^+, b_m^+, c_m^+ 是上行波的振幅系数, a_m^-, b_m^-, c_m^- 是下行波的振幅系数, 它们都是待定系数.

(115) 和 (116) 式又可写成

$$k\boldsymbol{U}^m = kU_B^m \boldsymbol{B}_m + kU_P^m \boldsymbol{P}_m + kU_c^m \boldsymbol{C}_m, \quad (119)$$

$$\dfrac{1}{\mu}\boldsymbol{S}(\boldsymbol{U}^m) \cdot \boldsymbol{e}_z = \dfrac{1}{\mu_1}\tau_B^m \boldsymbol{B}_m + \dfrac{1}{\mu_1}\tau_P^m \boldsymbol{P}_m + \dfrac{1}{\mu_1}\tau_c^m \boldsymbol{C}_m. \quad (120)$$

其中,

$$\begin{cases} kU_B^m = k^2 \mathrm{ch}\, az(a_m^+ + a_m^-) + k^2 \mathrm{sh}\, az(a_m^+ - a_m^-) + kb\mathrm{ch}\, bz(b_m^+ - b_m^-) + kb\mathrm{sh}\, bz(b_m^+ + b_m^-), \\ kU_P^m = ka\mathrm{sh}\, az(a_m^+ + a_m^-) + ka\mathrm{ch}\, az(a_m^+ - a_m^-) + k^2 \mathrm{sh}\, bz(b_m^+ - b_m^-) + k^2 \mathrm{ch}\, bz(b_m^+ + b_m^-), \\ \dfrac{1}{\mu_1}\tau_P^m = 2\dfrac{\mu}{\mu_1}\Omega\, \mathrm{ch}\, az(a_m^+ + a_m^-) + 2\dfrac{\mu}{\mu_1}\Omega\, \mathrm{sh}\, az(a_m^+ - a_m^-) + 2\dfrac{\mu}{\mu_1}kb\mathrm{ch}\, bz(b_m^+ - b_m^-) \\ \qquad\qquad + 2\dfrac{\mu}{\mu_1}kb\mathrm{sh}\, bz(b_m^+ + b_m^-), \\ \dfrac{1}{\mu_1}\tau_B^m = 2\dfrac{\mu}{\mu_1}ka\mathrm{sh}\, az(a_m^+ + a_m^-) + 2\dfrac{\mu}{\mu_1}ka\mathrm{ch}\, az(a_m^+ - a_m^-) + 2\dfrac{\mu}{\mu_1}\Omega\, \mathrm{sh}\, bz(b_m^+ - b_m^-) \\ \qquad\qquad + 2\dfrac{\mu}{\mu_1}\Omega\, \mathrm{ch}\, bz(b_m^+ + b_m^-), \\ kU_c^m = k k_\beta \mathrm{ch}\, bz(c_m^+ + c_m^-) + k k_\beta \mathrm{sh}\, bz(c_m^+ - c_m^-), \\ \dfrac{1}{\mu_1}\tau_c^m = \dfrac{\mu}{\mu_1}k_\beta b\mathrm{sh}\, bz(c_m^+ + c_m^-) + \dfrac{\mu}{\mu_1}k_\beta b\mathrm{ch}\, bz(c_m^+ - c_m^-). \end{cases} \quad (121)$$

上式给出了第 i 层中的位移谱和应力谱跟层中的波的振幅系数之间的线性变换. 为简明起见, 以 $\boldsymbol{H}^m$ 代表位移谱和应力谱构成的列向量, 其转置 $\boldsymbol{H}^{m\mathrm{T}}$ 是:

$$\boldsymbol{H}^{m\mathrm{T}} = \left[kU_B^m,\ kU_P^m,\ \dfrac{1}{\mu_1}\tau_P^m,\ \dfrac{1}{\mu_1}\tau_B^m,\ kU_c^m,\ \dfrac{1}{\mu_1}\tau_c^m \right], \quad (122)$$

以 $\boldsymbol{K}^m$ 代表波的振幅系数构成的列向量, 其转置 $\boldsymbol{K}^{m\mathrm{T}}$ 是:

$$\boldsymbol{K}^{m\mathrm{T}} = [a_m^+ + a_m^-,\ a_m^+ - a_m^-,\ b_m^+ - b_m^-,\ b_m^+ + b_m^-,\ c_m^+ + c_m^-,\ c_m^+ - c_m^-], \quad (123)$$

于是,

$$\boldsymbol{H}^m = \boldsymbol{D}(z)\boldsymbol{K}^m. \quad (124)$$

$\boldsymbol{D}(z)$ 如公式 (29) 所示. 从 (119) 到 (124) 式均略去了层的指标 i. 从 (121) 式或 (124) 和 (29) 式可以看到, 与向量 $\boldsymbol{B}_m$, $\boldsymbol{P}_m$ 相联系的振动和与向量 $\boldsymbol{C}_m$ 相联系的振动是不耦合的. 也就是, 对于体波来说, P 波、SV 波是耦合的, 而它们和 SH 波是不耦合的; 对面波来说, 瑞雷波和勒夫波是不耦合的.

完全重复上一节中从公式 (30)–(38) 的叙述, 就可以求得递推关系

$$\boldsymbol{H}_i^m = \boldsymbol{a}_i \boldsymbol{H}_{i-1}^m. \quad (125)$$

有了这个递推关系，就能导出多层弹性半空间中一般类型的点源产生的位移场的谱的表示式（形式解）.

设点源位于第 S 层中. 把这一层分成上、下两层，震源以下，以一撇表示，震源以上，以两撇表示. 因为

$$\mathbf{K}_n^m = \mathbf{E}_n^{-1} \mathbf{H}_{n-1}^m, \tag{126}$$

运用递推关系（125）式，可得

$$\mathbf{K}_n^m = \mathbf{E}_n^{-1} \mathbf{a}_{n-1} \mathbf{a}_{n-2} \cdots \mathbf{a}_{s+1} \mathbf{a}_s(d_s - h_s) \mathbf{H}_s^{m\prime}, \tag{127}$$

式中，$\mathbf{H}_s^{m\prime}$ 表示在震源所在平面上 S' 层中 $\mathbf{H}_s^m$ 的值. h_s 是震源至第 S 层顶面的距离.

$$\mathbf{H}_s^{m\prime\prime} = \mathbf{a}_s(h_s) \mathbf{H}_{n-1}^m. \tag{128}$$

运用递推关系（125）式，将上式化为

$$\mathbf{H}_s^{m\prime\prime} = \mathbf{a}_s(h_s) \mathbf{a}_{s-1} \mathbf{a}_{s-2} \cdots \mathbf{a}_2 \mathbf{a}_1 \mathbf{H}_0^m. \tag{129}$$

$\mathbf{H}_s^{m\prime\prime}$ 表示在震源所在平面上 s'' 层中 $\mathbf{H}_s^m$ 的值. 令

$$\mathbf{L} = \mathbf{E}_n^{-1} \mathbf{a}_{n-1} \mathbf{a}_{n-2} \cdots \mathbf{a}_{s+1} \mathbf{a}_s(d_s - h_s), \tag{130}$$

$$\mathbf{M} = \mathbf{a}_s(h_s) \mathbf{a}_{s-1} \mathbf{a}_{s-2} \cdots \mathbf{a}_2 \mathbf{a}_1. \tag{131}$$

那么因为

$$\mathbf{a}_s = \mathbf{a}_s(d_s - h_s) \mathbf{a}_s(h_s), \tag{132}$$

所以

$$\mathbf{J} = \mathbf{LM} = \mathbf{E}_n^{-1} \mathbf{a}_{n-1} \mathbf{a}_{n-2} \cdots \mathbf{a}_{s+1} \mathbf{a}_s \mathbf{a}_{s-1} \cdots \mathbf{a}_2 \mathbf{a}_1. \tag{133}$$

现在由（127），（129），（130），（131）式可知：

$$\mathbf{K}_n^m = \mathbf{L}\mathbf{H}_s^{m\prime}, \tag{134}$$

$$\mathbf{H}_s^{m\prime\prime} = \mathbf{M}\mathbf{H}_0^m. \tag{135}$$

令 $\mathbf{S}^m$ 代表 $\mathbf{H}_s^{m\prime}$ 与 $\mathbf{H}_s^{m\prime\prime}$ 之差：

$$\mathbf{H}_s^{m\prime} - \mathbf{H}_s^{m\prime\prime} = \mathbf{S}^m, \tag{136}$$

那么，

$$\mathbf{L}^{-1}\mathbf{K}_n^m - \mathbf{M}\mathbf{H}_0^m = \mathbf{S}^m, \tag{137}$$

$$\mathbf{K}_n^m - \mathbf{J}\mathbf{H}_0^m = \mathbf{L}\mathbf{S}^m. \tag{138}$$

因为第 n 层（半无限空间）中上行波的振幅系数 a_{mn}^+，b_{mn}^+，c_{mn}^+ 应当为零，所以 $\mathbf{K}_n^m$ 的转置 $\mathbf{K}_n^{m\mathrm{T}}$ 为：

$$\mathbf{K}_n^{m\mathrm{T}} = [a_{mn}^-, \; -a_{mn}^-, \; -b_{mn}^-, \; b_{mn}^-, \; c_{mn}^-, \; -c_{mn}^-]. \tag{139}$$

另外，在自由表面，应力为零，所以 $\mathbf{H}_0^m$ 的转置 $\mathbf{H}_0^{m\mathrm{T}}$ 为：

$$\mathbf{H}_0^{m\mathrm{T}} = [kU_B^m(0), \; kU_p^m(0), \; 0, \; 0, \; kU_c^m(0), \; 0]. \tag{140}$$

将（139），（140）两式代入（138）式，即得：

$$\begin{cases} (J_{11} + J_{21}) kU_B^m(0) + (J_{12} + J_{22}) kU_p^m(0) = -\sum_{i=1}^{4} (L_{1i} + L_{2i}) S_i^m, \\ (J_{31} + J_{41}) kU_B^m(0) + (J_{32} + J_{42}) kU_p^m(0) = -\sum_{i=1}^{4} (L_{3i} + L_{4i}) S_i^m, \\ (J_{55} + J_{65}) kU_c^m(0) = -\sum_{i=5}^{6} (L_{5i} + L_{6i}) S_i^m. \end{cases} \tag{141}$$

解上式，就求得在自由表面上 $kU_B^m(0)$, $kU_p^m(0)$ 和 $kU_c^m(0)$ 的表示式：

$$\begin{cases} kU_B^m(0) = F_R(k)^{-1} \left[(J_{12} + J_{22}) \sum_{i=1}^{4} (L_{3i} + L_{4i}) S_i^m - (J_{32} + J_{42}) \sum_{i=1}^{4} (L_{1i} + L_{2i}) S_i^m \right], \\ kU_p^m(0) = F_R(k)^{-1} \left[(J_{31} + J_{41}) \sum_{i=1}^{4} (L_{1i} + L_{2i}) S_i^m - (J_{11} + J_{21}) \sum_{i=1}^{4} (L_{3i} + L_{4i}) S_i^m \right], \\ kU_c^m(0) = F_L(k)^{-1} \left[- \sum_{i=5}^{6} (L_{5i} + L_{6i}) S_i^m \right]. \end{cases} \quad (142)$$

其中 $F_R(k)$, $F_L(k)$ 如（47），（48）式所示．

按照（113）式，在自由表面的位移谱就应当是

$$U(0) = \sum_m \int_0^\infty U^m(0) k \mathrm{d}k, \quad (143)$$

$$k\mathbf{U}^m(0) = kU_B^m(0) \mathbf{B}_m + kU_p^m(0) \mathbf{P}_m + kU_c^m(0) \mathbf{C}_m. \quad (144)$$

（143）式连同（144）式和（142）式就是多层弹性半空间中一般类型的点源辐射的地震波位移谱表示式（形式解）．将这个结果在圆柱坐标中表示出来，便是：

$$k\mathbf{U}^m(0) = kU_r^m(0) \mathbf{e}_r + kU_\varphi^m(0) \mathbf{e}_\varphi + kU_z^m(0) \mathbf{e}_z, \quad (145)$$

$$\begin{cases} kU_r^m(0) = \cos m\varphi \left[kU_B^{me}(0) \frac{\partial}{k\partial r} J_m(kr) - kU_c^{me}(0) \frac{m}{kr} J_m(kr) \right] \\ \qquad\qquad + \sin m\varphi \left[kU_B^{mo}(0) \frac{\partial}{k\partial r} J_m(kr) - kU_c^{mo}(0) \frac{m}{kr} J_m(kr) \right], \\ kU_\varphi^m(0) = -\sin m\varphi \left[kU_B^{me}(0) \frac{m}{kr} J_m(kr) - kU_c^{me}(0) \frac{\partial}{k\partial r} J_m(kr) \right] \\ \qquad\qquad + \cos m\varphi \left[kU_B^{mo}(0) \frac{m}{kr} J_m(kr) - kU_c^{mo}(0) \frac{\partial}{k\partial r} J_m(kr) \right], \\ kU_z^m(0) = \cos m\varphi \left[kU_p^{me}(0) J_m(kr) \right] + \sin m\varphi \left[kU_p^{mo}(0) J_m(kr) \right], \end{cases} \quad (146)$$

（146）式的形式还是比较整齐的，之所以能如此，是与 $\mathbf{C}_m$ 的定义有关的．在（146）式中 $kU_B^{me}(0)$, $kU_p^{me}(0)$, $kU_c^{me}(0)$ 和 $kU_B^{mo}(0)$, $kU_p^{mo}(0)$, $kU_c^{mo}(0)$ 是由（142）式中的 S_i^m 分别代入 S_i^{me} 和 S_i^{mo} 的值求得．如果 S_i^m 是实数，那么 S_i^{me} 就是 S_i^m，代入（142）式即得 $kU_B^{me}(0)$ 等；如果 S_i^m 是虚数，那么 S_i^{mo} 就是 $-iS_i^m$，代入（142）式即得 $kU_B^{mo}(0)$ 等．换言之，

$$\begin{cases} S_i^{me} = \mathrm{Re}\, S_i^m, & \text{当 } S_i^m \text{ 是实数}, \\ S_i^{mo} = -\mathrm{Im}\, S_i^m, & \text{当 } S_i^m \text{ 是虚数}. \end{cases} \quad (147)$$

令

$$\begin{cases} u_B^m(0) = (J_{12} + J_{22}) \sum_{i=1}^{4} (L_{3i} + L_{4i}) S_i^m - (J_{32} + J_{42}) \sum_{i=1}^{4} (L_{1i} + L_{2i}) S_i^m, \\ u_p^m(0) = (J_{31} + J_{41}) \sum_{i=1}^{4} (L_{1i} + L_{2i}) S_i^m - (J_{11} + J_{21}) \sum_{i=1}^{4} (L_{3i} + L_{4i}) S_i^m, \\ u_c^m(0) = -\sum_{i=5}^{6} (L_{5i} + L_{6i}) S_i^m, \end{cases} \quad (148)$$

那么在自由表面上的位移谱是

$$\mathbf{U}(0) = U_r(0) \mathbf{e}_r + U_\varphi(e) \mathbf{e}_\varphi + U_z(0) \mathbf{e}_z, \quad (149)$$

$$\begin{cases} U_r(0) = \sum_m \left\{ \cos m\varphi \int_0^\infty \left[\frac{u_B^{me}(0)}{F_R(k)} \frac{\partial}{k\partial r} J_m(kr) - \frac{u_c^{me}(0)}{F_L(k)} \frac{m}{kr} J_m(kr) \right] dk \right. \\ \qquad\qquad \left. + \sin m\varphi \int_0^\infty \left[\frac{u_B^{mo}(0)}{F_R(k)} \frac{\partial}{k\partial r} J_m(kr) - \frac{u_c^{mo}(0)}{F_L(k)} \frac{m}{kr} J_m(kr) \right] dk \right\}, \\ U_\varphi(0) = \sum_m \left\{ -\sin m\varphi \int_0^\infty \left[\frac{u_B^{me}(0)}{F_R(k)} \frac{m}{kr} J_m(kr) - \frac{u_c^{me}(0)}{F_L(k)} \frac{\partial}{k\partial r} J_m(kr) \right] dk \right. \\ \qquad\qquad \left. + \cos m\varphi \int_0^\infty \left[\frac{u_B^{mo}(0)}{F_R(k)} \frac{m}{kr} J_m(kr) - \frac{u_c^{mo}(0)}{F_L(k)} \frac{\partial}{k\partial r} J_m(kr) \right] dk \right\}, \\ U_z(0) = \sum_m \left\{ \cos m\varphi \int_0^\infty \left[\frac{u_p^{me}(0)}{F_R(k)} J_m(kr) \right] dk + \sin m\varphi \int_0^\infty \left[\frac{u_p^{mo}(0)}{F_R(k)} J_m(kr) \right] dk \right\}, \end{cases} \quad (150)$$

（150）式就是一般类型的点源在多层弹性半空间的自由表面上所引起的位移谱在圆柱坐标中的表示式（形式解）. 只要源系数 S_i^m 确定了，问题的解答就如同（150）式所示.

对于一般类型的位错源，源系数 S_i^m 可以这样求得：将表1中的系数 i_m，j_m，k_m 代入（93）和（94）式，求出 $\mathbf{H}_s^{m'}$ 和 $\mathbf{H}_s^{m''}$. 然后将结果代入（136）式就得到 S_i^m. 结果是：

$$\begin{cases} S_1^m = k(ki_m - bj_m)[1-(-1)^m], \\ S_2^m = k(-ai_m - kj_m)[1-(-1)^{m+1}], \\ S_3^m = 2\frac{\mu}{\mu_1}(\Omega i_m - bkj_m)[1-(-1)^m], \\ S_4^m = 2\frac{\mu}{\mu_1}(-kai_m - \Omega j_m)[1-(-1)^{m+1}], \\ S_5^m = kk_\beta k_m[1-(-1)^m], \\ S_6^m = -k_\beta bk_m \frac{\mu}{\mu_1}[1-(-1)^{m+1}]. \end{cases} \quad (151)$$

各种基本类型的位错源（偶极源）的源系数 S_i^m 的数值见表2.

如果把 P 点的元位错在自由表面上的 Q 点引起的位移谱记为 $\mathbf{U}_{kl}(P,Q)$，$\mathbf{U}_{kl}(P,Q)$ 即前面的 $\mathbf{U}(0)$. 那么任意形状的位错面 Σ 所引起的位移谱 $\mathbf{U}(Q)$ 为

$$\mathbf{U}(Q) = \iint_\Sigma \Delta U_k(P) \mathbf{U}_{kl}(P,Q) n_1(P) d\Sigma. \quad (152)$$

4 结语

哈斯克尔矩阵法无疑是解决多层介质中的波传播问题的方便工具. 在上述分析中，用元素是无量纲实数的 6×6 矩阵重新定义哈斯克尔层矩阵，研究了平面波在多层弹性半空间中的传播问题，导出了用元素是无量纲实数的层矩阵表示的一些结果，如多层弹性半空间对平面体波的响应谱，面波的频散方程，面波在不同深度处的相对激发程度以及瑞雷波的质点运动情况. 以往利用哈斯克尔矩阵分析和计算多层介质中波传播问题时，经常要遇到因为层矩阵的元素不完全是实数所招致的麻烦. 在这一工作中，因为用元素是无量纲实数的 6×6 矩阵定义哈斯克尔层矩阵，从而免去这一麻烦，便于计算. 吸取哈斯克尔矩阵法的这些优点以及用本征向量（汉森向量）表示圆柱坐标下运动方程的解答的优点，就可以比较方便地导出多层弹性半空间中一般类型的点源引起的位移谱表示式（形式解），特别是地震学中有意义的各种基本类型的偶极源引起的位移谱表示式（形式解）的具体形式. 在下面的工作中，将从这些结果出发，进一步

讨论多层弹性半空间中一般类型的震源辐射的地震波的传播问题.

表 2 源系数 S_i^m 的数值

S_i^m	kl						
	(11)	(22)	(33)	(23)	(31)	(12)	(11)+(22)+(33)
S_1^0	0	0	0	0	0	0	0
S_2^0	$-\dfrac{k}{2\pi}(1-2\gamma)$	$-\dfrac{k}{2\pi}(1-2\gamma)$	$\dfrac{k}{2\pi}$	0	0	0	$\dfrac{k}{2\pi}(4\gamma-1)$
S_3^0	0	0	0	0	0	0	0
S_4^0	$\dfrac{k}{2\pi}(4\gamma-1)\dfrac{\mu}{\mu_1}$	$\dfrac{k}{2\pi}(4\gamma-1)\dfrac{\mu}{\mu_1}$	0	0	0	0	$\dfrac{k}{\pi}(4\gamma-1)\dfrac{\mu}{\mu_1}$
S_5^0	0	0	0	0	0	0	0
S_6^0	0	0	0	0	0	0	0
S_1^1	0	0	0	$-\dfrac{ik}{2\pi}$	$\dfrac{k}{2\pi}$	0	0
S_2^1	0	0	0	0	0	0	0
S_3^1	0	0	0	0	0	0	0
S_4^1	0	0	0	0	0	0	0
S_5^1	0	0	0	$\dfrac{ik}{2\pi}$	$-\dfrac{k}{2\pi}$	0	0
S_6^1	0	0	0	0	0	0	0
S_1^2	0	0	0	0	0	0	0
S_2^2	0	0	0	0	0	0	0
S_3^2	0	0	0	0	0	0	0
S_4^2	$-\dfrac{k}{2\pi}\dfrac{\mu}{\mu_1}$	$\dfrac{k}{2\pi}\dfrac{\mu}{\mu_1}$	0	0	0	$\dfrac{ik}{2\pi}\dfrac{\mu}{\mu_1}$	0
S_5^2	0	0	0	0	0	0	0
S_6^2	$\dfrac{k}{2\pi}\dfrac{\mu}{\mu_1}$	$-\dfrac{k}{2\pi}\dfrac{\mu}{\mu_1}$	0	0	0	$-\dfrac{ik}{2\pi}\dfrac{\mu}{\mu_1}$	0

参 考 文 献

[1] N. A. Haskell, The dispersion of surface waves on multilayered media, *Bull. Seism. Soc. Amer.*, **43**, 1, 17–34, 1953.
[2] N. A. Haskell, Radiation pattern of surface waves from point sources in a multilayered medium, *Bull. Seism. Soc. Amer.*, **54**, 1, 377–393, 1964.
[3] K. Sezawa, On the transmission of seismic waves on the bottom surface of an ocean, *Bull. Earthq. Res. Inst. Tokyo Univ.*, **9**, 2, 115–143, 1931.
[4] W. W. Hansen, A new type of expansion in radiation problems, *Phys. Rev.*, **47**, 2, 139–143, 1935.
[5] A. Ben-Menahem and S. J. Singh, Multipolar elastic fields in a layered half-space, *Bull. Seism. Soc. Amer.*, **58**, 5, 1519–1572, 1968.
[6] T. Maruyama, On the force equivalents of dynamical elastic dislocations with reference to the earthquake mechanism, *Bull. Earthq. Res. Inst. Tokyo Univ.*, **41**, 3, 467–486, 1963.

（注：本文共分两部分，第二部分也收录在本书中）

Seismic waves in multilayered elastic half-space (I)

Chen Yun-Tai

Institute of Geophysics, Academia Sinica

Abstract The propagation problem of seismic waves radiated from general types of sources in a multilayered homogeneous isotropic and perfectly elastic half-space is studied. It is shown that the Haskell layer matrix can be defined in an alternative form with all its elements real and dimensionless for all real values of wave number, so as to facilitate theoretical analysis and numerical calculation. Using Haskell matrix method, the response spectra of multilayered elastic half-space to plane body waves, the dispersion equations for both Rayleigh and Love waves, the relative excitations of surface waves at any depth, and the particle motion condition for Rayleigh waves at free surface and at any depth are derived in detail. Finally, by using Hansen's expansion method and Haskell matrix formulation, the spectra (formal solution) for the displacement fields due to general types of sources in a multilayered elastic half-space, especially those due to various elementary types of dipolar sources, are given.

多层弹性半空间中的地震波（二）

陈运泰

中国科学院地球物理研究所

1 引言

地震面波的频散性质、地震波辐射的方向性等特性已经广泛地用于地壳和上地幔结构以及震源机制的研究中，并且取得了许多有用的成果．研究地震波如何从震源辐射出来、如何在实际介质中传播和衰减的这一问题，对于利用地震波确定地壳和上地幔结构以及震源的参数，是很有必要的．关于这一问题的研究，已经作过许多工作[1-6]．已往的工作中，为了分析方便，往往采用简单的地壳-上地幔模型或简单的震源模型，或两者都相当简单的模型．通过对简单介质模型或简单震源模型的地震波辐射、传播和衰减问题的分析，人们对地壳和上地幔结构以及震源的性质增进了认识．显然，进一步分析更接近于实际情况的震源模型辐射的地震波在复杂的地壳-上地幔系统中的传播和衰减问题，将有助于对地壳和上地幔结构以及震源机制的研究．

地震过程（至少是浅源地震过程）可以视为一种破裂过程．地震断层的两盘的块体相对运动（位错），释放积蓄在介质中的应变能．应变能一部分以弹性波的形式向周围传开，大部分转化为热．由于位错和力系的等效性[7-9]，原则上，如果知道集中力在多层弹性半空间引起的地震波位移场，便可以通过适当的微商导出位错源所引起的地震波位移场．实际上，涉及的计算颇为繁复，莫如另觅途径．

运用哈斯克尔（Haskell）[3,10]矩阵法计算多层介质中的波传播问题所导出的表达式，具有一定的规则性，便于检查、容易记住且便于数值计算．用汉森（Hansen）[11]向量表示圆柱坐标中运动方程的解答也相当有规则[12]．吸取哈斯克尔矩阵法和汉森展开的这些优点，可以比较方便地导出多层弹性半空间中一般类型的震源引起的位移场表示式（形式解），特别是地震学中所用到的各种基本类型的偶极源引起的位移场表示式（形式解）的具体形式[13]．从这些结果出发，可以进一步分析一般类型的震源辐射的地震波传播问题．

本文将在第一部分工作的基础上，从上述表示式（形式解）出发，计算被积函数的极点的留数的贡献，导出多层弹性半空间中地震震源辐射的面波位移场表示式，并指出所得结果对研究地震的震源深度、地壳和上地幔结构以及震源机制的意义．

2 多层弹性半空间中一般类型的点源引起的地震面波

2.1 自由表面上地震面波的位移谱

先研究多层弹性半空间中一般类型的点源引起的地震面波，然后研究一般类型的位错源引起的地震面波．凡在第一部分已采用过的符号，这里继续使用而不重新加以说明．凡提及第一部分的公式均冠以"I-"．

设 k_R 和 k_L 是频散方程（I-49）和（I-50）的根：

$$F_R(k_R) = 0, \tag{1}$$

---

* 本文发表于《地球物理学报》，1974 年，第 **17** 卷，第 3 期，173–185．

$$F_L(k_L) = 0, \tag{2}$$

那么，在积分解（I-150）中 k_R 和 k_L 是被积函数的极点．计算在极点 k_R 和 k_L 的留数，即得在多层弹性半空间的自由表面上地震面波的位移谱 $\widetilde{U}(0)$．结果是：

$$\widetilde{U}(0) = \widetilde{U}_r(0)\boldsymbol{e}_r + \widetilde{U}_\varphi(0)\boldsymbol{e}_\varphi + \widetilde{U}_z(0)\boldsymbol{e}_z, \tag{3}$$

$$\begin{cases} \widetilde{U}_r(0) = -\pi i \sum_m \left\{ \cos m\varphi \left[\dfrac{\widetilde{u}_B^{me}(0)}{F_R'(k_R)} \dfrac{\partial}{k_R \partial r} H_m^{(2)}(k_R r) - \dfrac{\widetilde{u}_C^{me}(0)}{F_L'(k_L)} \dfrac{m}{k_L r} H_m^{(2)}(k_L r) \right] \right. \\ \qquad\qquad\left. + \sin m\varphi \left[\dfrac{\widetilde{u}_B^{mo}(0)}{F_R'(k_R)} \dfrac{\partial}{k_R \partial r} H_m^{(2)}(k_R r) - \dfrac{\widetilde{u}_C^{mo}(0)}{F_L'(k_L)} \dfrac{m}{k_L r} H_m^{(2)}(k_L r) \right] \right\}, \\ \widetilde{U}_\varphi(0) = -\pi i \sum_m \left\{ -\sin m\varphi \left[\dfrac{\widetilde{u}_B^{me}(0)}{F_R'(k_R)} \dfrac{m}{k_R r} H_m^{(2)}(k_R r) - \dfrac{\widetilde{u}_C^{me}(0)}{F_L'(k_L)} \dfrac{\partial}{k_L \partial r} H_m^{(2)}(k_L r) \right] \right. \\ \qquad\qquad\left. + \cos m\varphi \left[\dfrac{\widetilde{u}_B^{mo}(0)}{F_R'(k_R)} \dfrac{m}{k_R r} H_m^{(2)}(k_R r) - \dfrac{\widetilde{u}_C^{mo}(0)}{F_L'(k_L)} \dfrac{\partial}{k_L \partial r} H_m^{(2)}(k_L r) \right] \right\}, \\ \widetilde{U}_z(0) = -\pi i \sum_m \left\{ \cos m\varphi \left[\dfrac{\widetilde{u}_P^{me}(0)}{F_R'(k_R)} H_m^{(2)}(k_R r) \right] + \sin m\varphi \left[\dfrac{\widetilde{u}_P^{mo}(0)}{F_R'(k_R)} H_m^{(2)}(k_R r) \right] \right\}. \end{cases} \tag{4}$$

上式中的 $H_m^{(2)}$ 是 m 阶汉克尔函数，

$$\begin{cases} \widetilde{u}_B^m(0) = u_B^m(0)|_{k=k_R}, \\ \widetilde{u}_P^m(0) = u_P^m(0)|_{k=k_R}, \\ \widetilde{u}_C^m(0) = u_C^m(0)|_{k=k_L}. \end{cases} \tag{5}$$

一般地说，频散方程的根 k_R 和 k_L 都有无穷多个，相应于基谐及高谐振型瑞雷面波和勒夫面波．面波的位移谱 $\widetilde{U}_r(0)$，$\widetilde{U}_\varphi(0)$，$\widetilde{U}_z(0)$ 是所有可能的极点的贡献之和．在（4）式以及下面的分析中，为简便起见，都略去了对所有可能的极点的贡献求和的记号．

为了把上面导出的结果化为物理意义明显和便于实际应用的形式，需要对矩阵 $\mathbf{a}_i$，$\mathbf{M}$，$\mathbf{A}$，$\mathbf{J}$ 的性质作一分析．

$\mathbf{a}_i$ 表示的是 $\mathbf{H}_{i-1}$ 和 $\mathbf{H}_i$ 之间的线性变换．从（I-37）式可以看出，层矩阵 $\mathbf{a}_i$ 具有如下性质：

$$\begin{cases} (\mathbf{a}_i)_{jk} = (-1)^{\delta_{2j}+\delta_{2k}+\delta_{3j}+\delta_{3k}} (\mathbf{a}_i)_{5-k,\,5-j}, & 1 \leq k, j \leq 4, \\ (\mathbf{a}_i)_{jk} = (\mathbf{a}_i)_{11-k,\,11-j}, & 5 \leq k, j \leq 6, \\ (\mathbf{a}_i)_{jk} = 0, & 1 \leq k \leq 4,\ 5 \leq j \leq 6, \\ & 5 \leq k \leq 6,\ 1 \leq j \leq 4. \end{cases} \tag{6}$$

这里，δ_{mn} 表示克罗内克尔（Kronecker）δ．还可以看出，$\mathbf{a}_i$ 的逆矩阵 $\mathbf{a}_i^{-1}$ 表示的是 $\mathbf{H}_i$ 和 $\mathbf{H}_{i-1}$ 之间的线性变换，因此，把 $(\mathbf{a}_i)_{jk}$ 中的 d_i 换成 $-d_i$，即得 $(\mathbf{a}_i^{-1})_{jk}$：

$$(\mathbf{a}_i^{-1})_{jk} = (-1)^{j+k} (\mathbf{a}_i)_{jk} \tag{7}$$

由（6）式和（7）式，立刻可以得出，层矩阵 $\mathbf{a}_i$ 具有下述性质：

$$\begin{cases} (\mathbf{a}_i^{-1})_{jk} = (-1)^{j+k+\delta_{2j}+\delta_{2k}+\delta_{3j}+\delta_{3k}} (\mathbf{a}_i)_{5-k,\,5-j}, & 1 \leq k, j \leq 4, \\ (\mathbf{a}_i^{-1})_{jk} = (-1)^{j+k} (\mathbf{a}_i)_{11-k,\,11-j}, & 5 \leq k, j \leq 6, \\ (\mathbf{a}_i^{-1})_{jk} = 0, & 1 \leq k \leq 4,\ 5 \leq j \leq 6, \\ & 5 \leq k \leq 6,\ 1 \leq j \leq 4. \end{cases} \tag{8}$$

如果矩阵 **a**，**b** 均具有上述性质，那么矩阵的积

$$\mathbf{c} = \mathbf{ab}$$

也具有上述性质. 这点可以直接验证.

既然矩阵 **M** 和 **A** 是具有上述性质的层矩阵的积，所以 **M** 和 **A** 均具有 (8) 式所示性质.

根据矩阵 **J** 的定义 (I-43) 式，可以求得：

$$\begin{cases} J_{11} + J_{21} = \dfrac{2}{k_\beta^2} A_{11} - \dfrac{2\Omega}{kak_\beta^2} A_{21} - \dfrac{\mu_1}{\mu k_\beta^2} A_{31} + \dfrac{\mu_1 k}{\mu a k_\beta^2} A_{41}, \\[4pt] J_{12} + J_{22} = \dfrac{2}{k_\beta^2} A_{12} - \dfrac{2\Omega}{kak_\beta^2} A_{22} - \dfrac{\mu_1}{\mu k_\beta^2} A_{32} + \dfrac{\mu_1 k}{\mu a k_\beta^2} A_{42}, \\[4pt] J_{13} + J_{23} = \dfrac{2}{k_\beta^2} A_{13} - \dfrac{2\Omega}{kak_\beta^2} A_{23} - \dfrac{\mu_1}{\mu k_\beta^2} A_{33} + \dfrac{\mu_1 k}{\mu a k_\beta^2} A_{43}, \\[4pt] J_{14} + J_{24} = \dfrac{2}{k_\beta^2} A_{14} - \dfrac{2\Omega}{kak_\beta^2} A_{24} - \dfrac{\mu_1}{\mu k_\beta^2} A_{34} + \dfrac{\mu_1 k}{\mu a k_\beta^2} A_{44}, \\[4pt] J_{31} + J_{41} = -\dfrac{2\Omega}{kbk_\beta^2} A_{11} + \dfrac{2}{k_\beta^2} A_{21} + \dfrac{\mu_1 k}{\mu b k_\beta^2} A_{31} - \dfrac{\mu_1}{\mu k_\beta^2} A_{41}, \\[4pt] J_{32} + J_{42} = -\dfrac{2\Omega}{kbk_\beta^2} A_{12} + \dfrac{2}{k_\beta^2} A_{22} + \dfrac{\mu_1 k}{\mu b k_\beta^2} A_{32} - \dfrac{\mu_1}{\mu k_\beta^2} A_{42}, \\[4pt] J_{33} + J_{43} = -\dfrac{2\Omega}{kbk_\beta^2} A_{13} + \dfrac{2}{k_\beta^2} A_{23} + \dfrac{\mu_1 k}{\mu b k_\beta^2} A_{33} - \dfrac{\mu_1}{\mu k_\beta^2} A_{43}, \\[4pt] J_{34} + J_{44} = -\dfrac{2\Omega}{kbk_\beta^2} A_{14} + \dfrac{2}{k_\beta^2} A_{24} + \dfrac{\mu_1 k}{\mu b k_\beta^2} A_{34} - \dfrac{\mu_1}{\mu k_\beta^2} A_{44}, \end{cases} \quad (9)$$

经过麻烦然而直接的运算可以证明：

$$\begin{aligned} & (J_{11}+J_{21})(J_{34}+J_{44}) - (J_{31}+J_{41})(J_{14}+J_{24}) \\ & = (J_{32}+J_{42})(J_{13}+J_{23}) - (J_{12}+J_{22})(J_{33}+J_{43}). \end{aligned} \quad (10)$$

矩阵 $\mathbf{a}_i$，**M**，**A** 和 **J** 的上述性质可以用来把地震面波的位移谱表示成物理意义明显和便于实际应用的形式.

根据 (I-133) 式，矩阵

$$\mathbf{L} = \mathbf{JM}^{-1}, \quad (11)$$

即

$$L_{jk} = \sum_{l=1}^{6} J_{jl}(\mathbf{M}^{-1})_{lk}. \quad (12)$$

考虑到矩阵 **M** 具有 (8) 式所示性质，可得

$$\begin{cases} L_{jk} = \displaystyle\sum_{l=1}^{4} J_{jl} M_{5-k,\,5-l}(-1)^{l+k+\delta_{2l}+\delta_{2k}+\delta_{3l}+\delta_{3k}}, & 1 \leq j,k \leq 4, \\[6pt] L_{jk} = \displaystyle\sum_{l=5}^{6} J_{jl} M_{11-k,\,11-l}(-1)^{l+k}, & 5 \leq j,k \leq 6. \end{cases} \quad (13)$$

从而

$$\begin{cases} \tilde{u}_B^m(0) = \left[(J_{12}+J_{22})(J_{34}+J_{44}) - (J_{32}+J_{42})(J_{14}+J_{24})\right]\left(K\chi_{R_4}S_1^m - \chi_{R_3}S_2^m - K\chi_{R_2}S_3^m + \chi_{R_1}S_4^m\right), \\ \tilde{u}_P^m(0) = -K\tilde{u}_B^m(0), \\ \tilde{u}_C^m(0) = (J_{56}+J_{66})(\chi_{L_2}S_5^m - \chi_{L_1}S_6^m). \end{cases} \quad (14)$$

式中，

$$\begin{cases} \chi_{R_1} = M_{11} - KM_{12}, \\ \chi_{R_2} = -\dfrac{M_{21}}{K} + M_{22}, \\ \chi_{R_3} = M_{31} - KM_{32}, \\ \chi_{R_4} = -\dfrac{M_{41}}{K} + M_{42}, \\ \chi_{L_1} = M_{55}, \\ \chi_{L_2} = M_{65}. \end{cases} \quad (15)$$

将上述结果代入（4）式，利用汉克尔函数在 $k_C r \gg 1$ 时的渐近展开式

$$H_m^{(2)}(k_C r) \sim \sqrt{\dfrac{2}{\pi k_C r}} e^{-ik_C r + \frac{im\pi}{2} + i\frac{\pi}{4}}, \quad C: R \text{ 或 } L, \quad (16)$$

以及当 $k_C r \gg 1$ 时，

$$\dfrac{\partial}{k_C \partial r} H_m^{(2)}(k_C r) \gg \dfrac{1}{k_C r} H_m^{(2)}(k_C r), \quad (17)$$

可将（4）式化为

$$\begin{cases} \widetilde{U}_r(0) \sim \mu_S A_R \chi_{R_1} \Phi_R, \\ \widetilde{U}_\varphi(0) \sim \mu_S A_L \chi_{L_1} \Phi_L, \\ \widetilde{U}_z(0) \sim -K\widetilde{U}_r(0), \end{cases} \quad (18)$$

式中，

$$\mu_S A_R = \dfrac{\mu_S}{\mu_1} \sqrt{\dfrac{k_R}{2\pi r}} e^{-ik_R r + i\frac{\pi}{4}} \dfrac{1}{F_R'(k_R)} \left[(J_{12} + J_{22})(J_{34} + J_{44}) - (J_{32} + J_{42})(J_{14} + J_{24}) \right], \quad (19)$$

$$\Phi_R = \dfrac{2\pi \mu_1}{k_R \mu_S} \sum_m e^{\frac{im\pi}{2}} \left[\cos m\varphi \left(K \dfrac{\chi_{R_4}}{\chi_{R_1}} S_1^{me} - \dfrac{\chi_{R_3}}{\chi_{R_1}} S_2^{me} - K \dfrac{\chi_{R_2}}{\chi_{R_1}} S_3^{me} + S_4^{me} \right) \right.$$
$$\left. + \sin m\varphi \left(K \dfrac{\chi_{R_4}}{\chi_{R_1}} S_1^{mo} - \dfrac{\chi_{R_3}}{\chi_{R_1}} S_2^{mo} - K \dfrac{\chi_{R_2}}{\chi_{R_1}} S_3^{mo} + S_4^{mo} \right) \right], \quad (20)$$

$$\mu_S A_L = \dfrac{\mu_S}{\mu_1} \sqrt{\dfrac{k_L}{2\pi r}} e^{-ik_L r + i\frac{\pi}{4}} \dfrac{1}{F_L'(k_L)} (J_{56} + J_{66}), \quad (21)$$

$$\Phi_L = \dfrac{2\pi \mu_1}{k_L \mu_S} \sum_m e^{\frac{im\pi}{2}} \left[\sin m\varphi \left(\dfrac{\chi_{L_2}}{\chi_{L_1}} S_5^{me} - S_6^{me} \right) + \cos m\varphi \left(\dfrac{\chi_{L_2}}{\chi_{L_1}} S_5^{mo} - S_6^{mo} \right) \right], \quad (22)$$

（18）式就是在多层弹性半空间中一般类型的点源引起的面波在自由表面上的位移谱. 当 χ_{R_1} 不等于零时，这个位移谱总是可以表示为三项的乘积. $\mu_S A_R$ 的意义是: 震源深度为零时由（kl）=（12）确定的单位面积单位错距的剪切位错元产生的瑞雷波在自由表面上震中距为 r 处的位移谱. Φ_R 是瑞雷波辐射图型因子，它和多层半空间的具体结构、震源深度和震源的特性有关. χ_{R_1} 表示震源深度对瑞雷波在自由表面上的位移谱的影响，其形式和瑞雷波的相对激发程度一样，$\mu_S A_L$, Φ_L, χ_{L_1} 的意义与以上各量完全相似，不过是相对于勒夫波.

2.2 接收深度对面波位移谱的影响

为研究接收深度对面波位移谱和应力谱的影响，先证明：面波的位移谱和应力谱在震源所在的平面

上是连续的. 当震源是爆炸源、垂直集中力和水平集中力等三种类型的震源时, 哈克莱德 (Harkrider)[5]曾经分别导出过这一结果. 这一结论对一般类型的震源依然成立. 这是因为, 按 (I-136) 式, 在震源所在平面上, 列向量 $\mathbf{H}_S^{m'}$ 和 $\mathbf{H}_S^{m''}$ 之差 $\mathbf{S}^m$ 中并不包含 $F_R^{-1}(k)$ 和 $F_L^{-1}(k)$ 这样的因子, 尽管列向量 $\mathbf{H}_S^m$ 在震源所在平面上不连续, 但 $\mathbf{S}^m$ 对 $\mathbf{H}_S^m$ 在极点 k_R, k_L 的留数的不连续并没有贡献. 所以, 面波的位移谱和应力谱在震源所在的平面上是连续的. 用式子表示, 即

$$\mathrm{Res}(\mathbf{H}_S^{m'}) - \mathrm{Res}(\mathbf{H}_S^{m''}) = 0. \tag{23}$$

这个性质可以用来很方便地导出任一层中面波引起的位移谱和应力谱. 按照 (I-84), (I-85), (I-89), (I-90) 式, 可以把第 i 层中的位移谱和应力谱表示为:

$$\mathbf{U}(D_i) = \sum_m \int_0^\infty \mathbf{U}^m(D_i) k \mathrm{d}k, \tag{24}$$

$$\mathbf{S}(\mathbf{U}(D_i)) \cdot \mathbf{e}_z = \sum_m \int_0^\infty \mathbf{S}(\mathbf{U}(D_i)) \cdot \mathbf{e}_z k \mathrm{d}k, \tag{25}$$

式中,

$$k\mathbf{U}^m(D_i) = kU_B^m(D_i)\mathbf{B}_m + kU_P^m(D_i)\mathbf{P}_m + kU_C^m(D_i)\mathbf{C}_m, \tag{26}$$

$$\frac{1}{\mu_1}\mathbf{S}(\mathbf{U}^m(D_i)) \cdot \mathbf{e}_z = \frac{1}{\mu_1}\tau_B^m(D_i)\mathbf{B}_m + \frac{1}{\mu_1}\tau_P^m(D_i)\mathbf{P}_m + \frac{1}{\mu_1}\tau_C^m(D_i)\mathbf{C}_m, \tag{27}$$

$D_i = z_{i-1} + h_i$, h_i 是接受点至第 i 层顶面的距离.

第 i 层中位移谱和应力谱列向量以 $\mathbf{H}_i^m(D_i)$ 表示. $\mathbf{H}_i^m(D_i)$ 的转置 $\mathbf{H}_i^{mT}(D_i)$ 是:

$$\mathbf{H}_i^{mT}(D_i) = \left[kU_B^m(D_i),\ kU_P^m(D_i),\ \frac{1}{\mu_1}\tau_P^m(D_i),\ \frac{1}{\mu_1}\tau_B^m(D_i),\ kU_C^m(D_i),\ \frac{1}{\mu_1}\tau_C^m(D_i) \right], \tag{28}$$

这里以 T 表示转置. 根据递推关系, 如果接收点在震源所在平面之上,

$$\mathbf{H}_i^m(D_i) = \mathbf{M}' \mathbf{H}_0^m, \tag{29}$$

$$\mathbf{M}' = \mathbf{a}_i(h_i)\mathbf{a}_{i-1}\mathbf{a}_{i-2}\cdots\mathbf{a}_2\mathbf{a}_1. \tag{30}$$

或者, 把位移谱写成明显的形式:

$$\begin{cases} kU_B^m(D_i) = M'_{11}kU_B^m(0) + M'_{12}kU_P^m(0), \\ kU_P^m(D_i) = M'_{21}kU_B^m(0) + M'_{22}kU_P^m(0), \\ kU_C^m(D_i) = M'_{55}kU_C^m(0). \end{cases} \tag{31}$$

如果接受点在震源所在平面之下,

$$\mathbf{H}_i^m(D_i) = \mathbf{a}_i(h_i)\mathbf{a}_{i-1}\mathbf{a}_{i-2}\cdots\mathbf{a}_{s+1}\mathbf{a}_s(d_s - h_s)\mathbf{H}_s^{m'}, \tag{32}$$

h_s 是震源距它所在那一层 (第 s 层) 的顶面的距离. 按照 (I-136) 式, 上式可化为

$$\mathbf{H}_i^m(D_i) = \mathbf{M}' \mathbf{H}_0^{m'} + \mathbf{M}' \mathbf{M}^{-1} \mathbf{S}^m. \tag{33}$$

把位移谱写成明显的形式, 则为:

$$\begin{cases} kU_B^m(D_i) = M'_{11}kU_B^m(0) + M'_{12}kU_P^m(0) + \sum_{i=1}^{4}(\mathbf{M}'\mathbf{M}^{-1})_{1i}S_i^m, \\ kU_P^m(D_i) = M'_{21}kU_B^m(0) + M'_{22}kU_P^m(0) + \sum_{i=1}^{4}(\mathbf{M}'\mathbf{M}^{-1})_{2i}S_i^m, \\ kU_C^m(D_i) = M'_{55}kU_C^m(0) + \sum_{i=5}^{6}(\mathbf{M}'\mathbf{M}^{-1})_{5i}S_i^m. \end{cases} \tag{34}$$

和 (I-149), (I-150) 式相仿, 在圆柱坐标下, 位移谱 $\mathbf{U}(D_i)$ 可以表示为:

$$\mathbf{U}(D_i) = U_r(D_i)\mathbf{e}_r + U_\varphi(D_i)\mathbf{e}_\varphi + U_z(D_i)\mathbf{e}_z, \tag{35}$$

$$\begin{cases} U_r(D_i) = \sum_m \left\{ \cos m\varphi \int_0^\infty \left[\frac{\mu_B^{me}(D_i)}{F_R(k)} \frac{\partial}{k\partial r} J_m(kr) - \frac{\mu_C^{me}(D_i)}{F_L(k)} \frac{m}{kr} J_m(kr) \right] dk \right. \\ \qquad\qquad \left. + \sin m\varphi \int_0^\infty \left[\frac{\mu_B^{mo}(D_i)}{F_R(k)} \frac{\partial}{k\partial r} J_m(kr) - \frac{\mu_C^{mo}(D_i)}{F_L(k)} \frac{m}{kr} J_m(kr) \right] dk \right\}, \\ U_\varphi(D_i) = \sum_m \left\{ -\sin m\varphi \int_0^\infty \left[\frac{\mu_B^{me}(D_i)}{F_R(k)} \frac{m}{kr} J_m(kr) - \frac{\mu_C^{me}(D_i)}{F_L(k)} \frac{\partial}{k\partial r} J_m(kr) \right] dk \right. \\ \qquad\qquad \left. + \cos m\varphi \int_0^\infty \left[\frac{\mu_B^{mo}(D_i)}{F_R(k)} \frac{m}{kr} J_m(kr) - \frac{\mu_C^{mo}(D_i)}{F_L(k)} \frac{\partial}{k\partial r} J_m(kr) \right] dk \right\}, \\ U_z(D_i) = \sum_m \left\{ \cos m\varphi \int_0^\infty \left[\frac{\mu_P^{me}(D_i)}{F_R(k)} J_m(kr) \right] dk + \sin m\varphi \int_0^\infty \left[\frac{\mu_P^{mo}(D_i)}{F_R(k)} J_m(kr) \right] dk \right\}, \end{cases} \quad (36)$$

其中，当接收点在震源所在平面之上时，

$$\begin{cases} u_B^m(D_i) = M'_{11} u_B^m(0) + M'_{12} u_P^m(0), \\ u_P^m(D_i) = M'_{21} u_B^m(0) + M'_{22} u_P^m(0), \\ u_C^m(D_i) = M'_{55} u_C^m(0); \end{cases} \quad (37)$$

当接收点在震源所在平面之下时，

$$\begin{cases} u_B^m(D_i) = M'_{11} u_B^m(0) + M'_{12} u_P^m(0) + F_R(k) \sum_{i=1}^{4} (\mathbf{M'M}^{-1})_{1i} S_i^m, \\ u_P^m(D_i) = M'_{21} u_B^m(0) + M'_{22} u_P^m(0) + F_R(k) \sum_{i=1}^{4} (\mathbf{M'M}^{-1})_{2i} S_i^m, \\ u_C^m(D_i) = M'_{55} u_C^m(0) + F_L(k) \sum_{i=5}^{6} (\mathbf{M'M}^{-1})_{5i} S_i^m. \end{cases} \quad (38)$$

计算（36）式的被积函数在极点 k_R，k_L 的留数，便求得在第 i 层中距该层顶面 h_i 的接收点的面波位移谱 $\widetilde{U}(D_i)$. 结果是：

$$\widetilde{U}(D_i) = \widetilde{U}_r(D_i) \boldsymbol{e}_r + \widetilde{U}_\varphi(D_i) \boldsymbol{e}_\varphi + \widetilde{U}_z(D_i) \boldsymbol{e}_z, \quad (39)$$

$$\begin{cases} \widetilde{U}_r(D_i) = -\pi i \sum_m \left\{ \cos m\varphi \left[\frac{\tilde{u}_B^{me}(D_i)}{F'_R(k_R)} \frac{\partial}{k_R \partial r} H_m^{(2)}(k_R r) - \frac{\tilde{u}_C^{me}(D_i)}{F'_L(k_L)} \frac{m}{k_L r} H_m^{(2)}(k_L r) \right] \right. \\ \qquad\qquad \left. + \sin m\varphi \left[\frac{\tilde{u}_B^{mo}(D_i)}{F'_R(k_R)} \frac{\partial}{k_R \partial r} H_m^{(2)}(k_R r) - \frac{\tilde{u}_C^{mo}(D_i)}{F'_L(k_L)} \frac{m}{k_L r} H_m^{(2)}(k_L r) \right] \right\}, \\ \widetilde{U}_\varphi(D_i) = -\pi i \sum_m \left\{ -\sin m\varphi \left[\frac{\tilde{u}_B^{me}(D_i)}{F'_R(k_R)} \frac{m}{k_R r} H_m^{(2)}(k_R r) - \frac{\tilde{u}_C^{me}(D_i)}{F'_L(k_L)} \frac{\partial}{k_L \partial r} H_m^{(2)}(k_L r) \right] \right. \\ \qquad\qquad \left. + \cos m\varphi \left[\frac{\tilde{u}_B^{mo}(D_i)}{F'_R(k_R)} \frac{m}{k_R r} H_m^{(2)}(k_R r) - \frac{\tilde{u}_C^{mo}(D_i)}{F'_L(k_L)} \frac{\partial}{k_L \partial r} H_m^{(2)}(k_L r) \right] \right\}, \\ \widetilde{U}_z(D_i) = -\pi i \sum_m \left\{ \cos m\varphi \left[\frac{\tilde{u}_P^{me}(D_i)}{F'_R(k_R)} H_m^{(2)}(k_R r) \right] + \sin m\varphi \left[\frac{\tilde{u}_P^{mo}(D_i)}{F'_R(k_R)} H_m^{(2)}(k_R r) \right] \right\}. \end{cases} \quad (40)$$

其中的 $\tilde{u}_B^m(D_i)$，$\tilde{u}_P^m(D_i)$，$\tilde{u}_C^m(D_i)$ 不管接收点是在震源所在平面之上还是之下，都是：

$$\begin{cases} \tilde{u}_B^m(D_i) = M'_{11} \tilde{u}_B^m(0) + M'_{12} \tilde{u}_P^m(0), \\ \tilde{u}_P^m(D_i) = M'_{21} \tilde{u}_B^m(0) + M'_{22} \tilde{u}_P^m(0), \\ \tilde{u}_C^m(D_i) = M'_{55} \tilde{u}_C^m(0), \end{cases} \quad (41)$$

这里的 $M'_{11}, M'_{12}, M'_{21}, M'_{22}$ 中的 k 要代入 k_R 的值,而 M'_{55} 中的 k 要代入 k_L 的值.

由 (14) 式,把上式化为

$$\begin{cases} \tilde{u}_B^m(D_i) = \chi'_{R_1} \tilde{u}_B^m(0), \\ \tilde{u}_P^m(D_i) = \chi'_{R_2} \tilde{u}_P^m(0) = -K' \tilde{u}_B^m(D_i), \\ \tilde{u}_C^m(D_i) = \chi'_{L_1} \tilde{u}_C^m(0), \end{cases} \quad (42)$$

K' 如 (I-77) 式所示.

当 $k_c r \gg 1$ 时,利用 (16),(17) 式可以求得:

$$\begin{cases} \tilde{U}_r(D_i) \sim \mu_S A_R \chi_{R_1} \chi'_{R_1} \Phi_R, \\ \tilde{U}_\varphi(D_i) \sim \mu_S A_L \chi_{L_1} \chi'_{L_1} \Phi_L, \\ \tilde{U}_z(D_i) \sim -K' \tilde{U}_r(D_i). \end{cases} \quad (43)$$

比较 (43) 和 (18) 式,可以得到下面的结论,即接收深度对面波位移谱的影响通过一个因子表达出来. 这个因子,对于 r 和 z 方向的位移来说,就是瑞雷波在深度 D_i 的相对激发程度;对于 φ 方向的位移来说,就是勒夫波在深度 D_i 的相对激发程度. z 方向的位移谱 $\tilde{U}_z(D_i)$ 和 r 方向的位移谱 $\tilde{U}_r(D_i)$ 之比,和平面波情形一样,都是 $-K'$.

3 多层弹性半空间中一般类型的地震震源引起的面波

3.1 破裂扩展方式对面波位移谱的影响

上面导出的结果 (43) 式是一般类型的点源引起的面波位移场表示式. 对于一般类型的地震震源(位错源)只要将[13]表 2 所列的 $S_i^m(kl)$ 的数值代入 (20),(22) 式,就可以得到单位面积单位错距的位错引起的面波位移谱 $\tilde{U}_{kl}(P, Q)$. 因此,任意形状的位错面 Σ 在分层弹性半空间中的 Q 点所引起的面波位移谱 $\tilde{U}(Q)$ 是:

$$\tilde{U}(Q) = \iint_\Sigma \Delta U_k(P) \tilde{U}_{kl}(P, Q) n_l(P) \mathrm{d}\Sigma. \quad (44)$$

设地震断层的断层面是一矩形,长为 L,宽为 D. 以 ξ, η 分别代表矩形的长和宽的方向 (图 1). 显然,ξ 即断层的走向,这里取和 x 轴方向一致;η 即其倾斜方向. 设破裂从矩形的一端 P_0 点开始,以速度 v_b 沿 ζ 方向扩展,ζ 和 ξ 的夹角以 α 表示. 断层面上的位错 $\Delta u_k(P)$ 可以表示成:

$$\Delta u_k(P) = U f_k g\left(t - \frac{\zeta}{v_b}\right), \quad (45)$$

式中,$g(t)$ 表示震源的时间函数,f_k 表示位错向量的方向余弦,U 表示错距. 由上式可知,位错 $\Delta u_k(P)$ 的谱 $\Delta U_k(P)$ 为

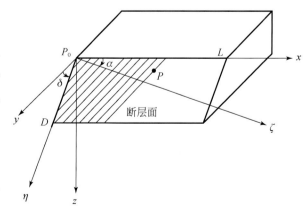

图 1 地震震源的模型

$$\Delta U_k(P) = Uf_k G(\omega) e^{-i\omega\frac{\zeta}{v_b}}, \tag{46}$$

$G(\omega)$ 表示震源的时间函数 $g(t)$ 的谱.

设 $\boldsymbol{R}_0$ 是 P_0 至观测点 Q 的向径，$\boldsymbol{R}$ 是断层面上的某一点 P 至 Q 的向径，显然，

$$\boldsymbol{R} = \boldsymbol{R}_0 - (\boldsymbol{\xi} + \boldsymbol{\eta}). \tag{47}$$

由此可以导出，当断层的线度 L 和 $\boldsymbol{R}_0$ 在水平面上的投影 r_0 之比 $\frac{L}{r_0} \ll 1$ 时，$\boldsymbol{R}$ 在水平面上的投影 r 为：

$$\begin{cases} r = r_0 \left\{ 1 - \frac{1}{r_0}(\xi\cos\varphi_0 + \eta\sin\varphi_0\cos\delta) + O\left[\left(\frac{L}{r_0}\right)^2\right] \right\}, \\ \varphi = \varphi_0 + \frac{1}{r_0}(\xi\sin\varphi_0 - \eta\cos\varphi_0\cos\delta) + O\left[\left(\frac{L}{r_0}\right)^2\right], \\ \cos m\varphi = \cos m\varphi_0 - \frac{m}{r_0}\sin m\varphi_0(\xi\sin\varphi_0 - \eta\cos\varphi_0\cos\delta) + O\left[\left(\frac{L}{r_0}\right)^2\right], \\ \sin m\varphi = \sin m\varphi_0 + \frac{m}{r_0}\cos m\varphi_0(\xi\sin\varphi_0 - \eta\cos\varphi_0\cos\delta) + O\left[\left(\frac{L}{r_0}\right)^2\right]. \end{cases} \tag{48}$$

把上式代入（44）式，便得到当 $\frac{L}{r_0} \ll 1$ 时，

$$\widetilde{\boldsymbol{U}}(Q) \sim G(\omega) U f_k n_l \widetilde{\boldsymbol{U}}_{kl}(P_0, Q) \iint_\Sigma e^{-i\omega\frac{\zeta}{v_b} + ik_C(\xi\cos\varphi_0 + \eta\sin\varphi_0\cos\delta)} d\Sigma, \tag{49}$$

式中，k_C 指的是 k_R 或 k_L. 如果是 $\widetilde{U}_r(Q)$，$\widetilde{U}_z(Q)$，则是 k_R；如果是 $\widetilde{U}_\varphi(Q)$，则是 k_L. 因为

$$d\Sigma = d\xi d\eta, \tag{50}$$

$$\zeta = \xi\cos\alpha + \eta\sin\alpha, \tag{51}$$

所以，

$$\widetilde{\boldsymbol{U}}(Q) \sim G(\omega) U f_k n_l \widetilde{\boldsymbol{U}}_{kl}(P_0, Q) \cdot SD_C, \tag{52}$$

其中，

$$S = DL, \tag{53}$$

$$D_C = \left(\frac{\sin X_C}{X_C} e^{-iX_C}\right)\left(\frac{\sin Y_C}{Y_C} e^{-iY_C}\right), \tag{54}$$

$$X_C = \frac{k_C L}{2}\left(\frac{v_C \cos\alpha}{v_b} - \cos\varphi_0\right), \tag{55}$$

$$Y_C = \frac{k_C D}{2}\left(\frac{v_C \sin\alpha}{v_b} - \sin\varphi_0\cos\delta\right), \tag{56}$$

$$v_C = \frac{\omega}{k_C}. \tag{57}$$

C 表示 R 或 L.

定义

$$m_0 = \mu_S U S, \tag{58}$$

m_0 称为地震矩. 那么

$$\widetilde{\boldsymbol{U}}(Q) \sim G(\omega) m_0 f_k n_l \frac{1}{\mu_S} \widetilde{\boldsymbol{U}}_{kl}(P_0, Q) D_C. \tag{59}$$

把上式中的 $\widetilde{\boldsymbol{U}}_{kl}(P_0, Q)$ 用（43）式代入，就可以得到：

$$\widetilde{\boldsymbol{U}}(Q) = \widetilde{U}_r(Q)\boldsymbol{e}_r + \widetilde{U}_\varphi(Q)\boldsymbol{e}_\varphi + \widetilde{U}_z(Q)\boldsymbol{e}_z, \tag{60}$$

$$\begin{cases} \widetilde{U}_r(Q) \sim Gm_0 D_R f_k n_l \Phi_R [S_i^m(kl)] \chi_{R_1} \chi'_{R_1} A_R, \\ \widetilde{U}_\varphi(Q) \sim Gm_0 D_L f_k n_l \Phi_L [S_i^m(kt)] \chi_{L_1} \chi'_{L_1} A_L, \\ \widetilde{U}_z(Q) \sim -K' \widetilde{U}_r(Q). \end{cases} \quad (61)$$

式中, 辐射图型因子 $\Phi_R [S_i^m(kl)]$, $\Phi_L [S_i^m(kl)]$ 要将由 (kl) 定义的源系数 $S_i^m(kl)$ 代入 (20), (22) 式求得. 从上式可见, 破裂扩展方式对面波位移谱的影响相当于乘上一个因子 D_R (对瑞雷波) 或 D_L (对勒夫波). D_R 或 D_L 称为有限性因子, 它们各包含两部分. 其中 X_R 或 X_L 表示破裂沿 ξ 方向传播对位移谱的影响, Y_R 或 Y_L 表示破裂沿 η 方向传播对位移谱的影响. 地震的强度通过地震矩 m_0 表现出来, 它与震源所在那层的介质的刚性系数 μ_s 和断层的规模 (面积) 成正比.

3.2 多层弹性半空间中一般类型的地震震源辐射的面波位移谱的表示式

综合上述结果, 可以把多层弹性半空间中一般类型的地震震源 (位错源) 辐射的面波位移谱表示为

$$\begin{cases} \widetilde{U}_r(Q) \sim Gm_0 D_R \Psi_R \chi_{R_1} \chi'_{R_1} A_R, \\ \widetilde{U}_\varphi(Q) \sim Gm_0 D_L \Psi_L \chi_{L_1} \chi'_{L_1} A_L, \\ \widetilde{U}_z(Q) \sim -K' \widetilde{U}_r(Q), \end{cases} \quad (62)$$

其中,

$$\begin{cases} \Psi_R = f_k n_l \Phi_R [S_i^m(kl)] = \Phi_R [f_k n_l S_i^m(kl)], \\ \Psi_L = f_k n_l \Phi_L [S_i^m(kl)] = \Phi_L [f_k n_l S_i^m(kl)]. \end{cases} \quad (63)$$

称为一般类型的位错源辐射的面波的辐射图型.

$\widetilde{U}(Q)$ 分解为七项的乘积. 头一项代表震源时间函数的谱. 第二项 m_0 代表地震的规模. 第三项代表破裂传播的方向性. 第四项 Ψ_R 或 Ψ_L 代表点源辐射的面波的方向性, 它与分层结构及震源的具体性质有关. 第五项 χ_{R_1} 或 χ_{L_1} 表示震源深度对面波位移谱的影响. 第六项 χ'_{R_1} 或 χ'_{L_1} 表示接收深度对面波位移谱的影响, 其形式和 χ_{R_1} 或 χ_{L_1} 是一样的. 第七项 A_R 或 A_L 仅与分层结构和传播路径有关, 而与震源的具体性质无关, 其意义是震源深度和接收深度均为零时由 $(kl)=(12)$ 确定的单位偶极矩的剪切位错元在震中距为 r 处所引起的 r 方向或 φ 方向的面波位移谱.

从所得结果中可以看到. 当 $S_1^m = S_2^m = S_3^m = 0$ 而 $S_4^m \neq 0$ 时, 对于 $\widetilde{U}_r(Q)$ 来说, 震源深度与接收深度是可互易的. 当 $S_1^m = S_2^m = S_4^m = 0$ 而 $S_3^m \neq 0$ 时, 对于 $\widetilde{U}_z(Q)$ 来说, 震源深度与接收深度也是可互易的. 当 $S_5^m = 0$ 而 $S_6^m \neq 0$ 时, 对于 $\widetilde{U}_\varphi(Q)$ 来说, 震源深度与接收深度也是可互易的. 对于一般情形, 即 S_i^m 不取上述值的情形, 震源深度和接收深度只是在形式上具有互易性.

作为一个例子, 根据 (63) 式, (20) 式, (22) 式以及文献 [13] 的表 2, 我们来推导地震学中有意义的一般滑动断层所引起的瑞雷波和勒夫波的辐射图型.

对于一般的滑动断层, 如图 1 所示. 错动方向与走向的夹角用 β 表示 ($0 \leq \beta \leq 2\pi$), 断层面与水平面的夹角 (倾角) 用 δ 表示 ($0 \leq \delta \leq \frac{\pi}{2}$), 那么 $\boldsymbol{f} = f_m \boldsymbol{e}_m$ 和 $\boldsymbol{n} = n_m \boldsymbol{e}_m$ 正交, 并且:

$$\begin{cases} f_1 = \cos\beta, \\ f_2 = \sin\beta\cos\delta, \\ f_3 = \sin\beta\sin\delta, \end{cases} \quad (64)$$

$$\begin{cases} n_1 = 0, \\ n_2 = \sin\delta, \\ n_3 = -\cos\delta. \end{cases} \quad (65)$$

由这两个式子和[13]表 2 可以求得 $f_k n_l s_i^m(kl)$。不等于零的 $f_k n_l S_i^m(kl)$ 计有：

$$\begin{cases} f_k n_l S_2^{0e}(kl) = \dfrac{k_R}{2\pi}(\gamma-1)\sin\beta\sin 2\delta, \\[4pt] f_k n_l S_4^{0e}(kl) = \dfrac{k_R}{2\pi}\left(\dfrac{4\gamma-1}{2}\right)\dfrac{\mu_S}{\mu_1}\sin\beta\sin 2\delta, \\[4pt] f_k n_l S_1^{1e}(kl) = -\dfrac{k_R}{2\pi}\cos\beta\cos\delta, \\[4pt] f_k n_l S_1^{10}(kl) = -\dfrac{k_R}{2\pi}\sin\beta\cos 2\delta, \\[4pt] f_k n_l S_5^{1e}(kl) = \dfrac{k_L}{2\pi}\cos\beta\cos\delta, \\[4pt] f_k n_l S_5^{10}(kl) = \dfrac{k_L}{2\pi}\sin\beta\cos 2\delta, \\[4pt] f_k n_l S_4^{2e}(kl) = \dfrac{k_R}{4\pi}\dfrac{\mu_S}{\mu_1}\sin\beta\sin 2\delta, \\[4pt] f_k n_l S_4^{20}(kl) = -\dfrac{k_R}{2\pi}\dfrac{\mu_S}{\mu_1}\cos\beta\sin\delta, \\[4pt] f_k n_l S_6^{2e}(kl) = -\dfrac{k_L}{4\pi}\dfrac{\mu_S}{\mu_1}\sin\beta\sin 2\delta, \\[4pt] f_k n_l S_6^{20}(kl) = \dfrac{k_L}{2\pi}\dfrac{\mu_S}{\mu_1}\cos\beta\sin\delta. \end{cases} \quad (66)$$

所以一般的滑动断层的瑞雷波和勒夫波辐射图型是：

$$\Psi_R = \left[\dfrac{\mu_1}{\mu_S}(1-\gamma)\dfrac{\chi_{R_3}}{\chi_{R_1}} + 2\gamma - \dfrac{1}{2}\right]\sin\beta\sin 2\delta - i\dfrac{\mu_1}{\mu_S}K\dfrac{\chi_{R_4}}{\chi_{R_1}}(\cos\varphi_0\cos\beta\cos\delta + \sin\varphi_0\sin\beta\cos 2\delta)$$
$$-\left(\dfrac{1}{2}\cos 2\varphi_0\sin\beta\sin 2\delta - \sin 2\varphi_0\cos\beta\sin\delta\right), \quad (67)$$

$$\Psi_L = i\dfrac{\mu_1}{\mu_S}\dfrac{\chi_{L_2}}{\chi_{L_1}}(\sin\varphi_0\cos\beta\cos\delta + \cos\varphi_0\sin\beta\cos 2\delta) + \left(\cos 2\varphi_0\cos\beta\sin\delta - \dfrac{1}{2}\sin 2\varphi_0\sin\beta\sin 2\delta\right), \quad (68)$$

这里，φ_0 是观测点相对于初破裂点的方位角 φ_S 和断层面的走向 φ_f 的差：

$$\varphi_0 = \varphi_S - \varphi_f, \quad 0 \leqslant \varphi_0 \leqslant 2\pi. \quad (69)$$

由 (67)，(68) 两式可见，多层弹性半空间的具体性质以及断层面的走向 φ_f，倾角 δ，错动方向 β 完全确定了面波的辐射图型。根据这些表示式，就可以计算出各种复杂介质结构及任意倾角和错动方向的滑动断层引起的面波辐射图型。而根据 (20)，(22) 式，则可以计算出任意多极源的面波辐射图型。

不难看出，

$$\begin{cases} \Phi_R(\varphi+\pi) = \Phi_R^*(\varphi), \\ \Phi_L(\varphi+\pi) = \Phi_L^*(\varphi), \end{cases} \quad (70)$$

这里，星号表示复共轭. 由上式可见，面波的辐射图型具有中心对称性，当两接收点的方位角相差 π 时，其 Φ_R 或 Φ_L 大小相等，位相角数值相等、符号相反. 哈斯克尔[4] 曾就偶极源指出瑞雷波辐射图型的中心对称性. 本·梅纳赫姆和哈克莱德[6] 也曾就单力偶和双力偶震源导出瑞雷波和勒夫波的这一性质. 上面就一般类型的点源（单力、偶极源乃至高级源）也导出了这一关系. 这就是说，辐射图型的中心对称性是一个普遍的性质，不只对偶极源成立.

(70) 式可以改写成

$$\begin{cases} |\Phi_R(\varphi+\pi)| = |\Phi_R(\varphi)|, \\ |\Phi_L(\varphi+\pi)| = |\Phi_L(\varphi)|, \end{cases} \quad (71)$$

$$\begin{cases} \arg \Phi_R(\varphi+\pi) + \arg \Phi_R(\varphi) = 0, \\ \arg \Phi_L(\varphi+\pi) + \arg \Phi_L(\varphi) = 0, \end{cases} \quad (72)$$

考虑到 $A_R \chi_{R_1}$ 和 $A_L \chi_{L_1}$ 的位相是 $\pm\pi + \frac{\pi}{4}$，所以

$$\begin{cases} \arg \widetilde{U}_r(0, \varphi+\pi) + \arg \widetilde{U}_r(0, \varphi) = \frac{\pi}{2}, \\ \arg \widetilde{U}_\varphi(0, \varphi+\pi) + \arg \widetilde{U}_\varphi(0, \varphi) = \frac{\pi}{2}, \\ \arg \widetilde{U}_z(0, \varphi+\pi) + \arg \widetilde{U}_z(0, \varphi) = -\frac{\pi}{2}, \end{cases} \quad (73)$$

这个结果可以看作是本·梅纳赫姆和哈克莱德[6] 类似结果的推广.

4 结语

上面从上一篇文章[13] 导出的多层弹性半空间中一般类型的震源引起的位移谱表示式（形式解）出发，进一步导出了点源辐射的瑞雷波和勒夫波的位移谱表示式. 证明了层矩阵及其他矩阵的一些性质，并利用这些性质把面波位移谱表示式化为物理意义明显和便于分析、计算的形式. 最后得到的面波位移谱表示式中清楚地反映了震源的特性、传播面波的介质特性以及接收点位置三者是如何影响面波位移场的. 表示式的第一项震源时间函数表示地震发生的时间进程，是震源性质的一种反映. 第二项地震矩与地震断层的规模和震源区介质的特性有关，是地震强度的一种合适的度量. 第三项有限性因子反映了破裂传播方式，也是震源性质的一种反映. 第四项辐射图型因子，反映了点状震源辐射面波的方向性，是震源性质和整个多层半空间性质的一种反映. 第五项不同深度的震源的面波的相对激发程度，反映了震源深度对面波位移场的影响，也是震源性质（深度）的一种反映. 第六项，不同深度接收点的相对激发程度，反映了接收点深度不同时面波位移场的差异. 第七项表示震源深度和接收深度均为零时由 $(kl) = (12)$ 确定的单位偶极矩的剪切位错元在震中距为 r 处所引起的面波位移谱，是整个多层弹性半空间的性质和传播路径的反映. 震源时间函数、地震矩、有限性因子、辐射图型因子、不同深度的震源辐射的面波的相对激发程度都是和震源性质有关的量，它们从不同的侧面反映了震源的性质. 采用适当的方法，从观测资料中将这些因素对面波位移场的影响分离出来，就能够确定震源的深度和研究地震的机制，从而可以对震源的性质有更多的了解. 面波的频散性质、多层弹性半空间对地震波的响应以及其他一些特性是由多层弹性半空间的介质的性质决定的，通过对这些性质的研究将有助于确定地壳和上地幔的结构. 因此，以上给出的一些

普遍结果可以应用于震源深度、震源机制和地壳及上地幔结构的研究中. 在进一步的工作中, 我们将给出有关的数值结果及其具体应用.

参 考 文 献

[1] A. Ben-Menahem, Radiation of seismic surface waves from finite moving sources, *Bull. Seism. Soc. Amer.*, **51**, 3, 401–435, 1961.

[2] A. Ben-Menahem, Radiation of seismic body waves from a finite moving source in the Earth, *J. Geophys. Res.*, **67**, 1, 345–350, 1962.

[3] N. A. Haskell, Radiation pattern of Rayleigh waves from a fault of arbitrary dip and direction of motion in a homogeneous medium, *Bull. Seism. Soc. Amer.*, **53**, 3, 619–642, 1963.

[4] N. A. Haskell, Radiation pattern of surface waves from point sources in a multilayered medium, *Bull. Seism. Soc. Amer.*, **54**, 1, 377–393, 1964.

[5] D. G. Harkrider, Surface waves in multilayered elastic media, Ⅰ. Rayleigh and Love waves from buried sources in a multilayered elastic half-space, *Bull. Seism. Soc. Amer.*, **54**, 2, 627–679, 1964.

[6] A. Ben-Menahem, and D. C. Harkrider, Radiation pattern of seismic surface waves from buried dipolar point sources in a flat stratified earth, *J. G. R.*, **69**, 12, 2605–2619, 1964.

[7] T. Maruyama, On the force equivalents of dynamical elastic dislocations with reference to the earthquake mechanism, *Bull. Earthq. Res. Inst.*, **41**, 3, 467–486, 1963.

[8] T. Maruyama, Statical elastic dislocations in an infinite and semi-infinite medium, *Bull. Earthq. Res. Inst.*, **42**, 2, 289–368, 1964.

[9] R. Burridge, and L. Knopoff, Body force equivalents for seismic dislocations, *Bull. Seism. Soc. Amer.*, **50**, 1, 117–134, 1960.

[10] N. A. Haskell, The dispersion of surface waves on multilayered media, *Bull. Seism. Soc. Amer.*, **43**, 1, 17–34, 1953.

[11] W. W. Hansen, A new type of expansion in radiation problems, *Phys. Rev.*, **47**, 2, 139–143, 1935.

[12] A. Ben-Menahem, and S. J. Singh, Multipolar elastic fields in a layered half-space, *Bull. Seism. Soc. Amer.*, **58**, 5, 1519–1572, 1968.

[13] 陈运泰, 多层弹性半空间中的地震波（一）, 地球物理学报, **17**, 1, 20–43, 1974.

(注: 本文共分两部分, 第一部分也收录在本书中)

Seismic waves in multilayered elastic half-space(II)

Chen Yun-Tai

Institute of Geophysics, Academia Sinica

Abstract On the basis of spectra expressions, obtained previously by using Hansen's expansion method and Haskell matrix formulation, for the displacement fields due to general types of sources in a multilayered homogeneous isotropic and perfectly elastic half-space, the spectra expressions for the displacement fields of surface waves radiated from the sources are derived by evaluating the contribution of the residues. It is shown that the displacement spectra for these surface waves can always be separated into seven factors representing source tine-function, seismic moment, propagation effect of fracture, radiation pattern, source depth, receiver depth and structure of elastic half-space respectively. For the multilayered elastic half-space, general expressions for both Rayleigh and Love wave radiation patterns are given explicitly in a form representing their dependence on the characteristics of the source, source depth as well as on the structure of the medium. Results here obtained can be used to determine focal depth, crust and mantle structure and source mechanism.

根据地面形变的观测研究 1966 年邢台地震的震源过程*

陈运泰　林邦慧　林中洋　李志勇

中国科学院地球物理研究所

摘要　本文以完整的形式给出拉梅常数不相等情形的半无限弹性介质中任意倾角的矩形滑动断层引起的地震位移场解析表示式．以一些数值结果说明介质的泊松比、断层面的倾角、上界和下界对地面的地震位移场的影响．在比较 1966 年邢台地震的地形变资料和计算得到的各种走向、倾向、倾角、断层面长度、宽度、震源深度和错距的单个的矩形滑动断层引起的地面位移之后指出，简单的滑动断层错动模式不能同时很好地解释观测到的邢台地震的水平和垂直形变．为了解释观测结果，提出了一个复合的断层模式．这个复合断层模式由六个简单的矩形滑动断层构成．运用网格尝试法，得到了基本上符合观测到的水平和垂直位移场的震源参数．结果是：第一、二、三部分，出露到地面，倾角 45°，宽度 15km；第四、五、六部分，倾角 82°，宽度 30km．其他参数是：断层面走向 N35°E，倾向 N125°E，断层总长度 50km，断层各部分的平均走向滑动错距依次为 −78，−134，−17，2，−3 和 1cm（负号表示右旋走向滑动，正号表示左旋走向滑动），平均倾向滑动错距依次为 2，50，88，24，−5 和 −23cm（负号表示逆断层，正号表示正断层），地震矩依次为 3.2，5.9，3.7，2.0，0.5，1.9×10^{25}dyn·cm，应力降依次为 22，42，33，5，1，4bar，应变降依次为 3.3，6.4，5.0，0.8，0.2，0.6×10^{-5}，释放的总能量的下限为 6.1×10^{22}erg．对比所得结果和 6 级以上的地震震中分布，表明邢台地震的两个较大的前震及主震和复合断层上部的错动关系较大，而大的余震则和复合断层下面那部分的端部的错动相联系．邢台地震区近半个世纪以来的水准测量资料表明，这个地区下降速率的上限约为 5mm/a．如果认为通过这次地震，积累在主要断层上的应变已基本释放完毕，那么，由这次地震的倾向滑动错距的大小可以估计这个地震断层上的地震复发周期的下限大约是 176a．

1　引言

大地震通常都伴随有地表面的断裂以及地面的大幅度水平和垂直运动，这种由地震引起的地面形变和地震震源的性质有密切的关系，因此，可以用它来研究震源的性质．如果在独立地运用体波、面波和地面形变等资料研究震源的基础上，对震源进行综合研究，无疑会使我们对震源过程的认识更为充分．

本文就是利用地面形变的观测资料研究 1966 年 3 月 22 日河北省邢台宁晋地震的一个结果．这次地震是群发型的，包括表 1 所示 5 个大于、等于 6 级的地震．主震群前后的地面形变资料表明[1]，邢台地震以后震中地区出现大范围、大幅度的水平形变和垂直形变（见图 1，图 2）．分析表明，专门埋设的水泥标石点的稳固性是可靠的，因此，这两幅图所表示的主要是地面本身的形变，而不是标石点相对于地面的移动．从这些观测结果中可以看到邢台地震极震区断裂带两盘块体的明显的相对运动．断裂带的走向大约是 N40°E，两盘的水平错动呈右旋性质，地面的水平位移在许多地点有 30cm 左右，最大达 40 多厘米．同

---

* 本文发表于《地球物理学报》，1975 年，第 **18** 卷，第 3 期，164–182.

时，还出现大范围的下降，下降带的走向和断裂带的走向一致，长约60km，宽约20km，下降幅度一般约为水平位移的一半，最大达44cm；下降带的外围是上升带，上升幅度远不及下降幅度，一般仅约为下降幅度的十分之一．这些观测事实强烈表明这次地震是由断层错动引起的．可以设想其震源为地球介质中的一个位移不连续的面——位错面．根据这个设想，可以运用弹性位错理论[2-6]分析地震断层错动引起的静力学和动力学变化过程．

表1

编号	日期	发震时刻（北京时间）	纬度	经度	地区	震级	震源深度/km
1	1966.3.8	05-29-14	37°21′N	114°55′E	河北隆尧	6.8	10
2	1966.3.22	16-11-36	37°30′N	115°05′E	河北宁晋	6.7	9
3	1966.3.22	16-19-46	37°32′N	115°03′E	河北宁晋	7.2	9
4	1966.3.26	23-19-04	37°41′N	115°16′E	河北束鹿	6.2	15
5	1966.3.29	14-11-59	37°21′N	115°02′E	河北巨鹿	6.0	25

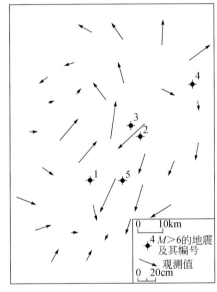

图1 邢台地震前后（1959年，1960年至1966年）
震中地区地面水平形变的向量图

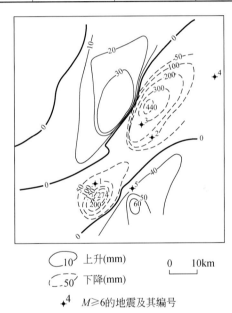

图2 邢台地震前后（1965年至1966年）
震中地区地面垂直形变的等值线图

弹性静力学位错理论已被成功地用以分析断层错动引起的地面位移场、应变场、倾斜场和应力场的变化[7-16]，借以讨论震源的性质．以往的工作中，为了分析的方便，通常采取简单的断层模式，即断层面和地面垂直的矩形的走向、倾斜或一般的滑动断层．这种模式虽能反映断层面和地面近于垂直的实际地震断层的许多主要图像，但对于大多数断层面和地面斜交的实际地震断层，它是不充分的，不能反映实际地震断层引起的位移场、应变场、倾斜场和应力场的若干重要图像，如位移场、应变场、倾斜场和应力场的不对称性等等．鉴于数学分析的繁琐，计算具有任意倾角的走向或倾向滑动断层引起的位移场通常运用丸山卓男（Maruyama）[4]的双重积分公式，这时，需要计算元位错面的贡献，然后，借助于二维数值积分求得位错面引起的位移场，如卡尼泽兹和托克索兹（Canitez and Toksöz）[15]、安藤雅孝（Ando）[16]等的作法就是这样．近来，曼新哈和斯迈里（Mansinha and Smylie）[17]给出了拉梅常数相等的、半无限介质中任意倾角的、矩形的、走向和倾向滑动断层的、地震位移场的精确的解析表示式．虽然精确的解析表示式颇为冗长，但只包括代数函数和初等的超越函数，易于数值计算．我们在这里将顺便给出拉梅常数不相等的半无限介质中任意倾角的矩形的走向和倾向滑动断层的地震位移场的精确的解析表示式．这些表示式，当拉梅常数相等时退化为曼新哈和斯迈里[17]的结果；当拉梅常数相等、倾角等于π/2并且观测点在地面时，退化为普雷斯（Press）[10]的结果．因此，下面将给出的地震位移场的精确的解析表示式可以看作是普雷斯以及曼新哈和斯迈里的结果的一个推广．

2 地震断层在地面引起的位移场

把地震断层视为介质中的一个位移向量不连续的面（位错面）．按照弹性位错理论，在均匀、各向同性和完全弹性的半无限介质中，任意形状的位错面 Σ 在介质中的某一点 Q（坐标 x_m, $m=1, 2, 3$）引起的位移是[4,10]：

$$\boldsymbol{u}(Q) = u_m(Q)\boldsymbol{e}_m, \tag{1}$$

$$u_m(Q) = \iint_\Sigma \Delta U_k(P) W_{kl}^m(P,Q) n_l(P) \mathrm{d}\Sigma, \tag{2}$$

这里，采用哑指标下的求和约定．$\boldsymbol{e}_m$（$m=1, 2, 3$）表示 x_m 方向的单位向量．$W_{kl}^m(P,Q)$ 是弹性半无限介质中由 (kl) 定义的、作用于某一点 P（坐标 ζ_m, $m=1, 2, 3$）的力系在 Q 点引起的沿 x_m 方向的位移（差一个量纲因子）．$\Delta U_k(P)$（$k=1, 2, 3$）是在 P 点的位错向量 $\Delta \boldsymbol{U}$ 的三个分量，$n_l(P)$（$l=1, 2, 3$）是在 P 点的面积元 $\mathrm{d}\Sigma$ 的法向 $\boldsymbol{n}$ 的方向余弦．$W_{kl}^m(P,Q)$ 由下式表示：

$$W_{kl}^m(P,Q) = \lambda \delta_{kl} \frac{\partial u_m^n}{\partial \zeta_n} + \mu \left(\frac{\partial u_m^k}{\partial \zeta_l} + \frac{\partial u_m^l}{\partial \zeta_k} \right). \tag{3}$$

式中，λ, μ 是拉梅常数，u_m^k 是弹性半无限介质中作用于 P 点的 x_k 方向的单位集中力在 Q 点引起的沿 x_m 方向的位移．u_m^k 的具体表示式见参考资料 [10], [18].

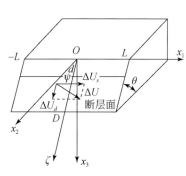

图 3 任意倾角的矩形断层错动模式

设断层面是一个矩形位错面，长为 $2L$, 宽为 ΔD. 将直角坐标系 (x_1, x_2, x_3) 的原点取在地面上，取和断层走向一致的方向为 x_1 方向，x_3 垂直于地面，向下为正（图3）．以 θ 代表断层面和地面的夹角（倾角），那么断层面上的法线方向 $\boldsymbol{n}$ 的方向余弦是

$$\boldsymbol{n} = \{0, \sin\theta, -\cos\theta\}. \tag{4}$$

对于走向滑动断层，若以 ΔU_s 表示走向滑动的错距，那么错位向量

$$\Delta \boldsymbol{U} = \{\Delta U_s, 0, 0, \}, \tag{5}$$

所以

$$u_m(Q) = \Delta U_s \iint_\Sigma \left(W_{12}^m \sin\theta - W_{31}^m \cos\theta \right) \mathrm{d}\Sigma. \tag{6}$$

以 ζ 代表断层面宽度方向的坐标，那么 $\mathrm{d}\Sigma = \mathrm{d}\zeta_1 \mathrm{d}\zeta$, 从而

$$u_m(Q) = \mu \Delta U_s \int_d^D \int_{-L}^L \left[\left(\frac{\partial u_m^1}{\partial \zeta_2} + \frac{\partial u_m^2}{\partial \zeta_1} \right) \sin\theta - \left(\frac{\partial u_m^1}{\partial \zeta_3} + \frac{\partial u_m^3}{\partial \zeta_1} \right) \cos\theta \right] \mathrm{d}\zeta_1 \mathrm{d}\zeta. \tag{7}$$

对于倾向滑动断层，若以 ΔU_d 表示倾向滑动的错距，那么

$$\Delta \boldsymbol{U} = \{0, \Delta U_d \cos\theta, \Delta U_d \sin\theta\}, \tag{8}$$

所以

$$u_m(Q) = \Delta U_d \iint_\Sigma \left[\frac{1}{2}(W_{22}^m - W_{33}^m)\sin 2\theta - W_{23}^m \cos 2\theta \right] \mathrm{d}\Sigma, \tag{9}$$

或者

$$u_m(Q) = \mu \Delta U_d \int_d^D \int_{-L}^{L} \left[\left(\frac{\partial u_m^2}{\partial \zeta_2} - \frac{\partial u_m^3}{\partial \zeta_3} \right) \sin 2\theta - \left(\frac{\partial u_m^2}{\partial \zeta_3} + \frac{\partial u_m^3}{\partial \zeta_2} \right) \cos 2\theta \right] d\zeta_1 d\zeta. \tag{10}$$

对于一般的滑动断层，若以 ΔU 表示其总错距，以 ψ 代表错动方向和断层面走向的夹角，顺时针为正，那么位错向量

$$\Delta U = \{\Delta U \cos\psi, \ \Delta U \sin\psi \cos\theta, \ \Delta U \sin\psi \sin\theta\}, \tag{11}$$

它所引起的位移是走向滑动断层和倾向滑动断层所引起的位移的叠加，其走向滑动错距 ΔU_s 和倾向滑动错距 ΔU_d 分别为：

$$\begin{cases} \Delta U_s = \Delta U \cos\psi, \\ \Delta U_d = \Delta U \sin\psi. \end{cases} \tag{12}$$

由参考资料 [10] 或 [18] 所给出的 u_m^k 的具体表示式，可将 (7) 和 (9) 式中对 ζ_1 和 ζ 的积分依次积出. 为此，引进以下几个量：$r_2, q_2, r_3, q_3, h, k, R, Q$，

$$\begin{cases} r_2 = x_2 \sin\theta - x_3 \cos\theta, \\ r_3 = x_2 \cos\theta + x_3 \sin\theta, \end{cases} \tag{13}$$

$$\begin{cases} q_2 = x_2 \sin\theta + x_3 \cos\theta, \\ q_3 = -x_2 \cos\theta + x_3 \sin\theta, \end{cases} \tag{14}$$

$$\begin{aligned} R^2 &= (x_1 - \zeta_1)^2 + (x_2 - \zeta_2)^2 + (x_3 - \zeta_3)^2 \\ &= (x_1 - \zeta_1)^2 + r_2^2 + (r_3 - \zeta)^2, \end{aligned} \tag{15}$$

$$\begin{aligned} Q^2 &= (x_1 - \zeta_1)^2 + (x_2 - \zeta_2)^2 + (x_3 + \zeta_3)^2 \\ &= (x_1 - \zeta_1)^2 + q_2^2 + (q_3 + \zeta)^2 \\ &= (x_1 - \zeta_1)^2 + h^2 = k^2 + (q_3 + \zeta)^2, \end{aligned} \tag{16}$$

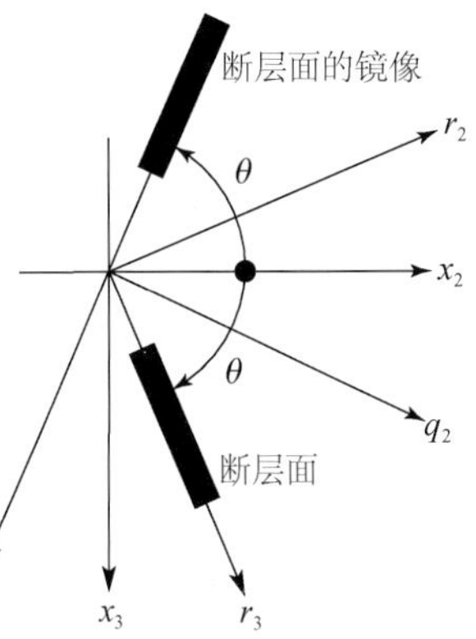

图 4 断层面和它的镜像

这些量的意义如图 4 所示.

略去繁琐的计算细节，下面给出以不定积分表示的结果，结果是：

对于走向滑动断层，

$$\begin{aligned} 8\pi \frac{u_1}{\Delta U_s} &= (x_1 - \zeta_1) \left[\frac{4\delta}{1+\delta} \frac{r_2}{R(R + r_3 - \zeta)} - \frac{1}{1+\delta} \frac{4q_2 - 4(1-\delta)x_3 \cos\theta}{Q(Q + q_3 + \zeta)} \right. \\ &\left. - 2\frac{1-\delta}{\delta} \frac{\tan\theta}{Q + x_3 + \zeta_3} + \frac{8\delta}{1+\delta} \frac{q_2 x_3 \sin\theta}{Q^3} - \frac{8\delta}{1+\delta} q_2 q_3 x_3 \sin\theta \frac{(2Q + q_3 + \zeta)}{Q^3 (Q + q_3 + \zeta)^2} \right] \\ &- 4\frac{1-\delta}{\delta} \tan^2\theta \tan^{-1} \left[\frac{(k - q_2 \cos\theta)(Q - k) + (q_3 + \zeta)k\sin\theta}{(x_1 - \zeta_1)(q_3 + \zeta)\cos\theta} \right] \\ &+ 2\tan^{-1} \frac{(x_1 - \zeta_1)(r_3 - \zeta)}{r_2 R} - 2\tan^{-1} \frac{(x_1 - \zeta_1)(q_3 + \zeta)}{q_2 Q}, \end{aligned} \tag{17}$$

$$\begin{aligned} 8\pi \frac{u_2}{\Delta U_s} &= \sin\theta \left[2\frac{1-\delta}{\delta} \tan\theta \sec\theta \ln(Q + x_3 + \zeta_3) - 2\frac{1-\delta}{1+\delta} \ln(R + r_3 - \zeta) \right. \\ &\left. - \left(2\frac{1-\delta}{1+\delta} + 2\frac{1-\delta}{\delta}\tan^2\theta \right) \ln(Q + q_3 + \zeta) \right] + \frac{4\delta}{1+\delta} \frac{r_2^2 \sin\theta}{R(R + r_3 - \zeta)} + \frac{4\delta}{1+\delta} \frac{r_2 \cos\theta}{R} \\ &- \frac{4\delta}{1+\delta} \sin\theta \frac{\left[2x_3(q_2 \cos\theta - q_3 \sin\theta) + q_2 \left(q_2 + \frac{1-\delta}{\delta} x_2 \sin\theta \right) \right]}{Q(Q + q_3 + \zeta)} \end{aligned}$$

$$-2\frac{1-\delta}{\delta}\tan\theta\frac{(x_2-\zeta_2)}{Q+x_3+\zeta_3}+\frac{4}{1+\delta}\frac{[\delta q_2\cos\theta-(1-\delta)q_3\sin\theta-(3\delta-1)x_3\sin^2\theta]}{Q}$$

$$+\frac{8\delta}{1+\delta}q_2x_3\sin\theta\frac{[(x_2-\zeta_2)+q_3\cos\theta]}{Q^3}-\frac{8\delta}{1+\delta}q_2^2q_3x_3\sin^2\theta\frac{2Q+q_3+\zeta}{Q^3(Q+q_3+\zeta)^2}, \quad (18)$$

$$8\pi\frac{u_3}{\Delta U_s}=\cos\theta\left[2\frac{1-\delta}{1+\delta}\ln(R+r_3-\zeta)+2\frac{1-\delta}{1+\delta}\left(1+\frac{1+\delta}{\delta}\tan^2\theta\right)\ln(Q+q_3+\zeta)\right.$$

$$\left.-2\frac{1-\delta}{\delta}\tan\theta\sec\theta\ln(Q+x_3+\zeta_3)\right]+\frac{4\delta}{1+\delta}\frac{r_2\sin\theta}{R}$$

$$+\frac{4}{1+\delta}\sin\theta\frac{[(2-3\delta)q_2+(3\delta-1)x_2\sin\theta]}{Q}-\frac{4\delta}{1+\delta}\frac{r_2^2\cos\theta}{R(R+r_3-\zeta)}$$

$$+\frac{4}{1+\delta}\frac{[(4\delta-1)\sin^2\theta-\delta]q_2x_3-(1-\delta)q_2q_3\sin\theta-\delta x_2x_3\sin\theta-\delta q_3x_2\sin^2\theta}{Q(Q+q_3+\zeta)}$$

$$+\frac{8\delta}{1+\delta}q_2x_3\sin\theta\frac{[(x_3+\zeta_3)-q_3\sin\theta]}{Q^3}-\frac{8\delta}{1+\delta}q_2^2q_3x_3\cos\theta\sin\theta\frac{(2Q+q_3+\zeta)}{Q^3(Q+q_3+\zeta)^2}. \quad (19)$$

对于倾向滑动断层，

$$8\pi\frac{u_1}{\Delta U_d}=(x_2-\zeta_2)\sin\theta\left[\frac{4\delta}{1+\delta}\frac{1}{R}+\frac{4}{1+\delta}\frac{1}{Q}-\frac{8\delta}{1+\delta}\frac{\zeta_3x_3}{Q^3}-2\frac{1-\delta}{\delta}\frac{1}{Q+x_3+\zeta_3}\right]$$

$$-\cos\theta\left[2\frac{1-\delta}{\delta}\ln(Q+x_3+\zeta_3)+\frac{4\delta}{1+\delta}\frac{(x_3-\zeta_3)}{R}+\frac{4}{1+\delta}\frac{(x_3-\zeta_3)}{Q}\right.$$

$$\left.+\frac{8\delta}{1+\delta}\frac{\zeta_3x_3(x_3+\zeta_3)}{Q^3}\right]+2\frac{1-\delta}{\delta}\frac{1}{\cos\theta}[\ln(Q+x_3+\zeta_3)$$

$$-\sin\theta\ln(Q+q_3+\zeta)]+4x_3\left[\frac{\cos\theta}{Q}-\frac{q_2\sin\theta}{Q(Q+q_3+\zeta)}\right], \quad (20)$$

$$8\pi\frac{u_2}{\Delta U_d}=\sin\theta\left\{-2\frac{1-\delta}{1+\delta}\ln(R+x_1-\zeta_1)+2\frac{1-\delta}{1+\delta}\ln(Q+x_1-\zeta_1)\right.$$

$$+\frac{8\delta}{1+\delta}\frac{\zeta_3x_3}{Q(Q+x_1-\zeta_1)}+2\frac{1-\delta}{\delta}\frac{(x_1-\zeta_1)}{Q+x_3+\zeta_3}+(x_2-\zeta_2)^2\left[\frac{4\delta}{1+\delta}\frac{1}{R(R+x_1-\zeta_1)}\right.$$

$$\left.\left.+\frac{4}{1+\delta}\frac{1}{Q(Q+x_1-\zeta_1)}-\frac{8\delta}{1+\delta}\zeta_3x_3\frac{(2Q+x_1-\zeta_1)}{Q^3(Q+x_1-\zeta_1)^2}\right]\right\}$$

$$-\cos\theta\left\{(x_2-\zeta_2)\left[\frac{4\delta}{1+\delta}\frac{(x_3-\zeta_3)}{R(R+x_1-\zeta_1)}+\frac{4}{1+\delta}\frac{(x_3-\zeta_3)}{Q(Q+x_1-\zeta_1)}\right.\right.$$

$$\left.+\frac{8\delta}{1+\delta}\zeta_3x_3(x_3+\zeta_3)\frac{(2Q+x_1-\zeta_1)}{Q^3(Q+x_1-\zeta_1)^2}\right]+4\frac{1-\delta}{\delta}\tan^{-1}\frac{(x_1-\zeta_1)(x_2-\zeta_2)}{(h+x_3+\zeta_3)(Q+h)}$$

$$-2\tan^{-1}\frac{(x_1-\zeta_1)(r_3-\zeta)}{r_2R}+\frac{2}{\delta}\tan^{-1}\frac{(x_1-\zeta_1)(q_3+\zeta)}{q_2Q}$$

$$+4\left[\frac{1-\delta}{\delta}\frac{1}{\cos\theta}\tan^{-1}\frac{(k-q_2\cos\theta)(Q-k)+(q_3+\zeta)k\sin\theta}{(x_1-\zeta_1)(q_3+\zeta)\cos\theta}\right]$$

$$\left.+4x_3\left[\frac{(\sin^2\theta-\cos^2\theta)(q_3+\zeta)+2q_2\cos\theta\sin\theta}{Q(Q+x_1-\zeta_1)}+\frac{(x_1-\zeta_1)\sin^2\theta}{Q(Q+q_3+\zeta)}\right]\right\}, \quad (21)$$

$$8\pi\frac{u_3}{\Delta U_d}=\sin\theta\left\{(x_2-\zeta_2)\left[\frac{4\delta}{1+\delta}\frac{(x_3-\zeta_3)}{R(R+x_1-\zeta_1)}+\frac{4}{1+\delta}\frac{(x_3-\zeta_3)}{Q(Q+x_1-\zeta_1)}\right.\right.$$

$$\left.-\frac{8\delta}{1+\delta}\zeta_3x_3(x_3+\zeta_3)\frac{(2Q+x_1-\zeta_1)}{Q^3(Q+x_1-\zeta_1)^2}\right]-4\frac{1-\delta}{\delta}\tan^{-1}\frac{(x_1-\zeta_1)(x_2-\zeta_2)}{(h+x_3+\zeta_3)(Q+h)}$$

$$+ 2\tan^{-1}\frac{(x_1-\zeta_1)(r_3-\zeta)}{r_2 R} - \frac{2}{\delta}\tan^{-1}\frac{(x_1-\zeta_1)(q_3+\zeta)}{q_2 Q}\Big\}$$

$$+ \cos\theta\Big\{2\frac{1-\delta}{1+\delta}\ln(R+x_1-\zeta_1) - 2\frac{1-\delta}{1+\delta}\ln(Q+x_1-\zeta_1)$$

$$- \frac{4\delta}{1+\delta}\frac{(x_3-\zeta_3)^2}{R(R+x_1-\zeta_1)} - \frac{4}{1+\delta}\frac{[(x_3+\zeta_3)^2-2\delta\zeta_3 x_3]}{Q(Q+x_1-\zeta_1)}$$

$$- \frac{8\delta}{1+\delta}\zeta_3 x_3(x_3+\zeta_3)^2\frac{(2Q+x_1-\zeta_1)}{Q^3(Q+x_1-\zeta_1)^2}\Big\}$$

$$+ 4x_3\Big\{\cos\theta\sin\theta\Big[\frac{2(q_3+\zeta)}{Q(Q+x_1-\zeta_1)}+\frac{(x_1-\zeta_1)}{Q(Q+q_3+\zeta)}\Big]$$

$$- q_2\frac{(\sin^2\theta-\cos^2\theta)}{Q(Q+x_1-\zeta_1)}\Big\}. \tag{22}$$

其中,常数 δ 和拉梅常数以及泊松比 σ 有如下的关系：

$$\delta = \frac{\lambda+\mu}{\lambda+3\mu} = \frac{1}{3-4\sigma}. \tag{23}$$

(17)-(22)式的右边诸项,均要代入二重积分的上、下限,即:

$$[f(\zeta_1,\zeta)]\| = f(L,D) - f(L,d) - f(-L,D) + f(L,d). \tag{24}$$

当拉梅常数 $\lambda=\mu$ 即 $\delta=1/2$ 时,(17)-(22)式退化为曼新哈和斯迈里的结果[17];当 $\lambda=\mu$ 且 $\theta=\pi/2$ 时,若观测点在地面上(即 $x_3=0$),则它们退化为普雷斯的结果[10].

3 断层参数对地震位移场的影响

现在根据(17)-(22)式的数值结果分析断层面的倾角、上界和下届对地震位移场的影响.实际地球介质可视为泊松体,即 $\lambda=\mu$ 的介质.计算表明,泊松比的变化对断层附近的位移场的影响不大.图5以断层面出露地面($d=0$)而 $D/L=1$、倾角等于 $60°$ 的情形为例,表示当泊松比分别为0.23和0.30两种情况下当 $x_1=x_3=0$ 时位移随 x_2 的变化.由该图可见,泊松比对断层附近的位移场的影响是不大的.考虑到这些情况,在以下的数值计算中,均按泊松体情形计算.

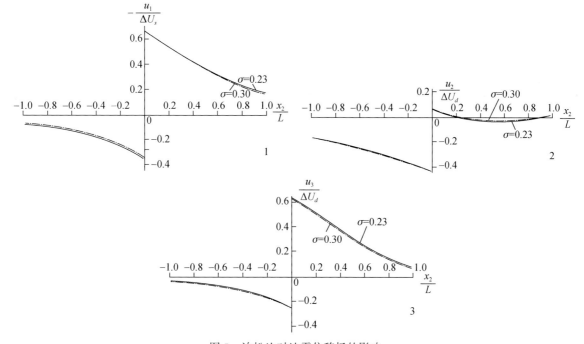

图5 泊松比对地震位移场的影响

$d=0$, $D/L=1$, $\theta=60°$

对于走向滑动断层,在不同的倾角的情况下,水平位移 u_1 沿 x_2 轴的变化可由图 6 和图 7 看出. 图 6 表示断层出露地面($d=0$)的情形,图 7 以 $d/L=0.1$ 为例,表示断层面上界离开地面有一定距离的情形. 这些曲线清楚地表明:在 x_2 轴上,水平位移 u_1 的不对称性取决于 θ. 由于这个原因,沿 x_2 轴方向的测线上的水平位移 u_1 的不对称性可以用来判断断层面的倾向和确定其倾角. 当 $D/L=1$ 时,对于断层出露地面的情形,断层面两边的位移之比 u_1^+/u_1^- 和倾角 θ 的关系如图 8 所示. u_1^+/u_1^- 随 θ 的变化比较显著,在这个例子中,大约 θ 增加 $5°$,u_1^+/u_1^- 减小 $0.1 \sim 0.2$.

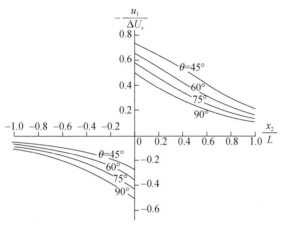

图 6　在不同倾角的情况下,当 $x_1=x_3=0$ 时,
u_1 随 x_2 的变化
$d=0$,$D/L=1$

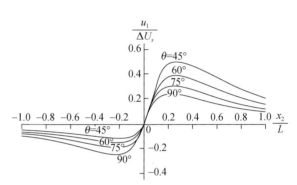

图 7　在不同倾角的情况下,当 $x_1=x_3=0$ 时,
u_1 随 x_2 的变化
$d/L=0.1$,$D/L=1$

图 9 表示断层面上界对地震位移场的影响. 当 $d/L=0$,0.1,0.2,0.3,0.4,0.5 时,断层面上界主要影响距离断层很近($x_2/L<0.2 \sim 0.5$)的位移场;对离开断层较远的位移场虽有影响,但不显著. 而断层面下界对位移场的影响则与上界的影响不同,它的影响范围较广,但不甚显著(见图 10). 在距离断层较远的地点,x_2 轴上的水平位移 u_1 的变化情况是断层面倾角 θ,下界 D 以及上界 d 的情况的反映. 对于 θ 和 d 固定的情形,u_1 随 x_2 的变化除了受 D 的不甚显著的影响外,还与断层的长度 $2L$ 有关. 因此,若是要从观测资料确定震源的参数,必须先设法确定断层长度,这样才有可能根据 u_1 随 x_2 的变化情况大致地估计 D. 否则,许多不同 L,D 的组合都可大致适合观测资料,结果将很不确定. 根据水平位移场、垂直位移场的形态,是有可能判断断层的类型和估计其长度的.

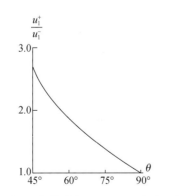

图 8　断层面两侧位移之比 u_1^+/u_1^- 和倾角 θ 的关系
$d=0$,$D/L=1$

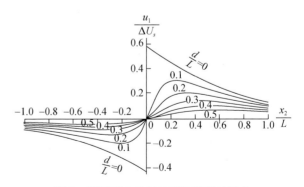

图 9　断层面上界对地震位移场的影响
$\theta=75°$,$D/L=1$

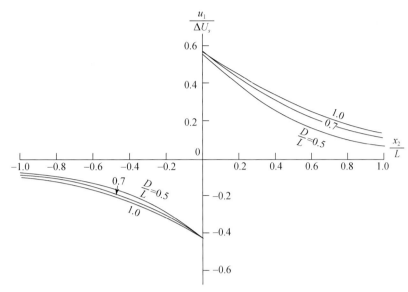

图 10 断层面下界对地震位移场的影响

$\theta = 75°$，$d = 0$

图 11 表示走向滑动断层的水平位移场和垂直位移场的一般图像. 由图可见：①走向滑动断层的垂直位移场幅度不大，仅约为水平位移的十分之一；②断层面和地面的交线（"断层线"）的两个端点，正是水平位移向量变化最急遽的地点，也是垂直位移幅度最大的地点（即垂直位移场的显著上升或下降的等值线位于"断层线"的端点附近）. 图 12 是倾向滑动断层的位移场的一般图像. 和走向滑动断层情形不同：①倾向滑动断层的水平位移场的幅度和垂直位移场的幅度大约相当；②"断层线"的两个端点，虽然也是水平位移场的位移向量变化最急遽的地点，但垂直位移场的显著上升或下降的等值线则是围绕着"断层线". 一般的滑动断层的位移场介于上述两种极端情况之间. 图 13 以 $\psi = 45°$（即 $\Delta U_s = \Delta U_d = \Delta U/\sqrt{2}$）为例，表示这种一般情况的位移场，它表示的是 ΔU_s，ΔU_d 均大于零的情形，即左旋-正断层的情

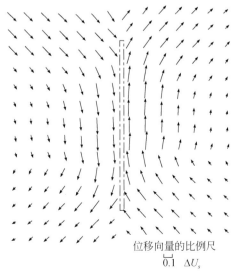

图 11-1 走向滑动断层的水平位移场

$\theta = 85°$，$d/L = 0.1$，$D/L = 0.7$

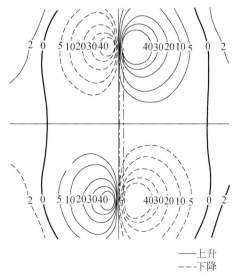

图 11-2 走向滑动断层的垂直位移场

（单位：$10^{-3} \Delta U_s$）. $\theta = 85°$，

$d/L = 0.1$，$D/L = 0.7$

形. 对于这种情形，水平位移场相对于 x_2 轴不对称，在水平位移幅度增大的一侧，垂直位移场的等值线朝着某一条等值线收拢，这条等值线距离断层大约等于断层的半长度 L；在水平位移幅度减小的一侧，则出现相反的情况：垂直位移的等值线稀疏. 右旋-逆断层的情形与图 13 所示的结果恰好反号. 右旋-正断层的情形见图 14. 与图 13 所示相仿，只不过其水平位移幅度增大的一侧是左旋-正断层情形的水平位移幅度减小的一侧. 至于左旋-逆断层，则与右旋-正断层情形反号.

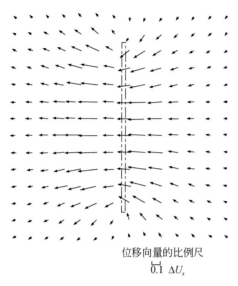

图 12-1 倾向滑动断层的水平位移场
$\theta=85°$，$d/L=0.1$，$D/L=0.7$

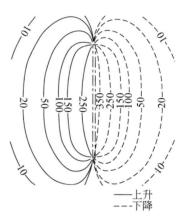

图 12-2 倾向滑动断层的垂直位移场
（单位：$10^{-3}\Delta U_d$）. $\theta=85°$，
$d/L=0.1$，$D/L=0.7$

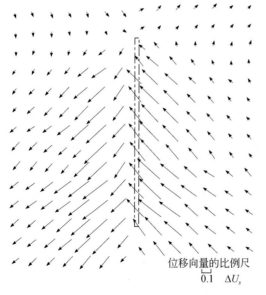

图 13-1 左旋-正断层的水平位移场
$\theta=85°$，$d/L=0.1$，$D/L=0.7$，$\psi=45°$

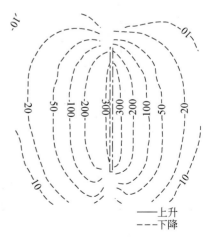

图 13-2 左旋-正断层的垂直位移场 u_3
（单位：$10^{-3}\Delta U_d$）. $\theta=85°$，
$d/L=0.1$，$D/L=0.7$，$\psi=45°$

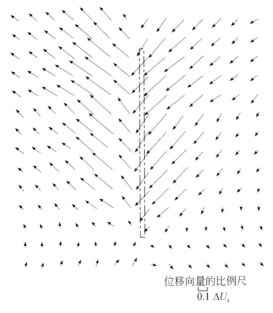

图 14-1　右旋-正断层的水平位移场

$\theta=85°$，$d/L=0.1$，$D/L=0.7$，$\psi=135°$

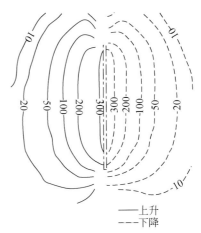

图 14-2　右旋-正断层的垂直位移场 u_3

（单位：$10^{-3}\Delta U_d$）．$\theta=85°$，

$d/L=0.1$，$D/L=0.7$，$\psi=135°$

4　邢台地震的断层错动模式

根据上面所得结果，我们来分析邢台地震的地面形变观测资料．试观察图 1 所示的邢台地震水平位移场，对照理论计算结果（图 11 至图 14），可以判断这个地震的断层具有右旋走向滑动性质，断层面接近于垂直，断层面的走向约束在 N30°E 至 N50°E 之间．既然垂直位移场显著上升或下降的等值线总是围绕着断层的端点（走向滑动断层）或"断层线"（倾向滑动断层），所以取下降幅度最大的两点之中点附近的一点为坐标原点．经过微小的调整，这个点取在图 16 和图 17 所示的 O 点．以 N35°E 的方向为 x_1 轴方向，以 N125°E 的方向为 x_2 轴方向，以铅垂方向为 x_3 轴方向，以显著下降地带（这里取为下降 200mm 以下的地带）的长度（约 50km）作为断层长度的估计值，即取 $L=25$km．

尽管邢台地震的水平位移场显示出这个地震的断层具有断层面接近于垂直的右旋走向滑动的性质，但其垂直位移场既不具备纯粹右旋走向滑动断层的特性（见图 11-2），也不具备一般的滑动断层的特性（见图 13-2）．具体地说，邢台地震垂直位移场虽然仿佛具有走向滑动断层的垂直位移场所特有的四象限分布的形态，但分析之后便可发现：①观测到的垂直位移场的四象限分布和走向滑动断层的垂直位移场四象限分布在形态上相差甚远（对比图 2 和图 11-2）；②若所观测到的垂直位移场是断层面接近于垂直的走向滑动断层所产生的，那么显著上升和显著下降的幅度应大体相等，并且其数量级应约为最大水平位移的十分之一（图 10-2）．但事实并非如此．尽管显著上升的幅度约为最大水平位移的十分之一，而显著下降的幅度却远远大于这个数值，和最大水平位移是同一数量级．这些观测事实表明，邢台地震的垂直位移场并不像是一个走向滑动断层的垂直位移场．

不仅如此，断层面接近于垂直的一般的滑动断层的垂直位移场在特点上也和它不同（见图 13-2）．根据一般的滑动断层的特点和观测到的邢台地震垂直位移场的特点，可知唯有倾角较小（$\theta=45°$）和宽度较窄的、出露到地面的正断层其垂直位移场具有实际观测到的两个特点，即：显著下降地带狭窄（20km 左右）；下降幅度远大于上升幅度（大一个数量级）．

一方面，水平位移场表明，邢台地震的震源应当是断层面接近于垂直的右旋走向滑动断层；另一方面，垂直位移场则表明，这个地震的震源应当是倾角较小和宽度较窄的正断层．这一表面上的矛盾意味着：邢台地震的震源具有更为复杂的特性，决非简单的断层模式所能解释，必须代之以更为复杂的断层模式．

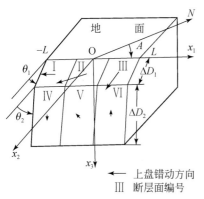

图 15　邢台地震的复合断层错动模式

为解决上述矛盾，更好地解释观测到的邢台地震位移场，这里提出一个复合断层模式．这个模式由六部分（六个简单的矩形断层）组成（图 15），其中，出露到地面的三部分是倾角较小、宽度较窄的断层，而它们下面的另外三部分则是倾角较大、宽度较大的断层．为简化起见，令每个简单断层的走向及其断层长度均相等，并令出露到地面的三部分简单断层的倾角及宽度均相等，它们下面的另外三部分简单断层的倾角及宽度也均相等．这样，待定的复合断层模式的震源参数便减少为 18 个，包括：断层面的走向 A，断层的半长度 L，上面三部分简单断层的倾角 θ_1 及宽度 ΔD_1，下面三部分简单断层的倾角 θ_2 及宽度 ΔD_2，还有每个简单断层的走向滑动及倾向滑动错距，它们总共有 $3 \times 2 \times 2 = 12$ 个．对于给定的一组参数 $(A, L, \theta_1, \theta_2, \Delta D_1, \Delta D_2)$，上述复合断层产生的位移场是 12 个未知的走向滑动或倾向滑动错距的线性函数，因此，可以用最小二乘法由观测资料确定它们．为断定断层面走向等六个参数，我们在六维参数空间 $(A, L, \theta_1, \theta_2, \Delta D_1, \Delta D_2)$ 中用网格尝试法求出使残差平方和极小的解答．残差均方根的极小值为 8.2 cm，相应的结果是：断层面走向为 N35°E，断层总长度为 50 km，θ_1 为 45°，θ_2 为 82°，ΔD_1 为 15 km，ΔD_2 为 30 km（参见图 15）．每个简单断层的走向滑动和倾向滑动错距的数值见表 2．

表 2　1966 年 3 月邢台地震群的地震断层各部分的参数

断层编号*	ΔU_s/cm	ΔU_d/cm	ΔU/cm	S/km^2	$M_0/10^{25}$ dyn·cm	$\Delta\sigma$/bar	$\Delta\varepsilon/10^{-5}$	$E/10^{22}$ erg
I	−78	2	78	250	3.2	22	3.3	0.21
II	−134	50	143	250	5.9	42	6.4	1.79
III	−17	88	90	250	3.7	33	5.0	3.54
IV	2	24	24	500	2.0	5	0.8	0.26
V	−3	−5	6	500	0.5	1	0.2	0.01
VI	1	−23	23	500	1.9	4	0.6	0.24

* 断层编号情况参见图 15.

图 16 是根据上面确定的断层参数计算出的水平位移向量的理论值，图中的六个矩形是符合断层的六个部分在地面上的投影．计算结果表明，理论值与观测值复合尚好，这意味着所采取的复合断层模式能够说明地面上观测到的水平位移场．

图 17 是按上面确定的断层参数计算出的垂直位移场的理论等值线图．与图 2 对比，可以看到理论计算结果和实际观测结果都具有以下特征：①有两个下降幅度最大的中心；②长约 60 km，宽约 20 km 的下降带；③垂直位移的等值线在形态上大体一样．这说明所采取的复合断层模式可以较好地解释地面上观测到的垂直位移场．

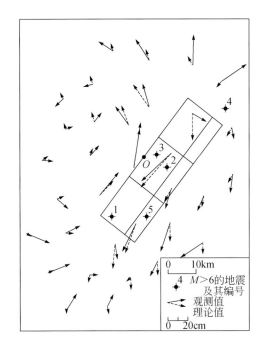
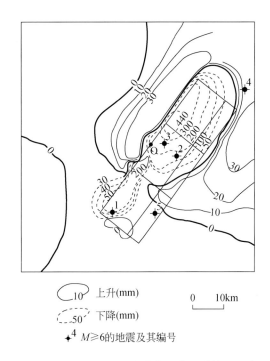

图 16　邢台地震前后地面水平形变的观测值和按图 15 所示的复合断层模式计算的理论值比较

图 17　按图 15 所示的复合断层模式计算的邢台地震前后地面垂直形变的理论等值线图

图 18 给出了残差均方根 Res 随 A，L，θ_1，θ_2，ΔD_1 和 ΔD_2 诸参数的变化情况．它表明，残差的均方根随断层面的走向 A，断层长度 $2L$，复合断层上部的倾角 θ_1 和宽度 ΔD_1 的变化较大，而随复合断层下部的倾角 θ_2 和宽度 ΔD_2 变化甚微．换句话说，地面形变场的观测资料可以用来有效地测定断层面走向、长

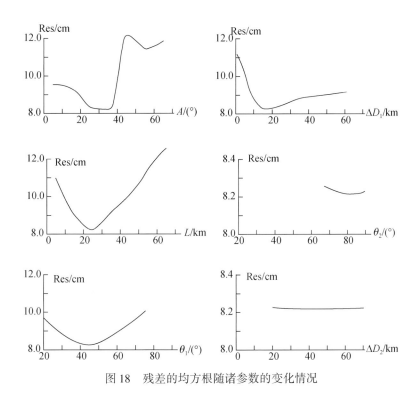

图 18　残差的均方根随诸参数的变化情况

度、近地面的那部分断层的倾角和宽度,而不适宜用来测定距地面较深的那部分断层的倾角和宽度.因此,这里确定出的诸震源参数中,走向、长度、复合断层上部的倾角及宽度是有一定代表意义的,而复合断层下部的倾角和宽度则只是一个极粗略的估计.

尽管由理论模式计算出的水平位移场和垂直位移场总的说来与观测结果符合尚好,但并不完全一致.主要表现为:①理论计算的垂直位移场虽然也有两个显著下降中心,并且其幅度与观测到的两个显著下降中心的幅度极为接近,但南边的那个下降中心的位置,计算结果和观测到的彼此距离较远.②理论计算结果和图1所示的观测到的水平位移场尽管大体符合,但在坐标原点东北边和西南边,理论值与观测值差异较大.东北边的观测点有明显的朝北东方向的位移,西南边的观测点也有明显的朝北东方向的位移,位移幅度远远超过测量误差.这两点不一致的地方是值得注意的,它表明,这里提出的复合断层模式虽然基本上能够反映邢台地震的地震断层的情况,但仍有一定的缺点,有待进一步改进,以求使理论结果更符合观测事实.

5 邢台地震的地震矩、应力降、应变降以及它所释放的能量

在上面所得结果的基础上,可以对邢台地震的地震矩、应力降、应变降以及它所释放的能量等震源参数作一估计.

地震矩由下式定义:

$$M_0 = \mu \Delta U S, \tag{25}$$

式中,μ 是刚性系数,取 $\mu = 3.3 \times 10^{11}\,\mathrm{dyn/cm^2}$;$\Delta U$ 是平均错距,S 是断层面总面积.断层各部分的地震矩见表2.

对于宽度为 $2a$ 的倾向滑动断层,应力降 $\Delta\sigma$ 为[19]:

$$\Delta\sigma = \mu \frac{(\lambda+\mu)}{(\lambda+2\mu)} \frac{U_m}{a}, \tag{26}$$

而对于走向滑动断层,应力降为[20]:

$$\Delta\sigma = \frac{\mu U_m}{2a}. \tag{27}$$

以上两式中,U_m 表示断层面上的最大错距.对于倾向滑动断层,U_m 和平均错距 ΔU_d 有如下关系[21]:

$$U_m = \frac{4}{\pi} \Delta U_d; \tag{28}$$

而对于走向滑动断层,由参考资料[20]可以求得 U_m 和 ΔU_s 也有和(28)式同样的关系.这样,根据以上几个公式,可以计算出复合断层各部分的应力降 $\Delta\sigma$,表2列出了计算结果,它们是走向和倾向方向应力降的向量和.

由应力降和应变降 $\Delta\varepsilon$ 的关系

$$\Delta\sigma = 2\mu\Delta\varepsilon \tag{29}$$

可以算出应变降,结果见表2.

长度为 $2L$ 的滑动断层在地震时释放的能量由参考资料[19],[20],[22]可以求得:

$$E = \frac{8}{\pi} \frac{\lambda+\mu}{\lambda+2\mu} \mu L \Delta U_d^2 f(\gamma), \quad (\text{倾向滑动}) \tag{30}$$

$$E = \frac{4}{\pi}\mu L \Delta U_s^2 f(\gamma), \quad (\text{走向滑动}) \tag{31}$$

其中,

$$f(\gamma) = \frac{1+\gamma}{1-\gamma}, \tag{32}$$

$1-\gamma$ 为分数应力降. 若这次地震是百分之百应力降, 那么 $\gamma=0$, 这样可以求得各部分释放的能量, 结果见表2. 一般情况下, $0<\gamma<1$, 所以表2所列的数值只能作为邢台地震时断层各部分所释放能量的下限的一个估计. 它们的总和就是这次地震所释放的总能量的下限, 估计为 6.1×10^{22} erg.

6 对邢台地震震源过程的几点认识

6.1 破裂扩展方式

1966年3月8日河北隆尧地震(表1所列的第一个地震), 其震源位置是在上面求出的地震断层南端, 而1966年3月22日河北宁晋6.7级和7.2级地震(表1的第二个和第三个地震), 其震源位置在断层的中部(图16), 这是值得注意的. 3月8日隆尧地震发生后, 立即进行了重复水准测量, 其垂直形变见图19. 按照前面叙述过的方法, 确定了这次地震的震源参数, 结果是: 断层面走向为N35°E, 断层长度为17km, θ_1 为45°, θ_2 为82°, ΔD_1 为15km, ΔD_2 为30km. 其他参数的数值如表3所示. 值得注意的是, 在第Ⅰ部分的简单断层上, 3月8日的走向滑动方向和整个复合断层的滑动方向相反, 而倾向滑动错距大于在第Ⅰ部分简单断层上的总错距, 显示出第Ⅰ部分断层的错动方向在3月8日地震时是左旋-正断层性质, 而在3月22日地震时则是右旋-逆断层性质. 图20是由上述震源参数计算得到的3月8日隆尧地震的垂直形变的理论等值线图. 对比图19和图20, 可见理论计算和观测结果符合尚好. 根据以上结果, 考虑到由地震记录确定的震源位置代表了初始破裂的地点, 那么, 可以推测, 3月8日隆尧地震是从地震断层的南端发动, 主要向北东方向扩展, 其破裂方式可近似地当作单侧破裂方式. 类似地, 1966年3月22日河北宁晋的6.7级和7.2级地震的破裂方式也是单侧破裂方式, 它们是在3月8日隆尧地震引起的应力场调整的基础上, 接着3月8日的破裂过程, 以更大的规模继续向北东方向破裂. 上述推测如果确切, 那么单侧破裂扩展方式(更确切点说, 可能是两侧不对称的双侧破裂方式)将对所辐射的地震波产生调制效应, 而这可以通过对地震图的分析加以证实, 并确定与此有关的震源参数.

表3 1966年3月8日河北隆尧地震的断层参数

断层编号	ΔU_s/cm	ΔU_d/cm	ΔU/cm	S/km²	$M_0/10^{25}$ dyn·cm	$\Delta\sigma$/bar	$\Delta\varepsilon/10^{-5}$	$E/10^{22}$ erg
Ⅰ	50	13	52	250	2.1	15	4.5	0.16
Ⅳ	-2	18	18	500	1.5	3	1.0	0.15

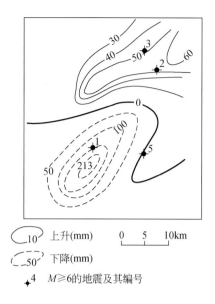 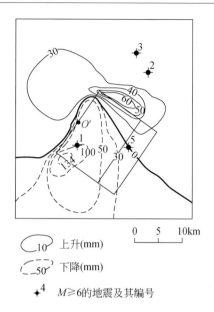

图 19 1966 年 3 月 8 日河北隆尧地震的垂直形变图

图 20 1966 年 3 月 8 日河北隆尧地震的垂直形变的理论等值线图

6.2 邢台地震群的时-空分布特点

邢台地震群的两个最大的前震和主震均发生在上面确定的断层上，而两个最大的余震则发生在其端部和附近，并具有比前震和主震略深的震源深度．这种空间分布的特点也许并非偶然．我们推测，这两个最大的余震可能是和上面确定的地震断层下面那部分的端部的错动相联系的．在两次前震和主震发生之后，震源区的应力场起了很大变化，促使或触发了端部的地震断层错动，发生余震．

上述邢台地震群的时-空分布特点究竟有无代表意义，需要在对其他地震进行广泛研究并对邢台地震进一步作研究之后才有可能看出．

6.3 形变速率和地震危险性的估计

邢台地震区从 1920 年起就开始进行水准测量，到 1966 年 3 月 8 日地震前，共进行过四次测量．测量结果表明这个地区在近半个世纪以来有一个长趋势的下降运动，下降速率约为 5mm/a．以这个数据作为这个地区地壳块体相对运动速度的垂直分量的估计值，设想这个地震群的发生是因为在胶结较牢的部位逐渐积累起来的应变的突然释放．以每年 5mm 的相对运动速率积累起 88mm 的倾向滑动错距，约需 176a．考虑到每年 5mm 的相对运动速率是这次大地震之前近半个世纪的平均速率，它可能比长期的平均运动速率大，可视为平均运动速率的上限的估计值．如果假定，通过这次地震，积累在主要断层上的应变已基本释放完毕，那么可以认为 176a 这个数值是这个地震断层的大地震复发周期的下限的一个粗略估计．

参 考 资 料

[1] 国家地震局地震测量队，1966 年邢台地震的地形变，地球物理学报，**18**，3，153–163，1975.

[2] J. A. Steketee, On Volterra's dislocations in a semi-infinite medium, *Can. Jour. Phys.*, **36**, 2, 192–205, 1958.

[3] J. A. Steketee, Some geophysical applications of the elasticity theory of dislocations, *Can. Jour. Phys.*, **36**, 9, 1168–1198, 1958.

[4] T. Maruyama, On the force equivalents of dynamical elastic dislocations with reference to the earthquake mechanism, *Bull. Earthq. Res. Inst., Tokyo Univ.*, **41**, 3, 467–486, 1963.

[5] T. Maruyama, Statical elastic dislocations in an infinite and semi-infinite medium, *Bull. Earthq. Res. Inst., Tokyo Univ.*, **42**, 2, 289–368, 1964.

[6] A. Ben-Menahem and S. J. Singh, Multipolar elastic fields in a layered half-space, *Bull. Seism. Soc. Amer.*, **58**, 5, 1519–1572, 1968.

[7] M. A. Chinnery, The deformation of the ground surface faults, *Bull. Seism. Soc. Amer.*, **50**, 3, 355–372, 1961.

[8] M. A. Chinnery, The stress changes that accompany strike-slip faulting, *Bull, Seism. Soc. Amer.*, **53**, 5, 921–932, 1963.

[9] M. A. Chinnery, The vertical displacements associated with transcurrent faulting, *J. G. R.*, **70**, 18, 4627–4532, 1965.

[10] F. Press, Displacements, strains, and tilts at teleseismic distances, *J. G. R.*, **70**, 10, 2395–2412, 1965.

[11] K. Kasahara, The nature of seismic origins as inferred from seismological and geodetic observations (1), *Bull. Earthq. Res. Inst., Tokyo Univ.*, **35**, 3, 473–532, 1957.

[12] K. Kasahara, Physical conditions of earthquake faults as deduced from geodetic data, *Bull. Earthq. Res. Inst., Tokyo Univ.*, **36**, 4, 455–464, 1958.

[13] J. A. Savage and L. M. Hastie, Surface deformation associated with dip-slip faulting, *J. G. R.*, **71**, 20, 4897–4904, 1966.

[14] J. A. Savage and L. M. Hastie, A dislocation model for the Fairview Peak, Nevada earthquake, *Bull. Seism. Soc. Amer.*, **59**, 5, 1937–1948, 1969.

[15] N. Cantiez and M. N. Toksöz, Static and dynamic study of earthquake source mechanism: San Fernando earthquake, *J. G. R.*, **77**, 14, 2583–2594, 1972.

[16] M. Ando, A fault origin model of the Great Kanto earthquake of 1923 as deduced from geodetic data, *Bull. Earthq. Res. Inst., Tokyo Univ.*, **49**, 1, 19–32, 1973.

[17] L. Mansinha and D. E. Smylie, The displacement fields of inclined faults, *Bull. Seism. Soc. Amer.*, **61**, 5, 1433–1440, 1971.

[18] R. D. Mindlin and D. H. Cheng, Nuelei of strain in the semi-infinite solid, *J. Appl. Phys.*, **21**, 9, 926–930, 1950.

[19] A. T. Starr, Slip in a crystal and rupture in a solid due to shear, *Proc. Camb. Phil. Soc.*, **24**, 489–500, 1928.

[20] L. Knopoff, Energy release in earthquakes, *Geophys. J. R. astr. Soc.*, **1**, 1, 44–52, 1958.

[21] K. Aki, Generation and propagation of G waves from the Niigata earthquake of June 16, 1964. Part 2. Estimation of earthquake moment, released energy, and stress-strain drop from the G wave spectrum, *Bull. Earthq. Res. Inst., Tokyo Univ.*, **44**, 1, 73–88, 1966.

[22] R. Burridge and L. Knopoff, The effect of initial stress or residual stress on elastic energy calculations, *Bull. Seism. Soc. Amer.*, **56**, 2, 421–424, 1966.

The focal mechanism of the 1966 Hsingtai (邢台) Earthquake as inferred from the ground deformation observations

Chen Yun-Tai, Lin Bang-Hui, Lin Zhong-Yang and Li Zhi-Yong

Institute of Geophysics, Academia Sinica

Abstract In this paper, analytical expressions for the earthquake displacement field produced by rectangular strike-slip and dip-slip faults of arbitrary dip in a semi-infinite elastic medium for the case of unequal Lamè constants are given in closed forms. Some numerical examples are presented to illustrate the effects of Poisson ratio of the medium, the dip angle, upper and lower boundary of the fault, to earthquake displacement field on surface. By comparing the geodetic data of the 1966 Hsingtai earthquake to the theoretical surface displacements that would be produced by a single rectangular slip fault of various strike, dip direction, dip angle, fault length, width, focal depth, and dislocations, it is shown that the simplified single slip fault model can not explain satisfactorily the observed horizontal and vertical ground deformation. In order to account for the observed data, a compounded fault model which consists of six simple rectangular slip faults, is proposed. By using the grid trial method, the source parameters appropriate to the observed horizontal as well as vertical displacement field are obtained. The results are as follows: the upper three simple rectangular slip faults breaking to the surface, have a dip angle of 45°, width 15 km, while the lower three simple rectangular slip faults have a dip angle of 82°, width 30 km. Other source parameters determined are as follows: strike, N35°E; dip direction, N125°E; total fault length, 50 km; average strike-slip dislocations respectively, −78, −134, −17, 2, −3, and 1 cm (minus stands for the right-lateral strike-slip while plus for the left-lateral strike-slip); average dip-slip dislocations respectively, 2, 50, 88, 24, −5, and −23 cm (minus stands for the reverse fault while plus for the normal fault); seismic moments respectively, 3.2, 5.9, 3.7, 2.0, 0.5, and 1.9×10^{25} dyne·cm, stress drops respectively, 22, 42, 33, 5, 1, and 4 bars, strain drops respectively, 3.3, 6.4, 5.0, 0.8, 0.2, and 0.6×10^{-5}. Lower limit of strain energy release is 6.1×10^{22} ergs. A comparison of the obtained results with the epicentral distribution of earthquakes of magnitudes greater than 6 indicates that, the main shock and two larger foreshocks are related more closely with the dislocation of the upper three portions of this compounded fault, and that the two larger aftershocks are related with the dislocation near the ends of the lower portions of this compounded fault. Results of leveling survey in Hsingtai area since 1920, indicate that the upper limit of the rate of subsidence is about 5 mm/a. If the strain accumulated in this area is assumed to be released entirely by the 1966 Hsingtai earthquake, the lower limit of the recurrence period of earthquakes of magnitudes as high as the 1966 Hsingtai earthquake can be estimated in terms of the average dip-slip dislocation of this compounded fault to be about 176 years.

巧家、石棉的小震震源参数的测定及其地震危险性的估计[*]

陈运泰[1] 林邦慧[1] 李兴才[1] 王妙月[1] 夏大德[2] 王兴辉[2] 刘万琴[1] 李志勇[1]

1. 中国科学院地球物理研究所；2. 国家地震局成都地震大队

摘要 根据巧家、石棉的小地震的观测资料，指出 P 波初动半周期在震级比较小时几乎是恒定的，在震级比较大时随震级的增大而增大；并指出 P 波初动振幅的对数也随震级的增大而增大. 以圆盘形均匀位错面作为中、小地震震源的理论模式，计算了它所辐射的地震波远场位移，从而导出了体波初动半周期及振幅与震源尺度及波速等物理量的定量关系，解释了 P 波初动半周期及振幅与震级之间的经验关系. 考虑到波在介质中的衰减和频散、地表面的影响以及地震仪器的频率特性，通过褶积方法合成了上述位错源产生的理论地震图，提出了直接由实际地震图上的初动半周期及振幅测定震源尺度、地震矩、应力降和错距等震源参数以及介质的品质因数 Q 的方法. 运用上述方法，测定了巧家、石棉两地区介质的品质因数 Q 和小震的震源参数. 这两个地区介质的品质因数 Q 分别为 620 和 560. 石棉地区小地震的应力降大约在 2~30bar 之间，巧家地区小地震的应力降比较接近，平均约 1.4bar. 将这个结果和 1962 年 3 月 19 日新丰江地震与 1975 年 2 月 4 日海城地震的前、主震的应力降作对比，我们看到，巧家地区小震的应力降的特征与上述两次大地震的前震的应力降的特征是类似的，因此不能排斥巧家地区的小震是一个较大地震的前震的可能性. 以测得的小震应力降的平均值（约 1.4bar）作为这个可能发生的较大地震应力降的下限估计值，从主震震级和主震应力降的经验关系可以推知，其震级的下限是 5.2 级.

1 引言

在地方震的地震图中，有许多值得注意的现象. 例如，体波初动的周期和震级有关；S 波初动的周期通常比 P 波的大，它们的比值与波速比有关；等等. 可是，体波初动的周期和震级的确切关系究竟是什么？S 波初动的周期和 P 波初动的周期之间的确切关系是什么？迄今仍缺乏系统的研究. 至于这些现象究竟反映震源和传播地震波的介质的什么特性？从这些现象中能否获得有关震源和它所处环境的更多的讯息？这些问题也很值得深究.

这里将试图从地方震的记录图中，寻找 P 波初动半周期及振幅和震级的经验关系，并进一步从震源理论的角度阐明这些关系，试验从这些现象推知一些有意义的震源参数（震源尺度、地震矩、应力降和错距）以及介质的特性（品质因数）的测量方法.

2 P 波初动半周期及振幅和震级的经验关系

为研究 P 波初动半周期及振幅和震级的关系，分析了四川省两个地震台的电流计记录地震仪在 1970 年—1973 年记下的震源位置相近的小地震. 图 1 是普格地震台记录的巧家附近的小震震中分布图，这些小地震震中距平均约 60km. 图 2 和图 3 分别是这些小地震的 P 波初动半周期 t_2 及振幅 A_m 和地方震震级

---

[*] 本文发表于《地球物理学报》，1976 年，第 **19** 卷，第 3 期，206–233.

M_L 的关系. 从图 2 可见, 在 $M_L<2.5$ 时, t_2 随 M_L 的变化不明显, 大约保持在 $1.0\sim0.3\mathrm{s}$ 之间; 而在 $M_L\geqslant 2.5$ 时, t_2 随 M_L 的增大而增大. 从图 3 可见, A_m 的对数也随 M_L 的增大而增大.

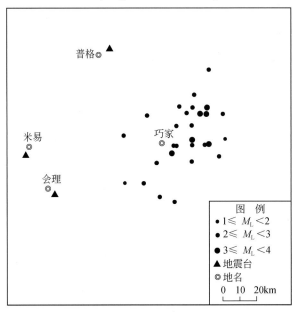

图 1 普格地震台记录的巧家地震的震中分布图

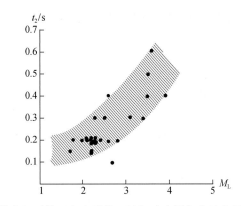

图 2 普格台记录的巧家地震的 P 波初动半周期和地方震震级的关系

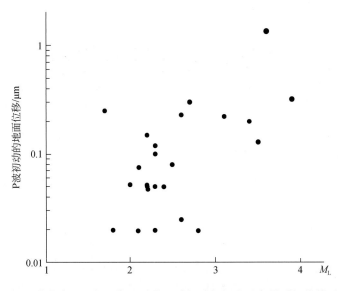

图 3 普格台记录的巧家地震的 P 波初动振幅和地方震震级的关系

石棉地震台记录的石棉附近的小地震也有类似的情况（图4—图6），所不同的是，在 $M_L<2$ 时，t_2 随 M_L 的变化不明显，大约保持在0.1s左右；而在 $M_L \geq 2$ 时，t_2 才随着 M_L 的增大而增大.

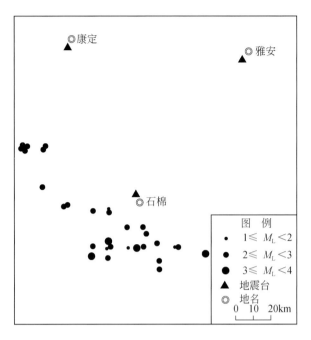

图4　石棉地震台记录的石棉地震的震中分布图

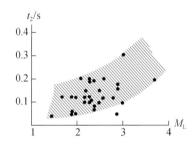

图5　石棉台记录的石棉地震的P波初动半周期和地方震震级的关系

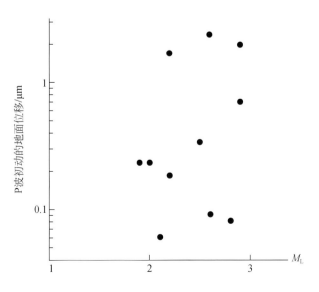

图6　石棉台记录的石棉地震的P波初动振幅和地方震震级的关系

3 定性解释

从以上例子看，虽然资料点有点分散，但总的看来，当震级比较小时，P波初动半周期几乎是恒定的；而当震级比较大时，它随着震级的增大而增大。此外，初动振幅也随着震级的增大而增大。这两个经验关系还是清楚的。从震源理论的角度看，这两个关系是可以得到理解的。

定性地说，根据对浅源大地震的断层长度的统计研究[1-5]，可知浅源大地震的断层长度的对数和震级成正比。就中、小地震而言，因为在地表一般无从发现其断层长度（或断层尺度）而从地震波资料得到的结果也不多，所以其断层尺度和震级的关系还不清楚。但是，如果浅源大地震的断层长度和震级的经验关系也适用于中、小地震，并且如果初动半周期和断层尺度成正比，那么就不难理解中、小地震的P波初动半周期和震级的经验关系。再者，如果中、小地震初动振幅的对数和断层尺度的对数成正比，那么也可以解释初动振幅的对数和震级成正比的关系。

地震时，由距离观测点最近的破裂点发出的扰动先到达观测点，距离它最远的点发出的扰动则后到，它们的时间差决定了初动的持续时间，也就是初动的半周期。很明显，这个时间差和震源的尺度成正比，而和震源所在处的体波速度成反比。

地震的强度跟断层面的面积与错距成正比，而断层面面积正比于震源尺度的平方、错距则正比于震源尺度，因此，地震的强度跟震源尺度的立方成正比。由于断层错动所引起的观测点的初动振幅不但和地震的强度成正比，还和振动的持续时间成反比。既然初动半周期正比于震源尺度，那么初动振幅便应当是和震源尺度的平方成正比。地方震震级是由最大地动位移的对数定义的。倘若认为，最大地动位移与初动振幅成正比，那么地方震震级便应当与初动振幅的对数成正比。

在地震波从震源传播到地震台时，由于它所通过的介质非完全弹性，对它有吸收作用，从而使得初动的半周期增大。显然，当震源尺度和震级较大时，介质吸收的影响可以忽略，因而初动半周期仍和震源尺度成正比。当震源尺度和震级较小，以至介质吸收所引起的初动半周期的变化比初动半周期本身还要大得多时，记录下来的初动半周期就不再由震源尺度和破裂速度之比决定，而是由介质的吸收性能决定。对于同一地震台记录的相同地区的地震，介质吸收的影响是一样的，因而，记录到的震级较小的地震，其初动半周期应当也是一样的。在这种情形下，初动半周期的数值应当和传播路径（震源距）的长短以及介质的吸收性能（品质因数）有关。震源距和品质因数的比值越大，吸收的影响就越大，即记录到的震级较小的地震的初动半周期越大。

以上简单的定性分析清楚地表明，初动半周期及振幅和震级的经验关系跟震源的尺度、错距、震源区的波速以及介质的吸收性能有关。这样，由初动半周期及振幅的测定便有可能推知震源参数（震源尺度、地震矩、应力降、错距等）及介质的性质（震源区的波速或波速比、波所通过的介质的品质因数）。为了做到这点，就需要建立初动半周期及振幅和上述震源参数及介质特性的定量关系。

4 圆盘形断层辐射的地震波

就浅源大地震而言，它的破裂面（断层面）的上界为地表面所限制，下界因为随深度而增加的摩擦应力，也受到限制，所以，以长度和宽度不一样的位错面（例如矩形或椭圆形位错面）模拟它是合适的。和浅源大地震的情况不同，对中、小地震来说，上述限制不那么显著；因此，以相等尺度的位错面（例如圆盘形或正方形位错面）模拟它则更合适一些。这里以圆盘形位错面作为中、小地震震源的理论模式。

图7表示完全弹性的无限介质中一个半径为a的圆盘形断层。断层面Σ的两侧分别以Σ_+，Σ_-表示。

以从 Σ_- 指向 Σ_+ 的方向为其法线方向 $\boldsymbol{n}$. 采用直角坐标系 (x_1, x_2, x_3)，原点与圆盘中心重合，x_3 轴与位错面的法向一致，x_1 轴与 Σ_+ 相对于 Σ_- 的错动方向一致.

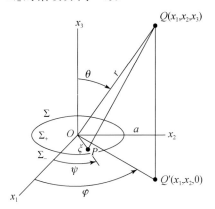

图7 圆盘形断层模式

在均匀、各向同性和完全弹性的无限介质中，圆盘形位错面 Σ 引起的地震波远场位移的频谱 $U(\omega)$ 是[6-13]：

$$\boldsymbol{U}(\omega) = U_r(\omega)\boldsymbol{e}_r + U_\theta(\omega)\boldsymbol{e}_\theta + U_\varphi(\omega)\boldsymbol{e}_\varphi, \tag{1}$$

$$U_j(\omega) = \frac{m_0}{4\pi\rho c^3 r}\mathscr{R}_j i\omega G(\omega) e^{-i\frac{\omega}{c}r} F_c(\omega),$$

$$j = r, \theta, \varphi,$$
$$c = \alpha, \quad \text{当} \quad j = r,$$
$$c = \beta, \quad \text{当} \quad j = \theta, \varphi. \tag{2}$$

式中，(r, θ, φ) 是观测点 Q 的球极坐标（见图7），$\boldsymbol{e}_r, \boldsymbol{e}_\theta, \boldsymbol{e}_\varphi$ 是球极坐标中的基向量；m_0 是地震矩：

$$m_0 = \mu \Delta \bar{u} S, \tag{3}$$

μ 是刚性系数，$\Delta \bar{u}$ 是平均错距，S 是断层面面积；ρ 是介质的密度，α，β 分别是 P 波和 S 波的波速；$\mathscr{R}_j$ 是辐射图型因子，对于 Σ_+ 相对于 Σ_- 沿 x_1 轴方向滑动的剪切错动，

$$\begin{cases} \mathscr{R}_r = \sin 2\theta \cos\varphi, \\ \mathscr{R}_\theta = \cos 2\theta \cos\varphi, \\ \mathscr{R}_\varphi = -\cos\theta \sin\varphi; \end{cases} \tag{4}$$

$G(\omega)$ 是震源时间函数 $g(t)$ 的频谱；$F_c(\omega)$ 是和断层面的几何形状、错距的分布以及破裂扩展方式有关的函数，在错距均匀分布的情况下，当破裂从圆心开始以有限的速度 v_b 向四周扩展时，它由以下的面积分表示：

$$F_c(\omega) = \frac{1}{S}\iint_\Sigma e^{-i\frac{\omega}{v_b}\xi + i\frac{\omega}{c}\xi\sin\theta\cos\psi} \xi \mathrm{d}\xi \mathrm{d}\psi, \tag{5}$$

(ξ, ψ) 是元位错面的平面极坐标（参见图7）.

在 $F_c(\omega)$ 的表示式中，对 ξ 的积分容易作出，结果是：

$$F_c(\omega) = \frac{iv_b}{\omega S}\int_0^{2\pi}\left[\frac{a}{q_c(\psi)}e^{-i\frac{\omega}{v_b}q_c(\psi)a} + \frac{v_b}{i\omega q_c^2(\psi)}(e^{-i\frac{\omega}{v_b}q_c(\psi)a} - 1)\right]\mathrm{d}\psi, \tag{6}$$

其中，

$$q_c(\psi) = 1 - \varepsilon_c \cos\psi, \tag{7}$$

$$\varepsilon_c = \frac{v_b}{c}\sin\theta. \tag{8}$$

返回时间域，便得到圆盘形位错面引起的地震波远场位移 $u(t)$ 的表示式：

$$\boldsymbol{u}(t) = u_r(t)\boldsymbol{e}_r + u_\theta(t)\boldsymbol{e}_\theta + u_\varphi(t)\boldsymbol{e}_\varphi, \tag{9}$$

$$u_j(t) = \frac{m_0}{4\pi\rho c^3 r}\mathscr{R}_j \dot{g}(t) * f_c\left(t - \frac{r}{c}\right), \quad j = r, \theta, \varphi,$$

$$c = \alpha, \quad \text{当} \quad j = r,$$
$$c = \beta, \quad \text{当} \quad j = \theta, \varphi. \tag{10}$$

式中，·表示对时间的微商，*表示褶积，$f_c(t)$ 是 $F_c(\omega)$ 的反演：

$$f_c(t) = \frac{v_b^2 t}{S}\left\{\frac{2\pi}{(1-\varepsilon_c^2)^{3/2}}[H(t) - H(t - t_2)]\right.$$
$$\left. - \left[\frac{2}{(1-\varepsilon_c^2)}\frac{\sqrt{(t - t_{1c})(t_{2c} - t)}}{t} + \frac{4}{(1-\varepsilon_c^2)^{3/2}}\tan^{-1}\sqrt{\frac{1+\varepsilon_c}{1-\varepsilon_c}\frac{t - t_{1c}}{t_{2c} - t}}\right][H(t - t_{1c}) - H(t - t_{2c})]\right\}, \tag{11}$$

其中，$H(t)$ 是单位函数，

$$t_{1c} = \frac{a}{v_b}(1 - \varepsilon_c), \tag{12}$$

$$t_{2c} = \frac{a}{v_b}(1 + \varepsilon_c). \tag{13}$$

若震源时间函数是单位函数，那么（10）式就简化为

$$u_j(t) = \frac{m_0}{4\pi\rho c^3 r}\mathscr{R}_j f_c\left(t - \frac{r}{c}\right). \tag{14}$$

$u_r(t)$ 就是 P 波的远场位移 $u_\alpha(t)$；$[u_\theta^2(t) + u_\varphi^2(t)]^{1/2}$ 就是 S 波的远场位移 $u_\beta(t)$；因此，

$$u_c(t) = \frac{m_0}{4\pi\rho c^3 r}\mathscr{R}_c f_c\left(t - \frac{r}{c}\right), \quad c = \alpha, \beta, \tag{15}$$

其中，$\mathscr{R}_\alpha$ 表示 P 波的辐射图型因子，也就是 $\mathscr{R}_r$；$\mathscr{R}_\beta$ 表示 S 波的辐射图型因子，它等于 $[\mathscr{R}_\theta^2 + \mathscr{R}_\varphi^2]^{1/2}$.

由（15）式可见，地震波远场位移的波形由 $f_c\left(t - \frac{r}{c}\right)$ 决定. 图 8 是 $f_c(t)$ 的图形. 当 $t < 0$ 时，位移为零；当 $0 < t < t_{1c}$ 时，位移随 t 线性增加；当 $t_{1c} < t < t_{2c}$ 时，它随 t 单调下降；最后，当 $t > t_{2c}$ 时，位移等于零.

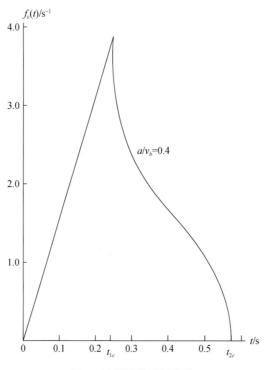

图 8 地震波的远场位移

5 初动半周期及振幅和震源尺度及波速等物理量的关系

前面得到的公式（13）表明，体波初动的半周期 t_{2c} 除了和震源尺度、体波速度及破裂速度有关外，还和观测点在震源球球面上的位置有关．对于在震源球球面上均匀分布的观测点，初动半周期在震源球球面上的平均值 $\langle t_{2c} \rangle$ 为：

$$\langle t_{2c} \rangle = \frac{a}{v_b}\left(1 + \frac{\pi}{4}\frac{v_b}{c}\right). \tag{16}$$

这相当于 $\theta = \theta_0 = \sin^{-1}\left(\frac{\pi}{4}\right) = 51°45'$．就是说，$\theta = \theta_0$ 处的初动半周期等于它在震源球球面上的平均值．

由（5）式容易证明 $F_c(0) = 1$，从而可以证明，体波初动的位移曲线下的面积 A_c 为：

$$A_c = \frac{m_0}{4\pi\rho c^3 r}\mathscr{R}_c \tag{17}$$

A_c 和观测点在震源球球面上的位置有关．对于在震源球球面上均匀分布的观测点，A_c 在震源球球面上的均方根为：

$$\langle A_c^2 \rangle^{1/2} = \frac{m_0}{4\pi\rho c^3 r}\langle \mathscr{R}_c^2 \rangle^{1/2}. \tag{18}$$

由（4）式容易求得：$\langle \mathscr{R}_\alpha^2 \rangle^{1/2} = \sqrt{4/15}$，$\langle \mathscr{R}_\beta^2 \rangle^{1/2} = \sqrt{2/5}$．

当观测点所在位置使得 $\mathscr{R}_c = \langle \mathscr{R}_c^2 \rangle^{1/2}$ 且 $t_{2c} = \langle t_{2c} \rangle$ 时，体波的位移表示式为

$$u_c(t) = \frac{m_0}{4\pi\rho c^3 r}\langle \mathscr{R}_c^2 \rangle^{1/2} f_{c0}\left(t - \frac{r}{c}\right), \tag{19}$$

其中，

$$f_{c0}\left(t - \frac{r}{c}\right) = f_c\left(t - \frac{r}{c}\right)\Big|_{\theta=\theta_0}. \tag{20}$$

表示式（19）有一定的代表性，它所给出的体波初动的半周期代表了初动半周期在震源球球面上的平均值；和地震矩成正比的初动位移曲线下的面积则代表了初动位移曲线下的面积在震源球球面上的均方根．因此，我们在下面将运用（19）式来分析初动半周期及振幅和震源尺度等物理量的关系．

如同（16）式所表明的，体波初动半周期和 a/v_b 成正比，比例系数和 v_b/c 有关（图9）．由（19）式可以得出，对于地震矩同样大小的震源，体波初动的振幅 u_{cm} 和 a/v_b 成反比，比例系数也和 v_b/c 有关（图10）：

$$u_{cm} = \frac{m_0}{4\pi\rho c^3 r}\langle \mathscr{R}_c^2 \rangle^{1/2} \Theta_c \frac{v_b}{a}, \tag{21}$$

$$\Theta_c = \frac{2}{\left(1 + \frac{\pi}{4}\frac{v_b}{c}\right)^{3/2}\left(1 - \frac{\pi}{4}\frac{v_b}{c}\right)^{1/2}}. \tag{22}$$

对于圆盘形断层面，平均错距和应力降的关系为[14,15]：

$$\Delta\bar{u} = \frac{16}{7\pi}\frac{\Delta\sigma a}{\mu}. \tag{23}$$

将上式代入（21）式，就得到初动振幅的另一个表示式：

$$u_{cm} = \frac{4}{7\pi}\langle \mathscr{R}_c^2 \rangle^{1/2}\frac{\Delta\sigma}{\rho r}\left(\frac{a}{v_b}\right)^2 \Theta_c', \tag{24}$$

$$\Theta_c' = \left(\frac{v_b}{c}\right)^2 \Theta_c. \tag{25}$$

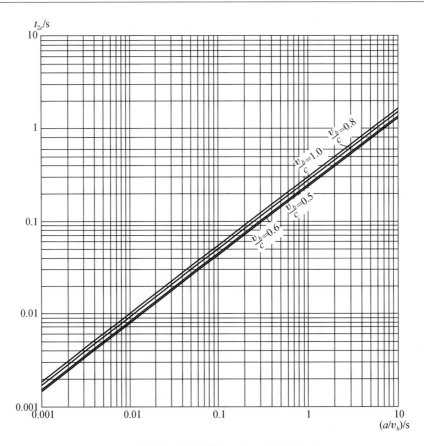

图 9 体波初动半周期和震源尺度的关系

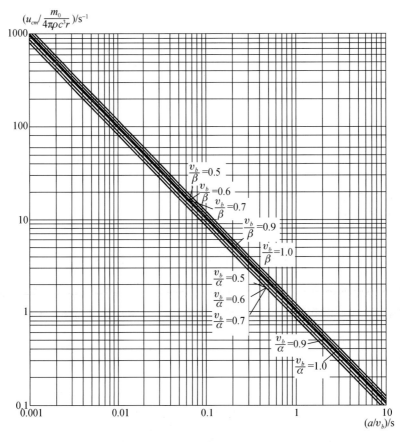

图 10 体波初动振幅和震源尺度的关系

(24)式表明，若 $\Delta\sigma$ 和 a^r 成正比，那么初动振幅就和 a^{r+2} 成正比；也就是说，初动振幅的对数和 a 的对数成正比.

6 震源参数的测定原理

运用上面得到的结果，可以测定断层面的尺度、地震矩、应力降和错距等震源参数. 下面分述这些参数的测定原理.

1. 断层面尺度

由（16）式可知

$$a = \frac{v_b}{\left(1 + \frac{\pi}{4}\frac{v_b}{c}\right)}\langle t_{2c}\rangle. \tag{26}$$

设 v_b 和 c 已知，那么由初动半周期 $\langle t_{2c}\rangle$ 的测定便可求得断层面的尺度 $2a$.

2. 地震矩

（21）式提供了由初动振幅求地震矩的方法. 设 ρ，c，r，v_b 已知，a 已由 $\langle t_{2c}\rangle$ 按上述方法测得，那么由初动振幅 u_{cm} 的测定便可求得地震矩 m_0.

3. 应力降

断层面尺度和地震矩的测定是基本的测定. 在测定这两个参数的基础上，可以由应力降和地震矩及震源尺度的关系式[14,15]计算应力降：

$$\Delta\sigma = \frac{7}{16}\frac{m_0}{a^3}. \tag{27}$$

4. 平均错距和最大错距

平均错距可以由（23）式计算，而最大错距 Δu_m 可以由下式[14,15]计算：

$$\Delta u_m = \frac{3}{2}\Delta\bar{u} = \frac{24}{7\pi}\frac{\Delta\sigma a}{\mu}. \tag{28}$$

7 理论地震图的合成

通过测定地震图上的 P 波或 S 波的初动半周期及振幅，可以得到许多有意义的震源参数. 然而，在实际应用之前，还必须考虑到地震波在传播过程中的衰减；地壳和上地幔的分层结构以及地表面的影响；最后，还要考虑到地震仪的频率特性的影响.

1. 体波的衰减和频散

观测和实验表明，地震体波的衰减系数在所观测的频段内与频率呈线性关系[16,17]. 考虑到体波的衰减，必须在前面得到的体波远场位移谱（公式（2））中乘上因子 $e^{-\frac{|\omega|r}{2cQ_0}}$ 才能得到远场地动位移谱. 这里，Q_0 表示介质的品质因数. 体波的频散总是和衰减成对出现（通常称为"吸收-频散对"）[16]，所以，与此同时，（2）式中的因子 $e^{-i\frac{\omega}{c}r}$ 里的体波速度 c 要代之以相速度 c_p. 对于弗特曼（W. I. Futterman[17]）的第一种形式的吸收-频散对，相速度 c_p 为：

$$c_P = \frac{c}{1 - \frac{1}{2\pi Q_0}\ln\left|\left(\frac{\omega}{\omega_0}\right)^2 - 1\right|}. \tag{29}$$

式中，ω_0 表示低频截止频率，在这里，ω_0 取超出地震仪频带的数值，即 $f_0 = \omega_0/2\pi = 10^{-3}$ Hz.

这样，在考虑了地震波的衰减和频散之后，地动位移谱为

$$U_c(\omega) = \frac{m_0}{4\pi\rho c^3 r}\mathcal{R}_c i\omega G(\omega) e^{-i\frac{\omega}{c}r} F_c(\omega) B_c(\omega). \tag{30}$$

其中，

$$B_c(\omega) = e^{-\frac{|\omega|r}{2cQ_0} + i\frac{\omega r}{2\pi c Q_0}\ln\left|\left(\frac{\omega}{\omega_0}\right)^2 - 1\right|}. \tag{31}$$

相应的地动位移为

$$u_c(t) = \frac{m_0}{4\pi\rho c^3 r}\mathcal{R}_c \dot{g}(t) * f_c\left(t - \frac{r}{c}\right) * b_c(t). \tag{32}$$

$b_c(t)$ 是 $B_c(\omega)$ 的反演，也就是衰减的脉冲响应。只考虑衰减不考虑频散就会出现违背因果律的情况，衰减和频散两者均加以考虑后就不会出现违背因果律的情况。图 11 是一个例子，说明衰减和频散均考虑后，衰减的脉冲响应。计算中，取 $r/Q_0 = 0.04$ km, $c = 6.06$ km/s.

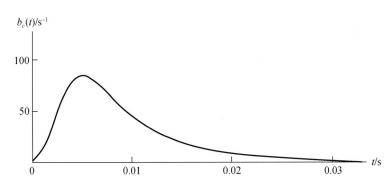

图 11　考虑了频散衰减的脉冲响应

2. 震源时间函数

这里设震源时间函数的形式为

$$g(t) = \frac{1}{2}\left(1 - \cos\frac{\pi t}{T_s}\right)[H(t) - H(t - T_s)] + H(t - T_s). \tag{33}$$

这个形式和观测结果及理论推算结果很接近[18,19]，因此可以将它作为震源时间函数的合理的一级近似。至于震源时间常数 T_s，则通过以下的考虑估算。

在断层面上的某一点刚发生破裂时，破裂点附近的质点运动速度

$$\dot{u}_1^+(t) = \frac{\beta_s}{\mu_s}\sigma_e(t), \tag{34}$$

式中，β_s 是震源处的横波速度，μ_s 是震源处的刚性系数，$\sigma_e(t)$ 是有效应力，它等于初始应力 σ_b 和摩擦应力 $\sigma_f(t)$ 之差：

$$\sigma_e(t) = \sigma_b - \sigma_f(t). \tag{35}$$

由 (33) 式可知，当 $t = T_s/2$ 时，破裂点附近的质点运动速度达到最大，今以 $\dot{u}_{1m}^+$ 代表它，那么

$$\dot{u}_{1m}^+ = \frac{\pi \Delta \bar{u}}{4T_s}, \tag{36}$$

从而

$$T_s = \frac{\pi \mu_s}{4\beta_s}\frac{\Delta \bar{u}}{\sigma_{em}}, \tag{37}$$

式中，σ_{em} 代表有效应力的最大值。

由于平均错距 $\Delta \bar{u}$ 和应力降 $\Delta \sigma$ 及震源半径 a 有一个简单关系（（28）式），所以

$$T_s = \frac{4}{7} \frac{a}{\beta_s} \frac{\Delta \sigma}{\sigma_{em}}. \tag{38}$$

这个结果意味着，震源时间常数 T_s 不仅和 a/β_s 有关，还和 $\Delta\sigma/\sigma_{em}$ 有关。如果剩余应力 σ_a 等于摩擦应力的最小值，那么 $\Delta\sigma = \sigma_{em}$，从而

$$T_s = \frac{4}{7} \frac{a}{\beta_s}. \tag{39}$$

在数值计算中，以上式估算震源时间常数。图12是个例子，表示当 $v_b/\beta = 0.9$ 而 a/v_b 分别为 0.2，0.4 和 1.0s 时的 $\dot{g}(t)$。

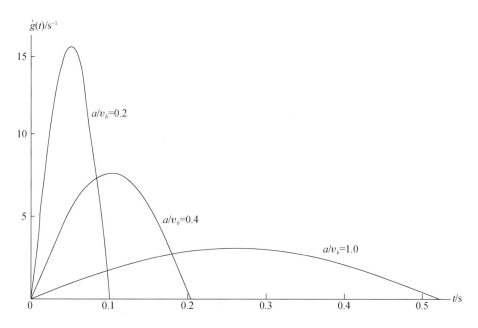

图 12 震源时间函数的时间微商

3. 地壳和上地幔的分层结构以及地表面的影响

这里没有考虑地壳和上地幔的分层结构的影响；至于地表面的影响，简单地用未考虑自由表面影响的地动位移谱乘上二倍的因子。

4. 地震仪的影响

为了直接从地震图上的初动半周期及振幅的测量求得震源参数，必须在前面得到的地动位移谱上乘以仪器的频率特性 $I_n(\omega)$ 才能得到观测地震图的频谱：

$$U_c(\omega) = \frac{m_0}{2\pi\rho c^3 r} \mathscr{R}_c i\omega G(\omega) e^{-i\frac{\omega}{c}r} F_c(\omega) B_c(\omega) I_n(\omega). \tag{40}$$

仪器的频率特性可以表示为：

$$I_n(\omega) = V_0 W(\omega) e^{-i\gamma(\omega)}. \tag{41}$$

式中，V_0 是静态放大倍数，它和频率无关；$W(\omega)$ 是振幅特性；$\gamma(\omega)$ 是相位特性。由于 V_0 和频率无关，且因台、因时而异，所以在测定振幅时均先归算到 $V_0 = 1$ 的情形，从而

$$I_n(\omega) = W(\omega) e^{-i\gamma(\omega)}. \tag{42}$$

普格地震台和石棉地震台所使用的维开克地震仪（配Fc6-10型振子），其频率特性的表示式是：

$$W(\omega) = \frac{2D_2/T_2}{\sqrt{\left(\frac{2\pi}{\omega}\right)^{-2} + a_1 + b_1\left(\frac{2\pi}{\omega}\right)^2 + c_1\left(\frac{2\pi}{\omega}\right)^4 + d_1\left(\frac{2\pi}{\omega}\right)^6}}, \tag{43}$$

$$\gamma(\omega) = \tan^{-1} \frac{s_1\left(\frac{2\pi}{\omega}\right)^4 - p_1\left(\frac{2\pi}{\omega}\right)^2 + 1}{q_1\left(\frac{2\pi}{\omega}\right)^3 - m_1\left(\frac{2\pi}{\omega}\right)}. \tag{44}$$

式中,

$$\begin{cases} a_1 = m_1^2 - 2p_1, \\ b_1 = p_1^2 - 2m_1q_1 + 2s_1, \\ c_1 = q_1^2 - 2p_1s_1, \\ d_1 = s_1^2, \end{cases} \tag{45}$$

$$\begin{cases} m_1 = 2\left(\dfrac{D_1}{T_1} + \dfrac{D_2}{T_2}\right), \\ p_1 = \dfrac{1}{T_1^2} + \dfrac{1}{T_2^2} + \dfrac{4D_1D_2}{T_1T_2}(1-\sigma^2), \\ q_1 = 2\left(\dfrac{D_1}{T_1T_2^2} + \dfrac{D_2}{T_2T_1^2}\right), \\ s_1 = \dfrac{1}{T_1^2T_2^2}. \end{cases} \tag{46}$$

摆的周期 $T_1 = 1s$, 阻尼系数 $D_1 = 0.5$; 电流计的周期 $T_2 = 0.1s$, 阻尼系数 $D_2 = 8$; 耦合系数 $\sigma^2 = 0.4$. 图13是维开克地震仪的振幅特性和相位特性. 图中, T 是周期.

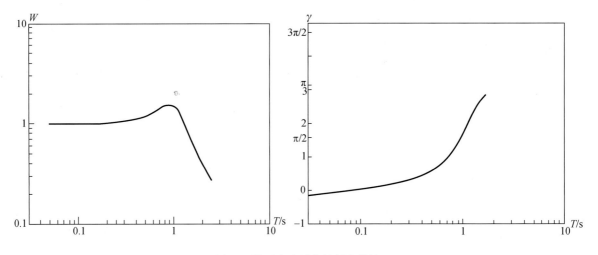

图 13 维开克地震仪的频率特性

由 (40) 式反演, 便可得到合成地震图:

$$u_c(t) = \frac{m_0}{2\pi\rho c^3 r}\mathscr{R}_c \dot{g}(t) * f_c\left(t - \frac{r}{c}\right) * b_c(t) * i_n(t), \tag{47}$$

$i_n(t)$ 表示 $I_n(\omega)$ 的反演, 即地震仪的脉冲响应.

8 数值计算结果

1. P 波的远场位移

在数值计算中, 取 $\beta = 3.50 \text{km/s}$, $\alpha : \beta : v_b = \sqrt{3} : 1 : 0.9$, 故 $\alpha = 6.06 \text{km/s}$, $v_b = 3.15 \text{km/s}$. 由这些数

据计算了 P 波的远场位移. 图 14 是当 a/v_b = 0.01, 0.02, 0.04 和 0.06s 四种情形下的 P 波远场位移, 由公式 (11) 可见, 当 a/v_b 增大 n 倍时, 只要把图 14 的横坐标的单位缩小 n 倍, 纵坐标的单位放大 n 倍, 就可以得到相应的 a/v_b 的地震波远场位移图.

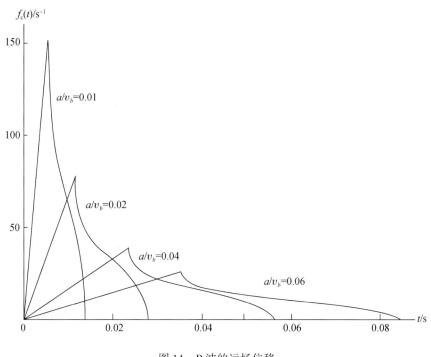

图 14 P 波的远场位移

2. 介质的吸收

图 15 是一个例子, 表明当震源距 r 和介质的品质因数 Q_0 的比值为 0.12, 0.20, 0.32km 时, 衰减的脉冲响应. 由图可以看出, 脉冲的宽度随着 r/Q_0 的增大而增大.

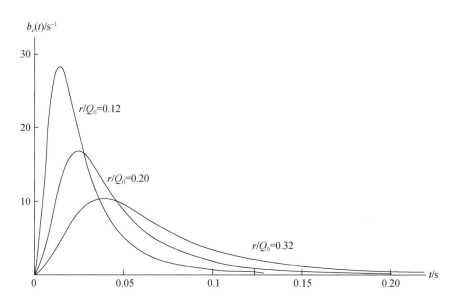

图 15 r/Q_0 取不同数值时衰减的脉冲响应

3. 地震仪的脉冲响应

根据公式 (42)—(44), 计算了维开克地震仪的脉冲响应, 结果如图 16 所示.

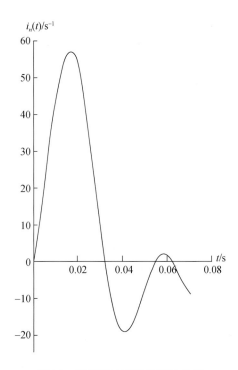

图 16 维开克地震仪的脉冲响应

4. 合成地震图

按照公式 (40)，依次对 P 波远场位移 $f_c(t)$、震源时间函数的时间微商 $\dot{g}(t)$、介质的吸收的脉冲响应 $b_c(t)$ 以及地震仪的脉冲响应 $i_n(t)$ 进行褶积，便得到合成地震图。图 17 是一个合成地震图的例子，$a/v_b=0.04\text{s}$，$r/Q_0=0.04\text{km}$。为便于作比较，逐次褶积的结果也表示于同一图上。图 18 和图 19 是另两个例子，其中图 18 的 $a/v_b=0.04\text{s}$，$r/Q_0=0.20\text{km}$；图 19 的 $a/v_b=1.0\text{s}$，$r/Q_0=0.04\text{km}$。图中也绘上逐次褶积的结果。从这些数值计算结果可以看到，介质的吸收和仪器的影响，使得初动的半周期、振幅和波形均发生畸变。

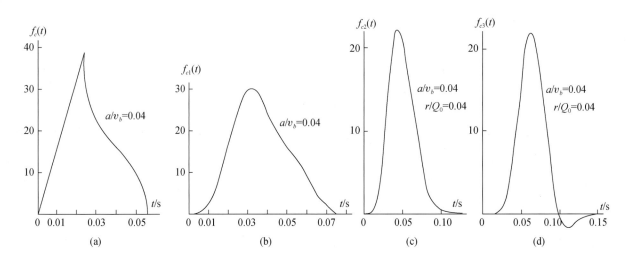

图 17 理论地震图

$f_{c1}(t)=f_c(t)*\dot{g}(t)$, $f_{c2}(t)=f_{c1}(t)*b_c(t)$, $f_{c3}(t)=f_{c2}(t)*i_n(t)$

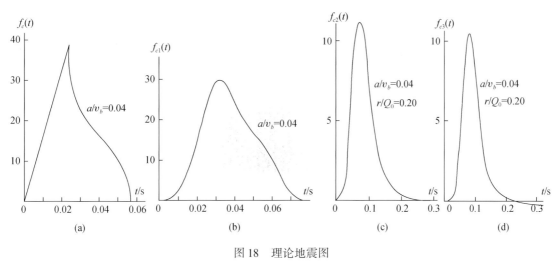

图 18 理论地震图

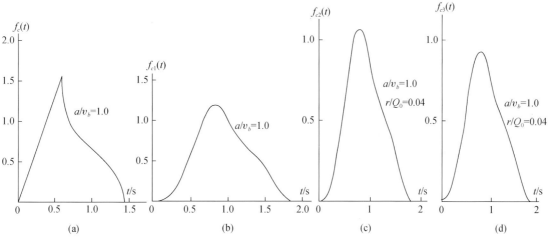

图 19 理论地震图

9 测定震源参数的理论曲线

由于介质的吸收作用、地表面的影响以及地震仪的影响，初动半周期和振幅跟震源尺度的关系不再如公式（16）和（21）所示．根据上节的数值计算结果，可以得到地震图上的初动半周期和震源尺度的关系．图 20 是 $r/Q_0 = 0.01$，0.04，0.08，0.12 和 0.20 km 时初动半周期 $t_{2\alpha}$ 和震源半径 a 的关系．它清楚地表明，当 a 较大时，$t_{2\alpha}$ 和 a 成正比；当 a 较小时，$t_{2\alpha}$ 趋近于和 a 无关的数值，这个数值随 r/Q_0 的增大而增大．以 $(t_{2\alpha})_{极小}$ 表示这个数值，它和 r/Q_0 的关系如图 21 所示．由观测到的 $t_{2\alpha}$-M_L 曲线的极小值便可确定 r/Q_0 的数值．r 是已知的，这样便可估算介质的品质因数 Q_0．在由 $(t_{2\alpha})_{极小}$ 确定了 r/Q_0 后，就可由相应于该 r/Q_0 值的 $t_{2\alpha}$-a 曲线测定震源尺度．

图 22 是 $u'_{\alpha m}$ 和 a 的关系曲线．由 a 可以读得相应于该 r/Q_0 值的 $u'_{\alpha m}$．$u'_{\alpha m}$ 表示 $u_{\alpha m}$ 和 $\dfrac{m_0}{2\pi\rho\alpha^3 r}$ 的比值．观测到的初动振幅 $u_{\alpha m}$ 与 $u'_{\alpha m}$ 的比值和地震矩有关．当 ρ，α，r 已知后，由图 23 所示的曲线便可读得该地震的地震矩．这种由初动振幅测地震矩的方法是和奥涅尔与希利（M. E. O'Neill and J. H. Healy）用 M_L 测地震矩的方法[20]不同的，他们需要先从理论上建立 M_L 和地震矩的关系，然后由 M_L 测地震矩，我们则不必．在我们的方法中，地震矩由初动振幅测得，而它和 M_L 的关系则靠观测数据来建立．

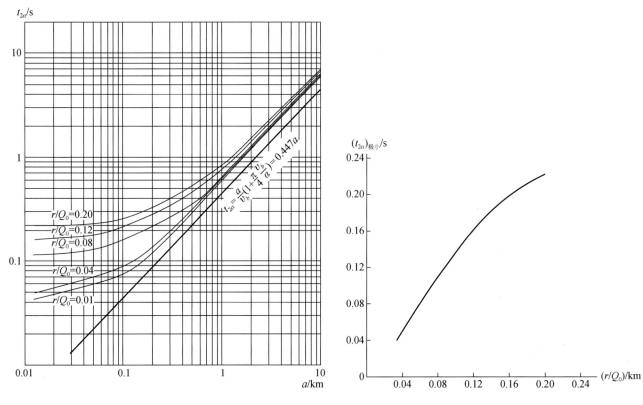

图 20　P 波初动半周期和震源半径的关系

图 21　初动半周期的极小值和震源距与品质因数的比值的关系

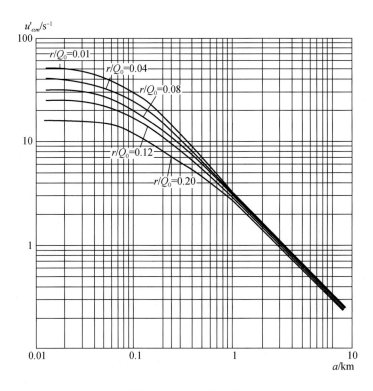

图 22　u'_{am} 和 a 的关系

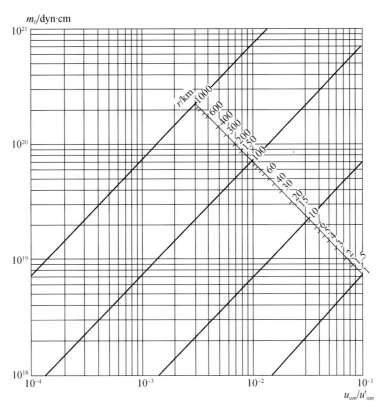

图 23　由 u_{am}/u'_{am} 和 r 测定 m_0 的理论曲线

在测得 a，m_0 后，由图 24 所示的曲线可以测得该地震的应力降，由图 25 所示的曲线可测得该地震的平均错距.

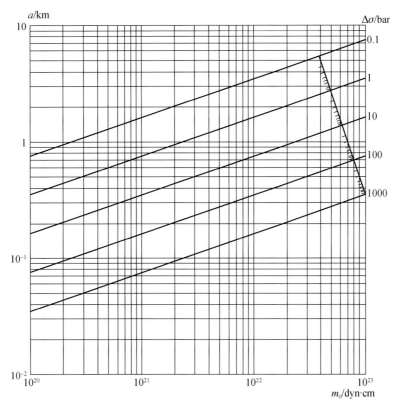

图 24　由震源半径和地震矩测定应力降的理论曲线

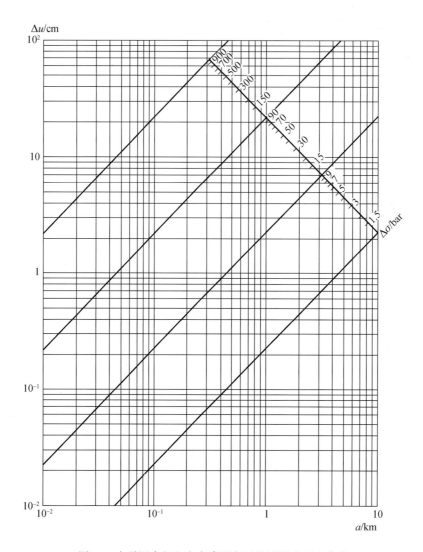

图 25　由震源半径和应力降测定平均错距的理论曲线

10　巧家、石棉的小震震源参数及其地震危险性的估计

按照上节叙述的步骤,运用计算得到的理论曲线,测定了巧家附近地区和石棉附近地区的小震震源参数. 表 1 和表 2 是测定结果.

由观测到的 P 波初动半周期的极小值 $(t_{2\alpha})_{极小}$ 测定了上述两个地区的 P 波的 Q 值. 巧家地区的 Q 值约为 620,石棉地区的约为 560,二者很接近. 这些结果和普雷斯(F. Press)[21]、萨顿(G. H. Sutton)[22] 对地壳中的 Q 值的估计一致.

图 26 是小震的震源尺度和震级的关系,它表明,震源尺度的对数虽然大体上和震级成正比,但资料的离散较大. 这一结果说明用一个简单的公式表示震源尺度和震级的关系看来是不可能的[11].

图 27 是地震矩的对数和震级的关系. 为了作比较,图中也绘上外斯和布龙(M. Wyss and J. N. Brune)[23]、道格拉斯和赖亚尔(B. M. Douglas and A. Ryall)[24],以及史密斯等(R. B. Smith et al.)[25] 的研究结果. 可以看到,本文得到的结果和上述作者的结果是一致的. 值得指出的是,其他作者的结果都是由频谱分析测得的,而这里的结果却是直接从地震图上测得的;所用的方法虽不同,而彼此的结果都比较一致. 另一个值得指出的事实是,本文测定的结果,其震级范围在 2.1 ~ 3.9 级之间,地震矩在 10^{19} ~ 10^{21} dyn·cm 之间,

表 1 由普格台的记录测定的巧家附近地区的小震震源参数

$(t_{2a})_{极小} \simeq 0.15s$ $r/Q_0 \simeq 0.12km$ $\bar{r} \simeq 74km$ $Q_0 \simeq 620$

编号	日期 年.月.日	发震时刻 时—分—秒	震中位置 东经	震中位置 北纬	震级 M_L	震中距 /km	震源深度 /km	震源距* /km	t_{2a}/s	$u_{am}/\mu m$	a/km	m_0/dyn·cm	$\Delta\sigma$/bar	$\Delta\bar{u}_1$/cm
1	1970.7.3	01–19–38	103°07′	27°10′	2.8	55			0.2	0.02				
2	8.9	12–39–11	102°41′	26°57′	2.3	50			0.2	0.12				
3	8.14	01–52–27	102°42′	26°42′	1.7	74			0.15	0.25				
4	10.5	07–41–08	103°18′	26°54′	1.8	85			0.2	0.02				
5	11.13	06–24–11	102°50′	27°03′	2.3	44			0.2	0.10				
6	12.7	21–13–59	102°54′	26°48′	2.2	71			0.15	0.05				
7	1971.3.29	21–54–13	103°06′	26°48′	2.2	81			0.2	0.05				
8	4.4	11–36–35	103°12′	27°18′	2.0	59			0.2	0.05				
9	4.28	09–36–29	103°09′	27°04′	3.6	65		(70)	0.6	1.35	0.76	1.9×10^{21}	2	0.32
10	4.28	10–32–19	103°06′	27°06′	2.2	59			0.2	0.15				
11	7.22	09–22–24	103°00′	26°54′	2.5	68		(73)	0.3	0.08	0.23	4.4×10^{19}	1.6	0.08
12	10.28	19–54–59	102°58′	26°52′	3.9	70		(75)	0.4	0.32	0.40	2.8×10^{20}	2	0.18
13	1972.5.2	06–39–10	103°00′	27°00′	2.6	58			0.2	0.23				
14	6.23	21–52–30	103°01′	27°07′	2.6	52		(58)	0.4	0.025	0.40	1.7×10^{19}	0.12	0.01
15	6.24	12–16–29	103°06′	26°56′	3.5	70		(75)	0.5	0.13	0.57			
16	8.22	16–30–47	103°12′	27°06′	3.5	66		(71)	0.4	0.20	0.40	1.1×10^{20}	0.7	0.06
17	1973.1.5	00–11–28	103°06′	27°04′	3.4	68		(73)	0.3	0.02	0.23	1.1×10^{20}	3	0.15
18	1.18	00–40–24	103°00′	26°36′	2.3	96		(100)	0.3	0.02	0.23	1.5×10^{19}	0.6	0.03
19	3.2	03–46–08	103°06′	26°54′	2.1	68			0.2	0.02				
20	3.2	04–34–02	103°12′	26°54′	2.2	78			0.15	0.05				
21	5.15	08–02–30	103°06′	26°54′	2.3	73			0.2	0.05				
22	11.16	09–56–33	102°48′	26°42′	2.2	80			0.2	0.075				
23	1974.3.26	21–01–37	103°06′	27°00′	2.1	65			0.2	0.22				
24	3.29	14–49–42	103°12′	26°54′	3.1	80			0.3	0.05				
25	4.4	07–05–07	103°17′	27°04′	2.4	76			0.2	0.30				
26	5.5	13–49–01	103°04′	27°04′	2.7	58	26		0.1					

* 因为缺乏可靠的震源深度资料，震源距系按震源深度为 26km 估算的.

表 2　由石棉台的记录测定的石棉附近地区的小震震源参数

$(t_{2a})_{极小} \doteq 0.1 \text{s}$　　$r/Q_0 \doteq 0.08 \text{km}$　　$\bar{r} \doteq 45 \text{km}$　　$Q_0 \doteq 560$

编号	日期 年.月.日	发震时刻 时-分-秒	震中位置 东经	震中位置 北纬	震级 M_L	震中距 /km	震源深度 /km	震源距* /km	t_{2a}/s	$u_{cm}/\mu m$	a/km	m_0/dyn·cm	$\Delta\sigma$/bar	$\Delta\bar{u}_1$/cm
1	1971.2.10	22-01-06	102°30′	28°53′	2.9	43			0.05	1.90				
2	5.9	00-49-03	102°12′	29°02′	3.7	30			0.2		0.16			
3	5.22	08-42-14	102°25′	29°04′	2.2	23			0.15					
4	6.2	12-58-09	102°18′	29°00′	1.7	30			0.12					
5	10.4	12-36-43	102°12′	29°12′	109	18			0.05					
6	10.20	19-05-23	102°22′	29°00′	3.0	30	16		0.1					
7	12.25	01-29-02	102°29′	29°02′	2.4	30			0.07					
8	1972.2.23	18-30-55	102°30′	28°54′	2.3	39			0.1					
9	2.28	09-07-01	102°06′	29°00′	1.9	34			0.12					
10	4.4	15-08-27	102°06′	29°00′	2.4	38			0.07					
11	5.20	01-51-43	101°54′	29°12′	2.3	42			0.12					
12	5.20	02-26-31	101°54′	29°12′	2.5	42			0.15					
13	5.24	04-14-22	102°18′	29°06′	2.3	19	15		0.1		0.36			
14	5.29	22-06-25	102°46′	28°58′	3.0	51			0.3		0.16			
15	6.25	16-51-35	101°48′	29°18′	2.3	54			0.2					
16	8.2	17-10-21	101°42′	29°30′	2.4	69			0.19					
17	8.2	18-08-59	101°42′	29°30′	2.3	69	17		0.19					
18	11.1	11-15-02	102°24′	29°00′	2.9	30			0.18					
19	2.12	02-45-34	102°24′	29°06′	2.2	19			0.1	1.65				
20	1973.4.11	02-37-55	102°36′	29°00′	1.9	37			0.05	0.23				
21	4.18	06-23-56	102°06′	29°12′	2.0	26			0.05	0.23				
22	4.24	13-25-02	102°36′	29°00′	2.2	38			0.12	0.18				
23	4.26	10-30-21	101°48′	29°30′	2.9	59			0.16	0.68				
24	5.13	18-26-57	102°12′	29°12′	2.6	19		(26)	0.2	2.30	0.16	2.8×10^{20}	28	0.92
25	5.17	04-57-44	101°42′	29°30′	2.1	69		(72)	0.2	0.06	0.16	2.1×10^{19}	2.3	0.08
26	5.19	05-40-38	101°42′	29°30′	2.6	69			0.12	0.09				
27	5.21	10-27-07	102°12′	28°54′	2.5	39			0.1	0.33				
28	5.31	10-21-33	101°48′	29°30′	2.8	59			0.12	0.08				
29	6.26	02-20-19	102°12′	29°00′	2.1	33			0.12					

* 震源距按震源深度 16km 估算.

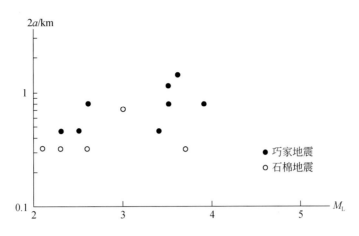

图 26 小地震的震源尺度和震级的关系

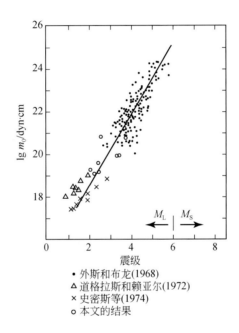

- • 外斯和布龙(1968)
- △ 道格拉斯和赖亚尔(1972)
- × 史密斯等(1974)
- ○ 本文的结果

图 27 地震矩和震级的关系

这个范围大部分是上述作者的工作未涉及的范围. 上述作者的工作加上本文的工作完整地显示了 $\lg m_0$ 和 M_L 的关系在 $1 \leq M_L \leq 6$ 的范围内大体上是线性的. 资料的离散也较大, 同样说明用一个简单的公式表示地震矩和震级的关系看来是不可能的[11].

图 28 是小地震的震源尺度和地震矩的关系. 从这幅图所表示的结果可以看到: 石棉附近的小地震, 应力降大约在 2~30bar 之间; 巧家附近的小地震的应力降较小, 且比较接近, 平均约 1.4bar.

吉博维茨 (S. J. Gibowicz)[26] 根据 18 个地震的资料, 统计出主震震级 M_L 和主震应力降 $\Delta\sigma$ 有如下关系:

$$M_L = 1.5\lg\Delta\sigma + 5.0. \tag{48}$$

前震系列的小震应力降和主震震级有无关系, 迄今仍缺乏系统的观测资料. 在 1962 年 3 月 19 日新丰江地震中, 可以看到[27], 其主震的应力降约 10bar, 和按 (48) 式估算的一个 6.1 级地震的应力降相近; 而其前震的应力降和主震的很一致. 这说明, 新丰江地震是发生在一种和 6.1 级地震相称的地质构造环境内. 与新丰江地震不同, 在海城地震中, 可以看到, 其主震的应力降低于按 (48) 式估算的一个 7.3 级地震

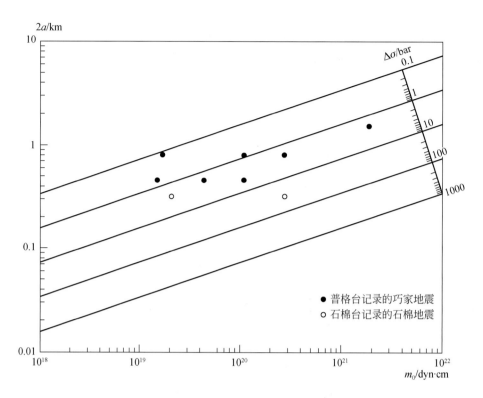

图 28 小地震的震源尺度和地震矩的关系

的应力降；而其前震的应力降虽然彼此相近，但低于主震的应力降．这说明，海城地震发生在一种比 7.3 级地震的平均地质构造环境要薄弱一些的构造环境内．单从巧家地区小震的应力降比较接近这一事实，目前还难以判断这些小震究竟是一个较大地震的前震还是一般的小地震．即便能够肯定这些小震是一个较大地震的前震，也还难以判断它究竟是属于新丰江地震这种情况还是属于海城地震这种情况．尽管如此，按目前的认识水平我们仍然不能排斥巧家地区的小震可能是一个较大地震的前震的可能性．果如其然，那么这个地震的应力降可能大于其前震的应力降（类似于海城地震），也可能等于其前震的应力降（类似于新丰江地震）．按照（48）式，可知这个地震的震级应大于或等于 5.2．因此，不能排斥巧家附近地区有发生 5.2 级地震的潜在危险性．

参 考 资 料

[1] D. Tocher, Earthquake energy and ground breakage, *Bull. Seism. Soc. Am.*, **48**, 2, 147–152, 1958.
[2] K. Iida. Earthquake energy and earthquake fault, *J. Earth Sci., Nagoya Univ.*, **7**, 2, 98–107, 1959.
[3] K. Iida. Earthquake magnitude, earthquake fault, and source dimensions, *J. Earth Sci., Nagoya Univ.*, **13**, 2, 115–132, 1965.
[4] C. Y. King and L. Knopoff, Stress drop in earthquakes, *Bull. Seism. Soc. Am.*, **58**, 1, 249–257, 1968.
[5] M. A. Chinnery, Earthquake magnitude and source parameters, *Bull. Seism. Soc. Am.*, **59**, 5, 1969–1982, 1969.
[6] M. A. Chinnery, Theoretical fault models, *Publ. Dominion Obs.*, **37**, 7, 211–223, 1969.
[7] В. И. 克依利斯-博罗克等，地震机制的研究，科学出版社，1961.
[8] T. Maruyama, On the force equivalents of dynamical elastic dislocations with reference to the earthquake mechanism, *Bull Earthq. Res. Inst., Tokyo Univ.*, **41**, 3, 467–486, 1963.
[9] J. C. Savage, Radiation from a realistic model of faulting, *Bull. Seism. Soc. Am.*, **56**, 2, 577–592, 1966.
[10] A. Ben-Menahem and S. J. Singh, Multipolar elastic fields in a layered half space, *Bull. Seism. Soc. Am.*, **58**, 5, 1519–1572, 1968.

[11] M. J. Randall, The spectral theory of seismic sources, *Bull. Seism. Soc. Am.*, **63**, 3, 1133–1144, 1973.

[12] T. Sato and T. Hirasawa, Body wave spectra from propagating shear cracks, *J. Phys. Earth*, **21**, 4, 415–432, 1973.

[13] F. A. Dahlen, On the ratio of P-wave to S-wave corner frequencies for shallow earthquake sources, *Bull. Seism. Soc. Am.*, **64**, 4, 1159–1180, 1974.

[14] V. I. Keils-Borok, On estimation of the displacement in an earthquake source and of source dimensions, *Ann. Geofis.*, **12**, 2, 205–214, 1959.

[15] J. D. Eshelby, The determination of the elastic field of an ellipsoidal inclusion and related problems, *Proc. Roy. Soc. (London) A*, **241**, 376–396, 1957.

[16] L. Knopoff, Q, *Rev. Geophys.*, **2**, 4, 625–660, 1964.

[17] W. I. Futterman, Dispersive body waves, *J. Geophys. Res.*, **67**, 13, 5279–5291, 1962.

[18] T. Usami, T. Odaka and Y. Sato, Theoretical seismograms and earthquake mechanism, *Bull. Earthq. Res. Inst., Tokyo Univ.*, **48**, 4, 533–579, 1970.

[19] Y. Ida and K. Aki, Seismic source time function of propagating longitudinal-shear cracks, *J. Geophys. Res.*, **77**, 11, 2034–2044, 1972.

[20] M. E. O'Neill and J. H. Healy, Determination of source parameters of small earthquakes from P-wave rise time, *Bull. Seism. Soc. Am.*, **63**, 2, 599–614, 1973.

[21] F. Press, Seismic wave attenuation in the crust, *J. Geophys. Res.*, **69**, 2, 4417–4418, 1964.

[22] G. H. Sutton, W. Mitronovas, and P. W. Poimeroy, Short-period seismic energy radiation patterns from underground nuclear explosions and small magnitude earthquakes, *Bull. Seism. Soc. Am.*, **57**, 2, 249–267, 1967.

[23] M. Wyss and J. N. Brune, Seismic moment, stress, and source dimension for earthquakes in the California-Nevada region, *J. Geophys. Res.*, **73**, 14, 4681–4694, 1968.

[24] B. M. Douglas and A. Ryall, Spectral characteristic and stress drop for microearthquakes near Fairview peak, Nevada, *J. Geophys. Res.*, **77**, 2, 351–359, 1972.

[25] R. B. Smith, P. L. Winkler. J. G. Anderson, and C. H. Scholz, Source mechanisms of microearthquakes associated with underground mines in eastern Utah, *Bull. Seism. Soc. Am.*, **64**, 4, 1295–1317, 1974.

[26] S. J. Gibowicz, Stress drop and aftershocks, *Bull. Seism. Soc. Am.*, **63**, 4, 1433–1446, 1973.

[27] 王妙月等, 新丰江水库地震的震源机制及其成因初步探讨, 地球物理学报, **19**, 1, 1–17, 1976.

The determination of source parameters for small earthquakes in Qiaojia (巧家) and Shimian (石棉) and the estimation of potential earthquake danger

Chen Yun-Tai[1], Lin Bang-Hui[1], Li Xing-Cai[1], Wang Miao-Yue[1], Xia Da-De[2], Wang Xing-Hui[2], Liu Wan-Qin[1] and Li Zhi-Yong[1]

1. Institute of Geophysics, Academia Sinica;
2. The Seismological Brigade of Chengdu, National Seismological Bureau

Abstract Using the observational data of small earthquakes which occurred in Qiaojia and Shimian, it is found that the half period of first P arrivals is nearly constant for the shocks of small magnitudes, and increases proportionally with magnitudes for larger earthquakes. It is also pointed out that the logarithm of the amplitudes of the initial motion of P- waves increases proportionally with the magnitudes. Taking a radially expanding circular shear crack with uniform dislocation as the theoretical model of the moderate and small earthquake sources, the far-field displacement of seismic body waves radiated from such a model is derived. By using the expression of far-field displacement, the quantitative relations between the source parameters, wave velocities and the half period, the amplitude of the first arrivals are inferred, and consequently, the empirical relationships between the half period as well as the amplitude of the first P arrivals and the magnitude of earthquakes are explained. Considering the attenuation and dispersion of waves in the medium, the effect of the free surface as well as the response characteristics of the seismograph instruments, the theoretical seismograms from this dislocation source are synthesized by the convolution technique, and a method by which the source parameters as well as Q value of the medium can be directly determined from the half period and the amplitude of the first P arrivals on the observed seismograms, is proposed. By applying the present method, the source parameters of small earthquakes as well as Q values of the medium of the two regions mentioned above are estimated. The Q values of the medium of the Qiaojia region and the Shimian region are 620 and 560, respectively. The stress drops of the small earthquakes in the Shimian region are estimated to be about 2 to 30 bars, and that of the small earthquakes in the Qiaojia region, are rather low and close to each other, the mean value of which is about 1.4bars. Comparing these results with the stress drops of the foreshocks and the main shocks of the Xinfengjiang(1962) and Haicheng(1975) earthquakes, we note the fact that the small earthquakes in the Qiaojia region are similar to the foreshocks of these two large shocks in the character of their stress drops, and so it may not be neglected that these small shocks in the Qiaojia region are foreshocks prior to a larger earthquake. Taking the mean value(about 1.4 bars)of the stress drops of the small earthquakes as an estimation of the lower limit of the stress drop of this hypothetical earthquake, it may be inferred that, from the empirical relationship between the stress drop and the magnitude of the mainshock, the lower limit of the magnitude of this hypothetical earthquake will be about 5.2.

1976年7月28日河北省唐山7.8级地震的发震背景及其活动性[*]

邱 群（陈运泰等）

中国科学院地球物理研究所

摘要 1976年7月28日河北省唐山7.8级地震发生在燕山地震带和河北平原地震带的交汇地区，认为它是华北地区北北东向构造体系和东西向构造体系复合构造共同作用的结果．它的发生更进一步说明华北地区的地震活动正处在其第四活动期的显著活动阶段中．

据P波初动符号的资料求得唐山地震的地震断层的走向为NE41°，倾向SE，倾角85°，它是发生在这个近乎直立的断层面上的右旋-正断层错动．观测表明，唐山地震的余震随时间迅速地、有起伏地衰减，表示余震的积累频度和震级的线性关系的斜率（b值）较高；而主震和强余震所释放的能量占整个地震系列能量的80%以上．这些事实说明唐山地震是一个主震型的地震．

1976年7月28日03点42分，我国河北省唐山—丰南一带发生了7.8级的强烈地震．伟大领袖毛主席和党中央对地震灾区人民极为关怀，地震发生后，党中央立即给灾区人民发出了慰问电．给了灾区人民巨大的鼓舞，激发了他们战胜地震灾害、重建家园的巨大力量．随之派出以华国锋总理为总团长的中央慰问团，深入地震灾区，对受灾群众和战斗在抗震救灾第一线的人民解放军指战员、各地支援人员进行慰问，转达了伟大领袖毛主席、党中央的亲切慰问和关怀．灾区人民坚决响应毛主席、党中央的战斗号召，认真学习毛主席的一系列重要指示，以阶级斗争为纲，联系各条战线阶级斗争、两条路线斗争的实际，继续批邓、反击右倾翻案风的伟大斗争，团结战斗，战胜灾害，在很短的时间内，夺取了一个又一个抗震救灾斗争的新胜利．

唐山地震的极震区在唐山市．当天07点17分，在主震的西南面发生了$M_S=6.5$的强余震，18点45分，在主震的东北面又发生了$M_S=7.1$的最大余震，主震和相继发生的强余震，使唐山、丰南及其附近地区遭受了严重的破坏．

尤其是唐山市的建筑物破坏极为严重．极震区内的工厂厂房也遭到严重破坏．强烈地震导致开滦煤矿坑道中的地下水急速上升，使正在井下作业的万名夜班工人的生命安全受到严重威胁．但是，由于井上、井下的工人和干部在战无不胜的毛泽东思想指引下共同奋战，在地震发生后几小时内，都胜利返回了地面，并在短短的十天时间，恢复了生产．地震使大部分工业烟囱折断，输电网、通信线路及地下管道也遭到损坏，但是，在毛主席和党中央的亲切关怀下，英雄的唐山电业工人同前来支援的各地电业工人紧密团结、顽强战斗，震后仅36小时就使唐山市区恢复了电力供应．唐山市部分市区的街道和郊区公路出现大规模的坍塌和大裂缝；京山铁路也遭到严重破坏，有的地段钢轨错位弯曲，呈波浪状起伏，有的地段钢轨被拉断，有的地段路基开裂、下沉．经过广大铁路工人和人民解放军铁道兵十天的奋战抢修，京山铁路已于8月7日下午胜利修复通车．整个唐山地震灾区呈现一派火热的战斗景象，广大军民在继续批邓、抗震救灾的斗争中奋勇前进．

---

[*] 本文发表于《地球物理学报》，1976年，第**19**卷，第4期，259-269．

1 唐山地震发震的构造背景

唐山地震发生在我国华北地区北北东向的河北平原的断层、断陷带与东西向的燕山构造带的交汇部位上. 华北地区是震旦纪末形成的一个古老地台, 震旦纪后经历了多次构造运动, 特别是中新生代以来, 由于现代板块运动的影响, 使这个地区的地质构造颇为复杂. 在华北地区, 有两个与地震活动密切相关的大构造体系 (图 1): 一个是东西向构造体系. 这是一个古生代的构造体系, 构造线总体呈东西向分布, 如燕山-阴山东西构造带、秦岭-大别山东西构造带以及伴随的山前拗陷或断陷带等. 这个构造体系对华北地区的地震的总体分布具有一定的控制作用. 另一个构造体系是北北东向构造体系. 这是一个中新生代的构造体系, 这个构造体系总体呈北北东向, 还有与北北东向共轭的北西向构造. 这一构造体系经历了印支、燕山和喜马拉雅三次构造运动. 特别是在燕山运动期间, 出现了大规模的断裂活动和以中酸性为主的岩浆活动, 形成了一系列与西太平洋板块和欧亚板块分界线平行的北北东向断裂所控制的大型拗陷和隆起. 这种隆起和拗陷往往相间出现, 从东向西平行地排列. 最外的一个隆起带是日本群岛、琉球群岛构造隆起带, 从东往西依次为日本海、黄海拗陷带, 辽东、山东隆起带, 松辽、河北、苏北拗陷带, 山西隆起带, 等等. 我们认为, 这可能是太平洋板块向欧亚板块俯冲、挤压的结果.

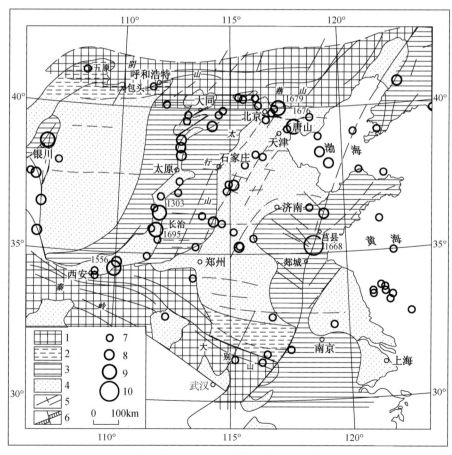

图 1 华北地区的地质构造与强震震中

图例: 1——东西向构造隆起带; 2——东西向山前拗陷 (断陷) 带; 3——北北东向构造隆起带; 4——北北东向拗陷带; 5——主要断裂带, 隐伏断裂带; 6——新生代断陷盆地; 7——$M=6$~6.9; 8——$M=7$~$7\frac{1}{2}$; 9——$M=7\frac{3}{4}$~8; 10——$M=8\frac{1}{2}$ (公元前 70—公元 1976 年地震震中)

从喜马拉雅运动开始、特别是第四纪以来,新华夏构造体系又进一步形成和发展.华北地区除了一些继承性断裂活动以外,在上述大型拗陷带和隆起带的内部又产生了一系列北北东向的具有水平剪切性质的断裂所切割的次一级隆起带和断陷盆地.这些活动断裂或断陷盆地自东向西平行地排列,它们包括郯城-庐江深断裂带,河北平原的一系列雁行状分布的构造隆起或断陷带,太行山山前断裂带,山西隆起区的锯齿状断陷盆地等.这些构造带反映了新第三纪和第四纪以来新华夏构造体系具有以走向滑动为主的右旋性质的运动.

根据地震震源机制的研究结果可知,华北地区主压应力方向以北东东方向占优势.这个方向和它东面的日本本州、伊豆、小笠原、马利亚纳岛弧一带的主压应力方向相一致.这可能是因为太平洋板块向欧亚板块推挤、并部分向欧亚板块下俯冲,在中朝地台产生了巨大的横向压力的结果.印度板块向北东方向推挤,可能也影响了华北地区.这两个方向的作用汇合在一起,构成了本区主压应力的优势方向为北东-南西向,造成了北北东方向的断裂带上的右旋走向滑动.

华北地区的强震,大都是沿着上述郯城-庐江深断裂带等四条北北东向的断裂分布,并且大都分布在北北东向构造和东西向构造的交汇部位附近(参见图1).

1976年7月28日唐山7.8级地震就发生在北北东向的河北平原内部的断层或断陷带和东西向的燕山构造带的交汇部位.在唐山一带,北北东向的构造带由于受到早期东西向构造带的限制,其走向逐渐转为北东向,向东到渤海一带甚至转为近东西向.唐山地震正位于这一转折部位.因此,我们认为唐山地震是北北东向构造体系和东西向构造体系复合构造共同作用的结果.

2 华北地区的历史地震

唐山地震发生在华北地震区的东北部.华北地震区是我国大陆东部的一个强震活动区,它的范围北起燕山、阴山以南,南到秦岭、大别山以北,西到银川盆地以东,东到渤海.这个地区的地震主要是浅源地震,震源深度一般是10～30km.地震大都聚集在上述活动构造带内,形成地震带.我们可以将它们分为渭河地堑地震带、山西地堑地震带、河北平原地震带、郯城-营口断裂地震带、燕山地震带等等.1966年河北邢台7.2级地震发生于河北平原地震带内,1969年渤海7.4级地震和1975年辽宁海城7.3级地震发生于郯城-营口断裂地震带上,而这次唐山地震则发生在燕山地震带和河北平原地震带的交汇地区.

华北地区早在公元前780年开始就有地震记载.当然,早期的地震记载缺失较多.自公元1000年以来,共发生了$8\frac{1}{2}$级地震1次,8级地震5次,7—7.9级地震12次,6—6.9级地震60余次.地震的频度不高,但强度很大.从地震震级和时间的分布图(图2)可以看出,这些地震集中在几段时间里连续发生,我们把一次相对平静过渡到显著活动称为一次地震活动期.华北地震区自公元1000年以来大体上经历了四次地震活动期,每期大致为300年左右.第一活动期大致在公元11世纪以前,第二活动期大致包括12至14世纪,第三活动期大致为15至18世纪初期,第四活动期大致为18世纪后期到现在.第一、第二活动期的资料缺失较多,从第三活动期开始,历史记载才较为详细.从图3可以看出,第三活动期从1369年开始到1739年,共370年,其中1369年至1476年为相对平静阶段,这个阶段没有6级以上地震,以能量积累为主、释放很少.自1477年以后,地震活动开始逐渐增加.1477年至1667年间共发生1次8级地震,3次7—$7\frac{1}{2}$级地震,22次6—6.9级地震,地震频度逐渐升高,强度增大.以后,1668年至1739年间接连发生了1668年山东郯城-莒县$8\frac{1}{2}$级大地震、1679年河北三河-平谷8级大地震、1695年山西临汾8级大地震,1739年宁夏银川8级地震和2次7级地震,我们把1477—1739年称为显著活动阶段,1740年以后地震活动开始逐渐衰减、趋于平静(见表1).

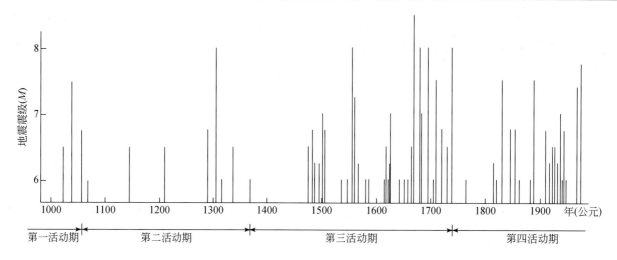

图 2　华北地震区震级-时间关系图

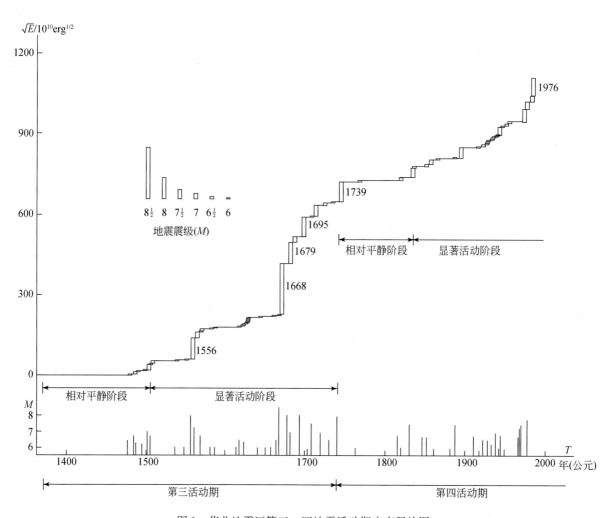

图 3　华北地震区第三、四地震活动期应变释放图

从1740年开始至今属第四活动期,它已持续了236年. 其中1740—1814年为相对平静阶段,没有记录到6级以上地震. 1815年以后地震活动逐渐增加、强度逐渐增大,从1815年到这次唐山地震前共发生6次7—7$\frac{1}{2}$级地震,20余次6—6.9级地震. 特别是1966年河北邢台7.2级地震以后,连续发生了1967年河北河间6.3级地震,1969年渤海7.4级地震,1975年辽宁海城7.3级地震,1976年4月内蒙古和林

格尔 6.3 级地震，其活动的频度和强度达到了 1815 年以来的最高峰．我们把 1815 年以后称为显著活动阶段．唐山 7.8 级大地震就发生在这个活动阶段．

表 1　华北地区的历史地震活动情况

活动期	阶段划分	各级地震发生次数			
		$8\frac{1}{2}$	8	7—7.9	6—6.9
第三活动期（1369—1739）	相对平静阶段（1369—1476）	0	0	0	0
	显著活动阶段（1477—1739）	1	4	5	25
第四活动期（1740—　　）	相对平静阶段（1740—1814）	0	0	0	0
	显著活动阶段（1815—　　）	(0)	(0)	(8)	(23)

3　京津唐张地区近期的地震活动

自 1966 年北京地震台网建立以来，测得的京津唐张地区 $M_L \geqslant 3.3$ 级的有感地震的分布一般比较均匀，如图 4 所示．从 1973 年开始，小地震有向东南方向的渤海集中的现象（见图 4 所示的 1973 年 1 月—1975 年 1 月的震中）．在 1975 年 2 月 4 日海城地震之后，这个现象依然存在，表明它不能用海城地震的影响予以解释．

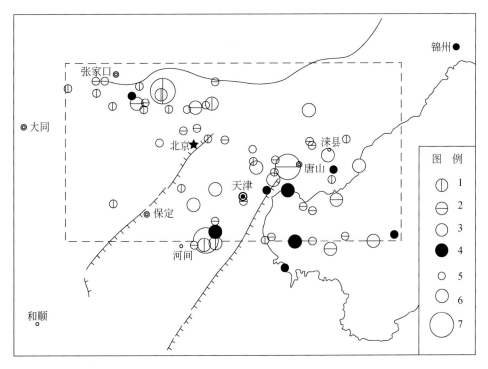

图 4　1966 年 4 月至 1976 年 6 月京津唐张地区 $M_L \geqslant 3.3$ 的地震震中分布

图例：1——1966 年 1 月至 1968 年 12 月；2——1969 年 1 月至 1972 年 12 月；3——1973 年 1 月至 1975 年 1 月；4——1975 年 2 月至 1976 年 6 月；5——$3.3 \leqslant M_L < 4.0$；6——$4.0 \leqslant M_L < 5$；7——$M_L \geqslant 5.0$

图 5 是 1973 年 12 月至 1976 年 7 月唐山 7.8 级地震之前的京津唐张地区震中分布图．从图 5 可以看出，1973 年 12 月后，在唐山东南面有一个以 2—4 级地震围起的空区．值得注意的是，这个空区在渤海

里,同这次唐山 7.8 级地震的余震区、极震区并不一致;并且从图 4 中我们还可以看到若干类似的空区. 这些情况说明了用地震填空性推测地震危险区时的复杂性.

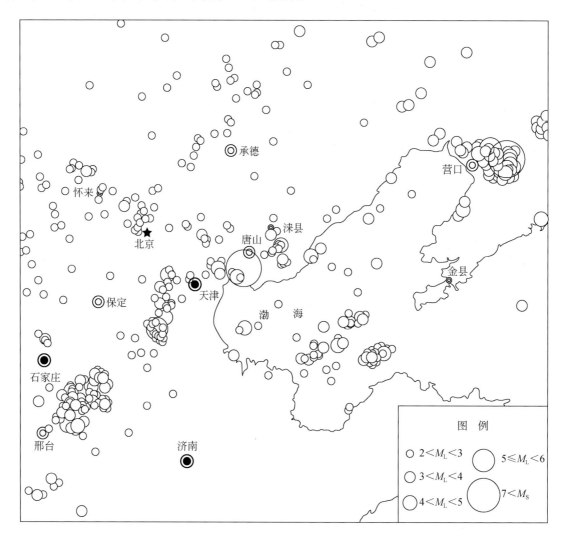

图 5　1973 年 12 月至 1976 年 7 月唐山 7.8 级地震前京津唐张地区的地震震中分布

4　唐山地震的震源参数

根据我国地震台网的测定,唐山地震的震中位置是 $39°.4N$, $118°.2E$,发震时刻是北京时间 03 时 42 分 53.8 秒,震级 M_S 为 7.8. 图 6 是主震和 $M_S \geqslant 4.5$ 的强余震的震中分布图. 由图可见,余震分布在 NE50°的方向上,余震区的长轴约 140km,短轴约 50km.

主震发生后,在三天半内,接连发生了 21 个 $M_S \geqslant 4.7$ 的余震. 最大余震在余震区的东北面,而主震的微观震中在余震区的西南面. 这种主震—最大余震—余震的空间分布特征和许多主余震型的地震的空间分布特征是类似的.

根据 P 波初动符号的资料,求得了主震的断层面解答. 图 7 是主震的断层面解. 主震的一个节面走向 NE41°,倾向 SE,倾角 85°;另一个节面走向 NW51°,倾向 NE,倾角 70°;主压应力轴的方位为 NE86°.5,

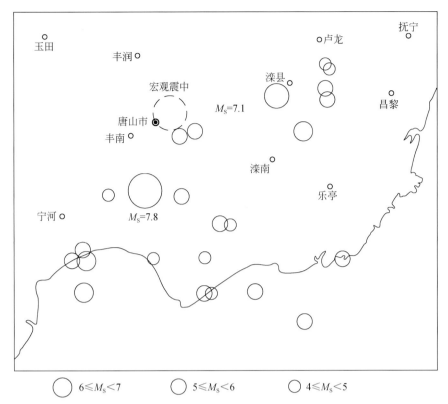

图 6 主震和 $M_S \geq 4.5$ 强余震的震中分布图（1976 年 7 月 28 日—1976 年 8 月 9 日）

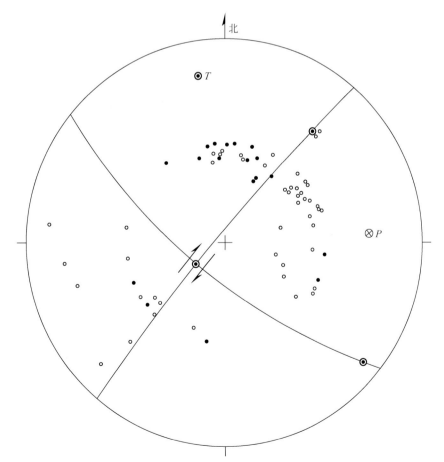

图 7 1976 年 7 月 28 日唐山 7.8 级地震的断层面解答

黑点表示初动为压缩，空心圆表示初动为膨胀．震源球上半球投影在乌尔夫网上

倾角 18°；主张应力轴的方位为 NW13°.5，倾角 10°；中间应力轴方位为 NW102°，倾角 69°. 北东向的节面的走向与余震区的走向基本一致，说明北东向的节面是这次地震的真正的断层面，这次地震是发生在走向为 NE41° 的近乎直立的断层上的右旋剪切错动，断层面略倾向东南，断层错动略具正断层分量. 根据主震的微观震中、宏观震中的分布特征，可以判断，主破裂开始于断层的西南面，破裂主要朝北东方向扩展，亦向南西方向扩展，是一种不对称的双侧破裂.

5 唐山地震系列的特征

唐山地震的余震随时间衰减很快. 图 8 是主震之后 $M_L \geq 3.8$ 的每日地震次数随时间变化图. 由图可见，余震随时间迅速地、有起伏地衰减的情况. 每日余震的频度 $n(t)$ 随着时间 t 大体上按

$$n(t) = \frac{A}{t^p}$$

的规律衰减. 式中，$A = 2.83$，$p = 1.30$.

图 9 是唐山地震余震系列的累积频度和震级的关系图，资料截至 1976 年 8 月 16 日 06 时. 余震累积频度 N 的对数和震级 M_L 呈线性关系：

$$\lg N = a - bM_L,$$

其中，表示大小地震比例关系的 b 值约为 1.05. 1975 年 2 月 4 日海城地震的 b 值是 0.87，1970 年 1 月 5 日通海地震的 b 值是 0.60. 同上述我国近年来的大地震的 b 值相比，唐山地震的 b 值是较高的. 这是表明唐山地震是主震型的一个显著标志.

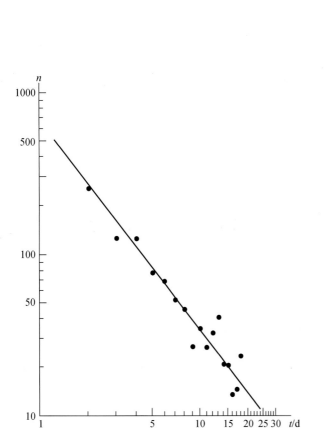

图 8 唐山地震主震后 $M_L \geq 3.8$ 的地震频度随时间的变化（1976.7.28—1976.8.15）

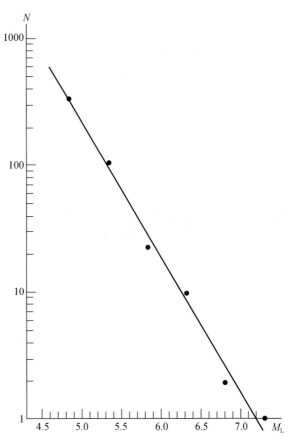

图 9 唐山地震余震系列的累积频度和震级的关系

图 10 是唐山地震余震系列的应变释放曲线. 由图 10 可见, 余震系列的大多数应变集中于主震发生后 1 小时. 主震之后半个月, 应变释放已渐趋缓和. 在整个地震系列中, 主震和 6 级以上余震所释放的能量占总能量的 80% 以上. 这一事实连同余震频度随时间的迅速衰减、余震系列的高 b 值等事实, 均表明唐山地震是主震型的地震.

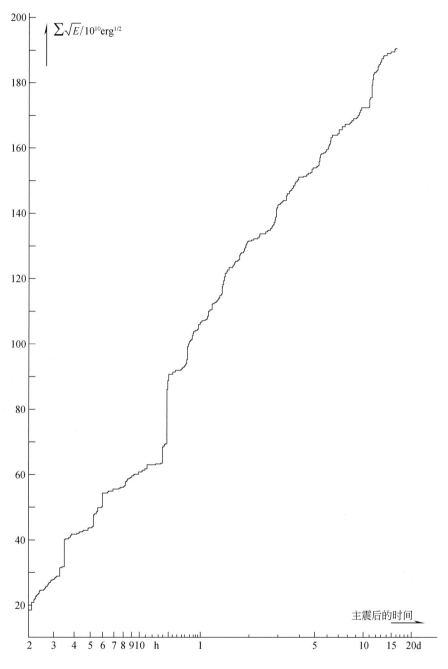

图 10　唐山地震余震系列的应变释放曲线

以上概述了唐山地震的发震背景和地震活动性. 关于这次地震的考察研究工作, 现正在继续进行中.

On the background and seismic activity of the $M=7.8$ Tangshan (唐山) earthquake, Hopei Province of July 28, 1976

Qiu Qun (Chen Yun-Tai et al.)

Abstract On July 28, 1976, an earthquake of magnitude 7.8 broke out in the vicinity of the city of Tangshan, Hopei Province, within a region corresponding to the conjunction of both the Yanshan (燕山) and the Hopei-Plain earthquake belts. It is thought of as the result of the combined action of north-northeast and east-west structural systems of North China. The occurrence of this earthquake also indicates that the seismic activity of North China has at present entered an active stage of the Fourth Period of seismic activity of North China.

Based on the signs of initial motions of P-waves, the strike of the earthquake fault is found to be NE41°, dipping to SE, dip angle, 85°. It behaves as a right-lateral normal fault dislocation, taking place on a nearly vertical fault plane. Observations show that the aftershocks of this earthquake sequence die down in certain fluctuations rather quickly with time. The inclination of the linear relation between the cumulated frequency and magnitudes of aftershocks (the b-value) is relatively high. More than 80% of the energy released by the entire earthquake sequence belong to the main shock and several stronger aftershocks. All such facts suggest that the aftershock sequence of the Tangshan Earthquake is the type beginning by the mainshock (the so-called mainshock-type).

1975年2月4日辽宁省海城地震的震源机制

顾浩鼎[1]　陈运泰[2]　高祥林[3]　赵　毅[4]

1. 辽宁省地震局；2. 中国科学院地球物理研究所；3. 宁夏回族自治区地震队；4. 国家地震局广州地震大队

摘要　由地震纵波初动符号的资料，求得了海城地震系列中 $M_S \geq 4.0$ 的24个地震的断层面解. 主震发生于1975年2月4日，它的一个节面走向N70°W，倾向NE，倾角81°；另一个节面走向N23°E，倾向SE，倾角75°. 根据余震的空间分布以及地面形变资料选取N70°W的节面为断层面，主震是发生在这个近乎直立的断层面上的左旋走向滑动，略具正的倾向滑动分量. 前震及大多数余震的震源机制和主震的相似，有四个 $M_S \geq 4.0$ 的余震的震源机制和主震的迥然不同，表现出滑动向量和主震的滑动向量相反的断层错动方式. 这种情况的一种可能的解释是主震时在断层的一些地段发生错动过头.

由野外资料及余震的空间分布资料计算了主震的震源参数. 主震断层长70km，宽20km，平均错距45cm，地震矩 2.1×10^{26} dyn·cm，应力降4.8bar，应变降 7.3×10^{-6}. 它是发生在不能积累起较高应力的薄弱地带的一次低应力降的地震.

由地震纵波初动的半周期和振幅的资料计算了81个前震和余震的震源尺度、地震矩、应力降和平均错距. 结果表明前震和余震的应力降都比较低，一般在0.1—1bar之间. 余震区中有两个应力降相对说来比较高（高于0.8bar）的地区，它们恰好对应于主破裂错动过头的部位. 这些结果意味着震前高应力、错动过头、相对高应力降和震源机制反向四者之间有内在联系，说明错动过头、相对高应力降和震源机制反向是震前高应力的表现和结果.

1　引言

震源机制的研究，可以给出大地震孕育和发生过程的重要信息，有助于阐明地震前兆的时-空分布特点. 因而在国内外有许多地震工作者做出种种努力，以确定地震的震源机制和测定越来越多的震源参数.

本文旨在探讨1975年2月4日海城7.3级地震的震源过程. 为此. 研究了整个地震系列中的较大地震的震源机制，估算了主震的震源参数，并测定了前震和余震的震源参数.

本文的一个结果是注意到了有些强余震的震源机制和主震的震源机制相反的现象，提出了错动过头的假设对这种现象给予力学上的解释. 另一个结果是，注意到了震前高应力、错动过头、震源机制反向和相对高应力降四者之间可能的内在联系. 错动过头，机制反向和相对高应力降是震前高应力的表现和结果.

2　主震的断层面解

为了了解海城地震的震源过程，由地震纵波初动符号的资料，求得了海城地震系列中 $M_S \geq 4.0$ 的24个地震的断层面解（见表1）. 其中，包括了主震、两个前震和21个余震. 图1是根据全国地震台网的记录和国外地震台网的资料得到的主震的断层面解. 图中，把震源球上半球投影在乌尔夫网上；黑点表示初

---

* 本文发表于《地球物理学报》，1976年，第**19**卷，第4期，270–285.

表 1 海城地震及其 $M_S \geq 4.0$ 的前震和余震的目录及断层面解

编号	日期 年.月.日	发震时刻 时-分-秒	震中位置 北纬	震中位置 东经	深度 /km	震级 M_S	节面 I 走向	节面 I 倾向	节面 I 倾角	节面 II 走向	节面 II 倾向	节面 II 倾角	P 轴 方位	P 轴 倾角	T 轴 方位	T 轴 倾角	B 轴 方位	B 轴 倾角	X 轴 方位	X 轴 倾角	Y 轴 方位	Y 轴 倾角	精度
1	1975.2.4	07-50-47	40°40′	122°45′	17	4.7	N70°W	SW	85°	N17°E	SE	65°	61°	15°	157°	22°	300°	64°	107°	26°	200°	5°	A
2	2.4	10-35-55	40	47	15	4.3	N71°W	SW	86	N17°E	SE	68	61	13	155	18	297	68	108	22	199	4	A
3	2.4	19-36-06	39	48	12	7.3	N70°W	NE	81	N23°E	SE	75	66	17.5	157	4	100	72.5	112	15	20	9	A
4	2.4	20-37	41	52		4.0																	
5	2.4	20-39-08	40	48		4.1																	
6	2.4	21-17-26	46	20		4.1																	
7	2.4	21-32-35	44	43		5.5																	
8	2.4	21-40-09	40	48		4.1																	
9	2.4	21-56-46	43	47		4.3																	
10	2.4	22-03-13	39	47		4.2																	
11	2.4	23-32-05	48	24		4.0																	
12	2.4	23-49-16	43	40		4.0																	
13	2.5	01-01-45	43	56	10	4.4	N62°W	NE	88	N28°E	NW	86	252	2	343	4	143	86	297	4	28	2	B
14	2.5	02-56-29	40	49	10	4.5	N68°W	SW	88	N22°E	SE	84	66	3	157	6	310	83	112	6	202	2	B
15	2.5	03-07-13	43	45	10	4.4																	
16	2.5	12-33-00	41	46	10	4.1	N54°W	SW	61	N41°E	NW	80	268	28	171	13	59	59	311	10	216	29	B
17	2.5	23-52-54	42	38	10	4.6	N55°W	SW	64	N37°E	NW	85	264	22	168	14	47	64	308	5	215	26	B
18	2.6	05-43-42	37	54	23	5.2	N68°W	NE	87	N22°E	SE	88	67	3	338	0	231	86	112	2	22	3	A
19	2.6	12-24-57	48	30	17	5.4	N18°W	SW	80	N79°E	NW	56	295	31	35	16	149	54	349	34	252	10	B
20	2.6	23-56-16	45	50	10	4.0	N64°W	SW	80	N14°E	SE	40	169	41	56	25	304	38	104	50	206	10	B
21	2.8	02-30-23	40°49′	122°28′	12	4.0	N19°W	SW	78	N82°E	NW	50	293	38	37	18	148	48	351	40	251	12	B
22	2.12	20-42-46	42	47	7	4.0	N37°E	SW	52	N3°E	SE	46	171	68	73	3	341	22	94	44	233	38	B
23	2.15	21-08-02	42	47	12	5.4	N69°W	SW	84	N18°E	SE	64	62	14	158	22	303	63	108	26	201	6	A
24	2.16	22-01-26	41	48	11	5.3	N57°W	SW	62	N30°E	SE	84	260	15	163	24	20	62	121	6	213	28	A
25	2.18	18-51-49	46	39	17	4.2	N66°W	SW	44	N14°E	SE	80	252	23	142	39	5	42	104	10	204	46	B
26	2.22	15-45-14	42	44	12	4.4	N67°W	SW	85	N19°E	NW	54	60	21	162	29	300	54	109	36	203	5	B
27	2.24	05-07-20	47	53	7	4.4	N48°W	SW	57	N78°E	SE	48	278	60	17	5	110	30	348	42	222	33	A
28	2.25	04-52-10	44	37	14	4.3	N46°W	SW	74	N46°E	SE	86	276	14	180	8	62	74	316	4	225	16	A
29	2.26	05-09-53	40	49	8	4.0	N59°W	SW	90	N30°E	NW	70	254	14	347	14	121	70	301	20	30	0	A
30	3.21	11-32-59	46	57	11	4.0	N80°W	NW	72	0°	E	62	38	34	132	7	231	56	90	28	350	18	B
31	3.29	23-16-36	46	36	6	4.1	N69°W	SW	85	N19°E	SE	66	63	13	158	20	303	65	109	24	201	5	A
32	4.10	03-55-37	43	29	10	4.6	N62°W	SW	85	N27°E	SE	81	253	3	163	10	357	80	118	5	208	9	B
33	4.21	00-17-06	46	27	8	4.0	N71°W	SW	84	N16°E	SE	60	59	17	157	25	299	59	106	30	199	6	B
34	7.4	07-06-29	43	40	10	4.1	N43°W	SW	56	N59°E	NW	74	283	36	185	12	80	52	329	16	227	34	B
35*	1974.12.22	12-46-18	41°25′	123°06′	6	4.8	N23°W	SW	40	N50°E	SE	76	179	45	291	22	39	37	140	14	247	50	A

* 辽阳参窝水库地震.

动向上，圆圈表示初动向下；大的符号表示国内台网的资料点，小的符号表示国外台网的资料点. 共有 162 个资料点，其中国内台网的 73 个，国外台网的 89 个. 矛盾的资料点共有 23 个，其中国内台网的 5 个，国外台网的 18 个. 考虑到国内台网记录的这次大地震的初动资料比较丰富和可靠，所以主要根据国内台网的初动资料确定主震的断层面解. 由外国地震台网的临时报告中得到的初动资料作为参考. 由图 1 可见，由纵波初动资料得出的两个可能的断层面可以确定得很好，可变动范围很小.

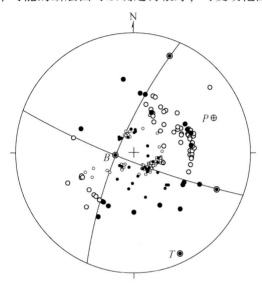

图 1　1975 年 2 月 4 日辽宁省海城 7.3 级地震的断层面解

表 1 的第三行是主震（编号 3）的断层面解. 为了从两个可能的断层面中，确定出真正的地震断层，可以借助其他资料，如余震的空间分布、地震引起的地面形变资料.

图 2 是余震的空间分布. 由图 2a 可以看出，余震震中分布在 N67°W 的长条形地带上，大体上呈椭圆形，其长轴沿北西向，长约 70km，走向同 N70°W 的节面（节面 I）几乎一致. 图 2b 是主震和余震在图 2a 所示的 AA' 剖面上的投影，它清楚地表明主震和余震在节面 I 上的分布情况，即：主震在余震区的下部，其震中靠近余震区的东南端，距此端点约 26 公里，在西北端，震源较深，而在东南端，震源较浅，由东南端至西北端，震源沿着与地面成 15°的轴向分布. 图 2c 是主震和余震在图 2a 所示的 BB' 剖面上的投影，它也清楚地表明，主震位于余震区的下部，还表明，余震大都分布在近乎直立的平面附近. 所有这些情况都表明由初动资料得出的节面 I 是真正的地震断层面.

地面形变的测量资料也支持北西向的节面是真正的断层面这一判断. 图 3 是震后对地震区的三角点复测的结果，它表示了海城地震引起的地面水平形变（1975 年相对于 1958 年的变化）图中的他山、大青山、海龙山三角形边长的变化，说明了主震是沿着北西西向断层的左旋剪切错动. 对于一个北西西向的左旋平移断层，靠近震中最近的大青山应当有最大的北西西向的水平位移，而离震中较远的他山的北西西向水平位移则应较小，因而他山—大青山的边长应当有明显的缩短. 否则，根据断层面解，这次大震就应当是沿北北东向断层的右旋剪切错动. 在这种情况下，大青山应当有最大的平行于北北东向节面的水平位移，而他山的位移则应较小；由于他山—大青山的测边几乎和北北东向节面垂直，因而他山—大青山的边长应当不会有明显的伸缩. 测量结果表明，他山—大青山边长的变化最大，缩短了 38cm. 这一事实说明，北西西向的节面 I 是主震的断层面.

余震的空间分布和地面形变的测量资料，一致地表明这次海城 7.3 级地震是发生在一个近乎直立的断层面上的剪切错动，其走向是 N70°W，倾向 NE，倾角 81°，走向滑动呈左旋性质，倾向滑动呈正断层性质，滑动向量与地面的夹角约 15°.

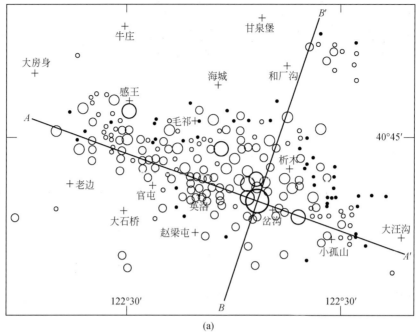

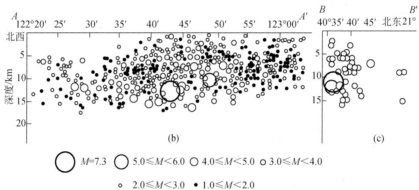

图2 余震的空间分布

图3 海城地震引起的地面水平形变

（1975年相对于1958年的变化）

海城地震释放的应力的主压应力轴方向是 N66°E，这同东北和华北地区的一些较大的浅源地震所释放的应力的主压应力轴方向是一致的. 表 2 列举了上述地区 6 个较大的浅源地震的断层面解. 由表 2 可见，发生这些较大的浅源地震时，释放的应力的主压应力轴的取向都是北东东—南西西向. 这一事实显示了在中国东北和华北地区，在这些地震所在范围内应力场的分布情况是很一致的. 鉴于这种情况，东北和华北地区的北北东向和北西西向的构造带或其交汇地区的地震危险性是应当首先给予注意的.

表 2 东北和华北地区一些较大的浅源地震的断层面解*

编号	日期	地点	震中位置		深度 /km	震级 M_S	节面 I			节面 II			主压应力轴		主张应力轴	
			北纬	东经			走向/°	倾向	倾角/°	走向/°	倾向	倾角/°	方位/°	倾角/°	方位/°	倾角/°
1	1960.4.13	吉林 土桥	44°39′	126°54′	8	5.7	55.3	NW	82	139	SW	49	285.5	34	180	21.5
2	1966.3.8	河北 隆尧	37°21′	114°55′	10	6.8	21	SE	75	114	SW	80	67	4	158	18
3	1966.3.20	河北 巨鹿	37°16′	114°58′	14	5.6	29	NW	65	311	NE	66	260	1	350	36
4	1966.10.2	吉林 怀德	43°47′	125°04′	24	5.2	42	NW	64	142.5	NE	69	271.5	3	3.5	34.5
5	1967.3.27	河北 河间	38°33′	113°36′	30	6.3	15	NW	61	287	NE	85	238	17	335	24
6	1969.7.18	山东 渤海	38°24′	119°36′	35	7.4	20	SE	80	292	SW	75	246	4	155	18

* 据国家地震局资料.

3 前震和余震的断层面解

表 1 给出了截至 7 月 4 日为止海城地震及其 $M_S \geq 4$ 的前震和余震的目录及断层面解. 在这个表中，包括了主震、两个前震和 21 个余震的断层面解. 有 10 个 4.0 级以上地震，因受主震的尾波或在它之前不久发生的地震的尾波的强烈干扰，未能求得其断层面解，所以在表中仅列出其目录. 作为比较，表 1 中也给出了 1974 年 12 月 22 日辽阳葠窝水库 4.8 级地震（编号 35）的断层面解. 这些结果主要是根据辽宁省地震台网的初动符号资料求得的. 在许多台站的记录图中，都可认出 P^*，P_n 和 $\bar{P}$ 等三个震相或其中的两个. 因此，在确定前震和余震的断层面解时，既用了这些地震的界面首波初动资料，也用了直达波初动资料，使有用的资料点的数目大大增加. 此外，由于许多台站的位置恰好有利于用来确定上述地震的节面（见图 4），因而大多数断层面解比较可靠. 在这些结果中，节面可变动范围较小（≤10°）、资料较丰富者属 A 类，节面可变动范围较大（>10°）、资料稍少者属 B 类.

图 5 是海城地震及其全部 $M_S \geq 4.0$ 的前震和截至 1975 年 7 月的全部 $M_S \geq 4.0$ 的余震的震源机制综合图. 可以看出，前震的断层面解几乎和主震的相同，都和葠窝水库地震的不同. 葠窝水库地震具有明显的倾滑分量，而海城地震则以走滑为主. 从 $M_S = 4.7$ 的最大的前震到主震，仅相隔 12 小时；而前震震中位置几乎和主震震中位置重合. 在短短 12 小时内，构造运动不会使震源区的应力场发生任何可觉察的变化；而由于前震的规模不大，也只能扰动局部范围的震源区应力场. 因此，前震的断层面解和主震的相同，说明在临近前震时，主震断层面附近的介质已处于应力相当集中的临界状态；还说明前震发生在应力较高的地点，它们的发生导致了破裂面边缘的应力集中，从而触发了沿此破裂面的主破裂，即主震. 这也就是说，海城地震的两个较大的前震的断层面和主震的断层面的取向应当是一致的. 根据这一分析，选取 N70°W 的节面为 4.7 级前震的断层面，选取 N71°W 的节面为 4.3 级前震的断层面.

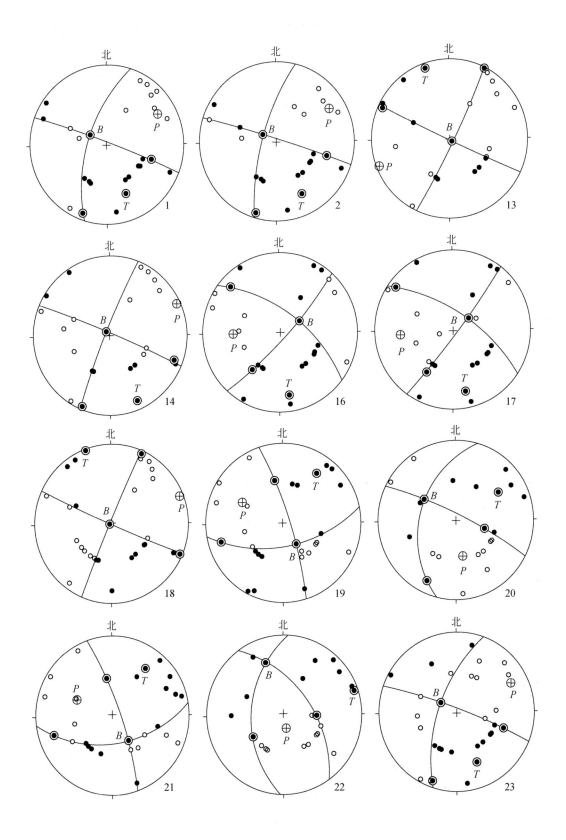

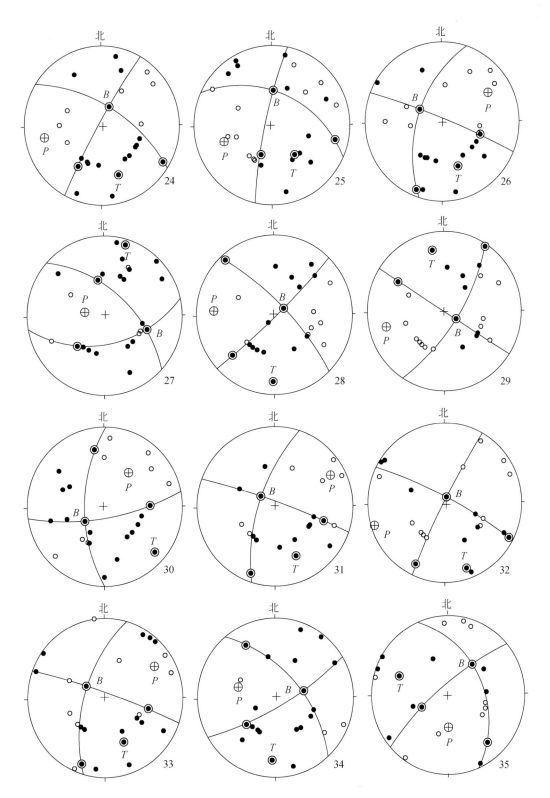

图 4 海城地震系列中 $M_S \geqslant 4.0$ 的前震和余震及 1974 年 12 月 22 日辽阳葠窝水库地震的断层面解
震源球上半球投影在乌尔夫网上. 黑点表示初动向上, 空心圆圈表示初动向下. 地震的编号和表 1 的地震的编号一致

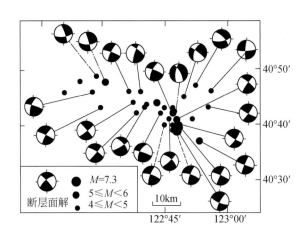

图 5 海城地震的 $M_S \geqslant 4.0$ 的前震和余震的震中分布及震源机制综合图

余震的情况和前震不同. 尽管多数余震的断层面解仍和前震、主震基本一致, 但主应力轴和节面的取向明显地偏离主震的主应力轴和节面的取向. 从图 5 可以看出, 那些断层面解基本上和主震一致的余震大都分布在主震断层上. 因此, 自然选取和主震断层面相近的那个节面为其断层面. 这些余震的主应力轴方向的离散说明, 主震显著地扰动了整个震源区的应力场. 图 6 综合了前、主、余震的主应力轴在乌尔夫网的投影. 图中, 大的符号表示主震的资料, 带一横的小的符号表示前震的资料, 不带一横的小的符号表示余震的资料. 这幅图清楚地显示, 尽管主压应力轴和主张应力轴的取向有些离散, 但零轴的分布则相当有规则, 它大体上分布在北西西向的平面上, 意味着前震、主震和大多数余震具有取向大体上一致的破裂面——北西西向的破裂面.

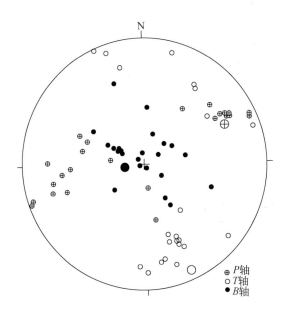

图 6 海城地震前、主、余震的主应力轴的综合图

在表 1 中, 编号 19, 20, 21, 22 等四个相继发生的 4.0 级以上地震, 它们的断层面解和前震、主震以及其他余震的断层面解迥然不同. 2 月 6 日, 在相隔很短的时间内发生了两个 5.0 级以上地震. 05 时 43 分, 在主震断层的东南端发生了 5.2 级地震 (编号 18), 它的断层面解和主震完全相同. 12 时 24 分, 在主震断层的西北端, 发生了 5.4 级地震 (编号 19), 它的断层面解和主震以及在它之前发生的 5.2 级余震完全不同. 这就是, 它的两个节面和主震以及 5.2 级余震的节面虽相接近, 但几乎全部台站记录的纵波

初动的极性都相反,主压应力轴从 N67°E 逆时针旋转了 132°,变到 N65°W. 2 月 6 日 23 时 56 分的 4.0 级地震(编号 20)、2 月 8 日 02 时 30 分的 4.0 级地震(编号 21)和 2 月 12 日 20 时 42 分的 4.0 级地震(编号 22)也有类似的情况. 出现这种情况的原因可以用主破裂在局部地段发生错动过头的假设来解释. 根据余震的空间分布以及由物探方法查明的地下隐伏断层的走向(图 7),可以确定 N79°E 的节面为 5.4 级余震的断层面. 主震震中靠近余震区的东南端,表明主破裂主要朝北西向传播,但亦朝南东向传播. 当它由东南向西北传播后,终止于 5.4 级余震的震源附近,它的终止可能和在它的西边的北东向的隐伏断层的存在有关. 在主破裂由东南传播至西北端时,由于已有的东西向隐伏断层的存在,破裂遂沿此断层进行. 在隐伏断层最初发生破裂的部位,即其东部,应力集中程度较高,因而在这里断层两盘相对错动冲过了平衡位置,即错动过头,将一部分动能转换为介质的应变能. 在余震期间的调整过程中,断层面的两盘回跳到其平衡位置,将这部分应变能释放出来,此时,两盘的错动方向正好和主震的相反. 与编号 19 的余震类似,编号 20,21 和 22 的余震的震源机制和主震的完全不同,同样可以根据上面提出的错动过头予以解释. 就震源位置而言,编号 19 和 21 的余震是在东西向隐伏断裂的东部;编号 21 和 22 的余震则和主震的初始破裂点靠近. 在发震的时间顺序上,这四个震源机制和主震相反的地震是在主震后不久相继发生的. 既然地震断层的初始破裂点是临近主震之前应力集中的地点,那么,前一个事实说明错动过头发生在应力集中的地点;后一个事实则说明,在震源区的介质内,由于错动过头导致的局部地区的应力集中状态在力学上是不稳定的,它将很快地以相反方向的错动恢复到平衡状态.

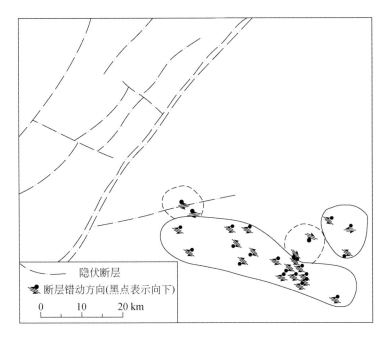

图 7 前、主、余震的断层错动方向和地下隐伏断裂的走向

4 一次低应力降的地震

这次地震的极震区中没有大规模的地震断裂,但在图 2a 所示的小孤山发现了一条近东西向的基岩裂缝带,在英洛及官屯一带的基岩中,也有一条北西西向、呈左旋性质的基岩裂缝,最大水平错距达 55cm. 今以此数据作为这次地震的断层面最大水平错距的估计值,由断层面解得知,错动向量和断层走向的夹角大约为 15°,由此可以估计这次地震的最大错距约 57cm. 根据余震的空间分布情况,分别取余震区的长度(70km)和它的深度(20km)作为断层长度(L)和宽度(a)的估计值. 由平均错距 $\Delta \overline{U}$ 和最大错距 U_m

的关系

$$\Delta \bar{U} = \frac{\pi}{4} U_m \tag{1}$$

以及地震矩 m_0 的定义

$$m_0 = \mu \Delta \bar{U} S, \quad S = La, \tag{2}$$

可以求得平均错距为45cm，地震矩为 2.1×10^{26} dyn·cm. 由应力降 $\Delta\sigma$ 和最大错距与断层宽度的关系

$$\begin{cases} \Delta\sigma = \mu U_m / 2a, & \text{对于走向滑动情形,} \\ \Delta\sigma = \mu \dfrac{(\lambda+\mu)}{(\lambda+2\mu)} \dfrac{U_m}{a}, & \text{对于倾向滑动情形.} \end{cases} \tag{3}$$

可以计算由于走向滑动引起的应力降和由于倾向滑动引起的应力降. 取拉梅常数 $\lambda = \mu = 3.3 \times 10^{11}$ dyn/cm^2 可以算出走滑和倾滑引起的应力降分别为4.5bar和1.6bar；从而得出应力降为4.8bar，应变降为 7.3×10^{-6}.

1966年3月22日7.2级邢台地震的应力降最大达42bar，应变降最大达 6.4×10^{-5}，总地震矩为 1.7×10^{26} dyn·cm[1]. 与此相比，海城地震的地震矩和邢台地震的地震矩接近，而应力降和应变降则比邢台地震的低一个数量级.

地面形变测量资料也支持这一结论. 在他山、海龙山、大青山三角形中，释放的水平应变比其他三角形都大，约为 2×10^{-5}；而邢台地震所释放的水平应变则约为 10^{-4}；海城地震所释放的水平应变也是比邢台地震低一个数量级.

对于一个7.3级地震，按照震级和应力降的经验关系[2]

$$M_L = 1.5 \lg \Delta\sigma + 5.0, \tag{4}$$

可知正常的应力降约为34bar. 由余震区的范围估计断层面的长度和宽度虽然粗略，但基本上反映了实际情况. 如果以震源最密集的区域的下部边界的深度（12km）作为断层宽度的估计值，那么，可算出应力降约8bar. 这充分说明，海城地震是发生在不能积累起较高应力的薄弱地带的一次低应力降的地震.

5　前震和余震的应力降和其他震源参数

利用参考资料[3]叙述的方法，由地震纵波初动的半周期（$T/2$）和振幅（A），测定了81个前震和余震的震源尺度（$2a$）、地震矩（m_0）、应力降（$\Delta\sigma$）和平均错距（$\Delta\bar{u}$）. 由草河掌地震台和鸡冠山地震台的记录图分别测定了上述地震的震源参数，结果见表3. 表中，以 S_1 和 S_2 分别表示草河掌台和鸡冠山台的测定结果.

图8表示海城地震的前震、余震的震源尺度、地震矩和应力降. 图中的斜线代表应力降的水平. 凡同一地震有两个台的测定结果者，均用细直线连结. 由图可见，测定结果的一致性是好的，说明了测定结果的可靠程度.

测定结果表明，前震和余震的应力降都比较低，一般在0.1—1bar这个数量级范围内，而且都低于主震的应力降.

图9是 $M_S \geq 4.0$ 的地震的应力降分布图，其中的小插图表示了草河掌台（S_1）和鸡冠山台（S_2）的位置. 从图9可以看见，有两个应力降相对说来比较高（高于0.8bar）的地区，在这两个区域以外，应力降低于0.8bar. 对比图9和图7，我们看到，高应力降地区对应于主破裂错动过头的部位，反映震前高应力、错动过头、震源机制反向和相对高应力降四者之间的内在联系. 海城地震这个震例说明后三者是震前高应力的表现和结果.

表 3 海城地震的前震和余震的震源半径、地震矩、应力降和平均错距

编号	日期 年.月.日	发震时刻 时-分-秒	震中位置 北纬	震中位置 东经	深度 /km	震级 M_L	$(T/2)/s$ S_1	$(T/2)/s$ S_2	$A/\mu m$ S_1	$A/\mu m$ S_2	a/km S_1	a/km S_2	m_0/dyn·cm S_1	m_0/dyn·cm S_2	$\Delta\sigma$/bar S_1	$\Delta\sigma$/bar S_2	$\Delta\bar{u}$/cm S_1	$\Delta\bar{u}$/cm S_2
1	1975.2.3	21-22-55	40°41′	122°50′	9	3.7	0.25	0.13	0.73	0.19	0.32		9.3×10^{19}		1.3		0.088	
2	2.4	04-31-26	42	48	12	3.8	0.19		0.83		0.20		7.2×10^{19}		4.1		0.18	
3	2.4	06-13-05	40	48	13	3.7	0.18	0.27	0.68	0.13	0.18	0.35	5.7×10^{19}	2.4×10^{19}	4.1	0.25	0.17	0.019
4	2.4	06-53-19	38	49		3.9	0.28	0.30	0.61	0.62	0.37	0.43	8.8×10^{19}	1.3×10^{20}	0.78	0.74	0.062	0.068
5	2.4	06-58-19	41	50	18	4.2	0.30	0.49	0.96	0.33	0.42	0.60	1.5×10^{20}	9.1×10^{19}	0.93	0.18	0.083	0.023
6	2.4	07-50-47	40	45	17	5.1	0.21	0.21	0.19	0.33	0.25	0.26	1.9×10^{19}	4.6×10^{19}	0.62	1.2	0.032	0.069
7	2.4	08-57-14	39	44	16	3.9	0.20		0.11		0.26		1.2×10^{19}		0.28		0.17	
8	2.4	09-04-39	38	49	11	2.9		0.19		0.022		0.20		2.6×10^{18}		0.15		0.062
9	2.4	10-28-31	48	39		3.3		0.15		0.037		0.13		3.1×10^{18}		0.66		0.019
10	2.4	10-35-55	40	47	15	4.7	0.22	0.25	0.94	0.38	0.27	0.32	9.8×10^{19}	6.4×10^{19}	2.3	0.83	0.013	0.058
11	2.4	10-47-19	40	50	7	2.7	0.18		0.082		0.18		6.7×10^{18}		0.52		0.020	
12	2.4	16-24-38	39	51	13	3.3	0.18	0.18	0.33	0.065	0.18	0.18	2.6×10^{19}	6.9×10^{18}	2.2	0.50	0.083	0.020
13	2.4	20-08	35	41		4.2		0.25		0.48		0.32		7.6×10^{19}		1.1		0.074
14	2.4	21-27-12	43	41		4.0		0.21		0.10		0.25		1.4×10^{19}		0.41		0.023
15	2.4	21-56-46	43	47	10	5.0	0.21	0.28	1.94		0.25	0.38	9.8×10^{19}		3.1		0.17	
16	2.4	22-03-13	39	47		4.4	0.18		0.14		0.18		5.7×10^{18}		0.49		0.015	
17	2.4	23-32-05	48	24		4.6	0.21		0.14		0.25		7.3×10^{18}		0.21		0.011	
18	2.4	23-49-16	43	40		4.2	0.19	0.30	0.24	0.22	0.20	0.42	1.1×10^{19}	4.5×10^{19}	0.60	0.27	0.025	0.033
19	2.5	01-01-45	43	56		4.8		0.30		0.074		0.41		1.5×10^{19}		0.083		0.0074
20	2.5	02-56-29	40	49		4.9	0.16	0.21	0.58	0.18	0.15	0.24	4.0×10^{19}	2.6×10^{19}	5.7	0.83	0.18	0.043
21	2.5	03-03-38	39	43		4.1	0.29		0.52		0.40		7.8×10^{18}		0.52		0.045	
22	2.5	03-07-13	43	45		4.7	0.39		0.35		0.58		7.2×10^{19}		0.17		0.022	
23	2.5	03-15-44	42	45		3.6	0.31	0.25	0.21	0.13	0.45	0.32	3.6×10^{19}	2.1×10^{19}	0.18	0.29	0.018	0.026
24	2.5	05-04-26	40	50		4.5	0.18	0.28	0.70	0.23	0.18	0.38	5.7×10^{19}	4.4×10^{19}	4.4	0.33	0.17	0.027
25	2.5	05-55-21	43	35		4.4	0.34		0.45		0.49		7.8×10^{19}		0.28		0.029	
26	2.5	06-02-29	42	44		3.7		0.22		0.093		0.26		1.3×10^{19}		0.33		0.019
27	2.5	06-55-33	44	42		3.6	0.18		0.17		0.18		1.5×10^{19}		1.3		0.050	

续表

编号	日期 年.月.日	发震时刻 时-分-秒	震中位置 北纬	震中位置 东经	深度 /km	震级 M_L	$(T/2)/s$ S_1	$(T/2)/s$ S_2	$A/\mu m$ S_1	$A/\mu m$ S_2	a/km S_1	a/km S_2	$m_0/dyn \cdot cm$ S_1	$m_0/dyn \cdot cm$ S_2	$\Delta\sigma/bar$ S_1	$\Delta\sigma/bar$ S_2	$\Delta\bar{u}/cm$ S_1	$\Delta\bar{u}/cm$ S_2
28	1975.2.5	09-13-39	40°43′	122°29′		4.0	0.28		0.12		0.37		1.7×10^{19}		0.16		0.013	
29	2.5	12-06-02	43	40		4.3	0.28	0.29	0.35	0.12	0.37	0.40	4.9×10^{19}	2.4×10^{19}	0.44	0.17	0.034	0.014
30	2.5	12-33-00	41	46		4.6	0.24	0.25	0.41	0.45	0.30	0.32	4.7×10^{19}	7.6×10^{19}	0.82	0.85	0.052	0.059
31	2.5	13-07-42	40	43		4.0	0.23		0.10		0.28		1.2×10^{19}		0.26		0.016	
32	2.5	14-39-02	40	45		3.6	0.23	0.25	0.12	0.084	0.28	0.32	1.5×10^{19}	1.4×10^{19}	0.31	0.18	0.019	0.012
33	2.5	15-39-48	40	39		4.0	0.21	0.25	0.16	0.16	0.25	0.32	1.7×10^{19}	2.5×10^{19}	0.46	0.33	0.025	0.023
34	2.5	18-41-23	42	29		4.1	0.31	0.25	0.076	0.21	0.44	0.32	1.3×10^{19}	3.4×10^{19}	0.067	0.45	0.0057	0.031
35	2.5	19-30-47	43	32		4.4	0.31	0.31	0.17	0.28	0.44	0.44	2.9×10^{19}	5.8×10^{19}	0.16	0.31	0.015	0.028
36	2.5	21-02-13	41	43		4.1	0.30		0.35		0.42		5.7×10^{19}		0.35		0.032	
37	2.5	23-52-54	42	38		5.0	0.25	0.30	0.11	0.43	0.32	0.41	1.4×10^{19}	8.7×10^{19}	0.21	0.50	0.015	0.045
38	2.5	05-43-43	37	54	23	5.5	0.31	0.38	0.090	0.39	0.44	0.54	1.5×10^{19}	9.7×10^{19}	0.088	0.25	0.0083	0.029
39	2.6	08-05-42	44	32		3.6	0.25		0.052		0.32		6.7×10^{18}		0.093		0.0067	
40	2.6	12-24-57	48	30	17	5.7	0.25	0.68	1.7		0.32	1.1	2.0×10^{20}		2.8		0.20	
41	2.6	23-56-16	45	50		4.5	0.16	0.25	0.21	0.24	0.15	0.32	1.4×10^{19}	4.1×10^{19}	2.1	0.58	0.067	0.041
42	2.7	18-31-46	41	50	15	4.2	0.25	0.20	0.39	0.27	0.32	0.23	4.7×10^{19}	3.5×10^{19}	0.67	1.3	0.048	0.066
43	2.8	02-30-04	49	40	12	4.5	0.25		0.035		0.32		4.4×10^{18}		0.062		0.0043	
44	2.8	02-30-23	49	30	2	4.5		0.22		0.17		0.27		2.4×10^{19}		0.54		0.032
45	2.8	18-10-48	38	47	10	2.2	0.15	0.22	0.044	0.20	0.13	0.26	2.8×10^{18}	2.9×10^{19}	0.62	0.74	0.017	
46	2.8	21-48-18	46	43	7	4.0	0.15	0.22	0.10	0.10	0.13	0.26	6.7×10^{18}	1.4×10^{19}	1.3		0.035	0.041
47	2.9	00-58-38	40	43	10	3.8	0.18	0.19	0.18	0.037	0.19	0.21	1.6×10^{19}	4.4×10^{18}	1.0	0.33	0.041	0.019
48	2.9	05-52-22	39	40	10	4.3	0.18	0.19	0.12	0.093	0.19	0.15	1.1×10^{19}	8.7×10^{18}	0.78	0.21	0.031	0.040
49	2.9	10-30-24	43	49	10	4.1	0.20	0.16	0.21		0.23		2.1×10^{19}	2.4×10^{19}	0.83	1.2	0.041	0.025
50	2.9	14-26-54	42	47	6	3.6		0.24		0.15		0.30				0.39		
51	2.9	21-35-09	37	52	2	4.1	0.18		0.035		0.19		3.0×10^{18}		0.18		0.072	
52	2.10	23-38-40	42	49	9	3.6	0.20		0.052		0.23		5.2×10^{18}		0.20		0.93	
53	2.11	02-51-51	44	46	7	4.1	0.18		0.070		0.19		6.2×10^{18}		0.41		0.017	
54	2.11	10-03-48	39	47	12	4.3	0.18		0.68		0.19		5.7×10^{19}		3.6		0.15	

续表

编号	日期 年.月.日	发震时刻 时-分-秒	震中位置 北纬	震中位置 东经	深度 /km	震级 M_L	$(T/2)/s$ S_1	$(T/2)/s$ S_2	$A/\mu m$ S_1	$A/\mu m$ S_2	a/km S_1	a/km S_2	$m_0/dyn \cdot cm$ S_1	$m_0/dyn \cdot cm$ S_2	$\Delta\sigma/bar$ S_1	$\Delta\sigma/bar$ S_2	$\Delta\bar{u}/cm$ S_1	$\Delta\bar{u}/cm$ S_2
55	1975.2.11	11-21-36	40°43′	122°42′	10	2.2	0.18		0.021		0.19		1.8×10^{18}		0.12		0.047	
56	2.11	15-04-04	40	45	16	3.8	0.18		0.073		0.19		6.2×10^{18}		0.41		0.017	
57	2.11	16-10-27	42	44	9	4.0	0.25		0.035		0.32		4.4×10^{18}		0.062		0.043	
58	2.12	19-11-00	41	33	11	3.6	0.19		0.056		0.21		5.2×10^{18}		0.26		0.012	
59	2.12	20-42-46	42	47	7	4.5	0.21	0.25	0.39	0.61	0.25	0.32	4.0×10^{19}	9.9×10^{19}	1.1	1.4	0.057	0.099
60	2.13	05-51-39	43	48	8	4.1	0.19	0.18	0.51	0.20	0.21	0.19	4.7×10^{19}	2.1×10^{19}	2.6	1.4	0.12	0.058
61	2.15	15-26-14	43	47	7	3.8	0.15	0.20	0.34	0.074		0.22		9.2×10^{18}		0.54		0.025
62	2.15	21-08-04	39	47	12	5.8		0.19		0.078		0.21		9.1×10^{18}		0.41		0.018
63	2.16	22-01-20	41	48	11	5.6	0.30	0.29	1.2	0.55	0.42	0.39	1.9×10^{20}	1.1×10^{20}	1.1	0.83	0.098	0.070
64	2.17	12-33-13	43	42	9	4.4		0.20		0.13		0.23		1.7×10^{19}		0.58		0.028
65	2.18	18-51-49	47	42	16	4.7	0.27	0.37	0.21	0.41	0.36	0.52	2.9×10^{19}	9.9×10^{19}	0.28	0.25	0.22	0.028
66	2.20	16-41-05	40	50	8	3.6		0.16		0.065		0.15		5.9×10^{18}		0.78		0.026
67	2.21	06-26-17	41	40	6	4.4	0.25		0.23		0.32		3.0×10^{19}		0.41		0.028	
68	2.22	15-45-14	42	44	12	4.8		0.41		0.63		0.60		1.7×10^{20}		0.31		0.041
69	2.24	05-07-20	47	53	7	5.0		0.35		0.28		0.51		6.4×10^{19}		0.21		0.023
70	2.24	06-01-47	41	49	5	4.0		0.18		0.10		0.19		1.2×10^{19}		0.74		0.030
71	2.25	04-52-10	44	37	14	5.0		0.30		0.59		0.40		1.2×10^{20}		0.83		0.073
72	2.26	05-09-53	40	49	8	4.8	0.31	0.38	0.56	0.40	0.43	0.54	8.8×10^{19}	1.0×10^{20}	0.52	0.25	0.049	0.028
73	2.26	07-06-56	39	41	3	4.2	0.22	0.19	0.33	0.13	0.27	0.21	3.7×10^{19}	1.6×10^{19}	0.93	0.74	0.052	0.033
74	2.27	09-19-57	42	57	7	4.0	0.18	0.30	0.27	0.61	0.19	0.42	5.7×10^{18}	3.7×10^{19}	0.36		0.015	
75	2.27	13-44-43	43	44	11	4.0	0.16	0.15	0.035	0.059	0.15	0.13	2.6×10^{19}	5.0×10^{18}	3.3	1.1	0.10	0.030
76	2.28	05-10-02	55	56	15	4.0	0.20	0.21		0.21	0.23	0.25		2.9×10^{19}		0.83		0.045
77	2.28	11-56-13	42	47	6	4.4		0.22		0.26		0.26		3.7×10^{19}		0.99		0.055
78	3.21	11-32-59	46	57	11	4.5		0.30		0.61		0.42		1.2×10^{20}		0.74		0.068
79	3.29	23-16-36	46	36	6	4.6		0.29		0.32		0.40		6.1×10^{19}		0.41		0.035
80	4.10	03-55-37	43	29	10	5.0		0.37		1.4		0.55		3.7×10^{20}		0.91		0.11
81	4.21	00-17-06	46	27	8	4.6		0.22		0.30		0.27		4.3×10^{19}		0.99		0.058

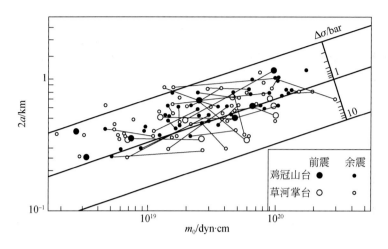

图 8　海城地震前震和余震的震源尺度、地震矩和应力降

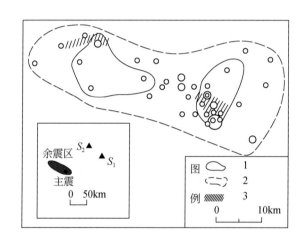

图 9　海城地震 $M_S \geq 4.0$ 的前震和余震的应力降的分布

1——应力降较大的地区（$\Delta\sigma \geq 0.8\text{bar}$）；2——应力降较小的地区（$\Delta\sigma < 0.8\text{bar}$）；3——错动过头的地区

6　结语

以上分析表明，海城地震是一个近乎直立的北西西向断层面上的左旋—正剪切错动．这个结果意味着，虽然在我国东北和华北地区，主压应力轴的取向都是北东东—南西西向，但破裂面的方向却不限于北北东方向；与北北东方向共轭的北西西方向的破裂同样是可能的．因此，在分析这个地区的地震危险性时，北北东向的构造带、北西西向的构造带及其交汇地区均要给予注意．

前震的机制与主震的相同，而余震的机制与主震的不尽相同．与主震机制相反的若干强余震可能是由于主破裂在局部地段发生错动过头以后的回跳引起的．

和震级相近的其他浅源大地震相比，海城 7.3 级地震的地震矩与它们相当，但应力降则低一个数量级，是一次低应力降的地震．前震和余震的应力降也较低．在应力降都相当低的背景下，有两个应力降相对高的地区．从所得结果看来，主震前的高应力地区与主震的初始破裂点、错动过头部位、强余震的滑动向量反向以及应力降可能有内在的联系．为了查明这些关系，有必要对其他地震系列进行类似的测定和分析．而为了鉴别前震和一般小地震，着重对大震的前震和一般小地震的震源参数进行系统的测定和分析也是很有意义的．本文涉及的结果仅是这方面工作的一个开始．

参 考 资 料

[1] 陈运泰等，根据地面形变的观测研究1966年邢台地震的震源过程，地球物理学报，**18**，3，164-182，1975.
[2] S. J. Gibowicz, Stress drop and aftershocks, *Bull. Seism. Soc. Amer.*, **63**, 4, 1433-1446, 1973.
[3] 陈运泰等，巧家、石棉的小震震源参数的测定及其地震危险性的估计，地球物理学报，**19**，3，206-233，1976.

Focal mechanism of Haicheng (海城), Liaoning Province, earthquake of February 4, 1975

Gu Hao-Ding[1], Chen Yun-Tai[2], Gao Xiang-Lin[3] and Zhao Yi[4]

1. The Seismological Bureau of Liaoning Province; 2. Institute of Geophysics, Academia Sinica;
3. The Seismological Brigade of the Ningxia Hui Nationality Autonomous Region;
4. The Seismological Brigade of Guangzhou, National Seismological Bureau

Abstract Fault plane solutions are obtained from data of the first motions for 24 earthquakes with $M_s \geqslant 4.0$ of the Haicheng earthquake sequence, the mainshock of which occurred on Feb. 4, 1975. One of the nodal planes of the mainshock strikes N 70°W, dipping 80° to the NE. while the other nodal plane strikes N 23°E, dipping 75° to the SE. Based on the data of the spatial distributions of the after shocks and the ground deformations, the N 70° W nodal plane is taken as the fault plane. The faulting is nearly a vertical, left-lateral strike-slip with a minor component of normal dip-slip movement. While the focal mechanisms of all the fore-shocks and most of the aftershocks are similar to that of the mainshock, those of 4 aftershocks with $M_s \geqslant 4.0$ are remarkably different from the mainshock. They represent a faulting with slip vector reversed in direction to that of the mainshock. One possible explanation for these exceptions is that during the mainshock the fault movement overshot along some segments of the fault.

The source parameters of the mainshock are calculated from the data of field observations and spatial distributions of the aftershocks. The fault length, width, average dislocation, seismic moment, and stress-drop, of the mainshock, are estimated as 70km, 20km, 45cm, 2.1×10^{26} dyne·cm, 4.8 bars, respectively. It is a low stress-drop earthquake occurring in a weak zone that is incapable of accumulating higher stresses.

For 81 foreshocks and aftershocks the source dimensions, seismic moments, stress-drops as well as average dislocations are calculated from the data of the first half cycles and amplitudes of the seismic P-waves. The results indicate that the stress-drops are rather low, and generally in the range of 0.1—1.0bars, for both the foreshocks and aftershocks. There are two regions with relatively higher stress-drops ($\geqslant 0.8$bars), which correspond to the overshooting portions on the main fracture. These results imply that there might be some intrinsic connections between higher initial stresses before the mainshock, overshooting of fault movement, relatively higher stress-drops, and reverse of the slip vectors of focal mechanisms. It seems that the latter three phenomena are the results of the relatively higher initial stress.

中、小地震体波的频谱和纵、横波拐角频率比[*]

陈运泰　王妙月　林邦慧　刘万琴

中国科学院地球物理研究所

由地震体波的频谱可以测定许多有意义的震源参数，如震源尺度、地震矩、应力降和错距等. 为了运用体波的频谱测定这些震源参数，需要计算地震体波的理论频谱. 它可以通过计算断层面上的预应力突然解除时的地震波辐射问题来求得. 一个简单的情形便是圆盘形断层面上的预应力突然解除时的地震波辐射问题. 与此相应的静力学解，Keilis-Borok[1]已经求出，可是动力学解迄未得到.

严格地处理这个动力学问题在数学上比较困难. 但从应用的角度，可以利用静力学解对动力学解作必要的、合理的限制. 根据静力学解所提供的圆盘形断层面上错距不均匀分布的特点，可计算当破裂速度有限时，圆盘形断层面所辐射的地震体波的理论频谱. 并由此导出测定震源参数的关系式，进而分析纵、横波拐角频率比，探讨由纵、横波拐角频率和初动半周期来测定震源所在处的纵、横波速度比的方法.

1 中、小地震震源的位错模式

以均匀、各向同性和完全弹性的无限介质中的圆盘形位错面表示中、小地震的断层面. 采用直角坐标系 $x_1 x_2 x_3$，原点与圆盘中心重合，x_3 轴垂直于断层面. 以 $u(Q; t)$ 代表坐标为 $x_i (i=1, 2, 3)$ 的 Q 点的位移向量，t 代表时间，以 $F(\omega)$ 代表任一时间函数 $f(t)$ 的频谱：

$$F(\omega) = \int_{-\infty}^{\infty} f(t) e^{-i\omega t} dt, \tag{1}$$

式中，ω 表示圆频率. 在无限介质中，任意形状的位错面 Σ 所辐射的地震波位移 $u(Q; t)$ 的频谱 $U(Q; \omega)$ 在远场近似地为[2]：

$$U(Q;\omega) = U_P(Q;\omega) + U_S(Q;\omega),$$

$$\begin{cases} U_P(Q;\omega) = \dfrac{m_0}{4\pi\rho\alpha^3 r} \mathscr{R}_P e^{-i\frac{\omega}{\alpha}r} F_\alpha(\omega), \\ U_S(Q;\omega) = \dfrac{m_0}{4\pi\rho\beta^3 r} \mathscr{R}_S e^{-i\frac{\omega}{\beta}r} F_\beta(\omega). \end{cases} \tag{2}$$

式中，$U_P(Q;\omega)$ 和 $U_S(Q;\omega)$ 分别表示 P 波和 S 波的远场位移谱；(r, θ, φ) 是观测点 Q 的球极坐标；ρ 是介质的密度；α, β 分别是纵、横波速度；m_0 是地震矩：

$$m_0 = \mu \langle \Delta U \rangle S, \tag{3}$$

μ 是介质的刚性系数，S 是断层面面积，$\langle \Delta U \rangle$ 是静力学位错在断层面上的平均值；$\mathscr{R}_P, \mathscr{R}_S$ 分别是 P 波和 S 波的辐射图型因子，对于滑动向量与 x_1 轴方向一致的剪切位错，它们的表示式分别是：

$$\begin{cases} \mathscr{R}_P = \{\sin 2\theta \cos \varphi, 0, 0,\}, \\ \mathscr{R}_S = \{0, \cos 2\theta \cos \varphi, -\cos 2\theta \sin \varphi,\}; \end{cases} \tag{4}$$

$F_\alpha(\omega), F_\beta(\omega)$ 是表示纵、横波频谱形状的函数. 当破裂从圆盘中心开始，以速率 v_b 沿径向扩展时，表

---

[*] 本文发表于《科学通报》，1976 年，第 21 卷，第 9 期，414–418.

示频谱形状的函数是下面形式的积分：

$$F_c(\omega) = \frac{1}{S} \iint_{\Sigma} \frac{i\omega \Delta U(P;\omega)}{\langle \Delta U \rangle} \times e^{-i\frac{\omega}{v_b}\xi + i\frac{\omega}{c}\xi\sin\theta\cos(\psi-\varphi)} d\Sigma, \tag{5}$$

式中，c 表示 α 或 β；(ξ, ψ) 是在 P 点的位错元 $d\Sigma$ 的平面极坐标；$\Delta U(P;\omega)$ 是 P 点的错距的频谱，可表示为：

$$\Delta U(P;\omega) = \Delta U(P) G(P;\omega). \tag{6}$$

(6) 式中 $\Delta U(P)$ 是静力学位错，$G(P;\omega)$ 是震源时间函数 $g(P;t)$ 的频谱.

对于半径为 a 的圆盘形剪切破裂面，它上面的静力学位错分布是[1]

$$\Delta U(P) = U_m [1 - (\xi/a)^2]^{1/2}, \quad \xi \leq a. \tag{7}$$

U_m 是圆盘中心处的位错，即最大错距. 由 (7) 式可知，$\langle \Delta U \rangle = \frac{2}{3} U_m$，从而

$$F_c(\omega) = \frac{3}{2S} \int_0^a d\xi \int_0^{2\pi} i\omega G(P;\omega) \times \xi [1 - (\xi/a)^2]^{1/2} e^{-i\frac{\omega}{v_b}\xi + i\frac{\omega}{c}\xi\sin\theta\cos(\psi-\varphi)} d\psi. \tag{8}$$

当 $g(P;t)$ 是单位函数时，$i\omega G(P;\omega) = 1$，此时通过积分变数的代换 $\zeta = \xi/a$，并利用 Bessel 函数的积分表示式，可将 $F_c(\omega)$ 化为：

$$F_c(\omega) = 3 \int_0^1 e^{-i\zeta y} J_0(\zeta x_c)(1-\zeta^2)^{1/2} \zeta d\zeta. \tag{9}$$

式中，J_0 是零阶 Bessel 函数，$y = \omega a/v_b$，$x_c = \omega a \sin\theta/c$.

(9) 式便是圆盘形位错源所辐射的体波远场位移谱的积分表示式.

2 圆盘形断层辐射的地震波远场位移谱

2.1 有限破裂速度情形

当破裂速度有限时，圆盘形断层辐射的地震波远场位移谱为：

$$F_c(\omega) = 3t\left(\frac{\partial K_c}{\partial y} + \frac{\partial^3 K_c}{\partial y^3}\right), \tag{10}$$

$$K_c(\omega) = \frac{\pi}{2} J_0(y) J_0^2(x_c/2) + \pi \sum_{k=1}^{\infty} (-1)^k J_{2k}(y) J_k(x_c/2) J_{-k}(x_c/2)$$

$$-i\pi \sum_{k=0}^{\infty} (-1)^k J_{2k+1}(y) J_{k+1/2}(x_c/2) \times J_{-k-1/2}(x_c/2). \tag{11}$$

这个结果是资料 [3] 中第 335 页的公式 (17) 和第 336 页的公式 (27) 的一个推广. 顺便指出，资料 [3] 第 335 页的公式 (17) 中，等号右边的 Bessel 函数 $J_{\frac{\nu}{2}-n-\frac{1}{2}}$ 的宗量应当是 $a/2$ 而不是 a.

分析 (10)，(11) 式，可以对体波频谱的特征有个了解. 首先，当 $|x_c| \gg 1$ 时，利用 Bessel 函数的渐近展开式以及其生成函数的展开式，可得

$$F_c(\omega) \sim 3x_c^{-3}(\sin x_c - x_c \cos x_c - ie_c x_c \sin x_c) e^{-iy}, \tag{12}$$

其中，$e_c = 3y/2x_c = 3c/2v_b \sin\theta$.

其次，当 $|x_c| \ll 1$ 时，根据 Anger-Weber 函数的定义，可将 $K_c(\omega)$ 近似地表示为

$$K_c(\omega) \doteq \frac{\pi}{2} [J_0(y) + i\mathbf{E}_0(y)], \tag{13}$$

式中，J_0 是零阶 Bessel 函数，$\mathbf{E}_0(y)$ 是零阶 Weber 函数.

借助 Anger-Weber 函数的级数展开式，可由 (10) 和 (13) 式得到 $F_c(\omega)$ 在 $|x_c| \ll 1$ 时的级数展

开式：

$$F_c(\omega) \doteq \frac{3\pi}{4}\sum_{m=0}^{\infty}\frac{e^{-\frac{m\pi i}{2}}}{\left[\Gamma\left(\frac{m+3}{2}\right)\right]^2}\times\left(\frac{m+1}{m+3}\right)\left(\frac{y}{2}\right)^m. \tag{14}$$

再次，在 $|x_c|\ll 1$ 的情形下，若 $|y|\gg 1$，可以利用 J_0 和 $\mathbf{E}_0$ 的渐近展开式，由（11）和（10）式得到 $F_c(\omega)$ 在这情形的渐近展开式：

$$F_c(\omega)\sim 3(\pi/2)^{1/2}y^{-3/2}e^{-i(y-3\pi/4)}. \tag{15}$$

$F_c(\omega)$ 作为 θ 的函数，对于 $\theta=\pi/2$ 是对称的，即 $F_c(\omega;\theta)=F_c(\omega;\pi-\theta)$，所以，下面仅对 $0\leq\theta\leq\pi/2$ 的情形进行讨论。由以上三种情况得到的结果可知，当 $\theta=0$ 时，振幅谱 $|F_c(\omega)|$ 的高频趋势和 $\omega^{3/2}$ 成反比。当 θ 很小但不为零时，若 $v_b/a\ll\omega\ll c/a\sin\theta$，则 $|F_c(\omega)|$ 和 $\omega^{3/2}$ 成反比。若 $\omega\gg c/a\sin\theta$，则和 ω^2 成反比。当 $\theta\neq 0$ 时，则 $|F_c(\omega)|$ 和 ω^2 成反比。

由（9）式可知，$F_c(0)=1$；这就是说，频谱的低频趋势为常数。由高频趋势和低频趋势的交点可以确定拐角频率 ω_c，结果是：

当 $\theta=0$ 时，它只与比值 a/v_b 有关，

$$\omega_c=\frac{3^{2/3}\pi^{1/3}}{2^{1/3}}\frac{v_b}{a}; \tag{16}$$

而当 $\theta\neq 0$ 时，它还与比值 $v_b\sin\theta/c$ 有关，

$$\omega_c=\frac{c}{a\sin\theta}\frac{3^{1/2}}{2^{1/4}}\left[1+\left(\frac{3c}{2v_b\sin\theta}\right)^2\right]^{1/4}. \tag{17}$$

当 θ 很小但不为零时，频谱有两个拐角，较低的拐角频率由（16）式表示，较高的拐角频率则为：

$$\omega_c=\frac{v_b}{\pi a}\left(\frac{c}{v_b\sin\theta}\right)\left[1+\left(\frac{3c}{2v_b\sin\theta}\right)^2\right]. \tag{18}$$

由于地震仪频带的限制，通常只有频率较低的第一个拐角频率可以被观测到，因此，着重分析第一个拐角频率是特别有意义的。由前面结果可知，当 $0\leq\theta\leq\theta_c$ 时，第一个拐角频率即如（16）式所示，当 $\theta_c\leq\theta\leq\pi/2$ 时，如（17）式所示。θ_c 是下列方程的解：

$$\frac{c}{v_b\sin\theta_c}\left[1+\left(\frac{3c}{2v_b\sin\theta_c}\right)^2\right]^{1/4}=\frac{3^{1/6}\pi^{1/3}}{2^{1/12}}. \tag{19}$$

对于在震源球球面上随机分布的观测点，第一个拐角频率的期望值是：

$$\langle\omega_c\rangle=(3^{1/2}\pi/2^{5/4})(c/a)\Theta_c \tag{20}$$

$$\Theta_c=\frac{2}{\pi}\int_0^{\theta_c}\frac{3^{1/6}\pi^{1/3}}{2^{1/12}}\frac{v_b\sin\theta}{c}d\theta+\frac{2}{\pi}\int_{\theta_c}^{\pi/2}\left[1+(3c/2v_b\sin\theta)^2\right]^{1/4}d\theta. \tag{21}$$

2.2 同时破裂情形

同时破裂，即破裂速度无穷大。就天然地震而言，这是不真实的。在有限破裂速度的情形下，只有所论及的波的周期比破裂过程所花费的时间大得多时，也就是 $\omega a/v_b\ll 1$ 时，才可以将它近似地当作同时破裂情形处理。这时体波远场位移谱可以令（9）式中的 $y=0$ 来求得。此时，（9）式可积出（资料［4］，第 329 页）：

$$F_c(\omega)=3(\pi/2)^{1/2}x_c^{-3/2}J_{3/2}(x_c), \tag{22}$$

$J_{3/2}$ 是半整数阶 Bessel 函数。以三角函数表示 $J_{3/2}$，则（22）式可表示为：

$$F_c(\omega)=3x_c^{-3}(\sin x_c-x_c\cos x_c). \tag{23}$$

Randall[5,6] 曾讨论过一个与这里讨论的完全不同的问题，即预应力球体突然置入无应力的介质中时，

它所辐射的体波位移谱问题. 有意思的是, 这里导出的 $F_c(\omega)$ 表示式和 Randall 讨论不同问题得到的表示式完全一样. 但是必须指出, 两者仅是形式上一样, 而问题的性质与结果的含义却有本质差别. 这里讨论的是半径为 a 的圆盘形位错面辐射的地震波远场位移谱, $F_c(\omega)$ 与观测点的方位有关, 在结果中通过 (22) 式或 (23) 式中的 $x_c = \omega a \sin\theta/c$ 表现出来; 而 Randall 讨论的是半径为 a 的预应力球体的辐射问题. 他得到的频谱形状的函数中的 $x = \omega a/c$, 和观测点的方位无关.

振幅谱的低频趋势为常数, 其高频趋势和频率的平方成反比. 由低频趋势和高频趋势的交点可以确定拐角频率 ω_c:

$$\omega_c = 3^{1/2} c/2^{1/4} a \sin\theta. \tag{24}$$

对于在震源球球面上随机分布的观测点, 拐角频率的期望值是

$$\langle \omega_c \rangle = (3^{1/2} \pi / 2^{5/4})(c/a) \doteq 2.29 c/a. \tag{25}$$

不言而喻, (24) 和 (25) 式分别是 (17) 和 (20) 式的特殊情形.

Brune[7] 最先用半理论、半经验的方法得到圆盘形位错源辐射的横波频谱的拐角频率为 $2.21\beta/a$, 后来改正为[8] $2.34\beta/a$. 他改正后的结果和我们算出的 $\langle \omega_\beta \rangle$ 很接近. 但是, Brune 求得的是横波的理论频谱, 他没有涉及纵波的理论频谱及拐角频率. 我们的结果既有横波情形, 也包括纵波情形, 所以由此可进一步分析纵、横波拐角频率比.

3 纵、横波拐角频率比

注意到 $\sin\theta_c/c$ 是方程 (19) 的解, 因此

$$\sin\theta_\alpha / \sin\theta_\beta = \alpha/\beta. \tag{26}$$

由于 $\alpha > \beta$, 所以 $\theta_\alpha > \theta_\beta$. 于是, 纵、横波的第一个拐角频率比为:

$$\begin{aligned}
\omega_\alpha/\omega_\beta &= 1, & &\text{当 } 0 \leq \theta \leq \theta_\beta, \\
&= 3^{1/6} 2^{-1/12} \pi^{1/3} \left(\frac{v_b \sin\theta}{\beta}\right)\left[1 + \left(\frac{3}{2}\frac{\beta}{v_b \sin\theta}\right)^2\right]^{-1/4}, & &\text{当 } \theta_\beta \leq \theta \leq \theta_\alpha, \\
&= \frac{\alpha}{\beta}\left[\frac{1+(3\alpha/2v_b \sin\theta)^2}{1+(3\beta/2v_b \sin\theta)^2}\right]^{1/4}, & &\text{当 } \theta_\alpha \leq \theta \leq \frac{\pi}{2}.
\end{aligned} \tag{27}$$

这意味着, 这个比不仅和波速比有关, 还和破裂速度以及观测点的方位有关.

Molnar 等[9] 根据合理假定的品质因数数值, 在对体波频谱作了衰减的校正后, 发现 P 波和 S 波的拐角频率之比为 1 到 3. 而 Furuya[10] 的观测表明 S 波和 P 波的优势周期之比也是 1 到 3. 看来, 这一观测结果是比较确切的.

倘若不考虑破裂传播的影响, 则拐角频率比为 α/β, 对于泊松体, 这个比值是 1.73, 它解释不了观测事实. 按 Sato 等[11] 的计算, 当 v_b/β 由 0.5 变至 0.9 时, 拐角频率比在 1.26 到 1.39 之间变化, 它无法说明这个比为什么能高于 1.39. 我们的理论计算表明, $\omega_\alpha/\omega_\beta$ 的下限为 1, 其上限可高达 $(\alpha/\beta)^{3/2}$, 对于泊松体, 这个数值是 2.28. 这比 Sato 等[11] 的理论计算要更符合观测事实.

4 由拐角频率比测定震源处纵、横波速度比的可能性

在地震预报实践中, 纵、横波速度比异常的研究是一个重要的问题, 目前, 波速比都是由地震波的走时测定的, 它依赖于 P 波到时 (T_P) 和 S 波与 P 波的走时差 (T_{S-P}) 的测定, 因而对时间服务的精度有较高的要求. 此外, 由地震波走时测定的波速比, 是从震源至台站的传播路径上的波速比.

本文得到的结果表明，纵、横波拐角频率比和纵、横波速度比有关（见（27）式）；当然，它还和破裂速度与观测点的方位有关．由（27）式可以看到，无须测定破裂速度与观测点的方位，只要设法测得比值 $v_b\sin\theta/\beta$ 或 $v_b\sin\theta/\alpha$，那么原则上由纵、横波拐角频率比就有可能测定波速比．用频谱的拐角频率比测定波速比不要求测定 P 波的到时，因此对时间服务的精度没有太高的要求．此外，和拐角频率相联系的波速，按其物理意义来说乃是震源所在处的波速．所以由纵、横波拐角频率比测得的波速比，和由地震波走时测得的波速比有不同的含意，在地震预报实践中，它们可以彼此印证，互相补充．

测定比值 $v_b\sin\theta/c$ 仿佛是很困难的．因为无论是测定每一个地震的破裂速度，还是确定断层面在空间的趋向（这样才有可能确定出 θ 角），都是困难的．这种困难在中、小地震尤为突出．然而，资料 [12] 导出的结果却提供了测定比值 $v_b\sin\theta/c$ 的可能性．我们在资料 [12] 中，导出了错距均匀分布的圆盘形断层所辐射的地震波远场位移表示式，和相应的初动四分之一周期与半周期的表示式．计算表明，对于错距按（7）式分布的圆盘形断层，其辐射的地震波远场位移表示式和均匀分布的圆盘形断层不同，但初动半周期 t_{2c} 的表示式也一样是：

$$t_{2c}=\frac{a}{v_b}\left(1+\frac{v_b\sin\theta}{c}\right). \tag{28}$$

所以，纵、横波初动半周期之比为：

$$\frac{t_{2\alpha}}{t_{2\beta}}=\frac{1+\dfrac{v_b\sin\theta}{\alpha}}{1+\dfrac{\alpha}{\beta}\dfrac{v_b\sin\theta}{\alpha}}. \tag{29}$$

把（27）式的第三式和（29）式联立，那么通过测定纵、横波拐角频率比和初动半周期之比，便有可能同时求得震源所在处的波速比以及比值 $v_b\sin\theta/\beta$．在实际的测定中．为了实现（27）式的第三式和（29）式联立的条件，必须避免采用 $0\leqslant\theta\leqslant\theta_\alpha$ 或 $0\leqslant\pi-\theta\leqslant\theta_\alpha$ 的台站，而只要避免采用节面附近的台站，就可以实现上述条件．

参 考 资 料

[1] Keilis-Borok, V., (Кейлис-Борок, В.) *Ann. Geofis.*, **12**(1959), 2, 205-214.
[2] Dahlen, F. A., *Bull. Seism. Soc. Amer.*, **64**(1974), 4, 1159-1180.
[3] Erdélyi, A. (ed.), *Tables of Integral Transforms*, 2, McGraw-Hill Book Company, New York, 1954.
[4] ———— (ed.), ibid., 1, McGraw-Hill Book Company, New York, 1954.
[5] Randall, M. J., *J. Geophys. Res.*, **71**(1966), 22, 5297-5302.
[6] ————, *Bull. Seism. Soc. Amer.*, **63**(1973), 3, 1133-1144.
[7] Brune, J. N., *J. Geophys. Res.*, **75**(1970), 26, 4997-5009.
[8] ————, ibid., **76**(1971), 20, 5002.
[9] Molnar, P. et al., *Bull. Seism. Soc. Amer.*, **63**(1973), 6, 2091-2104.
[10] Furuya, I. （古屋逸夫）, *J. Phys. Earth*, **17**(1969), 2, 119-126.
[11] Sato, T. （佐藤魂夫）& Hirasawa., T. （平泽朋郎）, ibid., **21**(1973), 4, 415-431.
[12] 陈运泰等，地球物理学报，**19**(1976), 3, 206-233.
[13] Savage, J. C., *Bull. Seism. Soc. Amer.*, **64**(1974), 6, 1621-1627.

由瑞雷波方向性函数研究1974年5月11日云南省昭通地震的震源过程[*]

刘万琴　陈运泰

国家地震局地球物理研究所

摘要 根据国内外100多个地震台的P波初动符号资料，确定了1974年5月11日昭通地震的断层面解．用面波方向性函数和广义的面波方向性函数，确定走向为N45°E的节面是这次地震的断层面，破裂传播方向是北东向，破裂速度为1.3km/s，破裂长度为53km．根据谱密度估算这次地震的地震矩为6.5×10^{25}dyn·cm．从上述结果，结合昭通地区的地震活动和区域构造特点，认为昭通地震是在东西向构造应力场的作用下，北东向的巧家-莲峰大断裂朝着北东方向继续破裂的结果，它把中断了的巧家-莲峰大断裂和华蓥山-宜宾大断裂贯通起来．基于上述发震模式，我们认为，在同一构造应力场的作用下，与北东构造共轭的北西向的彝良-水城断裂及峨眉-盐津构造带的地震危险性应予以注意．

1 引言

1974年5月11日云南昭通7.1级地震发生后，许多作者对它进行了研究[1-5]．这次地震发生在云南省昭通地区的永善县和大关县交界的山区中，所以有些作者称它为永善-大关地震或永善地震[1,2,4]．四川省地震台网测定这次地震的震中位置是北纬28.2°，东经103.9°，震源深度10km，发震时刻03时25分16秒（北京时间），震级M_s为7.1．成都地震大队由四川省地震台网和全国地震台网的一些台站资料，得到了这次地震断层面解[2]（见表1第1行）；昆明地震大队由云南省地震台网和全国地震台网的一些台站的资料，也得到了这次地震的断层面解[2]（见表1第2行）．两者的结果大体一致，都认为昭通地震P波的两个节面分别是北东向和北西向的几乎直立的平面．参考文献[3]的作者根据同样的资料，得到了和前面提到的两个结果差别较大的断层面解（见表1第3行）．按照文献[3]的结果，昭通地震P波的两个节面分别是北北东向和北北西向的平面，且倾角不大．这就引出一个问题，就是昭通地震是以走向滑动为主还是以倾向滑动为主？

表1　不同作者得到的昭通地震的断层面解

作者	节面Ⅰ			节面Ⅱ			P轴		T轴		B轴	
	走向	倾向	倾角	走向	倾向	倾角	方位	倾角	方位	倾角	方位	倾角
成都地震大队[2]	46°	北西	85°	320°	北东	80°	272°	3°	181.8°	11°	17°	79°
昆明地震大队[2]	57°	北西	81°	328°	北东	85.5°	102.5°	3.2°	192.5°	9.2°	352°	80°
朱成男、陈承照[3]	25°	北西	40°	349°	北东	63.0°	94°	71°	30.0°	15.8°	2°	9°
本文	45°	北西	86°	315°	南西	86°	90°	5°	179.0°	1.0°	268°	84°

在两个节面中，究竟哪一个是真正的断层面？参考文献[3]的作者根据震中区地壳形变的测量资料、地震烈度的衰减特点和余震区随时间推移等特征，认为北北西向的节面是真正的断层面．他们认为这

---

[*] 本文发表于《地震学报》，1979年8月，第1卷，第1期，25-37.

次地震的发震断层是走向为北北西的左旋逆断层.

这次地震发生在金沙江下游南岸的山区中,野外调查受到交通不便的影响,在足迹所能到达之处,都没有见到原生的地震裂缝.这给分析判断这次地震震源的性质增加了困难.为了确定这次地震的性质,我们利用国内和国外地震台的记录,重新确定这次地震的断层面解,并且运用地震面波的方向性函数,确定地震的断层面和地震矩,应力降等震源参数.

2 昭通地震的断层面解

参考文献[3]所得到的昭通地震的断层面解尽管和文献[2]所引用的两个结果有很大的差别,但所依据的资料大同小异.图1是不同作者得到的这次地震的断层面解.由图1(a)和图1(b)可见,受资料限制,在震源球球面上资料点的覆盖范围相当小.为了改善这种状况,我们在国内60个地震台的P波初动资料的基础上,增加了62个国外地震台的P波初动资料,得到了另外的结果(见表1最后一行和图1c).对比参考文献[2],[3]的结果和本文的结果,可知本文的结果和文献[2]所引用的两个结果

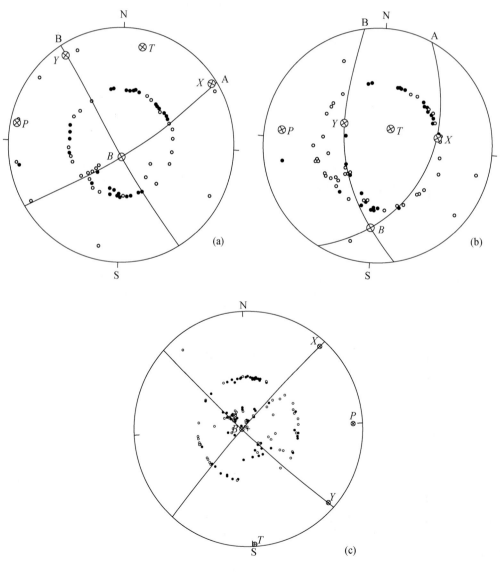

图1 不同作者得到的昭通地震的断层面解

震源球下半球投影在乌尔夫网上,黑点表示初动是压缩,圆圈表示初动是膨胀

是接近的,而与文献[3]的结果差别较大.若无本文所增补的资料,的确难以判断文献[2],[3]所给的断层面解哪一个比较合理.现在看来,[2]所给的结果和本文的结果可能合理些.本文的结果表明,昭通地震的发震断层或者是走向北东45°的右旋-正断层,或者是走向北西45°的左旋-正断层.两个可能的断层面都是几乎直立的.并且,无论哪一个节面是断层面,错动的性质都是以走向滑动为主.

3 由瑞雷波的方向性函数确定昭通地震的断层面和破裂扩展方向

3.1 面波的方向性函数

为了确定昭通地震的断层面和破裂扩展方向,我们利用面波的方向性函数的特性.对于无限介质中的一个长度为 b,宽度为 d 的垂直走向滑动断层,如果破裂以速度 v_b 沿着断层的走向扩展,则瑞雷波垂直分量的位移谱是[6]:

$$U_z^R = \frac{\sin 2\theta}{\sqrt{r}} g_z(\omega) \sqrt{k_\beta} \frac{\sin X_R}{X_R} \exp\left[i\left(\varphi_R + \frac{3\pi}{4}\right)\right], \tag{1}$$

式中,

$$\begin{cases} X_R = \frac{\pi b}{\lambda}\left(\frac{c_R}{v_b} - \cos\theta\right), \\ \varphi_R = \omega\left(t - \frac{r}{c_R}\right) - X_R, \end{cases} \tag{2}$$

c_R 是瑞雷波相速度,λ 是波长,r 是震中距,θ 是从断层走向逆时针测量的台站方位角,k_β 是横波的波数,$g_z(\omega)$ 是与震源时间函数、断层宽度、深度和圆频率 ω 有关的一个函数.

方向性函数 D 是从相反方向离开震源的两条射线上面波振幅谱的比值,也就是

$$D = \frac{|U_z^R(\theta)|}{|U_z^R(\theta+\pi)|} = \left|\frac{\sin\left[\frac{\pi bf}{c_R}\left(\frac{c_R}{v_b}-\cos\theta\right)\right]\left(\frac{c_R}{v_b}+\cos\theta\right)}{\sin\left[\frac{\pi bf}{c_R}\left(\frac{c_R}{v_b}+\cos\theta\right)\right]\left(\frac{c_R}{v_b}-\cos\theta\right)}\right| \tag{3}$$

式中,f 是频率.

公式(1)中,只有有限性因子 X_R 和辐射图型因子 $\sin 2\theta$ 中含有 θ,所以方向性函数可以容易地推广到夹角为 α 的任意两条射线上的面波振幅谱的比值:

$$D_\alpha = \frac{|U_z^R(\theta)|}{|U_z^R(\theta+\alpha)|} = \left|\frac{\sin\left[\frac{\pi bf}{c_R}\left(\frac{c_R}{v_b}-\cos\theta\right)\right]\left(\frac{c_R}{v_b}-\cos(\theta+\alpha)\right)\sin 2\theta}{\sin\left[\frac{\pi bf}{c_R}\left(\frac{c_R}{v_b}-\cos(\theta+\alpha)\right)\right]\left(\frac{c_R}{v_b}-\cos\theta\right)\sin[2(\theta+\alpha)]}\right| \tag{4}$$

式中,D_α 称为广义的方向性函数[7].

对于给定的 b,v_b 及 c_R,广义的方向性函数 D_α 有一系列的极大值和极小值,它们分别发生在频率为

$$f_{\max} = \frac{n}{b\left[\frac{1}{v_b} - \frac{\cos(\theta+\alpha)}{C_R}\right]}, \quad n=1,2,3,\cdots \tag{5}$$

和

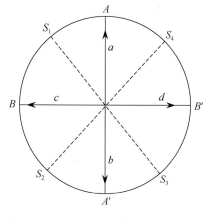

图 2 用面波方向性函数确定断层面及破裂传播方向的方法示意图

$$f_{min} = \frac{n}{b\left(\dfrac{1}{v_b} - \dfrac{\cos\theta}{C_R}\right)}, \quad n = 1, 2, 3, \cdots \quad (6)$$

处. 由 (5), (6) 式可见, 当 $\alpha = 180°$ 时, 若 $|\theta| < 90°$, 方向性函数的第一个极值是极大值; 若 $90° < |\theta| < 180°$, 它的第一个极值是极小值. 这个性质可以用来判断两个节面中哪一个是真正的断层面. 图 2 表示如何由观测资料判断断层面和确定破裂传播方向. 设 S_1, S_2, S_3, S_4 等四个台分别处于两个节面 AA' 和 BB' 所隔开的四个象限中, 如果用这四个台的面波振幅谱得到 $\alpha = 180°$ 时的方向性函数 $D_{13} = S_1/S_3$ 和 $D_{24} = S_2/S_4$, 那么对于图中所示的四种可能的破裂方向中的每一个方向, 比值 D_{13} 和 D_{24} 的第一个极值的性质 (极大或极小) 如表 2 所示.

表 2 破裂面、破裂方向与方向性函数的第一个极值性质的关系

破裂面	破裂方向	D_{13}	D_{24}
AA'	a	极大	极小
AA'	b	极小	极大
BB'	c	极大	极大
BB'	d	极小	极小

3.2 昭通地震的发震断层和破裂传播方向

为了运用面波的方向性函数确定昭通地震的断层面和破裂传播方向, 我们利用了世界标准地震台网 (WWSSN) 的六个台的长周期地震仪垂直向的瑞雷波记录. 这六个台站的名称、方位分布及所记录的垂直向瑞雷波如图 3 所示. 它们构成了三对方位角相差大约 180° 的台站对 (表 3).

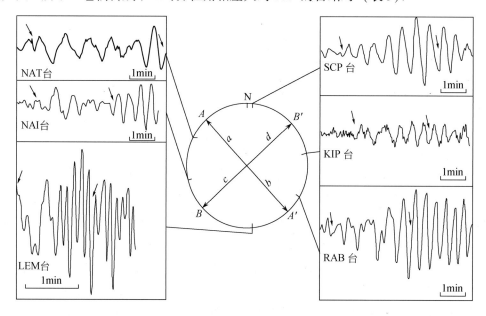

图 3 用以确定昭通地震的断层面和破裂传播方向的六个地震台的名称、方位分布及所记录的垂直向瑞雷波
图中 AA' 和 BB' 表示本文得到的昭通地震的两个节面 两个箭头分别表示瑞雷波波形数字化的起始和终止位置

表3 由三对地震台得到的瑞雷波方向性函数的第一个极值的性质

台站对	台站方位角之差	第一个极值
KIP/NAI	174°	极大
RAB/NAT	175°	极大
SCP/LEM	172°	极大

先将上述六个台的瑞雷波记录数字化（NAI，KIP台记录的采样间隔为0.7s，SCP，LEM，RAB，NAT台记录的采样间隔为0.27s），在DJS-6机上用快速傅里叶变换（FFT）的方法对它们作频谱分析．频谱分析时，采用矩形的时间窗，窗长度1—6min不等．然后对所得的频谱扣除地震仪器的影响（图4），这样便得到如图5所示的瑞雷波振幅谱．

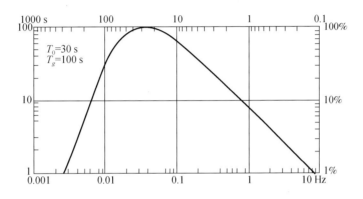

图4 世界标准地震台网（WWSSN）长周期地震仪的频率特性

T_0是摆的周期，T_g是电流计的周期

面波的方向性函数是震中距相同的两个台的面波振幅的比值，所以为求方向性函数，我们将表3所示的台站对的一个台站的振幅谱归算到另一个台站距离处的振幅谱：

$$A_2' = A_2 \left(\frac{\sin\Delta_2}{\sin\Delta_1} \right)^{1/2} \exp[-(\Delta_1' - \Delta_2)r], \tag{7}$$

然后将归算了的振幅谱相除便得到该台站对的瑞雷波方向性函数．在（7）式中，A_2表示第二个台站的振幅谱，Δ_1和Δ_2分别表示第一个和第二个台站的震中距，r表示衰减系数，A_2'表示第2个台站归一化后的振幅谱．计算中，取$r = 0.0002 km^{-1}$[8]．

图6是KIP/NAI，RAB/NAT和SCP/LEM等台站对的瑞雷波方向性函数，及相应的极值频率—序数关系图．图中的黑点表示作者辨认的第一个极值的位置，圆圈表示序数$n \geq 2$的极值的位置．公式（6）表明，作为序数n的函数，极值频率（f_{max}或f_{min}）是通过坐标原点的直线．由图6中的极值频率和序数的关系图可见，f_{max}-n图是通过坐标原点的直线，这说明图中的黑点所指示的极值是方向性函数的第一个极值．显而易见，上述SCP/LEM和KIP/NAI等两对台站的瑞雷波方向性函数的第一个极值都是极大值（见图6及表3）．

和SCP/LEM与KIP/NAI两对台站的情形不同，从RAB/NAT这对台站的方向性函数中难以直接找到它的第一个极值．RAB台相对于北东向节面BB'的方位角大约72°，如果BB'是断层面，那么根据公式（5），（6），同序数的极大值与极小值的位置应当十分接近图6中的极值频率——序数图所表明，尽管同序数的极大值频率与极小值频率十分接近，但极小值频率比极大值频率系统地偏高，这说明第一个极值应当是极大值．

对比图 3 和图 2 以及表 3 和表 2 可以判断，图 3 所示的 BB' 节面是断层面，而 d 是破裂传播方向．

关于昭通地震主震的发震断层，有许多不同的见解．国家地震局第一考察队认为这次地震的发震断层的走向是北西向，而国家地震局第二考察队则认为走向为北东．按照前面得到的结果，作者认为，昭通地震的断层面走向是北东 $45°$，倾向北西，倾角 $86°$，它的破裂传播方向是北东向．

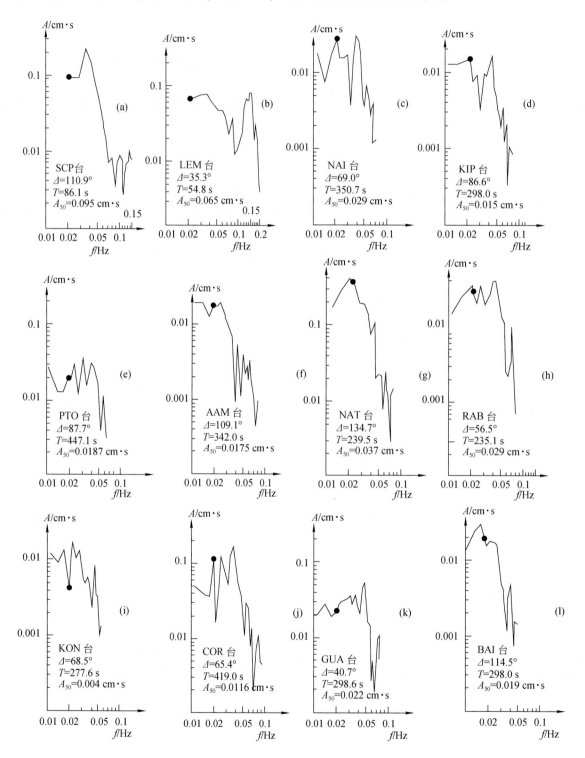

图 5　由世界标准地震台记录得到的昭通地震垂直向瑞雷波振幅谱

黑点表示周期 50 s 的振幅，Δ 表示震中距，T 表示时间窗的长度，A_{50} 表示周期为 50 s 的振幅谱

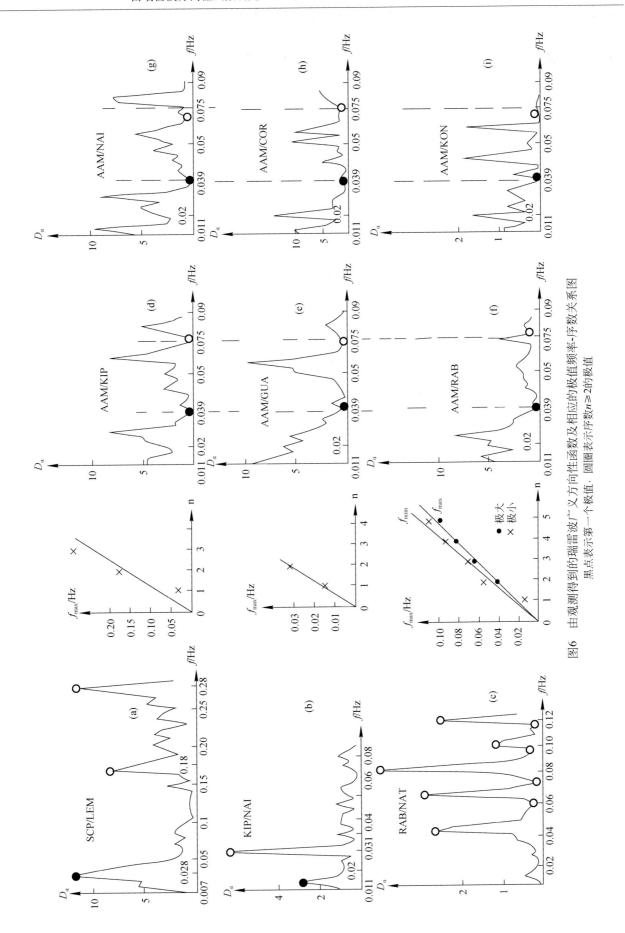

图6 由观测得到的瑞雷波广义方向性函数及相应的极值频率-序数关系图
黑点表示第一个极值，圆圈表示序数 $n \geq 2$ 的极值

3.3 昭通地震的破裂传播速度和破裂长度

从公式（6）可知，广义方向性函数的第一个极小点的周期 T_{min} 和 C_R，v_b 以及 θ 有关：

$$T_{min}=\frac{1}{f_{min}}=b\left(\frac{1}{v_b}-\frac{\cos\theta}{C_R}\right), \tag{8}$$

因此，可以用方向性函数第一个极小值的周期确定破裂传播速度 v_b 和破裂长度 b. 为此我们在上述六个台的资料的基础上再增加 3 个台的资料，这样共有 9 个台的资料. 它们的方位分布如图 7 所示.

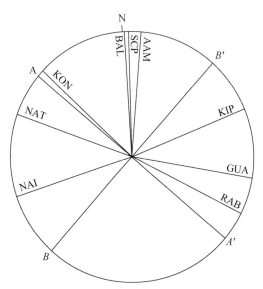

图 7 确定断层长度和破裂传播速度所用的台站的方位分布

$A'A$，$B'B$ 是两个节面

按照前面叙述过的方法，计算了这些台记录的垂直向瑞雷波谱（图 5）和广义方向性函数（图 6）. 然后利用广义方向性函数极小点只与方位角 θ 有关而与 α 无关的性质找出极小点，再根据 $f_{max}-n$ 图确定第一个极小点，表 4 列出了各个台站的名称、方位和极小点的周期.

表 4 用以确定昭通地震的震源参数的台站名称、方位和第一个极小点的周期

台名	RAB	GUA	KIP	AAM	SCP	BAL	KON	NAT	NAI
$\theta/(°)$	288.0	302.5	327.0	39.0	44.0	41.1	76.0	113.0	148.2
T_{min}/s	38.5	35.7	29.4	25.6	28.6	25.6	40.0	45.4	55.6

由图 8 可见，极小点周期和 $\cos\theta$ 呈线性关系. 取 $C_R=3.14$km/s，由最小二乘法可以求得破裂长度为 53km，破裂传播速度为 1.3km/s.

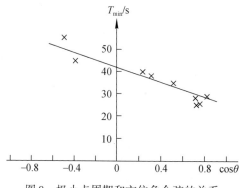

图 8 极小点周期和方位角余弦的关系

4 昭通地震的地震矩、平均错距和应力降

由瑞雷波的振幅谱可以计算地震矩[9]

$$M_0 = \frac{|U_z^R(\omega)|(2\pi r)^{1/2} c_R e^{\gamma r}}{N_{rz}|\chi(\theta)|} \tag{9}$$

假定震源时间函数是阶梯函数，$U_z^R(\omega)$ 表示圆频率为 ω 的瑞雷波垂直分量的振幅谱，r 表示震源距，c_R 表示相速度，N_{rz} 表示瑞雷波的单力转换函数，$\chi(\theta)$ 是辐射图型函数：

$$\chi(\theta) = d_0 + i(d_1\sin\theta + d_2\cos\theta) + d_3\sin 2\theta + d_4\cos 2\theta, \tag{10}$$

$$\begin{cases} d_0 = \dfrac{1}{2}\sin\lambda\sin 2\delta B(h), \\ d_1 = -\sin\lambda\cos 2\delta C(h), \\ d_2 = -\cos\lambda\cos\delta C(h), \\ d_3 = \cos\lambda\sin\delta A(h), \\ d_4 = -\dfrac{1}{2}\sin\lambda\sin 2\delta A(h), \end{cases} \tag{11}$$

λ 是滑动角，δ 是断层面的倾角，A，B，C 是和震源深度 h 有关的函数. 我们以周期为 50s 的振幅来计算地震矩. 结果得到昭通地震的地震矩 M_0 为 6.5×10^{25} dyn·cm.

平均错距可由下式求得：

$$M_0 = \mu \bar{u} b d, \tag{12}$$

式中，d 是断层的宽度，作为一种估计，取它为震源深度的二倍，即 $d = 20$ km，由前面可知，$b = 53$ km，若取 $\mu = 3.3\times 10^{11}$ dyn/cm^2，则可求得 $\bar{u} = 18.5$ cm.

最大错距 u_m 和平均错距有如下的简单关系[10,11]：

$$u_m = \frac{4}{\pi}\bar{u} \tag{13}$$

因此 u_m 为 23.6cm. 它沿走向和倾向的分量分别为

$$\begin{cases} u_m^{(s)} = u_m\cos\lambda \\ u_m^{(d)} = u_m\sin\lambda \end{cases} \tag{14}$$

由前面得到的结果可知 $\lambda = 4°$. 将这些数值代入（14）式，就得到 $u_m^{(s)}$ 和 $u_m^{(d)}$ 分别为 23.5cm 和 1.6cm.

应力降的走向分量 $\Delta\sigma_s$ 和倾向分量 $\Delta\sigma_d$ 为：

$$\begin{cases} \Delta\sigma_s = \mu\dfrac{u_m^{(s)}}{2d}, \\ \Delta\sigma_d = \mu\dfrac{\lambda+\mu}{\lambda+2\mu}\dfrac{u_m^{(d)}}{d}, \end{cases} \tag{15}$$

其数值分别为 1.9bar 和 0.2bar. 而总的应力降是走滑分量和倾滑分量的向量和，其数值 $\Delta\sigma = 2.1$bar.

由走向滑动和倾向滑动释放的能量为：

$$\begin{cases} E_s = \dfrac{4}{\pi}\mu b \bar{u}_s^2, \\ E_d = \dfrac{8}{\pi}\dfrac{\lambda+\mu}{\lambda+2\mu}\mu b \bar{u}_d^2, \end{cases} \tag{16}$$

其数值分别为 7.6×10^{20} 和 4.7×10^{18} erg，因而总能量 E 为 7.6×10^{20} erg.

为了清楚起见,表 5 列出了所得到的昭通地震震源参数.

表 5 昭通地震震源参数

M_0/dyn·cm	E/erg	b/km	d/km	$\bar{u}$/cm	u_m/cm	$\Delta\sigma$/bar	v_b/(km/s)
6.5×10^{25}	7.6×10^{20}	53	20	18.5	23.6	2.1	1.3

5 讨论

历史记载和二十世纪以来的仪器记录资料表明,滇东北地区的地震活动是相当频繁的(图 9). 自 1844 年 8 月大关县元亨地震以来,滇东北的较大地震的分布大体呈北东向,而这次昭通地震正好发生在这个北东向地带的东北端. 它的余震多数沿北东 35°方向分布(图 10). 余震区长约 45km,宽约 20km. 在余震区内,北东向构造具有重要地位(图 11). 巧家-莲峰大断裂长达 150km,其西南起于巧家与宁南之间,向北东以走向 40°一直延伸到余震区南端的高桥西北,断层面倾向西北. 它在晚古生代已有活动,燕山运动中发生了强烈的右旋-逆断层运动. 余震区东北的华蓥山-宜宾大断裂是划分川中劲儿川东两个构造区的重要断裂,这条大断裂在四川盆地内表现为梳状褶曲和间断分布的断裂,它延伸到余震区附近,和巧家-莲峰大断裂遥遥相对,在余震震中区间断了大约 30km 左右. 根据前面所得的结果,可以推断昭通地震是在东西向构造应力场作用下,北东向的巧家-莲峰大断裂朝着北东方向继续破裂的结果,它终止于华蓥山-宜宾大断裂的西南端,把上述两条北东向的大断裂贯通起来(图 11). 余震区附近有两条规模宏大的北西向构造带,其东南是彝良-水城断裂,西北端终止于大关一带,其北面是峨眉-盐津构造带,在绥江-盐津一段分布着同方向的断裂和褶皱,在盐津一带和华蓥山大断裂相交. 在近东西向构造应力场作用下,近年来北西向构造上陆续发生了一些中强震,如 1971 年 8 月 16 日马边 5.8 级地震,1973 年 4 月 22 日彝良 5 级地震和 1973 年 6 月 29 日马边 5.5 级地震. 在这些中强震活动的背景下发生了 1974 年 5 月 11 日昭通 7.1 级地震,它把北东向的巧家-莲峰大断裂和华蓥山-宜宾大断裂贯通起来(图 12). 考虑到昭通地震已使上述两条北东向大断裂贯通起来,减轻了滇东北地区北东向大断裂上的地震危险性,所以作

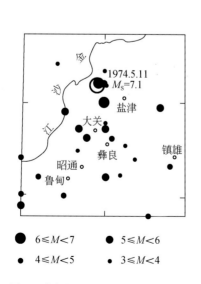

图 9 滇东北 4 级以上的历史地震和
近期 3 级以上地震的震中分布图

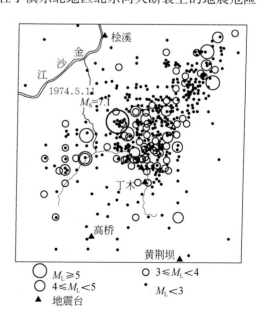

图 10 昭通地震余震震中分布图
(1974 年 5 月 11 日至 7 月 31 日)

者倾向于认为，滇东北的强震活动将从北东向构造带转移到与它共轭的北西—北北西向构造带．今后对北西向的彝良-水城断裂至峨眉-盐津构造带上的地震危险性更应予以注意．

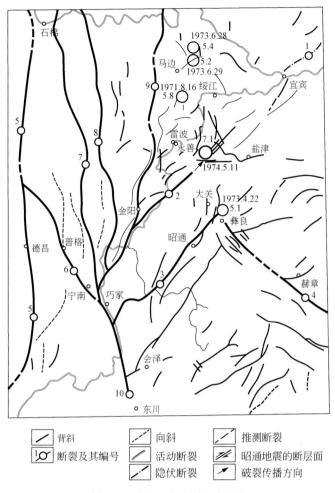

图 11　昭通地区区域构造略图

①华蓥山-宜宾断裂；②巧家-莲峰断裂；③洒鱼河断裂；④彝良-水城断裂；⑤安宁河断裂；⑥普格断裂；
⑦普雄河断裂；⑧汉源-甘洛断裂；⑨峨眉-金阳断裂；⑩小江断裂

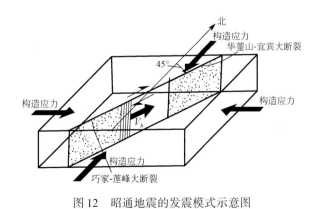

图 12　昭通地震的发震模式示意图

参 考 文 献

[1] 冯德益，1974 年 5 月云南省永善—大关 7.1 级强震前波速比的异常变化，地球物理学报，**18**，4，235-239，1975.
[2] 蜀水，震源应力场岩石膨胀性和水的扩散作用——云南省永善 7.1 级地震及其前兆孕育发展过程，地球物理学报，

19, 2, 74-94, 1976.
[3] 朱成男、陈承照, 1974年云南省昭通地震破裂机制, 地球物理学报, **19**, 4, 317-329, 1976.
[4] 刘正荣、雷素华、胡素华, 1974年5月11日云南省永善—大关地震, 地球物理学报, **20**, 2, 110-114, 1977.
[5] 曾融生, 师洁珊, 1974年5月10日云南永善—大关主震的多重性, 地球物理学报, **21**, 2, 160-173, 1978.
[6] A. Ben-Menahem, Radiation of seismic surface waves from finite moving sources, *Bull. Seism. Soc. Amer.*, **51**, 3, 401-435, 1961.
[7] A. Udias, Source parameters of earthquakes from spectra of Rayleigh waves, *Geophys. J. R. astr. Soc.*, **20**, 4, 353-375, 1971.
[8] E. Tryggvason, Dissipation of rayleiyh wave energy, *J. Geophys. Res.*, **70**, 6, 1449-1456. 1965.
[9] A. Ben-Menafem and D. G. Harkerider, Radiation patterns of seismic sarface waves from buried dipolar point sources in a flat stratified earth, *J. Geophys. Res.*, **69**, 12, 2605-2634, 1964.
[10] L. Knopoff, Energy release in earthquakes, *Geophys. J. R. astr. Soc.*, **1**, 1, 44-52, 1958.
[11] K. Aki, Generation and propagation of G waves from the Niigata earthquake of June 16, 1964. part 2, Estimation of earthquake moment, released energy, and stress-strain drop from the G wave spectrum, *Bull. Earthq. Res. Inst., Tokyo Univ.*, **44**, 1, 73-88, 1966.

用大地测量资料反演的 1976 年唐山地震的位错模式[*]

陈运泰[1]　黄立人[2]　林邦慧[1]　刘妙龙[2]　王新华[1]

1. 国家地震局地球物理研究所；2. 国家地震局测量大队

摘要　运用反演理论探讨了由"零频"资料反演大地震震源模式的基本原理和方法，并用大地测量资料反演了 1976 年唐山 7.8 级地震的位错模式. 得到的结果表明唐山地震的发震构造是一个总体走向为北东 49°的右旋–正断层，断层面倾向南东，倾角 76°. 这个地震的断层长 84km，宽 34km，走向滑动错距 459cm，倾向滑动错距 50cm，地震矩 4.3×10^{27}dyn·cm，应力降 29bar，应变降 4.3×10^{-5}，释放的能量 3.7×10^{23}erg. 由形变资料反演的平均错距和地震矩远大于由地震波资料定出的平均错距（270cm）和地震矩（1.8×10^{27}dyn·cm），它表明在地震区的地壳内震前可能已经发生了无震滑动——断层蠕动. 无震滑动的规模比主震还要大一些，它的矩估计约为 2.5×10^{27}dyn·cm. 唐山地震前虽然没有前震，但是却有规模这么大的"震前蠕动"，这可能是唐山地震与其他许多有前震的地震（如海城地震）的根本区别，它的许多与别的地震不同的前兆可能与此有关.

1　引言

自从弹性位错理论引进地震学领域以来[1-4]，位错就作为地震震源的一种理想化了的模式而得到相当广泛的研究和应用[5-14]. 根据这个理论，可以运用地面的持久位移、地倾斜和地应变等"零频"地震资料反演地震的震源参数[14-22]. 但是，迄今在运用观测资料反演震源参数时往往是采用试错法或网格搜寻法，有时也采用随机尝试法；并且往往是采用最简单的位错模式. 用试错法得到的结果，在相当大的程度上受主观判断的影响，故所得结果的可靠性较差. 用网格搜寻法或随机尝试法反演震源模式时，是在预先限定的参数空间内搜寻或尝试. 人们不可能寻遍或试遍所有的模式，所以这种搜寻或尝试总是欠周到的. 为了克服上述方法的缺点，我们运用近十年来发展起来的反演理论探讨由静力学形变资料反演大地震震源模式的原理和方法，并用大地测量资料反演 1976 年唐山 7.8 级地震的震源模式.

2　用大地测量资料反演震源模式的原理

2.1　经典的最小二乘估计

设在均匀、各向同性和完全弹性的半无限介质中有一个和地面斜交成 θ 角、长度为 $2L$、宽度为 W 的矩形断层. 这种断层引起的地震位移场可以由弹性静力学位错理论求得[8,14]. 将直角坐标系 (x_1, x_2, x_3) 的原点置于地面，以断层走向为 x_1 轴，以其倾向为 x_2 轴，以垂直地面向下的方向为 x_3 轴（图 1），则在这个坐标系(震源坐标系)中，任一观测点的位移分量 u_k ($k=1, 2, 3$) 都可以表示为[8,14]

$$u_k = u_k(x_1, x_2, x_3; \theta, L, d, D, \Delta U_S, \Delta U_d), \tag{1}$$

---

[*] 本文发表于《地球物理学报》，1979 年，第 22 卷，第 3 期，201–217.

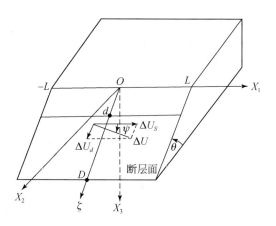

图 1 矩形断层模式

式中，(x_1, x_2, x_3) 是观测点的坐标，d 是断层面的上界，$D=d+W$，是断层面的下界，ΔU_s 是走向滑动错距，ΔU_d 是倾向滑动错距. 现在，另取一个坐标系 (X_1, X_2, X_3)，也将其原点置于地面，但 X_1 轴指向北，X_2 轴指向东，X_3 轴仍向下. 设震源坐标系的原点在这个坐标系（地理坐标系）中的坐标为 $(S_1, S_2, 0)$，并设 x_1 轴和 X_1 轴之间的夹角（断层的走向）为 α，那么在地理坐标系中坐标为 (X_1, X_2, X_3) 的观测点的位移分量 $u_k(k=1, 2, 3)$ 为：

$$\begin{cases} U_1 = u_1 \cos\alpha - u_2 \sin\alpha, \\ U_2 = u_1 \sin\alpha + u_2 \cos\alpha, \\ U_3 = u_3. \end{cases} \tag{2}$$

因为两个坐标系间有如下的坐标变换关系：

$$\begin{cases} x_1 = (X_1 - S_1)\cos\alpha + (X_2 - S_2)\sin\alpha, \\ x_2 = -(X_1 - S_1)\sin\alpha + (X_2 - S_2)\cos\alpha, \\ x_3 = X_3. \end{cases} \tag{3}$$

所以

$$U_k = U_k(X_1, X_2, X_3; \alpha, \theta, L, d, D, \Delta U_s, \Delta U_d, S_1, S_2). \tag{4}$$

为简便起见，以 $x_j(j=1, 2, 3, \cdots, m)$ 表示（4）式中的模式参数 α, θ, L, d, D, ΔU_s, ΔU_d, S_1, S_2. 以 y 表示位移的某一分量，则

$$y = f(x_j). \tag{5}$$

以 y_i^1 表示位移的第 i 个观测值，以 $y_i = f_i(x_j)$ 表示相应的理论计算值，则残差为

$$\epsilon_i^1 \equiv y_i^1 - f_i(x_j), \quad i = 1, 2, 3, \cdots, n. \tag{6}$$

式中，$f_i(x_j)$ 是 x_j 的函数. 我们将它在某一初始值 x_j^0 处作泰勒展开，然后略去二次和二次以上的高次项：

$$\epsilon_i^1 \doteq \epsilon_i \equiv y_i^1 - \left[f_i(x_j^0) + \sum_{j=1}^m \frac{\partial f_i}{\partial x_j} \Delta x_j \right], \tag{7}$$

式中，

$$\Delta x_j = x_j - x_j^0. \tag{8}$$

若以 Δy_i 表示观测的残差：

$$\Delta y_i \equiv y_i^1 - f_i(x_j^0), \tag{9}$$

以 A_{ij} 表示偏导数矩阵：

$$A_{ij} \equiv \left[\partial f_i / \partial x_j \right]_{x_j^0}, \tag{10}$$

则 ϵ_i 可以解释为观测的残差减去理论的残差（O-C 残差）：

$$\epsilon_i = \Delta y_i - A_{ij}\Delta x_j. \tag{11}$$

用矩阵符号，可以把上式表示成紧凑的形式：

$$\boldsymbol{\epsilon} = \Delta\boldsymbol{y} - \boldsymbol{A}\Delta\boldsymbol{x}. \tag{12}$$

若观测方程的个数 n 大于模式参数 x_j 的个数 m，则方程组（12）便是不相容的方程组. 在这种情形下，可以按最小二乘准则求解，这就是使残差平方和

$$S = \boldsymbol{\epsilon}^T\boldsymbol{\epsilon} = (\Delta\boldsymbol{y} - \boldsymbol{A}\Delta\boldsymbol{x})^T(\Delta\boldsymbol{y} - \boldsymbol{A}\Delta\boldsymbol{x}) \tag{13}$$

取极小值，以获得模式参数的改正向量 $\Delta\boldsymbol{x}$ 的估计值 $\Delta\hat{\boldsymbol{x}}$. 式中，T 表示转置. 按最小二乘准则不难得到下列的正则方程：

$$\boldsymbol{A}^T\boldsymbol{A}\Delta\hat{\boldsymbol{x}} = \boldsymbol{A}^T\Delta\boldsymbol{y}, \tag{14}$$

此时，如果 $\boldsymbol{A}^T\boldsymbol{A}$ 是非奇异矩阵，则可得到 $\Delta\hat{\boldsymbol{x}}$：

$$\Delta\hat{\boldsymbol{x}} = (\boldsymbol{A}^T\boldsymbol{A})^{-1}\boldsymbol{A}^T\Delta\boldsymbol{y}. \tag{15}$$

按广义反演理论的说法[23]，所谓反演问题就是：给定一个向量 $\Delta\boldsymbol{y}$，要求一个向量（"真实"参数）$\Delta\boldsymbol{x}$，使满足

$$\boldsymbol{A}\Delta\boldsymbol{x} = \Delta\boldsymbol{y}. \tag{16}$$

这也就是要求得一个算子 $\boldsymbol{H}$，使得它作用于 $\Delta\boldsymbol{y}$ 后得到 $\Delta\boldsymbol{x}$ 的估计值 $\Delta\hat{\boldsymbol{x}}$：

$$\Delta\hat{\boldsymbol{x}} = \boldsymbol{H}\Delta\boldsymbol{y}. \tag{17}$$

上式中的 $\boldsymbol{H}$ 称为广义反演算子. 对比（15）和（17）两式，可知在最小二乘估计情形下，

$$\boldsymbol{H} = (\boldsymbol{A}^T\boldsymbol{A})^{-1}\boldsymbol{A}^T. \tag{18}$$

将（16）式代入（17）式后得到：

$$\Delta\hat{\boldsymbol{x}} = \boldsymbol{R}\Delta\boldsymbol{x}, \tag{19}$$

式中，算子

$$\boldsymbol{R} = \boldsymbol{H}\boldsymbol{A} \tag{20}$$

是把"真实"参数 $\Delta\boldsymbol{x}$ 和估计的参数 $\Delta\hat{\boldsymbol{x}}$ 联系起来的"窗"或滤波器，因而可以称之为"分解度"算子. 显而易见，对于最小二乘估计来说，$\boldsymbol{R}$ 等于单位矩阵 $\boldsymbol{I}$：

$$\boldsymbol{R} = \boldsymbol{H}\boldsymbol{A} = (\boldsymbol{A}^T\boldsymbol{A})^{-1}\boldsymbol{A}^T\boldsymbol{A} = \boldsymbol{I}. \tag{21}$$

换言之，由最小二乘法得到的参数是完全分解了的. 若以 $\Delta\hat{\boldsymbol{y}}$ 表示 $\Delta\boldsymbol{y}$ 的理论计算值，即

$$\Delta\hat{\boldsymbol{y}} = \boldsymbol{A}\Delta\hat{\boldsymbol{x}}, \tag{22}$$

那么将（17）代入上式后可以得到

$$\Delta\hat{\boldsymbol{y}} = \boldsymbol{A}\boldsymbol{H}\Delta\boldsymbol{y} = \boldsymbol{S}\Delta\boldsymbol{y}. \tag{23}$$

上式中，算子

$$\boldsymbol{S} = \boldsymbol{A}\boldsymbol{H} \tag{24}$$

是将 $\Delta\hat{\boldsymbol{y}}$ 和 $\Delta\boldsymbol{y}$ 联系起来的、表示观测资料提供讯息情况的矩阵，通常称之为"讯息"分布矩阵.

以 E 表示数学期望算子，那么 $\Delta\hat{\boldsymbol{x}}$ 的协方差矩阵 $\boldsymbol{C}(\Delta\hat{\boldsymbol{x}})$ 是：

$$\boldsymbol{C}(\Delta\hat{\boldsymbol{x}}) = E\{(\Delta\hat{\boldsymbol{x}} - E\{\Delta\hat{\boldsymbol{x}}\})(\Delta\hat{\boldsymbol{x}} - E\{\Delta\hat{\boldsymbol{x}}\})^T\}. \tag{25}$$

将（17）式代入上式，得：

$$\boldsymbol{C}(\Delta\hat{\boldsymbol{x}}) = \boldsymbol{H}\boldsymbol{C}(\Delta\boldsymbol{y})\boldsymbol{H}^T. \tag{26}$$

如果 $\Delta\boldsymbol{y}$ 的分量 Δy_i 是彼此独立的、均值为 0、方差为 $\sigma^2_{\Delta y}$ 的随机变量，那么

$$\boldsymbol{C}(\Delta\hat{\boldsymbol{x}}) = \sigma^2_{\Delta y}\boldsymbol{H}\boldsymbol{H}^T. \tag{27}$$

可以证明，

$$\sigma^2_{\Delta y} = E\left\{\frac{S_{\min}}{n-m}\right\}, \tag{28}$$

式中，$S_{\min}$ 是 S 的极小值：

$$S_{\min} = (\Delta y - A\Delta\hat{x})^T(\Delta y - A\Delta\hat{x}). \tag{29}$$

由于（28）式的缘故，我们可以用 $\sigma^2_{\Delta y}$ 的无偏估计

$$\hat{\sigma}^2_{\Delta y} = \frac{S_{\min}}{n-m} \tag{30}$$

近似地代替 $\sigma^2_{\Delta y}$. 这样一来，

$$C(\Delta\hat{x}) \doteq \hat{\sigma}^2_{\Delta y} HH^T. \tag{31}$$

在最小二乘反演情形下，因为 A^TA 是对称矩阵，所以 $(A^TA)^{-1}$ 也是对称矩阵，从而

$$HH^T = (A^TA)^{-1}A^TA(A^TA)^{-1} = (A^TA)^{-1}, \tag{32}$$

$$C(\Delta\hat{x}) \doteq \hat{\sigma}^2_{\Delta y}(A^TA)^{-1}. \tag{33}$$

2.2 发散困难

用经典的最小二乘法估计 $\Delta\hat{x}$ 时，常因为 A^TA 是奇异矩阵或接近于奇异矩阵而出现发散困难. 出现这种困难的原因是在观测资料中没有或只有少量的某些模式参数或其线性组合的讯息. 为了说明这点，我们在这里引用兰乔士（Lanczos）的矩阵的基本分解定理[24]. 按照兰乔士的这个定理，任何一个 $n \times m$ 矩阵 A，都可分解成：

$$A = U_p \Lambda_p V_p^T, \tag{34}$$

式中，U_p 是 $n \times p$ 的半正交矩阵，V_p 是 $m \times p$ 的半正交矩阵，它们的行向量分别是资料空间和解空间的基向量. Λ_p 是 A 的非零本征值构成的对角线矩阵，p 是 A 的秩，将（34）式代入（15）式，可得：

$$\Delta\hat{x} = V_p \Lambda_p^{-1} U_p^T \Delta y, \tag{35}$$

式中，

$$\Lambda_p^{-1} = \begin{bmatrix} 1/\lambda_1 & 0 & \cdots & 0 \\ 0 & 1/\lambda_2 & \cdots & 0 \\ \vdots & \vdots & \ddots & \vdots \\ 0 & 0 & \cdots & 1/\lambda_p \end{bmatrix}, \tag{36}$$

$\lambda_1^2, \lambda_2^2, \cdots, \lambda_p^2$ 是正则方程（14）中的矩阵 A^TA 的本征值. 由（35）式可见，当 A^TA 的本征值中有一个或多个本征值很小时，$\Delta\hat{x}$ 的一个或多个分量便会很大. 当然，在对 A 进行分解时，已去掉了零本征值；可是在实际计算中，因为数值误差，一般会遇到本征值很小的情形. 这样，就不能以（13）式所示的残差平方和的近似值 S 代替残差平方和 s：

$$s = \epsilon^{1T}\epsilon^1, \tag{37}$$

因为使 S 取极小值的解 $\hat{x} = x^0 + \Delta\hat{x}$ 不一定也能使得 s 取极小值.

2.3 阻尼最小二乘法

为了克服上述困难，可以设法限制或阻止 $\Delta\hat{x}$ 的增长，以保证一级泰勒近似成立. 这可以通过同时使近似残差平方和以及解的增量取极小值来实现[25,26]. 现在，令

$$\bar{S} = \theta^{-2}S + Q, \tag{38}$$

其中，θ^{-2} 是一个权系数，而

$$Q = \Delta \hat{x}^{\mathrm{T}} \mathbf{W} \Delta x, \tag{39}$$

$\mathbf{W}$ 是一个权矩阵. 若设当 $\Delta x = \Delta \hat{x}$ 时 $\bar{S}$ 取极小值：$\bar{S} = \bar{S}_{\min}$, 则可以求得修正的正则方程：

$$(\mathbf{A}^{\mathrm{T}}\mathbf{A} + \theta^2 \mathbf{W}) \Delta \hat{x} = \mathbf{A}^{\mathrm{T}} \Delta y, \tag{40}$$

从而

$$\Delta \hat{x} = (\mathbf{A}^{\mathrm{T}}\mathbf{A} + \theta^2 \mathbf{W})^{-1} \mathbf{A}^{\mathrm{T}} \Delta y. \tag{41}$$

可以证明，由 $\bar{S} = \bar{S}_{\min}$ 得到的 $\hat{x} = x^0 + \Delta \hat{x}$ 能使 S 减少. 并且可以证明，解的增量在 x^0 的切线总是沿着 s 减少的方向. 另外还可以证明，如果 $\mathbf{W} = \mathbf{I}$，则解的增量在 x^0 的切线方向就是最陡下降方向. 在这种情形下，

$$\Delta \hat{x} = (\mathbf{A}^{\mathrm{T}}\mathbf{A} + \theta^2 \mathbf{I})^{-1} \mathbf{A}^{\mathrm{T}} \Delta y. \tag{42}$$

利用矩阵的基本分解定理，可将上式化为：

$$\Delta \hat{x} = \mathbf{V}_p \{ (\mathbf{\Lambda}_p^2 + \theta^2 \mathbf{I})^{-1} \mathbf{\Lambda}_p \} \mathbf{U}_p^{\mathrm{T}} \Delta y, \tag{43}$$

上式中

$$(\mathbf{\Lambda}_p^2 + \theta^2 \mathbf{I})^{-1} \mathbf{\Lambda}_p = \begin{bmatrix} \dfrac{\lambda_1}{\lambda_1^2 + \theta^2} & 0 & \cdots & 0 \\ 0 & \dfrac{\lambda_2}{\lambda_2^2 + \theta^2} & \cdots & 0 \\ \vdots & \vdots & \ddots & \vdots \\ 0 & 0 & \cdots & \dfrac{\lambda_p}{\lambda_p^2 + \theta^2} \end{bmatrix}. \tag{44}$$

经典的最小二乘法相当于 $\theta^2 \to 0$ 的极端情形. 对于这种情形，$\Delta \hat{x}_i$ 随着 $\lambda_i \to 0$ 趋于无穷大（图2）；对于阻尼最小二乘法，当 $\lambda_i \to 0$ 时，$\Delta \hat{x}_i \to 0$，这就保证了在本征空间中那些不提供讯息或只提供少量讯息的方向上本征谱是尖灭的（图2）.

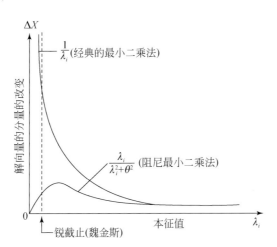

图2 解向量的分量的改变量随本征值的变化示意图

魏金斯（Wiggins）曾经用突然截断本征谱的办法来克服上述发散困难[27]. 这就是先按(34)式分解 $\mathbf{A}$，然后考察本征值很小时的本征谱，建立截断本征谱的准则以避免发散（图2）. 用这种方法可以得到和阻尼最小二乘法类似的结果，但方法本身比较麻烦，因为需要作分解矩阵的计算. 阻尼最小二乘法要简便的多，它只需要解一个修正的正则方程(40). 考虑到这种情况，我们在这里采用阻尼最小二乘法.

由(18)，(20)，(27)，(42)，(43)等公式，可得在阻尼最小二乘估计这种情况下，

$$\mathbf{H} = \mathbf{V}_p \{ (\mathbf{\Lambda}_p^2 + \theta^2 \mathbf{I})^{-1} \mathbf{\Lambda}_p \} \mathbf{U}_p^{\mathrm{T}}, \tag{45}$$

$$\mathbf{R} = \mathbf{V}_p\{(\mathbf{\Lambda}_p^2 + \theta^2 \mathbf{I})^{-1} \mathbf{\Lambda}_p^2\} \mathbf{V}_p^T, \tag{46}$$

$$\mathbf{C}(\Delta \hat{x}) \doteq \sigma_{\Delta y}^2 \mathbf{V}_p (\mathbf{\Lambda}_p^2 + \theta^2 \mathbf{I})^{-1} \mathbf{\Lambda}_p^2 (\mathbf{\Lambda}_p^2 + \theta^2 \mathbf{I})^{-1} \mathbf{V}_p^T. \tag{47}$$

2.4 收敛准则

按照上述方法，我们可以由某一初值 x^0 出发，求得新值 $x = x^0 + \Delta \hat{x}$。然后，以新值作为下次的初值，逐次迭代，直至模式参数满足预先给定的收敛准则，最后这一次的初值就取为问题的"最优解"。如果以 $\sigma_{\Delta \hat{x}_j}$ 代表 $\Delta \hat{x}_j$ 的标准误差，则当

$$|\Delta \hat{x}_j| \leqslant \sigma_{\Delta \hat{x}_j} \tag{48}$$

时，从讯号中就不能得到任何有意义的讯息，因为与资料相联系的随机噪声将完全淹没掉讯息。据此，我们可以用 (48) 式作为一条收敛准则。另一条准则是，相邻两次迭代得到的残差平方和 $s^{(k)}$ 和 $s^{(k+1)}$ 满足不等式：

$$r_1 \equiv \left| \frac{s^{(k)} - s^{(k+1)}}{s^{(k)}} \right| \leqslant \gamma, \tag{49}$$

式中，γ 是一个预先给定的小于 1 的正数。

2.5 系统的标准化

方程 (16) 并非标准形式，也就是说，观测资料和模式参数都是有量纲的量，它们在统计上也不独立。反演时必须将它们标准化。

位移资料的标准化问题比较简单，因为它们的单位都相同，只是各资料点的精度不同罢了。为简单计，我们忽略各资料点的相关性，只按各观测点位移的均方误差 σ_{U_i} 将位移 U_i 标准化，也就是

$$\widetilde{U}_i = \frac{U_i}{\sigma_{U_i}}, \tag{50}$$

$\widetilde{U}_i$ 表示观测点位移的某一分量的标准化形式。

至于模式参数，则应以 $\Delta \hat{x}_j$ 的标准误差 $\sigma_{\Delta \hat{x}_j}$ 将其标准化：

$$\widetilde{\Delta \hat{x}_j} = \Delta x_j / \sigma_{\Delta \hat{x}_j}, \tag{51}$$

其中，$\widetilde{\Delta \hat{x}_j}$ 是 $\Delta \hat{x}_j$ 的标准化形式。当然 $\sigma_{\Delta \hat{x}_j}$ 预先并不知道，所以，反演时先以模式参数的最大许可误差 t_j 近似地代替标准误差 $\sigma_{\Delta \hat{x}_j}$，待迭代初步成功求得 $\sigma_{\Delta \hat{x}_j}$ 后再改成以 $\sigma_{\Delta \hat{x}_j}$ 对 $\Delta \hat{x}_j$ 标准化。由 (51) 式，我们可将 (48) 式所示的收敛准则表示成：

$$|\widetilde{\Delta \hat{x}_j}| \leqslant 1. \tag{52}$$

为计算方便起见，我们不以 (48) 或 (52) 式作为收敛准则，而代之以

$$r_2 \equiv (\widetilde{\Delta \hat{x}}^T \widetilde{\Delta \hat{x}} / m)^{1/2} \leqslant 1. \tag{53}$$

这个准则是一个均方根准则，它比 (52) 所表示的准则要弱一些，允许有的模式参数超过其标准误差。

3 唐山地震地形变资料的反演

1976 年 7 月 28 日唐山 7.8 级地震前后在震中区及其周围广大地区进行了大地测量。测量结果表明，这次地震造成了地面明显的持久形变。我们在这里，根据反演理论，以大地测量资料反演这次地震的震源

模式，借以了解这次地震震源的性质.

3.1 精密水准测量

1975年在唐山地区施测的一等精密水准测量成果和1976年震后复测的一等精密水准测量成果是用以确定唐山地震造成的地面垂直形变的观测依据。重复测量的水准路线的分布见图3. 1975年的一等精密水准测量按自由网平差后求得单位权均方误差为±1.04mm/km$^{1/2}$；相对于起算点的最弱点高程均方误差为±11.9mm. 1976年的精密水准测量则分别为1.22mm/km$^{1/2}$和±14.9mm. 考虑到整个测区内各结点高程均方误差的差别不大（最大与最小之比为11∶7），因此我们对于所有的垂直形变资料采取一个统一的精度指标.

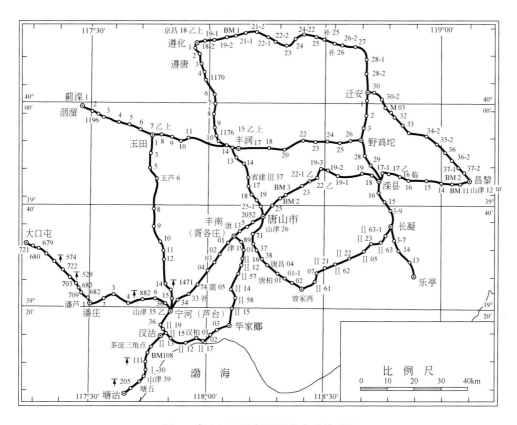

图3 唐山地区重复测量的水准路线图

考虑到反演时所有的位移向量都是相对于庙山三角点而言的，而庙山三角点的垂直位移则是从垂直形变图（图4）量取的，因而在反演中我们取各水准点相对于庙山三角点的垂直位移的均方误差为±22.3mm.

图4是唐山地震前后的垂直形变图. 测区的东南是滨海沉积层较厚的地区，这个部位的水准点受到较明显的局部地质条件的影响，我们在反演中舍去了这些点，共采用163个水准点的资料.

3.2 微波和三角测量

我们以1960年前施测的国家一、二、三等三角测量成果和1976年唐山地震后8—11月间用Dl-50微波测距仪复测的微波测量网的成果来确定这次地震引起的水平形变. 第一期的三角测量资料直接采用了国家三角测量成果表中的结果，其边长的相对精度为9×10^{-6}. 复测的微波测距网按自由网作了平差处理，其边长的相对精度约为2.2×10^{-6}.

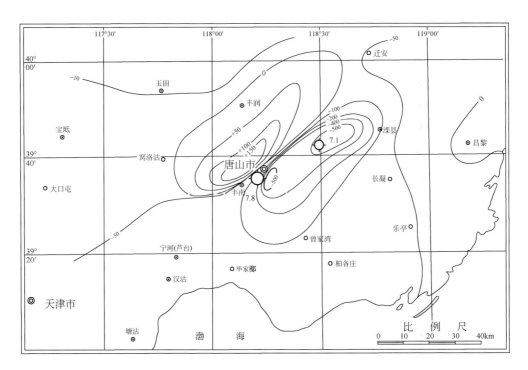

图 4　唐山地震前后的垂直形变（1975 年至 1976 年）图
垂直位移的单位是 mm，向上为正. 带十字的圆圈表示主震和最大余震的微观震中

图 5 是复测的微波测距网图. 图 6 为两期测量的公共三角点的位移向量图. 考虑到局部干扰因素的影响，我们舍去了图 6 中的福田村和皂甸两个三角点，反演时共采用 18 个观测点的水平形变资料.

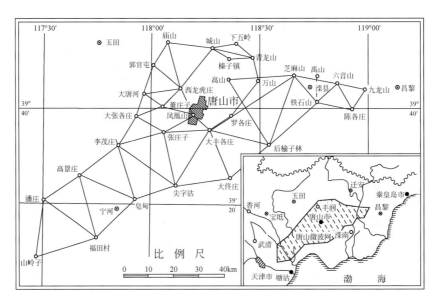

图 5　唐山地区复测的微波测距网图

所有的位移向量都是相对于庙山三角点计算的. 我们按文献 [28] 的公式 (166) 近似估算了这些水平位移的均方误差，并以与位移向量相同的比例尺用短线标在相应的位移向量（或其延长线）上.

3.3　不动点

在分析地震引起的地面形变时，通常假定参考点本身的位移为零，称之为"不动点". 当参考点位于

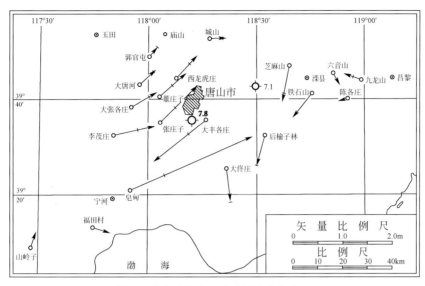

图 6 唐山地震前后的水平位移矢量图
带十字的圆点表示主震和最大余震的微观震中

远离震中区的构造上比较稳定的地区时,这样假定尚合理;但当它距离震中较近,仍然把它当作不动的观测点来处理就不适宜了. 为了克服这个缺点,我们在反演中都以观测点位移减去参考点位移作为该观测点位移的理论计算值.

3.4 计算结果

按照上述方法,同时运用水平和垂直形变测量资料反演唐山地震的震源参数. 计算时,取表1所示的模式参数的最大许可误差和初始值. 开始时,取阻尼因子 θ^2 为50;迭代时,按 0.7 的比例逐次降低它. 不等式(49)中的 γ 取为 0.01. 在 DJS-6 机上迭代一次约需 10min,共迭代了 19 次,全部计算时间约需 3.5h. 迭代结束时,(49)式中的 $r_1 = 0.001$,(53)式中的 $r_2 = 0.92$. 表1的第四行为迭代得到的结果,最后一行为标准误差. 现在定义分解度 r_j 为:

$$r_j = \left[\sum_{k=1}^{n} (R_{jk} - \delta_{jk})^2 \right]^{1/2}. \tag{54}$$

表 1 模式参数的最大许可误差、初值、终值和标准误差

模式参数	$\alpha/(°)$	$\theta/(°)$	L/km	d/km	D/km	ΔU_S/cm	ΔU_d/cm	S_1/km	S_2/km
最大许可误差	1.5	3	3	0.5	2.5	50	5	2	2
初值	45	85	55	0.5	30	−250	90	51.7	112.6
终值	49	76	42	0.0	34	−459	50	45	105
标准误差	1	3	2	0.3	2	48	5	2	2

若 r_j 愈接近于零,则意味着 **R** 愈接近于单位矩阵. 由表2可见,r_j 在迭代结束时大多数都很小,这说明 **R** 很接近于单位矩阵,也就是说所得结果的分解度是高的.

表 2 模式参数的分解度

模式参数	α	θ	L	d	D	ΔU_S	ΔU_d	S_1	S_2
分解度	0.0443	0.0415	0.0744	0.0218	0.0870	0.4953	0.1476	0.0592	0.0680

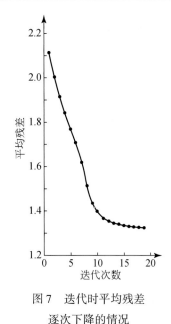

图 7 迭代时平均残差逐次下降的情况

图 7 表示平均残差

$$\bar{r} = (s/n)^{1/2} \tag{55}$$

在迭代时逐次下降的情况. 对于所采取的简单的矩形断层模式, 最优解的 $\bar{r}=1.323$. 由于观测资料总会有误差, 而以数学模式表示复杂的物理情况时模式本身总会有缺陷, 所以由反演得出的模式参数计算的理论位移值总不会完全符合观测值. 现在得到的最优解 $\bar{r}=1.323$, 意味着模式本身还有可以改进之处. 唐山地震可能是一个复杂的地震事件. 考虑到以一个如此简单的矩形断层模式竟能比较圆满地解释观测到的水平和垂直位移场, 应当认为这个模式在相当高的程度上客观地反映了唐山地震的发震构造.

图 8 表示唐山地震引起的地面垂直位移的理论等值线图, 图 9 表示水平位移的理论矢量图. 图中用虚线画的矩形表示反演得出的地震断层在地面上的投影; 断层迹线的中点以带十字的圆圈表示, 其坐标是 39°33.5′N, 118°07.7′E. 对比图 8 和图 4, 图 9 和图 6 可以看到, 如同 $\bar{r}=1.323$ 所意味的, 理论计算值与观测结果相当符合. 两者拟合程度最差的地方在震中东北部, 恰是唐山地震的最大余震 (滦县 7.1 级地震) 震中所在处. 这说明, 现在所采取的模式的最主要缺陷可能是未能考虑这个最大余震的发震构造. 可以预计, 倘若采取考虑了滦县 7.1 级地震的复杂的震源模式, 应当可以使理论值更佳地拟合观测值. 这部分工作放在下一步进行. 现在得到的结果已经可以使我们对唐山地震主震的震源过程有所了解.

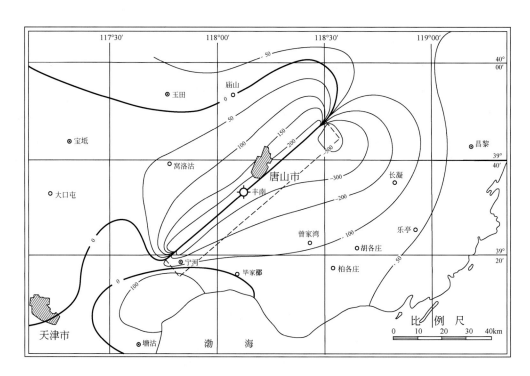

图 8 唐山地震引起的地面垂直位移的理论等值线图

垂直位移的单位是 mm, 向上为正

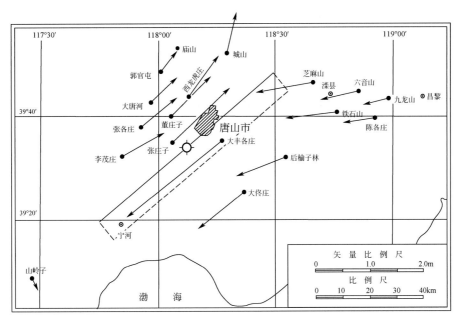

图 9　唐山地震引起的地面水平位移的理论矢量图

4　唐山地震震源过程的若干特点

4.1　总体破裂方向和初始破裂方向不一致

从地面形变的观测资料反演得到唐山地震的发震构造是一个走向为北东49°的近乎直立的右旋-正断层，断层面略向东南倾斜，倾角76°。这个结果和由P波初动资料得到的结果明显地不同[29-31]。文献[29]得到的结果是，断层面走向北东41°，倾向南东，倾角85°（图10中的1）；由地震体波频谱得到的结果表明，断层面是直立的，走向北东30°（图10中的2）。文献[30]由P波初动资料求得唐山地震的断层面走向是北东20°（图10中的3），而由合成面波地震图拟合观测到的面波地震图求得断层面走向是北东40°（图10中的4）。本文由形变资料得到的走向（图10中的5）和文献[30]由面波资料得到的比较接

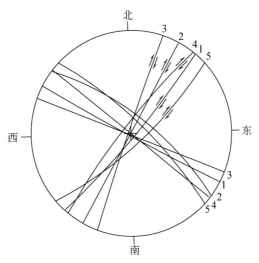

图 10　不同作者得到的唐山地震的断层面解答
震源球下半球投影

近. 既然面波的周期较体波的长, 而地面形变资料是"零频"资料, 它们两者都应当反映唐山地震总体破裂的性质, 所以由这两种资料得到的结果应当一致, 而不一定要与由 P 波初动资料得到的完全一致. 照此看来, 唐山地震不是一个简单的破裂过程, 它的总体的走向是北东49°, 而初始破裂的走向是北东30°.

唐山地震的余震分布在北东50°的地带上（图11）. 余震分布方向与面波和地形变资料定出的断层面走向一致, 而与 P 波初动资料定出的不一致. 这进一步说明了唐山地震初始破裂的走向和总体破裂的走向有明显的差异. 从 P 波初动资料得到的断层面走向与余震震中的分布方向可以有明显的差异, 这是值得注意的一个事实. 文献 [31] 曾报道过类似的结果. 图12是格林（Green）等得到的1969年9月29日南非西锐斯（Ceres）6.3级地震的断层面解和余震区的走向. 可以看出, 和唐山地震的情况类似, 由 P 波初动得到的断层面走向和余震区走向相差约20°. 这说明, 地震时的破裂过程要比通常设想的复杂, 过去人们常简单地将由 P 波初动解得到的断层面走向和余震区走向等同起来, 现在看来是不适宜的.

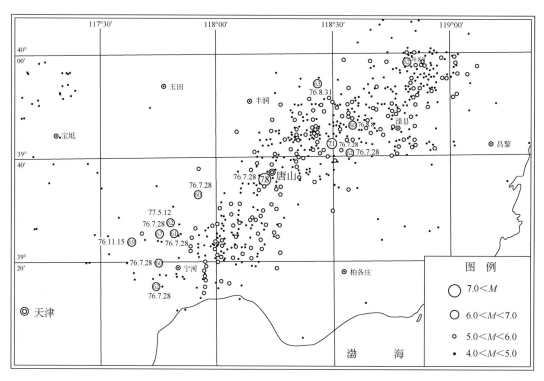

图 11　唐山地震及其余震震中分布图（1976.7.28—1978.6.30）

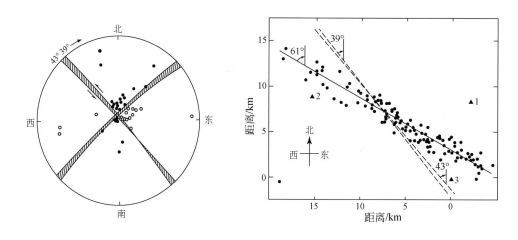

图 12　1969年9月29日南非西锐斯（Ceres）6.3级地震的断层面解和余震分布

黑色三角形表示地震台

4.2 唐山地震的破裂面从地表一直延伸到莫霍界面

唐山地震断层面的下界是34km（表1）．将这个结果和国家地震局物探队在极震区作的丰南-胡各庄地壳测深剖面作一比较是很有意思的，这条剖面大体上横穿余震区．它表明极震区下面的莫霍界面有大约4km的落差，东南深、西北浅．这与由地震波资料[30]和形变资料得到的唐山地震的发震构造一致，就是断层面微向东南倾斜，东南盘下降．按照地壳测深结果，可以推知极震区在地质年代里的构造运动的性质是东南下降，西北上升．由形变资料可知唐山地震继承了这个地区的构造运动，在从地面延伸到莫霍界面的长约84km的范围内发生了约50cm的正断层性质的倾向滑动，它是一个规模宏大的板内地震．

4.3 唐山地震没有前震，但震前可能有蠕动

唐山地震和海城地震明显地不同，没有发生前震．这两个震中相距不远的板内地震前的地震活动图像和前兆异常有很大的差异，充分地说明了地震前兆的多样性．为什么唐山地震没有前震？这个问题从本文的结果可以得到部分的回答．

由地面形变反演得到唐山地震的断层面长84km，宽34km，平均错距（462±50）cm，取刚性系数为3.3×10^{11}dyn/cm^2，则估计地震矩为4.3×10^{27}dyn·cm，应力降为29bar，应变降4.3×10^{-5}，释放的能量为3.7×10^{23}erg．断层比由地震波资料得到的要短和宽，平均水平错距和地震矩都大．由地震体波频谱资料得到，断层长度114km，宽31km，平均水平错距102cm，地震矩1.2×10^{27}dyn·cm．文献［30］的结果则是，断层长度140km，宽15km，平均错距270cm，地震矩1.8×10^{27}dyn·cm．由形变资料定出的平均错距和地震矩远大于由地震波资料定出的，其差别大大超过标准误差．地震波资料反映的是地震前后很短时间内的运动和变化，形变资料反映的是复测期间内的运动和变化，它既包含前者，也包含震前和震后的效应．两种资料所得结果的显著差异说明在震前和（或）震后，极震区的地壳内曾经发生过规模和主震相当的断层运动．从前面已经提到的情况我们知道唐山地震的最大余震（滦县7.1级地震）主要影响主震震中东北的位移，它对整个观测到的位移场影响不大．从文献［30］得到的结果看，这个地震的地震矩是8×10^{26}dyn·cm，还不到主震的一半．这些情况说明，上述差别主要应归因于震前地壳内的断层运动．自1945年9月23日滦县$6\frac{1}{4}$级地震以来，唐山及其附近就未曾有过6级以上地震．所以，上述断层运动必定是以无震滑动即蠕动的形式进行的．

由地震体波频谱资料求得的断层宽度和由形变资料求得的相近．如果接受这个数据，自然要认为唐山地震之前从地面直至莫霍界面的断层面上已经发生了大规模的无震滑动．如果发生这种情况而地面在震前又不出现可见断裂这是难以思议的．但是如果接受文献［30］关于断层宽度等于15km的结果；则必须设想在极震区地壳的下半部震前已经发生了断层蠕动．这种无震滑动的规模相当大，它的矩大约为2.5×10^{27}dyn·cm．这意味着在1960年至1976年间，在长约84km、深度15—34km的断层面上发生了滑距大约473cm的无震滑动．虽然唐山地震没有前震，但却有这种规模巨大的"震前蠕动"，许多观测到的前兆现象可能与此情况有关．前震和"震前蠕动"可能同是孕育地震过程中的重要现象——在有些情况下表现为前震，在另一些情况下表现为"震前蠕动"．在地震危险区监测无震滑动至少是和监测微震活动同样重要的．

4.4 华北及辽南地区的构造应力方向

由形变反演得到，唐山地区地震时释放的主压应力轴方向是北东104°，这个方向和近十年来华北及辽南地区的大地震如邢台、海城地震时释放的主压应力轴方向基本上一致[14,32]．统计地看[33]，这一事实说明华北及辽南地区的大地震都是在一个方向比较一致的构造应力场作用下发生的，这个构造应力场的主

压应力轴的方向是近东西向. 在近东西向的构造应力作用下, 这个地区的北东向及北西向构造是易于发震的, 如同岩石力学实验结果[33]及近十年来的地震活动性所表明的. 根据这种情况, 这个地区的北东向和北西向构造的潜在的地震危险性, 今后仍值得予以注意.

参 考 文 献

[1] А. В. Введенкая, Определение полей смещений при землетрясениях с помощью теории дислокаций, *Изв АН СССР, сер. Геофиз.*, No. 3, 277-284, 1956.

[2] J. A. Steketee, On Volterra's dislocation in a semi-infinite elastic medium, *Can. J. Phys.*, **36**, 2, 192-205, 1958.

[3] J. A. Steketee, Some geophysical applications of the elasticity theory of dislocations, *Can. J. Phys.*, **36**, 9, 1168-1198, 1958.

[4] T. Maruyama, Statical elastic dislocation in an infinite and semi-infinite medium, *Bull. Earthq. Res. Inst., Tokyo-Univ.*, **42**, 2, 289-368, 1964.

[5] M. A. Chinnery, The deformation of ground surface faults, *Bull. Seism. Soc. Amer.*, **50**, 3, 355-372, 1961.

[6] F. Press. Displacements, strains, and tilts at teleseismic distances, *J. Geophys. Res.*, **70**, 10, 2395-2412, 1965.

[7] J. C. Savage, and L. M. Hastie, Surface deformation associated with dip-slip faulting, *J. Geophys. Res.*, **71**, 20, 4897-4904, 1966.

[8] L. Mansinha, and D. E. Smylie, The displacement fields of inclined faults, *Bull. Seism. Soc. Amer.*, **61**, 5, 1433-1440, 1971.

[9] R. Sato, and M. Matsu'ura. Static deformations due to the fault spreading over several layers in a multi-layered medium. Part I. Displacement, *J. Phys. Earth*, **21**, 3, 227-249, 1973.

[10] R. Sato, Static deformations in an obliquely layered medium. Part I. Strike-slip fault, *J. Phys. Earth*, **22**, 4, 455-462, 1974.

[11] D. B. Jovanovich, M. I. Husseini, and M. A. Chinnery, Elastic dislocations in a layered half-space. I. Basic theory and numerical methods, *Geophys. J. R. astr. Soc.*, **39**, 2, 205-217, 1974.

[12] D. B. Jovanovich, M. I. Husseini, and M. A. Chinnery, Elastic dislocations in a layered half-space. II. The point source, *Geophys. J. R. astr. Soc.*, **39**, 2, 219-239, 1974.

[13] R. Sato, and T. Yamahita, Static deformations in an obliquely layered medium. Part II. Dip-slip fault, *J. Phys. Earth*, **23**, 2, 113-125, 1975.

[14] 陈运泰等, 根据地面形变的观测研究1966年邢台地震的震源过程, 地球物理学报, **18**, 3, 164-181, 1975.

[15] J. A. Savage, and L. M. Hastie, A dislocation model for the Fairview Peak, Nevada earthquake, *Bull. Seism. Soc. Amer.*, **59**, 5, 1937-1948, 1969.

[16] N. Canitez, and M. N. Toksöz, Static and dynamic study of earthquake source mechanism, San Fernando earthquake, *J. Geophys. Res.*, **77**, 14, 2583-2594, 1972.

[17] T. Mikumo, Faulting process of the San Fernando earthquake of February 9, 1971. inferred from static and dynamic near-field displacements, *Bull. Seism. Soc. Amer.*, **63**, 1, 249-269, 1973.

[18] M. Ando, A fault-origin model of the Great Kanto earthquake of 1923 as deduced from geodetic data, *Bull. Earthq. Res. Inst., Tokyo Univ.*, **49**, 1, 19-32, 1973.

[19] R. W. Alewine and T. M. Jordan, Generalized inversion of earthquake static displacement field (Abstract), *Geophys. J. R. astr. Soc.*, **35**, 1-3, 357-380, 1973.

[20] D. B. Jovanovich, An inversion method for estimating the source parameters of seismic and aseismic. events from static strain data, *Geophys. J. R. astr. Soc.*, **43**, 2, 347-365, 1975.

[21] M. Matsu'ura, Inversion of geodetic data. Part I. Mathematical formulation, *J. Phys. Earth*, **25**, 1, 69-90, 1977.

[22] M. Matsu'ura, Inversion of geodetic data. Part II. Optimal model of conjugate fault system for the 1927 Tango earthquake, *J. Phys. Earth*, **25**, 2, 233-255, 1977.

[23] R. S. Crosson, Crustal structure modeling of earthquake data. 1. Simultaneous, least-squares estimation of hypocenter and velocity parameters, *J. Geophys. Res.*, **81**, 17, 3036-3046, 1976.

[24] C. Lanczos. Linear Differential Operators, Chap. 3, D. Van Nostrand Co., London, 1961.

[25] K. Levenberg, A method for the solution of certain non-linear problems in least squares, *Quart. Appl. Math.*, **2**, 2, 164–168, 1944.

[26] D. W. Marquardt, An algorithm for the least squares estimation of nonlinear parameters, *J. Soc. Ind. Appl. Math.*, **11**, 2, 431–441, 1963.

[27] R. A. Wiggins, The general linear inverse problem: Implication of surface waves and free oscillations for earth structure, *Rev Geophys. Space Phys.*, **10**, 1, 251–285, 1972.

[28] К. Л. 普罗沃洛夫, 论连续三角网的精度, 1958.

[29] 邱群（陈运泰等），1976年7月28日河北省唐山7.8级地震的发震背景及其活动性, 地球物理学报, **19**, 4, 259–269, 1976.

[30] R. Butler, G. S. Stewart and H. Kanamori, The July 27, 1976 Tangshan, China Earthquake—a complex sequence of intraplate events, Manuscript, 1977.

[31] R. W. E. Green, and A. McGarr, A comparison of the focal mechanism and aftershock distribution of the Ceres, South Africa Earthquake of September 29, 1969, *Bull. Seism. Soc. Amer.*, **62**, 3, 869–871, 1972.

[32] 顾浩鼎、陈运泰、高祥林、赵毅，1975年2月4日辽宁省海城地震的震源机制, 地球物理学报, **19**, 4, 270–285, 1976.

[33] N. Yamakawa, Stress fields in focal regions, *J. Phys. Earth*, **19**, 4, 347–355, 1971.

A dislocation model of the Tangshan (唐山) earthquake of 1976 from the inversion of geodetic data

Chen Yun-tai[1], Huang Li-ren[2], Lin Bang-hui[1], Liu Miao-long[2] and Wang Xin-hua[1]

1. Institute of Geophysics, State Seismological Bureau;
2. The Geodetic Survey Brigade for Earthquake Research, State Seismological Bureau

Abstract The fundamental principles and method of deducing a source model of the earthquake from "zero frequency" data are developed, and a dislocation model of the Tangshan earthquake ($M_S = 7.8$) of 1976 is deduced from the inversion of geodetic data. The results obtained indicate that the source of the Tangshan earthquake is a right-lateral normal fault striking generally N 49°E, and diping 76°SW. The fault is of length, 84 km, width 34 km, average strike-slip dislocation 459 cm, average dip-slip dislocation 50 cm, seismic moment 4.3×10^{27} dyne·cm, stress-drop 29 bar, strain-drop 4.3×10^{-5}, and strain-energy release 3.7×10^{23} ergs. The fact that the average dislocation deduced from the geodetic data is much larger than that from seismic data, implies that before the earthquake, an aseismatic slip i.e. a precursory fault creep, had occurred within the crust beneath the epicentral area. The moment of the aseismic slip is estimated to be 2.5×10^{27} dyne·cm. It is concluded that although before Tangshan earthquake, no foreshock was observed, a large-scale pre-creep had occurred instead of the ordinary foreshock. This may be an essential characteristic of the Tangshan earthquake, and many peculiar precursory phenomena before the shock may be related to this pre-creep.

Variations of gravity before and after the Haicheng earthquake, 1975, and the Tangshan earthquake, 1976

Chen Yun-Tai[1], Gu Hao-Ding[2] and Lu Zao-Xun[2]

1. Institute of Geophysics, Academia Sinica; 2. The Seismological Bureau of Liaoning Province

Before and after the Haicheng earthquake of magnitude 7.3 which occurred on February 4, 1975, five repeated gravimeter surveys were carried out, three before and two after the earthquake, along a northwest-southeast profile of about 250km in length not far on the west of the epicenter. The mean-square error of the measurements of the gravity differences between two consecutive points on the profile is less than 40μGal. From June, 1972 to May, 1973, within a period of about one year, the results of three surveys indicated a clear decrease of the gravity values at points on the southeastern portion of the profile, amounting to about 352μGal. After the earthquake, the fourth survey, which was carried out in March, 1975, revealed that the gravity values had recovered to the levels of the first survey and continued to increase as was shown by a fifth survey carried out in July of the same year.

Variations of gravity were also observed before and after the Tangshan earthquake of magnitude 7.8 which occurred on July 28, 1976, but in this case, gravity was increasing instead of decreasing before the earthquake. Along an east-west profile of about 270km in length and not far on the north of the epicenter, two gravity surveys were made before and two after the earthquake. The results showed that after the main shock, the gravity values of the whole profile, especially at those points closer to Tangshan, tended to return gradually to their values of the first survey before the earthquake.

From these results, there seems to be a close relationship between these gravity variations and the occurrences of earthquakes. Based on results of repeated levelling work done in these regions, the estimated amount of gravity change caused by the change of elevation of the ground surface is far too small to account for the observed value. Therefore we speculate that some large earthquakes might be associated with some sort of mass transfer under ground, within the crust or in the upper mantle. This transfer would cause a large part of the gravity variation observed. We have made a theoretical analysis of this effect and attempted to obtain some estimate of the magnitude of this mass transfer, even though we are not yet clear about the physics of it.

1 Introduction

There was a change of gravity values before and after the Niigata earthquake of 1964 (Fujii, 1966). Similar change was also observed for the Alaska earthquake in the same year (Barnes, 1966). After that, and during the period 1965—1967 when the Matsushiro earthquake swarm occurred and before and after the Inangahua earthquake, 1968, and the San Fernando earthquake, 1971, changes of gravity were also observed (Hunt,

---

* 本文发表于 *Phys. Earth Planet. Interior*, 1979 年, 第 **18** 卷, 第 4 期, 330–338.

1970; Kisslinger, 1975; Oliver et al., 1975). These observations showed that major earthquakes might be accompanied by gravity changes. From the point of view of earthquake prediction, it is pertinent to examine more in detail the characteristics of these changes during an earthquake.

Both before and after the Haicheng earthquake, 1975, and the Tangshan earthquake, 1976, in China, repeated levelling surveys had been carried out. These surveys revealed some characteristics of the changes of gravity with time during an earthquake, and gave rise to this study of the cause of these changes. This paper is an attempt to analyze theoretically the cause of these gravity changes, in accordance with the above observations.

2 Gravity changes due to deformation and mass transfer

2.1 The change of gravitational potential at a fixed point in space due to deformation

What can cause a change of the value of gravity are the change of position of observation and the change of distribution of the attracting mass. Generally, under an observation point, not only could the medium undergo deformation, but also there might be a transfer of mass in the medium. These could cause a change of gravity. Therefore, in order to explain the change of gravity during an earthquake, it is necessary to analyze the changes due to both these causes. To analyze the effect of deformation, we assume that the deformed region is small compared with the whole Earth, so that the Earth may be approximately represented by a semi-infinite medium $z \geqslant 0$, z being positive downward (Fig. 1).

Let $P(\boldsymbol{r}_0)$ be the point of observation and $\boldsymbol{r}_0 = \boldsymbol{r}_0(x_0, y_0, z_0)$. We shall examine first the gravity change when P is referred to an inertial coordinate system and then the case when it is referred to a coordinate system which moves with the deforming medium. Let $Q(\boldsymbol{r})$ be any point in the deformed region V and $\boldsymbol{r} = \boldsymbol{r}(x, y, z)$. Let $\boldsymbol{R}$ be the vector $\overline{QP}$, that is,

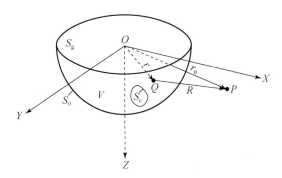

Fig. 1 Deformed region in a semi-infinite medium

$\boldsymbol{R} = \boldsymbol{r}_0 - \boldsymbol{r}$. Let the density at Q be $\rho(\boldsymbol{r})$ and the displacement be $\boldsymbol{u}(\boldsymbol{r})$. After deformation, the increase of mass in an element of volume dV which is fixed with respect to the inertial system is $-\nabla \cdot (\rho \boldsymbol{u}) dV$. This would cause an increase of the gravitational potential:

$$\frac{G \nabla \cdot (\rho \boldsymbol{u})}{R} dV$$

G being the gravitational constant. The flow of mass out of the surface element dS in the deformed region is $\rho \boldsymbol{u} \cdot \boldsymbol{n} dS$, $\boldsymbol{n}$ being an outward unit normal to dS. This out-flowed mass would cause an increase of the gravitational potential at P by an amount:

$$-\frac{G\rho \boldsymbol{u} \cdot \boldsymbol{n}}{R} \mathrm{d}S$$

The sum of the above increments is the total increase of the gravitational potential δU at a fixed point in space, i. e. :

$$\delta U = G \iiint_V \frac{\nabla \cdot (\rho \boldsymbol{u})}{R} \mathrm{d}V - G \oiint_S \frac{\rho \boldsymbol{u} \cdot \boldsymbol{n}}{R} \mathrm{d}S \tag{1}$$

This differs from the result of Walsh (1975) which is incorrect. The corrected result by Reilly and Hunt (1976) holds for the case when the surface of the deformed region is kept constant (see also Walsh, 1976), that is, $\boldsymbol{u}|_S = 0$. Our result (eq. 1) agrees with theirs if the second term, on the right-hand side of eq. 1 is put to zero.

2.2 Change of gravity due to deformation

The change of gravity corresponding to eq. 1 is:

$$\delta g = G \iiint_V \frac{(z_0 - z) \nabla \cdot (\rho \boldsymbol{u})}{R^3} \mathrm{d}V - G \oiint_S \frac{(z_0 - z)\rho \boldsymbol{u} \cdot \boldsymbol{n}}{R^3} \mathrm{d}S \tag{2}$$

The surface S consists of three parts: the Earth's surface $S_g(z=0)$; the external surface S_0, i. e. the surface of V with the exception of S_g; and the internal surface S_c of the deformed region, i. e. the surface of whatever cavities that region may contain. S_0 is the boundary between the deformed and the undeformed regions. Since the displacement must be continuous at the boundary, $\boldsymbol{u} = 0$ on S_0. The corresponding integral must therefore vanish. When P is on the Earth's surface before deformation, the surface integral over S_g is equal to $-2\pi G\rho h$, in which ρ is the density an the point of observation and h is the height of elevation of this point, in this case, the change of gravity due to deformation is given by:

$$\delta g = - G \iiint_V \frac{z \nabla \cdot (\rho \boldsymbol{u})}{R^3} \mathrm{d}V + G \oiint_S \frac{z \rho \boldsymbol{u} \cdot \boldsymbol{n}}{R^3} \mathrm{d}S - 2\pi G\rho h \tag{3}$$

2.3 Gravity change due to deformation and mass transfer at an observation point which moves with the deformed Earth's surface

In the case discussed above, the observation point is fixed with respect to an inertial system. Actually, it moves with the deforming Earth's surface, and is therefore approaching or going away from the Earth's center. Therefore, the observed gravity change contains two more terms, one is the free-air effect and the other corresponds to the difference of gravity caused by an equivalence of stratum of thickness h and density ρ at the observation point before and after deformation. The free-air effect depends principally on h, but less on latitude and is equal to (Grant and West, 1965):

$$\delta g_{FA} = -\frac{8\pi}{3} G\rho_E h \tag{4}$$

in which ρ_E is the mean density of the Earth. The constant coefficient is $(8\pi/3) G\rho_E = 0.3083 \mu\mathrm{Gal/mm}$. The effect due to the equivalence of stratum is equal to:

$$\delta g_e = 4\pi G\rho h \tag{5}$$

The mass transferred from distant or deep regions to the cavities near the observation point (e. g., cavities in the rock or magma chamber) leads to a gravity change:

$$\delta g_M = G \oiint_{S_c} \frac{z \rho_F \boldsymbol{u}_F \cdot \boldsymbol{n}}{R^3} \mathrm{d}S \tag{6}$$

where $\rho_F u_F \cdot n dS$ is the mass which flows into S_c through dS.

2.4 Change of gravity

Combining the above results, we obtain the change of gravity due to both deformation and mass transfer:

$$\delta g = -G\iiint_V \frac{z\nabla\cdot(\rho u)}{R^3}dV - 2\pi G\left(\frac{4}{3}\rho_E - \rho\right)h + G\oiint_{S_C}\frac{z n\cdot(\rho u + \rho_F u_F)}{R^3}dS \quad (7)$$

In the above expression, the first term represents the change of gravity due to the change of density after deformation; the second term, that due to the change of elevation caused by the deformation, i.e., the Bouguer effect δg_B. When $\rho = 2.67 \text{g/cm}^3$ the Bouguer gradient is $2\pi G\left(\frac{4}{3}\rho_E - \rho\right) = 0.1964 \mu\text{Gal/mm}$. The third term represents the effect on the gravity change due to the mass flow through the surfaces of the cavities (e.g., due to opening or closure of the pores or the expansion or contraction of the magmatic chambers) and due to fluids flowing through the cavities (e.g., water or magma).

3 Gravity effects of deformation and mass transfer

In order to assess the effects of deformation and mass transfer on the change of gravity in the processes of preparation and generation of earthquakes, let us analyze a simple example. Consider a cylinder which occupies the space, $0 \leq z \leq H$, $x^2 + y^2 \leq a^2$. Owing to the opening of the cavities, the surface of the earth is elevated by an amount h. If there is no change of density, the volume integral in eq. 7 is zero. Then the gravity effect δg_c due to the mass flow is equivalent to that of a cylinder of density $\rho(-h/H)$, height H and radius a, i.e.:

$$\delta g_c = -2\pi G\rho[1-F(H/a)]h \quad (8)$$

where $F(H/a)$ is given by:

$$F(H/a) = [\{1+(H/a)^2\}^{1/2} - 1]\frac{a}{H} \quad (9)$$

whose value is between 0 and 1 (Fig. 2). If there is mass transferring to the cavity, filling a space corresponding to an elevation h' of the Earth's surface, then the gravity effect δg_M would correspond to

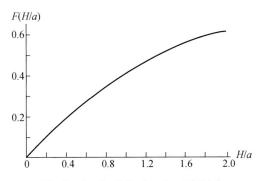

Fig. 2 Graph of the function $F(H/a)$

that of a cylinder of density $\rho_F(h'/H)$, height H and radius a, i.e.:

$$\delta g_M = 2\pi G\rho_F[1-F(H/a)]h' \quad (10)$$

Here, $h'>0$ corresponds to mass flowing in and $h'<0$ to mass flowing out. If $h'=h$, the mass flowing in would just fill the space left by the opening of the cavity. If $h'>h$, this means that a portion of the mass flowing in would occupy a part of the original void space of the cavity.

Summarizing the above results, we obtain the gravity change due to both deformation and mass transfer:

$$\delta g = -2\pi G\left[\frac{4}{3}\rho_E - \rho F(H/a)\right]h + 2\pi G\rho_F[1-F(H/a)]h' \tag{11}$$

in which, the first term represents the change due to deformation and its magnitude is between that of the Bouguer effect δg_B and that of the free-air effect δg_{FA} (Fig. 3). In the region below the straight line δg_D–h, mass flows out of the cavity, while in the region above it, the reverse is the case.

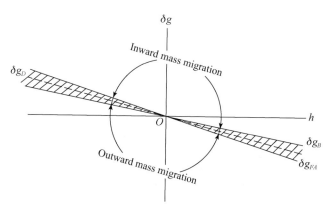

Fig. 3 Gravity effect of deformation and mass transfer

The above results indicate that when the inflow of the mass just fills the space generated by the opening of the cavity, then $h=h'$ and the gravity effect due to mass transfer will be of the same order of magnitude as that due to deformation. But when the mass flows into the cavity space which has already undergone an expansion such that h' is of a higher order of magnitude than h, then, the gravity effect due to mass transfer will be of a higher order of magnitude than that due to deformation.

4 Gravity changes before and after the Haicheng and Tangshan earthquakes and their interpretation

4.1 Gravity change before and after the Haicheng earthquake

Before and after the Haicheng earthquake of February 4, 1975, five gravity surveys were carried out along a NW-SE profile of about 250km long, not far on the west of the epicenter (Fig. 4). Three surveys were made before the shock, the first was in June, 1972, the second in November, 1972, and the third in May, 1973. Two surveys were made after the shock. One was in March, 1975 and the other in July, 1975.

All gravity points are marked and kept permanent by buried stones. The nearest gravity points to the epicenter are Yingkou and Gaixian, about 50km from it. Three quartz suspension type gravimeters, having a sensitivity of 20μGal, were used at the same time, among them, two are made in China, type ZS_2-67 and the other one, made in Canada, type Worldwide CG-2.

To reduce the effect of drift of the gravimeters in the field, a repetition method called "two-fold, three-way" had been adopted (Lu Zao-xun et al., 1978). The time interval between consecutive readings taken at the same point is generally less than 2h. The gravimeter readings have been corrected for the theoretical earth-tide effect.

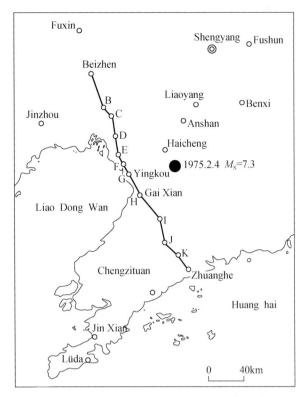

Fig. 4 Gravity traverse from Beizhen to Zhuanghe

The mean-square error of the gravity differences between two adjacent points in the profile is less than 40μGal. The three surveys before the shock indicate that there was a significant decrease of gravity in the southeast section of the profile, the maximum drop reaching 352μGal (Fig. 5). After the shock, in the fourth survey, it was discovered that the gravity value had recovered to the level of the first survey. It continued to rise, reaching a maximum of 382μGal, as was indicated by the fifth survey. From 1958 to 1971, the Liaotung Peninsular rose quickly, the southeast side with a higher speed than the northwest side. During this same period, the bench marks near Zhuanghe and Chengzituan rose about 60mm relative to that in Yingkou, the rising speed being about 5mm/a. From 1937 to 1958, the above two bench marks rose about 60mm, with a speed of about 3mm/a.

There was an evident subsidence in the epicentral area caused by the main shock. Subsequent levelling indicated that the maximum amplitude of the subsidence reached 140mm. On the other hand, on the northwest section of the profile, west of the epicentral area, the maximum subsidence of the Earth's surface reached 260mm.

Fig. 5 also shows the curve of the variation of the surface elevation along the gravity profile. In the years 1970 and 1971 before the earthquake the southeast of the profile rose relative to the northwest section. The rising area corresponded exactly to the area with a gravity drop during the period from June, 1972 to May, 1973. After the earthquake, in March, 1975, there was a significant surface subsidence along the northwest section of the profile relative to the elevations at 1970 and 1971, but the gravity values of the whole profile restored to the level at June, 1972. The north section, with a positive gravity anomaly, corresponds exactly with the area of significant depression.

The variations of gravity with surface elevation at the same point are shown in Fig. 6. Owing to limitations of observations, the data points do not correspond to variations at exactly the same time. For convenience of comparison, there are also shown the straight lines corresponding to the free-air and Bouguer effects. From this

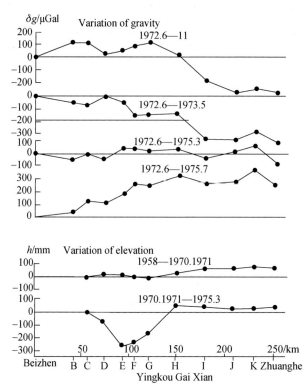

Fig. 5 Changes of gravity and elevation along the Beizhen-Zhuanghe profile

figure, it can be seen that the gravity change in about one-year period from June, 1972 to May, 1973 is nearly one order of magnitude higher than the gravity change during the 12-year period from 1958 to 1971 due to the change in elevation. Even though there are no data of elevation changes from June, 1972 to May, 1973, it can be inferred from the data presented above that if this change of gravity were due to the change in elevation, this latter would reach about 1000mm. This would at least be one order too high than is acceptable.

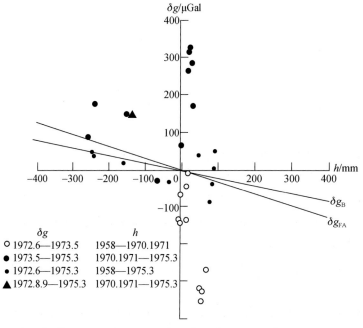

Fig. 6 Variations of gravity with surface elevation at the same point

The case is similar for the variation of gravity with elevation before and after the Haicheng earthquake. In Fig. 6, the black dots represent the variation of gravity with elevation before and after the earthquake and the black triangles those in the epicentral area. It is seen that from May, 1973 to March, 1975, there was a significant increase of gravity of the order of a few hundred microgals, but the change of elevation was quite inadequate to account for these, the gravity change being several times or even an order too large.

4.2 The gravity change before and after the Tangshan earthquake

Two gravity surveys were made before as well as after the Tangshan earthquake of magnitude 7.8 which occurred on July 28, 1976, along an east-west profile of about 270km long, not far on the north of the epicenter (Fig. 7). In this case, two gravimeters, type Worldwide CG-2, were used at the same time. A repetition method similar to the Haicheng case has been adopted, and the observations were also corrected for the theoretical earth tide effect. The mean-square error of the gravity differences between two adjacent points in the profile is less than 20μGal. The gravity changes referred to the values of the first survey, which was carried out in March 24, 1976, are shown in Fig. 8. It is seen that on July 3, 1976, before the earthquake, the gravity values on the east section of the profile increased, reaching a maximum of 165μGal. Just after the earthquake and on July 29, 1976, there was a slight decrease of gravity values along the whole profile and a continued decrease in August.

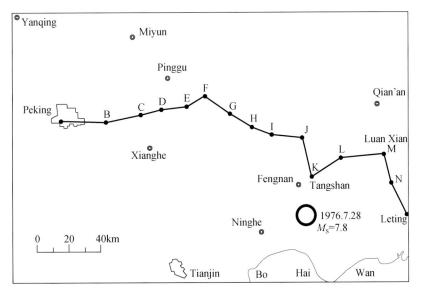

Fig. 7　Gravity traverse from Peking to Leting

Levelling data indicate that in the vicinity of Leting on the east of Tangshan, the earth's surface rose with a speed of about 5.0mm/a. in the period 1970—1975. In the Peking and Leting areas, we have no relevant data of elevation for the period from March to August, 1976. If we assume that the above rising speed also prevailed in this area before 1976, then the gravity change, both in sign and in magnitude, before the Tangshan earthquake cannot be explained in terms of change in elevation.

4.3 Mass transfer for the Haicheng and Tangshan earthquakes

The above facts show that even though the change of gravity is related to the change in elevation, the latter is, however, not the determining factor. There is yet another factor which would produce more important effects on gravity. According to the foregoing analysis, the effect on gravity due to deformation is between the Bouguer and

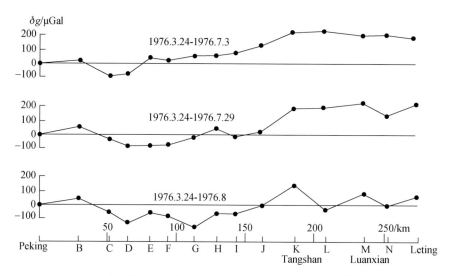

Fig. 8 Changes of gravity along the Peking-Leting profile

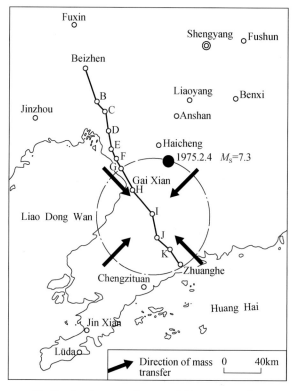

Fig. 9 A schematic diagram of the mass transfer

free-air effects and it is too small to account for the gravity changes before and after either the Haicheng or Tangshan earthquakes. It seems likely that mass transfer in the crust or mantle may give an adequate answer. Consider, for example, the Haicheng earthquake (Fig. 9). If we take the cavity aggregate region to be a circular cylinder of radius $a=60$km and height also $H=60$km. Let $\rho_F=\rho$. If we take h' to be -5.4m and $+5.8$m, we obtain from eq. 11 the gravity changes to be -352 and $+382\mu$Gal, respectively. This corresponds to an outflow of 61km^3 of mass from the cylindrical region during the period from June, 1972 to May, 1973. This would increase the porosity of the region by $9.0 \cdot 10^{-5}$. But during the period from May, 1973 to March, 1975, there will be an inflow of 66km^3 of mass, filling a space of $9.7 \cdot 10^{-5}$ porosity. This is, of course, only an order of magnitude estimate. If we set $\rho_F=$

1.5ρ, the volumes of the out-flowing and in-flowing masses will be 41 and 44 km^3, respectively. In the case of the Tangshan Earthquake, if we take the linear dimension of the region of maximum gravity change, say, 80 km, to be the diameter of the cylindrical region and let $H = a = 40$ km and $\rho_F = \rho$. Then, when $h' = 2.5$ m, the gravity change produced will be 165 μGal. This corresponds to an inflow of 12.5 km^3 of mass to the above cylindrical region, filling the space of $6.3 \cdot 10^{-5}$ porosity.

5 Conclusion

Gravity surveys before and after the Haicheng and Tangshan earthquakes indicate that a gravity change may accompany a major earthquake and the magnitude of the change far exceeds that which can be accounted for by a pure change of elevation. According to our analysis, certain major earthquakes may be related to some kinds of mass transfer in the crust or mantle and the observed change of gravity is due to this cause. We have calculated theoretically the gravity effect of this mass transfer, but we are not yet able to propose a physical mechanism for it.

Acknowledgements

The authors like to thank Professors Ku Kung-hsu and Fu Cheng-yi who critically read the manuscript and offered many valuable suggestions.

References

Barnes, D. F., 1966. Gravity changes during the Alaska earthquake. *J. Geophys. Res.*, **71**: 451–456.
Fujii, Y., 1966. Gravity changes in the shock area of the Niigata earthquake, 16 June 1964. *Zisin*, **19**: 200–216 (in Japanese).
Grant, F. S. and West, G. F., 1965. *Interpretation Theory in Applied Geophysics*. McGraw-Hill, New York, N. Y., 584 pp.
Hunt, T. M., 1970. Gravity changes associated with the 1968 Inangahua earthquake. *N. Z. J. Geol. Geophys.*, **13**: 1050–1051.
Kisslinger, C., 1975. Process during the Matsushiro, Japan, earthquake swarm as revealed by levelling, gravity, and spring-flow observations. *Geology*, **3**: 57–62.
Lu Zao-xun, Fang Chang-liu, Shi Zuo-ting, Zhang Mao-shu, Li Run-jiang and Yang Jun-zhen, 1978. Variation of the gravity field and the Haicheng earthquake. *Acta Geophys. Sin.*, **21**: 1–8 (in Chinese with English abstract).
Oliver, H. W., Robbins S. L., Grannell, R. B., Alewine, R. W. and Shawn Biehler, 1975. Surface and subsurface movements determined by remeasuring gravity. In: G. B. Oakeshott (Editors), *San Fernando, California, Earthquake of 9 February 1971*. Sacramento, Calif., pp. 195–211.
Reilly, W. I. and Hunt, T. M., 1976. Comment on An analysis of local changes in gravity due to deformation by J. B. Walsh. *Pure Appl. Geophys.*, **114**: 1131–1133.
Walsh, J. B., 1975. An analysis of local changes in gravity due to deformation. *Pure Appl. Geophys.*, **113**: 97–106.
Walsh, J. B., 1976. Reply to W. I. Reilly and T. M. Hunt. *Pure Appl. Geophys.*, **114**: 1135.

唐山地震引起的剩余倾斜场的空间分布和倾斜阶跃[*]

李兴才　陈运泰
国家地震局地球物理研究所

摘要　由倾斜断层的静态位错模式计算了唐山地震引起的剩余倾斜场的空间分布和泰安等倾斜仪台站处的理论倾斜阶跃，讨论了按不同震源参数计算得到的剩余倾斜场的空间分布特征和断层参数对倾斜变化的影响．各台站倾斜阶跃的观测值是：泰安，$1.4×10^{-8}$（北），$1.1×10^{-7}$（东）；沙城，$2.4×10^{-7}$（北）；易县，$-1.3×10^{-7}$（西）单位为 rad. 观测值与理论计算值的比较表明，观测的和计算的倾斜阶跃的方向是一致的，但数值上观测值较大．观测的和计算的倾斜阶跃方向的一致性说明了观测到的阶跃与地震的关系，而观测值较大则说明观测的倾斜阶跃仍包含其他局部因素的影响．观测倾斜阶跃的上升时间约 3h．分析表明，它可能是地震滑动和无震缓慢蠕动的综合效应．

1　引言

弹性静力学位错理论已成功地用于研究由断层错动引起的位移场、应变场和倾斜场的变化[1-3]，但迄今对应变及倾斜场的讨论还都限于矩形垂直断层．然而，一个实际的天然断层并非都是直立的，而且断层运动的方式通常也都是走滑和倾滑兼而有之，因此，研究倾斜断层的任意断层运动所引起的应变和倾斜变化，进而模拟实际观测的应变和倾斜变化对研究地震的震源过程无疑是必要的．陈运泰等[4]在用形变资料研究邢台地震的震源过程时已导出了拉梅常数不相等的半无限弹性介质中任意倾角的矩形走滑和倾滑断层的地震位移场的解析表达式．本文利用上述表达式做了数值计算，并以 1978 年 7 月 28 日唐山地震（$M_S=7.8$）为例，得到了沿唐山断层的错动所引起的倾斜场变化的空间分布图，讨论了倾斜变化的一些特征；在对泰安等倾斜仪台站观测到的阶跃和理论计算的阶跃进行比较后，作者分析了唐山地震的震源过程．

2　断层错动引起的倾斜场变化的空间分布

以埋在均匀、各向同性和完全弹性的半无限介质中的一个矩形断层为震源的模式[4]利用文献 [5-7] 中分别以不同方法给出的唐山地震的震源参数，计算得到了沿唐山断层的错动引起的倾斜变化的空间分布图像（图1和图2）．应该说明的是，由于按文献 [5] 和 [6] 给出的震源参数得到的倾斜变化的图像基本相同，所以本文只给出了其中的一种（图1）；另外，图2 只是由文献 [7] 中的震源参数计算得到的近场倾斜变化，因为它的远场图像和图1差别不大．

---

[*] 本文发表于《地球物理学报》，1982年，第 **25** 卷，第 3 期，219–226.

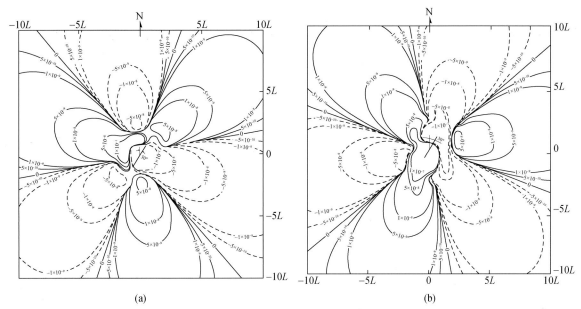

图 1 由文献 [5] 给出的震源参数计算的倾斜变化的空间分布图

(a) $\frac{\partial U_3}{\partial x_1}$，向北为正；(b) $\frac{\partial U_3}{\partial x_2}$，向东为正. L 为断层的半长度

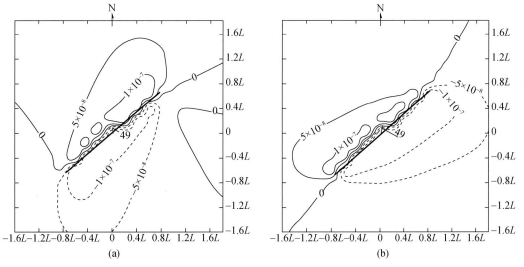

图 2 由文献 [7] 给出的震源参数计算的近场倾斜变化的空间分布图

(a) $\frac{\partial U_3}{\partial x_1}$，向北为正；(b) $\frac{\partial U_3}{\partial x_2}$，向东为正. L 为断层的半长度

表 1 唐山地震的震源参数

编号	断层走向/(°)	断层倾角/(°)	断层长度/km	断层上界/km	断层下界/km	走滑错距/cm	倾滑错距/cm	资料来源
1	N30E	直立	114	0	24	−136	0	文献 [5]
2	N40E	80	140	0	15	−270	0	文献 [6]
3	N49E	76	82	0	34	−459	50	文献 [7]

由图 1，2 所表示的倾斜变化的空间分布图可以得到如下几点看法：

(1) 与 Press 等人[3,8]的工作相比，本文采用的模式更具有一般性. 另外，由于倾斜变化的等值图是在地理坐标系中画出的，因此，便于和观测值进行对比．

（2）尽管在计算时采用的断层参数不大相同，但是，总体看来，除了数值上的差异之外，它们的远场图像是基本相同的，正、负倾斜变化相间分布，呈玫瑰叶状．

（3）断层附近的倾斜变化在方向上略有差别．由文献［5］和［6］给出的震源参数计算得到的断层附近的倾斜变化方向是断层两侧均向东南倾斜，而由文献［7］给出的震源参数计算得到的断层附近的倾斜变化方向是东南盘向西南倾斜，西北盘向东北倾斜．这种倾斜变化在方向上的差异或许可以归结为下面的原因：文献［5］和［6］得到的震源参数是用地震波资料得到的，因此，由这些参数计算的倾斜变化主要反映地震时的倾斜变化；而文献［7］中的震源参数是由大地测量资料反演得到的．我们知道，大地测量资料是在较长的时间尺度内获得的，所以，图2所表示的倾斜变化，可能不仅包括震时的变化，也包括震前和震后的倾斜效应，这种效应在断层附近的区域内比远处明显．另一方面，震时的倾斜变化方向与地震时断层两盘的相对运动方向（右旋）在符号上是一致的．

（4）唐山地震极震区的地壳测深结果（见文献［7］）表明，在地质年代里，沿唐山断裂的构造运动是东南盘下降，西北盘上升，所以，震时的倾斜变化方向说明地震时的断层运动是长期构造运动的继续．

（5）由文献［5］，［6］和［7］中的震源参数计算得到的倾斜变化在数值上的差异（特别是断层附近）也是比较大的．例如，在某些点上，由形变资料反演得到的震源参数计算得到的倾斜变化值可达 5×10^{-5} rad，比由地震波资料求得的震源参数计算的结果大两个量级．这种数值上的差异可能与方向的差异一样，可归因于在"（3）"中叙及的原因．

另外，我们还讨论了断层的几何参数对倾斜变化的影响，其中主要讨论了断层的倾角和埋藏深度特别是下界深度的影响．计算结果表明，走滑断层两侧的倾斜变化是对称的，而当断层运动有一定倾滑分量时，不对称性主要表现在断层附近，而且倾角越小，不对称性似乎也越明显，位于倾向方向上的那盘断层的倾斜变化稍大．

断层宽度的影响主要表现为，当增加断层下界深度时倾斜场空间分布的主节线将远离断层迹线，并使同一地点的倾斜变化的数值增大．这一对倾斜断层得出的结果与Rosenmen和Singh[8]对垂直走滑断层所得到的结果是一致的．因此可以说，在其他断层参数相同的情况下，断层的宽度越大，断层错动的影响范围也越大．

综上可以看出，研究断层运动引起的应变及倾斜场的空间分布，对于如何合理地布设台站以监视断层的运动（譬如，断层的蠕动等）无疑是有益的．因为它可以告诉我们哪些地点是灵敏点，哪些地点不是灵敏点．另外，如果观测点足够多，我们还可以用在地震时观测的倾斜变化反演震源参数．

3 唐山地震引起的倾斜阶跃

根据目前我们收集到的资料，唐山地震时至少有三个台站的倾斜仪，都不同程度地记到了倾斜的突然变化——阶跃．这三个台站是：泰安、沙城和易县台．其中泰安台的倾斜阶跃特别明显．

泰安台位于山东省，地理位置是东经117°07′17″.7，北纬36°12′43″.5，它在本文选用的地理坐标系中的坐标为 $(-334, -90.6, 0)$ km，距断层中点的距离为346km，方位角为190°．

泰安台的倾斜仪为SQ-70型石英水平摆式倾斜仪．仪器安装在长80m、上层覆盖约30m的闪长花岗岩山洞中，洞内温度年变化很小，仪器格值为0.002arc sec/mm，固体潮变化在记录上清晰可见．

唐山地震时，泰安台观测到的倾斜阶跃的方向为北东方向，以东西方向最明显，达 1.1×10^{-7} rad，南北方向的倾斜变化较小，约为 1.4×10^{-8} rad．如果忽略"地震图"或高频成分，则观测的倾斜阶跃的上升时间，即从倾斜开始变化至达到其剩余水平时的时间为3h左右．图3是泰安台地震前后的倾斜记录图．

沙城台和易县台均在河北省境内，安装在这两个台上的倾斜仪于唐山地震时也记到了阶跃．沙城台的仪器与泰安台相同，仪器安装在38m深的竖井内，基底为片麻花岗岩，仪器格值为0.01 arc sec/mm．该台距断层中点的距离为260km，方位角为302°，它在地理坐标系中的坐标为（138，−221，0）km．易县台距断层中点240km，在地理坐标系中的坐标为（19，−238，0）km，方位角为275°．安装在易县台的倾斜仪为金属丝水平摆倾斜仪，置放仪器的山洞为石灰岩，长约200m，上层覆盖200—300m，仪器格值为0.007 arc sec/mm．上述二台站在地震时记录到的倾斜阶跃方向是：沙城台为北东方向，其中南北方向的倾斜变化较明显，达 2.4×10^{-7} rad；易县台为北西方向，以东西方向较明显，达 -1.3×10^{-7} rad．

仔细观察沙城及易县台的倾斜记录就会发现，这两个台倾斜阶跃的上升时间也大约为3h，地震前后的记录复制件如图4和图5．

应该指出，单从图4所示的沙城台的倾斜记录看，似乎沙城的阶跃不如泰安的明显，但是，如果考虑到沙城台倾斜仪的灵敏度比泰安台低5倍这一事实，应当认为沙城的倾斜阶跃也是比较显著的．

另一方面，为了进一步分析观测到的倾斜阶跃，我们计算了理论的倾斜阶跃，因为空间任何一点的倾斜和应变的同震阶跃可以看成是把一个位错面引入半无限弹性空间所引起的倾斜和应变的变化[3,8,9]．为此，我们利用第2节中的位错模式和表1中列出的震源参数，分别计算了泰安、沙城和易县台的理论倾斜阶跃，计算结果与观测值一起列于表2．

比较表2所列观测到的和由不同震源参数计算得到的倾斜阶跃，我们可以得出两点结论：一是计算的理论倾斜阶跃的方向与观测的一致；二是观测值都比计算值大．各台站观测到的地震时倾斜阶跃的变化方向一致，说明观测到的阶跃与地震时的断层错动有关，或者说地震确实引起了剩余倾斜；而观测值比理论值大，则表明观测到的阶跃可能不只是由于地震这一种因素造成的，在其他诸因素中或许包括仪器的影响，以及局部条件的放大作用等，但无论如何，地震时的断层错动是主要的．

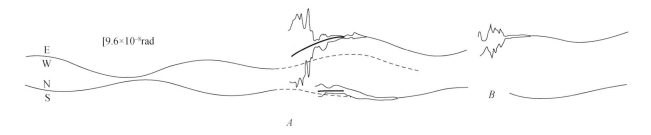

图3 泰安台倾斜仪唐山地震时的记录

A——主震；B——最大余震；虚线表示固体潮的正常变化趋势

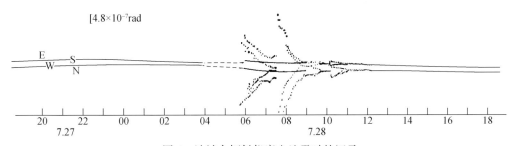

图4 沙城台倾斜仪唐山地震时的记录

虚线表示正常变化趋势

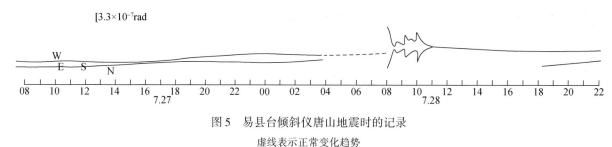

图 5 易县台倾斜仪唐山地震时的记录

虚线表示正常变化趋势

表 2 由位错模式计算得到的倾斜阶跃

编号	阶跃/10^{-8}rad						震源参数来源
	泰安		沙城		易县		
	北—南	东—西	北—南	东—西	北—南	东—西	
1	0.21	0.89	1.50	1.05	1.13	-1.31	[5]
2	1.15	1.60	3.60	1.24	2.77	-4.0	[6]
3	2.52	2.35	7.30	0.70	1.90	-8.0	[7]
观测值	1.4	11.0	24.0	向东	向北	-13.0	

地震引起了倾斜阶跃和其他形式的剩余形变还可以从唐山地震时地下水位的变化得到旁证. 唐山地震前后在震中及其周围地区有许多深井水位的观测数据, 其中许多井的含水层为封闭含水层, 所以这些井的水位较少受外界因素的影响, 对地应变的微小变化十分灵敏. 资料表明, 多数井的水位在地震时表现为大幅度地上升或下降. 例如, 唐山市人民公园内的一口井, 深 250m, 汉沽附近的双桥井, 深 64m, 它们于地震时均表现为突然上升, 甚至自流 (参见文献 [13] 第 108 页图 5—7). 如果我们把地震时井水水位的变化, 看成是由发震断层的错动所引起的水井所在地点体应变的变化造成的[9,10], 那么, 这种体应变的变化同样可通过前面给出的模式计算得到. 计算结果表明, 唐山市人民公园及双桥等处为压应变, 说明理论计算的体应变的变化与地震时井水水位的大幅度上升在符号上是一致的. 因此, 我们可以说, 唐山地震确实引起了剩余形变, 观测的倾斜阶跃是与地震有关的.

4 讨论

前面已经指出, 泰安等倾斜仪台站观测到的倾斜阶跃的上升时间约为 3h. 类似的观测结果还有: 日本的 ASO 台的倾斜仪, 记到了 1976 年 5 月 29 日我国云南龙陵地震引起的倾斜阶跃, 阶跃的上升时间有 20 多分钟[11]; Tomaschek[12] 于英国的温斯福也曾观测到 1951 年 10 月 21 日我国台湾省一次地震 ($M_S=7.3$) 时的倾斜阶跃, 上升时间约 5h. 因此, 地震时观测到的具有相当长上升时间的倾斜阶跃, 可能是某些地震震源过程的一种特征.

根据文献 [7] 得到的结果, 以及震前地下水和垂直形变的观测结果, 在唐山地震之前沿唐山断裂可能已经发生了断层蠕动, 断层的无震滑动必然对断层某些部位的物质起一种加载作用, 从而使应变进一步集中. 唐山地震前较短时期内震中区深井水位的加速变化[13], 可能是蠕动加速的一种表现; 蠕动加载作用的最终结果, 势必导致断层的突然错动, 即地震. 然而, 在地震时断层的地震滑动和无震滑动可能是并存的, 因此, 一些时间常数较长的仪器, 例如, 应变仪、倾斜仪及水位等就可能不仅记下了地震滑动的效应, 同时也记下了时间常数较长的无震断层蠕动的效应. 实际的记录可能是两者叠加的结果, 这种解释与金继宇[14]等关于地震具有慢开始和慢终结的说法是一致的.

感谢朱虎同志及沙城、易县台的同志提供资料.

参 考 文 献

[1] Chinnery, M. A., The deformation of the ground surface fault, *Bull. Seism. Soc. Am.*, Vol. **50**, 355–372, 1961.

[2] Chinnery, M. A., The stress changes that accompany strike-slip faulting, *Bull. Seism. Soc. Am.*, Vol. **53**, 921–932, 1963.

[3] Press, F., Displacement, strains and tilts at teleseismic distances, *J. Geophys. Res.*, Vol. **70**, 2395–2412, 1965.

[4] 陈运泰等, 根据地面形变的观测研究1966年邢台地震的震源过程, 地球物理学报, Vol. **18**, 164–181, 1975.

[5] 张之立等, 唐山地震的破裂过程及其力学分析, 地震学报, Vol. **2**, 111–129, 1980.

[6] Butler, R., Stewart, G. S. and Kanamori, H., The July 27, 1976 Tangshan China earthquake—A complex sequence of intraplate events, *Bull. Seism. Soc. Am.*, Vol. **69**, 207–220, 1979.

[7] 陈运泰等, 用大地测量资料反演的1976年唐山地震的位错模式, 地球物理学报, Vol. **22**, 201–216, 1979.

[8] Rosenmen, M. and Singh, S., Quasi-static strains and tilts due to faulting in a viscoelastic half-space. *Bull. Seism. Soc. Am.*, Vol. **63**, 1735–1752, 1973.

[9] Morenson, C. E., Lee, R. C. and Burford, R. O., Observations of creep-related tilt, strain and water-level changes on the central San-Andreas fault, *Bull. Seism. Soc. Am.*, Vol. **67**, 641–649, 1977.

[10] Wakita, H., Water wells as possible indicators of tectonic strain, *Science*, **189**, 4202, 553–555, 1975.

[11] Nagamune, T., Time characteristics of crustal deformation of the Yunnan earthquake of May 29, 1976, as inferred from tilt recording. *J. Phy. Earth.*, Vol. **35**, 209–218, 1977.

[12] Tomaschek, R., Earth tilts in the British Isles connected with far distant earthquakes, *Nature*, Vol. **176**, 24–25, 1955.

[13] 陈非比等, 唐山地震, 99–111, 地震出版社, 1979.

[14] King Chi-yu, Nason, R. D. and Burford, R. O., Coseismic steps reported on creep meters along the San-Andreas fault, *J. Geophys. Res.*, Vol. **82**, 1655–1662, 1977.

Spatial distribution of residual tilt field and tilt steps due to Tangshan earthquake

Li Xing-Cai and Chen Yun-Tai

Institute of Geophysics, State Seismological Bureau

Abstract Spatial distributions of residual tilt field and tilt steps at Tai-an, Sha-cheng and Yi-xian stations due to Tangshan earthquake were calculated by using the dislocation model for an inclined fault and characteristics of the spatial distribution as well as the effects of source parameters on the tilt are discussed here in detail. The observed tilt steps are as follows: Tai-an, 1.4×10^{-8}(N), 1.1×10^{-7}(E); Sha-cheng, 2.4×10^{-7}(N); Yi-xian, -1.3×10^{-7}(W) rad. A comparison of the observed and theoretical tilt steps indicates that the senses of the tilt steps at these stations are in good agreement. It shows that the observed steps are closely related to the earthquake, but on the other hand, the magnitudes of all the observed steps are larger than the calculated ones. It seems that there are still other effects which would affect the tilt steps observed. The rise-time of observed steps is nearly 3 hours; it seems to support the idea that the observed steps are caused by both seismic and aseismic slips during the process of earthquake.

地球自由振荡*

陈运泰　许忠淮

国家地震局地球物理研究所

地球自由振荡（free oscillations of the Earth）系指地球局部受到某种因素的激发时，地球整体产生的连续振动。地球在受到大地震、火山爆发或地下核爆炸的激发时，会发生整体的振动，并能持续一段时间。

由于地球很大，地球自由振荡的频率很低，振动周期一般为数十秒至数十分钟，通常振动亦很微弱，只有用灵敏的、可探测长周期振动的重力仪、应变地震仪和长周期地震仪等才能记录到。大地震激发的地球长周期自由振荡往往延续几天甚至几个星期才会逐渐消失。

1 研究简史

人类对地球自由振荡的认识是从理论研究开始的。1829年法国泊松（S. D. Poisson）最早研究了完全弹性固体球的振动问题。此后，英国的开尔文（Kelvin）和达尔文（G. H. Darwin）也有重要贡献。尽管理论工作延续多年，但只是在20世纪，地震学的发展使人类对地球内部构造的认识更加清楚以后，理论模式才比较接近真实地球。1952年11月4日堪察加大地震时，美国贝尼奥夫（H. Benioff）首次在他自己设计制作的应变地震仪上发现周期约为57min的长周期振动。1960年5月22日智利大地震时，贝尼奥夫和其他几个研究集体都观测到多种频率的谐振振型。地球长周期自由振荡的真实性遂被最后证实。至今已观测到的本征振荡频率已达1000多个，其中球型振荡约占三分之二，环型振荡约占三分之一。图1为由设在美国加利福尼亚伊沙贝拉台的应变地震仪记录得到的两个地震激发的地球自由振荡的功率谱密度曲线。δ是应变地震仪水平轴线同台站至震中大圆弧之间的夹角。

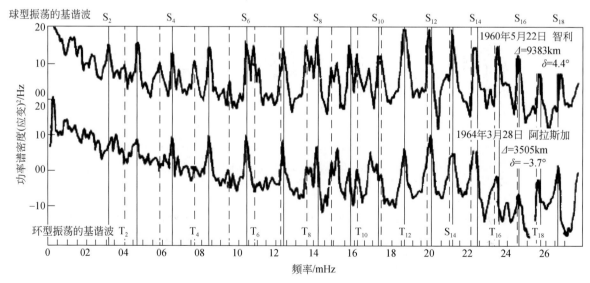

图1　地球自由振荡的功率谱密度曲线

---

* 本文发表于中国大百科全书出版社出版的《中国大百科全书·固体地球物理学卷》一书，1985年，115–116.

2 理论

地球自由振荡的理论是在适当的定解条件下求解确定地球振动的微分方程组. 方程组中包含 4 个微分方程式, 即: 表示牛顿定律的动量守恒方程; 表示质量守恒的连续方程; 表示万有引力定律的泊松方程; 表示介质弹性的弹性方程. 振动引起地球形变后必须满足的定解条件是: ①振动在地心处有限; ②地球外表应力为零; ③在地球表面和地球内部分界面上重力位及其梯度连续; ④在地球内部的固体和固体间分界面上位移和应力连续; ⑤在地球内部的固体和液体分界面上, 法向位移连续, 切向应力为零. 通常是在以地心为原点的球极坐标系中用驻波法求上述问题的解.

满足上述方程组和边界条件的振动只能取一些特定的频率, 称为地球的本征频率, 相应的本征角频率通常用 $_n\omega_l^m$ 来表示, 其数值取决于 3 个整数指标 n, l 和 m. 与本征频率相应的振动称做本征振荡. 每一种本征振荡都对应一种驻波, 是地球的一种谐振形式. n 代表某一振型振动位移沿地球半径方向的节点数; $l-|m|$ 表示位移在余纬方向的节点数 ($|m| \leq l$); $2|m|$ 表示位移在经度变化方向的节点数. n 最小时 (0 或 1) 的本征频率称基频, 其余称谐频.

本征振荡分成两类. 一类叫球型振荡, 通常用 $_nS_l^m$ 表示. 地球作球型振荡时, 其质点位移既有半径方向的分量, 也有水平分量. 这是一种无旋转振动. 重力仪、应变地震仪和长周期地震仪均可记录到这种振动. 另一类叫环型振荡, 通常用 $_nT_l^m$ 表示. 地球作环型振荡时, 各质点只在以地心为球心的同心球面上振动, 位移无径向分量, 地球介质只产生剪切形变, 无体积变化, 地球的重力场不受扰动, 重力仪记录不到这种振荡. 图 2 绘出 3 种最简单的振型 $_0S_0^0$, $_0S_2^0$ 和 $_0T_2^0$ 的振动方式. 振型 $_nT_0^0$ 无意义, 因为它表示地球各点位移恒为零. 此外, 地震及其他内力源激发不起 $_0S_1^m$ 和 $_0T_1^0$. $_0S_1^m$ 表示地球整体像刚体一样在太空中振动, 按动量守恒定律, 任何内力都不能激发这种振动. $_0T_1^0$ 表示地球自转角速度有变化, 按角动量守恒定律, 任何内力也都不可能激发这种振动. 理论上, 地球的谐振振型有无穷多个, 实际的振动就是这无穷多个振型叠加的总结果.

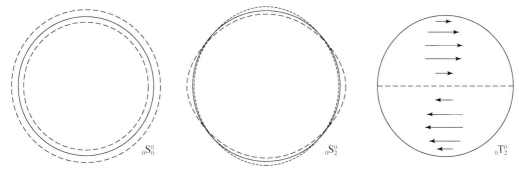

图 2 地球本征振荡的三种简单的振型

对于球对称的球体, n 和 l 相同而 m 不同 ($m=0$, ± 1, ± 2, $\cdots$, $\pm l$, 共 $2l+1$ 个) 的振型都有相同的谐振频率, 这种情形称为振型的简并. 地球的自转效应使地球的振荡频率对 m 不再简并. 在振动的频谱图上, 每条与某 n 和 l 相应的谐振谱线分裂为 $2l+1$ 条, 它们等间距对称地分布在 $m=0$ 谱线的两侧, 这与原子光谱线在磁场中发生分裂的塞曼效应十分相似. 自转还会使质点振动方向发生像傅科摆一样的变化, 从而导致球型振荡与环型振荡发生耦合. 真实地球并非球体, 而是接近于旋转椭球体. 地球的椭率效应使频谱线产生很微小的移动, 造成分裂谱线的不对称性. 对低频振型, 自转效应比椭率效应大得多; 高频振型反之. 实际观测中因有干扰, 不易发现椭率效应.

3 应用

计算不同地球模式产生的自由振荡频率，并与观测频率对比，可以检验并改善地球模式，从而研究地球内部的结构，与用地震体波研究地球内部结构的方法互为补充．测定相继时间间隔内地球自由振荡频谱谱峰的平均能量，或测定谐振谱峰的宽度（通常以能量降至谱峰能量的一半时相应的频率变化来量度），可以研究振动能量在地球内部的衰减情况，并进而研究地球介质的非弹性性质．此外，根据给定的地球模式和尝试的震源参数计算自由振荡的振幅和相位，然后与相应的观测值对比，可以确定地震的震源参数．

参 考 书 目

E. R. Lapwood and T. Usami, *Free Oscillations of the Earth*, Cambridge Univ. Press, London, 1981.

Static shear crack with a zone of slip-weakening*

Y. T. Chen and L. Knopoff

Institute of Geophysics and Planetary Physics, University of California, Los Angeles

Summary As a prologue to our investigation of the nature of sequences of earthquake events, we investigate the problem of a finite, 2-D, antiplane, static shear crack for two different models of a slip-weakening zone near the crack tip. The crack is imbedded in an infinite elastic medium. In the limit as the size of the slip-weakening zone vanishes, the problem becomes one considered by Knopoff for a finite shear crack in a uniform field. We solve for the displacement and the stress for representative values of dimensionless stress drop for the two models. Our results indicate that these quantities are not strongly dependent on the details of the constitutive relation in the transition zone. With decreasing dimensionless stress drop, the crack length increases monotonically and the size of the slip-weakening zone decreases monotonically. These results can be interpreted as the critical conditions for the further extension of an initial earthquake fault. In this interpretation, these results imply that the initiation of an earthquake is not only controlled by the stress drop on the fault plane, but also by the size of a pre-existing fracture: an earthquake can be initiated from a small pre-existing fracture if the stress drop on the initial fault is large enough; conversely it can be initiated under the influence of a lower stress drop, if the initial fault is large enough. No solution to the static problem exists if the dimensionless stress drop is greater than a certain value; this maximum dimensionless stress drop corresponds to a minimum crack length for which the slip-weakening zone occupies the entire crack. In the small dimensionless stress drop or small-scale yielding limit, the dimensionless half-length of the crack is inversely proportional to the square of the dimensionless stress drop; the dimensionless length of the slip-weakening zone has been calculated. We generalize the formula for the effective shear fracture energy obtained by Palmer & Rice and Andrews to include the effect of the size of the slip-weakening zone. Our results indicate that because the influence of the size of the slip-weakening zone has not been taken into account, earlier estimates of the effective shear fracture energy may have been overestimated by as much as 67 percent.

1 Introduction

Our interest in the model to be described in this and succeeding papers has as its genesis our effort to understand the mechanics of repetitive earthquake sequences such as aftershocks and earthquake swarms, as well as the circumstances under which isolated earthquakes may fail to be followed by aftershocks. In the course of these investigations we have been obliged to postulate, as others have before us, that the stress concentrations near the edges of fracture surfaces are relieved by redistribution on a time-scale of the order of the time between the events in the sequences. The mechanism for the redistribution is viscous flow. Our analysis of this process is tripartite: first, we

---

* 本文发表于 *Geophys. J. R. astr. Soc.*, 1986 年, 第 **87** 卷, 第 3 期, 1005–1024.

consider the nature of the stress concentration at the edge of a static fracture in an elastic medium, since we argue that the rates of extension of individual fractures are so high compared with the viscous relaxation rates, that the medium can be considered as elastic for these purposes. Secondly, we consider the problem of the relaxation of the stresses that are concentrated in the vicinity of the edge of the crack by taking the medium to be viscoelastic and calculating the nature of the slow extension of a crack against the viscoelastic drag forces; the identification of the rheological model that is appropriate to produce additional earthquakes in the sequence in finite time is crucial to this part of the argument. Third, we consider the details of the extension of cracks in viscoelastic media in the presence of a statistical distribution of those physical properties that control the acceleration and deceleration of the rate of extension of cracks. Rapid acceleration into the elastic regime generates identifiable earthquakes; deceleration into a subseismic or 'slow-earthquake' regime provides a description of the system in the time interval between earthquakes. We offer the complete discussion in three parts. Below we consider the first of these.

As a final introductory comment, we note that Yamashita (1980, 1981) has described a programme similar to ours, His point of departure was a description of the displacements on a crack if the stress drops are given, both on the fractured segment of the crack and its prolongation; this is the Dugdale (1960) model of the stresses and displacements in the vicinity of the crack tip. The model involves an inhomogeneous stress drop distribution on the crack plane. Yamashita (1981) comments that the diversity of earthquake processes that result may depend on the distributions of the inhomogeneity of the stress-drop and the critical shear displacement. In these papers we investigate the influence on these calculations if (1) we adopt a slip-weakening model for the zone at the edge of the crack within which the stresses are gradually relieved and (2) we assume that the stress-drop and cohesion distributions are both non-uniform.

2 Slip-weakening model

An earthquake is often modelled as a dynamic planar shear-crack in an otherwise homogeneous elastic continuum. In this model it is assumed that slip begins at that point on the plane of the crack where the local shear stress becomes equal to the breaking strength; the crack grows and on the ruptured part of the plane the stress decreases abruptly from its value before fracture began. If the stress drops abruptly at the edges of the crack, the stress just beyond the edges of the crack is singular and varies reciprocally as the square root of the distance from the edge. Since a real material cannot sustain an infinite stress within the framework of linear elasticity, the assumption that the stress drop is abrupt must be abandoned. This singularity of the shear stress can be removed by introducing a slip-weakening model for the constitutive relations (Barenblatt, 1959, 1962; Ida, 1972; Palmer & Rice, 1973; Andrews, 1976; Burridge, Conn & Freund, 1979; Kostrov & Das 1982). The slip-weakening model is an extension to the shear crack of Barenblatt's (1959, 1962) model of a zone of cohesion for a tensile crack, and is only one of a number of ways (Leonov & Panasyuk, 1959; Dugdale, 1960) in which we can mitigate the singularity at the edge of an abruptly terminated crack.

In the slip-weakening model it is assumed that the strength of the material in a slip-weakening zone is a function of the amount of slip. As a result, the zone of variation of strength is not confined to an infinitesimal region at the crack edges, but instead is distributed over a finite region near the edges of the crack. As before, the slip starts at a point in the plane where the local stress on the slip plane is equal to the breaking strength. However, in this case, the stress that resists slip decreases gradually to the dynamic friction, as the slip increases gradually up to a critical value. The stress required to sustain slip beyond the critical value is equal to the dynamic friction.

We can interpret this model either from the point of view of fracture of fresh rock or of a fracture on a pre-existing fault. In so far as earthquake events are concerned, we can assume that an earthquake fracture can be considered as having two parts, one being a main fault where the stress has dropped to its dynamic frictional stress, and the other, a slip-weakening zone adjacent to the main fault. Inspection of epicentral surface breakage of large earthquakes indicates that neither the main fault nor the fracture zone are planar. Instead, they are geometrically irregular. Since the fracture zone of the main fault and its extension are confined to a narrow band, we assume that they are planar for mathematical reasons and that the geometrical irregularities of the fracture surface are projected on to the plane in terms of some statistical variation of physical properties in the plane. The zone of variation of strength of the main fault is actually a region full of irregularly distributed microcracks, cracks, joints, secondary faults, etc. The slip-weakening zone is an appropriate model for representing the collective effect of these geometric heterogeneities, since the degree of microcracking, jointing, etc., decreases with distance from the main part of the fault.

Slip-weakening models, as models of earthquake instability, have been considered by many authors (Ida 1972; Palmer & Rice1973; Andrews 1976; Rice 1980; Kostrov & Das 1982). Palmer & Rice (1973) estimated the size of the slip-weakening zone for the case of small-scale yielding, i.e. the case in which the length of the main part of the fault can be taken to be large compared with the size of the slip-weakening zone. Palmer & Rice found that the size of the slip-weakening zone is proportional to a measure of the slip in the transition zone and inversely proportional to the difference between the peak and residual stresses. Since the model of Palmer & Rice is that of a semi-infinite crack, they have not calculated the influence of crack length on the size of the slip-weakening zone. Through the use of finite difference methods, Andrews (1976) calculated the distribution of the displacement and the stress on the crack plane for a dynamic antiplane shear crack. Andrews found that the size of the slip-weakening zone decreases as the length of the crack increases for the case for which the rupture velocity increases, but he did not indicate the form of the relationship, nor did he discuss the relationship between the size of the transition zone and the rupture velocity.

To investigate the relationship between the main part of the fault and its associated slip-weakening zone, we consider a 2-D antiplane static shear crack with a zone of slip-weakening in an otherwise homogeneous elastic medium. In a second paper, we will consider the interaction between this initial crack and the surrounding medium by taking the anelasticity of the surrounding material into account and study the early history of the subsequent earthquake rupture. In the present paper, we confine our analysis to the static condition of stress and displacement after an earthquake has occurred. In this condition the fault is in a critical equilibrium state. After a fracture event, there will be a slow redistribution of the stress due to the anelasticity of the surrounding material. Thus the rupture will proceed with very slow velocity at first, and then will accelerate up to near-seismic velocities; under the latter condition the medium is assumed to behave elastically. When this breakout phase is reached, the crack becomes a source which radiates seismic waves and a new earthquake is identified as occurring within the region of slip and the region of stress concentration outside it. Thus, the static conditions to be discussed in the present paper can also be considered to be the initial conditions for stress relaxation culminating in a subsequent earthquake event.

We will show that the size of the slip-weakening zone is a monotonically decreasing function of the length of the crack. These quantities are related through the dimensionless stress drop, which is defined as the ratio of the stress drop to the difference between the breaking strength and the sliding friction. As the dimensionless stress drop becomes smaller, the slip-weakening zone becomes shorter and the crack longer. These results imply that a larger fracture is usually associated with a smaller slip-weakening zone and, to initiate an earthquake from a smaller fault, a larger dimensionless stress drop on the crack plane is required. These results also imply that a large stress drop is not the only condition for the initiation of an earthquake. These statements are obvious for a model with a constant stress drop; they are not as obvious for the models of the slip-weakening zone we consider here. In principle, it is also possible to initiate an earthquake under the influence of lower stress drop on the crack, if a large enough fracture due to an earlier earthquake has taken place. As a consequence of these results, we show that the effective shear fracture energy as well as the critical displacement of the slip-weakening zone would probably be over-estimated if the effect of slip-weakening is not taken into account. The small-scale yielding case is a limiting case of our results: we find that the small-scale yielding approximation is applicable if the stress drop on the crack is very small compared to the difference between the static and dynamic shear stresses.

3 Slip-weakening shear crack with specified stress drop

We consider a 2-D antiplane shear crack (Fig. 1). The displacement $W(x, y)$ is parallel to the z-axis, and the crack is a segment of the x-axis, $-c<x<c$ in the plane $y=0$. The shear stress on the crack, $\tau(x)$ is symmetric about $x=0$. Because of symmetry, we only consider the region $x>0$. In the slip-weakening model we assume that a slip-weakening zone $a<x<c$, $y=0$ is located near the edge of the crack. Slip begins at a point on the plane $y=0$ where the local shear stress $\tau(x)$ on the slip plane is equal to τ_S (Fig. 2). The shear stress required to sustain slip decreases as the amount of slip on the crack, $W(x) \equiv W(x, 0)$, increases up to some critical amount δ_c; the shear stress required to sustain slip beyond the critical value is τ_d. The quantities τ_s, and τ_d are the static and dynamic frictional stresses. In the limit $(c-a) \to 0$, we have the usual model of an abrupt stress drop at the edge of the crack, with a correspondingly infinite shear stress just beyond the edge.

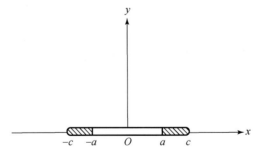

Figure 1 Schematic antiplane shear crack with slip-weakening zone $|a|<x<|c|$

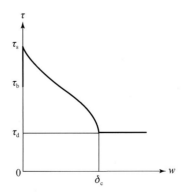

Figure 2 Stress versus displacement for the slip-weakening model. τ_b, τ_s and τ_d are the pre-stress, static and dynamic frictional shear stresses and δ_c is the critical displacement

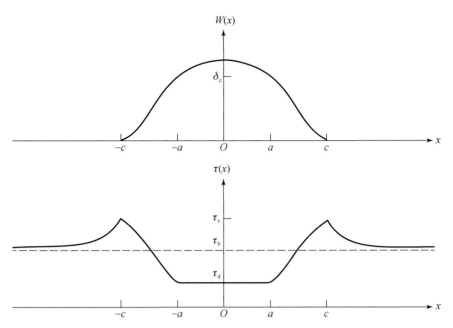

Figure 3 Schematic slip and stress distributions for the slip-weakening model

In the version of the slip-weakening model we consider in this section, we assume that the stress in the transition zone decreases linearly with distance from the edge of the crack. Thus, the stress distribution within the fracture zone on the crack plane $y=0$ is (Fig. 3)

$$\tau(x) = \tau_d, \qquad 0<x<a,$$
$$= \tau_s - (\tau_s - \tau_d)\left(\frac{c-x}{c-a}\right), \quad a<x<c, \tag{1}$$

where τ_d and τ_s are the dynamic and static frictional stresses in the plane $y=0$. In general, τ_s and τ_d are functions of position, but in the present case, we assume they and the pre stress τ_b are constants everywhere on the crack. In this problem, the slip $W(x)$ on the crack and the stress outside the crack are unknown. The stress drop $\Delta\tau(x)$ is

$$\Delta\tau(x) = \tau_b - \tau(x), \quad 0<x<c, y=0; \tag{2}$$

and beyond the edge of the crack, the stress after fracture is τ_b plus the redistributed stress.

The static displacement $W(x, y)$ satisfies Laplace's equation

$$\frac{\partial^2 W}{\partial x^2} + \frac{\partial^2 W}{\partial y^2} = 0, \tag{3}$$

and the excess of the shear stress over the initial stress is

$$\sigma_{yz}(x,y) = \mu \frac{\partial W}{\partial y}, \tag{4}$$

where μ is the rigidity. The cosine transform of (3) is

$$\frac{d^2 \hat{W}}{dy^2} - \xi^2 \hat{W} = 0, \tag{5}$$

which has as its solution

$$\hat{W}(\xi, y) = A(\xi)\exp(-\xi|y|). \tag{6}$$

The inverse transform of the latter expression is

$$W(x,y) = (2/\pi)^{1/2} \int_0^\infty A(\xi)\exp(-\xi|y|)\cos(\xi x)\,\mathrm{d}\xi, \tag{7}$$

and from (4) we have

$$\sigma_{yz}(x,y) = -\mu(2/\pi)^{1/2} \int_0^\infty \xi A(\xi)\exp(-\xi|y|)\cos(\xi x)\,\mathrm{d}\xi. \tag{8}$$

For the present problem, we have

$$W(x,0) = 0, \qquad x>c, \tag{9}$$

$$\sigma_{yz}(x,0) = -\Delta\tau(x), \quad 0<x<c. \tag{10}$$

As a consequence of equations (7) to (10), we can write

$$\int_0^\infty B(\xi) J_{-1/2}(\xi x)\,\mathrm{d}\xi = 0, \qquad x > c, \tag{11}$$

$$\int_0^\infty \xi B(\xi) J_{-1/2}(\xi x)\,\mathrm{d}\xi = \frac{\Delta\tau(x)}{\mu x^{1/2}}, \quad 0 < x < c, \tag{12}$$

where

$$B(\xi) = \xi^{1/2} A(\xi) \tag{13}$$

and $J_{-1/2}(\mu)$ is the usual Bessel function. The solution of equations (11) and (12) is (Sneddon 1951; Sneddon & Lowengrub 1969)

$$B(\xi) = \frac{1}{\mu}(2\xi/\pi)^{1/2} \int_0^c \eta J_0(\eta\xi)\,\mathrm{d}\eta \int_0^\eta \frac{\Delta\tau(x)}{(\eta^2 - x^2)^{1/2}}\mathrm{d}x. \tag{14}$$

From (13), $A(\xi)$ is

$$A(\xi) = \int_0^c g(\eta) J_0(\eta\xi)\,\mathrm{d}\eta, \tag{15}$$

$$g(\eta) = \frac{\eta}{\mu}(2/\pi)^{1/2} \int_0^\eta \frac{\Delta\tau(x)}{(\eta^2 - x^2)^{1/2}}\mathrm{d}x. \tag{16}$$

From (7), and after an exchange of the order of integration, we obtain an expression for the displacement on the crack:

$$W(x) \equiv W(x,0) = (2/\pi)^{1/2} \int_0^c g(\eta)\,\mathrm{d}\eta \int_0^\infty J_0(\eta\xi)\cos(\xi x)\,\mathrm{d}\xi. \tag{17}$$

Since

$$\int_0^\infty J_0(\eta\xi)\cos(\xi x)\,\mathrm{d}\xi = \frac{1}{(\eta^2 - x^2)^{1/2}}, \qquad \eta > x,$$
$$= 0, \qquad \eta < x, \tag{18}$$

we obtain

$$W(x) = (2/\pi)^{1/2} \int_x^c \frac{g(\eta)}{(\eta^2 - x^2)^{1/2}}\mathrm{d}\eta, \qquad x < c,$$
$$= 0, \qquad x > c. \tag{19}$$

In much the same manner, the redistributed stress, $\sigma(x) \equiv \sigma_{yz}(x,0)$, can also be written as

$$\sigma(x) = \mu(2/\pi)^{1/2} x \int_0^x \frac{g(\eta)}{(x^2 - \eta^2)^{3/2}}\mathrm{d}\eta, \qquad x < c,$$
$$= \mu(2/\pi)^{1/2} x \int_0^c \frac{g(\eta)}{(x^2 - \eta^2)^{3/2}}\mathrm{d}\eta, \qquad x > c, \tag{20}$$

where we have used the definite integral

$$\int_0^\infty \xi J_0(\eta\xi)\cos(\xi x)\,\mathrm{d}\xi = \frac{x}{(x^2 - \eta^2)^{3/2}}, \qquad 0 < \eta < x,$$
$$= 0, \qquad 0 < x < \eta. \tag{21}$$

The redistributed stress in the region $x>c$ is the difference between the total or final stress and the initial stress. If we substitute (16) into (19) and (20), we obtain

$$W(x) = \frac{1}{\mu}\int_0^c \Delta\tau(\zeta) Q(c,x,\zeta) \,d\zeta, \qquad x < 0, \tag{22}$$
$$= 0, \qquad x > c,$$

where

$$Q(c,x,\zeta) = \frac{1}{\pi}\ln\left|\frac{(c^2-x^2)^{1/2}+(c^2-\zeta^2)^{1/2}}{(c^2-x^2)^{1/2}-(c^2-\zeta^2)^{1/2}}\right|, \tag{23}$$

and

$$\sigma(x) = \int_0^c \Delta\tau(\zeta) R(c,x,\zeta) \,d\zeta, \qquad x > c, \tag{24}$$

where

$$R(c,x,\zeta) = \frac{2}{\pi}\frac{x}{(x^2-\zeta^2)}\left(\frac{c^2-\zeta^2}{x^2-c^2}\right)^{1/2}. \tag{25}$$

We have now completed the task defined at the beginning of this section. For a given stress drop on the entire crack, the displacement and stress distributions can be evaluated. For the case of constant stress drop, i.e. in the absence of a slip-weakening zone,

$$\Delta\tau(x) = \tau_b - \tau_d = \text{constant}, \qquad 0<x<c, \tag{26}$$

(22) and (24) can be integrated in closed form. The displacement and the stress are

$$W(x) = \frac{(\tau_b-\tau_d)}{\mu}(c^2-x^2)^{1/2}, \qquad 0<x<c, \tag{27}$$

$$\sigma(x) = (\tau_b-\tau_d)\left[\frac{x}{(x^2-c^2)^{1/2}}-1\right], \qquad x>c, \tag{28}$$

which are the same as the results of Knopoff (1958).

For the specific stress distribution (1), the result of the direct application of (2) and (22—25) is

$$W(x) = \frac{(\tau_s-\tau_d)}{\mu}[U(c,x,a)-(c^2-x^2)^{1/2}S(\epsilon,\gamma)], \qquad x<c, \tag{29}$$

$$\tau(x) = \tau_d + (\tau_s-\tau_d)\left[T(c,x,a) - \frac{x}{(x^2-c^2)^{1/2}}S(\epsilon,\gamma)\right], \qquad x>c, \tag{30}$$

where

$$U(c,x,a) = \frac{1}{(c-a)}\left[\frac{1}{\pi}(c^2-x^2)^{1/2}(c^2-a^2)^{1/2}-\frac{(x^2+a^2)}{2\pi}\ln\left|\frac{(c^2-x^2)^{1/2}+(c^2-a^2)^{1/2}}{(c^2-x^2)^{1/2}-(c^2-a^2)^{1/2}}\right|\right.$$
$$\left.+\frac{xa}{\pi}\ln\left|\frac{a(c^2-x^2)^{1/2}+x(c^2-a^2)^{1/2}}{a(c^2-x^2)^{1/2}-x(c^2-a^2)^{1/2}}\right|\right], \tag{31}$$

$$T(c,x,a) = \frac{2}{\pi(c-a)}\left[x\tan^{-1}\left(\frac{c^2-a^2}{x^2-c^2}\right)^{1/2} - a\tan^{-1}\frac{x}{a}\left(\frac{c^2-a^2}{x^2-c^2}\right)^{1/2}\right], \tag{32}$$

and

$$S(\epsilon,\gamma) = \frac{2[(1-\epsilon^2)^{1/2}-\epsilon\cos^{-1}\epsilon]}{\pi(1-\epsilon)}-\gamma, \tag{33}$$

with

$$\epsilon = \frac{a}{c}, \quad (34)$$

and γ is the dimensionless stress drop,

$$\gamma = \frac{\tau_b - \tau_d}{\tau_s - \tau_d}. \quad (35)$$

The quantities a and c are related by the condition for finiteness of the stress at the edge of the crack. Thus $S(\epsilon, \gamma) = 0$ and hence

$$\gamma = \frac{2[(1-\epsilon^2)^{1/2} - \epsilon \cos^{-1}\epsilon]}{\pi(1-\epsilon)}. \quad (36)$$

The solution to equation (36) is a single-valued, monotonic function.

Equation (36) is a condition that the sum of the stress intensity factors due to both the stress drop and the cohesive stress be zero. In the absence of the slip-weakening zone, the stress intensity factor, K_{III}, for the constant stress drop case is

$$K_{\mathrm{III}} = \lim_{x \to c^+} (2\pi)^{1/2} (x-c)^{1/2} \sigma(x). \quad (37)$$

By virtue of (28), we have

$$K_{\mathrm{III}} = (\pi c)^{1/2} (\tau_b - \tau_d). \quad (38)$$

The cohesive modulus for this slip-weakening model is

$$K_c = (\pi c)^{1/2} (\tau_s - \tau_d) \frac{2[(1-\epsilon^2)^{1/2} - \epsilon \cos^{-1}\epsilon]}{\pi(1-\epsilon)}. \quad (39)$$

Therefore, (36) is simply the fracture criterion $K_{\mathrm{III}} = K_c$.

The displacement at the inner edge of the slip-weakening zone, $x = a$, is the critical value δ_c. From (29) and (31)

$$\frac{\mu \delta_c}{\tau_s - \tau_d} = c\nu(\epsilon) = U(c, a, a), \quad (40)$$

where

$$\nu(\epsilon) = \frac{(1 - \epsilon^2 + 2\epsilon^2 \ln \epsilon)}{\pi(1-\epsilon)}. \quad (41)$$

From (36) we can solve for ϵ as a function of the stress drop γ; if we substitute this result into (40) we get $c(\gamma)$. Conversely, we can solve for ϵ as a function of the half-length c from (40), and then substitute this result into (36) to get $\gamma = \gamma(c)$. The equation $\gamma = \gamma(c)$ is the fracture criterion $K_{\mathrm{III}} = K_c$ restated in terms of the stress drop on the crack. The static crack of half-length c exists only when $\gamma = \gamma(c)$. If $\gamma < \gamma(c)$, no crack exists for this slip-weakening model; if $\gamma > \gamma(c)$, the crack begins to extend instantaneously.

Fig. 4 displays the numerical values of the crack dimension at its outer edge c and the size of the slip-weakening zone $c_1 = c - a$, as a function of the dimensionless stress drop γ; we have normalized the ordinates by the quantity $\lambda = \mu \delta_c / (\tau_s - \tau_d)$. For this slip-weakening model there is a critical stress drop γ_m above which no solution exists. This maximum stress drop corresponds to the limiting case in which the half-length of the crack is just equal to the size of the slip-weakening zone, i.e. $\epsilon = 0$. If we set $\epsilon = 0$ in equations (36) and (40) we obtain $\gamma_m = 2/\pi$ and $c(\gamma_m) = \pi\lambda$, respectively.

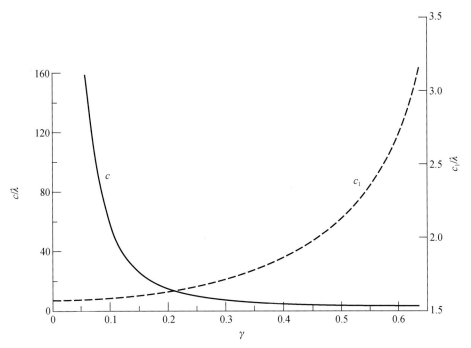

Figure 4 Half-length c and length of slip-weakening zone c_1 as functions of γ

For the static or equilibrium crack, we calculate $a(\gamma)$ and $c(\gamma)$ from (34), (36) and (40) and then $W(x)$ and $\tau(x)$ directly from (29) and (30). The distributions of the displacement and the stress on the crack plane are shown in Fig. 5 for selected values of the dimensionless stress drop.

In the 'small-scale yielding limit', the size of the slip-weakening zone is much smaller than the half-length of the crack $(c-a) \ll c$. From (36) and (40) we obtain

$$\epsilon = 1 - \frac{9\pi^2}{32}\gamma^2, \qquad (42)$$

and

$$c = \left(\frac{16}{9\pi}\right)\frac{\lambda}{\gamma^2}, \qquad (43)$$

in this limit. Therefore, the size of the slip-weakening zone is simply

$$c_1 = \frac{\pi}{2}\lambda. \qquad (44)$$

The small-scale yielding limit of this slip-weakening model has been discussed by Palmer & Rice (1973) for the case of a semi-infinite crack. Equations (43) and (44) are identical to the formulae for the fracture propagation criterion and the size of the end zone obtained by Palmer & Rice [1973, equations (41) and (45)]; they have been written here in our notation.

The antiplane problem we have considered in this section is similar to that solved by Kostrov & Das (1982) for the in-plane case. Our version differs from theirs in two additional respects: instead of considering a 2-D fault consisting of a periodic array of in-plane shear cracks, we have analysed a simple 2-D antiplane shear crack. In our study we have adopted the slip-weakening model for the transition zone, within which the stresses are gradually relieved, while Kostrov & Das (1982) adopted the Leonov-Panasyuk and Dugdale fracture criterion.

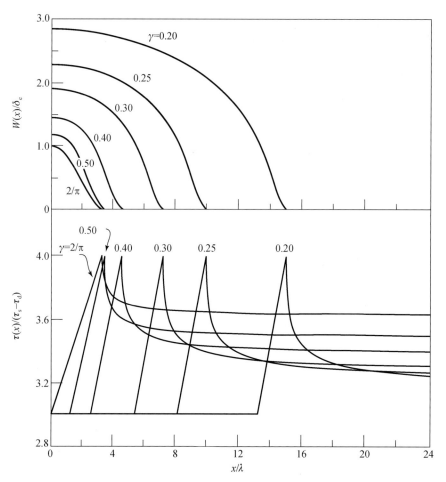

Figure 5 Displacement $W(x)$ and stress $\tau(x)$ as a function of position on the crack, for selected values of the stress drop, $\tau_d = 3.0$. The displacement is scaled by δ_c, the stress by $(\tau_s - \tau_d)$, and the distance along the crack by λ

4 Unspecified stress drop in the transition zone

The results obtained above, for the problem with specified stress drop everywhere on the crack, provide a basis for the analysis of the more difficult problem of the slip-weakening model with unspecified stress drop in the transition zone. In this case, we assume that there is a linear relationship between the stress $\tau(x)$ and the slip $W(x)$, measured from the initial equilibrium state in the transition zone. The stress distribution on the crack plane $y = 0$ is

$$\tau(x) = \tau_d, \qquad 0 < x < a,$$
$$= \tau_s - (\tau_s - \tau_d)\frac{W(x)}{\delta_c}, \qquad a < x < c, \qquad (45)$$

In this case the curve of Fig. 2 is straightened out. As before, we assume the stress in the plane $y = 0$ before fracture is τ_b. In this model, the stress drop is only given on the inner segment of the x-axis, $0 < x < a$. In this case, the displacement and hence the stress are both unknown in the slip-weakening zone. Substitution of (45) and (2) into (22) gives

$$\frac{\mu W(x)}{\tau_s - \tau_d} = \phi(c, x, a) - (c^2 - x^2)^{1/2} S(c, a, \gamma) + \int_a^c \frac{W(\zeta)}{\delta_c} G(c, x, \zeta) \, d\zeta, \qquad (46)$$

where

$$S(c,a,\gamma) = \chi(c,a) - \gamma + \int_a^c \frac{W(\zeta)}{\delta_c} K(c,\zeta) \, d\zeta, \qquad (47)$$

and, as before, γ is the dimensionless stress drop (35). The kernels $G(c,x,\zeta)$, $\phi(c,x,\zeta)$, $K(c,\zeta)$ and $\chi(c,\zeta)$ are defined as follows:

$$G(c,x,\zeta) = \frac{\partial \phi(c,x,\zeta)}{\partial \zeta} = \frac{1}{\pi} \ln \left| \frac{(c^2-x^2)^{1/2}+(c^2-\zeta^2)^{1/2}}{(c^2-x^2)^{1/2}-(c^2-\zeta^2)^{1/2}} \right| - \frac{2}{\pi} \left(\frac{c^2-x^2}{c^2-\zeta^2} \right)^{1/2}, \qquad (48)$$

$$\phi(c,x,\zeta) = \frac{\zeta}{\pi} \ln \left| \frac{(c^2-x^2)^{1/2}+(c^2-\zeta^2)^{1/2}}{(c^2-x^2)^{1/2}-(c^2-\zeta^2)^{1/2}} \right| - \frac{x}{\pi} \ln \left| \frac{\zeta(c^2-x^2)^{1/2}+x(c^2-\zeta^2)^{1/2}}{\zeta(c^2-x^2)^{1/2}-x(c^2-\zeta^2)^{1/2}} \right|, \qquad (49)$$

$$K(c,\zeta) = \frac{\partial \chi(c,\zeta)}{\partial \zeta} = -\frac{2}{\pi(c^2-\zeta^2)^{1/2}}, \qquad (50)$$

$$\chi(c,\zeta) = \frac{2}{\pi} \cos^{-1}\left(\frac{\zeta}{c}\right). \qquad (51)$$

The total stress outside the fractured region is

$$\tau(x) = \tau_b + \sigma(x), \qquad (52)$$

where $g(x)$ is the redistributed stress due to the fracture. Substitution of (45) and (2) into (24) gives the expression for the stress outside the broken region as

$$\frac{\tau(x)-\tau_d}{\tau_s-\tau_d} = \psi(c,x,a) - \frac{x}{(x^2-c^2)^{1/2}} S(c,a,\gamma) + \int_a^c \frac{W(\zeta)}{\delta_c} H(c,x,\gamma) \, d\zeta, \qquad (53)$$

where

$$H(c,x,\zeta) = \frac{\partial \psi(c,x,\zeta)}{\delta \zeta} = -\frac{2}{\pi} \frac{x}{(x^2-\zeta^2)} \left(\frac{x^2-c^2}{c^2-\zeta^2} \right)^{1/2}, \qquad (54)$$

$$\psi(c,x,\zeta) = \frac{2}{\pi} \tan^{-1}\left[\frac{x}{\zeta} \left(\frac{c^2-\zeta^2}{x^2-c^2} \right)^{1/2} \right]. \qquad (55)$$

In the limit as $x \to c^+$, the stress at the edge of the crack will have a reciprocal square root singular dependence on distance from the edge of the crack, if the factor $S(c,a,\gamma) \neq 0$. We set $S=0$ in (47) and obtain the condition for finiteness of the stress at the edge of the crack

$$\chi(c,a) - \gamma + \int_a^c \frac{W(\zeta)}{\delta_c} K(c,\zeta) \, d\zeta = 0. \qquad (56)$$

Equation (56) is a condition that the net stress intensity factor due to both the stress drop and the cohesive stress be zero and is formally similar to the corresponding condition (36) obtained for the first model. We can show this as follows. The contribution of the cohesive force in the slip-weakening zone to the total stress $\tau(x)$ is

$$\sigma_c(x) = (\tau_s - \tau_d) \left\{ \frac{x}{(x^2-c^2)^{1/2}} \left[\chi(c,a) + \int_a^c \frac{W(\zeta)}{\delta_c} K(c,\zeta) \, d\zeta \right] + \psi(c,x,a) + \int_a^c \frac{W(\zeta)}{\delta_c} H(c,x,\zeta) \, d\zeta \right\}, \qquad (57)$$

which introduces an additional singularity with cohesive modulus

$$K_c = \lim_{x \to c^+} (2\pi)^{1/2} (x-c)^{1/2} \sigma_c(x), \qquad (58)$$

i.e.

$$K_c = (\pi c)^{1/2} (\tau_s - \tau_d) \left[\chi(c,a) + \int_a^c \frac{W(\zeta)}{\delta_c} K(c,\zeta) \, d\zeta \right]. \qquad (59)$$

Comparison of (38), (56), and (59) yields

$$(\pi c)^{1/2} (\tau_s - \tau_d) S(c,a,\gamma) = K_c - K_{\text{III}}. \qquad (60)$$

Thus equation (56) is the fracture criterion, as in the first case.

$$K_{\mathrm{III}} = K_{\mathrm{c}}. \tag{61}$$

which is a statement that the net stress intensity factor due to both the stress drop and the cohesive stress vanishes.

From (56), the expressions for the displacement and the stress become

$$\frac{\mu W(x)}{(\tau_{\mathrm{s}} - \tau_{\mathrm{d}})} = \phi(c,x,a) + \int_{a}^{c} \frac{W(\zeta)}{\delta_{\mathrm{c}}} G(c,x,\zeta) \mathrm{d}\zeta, \tag{62}$$

$$\frac{\tau(x) - \tau_{\mathrm{d}}}{\tau_{\mathrm{s}} - \tau_{\mathrm{d}}} = \psi(c,x,a) + \int_{a}^{c} \frac{W(\zeta)}{\delta_{\mathrm{c}}} H(c,x,\zeta) \mathrm{d}\zeta, \tag{63}$$

respectively. As $x \to c^+$, $\tau(x) \to \tau_{\mathrm{s}}$ as a consequence of $H(c,x,\zeta) \to 0$ and $\psi(c,x,a) \to 1$ as $x \to c^+$.

For the slip-weakening model, the displacement at the inner edge of the slip-weakening zone, $x=a$, is equal to the critical value, δ_{c} and at the outer edge it is zero. Since $\phi(c,c,a) = 0$, and $G(c,c,\zeta) = 0$, the condition requiring that the displacement at the edge of the crack vanish is satisfied automatically. Therefore, to solve the integral equation for the displacement in the slip-weakening zone, we impose two constraints: one is that the stress be finite, which is $S = 0$ and gives (56), and the other that the displacement at $x = a$ be δ_{c}. The latter condition is

$$\frac{\mu \delta_{\mathrm{c}}}{\tau_{\mathrm{s}} - \tau_{\mathrm{d}}} = \phi(c,a,a) + \int_{a}^{c} \frac{W(\zeta)}{\delta_{\mathrm{c}}} G(c,a,\zeta) \mathrm{d}\zeta, \tag{64}$$

where

$$\phi(c,a,a) = \frac{2a}{\pi} \ln\left|\frac{c}{a}\right|. \tag{65}$$

Equations (56), (62) and (64) are integral equations for the determination of the displacement in the slip-weakening zone $a<x<c$ and the quantities a and c have been found, the displacement on the entire crack as well as the stress in the unbroken region may be calculated directly from (62) and (63).

5 Numerical results for the second model

We solve the non-linear integral equations numerically. We normalize the displacements on the crack by δ_{c}, the stresses by $(\tau_{\mathrm{s}} - \tau_{\mathrm{d}})$, and the coordinates by $\mu \delta_{\mathrm{c}}/(\tau_{\mathrm{s}} - \tau_{\mathrm{d}})$. In these units the simultaneous integral equations (56), (62) and (64) may be rewritten as

$$\phi(c,x,a) + \int_{a}^{c} W(\zeta) G(c,x,\zeta) \mathrm{d}\zeta = W(x), \tag{66}$$

$$\phi(c,a,a) + \int_{a}^{c} W(\zeta) G(c,a,\zeta) \mathrm{d}\zeta = 1, \tag{67}$$

$$\chi(c,a) + \int_{a}^{c} W(\zeta) K(c,\zeta) \mathrm{d}\zeta = \gamma \tag{68}$$

and the stress as

$$\tau(x) - \tau_{\mathrm{d}} = \psi(c,x,a) + \int_{a}^{c} W(\zeta) H(c,x,\zeta) \mathrm{d}\zeta. \tag{69}$$

These equations are solved by an iteration. We do not present the details of the numerical procedure here.

Some of the numerical results are illustrated in Figs 6–8. We have plotted the distributions of the displacement and the stress on the plane $y=0$ for several values of the dimensionless stress drop γ; γ is the only input parameter in the numerical calculation. As before a and c both increase monotonically as the dimensionless stress drop decreases, while the difference between a and c, i.e. the size of the slip-weakening zone c_1, decreases monotonically. The results of Fig. 7 can be written as

$$c_1 = 1.4626 \, F_1(\gamma) \frac{\mu \delta_c}{\tau_s - \tau_d} \tag{70}$$

and

$$c = \frac{2}{\pi} \frac{F(\gamma)}{\gamma^2} \frac{\mu \delta_c}{\tau_s - \tau_d}, \tag{71}$$

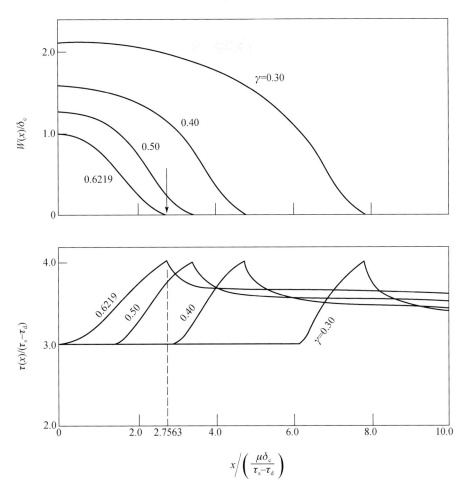

Figure 6 Displacement $W(x)$ and stress $\tau(x)$ as a function of position on the crack. γ is the stress drop. In the calculation, we arbitrarily chose $\tau_d = 3.0$. For scaling see Fig. 5

where $F_1(\gamma)$ and $F(\gamma)$ are monotonically increasing functions of γ (Fig. 8); $1 \leqslant F_1(\gamma) \leqslant 1.8845$, $1 \leqslant F(\gamma) \leqslant 1.6746$ as $0 \leqslant \gamma \leqslant 0.6219$. Equations (70) and (71) give a parametric relation between c_1 and c through γ. Since we simulate faulting by a crack with a slip-weakening zone, the results presented in (70) and (71) imply that a smaller main fault is usually accompanied by a larger slip-weakening zone and that an earthquake with a smaller fault length, requires a higher dimensionless stress drop on the crack. Equations (70) and (71) can also be interpreted as the critical conditions for the further extension of an initial fault under the influence of stress drop on the crack. These conditions imply that the initiation of an earthquake is not only controlled by the stress on the crack, but also by the size of the initial fault. In principle, it is also possible to initiate an earthquake under the influence of lower stress drop on the crack, for a larger initial fault. In our second paper, we will use this as a

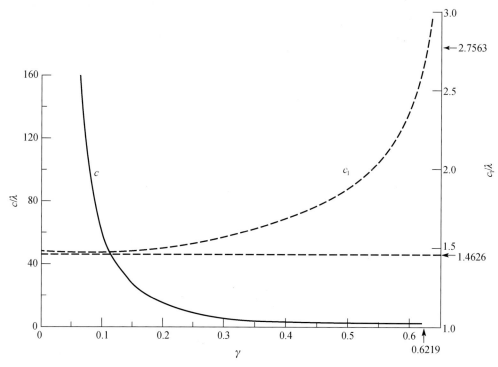

Figure 7 Half-length of the crack c and length of the slip-weakening zone c_1 as functions of the dimensionless stress drop. The lower dashed line represents the smaller-scale yielding limit of c_1

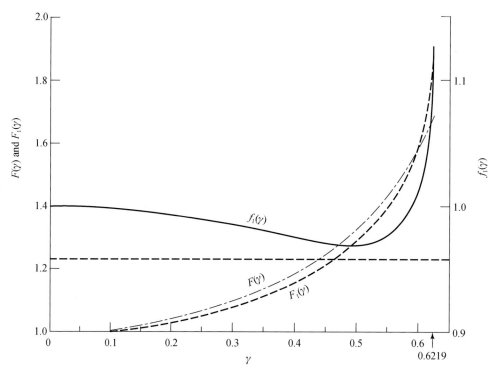

Figure 8 Scaling factors F, F_1, and f_1, as functions of the stress drop γ. These functions are defined in equations (70), (71) and (73)

stationary model for the initiation of a subsequent earthquake. These interpretations of the consequences of equation (71) are qualitatively in agreement with those for a model that assumes a constant critical stress-intensity factor at the crack tip; in fact these results are obvious for the constant critical stress-intensity factor model. However these conclusions could not have been drawn for the slip-weakening model without the prior calculation of the cohesive modulus (see equations 39 and 59). No statements regarding the size of the slip-weakening zone can be made from the stress-intensity factor model.

Finally, in Fig. 9, we have plotted the displacement W_1 in the slip weakening zone as a function of the distance from the inner edge $x_1 = x-a$, suitably scaled.

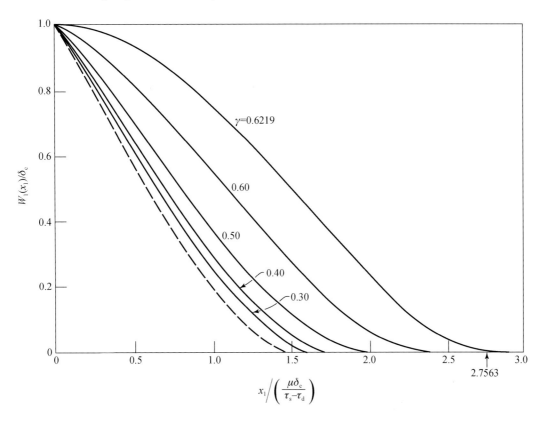

Figure 9 The distribution of the displacement in the slip-weakening zone as a function of position. γ is the stress drop. The dashed curve is the small-scale yielding limit

In the small-scale yielding limit, the quantities $F_1(\gamma)$ and $F(\gamma)$ are both equal to 1. Thus the coefficient 1.4626 in the present case is to be compared with $\pi/2$ obtained in the first case (see equation 44) and the coefficient $2/\pi$ here is to be compared with $16/(9\pi)$ in the first case (43). They are close to one another. We conclude that the size of the slip-weakening zone is not strongly dependent on the detailed form of the constitutive relation in the transition zone.

Equation (70) provides a method of estimating the size of the transition zone if γ, μ, δ_c and $(\tau_s - \tau_d)$ are known. As indicated above, δ_c is a quantity that represents the collective effect of the irregularly distributed geometric heterogeneities in the fracture zone. It can be deduced from the effective shear fracture energy, which in turn can be estimated by performing an integration over the experimental stress-slip diagram (Rice, 1980; Wong, 1982). Since a stress-slip diagram is not available for spontaneous earthquakes, an estimate is needed of the effective shear fracture energy G from seismological data; usually it is obtained by using a formula relating it to stress drop and fault length (Palmer & Rice, 1973, equation 40; Andrews, 1976, equation 11; Aki & Richards, 1980, equation 15, 80). A similar formula can be deduced as a consequence of (71). The quantity G is the excess of the energy required for fracture over the energy of frictional sliding. From (71) we have

$$G \equiv (\tau_s - \tau_d)\delta_c = \frac{\pi c (\tau_b - \tau_d)^2}{2F(\gamma)\mu}. \tag{72}$$

Palmer & Rice (1973) and Andrews (1976) obtained formula (72) without the factor $F(\gamma)$ in the denominator. The quantity $F(\gamma)$ describes the effect of the slip-weakening zone on the effective shear fracture energy; our result indicates that the above estimates of G may have been overestimated by as much as 67 percent. The case $F(\gamma) = 1$ gives an upper bound for the effective shear fracture energy, G. In the small-scale yielding limit, in which the size of the slip-weakening zone is very small compared to the length of the crack, this effect approaches a crack-size independent limit. We can derive this conclusion by taking the ratio c_1/c from (70) and (71):

$$\frac{c_1}{c} = 2.2975 f_1(\gamma)\gamma^2, \tag{73}$$

where $f_1(\gamma) \equiv F_1(\gamma)/F(\gamma)$ varies over the narrow range from 0.9707 to 1.1253 (Fig. 8). The ratio of c_1/c varies in a good approximation as γ^2; $c_1/c \ll 1$ if $\gamma \ll 1$. In other words, the small-scale yielding limit is just the case of small dimensionless stress-drop. In this limiting case, $F(\gamma) = 1$, and (72) reduces to the results given by Palmer & Rice (1973) and Andrews (1976).

If the value of $(\tau_s - \tau_d)$ is 500 bars (as is suggested by laboratory experiments), then the value of $F(\gamma)$ is 1.0408 for an earthquake with a fault length of 4km and a stress drop of 100 bars (we take $\mu = 3 \times 10^{11}$ dyne cm^{-2}). Thus the effective shear fracture energy is 1.1×10^7 J·m^{-2} (1.1×10^{10} erg cm^{-2}). This value seems to be consistent with various seismological estimates by diverse methods (Takeuchi & Kikuchi, 1973; Ida, 1973; Aki, 1979). If the above fracture were to break out and form a larger fault, we would have the same estimate of G as above. In this estimate, the value of the critical displacement δ_c is 0.20 m, and the size of the slip-weakening zone, c_1, is 0.18km. This latter estimate is only one-twentieth of the length of the initial fault. Thus the small-scale yielding approximation would seem to be appropriate.

The above result is dependent strongly on the value of $(\tau_s - \tau_d)$ for spontaneous earthquakes. If $(\tau_s - \tau_d)$ were smaller, say, 250 bars, then $F(\gamma)$ would be 1.1873, G would be smaller, 0.88×10^7 J·m^{-2}, but δ_c would be larger, 0.33 m, and c_1 would be much larger than before, 0.68km. In this case, c_1 is a non-negligible fraction of the length of the crack and the model of a finite crack with a size-comparable transition zone, as analysed by Andrews (1976) and in this paper, is more appropriate.

There is a critical stress drop above which no solution exists for the slip-weakening model (Fig. 7). This maximum stress drop corresponds to the limiting case in which the half-length of the crack is just equal to the length of the slip-weakening zone. If we set $a = 0$ in equations (66)—(68), the critical dimensionless stress drop is 0.6219 (as already noted), and the corresponding dimensionless half-length of the crack is 2.7563.

The dependence of the displacement and the stress on the critical displacement δ_c and the rigidity of the medium μ is given through the normalization. Thus a medium with a smaller rigidity, i.e. a 'soft medium' can only sustain a smaller crack, and as a consequence of the presence of a smaller crack, the stress will be redistributed into a smaller region outside the crack.

A comparison of the results obtained for the two cases is given in Table 1. The last two rows in the table are the results for the case of small-scale yielding. The results from the two models are very close to one another; the analysis is rather simpler in the first case.

Table 1 Comparison of two models of slip-weakening

	Stress drop proportional to distance	Stress drop proportional to displacement
Maximum stress drop γ_m	$2/\pi = 0.6366$	0.6219
Minimum crack dimension $c(\gamma_m)/\lambda$	$\pi = 3.1416$	2.7563
Small-scale yielding crack size $c\gamma^2/\lambda$	$16/(9\pi) = 0.5659$	$2/\pi = 0.6366$
Small-scale yielding transition zone size c_1/λ	$\pi/2 = 1.5708$	1.4626

$$\lambda = \frac{\mu\delta_c}{\tau_s - \tau_d}, \quad \gamma = \frac{\tau_b - \tau_d}{\tau_s - \tau_d}.$$

6 Conclusions

We have modelled an earthquake fracture as a 2-D crack with a zone of slip-weakening. We can also consider this crack as describing the critical state of an earthquake before breakout and a larger rupture ensues. We find that the results are not strongly dependent on the detailed form of the constitutive relation in the transition zone. For these cases of cracks of finite length, we have obtained the dependence of the size of the slip-weakening zone on the critical length of the crack. We find that a small initial fault will usually be accompanied by a large transition zone. The results also show that the initiation of an earthquake is controlled by the stress drop on the crack as well as its initial size. A small initial fault is able to develop into a larger earthquake if the dimensionless stress drop on the crack is large enough; and in principle, it is also possible to initiate an earthquake under the influence of lower stress drop on the crack, if a large enough initial fault is present. The second of these results can also be obtained as a consequence of a model that a constant critical stress-intensity factor at the crack tip can be used to determine the conditions of initiation of further fracture.

Our generalization of the effective shear fracture energy derived by Palmer & Rice (1973) is appropriate for a determination of the effective shear fracture energy as well as indicating how the size of the transition zone can be calculated from seismological information.

References

Aki, K., 1979. Characterization of barriers on an earthquake fault, *J. geophys. Res.*, **84**, 6140–6148.

Aki, K. & Richards, P. G., 1980. *Quantitative Seismology: Theory and Methods*, I and II, W. H. Freeman, San Francisco, California.

Andrews, D. J., 1976. Rupture propagation with finite stress in antiplane strain, *J. geophys. Res.*, **81**, 3575–3582.

Barenblatt, G. I., 1959. The formation of equilibrium cracks during brittle fracture. General ideas and hypothesis. Axially symmetric cracks, *J. appl. Math. Mech.*, **23**, 434–444.

Barenblatt, G. I., 1962. The mathematical theory of equilibrium cracks in brittle fracture, in *Advances in Applied Mechanics*, pp. 55–129, eds Dryden, H. L., Von Karman, T. & Kuerti, G., 7.

Burridge, R., Conn, G. & Freund, L. B., 1979. The stability of a rapid mode II shear crack, *J. geophys. Res.*, **84**, 2210–2222.

Dugdale, D. S., 1960. Yielding of steel sheet containing slits, *J. Mech. Phys. Solids*, **8**, 100–110.

Ida, Y., 1972. Cohesive force across the tip of a longitudinal shear crack and Griffith's specific surface energy, *J. geophys. Res.*, **77**, 3796–3805.

Ida, Y., 1973. The maximum acceleration of seismic ground motion, *Bull. seism. Soc. Am.*, **63**, 959–968.

Knopoff, L., 1958. Energy release in earthquakes, *Geophys. J. R. astr. Soc.*, **1**, 44–52.

Kostrov, B. V. & Das, S., 1982. Idealized models of fault behavior prior to dynamic rupture, *Bull. seism. Soc. Am.*, **72**, 679–703.

Leonov, M. Y. & Panasyuk, V. V., 1959. The development of very small cracks in a solid (in Ukrainian), *Prikladna Mekhanika (Applied Mechanics)*, **5**, 391–401.

Palmer, A. C. & Rice, J. R., 1973. The growth of slip surfaces in the progressive failure of overconsolidated clay, *Proc. R. Soc.*, **A332**, 527–548.

Rice, J. R., 1980. The mechanics of earthquake rupture, in *Physics of the Earth's Interior*, pp. 555–649, eds Dziewonski, A. M. & Boschi, E., Italian Phys. Soc., Bologna, Italy.

Sneddon, I. N., 1951. *Fourier Transforms*, McGraw-Hill, New York.

Sneddon, I. N., & Lowengrub, M., 1969. *Crack Problems in the Classical Theory of Elasticity*, John Wiley & Sons, Inc., New York.

Takeuchi, H. & Kikuchi, M., 1973. A dynamical model of crack propagation, *J. Phys. Earth*, **21**, 27–37.

Wong, T.-F., 1982. Shear fracture energy of westerly granite from post-failure behavior, *J. geophys. Res.*, **87**, 990–1000.

Yamashita, T., 1980. Quasistatic crack extensions in an inhomogeneous viscoelastic medium—A possible mechanism for the occurrence of aseismic faulting, *J. Phys. Earth*, **28**, 309–326.

Yamashita, T., 1981. Quasistatic crack extensions in an inhomogeneous viscoelastic medium—proposal of a unified seismic source model for the diverse earthquake rupture process, *J. Phys. Earth*, **29**, 283–304.

The quasistatic extension of a shear crack in a viscoelastic medium[*]

Y. T. Chen and L. Knopoff

Institute of Geophysics and Planetary Physics, University of California, Los Angeles

Summary We consider the quasistatic extension of an anti-plane shear crack with a zone of slip-weakening. The crack is imbedded in a medium that has a standard linear viscoelastic rheology ; the Kelvin and Maxwell solids are limiting cases. In general, three different crack histories can be identified for a crack of a given initial length: if the stress drop is smaller than some lower critical value, the crack does not extend. For stress drops larger than some upper critical value, the crack extends at velocities close to the S-wave velocity; we call this an infinite rupture velocity and the crack appears to grow instantaneously. For a stress drop in the interval between these two bounds, the crack extends quasistatically with finite velocities. For the case of the standard linear viscoelastic body in the quasistatic regime, the crack increases to a finite length and the rupture velocity becomes infinite after a finite time interval; after a finite time interval, both the crack length and the rupture velocity tend to infinity for a Kelvin body; no matter how small the stress drop is, a crack in a Maxwell solid will eventually extend to a critical length, at which point rupture occurs instantaneously. The breakout time, which is the time required for a growing crack to accelerate to infinite rates of extension, decreases monotonically with increasing stress drop.

1 Introduction

The time interval between successive earthquakes in an aftershock sequence in a given region often ranges from a few seconds to several months, is long in comparison to the rupture time of an individual earthquake and is short compared with the time constants for the accumulation of tectonic stress. The time scale of aseismic faulting before and/or after some large earthquakes is of the same order as the time intervals between the earthquakes of an earthquake sequence (Shimazaki 1974; Kasahara 1975). One can only attribute the cause of earthquake sequences such as aftershock sequences to the triggering of one earthquake by another, i.e. it is due to the redistribution of stress concentrations that are a consequence of the triggering event, by some process of stress relaxation or diffusion.

Under the conditions of temperatures and pressures at which earthquakes occur, crustal rocks deform non-elastically on long time scales. It is probably justified to take the medium to be elastic on the short time-scales of rapid rupture associated with fracture and seismic radiation. In the work summarized in this paper, we assume that the non-elastic rheology of the medium causes redistribution of the stresses concentrated after an earthquake. Furthermore, we assume that the rupture continues, albeit slowly at sub-seismic velocity rates. We find that the

---

[*] 本文发表于 *Geophys. J. R. astr. Soc.*, 1986 年, 第 **87** 卷, 第 3 期, 1025–1039.

rupture then accelerates, from these very slow velocities at first, up to a velocity comparable to elastic wave velocities. The acceleration may be so great that high velocities of deformation are reached catastrophically. When high rupture velocities are reached, a new earthquake is identified as occurring within the region of stress concentration outside of the slip zone as well as in the region of slip. An alternative consequence of the same set of initial conditions is that the redistributed stress outside the slowly growing fault may trigger another earthquake during the period of slow extension of the crack. The episode of slow quasistatic extension of the crack and the interval between two successive earthquakes have the same time constant because they are both controlled by the same rheological process.

From the above argument, we conclude that the rheological behaviour of the medium in the focal zone may be invoked to describe both slow aseismic faulting and the occurrence of earthquake sequences (Cohen 1978; Yamashita 1980, 1981). For this purpose. Yamashita (1980, 1981) considered the problem of the quasistatic extension of an anti-plane shear crack in a viscoelastic medium, with a fracture criterion based on the Dugdale model of the edge of the crack. A similar problem of a tensile crack with the Leonov-Panasyuk criterion as a fracture model was also considered by Kostrov & Nikitin (1970) and Kostrov, Nikitin & Flitman (1970). Both edge criteria take into account a zone of yielding near the edges of a crack. In the Leonov-Panasyuk model. the zone of yielding is assumed to be a region of weakened bonds or partially broken material, in which the slip increases under a constant stress until the slip reaches a critical value; the constant stress is the theoretical strength of the material. In the Dugdale model, the yielding zone is interpreted as a zone of plastic deformation. and the constant stress level is interpreted as the yield strength. These two models are formally equivalent and can be considered as a special case of the more general, slip-weakening model of Ida (1972) and Palmer & Rice (1973).

An earthquake fault surface is geometrically irregular since no natural surface is smooth. A study of the geometry of epicentres leads to the conclusion that these irregularities are fractally distributed (Kagan & Knopoff 1980). Since the irregularities are confined to a more or less narrow band, we take the fault zone to be planar for the purposes of this paper. The geometrical irregularities are projected onto this plane and reinterpreted as inhomogeneities of the physical properties in this plane. The influence of the spatial distribution of these inhomogeneities is considered in a separate paper; in this paper we are concerned with only one fracture event and its history under the influence of the rheology of the medium.

As indicated, we model an earthquake as an anti-plane shear crack with a zone of slip-weakening (Chen & Knopoff 1986). In this paper we investigate the quasistatic extension of a crack of this type imbedded in a viscoelastic medium with emphasis on understanding the processes that occur in the time interval between two successive earthquakes. By the adjective quasistatic we mean that the rates of extension are small compared with S-wave velocities and, hence, we shall be concerned with the early history of such extensions, i. e. before the catastrophic phase takes place.

In the quasistatic problem, the crack is assumed to grow so slowly that the inertial term in the equation of motion may be ignored. In the absence of an inertial term, infinite velocities of motion are possible. When the rate of extension of the rupture accelerates to 'very high velocities' we will describe the fracture in this phase as instantaneous. We scale the velocities of extension in units of the S-wave velocity; when the velocities of rupture approach the S-wave velocity, our solutions will no longer be valid. An equivalent statement is the following: if the time-scale of some part of the rupture history is very much less than the relaxation time of the viscoelastic material in which the fracture is imbedded. the rupture time of this phase will be taken to be zero. the rupture velocity will be taken to be infinite, and the rupture process will be considered as rapid or instantaneous. Put yet another way, in the regime for which the extension takes place with velocities of the order of sonic velocities, the inertial term in the equations of motion must be taken into account and the assumptions made in this paper are no longer valid. In the case of high sub-sonic or nearly sonic rupture velocities one encounters the problem of unstable propagation of cracks (Kostrov 1966; Burridge & Halliday 1971; Knopoff & Chatterjee 1982; Chatterjee & Knopoff 1983; and others). In this paper we focus on aspects of very slow growth of cracks prior to the sonic phase of rupture that are associated with the viscoelasticity of the medium. In our problem, the viscoelastic rheology is a linear one.

2 Quasistatic solution

The solution to the problem of a finite 2-D anti-plane static shear crack with a slip-weakening zone near the crack tip has been given by Chen & Knopoff (1986) for two cases of constitutive relationships in the transition zone. The results for the two cases are sufficiently close to one another, that we use only one of these models in the exposition below. The stress distribution for the viscoelastic problem is the same as for the elastic solid with the same geometry, except that the half-length of the crack and the size of the slip-weakening zone in the expression become functions of time; the displacement can be obtained from the solution for the equivalent elastic problem by replacing the elastic constant with a viscoelastic operator (Lee 1955; Shemyakin 1955; Kostrov & Nikitin 1970; Kostrov, Nikitin & Flitman 1970; Rice 1979). Thus, the static solution obtained by Chen & Knopoff (1986) can be used to deduce the solution for the viscoelastic problem of the same geometry.

We summarize the results derived in Chen & Knopoff (1986) for the case of a 2-D anti-plane shear crack in a uniform stress field at infinity. The crack occupies the region $0<|x|<c$ the region $a<|x|<c$ is the zone of slip-weakening. Because of symmetry, we only consider the region $x>0$. In the model in the above paper that we adopt here, the stress in the transition zone is assumed to be a linear function of the distance between coordinates a and c. The prestress τ_b, the dynamical friction τ_d and the static frictional shear stress τ_s are all assumed to be constants (cf. fig. 3 Chen & Knopoff 1986). We let ϵ be the ratio of the inner and outer dimensions of the transition zone and γ be the dimensionless stress drop (cf. equations (34) and (35) in Chen & Knopoff 1986). The stress on the crack play $y=0$ is expressed by equation (1) of Chen & Knopoff (1980). The displacements on the crack surface are

$$W(x) = \frac{(\tau_s - \tau_d)}{\mu} U(c, x, a), \quad x<c, \tag{1}$$

where $U(c,x,a)$ is given by equation (31) of Chen & Knopoff (1986). The relationship between the stress drop and the dimensions of the transition zone is

$$\gamma = \frac{2[(1-\epsilon^2)^{1/2} - \epsilon \cos^{-1}\epsilon]}{\pi(1-\epsilon)}. \tag{2}$$

The size of the static crack is fixed by the critical value of displacement at the inner edge of the slip-weakening zone, δ_c. This gives the relationship between c and ϵ (cf. equations (40) and (41) of Chen & Knopoff 1986).

$$\frac{\mu \delta_c}{\tau_s - \tau_d} = c\nu(\epsilon) = U(c, a, a). \tag{3}$$

where

$$\nu(\epsilon) = \frac{(1-\epsilon^2 + 2\epsilon^2 \ln \epsilon)}{\pi(1-\epsilon)}. \tag{4}$$

The stresses outside the crack are

$$\tau(x) = \tau_d + \frac{2(\tau_s - \tau_d)}{\pi(c-a)} \left\{ x \tan^{-1}\left(\frac{c^2-a^2}{x^2-c^2}\right)^{1/2} - a \tan^{-1}\left[\frac{x}{a}\left(\frac{c^2-a^2}{x^2-c^2}\right)^{1/2}\right] \right\}, \quad x>c. \tag{5}$$

In the viscoelastic case the quantities $\tau(x)$, a, c and $W(x, y)|_{y=0}$ are slowly varying functions of time. Kostrov, Nikitin & Flitman (1970) and Yamashita (1980) have suggested that the standard viscoelastic body

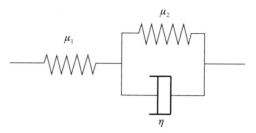

Figure 1 Three-element linear viscoelastic solid

shown in Fig. 1 be used to describe the rheology. In this case the relative displacement on the crack walls $W(x, t)$ is

$$\frac{W(x;t)}{\tau_s - \tau_d} = \frac{U[c(t), x, a(t)]}{\mu_1} + \int_{-\infty}^{t} U[c(\tau), x, a(t)] \frac{\exp[-\mu_2(t-\tau)/\eta]}{\eta} d\tau, \tag{6}$$

where $U(c, x, a)$ is given in equation (31) of Chen & Knopoff (1986), and μ_1 and μ_2 are the shear moduli of the elastic elements, and η the viscosity of the dashpot. The Kelvin and Maxwell bodies can be obtained as special cases of this model by letting $\mu_1 \to \infty$ or $\mu_2 \to 0$ respectively. Equation (5) represents the stress outside the edges of the crack and equation (36) of Chen & Knopoff (1986) represents the condition for finiteness of the stress at the edge of the crack, where the shear stress and a and c are functions of time.

Let the crack be stationary until $t=0$. Then, $a(t)$, $c(t)$ and $W(x; t)$ are constants for times $-\infty < t < 0$. We model this situation by imagining that $\eta = 0$ for $t<0$, and that the viscosity is suddenly switched on at $t=0$. The displacement on the crack is given by (1), which is now written as

$$\frac{\mu_0 W(x;t)}{\tau_s - \tau_d} = U(c, x, a), \quad -\infty < t < 0, \tag{7}$$

where

$$\mu_0 = \frac{\mu_1 \mu_2}{\mu_1 + \mu_2}, \tag{8}$$

and c and a are the values of these quantities prior to $t=0$, given in equations (2) and (3). The conceptual model of switching on the viscosity does no damage to one's physics expectations. Since the crack is at rest at $t=0$, the stresses are the same whether the viscosity is switched on or off. It is only when the crack begins to grow, which we assume takes place slowly at first, that the effective modulus begins to change. Below we shall see that the growth will become catastrophically fast at some later stage in the extension history; at this later stage, the effective modulus will be μ_1; but at $t=0$, the crack has zero velocity and, hence, it makes no difference whether the viscosity is switched in or out.

From the condition that the slip at the inner edge of the slip-weakening zone be δ_c, we obtain the relationship (3) between c_0 and $\epsilon = a/c$, where c_0 is the value of $c(t)$ for the stationary case:

$$\frac{\mu_0 \delta_c}{\tau_s - \tau_d} = c_0 v(\epsilon), \quad -\infty < t < 0. \tag{9}$$

Substitution of $\epsilon(\gamma)$ from (2) yields $c_0(\gamma)$.

Assume that the crack extends slowly for $t>0$. At any point beyond the initial position of the crack edge, $x > c_{(0)}$, the displacement begins to differ from zero at some time t_0. Therefore, (7) takes the form:

$$\frac{W(x;t)}{\tau_s - \tau_d} = \frac{U[c(t),x,a(t)]}{\mu_1} + \int_{t_0}^{t} U[c(\tau),x,a(\tau)] \frac{\exp[-\mu_2(t-\tau)/\eta]}{\eta} d\tau, \quad x > c(0). \quad (10)$$

From (3) we derive $c(t)$ by considering displacements for times when the inner edge of the crack $a(t)$ has moved beyond the initial location of the outer edge:

$$\frac{\delta_c}{\tau_s - \tau_d} = \frac{U[c(t),a(t),a(t)]}{\mu_1} + \int_{t_1}^{t} U[c(\tau),a(t),a(\tau)] \frac{\exp[-\mu_2(t-\tau)/\eta]}{\eta} d\tau, \quad a(t) > c(0), \quad (11)$$

where t_1 is the instant at which the displacement at the present position of the inner edge of the crack begins to differ from zero:

$$a(t) = c(t_1). \quad (12)$$

For slow motions of the crack we may approximate t_1 as

$$t_1 = t - \frac{c(t) - a(t)}{\dot{c}(t)}, \quad (13)$$

where $\dot{c}(t) \equiv dc(t)/dt$ is the rupture velocity.

If the rupture velocity is still so small that the inertial term in the equation of motion may be ignored, but the time-of-rupture $[c(t)-a(t)]/\dot{c}(t)$ is much smaller than the relaxation time of the rheological process η/μ_2, then the contribution from the integral on the rhs of (11) is negligible. In this case the crack extends instantaneously, i.e. with infinite rupture velocity (we remain in the regime of extension in which elastic wave propagation effects are ignored and hence infinite rupture velocities are not acausal). Let c_∞ be $c(t)$ for this limiting case. The position of the edge of the crack is given by

$$\frac{\mu_1 \delta_c}{\tau_s - \tau_d} = c_\infty v(\epsilon). \quad (14)$$

(14) is the same as (9) except that μ_1 is written in place of μ_0. As before, the function $\epsilon(\gamma)$ can be substituted from (2) into (14) to yield c_∞ as a function of γ. As the rupture velocity increases monotonically from zero to infinity, the integral in (11) decreases monotonically to zero, arid $c(t)$ increases monotonically from c_0 to c_∞.

Fig. 2 displays c_0 and c_∞ as functions of γ; we have scaled c_0 and c_∞ by $\mu_0 \delta_c/(\tau_s - \tau_d)$. In these units, for the stationary case, we have exactly the same curve as in fig. 4 of Chen & Knopoff (1986); for the case of instantaneous rupture, we have a similar curve with the ordinate the same as that for the stationary case but multiplied by $[1+(\mu_1/\mu_2)]$.

The curves c_0 and c_∞ in Fig. 2, divide the space $0<\gamma<\gamma_m$, $0<c<\infty$ into three sub-regions. For a given value of c, we can determine a 'safe' stress drop γ_0 from curve c_0; no rupture occurs for stresses less than γ_0, i.e. for stresses in region I. For the same value of c, we can find an ultimate stress drop γ_∞ from curve c_∞; for stresses greater than γ_∞, i.e. for stresses in region III, the rupture occurs instantaneously. For intermediate stress drops, $\gamma_0 < \gamma < \gamma_\infty$, rupture will accelerate gradually, at first slowly, and then at an increasing rate up to a velocity comparable to elastic wave velocities. Thus in region II the stress drop on the crack drives the crack at first into quasistatic extension and later into instantaneous fracture. The motion of the crack edge before instantaneous rupture takes place is given by (11).

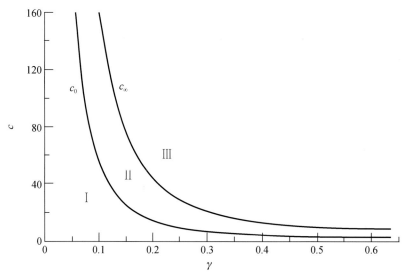

Figure 2 c_0 and c_∞ as functions of γ

$\mu_1/\mu_2 = 2.0$

3 The motion of the crack edge

In the case of slow extension of the crack. $c(\tau)$ in (11) may be approximated as
$$c(\tau) = c(t) - \dot{c}(t)(t-\tau), \tag{15}$$
and the quantity $U[c(\tau), a(t), a(\tau)]$, as
$$U[c(\tau), a(t), a(\tau)] = c(t)(1-s+\epsilon s) U\left(1, \frac{\epsilon}{1-s+\epsilon s}, \epsilon\right), \tag{16}$$
where s is a variable defined by
$$\tau = t \cdot \frac{c(t)-a(t)}{\dot{c}(t)} s. \tag{17}$$
Equation (11) becomes
$$\left[\frac{1}{\mu_1} + \frac{pF(p)}{\mu_2}\right]^{-1} \left(\frac{\delta_c}{\tau_s - \tau_d}\right) = c(t)v(\epsilon), \tag{18}$$
where
$$F(p) = \frac{1}{v(\epsilon)} \int_0^1 (1 - s + \epsilon s) U\left(1, \frac{\epsilon}{1 - s + \epsilon s}, \epsilon\right) \exp(-ps) \, ds, \tag{19}$$
and
$$p = \frac{\mu_2}{\eta} \frac{c(t)-a(t)}{\dot{c}(t)}. \tag{20}$$
Evidently (18) is the same as (3) if the time-dependent rigidity
$$\mu(t) = \left[\frac{1}{\mu_1} + \frac{pF(p)}{\mu_2}\right]^{-1} \tag{21}$$
is used in place of μ. If the rupture time is much greater than the relaxation time, p is infinite and the rupture process may be considered as stationary. If the rate of crack extension is very large so that the duration of the rupture is very much less than the relaxation time, the parameter p may be taken to be zero and the rupture process will be described as instantaneous; in the limit $p \to 0$, $pF(p) \to 1$. The time-dependent rigidity (21) is bounded by

its stationary limit μ_0 and its instantaneous rupture limit μ_1. For very slow crack growth rates the viscoelastic medium responds as an elastic solid with long-time rigidity μ_0, while for rapid growth the medium is also elastic and responds with instantaneous rigidity μ_1.

Substitution of (14) into (18) gives

$$\frac{c_\infty}{c(t)} = 1 + \frac{\mu_1}{\mu_2} pF(p). \tag{22}$$

Equations (20) and (22) are first order differential equations for $c(t)$ with the parameter p.

Integration yields

$$\frac{\mu_2(1-\epsilon)}{\eta} t = p_i - p + \int_p^{p_i} \frac{\mu_2/\mu_1 - p^2 F'(p)}{\mu_2/\mu_1 + pF(p)} dp, \tag{23}$$

where the p_i are the initial values.

For a Kelvin solid ($\mu_1 \to \infty$), $\mu_0 \to \mu_2$ and $c_\infty \to \infty$. Thus region III is absent. In this case, (22) and (23) simplify to

$$\frac{c_0}{c(t)} = pF(p) \tag{24}$$

and

$$\frac{\mu_2(1-\epsilon)}{\eta} t = p_i - p + \int_p^{p_i} \frac{pF'(p)}{F(p)} dp. \tag{25}$$

For a Maxwell solid ($\mu_2 \to 0$), $\mu_0 \to 0$ and $c_0 \to 0$. The curve c_0 degenerates into the c-axis and $0 < \gamma < \gamma_m$, i.e. region I disappears. In this case, the parametric equations (20) and (22) are replaced by the first order differential equation for $c(t)$,

$$\frac{c_\infty}{c(t)} = 1 + F(0) \frac{\mu_1(1-\epsilon)}{\eta} \frac{c(t)}{\dot{c}(t)}, \tag{26}$$

which has the closed form solution

$$\frac{\mu_1(1-\epsilon)}{\eta} t = \frac{1}{F(0)} \left\{ \ln\left|\frac{c_i}{c(t)}\right| + c_\infty \left[\frac{1}{c_i} - \frac{1}{c(t)}\right] \right\}, \tag{27}$$

where c_i is the initial value of $c(t)$.

In the case of small-scale yielding, $(1-\epsilon) \ll 1$. The quantity $U[c(\tau), a(t), a(\tau)]$ may be approximated as $U[c(t), a(t), a(t)]$, and (18) becomes a first-order differential equation for $c(t)$:

$$\left(\frac{\delta_c}{\tau_s - \tau_d}\right) \mu(t) = c(t) v(\epsilon), \tag{28}$$

where the time-dependent rigidity is

$$\mu(t) = \left\{ \frac{1}{\mu_1} + \frac{1}{\mu_2} - \frac{1}{\mu_2} \exp\left[-\frac{(1-\epsilon)\mu_2 c(t)}{\eta \dot{c}(t)}\right] \right\}^{-1}. \tag{29}$$

Integration of (28) yields the motion of the crack edges in the case of small-scale yielding

$$\frac{(1-\epsilon)\mu_2 t}{\eta} = \ln\left|1 + \frac{\mu_2}{\mu_1} \ln\left|\frac{c_i}{c(t)}\right|\right| - \int_{c_0/c(t)}^{c_0/c_i} \frac{\ln|1-z|}{z} dz \tag{30}$$

for the standard viscoelastic solid, and

$$\frac{(1-\epsilon)\mu_2 t}{\eta} = -\int_{c_0/c(t)}^{c_0/c_i} \frac{\ln|1-z|}{z} dz \tag{31}$$

for the Kelvin solid. For the Maxwell solid, in the case of small-scale yielding, the equation of motion of the edges of the crack, and its solution are easily obtained from equations (22) and (23) by setting $F(0) = 1$. In all three solutions the parameters ϵ, c_0, and c_∞, are

$$\epsilon = 1 - \frac{9\pi^2\gamma^2}{32}, \tag{32}$$

$$c_0 = \left(\frac{16}{9\pi\gamma^2}\right)\frac{\mu_0 \delta_c}{\tau_s - \tau_d}, \tag{33}$$

and

$$c_\infty = \left(\frac{16}{9\pi\gamma^2}\right)\frac{\mu_1 \delta_c}{\tau_s - \tau_d}, \tag{34}$$

(Chen & Knopoff, equation (43), 1986).

Fig. 3 shows the position of the crack edge and the rupture velocity as a function of time for the case of small-scale yielding, for several values of μ_1/μ_2 for an initial crack semilength c_0 for the standard viscoelastic solid as well as for the Kelvin body. Fig. 4 gives the same quantities for the Maxwell body for an initial crack semilength of $0.01c_\infty$. For the general viscoelastic body, the crack increases to some finite length and the rupture velocity increases to infinity in a finite time. For the special case of a Kelvin body, both the edge of the crack and its rupture velocity tend to infinity after a finite time interval; in other words a crack of finite length imbedded in a Kelvin solid does not rupture instantaneously. For a Maxwell body, the 'safe' stress drop is zero, i. e. no matter how small the stress drop is. the crack will eventually extend to its critical length, at which point rupture occurs instantaneously.

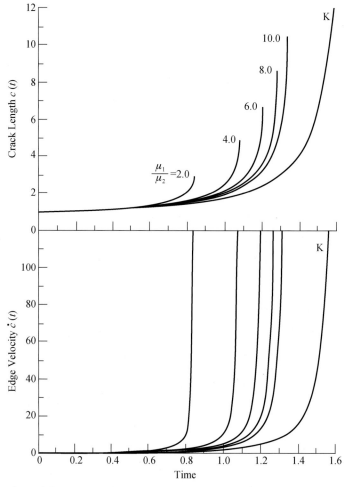

Figure 3 Position of the crack edge and the rupture velocity as functions of time for selected values of μ_1/μ_2 for the standard viscoelastic body as well as for the Kelvin body (K)

The units for t, $c(t)$ and $\dot{c}(t)$ are $\eta/(1-\epsilon)\mu_2$, c_0, and $c_0(1-c)\mu_2/\eta$, respectively

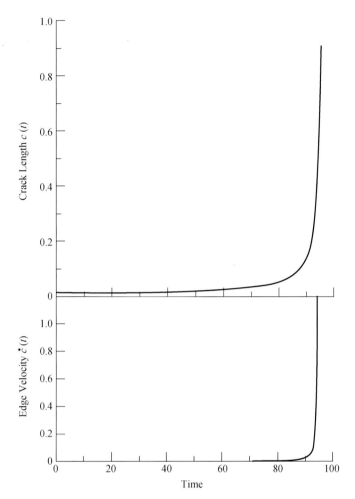

Figure 4   Same as Fig. 3 for Maxwell body

The units for t, $c(t)$ and $\dot{c}(t)$ are $\eta/(1-\epsilon)\mu_1$, c_∞ and $c_\infty(1-\epsilon)\mu_1/\eta$, respectively

Fig. 5 is a graph of the stress redistribution during extension of the crack for the case $\mu_1/\mu_2 = 2.0$, $\gamma = 0.1273$, and $c_i = c_0$. The rupture accelerates slowly at first and then at an increasing rate, and the region of stress concentration becomes broader and broader. If it continues to accelerate, a new earthquake occurs instantaneously both within the slipped region and the region of stress concentration outside it. An alternative possibility is that the elevated stress outside the growing crack may trigger another earthquake during the phase of slow extension of the crack. We do not consider the latter possibility here, i. e. we assume that the increment of stress outside the growing fault is too small to trigger an earthquake. We only discuss the first possibility.

4   Breakout time

If the stress drop γ on a crack lies between γ_0 and γ_∞, a crack of initial half-length c_i will extend quasistatically. After extending for a time t_m, the process will become unstable, where t_m, is the solution of the equation

$$c(t_m) = c_\infty. \tag{35}$$

We call t_m the breakout time of the crack. For the standard linear viscoelastic body,

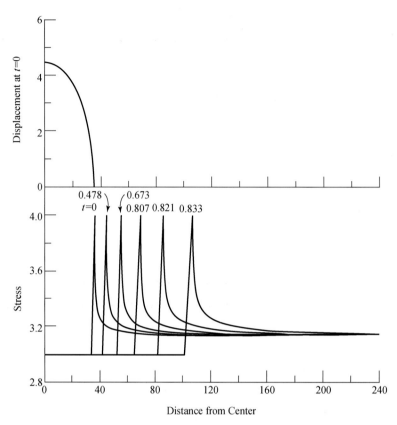

Figure 5 Stress distribution during extension of a crack imbedded in a standard linear viscoelastic body

Also shown is displacement at $t=0$, $\mu_1/\mu_2 = 2.0$, $\gamma = 0.1273$, and $c_i = c_0$.

The units of t, x, w and τ are $\eta/(1-\epsilon)\mu_2$, $\mu_0 \delta_c/(\tau_s - \tau_d)$, δ_c and $(\tau_s - \tau_d)$ respectively

$$t_m = \frac{\eta}{\mu_2(1-\epsilon)}\left(p_i - \int_0^{p_i} \frac{\mu_2/\mu_1 - p^2 F'(p)}{\mu_2/\mu_1 + pF(p)} dp\right), \tag{36}$$

while for a Kelvin body,

$$t_m = \frac{\eta}{\mu_2(1-\epsilon)}\left(p_i + \int_0^{p_i} \frac{pF'(p)dp}{F(p)}\right), \tag{37}$$

and for a Maxwell body,

$$t_m = \frac{\eta}{\mu_1(1-\epsilon)}\left(\ln\left|\frac{c_i}{c_\infty}\right| + \frac{c_\infty}{c_i} - 1\right). \tag{38}$$

Since the quantities c_0, c_∞ and ϵ are functions of the stress drop, the breakout time is a function of the stress drop as well as the initial length of the crack. The small-scale yielding limits of the quantities ϵ, c_0, and c_∞, are given in (32)–(34). The ultimate stress drop γ_∞ and the 'safe' stress drop γ_0 are determined by the equations

$$c_i = c_\infty(\gamma_\infty) \tag{39}$$

and

$$c_i = c_0(\gamma_0), \tag{40}$$

respectively. In the case of small-scale yielding, substitution of (35) into (36)–(38) yields

$$t_m = \frac{\eta}{\mu_2 \gamma_\infty^2}\left(\frac{32}{9\pi^2}\right)\left(\frac{\gamma_\infty}{\gamma}\right)^2\left[2\ln\left|1 + \frac{\mu_2}{\mu_1}\right|\ln\left|\frac{\gamma}{\gamma_\infty}\right| - \int_{z_1}^{z_2} \frac{\ln|1-z|}{z} dz\right], \tag{41}$$

for the general viscoelastic body, where $z_1 = (1+\mu_1/\mu_2)^{-1}$ and $z_2 = z_1(\gamma_\infty/\gamma)^2$,

$$t_m = -\frac{\eta}{\mu_2 \gamma_0^2}\left(\frac{32}{9\pi^2}\right)\left(\frac{\gamma_0}{\gamma}\right)^2 \int_0^{(\gamma_0/\gamma)^2} \frac{\ln|1-z|}{z}dz, \tag{42}$$

for the Kelvin body, and

$$t_m = -\frac{\eta}{\mu_1 \gamma_\infty^2}\left(\frac{32}{9\pi^2}\right)\left(\frac{\gamma_\infty}{\gamma}\right)^2\left[2\ln\left|\frac{\gamma}{\gamma_\infty}\right|+\left(\frac{\gamma_\infty}{\gamma}\right)^2-1\right] \tag{43}$$

for the Maxwell body.

Fig. 6 shows the breakout time as a function of the stress drop for the case of a small-scale yielding for a standard linear viscoelastic body for different values of μ_1/μ_2. For a given crack with initial half-length c_i, there is an ultimate stress drop γ_∞ above which fracture occurs instantaneously. On the other hand, for the same crack, there is a minimum stress drop γ_0 below which a crack of given length does not exist. For an intermediate stress drop γ, $\gamma_0 < \gamma < \gamma_\infty$, the breakout time required for a crack to develop into instantaneous fracture is finite, which decreases monotonically with increasing stress drop. We also display the breakout time for the case of the Maxwell body in Fig. 6; in the limiting case $\gamma/\gamma_\infty \ll 1$, the breakout time varies inversely with the fourth power of the stress drop. For the special case of a Kelvin body, a crack of finite length does not rupture instantaneously, but does so when it reaches infinite length; the breakout time in Fig. 7 is the time interval required for the crack to grow to infinite size.

The dependence of the breakout time on the stress level has been explored in the laboratory (Bridgman 1952) and its relevance to the problem of earthquake occurrence has been mentioned by many authors. The results presented here are qualitatively in agreement with the experimental results. Measurements by Lomnitz (1956) suggest the values $\eta/\mu_2 = 10^4$s, and $\mu_1/\mu_2 = 1.5$. The result presented in Fig. 6 shows that the breakout time for the case $\mu_1/\mu_2 = 1.5$ ranges from zero to $0.66\eta/\mu_2\gamma_\infty^2$, depending on the value of γ/γ_∞. If the ultimate dimensionless stress drop γ_∞ is chosen to be 0.2, the breakout time will range from zero to 1.6×10^5s, i.e. about 2 days. Low stress drops greatly prolong the breakout time. For instance, if the ultimate stress drop is as low as 10 bars, and the value of $(\tau_s - \tau_d)$ is 500 bars, the ultimate dimensionless stress drop will be 0.02; then the breakout time will range from zero to about 6 months. Thus, this result reasonably explains time delays between successive events in earthquake sequences ranging from a few seconds to several months.

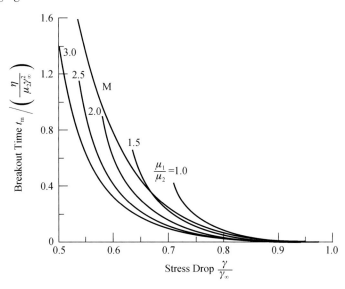

Figure 6 Breakout time as a function of stress drop for different values of μ_1/μ_2 for, the standard viscoelastic body as well as for the Maxwell body (M)

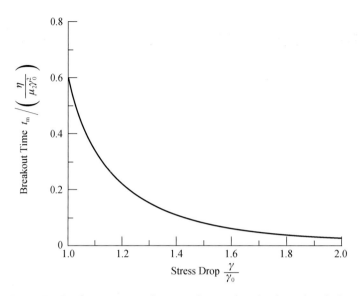

Figure 7 Breakout time as a function of stress drop for the Kelvin body

5 Comments

We have assumed that the 'quiet' period between successive events in an earthquake sequence is identified with the breakout time, i. e. the time required for a quasistatically growing crack to accelerate to infinite rates of extension. An anti-plane shear crack of a given initial length does not exist if the stress drop on the crack is smaller than some lower critical value. For stresses larger than some upper critical value, the crack grows instantaneously. For a stress drop in the interval between these two bounds. the crack extends quasistatically. The breakout time is estimated to range from a few seconds to several months and is reasonably in agreement with most observed time intervals between successive events in earthquake sequences.

In this paper we have confined our attention to the problem of the 'quiet' period between successive events in earthquake sequences. In our model, a crack will propagate dynamically once it becomes unstable. However, a dynamic crack will not stop under the simplified assumptions we have made; we have taken the distribution of the initial stress and the static and the dynamic frictional stresses to be uniform everywhere on the fault plane. Heterogeneity of the stress distributions will introduce complexity into the rupture process, but will also allow for the stopping of the crack. In a third paper in this series, we discuss the stopping of cracks as well as their restarting as a consequence of the non-uniform distribution of frictions and prestresses; we will be able to model prolonged sequences of earthquake events by invoking appropriate distributions of inhomogeneities of the frictions and prestresses.

References

Bridgman, P. W., 1952. *Studies in Large Plastic Flow and Fracture*, McGraw-Hill, New York.

Burridge. R. & Halliday, G. S., 1971. Dynamic shear cracks with friction as models for shallow focus earthquakes, *Geophys. J. R. astr. Soc.*, **25**, 261–283.

Chatterjee, A. K. & Knopoff, L., 1983. Bilateral propagation of a spontaneous two-dimensional anti-plane shear crack under the influence of cohesion, *Geophys. J. R. astr. Soc.*, **73**, 449–473.

Chen, Y. T. & Knopoff, L., 1986. Static shear crack with a zone of slip-weakening, *Geophys. J. R. astr. Soc.*, **87**, 1005–1024.

Cohen, S. C., 1978. The viscoelastic stiffness model of seismicity, *J. geophys. Res.*, **83**, 5425–5431.

Ida, Y., 1972. Cohesive force across the tip of a longitudinal shear crack and Griffith's specific surface energy, *J. Geophys. Res.*, **77**, 3796–3805.

Kagan, Y. Y. & Knopoff, L., 1980. Spatial distribution of earthquake: the two point correlation function, *Geophys. J. R. astr. Soc.*, **62**, 303–320.

Kasahara, K., 1975. Aseismic faulting following the 1973 Nemuro-Oki earthquake, Hokkaido, Japan (a possibility), *Pure Appl. Geophys.*, **113**, 127–139.

Knopoff, L. & Chatterjee, A. K., 1982. Unilateral extension of a two-dimensional shear crack under the influence of cohesive forces, *Geophys. J. R. astr. Soc.*, **68**, 7–25.

Kostrov, B. V., 1966. Unsteady propagation of longitudinal shear crack, *PMM*, **30**, 1241–1248.

Kostrov, B. V. & Nikitin, L. V., 1970. Some general problems of mechanics of brittle fracture, *Arch. Mech. Stosowanej*, **6**(22), 749–776.

Kostrov, B. V., Nikitin, L. V. & Flitman, L. M., 1970. The expansion of cracks in viscoelastic bodies, *Bull. Acad. Sci. USSR, Geophys. Ser.*, **7**, 413–420.

Lee, E. H., 1955. Stress analysis in visco-elastic bodies, *Quart. Appl. Math.*, **13**, 183–190.

Lomnitz, C., 1956. Creep measurements in igneous rocks, *J. Geol.*, **64**, 473–479.

Palmer, A. C. & Rice, J. R., 1973. The growth of slip surfaces in the progressive failure of overconsolidated clay, *Proc. R. Soc. A*, **332**, 527–548.

Rice, J. R., 1979. The mechanics of quasi-static crack growth, in *Proceedings of the 8th US National Congress of Applied Mechanics*, pp. 191–216. ed. Kelly, R. E., Western Periodicals, Hollywood, California.

Shemyakin, E, I., 1955. The Lamb problem for a medium with an elastic aftereffect (in Russian), *Doklady Akademiia Nauk SSR*, **104**, 193–196.

Shimazaki, K., 1974. Pre-seismic crustal deformation caused by an underthrusting oceanic plate, in eastern Hokkaido, Japan, *Phys. Earth Planet. Int.*, **8**, 148–159.

Yamashita, T., 1980. Quasistatic crack extensions in an inhomogeneous viscoelastic medium-possible mechanism for the occurrence of aseismic faulting, *J. Phys. Earth*, **28**, 309–326.

Yamashita, T., 1981. Quasistatic crack extensions in an inhomogeneous viscoelastic medium-proposal of a unified seismic source model for the diverse earthquake rupture processes, *J. Phys. Earth*, **29**, 283–304.

Simulation of earthquake sequences[*]

Y. T. Chen and L. Knopoff

Institute of Geophysics and Planetary Physics, University of California. Los Angeles

Summary An earthquake sequence can be considered to be a complex extension of a shear crack in a viscoelastic medium under the influence of non-uniform stresses. Because of these inhomogeneous stresses, the crack grows rapidly at rates comparable to seismic body-wave velocities during some intervals of time; during other intervals, which punctuate the episodes of rapid expansion, the major mode of slip and growth is one of creep. Under suitable pre-stress and frictional conditions, the post-seismic creep phase of one earthquake may become the pre-seismic creep phase of a succeeding earthquake. Thus inhomogeneity of the pre-stress and/or the static and dynamic frictional stresses combined with viscoelasticity of the medium provides a mechanism that accounts for not only the pre-seismic and the post-seismic creep and the stopping of the crack, but also the various types of earthquake sequences that occur in nature. By increasing the amplitude of the fluctuations in the spatial distribution of these stresses, the type of earth-quake sequence can be varied progressively from an 'isolated earthquake', to a sequence with foreshocks, main shock and aftershocks, and finally, to an earthquake swarm. As the wavelengths of the fluctuations in the stresses decrease, the frequency of earthquake occurrence increases. The type of earthquake sequence is also controlled by the general level of the stresses. A silent earthquake or aseismic creep event will occur if the pre-stress is sufficiently low and/or the breaking strength is sufficiently high.

1 Introduction

The most popular contemporary model of an earthquake source is that of a plane crack on which the local static frictional stress suddenly drops to the dynamic frictional stress. If the stress drop on the crack is finite up to the edges of the crack, then the stresses outside the crack at the edges are singular. These infinite stresses are mathematically correct consequences of the above assumption if the edge is defined to be the line in the plane separating the region of slip from the unbroken region. Real materials cannot support such singular stresses. The singularities at the edges of the crack can be removed if it is assumed that a transition zone exists near the edges of the crack (Barenblatt 1959, 1962; Leonov & Panasyuk 1959; Dugdale 1960), i. e. that the edge is not a mathematical line in space.

Although the model of a crack with a transition zone reasonably explains an isolated earthquake as an instability, it is incapable of accounting for the variety of different, repetitive earthquake sequences, such as aftershocks, swarms, isolated earthquakes, etc. with time intervals of the order of days or months between events. Nor does it account for the occurrence of aseismic faulting before and after earthquakes (Shimazaki 1974;

---

[*] 本文发表于 *Geophys. J. R. astr. Soc.*, 1987 年, 第 **91** 卷, 第 3 期, 693–709.

Kasahara 1975). During the 'quiet' period between two successive earthquakes in an earthquake sequence, the rheology of the medium allows for slow readjustment of the stresses that were concentrated or otherwise redistributed by the preceding event. In some models the rheological behaviour of the medium in the focal zone may not only account for slow aseismic faulting between two successive earthquakes, but also the repetitive character of the earthquakes themselves (Burridge & Knopoff 1967; Knopoff 1972; Cohen 1978; Yamashita 1980, 1981; Yin & Zheng 1983; Zheng & Yin 1983; Chen & Knopoff 1986b).

The model of a crack in a uniform stress field in a rheologically relaxing medium has some unsatisfactory properties. In a uniform stress field, a pre-existing crack will undergo accelerated extension quasistatically for a finite time interval; when its rate of extension reaches a velocity comparable to elastic wave velocities, it begins to radiate seismic waves. However, this crack never stops (Chen & Knopoff 1986b). Since it never stops, it is incapable of generating additional earthquakes in earthquake sequences, nor can it describe adequately the observed pre-seismic and post-seismic creep frequently associated with seismic events. Stated more dramatically, the first earthquake of this type would have sundered the Earth.

We must therefore assume that the earthquake process is governed by inhomogeneity of stresses: one or more of the initial pre-stress, the static and the dynamic frictional stresses are non-uniform. Surface breakage of a large earthquake exhibits a geometrical irregularity that is full of microcracks, cracks, joints, and secondary faults which appear as bifurcations, echeloning, parallelism, etc. Kagan (1982) has shown that real earthquake faults are not plane but are complex 3-D objects. Nevertheless, since the fracture zone of an earthquake is confined to a narrow band, we take it to be planar for mathematical reasons and assume that the geometrical irregularities of the fracture surface can be projected on to the plane in terms of some non-uniform distribution of stresses and breaking strengths in the plane. Where large deviations from planarity occur in the 3-D fault, large breaking strengths and/or large stress drops are to be expected in the plane of projection. Thus the projection of an irregular 3-D fault geometry is directly related to the barrier, or non-uniform strength, model (Das & Aki 1977a,b) and to the asperity, or non-uniform stress drop model (Kanamori 1978; Rudnicki & Kanamori 1981) of planar rupture.

In this paper we investigate the consequences of the assumption that a crack is both imbedded in a rheologically relaxing medium and the stresses and/or frictions are non-uniform. When a crack extends into a region of negative stress drop and/or a region with higher breaking strengths or cohesions, i.e. into appropriate regions of a non-uniform field, it decelerates and will finally come to rest. In the deceleration phase, the rheology of the medium becomes important, especially in the stages when the velocity of extension is far from sonic. If the crack extends in a highly inhomogeneous field, the process becomes very intricate and may consist of a number of events that alternate between instantaneous rupture and quasistatic extension. In this paper, we propose that both a viscoelastic rheology of the medium as well as inhomogeneity of the stresses are needed to explain the observed pre-seismic and post-seismic creep, and the stopping of the crack, as well as the variety of earthquake sequences.

The problem of the quasistatic extension of an anti-plane shear crack in a viscoelastic medium under the influence of inhomogeneous stresses, is similar to that considered by Cohen (1977, 1978) for a 1-D model and by Yamashita (1980, 1981) for a 2-D Dugdale model (1960). We model an earthquake as a 2-D anti-plane shear crack with a zone of slip-weakening at the outer edges (Ida 1972; Palmer & Rice 1973; Andrews 1976; Rice 1980; Burridge, Conn & Freund 1979; Kostrov & Das 1982; Chen & Knopoff 1986a, b) to avoid the singular stresses that are the consequence of the mathematically convenient assumption described above. Our rheological model is that of a three-element linear viscoelastic solid, a model also used by Kostrov, Nikitin &

Flitman (1970), Dieterich (1972), Cohen (1977, 1978), Yamashita (1980, 1981), Zheng & Yin (1983) and Chen & Knopoff (1986b); much earlier, Benioff (1951) made application of a similar model.

We invoke a model of linear viscoelasticity only for the purposes of mathematical tractability, recognizing fully that the description of creep prior to fracture is considerably more complicated. The minimum physical properties that are required of any rheological model are that the medium should be capable of storing energy elastically at slow rates of deformation, be capable of radiating energy in elastic waves at high frequencies, and that it should respond to increasing stress with increasing strain rates in the pre-shock creep state. A more accurate rheological simulation than the linear one is justified in a more quantitative exploration than is our purpose in this paper. We do not expect any loss in generality of our results in this paper due to this assumption.

2 Static shear crack with a zone of slip-weakening under the influence of non-uniform stresses

Assume that a plane 2-D crack occupies the strip in the x–y plane $-c<x<c$ $y=0$, with displacements W in the z-direction (Fig. 1). We assume the stress on the crack $\tau(x)$ is symmetric about $x=0$; hence we only consider the region $x>0$. We assume that a slip-weakening zone of length $(c-a)$ develops near the edge of the crack $a<x<c$, $y=0$; we shall only be concerned with the limiting case of small-scale yielding $(c-a)\ll c$. We have assumed that the stress distribution within the fracture zone on the crack plane is

$$\tau(x) = \tau_d(x),\ 0 < x < a,\ y = 0,$$

where τ_d is the dynamic friction. In the slip-weakening zone, the stress increases monotonically from $\tau_d(a)$ to $\tau_s(c)$, where $\tau_s(x)$ is the static friction (Fig. 2). The general form of the solution does not depend on the precise form of the increase; Chen & Knopoff (1986a) have considered two distinct models of the transition zone with only minor differences in the quantitative results.

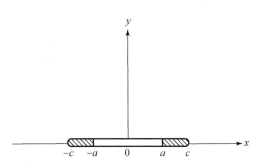

Figure 1 Schematic anti-plane shear crack with slip-weakening zone, $a<x<c$

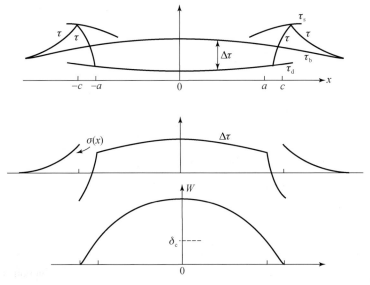

Figure 2 Schematic relationships among stress τ, pre-stress τ_b, dynamic friction τ_d, static friction τ_s, stress drop $\Delta\tau$, redistributed stress σ, and displacement W as functions of the crack coordinates

The critical dimensions of the crack are $(a,\ c)$ and critical displacement at the slip-weakening zone boundary is δ_c

The pre-stress $\tau_b(x)$ is assumed to be a function of position. In the plane of the crack the stress drop $\Delta\tau(x)$ within the fractured segment is

$$\Delta\tau(x) = \tau_b(x) - \tau(x), \quad 0 < x < c, \quad y = 0, \qquad (1)$$

and beyond the edge of the crack, the stress after fracture is

$$\tau(x) = \tau_b(x) + \sigma(x), \quad x > c, \quad y = 0, \qquad (2)$$

where $\sigma(x)$ is the redistributed stress. The solution of the problem of the stress distribution in relation to the critical displacement δ_c at the point $x = a$, $W(a) = \delta_c$ is somewhat cumbersome and its form depends on the form of the variation of stress in the slip-weakening zone; as indicated, these details are not critical to our further discussion. Particulars of the solution are to be found in Chen & Knopoff (1986a).

If the crack is to extend, the stress concentration at the edge of the crack must be greater than the strength of the material at the edge. We evaluate the critical state of equality between these two quantities as the fracture criterion,

$$K_{III} = K_c, \qquad (3)$$

where each of these quantities has the dimensions of a stress intensity factor. The stress concentration at the edge is measured by the stress intensity factor K_{III} for an anti-plane shear crack,

$$K_{III} = 2\left(\frac{c}{\pi}\right)^{1/2} \int_0^c \frac{\tau_b(\zeta) - \tau_d(\zeta)}{(c^2 - \zeta^2)^{1/2}} d\zeta \qquad (4)$$

which is evidently an appropriately weighted function of the stress drop on the entire crack. The cohesive modulus for the slip-weakening model is

$$K_c \cong 2\left(\frac{c}{\pi}\right)^{1/2} \int_a^c \frac{\tau(\zeta) - \tau_d(\zeta)}{(c^2 - \zeta^2)^{1/2}} d\zeta \cong C(c-a)^{1/2}[\tau_s(c) - \tau_d(c)] \qquad (5)$$

with a coefficient C of the order of unity. In the same small-scale yielding limit, the size of the slip-weakening zone $(c-a)$ is given by the solution to

$$\delta_c \simeq \frac{D(c-a)}{\mu}[\tau_s(c) - \tau_d(c)] \qquad (6)$$

with a coefficient D that is approximately $2/\pi$. Precise values of C and D are given in Chen & Knopoff (1986a) for particular models. Elimination of $(c-a)$ between (5) and (6) gives the cohesive modulus.

3 Quasistatic extension of the crack under the influence of non-uniform stresses

We use the result of the previous section for the static case to find the solution for the problem of the same geometry for the viscoelastic case by means of the correspondence principle (Lee 1955; Shemyakin 1955; Kostrov & Nikitin 1970; Kostrov, Nikitin & Flitman 1970; Rice 1979). We allow the quantities $W(x)$, a and c to be functions of time, and let the shear modulus become the appropriate viscoelastic operator. Assume that the rheology of the medium is that of the three element viscoelastic body shown in Fig. 3, with the shear moduli of the elastic elements being μ_1 and μ_2, and the viscosity of the dashpot η.

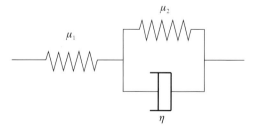

Figure 3 Three element linear viscoelastic solid

In the case of slow extension of the crack, for the limiting case of small-scale yielding, we find that the critical size of the slip-weakening zone $(c-a)$ is given by the solution to

$$\delta_c \simeq D(c-a)[\tau_s(c) - \tau_d(c)]\left\{\frac{1}{\mu_1} + \frac{1}{\mu_2} - \frac{1}{\mu_2}\exp\left[-\frac{\mu_2(c-a)}{\eta\dot{c}}\right]\right\}, \quad (7)$$

which is the same as (6) if the time-dependent rigidity

$$\mu(t) = \left\{\frac{1}{\mu_1} + \frac{1}{\mu_2} - \frac{1}{\mu_2}\exp\left[-\frac{\mu_2(c-a)}{\eta\dot{c}}\right]\right\}^{-1} \quad (8)$$

is used in place of μ. Eliminating $(c-a)$ between (5) and (7), we have the cohesive modulus for the viscoelastic case:

$$K_c(c, \dot{c}) = A^{-1}\left\{\frac{\delta_c[\tau_s(c) - \tau_d(c)]}{1/\mu_1 + 1/\mu_2 - (1/\mu_2)\exp[-(\mu_2/\eta cC^2)[K_c(c, \dot{c})/(\tau_s(c) - \tau_d(c))]^2]}\right\}^{1/2} \quad (9)$$

where the coefficient $A = CD^{-1/2}$ is approximately 3/4. From Fig. 3 and (8) we see that the time-dependent rigidity is bounded below by its stationary limit $(1/\mu_1 + 1/\mu_2)^{-1}$ and above by its instantaneous rupture limit μ_1; thus the cohesive modulus $K_c(c, \dot{c})$ has a lower bound

$$K_c(c, 0) = A^{-1}\left\{\frac{\mu_1\mu_2}{\mu_1 + \mu_2}\delta_c[\tau_s(c) - \tau_d(c)]\right\}^{1/2} \quad (10)$$

and an upper bound

$$K_c(c, \infty) = A^{-1}\{\mu_1\delta_c[\tau_s(c) - \tau_d(c)]\}^{1/2}. \quad (11)$$

From the distribution of the initial stress, and the static and dynamic frictional stresses, we may calculate the stress intensity factor $K_{\text{III}}(c)$ from (4), and the lower and upper bounds of the cohesive modulus $K_c(c, 0)$ and $K_c(c, \infty)$, from (10) and (11) respectively. If $K_{\text{III}}(c) \leq K_c(c, 0)$ for a particular model, a crack of semilength c does not extend. If $K_{\text{III}}(c) \gtrsim K_c(c, 0)$, a crack of semilength c begins to extend slowly. If $K_{\text{III}}(c) \geq K_c(c, \infty)$, a crack of semilength c extends instantaneously. For the intermediate cases, $K_c(c, 0) \leq K_{\text{III}}(c) \leq K_c(c, \infty)$, rupture will occur quasistatically. To determine the quasistatic motion of the crack edge in the intermediate case, we apply the fracture criterion

$$K_{\text{III}}(c) = K_c(c, \dot{c}), \quad (12)$$

where the stress intensity factor K_{III} is given by (4) with c a function of time. Substitution of the fracture criterion into (9) leads to an ordinary first-order differential equation for $c(t)$.

$$\frac{dc}{dt} = -\frac{\mu_2}{\eta C^2}\left[\frac{K_{\text{III}}(c)}{\tau_s(c) - \tau_d(c)}\right]^2 \frac{1}{\ln|1 + \mu_2/\mu_1 - \mu_2/\mu_1[K_c(c, \infty)/K_{\text{III}}(c)]^2|}. \quad (13)$$

A result equation similar to (13) is obtained for the Dugdale model [Yamashita 1981, equation (2.16)]. This is not unexpected since both formulations use the same rheological model. The slight differences are due to the differences in the fracture criteria.

As expected, from (13), the rupture velocity dc/dt approaches zero as $K_{\text{III}}(c) \to K_c(c, 0)$ and becomes infinite as $K_{\text{III}}(c) \to K_c(c, \infty)$. Therefore, the point of intersection of the curves $K_{\text{III}}(c)$ and $K_c(c, 0)$ determines the position of the crack edge at which rupture initiates or stops, and that of the curves $K_{\text{III}}(c)$ and $K_c(c, \infty)$ determines the point at which the nature of the rupture changes from quasistatic to instantaneous extension, or vice versa. For a given model, we calculate the position of these points of intersection numerically and determine the quasistatic rupture history by numerical integration of (13). In what follows we present some of the results without describing numerical details.

4 An 'isolated earthquake'

Before embarking on an analysis of models of spatially fluctuating stresses. we consider several examples of smoothly varying stresses. In the simplest case we take the initial stress, and the static and dynamic frictional stresses all to be uniform (Fig. 4a). The stress intensity factor varies as the square root of the crack size c and the cohesive moduli $K_c(c, 0)$ and $K_c(c, \infty)$ are constants (Fig. 4b) (see equations 4, 10 and 11). Thus, in a uniform stress field, a crack with a critical initial semilength extends at first slowly, and then at an increasing rate up to a velocity comparable to elastic wave velocities; once it reaches this catastrophic state, it continues to grow indefinitely. This crack never stops (Fig. 4c).

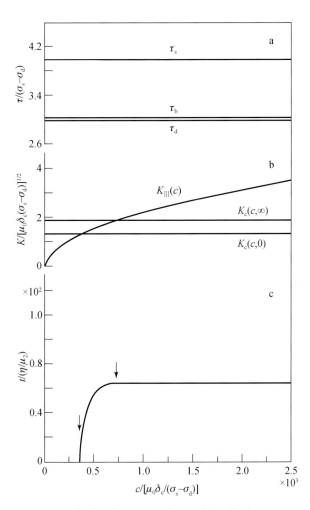

Figure 4 Crack extension in a uniform field
(a) The pre-stress $\tau_b(x)$, stat ci frictional stress $\tau_s(x)$ and dynamic frictional stress $\tau_d(x)$. (b) The stress intensity factor $K_{III}(c)$ and the lower and the upper bounds of the cohesive modulus $K_c(c, 0)$ and $K_c(c, \infty)$ as functions of the position of the crack edge. (c) The position of the crack edge c as a function of time. The arrows indicate the positions where the rupture phase changes. Numerical values are for the case $\tau_s = 4.0$, $\tau_b = 3.05$, $\tau_d = 3.0$, $\mu_1/\mu_2 = 1.5$ in the units indicated. The difference $(\tau_b - \tau_d)$ is small for graphical convenience, $\mu_0^{-1} = \mu_1^{-1} + \mu_2^{-1}$

In this as well as in all the succeeding examples, we have calculated the crack history in the high-velocity regime $K_{\mathrm{III}} > K_{\mathrm{c}}(c, \infty)$, $\dot{c} \to \infty$, as though the crack had continued to be quasistatic. This is evidently wrong, since loss of stress energy due to radiation is neglected. The problems of dynamical crack propagation fall outside the scope of this paper. Our neglect of the loss of energy due to radiation means that our estimates of crack lengths in the dynamical regime will be too great. We do not expect that our qualitative conclusions will be changed.

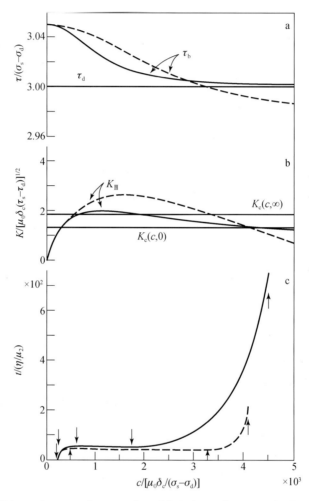

Figure 5 Decreasing positive (solid curve) or negative stress drop (dashed curve) as a mechanism of stopping the extension of a crack
In this example, we assume that $\tau_{\mathrm{s}} = 4.0$, $\tau_{\mathrm{d}} = 3.0$, $\tau_{\mathrm{b}}(x) = \dfrac{0.05}{1 + (x/1030)^2} + 3.0 \text{(solid)}$, and $\tau_{\mathrm{b}}(x) = \dfrac{0.08}{1 + (x/2500)^2} + 2.97 \text{(dashed)}$

As a second example (Fig. 5), we consider a case in which the crack is assumed to extend either into a region in which the stress drop decreases slowly to zero (solid curve) or into a region where the stress drop is negative (dashed curve). The inner region of large positive stress drop, can be considered to be an asperity. The initial episode of acceleration (starting at $c = 613$ and 480, respectively) is followed by one of acceleration due to the decreasing stress intensity factor starting at $c = 1730$ and 3268 in the two cases. This crack ultimately comes to a complete stop at $c = 4448$ and 4083 in the two cases.

For the case in which cohesion is absent (Burridge & Halliday 1971), entry into a region with a negative stress drop is necessary to stop the crack. In the presence of cohesion, a negative stress drop is not necessary for cessation of rupture; an alternative mechanism is an excessive cohesion in the path of extension of the crack

(Knopoff, Mouton & Burridge 1973; Chatterjee & Knopoff 1983). In this case (Fig. 6) we assume that the crack enters into a region with higher strength τ_s, but that the stress drop ($\tau_b - \tau_d$) is uniform. The outer region of large strength but uniform stress-drop, can be considered to be a barrier. As the crack extends, the resistance to crack extension becomes stronger, rupture slows down and finally stops just as in the case of the problem illustrated in Fig. 5. In both cases there is an episode of post-seismic creep between the time that the rupture velocity decreases below seismic velocities, and the time that the crack finally comes to rest.

In the examples of Figs 5 and 6 we have assumed that the dynamic shear stress or sliding friction τ_d is uniform for simplicity. Thus the non-uniform stress drop and the non-uniform cohesive modulus are the consequence of non-uniform pre-stress τ_b and of non-uniform static frictional stress, in the two cases. A non-uniform dynamic frictional stress introduces slightly more complexity into the crack histories since both the stress drop and the cohesion depend on this quantity. As the crack extends into a region with higher dynamic frictional stress, the stress intensity factor decreases but the cohesive modulus also decreases. Thus, the rate of growth of the crack may either decelerate or accelerate.

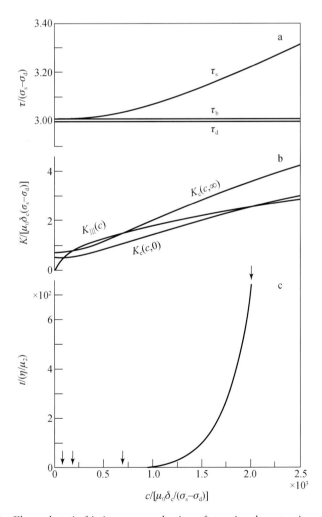

Figure 6 Elevated static friction as a mechanism of stopping the extension of a crack

In this example the quantities, scaled as shown, are $\tau_b = 3.008$, $\tau_d = 3.0$, $\tau_s = 4.1 - \dfrac{1.092}{1 + (x/4000)^2}$

We conclude that an 'isolated earthquake' can be simulated by the extension of a crack with a critical initial length in a non-uniform field; that the entry of the crack into a region with reduced stress drop and/or a higher cohesive modulus is necessary for the stopping of the crack, and that, in general, an 'isolated earthquake' is usually preceded by pre-seismic creep and is followed by post-scismic creep as well. We show below that a sufficient condition for an isolated earthquake to occur is that the stresses and frictions change smoothly, as shown in Figs 5 and 6; these are not necessary conditions since isolated earthquakes may occur in the presence of fluctuations in pre-stress and frictions; then, again, these fluctuations may give rise to more complex earthquake sequences. Finally. we note that the presence of pre-or post-shock creep is due to the smooth variation of the relevant quantities with distance; in the case of jump discontinuities that exceed the gap between the two bounds, one or both of the creep phases may not take place.

5 Earthquake sequences

In each of the cases considered in the preceding section, the rupture history was smooth which, as we now illustrate, is a consequence of the smooth variation of the frictional stresses and the pre-stress. If we add a fluctuating component to these stresses the character of the rupture history can change significantly, if the fluctuating component is large enough. To simulate complex sequences of earthquakes by introducing spatial fluctuations in the stress distributions into the calculation. we use a numerical procedure similar to that used by Cohen (1978) and Andrews (1980). The pre-stress is assumed to consist of two parts: a coherent part. which represents the general stress level and the largest scale inhomogeneities in the simulation, and a random part which represents a short-wavelength fluctuation about the coherent stress. We assume that the coherent pre-stress $S_b(x)$ varies monotonically in the same way as in the preceding section and that the random part has a cosine transform $A_b(k)$ that is a stochastic function of the wave number k. We also assume that the static and dynamic frictional stresses are of the same form as the pre-stress, with variables $S_s(x)$, $S_d(x)$, $A_s(k)$, $A_d(k)$. Thus

$$\frac{K_{\mathrm{III}}}{(\pi c)^{1/2}} = \frac{2}{\pi} \int_0^\infty \frac{S_b(\zeta) - S_d(\zeta)}{(c^2 - \zeta^2)^{1/2}} d\zeta + \frac{1}{\pi} \int_0^\infty [A_b(k) - A_d(k)] J_0(kc) dk. \tag{14}$$

In the numerical simulations, the functions $A_s(k)$, $A_b(k)$ and $A_d(k)$ are chosen to be uniformly distributed random numbers

$$0 \leqslant |A_s(k)| \leqslant \left(\frac{\lambda \Lambda}{2}\right)^{1/2} \alpha_s,$$
$$0 \leqslant |A_b(k)| \leqslant \left(\frac{\lambda \Lambda}{2}\right)^{1/2} \alpha_b, \tag{15}$$
$$0 \leqslant |A_d(k)| \leqslant \left(\frac{\lambda \Lambda}{2}\right)^{1/2} \alpha_d,$$

in the wave number interval $2\pi/\Lambda \leqslant k \leqslant 2\pi/\lambda$, where α_s, α_b and α_d are indices which describe the overall amplitude of the fluctuations and λ and Λ are the shortest and the longest wavelengths of the fluctuations. In the examples of Figs 7a-10a we use $\lambda/\Lambda = 64^{-1}$. The stress intensity factor (4) is more sensitive to small scale fluctuations in stress than is the cohesive modulus (5).

If the peak amplitudes of the fluctuations in the stresses are sufficiently small, the characteristics of the rupture and creep history are unchanged from the cases of no fluctuation discussed above, as might have been expected: there are only two intersections of the curve of stress intensity K_{III} with the upper and lower critical

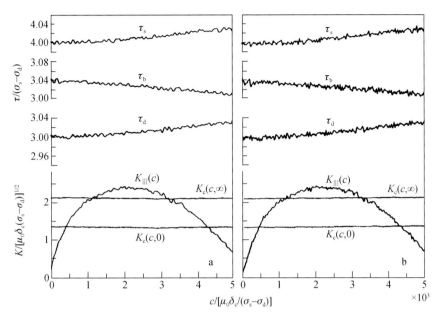

Figure 7　Simulation of an 'isolated earthquake'

$\alpha_s = \alpha_b = \alpha_d = 0.005$; (a) $\lambda/\Lambda = 64^{-1}$; (b) $\lambda/\Lambda = 128^{-1}$. The coherent parts of the stresses are $S_b(x) = \dfrac{0.074}{1 + (x/6400)^2} + 2.9635$,

$S_d(x) = \dfrac{-0.08}{1 + (x/6400)^2} + 3.08$, $S_s(x) = \tau_d(x) + 1.00$

thresholds of cohesive stress (Fig. 7a). There are small fluctuations in the pre-and post-seismic creep histories, but these do not influence the isolated character of the earthquake.

As the amplitudes of the stress fluctuations increase (Fig. 8a), the rupture history becomes more intricate and consists of a number of events that alternate between instantaneous rupture and quasistatic extension (curve 2, Fig. 10a). We identify the largest instantaneous rupture as a main shock, the small events before the main shock as foreshocks, and those after the main shock as aftershocks. We find that an earthquake sequence with foreshocks, main shock and aftershocks is usually associated with moderate fluctuations in the stresses; these are characterized by moderate values of α in our simulation.

If we increase the value of α still further (Fig. 9a), the complete rupture history consists of a series of size-comparable events that alternate between instantaneous rupture and quasistatic extension episodes (curve 3, Fig. 10a). We identify this sequence as an earthquake swarm.

In our simulations the amplitude of the short wavelength inhomogeneities is characterized by $A(k)$ or equivalently, by α. A medium with a larger value of α is more inhomogeneous than that with a smaller one. The results displayed in Figs 7a-10a are qualitatively in agreement with the experimental results of Mogi (1963).

We have tested the model to evaluate the influence of inhomogeneities of shorter wave-length than the cases described in the preceding paragraphs. In Figs 7b-10b we display the results for cases of reduction of the ratio between the short-and long-wavelength cut-offs from $\lambda/\Lambda = 64^{-1}$ to $\lambda/\Lambda = 128^{-1}$. In general, the change in the short-wavelength cut-off in the fluctuations of the stresses does not influence the type of earthquake sequence significantly; however, the inclusion of greater amounts of short wavelengths in the stress fluctuations does increase the frequency of occurrence of the earthquakes.

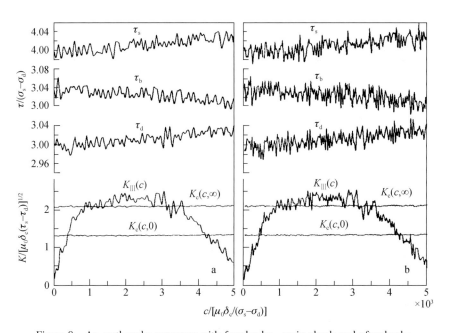

Figure 8 An earthquake sequence with foreshocks, main shock and aftershocks

$\alpha_s = \alpha_b = \alpha_d = 0.02$; (a) $\lambda/\Lambda = 64^{-1}$; (b) $\lambda/\Lambda = 128^{-1}$. The coherent parts of the stresses are the same as for Fig. 7

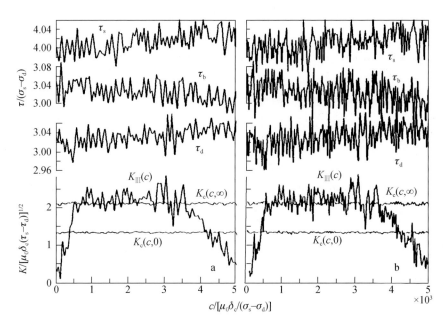

Figure 9 An earthquake swarm

$\alpha_s = \alpha_b = \alpha_d = 0.04$; (a) $\lambda/\Lambda = 64^{-1}$; (b) $\lambda/\Lambda = 128^{-1}$. The coherent parts of the stresses are the same as for Fig. 7

6 Effect of stress level

The cohesive modulus increases monotonically as the coherent part of the difference between the breaking strength and the sliding friction increases monotonically. The type of earthquake sequence that results through such a continuous change varies progressively from an 'isolated earthquake', to an earthquake sequence with foreshocks, main shock and after-shocks, then to an earthquake swarm, and finally to an aseismic event. Conversely, as the coherent part of the

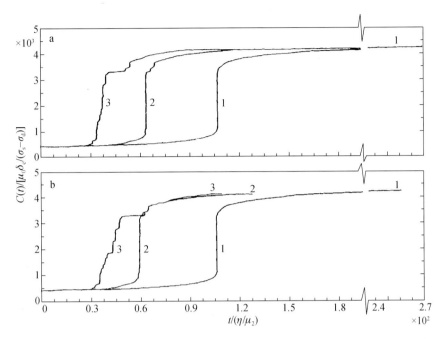

Figure 10   Crack edge locus as a function of time for the three cases shown in Figs. 7-9

Curve 1 is the 'isolated earthquake' presented in Fig. 7; curve 2, the earthquake sequence with foreshocks, main shock and aftershocks in Fig. 8; curve 3, the earthquake swarm in Fig. 9. The results for the two cases of isolated earthquake events are almost identical. The frequency of occurrence in the earthquake sequences increases as a result of the increase in the short-wavelength fluctuations in the stresses

stress drop increases, the type of earthquake sequence that results varies progressively in the reverse of the above order. In the examples of Figs 11 and 12 we assume that the breaking strength and the sliding friction as well as the fluctuation in the pre-stress are identical with those in Fig. 8a, while the coherent parts of the pre-stress are different. In these two figures, the result of Fig. 8a is reproduced as curve 2 for comparison. The results indicate that if the pre-stress levels are high, there is a reduction in the number of foreshocks and aftershocks as well as the amount of pre-stress and post-seismic creep (curves 1 and 2). In the extreme case that the pre-stress is very low (curve 4), a creep-event occurs without any accompanying instantaneous rupture. We identify this event as an aseismic creep event, i. e. as a silent earthquake. A silent earthquake is, on this model, an event for which the stress intensity in the creep phase never reaches the upper limit of the cohesive modulus $K_c(c, \infty)$.

7   Conclusions

We have modelled an earthquake sequence as a complex extension of a shear crack in a viscoelastic medium under the influence of non-uniform stresses. We can interpret the inhomogeneity of the physical properties in the fault plane as being due to the overall effect of geometrical irregularities such as microcracks, joints, secondary faults, etc. The crack stops when it penetrates deeply into a region with lower or negative stress drop and/or it encounters a region with higher cohesive modulus. We find that in general the crack ruptures instantaneously after experiencing pre-shock creep and comes to rest after a post-seismic creep episode. The simulation indicates that an isolated earthquake can be interpreted as the extension of a crack in a smooth non-uniform field; an earthquake sequence consisting of foreshocks, main shock and aftershocks is the result of the extension of a crack

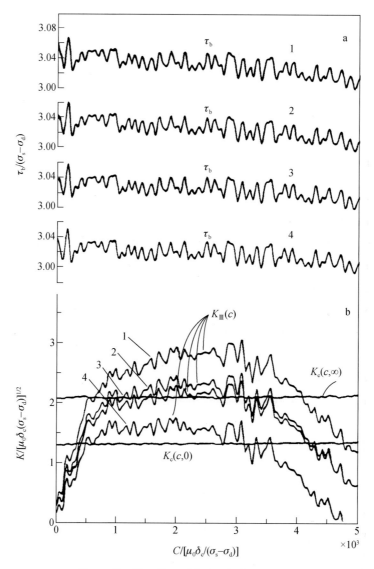

Figure 11 The effect of the level of pre-stress

As the coherent part of the pre-stress decreases, the type of earthquake changes from an 'isolated earthquake' (curve 1) to an earthquake sequence with foreshocks, main shock and aftershocks (curve 2) then to an earthquake swarm (curve 3) and finally to an aseismic creep event (or silent earthquake) (curve 4). All the parameters are identical with those in Fig. 8a and curve 2 in Fig. 10a except for the coherent part of the pre-stress:

$$S_b(x) = \frac{a}{1+(x/6400)^2} + b,$$

where

	a	b
Case 1	0.0830	2.9610
Case 2	0.0740	2.9635
Case 3	0.0620	2.9730
Case 4	0.0595	2.9685

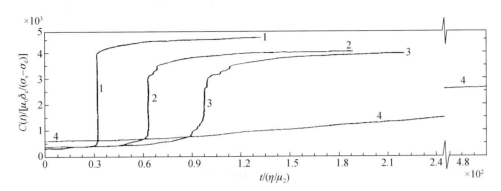

Figure 12   Crack edge locus as a function of time for the cases shown in Fig. 11

in a nonuniform field having moderate fluctuations; earthquake swarms arise as a consequence of the extension of a crack in a highly inhomogeneous field.

Our simulations show that an increase in the level of the short-wavelength components of the fluctuations in the stress tend to increase the frequency of occurrence of earthquakes, but do not cause a significant change in the type of earthquake sequence that is obtained.

We find that the types of earthquake sequences are controlled by both the inhomogeneity and the general level of the stresses. Usually isolated earthquakes occur in the presence of high pre-stress and/or low breaking stress. Earthquake sequences with foreshocks, main shock and aftershocks, occur in the presence of moderate pre-stress and breaking stress. Earthquake swarms occur in the cases of low pre-stress and/or high breaking strength. Very low pre-stress and/or very high breaking strength are preferred conditions for the occurrence of aseismic creep.

There can be little doubt that inhomogeneities play an important part in the earthquake process.

References

Andrews, D. J., 1976. Rupture propagation with finite stress in antiplane strain, *J. Geophys. Res.*, **81**, 3575–3582.

Andrews, D. J., 1980. A stochastic fault model: 1. Static case, *J. Geophys. Res.* **85**, 3867–3877.

Barenblatt, G, I., 1959. The formation of equilibrium cracks during brittle fracture: General ideas and hypothesis, axially symmetric cracks, *J. Appl. Math. Mech.*, **23**, 434–444.

Barenblatt. G. I., 1962. The mathematical theory of equilibrium cracks in brittle fracture, *Adv. Appl. Mech.*, **7**, 55–129.

Benioff. H., 1951. Earthquakes and rock creep. Part 1: Creep characteristics of rocks and the origin of aftershocks, *Bull. Seism. Soc. Am.*, **41**, 31–62.

Burridge, R. & Knopoff, L., 1967. Model and theoretical seismology. *Bull. Seism. Soc. Am.*, **57**, 341–371.

Burridge, R. & Halliday, G. S., 1971. Dynamic shear cracks with friction as models for shallow focus earthquakes, *Geophys. J. R. Astr. Soc.*, **25**, 261–283.

Burridge, R., Conn. G. & Freund, L. B., 1979. The stability of a rapid mode II shear crack. *J. Geophys. Res.*, **85**, 2210–2222.

Chatterjee. A. K. & Knopoff. L., 1983. Bilateral propagation of a spontaneous two-dimensional-anti-plane shear crack under the influence of cohesion. *Geophys. J. R. Astr. Soc.*, **73**, 449–473.

Chen, Y. T. & Knopoff, L., 1986a. Static shear crack with a zone of slip-weakening. *Geophys. J. R. Astr. Soc.*, **87**, 1005–1024.

Chen. Y. T. & Knopoff. L. . 1986b. The quasistatic extension of a shear crack with a zone of slip-weakening in a viscoelastic medium. *Geophys. J. R. Astr. Soc.*, **87**, 1025–1039.

Cohen. S. C., 1977. Computer simulation of earthquakes. *J. Geophys. Res.*, **82**, 3781–3796.

Cohen. S. C., 1978. The viscoelastic stiffness model of seismicity. *J. Geophys. Res.*, **83**, 5425–5431.

Das, S. & Aki. K., 1977a. A numerical study of two-dimensional spontaneous rupture propagation, *Geophys. J. R. Astr. Soc.*, **50**, 643–688.

Das. S. & Aki. K., 1977b. Fault plane with barriers: a versatile earthquake model. *J. Geophys. Res.*, **82**, 5658–5670.

Dieterich, J. H., 1972. Time-dependent friction as a possible mechanism for aftershocks, *J. Geophys. Res.*, **77**, 3771–3781.

Dugdale, D. S., 1960. Yielding of a steel sheet containing slits. *J. Mech. Phys. Solids*. **8**, 100–104.

Ida. Y., 1972. Cohesive force across the tip of a longitudinal shear crack and Griffith's specific surface energy, *J. Geophys. Res.*, **77**, 3796–3805.

Kagan. Y. Y., 1981, Spatial distribution of earthquakes: the four-point moment function. *Geophys. J. R. Astr. Soc.*, **67**, 719–733.

Kanamori. H., 1978. Use of seismic radiation to infer source parameters, *Proc. Conf. III Fault Mechanics arid its Relation to Earthquake Prediction*, U. S. Geol. Surv. Open File Rep., 78–380, 283–318.

Kasahara. K., 1975. Aseismic faulting following the 1973 Nemuro-Oki earthquake. Hokkaido, Japan (a possibility). *Pure Appl. Geophys.*, **113**, 127–139.

Knopoff, L., 1972. Model for aftershock occurrence, in *Flow and Fracture of Rock*, ed. Heard, H. C., *Monogr. Am. Geophys. Un.*, **16**, 259–263.

Knopoff. L., Mouton, J. O. & Burridge. R., 1973. The dynamics of a one-dimensional fault in the presence of friction. *Geophys. J. R. Astr. Soc.*, **35**, 169–184.

Kostrov. B. V. & Nikitin, L. V., 1970. Some general problems of mechanics of brittle fracture. *Arch. Mech. Stosowanej*, **6** (22), 749–776.

Kostrov. B. V. & Das, S., 1982. Idealized models of fault behaviour prior to dynamic rupture. *Bull. Seism. Soc. Am.*, **72**, 679–703.

Kostrov, B. V., Nikitin, L. V. & Flitman, L. N., 1970. The expansion of cracks in viscoelastic bodies, *Bull. Acad. Sci. USSR. Geophys. Ser.*, **7**, 413–420.

Lee. E. H., 1955. Stress analysis in visco-elastic bodies. *Quart. Appl. Math.*, **13**, 183–190.

Leonov, M. Y. & Panasyuk, V. V., 1959. The development of the smallest cracks in a solid (in Ukrainian), *Prikladna Mekhanika (Applied Mechanics)*, **5**, 391–401.

Mogi, K., 1963. Some discussions on aftershocks, foreshocks, and earthquake swarm. The fracture of a semi-infinite body caused by an inner stress origin and its relation to the earthquake (Third paper). *Bull. Earthq. Res. Inst. Tokyo Univ.*, **41**, 615–618.

Palmer. A. C. & Rice. J. R., 1973. The growth of slip surfaces in the progressive failure of over-consolidated clay, *Proc. R. Soc. A*, **332**, 527–548.

Rice. J. K.. 1979. The mechanics of quasi-static crack growth, in *Proc. 8th U. S. National Congr. Appl. Mech.*, 1978, pp. 191–216. ed. Kelly, R. E., Western Periodicals, Hollywood.

Rice, J. R., 1980. The mechanics of earthquake rupture, in *Physics of the Earth's Interior*, pp. 555–649, eds Dziewonski, A. M. & Boschi, E., North-Holland. Amsterdam.

Rudnicki, J. W. & Kanamori. H., 1981. Effects of fault interaction on moment. stress drop and strain energy release, *J. Geophys. Res.*, **86**, 1785–1793.

Shemyakin, E. I., 1955. Lamb's problem for a medium with an elastic aftereffect (in Russian), *Dokl. Akad. Nauk SSSR*, **104**, 193–196.

Shimazaki, K., 1974. Pre-seismic crustal deformation caused by an underthrusting oceanic plate, in eastern Hokkaido, Japan, *Phys. Earth Planet. Int.*, **8**, 148–159.

Yamashita, T., 1980. Quasistatic crack extensions in an inhomogeneous viscoelastic medium-a possible mechanism for the occurrence of aseismic faulting. *J. Phys. Earth*, **28**, 309–326.

Yamashita, T., 1981. Quasistatic crack extensions in an inhomogeneous viscoelastic medium-proposal of a unified seismic source model for the diverse earthquake rupture processes, *J. Phys. Earth*, **29**, 283–304.

Yin, X. C. & Zheng, T. Y., 1983. A rheological model for the process of preparation of an earthquake, *Sci. Sin. Beijing B*, 285–296.

Zheng, T. Y. & Yin, X. C., 1983. The subcritical extension of faulting and the process of preparation of earthquakes (in Chinese), *Kexue Tongbao (Science Bull.)*, **28**, 1325–1328.

Spontaneous growth and autonomous contraction of a two-dimensional earthquake fault[*]

Y. T. Chen[1], X. F. Chen[2] and L. Knopoff[1]

1. Institute of Geophysics and Planetary Physics, University of California, Los Angeles
2. Institute of Geophysics, State Seismological Bureau

Abstract We numerically study a model of an earthquake as a spontaneous rupture on a fault. We assume that a two-dimensional antiplane shear crack initiates at a point in an infinite, homogeneous and isotropic elastic medium and subsequently propagates with variable velocity under the influence of nonuniform stress drop on the crack and nonuniform cohesive resistance at the crack edges. To begin with, we analyze the dynamical rupture process immediately after nucleation. We determine the subsequent extension of each edge by solving an ordinary differential equation of the first order, as well as the dynamical stress in the regions between each edge and the nearest wave front. Application of this procedure to both edges of the crack in an alternating manner yields the complete history of extension of the crack. We use the critical stress-intensity fracture criterion to determine the conditions on propagation and arrest of the crack. To verify the accuracy and the stability of our numerical technique we compare it with some special cases in which analytical solutions are available. The comparison indicates that the numerical results are in good agreement with the analytical ones obtained previously by Knopoff and Chatterjee (1982) and Chatterjee and Knopoff (1983). We apply this technique to the computation of the dynamical extension of the crack for several representative cases. Among them are: (1) uniform stress drop but nonuniform cohesion (a barrier model), (2) nonuniform stress drop but uniform cohesion (an asperity model) and (3) nonuniform stress drop and nonuniform cohesion (a combination of both the barrier and the asperity models). The numerical results indicate that the heterogeneities of both the stress drop and the cohesion are the main factors which control the growth, cessation and healing of the crack, and that the complexities in the seismic radiation are caused by the complex healing process as well as by the complex rupture propagation. In contrast to the asperity model, the barrier model is characterized by abrupt changes in the slope of the source-time function and the theoretical seismogram. The collision of the stress pulses from each pair of barriers is the potential source of the crack fission. In the general model, the fission process occurs in a complex sequence, and the effects of this process have to be taken into account in the interpretation of observations.

1 Introduction

An earthquake source can be represented by a displacement discontinuity or a dislocation on an internal surface. If the slip on the internal surface of dislocation as a function of position and time (slip function) is known,

---

[*] 本文发表于 *Tectonophysics*, 1987 年, 第**144**卷, 第 1/3 期, 5–17.

the displacement at any point in the medium can be determined from a representation theorem (Maruyama, 1963; Burridge and Knopoff, 1964). Thus, one of the main problems in the mechanics of earthquake faulting is to find the slip function. The problems of earthquake faulting can be approached either kinematically or dynamically (Kostrov, 1975). In the kinematic dislocation model of the earthquake source, the slip function is specified a priori and the displacement field is deduced therefrom. In contrast, the slip function is not assigned a priori in the dynamical model of the earthquake source, but is a consequence of the redistribution of physically reasonable stresses specified a priori, and an appropriate fracture criterion. We call the latter the crack model.

In the crack model, earthquakes are assumed to be the consequence of a dynamical process in which the rupture on a fault occurs spontaneously by the sudden release of stress due to frictional instability. Once the rupture initiates, the slip motion on the fault surface develops and grows spontaneously, in a way that depends on the stress drop on the fault plane and on the strength of the material in the fault zone. This dynamical problem of earthquake faulting is difficult to approach, even for the simplest antiplane shear cases. However, since the pioneering work of Kostrov (1966), a number of dynamical crack problems have been solved analytically and numerically. Kostrov (1966) solved the problem of a semi-infinite, instantaneous, antiplane shear crack in an infinite medium. His method is also applicable to finite cracks, and has been successfully applied by many authors to study the dynamical problem of finite cracks. Burridge (1969) used a numerical technique to study the problem of in-plane as well as antiplane finite shear cracks with fixed rupture velocity. Using the Cagniard-De Hoop technique, Richards (1973, 1976) analytically solved the problem of the radiation from a self-similar, growing, elliptical shear crack. In his study, the crack grows with subsonic rupture velocities while preserving an elliptical shape. The crack never stops. By a finite difference calculation Madariaga (1976) solved a similar problem in which a circular shear crack grows at a fixed velocity and stops suddenly. In these studies, the rupture velocities are assigned a priori. Das (1976) and Das and Aki (1977a,b) used a numerical technique to study the problem of the propagation of two-dimensional shear cracks in an infinite, homogeneous medium. They used a critical stress-jump fracture criterion to determine the rupture history. The critical stress-jump is the finite difference approximation to the critical stress intensity factor used in Irwin's fracture criterion, and is useful for numerical calculations. Andrews (1976) used the finite difference method to solve the dynamical problem of a two-dimensional shear crack with a zone of slip-weakening. More recently, Mikumo and Miyatake (1978), Miyatake (1980a,b), Das (1980, 1981), Day (1982a, b) and Virieux and Madariaga (1982) have obtained numerical solutions to three-dimensional cases. Recent studies indicate that neither the distribution of the stress on the fault plane of an earthquake nor the strength of the material in the fault zone is uniform. Thus, there is a need for understanding how an earthquake rupture initiates, grows and is arrested and healed on a fault plane, as well as the way in which the seismic radiation is related to the process of the complex rupture of an earthquake fault. To solve this problem numerically, methods with negligible numerical instability and dispersion are required. In this study, we describe a numerical technique for solving the problem of the spontaneous growth and propagation of a two-dimensional earthquake fault, assuming that the stress drop on the crack and the cohesive resistance at the crack edges are both heterogeneous. This method is an alternative to the iteration method used by Knopoff and Chatterjee (1982) and Chatterjee and Knopoff (1983) for the same problem. Because the finiteness of the fault is one of the main factors which influences the nature of the radiation of seismic waves, these two-dimensional shear crack models are not expected to be satisfactory for providing a thorough understanding of the seismic radiation. Nevertheless, it is expected that radiation from three-dimensional models would preserve many of the features that are revealed in the two-dimensional cases. Thus, the solutions to the two-dimensional models to be described here may provide insights into the process of earthquake faulting and radiation.

2 Theory and method

We study the problem of the spontaneous rupture and subsequent growth of a two-dimensional earthquake fault. We assume that a two-dimensional antiplane shear crack in the plane $y=0$ initiates at $t=0$, $x=0$ (Fig. 1), and subsequently propagates bilaterally in the x-direction with variable velocity under the influence of nonuniform stress drop on the crack and nonuniform cohesive resistance at the crack edges. In this problem of an antiplane shear crack, the only component of the displacement $W(x, y, t)$, is in the z-direction. Let σ_0 be the prestress and τ the stress redistributed in the fracture. The total stress σ is given by:

$$\sigma = \sigma_0 + \tau \tag{1}$$

For the antiplane problem, the only nonvanishing components of the redistributed stress are:

$$\tau_{yz} = \mu \frac{\partial W}{\partial y} \tag{2}$$

and

$$\tau_{xz} = \mu \frac{\partial W}{\partial x} \tag{3}$$

The equation of motion is:

$$\frac{1}{\beta^2} \frac{\partial^2 W}{\partial t^2} = \frac{\partial^2 W}{\partial x^2} + \frac{\partial^2 W}{\partial y^2} \tag{4}$$

where $\beta = (\mu/\rho)^{1/2}$ is the shear wave velocity, μ the rigidity and ρ the density of the medium. The boundary conditions are that the stress drop on the crack is $T(x, t)$, a prespecified quantity:

$$-\mu \frac{\partial W(x, y, t)}{\partial y}\bigg|_{y=0} = T(x, t), \quad x_1(t) \leq x \leq x_2(t) \tag{5}$$

where $x_1(t)$ and $x_2(t)$ are the locations of the left (negative) and right (positive) edges of the crack (Fig. 1). The displacement W is continuous outside the fractures region of the crack across $y=0$, but discontinuous across the fractured region of the crack, and τ_{yz} is continuous across $y=0$.

We assume that initially, the displacements and velocities are zero everywhere in the medium:

$$W(x, y, t) = 0 \quad \frac{\partial W(x, y, t)}{\partial t} = 0 \quad t = 0 \tag{6}$$

The integral equation for the displacement on the crack is (Kostrov, 1966; Aki and Richards, 1980):

$$W(x_0, y_0, t_0) = -\frac{1}{\pi} \iint_S W_y(x, 0^+, t) \\ \times [(t-t_0)^2 - [(x-x_0)^2 \\ + y_0^2]/\beta^2]^{-\frac{1}{2}} dx dt \tag{7}$$

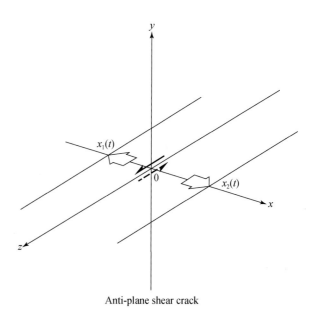

Anti-plane shear crack

Fig. 1 A bilaterally propagating crack at time t

$x_1(t)$ and $x_2(t)$ are the locations of the left (negative) and the right (positive) edges at time t, respectively

where S is the region that lies inside the characteristic cone in (x, y, t) space, defined by:

$$(x - x_0)^2 + y_0^2 \leq \beta^2 (t - t_0)^2, \quad t_0 \geq t \geq 0. \tag{8}$$

Since we do not know $W_y(x, 0^+, t)$ over the entire region of integration, eqn. (7) does not give the solution immediately. $W_y(x, 0^+, t)$ is known in the fractured region but not in the region between the edges of the crack and the wavefronts, i.e., in the disturbed region. We must determine $\tau_{yz}(x, 0^+, t)$ in the disturbed regions as well as the locations of the crack edges as functions of time. Once the trajectories of the crack edges in the (x, t) plane and the dynamic stresses in the disturbed regions are found, the slip function on the crack can be determined by carrying out the integration in eqn. (7). Many methods have been proposed to solve this problem; among them are those of Burridge (1969), Andrews (1976), Madariaga (1976), Das and Aki (1977a), Mikumo and Miyatake (1978), Miyatake (1980a, b), Das (1981), Knopoff and Chatterjee (1982), Chatterjee and Knopoff (1983) and Yoshida (1985). Below, we describe a numerical technique for solving this problem.

Following a procedure developed by Kostrov (1966), we deduce the simultaneous equations for the determination of the motion of the crack and the dynamical stress in both the positive and negative directions. Let the characteristic coordinates be:

$$\begin{cases} \xi = \dfrac{1}{\sqrt{2}}(\beta t - x), \\ \eta = \dfrac{1}{\sqrt{2}}(\beta t + x). \end{cases} \tag{9}$$

If the crack propagates subsonically, the trajectory of the crack edge lies above the characteristic line. If the trajectory is the characteristic line, the edge of the crack propagates with the shear wave velocity. In the undisturbed regions below the characteristics, the redistributed stress vanishes. The characteristics and the trajectories divide the upper $(x-t)$ half-plane into five regions. We denote the fractured region as S^{III}, the disturbed regions as S^{I} and S^{II}, and the undisturbed regions as S^{IV} and S^{V} (Fig. 2). In the new variables (ξ, η), the redistributed stress μW_y^{II} in the disturbed region S^{II} is:

$$\mu W_y^{\text{II}}(\xi_0, \eta_0) = \frac{1}{\pi \sqrt{\eta_0 - \eta_2(\xi_0)}} \left[\int_{\eta_1(\xi_0)}^{\eta_2(\xi_0)} T(\xi_0, \eta) \times \frac{\sqrt{\eta_2(\xi_0) - \eta}}{\eta_0 - \eta} d\eta \right.$$
$$\left. - \int_0^{\eta_1(\xi_0)} \mu W_y^{\text{I}}(\xi_0, \eta) \times \frac{\sqrt{\eta_2(\xi_0) - \eta}}{\eta_0 - \eta} d\eta \right] \tag{10}$$

where $\eta = \eta_1(\xi)$ and $\eta = \eta_2(\xi)$ are the loci of the crack edges $x = x_1(t)$ and $x = x_2(t)$, expressed in the characteristic coordinates, with ξ as an independent variable. Similarly, the redistributed stresses in the disturbed region S^{I} is:

$$\mu W_y^{\text{I}}(\xi_0, \eta_0) = \frac{1}{\pi \sqrt{\xi_0 - \xi_1(\eta_0)}} \left[\int_{\xi_2(\eta_0)}^{\xi_1(\eta_0)} T(\xi, \eta_0) \times \frac{\sqrt{\xi_1(\eta_0) - \xi}}{\xi_0 - \xi} d\xi \right.$$
$$\left. - \int_0^{\xi_2(\eta_0)} \mu W_y^{\text{II}}(\xi, \eta_0) \times \frac{\sqrt{\xi_1(\eta_0) - \xi}}{\xi_0 - \xi} d\xi \right] \tag{11}$$

where $\xi = \xi_1(\eta)$ and $\xi = \xi_2(\eta)$ are the loci of the crack edges $x = x_1(t)$ and $x = x_2(t)$, expressed in characteristic coordinates, with η as the independent variable.

Equations (10) and (11) describe the coupling between the dynamic stresses μW_y^{I} and μW_y^{II}, in the disturbed regions S^I and S^{II}. We can solve eqns. (10) and (11) for μW_y^I and μW_y^{II} provided we know the locations of both crack edges as functions of time, $\eta_1(\xi)$ and $\eta_2(\xi)$, or alternatively $\xi_1(\eta)$ and $\xi_2(\eta)$. Thus, we must determine the loci of the crack edges as functions of time by using an appropriate fracture criterion.

As the crack extends, there is a square root singularity at the advancing edge of the crack (Aki and Richards, 1980). Letting $\eta_0 \downarrow \eta_2(\xi_0)$, we find:

$$\mu W_y^{II}(\xi_0, \eta_0) \to \frac{k_2(\xi_0, \eta_2(\xi_0))}{\sqrt{\eta_0 - \eta_2(\xi_0)}} \qquad (12)$$

where:

$$k_2(\xi_0, \eta_2(\xi_0)) = \frac{1}{\pi}\left[\int_{\eta_1(\xi_0)}^{\eta_2(\xi_0)} \frac{T(\xi_0, \eta)}{\sqrt{\eta_2(\xi_0) - \eta}}d\eta - \int_0^{\eta_1(\xi_0)} \frac{\mu W^I(\xi_0, \eta)}{\sqrt{\eta_2(\xi_0) - \eta}}d\eta\right] \qquad (13)$$

which is related to the dynamical stress intensity factor $K_2(R)$ by:

$$K_2(R) = 2^{-1/4}\left[1 - \frac{\dot{x}_2(t_R)}{\beta}\right]^{1/2} k_2(R) \qquad (14)$$

where R represents $(\xi_p, \eta_2(\xi_p))$.

We employ the critical stress intensity fracture criterion (Irwin's fracture criterion) to determine the propagation and the arrest of the crack. The Irwin fracture criterion is extended to the dynamical problem assuming that during rupture propagation the relation:

$$K_2(R) = K_c(R) \qquad (15)$$

is satisfied where $K_2(R)$ is the dynamical stress intensity factor which depends on $\dot{x}_2(t_R)$, the instantaneous rupture velocity, and $K_c(R)$ is the prespecified critical dynamical stress intensity factor which is a measure of the strength of the material and is assumed to be a function of x_2 only. Substitution of (14) into (15) yields:

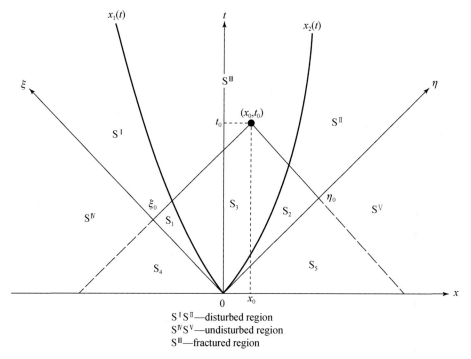

Fig. 2 The fractured regions S^{III}, the disturbed regions S^I and S^{II}, and the undisturbed regions S^{IV} and S^V in the (x, t) plane

$$2^{-1/4}g_2[(\eta_2(\xi_0)-\xi_0)/\sqrt{2}]\sqrt{1+\dot{\eta}_2(\xi_0)} = \int_{\eta_1(\xi_0)}^{\eta_2(\xi_0)}\frac{T(\xi_0,\eta)}{\sqrt{\eta_2(\xi_0)-\eta}}d\eta$$

$$-\int_0^{\eta_1(\xi_0)}\frac{\mu W_y^{\mathrm{I}}(\xi_0,\eta)}{\sqrt{\eta_2(\xi_0)-\eta}}d\eta, \quad (16)$$

where $g_2 = \pi K_c$ and is called the modulus of cohesion. Similarly, the dynamical stress condition near the left edge of the crack yields:

$$2^{-1/4}g_1[(\eta_0-\xi_1(\eta_0))/\sqrt{2}]\sqrt{1+\dot{\xi}_1(\eta_0)} = \int_{\xi_2(\eta_0)}^{\xi_1(\eta_0)}\frac{T(\xi,\eta_0)}{\sqrt{\xi_1(\eta_0)-\xi}}d\xi$$

$$-\int_0^{\xi_2(\eta_0)}\frac{\mu W_y^{\mathrm{II}}(\xi,\eta_0)}{\sqrt{\xi_1(\eta_0)-\xi}}d\xi. \quad (17)$$

Equations (16) and (17) are nonlinear differential integral equations for solving the motion of the bilaterally growing crack. Together with eqns. (10) and (11), they are simultaneous equations for the determination of the trajectories of the two crack edges and dynamical stresses in the two disturbed regions. Each pair of these eqns. (10) and (16), or (11) and (17) gives the motion of one edge and the dynamically redistributed stress in the region ahead of this edge from the dynamical stress drop on the growing crack, as well as the dynamically redistributed stress ahead of the opposite edge.

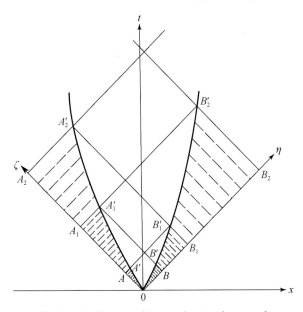

Fig. 3 A schematic diagram showing the procedures of the numerical technique for solving the motion of the crack

If we can derive the initial trajectories of both edges of the crack as well as the dynamically redistributed stresses in the disturbed region ahead of the edges, we can track forward, finite step by finite step, the motion of the crack edges and the dynamically redistributed stress ahead of the crack. This argument is in fact the basis for solving for the motion of the crack.

If the initial trajectories (OA' and OB') (Fig. 3) and the dynamically redistributed stress in the disturbed regions ($OA'A$ and $OB'B$) are known a priori, we can solve two first order independent ordinary differential equations to obtain the trajectories $A'A'_1$ and $B'B'_1$, for the motion of the crack edges in the intervals $0 \leq \eta \leq \overline{AB}$ and $0 \leq \xi \leq \overline{OA}$, as well as the dynamical redistributed stresses in the regions $AA_1A_1'A'$ and $BB_1B_1'B'$. Repeated application of this procedure yields the complete history of the growing crack and the dynamically redistributed stress in the disturbed regions.

3 Comparison with exact solutions

Chatterjee and Knopoff (1983) have shown that the bilateral extension of an antiplane shear crack in which the moduli of cohesion at the crack edges are proportional to the square root of the distance from the point of nucleation, and the dynamical stress drop on the crack is constant, has an exact solution. They found that the two edges of the crack propagate with constant velocities and hence that the crack is self-similar. They obtained the exact solution for the rupture velocities in closed form and also found the exact dynamical stress in the unfractured regions.

Assume that the moduli of cohesion in the neighborhood of the point of initiation can be approximated by the square root relations with coefficients of proportionality α_1 and α_2. We used the analytical solutions of Chatterjee and Knopoff (1983) to calculate the dimensionless ratios α_1/T and α_2/T (Table 1) for given values of the constant velocities of the edges, v_1 and v_2. From the numerical results, the trajectories for both edges have been plotted in Figs. 4 and 5 for two representative cases. There is good agreement between the exact and the numerical solutions for a crack with rupture velocities close to the shear wave velocity (Fig. 4). In the case of slow extension of the crack, the exact and the numerical solutions still agree, but with somewhat reduced accuracy (Fig. 5). This result is not surprising since in the latter case, there is stronger coupling between the dynamical stresses in the two disturbed regions ahead of the crack edges. which in turn produces larger errors in the numerical calculation due to the discretization. In the cases which we have tested against the exact solution, it is found that the numerical results are rather stable and that there are only minute fluctuations due to the discretization.

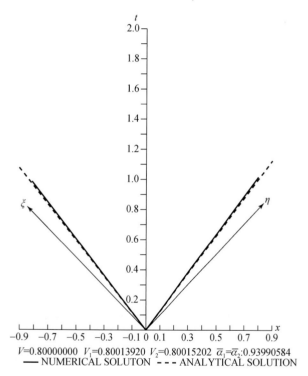

Fig. 4 A comparison of the numerical solutions with exact solutions for a bilaterally growing crack with rupture velocity close to shear wave velocity

The rupture velocities are scaled by shear wave velocity.

$\tilde{\alpha}_1 = \alpha_1/T$ and $\tilde{\alpha}_2 = \alpha_2/T$

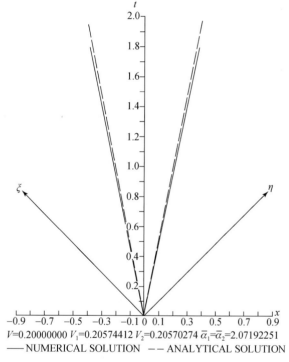

Fig. 5 A comparison of the numerical solutions with exact solutions for a bilaterally growing crack with low rupture velocity

For scaling, see Fig. 4

Table 1 A comparison between the numerical solutions and the exact solutions for the rupture velocities of self-similar growing cracks

α_1/T ($=\alpha_2/T$)	V_1/β ($=V_2/\beta$) (exact)	V_1/β (numerical)	V_2/β (numerical)
2.017192254	0.2	0.205744	0.205702
1.769412745	0.4	0.405142	0.404965
1.392371414	0.6	0.601576	0.601059
0.939905852	0.8	0.800139	0.800152

4 Nonuniform cracks

We now apply our numerical technique to the solution of the problem of the dynamical propagation of a two-dimensional antiplane shear crack under the influence of nonuniform stress drop and nonuniform cohesion.

4.1 Barrier model

We consider three models. In the first case, we assume that the stress drop on the crack is constant, but that the modulus of cohesion is nonuniform, with segments where the modulus of cohesion is very high (Fig. 6b, c). This is the barrier model of Das and Aki (1977a, b). In this example, the crack propagates symmetrically. We assume that there are two barriers at the scaled distance $x/L \doteq \pm 5.5$. In the present case, L is arbitrary, since there is no a priori scale imposed on the problem. Das and Aki (1977b) have studied several cases in which the areal extent of the barriers is small compared to the crack size at the encounter. They found that the crack edges and the barriers will interact differently depending on the magnitude of the barrier strength relative to the tectonic stress. When the areal extent is large, the growth of the crack ceases (Husseini et al., 1975). If the crack encounters an unbreakable barrier, it will be stopped abruptly (Knopoff and Chatterjee, 1982; Chatterjee and Knopoff, 1983). The examples we present here differ from the above in that the moduli of cohesion in the barriers are assumed to vary smoothly instead of discontinuously. In the case of Fig. 6, we find numerically that the edges of the case illustrated in Fig. 6a, both the left and the right edges of crack extend with almost constant velocity ($\doteq 0.8\beta$). When the crack encounters the barriers, it grows, but with a strong, smooth deceleration and finally comes to rest at the locations where $|x/L|$ is slightly less than 5.5. The crack advances into the barriers, but does not penetrate them nor generate slip in the regions beyond $|x/L| \doteq 5.5$.

The source-time functions on the crack can be calculated. The slip distributions on the crack at four successive times are shown in Fig. 6d, and the relative motions at several representative positions ($x/L = 2.8$, 0.0, -1.6 and -4.0) are shown in Fig. 6e. With regard to the healing condition, we assume that once the slip velocity becomes zero, the walls of the crack are frozen; reversal of the direction of relative particle motion on the crack faces is not allowed to occur (Burridge and Halliday, 1971). Freezing is therefore a kinematical rather than a dynamical condition (Knopoff, 1981).

In the case illustrated in Fig. 6, the crack edge decelerates when it encounters the barrier. At this time, a stress pulse is reflected from the barrier along the characteristic line. When the stress pulse arrives at an interior point, the slip on the crack is reduced to some nonzero value. When the stress pulse reflected from the second barrier arrives, the crack freezes. The collision of the stress pulses at the mid-point of the crack triggers fission of the crack into two segments, which shrink separately and ultimately disappear. Since the velocity of extension is close to the shear wave velocity, the retarding influence of the dynamically prestressed region ahead of the crack edge is too small to stop the slip. If the crack propagates with a lower rupture velocity because of low stress drop and high cohesion (Fig. 7) the first reflected stress pulse from the closer barrier will stop the slip on the crack. This is the case discussed in detail by Knopoff and Chatterjee (1982) and Chatterjee and Knopoff (1983). Once the slip as a function of time and coordinate along the crack is known, the displacement at any point (x_0, y_0, t_0) can be calculated from the expression (Burridge and Halliday, 1971; Chatterjee and Knopoff, 1983).

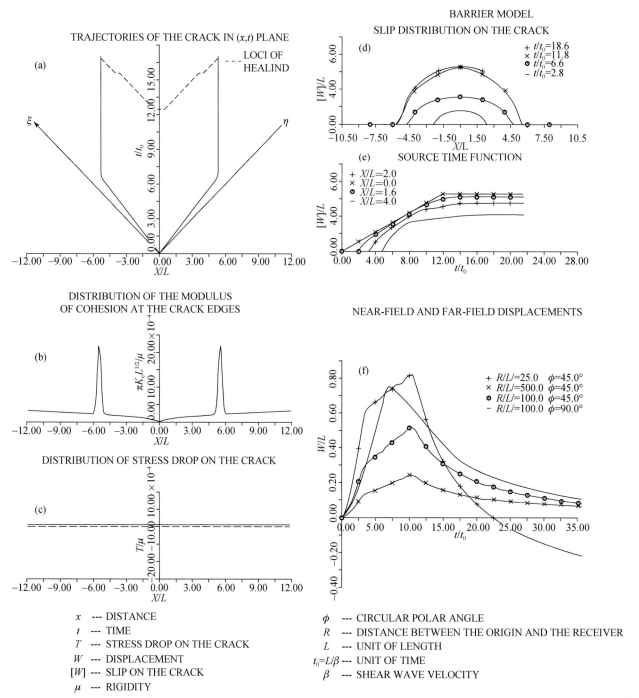

Fig. 6 (a) The extension and the arrest of the crack under the influence of non-uniform cohesion (b) and constant stress drop (c) (the barrier model). (d) The slip distribution on the crack at four successive times. (e) The source-time functions at several representative positions. (f) Near-field and the far-field theoretical displacement seismograms

In Fig. 6f we show the near-and far-field theoretical displacement seismograms. All of the seismograms show two abrupt changes in slope that correspond to the encounter of the crack with the barriers. In the case of locations $\phi = 90°$, the two disturbances originating at the edges arrive simultaneously and merge into a single strong event.

As indicated above, Fig. 7 displays the case for which the crack propagates symmetrically but with a lower rupture velocity under the influence of low stress drop and high cohesion. A comparison between the two cases illustrated in Figs. 6 and 7 shows that while the healing histories of these two cases are different, the theoretical displacement seismograms are quite similar in shape.

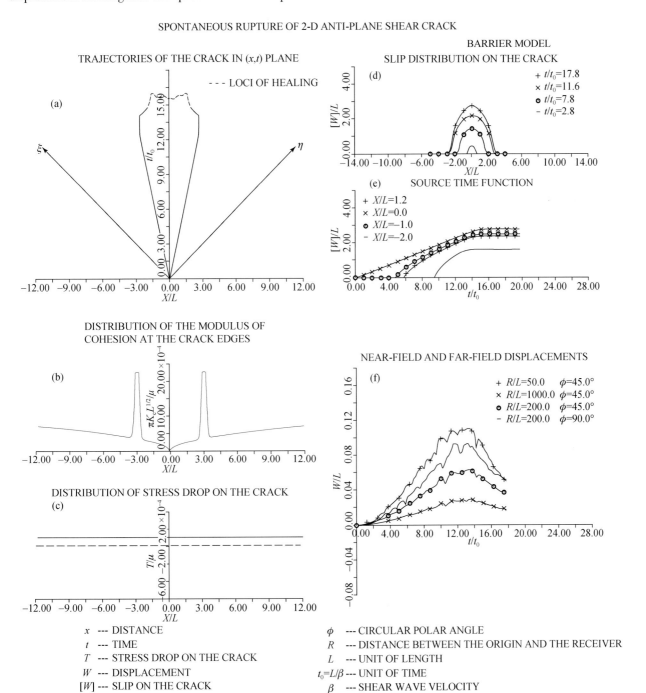

Fig. 7 The extension and the arrest of the crack under the influence of low stress drop and high cohesion

4.2 Asperity model

We consider a case (Fig. 8) in which the stress drop is very heterogeneous, with patches where the stress drop is high, called the asperities, as well as patches that may have already broken during previous events, and which we may take to have zero stress drop (Fig. 8c) (Kanamori, 1978; Rudnicki and Kanamori, 1981). For simplicity, we assume that the modulus of cohesion at the crack edges has a square root dependence on distance

(Fig. 8b). Thus, the crack will extend self-similarly until it encounters the region of zero stress drop (Fig. 8a). As the crack extends into the region of zero stress drop, the average stress drop on the crack becomes smaller. Thus the rate of growth of the crack will be reduced gradually, due to the smoothly decreasing average stress drop and the increasing cohesive resistance (Fig. 8a). For the asperity model, there is no abrupt change in the slope of the source-time function (Fig. 8e) and the displacement seismogram. It is interesting to note that in the example presented in Fig. 8, the crack extends deeply into the zero stress drop regions. This example indicates that in the case of great earthquakes along major plate boundaries, the aftershock areas of two great earthquakes would overlap by as much as half of the areas.

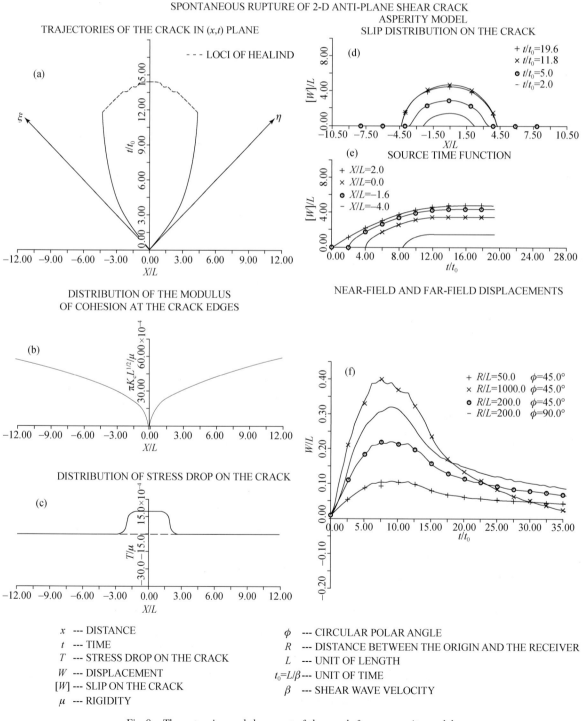

Fig. 8 The extension and the arrest of the crack for an asperity model

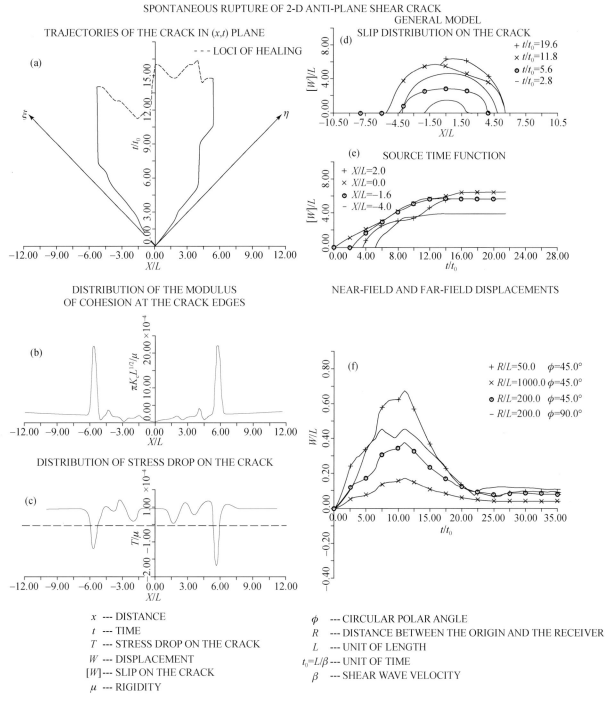

Fig. 9 The extension and the arrest of the crack for a general model

4.3 General model

It is unlikely that the process of earthquake rupture can be described purely by either the barrier or the asperity model. There are indications that the complexity of earthquake waveforms are attributable to both barriers and asperities. To get some idea of the effects of a complex model of earthquake rupture on the radiation, we consider a more general case (Fig. 9) in which the modulus of cohesion and the stress drop are both assumed to be nonuniform (Fig. 9b, c), i.e., we consider a composite of both the barrier and the asperity models. In this case, the stress pulses reflected from each barrier will arrive in complex sequence and will produce a very complex healing history (Fig. 9a). Inspection of the slip distribution on the crack at several times (Fig. 9d), the source-time function at different locations (Fig. 9e) and the near-and far-field displacement seismograms (Fig. 9f), shows that they are all quite different from those for the two simple cases. As before, the collision of the stress pulses from each barrier initiates fission of the crack. In the present case, a number of fission fragments is generated (Fig. 9a). The splitting of a dynamically propagating crack into fragments is peculiar to the general model and may be of special interest to the interpretation of observations. It is important to note that in the interpretation of near-field and far-field observations, the effects of the complex healing process have to be taken into account.

5 Summary

We have studied the problem of the spontaneous propagation of a two-dimensional earthquake fault. We have proposed a numerical technique for solving the dynamical crack problem. The method is based on the work of Kostrov (1966), and takes the analytical solutions of Chatterjee and Knopoff (1983) as a starting step. The numerical technique is verified in a number of experiments to be stable, and is useful and flexible in studying the complex dynamical crack problem of very heterogeneous stress drop and cohesive resistance. We have applied the method to several cases in which the stress drop on the crack and/or the modulus of cohesion at the crack edges are nonuniform. The preliminary results reveal that the heterogeneities of both the stress drop and the cohesion are the main factors which control growth, cessation and healing of the crack. The complexities in both the near-and far-field displacements are caused not only by instabilities in the propagation of the rupture but also by a complex healing history that results from the intricate interaction of the crack with the surrounding medium. The numerical examples also reveal differences between the influences of the barriers and the asperities on the source function and the seismic radiation. In the barrier model, abrupt changes in the modulus of cohesion cause abrupt changes in the slope of the source-time function and the theoretical seismograms, while in the asperity model, abrupt changes in the stress drop do not cause similar changes in the seismograms.

References

Aki, K. and Richards, P. G., 1980. *Quantitative Seismology: Theory and Methods*, Vol. 1 and 2. Freeman, San Francisco, 932 pp.

Andrews, D. J., 1976. Rupture velocity of plane-strain shear cracks. *J. Geophys. Res.*, **81**: 5679–5687.

Burridge, R., 1969. The numerical solution of certain integral equations with non-integrable kernels arising in the theory of crack propagation and elastic wave diffraction. *Philos. Trans. R. Soc. London*, *Ser. A*, **265**: 353–381.

Burridge, R. and Halliday, G. S., 1971. Dynamic shear cracks with friction as models for shallow focus earthquakes. *Geophys. J. R. Astron. Soc.*, **25**: 261–283.

Burridge, R. and Knopoff, L., 1964. Body force equivalents for seismic dislocation. *Bull. Seismol. Soc. Am.*, **54**: 1875–1888.

Chattejee, A. K. and Knopoff, L., 1983. Bilateral propagation of a spontaneous two-dimensional anti-plane shear crack under the influence of cohesion. *Geophys. J. R. Astron. Soc.*, **73**: 449–473.

Das, S., 1976. A numerical study of rupture propagation and earthquake source mechanism. ScD thesis, Mass. Inst. Technol., Cambridge, Mass.

Das, S., 1980. A numerical method for determination of source-time functions for general three-dimensional rupture propagation. *Geophys. J. R. Astron. Soc.*, **62**: 591–604.

Das, S., 1981. Three-dimensional rupture propagation and implications for the earthquake source mechanism. *Geophys. J. R. Astron. Soc.*, **67**: 375–393.

Das, S. and Aki, K., 1977a. A numerical study of two-dimensional spontaneous rupture propagation. *Geophys. J. R. Astron. Soc.*, **50**: 643–688.

Das, S. and Aki, K., 1977b. Fault plane with barriers: a versatile earthquake model. *J. Geophys. Res.*, **82**: 5658–5670.

Day, S. M., 1982a. Three-dimensional finite difference simulation of fault dynamics: rectangular faults with fixed rupture velocity. *Bull. Seismol. Soc. Am.*, **72**: 705–727.

Day, S. M., 1982b. Three-dimensional simulation of spontaneous rupture: the effect of non-uniform prestress. *Bull. Seismol. Soc. Am.*, **72**: 1881–1902.

Husseini, M. I., Jovanovich, D. B., Randall, M. J. and Freund, L. B., 1975. The fracture energy of earthquakes. *Geophys. J. R. Astron. Soc.*, **43**: 367–385.

Kanamori, H., 1978. Use of seismic radiation to infer source parameters. Proc. Conf. III, *Fault Mechanics and its Relation to Earthquake Prediction*. U. S. Geol. Surv. Open File Rep., 78-380: 283–318.

Knopoff, L. and Chatterjee, A. K., 1982. Unilateral extension of a two-dimensional shear crack under the influence of cohesive forces. *Geophys. J. R. Astron. Soc.*, **68**: 7–25.

Kostrov, B. V., 1966. Unsteady propagation of longitudinal shear cracks. *Prikl. Mat. Mekh*, **30**: 1042–1049 (in Russian).

Kostrov, B. V., 1975. *Mechanics of Tectonic Earthquake Sources*. Nauka, Moscow, 176 pp. (in Russian).

Madariaga, R., 1976. Dynamics of an extending circular fault. *Bull. Seismol. Soc. Am.*, **66**: 639–666.

Maruyama, T., 1963. On the force equivalents of dynamical elastic dislocations with reference to the earthquake mechanism. *Bull. Earthquake Res. Inst., Tokyo Univ.*, **41**: 467–486.

Mikumo, T. and Miyatake, T., 1978. Dynamical rupture process on a three-dimensional fault with non-uniform frictions and near-field seismic waves. *Geophys. J. R. Astron. Soc.*, **54**: 417–438.

Miyatake, T., 1980a. Numerical simulations of earthquake source process by a three-dimensional crack model. Part 1. Rupture process. *J. Phys. Earth*, **28**: 565–598.

Miyatake, T., 1980b. Numerical simulations of earthquake source process by a three-dimensional crack model. Part 2. Seismic waves and spectrum. *J. Phys. Earth*, **28**: 599–616.

Richards, P. G., 1973. The dynamic field of a growing plane elliptical shear crack. *Int. J. Solids Struct.*, **9**: 843–861.

Richards, P. G., 1976. Dynamic motions near an earthquake fault: a three-dimensional solution. *Bull. Seismol. Soc. Am.*, **66**: 1–32.

Rudnicki, J. W. and Kanamori, H., 1981. Effects of fault interaction on moment, stress drop, and strain energy release. *J. Geophys. Res.*, **86**: 1785–1793.

Vineux, J. and Madariaga, R., 1982. Dynamic faulting studied by a finite difference method. *Bull. Seismol. Soc. Am.*, **72**: 345–369.

Yoshida, S., 1985. Two-dimensional rupture propagation controlled by Irwin's criterion. *J. Phys. Earth*, **33**: 1–20.

Lancang-Gengma earthquake—a preliminary report on the November 6, 1988, event and its aftershocks*

Yuntai Chen[1] and Francis T. Wu[2]

1. Institute of Geophysics, State Seismological Bureau; 2. Department of Geological Sciences, State University of New York

On November 6, 1988, two earthquakes with magnitude >7 occurred within 15 minutes in southwestern Yunnan Province, China, near the Burmese border. The aftershock series in the next six weeks included three earthquakes with magnitude >6.0. Rapid deployment of accelerographs enabled us to record a large number of aftershocks, including two $M_S>6$ events, at near-source distances.

1 Seismicity

At 130314.5 UT on November 6 an earthquake with $M_S = 7.6$ (U.S. Geological Survey $M_S = 7.3$) occurred 40 km northwest of Lancang (Figure 1). Thirteen minutes later another large event with $M_S = 7.2$ (USGS $M_S 6.4$) occurred 60 km north-northwest of the first shock. By December 20 more than 600 aftershocks with $M_S > 3$ had occurred.

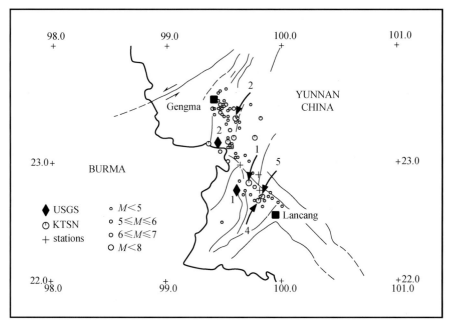

Fig. 1 Map showing locations of villages, main faults, and epicenters of $M_S>4$ events

The main shocks and the two largest aftershocks are marked by numbers corresponding to those in Table 1. The USGS main shock locations are indicated by diamonds. The aftershock monitoring stations set up by the ground motion study group are shown as crosses: the northernmost station is MUG, the middle one FUB and the southern one HGM. Note that the trend of the aftershock zone cuts across surface faults shown on the map. The faults on the map are those discernible on the LANDSAT image (ERTS E-2418-03055-7 01) and described in Yunnan Provincial Geological Bureau [1982]

---

* 本文发表于 EOS, Trans. Amer. Geophys. Union, 1989 年, 第 70 卷, 第 49 期, 1527, 1540.

The M_S used in this report refers to the surface wave magnitude or its equivalent as determined by the Kunming Telemetered Seismic Network (KTSN) unless otherwise noted. The magnitude scale is a modification of the Richter scale as defined for Southern California; it is generally higher than the teleseismic M_S assigned by USGS. The Chinese scale is also used as an "unified scale" for earthquakes. Thus M_S is also given to small events that do not produce many recognizable surface waves.

The hypocenters and magnitudes of the two main shocks and three largest aftershocks as determined by KTSN are listed in Table 1, together with magnitudes determined by USGS. The epicenters of the main shock as determined by KTSN and USGS (Figure 2) are very close; corresponding epicenters for the second shock are somewhat farther apart.

Table 1 Focal parameters of the mainshock and the largest aftershocks in the Lancang-Gengma area in November 1988

Number	Date	Origin Time	Latitude	Longitude	Depth/km	M_S	USGS m_b	USGS M_S
1	Nov. 6, 1988	13:03:14.5	22°50′	99°43′	18	7.6	6.1	7.3
2	Nov. 6, 1988	13:15:44.9	23°23′	99°36′	10	7.2	6.4	
3	Nov. 15, 1988	10:28:30.0	23°13′	99°35′	18	6.1	5.2	4.8
4	Nov. 27, 1988	04:17:53.4	22°41′	99°48′	16	6.3	5.0	5.1
5	Nov. 30, 1988	08:13:26.8	22°43′	99°50′	15	6.7	5.6	6.0

These earthquakes caused severe damage to houses in the meizoseismal area. The highest preliminary intensity assigned is X on the Chinese scale, which is very similar to the Modified Mercalli scale, in an area between the epicenters of the two largest earthquakes.

The preliminary intensity in Lancang county, in the epicentral area of the $M_S 7.6$ event, is only VII. This disparity is possibly related to site characteristics. The damage to structures was much more extensive in the vicinity of Gengma, because many villages were built on thick basin sediments. In contrast, Lancang county, closer to the epicenter of the first shock, is located in mountainous terrain; the villages are mostly in small intermontane basins. However, the directivity effect due a northwest-propagating rupture may also have played a role, since the epicenter is a likely initial rupture point of the fault and it lies in the southern part of the aftershock zone.

Although building damage was severe in Gengma, the $M_S 7.6$ mainshock near Lancang alerted the residents in the Gengma area and most of the residents rushed out of their houses. As a consequence, the $M_S 7.2$ earthquake, occurring 13 minutes later near Gengma and directly responsible for the damages, did not lead to as many casualties as it could have considering the higher population density in that county.

The shaking that resulted from the two closely spaced events was felt widely in the provincial capital of Kunming, 400 km from the meizoseismal area. Numerous landslides in the Gengma area limited access to Gengma county for many days. The official tally of casualties from this series of earthquakes stood at 733 by early December, 1988.

Figure 1 shows that many major faults are present in the area. Intensive efforts have been made by geologists of the Institute of Geology, State Seismological Bureau (SSB), PRC, to locate surface ruptures; preliminary results indicate right-lateral motions along some northwest-southeast trending faults. The overall trend of the epicenters shown in Figure 1 cuts across structures in the vicinity of Lancang and Gengma. Gengma county (Figure 1) is located in a large normal fault-controlled basin, a tectonic feature that commonly found in western and southwestern Yunnan [Wu and Wang, 1988].

2 Mainshock Focal Mechanism

The Global Digital Seismic Network (GDSN) recorded the main shocks at a number of stations. Moment tensor waveform inversion (for example, Langston [1981]) using P and SH waves from seven stations within 30°-90° of the earthquake was used to derive the focal parameters for the first main shock. As shown in Figure 2, the overall fitting of the waveforms is quite good. The P waves at KEV (Kevo, Finland) and GRFO (Grafenberg, Federal Republic of Germany) fix one of the P wave nodal planes well and the small amplitude SH wave at COL (College, Alaska) fixes the SH nodal plane. The CLVD (compensated linear vector dipole) component is relatively small ($<10\%$) and the parameters are well resolved. The time function is relatively simple, with a duration of about 10 seconds. The parameters for the solution are:

Strike 153.4°

Dip 86.7°

Rake 189°

M_0 4.5×10^{26}

M_W 7.0

Based on the overall trend of aftershock distribution (Figure 1), we conclude that the fault plane associated with the main shock is a N27°W-striking right-lateral strike-slip fault. Our solution shows a shallower depth (about 10 km) than that given by USGS in the November 1988 *Preliminary Determination of Epicenters*. The body waves of the second shock (M_S7.2, see Table 1) are not discernible on the long-period records because of interference due to the first shock.

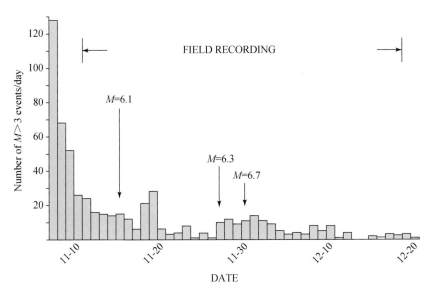

Fig. 2 Number of M_S>3 events in the Lancang-Gengma area as a function of time from November 6 to December 20, 1988

3 Aftershock Monitoring

The ground motion research group from the Institute of Geophysics, SSB, had been in Yunnan province since October 29, getting ready to deploy a temporary network around Xiaguan in northwestern Yunnan. Due to a delay in the shipping schedule, the instruments were still in Kunming when the earthquake occurred; they were subsequently shipped to the field area in Lancang.

The first portable digital accelerograph was deployed on November 11, 1988; between then and December 20, 1988, when the instruments were withdrawn, hundreds of events were recorded on four DCS-302 (Terra Technology) digital recorder-based accelerographs and six DTR-700 analog tape recorder-based seismographs.

Due to bad road conditions in the northern half of the meizoseismal area, the aftershock study was concentrated in the Lancang area. Figure 3 shows the number of $M_s > 3$ aftershocks from November 6 to December 20, 1988, determined by KTSN. Our monitoring effort ended when the overall aftershock activity tapered off to a relatively low level. The $M_s > 6$ events occurred in the Lancang area; they are listed in Table 1.

The accelerographs were purposely set at a relatively high threshold to reduce the number of triggers by smaller earthquakes. The tradeoff was that some events were recorded at only two or three sites. With relatively short recording time per tape (15 minutes), some events were missed because tapes became full between maintenance trips.

The accelerograms from the aftershock are being used, in conjunction with data from the more sensitive DTR-700 analog seismographs, to locate the aftershocks. They will also be used as empirical Green's functions in the synthesis for strong motions from larger earthquakes. The precisely located aftershock data will be of great importance in associating the earthquakes with active faults as well as in clarifying the tectonics of the area through a study of the focal mechanisms.

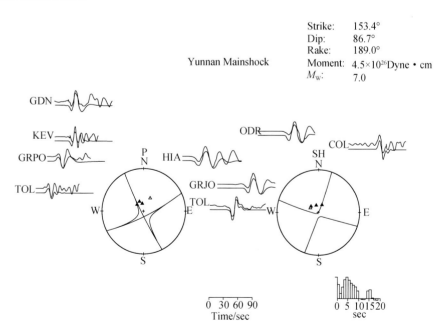

Fig. 3 Body waveform inversion result for the first main shock (1 in Table 1)

The time scale for the waves is shown at bottom center and the time function at bottom right

4 Strong Motion Records From Two $M_S > 6$ Aftershocks

Accelerograms of the November 27 and November 30, 1988, aftershocks (M_S 6.3 and 6.7, Table 1) are shown in Figure 4. They were recorded at the Fubang (FUB) station (Table 2, Figure 1). These two events are the largest aftershocks in the series. In addition to records at FUB, the November 27 event was also well recorded at Muga (MUG) and Haguoma (HGM). The November 30 event was well recorded at MUG (recording tape ran out at HGM). Distances to the stations and site conditions are listed in Table 2.

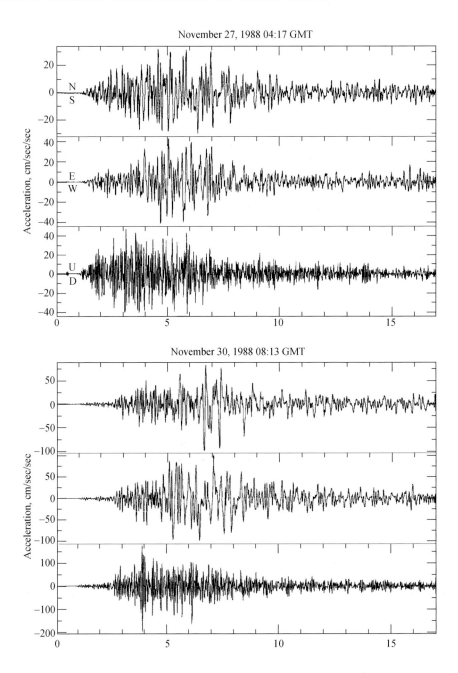

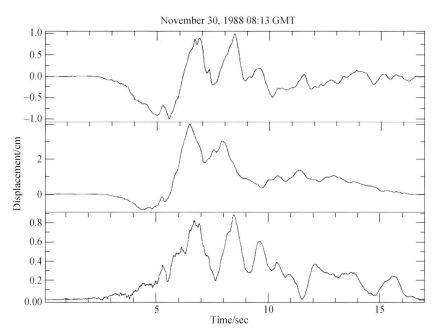

Fig. 4 top, Three-component accelerograms recorded for the November 27 event (4, Table 1) at FUB; middle, three-component accelerograms recorded for the November 30 event (5, Table 1) at FUB; bottom, displacement seismograms obtained from accelerograms of the November 30 event by double integration

Table 2 Stations and Site Conditions of the Aftershock Monitoring Stations

Code	Name	Site Condition	Event 4		Event 5	
			D/km	a_H/g	D/km	a_H/g
FUB	Fubang	Hard soil—edge of small basin	24	0.053	21	0.14
MUG	Muga	Soil—edge of basin	37	0.034	36	0.06
HGM	Haguoma	Soil—hill side	9.7	0.25		

Event numbers are the same as those in Table 1; a_H is the maximum horizontal acceleration.

The hypocentral locations of the November 27 and 30 events are only about 3 km apart and the paths from the hypocenters to FUB should be quite similar. However, the wave forms are noticeably different. The M_S6.3 accelerograms show a steady buildup (Figure 4, top) while the M_S6.7 accelerograms show an earlier small P wave train followed by a burst of large amplitude waves (Figure 4, middle). Furthermore, the clearly distinguishable P and S arrivals commonly seen on seismograms of small earthquakes are not found in these two accelerograms. Differences in source properties, in terms of space-time functions, most probably account for such differences. A dense local array of strong-motion instruments is needed for the study of detailed source dynamics.

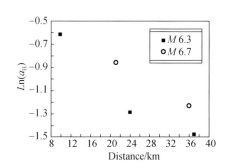

Fig. 5 Maximum peak horizontal accelerations for the two M_S>6 aftershocks as a function of distance

The horizontal components of the displacement seismograms (Figure 4, bottom), obtained by double integration of accelerograms of the November 27 event, show a continuous arrival of energy between P and the main S waves (between 4 and 5 seconds into the record). This observation is consistent with the theory of near-field elastic wave propagation [*Aki and Richards*, 1980, p. 87].

The three accelerograph stations that recorded the November 27 and 30 events (Figure 1, Table 2) range in epicentral distances from 9.7 to 37 km. The peak horizontal accelerations as a function of distance obtained from the accelerograms are listed in Table 2 and plotted in Figure 5. The more rapid attenuation of acceleration within the distance range of less than 20 km is suggested.

5 Discussion and Conclusion

The November 6, 1988, earthquakes near the China-Burma border are evidently associated with a north-northwest striking right-lateral strike-slip fault (hereby named Lancang-Gengma fault), which was not recognized as a major fault in the region [*Molnar and Tapponier*, 1975]. Several geomorphically prominent faults in the area did not participate in this tectonic event. For example, aftershock activity did not seem to extend to the vicinity of the left-lateral fault northwest of Gengma (Figure 1). As a consequence of this activity, however, the normal stress on the eastern part of the fault should be lessened and thus facilitate activity on that section of the fault. The Lancang-Gengma fault is sub-parallel to the Red River fault farther east [*Allen et al.*, 1984] and forms part of the escape system resulting from the collision of the Indian and Eurasian plates [*Tapponnier et al.*, 1982].

The epicentral distances to the largest aftershocks in the series range from 9.7 to 37 km and the accelerograms are of interest to near-source strong motion studies. In addition, the accelerograms for the small aftershocks (these events usually saturate the more sensitive instruments used in aftershock monitoring) can be used in more traditional aftershock location and focal mechanism determination. The wide dynamic range (112db) digital recorders coupled with force-balanced accelerometers are found to be ideal for both tasks.

The limitation in the present instruments is imposed by the small recording capacity (500 kbytes). New versions of digital recorders with several megabytes of memory or large-capacity tapes will augment dramatically the near-field acceleration data base and data needed for studying the dynamics of earthquakes.

Acknowledgments

The ground motion research group (Ming Wang, Peide Wang, Jianyu Zhou, Xiaoxi Ni, and Jiangchuan Ni) at the Institute of Geophysics, SSB, carried out the field work under difficult conditions and performed some of the analysis reported here.

The field work would not have been possible without the assistance of the Yunnan Seismological Bureau. We would like to acknowledge in particular the help of Jiang Kui, Tong Wonglian and Liu Zhuyin. David Salzberg assisted in moment tensor inversion work. The research was made possible with funds from the Seismological Research Foundation of China. Some of the instruments used were purchased under the Cooperative Research program of the U.S. National Science Foundation and the PRC State Seismological Bureau; Wu's travel to China was provided by the NSF Earth Science Program.

References

Aki, K. and P. G. Richards, *Quantitative Seismology*, Freeman and Co., San Francisco, Calif., 1980.

Langston, C, Source inversion of seismic waveforms: The Koyna, India, earthquakes of September 13, 1967, *Bull. Seism. Soc. Am.*, 71, **1**, 1981.

Molnar, P. and P. Tapponnier, Cenozoic tectonics of Asia: Effects of a continental collision, *Science*, **189**, 419, 1975.

Tapponnier, P., G. Peltzer, A. LeDain, R. Armijo, and P. Cobbold, Propagating extrusion tectonics in Asia: New insights from simple experiments with plasticine, *Geology*, **10**, 611, 1982.

Tong, W. L., *Statistics and Catalog of $M_S > 4$ Earthquakes in the Lancang-Gengma Area*, Kunming Telemetered Seismic Network, Yunnan Provincial Seismological Bureau, Kunming, Yunnan, December 14, 1988.

Wu, F. T., and P. D. Wang, Tectonics of western Yunnan Province, *Geology*, **16**, 153, 1988.

Yunnan Provincial Geological Bureau, *Report on Regional Geology of the Menglian Area (1: 2000000)*, Yunnan Provincial Geological Bureau, Kunming, Yunnan, 1982.

Inversion of near-source-broadband accelerograms for the earthquake source-time function

Y. T. Chen, J. Y. Zhou and J. C. Ni

Institute of Geophysics, State Seismological Bureau

Abstract A regularization method is applied to retrieve the far-field source-time function of some small and moderate aftershocks of the April 18, 1985, Luquan, Yunnan Province, China, $M_S = 6.1$ earthquake. Digital broadband accelerograms recorded at stations usually within 10 km of the epicentral distance are used in the inversion. To isolate the source effect from the effects of transmission path, recording site, anelastic attenuation and instrument response, the acceleration record of a smaller aftershock with the same hypocenter location and focal mechanism is treated as an empirical Green's function and incorporated in the inversion. The far-field source-time functions inverted independently from three components at all available stations of the deployed temporary accelerograph network are in good agreement. The results obtained show that the far-field source-time function of smaller aftershocks ($M_L \leq 3.0$) usually presents a simple spike-like pulse with a short rise time of approximately 0.1 s, while that of larger aftershocks ($M_L \approx 4.0$ or greater) has not only a longer rise time (0.3 s or longer), but also is fairly complex and consists of several prominent stages, exhibiting the heterogeneity of rupture process at the earthquake source.

1 Introduction

The earthquake source-time function is the time history of dislocation for a given point on the fault. It is an important parameter which manifests the nature of rupture process at the earthquake source. From theoretical and numerical modeling, it is inferred that the earthquake source-time function for any point on the fault is closely related to the entire rupture process of the earthquake source (e.g., Burridge and Halliday, 1971; Knopoff, 1981; Knopoff and Chatterjee, 1982; Chattejee and Knopoff, 1983; Madariaga, 1983; Chen et al., 1987; Das and Kostrov, 1988). In a numerical study, it is shown that the healing of a propagating fault, i.e., the cessation of relative particle motion on the fault walls, is either triggered by the stress pulse reflected from the encountered barrier, or initiated by the decreasing average stress-drop on the fault as it extends from an asperity into a low stress-drop region. Both the barrier and asperity models give dislocation rise times comparable to the overall rupture duration of the earthquake fault (Chen et al., 1987). In a study of the effect of the heterogeneity of the faulting process at the earthquake source on the complexity of the earthquake source-time function, Das and Kostrov (1988) found that the larger the density of barriers (ratio of barrier size to barrier spacing), the shorter the far-field pulse duration. The numerical models of Das and Kostrov (1988) also produce dislocation rise time comparable to the overall duration of earthquake faulting. These studies demonstrate the need to investigate the source complexity and its relation to the rupture process at the earthquake source, as well as the importance of determining the source-time function with high precision.

It is well known that the earthquake source-time function is one of the most difficult source parameters to extract from seismic records. In the past decade, with the development of high-quality seismological networks, techniques of retrieving source-time functions from seismic records have been developed, and significant progress has been made (Purcaru and Berckhemer, 1982; Kikuchi and Kanamori, 1982; Kikuchi and Sudo, 1984; Kikuchi and Fukao, 1985; Mueller, 1985; Niewiadomski and Meyer, 1986; Stavrakakis et al., 1987; Frankel et al., 1986; Frankel and Wennerberg, 1989; Mori and Frankel, 1990). The earthquake source-time function is determined by the use of the trial-and-error method, direct deconvolution method and iterative deconvolution method, from digital and/or analog, teleseismic and near-source seismograms. In the iterative deconvolution method, introduced by Kikuchi and Kanamori (1982) the earth response is calculated by assuming a source model of shear dislocation distribution with known focal mechanism, and a simple Earth structure model composed of homogeneous layers. However, situations continually arise where homogeneous layers are inappropriate or the focal mechanism is not known or poorly determined. Thus, the success of the iterative deconvolution method would largely depend on the knowledge of the actual focal mechanism and the true structure of the Earth. To determine the source-time function, Mueller (1985) applied a direct deconvolution technique with high-quality digital recordings of local earthquakes. In the direct spectral division deconvolution technique, small earthquake seismograms are used as empirical Green's functions (Hartzell, 1978). This technique does not require detailed knowledge of the transmission path, recording site, and instrument response. The only requirements are that the hypocenter location and focal mechanism are similar for both the larger and smaller earthquakes, the instrumentation is constant or at least that differences are well known, that the larger and smaller earthquakes are not greatly different in sizes so as the transmission path effect of the larger event can be approximately accounted for by a single Green's function (Mueller, 1985), and that the smaller earthquake be small enough that its far-field source-time function does not depart significantly from the Dirac δ-function. However, in practical use the direct spectral division deconvolution technique is limited by instability resulting from the zeros or low values occurring in the denominator as one calculates the quotient of the spectra of two earthquakes. The deconvolution can be stabilized by several techniques. One of these techniques is the so-called waterlevel technique. In the waterlevel technique, squared spectral amplitudes of the denominator are not allowed to fall below a fraction of the peak squared spectral amplitude (waterlevel). This is the technique proposed by Helmberger and Wiggins (1971). and used by Clayton and Wiggins (1976) and Mueller (1985). Another technique for stabilizing deconvolution is the regularization method (Tikhonov, 1963; Tikhonov et al., 1983). Niewiadomski and Meyer (1986) applied the regularization method to retrieve the far-field source-time function from teleseismic long-period records. To account for the Earth structure response, they calculated synthetic seismograms by assuming a simple half-space model. In the present study, we combine the regularization method with the empirical Green's function method to determine the far-field source-time function of small and moderate earthquakes, using the near-source digital broadband accelerograms recorded in a temporary seismological network that operated during the aftershock active period of the 1985 Luquan, Yunnan Province, China, $M_S = 6.1$ earthquake. The regularization method is applied to stabilize the deconvolution. The numerical and analytical difficulties of computing the responses of transmission path and instrument, and the effect of recording site, are avoided by treating the seismic record of a small earthquake as an empirical Green's function. The high-quality digital data recorded at short epicentral distances (usually less than 10 km) provide valuable opportunities to learn about the process of earthquake rupture. It is demonstrated that, with high-quality digital records and appropriate data processing techniques, the source-time history as well as the dislocation rise time can be determined precisely, and the complexity of rupture process of moderate, and even small earthquakes can also be retrieved.

2 Method

We model an earthquake source as a point shear dislocation. A point shear dislocation is defined as a shear dislocation source for which the smallest wavelength of interest is much larger than the linear dimension of the source (Stump and Johnson, 1977). For point shear dislocation, the recorded displacement at any point r and time t can be expressed as (Aki and Richards, 1980; Mueller, 1985):

$$u_i(\boldsymbol{r}, t) = M_{jk}(t) * G_{ij,k}(\boldsymbol{r}, t; \boldsymbol{r}_0, 0) * Q(t) * I(t) \tag{1}$$

where $u_i(\boldsymbol{r}, t)$ is the recorded displacement in the i direction, $M_{jk}(t)$ is the point moment tensor, $G_{ij}(\boldsymbol{r}, t; \boldsymbol{r}_0, 0)$ is the Green's function, $,k$ indicates the derivative with respect to the k direction source coordinate, $Q(t)$ is the attenuation factor, $I(t)$ is the instrument response, and $*$ represents temporal convolution. In the case that the temporal part of the point moment tensor can be separated and be written as:

$$M_{jk}(t) = M_{jk} \cdot s(t) \tag{2}$$

where M_{jk} stands for the limit of $M_{jk}(t)$ as $t \to \infty$ and $s(t)$, the source-time function, we may rewrite the recorded displacement as:

$$u_i(\boldsymbol{r}, t) = s(t) * B(t) \tag{3}$$

where:

$$B(t) = M_{jk} G_{ij,k}(\boldsymbol{r}, t; \boldsymbol{r}_0, 0) * Q(t) * I(t). \tag{4}$$

The equation (3) represents a Hilbert transform from Hilbert space S into Hilbert space U (Niewiadomski and Meyer, 1986):

$$\boldsymbol{A}s = u, \quad s \in S, \quad u \in U, \tag{5}$$

where u, s and the linear operator $\boldsymbol{A}$ represent $u_i(\boldsymbol{r}, t)$, $s(t)$ and $B(t)$, respectively. Our problem is to find s from the approximation of the operator $\boldsymbol{A}$ and the function u. In practice, we solve the following equation instead of eqn. (5):

$$\boldsymbol{A}_h s = u_\epsilon, \quad \|\boldsymbol{A}_h - \boldsymbol{A}\|_S \leq h, \quad \|u_\epsilon - u\|_U \leq \epsilon, \tag{6}$$

where $\boldsymbol{A}_h$ and u_ϵ are the approximation of $\boldsymbol{A}$ and u, respectively, and $\|\cdot\|$ represents the norm in the appropriate space.

To find the solution of the above equation by the regularization method (Tikhonov, 1963; Tikhonov et al., 1983; Niewiadomski and Meyer, 1986), we introduce a smoothing functional $M^\alpha[s]$:

$$M^\alpha[s] = \|\boldsymbol{A}_h s - u_\epsilon\|_U^2 + \alpha \|s\|_S^2, \quad \alpha \geq 0, \tag{7}$$

where α is a small parameter. The functional $M^\alpha[s]$ is in the form of the Fredhohn integral equation of the first kind with the kernel $B(t)$. Under the assumptions that U is a set of the square-integrable functions on the section $[a, b]$, i.e. $U = L^2[a, b]$ and that S is a set of continuous functions, having square-integrable first derivatives on $[a, b]$, i.e. $S = W_2^1[a, b]$, we can write:

$$M^\alpha[s] = \|B(t) * s(t) - u(t)\|_{L^2[a, b]}^2 + \alpha \|s(t)\|_{W_2^1[a, b]}^2. \tag{8}$$

Using the finite difference approximation of the integral operator in eqn. (8) and applying Parseval's theorem on

$B(t)$, $s(t)$ and $u(t)$, we can express $M^\alpha[s]$ by the Fourier coefficients of $B(t)$, $s(t)$ and $u(t)$, i.e. the coefficients $\tilde{B}_m$, $\tilde{s}_m$ and $\tilde{u}_m$:

$$M^\alpha[s] = \frac{1}{(b-a)} \sum_{m=0}^{n-1} [|\tilde{B}_m \tilde{s}_m - \tilde{u}_m|^2 + \alpha(1+\omega_m^2)|\tilde{s}_m|^2] \qquad (9)$$

where:

$$\omega_m = \frac{2\pi m}{b-a}, \quad m = 0, 1, 2, \cdots, n-1. \qquad (10)$$

The solution of eqn. (5), according to the regularization method, gives infimum (the smallest lower bound) $M^\alpha[s]$, expressed by eqn. (7) or (9). The value of the infimum can be calculated by means of the Frechet derivatives, which in the case of $M^\alpha[s]$ as expressed by eqn. (7) are given by:

$$\hat{M}^\alpha[s] = 2(A_h^* A_h s - A_h^* u_\epsilon + \alpha s) \qquad (11)$$

where A^* is the adjoint operator to A.

Substituting eqn. (9) into (11), we obtain:

$$\tilde{s}_m = \frac{\tilde{B}_m^* \tilde{u}_m}{\tilde{B}_m^* \tilde{B}_m + \alpha(1+\omega_m^2)}. \qquad (12)$$

To find the source-time function, we use the inverse Fourier transform of $\tilde{s}(\omega)$ and obtain:

$$s(t) = F^{-1}[\tilde{s}(\omega)]. \qquad (13)$$

It is interesting to note that the usual direct spectral division deconvolution is nothing but the regularization method when the parameter α equals zero.

The regularization technique provides a stable and straight-forward procedure to retrieve the source-time function. Its successful application for the determination of the source-time function is dependent upon detailed knowledge of the responses of the Earth's structure, anelastic attenuation, recording site, and instrument. In the first application of the regularization method, Niewiadomski and Meyer (1986) applied this method to determine the source-time function for the Gulf of Corinth, Greece, earthquake of February 24, 1981. They use teleseismic long-period S-wave records and calculate the operator $B(t)$ theoretically, using a simple half-space velocity model, well-determined focal mechanism, anelastic Futterman (1962) attenuation operator, and the instrument response for WWSSN stations. In the present study, we substitute the responses of the Earth's structure, anelastic attenuation, recording site and instrument by the ground motion record of a smaller earthquake with similar hypocenter location and focal mechanism. Stated another way, we assume that if two earthquakes with similar hypocenter location and focal mechanism are recorded by the same instrument in the same recording site, and that the smaller event is small enough that its source-time function can be assumed to be the Heaviside unit-step function (hence its far-field source-time function can be treated as the Dirac δ-function). Application of eqn. (3) to both the larger and smaller earthquakes yields:

$$u_i(r, t) = \frac{M_0}{M_0'} \dot{s}(t) * u'_i(r, t) \qquad (14)$$

where M_0 and M_0' are the scalar seismic moments of the larger and smaller earthquakes, respectively; $\dot{s}(t)$ is the derivative of the source-time function, i.e., far-field source-time function, and $u'_i(r, t)$ is the recorded displacement of the smaller earthquake. Thus the direct spectral division deconvolution of two earthquakes will yield the ratio of scalar seismic moments of the two events, M_0/M_0', as well as the far-field source-time function we wish to determine.

It is easy to verify that when applying the deconvolution method described by eqn. (14), the accelerogram as well as the velocity seismogram can be used instead of the displacement seismogram.

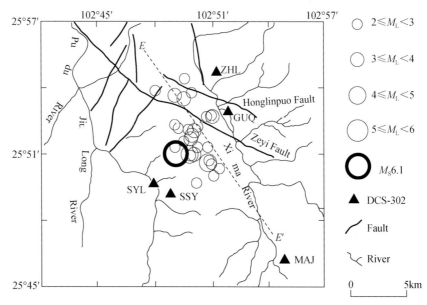

Fig. 1 Epicentral distribution of the well-located aftershocks of the 1985 Luquan, Yunnan Province, China, M_S = 6.1 earthquake

3 Far-field source-time function of four aftershocks of the Luqum Yunnan Province, China earthquake of April 18, 1985

3.1 1985 Luquan, Yunnan Province, China, earthquake

On April 18, 1985, a moderate earthquake with M_S = 6.1 occurred in Luquan, Yunnan Province, China, located about 100 km north of the provincial capital city Kumning (Fig. 1). The parameters for this earthquake are the following: origin time 5h 52m 53s, UTC (13h 52m 53s, BJUTC), epicenter coordinates 25°51′N, 102°49′E, focal depth 9 km. This earthquake was felt in Kunming. Six days after the occurrence of the main shock, we deployed a temporary small-aperture accelerograph network in the meizoseismal area. The temporary seismological network consisted of four stations, three components (vertical, E-W, N-S) each. The instruments used are the Terra Technology DCS-302 digital cassette seismograph which has a wide dynamic range (112 dB). The signal is digitized at 100 samples/s. The amplitude response curve is almost flat between 0 to 40 Hz (Fig. 2). The instruments are installed on limestone or sandstone basement with elevation ranging from 1980 to 2065 m (Table 1). During the operation period of April 24 to May 7, 1985, more than 400 events, including four events with magnitude M_L = 4.0 and larger, were recorded by the temporary seismological network.

The epicenters of the main shock and the well-located

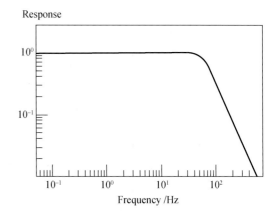

Fig. 2 Amplitude response curve of the Terra Technology DCS-302 digital cassette seismograph

aftershocks are shown in Fig. 1 (Wu et al., 1990). As shown in Fig. 1, the aftershocks are distributed in a small area confined by the Xima, Jiulong and Pudu Rivers, the Zeyi and Honglinpuo faults striking northwest, and faults striking northeast.

Table 1 Stations and site characteristics

Station	Code	Long. /°E	Lat. /°N	Elevation/m	Site
Zhuanlong	ZHL	102.854	25.911	2000	Limestone
Guiquan	GUQ	102.865	25.883	1980	Limestone
Majie	MAJ	102.911	25.772	2065	Limestone
Shayulang	SYL	102.800	25.831	1980	Sandstone
Shayulang[a]	SSY	102.810	25.822	2040	Sandstone

[a] Moved from Shayulang (SYL) station on May 3, 1985.

A well-constrained fault plane solution for the Luquan earthquake based upon the P-wave first motion data from the Kunming Telemetered Seismic Network (KTSN) is obtained (Wu et al., 1990). The focal mechanism solution indicates that one nodal plane strikes NE56.3° dipping 51.6°NW, and the other strikes NW76.1° dipping 49.6°SW, and that the P axis is nearly horizontal and in the N-S direction. From the spatial distribution of the well-located hypocenters, it is found that most aftershocks generally appear concentrated along the nodal plane striking NE56.3°, dipping NW; this nodal plane is then suggested to be the actual plane of faulting (Wu et al., 1990).

The ground motion recordings of the aftershocks of the Luquan earthquake, recorded at short epicentral distance ranging from a few to 10km or so, provide valuable opportunities to test the regularization method described in the previous section and to learn about the time history of particle motion on the fault. The aftershocks were well-located, with uncertainties of 0.5 km in epicenter location and 1 km in focal depth. In the present study, the near-source acceleration records of four aftershocks are selected for a test of the inversion method. In Table 2, we present the relevant parameters of these earthquakes as well as that of the correspondent smaller events used as empirical Green's functions. The smaller earthquakes are carefully selected to fulfil the requirement that the hypocenter location of each smaller event is close to that of the larger event (Table 2). In applying the regularization method, the parameter α introduced in eqn. (7) is chosen, through a number of tests, as 2% of the maximum spectral amplitude of the smaller event. A digital filter with a band width of 0.5 to 13.0 Hz is applied in the deconvolution, to depress the high frequency noise that contaminates the signal. Thus, the application of this bandpass filter enhances the ratio of signal to noise and facilitates the interpretation of the resultant far-field source-time function.

In frequency domain, a five-point sliding average is calculated to smooth the obtained far-field source-time function. It can be shown that this manipulation depresses the noise contaminated in the later portion of the signal but has no significant effect on the early portion of the signal (Weaver, 1983).

According to the procedures described in the previous section, we obtained the far-field source time functions of four larger aftershocks with magnitude ranging from $M_L=3.2$ to $M_L=4.8$ and the focal depths from near ground surface to about 9.6 km. The magnitude of smaller aftershocks used as empirical Green's functions ranges from $M_L=2.2$ to 3.0. As a by-product, we also obtained the ratio of scalar seismic moments of the larger to smaller earthquakes, M_0/M'_0 (Table 2). In the case that only the S-phase of the smaller event is recorded, we use the correspondent S-phase in the record of the larger event. In order to eliminate the cut-off effects generated from bandpass filtering, we use the Blackman filter both in time and spectral domains. The results of the four aftershocks are presented as follows.

Table 2 Information on the earthquakes investigated in this paper

Event No.[a]	Date yr-mo-d	Origin time h-m-s (UTC)	Epicenter location Long./°E	Epicenter location Lat./°N	Focal depth /km	M_L	M_0/M_0'	Rise time /s
20	85-05-01	13-08-28.1	102.831	25.865	9.6	3.5	8.4	0.12
8	85-04-29	04-42-46.8	102.824	25.860	8.5	2.2		
18	85-05-01	11-05-52.7	102.830	25.862	9.4	3.2	3.3	0.10
8	85-04-29	85-42-46.8	102.824	25.860	8.5	2.2		
13	85-05-01	08-59-04.8	102.829	25.849	4.1	4.8	46.0	0.51
*	85-05-01	08-42-02.0	102.829	25.849	4.1	2.9		
*	85-05-01	08-16-50.9	102.844	25.840	5.0	4.2	4.7	0.28
*	85-05-01	08-41-39.1	102.846	25.839	5.0	3.0		

[a] Numerated by Wu et al. (1990).

* Unnumerated in Wu et al. (1990).

3.2 The M_L = 3.5 aftershock (13h 08m 28.1s, May 1, 1985)

All of the acceleration records of the M_L = 3.5 aftershock at three stations have complete and clear P- and S-phases and are depicted in Fig. 3. The acceleration records of an M_L = 2.2 event used as empirical Green's functions are also shown in Fig. 3. It is easy to see the similarity between the wave trains of the two events. In Fig. 4, we plot all of the resultant far-field source-time functions from the three recording components at each of the three stations (Fig. 4) and the arithmetical averaged far-field source-time function (Fig. 4d). We note that the nine independent determinations in total from all components at all stations are consistent. The consistency is remarkable considering both the complexity of the input accelerograms and the striking difference in ground motion at the different stations. To verify the severity of the manipulations employed in stabilizing the deconvolution, we compare the synthetic seismograms (Fig. 4c) calculated by convolving the resultant far-field source-time functions (Fig. 4b) with the empirical function, with the high-cut-filtered observed seismograms (Fig. 4a). The synthetic seismograms are in good agreement with the observations. The far-field source-time function presented in this and the following figures is the derivative of the dislocation rise time, i.e. the derivative of the near-field source-time function. Inspection of the far-field source-time function of the M_L = 3.5 earthquake indicates that it is a simple spike-like impulse with a rise time of about 0.12 s. The azimuthal dependence of the resultant far-field source-time function is not as strong as predicted by Boatwright (1979). This implies that either the rupture process at the source of this small earthquake is quite simple, or the complexity of the rupture process at the source of an M_L = 3.5 earthquake can not be properly resolved only by the data presented in this paper. The surprisingly simplicity of the small earthquake exhibits that the striking difference in ground motion from this small earthquake at the different stations is mainly generated from the transmission path effect as well as focal mechanism.

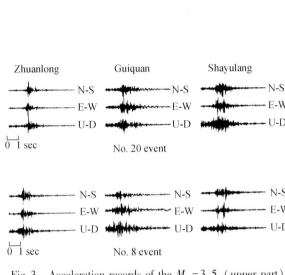

Fig. 3 Acceleration records of the $M_L = 3.5$ (upper part) and $M_L = 2.2$ (lower part) aftershocks, at Zhuanlong, Guiquan and Shayulang stations

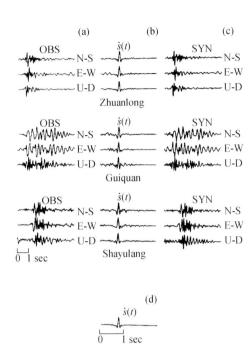

Fig. 4 Comparison of the observed (a) and synthetic (c) displacement seismograms at Zhuanlong, Guiquan and Shayulang stations, the far-field source-time function inverted from individual component (b) and the average far-field source-time function (d), for the $M_L = 3.5$ aftershock

3.3 The $M_L = 3.2$ aftershock (11h 05m 52.7s May 1, 1985)

Well-recorded accelerograms at three stations are obtained for this aftershock. From these records, the far-field source-time function is determined. As in the case of the $M_L = 3.5$ aftershock, the resultant far-field source-time functions from three components at all stations are all consistent. From Fig. 5, one can see that this small earthquake has a simple far-field source-time function similar to that of the $M_L = 3.5$ aftershock and a short rise time of 0.10 s. The results of the $M_L = 3.2$ event also indicate that either the rupture process at a smaller source is rather simple, or the present data set does not resolve the complexities of the source process of an $M_L \approx 3.0$ earthquake. The results demonstrate that one should attribute the striking difference in ground motion from an $M_L \approx 3.0$ earthquake mainly to the transmission path effect as well as focal mechanism.

3.4 The $M_L = 4.8$ aftershock (08h59m04.8s, May 1, 1985)

The $M_L = 4.8$ earthquake is the largest aftershock recorded by the temporary seismological network during the operation period in the field. The observed and synthetic seismograms at Zhuanlong station and the resultant far-field source-time function are shown in Fig. 6. A comparison between the far-field source-time function of this event and that of the smaller ones described above shows that this earthquake is characterized not only by a longer rise time (0.51s) but also a complicated time history in the source process. The resultant far-field source-time function reveals that the rupture process at this earthquake source consisted of two, perhaps three prominent stages. We attribute this characteristic to the complexity at the earthquake source. Due to the lack of available data for this event, we do not resolve the spatial dependence from the temporal dependence of this complexity.

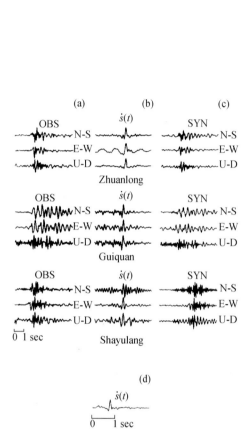

Fig. 5　Same as Fig. 4, for the $M_L = 3.2$ aftershock

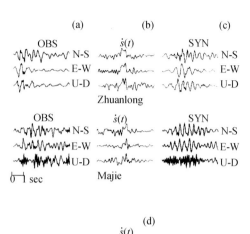

Fig. 7　Same as Fig. 4, but at Zhuanlong and Majie stations, for the $M_L = 4.2$ aftershock

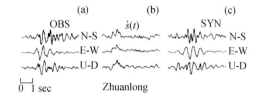

Fig. 6　Same as Fig. 4, but at Zhuanlong station, for the $M_L = 4.8$ aftershock

3.5　The $M_L = 4.2$ aftershock (08h 16m 50.9s, May 1, 1985)

The $M_L = 4.2$ aftershock is well-recorded at two stations. In Fig. 7, we present the observed and synthetic seismograms at Zhuanlong and Majie stations and the resultant far-field source-time function. All the results determined from the records from the two stations are in good agreement. These results reveal that the rupture process of this aftershock occurs in two distinct stages. As in the case of the $M_L = 4.8$ aftershock, we do not resolve the spatial dependence from the temporal dependence of this complexity.

From the results described above, we find that the far-field source time function of smaller aftershocks is nearly a single spike-like impulse with a short rise time, while that of larger events is more complex. The complexity of the source-time function for the larger earthquake may be generated from a small spatial extent but temporally in several distinct stages with a longer overall rise time, or from a larger spatial extent, but with a shorter rise time.

4 Summary

We have applied a regularization method to the determination of small and moderate earthquakes far-field source-time function, using the near-source digital broadband accelerograms registered on a temporary seismological network that operated during the aftershock active period of the 1985 Luquan, Yunnan Province, China, $M_S=6.1$ earthquake. In applying the regularization method to the high-quality digital data recorded at short epicenter distances, the advantages of the regularization method and the empirical Green's function method are manifested. It is demonstrated that the combination of these two methods enables one to use a straight-forward and stable procedure to retrieve the far-field source-time functions of small and moderate earthquakes, without computing the effects of transmission path and recording site and the responses of the anelastic attenuation and instrument. Since the complexities of the Earth's structure near the source and recording site, as well as that along the transmission path, are included automatically in the deconvolution procedure, the resultant far-field source-time functions are considered to be "neat" manifestations of the rupture process at the earthquake source. It is found that the rupture process of smaller earthquakes ($M_L \leq 3.0$) is rather simple and short (≈ 0.1 s), while that of larger earthquakes ($M_L \approx 4.0$ or greater) is more complex and longer (0.3 s or longer), and consisting of several prominent stages. We do not resolve the spatial dependence from the temporal dependence of this complexity.

Acknowledgement

This study is supported jointly by the Western Yunnan Experimental Site for Earthquake Prediction and Seismological Joint Foundation, State Seismological Bureau, P. R. of China. The authors would like to acknowledge in particular the help of Messrs. Kui Jiang and Jinhai Chen, of the Yunnan Province Seismological Bureau, and Messrs. Shiyu Li and Zuqiang Fu and Ms. Lianqiang Zhang, of the Institute of Geophysics, State Seismological Bureau, P. R. of China.

References

Aki, K. and Richards, P. G., 1980. *Quantitative Seismology: Theory and Methods*, Vols. 1 and 2. Freeman, San Francisco, Calif., 932 pp.

Boatwright, J., 1979. The Radon transform and the inversion of body wave pulse shapes. *Earthquake Notes*, **50**: 31–32.

Burridge, R. and Halliday, G. S., 1971. Dynamic shear cracks with friction as models for shallow focus earthquakes. *Geophys. J. R. Astron. Soc.*, **25**: 261–283.

Chatterjee, A. K. and Knopoff, L., 1983. Bilateral propagation of a spontaneous two-dimensional anti-plane shear crack under the influence of cohesion. *Geophys. J. R. Astron. Soc.*, **73**: 443–473.

Chen, Y. T., Chen, X. F. and Knopoff, L., 1987. Spontaneous growth and autonomous contraction of a two-dimensional earthquake fault. In: R. L. Wesson (Editor), Mechanics of Earthquake Faulting. *Tectonophysics*, **144**: 5–17.

Clayton, R. W. and Wiggins, R. A., 1976. Source shape estimation and deconvolution of teleseismic body waves. *Geophys. J. R. Astron. Soc.*, **47**: 151–177.

Das, S. and Kostrov, B. V., 1988. An investigation of the complexity of the earthquake source-time function using dynamic faulting models. *J. Geophys. Res.*, **93**: 8035–8050.

Frankel, A. and Wennerberg, L., 1989. Rupture process of the M_S6.6 Superstition Hills, California, earthquake determined from strong-motion recordings: application of tomographic source inversion. *Bull. Seismol. Soc. Am.*, **79**: 515–541.

Frankel, A., Fletcher, J., Vernon, F., Haar, L., Berger, J., Hanks, T. and Brune, J., 1986. Rupture characteristics and tomographic source imaging of $M_L \approx 3$ earthquakes near Anza, Southern California. *J. Geophys. Res.*, **91**: 12633–12650.

Futterman, W. I., 1962. Dispersive body waves. *J. Geophys. Res.*, **67**: 5279–5291.

Hartzell, S. H., 1978. Earthquake aftershocks as Green's function. *Geophys. Res. Lett.*, **5**: 1–4.

Helmberger, D. V. and Wiggins, R. A., 1971. Upper mantle structure of midwestern United States. *J. Geophys. Res.*, **76**: 3229–3245.

Kikuchi, H. and Fukao, K., 1985. Iterative deconvolution of the complex body waves from great earthquake: the Tokachi-Oki earthquake of 1986. *Phys. Earth Planet. Inter.*, **37**: 235–248.

Kikuchi, H. and Kanamori, H., 1982. Inversion of complex waves. *Bull. Seismol. Soc. Am.*, **72**: 491–506.

Kikuchi, H. and Sudo, K., 1984. Inversion of teleseismic P waves of the Izu-Oshima, Japan, earthquake of January 14, 1982. *J. Phys. Earth*, **32**: 161–171.

Knopoff, L., 1981. The nature of the earthquake source. In: E. S. Husebye and S. Mykkeltveit (Editors), *Identification of Seismic Sources — Earthquake or Underground Explosion*. Reidel, Dordrecht. pp. 49–69.

Knopoff, L. and Chatterjee, A. K., 1982. Unilateral extension of a two-dimensional shear crack under the influence of cohesive forces. Geophys. *J. R. Astron. Soc.*, **68**: 7–25.

Madariaga, R., 1983. Earthquake source theory: a review. In: H. Kanamori and E. Boschi (Editors), *Earthquake: Observation, Theory and Interpretation*. Soc. Ital. di Fisica, Bologna. pp. 1–44.

Mori, J. and Frankel, A., 1990. Source parameters for small events associated with the 1986 North Palm Springs, California, earthquake determined using empirical Green's functions. *Bull. Seismol. Soc. Am.*, **80**: 278–295.

Mueller, C. S., 1985. Source pulse enhancement by deconvolution of an empirical Green's function. *Geophys. Res. Lett.*, **12**: 23–36.

Niewiadomski, J. and Meyer, K., 1986. Application of the regularization method for determination of seismic source-time functions. *Acta Geophys. Pol.*, **34**: 137–144.

Purcaru, G. and Berckhemer, H., 1982. Quantitative relations of seismic source parameters and an attempt for the classification of earthquakes. *Tectonophysics*, **84**: 57–128.

Stavrakakis, G. N., Tselentis. A. G. and Drakopoulos. J., 1987. Iterative deconvolution of teleseismic P waves from the Thessaloniki (N. Greece) earthquake of June 20. 1978. *Pure Appl. Geophys.*, **124**: 1039–1050.

Stump. B. W. and Johnson. L. R., 1977. The determination of source properties by the linear inversion of seismograms. *Bull. Seismol. Soc. Am.*, **67**: 1489–1502.

Tikhonov, A. N., 1963. On the solution of ill-posed problems and the method of regularization. *Dokl. Akad. Nauk S. S. S. R.*, **3**: 501–504 (in Russian).

Tikhonov, A. N., Goncharskiy, A. V., Stepanov, V. V. and Yagola, A. G., 1983. *Regularization Algorithms and Apriori Information*. Nauka, Moscow (in Russian).

Weaver, H. J., 1983. *Applications of Discrete and Continuous Fourier Analysis*. Wiley, New York, pp. 90–109.

Wu, M. X., Wang, M., Sun, C. C., Ke, Z. M., Wang, P. D., Chen, Y. T. and Wu, F. T., 1990. Accurate hypocenter determination of aftershocks of the 1985 Luquan earthquake. *Acta Seismol. Sin.* (Chinese edition), **12**: 121–129.

Seismological studies of earthquake prediction in China: a review[*]

Y. T. Chen[1], Z. L. Chen[2] and B. Q. Wang[1]

1. Institute of Geophysics, State Seismological Bureau; 2. State Seismological Bureau

Abstract In the studies of earthquake prediction, the seismic precursory phenomena have received much attention of the Chinese seismologists and the seismological approach has played an important role in the practice of the earthquake prediction made by the Chinese seismologists. This paper selectively and briefly reviews the progress in seismological studies of earthquake prediction in China, with emphasis on the methods widespreadly employed by the Chinese seismologists. Among these approaches are: (1) historical seismicity; (2) correlation of strong earthquake activity in different seismic belts; (3) seismic gap; (4) stripe-like pattern of seismicity; (5) recurrence of earthquakes; (6) migration of seismicity; (7) anomalously high seismicity and seismic quiescence; (8) spatial and temporal change of b value in the magnitude-frequency distribution; (9) earthquake sequence; (10) foreshock; (11) aftershock prediction; (12) pattern recognition.

1 Introduction

Earthquake is one of the most destructive natural disasters. From 1904 to 1980, there were 53 large earthquakes with magnitude 8.0 and larger occurred in the world (Abe, 1981; 1984; Kananori, 1983). Lives lost due to these earthquakes were as high as 2.6 million, a figure about 58% of the total number caused by various natural disasters in the same period.

China is one of the seismic active countries in the world. In China large earthquakes occur frequently. Earthquakes in China's mainland as well as in Taiwan Province are considerably active and frequent. The earthquakes in China's mainland, referred to as intraplate or continental earthquakes, are affected by the activities of the Circum-Pacific seismic belt and the Alps-Himalaya seismic belt. The destructive earthquakes pose a continuing major threat to lives and property in almost all province, autonomous regions and municipalities of China. Historically, China has been struck by a number of major destructive earthquakes. For example, the 1556 Guanzhong earthquake is responsible for 830 thousand deaths. In 1920, the Haiyuan earthquake caused 230 thousand deaths. On February 4, 1975, an earthquake of magnitude $M_S = 7.3$ occurred in Haicheng, Liaoning Province. Although this earthquake had been successfully predicted by the Chinese seismologists, yet it still caused major damage and 1,328 fatalities in the densely populated meizoseismal area. On July 28, 1976, the Tangshan $M_S = 7.8$ earthquake caused 242 thousand people died and another 164 thousand people seriously injured. A statistic indicates that in the 37 years period from 1949 to 1986, China's mainland experienced over 37 destructive earthquakes. These earthquakes caused 273 thousand deaths and 763 thousand injured (Guo and Chen, 1986).

---

[*] 本文刊载于意大利罗马 Il Cigno Galileo Galilei 出版社出版的 *Earthquake Prediction* 一书中，1992 年，71–109.

Earthquake prediction has been a matter of common concern in China and a challenge to the Chinese seismologists since the occurrence of the March 22, 1966 Xingtai $M_s = 7.2$ earthquake. This earthquake occurred in a densely populated area of North China plain, and caused 8,064 fatalities and 9,492 seriously injured. It has a great influence on the earthquake research in China and became an important turning point in the development of the earthquake research in China. Since the occurrence of the great Xingtai earthquake, extensive earthquake research program has been made and carried out (Chen, 1986). During the period of 1966 to 1976, a number of major earthquakes have struck China's mainland. The earthquakes occurred in this seismic active period provided valuable research opportunities for the Chinese seismologists and greatly promoted the quiescent period of 1977 to 1984. The seismic activity in China's mainland revived in 1985 and continues up to the present with fluctuations. In the last 23 years, by actual experience the Chinese seismologists gained some knowledge about the regularities of earthquake occurrence and the characteristics of earthquake precursors, and acquired some experience in the intermediate-and long-term earthquake predictions. In the meantime, more and more Chinese seismologists as well as the public recognize the difficulties encountered in the studies of earthquake prediction, especially in the studies of the short-term and imminent earthquake predictions. While the occurrence of earthquake is predictable, the appearance of the earthquake precursor displays far more complicatedly than one conceived before. Earthquake prediction as a science is still in its infant and exploratory stage, the capability to predict earthquake accurately remains very poor. To bring about the success of earthquake prediction, a better understanding of the regularities of earthquake occurrence, and of the characteristics of earthquake precursors is needed, and much efforts in a number of aspects such as the improvement of the observation conditions, the researches on earthquake precursors, etc. Should be intensified.

Earthquakes prediction means the forecasting of a specified time and place of an earthquake of specified magnitude. In China, these predictions are classified into long-term (decades to years), intermediate-term (years to months), short-term (months to weeks) and imminent (weeks to days and even hours). Since at present our understanding of the cause and the regularities of earthquake occurrence is still very preliminary, several empirical techniques are explored. Those techniques mainly adopted in China include seismological method, crustal deformation, level of ground water, hydrochemical change, geoelectricity, geomagnetism, crustal stress, gravity, anomalous behavior of animals, etc. There are about 20 items adopted for observation. Considering the lack of a thorough understanding of the physics of earthquake occurrence, earthquake prediction research in China is mainly empirical.

In the studies of earthquake prediction, as well as in the monitoring of earthquake regime in China, the seismological approach plays an important role and attracts much attention of the Chinese seismologists. This paper present the progress in seismological studies of earthquake prediction in China, with emphasis on the widespreadly employed methods such as seismicity, seismic gap, seismic stripe-like pattern, b-value, migration of seismicity, etc.

2 Studies of historical seismicity and long-term earthquake prediction

2.1 Historical materials of earthquakes in China

As an early developed country in science and culture, China has long recorded history. The abundant historical materials are indispensable for the prolongation of time span in the study of historical seismicity. In 1956, the Seismological Committee of Academia Sinica (1956) compiled the *Chronological Table of Chinese Earthquakes*. In 1978, Xie and Cai (1983-1987) organized more than 1,000 Chinese seismologists, historians, archaeologists, archivists and librarians from various institutions and localities to revise it. As a result, a series of five volumes of *Compilation of Historical Materials of Chinese Earthquakes* were published successively in the period of 1983-1987. These volumes include the first entry about earthquake dating back to the 23rd century B. C. and continue through 1980 A. D. In the *Chronological Table of Chinese Earthquakes*, more than 1600 counties among the 2000 and more counties have materials of historical earthquakes. Most of them are distributed in the eastern part of China's mainland. For the bordering regions, historical records are rare. In the new *Compilation of Historical Materials of Chinese Earthquakes*, the number of earthquakes collected increased more than ten times. Even for the densely populated eastern China, where culture and economy were well developed, many new materials were found and include in the new *Compilation of Historical Materials of Chinese Earthquakes*.

Based on the *Chronological Table of Chinese Earthquakes* (the Seismological Committee of Academia Sinica, 1956) Lee (1960) published a *Catalogue of Chinese Earthquakes*. This catalogue contains about 10,000 earthquakes occurred between 1189 B. C. and 1955 A. D. Among them, 1,180 events have magnitudes 4.7 and larger. In its revised and enlarged edition (Central Seismological Group, 1971), 2,257 earthquakes with magnitude 4.7 and larger occurred between 1177 B. C. and 1969 A. D. were included. Several concise catalogues with focal parameters such as location, focal depth, origin time and magnitude only (State Seismological Bureau, 1977; Lee et al., 1978; Shi et al., 1986) and a *Collection of Isoseismals of Chinese Earthquakes* (Seismic Intensity Regionalization Group, State Seismological Bureau, 1979) were published later. The latest edition of *Catalogue of Chinese Earthquakes* (Gu, 1983) was published by the end of 1983. In the *Compilation of Historical Materials of Chinese Earthquakes* edited by Xie and Cai (1983-1987), a catalogue of instrumentally determined focal parameters of earthquakes with magnitude 4.7 and larger, of the 20th century is included. These comprehensive earthquake catalogues, along with the other observation reports of the nationwide and local seismic networks, constitute the important database, and are widely used in the studies of seismicity and its regularities in China.

2.2 Seismicity of China

The seismicity activity in China is characterized by its widely distributed area, large magnitude, high frequency of earthquake occurrence and shallow focal depth. In the epicentral map of large earthquakes (magnitude $\geqslant 6.0$) occurred in China and its adjacent areas shown in Figure 1, the wide distribution of strong earthquakes in China can be seen at a glance. The strong earthquakes of magnitude of magnitude 6.0 and larger occurred in nearly all the administrative provinces, autonomous regions and municipalities. In the 20th century, 104 earthquakes of 7.0 and larger have stricken 21 of the 30 administrative provinces, autonomous regions and municipalities. The earthquake distribution in China is quite different from that in USSR, where earthquakes mainly occur in its southern border and the Far-East region, nor from that in the United States, where as is well known, most major earthquakes occur along the San Andres fault zone in the western United States.

Seismicity in China varies in different geographical locations (Figure 1). In Taiwan Province earthquake activity is the most extensive. In China's mainland most of the large earthquakes occur in Southwest and Northwest China, i. e., Xizang (Tibet), Yunnna, Sichuan, Qinhai, Gansu, Ningxia and Xinjian. Large earthquakes also occur in North and Southeast China, including Hebei, Shan'xi, Shandong, Liaoning, Fujian and Guangdong Provinces.

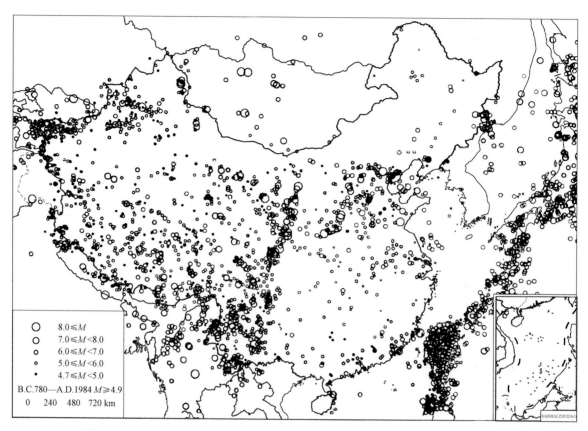

Fig. 1 Epicentral distribution in China and its adjacent areas (780 B. C. —1984 A. D.)

There are 291 major earthquakes with magnitude 6. 0 and larger occurred in China's mainland, within the period of 1900—1987. In the same period, the total number of earthquakes with magnitude 8. 0 and larger is 7.

The earthquakes in China are generally shallow-focus events, except in Helongjiang and Jilin Provinces, Northeast China, where some intermediate-and deep-focus earthquakes occurred. The focal depth of the earthquakes in East China ranges from 10 to 25 km, while that in West China, from 15 to 30 km. It is these shallow events that cause high intensity and serious damage. The destruction caused by these earthquakes, or the earthquake risk in China is considerably serious and complex because besides their shallow focal depth, many factors that enhance the risk, such as large magnitude, high frequency, local geological and soil conditions, and poor earthquake-resistant ability of buildings enter into the problem.

As is shown in Figure 1, the epicentral distribution of the strong earthquakes occurred in China from 780 B. C. to 1984 A. D. clearly depicts the characteristics of the geographical distribution of the seismicity in China. According to the earthquake distribution (including magnitude, frequency of occurrence, cyclic characteristics of earthquake occurrence, recurrence period of paleo-and historic earthquakes, characteristics of earthquake energy accumulation and release, focal depth, etc.), the characteristics of geological structure (especially that of the neotectonic and recent tectonic movements) and the structure of the Earth's crust and other geophysical data, 23 seismic belts can be identified (Figure 2). Along these seismic belts, the epicenters are rather concentrated, and in general they coincide with the boundaries of the major crustal blocks or the structural faults inside the blocks.

In most seismic belts, during the long time process of earthquake activity, the seismic active and quiescent periods occurred alternatively (Figure 3). In general, the seismic cycle in individual seismic belt can be identified. In most seismic cycles, four periods can be distinguished. These periods are characterized by strain accumulation, accelerated strain release, catastrophic strain release and strain adjustment, respectively (Shi et al., 1974).

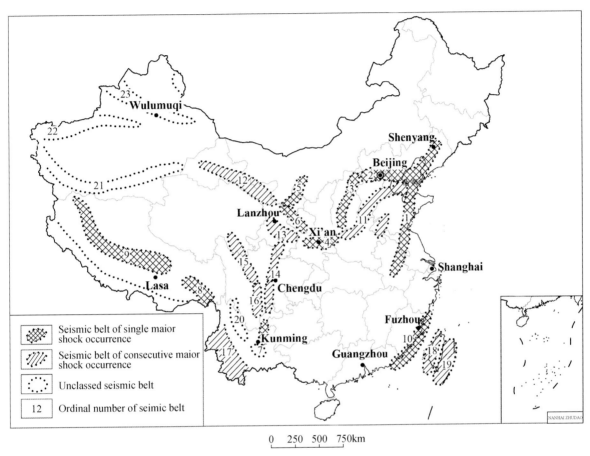

Fig. 2 Seismic belts in China

In general, high seismicity appears either in the way characterized by the occurrence of a few very strong earthquakes, or in the way characterized by the occurrence of a number of strong earthquakes. Thus two parameters are adopted to describe the seismicity in these cases. The first parameter employed to describe seismicity is the occurrence frequency of earthquakes with magnitude larger than a specified threshold magnitude. The second one is the total energy or strain release of earthquakes. Usually these two descriptions of seismicity give similar results. As an example, Figure 4a presents the temporal distribution of earthquakes with $M_S \geq 6.0$ since the 15th century in North China. It is obvious that the earthquake activity in different time intervals presents distinctly different characteristics. During the periods of 1484—1730 and 1815 up to present, both the occurrence frequency and the magnitude of earthquakes were considerably high. On the contrary, during the period of 1731—1814, the seismicity was rather low. The 1765 Huanhai $M_S = 6.0$ earthquake was the only $M_S \geq 6.0$ earthquake occurred in this period. Thus the periods of 1484—1730 and 1815 up to present are recognized as the seismic active periods in North China. The recognition of the seismic active and quiescent periods provides a lead in forecasting the long-term earthquake activity.

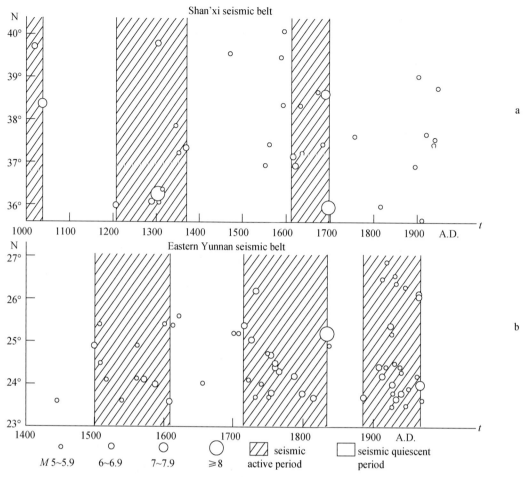

Fig. 3 Seismic active and quiescent periods occurred alternatively in Shan'xi (a) and eastern Yunnan (b) seismic belts

2.3 Comparison between historical and present seismicities

One of the most important methods in the long-term earthquake prediction is by comparing phenomenologically the characteristics of historical seismic activity in individual seismic belt or zone, with the present status of the same belt or zone, to judge and infer the seismic tendency in this belt or zone. The comparison includes the level of seismicity, duration time of earthquake activity, evolution process, among others. It is found that in a seismic belt or zone, usually there is a rather long preparation period prior to the catastrophic strain release period. This characteristic is especially prominent in North China seismic zone (Figure 4a). In this seismic zone the preparation period prior to the seismic active period between 1484 and 1730 is more than 180 years long. It consists of three relatively active sub-periods. The occurrence frequency of earthquake with $M_S \geqslant 6.0$ in the sub-period of 1614—1626 is the highest. Its annual average is about 0.5. On the contrary, in the other two sub-periods the annual average of the occurrence frequency of earthquake with $M_S \geqslant 6.0$ is only about 0.2. In the catastrophic strain release period, not only the occurrence frequency of earthquake with $M_S \geqslant 7.0$ is very high, but also the magnitude of earthquake is very large. A comparison between the seismicity of the periods since 1815 and that of the last active period shows the similarity between the seismicities in these two periods. During the 47 years of the period 1906—1952, there were 17 earthquakes with $M_S \geqslant 6.0$ occurred in North China, while in the nearly 100 years period prior to 1906, there were only 11 earthquakes with $M_S \geqslant 6.0$ occurred. The similarity between the seismicities in the period of 1906–1952 and that in the period of 1614—1626 is striking. Thus these two periods are recognized as the accelerated strain release periods before catastrophic strain release. All of the recent large earthquakes occurred in North China, i. e., the 1966 Xingtai $M_S = 7.2$ earthquake, the 1969 Bohai $M_S = 7.4$ earthquake, the 1975 Haicheng $M_S = 7.3$ earthquake and the 1976 Tangshan $M_S = 7.8$ earthquake, mark that the seismic activity in this region is in the period of catastrophic strain release.

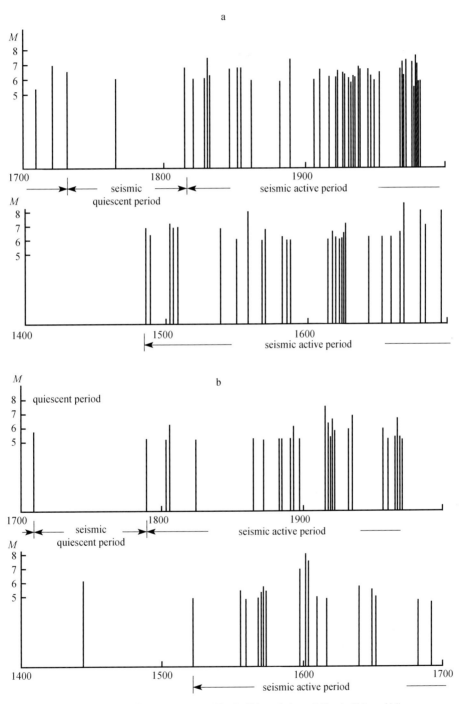

Fig. 4 Historical seismicities in North China (a) and South China (b)

It is worth to note that while the method of comparing the historical seismicity provides a practical clue in predicting the general tendency of the seismicity in long-term sense, the successful application of this method is limited by the short time span of available historical records of earthquakes, as well as by the difficulties in the recognition of the earthquake activity arising from the intrinsic complexities.

2.4 Correlation of strong earthquake activities in different seismic belts

It seems that some correlations exist in the seismicities of different seismic belts (zones). These correlations provide useful lead for inferring the tendency of earthquake activity. Correlation phenomena mainly include: (1) the similarity between two time sequences of earthquakes; (2) the time lag between the seismic active periods of two seismic belts (zones); and (3) the negative correlation between the seismicities of two seismic belts (zones).

Figure 5 shows the similarity between the global seismicity ($M_S \geq 7.0$) and the seismicity of China's mainland and its adjacent areas ($M_S \geq 6\frac{3}{4}$) occurred in the active period of Mediterranean-South Asia seismic belt. Earthquakes of $M_S \geq 6.0$ since the 20th century in North China seismic zone mostly occurred in the active period of the West Pacific seismic belt.

Figure 4b shows the temporal distribution of earthquakes with $M_S \geq 5.0$ in the period of 1400 to 1979 A.D. A comparison between Figure 4a and 4b shows that the seismic active periods of both South China and North China seismic zones are about 100 to 200 years, and that both of them started at nearly the same time but the catastrophic strain release period of South China seismic zone stared remarkably in advance of that of the North China seismic zone. Since the 20th century, the seismic active period of earthquakes with $M_S \geq 6.0$ in the North China seismic zone also started after the activity of $M_S \geq 6.0$ earthquakes in the South China seismic zone started to increase.

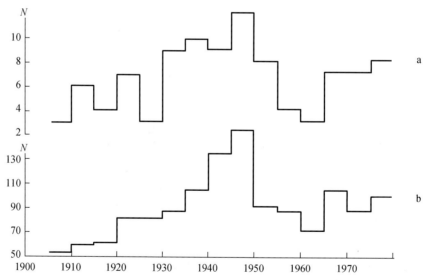

Fig. 5 Correlation between the seismicity of China's mainland (a) and the global seismicity (b)

N is the cumulative number of earthquakes in five years

In Liupanshan-Qilianshan-Aerjin seismic belt, as well as in North China-Northeast China plain, earthquakes of $M_S \geq 6.0$ usually occur alternatively: the active seismic period in the Liupanshan-Qilianshan-Aerjin seismic belt just correspond to the seismic quiescent period of the North China-Northeast China plain, and vice versa.

3 Seismicity pattern

During the preparation process of a strong earthquake, in the focal region and its surrounding regions, the activity of moderate and small earthquakes appears anomalously, and can be distinguished from the normal level of seismicity. These anomalous patterns include seismic gap, strike-like pattern of seismicity, recurrence of earthquake, migration of seismicity, anomalously high seismicity and quiescence prior to large earthquake, and spatial and temporal variation in b-value.

3.1 Seismic gap

Seismic gap is one of the phenomena in spatial distribution of seismicity. This phenomenon has received the Chinese seismologists attention very early (Mei, 1960). It is found that, before the occurrence of some major earthquakes in China's mainland, a seismic gap appears in the aftershock region of the forthcoming major earthquake; seismicity outside the gap is rather high, while that inside the gap is low.

In a systematic searching for the seismic gap as a precursor of earthquake, Chen et al. (1984) and Xue and Chen (1989) found 28 seismic gaps according to the temporal variation of the epicentral distribution of earthquakes occurred in the period of 1970 to 1973. In the criteria proposed by Chen et al. (1984), 14 among the 28 seismic gaps are recognized as precursors of forthcoming major earthquakes. Twelve major earthquakes among

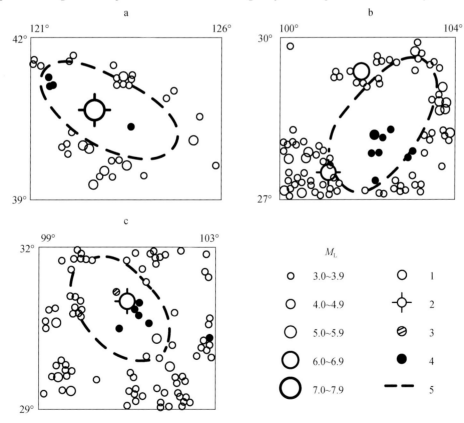

Fig. 6 Seismic gaps recognized as the precursors of the forthcoming major earthquakes

(a) The Feb. 4, 1975 Haicheng M_S =7.3 earthquake. (b) The Aug. 16, 1976 Songpan M_S =7.2 earthquake. (c) The Jan. 24, 1981 Daofu M_S =6.9 earthquake. 1-earthquake outside seismic gap; 2-main shock; 3-earthquake inside seismic gap before the occurrence of the main shock; 4-earthquake inside seismic gap after the occurrence of the main shock; 5-boundary of the seismic gap

these 14 cases did occur and filled the respective seismic gaps. Figure 6 presents three seismic gaps as the precursors of the forthcoming main shocks. In the same criteria, the other 14 seismic gaps are excluded to be precursors of any forthcoming major earthquake. While there is no any large earthquake occurred inside the 12 among the other 14 seismic gaps, two major earthquakes occurred inside the respective seismic gaps.

Early studies of seismic gap laid emphasis on the identification of earthquake-prone area. Based on the empirical formula relating the magnitude of the forthcoming strong earthquake with the area or linear dimension of the seismic gap, the strength of the forthcoming strong earthquake can be estimated (Wei et al., 1978; Lu et al., 1984). Lately, the evolution characteristics of the seismic gap are used to predict the occurrence time of the forthcoming strong earthquake. As Xue and Chen (1989) noted that, while the appearance of seismic gap is relatively common and is usable as an important clue in long-term earthquake prediction, caution must be taken when applying this characteristic.

3.2 Stripe-like pattern of seismicity

Before the occurrence of a large earthquake, the regional seismicity changes from a more or less random distribution to a concentrated stripe-like or band-like area, outside this area, the seismicity in a broad area is rather quiescent. This distinct stripe-like pattern in seismicity is one of the methods extensively used in the long-term earthquake prediction in China.

Chen et al. (1984) and Liu and Chen (1989) studied the stripe-like pattern of seismicity before the occurrence of 22 moderate to large earthquakes of $M_S \geqslant 5.5$ in China's mainland and its adjacent area (Figure 7). Among these events, 18 earthquakes were preceded by a distinct stripe-like pattern of seismicity (Figure 8). The length of the stripe ranges from 200 km to about 900 km, a figure much longer than the dimension of the focal region.

Liu and Chen (1989) proposed an empirical relationship between the duration time ΔT (in *month*) of the stripe-like pattern and the magnitude of the forthcoming main shock M (Figure 9a):

$$\Delta T = 7.8M - 39. \tag{1}$$

They also proposed an empirical formula describing the relationship between the precursor time T (in *month*) and the magnitude of the forthcoming main shock (Figure 9b):

$$T = 8.7M - 43. \tag{2}$$

In general, the strike of the stripe is in accordance with the main fracture plane of the forthcoming main shock. Thus the appearance and activity of the stripe-like pattern of seismicity are considered to be the manifestation of the activity of the seismogenic zone which directly relates to the main fracture.

3.3 Recurrence of earthquake

The anomalous pattern of the spatial distribution of seismicity can be considered as one promising precursor prior to the occurrence of a main shock; besides, the recurrence of earthquake activity can also be considered as another promising precursor. The recurrence of earthquake activity occurs either in a seismic belt (zone) or in the same location of a seismic belt (zone). For example, in 1923, an $M_S = 7\frac{1}{4}$ earthquake occurred in Luhuo along the Xianshuihe fault. Fifty years later after the occurrence of this earthquake, in the same location an earthquake of $M_S = 7.9$ occurred in 1973.

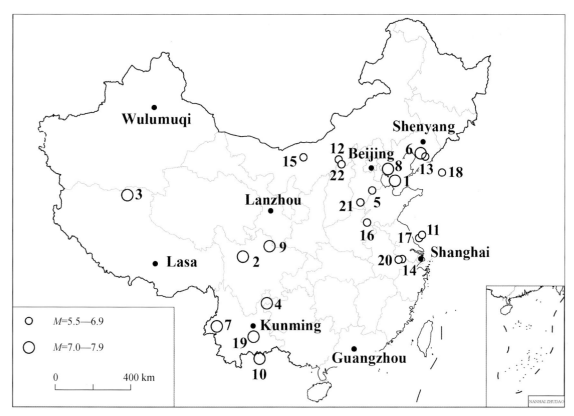

Fig. 7 Map showing the locations of 22 earthquakes with $M_S \geq 5.5$, used in the studies of the stripe-like pattern of seismicity

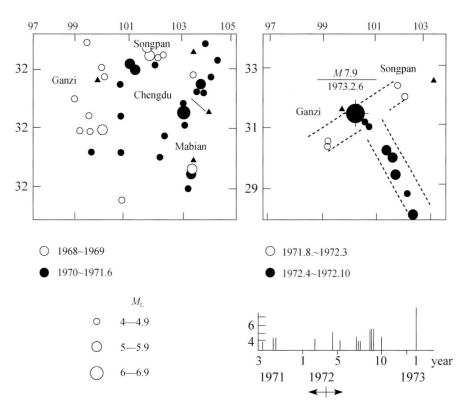

Fig. 8 Stripe-like pattern of seismicity before the 1973 $M_S = 7.9$ Luhuo earthquake

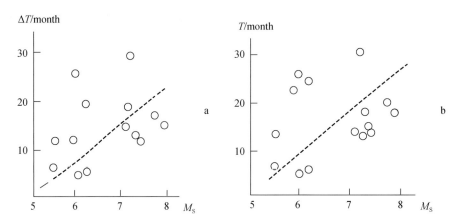

Fig. 9 (a) Relationship between the duration time ΔT of the stripe-like pattern and the magnitude of the forthcoming main shock, M_S.
(b) Relationship between the precursor time T of the stripe-like pattern and the magnitude of the forthcoming main shock, M_S

3.4 Migration of seismicity

Migration of seismicity is well-known in seismology (Båth, 1966; Guo and Chin, 1966; Mogi, 1969). It describes the consecutive occurrence in 3-D space, of strong earthquake or seismicity following some regularities. The migration of seismicity is important in forecasting roughly the location and occurrence time of the forthcoming major earthquake.

In West China, there is a tendency that the epicenter of strong earthquake usually migrates along a specific route. As an example, Figure 10 shows the migration of strong earthquake along the NW strike Xianshuihe fault (Liu, 1976). The first migration occurred during the period of 1768 to 1811. The second one occurred during the period of the 1892 Qianning earthquake to the 1910 Ganzi earthquake. The third one started from the 1910 Ganzi earthquake and ended in the 1935 Shimian earthquake. The fourth one started from the 1935 Shimian earthquake and ended in the 1967 Zhuowo earthquake. The fifth one started from the 1967 Zhuowo earthquake up to the present. Since the occurrence of the 1967 Zhuowo earthquake, along the Xianshuihe fault two earthquakes of $M_S = 7.9$ and 6.7 occurred consecutively in Luhuo and Daofu, respectively.

Historical seismicity data of about 200 years long indicate that the earthquakes of $M_S \geqslant 7.0$ occurred in the North-South seismic belt migrate regularly from north to south, and then from south to north (Figure 11). The North-South seismic belt, located in 24°—39°N and 102°—127°E is an important and major seismic belt in China. This seismic belt is a tectonic boundary which separates China's mainland into eastern and western parts. Along this boundary, tectonic movement is strong and major earthquakes occur very frequently. During the course of epicenter migration, major earthquakes mainly occurred in the locations where the tectonic structures with different strikes intersect each other or the strike of tectonic structures deflects. The major earthquakes migrated along the North-South seismic belt did not recur in the same location. In most cases they filled the seismic gaps distributed along the North-South seismic belt (Chengdu Seismological Brigade, 1972).

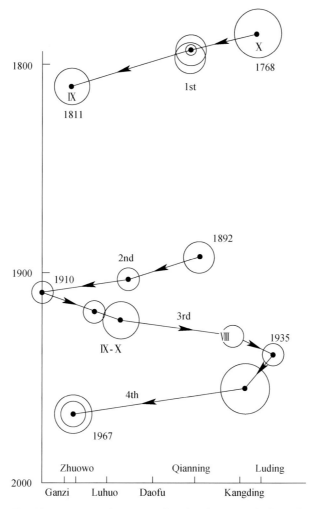

Fig. 10 Migration of strong earthquake along Xianshuihe fault

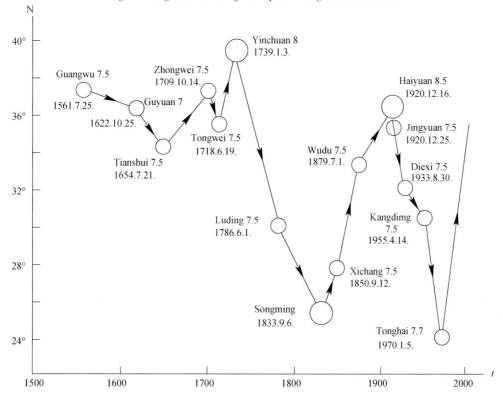

Fig. 11 Migration of strong earthquake along the North–South seismic belt

It is found that besides major earthquake, seismic active region also migrates along some seismic belts. Figure 12 depicts the migration of seismic active region along the Shan'xi seismic belt from 231 B. C. to 1695 A. D. This seismic belt experienced six relatively active periods, during which the seismic active region straightaway migrated from north toward south.

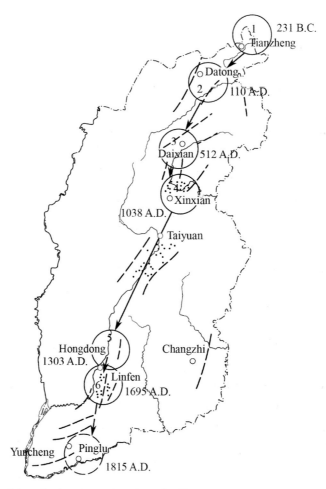

Fig. 12 Migration of the seismic active region along the Shan'xi seismic belt since 231 B. C. to 1695 A. D.

3.5 Anomalously high seismicity and seismic quiescence prior to major earthquake

Mei (1960) has studied the anomalously high seismicity and seismic quiescence prior to major earthquake. She reported that prior to the 1668 Tancheng and the 1679 Sanhe earthquakes the frequency of belt earthquakes increased at first, and then decreased (Figure 13).

Liu and Chen (1989) examined ten large earthquakes with magnitude 7.0 and larger and concluded that these large earthquakes were all preceded by an increase in activity followed by a decrease before the main shock (Figure 14). In general the anomalously high seismicity prior to a large earthquakes appeared either in the epicentral region and its adjacent regions of the forthcoming large earthquake (Figure 14e), or in a broader region (Figure 14d, f, g).

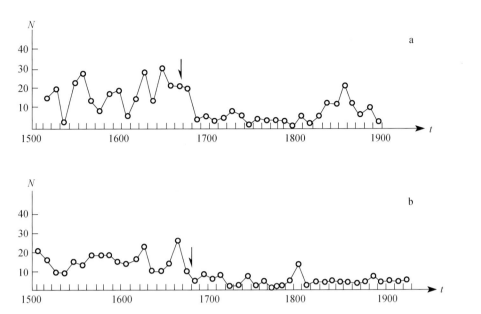

Fig. 13 Anomalously high seismicity and seismic quiescence prior to the 1668 Tancheng (a) and the 1679 Sanhe (b) earthquakes

N represents the number of the felt earthquakes in ten years

It is worth to note that two kinds of seismic activity pattern are often observed prior to a large earthquake. As shown in Figure 14, most of the large earthquakes such as the Tonghai, Luhuo, Bakeharu, Zhaotong and Tangshan earthquakes were preceded by an increase in seismic activity followed by a quiescence before the burst of the main shock. There are also examples, e. g. the Xingtai and Longling earthquakes, in which the quiescence was followed by foreshock activity shortly before the burst of the main shock.

3.6 Spatial and temporal change of b value in the magnitude-frequency distribution

The magnitude-frequency relationship is given by the Gutenberg-Richter formula:

$$\lg N = a - bM, \tag{3}$$

where N is the frequency of earthquake in the magnitude interval $M \pm dM$ and a and b are constants. The b value is a parameter expressing the proportionality between the frequency of larger and smaller events. Normally it is close to 1.0. The value a represents the overall level of seismic activity.

In a systematic search for the spatial and temporal change of b value as a possible precursor of earthquake, Li and Chen (1989) through long-time average of b value found that the normal or background b value in East China is about 0.9. They empirically defined 0.65 as the lower threshold and 1.10 as the higher threshold, of the normal level of b value, and found that for the seven major earthquakes with magnitude 7.0 and larger occurred in China's mainland in 1969–1976 (Figure 15), with the epicentral region and its surrounding regions of the forthcoming earthquake, in all cases the b value changes remarkably below the normal level before the occurrence of the forthcoming main shocks (Figure 16). In the cases of the 1974 Zhaotong $M_S = 7.1$ earthquake and the 1976 Tangshan $M_S = 7.8$ earthquake, it seems that the main shocks were preceded by a time interval with abnormally higher b value followed by a time interval with abnormally lower b value (Figure 16c, f).

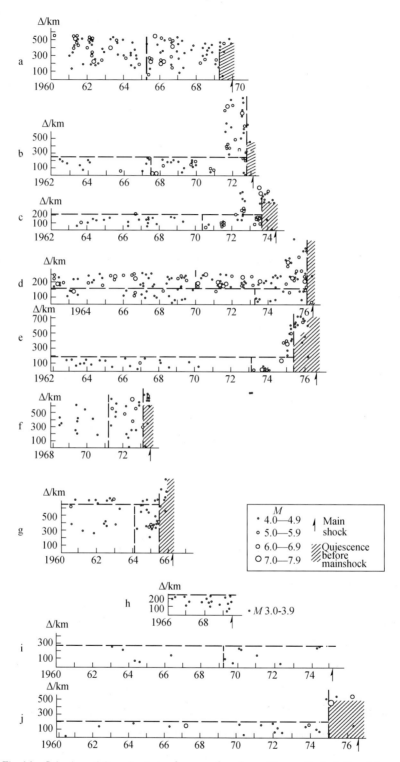

Fig. 14 Seismic activity prior to ten large earthquakes with magnitude 7.0 and larger

(a) Tonghai; (b) Luhuo; (c) Zhaotong; (d) Longling; (e) Songpan; (f) Bakehalu; (g) Xingtai; (h) Bohai; (i) Haicheng; (j) Tangshan

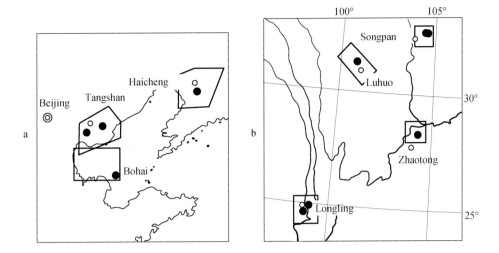

Fig. 15 Map showing the epicentral regions and their surrounding regions of the seven major earthquakes with magnitude 7.0 and larger occurred in China's mainland in 1969—1976, for which the spatial and temporal change of b value are studied.

(a) North China; (b) Southwest China

3.7 Characteristics of earthquake sequence

Besides the regional seismicity pattern can be considered as one important approach in earthquake prediction, the characteristics of earthquake sequence clustering in space and time can be considered as another important approach.

The earthquake sequence is defined as a series of earthquakes occurred within a few days to years in time and clustered in the same volume of the seismic source. The spatial and temporal clustering of earthquakes in a sequence is thought to be the manifestation of the evolution process at the earthquake source. Thus it is expected that the studies of the characteristics of earthquake sequence will promote the under the understanding of seismogenic process and provide one clue to predicting the occurrence of major earthquake. The studies of earthquake sequence in China mainly concentrated on the catalogue and identification of earthquake sequence, identification of foreshock, and prediction of the aftershock activity, especially that of the late stronger aftershock.

3.8 Types of earthquake sequence

Based on their characteristics, the earthquake sequences can be broadly catalogued into main shock, swarm and isolated types. At present, the type of the earthquake sequence is mainly catalogued and identified by a number of statistical characteristics such as energy release, change of earthquake frequency, and b-value and its change with time.

The earthquake sequence of the main shock type can be further divided into the foreshock-main shock-aftershock and the main shock-aftershock types. In the earthquake sequence of the main shock type, the energy release of the main shock is more than 90% of the total energy release in the whole earthquake sequence, and the difference between the magnitude of the main shock and that of the second largest earthquake, ΔM, usually ranges from 0.6 to 2.4. In most cases, for the aftershocks of the main shock type, the energy release in aftershock activity decreases rapidly within the first two or three days. The earthquake sequences of the 1967 Hejiang $M_S = 6.3$ earthquake, the 1969 Bohai $M_S = 7.4$ earthquake and the 1976 Tangshan $M_S = 7.8$ earthquake are catalogued into the main shock type.

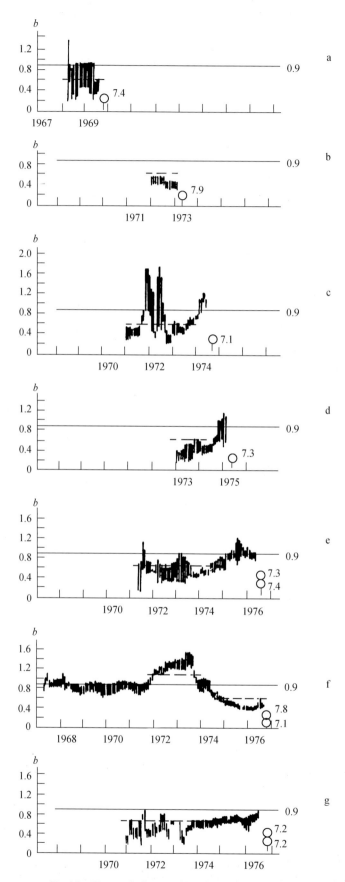

Fig. 16 Temporal change of b value in seven regions

(a) Bohai; (b) Luhuo; (c) Zhaotong; (d) Haicheng; (e) Longling; (f) Tangshan; (g) Songpan

In the earthquake sequence of the swarm type, the energy release of the largest earthquake is usually less than 80% of the total energy release in the whole sequence, and the difference between the magnitude of the largest shock and that of the second largest shock, ΔM, is usually less than 0.6. The earthquake swarm is also characterized by its high frequency of earthquake occurrence, fluctuation in energy release, slow attenuation rate of the earthquake occurrence frequency, and long-time duration of the earthquake sequence.

Before the occurrence of the 1975 Haicheng $M_S = 7.3$ earthquake, there were more than 500 foreshocks recorded at the Yingkou seismic station, about 20km away from the epicenter region of the main shock. The occurrence the main shock immediately before its occurrence (Figure 17). Similar example is the 1976 Longling, Yunnan, $M_S = 7.3$ and 7.4 earthquakes. The imminent prediction had been issued based on the analysis of the foreshock sequence which started half an hour before the occurrence of the $M_S = 7.3$ earthquake. Figure 18 shows the Longling foreshock activity recorded at a seismic station about 50km away from the epicentral region.

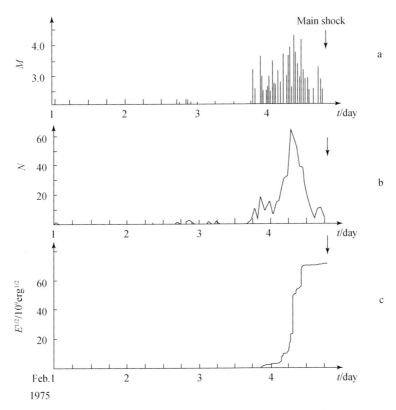

Fig. 17 Foreshocks before the 1975 Haicheng $M_S = 7.3$ earthquake

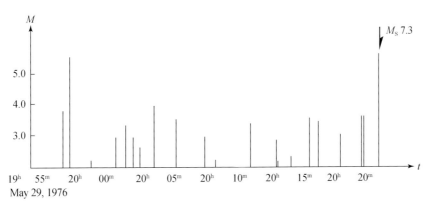

Fig. 18 Foreshocks before the 1976 Longling $M_S = 7.3$ earthquake

Among the nine $M_S \geq 7.0$ earthquakes occurred in the period 1966—1976 in China's mainland, three earthquakes (the 1966 Xingtai earthquake, the 1975 Haicheng earthquake and the 1976 Longling earthquake) were accompanied by foreshocks occurred a few days or a few hours before the occurrence of the main shock, and one earthquake (the 1974 Zhaotong $M_S = 7.1$ earthquake) has been preceded by three foreshocks with magnitude $M_L = 1.2$, 1.7, 2.3, respectively. No foreshock in a strict sense was observed before the occurrence of the other major earthquakes (the 1969 $M_S = 7.4$ Bohai earthquake, the 1970 Tonghai $M_S = 7.7$ earthquake, the 1976 Tangshan $M_S = 7.8$ earthquake and the 1976 Songpan $M_S = 7.2$ earthquake).

Foreshock activity or foreshock sequence also occurred before many earthquakes with magnitude smaller than 7.0. For example, prior to the occurrence of the November 7, 1976 Yanyuan-Ninglang $M_S = 6.7$ earthquake, the foreshock activity commenced with an $M_L = 4.0$ foreshock 9 days before the occurrence of the main shock, followed by at least 14 small events detected and located by the local seismic network. Other examples of earthquake with $M_S \leq 6.0$ accompanied by foreshocks are the 1962 Xinfengjiang $M_S = 6.1$ earthquake, the 1967 Luhuo $M_S = 6.8$ earthquake, the 1971 Dayi $M_S = 6.2$ earthquake and the 1973 Nanping $M_S = 6.5$ earthquake.

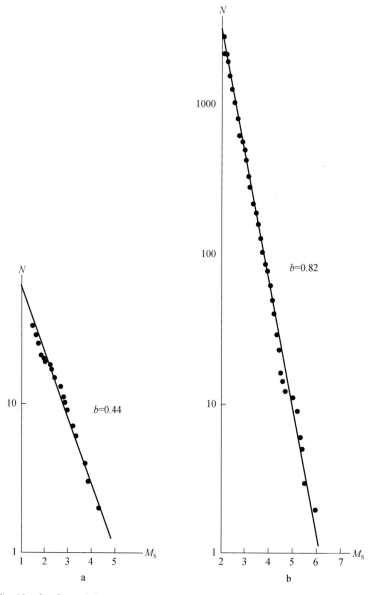

Fig. 19 b values of the Haicheng foreshock (a) and aftershock (b) sequences

A number of authors have examined the b values of the foreshock and aftershock sequences. They found that usually the former is less than the later. Wu et al. (1976) noted that the b value of the 1975 Haicheng foreshock sequence is 0.56, which sharply contrasts with the b value of the Haicheng aftershock sequence (Figure 19).

The frequency attenuation coefficient, p value, in the modified Omori's law is determined for distinguishing between foreshock sequence and swarm that occur immediately prior to a major event. It is found that in general the p value for the earthquake swarm is larger than 1.0, while for foreshock sequence this value is less than 1.0 (Liu et al., 1979).

3.9 Foreshock in a broad sense

Prior to a large earthquake, seismic activity usually occurs over a wide area (0.5°—1.0° in dimension) and a long period (years to decades). This kind of seismicity is referred to as foreshock in a broad sense (Evison, 1977). The activity of foreshock in a broad sense can be divided into four stages. The first stage is characterized by the normal seismic activity. Following this stage is the stage of the precursory swarm activity. In the third stage there is a reduction in seismicity prior to the main shock. This stage is referred to as the precursory seismic gap. The fourth stage is the stage of the main shock activity during which the foreshocks in the strict sense, main shock and the aftershocks occur.

Zhao (1980) studied the temporal and spatial distribution of 31 major earthquakes occurred since 1960 in China's mainland, based on the seismic observation data and the historical earthquakes data. He found that except a few earthquakes for which the data are insufficient, precursory seismic activities, i. e., foreshocks in broad sense, can be recognized. As an example, Figure 20 shows the precursory seismic activity before the 1976 Tangshan earthquake. Zhao (1980) obtained the following empirical relationship between the time interval T_1 (in day), from when the precursory seismic activity begins to occur first time until the main shock occurs, and the magnitude of the main shock, M:

$$\lg T_1 = 0.80M - 0.25. \tag{4}$$

He also obtained an empirical relationship between the time interval T_2 (in day), from when the precursory seismic activity begins to occur once again until the main shock, and M:

$$\lg T_2 = 0.82M - 3.04. \tag{5}$$

The magnitude difference between the main shock and the largest precursory event ΔM is found to be

$$\Delta M = 2.02 \pm 0.5. \tag{6}$$

3.10 Characteristics of aftershock sequence and aftershock prediction

Aftershock activity reflects the physical process in the regions of earthquake source. It provides a valuable opportunity for studying the properties of the medium and the physical process in the regions of earthquake source. It is also very important for the estimation of the tendency of the aftershock activity and the prediction of the larger aftershocks soon after the occurrence of a major earthquake. Thus the studies of the characteristics as well as the physics of the aftershock sequence have received extensive attention in China. Early studies of the aftershock sequence focused on the statistics of the spatial and temporal distribution of the aftershock sequence and the mechanism of the aftershock activity. Recently, the prediction of the aftershock activity and the forecasting of the stronger aftershock have received more and more attention.

Figure 20 shows the epicentral distribution of aftershocks for the nine major earthquakes occurred in the period

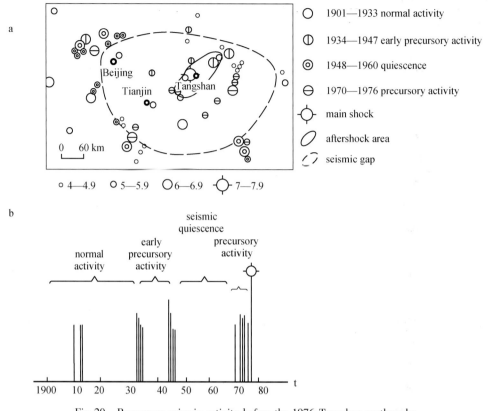

Fig. 20 Precursory seismic activity before the 1976 Tangshan earthquake

1966—1976 in China's mainland (Ma et al., 1989). In general, most aftershocks distribute in an elongated area with aspect ratio being about 2 to 3. Usually the elongated direction is parallel with the strike of the seismogenic fault. Lu et al. (1985) proposed an empirical relationship between the linear dimension, L (in km), of the aftershock area, and the magnitude of the main shock, M:

$$\lg L = 0.51M - 1.87. \tag{7}$$

Lu et al. (1985) also proposed an empirical relationship between the aftershock area, A (in km^2), and the magnitude of the main shock, M:

$$\lg A = 0.85M - 2.93. \tag{8}$$

It seems that there are some correlations between the main shock distribution and the mode of earthquake rupture. For unilateral propagating rupture, the epicenter of the main shock usually locates at one end of the elongated aftershock area (Figure 21e). For bilateral propagating rupture, the epicenter of the main shock mostly locates the center of the elongated aftershock area, instead of being near the ends of the elongated aftershock area (Figure 21c). For the earthquake swarm, without exception the strongest events distribute along the rupture propagating direction (Figure 21a, i).

It is also found that the aftershock activity closely correlates with the location of the stronger aftershocks. In the case that the stronger aftershocks distribute close to the trace of the seismogenic fault, the frequency of earthquake occurrence is very low and the total number of earthquakes, very small. For the case that the stronger aftershocks locate near the ends of the main fault, the aftershocks are abundant and stronger, and last for a longer time.

The focal depth for the most aftershock sequences ranges from 10 to 20 km. In many cases, the focal depths of the foreshock and main shock are deeper than that of the most aftershocks.

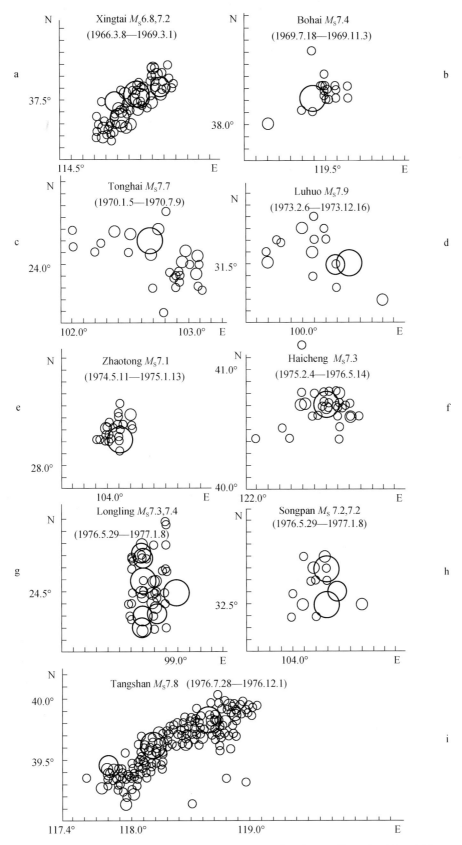

Fig. 21 Epicentral distribution of the nine major earthquakes in China (1966—1976)

The occurrence frequency of the aftershock of the major earthquake follows the modified Omori's law with the frequency attenuation coefficient p ranging between 1.0—1.6. The average of the p value is about 1.3.

For most earthquake sequences, the b value of the aftershocks is higher than that of the foreshocks, and is close to or slightly higher than the b value in the normal activity period.

3.11 Pattern recognition

Pattern recognition (Gelfand et al., 1972; 1976; Gvishiani et al., 1981) is one of the most important contemporary methods for recognizing earthquake-prone regions. This method allows of comprehensive analysis of factors (features) of geology, geomorphology, gravity, geomagnetism and seismicity, and has the merit of choosing good features and abandoning other features that provide less information, and the merit of recognizing earthquake-prone regions quantitatively. Wang (1984) and Wang et al. (1984; 1985; 1988; 1989) proposed an improved consecutive Hamming method (ICHAM) and applied the ICHAM to recognize earthquake-prone regions in Beijing area and its vicinity (Figure 22). Thirteen important features, including features of active fault, gravity and geomagnetism anomalies, elevation, Cenozoic graben basin, Tertiary and Quaternary sediments, etc. Were selected for pattern recognition. Thirty-nine specimens described by thirteen features are recognized by using the ICHAM (Wang et al., 1989). According to the criteria for discriminating the earthquake-prone regions, Wang et al. (1989) recognized the area around Wen'an and Renqiu (specimen 29 in Figure 22) as the first earthquake-prone region, i.e. the region where an earthquake with $M \geqslant 6.0$ is likely to occur in the future. They also recognized two other earthquake-prone regions, i.e. Huailai (specimen 2) and Fuping (specimen 27). As Wang et al. (1989) noted, in each of these three regions, depicted with shadow in Figure 22, a major earthquake with $M \geqslant 6.0$ is expected to occur in the future.

4 Summary

We have briefly reviewed the progress in seismological studies of earthquake prediction in China, with emphasis on the widespreadly employed methods such as seismicity, etc. We have pointed out that earthquake prediction is a complicated and difficult scientific problem to attack. At present, our understanding of the regularities and physics of earthquake occurrence and the precursory phenomena is still in the preliminary stage. The capability to predict earthquake accurately remains very poor. There is no any anomalous phenomenon which has been observed without exception prior to all the major earthquakes, nor any major earthquake which has followed to the appearance of an anomalous precursor without exception. Complexities of the anomalous precursory change from individual local regions, and the differences existed in the individual anomalous phenomenon prior to different earthquakes, make the practice in earthquake prediction difficult. To attack this important and difficult scientific problem successfully, the observation conditions should be improved and the search for the earthquake precursors and the studies of the underlying principles should be intensified. In this aspect, the seismological approach promises to be a perspective one to the earthquake prediction.

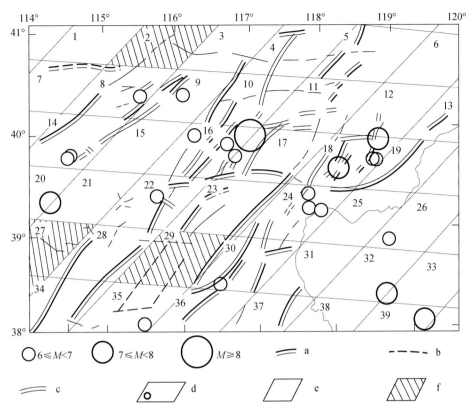

Fig. 22 Earthquake-prone regions of strong earthquakes with $M \geq 6.0$ in Beijing and its surrounding areas, predicted by pattern recognition

References

Abe K., 1981. Magnitude of large shallow earthquakes from 1904 to 1980. *Phys. Earth. Planet. Inter.*, **27**, 72–92.

Abe K., 1984. Complements to Magnitudes of large shallow earthquakes from 1904 to 1980. *Phys. Earth. Planet. Inter.*, **34**, 17–23.

Båth M., 1966. Earthquake prediction. *Scientia* (Milan), Mai-Juin, 1–10.

Central Seismological Group, 1971. *Catalogue of Chinese Earthquakes*, Vols. 1 and 2. Science Press, Beijing, 664 pp (in Chinese).

Chen Z. L., Liu P. X., Huang D. Y., Zheng D. L., Xue F. and Wang Z. D., 1984. Characteristics of regional seismicity before major earthquakes. In: *Proceedings of the International Symposium on Earthquake Prediction*, Terra Scientific Publishing Company, Tokyo. UNESCO, Paris. 505–521.

Chen Z. L., 1986. Earthquake prediction research in China: status and prospects. *J. Phys. Earth*, **34**, (Suppl.): S1–S11.

Chengdu Seismological Brigade, 1972. Temporal and spatial migration characteristics of earthquakes in the North-South seismic belt. *Earthquake Front*, **2**, 17–23 (in Chinese).

Evison F. F., 1977. Precursory seismic sequences in New Zealand. *New Zealand J. Geol. Geophys.*, **20**, 129–141.

Fu Z. X., 1982. Migration of regional seismicity before some great earthquakes. *Acta Geophysica Sinica*, 25, 509–515 (in Chinese).

Gelfand I. M., Guberman Sh. A., Izvekova M. L., Keilis-Borok V. I. and Ranzman E. Ya., 1972. Criteria of high seismicity, determined by pattern recognition. *Tectonophysics*, **13**, 415–422.

Gelfand I. M., Guberman Sh. A., Keilis-Borok V. L., Knopoff L., Press F., Ranzman E. Ya., Rotwain I. M. and Sadovsky A. M., 1976. Pattern recognition applied to earthquake epicenters in California. *Phys. Earth Planet. Inter.*, **11**, 277–283.

Gu G. X. (Chief Editor), 1983. Catalogue of Chinese Earthquakes, Vols. 1 and 2. Science Press, Beijing, **1**, 228 pp (in Chinese).

Kanamori H., 1983. Global seismicity. In: Kannmori, H. and Boschi, E. (Editors), *Earthquakes: Observation, Theory and Interpretation*, 596–608.

Guo Z. J. and Chin B. Y., 1966. The migration of earthquakes in Gansu Province. *Chinese Science Bulletin*, **17**(5), 238–240 (in Chinese).

Guo Z. J. and Chen X. L. (Chief Editors), 1986. Earthquake Countermeasures. *Seismological Press*, 503 pp (in Chinese).

Lee S. P. (Chief Editor), 1960. Catalogue of Chinese Earthquakes, Vols. **1** and **2**. *Science Press*, Beijing, 790 pp (in Chinese with English abstract).

Lee W. H. K., Wu, F. T. and Wang, S. C., 1978. A catalogue of instrumentally determined earthquakes in China (magnitude ≥ 6) compiled from various sources. *Bull. Seism. Soc. Am.*, **68**, 383–398.

Li Q. L., Chen J. B., Yu L. and Hao B. L., 1978. Time and space scanning of the b-value — a method for monitoring the development of catastrophic earthquakes. *Acta Geophysica Sinica*, **21**(2), 101–124 (in Chinese).

Li Q. L. and Chen J. B., 1989. Studies on b-value. In: *Compilation of Methods in Earthquake Surveillance and Prediction: Seismological Approaches*. Seismological Press, Beijing, 224–242 (in Chinese).

Liu P. X., Huang D. Y., Wang L. P., Wang Z. D., Zheng D. L. and Feng H., 1984. Seismicity pattern over the preparation process of strong earthquakes. In: Gu, G. X. and Ma, X. Y. (Editors), *On Continental Seismicity and Earthquake Prediction*, Seismological Press, Beijing, 100–110.

Liu P. X. and Chen Z. L., 1989. Stripe-like pattern of seismicity and its potential in earthquake prediction. In: *Compilation of Methods in Earthquake Surveillance and Prediction: Seismological Approaches*. Seismological Press, Beijing, 89–99 (in Chinese).

Liu Z. R., 1976. *Fundamentals of Seismology*. Science Press, Beijing, 256 pp (in Chinese).

Liu Z. R., Qian Z. X. and Wang W. Q., 1979. An indication of the foreshock-attenuation of earthquake frequency. *J. Seismol. Res.*, **2**(4), 1–9 (in Chinese).

Lu Y. Z., Shen J. W. and Wang W., 1984. Seismic gaps in the mainland of China. In: Gu, G. X. and Ma, X. Y. (Editors), *On Continental Seismicity and Earthquake Prediction*, Seismological Press, Beijing, 111–131.

Lu Y. Z., Chen Z. L., Wang B. Q., Liu P. X., Liu W. R. and Dai W. L., 1985. *Seismological Methods in Earthquake Prediction*. Seismological Press, Beijing, 268 pp (in Chinese).

Ma Z. J., Chen Z. L., Zhu Y. Q., Wang L. P. and Xue F., 1984. The basic characteristics of the continental earthquakes. In: Gu, G. X. and Ma, X. Y. (Editors), *On Continental Seismicity and Earthquake Prediction*, Seismological Press, Beijing, 299–311.

Ma Z. J., Fu Z. X., Zhang Y. Z., Wang C. M., Zhang G. M. and Liu D. F., 1989. *Earthquake Prediction: Nine Major Earthquakes in China (1966–1976)*. Seismological Press, Beijing, 332 pp.

Mei S. R., 1960. Seismicity of China. *Acta Geophysica Sinica*, **9**(1), 1–19.

Mogi K., 1969. Some features of recent seismic activity in and near Japan. *Bull. Earthq. Res. Inst.*, **47**, 395–417.

Seismic Intensity Regionalization Group, State Seismological Bureau, 1979. *Collection of Isoseismals of Chinese Earthquakes*. Seismological Press, Beijing (in Chinese).

Seismological Committee of Academia Sinica, 1956. *Chronological Table of Chinese Earthquakes*, Vols. **1** and **2**. Science Press, Beijing, **1**, 653 pp (in Chinese).

Shi Z. L., Zhao R. G., Wang S. Z. and Wang J. X., 1986. *World Catalogue of Earthquakes (1900–1986, $M_S \geq 6$)*. Cartographic Publishing House, Beijing, 412 pp (in Chinese).

State Seismological Bureau, 1977. *Concise Catalogue of Chinese Earthquakes*. Seismological Press, Beijing (in Chinese).

Wang B. Q. 1984. Study of the preparatory process of strong earthquakes using the cluster analysis method. *Acta Seismologica Sinica*, **6**(2), 121–128 (in Chinese).

Wang B. Q., Chen Z. Y. and Ma X. F., 1984. A nonparametric clustering method for the order samples and its application. In: *Proceedings of the 7th International Conference on Pattern Recognition*, Montreal, Canada, July 30-August 2, 1984. 566–568.

Wang B. Q. and Ma X. F., 1985. Application of pattern recognition to the study of the dynamic factor of occurrence of large earthquakes. In: Institute of Geophysics, State Seismological Bureau (Editor), *Geophysical Research*, **1**, 25–39. China Academic Publisher, Beijing.

Wang B. Q., Chen Z. Y., Tong G. B. and Wang C. Z., 1988. Order-clustering methods in the study of the seismogenic processes of strong earthquakes. *Earthquake Research in China*, **1**(4), 511–528.

Wang B. Q. and Wang C. Z., 1989. An improvement on the consecutive Hamming method and its application to the identification of earthquake-prone regions. *Acta Seismologica Sinica*, **2**(2), 163–174.

Wei G. X., Lin Z. X., Zhu X. J., Zhao Y. H., Zhao X. L. and Hou H. F., 1978. On seismic gaps previous to certain great earthquakes occurred in North China. *Acta Geophysica Sinica*, **21**(2), 213–217.

Wu K. T., Yue M. S., Wu H. Y., Cao X. L., Chen H. T., Huang W. Q., Tian K. Y. and Lu S. D., 1976. Certain Characteristics of Haicheng earthquake ($M_S=7.3$) sequence. *Acta Geophysica Sinica*, **19**(2), 95–109.

Xie Y. S. and Cai M. B., 1983–1987. *Compilation of Historical Materials of Chinese Earthquakes*, Vols. **1** to **5**. Science Press, Beijing, 4,471 pp (in Chinese).

Xue F. and Chen Z. L., 1989. Seismic gap and its potential in earthquake prediction. In: *Compilation of Methods in Earthquake Surveillance and Prediction: Seismological Approaches*, Seismological Press, Beijing, 15–33 (in Chinese).

卢龙地区 S 波偏振与上地壳裂隙各向异性

姚 陈[1]　王培德[2]　陈运泰[2]

1. 国家地震局地质研究所；2. 国家地震局地球物理研究所

摘要　由三分量数字地震仪组成的小孔径流动台网记录了 1982 年 10 月 19 日河北卢龙 $M_S = 6.1$ 级地震的部分余震. 用质点运动图的方法对横波的偏振进行了分析. 研究结果表明, 在横波窗内的各观测点都存在横波的分裂现象. 不同离源角和方位角快波偏振的水平投影都具有近 NE40° 方向的优势取向, 与根据卢龙地震两组断层错动在各向同性介质中所辐射的横波的偏振方向不一致. 这可以由传播介质中应力所导致裂隙的定向排列来解释. 这一观测结果提供了卢龙地区脆性上地壳大范围膨胀各向异性（EDA）的证据, 并表明这一地区直立平行排列裂隙取向和水平主压应力的方向为 NE40°.

1　引言

近十几年来, 对各向异性介质中地震波传播的理论研究与观测迅速发展. Crampin 对横波分裂现象的研究最引人瞩目, 他指出横波分裂是地震体波在各向异性介质中传播的最显著特征[1]. 对土耳其北安纳托利亚断层区的地震研究以及在世界其他一些地区对横波分裂的观测和研究, 证实了脆性上地壳存在着大范围膨胀各向异性[2]. 横波携带的源和波传播所经介质的信息量约比纵波携带的高 4 倍, 由横波的特征可以得到应力和裂隙的几何图像及动态变化, 对于地震预报也有重要意义, 所以得到地球物理学家和地质学家的广泛关注[3].

迄今在国内尚未见到由横波分裂给出深部介质各向异性的证据. 随着观测技术水平的提高, 三分量数字地震记录逐渐增多, 为对横波震相做深入的分析提供了可能性, 特别是小孔径流动数字化台网的地震近场记录对横波分裂研究极其有利. 本文利用河北卢龙地震的近场记录研究横波的特征, 着重分析横波分裂的偏振异常, 讨论源的辐射、介质特性和自由表面对横波的作用. 最后由快波偏振水平投影方向给出直立平行排列裂隙取向和地壳内水平主压应力的方向.

若地震波传播所经过的介质为各向同性介质, 那么从位错源发出的横波的偏振方向与接收点的方位角、离源角有关. 由于横波在均匀各向同性介质中的传播速度与其偏振无关, 所以横波在传播过程中将始终保持线性偏振的特征. 当横波在临界角（对于泊松比为 0.25 的介质, 约为 35°）以内入射自由表面时, 仪器记录到的地表位移的振动特征与入射波位移的振动特征基本一致, 通常把这一入射范围称为横波窗. 在横波窗外, 自由表面产生的反射波及其相移使地表位移的特征不同于入射波位移的特征, 且地表位移记录中的横波偏振也不同于入射波的偏振. 在各向异性介质中横波分成两个以不同的速度和偏振方向传播的准横波[4], 这就是横波分裂现象. 横波分裂表现为横波的偏振异常, 这种异常分布的特性依赖于各向异性介质的性质及其空间取向, 除了在一些特殊的传播方向上, 分裂开的快波 qS1 和慢波 qS2 的偏振一般不能再区分成 SV 波和 SH 波[5]. 当两准横波由源至接收点的传播路径相近时, 快波和慢波的偏振近似正交. 近震源观测对应着由源激发地震波以曲面波前传播至接收点. 波群传播与相位传播的不同会导致附加的震相[6]. 实际观测接收的是

---

* 本文发表于《地球物理学报》, 1992 年, 第 35 卷, 第 3 期, 305–315.

波群的传播，与相位传播给出的横波窗半径比较，波群传播所给出的稍有扩大或缩小[7,8]. 在临界角附近，会出现径向分量很强的沿地表滑行的 SP 波，容易被误认为横波分裂[8,9]. 为了避免横波窗外及临界角附近问题的复杂性，和许多研究者的通常做法一样，我们限于在横波窗内讨论横波的偏振特性.

2 卢龙余震横波分裂特征

由 5 台 DCS-302 数字磁带记录地震仪组成的小孔径流动台网对 1982 年 10 月 19 日河北卢龙地震（M_S = 6.1）的部分余震作了近场观测，得到了一批近场数字记录[10]. DCS-302 是美国 Terra Technology Corp. 制造的三分向数字地震仪，与 SSA-312 伺服加速度计配合使用记录垂直、东西和南北三个方向的地面运动加速度. SSA-312 加速摆的频率响应范围为 0—100Hz，摆体检测到的加速度转换成电信号后送入 DCS-302 地震仪，信号经截止频率为 30Hz 的 5 阶 Butterworth 低通滤波器滤波后，由模−数转换电路变成数字量. 该地震仪的采样率为 100sps，字长 12 位二进制数，具有 4 档自动量程转换，总动态范围为 112dB，最大量程为一个重力加速度 g，分辨率为 $5.0 \times 10^{-6}g$. 仪器采用了带有内部存储单元的延时触发记录，触发阈值可按工作地点的本底噪声背景和工作的需要设定，内部存储单元可记忆 1.92s 的三分量数据.

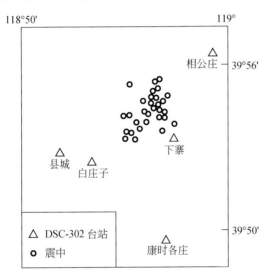

图 1 1982 年 10 月 19 日卢龙地震（M_S = 6.1）部分余震的震中位置和观测台站分布图

据北京台网测定，卢龙地震主震震中的位置在北纬 39°57′，东经 119°04′. 流动台网在 1982 年 10 月 21—29 日之间进行了观测并获得了大量余震的近场数字记录，到时数据足以定位的地震有 24 个. 定位结果表明，平均震源深度为 7—8km，即大多数卢龙余震发生于脆性的上地壳内. 记录台站为县城、白庄子、下寨、康时各庄和相公庄，震中与台站分布如图 1 所示. 还有一些地震只有一个或两个台获得记录，由于到时数据不够而未能用小孔径网内定位.

2.1 观测记录的选取和预处理

由于横波窗的限制，资料选取考虑如下因素：

小孔径台网的定位精度要高于区域台网给出的精度，但小孔径台网的定位仍基于介质各向同性假设，在各向异性介质中的定位则要求有更高的观测台网密度. 我们以小孔径网在各向同性介质条件定位为基准，并考虑到由此会产生一些震中位置和震源深度的误差.

我们对横波窗内的地震记录进行研究，当台基下面有一较薄的低速盖层时，实际横波窗的半径会有所扩大. 由于缺乏各接收点下面介质的速度结构资料，对已定位和没有定位的地震如符合以下条件的也加以研究：（1）P波和S波之间到时差小于1.5s，根据平均震源深度估算这类地震也可能在横波窗内；（2）原始加速度记录横波初动为线性偏振，各分量无超临界入射引入的相移．有的地震虽在横波窗内，由于位错源的辐射图型的影响，横波过于微弱，对这种记录不做进一步分析．

本文以三分量位移记录对横波分裂进行研究．原始记录为加速度数字记录，经两次积分得到位移．考虑卢龙余震记录的主要频率成分在10Hz左右，用中心频率为10Hz带宽为0.2—20Hz的高斯窗做带通滤波，以压制低频成分．图2是康时各庄台记录的1982年10月22日 $M_L=1.5$ 级地震的加速度图和经滤波处理后的位移图．

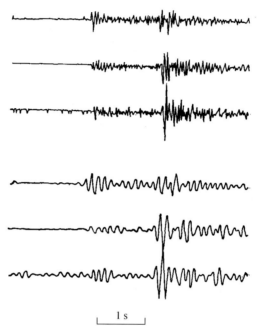

图2 康时各庄台记录的1982年10月22日 $M_L=1.5$ 级地震的加速度图和经滤波处理后的位移图

由横波分裂导致的快、慢波的相对时间差较小时，高频波比低频波有更高的分辨率．我们选取位移时间序列的主频率为10Hz，在这一频率成分的三分量位移记录图上容易直接看到快波和慢波偏振方向的变化．

2.2 横波的偏振与质点运动图

我们用未经坐标转换的三分量位移图分析横波的偏振，并由沿南北向的垂直面和水平面上的质点运动图来显示偏振方向的变化．这样做需要地表接收条件比较简单，能直接判断出SP波的尖锐波至和S波超临界入射在两水平向的到时差．图3a为相公庄台记录的1982年10月22日0时48分 $M_L=1.9$ 级地震的位移图，从图上可以见到沿地表滑行SP波的尖锐波至．图3b为相公庄台记录的1982年10月28日15时30分 $M_L=1.5$ 级地震的位移图，从图上可以见到由于横波超临界入射而引起横波在两个水平方向上具有不同的到时的震相．

对下寨台而言，大多数经流动台网定位地震都落在横波窗内．图4为该台对3个地震的记录，可以看出直达P波和S波之间其他震相较弱，横波的偏振主要在水平向；尽管3个地震横波的波列不同，但都可以看出其中有两种不同偏振方向的波．在水平面上的质点运动图中尤为明显．在接近线性偏振的快波 qS1 初动后的几个周期，两水平道振动极性反向，偏振方向出现近90°的变化，即慢波 qS2 的振动．qS2 波的初动不易直接看出，但其开始振动部分往往对应着两水平道位移波形或视周期的变化．如不区分快波 qS1 初动的极性，可见图4中快波偏振的水平投影为北东向．这3个地震记录的区别是：图4a中快波振幅很弱；图4b中慢波振幅弱；图4c中快波与慢波的振幅相当．从水平位移图中可见，快波初动后到极性发生反转出现波峰的次数不同,即快慢波的相对时间延迟是有区别的.尽管其他接收点的场地条件不同,但都可见到类似的横波偏振方向的突然变化.

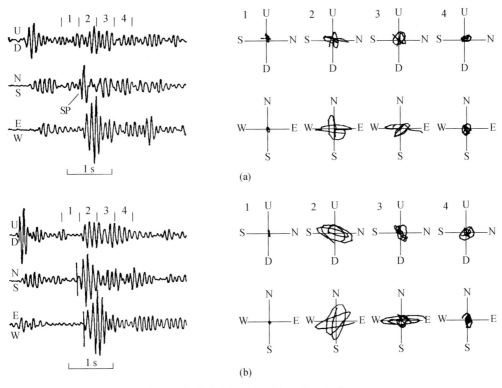

图 3　相公庄台记录的两个地震的位移图

(a) 1982 年 10 月 22 日 0 时 48 分 M_L = 1.9 级地震的位移图;(b) 1982 年 10 月 28 日 15 时 30 分 M_L = 1.5 级地震的位移图

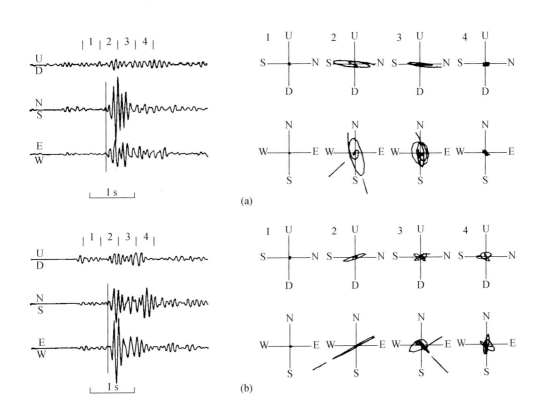

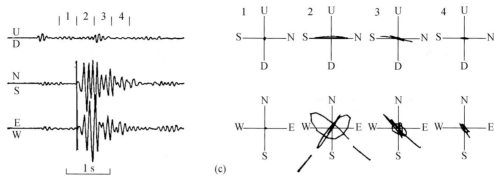

图 4 下寨台记录的 3 个余震的位移记录和质点运动轨迹图

图 5 显示了白庄子、县城、康时各庄和相公庄等台记录到的横波分裂,且快波的偏振呈北东向,慢波的偏振呈北西向.

有些直达横波在横波窗外,但深界面反射的横波和盖层的 PS 透射转换横波及多次波却在横波窗内,也可以看到上述横波分裂的特征,如图 5f,g.

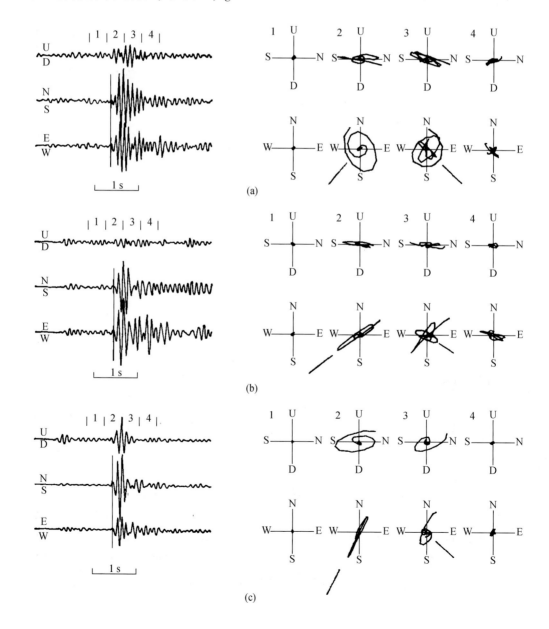

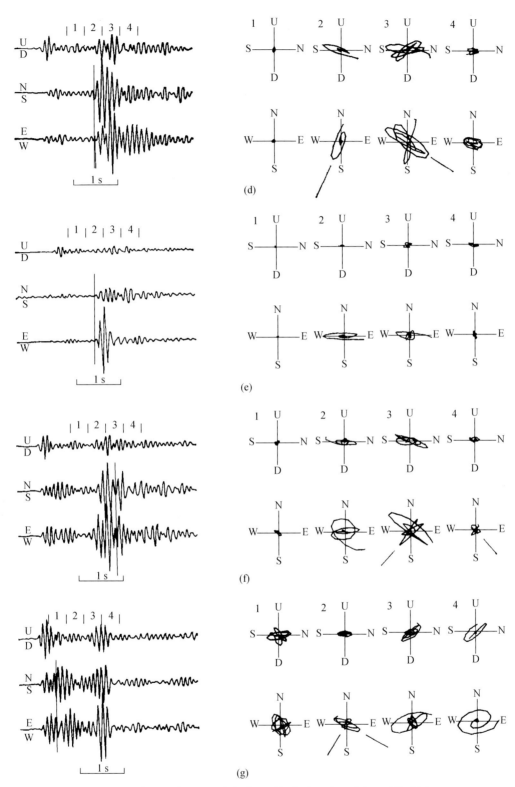

图 5 卢龙地震余震观测中部分地震的位移记录和质点运动轨迹图

(a) 白庄子台 (10月26日18时07分, $M_L = 2.0$); (b) 县城台 (10月22日01时55分, $M_L = 1.5$); (c) 康时各台 (10月22日13时55分, $M_L = 2.4$); (d) 相公庄台 (10月23日04时46分, $M_L = 2.0$); (e) 下寨台 (10月27日10时25分, $M_L = 1.5$); 图中表现横波窗内S波分裂奇异性; (f) 白庄子台 (10月27日01时43分, $M_L = 4.0$), 该记录示出深界面反射波的S波分裂; (g) 白庄子台 (10月22日16时59分, $M_L = 2.0$), 该图示出沿沉积层的P-S转换波的分裂

2.3 快波偏振的等面积投影

各个地震辐射的直达横波至不同的接收台站有不同的离源角和方位角,图 6 用下半球的等面积投影表示以离源角和方位角为参数的快波偏振方向的分布. 快波偏振方向根据质点运动图椭圆偏振长轴方向确定. 鉴于白庄子、相公庄和康时各庄的台基为沉积层,我们将横波窗半径扩大到离源角 50°,尽管由于没有校正沉积薄层引起的离源角的计算误差,以及如前所述的震源定位不够准确会给快波偏振方向在等面积投影图上的位置带来的偏差,并且有些台记录到的横波窗内的地震数目也比较少,但不同方位、不同离源角快波偏振的水平投影都在北东方向呈优势取向. 为表示各台快波偏振水平方向投影的整体特征,我们将 5 个台记录的 S 波分裂快波偏振的等面积投影示于图 6f,可见偏振方向的一致性是很明显的.

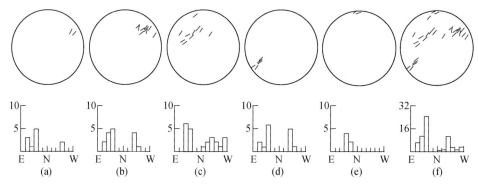

图 6 5 个观测台记录的 S 波分裂快波偏振等面积投影和快慢波偏振方向分布统计直方图

在有些地震记录中只能看到清楚的慢波偏振,还有些地震记录中横波分裂的现象看得很清楚,但没有能对地震定位. 为弥补这一不足,我们在图中给出快波偏振方位分布的直方图. 由于没有区分极性,只归并为北东和北西两个方位范围. 图中的两个峰值对应着横波初动偏振方位的优势取向. 结合快波偏振等面积投影图我们得到快波的水平投影较一致地呈近 NE40°的优势方向.

3 对快波偏振方向的解释

吴大铭等对卢龙地震余震的研究表明,卢龙地震序列是由两组断层面的活动构成的,分别来自走向北北东和走向近东西两组非共轭平面断层活动[11]. 在介质是各向同性介质假设条件下,由这两组断层的远场辐射图型作出的横波偏振方向的等面积投影图(图 7)[12],与实际观测的结果有较大的差别,因此不能用各向同性介质条件下位错源辐射的横波偏振来解释实际观测中见到的快波偏振的优势取向和慢波的偏振.

卢龙余震发生在 10km 深度内的上地壳中,在小孔径台网内直达波是强震相. 在横波出现以前到达的其他次生震相很弱,这在台基为基岩的情况下特别明显. 当台基下面有沉积层时,在离源角较大的情况下,在直达 P 波和 S 波之间会出现转换波或层内多次波,根据其偏振特征可辨认出这种波为 P-S 或 S-P 转换波. 我们没有使用横波不发育的记录,直达横波作为记录中的强震相,保留了记录中其初动的主要偏振特征,不能用界面次生震来解释快慢波偏振近似正交的现象. 来自深层界面的反射波也不能解释慢波的偏振和相对快波的相位延迟. 界面的倾斜或弯曲可以改变波的偏振特征,但是难以同时解释不同方位不同离源角横波的分裂,特别是快波偏振水平投影的一致性.

过去对于华北地区应力场的研究已经证明华北地区水平向主压应力的方向为北东向. 与吴大铭等确定

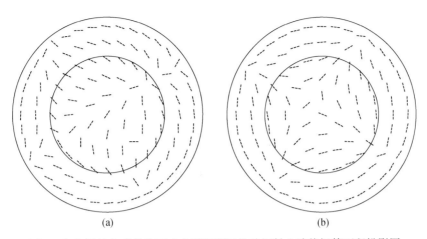

图 7 各向同性介质条件下卢龙两组断层错动辐射 S 波偏振等面积投影图

(a) 第 1 条断裂, 倾角 65°, 滑动角 135°, 走向角 270°; (b) 第 2 条断裂, 倾角 20°, 滑动角 165°, 走向角 195°

的卢龙地震第 2 条断裂综合断层面解所确定的主压应力轴的方向相近, 而与第 1 条断裂所确定的主压应力轴方向不同[11]. 地震的发生特别是大地震以后的余震可以是已有断层在应力作用下再次错动的结果. 根据余震的综合断层面解得到的应力方向, 并不能完全代表区域应力的方向, 据此我们认为卢龙地区水平向主压应力的方向为北东向. 我们以裂隙介质对横波传播的作用来解释观测到的分裂横波的快波偏振的优势取向与区域应力场主压应力方向的一致性. Crampin 提出了大范围膨胀各向异性 (EDA) 假设. 这个假说认为脆性地壳弥漫着微裂隙和介质水. 在地壳内水平向主压应力的作用下, 可产生沿主压应力方向的直立、平行排列的裂隙, 这种裂隙介质导致地震波传播的各向异性[13,14].

使用各向异性介质中理论地震图的计算证明, 穿过裂隙各向异性介质的地震体波与各向同性介质中体波相比, 最突出的特征是横波分裂. 对于直立平行排列裂隙, 虽横波分裂的特征随传播方向变化, 但横波分裂的快波偏振水平方向却平行于裂隙取向. 使用由 Hudson 推导、并经 Crampin 调整的裂隙介质 4 阶弹性张量进行理论计算[9,15], 设直立平行排列裂隙的走向为 NE40°, 裂隙密度为 0.03, 裂隙纵横比为 0.025, 介质在不含裂隙时的纵波速度为 5.8km/s, 横波速度为 3.349km/s, 密度为 2.6g/cm^3, 其快、慢波偏振的方向等面积投影如图 8. 图 8a 对应含水的裂隙介质, 图 8b 对应干裂隙. 理论计算的结果与实际观测的一致性, 表明所作假设是合理的. 由于 EDA 现象是低应力现象, 从物理机制上我们更倾向于实际介质是含水的裂隙介质. 上述分析也同时确认了卢龙地区水平主压应力的方向为 NE40°.

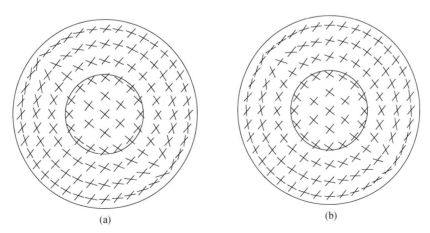

图 8 各向异性介质中横波分裂快、慢波偏振方向的下半球等面积投影图

4 讨论

快波和慢波的偏振方向与源的辐射无关,仅取决于裂隙介质的性质和波传播的路径[9],但快慢波能量的变化取决于各个地震位错源的辐射图型.裂隙散射导致3类体波吸收的各向异性,慢波的衰减比快波的衰减更快一些[15].在实际观测中发现了比快波振幅大得多的慢波振幅,这说明快慢波的振幅比还应包含有位错源的影响.目前的理论对震源做了大量假设和简化,裂隙介质中源的问题有待进一步研究,它关系着用研究吸收获取比从偏振和到时差的研究中得到有关裂隙介质的更详细的信息.

裂隙介质中快、慢波的传播速度随方向变化,其相对到时不仅依赖于传播路径的长度,也与裂隙参数有关.当裂隙参数发生随时间的变化时,有可能从相对到时差的变化中发现它们的变化.由于地震记录数量的不足,也没有在横波窗内得到一个包括较大地震和它前后的多个较小地震构成的序列,以便于对到时差的变化进行分析.从图5e可以看到在 $M_L=4.2$ 级地震的前后下寨台记录的横波运动特性有所变化,说明有可能通过横波分裂的观测来得到介质裂隙参数动态变化的信息.

近年来国外有些关于近场地震记录的研究注意到场地效应对记录频谱的影响[16,17].我们在横波分裂的研究中讨论的频率范围为10Hz左右,波长为几百米左右,一般不会由于场地效应而影响横波的偏振特性.如何从实际记录中消除场地效应的影响也是一个正在研究的课题,因此详细的讨论场地效应对横波偏振的影响还需要进一步工作.

我们对卢龙地区近场记录的波形分析主要是对横波的分析,还是比较初步的,但是从这种分析中得到了一些有意义的结果.确认了在卢龙地区的近场记录中存在横波分裂现象,并得到了快波偏振水平投影的优势取向.由此可推断在该地区脆性上地壳内存在直立平行排列微裂隙,其方向为NE40°,即华北地区水平主压应力场的方向.这一结果与以往测震学和地震地质学所得出的华北地区北东向主压应力场的方向一致[18].为加强对震源过程和震源区介质性质及状态的研究,需要有更密集的三分量数字化记录的台网对地震做更长时间的记录,同时也需要进一步发展波形的理论研究.

参 考 文 献

[1] Crampin, S., Seismic wave propagation through a cracked solid: polarization as a possible dilatancy diagnostic, *Geophys. J. R. astr. Soc.*, **53**, 467-496, 1978.

[2] Crampin, S., Geological and industrial implications of extensive-dilatancy anisotropy, *Nature*, **328**, 491-496, 1987.

[3] 林蓉辉,地震各向异性与S波分裂研究,地震科技情报,**3**,1-32,1989.

[4] Crampin, S., An introduction to wave propagation in anisotropic media, *Geophys. J. R. astr. Soc.*, **76**, 17-28, 1984.

[5] 姚陈,穿透裂隙介质远震PS波的分裂,中国地震,**5**,1,38-47,1989.

[6] Booth, D. C. and Crampin, S., The anisotropic reflectivity technique theory, *Geophys. J. R. astr. Soc.*, **72**, 755-765, 1983.

[7] Evans, R., Effects of free surface on shear wavetrains, *Geophys. J. R. astr. Soc.*, **76**, 165-172, 1984.

[8] Booth, D. C. and Crampin, S., Shear-wave polarizations on a curved wavefront at an isotropic surface, *Geophys. J. R. astr. Soc.*, **77**, 31-45, 1985.

[9] 姚陈,地方震各向异性合成地震图研究,地球物理学报,待发表.

[10] 王培德,近震源强地面运动研究,博士研究生毕业论文,国家地震局地球物理研究所,1987.

[11] 吴大铭、王培德、陈运泰,用SH波和P波振幅比确定震源机制解,地震学报,**11**,275-281,1989.

[12] Aki, K. and Richards, P. G., *Quantitative Seismology: Theory and Methods*, W. H. Freeman, San Francisco, 1980.

[13] Crampin, S., Evans, R. and Atkinson, B. K., Earthquake prediction: a physical bases, *Geophys. J. R. astr. Soc.*, **76**, 147-156, 1984.

[14] Crampin, S. and Booth, D. C., Shear-wave polarizations near the North Anatolian Fault II. Interpretation in terms of crack-induced anisotropy, *Geophys. J. R. astr. Soc.*, **83**, 75–92, 1985.

[15] Crampin, S., Effective elastic constants for wave propagation through cracked solids, *Geophys. J. R. astr. Soc.*, **76**, 135–145, 1984.

[16] Castero, R., Anderson, J. G. and Singh, K., Site response, attenuation and source spectra of S waves along the Guerrero, Mexico subduction zone, *Bull. Seism. Soc. Amer.*, **80**, 1481–1503, 1990.

[17] Blakeslee, S. and Malin, P., High-frequency site effects at two Parkfield downhole and surface stations, *Bull. Seism. Soc. Amer.*, **81**, 332–345, 1991.

[18] 国家地震局《一九七六年唐山地震》编辑组，一九七六年唐山地震，地震出版社，1982.

Shear-wave polarization and crack induced anisotropy of upper crust in Lulong, North China

Yao Chen[1], Wang Pei-De[2] and Chen Yun-Tai[2]

1. Institute of Geology, State Seismological Bureau; 2. Institute of Geophysics, State Seismological Bureau

Abstract We have used a temporal network consisting of three-component digital seismographs to record aftershock sequence of the Lulong, Hepei Province, China, earthquake of October 19, 1982, ($M_S = 6.1$). Polarization of shear-wave is analysed by using particle motion diagram. It is demonstrated that the shear wave splitting exists at all seismic stations within shear-wave window. For different azimuths and take-off angles, the horizontal projections of faster shear-wave polarization present a dominant azimuth of polarization of about NE 40°. This azimuth is different from the shear-wave polarization pattern determined from two groups of fault planes in Lulong earthquake sequence if the anisotropy is not taken into account. The stress-induced crack alignments in the propagation media can be used to explain this discrepancy. The observation presented in this study exhibits the existance of extensive dilatancy anisotropy (EDA) in upper crust of the Lulong, North China, area and demonstrates that the orientation of the vertical parallel-aligned cracks and the regional horizontal maximum compression is about NE 40°.

平面内剪切断层的超 S 波速破裂*

李世愚　陈运泰

国家地震局地球物理研究所

摘要　研究了平面内剪切断层的自然破裂速度，特别是超 S 波速自然破裂是否存在的问题．采用经典的线弹性断裂力学模型，用 1 个平面内剪切裂纹沿自身所在的平面扩展，作为平面内剪切断层的模型．通过理论推导，把 Kostrov（1975）的解从低于瑞利波速度 v_R 推广到高于 S 波速度，得出了应力强度因子 K_2 在 $\alpha > v > \beta$ 的条件下的解析式．对于泊松介质，K_2 在（β, 1.70β）为正实数，其中 v 为破裂速度，α 为 P 波速度，β 为 S 波速度．这表明（β, 1.70β）这个范围是满足自发破裂条件的 v 的解的存在范围．对 v 在不同区间内 K_2 的存在性、收敛性以及取值的正负进行了总结，得出以下结论：（1）平面内剪切裂纹的自然破裂速度 v 有 3 个物理区间，第 1 个在 0 和 v_R 之间，第 2 个在 β 和 1.70β 之间，第 3 个为 α；（2）v 有两个物理"禁区"，第 1 个在 v_R 和 β 之间，第 2 个在 1.70β 和 α 之间，它们分别构成了破裂速度的屏障．

导出的解析式不仅适用于经典模型，也适用于平面内剪切裂纹自然失稳扩展的各种其他派生模型（例如，滑动弱化模型、重整化模型等），所采用的模型比起前人的稳态模型更接近实际情况．

1 引言

对于平面内剪切断层（裂纹）的自然破裂来说，是否存在超 S 波速的破裂速度？Burridge（1973）曾经根据稳态解（破裂速度 v 为常数）的分析结果从理论上预言：在介质抗剪切强度为有界的条件下，裂纹的破裂速度可能超越这个"禁区"，以 P 波速度扩展．Burridge 的模型并没有考虑破裂能的问题．后来 Andrews（1976）指出，Burridge 处理的实际是内聚带长度为零的极限情形，因此他的模型中的表面能为零．Burridge 的稳态模型和实际的自然的非稳定破裂有差距，因此这个问题还不能说已经从理论上解决了．Andrews（1976，1985）采用数值解法验证了 Burridge 的预言，发现在某种滑动弱化模式的假定下，当破裂速度逐渐接近瑞利波速时，裂纹前缘会出现隧道效应，产生的副裂纹以大于 S 波的速度扩展，逐渐达到 P 波的速度．破裂速度是通过隧道效应越过物理禁区的．Das 等（1977）也采用数值方法计算过平面内剪切裂纹自然扩展问题，并肯定了破裂速度超出瑞利波速的可能性．Das 和 Aki 采用了 Hamano（1974）提出的临界应力判据，计算的结果表明破裂速度最大可以达到 P 波速度，但是并没有给出破裂过程细节，更没有说明破裂速度是怎样超越瑞利波速度的．

为了从理论上证明数值计算的结果，我们必须得到这个问题在非稳定自然破裂过程中的解析解，为此，我们需要找到一个对于各种模型都普遍适用的基本物理参数，在破裂动力学里，这个参数就是动态应力强度因子 $K_2(t)$，$K_2(t)$ 是破裂速度 $v(t)$ 的函数，它代表了破裂的驱动力．是否能取某一数值上的 v，要看 v 取该值时 K_2 是否为收敛且大于（或等于）零的实数（如果零速度起始破裂的 K_2 为正）．Kostrov（1975）在经典模型（线弹性断裂力学模型）内导出过 $K_2(t)$ 的解析表示式，然而他的解在物

---

*　本文发表于《地震学报》，1993 年，第 **15** 卷，第 1 期，9–14．

理上只适用于 $v < v_R$ 的情形. 至于在 $\beta < v < \alpha$ 情况下的 $K_2(t)$ 的解析式, 迄今我们还没有在文献上见到过. 这里 v_R 为瑞利波速度, α 为 P 波速度, β 为 S 波速度. 现在, 我们需要研究的问题就归结为: 在经典模型内, 超 S 波速破裂的 $K_2(t)$ 的解析式是否存在? 是否收敛? 在什么区间内大于 (或等于) 零?

2 基础公式

我们用一个平面内剪切裂纹沿自身所在的平面扩展, 作为平面内剪切断层的模型. 令二维裂纹位于 $x_1 = 0$ 的平面上, 端部位于 $l_-(t) < x_2 < l_+(t)$, $-\infty < x_3 < +\infty$. 和 Kostrov (1975) 以及许多研究者一样, 我们假定裂纹是在 $t = 0$ 时刻突然出现的. 假定体力为零, 并假定所有的物理量不随 x_3 变化, 应力分量 σ_{ij} 满足运动方程

$$\sigma_{ij,j} = \rho \ddot{u}_i \qquad i, j = 1, 2, 3. \tag{1}$$

在弹性介质中, 应力–应变的本构关系为

$$\sigma_{ij} = c_{ijpq} e_{pq} \qquad p, q = 1, 2, 3. \tag{2}$$

在均匀、各向同性的完全弹性介质中, 应力、应变满足本构关系

$$\sigma_{ij} = \lambda \delta_{ij} e_{kk} + 2\mu e_{ij}, \tag{3}$$

其中应变张量

$$e_{pq} = \frac{1}{2}(u_{p,q} + u_{q,p}). \tag{4}$$

设本问题为平面应变问题, 则运动方程成为

$$\sigma_{pq,q} = \rho \ddot{u}_p, \qquad \sigma_{3q,q} = \rho \ddot{u}_3, \qquad p, q = 1, 2, \tag{5}$$

其中, ρ 为介质密度. 这个问题的边界条件为

$$\sigma_i(x_2, t) \equiv \sigma_{i1}(0, x_2, t) = -p_i(x_2, t), \quad i = 1, 2, 3, \quad x_1 = 0, \quad l_-(t) < x_2 < l_+(t), \tag{6}$$

其中, $p_i(x_2, t)$ 为已破裂部分上的载荷, 对于本文的问题来说, $p_1(x, t) = p_3(x, t) = 0$; 以及在实际断层面内还存在压力, 即

$$\sigma_1(x_2, t) \equiv \sigma_{11}(0, x_2, t) = -\sigma_N(x_2, t) \leq 0,$$

但是它不会改变断层面上的剪切应力. 我们可以把本文的问题看成为上式 (相当于没有断层) 和 (6) 式的两个问题的叠加. 记 α, β 为弹性波速, $\alpha = \sqrt{(\lambda + 2\mu)/\rho}$, $\beta = \sqrt{\mu/\rho}$.

3 广义应力和广义应力强度因子

Kostrov (1975) 采用 Cagniard 方法, 结合 Weiner-Hopf 的分解因式法, 导出了边界积分方程. Kostrov 引进了广义应力 $F_i(i = 1, 2)$. 在本问题里, $F_3 = 0$, 而 F_1 和 σ_1 不直接参与剪切破裂判据的计算, 因此我们只需要考察 F_2. F_2 与真实应力 σ_2 的相互变换公式为 (Kostrov, 1975)

$$\sigma_2(x,t) = F_2^+(x,t) + \frac{1}{2\pi}\frac{\partial}{\partial t}\int_{1/\beta}^{1/\alpha}\{S(-s)\}\frac{1/v_R - s}{(s - 1/\alpha)^{1/2}(1/\beta - s)^{1/2}}$$

$$\cdot \int_0^{t/s} F_2^+(x - \eta, t - s\eta) \, d\eta \, ds, \tag{7}$$

式中,

$$S(s) = \exp\left\{-\frac{1}{\pi}\int_{1/\alpha}^{1/\beta} \tan^{-1}\frac{4\xi^2(\xi^2 - 1/\alpha^2)^{1/2} \cdot (1 - \beta^2 - \xi^2)^{1/2}}{(2\xi^2 - 1/\beta^2)^2}\frac{d\xi}{\xi + s}\right\}. \tag{8}$$

在复 s 平面上的支点 $s = 1/\beta$ 和 $s = 1/\alpha$ 之间做割线（图1）.

$$\{S(-s)\} = S(-s+i0) + S(-s-i0) \tag{9}$$

其中，$S(s)$，$S(-s)$ 满足关系式（Kostrov，1975；Das et al.，1977）

$$R(s) = 2\left(\frac{1}{\beta^2} - \frac{1}{\alpha^2}\right)\left(\frac{1}{v_R^2} - s^2\right) \cdot S(s) \cdot S(-s), \tag{10}$$

$R(s)$ 为瑞利函数

$$R(s) = \left(2s^2 - \frac{1}{\beta^2}\right)^2 + 4s^2\left(\frac{1}{\alpha^2} - s^2\right)^{1/2}\left(\frac{1}{\beta^2} - s^2\right)^{1/2}.$$

相应地还有 F_2^-. 我们在这里只涉及 F_2^+，以下将 F_2^+ 简记为 F_2. 在 $x > l_+(t)$ 部分，计算广义应力的公式为

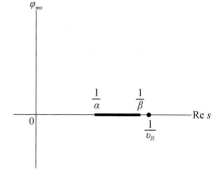

图1 复 s 平面上的割线

$$F_2(x, t) = \frac{1}{\pi [x - l_+(t^*)]^{1/2}} \int_{x-\alpha t}^{l_+(t^*)} f_2\left(x_1, t - \frac{x - x_1}{\alpha}\right) \frac{[l_+(t^*) - x_1]^{1/2}}{x - x_1} dx_1, \tag{11}$$

其中，t^* 为方程 $x - l_+(t^*) = \alpha(t - t^*)$ 的根，f_2 为广义载荷，它和已破裂部分的真实载荷的关系满足（7）式. 由 Kostrov（1975）应力强度因子 $K_2(t)$ 定义为

$$\sigma_2(x, t) \approx \frac{K_2(t)}{[2(x - l_+(t))]^{1/2}} + O(1), \quad x \to l_+(t) + 0. \tag{12}$$

广义应力强度因子 $m_2(t)$ 定义为

$$F_2(x, t) \approx \frac{m_2(t)}{[2(x - l_+(t))]^{1/2}} + O(1), \quad x \to l_+(t) + 0. \tag{13}$$

由（11）和（13）式可得（Kostrov，1975）

$$m_2(t) = \frac{\left[2\left(1 - \frac{v(t)}{\alpha}\right)\right]^{1/2}}{\pi} \int_{x-\alpha t}^{l(t)} f_2\left(x_1, t - \frac{l - x_1}{\alpha}\right) \frac{dx_1}{[l(t) - x]^{1/2}}. \tag{14}$$

在导出以上公式之后，Kostrov（1975）只考虑了 $s = 1/v$ 位于复 s 平面的实轴上 $\text{Re}\, s > 1/\beta$，也就是 $v < \beta$ 的情况，因此他所得到的 $K_2(t)$ 的解析表示式（Kostrov，1975），只是在 $0 < v < \beta$ 时为实数，在 $v < v_R$ 时大于零，在 $v_R < v < \beta$ 时则变为小于零. 这就是为什么他的解在物理上只适用于 $v < v_R$ 的情况. 事实上，(1)—(14)式的推导过程并没有受到 $v < v_R$ 的约束. 更详尽的推导可参见李世愚的文章（李世愚，1991）. $s = 1/v$ 位于实轴上支点 $s = 1/\beta$ 和 $s = 1/\alpha$ 之间的情况也适用于（1）—（14）式，只不过由于这一段是割线，必须采用另外的方法去进一步推导应力强度因子，这是本文的工作和 Kostrov（1975）所不同的地方. 因此，在本文后面的推导中，我们可以把 v 的取值范围延拓到区间 $[0, \alpha]$，仍然以（1）—（14）式为基础，分析平面内剪切断层的自然破裂速度问题.

4 动态应力强度因子的广义解

假定 $v(t) > \beta$，将（12），（13）式代入（7）式，可得到平面内剪切断层自然破裂的动态应力强度因子的广义解. 在推导过程中，需要区分 $\text{Re}\, s < 1/v$ 和 $\text{Re}\, s > 1/v$ 两种不同情况，并证明 Cauchy 型积分 $\int_{1/\alpha}^{1/\beta} \frac{Q(s)}{s - 1/v(t)} ds$ 有意义，其中 $Q(s)$ 为在区间 $(1/\alpha, 1/\beta)$ 内连续、可积的解析函数. 动态应力强度因子的广义解可以写成

$$K_2(t) = m_2(t)\left\{1 + \frac{1}{2\pi}\int_{1/\alpha}^{1/\beta} \frac{1}{s - \frac{1}{v(t)}} \{S(-s)\} \frac{1/v_R - s}{(s - 1/\alpha)^{1/2} \cdot (1/\beta - s)^{1/2}} ds\right\},$$

$$0 < v(t) < \alpha, \tag{15}$$

式中，$m_2(t)$ 由 (14) 式表示. 由上面的推导，(15) 式不仅适用于 $v(t) \in (\beta, \alpha)$，也适用于 $v(t) \in (0, \beta)$. 这就是为什么 (15) 式的条件标注为 $0 < v(t) < \alpha$. 记 (15) 式的大括号部分为 $C_k(v)$

$$C_k(v) = 1 + \frac{1}{2\pi} \int_{1/\alpha}^{1/\beta} \frac{1}{s - \frac{1}{v(t)}} \{S(-s)\} \frac{1/v_R - s}{(s - 1/\alpha)^{1/2} \cdot (1/\beta - s)^{1/2}} \mathrm{d}s, \tag{16}$$

于是

$$K_2(t) = m_2(t) \cdot C_k(v). \tag{17}$$

图 2 给出了 $C_k(v)$ 曲线在区间 $v(t) \in (\beta, \alpha)$ 的部分. 可以看出，在 $v(t) \in (\beta, \alpha)$ 内，$K_2(t)$ 还有一个零点，它是方程 $C_k(v) = 0$ 的根. 在泊松介质 ($\lambda = \mu$) 的假定下，这个根位于 $v = 1.70\beta$. 在 $v(t) \in (\beta, 1.70\beta)$ 内，$C_k(v) > 0$，而在区间 $v(t) \in (1.70\beta, \alpha)$ 内，$C_k(v) < 0$. 区间 $v(t) \in (1.70\beta, \alpha)$ 似乎成了破裂速度的第二物理"禁区".

5 $v(t) = \beta$ 和 $v(t) = \alpha$ 时的应力强度因子

(1) $v(t) = \beta$. 此时 (15) 式积分的结果为 $-\infty$，$K_2(t)$ 在 $v(t) = \beta$ 点也应为 $-\infty$，$v(t) = \beta$ 成为 $K_2(t)$ 的一类间断点.

(2) $v(t) = \alpha$. 由 (14) 式可知，此时 $m_2(t)|_{v=\alpha} = 0$. 代入 (15) 式，立即得到 $K_2(t)|_{v=\alpha} = 0$. 前面我们提到过，在区间 $v(t) \in (1.70\beta, \alpha)$ 内，$\lim_{v \to \alpha^-} K_2(t) = -\infty$，但是这是从极限的意义上讲的，现在，我们进一步指出，$v(t) = \alpha$ 时 $K_2(t)$ 本身并不是 $-\infty$，而是零.

6 次 S 波速破裂的动态应力强度因子

前面已经提到，(15) 式是应力强度因子在区间 $v(t) \in [0, \alpha]$ 内的普遍表达式，只不过在 $v(t) \in (0, \beta)$ 的条件下，点 $s = 1/v(t)$ 不在图 1 复 s 平面的割线上，我们还可以进一步把它积出来. 把 (15) 式中的积分延拓到复 s 平面上，积分回路包围的是割线 $1/\alpha < \mathrm{Res} < 1/\beta$，$s = 1/v(t)$ 是回路 Γ_s 外的一个单极点，可以应用残数定理. Kostrov (1975) 已给出 (15) 式的积分结果是

$$K_{2d}(x, t) = m_2(t) \left\{ S(-s) \frac{1 - \frac{v(t)}{v_R}}{\left[1 - \frac{v(t)}{\alpha}\right]^{1/2} \left[1 - \frac{v(t)}{\beta}\right]^{1/2}} \right\}. \tag{18}$$

(18) 式实际是 (15) 式在 $v < v_R$ 情况下的特解.

7 破裂速度的"禁区"与非"禁区"

图 2 显示了变换因子 $C_k(v)$ 的函数曲线. $K_2(t)$ 在区间 $v(t) \in (\beta, \alpha)$ 内虽然处处收敛，却并不一致收敛，这些结果的收敛性只具有数学意义，加上物理约束条件才构成破裂速度的实际判断. 如果我们规定 $m_2(t) \geq 0$，则物理约束条件就是 $K_2(t) \geq 0$，下面我们将对不同区间的分析结果列于表 1.

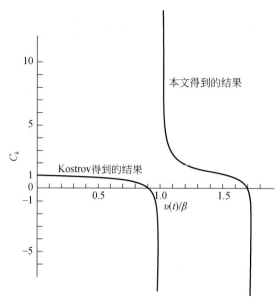

图2 $0 < v(t) < \alpha$ 的全部应力强度因子变换系数曲线

表1 破裂速度在不同区间的存在判断

$v(t)$	收敛性	$K_2(t)$	$v(t)$
$0 - v_R$	收敛	≥ 0	有非稳定解（第一区间）
$v_R - \beta$	收敛	≤ 0	第一物理禁区
β	发散	间断	无解
$\beta - 1.70\beta$	收敛	> 0	有非稳定解（第二区间）
$1.70\beta - \alpha$	收敛	< 0	第二物理禁区
α		$= 0$	有解（破裂速度上限）

8 结语

我们已经完成了本文开头提出的任务，在经典模型中导出了超 S 波速破裂的应力强度因子．本文导出的解析式（15）适用于平面内剪切裂纹自然失稳扩展的各种模型，因此它可以作为以往的数值结果的理论依据．本文所采用的模型比起 Burridge（1973）的稳态模型更具有实际意义．另外，它具有更高的分辨率，把 $v(t) \in (\beta, \alpha)$ 区间分成一段非稳定区和一段狭窄的物理"禁区"．本文所得到的解实际上是平面内剪切断层超 S 波速破裂存在的必要条件，而不是充分条件，关于它的充分条件，还需要进一步的考虑．事实上我们已经看到，超 S 波速破裂的应力强度因子虽然也是经典方程的解却在经典模型中单一裂纹的情况下无法实现，原因就是经典模型中存在的应力奇异性使得它的破裂速度无法超越第一物理禁区所形成的壁垒．在共线裂纹的情况下，这个问题也许可以考虑用隧道效应（李世愚，1991）去解决．不过在分析实际断层破裂问题时，人们往往借助于派生模型（例如滑动弱化模型、重整化模型等）去理解这个速度超越问题．

作者曾与吴明熙、周家玉、李大鹏、张洪魁进行过有益的讨论，在此谨表谢意．

参 考 文 献

李世愚, 1991. 平面内剪切断层的自然扩展及地震破裂机制、破裂判据的研究. 博士学位论文, 国家地震局地球物理研究所, 北京.

Andrews, D. J., 1976. Rupture velocity of plane-strain shear cracks. *J. Geophys. Res.*, **81**, 5679–5687.

Andrews, D. J., 1985. Dynamic plane-strain shear rupture with a slip-weakening friction law calculated by a boundary integral method. *Bull. Seism. Soc. Amer.*, **75**, 1–21.

Burridge, R., 1973. Admissible speeds for plane strain self similar shear cracks with friction but lacking cohesion. *Geophys. J. R. astr. Soc.*, **35**, 439–455.

Das, S. and Aki, K., 1977. A numerical study of two-dimensional spontaneous rupture propagation, *Geophys. J. R. astr. Soc.*, **50**, 643–668.

Hamaro, Y., 1974. Dependence of rupture time history on the heterogeneous distribution of stress and strength on the fault plane (abstract). *Eos Trans. Amer. Geophys. Un.*, **55**, 352.

Kostrov, B. V., 1975. On the crack propagation with variable velocity. *Int. J. Frac.*, **11**, 47–56.

The China Digital Seismograph Network*

Y. T. Chen, Q. D. Mu and G. W. Zhou

Institute of Geophysics, State Seismological Bureau

The China Digital Seismograph Network (CDSN) program was initiated in May 1983. On October 1, 1986, the CDSN began to distribute the network-day tapes to the research community. On October 22, 1987, the CDSN began full operation. The CDSN are supported by the State Seismological Bureau, People's Republic of China and the United States Geological Survey. The operation and maintenance of the network are taken by the staffs at the 10th Division of the Institute of Geophysics, State Seismological Bureau (IGSSB). At present the CDSN includes ten field stations, i.e. Beijing (BJI), Lanzhou (LZH), Enshi (ENH), Kunming (KMI), Qiongzhong (QIZ), Shanghai (SSE), Urumqi (WMQ), Hailar (HIA), Mudanjiang (MDJ), and Lhasa (LSA), two national centers, i.e. the Network Maintenance Center (NMC) and the Data Management Center (DMC), both at the Institute of Geophysics, State Seismological Bureau, Beijing. Figure 1 shows the distribution of CDSN stations and support facilities. Table I lists the station parameters of CDSN.

1 Operating status of the CDSN

Operation and data availability of the CDSN have improved significantly and kept at a high level during 1992 and 1993. Data from the CDSN are used widely by research institutions. The Data Management Center (DMC) of the CDSN has provided users with 1000 MB-CDSN event data and 150 MB-NEIC CD-ROMs data, for studying the mechanism of earthquake source and the structure of crust and upper mantle. The CDSN data availability is 98.6% for 1992, and 98.1% for 1993 (table II).

2 New CDSN program

In order for the CDSN to become a fully participating partner in the GSN program, it will be necessary to update the CDSN station and the DMC equipment to meet IRIS standards. Since November 1992, the new CDSN program has been carried out jointly by the Institute of Geophysics, State Seismological Bureau and USGS Albuquerque Seismological Laboratory.

2.1 Major goals of the second-phase technical upgrade of the CDSN

According to the 《agreement-in-principle》 between the State Seismological Bureau, People's Republic of China and the United States Geological Survey for upgrades to the CDSN, the second-phase technical upgrade of the CDSN commenced in November of 1992 and is planned to be completed by the end of 1995. Its major goals are:

---

* 本文发表于 Annali di Geofisica, 1994 年, 第 37 卷, 第 5 期, 1049–1053.

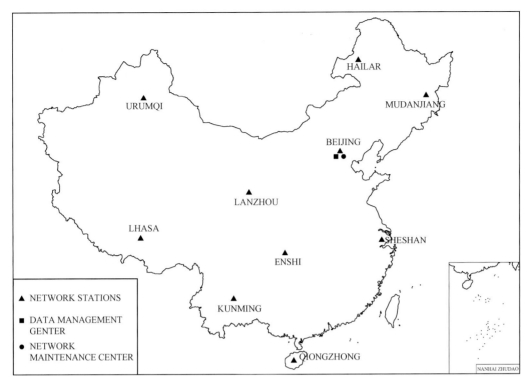

Fig. 1 Map showing distribution of CDSN stations and support facilities

Table I Station parameters of CDSN

Serial N.	Station	Code	Location		Elevation/m
			Lat. /°N	Long. /°E	
101	Baijiatuan (Beijing)	BJI	40.0403	116.1750	43
102	Lanzhou	LZH	36.0867	103.8444	1560
103	Enshi	ENH	30.2717	109.4868	487
104	Kunming	KMI	25.1232	102.7400	1945
105	Qiongzhong	QIZ	19.0293	109.8432	230
106	Sheshan (Shanghai)	SSE	31.0956	121.1867	10
107	Urumqi	WMQ	43.8136	87.7047	970
108	Hailar	HIA	49.2666	119.1666	610
109	Mudanjiang	MDJ	44.6163	129.5918	250
110	Lhasa (Tibet)	LSA	29.7008	91.1167	3789

Table II Availability of CDSN data

Station	Code	Data availability/%	
		1992	1993
Baijiatuan (Beijing)	BJI	100	96.56
Lanzhou	LZH	98.9	100
Enshi	ENH	98.6	91.22

Continued

Station	Code	Data availability/%	
		1992	1993
Kunming	KMI	100	100
Qiongzhong	QIZ	94	100
Sheshan (Shanghai)	SSE	98.6	100
Urumqi	WMQ	100	95
Hailar	HIA	96.7	100
Mudanjiang	MDJ	99.45	100
Lhasa (Tibet)	LSA	100	98.43

(1) expansion of the network to include additional stations to provide denser area coverage (for this purpose, the Lhasa station in Xizang, Tibet, was installed in December 1991);

(2) broadening the frequency band from BB to VBB, adoption of continuous recording mode, adding very-short period and/or low-gain records at some stations;

(3) change of the present instrumentation at 10 field stations and Network Maintenance Center (NMC) for newly-produced GTSN digital instruments;

(4) change of the PDP-11/44 computer system at DMC for update SUN 4 Sparkserver 490 and SUN 4/65 workstations, adopting new software for data management;

(5) realization of data transmission for domestic stations to DMC and international satellite link.

2.2 Progress of the second-phase technical upgrade of the CDSN

At present the second-phase technical upgrade of the CDSN is progressing successfully.

A set of the CDSN equipment has been installed at Baijiatuan(Beijing)Station(BJI). Presently this station is operating in the VBB configuration and has been producing the new data in SEED format since June 2, 1993. The site of Mudanjiang station(MDJ) has been improved installing GTSN equipment. The establishment and operation (first-phase equipment) of Lhasa station(LSA) have been accepted jointly by PRCSSB and USGS. Both sides had highly evaluated the operation status and data quality.

The technical upgrade of DMC has now been basically accomplished, by configuring SUN computer system to replace PDP-11/44 computer system and adopting new software. A block diagram of DMC of NCDSN hardware configuration is shown in fig. 2. The plans for data transmission from station to DMC have been made. The international satellite link has been operational via installing the satellite track system since December 8, 1993. It is planned that a set of SUN work-station with Analyst Review Station(ARS) software will be installed in each station of NCDSN so that station staff can analyse and use data. At present, DMC staff are getting familiar with using ARS software. A set of Seimic Analysis Code(SAC) software is being used for earthquake analysis, especially for fast determination of large earthquake parameters.

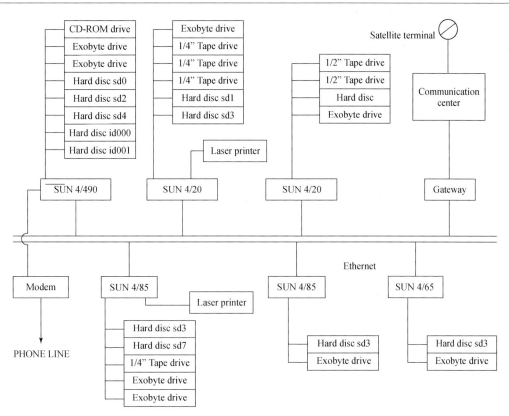

Fig. 2 Block diagram of DMC/NCDSN hardware

3 Plan for 1994 and task groups

3.1 Plan of the CDSN for 1994

We will install GTSN instruments at Mudanjiang (MDJ), Lanzhou (LZH), Urumqi (WMQ), Enshi (ENH) and Kunming (KMI) stations. Staff at the CDSN and staff at the Albuquerque Seismological Laboratory will continually improve DMC software for data managements and processing. During the first half-year, 1994, we will determine and test data transmission from stations to DMC, and realize data transmission via international satellite link and access to event data.

In April 1994, we will draw up a plan for the installation of four additional stations. These new stations will be installed with the equipment used in the first-phase of the CDSN.

3.2 Task groups

Several task groups have been formed in order for the CDSN to operate properly and upgrade successfully. These groups include:

(1) CDSN quality control group for checking and quality control of the CDSN data;
(2) NCDSN station group for installation of the GTSN equipment in 10 stations;
(3) Data communication group for management of the satellite communication system;
(4) SUN system management group for daily management of SUN system and data base resources;
(5) CDSN data application group for events detection, determination of earthquake parameters and study of broadband seismology.

Source process of the 1990 Gonghe, China, earthquake and tectonic stress field in the northeastern Qinghai-Xizang (Tibetan) Plateau

Y. T. Chen[1], L. S. Xu[1], X. Li[1] and M. Zhao[1]

Institute of Geophysics, State Seismological Bureau

Abstract The M_S = 6.9 Gonghe, China, earthquake of April 26, 1990 is the largest earthquake to have been documented historically as well as recorded instrumentally in the northeastern Qinghai-Xizang (Tibetan) Plateau. The source process of this earthquake and the tectonic stress field in the northeastern Qinghai-Xizang Plateau are investigated using geodetic and seismic data. The leveling data are used to invert the focal mechanism, the shape of the slipped region and the slip distribution on the fault plane. It is obtained through inversion of the leveling data that this earthquake was caused by a mainly reverse dip-slipping buried fault with strike 102° dip 46° to SSW, rake 86° and a seismic moment of 9.4×10^{18} Nm. The stress drop, strain and energy released for this earthquake are estimated to be 4.9 MPa, 7.4×10^{-5} and 7.0×10^{14} J, respectively. The slip distributes in a region slightly deep from NWW to SEE, with two nuclei, i.e., knots with highly concentrated slip, located in a shallower depth in the NWW and a deeper depth in the SEE, respectively.

Broadband body waves data recorded by the China Digital Seismograph Network (CDSN) for the Gonghe earthquake are used to retrieve the source process of the earthquakes. It is found through moment-tensor inversion that the M_S =6.9 main shock is a complex rupture process dominated by shear faulting with scalar seismic moment of the best double-couple of 9.4×10^{18} Nm, which is identical to the seismic moment determined from leveling data. The moment rate tensor functions reveal that this earthquake consists of three consecutive events. The first event, with a scalar seismic moment of 4.7×10^{18} Nm, occurred between 0–12s, and has a focal mechanism similar to that inverted from leveling data. The second event, with a smaller seismic moment of 2.1×10^{18} Nm, occurred between 12–31s, and has a variable focal mechanism. The third event, with a scalar seismic moment of 2.5×10^{18} Nm, occurred between 31–41s, and has a focal mechanism similar to that inverted from leveling data. The strike of the 1990 Gonghe earthquake, and the significantly reverse dip-slip with minor left-lateral strike-slip motion suggest that the pressure axis of the tectonic stress field in the northeastern Qinghai-Xizang plateau is close to horizontal and oriented NNE to SSW, consistent with the relative collision motion between the Indian and Eurasian plates. The predominant thrust mechanism and the complexity in the tempo-spatial rupture process of the Gonghe earthquake, as revealed by the geodetic and seismic data, is generally consistent with the overall distribution of isoseismals, aftershock seismicity and the geometry of intersecting faults structure in the Gonghe Basin of the northeastern Qinghai-Xizang plateau.

---

* 本文发表于 *Pure Appl. Geophys.*, 1996 年, 第 **146** 卷, 第 3/4 期, 697–715.

1 Introduction

On April 26, 1990, a destructive earthquake occurred in Gonghe, Qinghai Province of western China (Fig. 1). The epicentral location of the earthquake is latitude $\phi = 35.986°$ N, longitude $\lambda = 100.245°$ E. The focal depth, $h = 8.1$ km. The origin time, $O = 09$h 37 min 15s UTC (17 h 37 min 15s BTC). The magnitude, $M_S = 6.9$. The earthquake occurred beneath the Gonghe Basin of the northeastern part of Qinghai-Xizang (Tibetan) plateau near the south margin of the Gonghe Basin. This was the largest earthquake to have occurred in the northeastern part of the Qinghai-Xizang plateau, and the first significant earthquake in this region to be recorded by the modern digital broadband seismograph network. No earthquake larger than 6.0 was documented historically (780 B. C. to 1900 A. D.), nor recorded instrumentally (1900 to 1989). The microearthquake seismicity in this region, as compared to the overall level of seismicity in the Qinghai Province of China, is very low in the last 20 years since 1970 (Fig. 1). The maximum intensity in this region, given by the seismic zoning map of the Qinghai Province of China, as assessed largely on the basis of surficial geologic observations and historic seismicity, is grade IV, a numeral even lower than that in the regions surrounding the Gonghe Basin (Zeng, 1990, 1991, 1995).

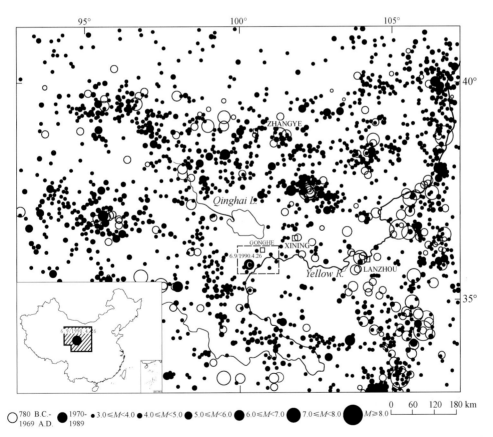

Figure 1   Map showing the epicenters of the $M_S = 6.9$ Gonghe, China, earthquake of April 26, 1990 and seismicity from 780 B. C. to 1989 A. D. in the northeastern Qinghai-Xizang (Tibetan) plateau

The earthquake caused 126 casualties, 2,049 injuries, the collapse of 21,200 houses, and damage to 66,800 houses. During the earthquake, buildings in the meizoseismal region of intensity grade IX totally collapsed. The earthquake was felt in the entire Qinghai Province (~ 93° to 108° E, 33° to 41° N), as far east as Lanzhou (about 300km to the east), and as far north as Zhangye (about 360km to the north) (Fig. 2).

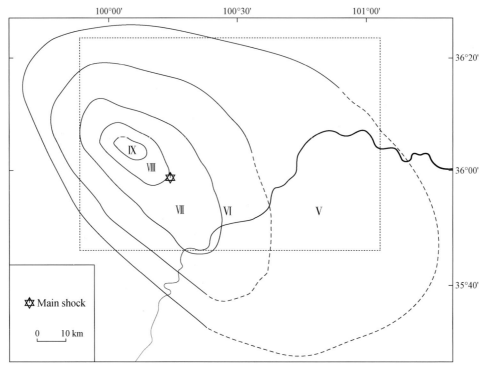

Figure 2　Isoseismals of the 1990 Gonghe earthquake

Dotted rectangle shows the area depicted in Figure 3

The focal mechanism and seismogenic structure of the Gonghe earthquake have been studied by Zeng(1990) and Tu(1990) using mainly field geologic data, and routinely determined by Person(1991), and Dziewonski et al. (1991) using seismic data, among others. While considerable effort has been expended in the studies of the focal mechanism and seismogenic structure of the earthquake, the answer regarding these questions remains controversial.

The relatively low level of historical as well as present seismic activity prevents our understanding of the tectonic stress field in the Gonghe basin and its neighbouring region, using the seismological method. Using P-wave first motion data from a large number of small earthquakes, Xu et al. (1992) found that the pressure axis of the tectonic stress field in the northeastern Qinghai-Xizang plateau is close to horizontal and trends NE-SW to NNE-SSW. Nevertheless, as Xu et al. (1992) noted, the data available for the determination of this region are relatively rarer and less reliable.

To understand the focal mechanism and seismogenic structure of the Gonghe earthquake and the tectonic stress field in the northeastern Qinghai-Xizang plateau, we study the source process of the Gonghe earthquake, using geodetic and seismic data. We invert leveling data obtained before and after the occurrence of the Gonghe earthquake to determine the focal mechanism, fault shape, and spatial distribution of variable slip on the fault. We also invert broadband body waves data obtained from the China Digital Seismograph Network (CDSN) to determine the moment release history, using the moment-tensor inversion technique. We then apply the empirical Green's function (EGF) technique to seismic waves recorded at stations of the CDSN to invert the records for the tempo-spatial rupture process of this earthquake. Finally, we compare the results obtained with geological, meizoseismal and aftershock data obtained by other investigators.

2 Focal mechanism from leveling data

Gong and Guo (1992) and Gong et al. (1993) at the No. 2 Crustal Deformation Monitoring Center, State Seismological Bureau, Xi'an, China, have conducted a leveling survey in an area of 100km × 60km which enclosed the meizoseismal regions, in 1978 to 1979. Following the main shock, in May to June, 1990, they carried out leveling remeasurements along the same leveling route, and obtained valuable geodetic data of this earthquake (Fig. 3). These data provide a good opportunity to study the focal mechanism and seismogenic structure of this earthquake. In this article, the leveling data obtained by Gong and Guo (1992) and Gong et al. (1993) are used to invert the focal mechanism of the Gonghe earthquake.

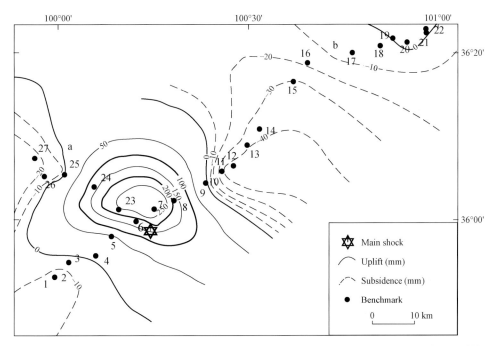

Figure 3 Vertical displacement distribution of the M_S = 6.9 Gonghe earthquake of April 26, 1990

There are 27 benchmarks in total, distributed in two leveling routes of 162.6 km in total length. The leveling route (a) runs northwestward of the epicenter. The leveling route (b) runs southwestward from benchmark No. 22, crossing the epicentral area, and ending about 25km southwest of the epicenter of the main shock. Relative to the benchmark No. 22 the largest vertical deformation located in an uplift region of 30km-wide, is 358mm. The random survey error σ_1 is 0.43mm/km$^{1/2}$, and the root-mean-squared errors of the observation data σ is 3.9mm. Considering that the ground deformation in the northeastern part of the epicentral area is substantially influenced by the tectonic movement of the Nanshan mountain, which is close to the benchmark No. 22, in this study we exclude the data of the benchmarks Nos. 16 to 22 from inversion, and take benchmark No. 15 as a reference point.

In the first step of the leveling data inversion, we model the observed coseismic deformation by a single rectangular fault buried in an isotropic, homogeneous and perfect elastic half-space, with uniform oblique slip. We define fault length $2L$ along the fault-strike direction α, width W along perpendicular direction to the strike, with upper edge d and lower edge D, dip angle θ, strike-slip ΔU_s and dip-slip ΔU_d. In the Cartesian coordinate

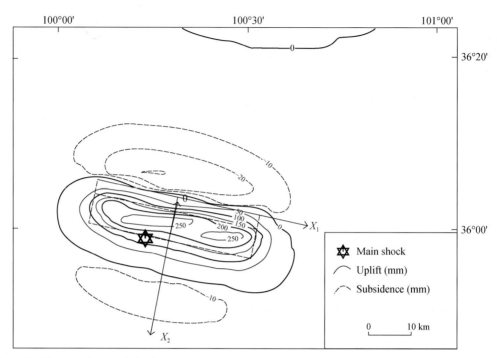

Figure 4 Theoretical vertical displacement distribution contours for the single, rectangular uniform slip model

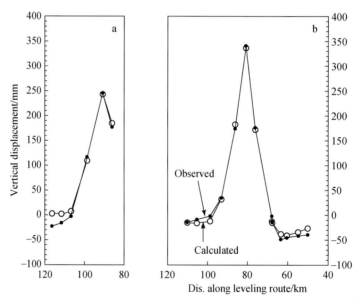

Figure 5 Comparison of theoretical (open circles with dotted lines) and observed (solid circles with solid lines) vertical displacements for the single, rectangular uniform slip model

Left part (a) and right part (b) refer to leveling routes shown in Figure 3, respectively

system, as shown in Figure 4, the i-th component of the displacement, $u_i(i = 1, 2, 3)$ at an observation point (x_1, x_2, x_3), due to this fault, is given by Mansinha and Smylie (1971) for the Poisson medium, and is given by Chen et al. (1975) for the case of unequal Lamé constants.

If we denote the geographical coordinates of the origin of the source coordinates system as $(s_1, s_2, 0)$, the vertical component of the displacement at a given point on the ground surface $(X_1, X_2, 0)$ can be expressed as follows (Chen et al., 1979):

$$u_3 = u_3(\alpha, \theta, L, d, D, \Delta U_s, \Delta U_d, s_1, s_2; X_1, X_2). \tag{1}$$

The displacement at a given point on the ground surface is a nonlinear function of the source parameters (α, θ, L, d, D, ΔU_s, ΔU_d, s_1, s_2). To invert the data of the vertical displacement on the ground surface for the source parameters, the function u_3 was linearized by making Taylor series expansion of the function u_3 about an initial model, and neglecting the higher order terms. Starting from initial model parameters, the estimation of the model parameters which satisfy convergence criteria described in Chen et al. (1979), was obtained as $\alpha = 102°$, $\theta = 46°$, $2L = 2 \times 20$km, $d = 5$km, $D = 14$km, $\Delta U_s = 5$cm, $\Delta U_d = -79$cm, $s_1 = 93.0$km, and $s_2 = 34.5$km. For the single rectangular uniform slip model, this final model is the best fit to the observed leveling data. The root-mean-squared residual r for the final model is 14.2mm.

Figure 4 shows the projection of the inverted fault model on the ground surface and the contour of the theoretical vertical displacement due to the inverted fault model. Figure 5 shows the fitness of the theoretical vertical displacement due to the inverted fault model and the observation data. The overall focal mechanism we obtained for this earthquake from leveling data indicates that this earthquake is mainly reverse faulting with a small left-lateral component on a dipping 46° to SSW striking 102° fault plane.

We also calculated seismic moment, stress drop, strain drop and energy released by the Gonghe earthquake, using the inverted fault parameters. Using a rigidity of 3.0×10^{11} dyn/cm^2, the seismic moment M_0 is estimated to be 9.4×10^{18} Nm. The stress drop $\Delta\sigma = 4.9$ MPa. The strain drop $\Delta\varepsilon = 7.4 \times 10^{-5}$, and the energy released $\Delta E = 7.0 \times 10^{14}$ J. It can be seen that the stress drop of the Gonghe earthquake is low (4.9 MPa) as compared to the usual intraplate earthquake (normally several ten MPa) (Kanamori and Anderson, 1975).

While the single rectangular fault model inverted from leveling data accounts well for the observed vertical deformation, it is worthwhile to note that the misfit for this model is still several times larger than the root-mean-squared errors of the observation data ($\sigma = 3.9$mm). These discrepancies are probably caused by the inadequacy of the single rectangular dislocation model and one would therefore not expect the oversimplified model to fit the observation data quite well.

In the second step of the leveling data inversion, following Ward and Barrientos (1986), we use a variable slip fault model in which we initially fix the focal mechanism, i.e., fault strike (α), dip angles (θ), depth and width parameters (d and D) and rake (λ), and allow the slip amount to change on the fault plane. The fault plane is then divided into subfaults of equal size, and point dislocation sources are distributed uniformly across each of the subfaults. At individual observation points, displacement is computed by summing the contribution of each point source:

$$u_i = K_{ij} D_j, \quad i = 1, 2, \cdots, n, \tag{2}$$

where u_i is the displacement at the i-th observation point, D_j, $j = 1, 2, \cdots, m$ is the slip of the j-th subfault, and K_{ij} the displacement at the i-th observation point from the j-th subfault of unit slip.

Equation (2) represents a linear relation between the displacements at observation points on the ground surface and the slips of the subfaults. The theoretical and observed displacements describe an underdetermined system of linear equations. We solve for D_j using a gradient inversion scheme proposed by Ward and Barrientos (1986), which invokes a positivity constraint on the solutions. The fault plane of 100km × 40km is divided into 20 × 8 subfaults with dimensions of 5km × 5km. The fault strike α, dip angles δ and rake λ are taken from the inversion result for the single rectangular fault model, i.e., $\alpha = 102°$, $\delta = 46°$, and $\lambda = 86°$.

The inversion yields a variable slip distribution on the fault plane (Fig. 6) and an estimation of the seismic

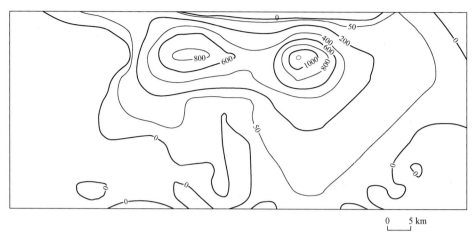

Figure 6 Contoured distribution of variable slip on a fault plane

The figure in the contour shows the slip in mm

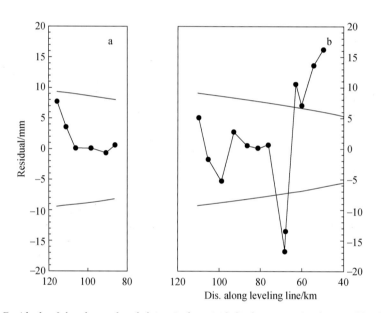

Figure 7 Residuals of the observed and theoretical vertical displacements for the variable slip models

Left part (a) and right part (b) refer to leveling routes shown in Figure 3, respectively. Ordinate represents residuals in mm. Abscissa represent distance along leveling route. Dashed lines represent twice standard deviation along the leveling route

moment of 9.8×10^{18} Nm, which is very close to the estimation of the seismic moment from the single, rectangular fault model of uniform slip. The slip distributes in a region slightly deep from NWW to SEE with two nuclei, i. e., knots with highly concentrated slip (equal to and larger than 60cm), located in a shallower depth (4 to 12km) at the NWW and, about 25km away, in a deeper depth (5 to 16 km) at the SEE, respectively. At most observation points the theoretical vertical displacements for the variable slip fault model fit the observed vertical displacements quite well. Figure 7 illustrates the residuals at individual observation points versus distance along the leveling routes. Dashed lines represent an envelope of two times standard deviation. The standard deviation at distance l along the leveling route, $\sigma(l)$, is expressed as (Ward and Barrientos, 1986)

$$\sigma(l) = \sigma_1 \sqrt{l} \qquad (3)$$

where σ_1 is the random survey error. It can be seen that the residuals at most observation points are within one standard deviation uncertainty. The root-mean-squared residuals for the variable slip models is 7.8mm, which is twice the root-mean-squared errors of 3.9mm.

3 Source process from moment-tensor inversion

Moment-tensor inversion technique is used to study the source process of the Gonghe earthquake (Dziewonski et al., 1991, among others).

If the dimension of the source is considerably smaller than the dominant wavelength, the relationship between the seismic displacement $u_i(r, t)$ and the seismic moment $M_{jk}(\mathbf{0}, t)$ is linear:

$$u_i(r,t) = G_{ij,k}(r,t; \mathbf{0},0) * M_{jk}(\mathbf{0},t) \qquad (4)$$

where $G_{ij,k}(r,t; \mathbf{0},0)$ is the partial derivative with respect to source coordinates x'_k, of the Green's function $G_{ij}(r, t; \mathbf{0},0)$. The Green's function $G_{ij}(r,t; \mathbf{0},0)$ denotes the i-th component of displacement at the position r and at the time t due to a unit impulse applied at the origin $\mathbf{0}$ and at the time 0 and in the j-th direction.

In the frequency domain, equation (4) can be written

$$u_i(r,\omega) = G_{ij,k}(r,\mathbf{0};\omega) \cdot M_{jk}(\mathbf{0},\omega). \qquad (5)$$

In this study the generalized reflection-transmission coefficient matrix method (Kennett, 1979, 1983) and the discrete wave slowness integration method are used to synthesize the partial derivatives of the Green's function $G_{ij,k}$. In our calculation, the wave slowness interval is taken to be 0.09 to 0.32s/km and the frequency interval, 0.01 and 0.5 Hz. We use the broadband P and S body wave phases of the CDSN to invert the moment tensor of the Gonghe earthquake. The original seismograms were preprocessed for N-E-U to Z-R-T and filtered in the passband between 0.05 and 0.2 Hz.

Figure 8 shows the observed (thick lines) and synthetic (thin lines) broadband displacement wave forms at seven stations of the CDSN, for vertical (upper traces), radial (middle traces) and tangential (lower traces) components, respectively, and the lower hemisphere projection of the moment-tensor solution. In the lower hemisphere projection of the focal sphere, P nodals for the moment tensor and the best double-couple are represented by thick and thin lines, respectively. The best double-couple solutions for the Gonghe earthquake have one nodal plane (N.P.1) with strike 113°, dip 68°, rake 89°, and another nodal plane (N.P.2) with strike 294°, dip 22°, rake 91°. Considering that the nodal plane 1 is so close to the fault plane inverted from the leveling data, we prefer the nodal plane 1 to the fault plane. The focal mechanism of the Gonghe earthquake obtained from moment-tensor inversion is a mainly reverse dip slip fault with strike 113°, dip 68° to SSW and rake 89°. The pressure axis P is close to horizontal (plunge 23°) and oriented NNE to SSW (azimuth 23°). The plunge and the azimuth of the tension axis T are 67° and 22° respectively. The null axis B is almost horizontal (plunge 1°) and oriented NWW to SEE (azimuth 113°).

Figure 9 shows the moment rate tensor (left part), isotropic component (EP), best double-couple (DC) and compensated linear vector dipole (LD) (central part), and the azimuth (thick lines) and plunge (thin lines) of the T, P, and B axes (right part), of the Gonghe earthquake, versus time. The figures in the lower right are the integrations of individual quantities. It is evident that the Gonghe earthquake is a rupture process dominated by shear faulting. The scalar seismic moment of the best double-couple is 9.4×10^{18} Nm, which is identical to that determined from leveling data. The moment of the isotropic part is only 7.8×10^{17} Nm, about 8.3% of that of the best double-couple, and the moment of the compensated linear vector dipole is vanishing.

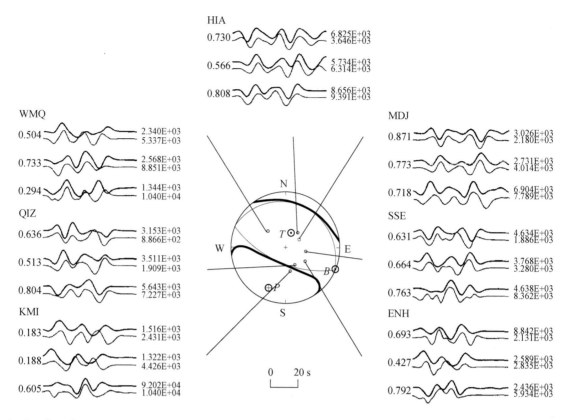

Figure 8　Focal mechanism of the M_S = 6.9 Gonghe earthquake of April 26, 1990 inverted from broadband body waves data, and observed and synthetic seismograms at seven stations of the CDSN

For detail refer to text

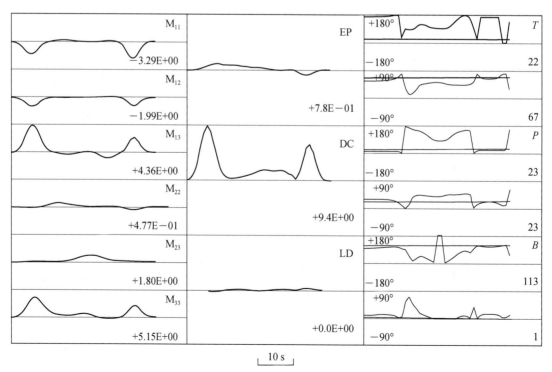

Figure 9　Moment rate tensor of the M_S = 6.9 Gonghe earthquake of April 26, 1990 inverted from broadband body waves data

For detail refer to text

The central part of Figure 9 clearly depicts that the Gonghe earthquake consists of three consecutive events. The first event which occurred between 0—12s, with a scalar seismic moment of 4.7×10^{18} Nm, has a focal mechanism of strike 96° dip 73°, and rake 77°. The second event which occurred between 12—31s with a smaller seismic moment of 2.1×10^{18} Nm, has a variable focal mechanism with average strike 107°, dip 8°, and rake 169°. The third event which occurred between 31—41s, with a seismic moment of 2.5×10^{18} Nm, has a focal mechanism of strike 100°, dip 67°, and rake 75°. As the right part of Figure 9 indicated, the principal axes in the time intervals for the first and the third events are stable and almost identical. The principal axes in the time interval for the second event (12—31s) are different from that for the first and the third events, and evidently change with time.

4 Tempo-spatial rupture process from inversion of source time functions

The time domain inversion technique proposed by Hartzell and Iida (1990) and Dreger (1994) are used to image the rupture process of the Gonghe earthquake. The source time functions (STF) retrieved from the recordings at a station is the summation of the STFs of subfaults,

$$S_i(t) = \sum_{j=1}^{N} m_j(t - \tau_{ij}) \tag{6}$$

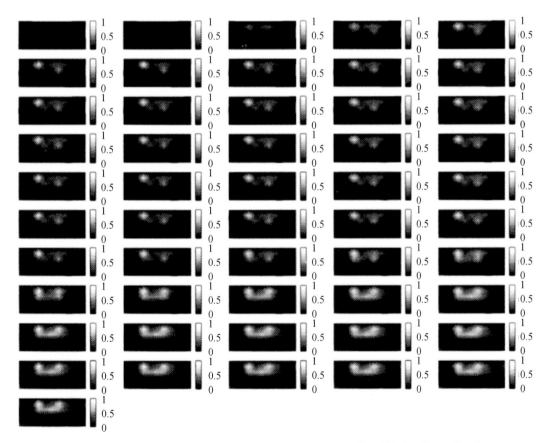

Figure 10 Snapshots showing tempo-spatial rupture process of the 1990 Gonghe earthquake

From top to bottom and from left to right, these pictures indicate variable slip on the fault plane of 88km long and 40km wide between 0—41s at time interval of 1s

where S_i is the STF observed at the i-th station, N is the number of subfaults, m_j is the STF of the j-th subfault, and τ_{ij}, the time delay of the j-th subfault to the reference point, determined by

$$\tau_{ij} = \frac{R_{ij}}{V} \tag{7}$$

where R_{ij} is the distance of the i-th station to the j-th subfault, and V, the wave velocity at the earthquake source.

Equation (6) can be rewritten as convolution of the STF of the j-th subfault with Dirac δ-function:

$$S_i(t) = \sum_{j=1}^{N} \delta(t - \tau_{ij}) * m_j(t). \tag{8}$$

Equation (8) represents a linear relation between the STF observed at a station and the STFs of subfaults.

Solving the $m_j(t)$ is an underdetermined problem. In order to obtain a stable solution, some appropriate constraints should be imposed. In our study, the positivity constraint, which physically means that no backward slip is allowed to take place, is used in the inversion.

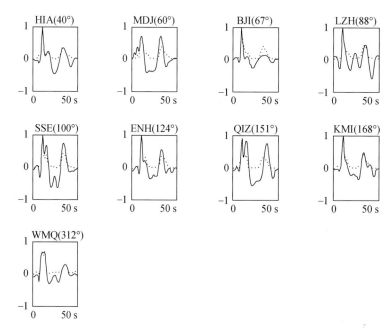

Figure 11 Comparison of theoretical (dotted lines) and observed (solid lines) source time functions of the Gonghe earthquake at nine stations of the CDSN

Figure enclosed in parenthesis represents the azimuth of the station

In the study of Gonghe earthquake, the fault plane of 88km long and 40km wide is divided into 11×5 subfaults with dimensions of 8km × 8km. The STFs retrieved from Love waves are used to invert for the tempo-spatial rupture process of the Gonghe earthquake. In the inversion, the wave velocity is taken to be 6.2 km/s. The inversion yields a result with final slip distribution similar to the static variable slip distribution inverted from leveling data. The slip distribution on the fault plane inverted from the STFs (Fig. 10) also exhibits two slip concentrated regions about 25km distant in the NWW to SEE direction. Using a total seismic moment of 9.4×10^{18} Nm and rigidity of 3.0×10^{11} dyn/cm^2, the maximum slip on the fault is estimated to be about 50cm, which is very close to that (about 60cm) obtained using leveling data.

The advantage of the technique described above is that not only the static, final slip distribution could be inverted but also the tempo-spatial variable slip on the fault plane could be obtained. The snapshots depicted in Figure 10 demonstrate that the rupture of the Gonghe earthquake is complex, both temporally and spatially. Two noticeable events, the first event and the third event as revealed by moment-tensor inversion, do not simply respectively correspond to the nuclei at the NWW and at the SEE ends of the final, variable slip model inverted from leveling data (Fig. 6). The tempo-spatial variable slip from STF inversion, shows that both nuclei were involved in the entire rupture process. At the onset of the earthquake, rupture initiated at the NWW end of the fault plane, and then expanded mainly toward the SEE end of the fault plane and triggered rupture at the SEE end of the fault plane. During the time period of the first event, rupture involved most of the fault plane but mainly concentrated at the NWW end of the fault plane, and during the time period of the third event, rupture also involved most of the fault plane but mainly concentrated at the SEE end of the fault plane.

We also calculate the theoretical STFs at different stations from the inverted tempo-spatial slip distribution (Fig. 11). In general, the theoretical STFs at individual stations are in good agreement with the observed STFs.

5 Conclusions and discussion

Table 1 summarizes the focal mechanisms from geodetic and seismic data. Comparing the fault-plane solutions from geodetic and seismic data, it is concluded that the seismogenic structure of the 1990 Gonghe earthquake is a buried fault with strike NWW to SEE ($\alpha = 102°$ to $113°$). The P axis of the earthquake shown is close to horizontal (plunge $23°$) and oriented NNE to SSW (azimuth $23°$). The seismogenic fault of the Gonghe earthquake is a mainly reverse dip-slip fault with a minor left-lateral strike component. The scalar seismic moments determined from geodetic data, and broadband data range from 9.4×10^{18} to 9.8×10^{18} Nm. By using rigidity $\mu = 3.0 \times 10^{11}$ dyn/cm^2 the stress drop, $\Delta\sigma$, is calculated to be 4.9 to 5.1 MPa. The slip distributes in a region slightly deep from NWW to SEE, with two nuclei located in a shallower depth in the NWW and a deeper depth in the SEE, respectively. The earthquake consists of three consecutive events. The first event which occurred between 0—12s, with a scalar seismic moment of 4.7×10^{18} Nm, has a focal mechanism with strike $96°$, dip $73°$, and rake $77°$, which is similar to that inverted from leveling data. The second event which occurred between 12–31s, with a smaller scalar seismic moment of 2.1×10^{18} Nm, has a variable focal mechanism with average strike $107°$, dip $8°$, and rake $169°$. The third event which occurred between 31–41s, with a scalar seismic moment of 2.5×10^{18} Nm, has a focal mechanism almost identical with that of the first event. Both slip concentrated nuclei were involved in the entire rupture process, but during the first 12s, rupture mainly occurred at the NWW end, and during the last 10s, mainly at the SEE end, of the fault plane.

Table 1　Focal mechanisms from geodetic and seismic data

Data	strike / (°)	dip / (°)	rake / (°)	M_0 /10^{18} Nm	Remarks
Leveling (1)	102	46	86	9.4	a single rectangular uniform-slip fault
Leveling (2)	102	46	86	9.8	a variable-slip fault with two nuclei
Broadband body waves (1)					one source
	113	68	89	9.4	total rupture duration 41s
Broadband body waves (2)					three events
	96	73	77	4.7	occurred between 0—12s
	107	8	169	2.1	occurred between 12—31s
	100	67	75	2.5	occurred between 31—41s

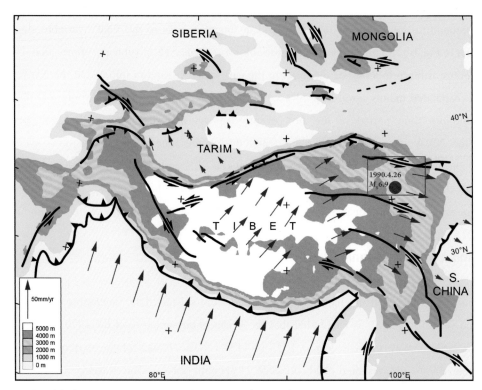

Figure 12　Simplified map of tectonics of Qinghai-Xizang (Tibetan) plateau after Avouac and Tapponnier (1993)

Inner solid rectangle represents the area shown in Figure 13

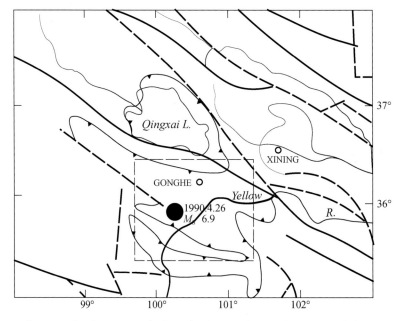

Figure 13　Map of the northeastern Qinghai-Xizang (Tibetan) plateau showing the epicentral location of the $M_S = 6.9$ Gonghe earthquake of April 26, 1990 and the faults (heavy solid lines), inferred faults (heavy dashed lines) and basins after Institute of Geology, State Seismological Bureau (1981)

Inner thin dash rectangle represents the area shown in Figure 14

The 1990 Gonghe earthquake occurred in the south margin of the Gonghe basin (Figs. 12 and 13). The Gonghe, Qinghai Lake, Xining-Minghe and Qaidam basins are known as large Cenozoic basins (Ma, 1987).

The thickness of the Quaternary accumulation beneath these basins is more than 1000m. The thickness of the Quaternary accumulation beneath the Qaidam and Gonghe basins is 2800m and 1200m, respectively. The block of Qaidam and Gonghe basins, separated by the left-lateral strike-slip faults such as Qilianshan, Haiyuan, Altyn Tagh and Kunlun fault zones is a region of strong differential tectonic movement. As Avouac and Tapponnier (1993) noted, this curved block rotates clockwise with respect to the Siberia block. Also, Zeng et al. (1993) and Zeng and Sun (1993) proposed a new model of the continental collision from the Indian to Eurasian plates and an eastward transfer of the crustal material underneath the Qinghai-Xizang plateau. The 1990 Gonghe earthquake strike, and the significantly reverse dip-slip with minor left-lateral strike-slip motion obtained in this study, confirm that the pressure axis of the tectonic stress field in the northeastern Qinghai-Xizang plateau is close to horizontal and oriented NNE to SSW, consistent with the relative collision motion between the Indian and Eurasian plates. This result gives support to the new model proposed by Zeng et al. (1993) and Zeng and Sun (1993), of the eastward transfer of the crustal material beneath the Qinghai-Xizang plateau, and to the kinematic model proposed by Avouac and Tapponnier (1993), of the clockwise rotation of the Qaidam basin-Gonghe basin block with respect to the Siberian block as a whole.

Figure 14 summarizes the pressure axis of the tectonic stress field in the northeastern Qinghai-Xizang plateau, isoseismals, aftershock activity and the projection of the inverted fault model on the ground surface. Comparing these results, it is believed that the occurrence of the Gonghe earthquake is mostly related to a buried fault striking, NW to SE, near the south margin of the Gonghe basin, inferred by the Institute of Geology, State Seismological Bureau (1981). The predominantly thrust mechanism and the complexity in the tempo-spatial rupture process of the Gonghe earthquake, as revealed by the geodetic and seismic data, is generally consistent with the overall distribution of isoseismals, aftershock seismicity and intersecting faults structure in the Gonghe basin of the northeastern Qinghai-Xizang plateau.

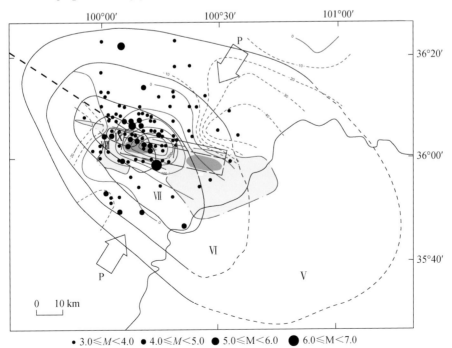

Figure 14 Map showing the pressure axis in the northeastern Qinghai-Xizang (Tibetan) plateau, vertical ground deformation, surface projection of the uniform slip and variable slip model inverted from leveling data, epicenters of aftershocks and isoseismals, of the Gonghe earthquake

References

Avouac, J. P., and Tapponnier, P. (1993), Kinematic model of active deformation in central Asia, *Geophys. Res. Lett.* **20**, 895–898.

Chen, Y. T., Lin, B. H., Lin, Z. Y., and Li, Z. Y. (1975), The focal mechanism of the 1966 Hsingtai earthquake as inferred from the ground deformation observations, *Acta Geophysica Sinica* **18**, 164–182 (in Chinese with English abstract).

Chen, Y. T., Lin, B. H., Wang, X. H., Huang, L. R., and Liu, M. L. (1979), A dislocation model of the Tangshan earthquake of 1976 from the inversion of geodetic data, *Acta Geophysica Sinica* **22**, 201–217 (in Chinese with English abstract).

Dreger, D. S. (1994), Investigation of the rupture process of the 28th June 1992 Landers earthquake utilizing TERRA Scope, *Bull. Seismol. Soc. Am.* **84**, 713–724.

Dziewonski, A. M., Ekström, G., Woodhouse, J. H., and Zwart, G. (1991), Centroid-moment tensor solutions for April-June in 1990, *Phys. Earth Planet. Inter.* **66**, 133–143.

Gong, S. W., and Guo, F. Y. (1992), Vertical ground deformation in the earthquake of Gonghe, Qinghai Province, *Acta Seismologica Sinica* (Chinese edition) **14**(Supplement), 725–727 (in Chinese).

Gong, S. W., Wang, Q. L., and Lin, J. H. (1993), Study of dislocation model and evolution characteristics of vertical displacement field of Gonghe $M_S = 6.9$ earthquake, *Acta Seismologica Sinica* (English edition) **6**(3), 641–648.

Hartzell, S., and Iida, M. (1990), Source complexity of the 1987 Whittier Narrows, California, earthquake from the inversion of strong motion records, *J. Geophys. Res.* **98**(B12), 22123–22134.

Institute of Geology, State Seismological Bureau (ed.), *Seismotectonic Map of Asia and Europe* (China Cartography Press, Beijing 1981) (in Chinese).

Kanamori, H., and Anderson, D. L. (1975), Theoretical basis of some empirical relations in seismology, *Bull. Seismol. Soc. Am.* **65**, 1073–1095.

Kennett, B. L. N., *Seismic Wave Propagation in Stratified Media* (Cambridge University Press, 1983), 342 pp.

Ma, X. Y. (ed.), *Outlines of Lithospheric Dynamics of China* (Seismological Press, Beijing, 1987) (in Chinese).

Mansinha, L., and Smylie, D. E. (1971), The displacement fields of inclined faults, *Bull. Seismol. Soc. Am.* 61, 1433–1400.

Person, W. J. (1991), Seismological Notes—March-April 1990, *Bull. Seismol. Soc. Am.* **81**, 297–302.

Tu, D. L. (1990), Geological structure background of gonghe earthquake $M_S = 6.9$, on April 26, 1990, *Plateau Earthq. Res.* **2**(3), 15–20 (in Chinese with English abstract).

Velasco, A. A., Ammon, C. J., and Lay, T. (1994), Empirical Green function deconvolution of broadband surface waves: rupture directivity of the 1992 Landers, California ($M_W = 7.3$) earthquake, *Bull. Seismol. Soc. Am.* **84**(3), 735–750.

Ward, S. N., and Barrientos, S. (1986), An inversion for slip distribution and fault shape from geodetic observations of the 1983, Borah Peak, Idaho, earthquake, *J. Geophys. Res.* **91**, 4909–4919.

Xu, Z. H., Wang, S. Y., Huang, Y. R., and Gao, A. J. (1992), Tectonic stress field of China onferred from a large number of small earthquakes, *J. Geophys. Res.* **97**(B8), 11867–11877.

Zeng, Q. S. (1990). Survey of the earthquake $M_S 6.9$ between Gonghe and Xinghai on April 26, 1990, *Plateau Earthq. Res.* **2**(3), 3–12 (in Chinese with English abstract).

Zeng, Q. S. (1991). Seismicity and earthquake disaster of Qinghai province, *Plateau Earthq. Res.* **3**(1), 1–11 (in Chinese with English abstract).

Zeng, Q. S. (1995), Earthquake resistance and disaster reduction and short-impending prediction of Qinghai Province, *Plateau Earthq. Res.* **7**(1), 42–51 (in Chinese with English abstract).

Zeng, R. S., and Sun, W. G. (1993), Seismicity and focal mechanism in Tibetan Plateau and its implications to lithospheric flow, *Acta Seismologica Sinica* (English edition) **6**(2), 261–287.

Zeng, R. S., Zhu, J. S., Zhou, B., Ding, Z. F., He, Z. Q., Zhu, L. R., Luo, X., and Sun, W. G. (1993), Three-dimensional Seismic Velocity Structure of the Tibetan Plateau and its eastern neighboring areas with implications to the model of collision between continents, *Acta Seismologica Sinica* (English edition) **6**(2), 251-260.

Zhao, M., Chen, Y. T., Gong, S. W., and Wang, Q. L. (1993), Inversion of focal mechanism of the Gonghe, China, earthquake of April 26, 1990, using leveling data. In *Continental Earthquakes* (eds. Ding, G. Y. and Chen, Z. L.) IASPEI Publication Series for the IDNDR **3**, 246-252.,

Delineation of potential seismic sources for seismic zoning of Iran[*]

Noorbakhsh Mirzaei[1,2], Mengtan Gao[1] and Yun-tai Chen[1]

1. Institute of Geophysics, China Seismological Bureau; 2. Institute of Geophysics, Tehran University

Abstract A total of 235 potential seismic sources in Iran and neighboring regions are delineated based on available geological, geophysical, tectonic and earthquake data for seismic hazard assessment of the country. In practice, two key assumptions are considered: first, the assumption of earthquake repeatedness, implying that major earthquakes occur preferentially near the sites of previous earthquakes; second, the assumption of tectonic analogy, which implies that structures of analogous tectonic setting are capable of generating same size earthquakes. A two-step procedure is applied for delineation of seismic sources: first, demarcation of seismotectonic provinces; second, determination of potential seismic sources. Preferentially, potential seismic sources are modeled as area sources, in which the configuration of each source zone is controlled, mainly, by the extent of active faults, the mechanism of earthquake faultings and the seismogenic part of the crust.

1 Introduction

Demarcation of potential seismic sources is often the major stage of seismic hazard analysis. In the classic form, earthquake sources range from clearly understood and well defined faults to less well understood and less well defined geologic structures to hypothetical seismotectonic provinces extending over many thousands of square kilometers whose specific relationship to the earthquake generating process is not well known (Reiter, 1990). It is known that the combination of seismological and tectonic data presents a better understanding of the complex processes governing the origin of earthquakes over long periods and in geologically different structural units (Karnik, 1969).

In many areas, even in highly seismic regions of the world, there is not enough information available to delineate potential seismic sources with a sufficient degree of reliability. Although during a quarter of century, from pioneering works of Nowroozi (1971, 1972) and McKenzi (1972), considerable efforts have been made by Tchalenko (1975), Berberian (1976a, 1977a, 1981, 1983, 1995), Jackson (1980a, b), Jackson and McKenzie (1984, 1988), Kadinskey-Cade and Barazangi (1982), Ni and Barazangi (1986), Byrne et al. (1992), Baker et al. (1993), Priestley et al. (1994) and Jackson et al. (1995) to understand active tectonics of Iran and neighboring regions, knowledge about earthquake generating structures is still inadequate for detail seismic source delineation. For example, the main surface rupture of the Rudbar-Tarum catastrophic earthquake of 1990.06.20 [M_S 7.7 (USGS/NEIC) 7.4 (ISC), M_W 7.2 (Priestley et al., 1994)] in northwest Iran, occurred on a fault that had not previously been recognized (Berberian et al., 1992); the Tabas earthquake fault

---

[*] 本文发表于 J. Seismol., 1999 年, 第 3 卷, 第 1 期, 17–30.

which caused the destructive earthquake of 1978. 09. 16, M_s7.4, was not known as active prior to this event and the region had not been recognized as a high seismic zone (Berberian, 1981). Similarly, the Ipak Fault responsible for the catastrophic earthquake of 1962. 09. 01, M_s7.2, in Buyin Zahra, and Dasht-e-Bayaz Fault which caused the major earthquake of 1968. 08. 31, M_s7.3, was not recognized as an active fault before the earthquake occurrences (Berberian, 1983). In the Zagros region of southwest Iran, most of the earthquakes nucleate on blind (buried) thrust faults, with not well known characteristics. Moreover, because of inadequate local networks and regional seismographic stations, the Iranian earthquake catalog is mainly based on tele-seismic earthquake records, located by international agencies. Therefore, there are large uncertainties in different earthquake parameters and a considerable number of small events are left out. Consequently, earthquake information is severely incomplete, especially for the time period before installation of World-Wide Seismographic Station Network (WWSSN) in 1963. Our knowledge about historical seismicity (before 1900), is even more incomplete. Clearly, available earthquake data can not fully represent seismic activity in Iran and there is not any additional information, like paleoseismic investigations to show evidences for repeated ground breaking earthquakes with recurrence intervals of hundreds to tens of thousands of years, and geodetic measurements to help us delineate potential seismic sources with a high degree of reliability.

A two step procedure for demarcation of potential seismic sources is applied:

(1) Demarcation of seismo-tectonic provinces. The seismotectonic province is considered to be an area that, under the present-day geodynamic regimes, has a comparable tectonic setting and unified seismicity pattern. (Ye et al., 1993, 1995)

(2) Determination of potential seismic sources. Seismicity level and the magnitude of maximum earthquake varies from place to place in a seismotectonic province due to the variability of local tectonic setting. Therefore, potential seismic sources with different magnitudes of maximum earthquake and different activity rates should be delineated within each seismo-tectonic province to represent a region with uniform seismic potential.

A total of 235 potential seismic source zones in Iran and adjacent regions are delineated (Figure 1), based on all available geological, tectonic, geophysical and earthquake information, for seismic hazard assessment of the country.

2 Seismotectonic provinces

Five major seismotectonic provinces in Iran are delineated based on all available seismicity, geological and tectonic, as well as geophysical information. Continental-continental collision zone of Zagros in southwest Iran, is one of the youngest and most active continental collision zones on the Earth (Synder and Barazangi, 1986), resulting from the continuing northeastward drift of the Arabian Plate relative to the Eurasia and Central Iranian Microcontinent. The highly seismic region of Alborz-Azarbayejan covering north and northwest of Iran, constitutes a part of the northern limit of Alpine-Himalayan orogenic belt. The continental collision zone of Kopeh Dagh in the northeast represents a northern segment of the Alpine-Himalayan orogenic belt, which faces the stable Turan Platform (Eurasia) to the north. The oceanic-continental subduction zone of Makran, where the consumption of oceanic crust of Arabian Plate has occurred continuously since the Early Cretaceous along a north dipping subduction zone underneath the Eurasia-Central Iranian Microcontinent, covers the southeast of the country. Central-East Iran represents an intraplate environment which is surrounded by the foregoing major seismotectonic provinces [for detail discussion on the major seismotectonic provinces of Iran, see Mirzaei et al. (1997c)].

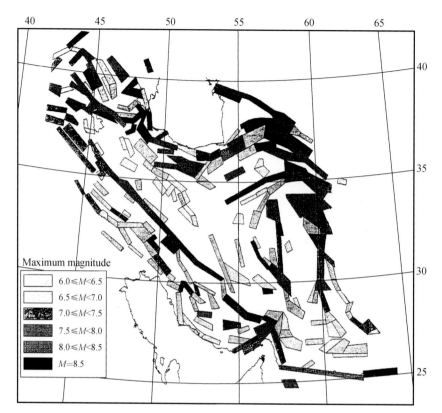

Figure 1 Potential seismic sources in Iran

3 Potential seismic sources

The goal of potential seismic source identification is to identify and include in the analysis all structures (mainly faults) that are believed to be tectonically active; even faults with low rates of activity (EERI Committee on Seismic Risk, 1989). The description of configuration of potential seismic sources requires the consideration of many variables; in practice, tectonic maps and earthquake epicenter distribution maps are used as guidance. There is no standard method for delineating seismic sources; and the final decision on how to model the geometry of seismic sources relies largely on personal judgment and expert opinion (e. g., Yucemen and Gulkan, 1994).

Two key assumptions are considered in most practices of potential seismic source delineation. First, the assumption of earthquake repeatedness; second, the assumption of tectonic analogy (e. g., Zhang, 1993; Ye et al., 1993; He et al., 1994). However, common to all source zone models is the assumption that observed seismic activity is due to reactivation of structures (faults) formed during prior tectonic regimes (Thenhaus et al., 1987).

The first assumption implies that major earthquakes occur preferentially near the sites of previous earthquakes, so that, when an earthquake occurs on a fault, the probability of future earthquakes is larger, at a given distance along that fault, and relatively smaller off the fault (Kagan and Jackson, 1994). The accurate location of recorded events is of particular importance to the identification and characterization of earthquake sources (e. g., Reiter, 1990). However, there are considerable errors in teleseismically assigned locations of earthquakes in

Iran, so that earthquakes often can not be correlated to a particular capable structure; therefore, available data are not useful for very detail seismotectonic analysis. To use available data, considerable efforts were made to include the most reliable earthquake records based on the quality of earthquake records in the analysis (e. g. Mirzaei et al., 1997a). While epicenters in Iran are often mislocated by a few tens of kilometers, the depth of earthquakes determined by international agencies are even less accurate (e. g., Ambraseys, 1978; Berberian, 1979a; Asudeh, 1983; Mirzaei et al., 1997a). Although any inability to accurately determine the depths of crustal earthquakes does not usually cause large errors in epicentral or surface location, it can be a serious problem when attempting to correlate seismicity with the geologic features at depth, which may be quite different from those observed at the surface (Reiter, 1990).

The second key assumption implies that structures with analogous tectonic setting are capable of generating the same size earthquakes, and absence of earthquake record for a structure is not evidence for absence of earthquake occurrence on it. Therefore, an area without record of strong earthquake, may be modeled as a potential seismic source with the upper bound magnitude similar to the region of comparable structural and tectonic characteristics which has the record of strong earthquakes.

A total of 235 potential seismic source zones in Iran and neighboring regions are delineated for seismic hazard analysis of the country (Figure 1). Boundaries of the potential seismic sources inside the country are delineated using 1 : 1000000 base maps while boundaries for sources outside the country are based on 1 : 2500000 and some times smaller scale maps.

4 Fundamental data

Following documentary materials are the principal sources used to model potential seismic sources in and around the country.

4.1 Earthquake catalog

The uniform catalog of earthquakes in Iran and neighboring regions (Mirzaei et al., 1997d) including historical and instrumental data from 400 B. C. through 1994 covering the area bounded by $22° \sim 42°$N, $42° \sim 66°$E. The catalog is extended to May 1997, using USGS/NEIC Preliminary Determination of Epicenters (PDE) data file.

4.2 Geological, tectonic and geophysical maps

(1) Seismotectonic map of Iran, scale 1 : 1000000, Compiled by Nogole Sadat (1993).

(2) Seismotectonic map of Iran, scale 1 : 2500000, Compiled by Berberian (1976b).

(3) Seismotectonic map of the Middle East, scale 1 : 5000000, Compiled by Haghipour (1992).

(4) Map of active faults and seismotectonics of Armenian upland territory and adjacent regions of the West Asia, scale 1 : 2500000, compiled by Karakhanian (1993).

(5) Isoseismal map of Iran (1900–1977), scale 1 : 5000000, compiled by Berberian (1977b).

(6) Total (magnetic) intensity map (of Iran), scale 1 : 1000000, compiled by Yousefi (1989).

(7) Geological and tectonic maps of Iran published by the Geological Survey of Iran and National Iranian Oil Company.

(8) In the Zagros region of southwest Iran, most seismogenic structures are blind thrust faults, the recent work of Berberian (1995), in which mainly based on geomorphic indicators some blind thrust faults are mapped, also, utilized to model potential seismic sources.

4.3 Focal mechanism solutions

The map of focal mechanism solutions of earthquakes in Iran is not published, but considerable sources of regional and case studies, using first P motion and waveform modelling techniques, are available which are used in delineation of potential seismic sources; e. g., McKenzi, (1972); Chandra (1981, 1984); Jackson and Fitch (1981); Berberian (1982, 1983); Kadinsky-Cade and Barazangi (1982); Jackson and McKenzi (1984); Ni and Barazangi (1986); Byrne et al. (1992); Baker et al. (1993); Priestley et al. (1994); Gao and Wallace (1995). For recent earthquakes for which results of special studies are not available results of Centroid Moment Tensor (CMT) solutions are used, e. g., Dziewonski et al. (1994, 1995).

It is worth noting, that the seismotectonic map of Iran (Nogole Sadat, 1993), scale 1 : 1000000, is utilized as the principal base map for demarcation of potential seismic sources inside the country.

5 Delineation of potential seismic sources

Traditionally, the configuration of individual seismic sources could be points, lines, areas or volumes, depending upon the type of source chosen and the ability to define it in geologic space (Reiter, 1990). If possible it is preferred to model the seismic sources in three dimensions (EERI Committee on Seismic Risk, 1989). The point source model is used when there is no ability to define a causative structure for recorded events; and line sources are modeled for identified active faults. Consideration of line sources imply that, either causative structure is vertical or we are dealing with a dip slip fault which generates earthquakes of zero depth along its length, which rarely is the real case. We prefer to model potential seismic sources as areas in which configuration of each source zone is controlled, mainly, by the fault extent, seismogenic crust (a part of the earth crust in which large earthquakes usually nucleates), and mechanism of earthquake faultings or a type of active faults. In the cases in which available information does not permit to reliably specify earthquake-generating faults, a concentration of known and/or unknown active faults [localizing structure (Reiter, 1990)], are included in a unique source zone.

Along each seismogenic zone [over a broad seismotectonic province, major earthquakes usually tend to cluster in particular zones with particular tectonic settings which are structurally related to a group of tectonic features; these are major seismic energy release zones under the present-day geodynamic regime, called seismogenic zones (Ye et al., 1993, 1995)] segmentation analysis is applied. In the segmentation concept, large earthquakes, commonly of a characteristic size, repeatedly rupture the same part or segment of a fault, and less commonly extend into adjacent segments (Schwartz and coppersmith, 1984; McCalpin, 1996). Possible rupture segments may be reflected by structural discontinuities of the fault, especially where strike-slip tear faults intersect and offset the thrust, or at the ends of overlapping imbricate faults, changes in orientations of the fault, intersections with branch faults, abrupt changes in dip, changes in net slip (Knuepfer, 1989), and changes in spatial pattern of seismicity (Ye et al., 1995). Boundaries of potential seismic source zones are delineated mainly based on the mapped or inferred extents of corresponding tectonic features. The aftershock distribution and the clustering of small earthquakes are used as well, in making judgment on the boundaries of seismic sources. For the width of the seismic source, we take 30km for thrust fault zones and 20km for strike-slip faultings. Since crustal thrust and reverse faults seldom occur individually, but instead are generally part of imbricate or overlapping systems made up of multiple faults and folds (Carver and McCalpin, 1996), the width of the seismic source zone is wider if individual active faults could not be identified in the system.

5.1 Configuration of potential seismic source zones

(1) If potential seismic source zones include a single structure, configuration of the source zone is controlled by the type of structure (fault, fold or basin) and its extent. If because of complexity of the region, several structures are included in a same potential seismic source, configuration of source zone depends on the prominent extent of involved structures. In the later case sometimes there is no preferred orientation in the source zone.

(2) Spatial distribution of main shocks and related aftershocks is the most convenient approach for determination of extent and direction of different segments of active faults. For the cases in which causative structure does not rupture the surface, spatial distribution of aftershocks is of particular importance to search active structures and model the source zone.

(3) Clustering of small earthquakes can be used to delineate potential seismic source zone especially if it is supported by geomorphic indicators or geophysically inferred structures and anomalies.

(4) In an area where an isolated moderate to strong earthquake is located, the potential source zone may have no preferred orientation.

(5) A great number of earthquakes in highly seismic region of Zagros, occur on hidden faults. There is considerable uncertainty about their extent, geometry and the mechanism. Many potential seismic sources in this region are based on geomorphic evidences of blind thrusts [blind thrusts are dip-slip faults (low angle thrust ramps and decollements as well as high angle reverse faults) that do not cut the Earth's surface and often have overlying, cogenetic fold trends (Shaw and Suppe, 1996)].

6 Estimation of M_{max} in potential seismic sources

The knowledge of the largest magnitude which may occur within a potential seismic source is of practical importance. However, its determination encounters serious difficulties. Destructive shocks are rare, therefore their recurrence interval is long and cannot be determined based on short time available earthquake history. Moreover, we do not know the physical conditions of focal zones; and we are not able to estimate in advance the capacity of a region to store the energy. The estimation of M_{max} in potential seismic source zones is usually based on the features of seismic activity and geological analogy. It is often estimated from empirical correlation between magnitude and various fault parameters, such as rupture length, rupture area, maximum surface displacement and rate of seismic moment release (e.g., Wyss, 1979; Bonilla et al., 1984; Wells and Coppersmith, 1994); paleoseismic analysis (e.g., Slemmons et al., 1987) and characteristic earthquake model (Schwartz and Coppersmith, 1984). However, determination of maximum magnitude for a source zone is a subjective problem and encounters considerable scientific judgment. Our approach to estimate maximum magnitude for potential seismic source zones in Iran is mainly based on the concept of tectonic analogy.

(1) For the sources with the record of great earthquakes ($M_S 7.5$); if the largest earthquake has occurred in

a historical time-period, the observed largest magnitude is taken as the upper bound magnitude directly, or an increment of up to 0.5 magnitude unit is added based on frequency and accuracy of earthquake records in the source zone. If the largest earthquake occurred in an instrumental time-period with good accuracy of the recorded event, an increment of 0.3~0.5 magnitude unit is added to the observed magnitude to estimate maximum possible magnitude in the potential seismic source.

(2) For the sources in which major earthquakes ($6.0 < M_S < 7.5$) are recorded, based on the accuracy of earthquake record, time span of the catalog and tectonic setting of the causative structure, an increment of 0.5~1 is added to the largest observed magnitude, and/or, directly it is taken as the maximum magnitude.

(3) Since the available earthquake records can not be representative of recurrence time for large earthquakes in most of the regions, the tectonic analogy concept is preferred to other approaches. According to the assumption of tectonic analogy, for the areas with no long history of earthquake record, maximum magnitude of the analogous potential seismic source may be adopted directly.

It is worth noting that in any case, the maximum magnitude for potential seismic source does not exceed the estimated maximum magnitude (Mirzaei et al., 1997b) for the relevant major seismotectonic province.

7 Background seismicity

In the regions in which scarcity and/or lack of information does not permit to delineate potential seismic sources, and even in the areas where active faults are defined, it is needed to model the background seismicity. In the background seismicity concept, small and moderate sized earthquakes may occur in the defined area randomly. Its value is usually defined as earthquakes of magnitude 6.5 and smaller, based on the activity level of the region.

There are many reports of the occurrence of moderate and major earthquakes on previously unknown faults in all over the world; the magnitude 7.4 Tabase-Golshan earthquake of 16 September 1978 in central Iran (Berberian, 1979b), and magnitude 6 Whittier Narrows earthquake of October 1, 1987, in southern California (EERI Committee on Seismic Risk, 1989) are good examples to show importance of consideration of background seismicity. For seismic hazard assessment in Iran, background earthquake of magnitude 6.0 is considered for the seismotectonic provinces of Zagros and Alborz-Azarbayejan, and for Central-East Iran, Kopeh Dagh and Makran seismotectonic provinces, magnitude 5.5 is taken as background earthquake.

8 Examples

In Zagros major seismotectonic province, the Main Recent Fault (Tchalenko and Braud, 1974) is a prominent right-lateral strike-slip fault system with northwest-southeast trend, in northern part of northeast Zagros border zone. It is a major structure broadly parallel but quite distinct from and younger than the Main Zagros Reverse Fault which transects it in several places. The Main Recent Fault is a major seismogenic structure constituted of several segments with different levels of seismicity. More intense seismic activity is concentrated between 33°~35°N on the Dorud, Nahavand, Sahneh and Dinavar segments, while to the north, Sartakht, Morvarid, Marivan, and Piranshahr segments show relative seismic quiescence (Berberian, 1995). A total of five potential seismic source zones are delineated in this siesmogenic zone. Each potential seismic source includes one or more fault segments based on interpretation of all available information.

The Dorud segment of Main Recent Fault which has nucleated the largest recorded earthquake in Zagros [the Silakhor (Dorud) earthquake of 23 January 1909, M_S7.4, with 45km documented surface faulting (Ambraseys and Melville, 1982)], is distinct from the Nahavand Fault to the north, by structural discontinuity as well as more intense seismic activity (Figure 2, zone 53). The upper limit magnitude for this potential seismic source zone estimated to be 7.7 (the maximum expected magnitude in Zagros) based on the statistical analysis in seismotectonic province and uncertainty in earthquake magnitude estimation for its relevant events.

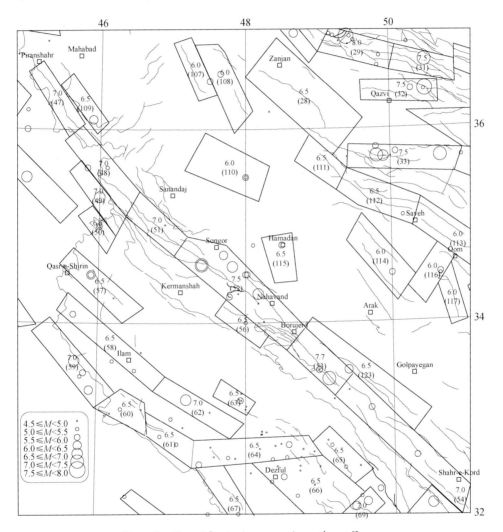

Figure 2　Potential seismic sources in northwest Zagros

The Sahneh Fault segment of the Main Recent Fault is contributed to the Farsinaj (Farsineh) earthquake of 13 December 1957, M_S6.7, by some authors (e.g., Berberian, 1995). Indeed, the Farsinaj earthquake fault did not break the surface (e.g., Tchalenko and Braud, 1974). Field investigations and interviews with local people show that the earthquake occurred on a buried fault, probably in continuation of the Nahavand segment. The Sahneh and Nahavand Faults are responsible for several documented destructive earthquakes in historical and instrumental time periods (e.g., Tchalenko and Braud, 1974; Ambraseys and Melville, 1982). Moreover, Archeological excavations at Kangavar, Gowdin Tapeh and at other localities, provide some indications of much earlier destructive earthquakes in the region (Tchalenko and Braud, 1974; Ambraseys, 1974; Moinfar, 1976). It is very possible that the Sahneh Fault and Nahavand Fault are surface expressions of some more complicated buried faults in the region. Therefore, the area bounded by these faults are included in a unique potential seismic source zone, even though the Sahneh Fault (including Dinavar segment) has more frequent earthquake records during history (Figure 2, zone 52). The upper bound magnitude for this potential seismic source is estimated to be 7.5, which is 0.5 magnitude unit larger than the not well documented largest recorded event of 27 April 1008 A.D. in Dinavar, which caused more than 16000 loss of human life and considerable geomorphic changes and ground rupture (e.g., Ambraseys and Melville, 1982).

In Central-East Iran major seismotectonic province, a prominent zone of seismicity extends from Tabas southward along Nayband-Gowk Fault System (Figure 3, zones 142, 141, 139 and 132). It is a major seismogenic zone, which borders the western edge of the Dasht-e-Lut desert (e.g., Berberian et al., 1979, 1984; Ambraseys et al., 1979).

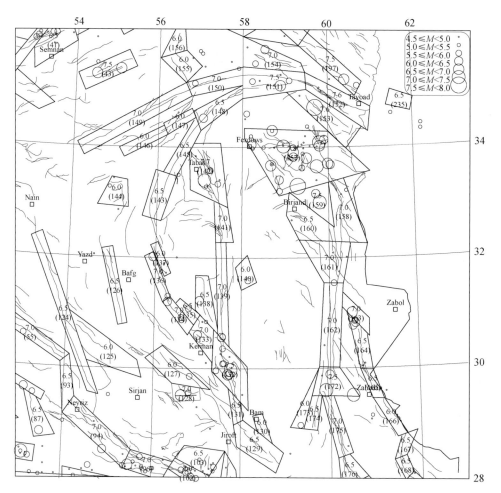

Figure 3 Potential seismic sources in East Iran

The Gowk Fault system is a major structure which extends for about 180km from northeast Kerman to southwest Bam in a NNW trend. It is clearly recognizable on landsat imagery and aerial photographs, and is made up of several distinct en echelon segments containing individual faults that dip steeply (60°~90°) to both the east and the west (Berberian et al., 1984). Observation of uplift, offset escarpments in Quaternary alluvium, and slickensides on exposed faults in the Gowk Fault Zone show evidence of recent motion involving both reverse (thrust) and right-lateral strike-slip components (Berberian et al., 1984).

Destructive shocks occurred in the northern part of the Gowk Fault System in 1877, M_S5.6, 1909.10.27,

M_S 5.5, 1911.04.29, M_S 5.6, 1948.07.05, M_S 6.0, 1981.06.11, M_S 6.6 and 1981.07.28, M_S 7.0 (Ambraseys and Melville, 1982; Berberian et al., 1984). The earthquake of 1989.11.20, M_S 5.7, was associated with a surface deformation that followed the mapped traces of the southern Gowk active fault system (Berberian and Qorashi, 1994). More intense seismic activity is concentrated on the Sirch-Golbaf segment, where the north-south Nayband Fault System joins the northwest-southeast trending seismically active Kuhbanan Fault System, representing a relatively wide zone of seismic activity (Figure 3, zone 132).

One of the interesting characteristics of seismicity in Central-East Iran is that, intense seismic activity is concentrated in the special areas on patches of active faults, which in some cases show very complicated local tectonic setting. The Tabas-e-Golshan destructive earthquake of 1978.09.16, M_S 7.4, caused multiple frontal thrust faulting at the surface and bedding-plane slips of thrust mechanism in the hanging-wall block along Tabas Fault (Berberian, 1979b, 1982). The surface rupture extended for about 85km in ten discontinuous thrust segments with a NNW-SSE trend. The fault segments were separated by gaps in the surface ruptures in a complicated pattern along curved lines. The main fault was associated with a zone of bedding plain slips of thrust mechanism extending for 6 km east of the main surface break (Berberian, 1979b). The Tabas Fault segments together with the Shotori Fault and Esfandiar Fault to the east, have concentrated many shocks in a limited area and permit to model the area as an unique potential seismic source zone located north of Nayband Fault (Figure 3, zone 142). The east-west Dasht-e-Bayaz fault, demonstrated by 80km of left-lateral strike-slip surface faulting in Dasht-e-Bayaz earthquake of 1968.08.31, M_S 7.3 (Ambraseys and Tchalenko, 1969; Tchalenko and Berberian, 1975; Jackson and McKenzi, 1984) and 60km in Kowli-Bonyabad earthquake of 1979.11.27, M_S 7.1 (Haghipour and Amidi, 1980). In the south of the east-west Dasht-e-Bayaz fault, there are a number of earthquakes, in which north-south right-lateral faulting was observed (Jackson and McKenzi, 1984). In the earthquakes of 1941.02.16, M_S 6.1, and 1947.09.23, M_S 6.8, such faulting with lengths of 10 and 20km, respectively, is known from field studies and interviews with the local population (Ambraseys and Melville, 1977, 1982; Jackson and McKenzi, 1984). The Korizan earthquake of 14 November 1979, M_S 6.7, involved a curiously shaped fault of 20km with predominantly north-south right-lateral strike-slip mechanism, but curving to the east at each end (Haghipour and Amidi, 1980; Jackson and McKenzi, 1984). The 20km Korizan fault rupture extended for another 10km northeastward (preserved its character of movement) by the Kowli-Bonyabad earthquake of 27 November 1979 (Haghipour and Amidi, 1980). Preliminary studies on the most recent earthquake of 1997.05.10 in the Qaen-Birjand region, show at least 80km of right-lateral strike-slip surface faulting on the Korizan Fault. The 1976.11.07, M_S 6.2, south of the Dasht-e-Bayaz Fault, also involved right-lateral strike-slip in the north-south direction (Jackson and McKenzi, 1984). To the west, the Ferdows Fault is a northwest-southeast trending reverse fault of 100km length which reactivated by the occurrence of Dashte-Bayaz earthquake of 31 August 1968 (Berberian, 1981). It passes through unconsolidated Quaternary deposits and cataclastics, therefore, the fault trace is inferred from geomorphic indications, as a sharp border between Quaternary deposits in the southwest and older cataclastics and conglomerates in the northeast. The Ferdows earthquakes of 1968.09.01, M_S 6.3, and 1968.09.04, M_S 5.2, involved thrust faulting with a northwest strike, as shown by their fault plane solutions, aftershock locations and tentative reports of ground displacement on a reverse fault dipping east (Jackson and Fitch, 1979; Berberian, 1979b; Jackson and McKenzi, 1984). Concentration of several earthquake faults of different mechanisms and different orientations in the limited described area do not permit to model each active fault as a single seismic source; therefore, they are included in a relatively large potential source zone (Figure 3, zone 157).

In Alborz-Azarbayejan major seismotectonic province, the North Tabriz Fault, extends for over 180km in a NW-SE direction from northwest Marand to northeast Hashtrud (Nogole Sadat, 1993). It is responsible for many earthquakes in historical time-period; for example, two catastrophic earthquakes of 1721.04.26 M_S7.7, and 1786.10, M_S6.3, responsible for the destruction of Tabriz, involved substantial motion on North Tabriz Fault. These earthquakes have been studied by Berberian and Arshadi (1976) and Ambraseys and Melville (1982), who agree that the southwest side of this fault was downthrown, but disagree on whether it was accompanied by right lateral strike-slip motion (Jackson and McKenzi, 1984). Two potential seismic sources are delineated on this fault (Figure 4, zones 11 and 12) to reflect changes in seismic activity rate as well as changes in fault orientation. The potential seismic source 11 has more frequent earthquake record, but maximum possible magnitude in both potential seismic sources is estimated to be the same, based on the assumption of tectonic analogy. To the west, the potential seismic source 10 is an area source without preferred elongation. This includes surface faults of different orientation and earthquake epicenters are, spatially, scattered.

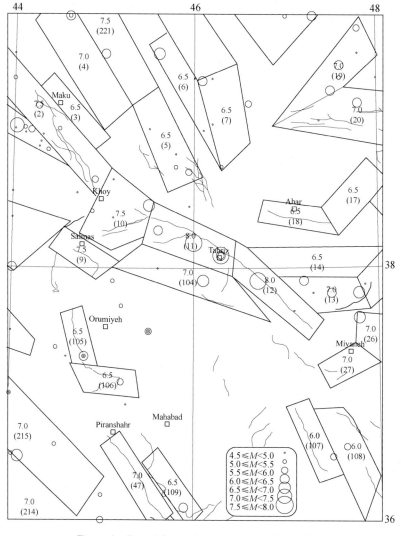

Figure 4　Potential seismic sources in northwest Iran

The Buyin Zahra earthquake of 1962.09.01, M_S7.2, caused 85km surface faulting which had a mechanism involving thrusting and east-west left-lateral strike-slip with NE slip vector on the Ipak Fault in the border zone of

Alborz-Azarbayejan and Central-East Iran major seismotectonic provinces. To the east, between Eshtehard and Tehran, there exists a record of a large historic earthquake but no mapped surface rupture is known to be related to this event. Geomorphic evidences and tectonic setting of this region, strongly permit to include these events in a same potential seismic source zone, in which elongation of the source zone follows the orientation of Ipak Fault and its width is approximately 20km which is the case for strike-slip faulting (Figure 5, zone 33). Further to the east, several historical earthquakes lie in a northwest direction between Tehran and Garmsar. There is no documented fault related to these events, but alignment of earthquake epicenters on a clear geomorphic lineament which separates unfolded Quaternary deposits from Oligo-Miocene shallow marine Qom Formation and synorogenic sediments deposited during Late Alpine event, helped to model them in a same potential seismic source zone (Figure 5, zone 36).

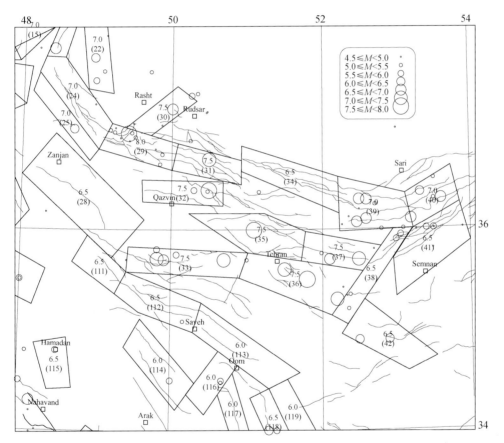

Figure 5 Potential seismic sources in north-central Iran

In Makran, the Makran subduction zone exhibits strong segmentation between east and west in its seismic behavior. Eastern Makran has experienced large and great thrust earthquakes and currently experiences small and moderate size earthquakes. In contrast, western Makran exhibits no well documented great earthquakes in historic times and modern instrumentation has not detected any shallow events along the plate boundary (Byrne et al., 1992). There is no documented fault rupture following any earthquake in Makran even in the great Pasni-Ormara earthquake of 1945.11.27, $M_S 8.0$ (Ambraseys and Melville, 1982) $M_W 8.1$ (Byrne et al., 1992). Based on the tectonic geomorphology and damage area, Jackson and McKenzi (1984) suggested that the mechanism of great Pasni-Ormara earthquake is consistent with a shallow thrust dipping northward, which is confirmed by the recently published focal mechanism solution for this event by Byrne et al. (1992). On the other hand, catalog of earthquakes in Makran is severely incomplete; so that, there is no reliable record of earthquake until 1919 A. D. ; therefore, to delineate potential seismic sources in this region we should be careful not to downgrade potential seismic risk which threatens the region. Based on earthquake epicenter distribution and geomorphic indications, the region in which the great Pasni-Ormara occurred is delineated as a potential seismic source along shoreline of Oman Sea, and the western Makran shoreline is modeled as a potential seismic source based on tectonic analogy as well as for safety (Figure 6).

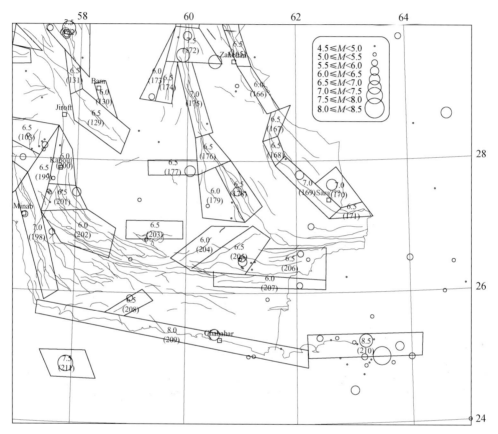

Figure 6 Potential seismic sources in Makran

9 Conclusions

Potential seismic sources in Iran and adjacent regions are delineated, using all available geological, geophysical, tectonic and earthquake data for seismic hazard assessment of the country. Potential seismic sources are modeled as area sources, in which configuration of each source zone is controlled, mainly, by the extent of active faults, mechanism of earthquake faulting and seismogenic part of the crust. A great number of earthquakes in the highly seismic region of Zagros, occur on hidden faults, therefore, many potential seismic sources in this region are modeled based on geomorphic evidences of blind thrust faults and geophysically inferred lineaments and anomalies. In Makran, there is no documented fault rupture following any earthquake, even in the great earthquake of 1945, $M_S 8.0$. On the other hand, catalog of earthquakes in Makran is severely incomplete; to delineate potential seismic sources, we were careful not to downgrade potential seismic risk which threatens the region.

Since, the available earthquake records can not be representative of recurrence time for large earthquakes in most of the regions, estimation of maximum magnitude for potential seismic source zones is mainly based on the concept of tectonic analogy, which implies that structures of analogous tectonic setting are capable of generating the same size earthquakes.

Acknowledgements

The authors would like to thank Professor Zhang Yuming from the Institute of Geology, China Seismological Bureau; Professor Yan Jiaquan and Professor Jin Yan from the Institute of Geophysics, China Seismological Bureau, for their helpful discussions and suggestions.

References

Ambraseys, N. N., 1974, The historical seismicity of North-Central Iran, *Geol. Surv. Iran*, Rep. **29**, 47–96.

Ambraseys, N. N., 1978, The relocation of epicenters in Iran, *Geophys. J. R. Astr. Soc.* **53**, 117–121.

Ambraseys, N. N. and Melville, C. P., 1977, The seismicity of Kuhistan, Iran, *Geogr. J.* **143**, 179–199.

Ambraseys, N. N. and Melville, C. P., 1982, *A History of Persian Earthquakes*, Cambridge University Press, Cambridge, 219 pp.

Ambraseys, N. N. and Tchalenko, J. S., 1969, The Dasht-e-Bayaz (Iran) earthquake of August 31, 1968: a field report, *Bull. Seism. Soc. Am.* **59**, 1751–1792.

Ambraseys, N. N., Arsovski, M. and Moinfar, A., 1979, The Gisk earthquake of 19 December 1977 and the seismicity of the Kuhbanan Fault-zone, UNESCO Publ. No. FMR/SC/GEO/79/192, Paris.

Asudeh, I., 1983, ISC mislocation of earthquakes in Iran and geometrical residuals, *Tectonophysics* **95**, 61–74.

Baker, C., Jackson, J. and Priestley, K., 1993, Earthquakes on the Kazeron line in the Zagros mountains of Iran: strike-slip faulting within a fold and-thrust belt, *Geophys. J. Int.* **115**, 41–61.

Berberian, M., 1976a, Contribution to the seismotectonics of Iran (part II), *Geol. Surv. Iran*, Rep. 39.

Berberian, M., 1976b, Seismotectonic map of Iran (1: 2500000), *Geol. Surv. Iran*, Rep. 39.

Berberian, M., 1977a, Contribution to the seismotectonics of Iran (part III), *Geol. Surv. Iran*, Rep. 40.

Berberian, M., 1977b, Isoseismal map of Iran (1900–1977), 1: 5000000, *Geol. Surv. Iran*, Rep. 40.

Berberian, M., 1979a, Evaluation of the instrumental and relocated epicenters of Iranian earthquakes, *Geophys. J. R. Astr. Soc.* **58**, 625–630.

Berberian, M., 1979b, Earthquake faulting and bedding thrust associated with the Tabas-e-Golshan (Iran) earthquake of September 16, 1978, *Bull. Seism. Soc. Am.* **69**, 1861–1887.

Berberian, M., 1981, Active faulting and tectonics of Iran. In: Gupta, H. K. and Delany, F. M. (eds), *Zagros-Hindukush-Himalaya Geodynamic Evolution*, Am. Geophys. Union and Geol. Soc. Am., Geodyn. Ser. **3**, 33–69.

Berberian, M., 1982, Aftershock tectonics of the 1978 Tabas-e-Golshan (Iran) earthquake sequence: a documented active 'thin- and thick-skinned tectonic' case, *Geophys. J. R. Astr. Soc.* **68**, 499–530.

Berberian, M., 1983, Continental deformation in the Iranian plateau (contribution to the seismotectonics of Iran, part IV), *Geol. Surv. Iran*, Rep. 52.

Berberian, M., 1995, Master 'blind' thrust faults hidden under the Zagros folds: active basement tectonics and surface morphotectonics, *Tectonophysics* **241**, 193–224.

Berberian, M. and Arshadi, S., 1976, On the evidence of the youngest activity of the North Tabriz Fault and the seismicity of Tabriz city, *Geol. Surv. Iran*, Rep. 39, 397–418.

Berberian, M. and Qorashi, M., 1994, Coseismic fault-related folding during the South Golbaf earthquake of November 20, 1989, in southeast Iran, *Geology* **22**, 531–534.

Berberian, M., Asudeh, I. and Arshadi, S., 1979, Surface rupture and mechanism of the Bob-Tangol (south-eastern Iran) earthquake of 19 December 1977, *Earth Planet. Sci. Lett.* **42**, 456–462.

Berberian, M., Jackson, J. A., Qorashi, M. and Kadjar, M. H., 1984, Field and teleseismic observations of the 1981 Golbaf-Sirch earthquakes in SE Iran, *Geophys. J. R. Astr. Soc.* **77**, 809–838.

Berberian, M., Qorashi, M., Jackson, J. A., Priestley, K. F. and Wallace, T., 1992, The Rudbar-Tarom Earthquake of 20 June 1990 in NW Persia: Preliminary field and seismological observations and its tectonic significance, *Bull. Seism. Soc. Am.* **82**, 1726–1755.

Bonilla, M. G., Mark, R. K. and Lienkaemper, J., 1984, Statistical relations among earthquake magnitude, surface rupture length and surface fault displacement, *Bull. Seism. Soc. Am.* **74**, 2379–2411.

Byrne, D. E., Sykes, L. R. and Davis, D. M., 1992, Great thrust earthquakes and aseismic slip along the plate boundry of Makran subduction zone, *J. Geophys. Res.* **79**, 449–478.

Carver, G. A. and McCalpin, J. P., 1996, Paleoseismology of compressional tectonic environments. In: McCalpin, J. P. (ed.), *Paleoseismology*, Academic Press, Inc., San Diego, California, pp. 183–270.

Chandra, U., 1981, Focal mechanism solutions and their tectonic implication for the eastern Alpine-Himalayan region. In: Gupta, H. K. and Delany, F. M. (eds), *Zagros-Hindukush-Himalaya Geodynamic Evolution*, Am. Geophys. Union and Geol. Soc. Am., Geodynamic Ser. **3**, 243–271.

Chandra, U., 1984, Focal mechanism solutions for earthquakes in Iran, *Phys. Earth Planet. Inter.* **34**, 9–16.

Dziewonski, A. M., Ekstrom, G., Salganik, M. P., 1994, Centroid-moment tensor solutions for January-March 1994, *Phys. Earth Planet. Inter.* **86**, 253–261.

Dziewonski, A. M., Ekstrom, G., Salganik, M. P., 1995, Centroid-moment tensor solutions for April-June 1994, *Phys. Earth Planet. Inter.* **88**, 69–78.

EERI Committee on Seismic Risk, 1989, The basics of seismic risk analysis, *Earthquake Spectra* **5**, 675–702.

Gao, L. and Wallace, T. C., 1995, The 1990 Rudbar-Tarum Iranian earthquake sequence: evidence for slip partitioning, *J. Geophys. Res.* **100**, 15317–15332.

Haghipour, A., 1992, Seismotectonic map of Middle East (1:5000000), *Geol. Surv. Iran* and CGMW.

Haghipour, A. and Amidi, M., 1980, The November 14 to December 25, 1979 Ghaenat earthquakes of northeast Iran and their tectonic implications, *Bull. Seism. Soc. Am.* **70**, 1751–1757.

He, H., Zhang, Y. and Zhou, B., 1994, Delineation of potential seismic sources in continent of China. In: Proc. Workshop on *Implementation of GSHAP in Central and Southern Asia*, Beijing, China, pp. 169–183.

Jackson, J. A., 1980a, Errors in focal depth determination and the depth of seismicity in Iran and Turkey, *Geophys. J. R. Astr. Soc.* **57**, 209–229.

Jackson, J. A., 1980b, Reactivation of basement faults and crustal shortening in orogenic belts, *Nature* **283**, 343–346.

Jackson, J. A. and Fitch, T. J., 1979, Seismotectonic implications of relocated aftershock sequences in Iran and Turkey, *Geophys. J. R. Astr. Soc.* **57**, 209–229.

Jackson, J. A. and Fitch, T. J., 1981, Basement faulting and the focal depth of the large earthquakes in the Zagros mountains (Iran), *Geophys. J. R. Astr. Soc.* **64**, 561–586.

Jackson, J. A. and McKenzi, D. P., 1984, Active tectonics of the Alpine-Himalayan belt between Western Turkey and Pakistan, *Geophys. J. R. Astr. Soc.* **77**, 185–264.

Jackson, J. A. and McKenzi, D. P., 1988, The relation between plate motions and seismic moment tensors, and the rates of active deformation in the Mediterranean and the Middle East, *Geophys. J. R. Astr. Soc.* **93**, 45–73.

Jackson, J., Haines, J. and Holt, W., 1995, The accommodation of Arabia-Eurasia plate convergence in Iran, *J. Geophys. Res.* **100**, 15205–15219.

Kadinsky-Cade, K. and Barazangi, M., 1982, Seismotectonics of southern Iran: The Oman line, *Tectonics* **1**, 389–412.

Kagan, Y. Y. and Jackson, D. D., 1994, Long-term probabilistic forecasting of earthquakes, *J. Geophys. Res.* **99**, 13685–13700.

Karakhanian, A. S., 1993, Map of active faults and seismotectonics of Armania Upland territory and adjacent regions of the West Asia, National survey of seismic protection, Armania.

Karnik, V., 1969, *Seismicity of the European Area*, Part 1, D. Reidel Publ. Co., Dordrecht-Holland, 364 pp.

Knuepfer, P. L. K., 1989, Implications of the characteristics of endpoints of historical surface fault ruptures for the nature of fault segmentation. In: Schwartz, D. P. and Sibson, R. H. (eds), *Fault Segmentation and Controls of Rupture Initiation and Termination*, U. S. Geol. Surv. Open File Rep. pp. 89-315, 193-228.

McCalpin, J. P., 1996, Application of paleoseismic data to seismic hazard assessment and neotectonics research. In: McCalpin, J. P. (ed.), *Paleoseismology*, Academic Press, Inc., San Diego, California, pp. 439-493.

McKenzi, D. P., 1972, Active tectonics of Mediterranean region, *Geophys. J. R. Astr. Soc.* **30**, 109-185.

Mirzaei, N., Gao, M. and Chen, Y. T., 1997a, Evaluation of uncertainty of earthquake parameters for the purpose of seismic zoning of Iran, *Earthquake Research in China* **11**, 197-212.

Mirzaei, N., Gao, M. and Chen, Y. T., 1997b, Seismicity in major seismotectonic provinces of Iran, *Earthquake Research in China* **11**(4), in press.

Mirzaei, N., Gao, M. and Chen, Y. T., 1997c, Seismic source regionalization for seismic zoning of Iran: Major seismotectonic provinces, Submitted to *Earthquake Research in China*.

Mirzaei, N., Gao, M., Chen, Y. T. and Wang, J., 1997d, A uniform catalog of earthquakes for seismic hazard assessment in Iran, *Acta Seismol. Sinica* **10**, 713-726.

Moinfar, A., 1976, The importance of macroseismic studies of past earthquakes. In: *Proc. CENTO Semin. Recent Advances in Earthq. Hazard Minimization*, Tehran, pp. 65-69.

Ni, J. and Barazangi, M., 1986, Seismotectonics of the Zagros continental collision zone and a comparison with the Himalayas, *J. Geophys. Res.* **91**, 8205-8218.

Nogole Sadat, M. A. A., 1993, Seismotectonic map of Iran (scale 1: 1000000), Geol. Surv. Iran (proof print).

Nowroozi, A. A., 1971, Seismotectonics of the Persian plateau, Eastern Turkey, Caucasus and Hindu-Kush region, *Bull. Seism. Soc. Am.* **61**, 317-341.

Nowroozi, A. A., 1972, Focal mechanism of earthquakes in Persia, Turkey West Pakistan, and Afganistan and plate tectonics of the Middle East, *Bull. Seism. Soc. Am.* **62**, 823-850.

Priestley, K., Baker, C. and Jackson, J., 1994, Implications of earthquake focal mechanism data for the active tectonics of the south Caspian basin and surrounding regions, *Geophys. J. Int.* **118**, 111-141.

Reiter, L., 1990, *Earthquake Hazard Analysis*, Colombia University Press, New York, 254 pp.

Schwartz, D. P. and Coppersmith, K. J., 1984, Fault behavior and characteristic earthquakes: Examples from the Wasatch and San Andreas fault zones, *J. Geophys. Res.* **89**, 5681-5698.

Shaw, J. H. and Suppe, J., 1996, Earthquake hazards of active blind-thrust faults under the central Los Angles basin, California, *J. Geophys. Res.* **101**, 8623-8642.

Slemmons, D. B., Bodin, P. and Zhang, X., 1987, Determination of earthquake size from surface faulting events. In: *Proc. International Seminar on Seismic Zonation*, Guangzhou, China, 157-169.

Synder, D. B. and Barazangi, M., 1986, Deep crustal structure and flexure of the Arabian plate beneath the Zagros collisional mountain belt as inferred from gravity observation, *Tectonics* **5**, 361-373.

Tchalenko, J. S., 1975, Seismicity and structure of the kopet Dagh (Iran, USSR), *Phil. Trans. R. Soc. Lond. A* **278**, 1-28.

Tchalenko, J. S. and Berberian, M., 1975, Dasht-e-Bayaz fault, Iran: earthquake and earlier related structures in bedrock, *Geol. Soc. Am. Bull.* **86**, 703-709.

Tchalenko, J. S. and Braud, J., 1974, Seismicity and structure of the Zagros (Iran): the main recent fault between 33° and 35°N, *Phil. Trans. R. Soc. Lond. A* **227**, 1-25.

Thenhause, P. C., EERI, M., Perkins, D. M., Algermissen, S. T., EERI, M. and Hanson, S. L., 1987, Earthquake hazard in the Eastern United States: consequences of alternative seismic source zones, *Earthquake Spectra* **3**, 227-261.

Wells, D. L. and Coppersmith, K. J., 1994, New empirical relationship among magnitude, rupture length, rupture width, rupture area, and surface displacement, *Bull. Seism. Soc. Am.* **84**, 974–1002.

Wyss, M., 1979, Estimating maximum expectable magnitude of earthquakes from fault dimensions, *Geology* **7**, 336–340.

Ye, H., Chen, G. and Zhou, Q., 1995, Study on the intraplate potential seismic sources. In: *Proc. Fifth International Conf. Seismic Zonation*, **2**, Nice, France, pp. 1424–1430.

Ye, H., Zhou, Y., Zhou, Q., Yang, W., Chen, G. and Hao, C., 1993, Study on potential seismic sources for seismic zonation and engineering seismic hazard analysis in continental areas. In: *Continental Earthquakes*, IASPEI Publication Series for the IDNDR **3**, pp. 473–478.

Yucemen, M. S. and Gulkan, P., 1994, Seismic hazard analysis with randomly located sources, *Natural Hazards* **9**, 215–233.

Yousefi, E., 1989, *Total Intensity Map (1:1000000)*, Geol. Surv. Iran.

Zhang, Y., 1993, Principles and methods on delineation of potential earthquake source area. In: *Proc. PRC/USSR Workshop on Geodynamics and Seismic Risk Assessment*, Beijing, China, pp. 201–207.

A time-domain inversion technique for the tempo-spatial distribution of slip on a finite fault plane with applications to recent large earthquakes in the Tibetan Plateau[*]

Chen Y. T. and Xu L. S.

Institute of Geophysics, China Seismological Bureau

Summary A time-domain inversion technique is proposed to invert for the tempo-spatial distribution of slip on an earthquake fault plane. This technique is based on the idea that a finite fault plane can be divided into several subfaults, each of which can be treated as a point source, and that the source time function (STF) of the finite fault is the weighted sum of the STFs of all the subfaults. This technique is applied to deduce the source processes of three recent large earthquakes in the Qinghai-Xizang (Tibetan) Plateau. The 1990 April 26 Gonghe, Qinghai, earthquake, $M_S 6.9$, ruptured on a fault plane with a strike of 113°, a dip of 68° and a rake of 89°. The fault has a length of 45 km, a width of 15 km and a total moment release of 9.4×10^{18} Nm. The rupture initiated at the WNW end of the fault and propagated unilaterally towards the ESE end of the fault. In total, the rupturing lasted for 45 s and was divided into two episodes (the first 9 s long and the second 16 s long with 20 s quiescence in between). There are two distinct rupture regions (nuclei) on the fault plane. These two nuclei, with maximum slips of about 62 and 55 cm, are about 25 km apart in the WNW-ESE direction. The 1996 February 3 Lijiang, Yunnan, earthquake, $M_S 7.0$, has a focal mechanism with a strike of 157°, a dip of 48° and a rake of −102°. The earthquake fault is about 45 km with a width of 20 km and a total moment release of 9.8×10^{18} Nm. The total duration of the rupture was 16 s, is divided into two episodes (the first 8.5 s long and the second 7.5 s long). The inverted final slip distribution of the Lijiang earthquake also shows two nuclei. One nucleus, with a maximum slip of 40 cm, is located on the NNW portion of the fault at a shallower depth and the other, with a maximum slip of 50 cm, is located on the SSE portion of the fault at a greater depth. The 1997 November 8 Mani, Xizang (Tibet), earthquake, $M_S 7.9$, has a focal mechanism with a strike of 250°, a dip of 88° and a rake of 19° and a moment release of 3.4×10^{20} Nm. The total duration of the rupture was about 15 s. The inverted final slip distribution of the Mani earthquake shows three nuclei on the fault plane. The first nucleus, with a maximum slip of 956 cm, is located at the WSW end of the fault at 10 km in depth. The second, with a maximum slip of 743 cm, is located at the ENE end of the fault, 55 km away from the WSW end of the fault and at 35 km depth. The third, with a maximum slip of 1060 cm, is about 30 km away from the WSW end of the fault and at about 40 km depth. These three nuclei form a total rupture area of about 70 km in length and 60 km in depth.

[*] 本文发表于 *Geophys. J. Int.*, 2000 年, 第 **143** 卷, 第 2 期, 407–416.

1 Introduction

In the 1980s, one of the important seismological subjects was investigating the time history of a composite earthquake by modelling waveform recordings using empirical or synthetic Green's functions (Mueller 1985; Hartzell & Heaton 1985; DiBona & Boatwright 1989). Since the late 1980s and early 1990s, seismologists have begun to analyse the spatial complexity along with the temporal complexity of earthquakes (Frankel et al. 1986; Velasco et al. 1994; Dreger 1994a; Mori 1996). The investigation of the rupture process both temporally and spatially has become one of the most active fields. Different data and different techniques have been used in this study. Data such as near-field (Mendez & Anderson 1991; Fletcher & Spudich 1998), broad-band (Velasco 1994) and long-period (Chen et al. 1996) recordings as well as geodetic data have been used (Ward & Barrientos 1986; Wald & Somerville 1995). Time-and frequency-domain inversion techniques were used, based on the idea that an earthquake is composed of several temporally and spatially simple quakes. Based on the same idea, we propose a time-domain inversion technique based on the technique used by Dreger (1994b), similar to the technique used by Fletcher & Spudich (1998). The common element is in the modelling of the shape of source time functions (STFs) that have been retrieved from the waveforms at different stations. The differences are twofold. First, not only the spatial distribution of the final slip but also the slip variation with time are inverted. Second, it is unnecessary to assume a rupture velocity *a priori*; it actually becomes a parameter to be determined by inversion. The technique and the corresponding program were tested for their practicability and reliability (Xu 1995).

We have applied this inversion technique to three recent large earthquakes in the Qinghai-Xizang (Tibetan) Plateau, which are shown in Fig. 1, that occurred on 1990 April 26, 1996 February 3 and 1997 November 8. There are three smaller events chosen as empirical Green's functions (EGFs) for the 1990 April 26 earthquake and two smaller events for the 1996 February 3 earthquake. No appropriate aftershock, however, can be used as an EGF for the 1997 November 8 earthquake. For this reason, instead of EGFs, synthetic Green's functions were used in retrieving STFs. The techniques of moment tensor inversion, STF retrieval and slip distribution inversion used in these studies have been described in Xu (1995), Chen et al. (1996), Xu & Chen (1996, 1997, 1999) and Xu et al. (1997a, b). The purpose of this paper is to illustrate the practicability and effectiveness of the inversion technique of slip distribution by applying this technique to three recent large earthquakes in the Tibetan Plateau. To facilitate the retrieval of total moment release and details of the source complexities, both the long-period recordings with 1 sample s^{-1} sampling rate and the broad-band recordings with 20 sample s^{-1} sampling rate are used. The success in imaging the tempo-spatial slip distribution of the three recent large earthquakes in the Tibetan Plateau strongly suggests (1) that this technique is practicable and robust, (2) that high-resolution data lead to a high-resolution image, showing a detailed rupture process temporally and spatially, and (3) that synthetic Green's functions can also be adopted if no appropriate aftershock can be used as an EGF. The present inversion technique is robust even in the case of using only a few station data, but more robust results are obtained using more station data with a wide-range azimuthal coverage. The data we have used in this study come mainly from the China Digital Seismograph Network (CDSN) (Fig. 1), which presently has 11 stations, but on average had 7 – 9 stations, which produced good-quality recordings for $M_s \geqslant 6.0$ earthquakes that occurred in China.

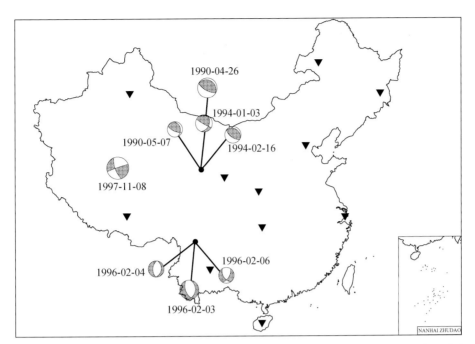

Figure 1 The distribution of China Digital Seismograph Network (CDSN) stations (inverted triangles) and three recent large earthquakes in the Tibetan Plateau

The M_S6.9 Gonghe earthquake occurred on 1990 April 26, with three aftershocks that were used as EGFs. The 1996 February 3 M_S7.0 Lijiang earthquake had two aftershocks that were used as EGFs. The 1997 November 8 M_S7.9 Mani earthquake had no smaller events that could be taken as EGFs

2 Theoretical background

In order to construct an image showing the tempo-spatial distribution of slip on the fault plane, one has to get an idea about the parameters of the fault plane, the scalar seismic moment and the STFs retrieved from different stations. The procedure is divided into three steps. In the first step the parameters of the fault plane are obtained by moment tensor inversion. In the second step the azimuth-dependent STFs are retrieved from waveforms at different stations by deconvolving empirical or synthetic Green's functions from the recordings of the earthquake under study. In the third step the slip distribution is imaged by inverting the observed STFs obtained in the second step.

2.1 Moment tensor inversion

Assuming that the earthquake source is at the origin of the coordinate system, as the linear dimension of the earthquake source is small compared to the wavelength of interest, the displacement at the field point r can be written as (Aki & Richards 1980)

$$u_i(r, t) = G_{ij,k}(r, t) * M_{jk}(t), \qquad (1)$$

where the asterisk represents the convolution in the time domain, $u_i(r, t)$ is the observed displacement, $M_{jk}(t)$ is the moment tensor function and $G_{ij,k}(r, t)$ is the Green's function. Taking the Fourier transform, eq. (1) in the frequency domain becomes

$$u_i(r, \omega) = G_{ij,k}(r, \omega) M_{jk}(\omega), \qquad (2)$$

where ω is the angular frequency. The moment tensor functions are obtained by frequency-domain inversion (Xu et al. 1997a, b).

2.2 Retrieval of the source time function

In general, the recorded displacement generated by an earthquake with scalar seismic moment M_0 may be expressed as (Hartzell 1978; Mueller 1985)

$$u(t) = M_0 S(t) * P(t) * I(t), \tag{3}$$

where $S(t)$ is the normalized far-field STF (the time derivative of the normalized near-field STF), $P(t)$ is the impulse response of the transmitting path and $I(t)$ is the impulse response of the instrument. In the frequency domain, eq. (3) becomes much simpler, i.e.

$$u(\omega) = M_0 S(\omega) P(\omega) I(\omega). \tag{4}$$

By analogy with eq. (4), we have a similar equation for a second earthquake:

$$u'(\omega) = M'_0 S'(\omega) P'(\omega) I'(\omega), \tag{5}$$

where u' M'_0 S' P' and I' are the respective quantities for this earthquake. For an earthquake with the same hypocentre and focal mechanism as the first one, recorded by the same instrument at the same site and small enough so that its far-field STF can be regarded as a Dirac δ function in time, the seismogram of the smaller event is the impulse response of the transmitting path and instrument and can be regarded as an EGF of the first event. The relationship between the displacements of these two earthquakes is

$$\frac{u(\omega)}{u'(\omega)} = \frac{M_0}{M'_0} S(\omega) \tag{6}$$

in the frequency domain, since $P(\omega) = P'(\omega)$, $I(\omega) = I'(\omega)$ and $S'(\omega) = 1$.

As eq. (6) indicates, the STF of the larger earthquake is scaled by the ratio of the scalar seismic moment of the larger earthquake to that of the smaller earthquake and is a relative STF. In some cases, there is no smaller event available satisfying these requirements, and synthetic seismograms have to be adopted. If the synthetic seismograms are used to retrieve the STF of the larger earthquake, the ratio of the scalar seismic moments is simply the scalar seismic moment of the larger earthquake. The shortcoming of using a synthetic Green's function is that the retrieved STFs are affected by the inaccuracy of the synthetic Green's function. For this reason, EGFs are preferred.

The STFs of the larger earthquake used in the next step of this study are obtained by spectral division in the frequency domain (DiBona & Boatwright 1989; Dreger 1994a, b; Xu & Chen 1996).

2.3 Construction of slip distribution

If the final area of an earthquake fault is A and the scalar seismic moment as a function of time is $M_0(t)$, the average slip as a function of time, $D(t)$, on the fault plane is

$$D(t) = \frac{M_0(t)}{\mu A}, \tag{7}$$

where μ is the rigidity of the material in the earthquake source region. For a finite earthquake fault, the slip is generally nonuniform and variable with time and space, and the rupture area is irregular in shape. In this case, the finite fault can be divided into several subfaults, each of which can be regarded as a point source. The slip as a function of time, $D_j(t)$, of the jth subfault is expressed as

$$D_j(t) = \frac{M_j(t)}{\mu A_j}, \tag{8}$$

where $M_j(t)$ and A_j are the scalar seismic moment and the area of the jth subfault, respectively. The far-field STF observed at the ith station, $S_i(t)$, is the weighted sum of the far-field STF of the jth subfault, $s_j(t)$,

$$S_i(t) = \sum_{j=1}^{J} w_j s_j(t - \tau_{ij}), \qquad (9)$$

where J is the number of the subfaults, τ_{ij} is the time delay associated with the wave propagation and w_j is a weight defined by the ratio of the scalar seismic moment of the jth subfault, M_j, to that of the total scalar seismic moment, M_0,

$$w_j = \frac{M_j}{M_0}. \qquad (10)$$

Let $m_j(t)$ denote the weighted far-field STF of the jth subfault,

$$m_j(t) = w_j s_j(t). \qquad (11)$$

The weight for each subfault is obtained by normalizing the weighted far-field STF of the subfault,

$$w_j = \int_0^\infty m_j(t)\,\mathrm{d}t. \qquad (12)$$

Eq. (9) can be rewritten as

$$S_i(t) = \sum_{j=1}^{J} m_j(t - \tau_{ij}). \qquad (13)$$

Hence, the time derivative of the slip as a function of time of the jth subfault is determined by

$$\dot{D}_j(t) = \frac{M_0}{\mu A_j} m_j(t). \qquad (14)$$

The STFs observed at different distances and/or different azimuths are identical if the linear dimension of a fault is small enough to be considered as a point source. However, the shapes of the STFs vary with azimuth for a finite fault. If a finite fault is considered as a combination of several subfaults, as indicated by eq. (13), the observed STF for the finite fault is the weighted sum of all the STFs of the subfaults. Eq. (13) can be rewritten as

$$S_i(t) = \sum_{j=1}^{J} \delta(t - \tau_{ij}) * m_j(t), \qquad (15)$$

$$\tau_{ij} = \frac{r_j}{v_i}, \qquad (16)$$

where r_j is the distance between the jth subfault and the reference point on the fault plane and v_i is the apparent velocity, which depends on the wave velocity and propagation direction and the location of the jth subfault.

In matrix form eq. (15) can be written as

$$\mathbf{S} = \mathbf{KM}, \qquad (17)$$

where $\mathbf{S}$ is the data vector, consisting of the observed STFs, $\mathbf{M}$ is the unknowns vector made up of the weighted STFs of all the subfaults and $\mathbf{K}$ is the coefficient matrix determined by the time delay associated with the wave propagation.

Usually, this problem is underdetermined since the number of unknowns is larger than the number of observation equations and there is no constraint imposed on the STFs of subfaults. In order to stabilize the solution, the following condition is imposed:

$$m_j(t) \geq 0, \qquad (18)$$

which means physically that no backward slip occurs during the rupture process. This condition is very easy to apply if the conjugate gradient method is employed in solving eq. (17) (Ward & Barrientos 1986). The conjugate gradient method is very helpful and powerful in solving underdetermined problems such as eq. (17). This method has the advantages that it is not necessary to find the inverse matrix and that any non-negative model can be used as an initial model.

3 Applications to three recent large earthquakes in the Qinghai-Xizang (Tibetan) Plateau

3.1 The 1990 Gonghe, Qinghai Province of China, $M_S 6.9$ earthquake

On 1990 April 26, a destructive earthquake occurred in Gonghe, Qinghai Province of China. The epicentral location of the earthquake was longitude $\lambda = 100.245°$ E, latitude $\phi = 35.986°$ N, the focal depth $h = 8.1$ km, the origin time O = 08 hr 37 min 15s (UTC) and the magnitude $M_S = 6.9$. Later, several moderate-sized aftershocks occurred approximately at the same location (Fig. 1), and they were been used to retrieve the STFs of the $M_S 6.9$ event. Many studies have been conducted concerning its focal mechanism, seismogenic structure and moment tensor using field geological data and long-period and broad-band waveform data (Tu 1990; Zeng 1990; Zhao et al. 1993; Xu 1995; Xu & Chen 1996, 1997; Chen et al. 1996). The slip distribution on the fault plane was obtained by inverting the levelling data and the waveform data (Zhao et al. 1993; Chen et al. 1996). Here we just cite the results from the imaging of the tempo-spatial rupture process to illustrate the technique described

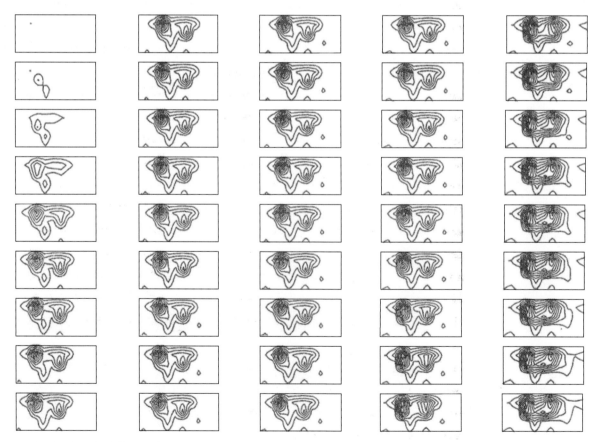

Figure 2 Snapshots of the rupture process on the fault plane of the 1990 April 26 $M_S 6.9$ Gonghe earthquake

The time interval between two neighbouring shots is 1s from top to bottom and from left to right. Each shot covers a fault plane 88km long in the strike direction (113°) and 40km wide in the dip direction

above.

In this study, the fault plane used for imaging of the tempospatial rupture process is obtained by moment tensor inversion of long-period digital waveform data and has a strike of 113°, a dip of 68° and a rake of 89° (Fig. 1). The scalar seismic moment obtained by moment tensor inversion is 9.4×10^{18} Nm. A rectangular fault, which is 88km long in the strike direction and 40km wide in the dip direction, is divided into 11×5 subfaults of dimension 8km × 8km. The inverted result is smoothed by bilinear interpolation to give a smoother image. The snapshots in Fig. 2 show how the rupture evolved during the source process. The time interval between two neighbouring shots from top to bottom and left to right is 1s. Note that the rupture initiated at the WNW end and propagated towards the ESE, suggesting that rupture was unilateral. However, the rupture is heterogeneous in strength, velocity and direction. The rupture initiated at the WNW end and propagated towards the ESE end very

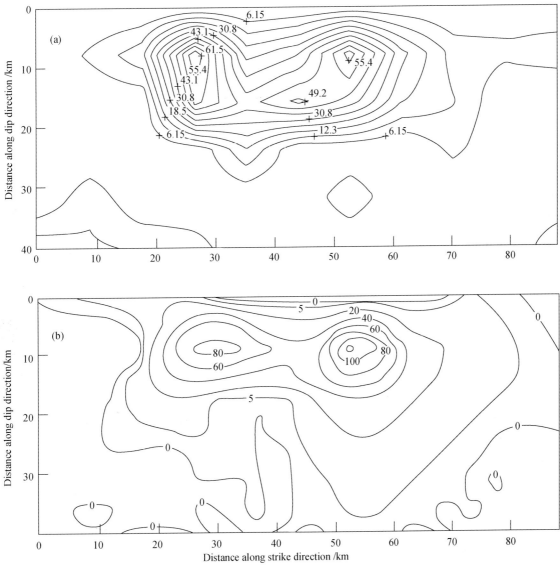

Figure 3 (a) The static slip distribution (in cm) inverted from waveform data for the 1990 April 26 M_S6.9 Gonghe earthquake. The slip was mainly concentrated on the two nuclei in the strike direction (113°) with maximum slips of about 62 and 55cm, which unexpectedly agrees with the result from the levelling data inversion shown in (b). (b) The static slip distribution (in cm) inverted from levelling data for the same earthquake. The slip was mainly concentrated on the two nuclei in the strike direction (113°) with maximum slips of about 80 and 100cm (Zhao et al. 1993)

quickly in the first 9s. After this the rupturing region remained almost unchanged for about 20s, but the slip within the ruptured region still varied in strength, especially in the WNW portion. At 30s, the earthquake rupture restarted in the ESE portion of the fault plane, but it varies mainly in strength, not in area. Later the rupture varied in area and direction slightly until the termination of the rupture process. The second episode of rupturing lasted for about 16s. The total period of rupture was about 45s.

3.2 The 1996 Lijiang, Yunnan Province of China, $M_S 7.0$ earthquake

On 1996 February 3, an $M_S 7.0$ earthquake occurred in Lijiang, northwestern Yunnan Province of China. The origin time was O = 11 hr 14 min 18.1s (UTC). The epicentre was located at longitude λ = 110.13° E, and latitude ϕ = 27.18° N. The focal depth was determined to be 10km. After the occurrence of the main shock, a number of aftershocks followed, but only two larger aftershocks were chosen as EGFs in retrieving the STFs of the main shock (Fig. 1). The time history of the earthquake rupture process was analysed using these two aftershocks as EGFs (Xu et al. 1997a). Other than this, many studies were carried out about the focal mechanism, seismogenic structure and rupture process of the Lijiang earthquake using field geological survey data and aftershock data (Yan 1998). What is given in the following is just the tempo-spatial distribution of slip on the fault plane inverted using STFs retrieved from the CDSN broad-band waveform data.

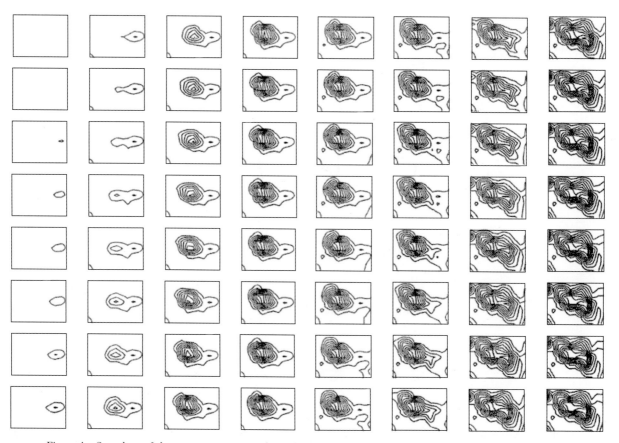

Figure 4   Snapshots of the rupture process on the fault plane for the 1996 February 3 $M_S 7.0$ Lijiang earthquake

The time interval between two neighbouring shots is 0.25s from top to bottom and from left to right. Each shot covers a fault plane 66km long in the strike direction (157°) and 49km wide in the dip direction

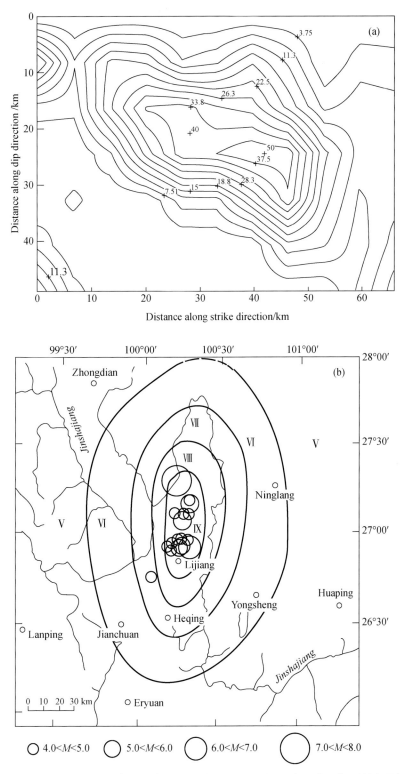

Figure 5 (a) The static slip distribution (in cm) inverted from waveform data for the 1996 February 3 $M_S 7.0$ Lijiang earthquake. The slip was mainly concentrated on the two nuclei in the strike direction (157°) with maximum slips of about 40 and 50cm. (b) The isoseismals of the same earthquake obtained from a field survey after the main shock. The two distinct mesoseismal regions agree well with the overall two-nucleus rupture pattern of the static slip distribution shown in Fig. 4

The fault plane with strike 157°, dip 48° and rake −102° and the scalar seismic moment $M_0 = 9.8 \times 10^{18}$ Nm are used in studies of this earthquake (Yan 1998). A rectangular fault 66km long in the strike direction and 49km wide in the dip direction was divided into 11 × 7 subfaults with dimensions 6km × 7km in the strike and dip directions, respectively. The inverted result was smoothed spatially by bilinear interpolation.

The slip of the $M_S 7.0$ Lijiang earthquake mainly concentrated at two loci, one on the NNW portion of the fault, with a maximum slip of about 40cm and a shallower depth, and another on the SSE portion of the fault, with a maximum slip of about 50cm and a greater depth (Fig. 5). The isoseismals of the Lijiang earthquake obtained from a field survey after the main shock are shown in Fig. 5(b). The aftershocks with $M_L \geqslant 4.0$ were distributed in a nearly N-S direction and clustered in central and southern parts of the mesoseismal regions. It is observed that there are two distinct mesoseismal regions. One is located in the area between the epicentre of the main shock and the central group of aftershocks; the other is located in the area between the central and southern groups of aftershocks. It is worth noting that the two-nucleus pattern of the final slip distribution corresponds to the two distinct mesoseismal regions observed from the field survey after the main shock.

Fig. 4, in which two neighbouring shots have a time interval of 0.25s from top to bottom and left to right shows how the rupture initiated and developed on the fault plane. The total duration of the earthquake rupture was about 16s, divided into two episodes. The rupturing propagated unilaterally from the SSE end to the NNW end of the fault. In the first episode of 8.5s, the rupture initiated from the SSE end of the fault, but moment release mainly occurred in the NNW portion of the fault. In the second episode of 7.5s, rupture began at the deeper depth of the SSE end of the fault, and then expanded and spread out at the ground surface.

3.3 The 1997 Mani, Xizang (Tibet) of China, $M_S 7.9$ earthquake

An $M_S 7.9$ earthquake occurred in Mani, Xizang (Tibet), on 1997 November 8. This earthquake is the largest earthquake to have occurred in the Qinghai-Xizang (Tibetan) Plateau since the 1950 August 15 Chayi, Xizang (Tibet), earthquake of $M_S 8.6$. The moment tensor solution was obtained by inverting the long-period body wave data (Xu & Chen 1998, 1999). The results for the best double couple (DC) are NP1: strike 250°/dip 88°/rake 19°, NP2: 159°/71°/178° and scalar seismic moment $M_0 = 3.4 \times 10^{20}$ Nm. Synthetic Green's functions were calculated by using the generalized reflection/transmission coefficient matrix method (Kennett 1983), the STFs were retrieved by deconvolving a synthetic Green's function instead of the EGF from the observed P and S waves, and the image of slip distribution on the fault plane was constructed by inverting the azimuth-dependent STFs (Xu & Chen 1999). In the following we present the tempo-spatial image of slip obtained using the above technique.

The thickness of the crust underneath the Tibetan Plateau is about 70km (Zeng & Sun 1993). In the inversion, we set the fault plane to be 70km in the dip direction and 110km in the strike direction and divided the rectangular fault (strike 250°, dip 88° and rake 19°) into 11 segments in the strike (length) and 7 segments in the dip (width) directions.

Snapshots of the slip distribution show that the rupture initiated at the WSW end of the fault and propagated towards the ENE end (Fig. 6a), indicating a unilateral rupture in the overall trend. The total duration of the earthquake rupture was about 15s. The rupture area is actually concentrated in three subregions or nuclei, which are more visible in Fig. 6(b). The first nucleus is located at the WSW end of the fault at about 10km depth with a maximum slip of 956cm. The second nucleus is located at the ENE end of the fault at about 35km depth with a

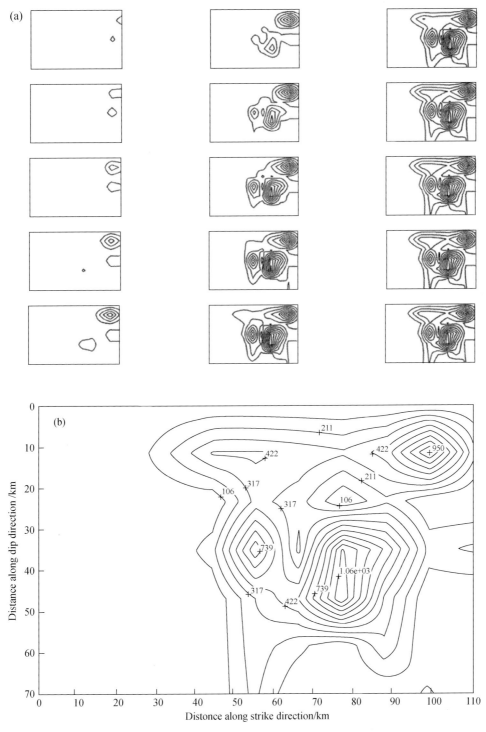

Figure 6 (a) Snapshots of the rupture process on the fault plane for the 1997 November 8 M_s7.9 Mani earthquake. The time interval between two neighbouring shots is 1s from top to bottom and from left to right. Each shot covers a fault plane 110km long in the strike direction (250°) and 70km wide in the dip direction. (b) The static slip distribution (in cm) inverted using waveform data for the same earthquake. The rupture area is about 70km × 60km and is separated into three subareas or nuclei with maximum slip values of 956, 743 and 1060cm

maximum slip of about 743cm. The third nucleus, with a maximum slip of 1060cm, is about 30km from the WSW end of the fault and at about 40km depth. The entire rupture area is about 70km in length and 60km in width and it is to the east of the epicentre. However, the aftershocks were clustered in the inverted rupture area and are denser to the east of the epicentre but very sparse to the west of the epicentre of the main shock. It is inferred that the M_S7.9 Mani earthquake resulted from the nearly eastward extension of the ENE-WSW-to nearly E-W-striking fault in the northwestern Tibetan Plateau.

4 Discussion and conclusions

We have proposed a time-domain inversion technique for constructing an image of the tempo-spatial distribution of slip on a finite fault plane based on (1) the idea that a finite fault is a combination of several subfaults, each of which can be regarded as a point source, (2) the fact that the STF of a finite fault is formed from the weighted STFs of all the subfaults, and (3) the fact that the azimuthal dependence of the observed STFs of the finite fault can be traced back to the different time delays of the STFs of the subfaults. Because of the fact that the rupture may occur more than once on the same subfault, no constraint is imposed on the time window of the STFs of subfaults or on the rupture velocity. However, in order to stabilize the solution, an inequality condition is imposed such that the slip is always equal to or greater than zero. This condition implies that no backward slip is allowed, which is physically a kinematic condition rather than dynamic condition (Knopoff et al. 1973). The conjugate gradient method is employed in solving this underdetermined problem since it is not necessary to solve the inverse matrix, and any non-negative model can be taken as an initial model. It will hopefully become a powerful tool for obtaining the tempo-spatial image of slip on a finite fault plane.

The inversion technique presented in this paper has turned out to be practicable and effective. The Gonghe, Qinghai Province of China, M_S6.9 earthquake has been analysed for its tempo-spatial rupture process, and the result shows the complexity of the earthquake rupture temporally as well as spatially. Although there is no other evidence to support our rupture pattern, the agreement of the static slip distribution with that inverted from the levelling data gives a positive confirmation. Also, the inversion of the Lijiang, Yunnan Province of China, M_S7.0 earthquake gave an image with higher resolution, showing how the rupture developed and what the tempo-spatial complexity was like during the quake. The overall rupture feature obtained using our technique coincided with the isoseismals obtained from a field survey carried out after the main shock, thus giving support to our results. The inverted result of the Mani, Xizang (Tibet) of China, M_S7.9 earthquake did not have any evidence from other research or investigation (Xu & Chen 1998, 1999). We note that the main feature we have obtained is consistent with the surface fault ruptures from a field investigation made in 1999 (X. W. Xu, personal communication, 1999) and from satellite synthetic aperture radar (SAR) interferometry (Peltzer et al. 1999).

We have used only the CDSN data since the results presented in this paper were obtained within a few days to a few months after the main shocks; at that time, data from other broadband or long-period stations were not available. As mentioned above, on average only 7–9 stations have good-quality data available for M_S6.0 earthquakes occurring in China. Even for this number of stations a robust result can be obtained. This means that the inversion technique proposed in this paper is not particularly limited by the number of stations. However, it will not work if the azimuthal coverage of stations is too limited because the information of azimuthal dependence is not enough to build up a reliable tempo-spatial image of slip.

Acknowledgments

We wish to express our sincere thanks to Drs Anthony F. Gangi, Amiya Chatterjee and Angela Sarao for their critical reviews that greatly helped us to improve the manuscript. This work was supported by the SSTCC Climb Project 95-S-05. Publication no. 00AE1001, Institute of Geophysics, China Seismological Bureau.

References

Aki, K. & Richards, P. G., 1980. *Quantitative Seismology. Theory and Methods*, Vols 1 & 2, W. H. Freeman, San Francisico.

Chen, Y. T., Xu, L. S., Li, X. & Zhao, M., 1996. Source process of the 1990 Gonghe, China earthquake and tectonic stress field in the northeastern Qinghai-Xizang (Tibetan) plateau, *PAGEOPH*, **146**, 697-715.

DiBona, M. & Boatwright, J., 1989. Single-station decomposition of seismograms for subevent time historries, *Geophys. J. Int.*, **105**, 103-117.

Dreger, D. S., 1994a. Empirical Green's function study of the January 17, 1994 Northridge, California earthquake, *Geophys. Res. Lett.*, **21**, 2633-2636.

Dreger, D. S., 1994b. Investigation of the rupture process of the 28 June 1992 Landers earthquake utilizing TERRAscope, *Bull. seism. Soc. Am.*, **84**, 713-724.

Fletcher, J. B. & Spudich, P., 1998. Rupture characteristics of the three $M \approx 4.7$ (1992 - 1994) Parkfield earthquakes, *J. geophys. Res.*, **103**(B1), 835-854.

Frankel, A., Fletcher, J., Vernon, F., Haar, L., Berger, J., Hanks, T. & Brune, J., 1986. Rupture characteristic and tomographic source imaging of $M_L \approx 3$ earthquakes near Anza, Southern California, *J. geophys. Res.*, **91**, 12633-12650.

Hartzell, S., 1978. Earthquake aftershocks as Green's functions, *Geophys. Res. Lett.*, **5**, 1-4.

Hartzell, S. H. & Heaton, T. H., 1985. Teleseismic time functions for large, shallow subduction zone earthquakes, *Bull. seism. Soc. Am.*, **75**, 965-1004.

Kennett, B. L. N., 1983. *Seismic Wave Propagation in Stratified Media*, Cambridge University Press, Cambridge.

Knopoff, L., Mouton, J. O. & Burridge, R., 1973. The dynamics of a one-dimensional fault in the presence of friction, *Geophys. J. R. astr. Soc.*, **35**, 169-184.

Mendez, A. J. & Anderson, J. G., 1991. The temporal and spatial evolution of the 19 September 1985 Michoacan earthquake as inferred from near source ground-motion records, *Bull. seism. Soc. Am.*, **81**, 844-861.

Mori, J., 1996. Rupture directivity and slip distribution of the M4.3 foreshock to the 1992 Joshua Tree earthquake, Southern California, *Bull. seism. Soc. Am.*, **86**, 805-810.

Mueller, C. S., 1985. Source pulse enhancement by deconvolution of an empirical Green's function, *Geophys. Res. Lett.*, **12**, 33-36.

Peltzer, G., Crampe, G. & King, G., 1999. Evidence of nonlinear elasticity of the crust from the M_W7.6 Manyi (Tibet) earthquake, *Science*, **286**, 272-276.

Tu, D. L., 1990. Geological structure background of Gonghe earthquake $M_s = 6.9$, on April 26, 1990, *Plateau Earthq. Res.*, **2**(3), 15-20 (in Chinese with English abstract).

Velasco, A. A., Ammon, C. J. & Lay, T., 1994. Recent large earthquakes near Cape Mendocina and in the Gorda plate: broadband STFs, fault orientations and rupture complexities, *J. geophys. Res.*, **99**, 711-728.

Wald, D. J. & Somerville, P. G., 1995. Variable-slip rupture model of the great 1923 Kanto, Japan, earthquake: geodetic and body-waveform analysis, *Bull. seism. Soc. Am.*, **85**, 159-177.

Ward, S. N. & Barrientos, S. E., 1986. An inversion for slip distribution and fault shape from geodetic observations of the 1983, Borah Park, Idaho, earthquake, *J. geophys. Res.*, **91**(B5), 4909-4919.

Xu, L. S., 1995. Complexities of the Gonghe, Qinghai Province of China, earthquake sequence, *PhD thesis*, Institute of Geophysics, China Seismological Bureau, Beijing (in Chinese with English abstract).

Xu, L. S. & Chen, Y. T., 1996. Source time functions of the Gonghe, China earthquake retrieved from long-period digital waveform data using empirical Green's function technique, *Acta Seism. Sinica*, **9**, 209–222.

Xu, L. S. & Chen, Y. T., 1997. Source parameters of the 1990 Gonghe Qinghai, China earthquake determined from digital broadband seismic waveform data, *Acta Seism. Sinica*, **10**, 143–159.

Xu, L. S. & Chen, Y. T., 1998. Source process of the November 8, 1997 Mani earthquake ($M_S = 7.9$), northern Tibetan plateau of China, *2nd Mtng Asian Seismological Commission and Symposium on Earthquake Hazard Assessment and Earths Interior Related Topics*, Hyderabad, India, p. 77 (abstract).

Xu, L. S. & Chen, Y. T., 1999. Tempo-spatial rupture process of the 1997, Mani, Xizang (Tibet), China earthquake of $M_S = 7.9$, *Acta Seism. Sinica*, **12**, 495–506.

Xu, L. S., Chen, Y. T. & Fasthoff, S., 1997a. Temporal and spatial complexity in the rupture process of the February 3, 1996 Lijiang, Yunnan, China, in *Advances in Seismology in China*, pp. 91–105, ed. Chen, Y. T., Seismological Press (in Chinese with English abstract).

Xu, L. S., Fasthoff, S., Duda, S. J. & &. Chen, Y. T., 1997b. *MomTen User's Guide*, Institute of Geophysics, China Seismological Bureau/Institute of Geophysics, University of Hamburg.

Yan, F. T., ed., 1998. *The 1996 Lijiang Earthquake*, Seismological Press, Beijing (in Chinese with English abstract).

Zeng, Q. S., 1990. Survey of the earthquake $M_S 6.9$ between Gonghe and Xinghai on April 26, 1990, *Plateau Earthq. Res.*, **2**(3), 3–12 (in Chinese with English abstract).

Zeng, R. S. & Sun, W. G., 1993. Seismicity and focal mechanism in Tibetan Plateau and its implications to lithospheric flow, *Acta seism. Sinica*, **6**, 261–287.

Zhao, M., Chen, Y. T., Gong, S. W. & Wang, Q. L., 1993. Inversion of focal mechanism of the Gonghe, China, earthquake of April 26, 1990, using leveling data, in *Continental Earthquakes*, pp. 246–252, eds Ding, G. Y. & Chen, Z. L., IASPEI Publication Series for the IDNDR, Seismological Bureau, Beijing.

Source parameter determination of regional earthquakes in the Far East using moment tensor inversion of single-station data[*]

Kim S. G.[1], Kraeva N.[2] and Chen Y. -T.[3]

1. The Seismological Institute, Hanyang University;
2. Institute of Marine Geology and Geophysics, Far Eastern Division of Russian Academy of Sciences;
3. Institute of Geophysics, China Seismological Bureau

Abstract The source mechanisms of the Kobe earthquake on 16 January 1995 and the Neftegorsk on 27 May 1995, including their strongest aftershocks as well as two recent South Korean earthquakes of Youngwol on 13 December 1996 and Kyongju on 25 June 1997 are calculated using moment tensor inversion technique (Green's functions are calibrated for path of source to receiver for these earthquakes). Results derived from data of only one or two stations for Kobe and Neftegorsk mainshocks show a good agreement with those of other authors. The inversion solutions for aftershocks of the Kobe and the Neftegorsk earthquakes turn out to be very similar to their mainshocks. The focal mechanism for the Yongwol earthquake on 13 December 1996 is found to be a right-lateral slip event with a NE strike, and the Kyongju earthquake on 25 June 1997 is found to be an oblique reverse fault with a slight component of left-lateral slip in the NE direction (Kim and Kraeva, 1999).

1 Introduction

During the last decade global seismographic networks such as IRIS systems were installed in Far East Asia. The information provided by digital IRIS systems makes it possible to solve various problems of seismology on a qualitatively higher level than was possible previously.

In particular, new prospects are appearing in research of source mechanisms of shallow, moderate sized earthquakes, whose magnitudes are too small to be registered by teleseismic stations. Earthquakes of this type have widespread geographic occurrence, and in some cases provide the only clue to the active tectonics of a region.

In this paper we attempted to use the moment tensor inversion routine TDMT_INV by Douglas Dreger (e. g. Pasyanos et al., 1996). This program is a time domain inverse code designed to invert regional long-period waveform data incorporating both body and surface waves. It has been shown to constrain focal parameters using the data of a single-station, but more robust results are obtained using multiple stations.

For our investigations we have chosen two recent South Korean earthquakes and two large earthquakes and their strong aftershocks that occurred in Kobe and Neftegorsk. The focal mechanisms of all these earthquakes are very important for understanding of the local and regional tectonic process in and near the Korean Peninsula.

---

[*] 本文发表于 *Tectonophysics*, 2000 年, 第 **317** 卷, 125–136.

Appropriate Green functions that are used for these particular regions to fit the source parameters for the data are the most important components for the sake of methodology. Fortunately at low frequencies on which point-source assumptions are feasible (Helmberger and Engen, 1980), relatively simple velocity models may be used to effectively model the data. In this study we carry out all follow-ups to test and calibrate the Green's function.

The source depth of the event is determined iteratively by performing inversions with Green's functions computed for a suite of source depths. The best fitting source depth may be determined from the variance reduction,

$$VR = \left[1 - \frac{\sum (d_i - s_i)^2}{\sum (d_i)^2} \right] \times 100$$

where d_i are data and s_i is a synthetic seismogram. A variance reduction value of 100 means the perfect correlation of data and synthetics in given frequency bands.

In the fitting of the waveforms the consequences of a 'wrong' crustal model are usually compensated for by incorrect focal depth if the velocity perturbation is small (Dreger and Helmberger, 1993; Walter, 1993). Thus, there is a trade-off between structure and a solution with focal depth. An incorrect focal depth can accompany a reasonably recovered source (Fan and Wallace, 1991). The knowledge of the velocity structure is the most important a priori information using complete waveforms.

2 Inversion results

In this study, we determined focal mechanism solutions of the major two earthquakes which occurred in the Far East, their strongest aftershocks and, locally, two large earthquakes that occurred in the Korean Peninsula (Table 1 and Fig. 1) using very broadband single-station data. Using the moment tensor inversion algorithm of Dreger, we had results good enough to estimate a focal mechanism solution of the events and to do that, we used average plane-layered crustal models from other researchers. Figs. 2–4 show displacement seismograms for Kobe, Neftegorsk, and South Korean earthquakes, respectively.

Table 1 The events used in this study (IRIS Data Center catalogue)

Region	Event ID	Date mm/dd/yy	O.T. (UTC) hh:mm:ss	Lat. /°N	Long. /°E	Depth /km	Mag.	Station[b]	Distance /km	BAZ
Kobe	1[a]	01/16/95	20:46:52.1	34.583	135.018	22.0	6.9	MAJO	361.0	234.0
	2	01/17/95	04:05:21.6	34.488	135.179	10.0	4.7	MAJO	357.0	231.0
	3	01/17/95	13:18:36.4	34.520	134.801	10.0	4.6	MAJO	381.0	235.0
	4	01/17/95	15:51:28.5	34.583	135.180	10.0	4.6	MAJO	353.0	232.0
	5	01/17/95	21:50:17.5	34.639	135.072	10.0	4.7	MAJO	354.0	234.0
	6	02/18/95	12:37:32.3	34.444	134.757	10.0	4.7	MAJO	390.0	234.0
Neftegorsk	1[a]	05/27/95	13:03:52.6	52.629	142.827	11.0	7.5	YSS MA2	630.0 915.00	0.0 216.0
	2	05/28/95	02:02:53.6	52.909	142.879	33.0	5.0	YSS MA2	662.0 886.0	1.0 1.0
	3	05/29/95	10:21:34.2	52.686	142.850	33.0	5.3	YSS MA2	637.0 908.0	1.0 216.0

Continued

Region	Event ID	Date mm/dd/yy	O. T. (UTC) hh：mm：ss	Lat. /°N	Long. /°E	Depth /km	Mag.	Station[b]	Distance /km	BAZ
Neftegorsk	4	06/13/95	10：42：39.8	53.098	142.908	13.0	5.2	YSS MA2	683.0 867.0	1.0 217.0
	5	06/13/95	21：35：00.0	53.000	142.748	11.0	5.3	YSS MA2	672.0 882.0	0.0 218.0
	6	12/18/95	02：05：57.9	52.651	142.720	33.0	5.5	YSS MA2	633.0 916.0	0.0 216.0
Korea	1[c]	12/13/96	04：10：16.5	37.141	128.764	10.0	4.8	INCN	192.0	101.0
	2[d]	06/25/97	18：50：21.1	35.820	129.189	10.0	4.7	INCN	294.0	128.0

[a] Mani shock.

[b] Stations used for the inverse calculation.

[c] Youngwol Earthquake.

[d] Kyongju Earthquake.

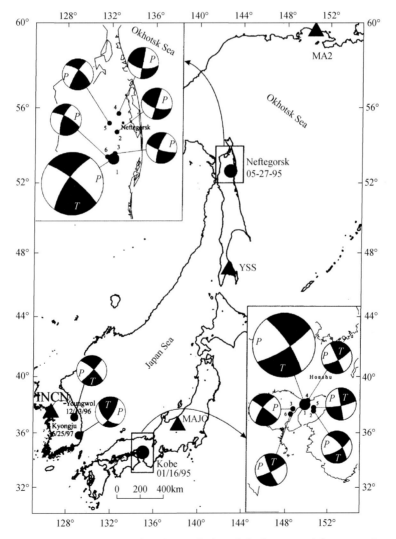

Fig. 1 Map showing the locations of IRIS stations (black triangles) and the locations of the events (gray circles) in this study. The numbers near earthquake locations correspond to numbers of aftershocks listed in Table 1. The mechanisms (Table 5) were determined by inversion of three-component data recorded by the shown stations. Black and open quadrants correspond to compression and dilatation. The characters P and T represent the orientation of maximum compression and tension axes, respectively

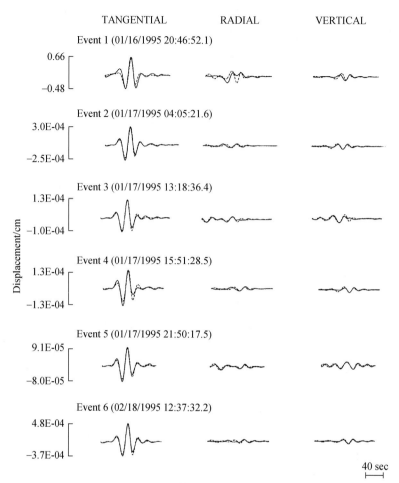

Fig. 2 Displacement fields of the Kobe mainshock and its aftershock in the source coordinate system (radial, transverse and vertical) at the MAJO station area

The displacements are filtered in the broadband channel. The solid line represents an instrumental displacement and the dotted line corresponds to the synthetic displacement calculated by the TDMT_ INV program of Dreger. Event numbers are shown in Table 1

2.1 The Kobe earthquake

The best solution of the mainshock moment tensor inversion was obtained using an average plane-layered crust structure model representing the Honshu between the station MAJO and Kobe (Fig. 5a) in accordance with studies of Tuezov and Jiltsov (1972), Tuezov (1975) and Gnibidenko (1976). Values of attenuations Q_P and Q_S were taken as mean values for the earth's crust from Lay and Wallace (1995).

Results of moment tensor inversions for this earthquake sequence are given in Table 5 and Fig. 6. For the mainshock our result is very similar to other authors results (Tables 2 and 5). Comparison of all these results shows some variations relative to the dip of the focal plane extending in a NE-SW direction. It is clear that this plane is near vertical and the fault mechanism is a right-lateral slip along this plane. Choice of this plane as the main focal plane was made according to Ide et al. (1996).

Our result shows that the damage of the event in the Kobe city may have been due to its close location to the fault plane and also the fault type which was nearly pure strike-slip. For the main shock, the source duration was about 11s, and the total length of the fault dip reaches about 40km in the NE-SW direction with a fault slip of 1–2 m along the Nojima Fault on Awaji Island (Kikuchi and Kanamori, 1996). It is also in agreement with our result.

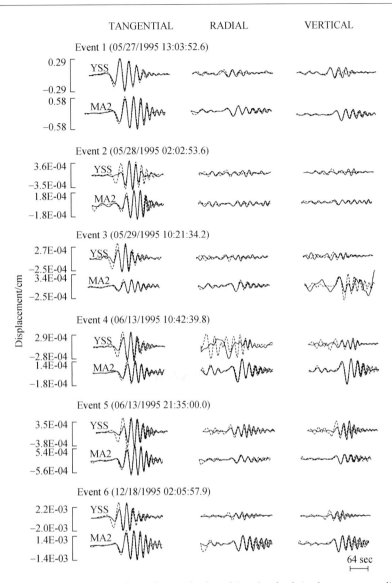

Fig. 3 Displacement fields of the Neftegorsk mainshock and its aftershock in the source coordinate system
(radial, transverse and vertical)

Upper and lower traces of each event correspond to the station YSS and MA2, respectively. The displacements are also filtered in the broadband channel. The solid line represents an instrumental displacement and the dotted line corresponds to the synthetic displacement calculated by the TDMT_INV program of Dreger. See also Table 1 for the number of the earthquake sequence

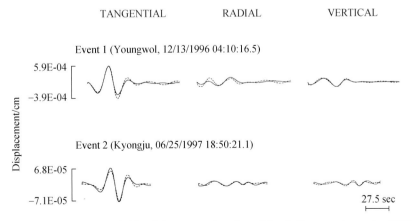

Fig. 4 The displacement fields of the two earthquakes of South Korea at the station INCN
Solid and dotted lines has the same meaning as in Figs. 2 and 3

The focal mechanisms of the strongest five aftershocks appear to be very similar to fault plane solutions of the mainshock. All these earthquakes are right-lateral slips, caused by near horizontal compression in the NW-SE direction.

2.2 The Neftegorsk earthquake

For the Neftegorsk earthquake and its aftershocks, we used two stations of Yuzhno-Sakhalinsk (YSS) and Magadan (MA2) at which the event was regionally recorded (Fig. 3). Here, we can see that even if the distance of station MA2 from the epicenter is one and half times the distance of station YSS, the displacement field at station MA2 is potentially twice as large as that at station YSS. This can be explained only by northward radiation directivity (Bath, 1974).

Recently, crustal structure information beneath the event-station pairs of the North Sakhalin-YSS and the North Sakhalin-MA2 was not well constrained; data regarding P-wave velocities in the crust and upper mantle in and near these area were previously obtained using Deep Seismic Soundings (Zverev, 1971; Tuezov, 1975; Gnibidenko, 1976). These results show that the crustal thickness of the Sakhalin Island varies from 30km to 37.5km from North to South across the Sakhalin.

Using trial-and error-methods based on these data, the average cross-sections for the two paths were fitted (Fig. 5a). For the Neftegorsk-YSS pair, depth distribution attenuation parameters Q_P and Q_S taken analog for the Korean Peninsula crust (Kim and Lee, 1994, 1996) proved to fit synthetics to observations much better than average crust values.

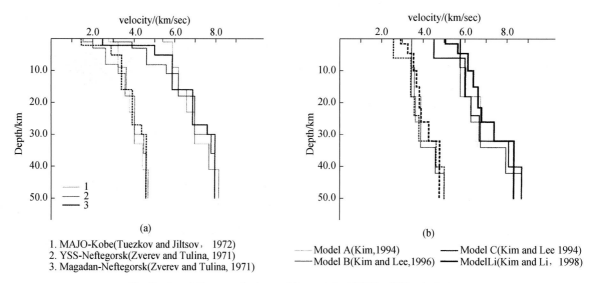

Fig. 5 P- and S-wave velocity models used for TDMT_INV calculations

(a) shows the velocity structure between the event-station pairs MAJO-Kobe, Yuzhno-Sakhalinsk-Neftegorsk, and Magadan-Neftegorsk. The four different crustal models of the Korean Peninsula are shown in (b). In each figure, dotted lines represent the corresponding velocity structure of the S-wave

For the mainshock three different inverse calculations were tested using two stations YSS and MA2 stations (Table 3). Note that the seismic moment value M_0 determined from the MA2 record is three times as large as that from the YSS record; the parameters of a rake and dip estimated from coupled station inversion (YSS and MA2) were almost equal to Harvard University estimations (Tables 2 and 3). For the strike, however, each single-station inverse calculation gives results similar to the others; for the coupled inversion the value is rotated clockwise at between 15° and 19° (see Tables 2 and 3). The coupled station inversion shows that the MA2 station is located in the nodal plane, and it brings the difference of a displacement field around the stations, as a mentioned previously.

Fig. 7 shows a variable reduction in inversion calculations as a function of the source depth. Maximum values of variance reductions were obtained for single-station inversions at a depth equal to 35km for every station and for the couple stations inversion, at a depth of 25km (Table 3). Note that the results of Ivashcenko et al. (1995) give an average depth of mainshock as 18±3km.

The focal mechanism of the mainshock is almost a pure right-lateral strike-slip with a dip of between 73° and 78° to the NW and strike between 198° and 199° with a small reverse component. These results agree well with those of the field survey (Arephyev et al., 1995; Shimamoto et al., 1995): the NW side of the fault moved to the NE and up, relative to the SE side (the average shift is 3.8 m, maximum 8.1 m; the average thrust is about 0.3 m).

A focal mechanism of the five strongest aftershocks is also a right-lateral nearly pure strike-slip along a fault plane slightly inclined to the west with strike about 200° (Table 5). These results imply that all these earthquakes were the prolongation of the tectonic event started by the main, very destructive, shock under conditions of a near horizontal compression in the NE-SW direction. Here we note that owing to a high level of long-period noise on the seismograms resulting from aftershocks numbers 3 and 4 (Table 1; Fig. 3) a variance reduction of combined solutions is very low.

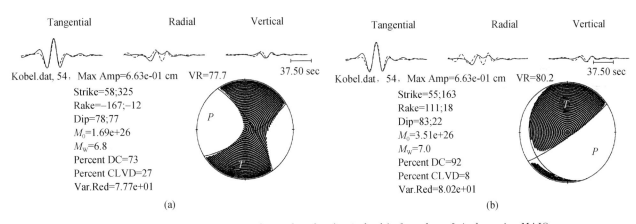

Fig. 6 Two inverse solutions of Kobe earthquake (mainshock) from data of single-station MAJO

The solution (a) (correct) is computed for the zero off set equal to 25, the solution (b) ('degenerate') for the zero off set equal to 29. The solid line corresponds to observed seismograms and the dotted line to synthetics

2.3 Two earthquakes of the South Korea

Although the magnitude of the Youngwol on 13 December 1996 and the Kyongju on 25 June 1997 earthquakes (Table 1) was much smaller than the magnitude of the major two earthquakes which occurred in Far East, locally these are the largest events in the Korean Peninsula, which is a low seismicity area close to the SW Japan. There has been much information obtained about the crustal structure of south of the Korean Peninsula (Kim, 1994; Kim and Lee, 1994, 1996; Kim and Li, 1998; Fig. 5b).

Using the Inchon (INCN) station, we calculated the focal mechanism of the two events for the four different crustal structure models. The model of Li gives a best fit result for both earthquakes, and for a low-velocity model, Model C, the change in compression and dilatation axis was more impressive (Table 4).

For the Youngwol earthquake the inverse calculation gave a maximum variance reduction at a depth of 10km and for the Kyongju earthquake it was at a depth of 7km. A variance reduction of models A and Li as a function of source depth is shown in Fig. 8 for the Youngwol earthquake. The maximum for the model Li is higher and narrower than that for three-layered model A, so model Li in Fig. 9 is more sensitive to the change in depth and gives a better approximation to seismograms. We obtained the same results for the Kyongju earthquake.

Table 2 Focal mechanism solutions of the mainshocks of the major earthquakes from other researchers

Region	$M_0/10^{26} \text{dyn} \cdot \text{cm}$	Strike	Rake	Dip	Authors
Kobe	1.8	65	−169	81	USGS
	2.5	230	162	79	Harvard University[b]
	1.7	231	173	84	Kawakatsu
	1.8	323, 233	0, 163	73, 90	S. Sipkin
	−	322, 230	7, 163	73, 82	Katao et al., 1997
Neftegorsk	4.3	200, 299	160, 29	63, 73	Sipkin[a]
	4.2	196, 289	170, 18	73, 80	Harvard University[b]

[a] Personal communication.

[b] Harvard University quick CMT solution (Salganik et al., 1995).

Table 3 Moment tensor inversion results of the Neftegorsk mainshock for the single and composite stations

Stations	Strike	Rake	Dip	$M_0/10^{26}\text{dyn} \cdot \text{cm}$	M_W	Depth /km	VR/%	DC/%
YSS	198, 290	168, 9	81, 78	2.7	6.9	35	80.3	83
MA2	199, 289	179, 16	74, 89	8.9	7.3	35	84.4	62
YSS+MA2	215, 308	168, 17	73, 78	5.3	7.1	25	73.7	37

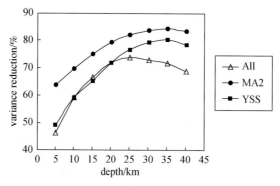

Fig. 7 Variance reductions for the Neftegorsk mainshock as a function of source depth from single-station inversion (YSS and MA2 separately) and the two station inversion (YSS and MA2 together)

Table 4 Moment tensor inversion results of the Youngwol earthquake for the different crustal models

Model	Strike	Rake	Dip	VR/%	DC
A	301, 209	5, 157	67, 86	85.6	34
B	310, 42	−19, −172	83, 71	85.3	68
C	311, 45	148, 6	85, 58	85.8	97
Li	311, 43	−6, −167	77, 84	89.1	70

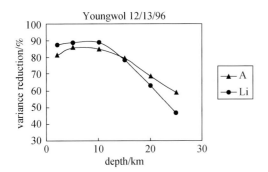

Fig. 8 Youngwol earthquake variance reduction as function of source depth in the case of two different crust models of Korean peninsula

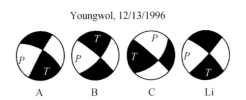

Fig. 9 Moment tensor solution of Youngwol earthquake calculated for different crustal structure models (Table 4)

Table 5 Focal mechanism solutions of the earthquakes in this study

%	Event ID	Strike	Rake	Dip	M_0 /10^{22} dyn·cm	M_W	Depth /km	VR /%	DC /%
Kobe	1[a]	58, 325	−167, −12	78, 77	1.7e+04	6.8	20.0	77.7	73.0
	2	59, 319	−154, −23	69, 66	6.2	4.5	5.0	95.2	91.0
	3	307, 216	18, 175	85, 72	3.2	4.3	10.0	93.3	89.0
	4	153, 63	15, 178	88, 75	2.6	4.2	5.0	88.5	89.0
	5	169, 79	10, 180	90, 80	3.1	4.3	15.0	93.7	98.0
	6	155, 64	10, 172	82, 81	11.6	4.7	15.0	95.6	72.0
Neftegorsk	1[a]	215, 308	168, 17	73, 78	5.3e+04	7.1	25.0	73.7	37.0
	2	198, 107	−154, −2	88, 64	13.0	4.7	15.0	61.7	94.0
	3	200, 107	−166, −15	76, 76	23.0	4.9	35.0	30.0	29.0
	4	190, 99	−149, −2	88, 59	24.0	4.9	35.0	40.5	71.0
	5	212, 307	155, 12	79, 65	49.0	5.1	25.0	67.8	83.0
	6	201, 298	162, 22	69, 73	176.0	5.5	30.0	58.2	49.0
Korea	1	311, 43	−6, −167	77, 84	7.1	5.2	10.0	89.1	70.0
	2	135, 31	22, 140	52, 72	1.7	4.8	7.0	91.6	99.0

[a] Main shock.

The final results of a moment tensor inversion for these two earthquakes are represented in Fig. 1 and Table 5 (a comparison of observed and synthetic displacements in the frequency range 0.02–0.05 Hz). So our focal mechanism solution from data of only one broad band IRIS station INCN is the pure strike-slip along a right-lateral fault plane with a dip of 85° to the SE and the strike 41° for Youngwol earthquake. The results here are in agreement with the data of Baag et al. (1997) regarding the fault plane solution obtained from P-wave first motion signs recorded at the KIGAM, KSRS and JAPAN stations. For Kyongju earthquake our solution is a left-lateral slip-reverse faulting along an oblique (57° to the SW) fault plane with strike of 131° (Kim and Kraeva, 1999). These nodal planes were chosen as fault planes by studying the tectonics in the region. Notice that both earthquakes were caused by the near horizontal compression in the near latitude direction, reflecting a general tectonic peculiarity of this region.

3 Discussion and conclusions

The case of change in a compression and dilatation axis of the Youngwol earthquake mechanism solution, calculated using the model C at a sufficiently high value of a variance reduction (85.8%), is one of the unstable examples sometimes arising at inversions using data from only one station. We need to reiterate that the knowledge of the crust structure is the most important a priori information for this discussed method.

The next unstable example of inverse solutions from data using a single-station is when this single-station is in a nodal plane of the earthquake. In this case radial and vertical components are relatively small for P-waves and relatively large for S-waves in comparison with a tangential component. The solution can be optimized by manually sliding the data about the synthetic to find the zero offset, which maximizes the variance reduction (Dreger and Langston, 1995). However, in practice such manual 'optimization' for stations being in one of the nodal planes of an earthquake very often gives a false 'degenerate' solution (Fig. 6). All these solutions always have near vertical dip-slip (so the station is found near null-vector while the value of seismic moment M_0 is accelerating) and have a variance reduction a few percent higher than the correct solution. The value of zero offset shift is usually 4 to 6s. Fortunately, the fact that we never see such instability in automatic solutions that are cross-correlation procedures in the inverse program is proof enough against such mistakes.

The important limitation of the given method is the plain-layered approach of the crust structure. In the regions of an ocean-continent transfer zone, a crust thickness varies very strongly. For example we could not calibrate Green's function for paths Japan Sea-MAJO station (Honshu) because the crustal thickness changes very quickly here from 10 to 40km (Tuezov and Jiltsov, 1972; Tuezov, 1975). This is disappointing because, if a ray path was under Japan or the Sakhalin islands, we could calculate Green's functions from stations located in or near nodal plane. If a ray path passes a cross tectonic-geomorphologic structures of the Far East Asia region we are not able to fit an adequate plain-layered crust model.

Nevertheless, in many cases, even in such a complex region, the discussed method is a very useful new tool for earthquake research. This method is very important for focal mechanism determination of small earthquakes ($M \leq 5.0$), using three component single-station broadband records. The possibilities of its applications here will grow as more broadband digital stations are installed in the Far East Asia region.

Acknowledgement

This work was supported by the Korean Ministry of Science and Technology (1-3-090).

References

Arephyev, S. K., Pletnev, R., Tatevosyan, P., Alexin, B., Borisov, S., Lukyanenko, I., Matveev, S., Molotkov, E., Rogojin, J., Aptekman, B., Osher, A., Petrosyan, O., Erteleva, A., Kojurin, A., Ivaschenko, D., Kuznetsov, S. R., Se, M., Streltsov, K. C., Un, M., Kasahara, A., Katsumata, K., 1995. Preliminary results of epicenter observations of May 27 (28) 1995 Neftegorsk earthquake. Federal System of Seismological Observations and Earthquake Prediction. Information-Analytical Bulletin. MES of Russia, Moscow. in Russian.

Baag, C. E., Chi, H. C., Kang, I. B., Shin, J. S., 1997. Source parameters of December 13 1996 Youngwol earthquake. Study of Youngwol Earthquake. Open Report of the Association of Korean Earthquake Engineering Society, Seoul. (in Korean).

Bath, M., 1974. *Spectral Analysis in Geophysics*. Elsevier Science, Amsterdam.

Dreger, D. S., Helmberger, D. V., 1993. Determination of source parameters at regional distances with three-component sparse network data. *J. Geophys. Res.* **98**, 8107–8125.

Dreger, D. S., Langston, C., 1995. Distributed by Incorporated Research Moment Tensor Inversion Workshop (an IRIS DMS short Course) December 15–16 1995. Institutions for Seismology. Data Management Center, Seattle, Washington.

Fan, G., Wallace, T. C., 1991. The determination of source parameters for small earthquakes from a single very broadband seismic station. *Geophys. Res. Lett.* **18**, 1385–1388.

Gnibidenko, G., 1976. Earth Crust and Upper Mantle Structure in the Asia-Pacific Ocean Transition Zone. Novosibirsk, *Nauka*. in Russian.

Helmberger, D. V., Engen, G. Q., 1980. Modeling the long-period body waves from shallow earthquakes at regional distances. *Bull. Seismol. Soc. Am* **70**, 1699–1714.

Ide, S., Takeo, M., Yoshida, Y., 1996. Source process of the 1995 Kobe earthquake: determination of spatio-temporal slip distribution modeling. *Bull. Seismol. Soc. Am.* **86**, 547–566.

Ivashcenko, A. I., Kuznetsov, D. P., Un, K. C., Oskorbin, L. S., Poplavskaya, L. N., Poplavskiy, A. A., Burymskaya, R. N., Mihaylova, T. G., Streltsov, M. I., Sholohova, A. A., Davydova, N. A., Nagornyh, T. V., Ovchinnikov, V. V., Sadchikova, A. A., Se, S. R., Charlamov, A. A., Shalgin, S. V., Fokina, T. A., Ephimov, S. A., Chritova, L. I., Koykova, L. F., Levin, Y. N., 1995. Neftergorsk earthquake in the Sakhalin. Federal System of Seismological Observations and Earthquake Prediction. *Information-Analytical Bulletin*. MES of Russia, Moscow. in Russian.

Katao, H., Maeda, N., Hiramatsu, Y., Iio, Y., Nakao, S., 1997. Detailed mapping of focal mechanisms in/around the 1995 Hyogo-ken Nanbu earthquake rupture zone. *J. Phys. Earth* **45**, 105–119.

Kikuchi, M., Kanamori, H., 1996. Rupture process of the Kobe, Japan, earthquake of Jan 17, 1995, determined from teleseismic body waves. *J. Phys. Earth* **44**, 429–436.

Kim, S. G., 1994. The applicability of seismic waves to detect a low velocity body of the geothermal area. *Kor. J. Eng. Geol* **4**(3), 333-341.

Kim, S. G., Kraeva, N., 1999. Source parameter determination of local earthquakes in Korea using moment tensor inversion of single station data. *Bull. Seismol. Soc. Am.* **89**(4), 1077–1083.

Kim, S. G., Lee, S. K., 1994. Crustal modeling for Southern Parts of the Korean Peninsula using observational seismic data and ray method. *J. Korea Inst. Mineral and Energy Resources Engineers* **31**, 549–558.

Kim, S. G., Lee, S. K., 1996. Seismic velocity structure in the Central Korean Peninsula using the artificial explosions. *BSA FE* 2(1), 4–17.

Kim, S. G., Li, Q., 1998. 3-D crustal velocity tomography in the southern part of the Korean Peninsula. *Kor. Econ. Environ. Geol.* **31**(2), 127–139.

Lay, T., Wallace, T., 1995. *Modern Global Seismology*. International Geophysics Series Vol. **58**. Academic Press, San Diego.

Pasyanos, M. E., Dreger, D. S., Romanovwicz, B., 1996. Forward real-time estimation of regional moment tensors. *Bull. Seismol. Soc. Am* **86**, 1255–1269.

Salganik, M., Eksron, G., Sianissian, S., 1995. Quick CMT Determination, May 27, 1995, Sakhalin Islands. *Harvard Event-File*, C052795X.

Shimamoto, T., Watanabe, M., Sudzuki, Y., 1995. Neftergorsk earthquake. Federal System of Seismological Observations and Earthquake Prediction. *Information-Analytical Bulletin. MES of Russia*, Moscow. in Russian.

Tuezov, I., 1975. Litosphere of the Asia-Pacific Ocean Transition Zone. Nauka, Novosibirsk. in Russian.

Tuezov, I., Jiltsov, E., 1972. Deep structure of Japan from seismic data. Methods and Results Researches of the Earth's Crust and Upper Mantle, N 8. From: Results of Researches on the International Geophysical Projects. *Nauka*, Moscow. in Russian.

Walter, W. R., 1993. Source parameters of the June 29, 1992 Little Skull Mountain earthquake from complete regional waveforms at a single station. *Geophys. Res. Lett.* **20**, 403–406.

Zverev, S., Tulina, Y., 1971. Deep Seismic Sounding of the Crust of Sakhalin-Hokkaido-Primorie Zone. *Nauka*, Moscow. in Russian.

Source process of the 4 June 2000 southern Sumatra, Indonesia, earthquake

Zhou Y. H., Xu L. S. and Chen Y. T.

Institute of Geophysics China, Seismological Bureau

Abstract Source parameters of the 4 June 2000 southern Sumatra, Indonesia, earthquake (M_S 8.0) were estimated from teleseismic body waves recorded by long-period seismograph stations of the global seismic network. The analysis shows that the earthquake had a mechanism with a strike of 199°, a dip of 82°, and a rake of 5° and ruptured unilaterally from the northeast to the southwest, nearly perpendicular to the Java Trench. The focal mechanism of mainly left-lateral strike slip, with a small thrust component, explains well the reason why no tidal waves were generated by this great earthquake. The double-couple component of this earthquake was 1.5×10^{21} Nm, the compensated linear-vector-dipole component was 1.2×10^{20} Nm, and the explosion component was 5.9×10^{19} Nm. The analysis indicates that the source duration was about 16 sec. The fault area is about 95km in length and 60km in width. The average static slip is estimated to be about 11 m, and the average static stress drop, 90 MPa.

1 Introduction

The 4 June 2000 southern Sumatra, Indonesia, earthquake (16h28m26.2sec coordinated universal time; 4.72°S, 102.09°E, 33km, M_S 8.0 [IRIS]) occurred under the Indian Ocean, near the Mentawai Fault, along the well-known Sumatran subduction zone and the Great Sumatran Fault, all of which trend NW-SE. One thousand eight hundred houses were totally destroyed, 10,196 were heavily damaged, and 18,378 were slightly damaged by the earthquake. At least 97 people were killed, 1,900 were injured, and 122,000 were left homeless. Extensive damage and landslides occurred in the Bengkulu area (intensity V – VI). Injures and damage were also caused on Enggano. This earthquake was felt at Lampung and Palembang (intensity IV) and at Jakarta (intensity II – III). It was also felt in much of southern Sumatra, Indonesia, and Singapore (intensity III – V), and at Johor Bahru, Kuala Lumpur, and Petaling Jaya, Malaysia, 875km north of the epicenter (IRIS 2000_QED_ daily; Pan et al., 2001).

Historically, both the Sumatran subduction zone and the Great Sumatran Fault are seismically very active (Zachariasen et al., 1999). In the past two centuries, numerous large and small earthquakes have occurred on both structures (Katili and Hehuwat, 1967; Fitch, 1972; Newcomb and McCann, 1987; Fauzi et al., 1996). Among these earthquakes, the two greatest occurred in 1833 and 1861. The 1833 event released a moment of $12-21 \times 10^{21}$ Nm (M_W 8.7-8.8), with a fault area of 500km in length and 100km in width and an average slip of 4.3-7.5m (Newcomb and McCann, 1987). The 1861 event released a moment of $4.1-7.5 \times 10^{21}$ Nm (M_W 8.3-8.5), with a fault area of 300km in length and 100km in width and an average slip of 2.8-5.0 m (Newcomb and

---

* 本文发表于 Bull. Seism. Soc. Amer., 2002 年, 第 **92** 卷, 第 5 期, 2027-2035.

McCann, 1987). Both the 1833 and 1861 earthquakes caused severe damage and generated huge tidal waves that extended over 500km along the coast of Sumatra (Newcomb and McCann, 1987).

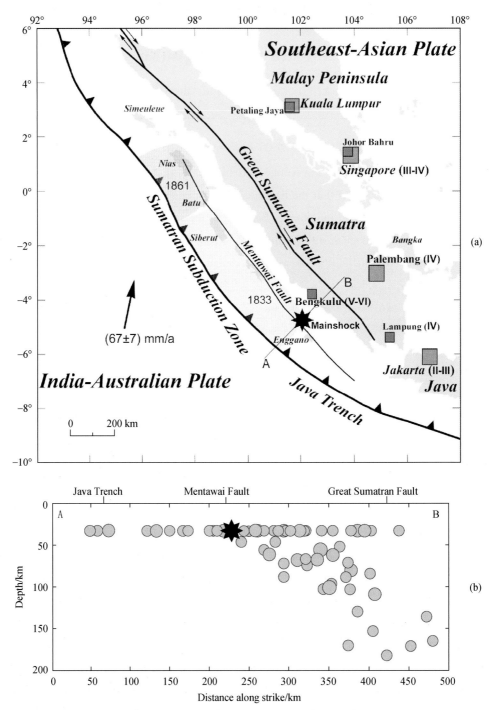

Figure 1 (a) Epicenter of the 4 June 2000 southern Sumatra great earthquake (the black star), its intensities for some cities in the nearby region, rupture zones of the 1833 and 1861 great earthquakes (the darkened zones), and tectonic setting of Sumatra. (b) Projection, in vertical cross section along line A-B in (a), of earthquakes $M_B \geqslant 4.0$ that occurred in the 0°–10° S, 96°–107° E region within 1 year before the 2000 great earthquake, which shows clearly that along the Sumatran subduction zone, the India-Australian plate thrusts below the Southeast-Asian plate. The smaller gray circles represent earthquakes of $4.0 \leqslant M < 5.0$; the larger gray circles, $5.0 \leqslant M < 6.0$; the still-larger gray circles, $6.0 \leqslant M < 7.0$

Figure 1a shows the epicenter of the 4 June 2000 great earthquake (the black star), the intensities for some cities in the nearby region, the rupture zones of the 1833 and 1861 great earthquakes (the darkened zones), and the tectonic setting of Sumatra. Figure 1b shows the projection, in vertical cross section along line A-B striking 45° in Figure 1a, of earthquakes (M_B 4.0) that occurred in the 0°–10°S, 96°–107°E region within 1 year before the 4 June 2000 great earthquake. The earthquake distribution shows clearly that, along the Sumatran subduction zone, the India-Australian plate thrusts below the Southeast-Asian plate. Earlier research showed that the relative convergence rate between the India-Australian plate and the Southeast-Asian plate, measured between west Java and Christmas Island, is 67 ± 7mm/yr in a direction N11° ± 4°E, nearly orthogonal to the trench axis near Java but highly oblique near Sumatra (Tregoning et al., 1994). The 4 June 2000 earthquake is the strongest one since 1861. It is worth noting that unlike the last two great earthquakes, this earthquake had not generated tidal waves (Pan et al., 2001), and that big differences exist between moment-tensor solutions of the earthquake reported by three institutions (U.S. Geological Survey [USGS], Harvard CMT [CMT], and ERI [ERI] at the University of Tokyo). Much research needs to be done to understand the source mechanism.

In this short note, we report on a study of source properties of the 4 June 2000 earthquake. Source parameters and general features of the source-rupture process are derived from teleseismic body waves recorded by long-period seismographs of the IRIS global seismic network.

2 Data and methods

To estimate the source parameters and the source-rupture process, we used the body-wave-inversion technique developed by Xu and Chen (Xu and Chen, 1997, 1999; Chen and Xu, 2000). Waveform data from 28 long-period seismic stations of the IRIS network in the epicentral distance ranging from 30° to 95° were used. In this range, the simplicity of direct P and S waves makes them useful for retrieving the source information. Generally speaking, outside this distance range, the direct phases become complicated owing to the complex structure of the mantle and the Earth's core, in addition to the interactions with the crust-mantle boundary, the discontinuities in the upper mantle, and the core-mantle boundary (Langston and Helmberger, 1975). Only two additional stations, at epicentral distances of 29° and 23°, respectively, with simple, direct P and S phases, were included in the moment-tensor inversion, the rupture-direction, and the slip-distribution analyses. In computing Green's functions, we used a flattened IASPEI 91 model (Kennett and Engdahl, 1991), partitioning 183 layers in the Earth's crust and the mantle by linear interpolation. All observed data and Green's functions were restituted to displacement and bandpass-filtered with an eighth-order Butterworth filter, with a bandwidth of 0.005–0.05 Hz for the moment-tensor inversion and a bandwidth of 0.005–0.1 Hz for the slip-distribution analysis. In these frequency ranges, the amplitude-frequency response curves of all velocity-recording instruments are flat. The lower-frequency threshold of 0.005 Hz is necessary to filter out the long-period Earth vibration and the long-period excursion caused by integrating velocities to obtain displacements. Teleseismic waves in the frequency range higher than 0.1 Hz are sensitive to the Earth media; therefore, the source information in this frequency range was "contaminated" and was filtered out.

3 Moment-tensor solution and rupture-propagation direction

To perform the moment-tensor inversion, a total of 55 P phases of vertical (z) and radial (r) components and 20 SH phases of the transverse (t) component from 30 teleseismic stations were used. In order to find the stabilized moment-tensor solution by a trial-and-error method, we applied a constraint of specifying the source-time function to be a half-period cosine function, with a duration of 16 sec in the inversion. The estimated double-couple component of the moment-tensor solution of this earthquake was 1.5×10^{21} Nm (USGS, 4.0×10^{20} Nm; HRV, 7.5×10^{20} Nm). The compensated linear-vector-dipole component was 1.2×10^{20} Nm, 8% of the double-couple component. The explosion component was 5.9×10^{19} Nm, <4% of the double-couple component (Table 1).

Table 1 Moment-Tensor Solution of the 4 June 2000 Southern Sumatra Earthquake

$M_{ij}/10^{21}$ Nm	
M_{11}	-1.1
M_{12}	1.2
M_{13}	0.15
M_{22}	0.74
M_{23}	0.15
M_{33}	0.20
M_{dc}	1.5
M_{clvd}	0.12
M_{ep}	-0.059

Principal Axes		
	Azimuth	Plunge
P	154°	2°
T	64°	10°
B	256°	80°

Best Double Couple			
	Strike	Dip	Rake
Nodal plane I	199°	82°	5°
Nodal plane II	109°	85°	172°

dc, double couple; clvd, compensated linear-vector dipole; ep, explosion (component).

Figure 2 shows a comparison of observed and synthetic seismograms in the moment-tensor inversion. The synthetics fit the data very well. Of the 75 phases used, >97% of the correlation coefficients between the synthetics and the data were >0.6, 92% of them were >0.7, 84% of them were >0.8, and >45% of them were >0.9. Figure 3 shows a comparison of the derived best-double-couple solution in this research with those given by the USGS, Harvard CMT, and the ERI at the University of Tokyo. The solution given by ERI at the University of Tokyo is obviously contradictory to the compression-dilatation distribution of the P-wave first motion. While the solutions given by the USGS and Harvard CMT are similar, they still cannot explain the clearly dilatational polarities of the P-wave first motion at LSA, CHTO, and VNDA stations, which seem to be away from, rather than near, the nodal planes in the dilatational quadrants, nor can they explain the polarities of the P-wave first motion at SNZO, TRTE, and GRFO stations, which are uncertain because of the small amplitudes of P-wave onsets that could be attributed to nodal-plane behavior. Therefore, the mainly strike-slip mechanism solution obtained in this research (see bottom of Table 1) is considered more reasonable.

Figure 4 shows the moment-tensor solution and z component of P waveforms of the 30 stations used in this study. The moment-tensor solution is in good agreement with the compression-dilatation distribution of the P-wave first motion. A prominent feature in Figure 4 is that the P waveforms show directivity. The P waves recorded at northeastern stations are of longer period and decrease slowly, while the P waves recorded at southwestern stations are of shorter period and decrease rapidly. This directivity effect provides strong support for the conclusions that the nodal plane of strike 199°, rather than the nodal plane of strike 109°, is the fault plane and that the earthquake ruptured unilaterally from the northeast to the southwest.

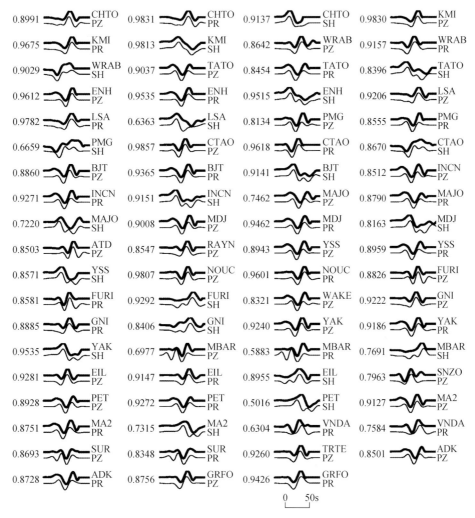

Figure 2 Comparison of observed and synthetic seismograms in the moment-tensor inversion

Top trace of each seismogram pair is the observed, and bottom trace is the synthetic. Number on the left shows correlation coefficient of each seismogram pair. Station code is written on the right side, and phase name is below the station code. PZ means z component of P phase, PR means radial component of P phase, and SH means SH phase

Based on IRIS EHDF_ monthly, Figure 5 shows the epicentral distribution of the 4 June 2000 southern Sumatra earthquake sequence. It is worth notice that the 1-day aftershock zone is SE trending, nearly perpendicular to the fault plane inferred from P-wave directivity, and that the 1-month aftershock zone appears quite complex, with both SE-trending and SSW-trending after shock distributions. Considering the P-wave directivity effect and the tectonic setting of this region, we infer that the aftershocks of the 4 June 2000 southern Sumatra earthquake occurred on or near the SSW-striking fault plane of the mainshock and on or near the pre-existing SE-trending Mentawai Fault.

4 Slip and stress-drop distributions

To study the slip distribution on the fault plane, we used the inversion technique of Xu and Chen (Xu and Chen, 1997, 1999; Chen and Xu, 2000). At first, similar to the empirical Green's function method (Hartzell, 1978; Mueller, 1985), we calculated synthetic seismograms of a small earthquake with the same hypocentral location and focal mechanism as the earthquake under study and retrieved the source-time functions of this earthquake "observed" at individual stations by deconvolution. By comparing the retrieved source-time functions to identify the common features due to the source signal and the individual features due to an abnormal medium or other reasons, we selected a total of 42 source-time functions, 23 from the z component of P phases and 19 from SH phases, for the inversion. Figure 6 shows the directivity exhibited by the retrieved source-time functions. Both of the source-time functions retrieved from P phases (Fig. 6a) and those from SH phases (Fig. 6b) show directivity in accordance with the P-waveform directivity shown in Figure 4. The directivity of the source-time functions strengthens the conclusion that the nodal plane of strike 199° is the fault plane and that the earthquake ruptured unilaterally from the northeast to the southwest. On the basis of this result, we selected a rectangular plane, 165km in length and 60km in width, with a strike of 199° and a dip of 82° as the fault plane to invert for the temporal-spatial rupture process of the earthquake. We divided the fault plane into 11×4 subfaults, each of which was a square of 15×15km. The 15km was selected by the resolving power of this study: because the seismic waves were bandpass-filtered with a passband of 0.005–0.1 Hz, if the slowest P-wave velocity is assumed to be about 6km/sec, then the shortest P wavelength λ would be about 60km. Thus, the resolving power turns out to be $\lambda/4$, or 15km.

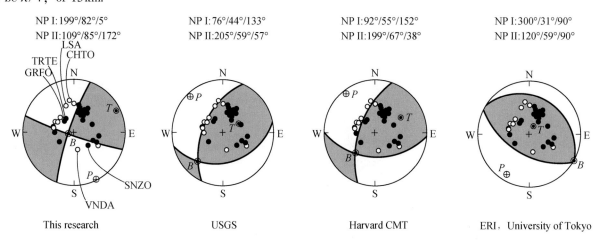

Figure 3　Comparison of best-double-couple solutions given by USGS, Harvard CMT, ERI at the University of Tokyo, and this research

Polarities of the P-wave first motion recorded at 30 stations used in this research are ploted in all of the equal-area projections of the geometrical representation of the solutions. The closed and open circles represent compressional and dilatational motions, respectively

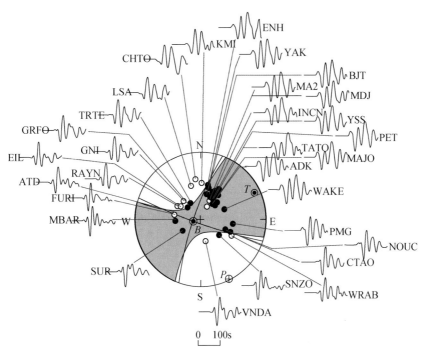

Figure 4 Moment-tensor solution and displacement P waveforms (z component) observed at 30 stations used in this study

The P waveforms show directivity, which indicates that the source ruptured from the northeast to the southwest. The amplitudes of the P waveforms are normalized

Figures 7 and 8 show "snapshots" of the slip-rate distribution and the static (final) slip distribution of the earthquake, respectively. It can be seen that the main rupture of the earthquake initiated at the hypocenter of the main shock, which was located in the central part of the fault plane, propagated unilaterally to the southwest, and terminated at the southwestern end of the fault. The rupture lasted for about 16 sec. This result is in good agreement with the source duration applied in the moment-tensor inversion. The ruptured area is about 95km in length and 60km in width. The maximum static (final) slip is 27 m, and the average static slip is about 11 m.

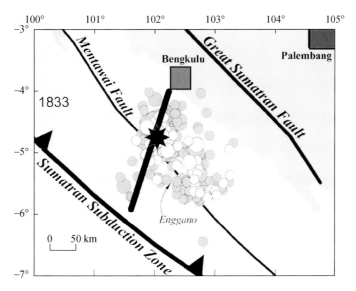

Figure 5 Epicentral distribution of the southern Sumatra earthquake sequence

The black star represents the mainshock. The aftershocks of $M_B \geq 4.0$ that occurred within 1 day after the mainshock are represented by the open circles. The aftershocks of $M_B \geq 4.0$ that occurred in the period of 16:30, 5 June, 1 day after the mainshock, to 16:30, 4 July, 1 month after the mainshock, are represented by the gray circles. The thick black bar shows the strike of the fault plane of the mainshock

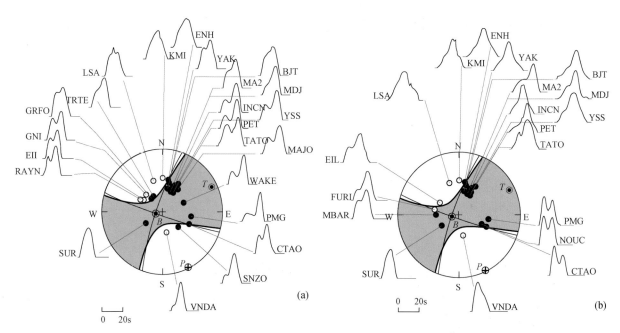

Figure 6 (a) Directivity of source-time functions retrieved from z component of P phases. (b) Directivity of source-time functions retrieved from SH phases. The directivities exhibited by the source-time functions are coincident with the directivity presented by of observed P-waves (Figure 4)

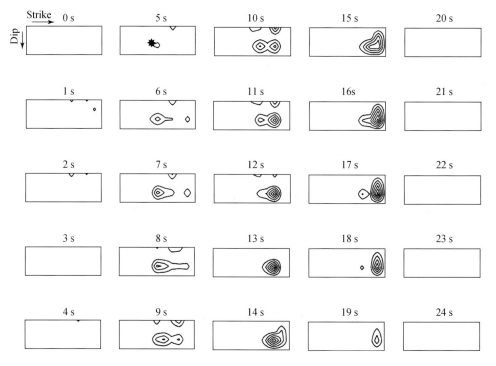

Figure 7. "Snapshots" of the rupture process of the mainshock on the fault plane, 199° in strike direction and dipping 82°

The contours are the slip-rate distribution. The time interval between two neighboring shots is 1 sec. Each shot covers a fault plane 165 km long in the strike direction and 60 km wide in the dip direction. The snapshots show that the significant rupture initiated at the central part of the fault plane at the fifth second, propagated unilaterally to the southwest, and terminated at the southwestern side of the fault plane. The rupture lasted for 20 sec, the main energy release occurring during the last 16 sec. The star represents the location of the hypocenter of the mainshock

Assuming each subfault to be an equivalent circular fault, we estimated the static stress drop of the earthquake. The highest static stress drop was 220 MPa, and the average static stress drop was 90 MPa.

5 Discussion and conclusions

For most undersea large earthquakes in subduction zones, their sources are generally thrust slip or dip slip, and they generate tidal waves. However, in this study we found that the 4 June 2000 southern Sumatra $M_S 8.0$ earthquake was mainly due to strike-slip faulting; this explains why no tidal waves were generated by this event. In many studies, the lineation of aftershock epicenters is used to determine the fault plane of the mainshock. In this study, we found that the application of this rule was not straightforward. The aftershocks of the 2000 southern Sumatra $M_S 8.0$ earthquake occurred on both the fault plane of the main shock and its conjugate plane. Another unusual feature of the 2000 southern Sumatra $M_S 8.0$ earthquake was that its fault plane was NNE-SSW trending, nearly perpendicular to the strikes of the fault plane of the 1833 and 1861 earthquakes and the strikes of the Great Sumatran Fault, the Mentawai Fault, and the Java Trench, all of which are NW-SE trending. To explain the features of the focal mechanism and aftershock distribution of this earthquake, we propose a model as shown in Figure 9. In this model, we assumed that before the occurrence of the 4 June 2000 earthquake, segments A and B were one unbroken segment. On a long time scale, the subducting plate, the nonseparated A and B, moves slowly downward to the overriding plate. Subsequently, the overriding plate encounters barriers on the subducting plate, on side A. The barriers resist the motion of the subducting plate, but, drawn by the gravitational force, the front part of the subducting slab keeps moving slowly downward. Consequently, this motion causes shear-stress accumulation in the neighborhood of the octagonal area between side A and side B near the subduction interface. Once the accumulated shear stress increases to the strength of the subducting plate, left-lateral strike faulting will occur abruptly, with the strike of the rupture plane nearly perpendicular to the subduction interface. The slip of segment B causes stress redistribution in the area near the rupture plane of the main shock and the pre-existing Mentawai Fault. The elevated stress then triggers aftershocks, which occur either on or near the fault plane of the main shock or on or near the Mentawai Fault. To answer the question of unilateral-rupture mode of the mainshock, we assumed that barriers existed in the locked segment of the fault plane of the mainshock. This is a very preliminary model, and more research needs to be done.

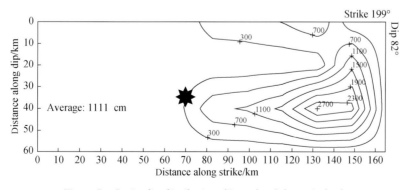

Figure 8   Static slip distribution (in cm) of the mainshock

The ruptured area is about 95km in length and 60km in width, with an average static slip of about 11 m and maximum static slip of 27 m. The star represents the location of the hypocenter of the mainshock

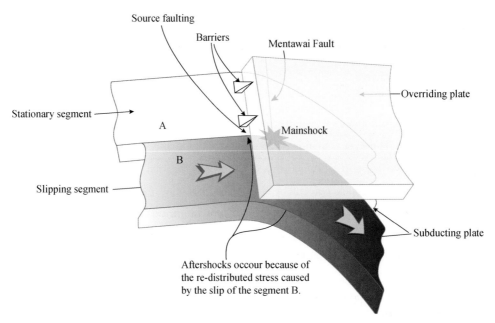

Figure 9　A source model proposed for the 4 June 2000 southern Sumatra, Indonesia, earthquake

In conclusion, this study demonstrates that the 4 June 2000 southern Sumatra, Indonesia, earthquake (M_S 8.0) had a focal mechanism with a strike of 199°, a dip of 82°, and a rake of 5° and ruptured unilaterally from northeast to southwest, nearly perpendicular to the Java Trench. The focal mechanism of mainly left-lateral strike slip, with a small thrust component, explains well the reason why no tidal waves were generated by this great earthquake. The double-couple component of this earthquake was 1.5×10^{21} Nm. The strike-slip faulting was caused by an asymmetrical distribution of barriers on the subducting plate. The NNE-SSW-distributed aftershocks occurred on or near the fault plane of the main shock, while the NW-SE-distributed aftershocks occurred on or near the Mentawai Fault.

Acknowledgments

We would like to express our gratitude to Dr. S. Y. Li for stimulating discussions and to Dr. D. I. Doser and Dr. R. McCaffrey for very helpful comments and suggestions. This work was supported by the National Natural Science Foundation of China under Grant No. 49904004 and the SSTCC Climb Project 95-S-05, publication No. 01AE1001, Institute of Geophysics, China Seismological Bureau.

References

Chen, Y. T., and L. S. Xu (2000). A time domain inversion technique for the tempo-spatial distribution of slip on a finite fault plane with applications to recent large earthquakes in Tibetan Plateau, *Geophys. J. Int.* **143**, 407–416.

Fauzi, R., McCaffrey, D., Wark, Sunaryo, and P. Y. P. Haryadi (1996). Lateral variation in slab orientation beneath Toba Caldera, northern Sumatra, *Geophys. Res. Lett.* **23**, 443–446.

Fitch, T. J. (1972). Plate convergence, transcurrent faults, and internal deformation adjacent to Southeast Asia and the Western Pacific, *J. Geophys. Res.* **77**, 4432–4460.

Hartzell, S. (1978). Earthquake aftershocks as Green's functions, *Geophys. Res. Lett.* **5**, 1–4.

Katili, J. A., and F. Hehuwat (1967). On the occurrence of large transcurrent earthquakes in Sumatra, Indonesia, Osaka Univ. *J. Geosci.* **10**, 5–7.

Kennett, B. L. N., and E. R. Engdahl (1991). Travel times for global earthquake location and phase identification, *Geophys. J. Int.* **105**, 429–465.

Langston, C. A., and D. V. Helmberger (1975). A procedure for modeling shallow dislocation sources, *Geophys. J. R. Astr. Soc.* **42**, 117–130.

Mueller, C. S. (1985). Source pulse enhancement by deconvolution of an empirical Green's function, *Geophys. Res. Lett.* **12**, 33–36.

Newcomb, K. R., and W. R. McCann (1987). Seismic history and seismotectonics of the Suda Arc, *J. Geophys. Res.* **92**, 421–439.

Pan, T. C., K. Megawati, J. M. W. Brownjohn, and C. L. Lee (2001). The Bengkulu, southern Sumatra, earthquake of 4 June 2000 ($M_W = 7.7$): another warning to remote metropolitan areas, *Seism. Res. Lett.* **72**, 171–185.

Tregoning, P., F. K. Brunner, Y. Bock, S. S. O. Puntodewo, R. McCaffrey, J. F. Genrich, E. Calais, J. Rais, and C. Subarya (1994). First geodetic measurement of convergence across the Java Trench, *Geophys. Res. Lett.* **21**, 2135–2138.

Xu, L. S., and Y. T. Chen (1997). Source parameters of the Gonghe, Qinghai Province, China, earthquake from inversion of digital broadband waveform data, *Acta Seism. Sin.* **10**, 143–159.

Xu, L. S., and Y. T. Chen (1999). Tempo-spatial rupture process of the 1997 Mani, Xizang (Tibet), China earthquake of $M_S = 7.9$, *Acta Seism. Sin.* **12**, 495–506.

Zachariasen, J., K. Sieh, F. W. Taylor, R. L. Edwards, and W. S. Hantoro (1999). Submergence and uplift associated with the giant 1833 Sumatran subduction earthquake: evidence from coral microatolls, *J. Geophys. Res.* **104**, 895–919.

Temporal-spatial rupture process of the 1999 Chi-Chi earthquake from IRIS and GEOSCOPE long-period waveform data using aftershocks as empirical Green's functions[*]

Xu L. S.[1], Chen Y. T.[1], Teng T. L.[2] and Patau G.[3]

1. Institute of Geophysics, China Seismological Bureau; 2. Department of Earth Science, University of Southern California; 3. Institut de Physiques du Globe de Paris

Abstract A large earthquake (M_W7.6) occurred near Chi-Chi, Taiwan, on 20 September 1999 (UTC) and was followed by many moderate-size aftershocks in the following days. The two largest aftershocks with magnitudes M_W6.1 and 6.2, respectively, were used as empirical Green's functions (EGFs) to retrieve the source time functions (STFs) and further image the temporal-spatial rupture process of the mainshock. For each station, two types of STFs were retrieved, one from P phases and another from S phases. A total of 178 STF individuals were retrieved for source-process analysis of the event. From the STFs retrieved, firstly, similarities appeared on the STFs from most stations except that several STFs in special azimuths looked different or odd because of the focal mechanism difference between the mainshock and the EGF aftershocks; and secondly, systematic shape-variation with azimuth appeared. The analysis of the STFs indicated that this event consisted of two subevents; on the average, the second event was about 7 sec later than the first one, and the first event was about 15% larger than the second one in the moment-release rate. The total duration time of earthquake rupture process was about 26 sec. From the image of the static slip distribution, there were two slip-concentrated areas on the fault plane, and their centers were about 45km away from each other. The maximum slip of about 6.5 m appeared on the northern one. The rupture area, with the slip greater than 0.5 m, was about 80km long and 60km wide. The maximum stress drop was about 25 MPa (250 bar), and the average stress drop on the entire fault was 9.2 MPa (92 bar). From the snapshots of the temporal-spatial variation of slip and slip rate, the rupture initiated at the southern end and stopped at the northern end of the fault, which suggested an overall unilateral rupture at a rupture velocity of about 2.5km/sec. The depth of the initiation locus was about 20km. The rupture duration times on the subfaults were estimated. A good correlation was noticed between the rupture duration time and slip amplitude of the subfaults: the larger the slip amplitude, the longer the rupture duration time of the subfault. The maximum duration time for the subfaults was 16 sec, which was about two-thirds the total duration time of the earthquake rupture process. In the beginning and near the end of the source process, the rupture propagated like a self-healing pulse of slip.

1 Introduction

On 20 September 1999 (UTC), a great earthquake occurred near the town of Chi-Chi in Nantou County,

---

[*] 本文发表于 *Bull. Seism. Soc. Amer.*, 2002 年, 第 **92** 卷, 第 8 期, 3210–3228.

Taiwan. The origin time was determined to be 17h47m15sec on 20 September (UTC) or 1h47m15sec on 21 September (local time) although different agencies offered slightly different results (Table 1). The epicenter was finally determined to be longitude 120.82° E and latitude 23.85° N (Chang et al., 2000a, b; Kao et al., 2000; Shin et al., 2000) though some difference existed in the quick results given by different agencies (Table 1). As usual, the focal depth is a difficult parameter to determine. In the quick results, it ranged from 1 to 33km (Table 1). However, several detailed studies strongly suggested the focal depth be from 7 to 10km (Chang et al., 2000a, b; Kao et al., 2000; Ma et al., 2000; Shin et al., 2000; Wang et al., 2000; Yeh, 2000).

Table 1 Hypocentral locations and origin time of the 1999 Chi-Chi earthquake

Agency	Epicentral Location		Depth/km	Origin Time (UTC) /hr: min: sec
	Latitude/°N	Longitude/°E		
CWB, RTD Network	23.87	120.75	10	17: 47: 15.89
CWB, S13 Network	23.87	120.81	7	17: 47: 15.85
CWB, Final Report	23.85	120.82	8	17.47.15.85
PDE, Daily Report	23.73	121.06	33	17.47: 18.40
PDE, Weekly Report	23.77	120.98	33	17.47: 18.40
NEIC Homepage	23.78	121.09	33	14.47: 19.01
Harvard CMT	24.15	120.80	21	17.47: 35.30

S13, Short-Period Seismographic Network; RTD, Taiwan Rapid Earthquake Information Release System; PDE, Preliminary Determination of Epicenter.

The focal mechanism of the Chi-Chi earthquake was determined by moment-tensor inversion and with first motions of P waves soon after the mainshock (Table 2). Subsequently detailed investigations were made, from which similar results were obtained (Chang et al., 2000; Shin et al., 2000). The hypocenters of the aftershocks were located (see IRIS Data Center's Website). Iwata et al. (2000) noticed that aftershocks distributed on a dipping plane of about 30° toward east in the southern part of the source region, namely, near the hypocenter. On the contrary, there was no significant dipping plane of the aftershocks in the northern part of the source region. From field investigation, it was found that the Chi-Chi earthquake was caused by an existing active fault, the Chelungpu fault. The fault plane extended from south to north with a strike of 5° and a dip of 30° and bent 20° eastward on the northern part (Wu et al., 2000). The dip angle of the fault also changed from 25° at southern and central part to 20° at the northern part (Yeh, 2000). Ma et al. (2000) observed that the surface rupture along the Chelungpu fault had a strike of about 7° to the east and a dip angle of about 25° to 30°. Lee et al. (2000) concluded by waveform inversion that the average rake angle in fault plane was about 60°.

The fault length of the earthquake was estimated by a few investigators. For example, Wang et al. (2000) proposed that the fault length was 90km, the southern 70km of the fault trended in the south-north direction with pure reverse thrusting at a low dip angle of 20° to 30°, and the northern 20km made a 70° turn toward the east but was also dominated by the reverse displacement on the fault surface. Wang (2000) estimated that the fault length was 105km and the fault trace bent toward the northeast at its northern tip, 40km away from the epicenter of the mainshock. Anyhow, it is reasonable to constrain the fault length in 80–105km (Chang and Lee, 2000; Lee et al., 2000; Ma et al., 2000; Ota et al., 2000; Wang, 2000; Wang et al., 2000; Yeh, 2000; Yuan et al., 2000).

Table 2 The quick source parameters of the 1999 Chi-Chi earthquake

Nodal Plane I			Nodal Plane II			M_0 / (10^{20} Nm)	M_W	Epicentral Location		Depth /km	Duration /sec	Agency
Strike /(°)	Dip /(°)	Rake /(°)	Strike /(°)	Dip /(°)	Rake /(°)			Latitude /°N	Longitude /°E			
26	27	82	215	64	94	4.1	7.7	23.94	120.71	21	37.6	Harvard
357	29	67	202	63	102	2.4	7.5	23.78	121.09	5		USGS
44	38	114	194	55	72	2.5	7.5	23.72	121.12	24		Tokyo
17	28	87	201	62	92	3.5	7.6					Xu et al. (1999)
						5.7						Lee et al. (2000)
						3.5						Yagi et al. (2000) *
20	30	85										Shin (2000) †
5	34	65										Chang et al. (2000)

* GPS.

† First P motion.

Taking into account the solutions of moment tensor and first motions of P waves, field investigations, and other studies, as mentioned previously, we conclude that the mainshock initiated at 120.82° E and 23.85° N and then propagated northward, the focal depth was about 8km, and the fault length was about 90km, and was mainly a thrust movement (with rake of around 80°) on the fault plane striking about 5°, parallel to the Chelungpu fault, and dipping about 30°.

Despite the studies mentioned previously, many issues remain and no clear concensus has been reached regarding the source process of the mainshock. The estimate of total duration time of the mainshock rupture is important for a variety of reasons, particularly for earthquake engineering purpose. In Yagi and Kikuchi's determination (Yagi and Kikuchi, 1999), it was 28 sec. By imaging source process using only long-period data from China Digital Seismograph Network (CDSN), Xu et al. (1999) estimated the total rupture duration time was about 27 sec. Huang (2000) claimed that it was 38 sec, that within the first 10 sec, seismic energy radiation was concentrated at the southern part of the fault, and that the major rupture propagation from south to north began at the second 15, when major energy was radiated at the northern part of the fault. Yagi and Kikuchi (2000) pointed out that the total rupture duration time was about 35 sec; Wu et al. (2000) noted that a large seismic moment was released in about 40 sec. Shin (2000) examined the strong-motion data and concluded that three separate events were induced during the first 20 sec of the main wave train, among them, two in the southern part and one in the northern part. From totally 178 source time functions (STFs) retrieved from different phases of different IRIS and GEOSCOPE stations using two largest aftershocks as empirical Green's functions (EGFs), Xu et al. (2002) concluded that the rupture duration time was 26 sec.

There have been a number of studies in determining the source complexity of the Chi-Chi earthquake. A few days after the occurrence of the Chi-Chi earthquake, Yagi and Kikuchi (1999) quickly obtained the rupture process and static slip distribution of the mainshock and posted it on their Website. They concluded that the source process of the Chi-Chi earthquake was characterized mainly by a unilateral rupture propagation. The rupture propagated 45km to north, 25km to south, and 40km in dip direction, forming a 70km × 40km of rupture area. The maximum dislocation was about 6 m, and the averaged dislocation over the entire rupture area was about 2.5 m. The dislocation was nonuniformly distributed on the fault plane. The rupture area was isolated into two parts: one in the south of the epicenter and another in the north of the epicenter. The southern part was relatively weak, which, furthermore, had three areas of large slip located at different depths; the northern area of large slip, the main rupture area of this event, was much stronger than the southern one. The locus of the maximum slip was about 35km away from the epicenter and about 12km deep. Xu et al. (1999) obtained the preliminary image of source process using only long-period data from the CDSN. The image showed that the rupture was nonuniform on the fault plane. There were three slip-concentrated areas (nuclei) with the maximum slips greater than 4 m. The first nucleus was 15km south of the epicenter and about 28km deep, with maximum slip of about 4.5 m; the second nucleus was about 30km north of the epicenter and about 15km deep, with maximum slip of about 4.4 m; the third nucleus was about 30km north of the epicenter and close to the ground surface, with maximum slip of about 6 m. The total rupture area was around 100km long from south to north and 75km wide along the down-dip direction. Chen and Zeng (personal comm., 1999) inverted the source process of the mainshock. Their result indicated that the rupture area consisted of two subareas with large slips, and these two areas had different focal mechanisms. The first was south of the epicenter, with a left-lateral strike-slip mechanism, and the second was north of the epicenter, with a thrusting mechanism. The slips of the two subareas were all greater than 10 m. Subsequently more detailed studies were made. Ma et al. (2000) used 13 telesesimic P-wave waveforms of the M_W6.4, 26 September 1999, aftershock as the EGF and the Haskell model (100km long and 20km wide) as the fault model with an assumed rupture velocity of 2.5km/sec to analyze slip along the seismogenic fault. They concluded that in the first 25km, the slip was quite small, then the significant slip of 3–7 m occurred near 25km and 50km north of the mainshock. Lee et al. (2000) used 22 broadband teleseismic records from the stations within epicentral distances of 40° and 100° and the generalized ray method in calculation of synthetic Green's function to determine its temporal and spatial slip distribution. In their inversions, the STF for each subfault was fixed to be a triangle function of the total duration of 2 sec, with a rising time of 1 sec and a falling time of 1 sec. The rupture velocity was assumed to be 2.5km/sec, which is about 80% of the shear-wave velocity. Their results showed an anomalous large slip region centered about 40–50km north of the hypocenter at a shallow depth. The largest slip was about 10 m. The slip in the vicinity of the hypocenter was relatively smaller. In the largest slip area, the slip was dominated by thrust dip slip. Some strike-slip component was found in the middle segment of the fault. Ma et al. (2000) and Lee et al. (2000) used the multiple-time-window method (Wald and Heaton, 1994) to 22 teleseismic P-wave displacement waveforms from IRIS between 35° and 90° of epicentral distances to obtain temporal-spatial distribution of slip on the fault plane. In the inversion a fault geometry with length of 110km and width of 40km was considered, and it was divided into 33 elements along the strike direction and 10 elements along the dip direction. The generalized ray method was employed in Green's function calculation, and a velocity of 2.5km/sec was assumed for the initial rupture velocity. The spatial distribution of the slip revealed a significant asperity with a dimension of about 45km in length and about 15km in width in the region of about 15 to 60km to the north and above the hypocenter of the earthquake. The maximum slip of about 8 m was located at about 25km north of the epicenter. In addition, Yagi and Kikuchi (2000) proposed that toughly three asperities or areas of large slip could be identified: the first was located about 10km west of the epicenter; the second about 20km north, and the third about 40km northwest. Chung and Shin (1999) and Ma et al. (2000) noted that a larger amount of slip occurred at 10–20km north of the fault and the largest surface break occurred at about 45–50km north of the epicenter.

Various estimates about the maximum slip on the fault plane have existed. Ho et al. (2000) analyzed the displacements on the seismogenic fault by dividing the fault into five segments along the strike direction and concluded that on each segment the displacements were 5 to 12 m and 1 to 9 m on dip and strike directions, respectively. Yuan et al. (2000) suggested that the vertical displacement was 11 m, the horizontal displacement was 7.9 m, and the resultant slip was 13.5 m. Yu et al. (2000) suggested that at the northern part of the fault, the slip was 13 m on dip direction and 4 m on strike direction. Yoshioka (2000) claimed that the average slip was 2.1 m and the maximum slip was 3.0 m. Azuma et al. (2000) showed that the amount of the total slip ranged from 3 to 5 m except in the northernmost part, where the largest slip of about 16 m occurred. Chang and Lee (2000) noted that the northern segment struck east-northeast and had a slip direction of north-northwest with horizontal displacement of about 8 to 9 m and uplift of 5 to 15 m; the central segment struck north with horizontal displacement of 3 to 4 m and uplift of less than 6 m; the southern segment struck north with horizontal displacement of 3 to 4 m and uplift of less than 4 m. Wu et al. (2000) suggested that largest slip was as large as 10–20 m. Yagi and Kikuchi (2000) estimated that the maximum slip amounted to more than 11m.

Different studies have shown that overall, the mainshock ruptured unilaterally from south to north but the rupture propagation direction as well as the rupture velocity were variable (Ma et al., 2000). Wu et al. (2000) showed that at the beginning of the earthquake rupture there was some propagation southward, with average velocity less than 2km/sec, whereas the northward propagation was at a velocity of about 2.5km/sec. Chen and Zeng (personal comm., 1999) proposed that the rupture velocities varied from 2.3 to 2.7km/sec, with an average rupture velocity of 2.5km/sec. In contrast, Lee et al. (2000) proposed after multiple-time-window analysis that the rupture velocities were initially about 2.5km/sec near the hypocenter, similar to common observations of about 85% of the shear-wave velocity, but decreased substantially to only about 1.2km/sec in the largest slip region. For the entire process, the rupture velocity ranged from 1.2 to 4.0km/sec.

As mentioned previously, the overall characteristics associated with the Chi-Chi earthquake source, according to the different investigators, are similar, such as the seismogenic fault of near north-south to north-northeast strike and dip of about 30°, the overall thrusting mechanism, the overall characteristics of rupture propagating from south to north, and the total rupture area. However, differences in some aspects still exist, such as the temporal and spatial complexity of the rupture process, rupture velocity, and maximum slip. In this article we present our results regarding temporal-spatial complexity of the Chi-Chi earthquake rupture. They were obtained from long-period waveform data from IRIS and GEOSCOPE stations using two largest aftershocks as EGFs, instead of synthetic Green's function as used by Lee et al. (2000) and Ma et al. (2000), and the time domain inversion technique of Chen and Xu (2000).

2 Data and Data Preprocessing

For study of STFs using EGF technique, it is important to find a smaller earthquake that can be used as EGF. The EGF must have similar focal mechanism and hypocentral location to the mainshock and its source process must be simple enough compared with the source process of the mainshock so that its far-field STF can be treated as a Dirac δ function. Usually, the smaller the event, the simpler its source process. However, the signal of an excessively small event cannot go far; that is, some distant stations cannot receive its signal, which prohibits use of the EGF technique. Therefore, the size of the smaller event to be used as the EGF should not be too small in order to have as much data as possible. According to these principles, we selected two aftershocks from all the aftershocks until 31 December 1999, whose focal mechanisms have been determined by Harvard (Table 3). One is the $M_W 6.1$ aftershock (aftershock E4, the fourth event in Table 3, or aftershock 99-09-22a in Fig. 1) occurring at 00h14m39.20sec on 22 September 1999 (UTC), another is the $M_W 6.2$ aftershock (aftershock E9, the ninth event in Table 3, or aftershock 99-09-25b in Figure 1) occurring at 23h52m48.70 of 25 September 1999 (UTC). The focal mechanisms of the mainshock and these two aftershocks are similar to each other. The epicenters of these two aftershocks seem to be a slightly distant in view of centroid positions offered by Harvard, but the epicentral location (longitude 120.82° E, latitude 23.85° N) of the mainshock is quite close to them (120.84 E, 23.83° N for E4 and 120.87° E, 23.88° N for E9, respectively). Moreover, there are a large number of clear signals with high signal-to-noise ratio available for both aftershocks. Tables 4 and 5 list the stations with data available and the phases picked to be used, where the asterisk means that no phase is usable because of special epicentral distances or azimuths. As the Tables 4 and 5 show, 26 three-component stations were selected across the world for the aftershock E4, and 24 three-component stations were selected for the aftershock

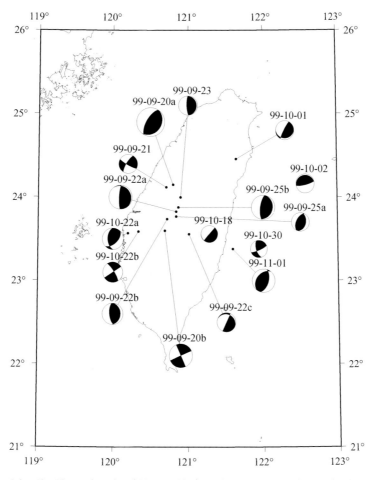

Figure 1 The distribution of the Chi-Chi earthquake (99-09-20a) and its major aftershocks (full circles) from September 20 to December 31, 1999 and the beach-balls of their focal mechanisms offered by Harvard

The size of the beach-ball is proportional to the magnitude (see Table 3). The positions indicated in the figure are centroid locations of the earthquakes

E9. A total of 57 P phases and 40 S phases were adopted for E4, and 51 P phases and 30 S phases for E9. For those stations with epicentral distances greater than 90°, the phases used are diffracted P or PKP or diffracted S (see Tables 4 and 5 for details).

In data processing three steps were performed. First, three-component recordings were rotated from up-down, north-south, and east-west to Z, R, and T (vertical, radial and tangential components, respectively). Second, all the original velocity recordings of the mainshock and the aftershocks were integrated to obtain the displacement recordings. Finally, all the integrated recordings were filtered using a third-order butterworth filter with a frequency band of 0.01–0.1 Hz.

Table 3 The source parameters of the mainshock and larger aftershocks until 31 December 1999

No.	Date yr/mo/day	Origin Time (UTC) hr: min: sec	Epicentral Location Latitude / (°N)	Epicentral Location Longitude / (°E)	Depth /km	Nodal Plane I Strike / (°)	Nodal Plane I Dip / (°)	Nodal Plane I Rake / (°)	Nodal Plane II Strike / (°)	Nodal Plane II Dip / (°)	Nodal Plane II Rake / (°)	M_0 /10^{19} Nm	M_S
1	1999/09/20	17: 47: 18.50	24.150	120.800	21	37	25	96	211	65	87	34.0	7.4
2	1999/09/20	21: 46: 42.90	23.600	120.690	20	246	89	179	336	89	1	4.8	6.1
3	1999/09/21	18: 18: 40.00	24.120	120.710	33	212	74	9	119	81	164	7.1	4.9
4*	1999/09/22	00: 14: 39.20	23.830	120.840	28	327	12	55	183	80	97	5.0	6.1
5	1999/09/22	00: 49: 42.80	23.740	120.720	37	3	25	97	175	65	87	6.3	5.5
6	1999/09/22	12: 17: 19.30	23.560	121.010	33	294	18	0	24	90	−108	9.3	5.0
7	1999/09/23	12: 44: 34.70	24.000	120.900	33	350	13	83	178	77	92	8.8	5.0
8	1999/09/25	08: 43: 31.60	23.770	120.840	25	12	38	80	205	52	98	5.1	4.8
9*	1999/09/25	23: 52: 48.70	23.880	120.870	17	12	20	95	187	70	88	6.0	6.2
10	1999/10/01	12: 54: 12.40	24.460	121.640	31	91	18	153	207	82	74	7.0	4.9
11	1999/10/02	17: 14: 17.80	24.160	122.570	21	259	12	94	75	78	89	6.1	4.9
12	1999/10/18	16: 00: 44.50	23.560	121.280	25	90	3	140	220	88	87	1.3	5.1
13	1999/10/22	02: 18: 58.60	23.570	120.200	15	178	49	54	46	52	125	7.0	5.6
14	1999/10/22	03: 10: 19.00	23.590	120.340	15	327	78	0	237	90	168	2.5	5.3
15	1999/10/30	08: 27: 53.80	23.380	121.940	56	162	64	13	66	78	154	1.3	5.1
16	1999/11/01	17: 53: 00.10	23.380	121.590	46	218	38	108	15	54	76	3.3	6.0

* Events used as EGFs in this study.

Table 4 The parameters of stations and their phases used as E4 is used as EGF

Station Code	Latitude / (°N)	Longitude / (°E)	Elev. /km	Azi. / (°)	Epi distance /km	Epi distance / (°)	Component Z	Component R	Component T
ATD	11.53	42.85	0.61	275.99	8211.59	74.51	P and S	P and S	P and S
BJT	40.02	116.17	0.20	347.69	1856.50	16.84	P	P	*
CHTO	18.79	98.98	0.32	260.93	2300.33	20.87	P	P	*
DAV	7.088	125.58	0.09	163.55	1877.05	17.03	P	P	*
ENH	30.28	109.49	0.49	305.92	1330.70	12.07	P	P	*
GUMO	13.59	144.87	0.01	109.29	2739.87	24.86	P	P	*
HYB	17.42	78.55	0.51	269.26	4398.99	39.92	P and S	P and S	P and S
KEV	69.76	27.01	0.08	338.22	7647.61	69.39	P and S	P and S	P and S
KIEV	50.69	29.21	0.16	318.36	8053.97	73.08	P and S	P and S	P and S
KMI	25.12	102.74	1.98	279.02	1819.73	16.51	P	P	P
KOG	5.21	-52.73	0.01	346.71	16606.70	150.69	PKP_{df} and SKS_{df}	SKS_{df}	SKS_{df}
KONO	59.65	9.60	0.22	331.26	8811.72	79.96	P and S	S	P and S
LSA	29.70	91.15	3.79	289.39	2989.89	27.13	P	P	P
MA2	59.58	150.77	0.34	22.49	4593.54	41.68	P and S	P and S	S
MAJO	36.55	138.21	0.41	45.34	2184.00	19.81	P	P	P
MAKZ	46.81	81.98	0.60	316.51	4266.94	38.72	P and S	P and S	*
MDJ	44.62	129.60	0.25	16.84	2445.25	22.18	P and S	P and S	S
NWAO	-32.93	117.23	0.27	183.58	6203.39	56.29	P and S	P and S	*
PMG	-9.41	147.15	0.07	138.79	4596.98	41.71	P and S	P and S	S
SSB	45.28	4.54	0.70	320.71	9947.62	90.26	P and S	S	P and S
SSE	31.10	121.19	0.02	2.67	825.905	7.49	P	P	*
TAM	22.79	5.53	1.37	301.43	11241.00	102.00	P_{diff}	P_{diff}	P_{diff}
TIXI	71.65	128.87	0.05	3.42	5324.48	48.31	P	S	S
WMQ	43.82	87.70	0.91	314.76	3725.95	33.81	P and S	P and S	*
XAN	34.03	108.92	0.63	317.74	1619.80	14.69	P	P	*
YSS	46.96	142.76	0.10	31.65	3222.31	29.24	P and S	*	S

Table 5 The parameters of stations and their phases used as E9 is used as EGF

Station Code	Latitude / (°N)	Longitude / (°E)	Elev. /km	Azi. / (°)	Epi distance /km	Epi distance / (°)	Component Z	Component R	Component T
ATD	11.53	42.85	0.61	275.99	8211.59	74.52	P and S	P and S	P and S
BJT	40.02	116.17	0.20	347.69	1856.50	16.85	P	P	*
CCM	38.06	-91.25	0.22	26.78	12310.30	111.71	P_{diff}	P_{diff} and S_{diff}	P_{diff} and S_{diff}
CHTO	18.79	98.98	0.32	260.93	2300.33	20.87	P	P and S	P and S
COLA	64.88	-147.85	0.19	27.11	7658.02	69.49	P	*	*
DAV	7.09	125.58	0.09	163.55	1877.05	17.03	P	P	*
FURI	8.90	38.69	2.55	275.08	8737.42	79.29	P and S	P and S	S
GUMO	13.59	144.87	0.01	109.29	2739.87	24.87	P	P	*
HIA	49.27	119.74	0.61	358.43	2826.64	25.65	P and S	P and S	P and S
HKT	29.96	-95.84	-0.41	35.23	12804.40	116.19	P_{diff} and S_{diff}	S_{diff}	S_{diff}
HYB	17.42	78.55	0.51	269.26	4398.99	39.92	P and S	P and S	S
KEV	69.76	27.01	0.08	338.22	7647.61	69.39	P and S	P and S	S
KIP	21.42	-158.02	0.07	73.29	8158.00	74.03	*	P	*
KMI	25.12	102.74	1.98	279.02	1819.73	16.51	P	P	*
KOG	5.21	-52.73	0.01	346.71	16606.70	150.69	P	*	*
LSA	29.70	91.15	3.79	289.40	2989.89	27.13	P	P and S	*
MAJO	36.54	138.21	0.41	45.35	2184.00	19.82	P	P	*
MDJ	44.62	129.59	0.25	16.84	2445.25	22.19	P and S	P	P and S
PEL	-33.15	-70.68	0.66	135.95	18308.10	166.14	PKP_{diff}	PKP_{diff}	PKP_{diff}
SSE	31.10	121.19	0.02	2.67	825.91	7.49	P	P	*
TAM	22.79	5.53	1.38	301.43	11241.00	102.01	P_{diff} and S_{diff}	P_{diff}	P_{diff} and S_{diff}
WMQ	43.82	87.70	0.90	314.76	3725.95	33.81	P and S	P and S	S
XAN	34.03	108.92	0.63	317.74	1619.80	14.69	P	P	*
YSS	46.96	142.76	0.10	31.66	3222.31	29.24	P	P	*

3 Source Time Functions

The EGF technique (Hartzell, 1978; Mueller, 1985; Dreger, 1994; Velasco et al., 1994; Hough and Dreger, 1995; Xu, 1995; Xu et al., 1996a, b, 1999) was used in retrieval of STFs. In frequency domain, the recorded displacement (velocity, acceleration) spectrum of the mainshock $U(\omega)$ is equal to the recorded displacement (velocity, acceleration) spectrum of the EGF aftershock $U'(\omega)$ multiplied by the spectrum of the far-field STF of the mainshock $S(\omega)$ and the ratio of the scalar seismic moments of the mainshock to that of the aftershock:

$$U(\omega) = \frac{M_0}{M'_0} S(\omega) U'(\omega). \tag{1}$$

Therefore, obtaining $S(\omega)$ and the ratio of the scalar seismic moment M_0/M'_0 is straight forward if $U(\omega)$ and $U'(\omega)$ are given. In practice, the following scheme is adopted to stabilize the spectral division (Chen et al., 1991),

$$\frac{M_0}{M'_0} S(\omega) = \frac{U(\omega) U'^*(\omega)}{U'(\omega) U'^*(\omega) + \alpha}, \tag{2}$$

where the asterisk refers to conjugate and α is a small constant called "water level," which usually is a small percentage of the maximum value of the $|U'(\omega)|$. The principle of selecting α is to stabilize the spectral division. It should be determined by the comprehensive analysis of the noise level of the EGF displacement recordings. In this study, α was adopted as 0.1. As shown in the subsequent paragraphs, this value proved to be appropriate for getting the stable results. For each pair of displacement recordings or phases, we obtained $S(\omega)$ and then transformed it into time domain to get the far-field STF, $S(t)$ and a low-pass butterworth filter of 0.1 Hz was applied to remove the unreliable high-frequency component. We retrieved the STF from each mainshock-EGF pair of phases first and then averaged those from the same phases to get an STF from that type of phases in that station.

Figure 2a briefly illustrates the procedure of retrieving an STF. Specifically, the selected phases, which may be P or S waves, are transformed into a frequency domain, and the spectrum of the mainshock phase is divided by the spectrum of the corresponding phase of the aftershock, and the water-level technique as described by equation (2) is used to stabilize the spectral division. Finally the quotient spectrum is transformed back into the time domain to get the STF. The first column in Figure 2a shows the P and S waves of the mainshock recorded at station HYB. They were rotated into vertical (Z), radial (R), and tangential (T) components from up-down, north-south, and east-west components. In the second column of the Figure 2a are the corresponding P and S waves of the aftershock E4 recorded at the same station. In the third column of the Figure 2a are the STFs retrieved from the corresponding waveforms of the mainshock-EGF pair. The STF is named P-STF if it is obtained from the P wave. Similarly, it is named S-STF if it is from the S wave. The synthetic waveforms of the mainshock are shown in the fourth column of Figure 2a for comparison, which were calculated using the corresponding STF retrieved and the waveform of the EGF aftershock. For each station, we obtain an STF from the waveforms of the mainshock-EGF pair, and then we average the P-STFs and the S-STFs, respectively, to get an averaged P-STF and an averaged SSTF, respectively, for that station. In Figure 2a and other figures, the STFs retrieved were all normalized to emphasize the similarities of the their shapes.

Figure 2b and c shows the error evaluation of the retrieved STFs in general; the thick lines are the average P-STF and average S-STF from station HYB, and the thin lines form the error envelopes. The upper limit of error was determined with the maximum values of all the normalized STFs at the same times, and the lower limit of error was determined with the minimum values of all the normalized STFs at the same times. As Figure 2b and c shows the error range is smaller in the time windows of the robust signal than that outside these windows.

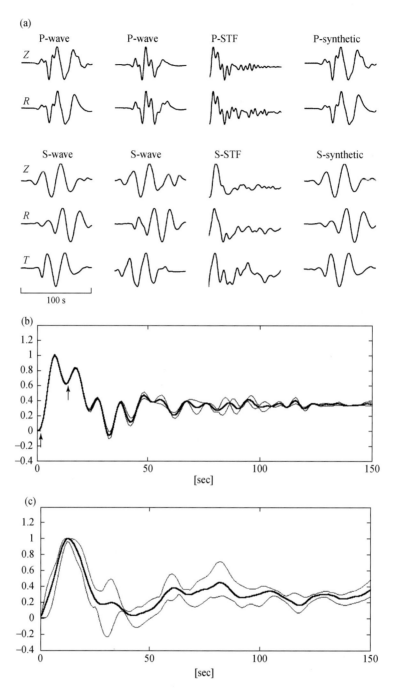

Figure 2 (a) An example to illustrate the procedure of retrieving STF using P and S phases of the HYB station. In the first column are the P and S waves for the main shock; in the second column are the P and S waves for the E4 aftershock; in the third column are the normalized STFs obtained from the corresponding pairs of phases; in the fourth column are the synthetic phases by convolution of the related STFs and EGF waves; (b) P-STF(thick line) with its error range(thin line). The two arrows denote the two large subevents; (c) S-STF (thick line) with its error range (thin line)

Figure 3a depicts the P-STFs of 26 stations across the world from which the P-STFs were retrieved and the epicenter of the mainshock, and Figure 3b depicts the S-STFs of 15 stations from which the S-STFs were retrieved and the epicenter of the mainshock, as E4 was used as EGF. In these figures the STFs are divided into two groups according to their azimuths. The STFs within azimuths of 0°–180° are put in one group and plotted on the right of the figure, and their azimuths increase from top to bottom; the STFs within azimuths of 180°–360° are put in another

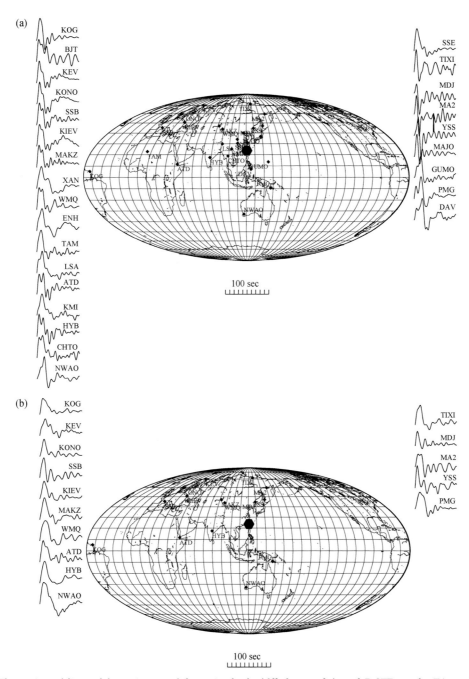

Figure 3 (a) The stations (diamonds), epicenter of the main shock (filled sexangle) and P-STFs as the E4 was EGF; (b) The stations, epicenter of the main shock and S-STFs as the E4 was EGF. The STFs are arranged according to their azimuths. On the left are the STFs the azimuths of which increase from 0° to 180° from top to bottom; on the right are the STFs the azimuths of which decrease from 360° to 180° from top to bottom. Each of the STFs is obtained by averaging the STFs from P-phases of different components

group and plotted on the left, and their azimuths decrease from top to bottom. Similarly, Figure 4a shows the P-STFs of 24 stations from which the P-STFs were retrieved and the epicenter of the mainshock, and Figure 4b shows the S-STFs of 12 stations from which the S-STFs were retrieved, as E9 was used as EGF. They are also grouped and plotted according to the same method as Figure 3. This arrangement is helpful for finding not only the similarities of the STFs obtained at different stations but also their variations with azimuth. The analysis of these figures led us to at least three conclusions.

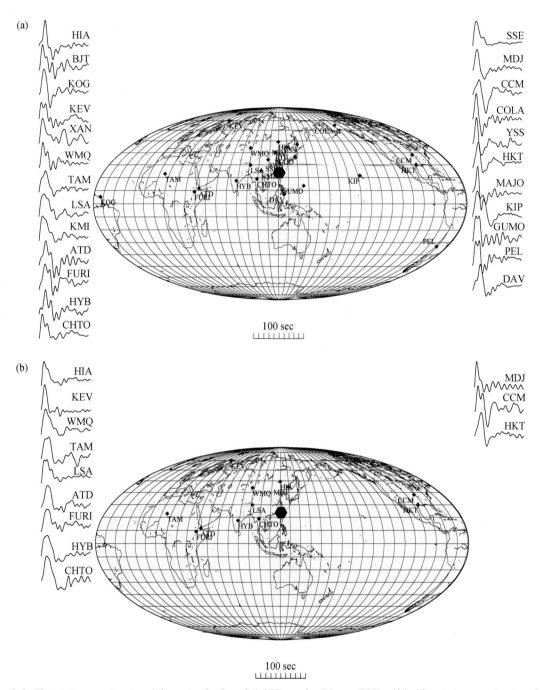

Figure 4 (a) The stations, epicenter of the main shock and P-STFs as the E4 was EGF; (b) The stations, epicenter of the main shock and S-STFs as the E9 was EGF. Plotted in the same way as in Figure 3

First, overall, the rupture process of the Chi-Chi earthquake had no interruption in time. This is unlike the Gonghe, Qinghai, $M_S 6.9$ earthquake (Xu and Chen, 1996) and the Lijiang, Yunnan, $M_S 7.0$ earthquake (Xu et al., 1998b), both of which had an obvious quiescent time interval between two subevents (Chen and Xu, 2000). The entire rupture duration of the Chi-Chi earthquake was about 26 sec, provided that a reasonable noise level of 10% was taken into account.

Second, it is very possible that the event consisted of at least two large subevents. The double-event feature can be clearly distinguished in most of stations such as GUMO, PMG, DAV, NWAO, CHTO, HYB, ATD,

LSA, WMQ, XAN, MAKZ, KIEV, SSB, KONO, KEV, and BJT in Figure 3a; and KIP, GUMO, PEL, DAV, CHTO, HYB, FURI, ATD, WMQ, KEV, and BJT in Figure 4a; and weakly recognized in some stations such as MDJ, MA2, YSS, ATD, MAKZ, KIEV, SSB, and KEV in Figure 3b; and LSA, TAM, WMQ and HIA in Figure 4b. The time difference between the two subevents is azimuth dependent and is estimated to be about 7 sec in the sense of average because the time intervals between two peaks reflecting subevents on the STFs vary with azimuth. Their relative amplitudes seem to be azimuth dependent too. But, on the average, the amplitude of the far-field STF, that is, the moment-release rate, of the first event is about 15% larger than that of the second one. As an example, the two large subevents are denoted with two arrows in Figure 2b.

Third, this event occurred on a quite large fault plane, and the rupture propagation was of clear directivity. Because the shapes of the STFs vary with azimuth, that is, there exists an azimuthal dependence. The STFs in the different azimuths have different features; see, for example, in Figure 3a, the STFs in stations TIXI, MDJ, MA2, YSS, and MAJO; the STFs in stations GUMO, PMG, DAV, and NWAO; the STFs in stations CHTO, HYB, and ATD; and the STFs in stations WMQ, MAKZ, KIEV, SSB, KONO, and KEV; and in Figure 4a, the STFs in stations CCM, COLA, YSS, HKT, and MAJO; the STFs in stations GUMO, PEL, and DAV; the STFs in stations CHTO, HYB FURI, and ATD; the STFs in stations KMI, LSA, and TAM; and the STFs in stations WMQ, XAN, KEV, and KOG.

4 Discussion of Source Time Function

STFs can be retrieved in the time domain by deconvolution of recordings of a smaller earthquake as EGFs from recordings of the mainshock and usually, with a positivity constraint imposed, can also be obtained in the frequency domain by spectral division of the recordings of the mainshock to a smaller shock, as used in this study. For time-domain deconvolution, the advantage is that physically unreasonable negative points of solution are easily discarded, but the disadvantage is that excessive subevents are obtained, which seem to be unreasonable (Hartzell et al., 1983; Ihmle, 1998). The frequency-domain spectral division often produces a relatively smooth STF reflecting a more reasonable source process, but it causes quite large oscillations and negative points in the resultant STF (Xie et al., 1991; Dreger, 1994; Hough and Dreger, 1995), which obscures interpretation of results. In both cases, the researcher's analysis and judgment are needed. The usual solution is usually to keep the robust signal and throw away the blurry part as noise (Dreger, 1994; Xu and Chen, 1996; Lanza et al., 1999). Our analysis and decisions followed these rules.

An STF is the function that describes the time history of seismic moment released at a point on the seismogenic fault. It is easily understood that the STF does not vary with one's viewpoint if the earthquake source is small enough to be considered as a point, but in most cases, or in realistic cases, the earthquake source is finite in extent and the observed STF is not only a function of time but also of location or space, so one has to consider the viewpoint when one interprets an observed STF. An observed STF or an STF retrieved from a specific station only describes the moment release watched just at that distance and orientation to the earthquake. That is why the individual STFs are often investigated from different phases at different stations (Dreger, 1994; Hough et al., 1995; Lanza, 1999), as done previously; an STF averaged over the STFs retrieved from different stations describes overall features of the moment release on the fault plane. Understanding the above points is very helpful for interpreting the STFs processed in different ways.

As we know, it is almost impossible to find two events with fully identical focal mechanisms and hypocenters, so the effects caused by the differences in focal mechanism and locations are often involved in results. Fortunately, the effect of location difference, including depth difference, depends on the wavelength of interest (Velasco et al., 1994; Xu and Chen, 1996, 1999). The wavelength of data used in this study is larger than about 50km; thus, the distance difference less than one wavelength does not cause observable effects on STFs. The focal depths of the main shock and aftershocks E4 and E9, which are the loci of their primary moment release, are 21, 28, and 17km, respectively, as shown by Harvard's results. They are close to each other and hardly affect the resultant STFs. Xu (1995) investigated the effect of focal mechanism difference between the mainshock and the EGF on the retrieved STF using a numerical test and noticed that focal mechanism difference can result in a phase shift on STFs at those stations whose azimuths are around the strike of fault plane. Thus, the abnormal STFs in stations TIXI, MDJ, MA2, YSS, MAJO, PMG, DAV, and NWAO in Figure 3a and in stations MDJ, CCM, COLA, YSS, HKT, MAJO, and KIP in Figure 4a can be attributed to the focal mechanism difference between the mainshock and the EGFs.

As we discussed previously, the rupture duration time obtained by the EGF technique is only an averaged or statistical estimate. Besides, error caused by the rupture duration time of the EGF is possibly involved. If the EGF were exactly an impulse response of the transmitting medium, the estimated rupture duration time of the mainshock would be the true rupture duration time, if not, the true rupture duration time would include the rupture duration time of the EGF as error. This may be one of the reasons why the width of the STF obtained by Yagi and Kikuchi (IRIS Data Center, 1999) was about 40 sec and the half-duration time estimated by Harvard was 18.8 sec (see Harvard CMT catalogue), while the rupture duration time obtained in this study was about 26 sec. It is interesting to note that the rupture duration time obtained in this study is in a good agreement with that obtained by Lee et al. (2000).

5 Temporal-Spatial Distribution of Slip on the Fault Plane

Many studies indicated that rupture propagation on the fault plane is not at a constant speed and that slip or rupture duration time at a certain point on fault plane is not constant (Mendoza and Hartzell, 1999; Chen and Xu, 2000; Lee et al., 2000; Ma et al., 2000; Wu et al., 2000). Thus it seems unreasonable to fix a shape for STFs of the subfaults and/or a rupture duration time *a priori* for a certain point on the seismogenic fault plane. Considering this, a time-domain inversion technique without constraints on rupture velocity or slip duration time was used for imaging the temporal-spatial distribution of slip on the fault plane (Xu et al., 1998a, b; Chen and Xu, 2000). Briefly speaking, an observed STF at the ith station, $S_i(t)$, is equal to summation of all the weighted STFs of subfaults $m_j(t)$ with different time delays τ_{ij}

$$S_i(t) = \sum_{j=1}^{J} m_j(t - \tau_{ij}) \tag{3}$$

or

$$S_i(t) = \sum_{j=1}^{J} \delta(t - \tau_{ij}) * m_j(t) \tag{4}$$

$$m_j(t) = \frac{M_j}{M_0} S_j(t) \tag{5}$$

$$\tau_{ij} = \frac{r_j}{v_i}, \tag{6}$$

where the asterisk represents time convolution, $S_j(t)$ is the far-field STF of the jth subfault, M_j is the scalar seismic moment of the jth subfault, M_0 is the total scalar seismic moment, r_j is the distance between the jth subfault and the reference point on the fault plane, v_i is the apparent velocity depending on the wave type, from which wave the observed STF is retrieved.

Slip rate $\dot{D}_j(t)$ of the jth subfault is determined with weighted STF $m_j(t)$, total scalar seismic moment M_0, area of the subfault A_j, and the rigidity μ at the source by

$$\dot{D}_j(t) = \frac{M_0 m_j(t)}{\mu A_j}. \tag{7}$$

In matrix form, equation (4) can be written as

$$\mathbf{S} = \mathbf{KM} \tag{8}$$

where $\mathbf{S}$ is the data vector consisted of the observed STFs, $\mathbf{M}$ is the unknown vector made up of the weighted STFs of all the subfaults, and $\mathbf{K}$ is the coefficient matrix determined by the time delay associated with the wave propagation.

Equation (8) is generally underdetermined because the number of the unknowns is larger than the number of the observation equations and there is no constraint imposed on the STFs of subfaults. In order to stabilize the solution, a constraint condition is imposed, that is,

$$m_j(t) \geq 0, \tag{9}$$

which physically means that no backward slip occurs during the rupture process. This condition is very easy to apply if the conjugate gradient method is employed in solving equation (8) (Ward and Barrientos, 1986). The conjugate gradient method is very helpful and powerful in solving an underdetermined problem such as equation (8). This method is advantageous because it is not necessary to find the inverse matrix and any nonnegative model can be used as an initial model. Thus, the locus of the rupture initiation, the slip amplitude, the rupture duration times of the subfaults, and the rupture velocity are derived from the inversion.

As we know, the seimogenic fault has been determined to be that striking toward north, dipping nearly toward east (Table 2). According to this conclusion, we choose a fault plane 170km long along the strike direction and 90km wide along the dip direction as an initial model, whose area should be large enough to encompass the ruptured area and divided it into 15 segments in the strike direction and 7 segments in the dip direction, respectively. In other words, a fault of 170km × 90km was divided into 15 × 7 equal-area subfaults. In order to reduce the effect on the mainshock's STFs caused by the differences in the focal mechanisms and the hypocentral locations between the mainshock and the two aftershocks whenever possible, we averaged the two STFs retrieved from the same phases of the same station as the different aftershocks were used as EGFs. The averaged one was taken as the STF at that station to be adopted in the inversion. However, a few of STF individuals were used in the inversion for those stations without the averaged ones. In this way, a total of 33 P-STFs and 22 S-STFs were produced for the inversion. The apparent velocities were calculated based on the IASPEI 91 model (Kennett and Engdahl, 1991) (Table 6).

Table 6 Apparent velocities at those stations used in the inversion

Station Code	V_P / (km/sec)	V_S / (km/sec)	Station Code	V_P / (km/sec)	V_S / (km/sec)	Station Code	V_P / (km/sec)	V_S / (km/sec)
BJT	7.8182		CHTO	8.2547	4.4840	DAV	7.8300	
ENH	7.7722		GUMO	8.4215		KEV	11.4652	6.2990
KIEV	11.7044	6.4190	KMI	7.8796		KONO	12.1072	6.6197
LSA	8.7399	4.8564	MA2	9.6724	5.4122	MAJO	8.0185	
MAKZ	9.6560	5.3439	MDJ	8.2283	4.4709	NWAO	5.9204	10.8566
PMG	9.7971	5.4245	SSE	7.2416		TIXI	10.2081	5.6491
WMQ	9.3191	5.1540	XAN	7.7290		YSS	8.8569	4.8873
ATD	11.8215	6.4763	HYB	9.7645	5.4076	KOG	14.0056	10.2943
SSB	12.7764	6.9510	TAM	13.5326	7.3449	CCM	14.0121	7.5996
COLA	11.4280		FURI	12.1099	6.6208	HIA	8.5634	4.7138
HKT	14.2493	7.7254	KIP	11.6773		PEL	15.1685	

The algorithm designed by Xu et al. (1998a) was used to image temporal-spatial slip distribution. Figure 5a shows the static slip distribution inverted. In Figure 5a two slip-concentrated areas with slip value greater than 3 m can be identified. Their centers are about 45km away from each other. The location of the maximum slip of the southern one is deeper than that of the northern one. The maximum slip of about 6.5 m occurred on the northern one, which is similar to the result obtained by Yagi and Kikuchi (1999). If we defined the area with the slip amplitude greater than 0.5 m as the rupture area, the rupture area of the Chi-Chi earthquake was estimated to be about 80km long and 60km wide, which was slightly larger than the 70km × 40km rupture area obtained by Yagi and Kikuchi (1999). The centroid depths of the southern and northern slip-concentrated areas are about 10 and 15km, respectively, which are different from the Harvard's centroid depth of 21km. From this study the fault length was estimated to be about 80km, which was in good agreement with the conclusion from Wang et al. (2000). The spatial distribution of the static stress drop on the fault plane was calculated assuming each subfault was of circular crack model (Keilis-Borok, 1959). The maximum stress drop was about 25 MPa (250 bar), and the average stress drop on the entire fault was 9.2 MPa (92 bar). In performing the inversion, we used source models of different fault strike and dip angle and found the fit of the synthetic STFs and the retrieved STFs was better for the fault model of strike 5° and dip angle 30°.

Figure 5b shows the snapshots of slip variation with time and space. The rupture initiated at the southern part and overall propagated northward unilaterally. The rupture process appeared to be complex. On the average, the rupture velocity was estimated to be about 2.5km/sec although it was variable during the rupture process. The initiation locus shown in Figure 5b was about 55km away from the epicenter, determined by Shin et al. (2000), and about 20km deep. This result differed from that of Shin et al. (2000) because our result reflects long-period energy release over the frequency range from 0.01 to 0.1 Hz, whereas that of Shin et al. (2000), determined from the seismic network including near-source and local stations, reflects broadband energy release. Figure 5c shows the snapshots of slip-rate variation on the fault plane. From this figure, the so-called "self healing pulse" (Heaton, 1990) can be seen in the beginning and near the end of source process; only a small portion of the fault was slipping within a relatively short time period, whereas a large portion of the fault was slipping within a relatively long time period in the middle of the source process.

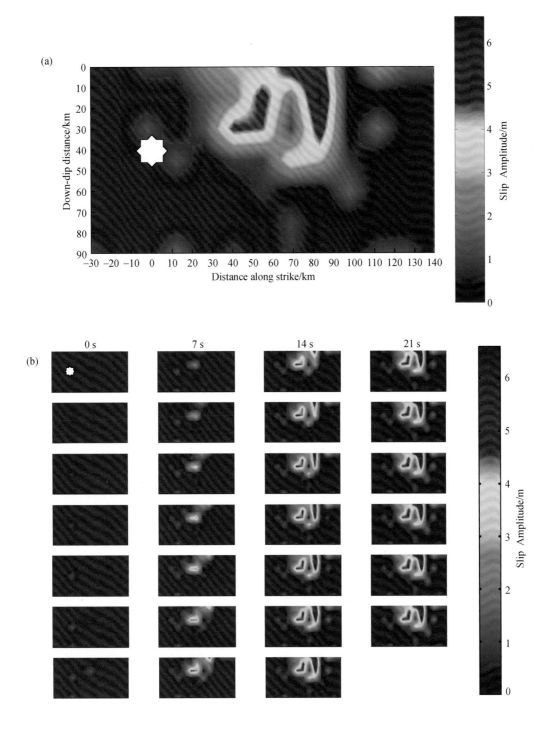

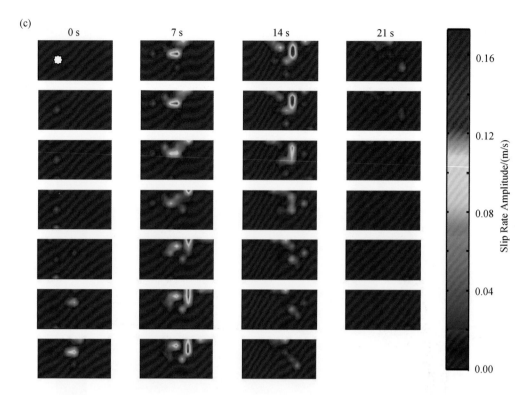

Figure 5 (a) The static slip distribution of the main shock rupture obtained by inversion of the retrieved STFs. White octagon represents the initiation locus of the 1999 Chi-Chi earthquake rupture; (b) The snapshots of the slip evolution on the fault plane, which shows the propagation of rupture on the fault plane at 7-sec intervals; (c) The snapshots of the slip rate evolution on the fault plane, which shows when and where there was slip during the earthquake process. White octagon represents the initiation locus of the 1999 Chi-Chi earthquake rupture

In some studies, the slip function for each subfault was fixed to be a prescribed function, such as box-car function, or a common time window was used for each subfault. Doing this is very likely to distort the results. Therefore, in this study, we did not constrain the shapes of slip function or the width of time window for each subfault *a priori*, instead, the rupture duration time of each subfault was obtained on the basis of the inversion. Figure 6 shows the spatial distribution of the rupture duration time, which was derived from the widths of the far-field STFs, that is, the moment-release rate of the subfaults. We noticed that the maximum rupture duration time for individual subfaults was about 16 sec, which was about two-thirds the total rupture duration time (26 sec) of the earthquake rupture process, and that there is a good correlation between the rupture duration times and the slip amplitudes of the subfaults. Grossly, the larger the slip amplitudes, the longer the duration times of the subfault (Fig. 7). Probably these phenomenon is challenging the way of prescribing time window or even a common slip function.

Figure 8 shows the comparison of the retrieved STFs (thick lines) used in the inversion and the synthetic STFs (thin lines) calculated using the inverted kinematical model of the Chi-Chi earthquake rupture. As seen in Figure 8, the overall features of almost all the observed STFs were well interpreted with the inverted source model. The correlation coefficients between the synthetic and the retrieved STFs in most stations (78% of the 55 retrieved STFs) are larger than 0.90.

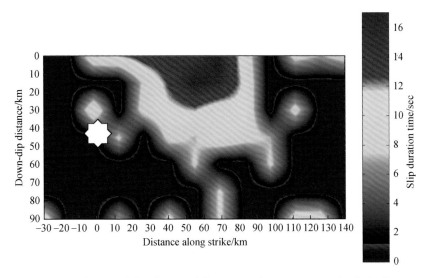

Figure 6 The spatial distribution of the rupture duration time on the fault plane

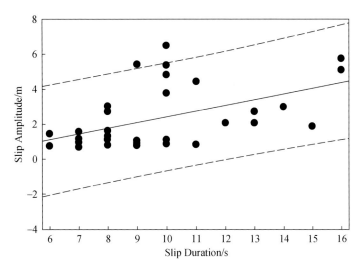

Figure 7 The correlation between the rupture duration time and static slip amplitude for subfauts on the fault plane

The filled circles were determined by the slip amplitudes and the slip duration times of the subfaults with the slip amplitude greater than 0. 5 m. The thick line was built by the least square method. The dashed lines denote the confidence interval of 95%

Stability and reliability of the result are concerned all the time for inversion. In this case, following Wald and Heaton (1994), numerous inversions were performed to find a stable and reasonable solution. It is worthy to note that the results above are stable to varieties of initial models and are reasonable as comparing with the results of others.

The slip distribution with time and space on a finite-fault plane could be obtained through various ways, which have advantages as well as disadvantages (Heaton, 1990; Dreger, 1994; Mendoza et al., 1999; among others). The technique adopted in this study (Xu et al., 1998a) is similar to that of Dreger (1994), which inverts the STFs extracted from different observation points instead of the original waveforms. In this technique, the STFs are allowed to be different from subfault to subfault and have no duration limit imposed, as mentioned in the previous sections. The effectiveness of this technique has been demonstrated by previous studies on some large

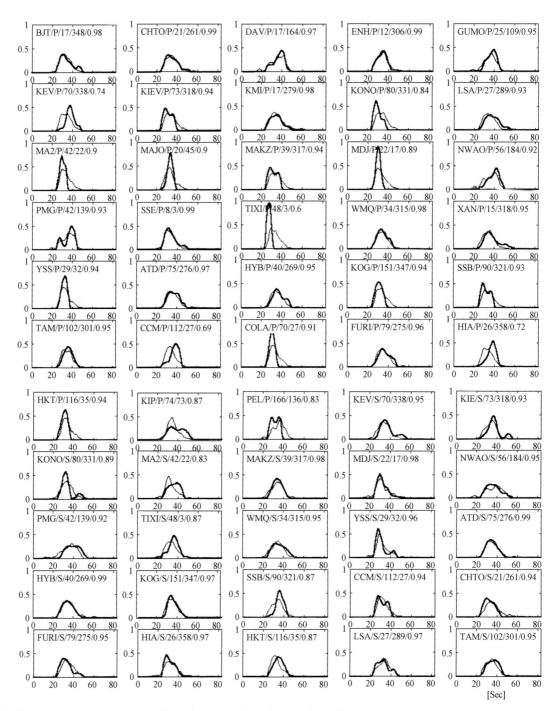

Figure 8 The comparison of the retrieved STFs (thick lines) and the synthetic STFs (thin lines) calculated using the rupture model of the Chi-Chi main shock obtained in this study

Each pair of STFs are labeled in order of station, phase, epicentral distance in degrees, station azimuth in degrees and correlation coefficient, respectively

earthquakes such as the 26 April 1990, Qinghai, China, $M_S6.9$ earthquake (Chen et al., 1996), the 3 February 1996, Lijiang, China, $M_S7.0$ earthquake, and the 7 November 1997, Mani, Tibet, $M_S7.9$ earthquake (Chen and Xu, 2001). In this study, the result we obtained independently using only teleseismic long-period waveforms offered the main features on the source process, as presented by other researchers using different techniques, different models, and different datasets. For example, Wang et al. (2001) used the surface displacement from GPS and the integrations of near-field strong motions, as well as 2D and 3D fault models; Chi et al. (2001) used the velocity waveform of strong motion, multiple time windows for subfaults, and a planar fault model; Wu et al. (2001) used the strong ground motion and GPS data, a multiple time window of 12 sec for subfaults, and a multifault model; Zeng et al. (2001) used a dataset of near-field strong ground motion and GPS, a single time window for subfaults, and a 3D curved model; Ma et al. (2001) used a dataset of near-source strong-motion records, broadband teleseismic displacement waveforms and GPS, a multiple time window of 12 sec, and both a single planar fault and a two-segment fault models. The common features exhibited by the inverted results justified our inversion. The discrepancies among the inverted results may be attributed to the differences of the inversion techniques adopted or datasets used.

6 Conclusions

Using long-period waveform data from IRIS stations and GEOSCOPE stations, we studied the temporal-spatial rupture process of the M_W7.6 earthquake, which occurred near Chi-Chi on 20 September 1999 (UTC).

We retrieved the STFs using EGF technique, including 178 STF individuals from P and S phases. Some similarities were found on the STFs of most stations except that a few of STFs in special azimuths looked different or odd due to the focal mechanism difference between the mainshock and the aftershocks. And a systematic shape-variation with azimuth was also observed. Both of these characteristics reflected the stability and reliability of the retrieved STFs. The analysis of them indicated that the Chi-Chi earthquake consisted of two large subevents, the second event was about 7 sec later than the first one, and the first event was 15% larger than the second one in the moment-release rate. The total duration time of the mainshock rupture was about 26 sec, which was in good agreement with that obtained by Yagi and Kikuchi (1999).

The temporal-spatial rupture process of the Chi-Chi earthquake was imaged using the STFs obtained by means of the time-domain inversion technique (Xu, 1995; Chen et al., 1996; Xu and Chen, 1996, 1999; Xu et al., 1998a, b; Chen and Xu, 2000). From the image of the static slip distribution, two slip-concentrated areas on which slip was greater than 3 m were identified, and their centers were about 45km away from each other. The maximum slip, which occurred on the northern one, was about 6.5 m. The rupture area, on which the slip was greater than 0.5 m, was about 80km long and 60km wide. The centroid loci of the two slip-concentrated areas were about 10 and 15km deep, respectively. The maximum stress drop was about 25 MPa (250 bar), and the average stress drop on the entire fault plane was 9.2 MPa (92 bar). From the snapshots of the temporal-spatial variation of slip, the rupture initiated at southern end and stopped at northern end, which suggested an overall unilateral rupture propagation. The propagation of rupture was quite complicated. The rupture velocities of the subfaults were variable. However, on the average, the rupture velocity was about 2.5km/sec. From the snapshots of the temporal-spatial variation of slip-rate, only a small portion of the fault was slipping within a relatively short time period in the beginning and near the end of the source process, which is like the self-healing pulse, whereas a large portion of the fault was slipping within a relatively long time period in the middle of the source process. From the spatial distribution of the rupture duration times, there is a good correlation between the rupture duration time and the slip amplitude of the subfault. Grossly, the larger the slip amplitudes, the longer the rupture duration times of the subfaults. The maximum rupture duration time for subfaults was 16 sec, about two-thirds the total rupture duration time of the earthquake rupture process.

Acknowledgments

This research was supported by the 973 Project (G1998040705), Ministry of Science and Technology, P. R. China; the National Science Foundation of China under Grant Nos. 49904004 and 40074012, and the Institut de Physiques du Globe, Paris, France. The authors would like to express their sincere thanks to Professors J. P. Montagner and P. Bernard for their kind help and to Dr. P. Spudich of the U. S. Geological Survey, Dr. D. D. Oglesby of University of California-Riverside and an anonymous reviewer, whose constructive comments and suggestions greatly improved this article. This is publication No. 02A1004, Institute of Geophysics, China Seismological Bureau.

References

Azuma, T., Y. Sugiyama, Y. Kariya, Y. Awata, Y. H. Lee, T. S. Shin, and T. Nagata (2000). Earthquake segments of the Chelungpu Fault based of slip distribution on the fault scarp of the 1999 Chi-Chi earthquake, 2000 Western Pacific Geophysics Meeting, Tokyo, Japan, 27 June-1 July 2000.

Chang, H. C. and Y. H. Lee (2000). The surface rupture of the 1999 ChiChi, central Taiwan, earthquake, 2000 Western Pacific Geophysics Meeting, Tokyo, Japan, 27 June-1 July 2000.

Chang, C. H., Y. M. Wu, T. C. Shin, and C. Y. Wang (2000a). Relocated Chi-Chi, Taiwan earthquake and its characteristics, 2000 Western Pacific Geophysics Meeting, Tokyo, Japan, 27 June-1 July 2000.

Chang, S. H., C. H. Chen, and W. H. Wang (2000). Fault slips inverted from surface displacement during Chi-Chi earthquake, 2000 Western Pacific Geophysics Meeting, Tokyo, Japan, 27 June-1 July 2000.

Chang, C. H., Y. M. Wu, T. C. Shin, and C. Y. Wang (2000b). Relocation of the 1999 Chi-Chi earthquake in Taiwan, *TAO* **11**, 581–590.

Chen, Y. T., and L. S. Xu (2000). A time-domain inversion technique for the tempo-spatial distribution of slip on a finite fault plane with applications to recent large earthquakes in the Tibetan Plateau, *Geophys. J. Int.* **143**, 407–416.

Chen, Y. T., J. Y. Zhou, and J. C. Ni (1991). Inversion of near-source broadband accelerograms for the earthquake source-time function, *Tectonophysics* **197**, 89–98.

Chen, Y. T., L. S. Xu, X. Li, and M. Zhao (1996). Source Process of the 1990 Gonghe, China Earthquake and Tectonic Stress Field in the Northeastern Qinghai-Xizang (Tibetan) Plateau, *Pageoph* **146**, no. 314, 97–105.

Chi, W. C., D. Dreger, and A. Kaverina (2001). Finite-source modeling of the 1999 Taiwan (Chi-Chi) earthquake derived from a dense strongmotion network, *Bull. Seism. Soc. Am.* **91**, 1144–1157.

Chung, J. K., and T. C. Shin (1999). Implications of the rupture process from the displacement distribution of strong ground motions recorded during the 21 September 1999 Chi-Chi, Taiwan earthquake, *TAO* **10**, 777–786.

Dreger, D. S. (1994). Emrirical Green's function study of the January 17, 1994 Northridge, California earthquake, *Geophys. Res. Lett.* **21**, 2633–2636.

Hartzell, S. H. (1978). Earthquake aftershocks as Green's functions, *Geophys. Res. Lett.* **5**, 1–4.

Hartzell, S. H., and T. H. Heaton (1983). Teleseismic time functions for large, shallow subduction zone earthquakes, *Bull. Seism. Soc. Am.* **75**, 965–1004.

Heaton, T. H. (1990). Evidence for and implications of self-healing pulses of slip in earthquake rupture, *Phys. Earth. Planet. Int.* **64**, 1–20.

Hough, S. E., and D. S. Dreger (1995). Source parameters of the 23 April 1992 M6.1 Joshua Tree, California, earthquake and its aftershocks: empirical Green's function analysis of GEOS and TERRAscope data, *Bull. Seism. Soc. Am.* **85**, 1576–1590.

Ho, C. C., R. J. Rau, T. T. Yu, J. Y. Yu, M. Yang, and C. L. Tseng (2000). The co-seismic slip distribution of the 1999 Chi-Chi, Taiwan, earthquake sequence, 2000 Western Pacific Geophysics Meeting, 27 June-1 July 2000. Tokyo, Japan.

Huang, B. S. (2000). Two-dimensional reconstruction of the surface ground motions of an earthquake: the September 21, 1999, Chi-Chi, Taiwan earthquake, 2000 Western Pacific Geophysics Meeting, 27 June-1 July 2000. Tokyo, Japan.

Ihmle, P. F. (1998). On the interpretation of subevents in teleseismic waveforms: the 1994 Bolivia deep earthquake revisited, *J. Geophys. Res.* **103**, 17, 919–17, 932.

IRIS Data Center (1999). Spatiotemporal distribution of source rupture process for Taiwan earthquake ($M_S = 7.7$), EIC Seismological Note No. 66, 21 September 1999.

Iwata, T., H. Sekiguchi, and K. Irikura (2000). Source process of the 1999 Chi-Chi, Taiwan, earthquake and its near-fault strong ground motions, 2000 Western Pacific Geophysics Meeting, Tokyo, Japan, 27 June-1 July 2000.

Kao, H., R. Y. Chen, and C. H. Chang (2000). Exactly where does the 1999 Chi-Chi earthquake in Taiwan nucleate? Hypocenter relocation using the master station method, *TAO* **11**, 567–580.

Keilis-Borok, V. I. (1959). On estimation of the displacement in an earthquake source and of source dimension, *Ann. Geofis.* **12**, 205–214.

Kennett, N. L. N., and E. R. Engdahl (1991). Traveltimes for global earthquake location and phase identification, *Geophys. J. Int.* **105**, 429–465.

Lanza, V., D. Spallarossa, M. Cattaneo, D. Bindi, and P. Augliera (1999). Source parameters of small events using constrained deconvolution with empirical Green's functions, *Geophys. J. Int.* **137**, 651–662.

Lee, S. J., and K. F. Ma (2000). Rupture process of the 1999 Chi-Chi, Taiwan, earthquake from the inversion of teleseismic data, *TAO* **11**, 591–608.

Lee, S. J., K. F. Ma, J. Mori, and S. B. Yu (2000). Teleseismic and GPS data analysis of the 1999 Chi-Chi, Taiwan, earthquake, 2000 Western Pacific Geophysics Meeting, Tokyo, Japan, 27 June-1 July 2000.

Ma, K. F., and J. Mori, (2000). Rupture process of the 1999 Chi-Chi, Taiwan earthquake from direct observations and joint inversion of strong motion, GPS and teleseismic data, 2000 Western Pacific Geophysics Meeting, Tokyo, Japan, 27 June-1 July 2000.

Ma, K. F., J. Mori, S. J. Lee, and S. B. Yu (2001). Spatial and temporal distribution of slip for the 1999 Chi-Chi, Taiwan, earthquake, *Bull. Seism. Soc. Am.* **91**, 1069–1087.

Ma, K. F., T. R. A. Song, S. J. Lee, and H. I. Wu (2000). Spatial slip distribution of the September 20, 1999, Chi-Chi, Taiwan, earthquake ($M_W 7.6$): inverted from teleseismic data, *Geophys. Res. Lett.* **27**, 3417–3420.

Mendoza, C., and S. Hartzell (1999). Fault-slip distribution of the 1995 Colima-Jalisco, Mexico, earthquake, *Bull. Seism. Soc. Am.* **89**, 1338–1344.

Mueller, C. S. (1985). Source pulse enhancement by deconvolution of an empirical Green's function, *Geophys. Res. Lett.* **12**, 23–36.

Ota, Y., M. Watanable, Y. Suzuki, C. Y. Huang, and H. Sawa (2000). Earthquake fault associated with the 921 Chi-Chi earthquake with special reference to coincidence with pre-existing fault and progressive deformation, 2000 Western Pacific Geophysics Meeting, Tokyo, Japan, 27 June-1 July 2000.

Shin, T. C. (2000). Some seismological aspects of the 1999 earthquake in Taiwan, *TAO* **11**, 555–566.

Shin, T. C., K. W. Kuo, W. H. K. Lee, T. L. Teng, and Y. B. Tsai (2000). A preliminary report on the 1999 Chi-Chi (Taiwan) earthquake, *Seism. Res. Lett.* **71**, 24–30.

Velasco, A., C. Ammon, and T. Lay (1994). Empirical Green's function deconvolution of broadband surface waves: rupture directivity of the 1992 Landers, California ($M_W = 7.3$) earthquake, *Bull. Seism. Soc. Am.* **84**, 735–750.

Wald, D., and T. Heaton (1994). Spatial and temporal distribution of slip for the 1992 Landers, California earthquake, *Bull. Seism. Soc. Am.* **84**, 668–691.

Ward, S. N., and S. E. Barrientos (1986). An inversion for slip distribution and fault shape from geodetic observations of the 1983, Borah Park, Idaho, earthquake, *J. Geophys. Res.* **91**, 4909–4919.

Wang, C. Y. (2000). An interpretation of 1999 Chi-Chi earthquake, Taiwan based on the thin-skinned thrust model, 2000 Western Pacific Geophysics Meeting, Tokyo, Japan, 27 June-1 July 2000.

Wang, C. Y., C. H. Chang, and H. Y. Yen (2000). An interpretation of the 1999 Chi-Chi earthquake in Taiwan based on the thin-skinned thrust model, *TAO* **11**, 609–630.

Wang, W. H. (2000). Static stress transfer and aftershock triggering by the 1999 Chi-Chi earthquake in Taiwan, *TAO* **11**, 631–642.

Wang, W. H., S. H. Chang, and C. H. Chen (2001). Fault slip inverted from surface displacements during the 1999 Chi-Chi, Taiwan, earthquake, *Bull. Seism. Soc. Am.* **91**, 1167–1181.

Wu, C. J., M. Takeo, and S. Ide (2000). Source process of the Chi-Chi earthquake, 2000 Western Pacific Geophysics Meeting, Tokyo, Japan, 27 June-1 July 2000.

Wu, C. J., M. Takeo and S. Ide (2001). Source process of the Chi-Chi earthquake: a joint inversion of strong motion data and global positioning system data with a multifault model, *Bull. Seism. Soc. Am.* **91**, 1128–1143.

Xie, J., Z. Liu, R. B. Herrmann, and E. D. Cranswick (1991). Source parameters of three aftershocks of the 1983 Goodnow, New York, earthquake: high-resolution image of rupturing cracks, *Bull. Seism. Soc. Am.* **81**, 818–843.

Xu, L. S. (1995). Tempo-spatial complexity of the Gonghe, Qinghai, China, Ph. D. Thesis, Institute of Geophysics, China Seismological Bureau, Beijing, P. R. China.

Xu, L. S., and Y. T. Chen (1996). Source time functions of the Gonghe, China earthquake retrieved from long-period digital waveform data using empirical Green's function technique, *Acta Seismologica Sinica* **9**, 209–222.

Xu, L. S., and Y. T. Chen (1999). Tempo-spatial rupture process of the 1997, Mani, Xizang (Tibet), China earthquake of $M_S = 7.9$, *Acta Seismologica Sinica* **12**, 495–506.

Xu, L. S., Y. T. Chen, and S. Fasthoff (1998a). *TempSpac User's Guide*, Institute of Geophysics, China Seimologica Bureau, Beijing, P. R. China, 1–47.

Xu, L. S., Y. T. Chen, and S. Fasthoff (1998b). Inversion for rupture process of the 1996 Lijiang, Yunnan, China $M_S 7.0$ earthquake by empirical Green's function technique, in *1996 Lijiang Earthquake*, Seismological Press, Beijing, P. R. China, 79–81.

Xu, L. S., G. Patau, and Y. T. Chen (2002). Source time functions of the 1999, Taiwan, $M_S 7.6$ earthquake retrieved from IRIS and GEOSCOPE long period waveform data using aftershocks as empirical Green's functions, *Acta Seismologica Sinica* **15** (2), 121–133.

Xu, L. S., Z. X. Yang, and Y. T. Chen (1999). A preliminary analysis of the Chi-Chi earthquake sequence and the rupture process of the main shock, in *Proc. of the 20th Anniversary of the Foundation of the Seismological Society of China*, Y. T. Chen (Editor), 97–112 (in Chinese with English abstract).

Xu, L. S., Z. X. Yang, and Y. T. Chen (2000). Tempo-spatial rupture process of the 1999 Jiji (Chi-Chi) earthquake, *in Extended Abstracts with Programs: The Second World Chinese Conference on Geological Sciences*, Stanford, California, 2–4 August 2000, A. 39.

Yagi, Y., and M. Kikuchi (2000). Source rupture process of the Chi-Chi, Taiwan, earthquake of 1999, obtained by seismic wave and GPS data, 2000 Western Pacific Geophysics Meeting, Tokyo, Japan, 27 June-1 July 2000.

Yagi, Y. and M. Kikuchi (1999). Earthquake Research Institute, University of Tokyo. October 1999. http://www.eri.u-tokyo.ac.jp.

Yeh, Y. H. (2000). The Chi-Chi earthquake in Taiwan, 2000 Western Pacific Geophysics Meeting, Tokyo, Japan, 27 June-1 July 2000.

Yoshioka, S. (2000). Coseismic slip distribution of the 1999 Chi-Chi earthquake, Taiwan, deduced from inversion analysis of GPS data. 2000 Western Pacific Geophysics Meeting, Tokyo, Japan, 27 June-1 July 2000.

Yu, T. T., C. C. Ho, and C. L. Tseng (2000). Model of static deformation produced by the Chi-Chi earthquake, Taiwan on September 20, 1999, 2000 Western Pacific Geophysics Meeting, Tokyo, Japan, 27 June-1 July 2000.

Yuan, B. D., C. W. Lin, C. Y. Lai, M. L. Huang, and Y. C. Liu (2000). Bonding mechanism of surface rupture associated with the 1999 Chi-Chi earthquake, central Taiwan, 2000 Western Pacific Geophysics Meeting, Tokyo, Japan, 27 June-1 July 2000.

Zeng, Y. H., and C. H. Chen (2001). Fault rupture process of the 20 September 1999 Chi-Chi Taiwan, earthquake, *Bull. Seism. Soc. Am.* **91**, 1088–1098.

Realistic modeling of seismic wave ground motion in Beijing City[*]

Ding Z.[1], Romanelli F.[2], Chen Y. T.[1] and Panza G. F.[2,3]

1. Institute of Geophysics, China Seismological Bureau, Beijing, 100081, China;
2. Dipartimento di Scienza della Terra, Via Weiss 4, 34127 Trieste, Italy;
3. SAND Group, ICTP, Strada Costiera 11, 1-34100 Trieste, Italy

Abstract Algorithms for the calculation of synthetic seismograms in laterally heterogeneous anelastic media have been applied to model the ground motion in Beijing City. The synthetic signals are compared with the few available seismic recordings (1998, Zhangbei earthquake) and with the distribution of observed macroseismic intensity (1976, Tangshan earthquake). The synthetic three-component seismograms have been computed for the Xiji area and Beijing City. The numerical results show that the thick Tertiary and Quaternary sediments are responsible for the severe amplification of the seismic ground motion. Such a result is well correlated with the abnormally high macroseismic intensity zone in the Xiji area associated with the 1976 Tangshan earthquake as well as with the ground motion recorded in Beijing city in the wake of the 1998 Zhangbei earthquake.

1 Introduction

Beijing City is situated in an active seismic zone, oriented in the NW-SE direction, stretching from Bohai Sea to the city of Zhangjiakou along the northern margin of the North China Plain. Historically, Beijing City has been rocked by destructive earthquakes in the past (see Fig. 1). The last great event was the 1697 Sanhe-Pinggu earthquake ($M = 8$), which occurred approximately 50km from the city. The maximum observed macroseismic intensity in Beijing, caused by that earthquake, was XI, on the China Seismic Intensity Table (Xie, 1957), which is close to the MSK scale. The 1976 Tangshan earthquake ($M = 7.8$) caused a maximum intensity of VIII in Beijing City. The latest strong event felt in Beijing was in 1998 caused by the $M_S 6.2$ Zhangbei earthquake.

Estimation of the expected seismic ground motion is a key issue in the design of rational measures for mitigation impact of seismic hazard. For a given study area, a possible solution to the seismic microzonation problem is to assemble a comprehensive set of recorded strong ground motions and to group those seismograms that represent similar source, path and site effects. However, such a database is ordinarily not available in practice. An alternative and complementary way is based on the use of computer codes, which take into account the seismic waves propagation in anelastic laterally heterogeneous media, the complexity of seismic sources and the site effects. Realistic models thus formulated are expected to simulate ground motions at a given site arising from a given earthquake scenario.

---

[*] 本文发表于 *Pure Appl. Geophys.*, 2004年, 第**161**卷, 第5-6期, 1093-1106.

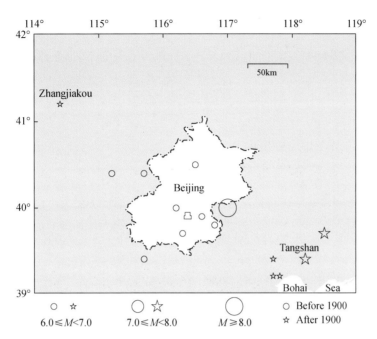

Figure 1 Epicenters in and around the Beijing area
stars represent the events after 1900; circles represent the events before 1900

In the present paper, we use the modal summation method (Panza, 1985; Panza and Suhadolc, 1987; Florsch et al., 1991; Panza et al., 2000) to compute the synthetic broadband seismogram in the reference bedrock model, using two different approaches to calculate the synthetic seismograms in laterally heterogeneous anelastic structures. The analytical coupling coefficient algorithm (Levshin, 1985; Vaccari et al., 1989; Romanelli, et al., 1996, 1997) uses several contiguous 1-D models with vertical discontinuities in welded contact to mimic a 2-D model. The hybrid method uses the results of the modal summation as the input signal, and calculates the seismograms in the local laterally heterogeneous structures by using a finite-difference algorithm (Fäh et al., 1993, 1994; Fäh and Suhadolc, 1995).

In this work, we compute seismic ground motions in Beijing in respect of two strong earthquakes at epicentral distances in the range 110–160km (the 1976 M_S7.8 Tangshan earthquake) and 200–235km (the 1998 M_S6.2 Zhangbei earthquake), and compare these with the observed records and macroseismic intensities.

2 Structures and modeling method

Beijing City is situated in the Beijing Tertiary Depression Zone, the thickness of the Tertiary sediments reaching 2km in the southwestern part of the city (Figs. 2, 3). Quaternary sediments whose thickness increases from the mountain-plain boundary to the southeastern direction cover all of the plain area. Whilst most of the city is covered by Quaternary sediments which average about 0.1km in thickness (Fig. 4), there are two abnormally thick Quaternary sediment zones near the city, one in the northwest, and the other in the northeast suburbs reaching a maximum thickness of about 0.8km.

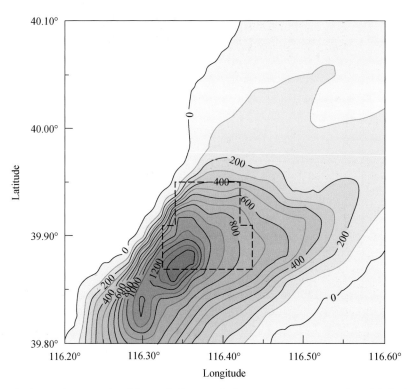

Figure 2 Thickness (in meters) of the Early Tertiary Sediment Layer in Beijing City (dashed polygon)

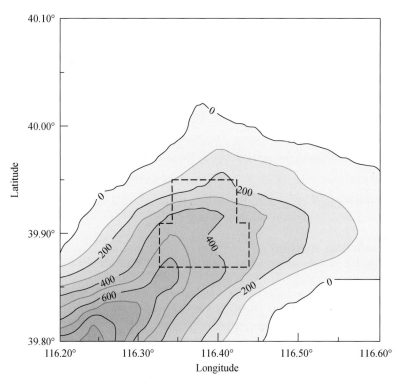

Figure 3 Thickness (in meters) of the Late Tertiary Sediment Layer in Beijing City (dashed polygon)

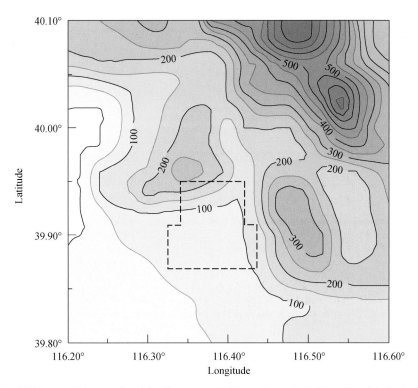

Figure 4 Thickness (in meters) of the Quaternary Sediment Layer in Beijing City (dashed polygon)

The seismic ground motion in Beijing City is computed in accordance with two great earthquakes, the 1976 Tangshan earthquake and the 1998 Zhangbei earthquake. The hybrid algorithm is applied to model the seismic ground motion in the Beijing City area, while the coupling coefficient algorithm is applied in the Xiji area, on the eastern border of Beijing City. In both cases, the effects of source, path and site have been taken into account.

To provide for the source finiteness for different magnitudes, we use the scaling laws of Gusev (1983) by properly weighting the source spectrum in the frequency domain, as reported in Aki (1987).

Comparison of the observed data with the synthetic results represents the quality check of our modeling.

3 Modeling for the Tangshan earthquake

The July 28, 1976 Tangshan earthquake claimed at least 230, 000 victims and greatly disrupted the political and economic life of China. The earthquake, located about 160km east of Beijing City, had a magnitude $M_S = 7.8$ (Chinese Seismic Network Report). According to the general trend of isoseismals, Beijing City lies in the zone of macroseismic intensity VI, but abnormally high intensities are reported as well. More specifically, in the northwest, the observed macroseismic intensity is VII, and in the Xiji area, on the eastern border of Beijing City, as high as VIII (Fig. 5).

Based on previous studies (Zhang et al., 1980), the parameters of the Tangshan earthquake adopted in this study are:

Location: 39.4°N, 118.2°E

Depth = 11km

$M_0 = 1.8 \times 10^{27}$ dyn · cm

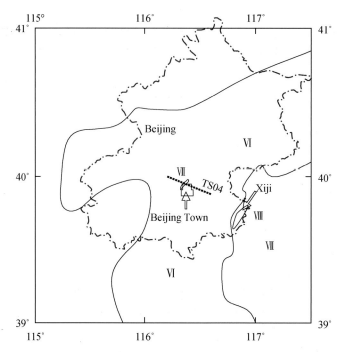

Figure 5　Intensities of the 1976 Tangshan Earthquake reported in the Beijing Area

The intensity in most areas of Beijing City is Ⅵ. While intensities in Xiji (Ⅷ) and the northwestern part of Beijing City (Ⅶ) are characterized by anomalously high intensity values

Source Mechanism: Strike = 30°, Dip = 90°, Rake = 180°.

The Xiji area, located over the Dachang Tertiary depression zone, has a specially shallow structure. The thickness of the Quaternary sediment is about 0.4km and the depth of the Tertiary sediments in the Xiji area reach 3.0km. The Tertiary sediments in Xiji are substantially thicker compared with that in the neighboring areas. Sun et al. (1998) simulated the seismic ground motion for SH waves in the Xiji area, using the hybrid method. Here we extend their work using the coupling coefficient algorithm (Levshin, 1985; Vaccari et al., 1989; Romanelli et al., 1996, 1997) to compute the synthetic seismograms for SH and P-SV waves.

Figure 6 shows the local structure and the synthetic three-component ground accelerations in the Xiji area in respect of the event parameters representing the 1976 Tangshan earthquake, the largest amplitudes being obtained in the transverse component of motion (SH waves).

To estimate site effects, we use as reference signals the synthetic seismograms computed for the average one dimension (1-D) bedrock model defined by Sun et al. (1998), which includes the crust and upper mantle structures and has no sediments at the surface. Figure 7 shows the comparison between the peak values obtained with the bedrock (1-D) and the realistic laterally varying (2-D) structures. There are abnormally large amplitudes at epicentral distances in the range between 109 and 112km. At approximately 110km, the maximum amplitude is larger than 200 (cm/s * s), quite compatible with the observed anomalous intensity (Ⅷ). The ratios of maximum amplitude and Arias intensity (Arias, 1970), A_{max} (2D) /A_{max} (1D) and W (2D) / W (1D), reach the highest values in the Xiji area. There are two peaks in the Arias intensity ratio W (2D) / W (1D): more than 5 at a distance close to 110km, due to the large peak values of the seismograms in the laterally varying model, and over 4 at a distance close to 117km, mainly due to the long duration of the ground motion in the laterally varying model. These results explain quite naturally how the local mechanical properties

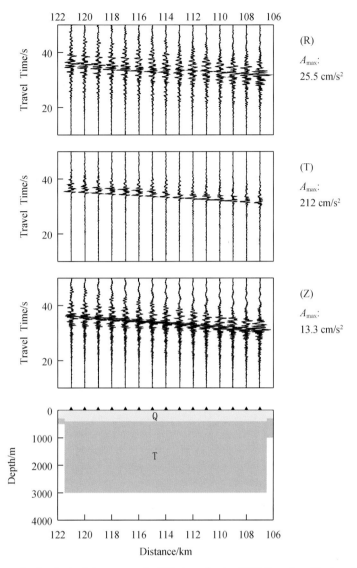

Figure 6 Synthetic Seismograms in Xiji Area for the 1976 Tangshan Earthquake

Q and T represent the Quaternary and Tertiary sediments, which cover the bedrock. The geophysical properties of the sediments are given in Table 1 of Ding et al. (2004)

amplify the seismic ground motion and cause the abnormally high intensity value, observed in the Xiji area during the 1976 Tangshan earthquake. Our results confirm the results obtained by Sun et al. (1998) who used the 2-D finite-difference algorithm.

An abnormally high intensity (VII) area is observed in the northern part of Beijing town as well. To study this phenomenon we consider a profile TS04 (see Fig. 5) that points from the epicenter of the 1976 Tangshan earthquake towards Beijing. The profile penetrates the northern part of Beijing town, and crosses the abnormally high intensity (VII) area. The distance along the profile is measured from the epicenter of the 1976 Tangshan earthquake.

Along this profile, two thick Quaternary sediment areas are encountered at distance of 153–163km and 170km. The synthetic three-component acceleration seismograms along the profile TS04 were obtained (Fig. 8) by using the hybrid method (Fäh et al., 1993, 1994; Fäh and Suhadolc, 1995). In the simulated seismograms,

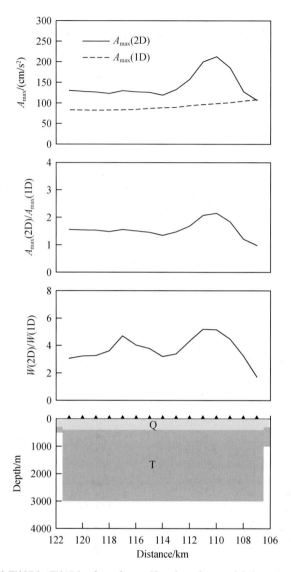

Figure 7　A_{max}(2D)/A_{max}(1D) and W(2D)/W(1D) along the profile adopted to model the seismic ground motion in the Xiji Area

Q and T represent the Quaternary and Tertiary sediments, which cover the bedrock

the amplitude of the transverse component is found to be about 10 times the radial and vertical components, as in the Xiji area. The waveform variations of all three components along the profile are strongly correlated to the thickness of the Quaternary sediments. At distances of 153–160km and 168–171km, the amplitudes and the durations of the seismograms are enhanced. The values of A_{max} around 150cm/s ∗ s are quite compatible with intensity Ⅶ. The ratios A_{max} (2D)/A_{max} (1D) and W (2D)/W (1D), obtained using the bedrock model of Sun et al. (1998) reach their peaks over the two thick Quaternary sediment areas (Fig. 9). In the northwestern part of Beijing town (about 170km in TS04), the ratios A_{max} (2D)/A_{max} (1D) and W (2D)/W (1D) can be as large as 2.5 and 8.0, respectively. This site coincides with the narrow, abnormal (one-degree higher) intensity zone, observed during the 1976 Tangshan earthquake.

4　Modeling for the Zhangbei earthquake

Next, the source parameters of the 1998 Zhangbei earthquake (Havard CMT solution), given below were

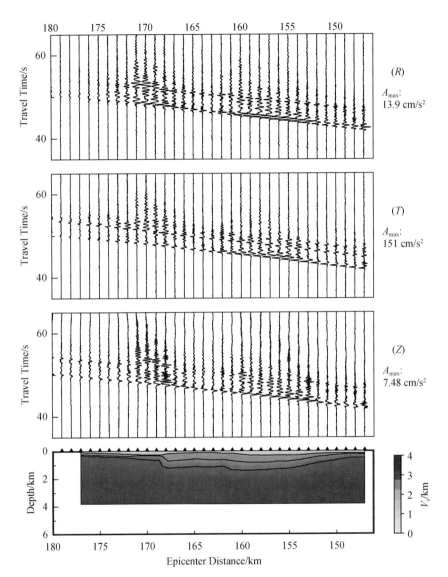

Figure 8 Cross section and the synthetic acceleration (radial, R; transverse, T, and vertical, Z) along the profile TS04
The contour lines shown in the model represent the boundaries of the three sedimentary layers, which cover the bedrock

used to computer ground accelerations in Beijing.

Date: January 10, 1998

Location: (41.2°N, 114.4°E)

Depth = 15km

$M_0 = 6.1 \times 10^{24}$ dyn·cm

NP1: Strike = 200°; Dip = 44°; Slip = 136°

NP2: Strike = 324°; Dip = 61°; Slip = 55°

Because of the NE orientation of the distribution of aftershocks, we selected NP1 as the preferred fault plane solution, and Rake = 180°.

The Zhangbei earthquake was the latest strong earthquake that occurred near Beijing City. Since several digital seismic stations had been deployed in the Beijing area before the earthquake (Fig. 10), we had an opportunity to compare our synthetic seismograms computed in this area with real seismic records.

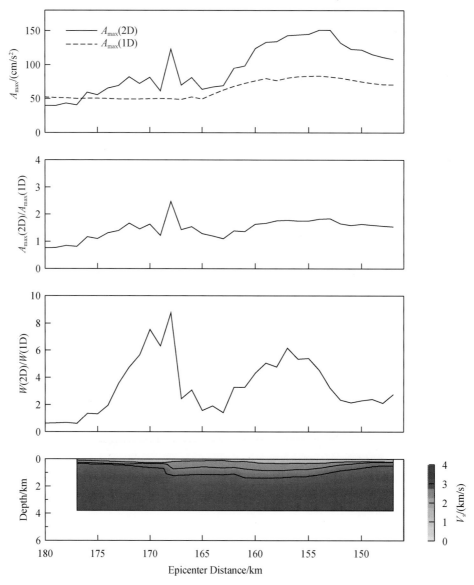

Figure 9 A_{max}, A_{max} (2D) $/A_{max}$ (1D) and W (2D) $/W$ (1D) along profile TS04

For the main shock, only one station, FHSD, recorded a good seismograms, others being saturated. Some aftershocks with magnitude $M_L > 4$ were well recorded by most of these stations. Figure 11 shows the records (transverse component—the largest one) for the aftershock that occurred on January 17, with magnitude $M_L =$ 4.6. The four stations, QSDD, LQSD, FHSD, and ZKDD are almost from the same azimuth with respect to the epicenter. In the records of QSDD, LQSD, and ZKDD, located in the mountain area, the amplitude of the seismic waves decreases with increasing epicentral distance. The station FHSD, located over thick sediments, does not follow the rule showing relatively large amplitudes (comparable to the ones of QSDD).

Seismograms at stations QSDD, LQSD, FHSD, and ZKDD were also simulated (Fig. 12), at QSDD, LQSD, and ZKDD, using the bedrock model and the modal summation method. At FHSD, we used hybrid method with the laterally heterogeneous model, shown in Figure 8. The synthetic signals show that the thick sediments beneath FHSD amplify the peak values of seismic ground motion, as seen in the actual records.

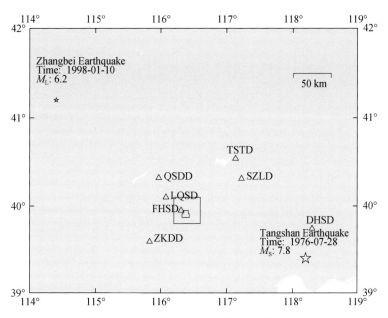

Figure 10 Beijing Seismic Network Stations

Triangles represent the location of seismic stations and stars the epicenters of the 1976 Tangshan and the 1998 Zhangbei earthquakes

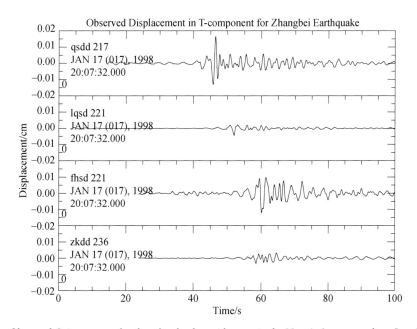

Figure 11 Observed Seismograms for the aftershock, with magnitude $M_L = 4.6$, occurred on Jan. 17, 1998,

of the 1998 Zhangbei earthquake

The epicenter distances of the recording stations, QSDD, LQSD, FHSD, and ZKDD, are 164, 187, 213 and 216km, respectively. Station FHSD is located in the thick sediments area, while the other three stations lie on bedrock. The signals, after deconvolution for the instrument response, are low-pass filtered with a cut-off frequency of 1 Hz

5 Conclusions

Three-component broadband synthetic accelerograms have been calculated for sites in Beijing City in respect of

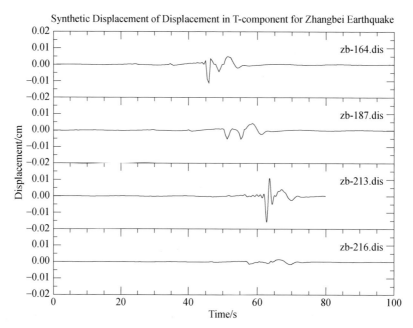

Figure 12 Synthetic seismograms for the aftershock, with magnitude $M_L = 4.6$, occurred on Jan. 17, 1998, of the 1998 Zhangbei earthquake

The synthetic seismograms are calculated at stations QSDD, LQSD, FHSD and ZKDD. The 1st, 2nd and 4th seismograms are calculated using the modal summation method applied to the bedrock model. The 3rd seismogram is the result of the finite-difference method applied to the model shown in Figure 8. The maximum frequency in the calculation is 1 Hz

the 1976 Tangshan and the 1998 Zhangbei earthquake, to be regarded as calibration events.

Both the synthetic seismograms and observed data show that the thick Quaternary sediment in the northwest of Beijing town and in Xiji area amplify the seismic wave ground motion.

This satisfactory comparison between observed and synthetic waveforms represents a sound justification for extending ground motion modeling to sites where observations are not available. Thus, it would be possible to study expected seismic ground motions at close sites covering all of Beijing City, thereby generating a seismic microzonation map as a basic map for land-use planning and specification of building codes and practices by local authorities, city planners, land-use specialists and civil engineers.

Acknowledgements

This work is a contribution to the UNESCO-IUGS-IGCP Project 414 "Realistic Modeling of Seismic Input for Megacities and Large Urban Areas." This research has been carried on in the framework of the bilateral project "Geophysical Studies for the Deterministic Evaluation of Seismic Risk," with the contribution of the Italian Ministry of Foreign Affairs (MAE), Directorate General for Cultural Promotion and Cooperation. The research received support from the Chinese National Key Basic Research and Development Program (973 Program) No. 2002CB412709, MOST of China and China Seismological Bureau. We acknowledge Dr. Franco Vaccari and Dr. Francesco Marrara for their kind help in the calculations. Dr. Zhang Wenbo supplied some seismic recordings. This is the contribution No. 02A10001, Institute of Geophysics, China Seismological Bureau. We have used the public domain graphics software (Wessel and Smith, 1995a, 1995b).

References

Aki, K., Strong motion seismology. In *Strong Ground Motion Seismology* (eds. Erdik, M. Ö. and Toksöz, M. N.) (NATO Advanced Study Institute Series, Series C: Mathematical and Physical Sciences, D. Reidel Publishing Company. The Netherlands 1987), 204, pp. 3–39.

Arias, A. (1970), A measure of earthquake intensity. In *Seismic Design for Nuclear Power Plants* (ed. R. Hansen), Cambridge, Massachusets.

Ding, Z., Chen, Y. T., and Panza, G. F. (2004), Estimation of site effects in Beijing City, *Pure Appl. Geophys.* **161**(5/6), 1107–1123.

Florsch, N., Fäh, D., Suhadolc, P., and Panza, G. F. (1991), Complete synthetic seismograms for high-frequency multimode SH-waves, *Pure Appl. Geophys.* **136**, 529–560.

Fäh, D., and Suhadolc, P. (1995), Application of numerical wave-propagation techniques to study local soil effects: The Case of Benevento (Italy), *Pure Appl. Geophys.* **143**, 513–536.

Fäh, D., Iodice, C., Suhadolc, P., and Panza, G. F. (1993), A new method for the realistic estimation of seismic ground motion in megacities: The Case of Rome, *Earthquake Spectra* **9**(4), 643–668.

Fäh, D., Suhadolc, P., Mueller, St., and Panza, G. F. (1994), A hybrid method for the estimation of ground motion in sedimentary basin: quantitative modelling for Mexico City, *Bull. Seismol. Soc. Am.* **84**, 383–399.

Gusev, A. A. (1983), Descriptive statistical model of earthquake source radiation and its application to an estimation of short-period strong motion, *Geophys. J. R. Astr. Soc.* **74**, 787–800.

Panza, G. F. (1985), synthetic Seismograms: The Rayleigh waves modal summation, *J. Geophysics* **58**, 125–145.

Panza, G. F., Romanelli, F., and Vaccari, F. (2000), Seismic wave propagation in laterally heterogeneous anelastic media: theory and applications to seismic zonation, *Advances in Geophysics* **43**, 1–95.

Panza, G. F., Schwab, F. A., and Knopoff, L. (1973), Multimode surface waves for selected focal mechanisms, I. Dip-slip sources on a vertical fault plane, *Geophys. J. R. Astr. Soc.* **34**, 265–278.

Panza, G. F. and Suhadolc, P., Complete strong motion synthetics. In *Seismic Strong Motion Synthetics* (ed. B. A. Bolt) (Academic Press, Orlando, 1987), *Computational Techniques* **4**, 153–204.

Romanelli, F., Bekkevold, J., and Panza, G. F. (1997), Analytical computation of coupling coefficients in Non-Poissonian media, *Geophys. J. Int.* **129**, 205–208.

Romanelli, F., Zhou, B., Vaccari, F., and Panza, G. F. (1996), Analytical computation of reflection and transmission coupling coefficients for Love waves, *Geophys. J. Int.* **125**, 132–138.

Sun, R., Vaccari, F., Marrara, F., and Panza, G. F. (1998), The main features of the local geological conditions can explain the macroseismic intensity caused in Xiji-Langfu (Beijing) by the Tangshan 1976 earthquake, *Pure Appl. Geophys.* **152**, 507–522.

Vaccari, F., Gregersen, S., Furlan, M., and Panza, G. F. (1989), Synthetic seismograms in laterally heterogeneous, anelastic media by modal summation of P-SV waves, *Geophys. J. Int.* **99**, 285–295.

Wessel, P., and Smith, W. H. F. (1995), The Generic Mapping Tools (GMT) Version 3.0 Technical Reference and Cookbook, SOEST/NOAA.

Wessel, P. and Smith, W. H. F. (1995), New version of the generic mapping tools released, *EOS Trans. AGU* **76**, 329.

Xie, Y. (1957), A new scale of seismic intensity adapted to the conditions in Chinese Territories, *Acta Geophysica Sinica* **6**, 35–48.

Zhang, Z., Li Q., Gu, J., Jin, Y., Yang, M., and Liu, W. (1980), The fracture processes of the Tangshan earthquake and analyses of mechanics, *Acta Seismologica Sinica* **1**, 111–129.

Estimation of site effects in Beijing City

Ding Z.[1], Chen Y. T.[1] and Panza G. F.[2,3]

1. Institute of Geophysics, China Seismological Bureau, Beijing, 100081, China;
2. Dipartimento di Scienza della Terra, Via Weiss 4, 34127 Trieste, Italy;
3. SAND Group, ICTP, Strada Costiera 11, I-34100 Trieste, Italy

Abstract For the realistic modeling of the seismic ground motion in lateral heterogeneous anelastic media, the database of 3-D geophysical structures for Beijing City has been built up to model the seismic ground motion in the City, caused by the 1976 Tangshan and the 1998 Zhangbei earthquakes. The hybrid method, which combines the modal summation and the finite-difference algorithms, is used in the simulation. The modeling of the seismic ground motion, for both the Tangshan and the Zhangbei earthquakes, shows that the thick Quaternary sedimentary cover amplifies the peak values and increases the duration of the seismic ground motion in the northwestern part of the City. Therefore the thickness of the Quaternary sediments in Beijing City is the key factor controlling the local ground effects. Four zones are defined on the base of the different thickness of the Quaternary sediments. The response spectra for each zone are computed, indicating that peak spectral values as high as 0.1 g are compatible with past seismicity and can be well exceeded if an event similar to the 1697 Sanhe-Pinggu occurs.

1 Introduction

China is one of the countries exposed to the largest seismic hazard in the world. Death caused by seismic activity exceeds the sum of victims caused by other natural hazards. Most provinces in China have historical records of destructive earthquakes.

China is located at the intersection of the Pacific Ocean seismic belt with the Euro-Asian seismic belt, and it is affected by the strongest continental seismic activity in the world. The historical records contain thousands of destructive earthquakes which occurred in China. They include eight earthquakes with magnitude greater than or equal to 8 before 1900. The 1556 ShanXi earthquake ($M=8.0$) claimed 830,000 victims. Past 1900, there were nine earthquakes with magnitude greater than or equal to 8, seven of which occurred in the continental area. The most immense was the 1950 Tibet earthquake with a magnitude of 8.6. The 1976 Tangshan earthquake ($M=7.8$) claimed at least 230,000 victims and destroyed a modern city in a few seconds. Based on the high level of seismic activity in China, the loss of human life and property due to seismic hazard is important and urgent to mitigate, especially in the megacities and large urban areas.

Beijing, the capital city of China, has a large population (about 12 million) and hosts many important economic and political centers. The city recurrently suffered from earthquakes. The latest great event was the 1697 Sanhe-Pinggu earthquake ($M=8$), sited 50km from the city. The maximum observed macroseismic intensity in

---

* 本文发表于 *Pure Appl. Geophys.*, 2004 年, 第 **161** 卷, 第 5-6 期, 1107-1123.

Beijing was XI. The intensity was scaled on the China Seismic Intensity Table (Xie, 1957). The 1976 Tangshan earthquake ($M = 7.8$) registered in Beijing a maximum intensity of VIII. The 1998 Zhangbei earthquake ($M = 6.2$) was the latest strong event felt in Beijing. The spatial distribution and the physical properties of the local structures are often correlated to the seismic damage distribution.

The estimation of the seismic ground motion produced by possible strong earthquakes (earthquake scenarios) is useful to reduce the seismic damage. In a companion paper Ding et al. (2004) have modeled digital recordings of the 1998 Zhangbei earthquake. One of the stations, FHSD, is located on the thick Quaternary sediment area, approximately 200km from the epicenter. The observed seismograms show that the seismic waves are substantially stronger at the station FHSD than at other stations located in the mountain area, at similar epicentral distances. The modeling of Ding et al. (2004) explains quite naturally the observations, i.e. that the Quaternary sediments enlarge the peak values and the duration of the seismic waves at the station FHSD. Using the same modeling technique validated by Ding et al. (2004) we can immediately compute the ground motion due to any scenario earthquake. In this manner, we obtain the seismic response at any place for any potential strong earthquake, and can estimate the distribution of future earthquakes, effects and damage in the research area.

For such a purpose, we built the data set of the physical properties of the local 3-D underground structure in Beijing City. With this data set, the geophysical structure along an arbitrary cross section can be obtained, and the latest computer codes developed at the Department of Earth Science, University of Trieste, Italy are used for the calculation of realistic ground motion (Panza, 1985; Panza and Suhadolc, 1987; Florsch et al., 1991; Fäh et al., 1993, 1994; Panza et al., 2000).

2 The structures database

The research area for Beijing City is defined latitudinally 39.8°N to 40.1°N and longitudinally 116.2°E to 116.6°E. The data from local dense drilling wells and geological survey results (Gao and Ma, 1993) are employed to define the distribution of the Quaternary and Tertiary sediment properties and thickness. In the research area there is no sedimentary cover in the mountain area, while the Quaternary sediments cover the entire plain district where the thickness increases from the northwestern mountain-plain boundary to the southeast. There are two abnormally thick Quaternary sediment zones near the city; one at the northwest margin of the city, the other within the northeast suburbs. In the latter the thickness reaches 800 meters.

The parameters in the database constructed to study the seismic ground motion in Beijing City include the density, the seismic velocities of P and S waves, and the attenuation parameter Q values for the different sedimentary units (Quaternary, Late Tertiary and Early Tertiary). The density values are obtained from local geophysical surveys and gravity inversion results (Group of Results of Deep Geophysical Prospecting, 1986). The S wave velocity is derived from shallow seismic exploration and drilling well data.

Table 1 lists the available ranges of the considered parameters of the sediments. In the local geophysical data set the thickness of the three sedimentary layers is specified on a 0.02°×0.02° horizontal grid. At any point in the volume, the density, the P and S-wave velocity and the Q value can be obtained from the data set. These parameters are then used to build-up the input structural model for the calculation of the synthetic seismograms in the laterally heterogeneous anelastic media.

The synthetic seismograms are calculated along the profiles by using the hybrid method, which combines the modal summation (1-D) and the finite-difference (2-D) algorithms (Panza, 1985; Panza and Suhadolc, 1987; Florsch et al., 1991; Fäh et al., 1993, 1994; Panza et al. 2000). In the calculation, the seismic source, the travel path from the source to the research area and the site structure are all taken into account. The bedrock reference 1-D structure is taken from Sun et al. (1998).

In the hybrid calculation the finite-difference method requires that there are at least ten grid points inside the shortest wavelength (Fäh, et al., 1994). In our model the minimum S-wave velocity is 0.4 km/s. If we consider an upper frequency limit of 4Hz, the minimum wavelength is 100meters, thus a grid size of 10meters is appropriate to describe the structural model. The dimensions of the local laterally heterogeneous models used in Beijing City are 4 km (depth) by about 40km (horizontal).

To provide for the source finiteness we use the scaling laws of Gusev (1983) by properly weighting the source spectrum in the frequency domain, as reported in Aki (1987). By using the scaled signals in the frequency domain, the response spectral ratio (RSR) corresponding to the laterally varying model and to the bedrock model, versus frequency and distance along the profiles, have been calculated to estimate the local response.

Table 1 Geophysical properties of the sediments

	Density/(g/cm^3)	V_P/(km/s)	V_S/(km/s)	Q_S
Quaternary	1.8—2.2	1.0—3.5	0.4—2.0	40—60
Later Tertiary	2.4—2.5	4.0—4.6	2.35—2.65	100—130
Early Tertiary	2.6—2.8	5.0—5.6	2.9—3.3	150—175
Bedrock	2.85	5.9—6.1	3.44—3.54	200

The seismic ground motion in Beijing City is simulated for two selected large earthquakes: the 1976 Tangshan and the 1998 Zhangbei earthquakes, which lie to the southeast and the northwest of the City, respectively. These events can be considered representative of the most dangerous seismogenic areas around Beijing. The three-component broadband synthetic seismograms along five profiles crossing the city area have been calculated, with the maximum frequency of 4Hz. The ground motion in this frequency band is effective for various buildings existing inside the research area. Three profiles, TS02, TS03, TS04, point towards the epicenter of the 1976 Tangshan earthquake, two profiles, ZB05 and ZB06, point towards the epicenter of the 1998 Zhangbei earthquake.

3 Synthetic results

The 1976 Tangshan earthquake produced damage in Beijing City. The observed macroseismic intensity was VI in most of Beijing City. However an abnormal intensity was observed in the northwestern part of the city with a value of VII. The source parameters of the earthquake can be referred to Ding et al. (2004). The seismic ground motion for the 1976 Tangshan earthquake is modeled along the three profiles, TS02, TS03 and TS04, and shown in Figure 1. The distance along the profile is measured from the epicenter of the 1976 Tangshan earthquake.

The profile TS02 passes south of the city. The Quaternary sediment is thinner westward (toward the mountain area) with increasing epicenter distance. The Tertiary Beijing depression is located between the distance of 160 and 172km (Fig. 2). The synthetic seismograms reveal that the thick Tertiary sediments amplify only the seismic ground motion of Rayleigh and P-SV waves. The Love and SH waves which, due to the source orientation with respect to the considered profiles in the city are the strongest waves in this case, do not show significant unusual phenomena.

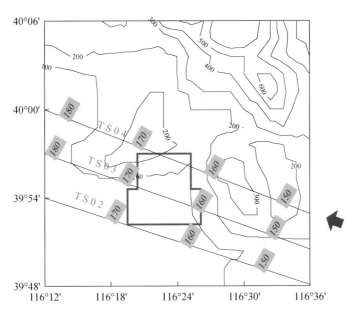

Figure 1　Profiles for the 1976 Tangshan Earthquake

The background contours represent the Quaternary sediment depth in meters. The polygon represents the city of Beijing. Three profiles, TS02, TS03 and TS04 are shown in the figure. The profiles point towards the epicenter of the 1976 Tangshan earthquake, which is located in the southeast. The numbers along the profiles are the distances from the epicenter, in km

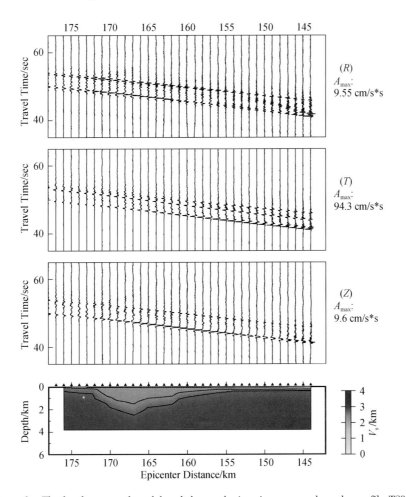

Figure 2　The local structural model and the synthetic seismograms along the profile TS02

The lines in the bottom figure outline the three sediment layers. Radial, transverse and vertical components of the synthetic ground acceleration

The profile TS03 passes through the center of the city. The seismic waves are mainly controlled by the thickness of the Quaternary sediments (Fig. 3). The acceleration amplitudes are enlarged at the distance of 148-160km and 175km, where thicker Quaternary sediments exist.

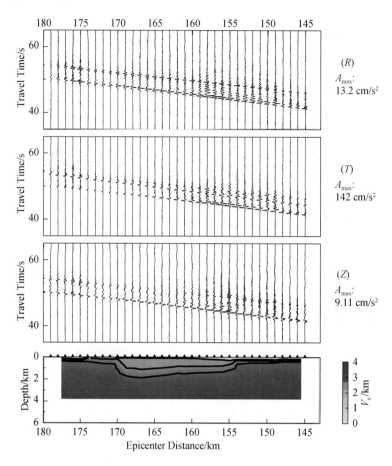

Figure 3 The local structural model and the synthetic seismograms along the profile TS03

The lines in the bottom figure outline the three sediment layers. Radial, transverse and vertical components of the synthetic ground acceleration

The profile TS04 passes through the northern part of the City. The cross section cuts across the Tertiary depression zone between the distance of 157 and 168 km, and the thickness of the Tertiary sediments along TS04 is not as large as along TS02 and TS03. Two thick Quaternary sediment areas are situated around 153—163km and 170km (see Fig. 4). The synthetic acceleration time history of profile TS04 is shown in Figure 4. The amplitude of the transverse component is 10 times larger than the radial and vertical components, and the waveforms of all three components along the profile are mainly controlled by the thickness of the Quaternary sediment. Enlarged amplitudes and longer durations of seismic ground motion characterize the northwestern part of the city, at the distance of 170km. This is the district of Beijing where, during the 1976 Tangshan earthquake, abnormally high—one degree higher than the value in the surrounding area—macroseismic intensity has been observed.

In Figure 5 the RSR versus epicentral distance and frequency reaches the large values corresponding to of the thick Quaternary sediment at about 170km and at the frequency of 1-2Hz, which is the fundamental resonant frequency of the Quaternary sediment layer there. For the transverse component, which is the dominant one, the RSR for a set of selected sites is shown in Figure 6, to illustrate the variation of the dominant frequency along the profile.

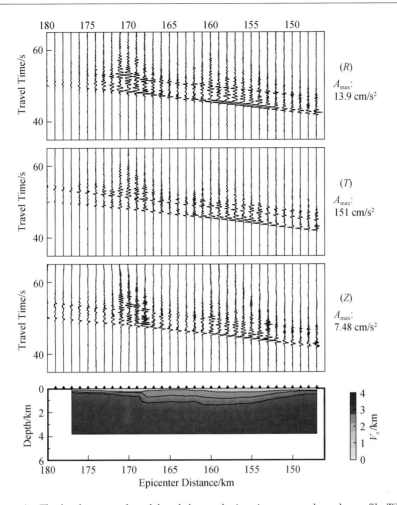

Figure 4 The local structural model and the synthetic seismograms along the profile TS04

The lines in the bottom figure outline the three sediment layers. Radial, transverse and vertical components of the synthetic ground acceleration

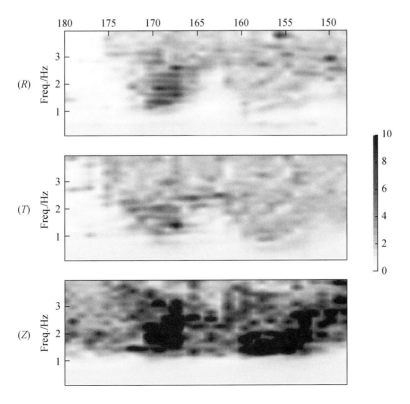

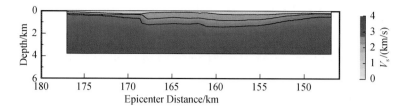

Figure 5　RSR versus frequency and distance along profile TS04

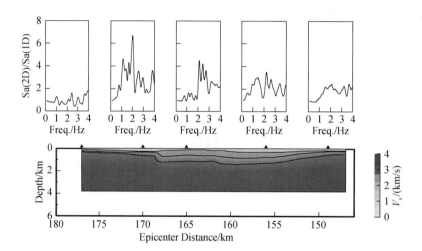

Figure 6　RSR of SH waves at selected sites along profile TS04

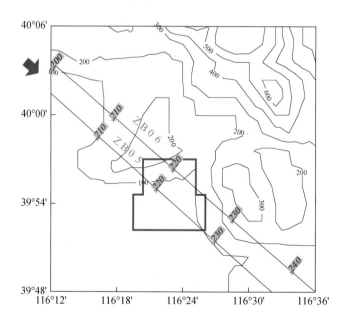

Figure 7　Profiles for Zhangbei Earthquake

The background contours represent the Quaternary sediment depth in meters. The polygon represents the city of Beijing. Two profiles, ZB05 and ZB06 are shown in the figure. The profiles point towards the epicenter of the 1998 Zhangbei earthquake, which is located in the northwest. The numbers along the profiles are the distances from the epicenter, in km

On January 10, 1998 an earthquake occurred in Zhangbei County, which is located to the northwest of Beijing City at a distance of about 200km. The source parameters of the earthquake can be referred to Ding et al.

(2004). The scale of the profiles is the distance from the epicenter.

The profile ZB05 passes through the northwest and southeast corners of the City. There are thick Quaternary sediments between 212km and 217km, while the Tertiary depression zone is located between 217km and 231km. Figure 8 shows the local structure and the synthetic seismograms along the profile. The seismic waves are obviously amplified at about 215km, where the thick Quaternary sediments are located.

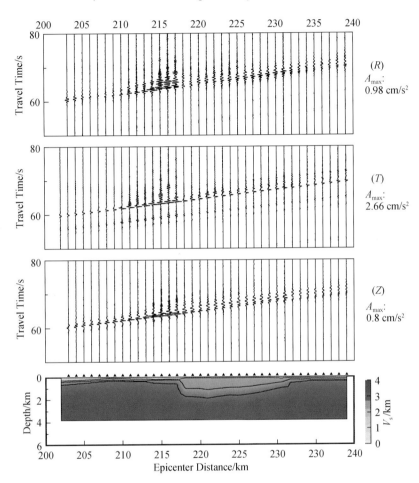

Figure 8   The local structural model and the synthetic seismograms along the profile ZB05

The lines in the bottom figure outline the three sediment layers. Radial, transverse and vertical components of the synthetic ground acceleration

Figure 9 shows the maximum acceleration (A_{max}) and the ratios $A_{max}(2D)/A_{max}(1D)$ and $W(2D)/W(1D)$ versus epicenter distance. Here (2D) represents the local lateral 2-D inhomogeneous structure, (1D) the reference 1-D bedrock structure and W is the so-called relative Arias intensity (Arias, 1970) defined as:

$$W = \frac{\pi}{2g}\int_0^\infty [\ddot{x}(\tau)]^2 d\tau,$$

where x is the ground displacement and g is the gravity acceleration. These parameters reach peak values at an epicenter distance of approximately 215km, where the thick Quaternary sediments are located.

From the RSR versus frequency and distance along this profile (Fig. 10) the dominant part at 213 km is mainly at high frequency, from 1.7 to 3Hz. This is the fundamental resonant frequency for the thick Quaternary sediment. For other places with thinner Quaternary sediments the dominant frequency is higher, exceeding 2.5Hz.

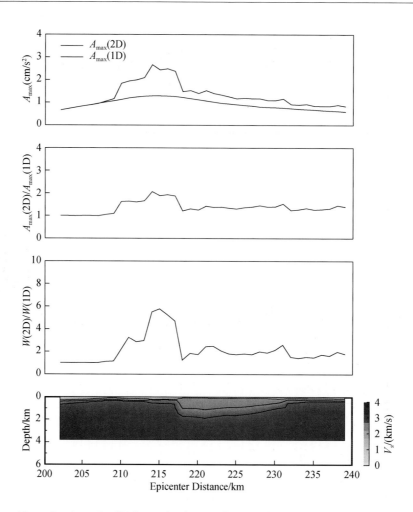

Figure 9 A_{max}, $A_{max}(2D)/A_{max}(1D)$ and $W(2D)/W(1D)$ along the profile ZB05

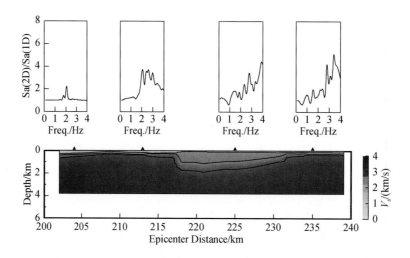

Figure 10 RSR of SH waves at selected sites along the profile ZB05

Profile ZB06 crosses the north and east suburbs and the northeastern part of the City. Two thick Quaternary sediment zones lie on the profile at about 215km and 232km. Figure 11 presents the synthetic three-component acceleration time histories along the profile: the amplification of the seismic waves is controlled by the thickness of the Quaternary sediments.

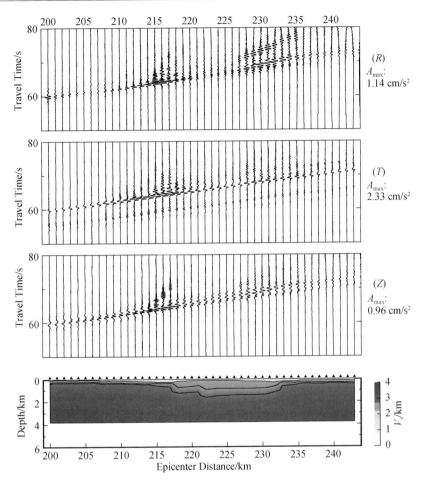

Figure 11   The structural model and the synthetic seismograms along the profile ZB06

4   Site effects in Beijing City

From the synthetic acceleration time histories used to model the ground motion arising from the 1976 Tangshan and the 1998 Zhangbei earthquakes, it can be concluded that mainly the thickness of the Quaternary sediments controls the seismic ground motion variations in Beijing City. The two earthquakes are at different azimuths, therefore it is reasonable to extend such a conclusion to the entire research area (Panza et al., 2000). Four zones can be defined accordingly to the thickness of the Quaternary sediments: Zone 1 includes the areas with a thickness of Quaternary sediments less than 50m, Zone 2 between 50m and 100m, Zone 3 between 100m and 200m, and Zone 4 greater than 200m. The distribution of the four zones is shown in Figure 12.

For the Tangshan earthquake the RSR have been computed for all sites located in each of the four zones along the profiles TS02, TS03 and TS04. From these values the average and the maximum RSR for 0% and 5% damping of the oscillator are determined and shown in Figure 13. Such spectral properties can be considered representative of the four zones shown in Figure 12.

Figure 14 displays the absolute response spectra for each zone. As in the case for the EC8 design spectrum, the DGA can be obtained by dividing the largest spectral value by 2.5 in each zone. Therefore the DGA in Beijing City for the 1976 Tangshan earthquake could be obtained as 0.03g, 0.04g, 0.07g and 0.1g for Zone 1, Zone 2, Zone 3 and Zone 4, respectively.

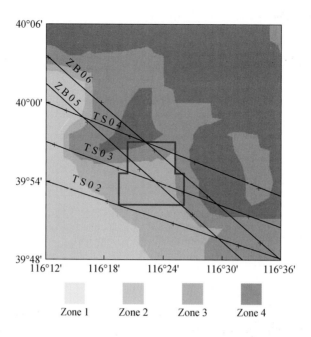

Figure 12　Different site effect zones in Beijing City

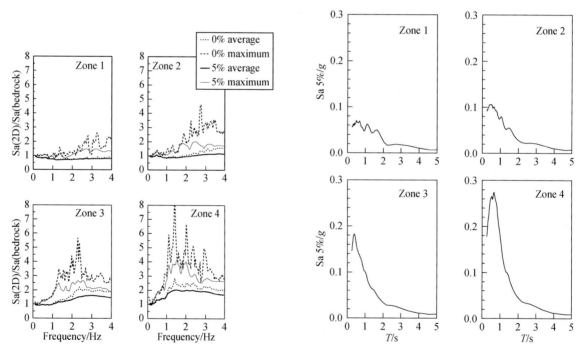

Figure 13　Maximum and average RSR for the four zones (for 0% and 5% damping)

Figure 14　Absolute spectral acceleration (5% damping) for the four zones

5　Conclusions

3-D geological and geophysical models have been built for Beijing City. With this database realistic three-component broadband synthetic seismograms have been calculated with a cutoff frequency of 4Hz. The synthetic acceleration time histories are computed along five selected profiles.

Along all profiles the thick Quaternary sediments effect a large amplitude and long duration of the ground motion due to resonance effects and to the excitation of local surface waves.

Four zones are defined in accordance with the thickness of the Quaternary sediments in the area of Beijing City; each characterized by different spectral responses. The response spectra indicate that peak spectral values as high as 0.1g are compatible with the past seismicity and can be well exceeded if an event similar to the 1697 Sanhe-Pinggu occurs.

Acknowledgements

This work is a contribution to the UNESCO-IUGS-IGCP Project 414 "Realistic Modeling of Seismic Input for Megacities and Large Urban Areas." This research has been carried on within the framework of the bilateral project "Geophysical Studies for the Deterministic Evaluation of Seismic Risk," with the contribution of the Italian Ministry of Foreign Affairs (MAE), Directorate General for Cultural Promotion and Cooperation. The research received support from the Chinese National Key Basic Research and Development Program (973 Program) No. 2002CB412709, MOST of China and China Seismological Bureau. We acknowledge Drs. Fabio Romanelli, Franco Vaccari and Francesco Marrara for their kind assistance in the calculation. This is contribution No. 02A10002, Institute of Geophysics, China Seismological Bureau. We have used public domain graphics software (Wessel and Smith, 1995a, 1995b).

References

Aki, K., Strong motion seismology. In *Strong Ground Motion Seismology* (eds. Erdik, M. Ö. and Toksöz, M. N.) (NATO Advanced Study Institute Series, Series C: Mathematical and Physical Sciences, D. Reidel Publishing Company. The Netherlands 1987), **204**, pp. 3-39.

Arias, A. (1970), A measure of earthquake intensity. In *Seismic Design for Nuclear Power Plants* (ed. R. Hansen), Cambridge, Massachusets.

Ding, Z., Romanelli, F., Chen, Y., and Panza, G. F. (2004), Realistic modeling of seismic strong ground motion in Beijing Area, *Pure Appl. Geophys.* **161**, 1093-1106.

Fäh, D., Iodice, C., Suhadolc, P., and Panza, G. F. (1993), A new method for the realistic estimation of seismic ground motion in megacities: The case of Rome, *Earthquake Spectra* **9**(4), 643-668.

Fäh, D., Suhadolc, P., Mueller, St., and Panza, G. F. (1994), A hybrid method for the estimation of ground motion in sedimentary basin: Quantitative modelling for Mexico City, *Bull. Seismol. Soc. Am.* **84**, 383-399.

Florsch, N., Fäh, D., Suhadolc, P., and Panza, G. F. (1991), Complete synthetic seismograms for high-frequency multimode SH waves, *Pure Appl. Geophys.* **136**, 529-560.

Gao, W., MA, J., *Seismo-geological Surroundings and Seismic Hazards in Capital Area* (Seismological Press, Beijing, 1993).

Group of Results of Deep Geophysical Prospecting, State Seismological Bureau, *Results of Deep Exploration of the Crust and Upper Mantle of China* (Seismological Press, Beijing, 1986).

Gusev, A. A. (1983), Descriptive statistical model of earthquake source radiation and its application to an estimation of short-period strong motion, *Geophys. J. R. Astr. Soc.* **74**, 787-800.

Panza, G. F. (1985), Synthetic seismograms: The Rayleigh waves modal summation, *J. Geophysics* **58**, 125-145.

Panza, G. F. and Suhadolc, P., Complete strong motion synthetics. In *Seismic Strong Motion Synthetics* (ed. B. A. Bolt) (Academic Press, Orlando, 1987), Computational Techniques **4**, 153-204.

Panza, G. F., Romanelli, F., and Vaccari, F. (2000), Seismic wave propagation in laterally heterogeneous anelastic media:

Theory and applications to seismic zonation, *Advances in Geophysics* **43**, 1–95.

Sun, R., Vaccari, F., Marrara, F., and Panza, G. F. (1998), The main features of the local geological conditions can explain the macroseismic intensity caused in Xiji-Langfu (Beijing) by the Tangshan 1976 earthquake, *Pure Appl. Geophys.* **152**, 507–522.

Wessel, P. and Smith, W. H. F. (1995a), New version of the Generic Mapping Tools Released, *EOS Trans. AGU* **76**, 329.

Wessel, P. and Smith, W. H. F. (1995b), *The Generic Mapping Tools (GMT) Version 3.0 Technical Reference and Cookbook*, SOEST/NOAA.

Xie, Y. (1957), A new scale of seismic intensity adapted to the conditions in Chinese Territories, *Acta Geophysica Sinica* **6**, 35–48.

Decade-scale correlation between crustal deformation and length of day: implication to earthquake hazard estimation[*]

Wang Q. L.[1,2], Chen Y. T.[3], Cui D. X.[2] and Wang W. P.[2]

1. Second Crustal Monitoring Center, China Earthquake Administration, Xi'an 710054 (China);
2. College of Geological Engineering and Survey Engineering, Changan University, Xi'an 710054 (China);
3. Institute of Geophysics, China Earthquake Administration, Beijing 100081 (China)

Abstract Crustal deformations are measured to fluctuate irregularly on different time scales, but we know little about its mechanism. In recent years' comparison, we found a popular decade-scale correlation phenomenon between crustal deformation and length of day (l.o.d.) of the Earth. The decade correlations mainly display as the following characteristics: (1) Original or de-trended crustal deformations fluctuate synchronously along with changes in length of day; (2) Crustal deformation and length of day share the same or close turning points; (3) The correlations are location dependent and direction dependent.

The finding of the decade-scale correlation is of importance to geodynamics and earthquake hazard estimation: First, it reveals that core-mantle coupling is an important modulation for crustal deformations; Second, the decade-scale correlations can be used to improve the reliability of earthquake precursor identification.

1 Introduction

Earthquake prediction is an important topic in science. Based on rock mechanical experiments and quasi-linear movements of tectonic plates, it is generally believed that some kinds of precursors or anomalies, traditionally classified into long-term anomaly, medium-term anomaly and short-term anomaly, should exist before a strong earthquake, and the anomaly itself is defined as a deviation from linear trend or tidal fluctuation of the observations. Estimation of earthquake hazard at decade scale demonstrates that, relative to the very limited short-term and imminent precursors detected in or around the earthquake region, most of the fluctuating anomalies seem to have little thing to do with earthquakes although they often synchronously appear in a large area.

Although several mechanisms, for example, the earthquake energy release and stress propagation mechanism (Ishii et al., 1978, 1980), have been proposed to explain the background variation of premonitory observations, the reason why crustal deformations and geophysical observations often change irregularly on different time scales has not been clearly understood yet, which, in turn of course, restricts the proper identification of real earthquake precursors.

In early 1999, in comparing time series data of fault deformations in China with Earth orientation parameters (EOP) downloaded from IERS web site, we surprisingly found that fault deformations correlate well to length of day on decade time scales at many monitoring sites (Wang et al., 2000). In the following years, we further

---

[*] 本文刊载于地震出版社出版的 *Earthquake: Hazard, Risk, and Strong Ground Motion* 一书中, 2004 年, 151–163.

extended our comparison to the monitoring data of tilt, strain, stress and gravity, and also found lots of similar decade correlation phenomena to l.o.d (Wang, 2003). In this paper, we will introduce some representative correlation evidences in crustal deformation observations, and then discuss its implications to earthquake hazard.

2 Earth rotation variation

Paleontological clock evidence, ancient and modern astronomical observations demonstrate that, the Earth does not spin at a stable rate. The variations in rotation rate or l.o.d can be roughly classified into three types: secular deceleration, periodic variations and irregular variations. The secular variation, about $1 \sim 2$ ms in l.o.d per century, is primarily due to friction of ocean tides on the surface of the Earth. Periodic l.o.d fluctuations, typically on annual and semi-annual time scales (Figure 1), are mainly excited by global atmospheric circulation and solar tide (Lambeck, 1980). The reason for irregular l.o.d fluctuations has not been well understood yet, but the irregular variations with characteristic time scale of $10 \sim 20$ years (Figure 1), are generally attributed to angular momentum exchanges between the core and the mantle (Lambeck, 1980). Three possible core-mantle mechanisms, topographic coupling, electromagnetic coupling and gravitational coupling, have been proposed to explain the decade-scale variations of length of day (Hide and Horai, 1968; Hide, 1995; Jault and Le Mouel, 1989, 1990; Rochester, 1968, 1984; Holme, 1998a, b; Buffett, 1996a, b), but no one was thoroughly verified or ruled out up to now.

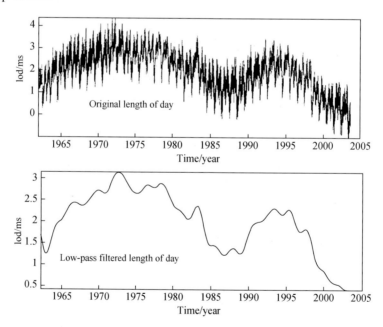

Figure 1 Time series of length of day (based on IERS product)
Upper subplot: Original observations of length of day with typical annual and semi-annual variations;
Lower subplot: Low-pass filtered length of day with tow typical decade-scale fluctuations

3 Decade-scale correlation characteristics

In order to capture possible earthquake precursors, some comprehensive premonitory networks have been established in China, Japan, western United States and other seismic dangerous areas of the world. Crustal

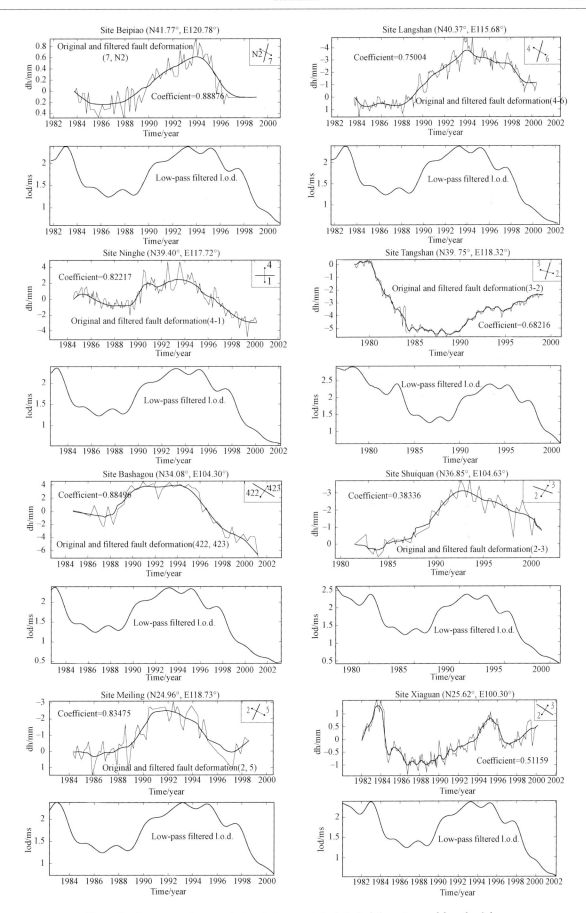

Figure 2 Representative correlations between original fault deformation and length of day

deformation, owing to its definite constitutive relation to stress, has all along been a major useful observation method in the premonitory networks. Our correlation comparisons are mainly base on the deformation data collected in China, California, U. S. A and Friuli, NE Italy.

According to our investigations, as long as precipitation and artificial effects are small enough, and the observation precisions are good enough, long-term deformations would appear correlation property more or less to l. o. d. on decade time scales, and the correlations mainly present the following characteristics:

(1) Synchronous fluctuation: Along with decade fluctuations in length of day, crustal deformation also changes its process synchronously, both of the variation patterns are very similar. Figure 2 shows some representative correlation examples between length of day and fault deformations in China, the observation sites are located respectively in northern China, southern China, north-western China and south-western China.

(2) De-trended correlation: At some monitoring sites, the original crustal deformations display as quasi-linear changes (plate-movement dominated or instrument-drift dominated), after the de-trende process. However, the correlations between length of day and the residual deformations become more clear. Figure 3 shows some detrended correlation examples of the fault deformation and tilt observations in China, the sites are also located in different areas of China.

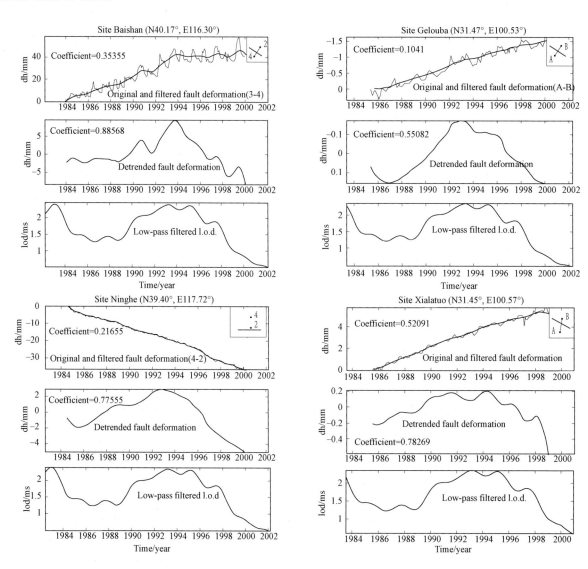

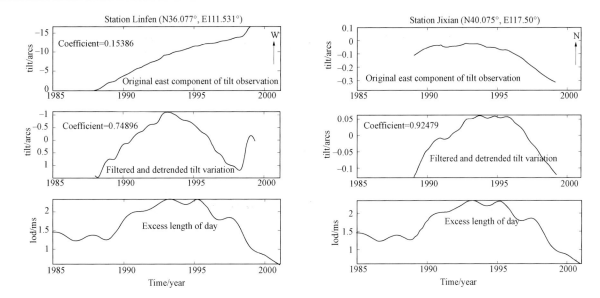

Figure 3 Representative correlations between crustal deformation and length of day

The upper four plots are the correlation examples between fault deformation and length of day; The lower two figures are the correlation examples between tilt observation and length of day

(3) With the same or close turning points: This is the most common case of the correlations. In this case, crustal deformation does not change very synchronously along with variation in length of day. The turning points of the both, however, are nearly the same or close. Figure 4 shows some examples of the synchronous turning points of l.o.d and tilt observations of China, it can be seen that the turning points of l.o.d round the year 1990 and 1995 are also the turning points of tilt variations.

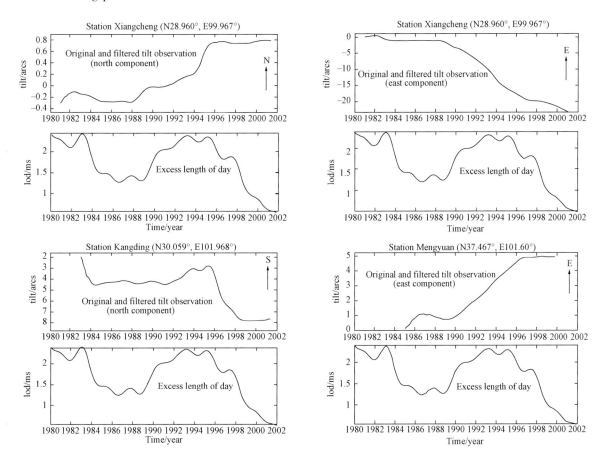

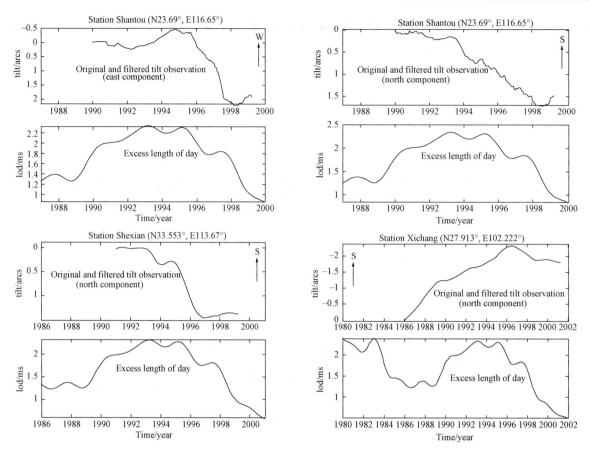

Figure 4 The same or close turning points between tilt observation and length of day

(4) Direction dependence: For the same station, different directional observations or different component observations often display as different correlation properties to l.o.d Figure 5 presents two examples of the direction

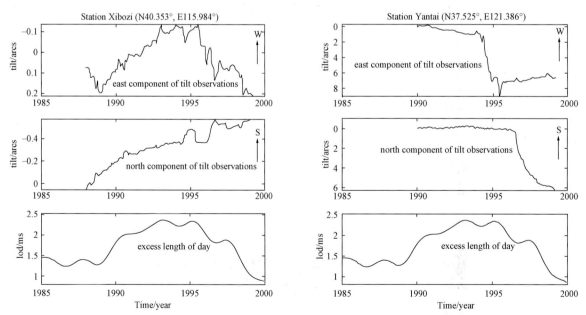

Figure 5 Direction-dependent correlation between tilt observation and length of day

Left: Tilt gauge Xibozi located near Beijing, where the whole process of the east component correlate well to length of day; for the north component, only the first half process correlate well to length of day.

Right: Tilt gauge Yantai located in eastern China, where the turning point of north component lags about two years after the east component

dependence of the tilt observations in China. We can see that, for station Xibozi at Beijing, the east-west component fluctuates synchronously along with decadal variation of length of day, while the north-south component dose not. For station Yantai at Shandong Province, the turning point of north-south component lags about two years after that of east-west component.

(5) Location dependence: There are two meanings for the location dependence: first, different stations at different places usually appear different correlation pattern between crustal deformation and l.o.d. Second, even within the same monitoring station or site, the correlation is also geological location dependent, Figure 6 shows two examples of the geological location dependence in fault deformations.

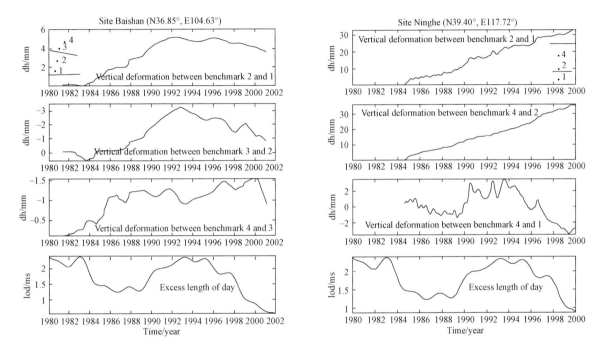

Figure 6 Location-dependent correlation between fault deformation and length of day

Left: Leveling monitoring site Shuiquan across Haiyuan fault at northeastern margin of Tibetan plateau, where the vertical deformation between benchmark 2 and 3 at both sides of the fault possesses the best correlation to length of day;

Right: Leveling monitoring site Ninghe across Ninghe fault in northern China, where the vertical deformation between benchmark 1 and 4 possesses the best correlation to length of day, other two deformations between benchmark 1 and 2 as well between 2 and 4 change quasi-linearly along with time

(6) Global correlation: Decade-scale correlation phenomena between crustal deformation and length of day not only appears in China, but also exists in other country or areas. Figure 7 is the two-color geodimeter network at Parkfield, which across the right-lateral San Andreas fault, California. For this ranging network, the cross-fault de-trended baselines change synchronously along with decade variation in length of day (Figure 8), where for the shortening baselines between the instrument station CARR and mirror stations BARE, BUCK, CANN and MIDE at the northwestern side, the correlations appear to be positive, and for the lengthening baselines between center station CARR and mirror stations HUNT, TURK, MELV and GOLD at the southeastern side, the correlations appear to be negative. Langbein et al. (1999) once found the slip-rate increase in 1993, our comparison here implies that the slip-rate increase in 1993 may has something to do with the Earth's rotation variation.

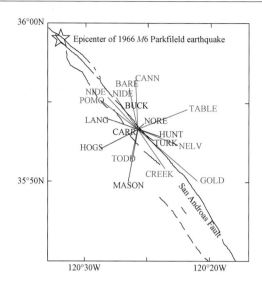

Figure 7 Two-color Geodimeter network at Parkfield, California

where center station CARR at west side of San Andreas fault is the instrument station, other stations around CARR are the mirror stations

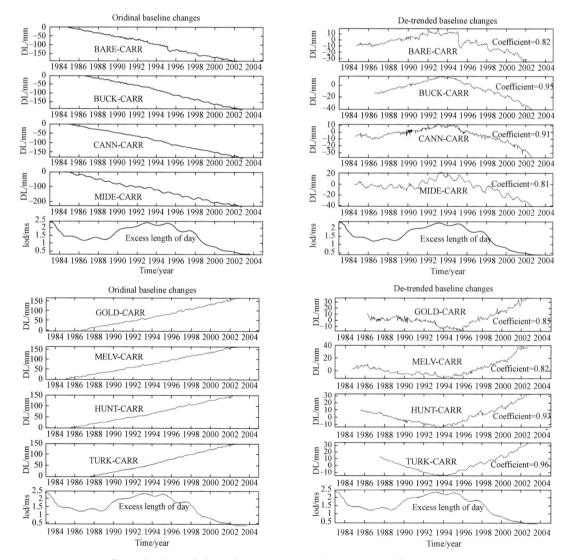

Figure 8 Detrended correlation between baseline change and length of day

The original cross-fault baselines change quasi-linearly and do not appear any correlation to length of day; The detrended cross-fault baselines change synchronously along with decade variations in length of day

Figure 9 is the tilt/strain gauge network of Friuli area, NE Italy, which was established mainly after the 1976 Friuli 6.4 earthquake. Rossi and Zadro (1996) once noticed the long-term variations in the tilt/strain observations, but they attributed the long-term variations to deep crust. Our comparison work reveals that the secular variations in tilt/strain observations in Friuli, Italy are also actually l.o.d correlated on decade time scales (Figure 10), especially for the station IN (Invillino), VI (Villanova) and Trieste.

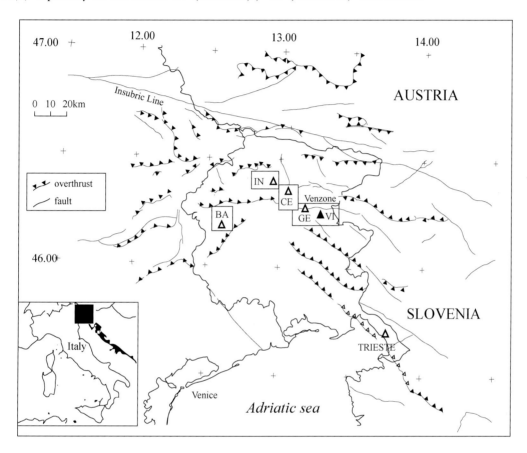

Figure 9 Distribution map of faults and tilt/strain gauges in Friuli, NE Italy

△ tilt station; ▲ tilt/strain station

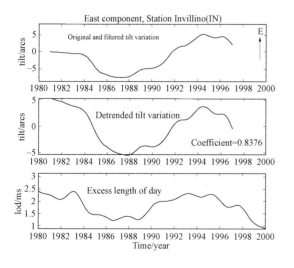
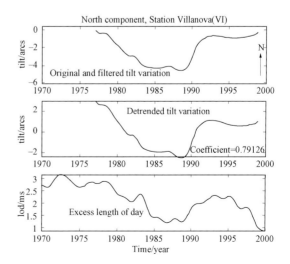

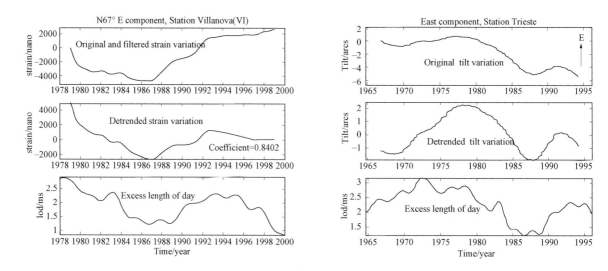

Figure 10 Representative correlations between tilt/strain observations and length of day

4 Correlation Mechanism and Implication to Earthquake Hazard

There are four possible mechanisms for the decade-scale correlation between crustal deformation and length of day: first, decade fluctuations in l.o.d modulate decade fluctuations in crustal deformations; second, decade fluctuations in crustal deformations induce decade fluctuations in the Earth's rotation rate; third, both decade fluctuations in crustal deformation and length of day have the same origin; forth, crustal deformation and length of day are only mathematically correlative, but not physically relative.

In his doctoral dissertation, Wang (2003) has discussed the possibility of the four candidates, and inferred that the most acceptable mechanism is that both decade fluctuations in crustal deformation and l.o.d originate from the same core-mantle coupling process (topographic or electromagnetic coupling). Wang (2003) further pointed out that, decade variations of the Earth's rotational rate are only sensitive to the change of latitudinal torsion torque acting on bottom of the mantle. Crustal deformations, however, are dependent not only on the change of latitudinal traction force, but also on the changes of longitudinal traction force and radial normal force on the core-mantle boundary (CMB). Moreover, inhomogeneity of the mantle and the crust also exerts great influence on crustal deformation property. Therefore, correlations between crustal deformation and length of day present very complicated patterns station by station, direction by direction, and time by time.

The finding of decade-scale correlation between crustal deformation and length of day is of importance to earthquake prediction:

First, it widens and deepens our understanding of the premonitory observations, for instance, from the decade-scale correlation evidences introduced in this paper, we realize that large part of the background anomalies in premonitory observations are actually l.o.d correlated and deep core-mantle coupling originated. Therefore, by employing the Earth rotation data downloaded from the IERS web site, we might improve the reliability of earthquake precursor identification to certain extent, for example, if some moderate anomalies come out in a large area and change quasi-synchronously along with length of day, they most possibly are the core-mantle coupling modulated variations, and should not be taken as earthquake-related precursors.

Second, global and regional seismic activities are investigated to correlate with variation in the Earth's rotation rate change to certain extent (Xu, 1980; Du and Xu, 1989; Anderson, 1974), but the reason was not clear yet. Centrifugal force induced by changes in the Earth's rotation rate is only an order of $10^{-1} \sim 10^{0}$ Pa, which is too small to trigger an earthquake (Wang, 2003), the decade-scale correlation between crustal deformation and length of day reveals that core-mantle coupling process is actually an important trigger contribution to earthquake occurrence.

Third, complexity of decade-scale correlation between crustal deformation and length of day provides an excuse for the difficulty of earthquake prediction. Owing to inhomogeneity of the crust and complexity of core-mantle coupling forces at CMB, crustal deformation is not always strong correlated to length of day, and in most cases, they only share the same or close turning points. Moreover, l.o.d correlated crustal deformation usually surpass the possible medium-term anomaly of an earthquake, so on this strong 'noisy' background (together with other factors such as precipitation and artificial activity), it is very difficult or nearly impossible to pick up any reliable medium-term precursor of an earthquake. The short-term or impending precursor, however, might be identified before an earthquake just because its bursting characteristics, so we should not be over-pessimistic to earthquake prediction work.

5 Conclusions

Crustal deformations are measured to fluctuate on different time scales, and on decade time scales, they appear to correlate well to great extent with decade fluctuations in length of day of the Earth. Core-mantle coupling is the major mechanism for the decade-scale correlation.

The finding of decade-scale correlation between crustal deformation and length of day is of importance for earthquake prediction, because it might help to improve the reliability of precursor identification.

References

Anderson, D. L., 1974. Seismic activity and changes in the Earth's rotation rate correlated by CalTech seismologist. *Earthquake Information Bulletin* **6**, 2.

Buffett, B. A., 1996a. Gravitational oscillations in the length of day. *Geophys. Res. Lett.*, **23**, 2279-2282.

Buffett, B. A., 1996b. A mechanism for decadal fluctuations in the length of day. *Geophys. Res. Lett.*, **23**, 3803-3806.

Du, P.-R. and Xu, D.-Y., 1989. *An Introduction to Astroseismology*. Seismology Press, Beijing (in Chinese).

Hide, R. and Horai, K. I., 1968. On the topography of the core-mantle interface. *Phys. Earth Planet. Inter.* **1**, 305-308.

Hide, R., 1995. The topographic torque on a bounding surface of a rotating gravitating fluid and the excitation by core motions of decadal fluctuations in the Earth's rotation. *J. Geophys. Res.* **22**, 3561-3565.

Holme, R., 1998a. Electromagnetic core-mantle coupling I, Explaining decadal changes in length of day. *Geophys. J. Int.* **132**, 167-180.

Holme, R., 1998b. Electromagnetic core-mantle coupling II, Probing deep mantle conductance. Gurnis, M., Wysession, M. E., Knittle, E. and Buffett, B. A. (Eds.), The Core-Mantle Boundary Region, *Geodynamics Series* **28**(AGU), 139-151.

Ishii, H., Sato, T. and Takagi, A., 1978. Characteristics of strain migration in the northeastern Japan arc (I)—propagation characteristics. *Sci. Rep. Tohoku Univ. Ser. S. Geophysics* **25**, 83-90.

Ishii, H., Sato, T. and Takagi, A., 1980. Characteristics of strain migration in the northeastern Japan arc (II)—amplitude characteristics. *J. Geod. Soc. Japan* **26** (1), 17-25.

Jault, D. and Le Mouel, J. L., 1989. The topographic torque associated with a tangentially geostrophic motion at the core surface and inferences on flow inside the core. *Geophys. Astrophys. Fluid Dyn.* **48**, 273-296.

Jault, D. and Le Mouel, J. L., 1990. Core-mantle boundary shape: constraints inferred from the pressure torque acting between the core and the mantle. *Geophys. J. Int.* **101**, 233–241.

Lambeck, K., 1980. *The Earth's Variable Rotation*, Cambridge Univ. Press, New York.

Langbein, J., Gwyther, R. and Gladwin, M. T., 1999. Slip-rate increase at Parkfield in 1993 detected by high-precision EDM and borehole tensor strainmeters. *Geophys. Res. Lett* **26**, 2529–2532.

Rochester, M. G., 1968. Perturbations in the Earth's rotation and geomagnetic core-mantle coupling. *J. Geomagnetism and Geoelectricity* **20**, 387–402.

Rochester, M. G., 1984. Causes of fluctuations in the rotation of the Earth. *Philos. Trans. R. Soc. London* **A313**, 95–105.

Rossi, G. and Zadro, M., 1996. Long-term crustal deformations in NE Italy revealed by tilt-strain gauges. *Phys. Earth Planet. Interi.* **97**, 55–70.

Wang, Q. L., Chen, Y. T., Cui, D. X., Wang, W. P. and Liang, W. F., 2000. Decadal correlation between crustal deformation and variation in length of day of the Earth. *Earth, Planets, Space* **52**, 989–992.

Wang, Q. L., 2003. Decade-scale correlation between crustal deformation, geophysical variation and length of day. Ph. D. dissertation, Institute of Geophysics, China Seismological Bureau (in Chinese).

Xu, D. Y., 1980. *Planet Movement and Earthquake Prediction*. Seismological Press, Beijing (in Chinese).

Deterministic seismic hazard map in North China[*]

Ding Z.[1], Vaccari F.[2], Chen Y. T.[1] and Panza G. F.[2,3]

1. Institute of Geophysics, China Earthquake Administration, Beijing 100081 (China);
2. Department of Earth Science, Trieste University, Trieste (Italy);
3. The Abdus Salam International Center for Theoretical Physics, SAND Group, Trieste (Italy)

Abstract The seismic hazard map in North China is assessed using a deterministic approach. The seismic sources are represented at a set of grid points located at distances of 0.2° from each other. The main inputs for this computation are earthquake catalogue, source mechanisms, the level of seismic activity and the structural models. Synthetic seismograms are generated by different sources, and are used to obtain the distribution map of the maximum ground motion and Design Ground Acceleration (DGA). In the research area, the highest value of DGA is at east of Beijing, with the value as high as 0.75g.

1 Introduction

The research area in this study is part of Northern China. It includes the Northern China Plain, part of Bohai Sea, Taihang Mountain and Fengwei Rift. Some large cities, such as Beijing, Tianjin, Taiyuan, Jinan, Tangshan, are located in this region.

The seismicity is very active in recent centuries in this region. Some great earthquakes of magnitude 7.0 and above have occurred. The event of Sep. 2, 1679 is the strongest one, with the magnitude of 8.0, which was very close to the capital city Beijing. The events of March 22, 1966 in Xingtai and July 28, 1976 in Tangshan destroyed many buildings and caused a high number of deaths.

The economic and social effects of earthquake disasters can be reduced through a comprehensive assessment of seismic hazard and risk. Such a study leads to an increased public awareness, with a consequent upgrading of the existing buildings and engineering works as well as reliable earthquake-resistant designs for new structures. With the knowledge of the geological and geophysical structure and probable earthquake source mechanisms, realistic modeling for the ground motions at all sites of interest can be determined. In this way, we can map the seismic ground motion before the installation of the dense seismic network and the occurrence of earthquakes.

This is the first step to obtain the seismic hazard map of China. Based on the previous work in the Beijing area (Sun et al., 1998; Ding et al., 2003, Ding et al., 2003), we extend our research area to the North China, to obtain a first-order of deterministic seismic hazard map.

The main aim of the present study is to obtain the map of possible seismic ground motion parameters distribution. It includes the parameters of peak value of displacement (D_{max}), velocity (V_{max}), acceleration (A_{max}), and the design ground acceleration (DGA) over the investigated area.

---

[*] 本文刊载于地震出版社出版的 *Earthquake: Hazard, Risk, and Strong Ground Motion* 一书中, 2004 年, 351–359.

2 Method

Based on the model summation method for the layered structures (Panza, 1985), a deterministic approach to estimate the seismic hazard was developed at the Department of Earth Science, University of Trieste, Italy (Costa et al., 1992, 1993; Panza et al., 1999), and subsequently applied to several regions of the world (e.g. Orozova-Stanishkova et al., 1996; Alvarez et al., 1999; Panza et al., 1999; Aoudia et al., 2000; AA VV, 2000; El-Sayed et al., 2001; Parvez et al., 2002). The procedure uses the available information on the Earth structure parameters, seismic sources, the level of seismicity of the research area, and wave propagation in an elastic media to compute the synthetic seismograms. The outcome of the procedure is maps of the peak displacement (D_{max}), peak velocity (V_{max}), acceleration (A_{max}), and design ground acceleration (DGA) over the investigation territory.

3 Input

The input data needed to compute synthetic seismograms consists of four main parts: earthquake catalogue, seismogenic zones, focal mechanisms and structural models.

Earthquake catalogue within the research area, between latitude 36°N and 42°N, longitude 112°E and 120°E, have been assembled for the time period 1500～2002, with the magnitude $M \geqslant 5.0$. The catalogue contains both historical (1500～1900) and instrumental (1900～2002) ones. Since 1970, the North China Telemetry Seismic Network has been built. It is powerful to monitor the seismic activity in this region. Because of the dense population and the continuous culture, it is assured the catalogue for strong historical events are complete. The epicenter distribution in and around the research area is shown in Figure 1.

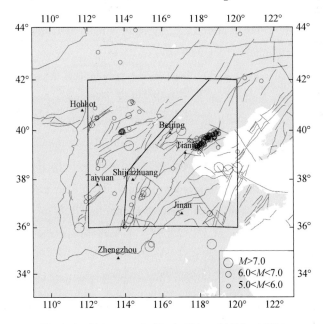

Figure 1　Epicenters in North China (1500～1992)

The spatial distribution of the collected earthquakes are mainly in three regions: (1) the Beijing-Tianjin-Bohai Sea region; (2) the south of the North China Plain; and (3) the western mountain area and the Fengwei rift region. Figure 2 shows the major seismogenic zones used in this study.

The focal mechanism results downloaded from the Harvard centroid moment tensor (CMT) consist of six events. Three of them belong to the 1976 Tangshan earthquake series and the mechanism of main Tangshan earthquake ($M=7.8$) is selected as the representative of this zone. Only one CMT result is in the south of North China Plain. The other two CMT results locate in the Fengwei rift and the western mountain area. The mechanism of 1998 Zhangbei earthquake is chosen as the typical one in this zone. The selected focal mechanisms are shown also in Figure 2.

The structure model of the crustal and mantle in the research area is from the deep seismic survey profile and seismic tomographic results (Zeng et al., 1985; Sun et al., 1998). Q values are from Liu et al. (1997). The research area is divided into two parts by the thickness of the crust. The border of structure I and II is the contour line of the Moho depth of 42km. It coincides with the gravity anomaly results and the transition zone of the plain and mountain areas (Editorial Board for the Lithospheric Dynamics Atlas of China, State Seismological Bureau, 1989). The selected structures for region I and II are shown in Figure 3.

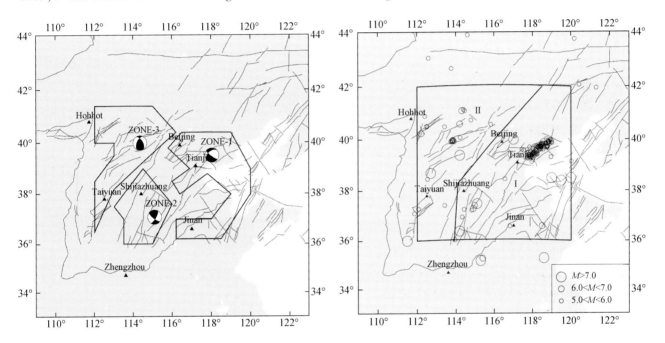

Figure 2 Focal mechanism in seismogenic zones   Figure 3 The distribution for structure zones I and II

4 Calculations

Starting from the available information on the Earth's structure, seismic sources and the level of seismicity in the investigated area, the synthetic seismograms are computed as follow:

(1) Definition of seismic sources. For discretized seismicity, the sturdy area is subdivided into cells (0.2°× 0.2°). The magnitude of the strongest earthquake that occurred within the cell is assigned to that cell. The smooth procedure then is applied to account for the source dimension of the largest earthquake and for the location errors, which may be particularly severe for historical events (Costa et al., 1993; Panza et al., 1999). To each cell we assign the magnitude value. We use the centered smoothing window with the radius of 0.6°. After smoothing, only the sources falling within the seismogenic zones are taken into account for the computations of synthetic seismograms.

(2) Definition of observation points. The observation points are placed in a grid with the dimension of 0.2°× 0.2°. To reduce the computations, the source-receiver distance is kept below an upper threshold (90km at most)

that depends on magnitude (Costa et al., 1992; Panza et al., 1999). At each observation point, all seismograms are generated by different sources. The largest component of ground motion is selected for further analysis.

(3) The synthetic seismograms are computed to obtain the peak ground displacement (D_{max}), velocity (V_{max}) and acceleration (A_{max}) up to the maximum frequency of 1Hz. The synthetic signals are properly scaled according to the magnitude associated with the cell of the source, using the moment-magnitude relation given by Kanamori (1977) and the spectral scaling law proposed by Gusev (1983) as reported in Aki (1987). The design ground acceleration is obtained through extrapolation using standard code response spectra following the procedure described by Panza et al. (1996). Here, EC8 for soil A (European code) is used. The choice of soil A is for the hard soil topmost structure layer with the S-wave velocity greater than 0.8 km/s.

5 Results and discussion

The spatial distribution of D_{max}, V_{max}, A_{max} and DGA due to the source located at an epicenter distance of 90km at most is shown in Figures 4, 5, 6 and 7.

The strongest synthetic ground motion is at east of Beijing. This is mainly caused by the 1679 Sanhe-Pinggu earthquake. The peak ground acceleration and DGA can be as high as 0.5g and 0.75g, separately in this area. The DGA in south of the North China Plain (south of Shijiazhuang City) is 0.4g. In Fengwei rift, north of Taiyuan City, the DGA can be as high as 0.25g.

The synthetic results are from the seismic activities from 1500 to 2005 in North China. The deterministic seismic hazard maps are given with the maximum ground motion we obtained. It can be found that the cities of Beijing, Shijiazhuang and Taiyuan are located in the area with strong seismic ground motion.

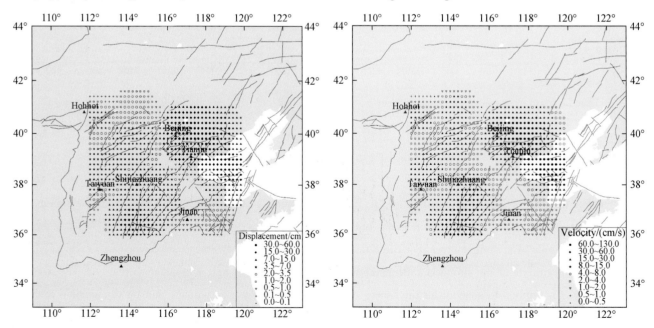

Figure 4 Synthetic Maximum Displacement in North China   Figure 5 Synthetic Maximum Velocity in North China

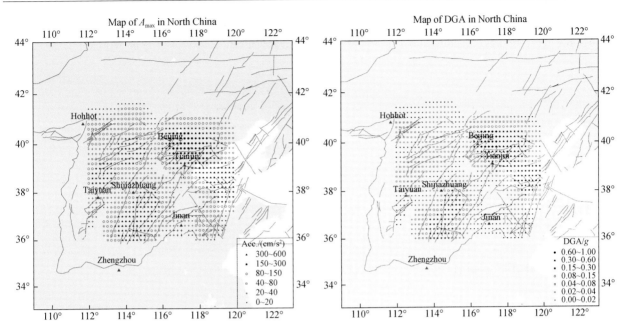

Figure 6　Synthetic Maximum Acceleration in North China

Figure 7　Map of Designed Ground Acceleration (DGA) in North China

Acknowledgements

This research has been carried on in the framework of the bilateral project "Geophysical Studies for the Deterministic Evaluation of Seismic Risk", with the contribution of the Italian Ministry of Foreign Affairs (MAE), Directorate General for Cultural Promotion and Cooperation. The research got support from Ministry of Science and Technology (MOST) of China and China Earthquake Administration. We have used public domain graphics software (Wessel and Smith, 1995a, b).

References

AA. VV., 2000, Seismic hazard of the Circum-Pannonian Region. *Pure Appl. Geophys.* **157**, 171–247.

Aki, K., 1987. Strong motion seismology. Erdik, M. Ö. and Toksöz, M. N. (Eds.), *Strong Ground Motion Seismology*, D. Reidel Publishing Company, Dordrecht, 3–39.

Alvarez, L., Vaccari, F. and Panza, G. F., 1999. Deterministic seismic zoning of eastern Cuba. *Pure Appl. Geophys.* **156**, 469–486.

Aoudia, A., Vaccari, F., Suhadolc, P. and Meghraoui, M., 2000. Seismogenic potential and earthquake hazard assessment in the Tell Atlas of Algeria. *J. Seism.* **4**, 79–98.

Costa, G., Panza, G. F., Suhadolc, P. and Vaccari, F., 1992. Zoning of the Italian region with synthetic seismograms computed with known structural and source information. *Proc. 10th World Conference on Earthquake Engineering*, 435–438, Madrid.

Costa, G., Panza, G. F., Suhadolc, P. and Vaccari, F., 1993. Zoning of the Italian territory in terms of expected peak ground acceleration derived from complete synthetic seismograms. *J. Appl. Geophys.* **30**, 149–160.

Ding, Z. F., Chen, Y. T. and Panza, G. F., 2003. Estimates of site effects in Beijing City. *Pure Appl. Geophys.* **161**, 1107–1123.

Ding, Z. F., Romanelli, F., Chen, Y. T. and Panza, G. F., 2003. Realistic modeling of seismic wave ground motion in Beijing City. *Pure Appl. Geophys.*, **161**, 1093–1106.

Editorial Board for the Lithospheric Dynamics Atlas of China, State Seismological Bureau (Ed.), 1989. *Lithospheric Dynamics Atlas of China*. China Cartographic Publishing House, Beijing.

El-Sayed, A., Vaccari, F. and Panza, G. F., 2001. Deterministic seismic hazard in Egypt. *Geophys. J. Int.* **144**, 555–567.

Gusev, A. A., 1983. Descriptive statistical model of earthquake source radiation and its application to an estimation of short period strong motion. *Geophys. J. R. astr. Soc.* **74**, 787–800.

Kanamori, H., 1977. The energy release in great earthquakes. *J. Geophys. Res.* **82**, 2981–2987.

Liu, J., Liu, F., Yan, X. and He, J., 1997. Characteristics of Q distribution in and around Beijing from Lg coda waves. Chen, Y., Wang, S., Qin, Y. and Chen, B. (Eds.), Chun Dan Ji, 777–787, Science Publishing House of China, Beijing.

Orozova-Stanishkova, I. M., Costa, G., Vaccari, F. and Suhadolc, P., 1996. Estimates of 1Hz maximum acceleration in Bulgaria for seismic risk reduction purposes. *Tectonophysics* **258**, 263–274.

Panza, G. F., 1985. Synthetic seismograms: the Rayleigh waves model summation. *J. Geophys.* **58**, 125–145.

Panza, G. F., Vaccari, F., Costa, G., Suhadolc, P. and Fäh, D., 1996. Seismic input modeling for zoning and microzoning. *Earthq. Spectra* **12**, 529–566.

Panza, G. F., Vaccari, F. and Romanelli, F., 1999. Deterministic seismic hazard assessment. Wenzel, F. and Lungu, D. (Eds.), *Vrancea Earthquakes: Tectonics and Risk Mitigation*, Kluwer Academic Publishers, The Netherland, 269–286.

Parvez, I. A., Vaccari, F. and Panza, G. F., 2002. A deterministic seismic hazard map of India and adjacent areas (submitted to *Geophys. J. Int.*).

Sun, R., Vaccari, F., Marrara, F. and Panza, G. F., 1998, The main features of the local geological conditions can explain the macroseismic intensity caused in Xiji-Langfu (Beijing) by the Tangshan 1976 earthquake. *Pure Appl. Geophys.* **152**, 507–522.

Wessel, P., and Smith, W. H. F., 1995a. New version of the generic mapping tools released. *EOS Trans. AGU* **76**, 329.

Wessel, P., and Smith, W. H. F., 1995b. *The Generic Mapping Tools (GMT) Version 3.0 Technical Reference and Cookbook*, SOEST/NOAA.

Zeng, R., Zhang, S., Zhou, H. and He, Z., 1985. Crustal structure in Tangshan seismogenic zone and the discussion on the formation of continent earthquake. *Acta Geophysica Sinica* **36**(2), 125–142.

从全球长周期波形资料反演2001年11月14日昆仑山口地震时空破裂过程*

许力生　陈运泰

中国地震局地球物理研究所，北京100081

摘要　利用从全球数字地震台网记录的资料中选择出的震中距小于90°且震相清晰的20个台站的长周期垂直分量P波震相，通过反演得到了2001年11月14日昆仑山口地震的震源时空破裂过程. 结果表明, 这次地震由三次子事件构成. 第一次子事件的破裂从震中位置(35.97°N, 90.59°E)开始并向东西两侧扩展, 向西以4.0km/s的破裂速度扩展了140km, 向东以2.2km/s的破裂速度扩展了80km, 表现为以自东向西为主的不对称双侧破裂, 形成了约220km长的断层. 在第一次子事件的破裂开始后大约52s, 在震中以西约220km的地方, 第二次子事件的破裂开始. 此时, 第一次事件没有结束, 但已进入愈合阶段. 第二次子事件的破裂向东西两侧扩展, 向西以2.2km/s的破裂速度扩展了50km, 向东以5.8km/s的破裂速度扩展了70km, 表现为以自西向东为主的不对称双侧破裂, 形成了约120km长的断层. 在第二次子事件开始后大约12s, 第二次子事件的破裂与第一次子事件的破裂在震中以西约140km处发生了聚合. 在第一次子事件的破裂开始后大约56s, 在震中以东约220km的地方, 第三次子事件开始. 此时, 第一次事件仍未结束, 但已进入愈合阶段的尾声. 第三次子事件的破裂向东西两侧扩展, 向西以4.0km/s的破裂速度扩展了140km, 向东以3.7km/s的破裂速度扩展了130km, 基本上是一次不对称双侧破裂, 形成了约270km长的断层. 在第三次子事件开始后大约36s, 第三次子事件的破裂与第一次子事件的破裂在震中以东约80km处发生了聚合. 此后, 震源过程主要是第一次子事件与第三次子事件聚合后的断层运动过程.

2001年11月14日09∶26∶10协调世界时(UTC), 在昆仑山断裂西部(36.2°N, 90.9°E)发生了一次矩震级M_W=7.8的地震. 地震之后, 国内外一些地震机构很快确定了这次地震的发震时刻、震中位置、震源深度、震级、地震矩以及震源机制等参数(表1). 不同机构测定的震源参数有所不同, 尤其是在震源机制的测定结果之间存在着明显的差异, 这一情况在一定程度上反映了这次地震震源破裂的复杂性.

表1　2001年11月14日昆仑山口地震震源参数测定结果

发震时刻(UTC)	震中位置		深度/km	M_0/10^{20}Nm	震级			节面 I			节面 II			资料来源[a]
	北纬/(°)	东经/(°)			M_S	M_W	m_b	走向/(°)	倾角/(°)	滑动角/(°)	走向/(°)	倾角/(°)	滑动角/(°)	
9∶27∶10.8	35.54	92.25	15	5.8		7.8		96	54	−7	190	84	−144	Harvard
9∶26∶10.0	35.95	90.54	10	3.5	8.0	7.7	6.1	32	4	−76	198	86	−91	USGS
	36.10	90.50	20	3.9		7.7		89	88	−2	179	88	−178	ERI, Tokyo Univ
	35.97	90.59	11	3.2	7.6	7.6	5.9	113	68	49	0	46	149	IGCSB
	36.60	90.80		2.2		7.5		144	83	−81	273	11	−140	APCSB
	35.97	90.59	11	3.2		7.6		290	85	−10	21	80	−175	本文

a) IGCSB, 中国地震局地球物理研究所; APCSB, 中国地震局分析预报中心; ERI, Tokyo Univ, 日本东京大学地震研究所.

---

* 本文发表于《中国科学: D辑 地球科学》, 2004年, 第**34**卷, 第3期, 256-264.

青海省地震局的地方性地震台网对这次地震的余震进行了监测,截至 11 月 27 日测定出震级和震中位置的事件达 2000 余次,其中,在位于 88°~96°E 及 34°~37°N 区域内,震级 $M_L>1.0$ 的事件 1288 次,震级 $M_L1.0~2.0$ 的事件 907 次,震级 $M_L2.0~3.0$ 的事件 270 次,震级 $M_L3.0~4.0$ 的事件 57 次,震级 $M_L4.0~5.0$ 的事件 8 次,震级 $M_S5.0~6.0$ 的事件 6 次. 余震最大震级为 $M_S5.7$. 主震和余震主要沿昆仑山断裂带分布(图1).

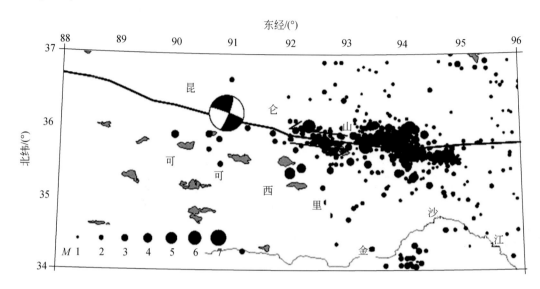

图 1 昆仑山口地震的震源机制、余震的震中分布与昆仑山断层图

"沙滩球"表示用全球长周期波形资料反演确定的 2001 年 11 月 14 日在昆仑山口发生的 $M_W7.8$ 地震的震源机制,实心圆表示截至 2001 年 11 月 27 日由青海省地震局地方台网确定 2000 余次余震的震中位置,粗实线表示昆仑山断层

2001 年 11 月 16 日中国地震局派出了考察组,对地震现场的地质与震害进行了考察. 现场考察发现,同震地表破裂沿昆仑山断裂从西向东展布,长达 350km,断裂活动主要表现为左旋走滑运动,同震断裂带的大部分地段的滑动量为 2~4m,最大滑动量达 6m[1]. 2002 年 2 月中国地震局派出综合地球物理考察团,再次对地震现场进行了综合地球物理考察. 考察发现在仪器震中(35.97°N,90.59°E)以西还有约 90km 的地表破裂存在,而且在仪器震中附近的最大水平滑动量达 5m 之多. 综合以上考察结果,可以断定昆仑山口地震地表破裂至少长 440km.

日本东京大学地震研究所在网上发布了这次地震的震源时间函数和断层面滑动量分布的快速测定结果①. 根据他们的结果,这次地震的震源时间函数比较复杂,总持续时间约 120s,在前 45s 内信号相当弱,在第 45s 之后,地震信号比较强. 从他们得到的滑动量分布的图像看,震源破裂区分为两部分,一部分在破裂的起始点周围,面积较小,滑动量较小,另一部分在破裂起始点的东面 35~200km 之间,面积较大,滑动量也较大. 地震断层的总长度达 200km. 地震发生后不久(2001 年 11 月 25 日),许力生[2]利用中国数字地震台网(CDSN)数量非常有限的资料得到了关于断层面上滑动时空分布的初步结果. 该结果表明,地震断层总体上具有自西向东扩展的单侧破裂特性,断层总长度约 130km,破裂持续时间约 25s.

根据昆仑山口地震的震源机制[2]、余震的空间分布(图1)和震后的现场考察[1],可以确信这次地震是昆仑山断裂活动的结果[3]. 但是,为什么不同的研究机构测定的震源机制存在着比较大的差异?为什么地表破裂长度远大于从地震波形反演得到的断层长度?这次地震震源的时空破裂过程究竟如何?要回答这些问题,需要做深入细致的研究工作. 更为重要的是,分析研究大地震震源断层的运动学和几何学参

---

① http://www.eic.eri.u-tokyo.ac.jp.2001.

数对了解地震的成因与机制有着十分重要的意义. 我们将着重分析用全球地震台网的长周期波形资料获得的昆仑山口地震震源的时空破裂过程.

1 资料与处理

昆仑山口地震发生后, 我们从 IRIS 数据中心获得了这次地震的波形数据的 SEED 卷, 从中提取出了 52 个台站的数据. 从这些数据中选用了震中距小于 90°且震相清晰的 20 个台站的垂直分量 P 波波形资料. 首先, 将原始速度记录用 0.01~0.1 Hz 的三阶 Butterworth 滤波器进行滤波; 然后, 经积分变换成位移记录; 最后, 用拐角频率为 0.03 Hz 的三阶低通 Butterworth 滤波器对积分后的位移记录进行滤波. 图 2 表示了初动前 50 s 和初动后 200 s 的 P 波波形.

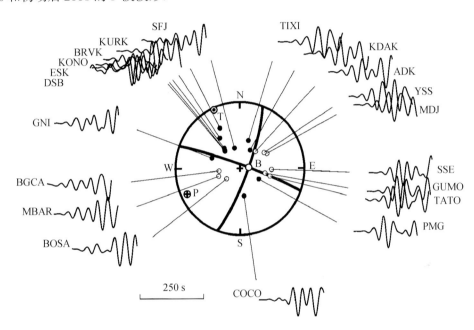

图 2 长周期垂直分量的 P 波位移记录、相关台站分布和震源机制解

为了获得震源断层面上滑动随时间和空间变化的图像, 首先需要确定地震的震源机制解. 我们用 IASPEI91 模型作为地球的近似模型[4], 采用反射-折射率方法计算格林函数[5], 借助频率域矩张量反演方法[6]确定了这次地震的矩张量解. 相应的最佳双力偶解如图 2 所示, 节面 I 走向 290°, 倾角 85°, 滑动角 −10°; 节面 II 走向 21°, 倾角 80°, 滑动角 −175°. 通过矩张量反演得到的标量地震矩 $M_0 = 3.2 \times 10^{20}$ Nm, 矩震级 $M_W = 7.6$. 根据震后的宏观考察[1]、余震分布和昆仑山断裂带的空间分布(图 1), 可以确认这次地震的发震断层为节面 I. 换句话说, 昆仑山口地震是发生在近东-西走向、几乎直立的断层上的左旋运动, 南盘相对于北盘向东运动.

获得断层面上滑动量随时间和空间变化图像的重要环节之一是从不同台站的波形资料中用反褶积方法提取震源时间函数[7~9]. 我们以这次地震的断层面, 即上述最佳双力偶解的节面 I(走向 290°, 倾角 85°, 滑动角 −10°)作为震源模型, 以 IASPEI91 模型作为地球介质的近似模型[4], 采用反射-折射率方法计算格林函数[5], 然后, 在对震源时间函数加非负约束的条件下, 在时间域中用合成格林函数通过反褶积提取了震源时间函数[10]. 最后, 我们用 0.02 Hz 的三阶低通 Butterworth 滤波器对直接反褶积之后的震源时间函数进行滤波, 得到如图 3 所示的 20 个地震台站接收到的震源时间函数. 这组震源时间函数具有很好的方位覆盖, 适宜用作震源时空破裂过程成像的资料.

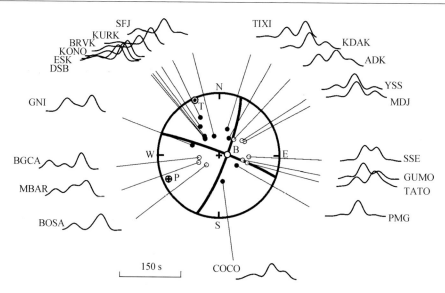

图3 从不同台站的垂直分量P波段提取的昆仑山口地震的震源时间函数、相关台站分布和震源机制

由观测资料提取得到的震源时间函数是反演震源时空破裂过程的基础,所以必须恰当地评价其可靠性. 为此, 利用直接反褶积的结果计算了合成地震图, 并和相应的观测地震图进行了比较. 我们注意到合成地震图能够很好地解释观测地震图, 表明了由观测资料提取得到的震源时间函数的可靠性. 此外, 从图3可以看到, 由不同台站获得的震源时间函数的形状随着台站的方位呈现系统的变化, 这从另一个角度说明所得到的震源时间函数是非常稳定而可靠的, 同时又反映了这次地震震源破裂过程的复杂性.

2 昆仑山口地震震源破裂时空过程

对昆仑山口地震震源过程的成像方法是用不同观测点获得的震源时间函数在时间域反演断层面上滑动量时空分布的方法[6].

上面的分析表明, 这次地震的发震断层为走向、倾角和滑动角分别为290°/85°/−10°的断层. 在本研究所用的反演方法中, 无须先验地假定破裂区域的范围和形状, 实际的破裂区域是由反演的结果得到的. 因此, 为了客观地反演出实际破裂区域, 在反演时, 总是将这个范围选取得足够大, 使之大于实际破裂区域. 为此, 以仪器震中 (35.97°N, 90.59°E) 为中心点, 沿断层走向在震中以东和以西分别取400km, 沿断层倾斜方向 (倾向N20°E, 倾角85°) 取40km, 将此长800km, 宽40km的矩形作为反演滑动量时空分布的范围, 并把此矩形在长度方向上40等分, 在宽度方向上4等分. 换句话说, 把800km×40km的矩形断层面分成40×4=160个20km×10km的矩形子断层.

根据前面所述, 我们从20个台站获取了依赖于台站方位的20个震源时间函数, 而滑动量分布成像的断层却被分成160个子断层. 对于每一时刻而言, 我们拥有的观测方程的数目为20, 而需要确定的未知数却为160. 所以, 这个反演问题为一欠定问题. 为了使反演稳定, 我们采用了三个约束条件. 第一个约束条件是"解大于或等于零"[9]. 这个约束条件的物理意义是, 在地震破裂过程中地震断层不发生反向运动. 尽管这一约束条件是一个运动学条件, 但是迄今为止的大量实测表明, 至少在目前精度所及的情况下, 在地震破裂过程中, 地震断层不发生反向运动. 第二个约束条件是破裂速度不大于震源区P波速度的约束 (这里使用的P波速度为6km/s). 这个约束条件是因果性关系. 第三个约束条件是同一时刻相邻子断层上的滑动率的梯度不大于某个常数, 这个约束条件是光滑性约束条件. 通过试验, 我们在反演中使用的光滑常数为0.00015.

我们使用的反演方法是共轭梯度法[11]. 用共轭梯度法得到的反演结果对初始模型具有一定的依赖性. 为了得到稳定而可靠的反演结果, 我们使用随机函数建立初始模型. 数值试验结果表明, 虽然利用不同的初始模型会得到不同的反演结果, 但是反演结果的主要特征都很稳定. 仅在震源过程临近结束时和用作反演范围的断层面的外围区域, 反演结果的残差相对较大. 我们运用随机产生的初始模型进行了 50 次反演, 这 50 次反演结果之间只存在微小的差别. 下面介绍的结果是从 50 次反演中得到的平均结果, 该结果反映了不依赖于初始模型的破裂过程的稳定特征.

图 4 展示了反演得到的断层面上滑动量的静态分布. 从滑动量的静态分布图像看, 滑动量大于 0.5m 的区域长达 610km, 分布在震中以西约 260km 与震中以东约 350km 之间. 滑动量大于 1.0m 的区域长达 420km, 分布在震中以西约 120km 与震中以东约 300km 之间. 滑动比较集中的区域 (滑动量大于 1.5m 的区域) 分为两个, 一个在震中以东 30km 至震中以西 80km 之间, 长度为 110km; 另一个在震中以东 150～280km 之间, 长度为 130km. 地震震源破裂在空间上的分布是非均匀的. 断层面上的最大滑动量达 2.2m (按其意义, 这个数值实际上是滑动量在 20km×10km 破裂面上的平均值), 平均滑动量约为 1.2m. 基于圆盘形位错模型[12～14], 我们计算了每个子断层的应力降, 得出断层面上最大应力降为 7.9MPa, 平均应力降为 4.0MPa.

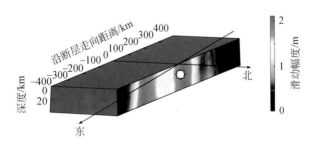

图 4 昆仑山口地震断层面上的静态位错分布
空心八角星表示破裂的起始点 (震源)

图 5 展示了地震震源破裂随时间和空间变化的过程. 图 5 (a) 反映滑动率随时间和空间的变化过程, 而图 5 (b) 反映滑动量随时间和空间的变化过程. 由滑动率的时空分布可以看出, 昆仑山口地震由三次子事件构成. 第一次破裂子事件, 即图 5 (a) 折线 Q1-W1-S1-E1 所围的事件, 时间上发生于发震以后最初 72s, 空间上发生于仪器震中以东 80km 至以西 140km 之间. 第二次破裂子事件, 即图 5 (a) 折线 Q2-W3-E2-S1-W2 所围限的事件, 时间上发生于发震以后的第 52～92s 之间, 空间上发生于仪器震中以西 270km 至以东 80km 之间. 第三次破裂子事件, 即图 5 (a) 折线 Q3-E2-S2-S3-E3 构成的事件, 时间上发生于发震以后的第 56～140s 之间, 空间上发生于仪器震中以西 100km 至以东 360km 之间. 根据图 5 (b) 所示的滑动量的时空分布, 可以清楚地看出构成昆仑山口地震的三次子事件的 "外貌". 这三次子事件本身规模都较大, 且具有不同的起始时间和起始位置, 但最终连接在一起形成一次更大的事件. 如前已述, 第一次子事件起始于仪器震中位置, 在图 5 (a) 与 (b) Q1 所示的位置. 第二次子事件起始位置位于仪器震中以西 220km 的地方, 即 Q2 所示的位置, 起始时刻为第 52s; 在约第 64s 时与第一次子事件几乎完全连接在一起. 第三次子事件起始位置为仪器震中以东 220km 的地方, 即 Q3 所示的位置, 起始时刻为第 56s; 在约第 92s 时与第一次子事件几乎完全连接在一起.

迄今常用于表示震源破裂的最简单的运动学模式有两种: 一种是单侧破裂模式, 即破裂起始于某一点, 然后向一侧传播; 另一种是双侧破裂模式, 即破裂起始于某一点, 然后向起始点的两侧传播. 然而, 这两种模式只是震源实际破裂过程的十分几近的描述. 我们的反演结果表明, 昆仑山口地震既不是一次纯粹的单侧破裂事件, 也不是一次纯粹的双侧破裂事件, 而是一次以从西向东为主的不对称的双侧破裂事件. 从几何学的特征看, 总体上, 震中以东的破裂长度大于震中以西的破裂长度. 如果考虑滑动量大于

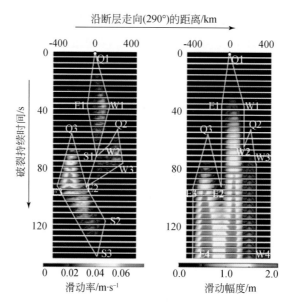

图 5　昆仑山口地震断层面上的滑动率（左）和滑动量（右）随时间和空间的变化图像
每个矩形表示 800km 长、40km 宽的断层面．空心八角星表示破裂的起始位置（参看图 4）

0.5m 的区域，仪器震中以西的破裂长度为 260km，而仪器震中以东的破裂长度为 360km；如果考虑滑动量大于 1m 的区域，仪器震中以西的破裂长度为 120km，而仪器震中以东的破裂长度为 300km；如果考虑滑动量大于 1.5m 的区域，区域西端距震中 80km，而区域东端距震中 280km（图 4）．从运动学特征看，总体上，向东的破裂速度明显大于向西的破裂速度．在整个震源破裂过程中，向西的平均破裂速度为 3.5km/s，而向东的平均破裂速度为 3.9km/s．如果更详细地分析一下震源断层面上滑动的时空变化过程，便可发现，组成这次昆仑山地震的三次子事件的破裂速度和破裂方向各不相同．如图 5（b）所示，第一次子事件是一次以从东向西为主的不对称双侧破裂事件，在起始破裂点 Q1 以西的破裂长度为 140km，而在起始破裂点以东的破裂长度为 80km；而且，从起始破裂点向西的破裂速度为 4.0km/s，从起始破裂点向东的破裂速度为 2.2km/s．第二次子事件是一次以从西向东为主的不对称双侧破裂事件，在起始破裂点 Q2 以西的破裂长度为 50km，而在起始破裂点以东的破裂长度为 70km；而且，从起始破裂点向西的破裂速度为 2.2km/s，从起始破裂点向东的破裂速度为 5.8km/s．第三次子事件则是一次基本上对称的双侧破裂事件，在起始破裂点 Q3 以西的破裂长度为 140km，而在起始破裂点以东的破裂长度为 130km；而且，从起始破裂点向西的破裂速度为 4.0km/s，从起始破裂点向东的破裂速度为 3.7km/s．

昆仑山口地震的震源破裂过程，既不是一次简单的单侧破裂，也不是一次简单的双侧破裂，而是由上述三次破裂子事件构成的复杂的破裂过程，在长达 140s 的破裂时间中，在长达 610km 的断层面上，经历了错综复杂的破裂起始、传播、聚合、愈合和停止的过程．特别是从滑动幅度和滑动率的时空分布，可以看到地震破裂过程中复杂的"聚合（fusion）"和"愈合（healing）"现象．"聚合"是裂纹发育的一种重要机制[15]．两个裂纹的发育有两种典型的情况．在第一种情况下，两个裂纹在不同的时刻独立发育，每个裂纹都受到高强度障碍体的阻止而不能扩展到另外一个裂纹的裂隙区．在第二种情况下，第二个裂纹突破二者之间的障碍体与第一个裂纹连在一起，这就是所谓的裂纹的聚合[15]．在聚合情况下，裂纹扩展辐射的长周期地震能量要比非聚合情况下大得多，即使在第一个裂纹区的应力已经降到很低水平的情况下也是如此[15]．我们的反演结果（图 5）表明，在昆仑山口地震过程中发生了裂纹的聚合．地震开始于仪器震中位置（在图 5 中以 Q1 标记，定义为第一个裂纹或事件），在大约第 52s，震中西约 220km 的地方（在图 5 中以 Q2 标记）有一个新的裂纹开始启动（定义为第二个裂纹或事件）．此时，第一个裂纹的扩展并没有结束，但已进入愈合阶段（参看图 5（a））．第二个裂纹迅速扩展，在大约 12s 后与第一个裂纹聚合,即将破裂过程扩展至第一个裂纹的已破裂区域(聚合点在图 5 中以 W2 标记). 在第一个裂纹扩展开始后大约 56s 时,在震中东约 220km 的地方(在图 5 中以 Q3 标记)又有一个新的裂纹开始启动(定义为第三个裂纹或事件),此时,第一个裂纹的扩展仍未结束,但已进入愈合阶段的尾声(参看图 5 (a)). 第三个裂纹迅速扩展,大约 36s 后与第一个裂纹聚合(聚合点在图 5 中以 E2 标记). 此后,震源过程主要是第三个裂纹与第一个裂纹聚合后的断层运动过程(参看图 5 (a)). 最后,在发震时刻之后的 140s,全部破裂过程结束. 因此,在昆仑山口地震的情况下,我们所说的三次子事件指的是三个"裂纹"发生聚合之前,即在表示滑动幅度时空变化的图 5 (b) 中的三个"尖峰".

3 反演得到的断层面上滑动量与同震地表位错量的比较

根据中国地震局第一次震后考察结果,地表破裂带西起布喀达板峰东缘(36°01′N, 91°08′E),东止于青藏公路东 70km 附近(35°33′N, 94°48′E),全长约 350km,地表滑动量很不均匀,最大水平滑动达 6m(图 6 (a))[1]. 根据中国地震局第二次震后考察结果,在布喀达板峰以西也有长约 90km 的地表破裂,而且在布喀达板峰附近的最大滑动量达 5m. 但是,到目前为止,布喀达板峰以西地表滑动的确切情况尚不清楚. 现将徐锡伟[1]实地考察获得的震中以东的地表滑动资料和我们得到的反演结果作一比较. 图 6 (a)表示徐锡伟[1]获得的实测地表滑动量,图 6 (b) 中的粗实线表示反演得到的断层面上的滑动量沿断层走向的平均值. 为了使二者具有可比性,在作比较前,对实测地表滑动量作如下处理. 首先,因为实测地表滑动量在空间上分布很不均匀,分辨率也很不一致,所以,我们对实测地表滑动量进行了滤波处理,使其与我们的反演结果具有大体相当的空间分辨率. 其次,我们假设这次地震造成的滑动在地表最大而在断层最深处(40km)为 0,并假定从地表到 40km 深处的滑动量线性地递减. 在此假定下,我们计算了沿断层走向断层面上的平均滑动(图 6 (b)). 可以看出,在具有实测资料的地段,反演结果和实测结果大体上是一致的.

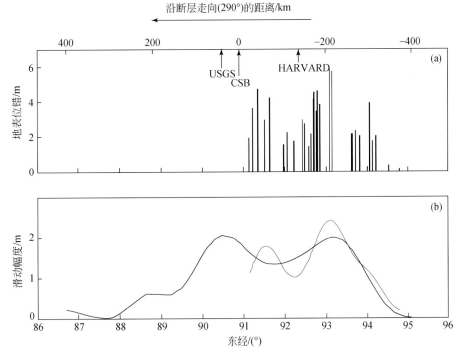

图 6 昆仑山口地震断层的实测地表位错和地震波反演结果的比较

(a) 实测地表位错量. 图中箭头所指的位置分别是美国地质调查局(USGS),中国地震局(CSB)与美国哈佛大学(HARVARD)测定的昆仑山口地震震中在该地震断层走向上的投影. (b) 由实测地表位错量推断的沿断层走向方向断层面上的平均位错量(细线)和由地震波资料反演得到的断层面上沿断层走向的平均滑动量(粗线)

4 讨论与结论

利用从全球数字地震台网记录的资料中选择的震中距小于 90°且震相清晰的 20 个台站的长周期垂直分量 P 波震相, 通过反演得到了 2001 年 11 月 14 日昆仑山口地震的震源时空破裂过程.

从滑动量的静态分布图像看, 滑动量大于 0.5m 的区域长达 610km, 震中以西约 260km, 震中以东约 350km. 这就是说, 由波形反演得到的这次地震的破裂长度 (610km) 大于迄今为止发现的地表破裂带的长度 (440km), 更大于早期野外考察发现的地表破裂带的长度 (350km) 以及用少量地震波资料反演得到的断层长度[2],①. 早期的结果或因考察范围的局限, 或因用作分析的波形资料数量的不足以及时间窗偏窄, 不能较全面地反映整体破裂的全貌. 这里的反演结果表明, 滑动量大于 1.0m 的破裂区域长达 420km, 在震中以西约 120km, 震中以东约 300km. 这个长度和震后宏观考察的最新结果非常吻合, 表明了地表破裂带与地震断层面上滑动量较大的区域 (滑动量大于 1.0m 的区域) 是相对应的. 如果考虑滑动量大于 1.5m 的区域, 可以从空间上分辨出两个滑动比较集中的区域, 一个在震中以东 30km 至震中以西 80km 之间, 长度约 110km; 另一个在震中以东 150~280km 之间, 长度约 130km. 后者恰好是地表破裂最严重的库赛湖及其以东地段. 反演结果表明, 昆仑山口地震断层面上的滑动分布是非均匀的, 断层面上的最大滑动量达 2.2m, 平均滑动量约为 1.2m. 基于圆盘形位错模型计算得到的最大应力降为 7.9MPa, 平均应力降为 4.0MPa.

我们注意到, 通过波形反演得到的滑动量大于 0.5m 的区域内 610km 的破裂长度比地表破裂带 440km 的长度大得多; 断层面上的最大滑动量 (2.2m) 和平均滑动量 (1.2m) 相差 1.0m, 但比地表破裂的最大位错量 (6.0m) 小许多; 断层面上的最大应力降 (7.9MPa) 和平均应力降 (4.0MPa) 也相差不大. 作者认为, 造成这些差别的共同原因可能是波形反演所使用的资料的波长较长, 降低了空间分辨率并使得反演结果过于"光滑". 此外, 由现在使用的长周期体波资料反演得到的结果几乎无法分辨滑动在深度方向上的变化. 需要说明的是, 对于昆仑山口地震的情形, 断层的长度和宽度相差甚大, 用本文采用的方法和所使用的资料, 在反演中难于同时兼顾沿走向和沿倾向的空间分辨率. 因此, 在本研究中, 首先关注的是沿断层走向方向的空间分辨率. 滑动量沿倾向方向的空间分布尚需要进一步研究.

根据对破裂在震源断层面上的时空变化过程的分析, 昆仑山口地震显示了一次大地震所具有的复杂震源过程. 震源过程的复杂性不仅表现在一次大的地震一般是由若干个子事件构成, 而且还表现在地震破裂过程中发生了错综复杂的聚合与愈合. 昆仑山口地震的震源过程可以概述如下: 第一次子事件的破裂从震中位置 (35.97°N, 90.59°E) 开始, 并向东西两侧扩展, 向西以 4.0km/s 的破裂速度扩展了 140km, 向东以 2.2km/s 的破裂速度扩展了 80km, 表现为以自东向西为主的不对称双侧破裂, 形成了约 220km 长的断层. 在第一次子事件破裂开始后大约 52s, 在震中以西约 220km 的地方, 第二次子事件的破裂开始. 此时, 第一次事件没有结束, 但已进入愈合阶段. 第二次子事件的破裂向东西两侧扩展, 向西以 2.2km/s 的破裂速度扩展了 50km, 向东以 5.8km/s 的破裂速度扩展了 70km, 表现为以自西向东为主的不对称双侧破裂, 形成了约 120km 长的断层. 在第二次子事件开始后大约 12s, 第二次子事件的破裂与第一次子事件的破裂在震中以西约 140km 处发生了聚合. 在第一次子事件的破裂开始后大约 54s, 在震中以东约 220km 的地方, 第三次子事件开始. 此时, 第一次事件仍未结束, 但已进入愈合阶段的尾声. 第三次子事件的破裂向东西两侧扩展, 向西以 4.0km/s 的破裂速度扩展了 140km, 向东以 3.7km/s 的破裂速度扩展了 130km, 基本上是一次对称的双侧破裂, 形成了约 270km 长的断层. 在第三次子事件开始后大约 36s, 第

---

① http://www.eic.eri.u-tokyo.ac.jp.2001.

三次子事件的破裂与第一次子事件的破裂在震中以东约80km处发生了聚合. 此后, 震源过程主要是第一次子事件与第三次子事件聚合后的断层运动过程. 这三次子事件规模不同, 破裂方式和速度也彼此不同, 但由这三次事件组成的昆仑山口地震总体上是一次以由西向东扩展较强的不对称的双侧破裂, 破裂时间达140s.

2001年昆仑山口M_W7.8地震和1997年西藏玛尼M_W7.6地震是1950年8月15日西藏察隅M_W8.6地震以来发生于青藏高原的重要地震事件[6,16], 断层走向都是沿近东-西方向, 而且断层面几乎直立、以左旋走滑为主, 破裂扩展的主体方向都是自西向东. 这表明这两次地震都受同一构造应力场作用, 是青藏高原及其周边地区的物质"差速"东流的结果; 它们的发生进一步支持了青藏高原及其周边地区的物质在向东运动的见解[17,18].

致谢 作者谨向为本项研究提供波形资料的美国IRIS数据中心、提供余震资料的中国国家数字地震台网分中心刘瑞丰博士表示感谢.

参 考 资 料

[1] 徐锡伟. 2001年11月14日昆仑山库塞湖地震(M_S8.1)地表破裂带的基本特征. 地震地质, 2001, **24**(1): 1-13

[2] 许力生, 陈运泰. 2001年11月14日昆仑山口西8.1级大地震简介. 震情研究, 2001, **51**(4): 80-82

[3] Woerd J V D, Meriaux A S, Klinger Y, et al. The 14 November 2001, M_W = 7.6 Kokoxili Earthquake in Northern Tibet (Qinghai Province, China). *Seism Res Lett*, 2002, **73**(2): 125-135

[4] Kennett B L N, Engdahl E R. Travel times for global earthquake location and phase identification. *Geophys J Int*, 1991, **105**: 429-465

[5] Kennett B L N. *Seismic Wave Propagation in Stratified Media*. Cambridge: Cambridge University Press, 1983. 1-339

[6] Chen Y T, Xu L S. A time-domain inversion technique for the tempo-spatial distribution of slip on a finite fault plane with applications to recent large earthquakes in the Tibetan Plateau. *Geophys J Int*, 2000, **143**(2): 407-416

[7] Dreger D S. Empirical Green's function study of the January 17, 1994 Northridge, California earthquake. *Geophys Res Lett*, 1994, **21**: 2633-2636

[8] Hough S E, Dreger D S. Source parameters of the 23 April 1992 M6.1 Joshua Tree, California, earthquake and its aftershocks: empirical Green's function analysis of GEOS and TERRAscope data. *Bull Seism Soc Amer*, 1995, **85**: 1576-1590

[9] 许力生, 陈运泰. 用经验格林函数方法从长周期波形资料中提取共和地震的震源时间函数. 地震学报, 1996, **18**(2): 156-169

[10] Ihmle P F. On the interpretation of subevents in teleseismic waveforms: the 1994 Bolivia deep earthquake revisited. *J Geophys Res*, 1998, **103**: 17919-17932

[11] Ward S N, Barrientos S E. An inversion for slip distribution and fault shape from geodetic observations of the 1983, Borah Park, Idaho, earthquake. *J Geophys Res*, 1986, **91**(B5): 4909-4919

[12] Keylis-Borok V I. On estimation of the displacement in an earthquake source and of source dimensions. *Ann Geofis*, 1959, **12**: 205-214

[13] Brune J N. Tectonic stress and the spectra of seismic shear waves from earthquakes. *J Geophys Res*, 1970, **75**: 4997-5009

[14] Brune J N. Correction. Tectonic stress and the spectra of seismic shear waves from earthquake. *J Geophys Res*, 1971, **76**: 5002

[15] Chatterjee A K, Knopoff L. Crack breakout dynamics. *Bull Seism Soc Amer*, 1990, **80**(6): 1571-1579

[16] 许力生, 陈运泰. 1997年中国西藏玛尼M_S7.9地震的时空破裂过程. 地震学报, 1999, **21**(5): 449-459

[17] 曾融生, 孙为国. 青藏高原及其邻近地区的地震活动性和震源机制以及高原物质东流的讨论. 地震学报, 1992, **14**(增刊): 534-563

[18] Avouac J P, Tapponnier P. Kinematic model of active deformation in Central Asia. *Geophys Res Lett*, 1993, **20**: 895-898

Double-difference relocation of earthquakes in central-western China, 1992—1999

Yang Z. X.[1], Waldhauser F.[2], Chen Y. T.[1] and Richards P. G.[2]

1. Institute of Geophysics, China Seismological Bureau, Beijing 100081, China;
2. Lamont-Doherty Earth Observatory, Columbia University, NY 10964, USA

Abstract The double-difference earthquake location algorithm was applied to the relocation of 10,057 earthquakes that occurred in central-western China (21°N to 36°N, 98°E to 111°E) during the period from 1992 to 1999. In total, 79,706 readings for P waves and 72,169 readings for S waves were used in the relocation. The relocated seismicity (6,496 earthquakes) images fault structures at seismogenic depths that are in close correlation with the tectonic structure of major fault systems expressed at the surface. The new focal depths confirm that most earthquakes (91%) in this region occur at depths less than 20km.

1 Introduction

Our study area is the central-western China, extending roughly from 21°N to 36°N and 98°E to 111°E (Figure 1). As the tectonic map indicates (see Figure 1a), the Chinese mainland and its surrounding area are located in the southeastern part of the Eurasian plate bounded by the Indian, the Philippine Sea and the Pacific plates. The region is characterized by strong northeastward motion of the Indian plate with respect to southwestern China, by the westward subduction of the Pacific plate beneath eastern China, and by the northwestward impact of the Philippine Sea plate (Molnar and Tapponier, 1975; Teng et al., 1979; Zhou et al., 1998; Wang et al., 2001; Qin et al., 2002). The study area can be broadly divided into a western part and an eastern part along 105°E longitude (see Figure 1b). The western part has a complicated tectonic structure and is characterized by a high rate of seismic release, while the eastern part has a lower level of seismic activity (Chen et al., 1992; Min, 1995; Wang et al., 1999). In the western part, several major active faults, including the Honghe (Red River) Fault (F1 in Figure 1b), the Xiaojiang Fault (F2), the Xianshuihe Fault (F3) and the Jinshajiang Fault (F4), and seismic belts, including the Lijiang seismic belt (R3) and the Yongsheng-Ninglang-Muli-Jiulong seismic belt (R4), outline a diamond-shaped tectonic block, first noted and named as the Sichuan-Yunnan rhombic tectonic block by Kan et al. (1977). To the northeast of this structure is the 470km long, NE striking Longmenshan Fault (F5), and to the southwest the NNW striking Tengchong-Longling (R1) and Lancang-Gengma seismic belts (R2). Previous studies (Kan et al., 1977; Li, 1993; Wen, 1998) indicate that some of the larger earthquakes of the region have occurred along these major fault systems, both historically and recently (Figure 1b, Table A1). In contrast, to the east of 105°E longitude the seismicity is low, in particular to the south of 34°N where no major earthquake has been documented and current micro-seismicity is low. An exception is the area of increased seismicity around the Baise-Hepu Fault (F8). The northeastern part is dominated by the E-W striking Northern Qinling Fault (F6 in Figure 1b) and the NE striking Weihe fault (F7), along which several larger earthquakes occurred historically.

The large earthquakes with $M \geqslant 7$ shown in Figure 1b have taken place mainly along active faults expressed at the surface, and have in fact contributed, in many cases, to their formation (Ma, 1987, 1989; Ding, 1991, 1996; Deng et al., 1994). But the relationship between the thousands of microearthquakes recorded with modern instruments (Figure 2) and the active faults and associated large earthquakes has never previously been studied in detail in this area- a fact which has motivated our work. The instrumental seismicity is diffuse, and in previous studies has not indicated a clear association to the faults at the surface, or to the large events that occurred along these structures. A main problem in studying the temporal and spatial distribution of the seismicity is the large error associated with published locations of these earthquakes. These errors are mainly due to oversimplified velocity models used in the routine location process, errors in arrival time readings, and lack of suitable station coverage. The latter source of error is especially common for events near the boundaries of individual provinces, resulting in large azimuthal gaps in the station distribution, unless as in this study, the data from different networks are combined.

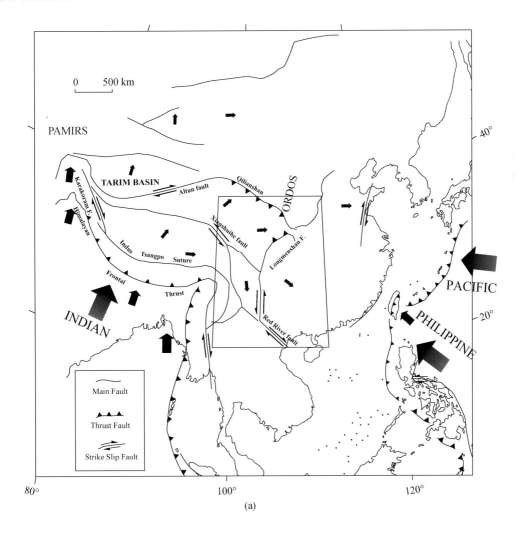

(a)

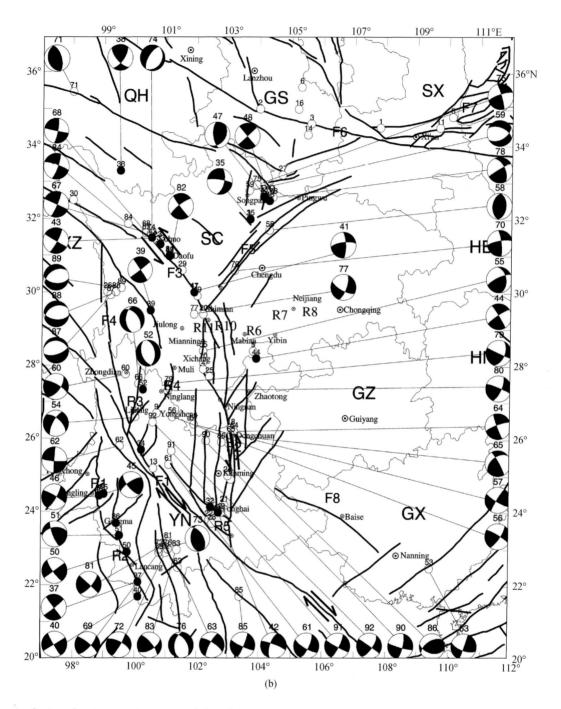

Figure 1 Setting of active tectonic structure (a) and seismicity of major earthquakes of $M \geq 7$ from 780 B. C. to April 2003 in central-western China (b)

Thick black lines represent surface traces of active faults (Deng et al., 1994), thin lines represent provincial boundaries; gray and black circles indicate major events with $M \geq 7$ before 1911 (Min, 1995) and after 1911 (Wang et al., 1999), respectively; open circles represent some events of $6 \leq M < 7$ with available fault plane solutions. The identification number, above each circle and beach ball representation of lower hemisphere projection of focal mechanism, refers to the earthquake numeral given in Table A1 and cited in the text. The thick black arrow represents the NNW to SSE motion of the Sichuan-Yunnan diamond-shaped tectonic block

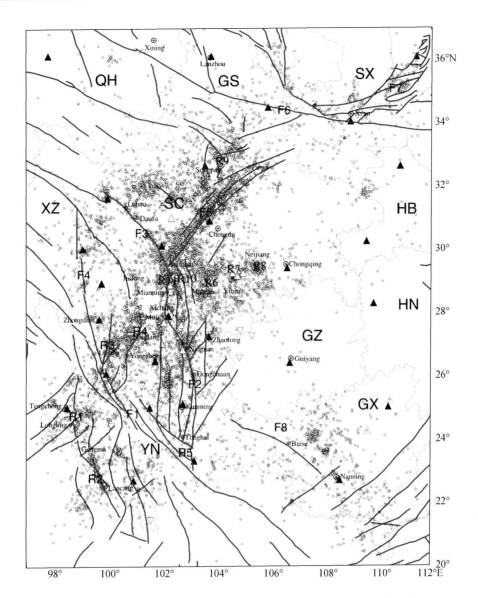

Figure 2 Epicentral distribution of the all 15,092 routinely located earthquakes
(gray circles) for the period from 1992 to 1999

Thick black lines represent surface traces of active faults, thin lines represent provincial boundaries. Gray triangles represent stations of Sichuan (SC) Province Seismic Network, gray upside-down triangles represent stations of the Yunnan (YN) Province Seismic Network (YNSN), open upside-down triangles represent stations of the Shaanxi (SX) Province Seismic Network, open triangles represent stations of the Guangxi (GX) Province Seismic Network and black triangles represent stations of the China National Seismic Network (CNSN)

The study area includes all of four provinces, namely Sichuan (hereafter referred to as SC), Yunnan (YN), Guizhou (GZ) and Guangxi (GX); and parts of at least six additional provinces, namely Qinghai (QH), Gansu (GS), Shaanxi (SX), Hubei (HB), Hunan (HN) and Xizhang (XZ).

In this study we combine all the available phases from the four provincial (SC, YN, SX and GX) seismic networks and the China National Seismic Network (CNSN). We relocate the events using the double-difference algorithm (DD algorithm) of Waldhauser and Ellsworth (2000). The relocated seismicity distribution is investigated in terms of its relationship to faults mapped at the surface and the location of large recent and historical earthquakes.

2 Seismotectonic setting

While the Honghe Fault (F1 in Figure 1b), marking the southwestern end of the diamond-shaped tectonic block, is a right-lateral strike-slip fault, both the Xianshuihe Fault (F3) at the northern end and the Xiaojiang Fault (F2) at the eastern end are left-lateral strike-slip faults. Focal mechanisms (Kan et al., 1977; Xu et al., 1989) of earthquakes bounding the tectonic block (Figure 1b) reveal a NNE horizontal maximum compressive stress for the region, which suggests a relative movement of the block to the SSE (Kan et al., 1977). Along the Honghe Fault, the rate of slip since the Quaternary is about 4-6 mm/yr on the northern segment, and 7-9 mm/yr on the southern segment of the fault (Guo et al., 1984; Su and Qin, 2001). Strong earthquakes have occurred frequently along the Honghe Fault in historic times. Among the most important historical events are the 1925 Dali earthquake of $M_S 7.0$ (No. 34 in Figure 1b and Table A1- hereafter a numeral cited in the text refers to the earthquake identification number, above each circle and beach ball representation of lower hemisphere projection of focal mechanism in Figure 1b and Table A1), and the instrumental 1970 Tonghai earthquake of $M_S 7.8$ (No. 42) that ruptured the northwestern and southeastern parts of the Honghe fault, respectively. Along the Xiaojiang Fault (F2), the rate of slip since the Quaternary is about 6.4-8.8 mm/yr (Su and Qin, 2001; Li, 1993) and several strong earthquakes, including the 1833 Songming earthquake of $M8$ (No. 24) and the 1966 Dongchuan earthquake of $M_S 6.5$ (No. 64) have occurred. Along the Xianshuihe Fault (F3), the annual rate of slip since the Quaternary is about 8-15 mm/yr (Su and Qin, 2001). Historically, a series of large earthquakes have taken place along this fault, in particular on the Luhuo-Daofu segment (Figure 1b) as follows: the 1816 earthquake of $M7.5$ (No. 23), the 1893 earthquake of $M7$ (No. 29), the 1904 earthquake of $M7$ (No. 31), the 1923 earthquake of magnitude $M7.3$ (No. 33), the 1973 earthquake of $M_S 7.6$ (No. 43), and the 1981 earthquake of $M_S 7.0$ (No. 49).

Several structures with a high rate of seismic activity exist outside and inside the Sichuan-Yunnan diamond-shaped tectonic block. The Longmenshan Fault (F5) northeast of the block frequently ruptures in small to moderate size earthquakes, but no earthquakes with $M \geqslant 7$ have been documented along this fault. As shown in Figure 2, about 2,400 events with magnitudes between 1.0 and 5.0 occurred on or near the Longmenshan Fault (F5) during 1992 to 1999. Some earthquakes with magnitudes around 6 occurred here historically, such as the 1657 Wenchuan earthquake of $M6.5$ (epicentral location: 31.3°N, 103.5°E) and the 1970 Dayi earthquake of $M_S 6.2$ (No. 70) (Gu, 1983; Xie and Cai, 1983-1987). Along the Tengchong-Longling seismic belt (R1) in the southwestern corner of the study area, two large earthquakes with $M_S 7.3$ and $M_S 7.4$ (Nos. 45 and 46 in Figure 1b and Table A1) occurred within about 100min on May 29, 1976. Similarly, two large earthquakes with $M_S 7.4$ and $M_S 7.2$ (Nos. 50 and 51) occurred within about 12 min on the Lancang-Gengma seismic belt (R2) on November 6, 1988. The two parallel running, N-S striking Lijiang (R3) and Yongsheng-Ninglang seismic belts (R4) within the Sichuan-Yunnan diamond-shaped tectonic block ruptured in a $M_S 7.0$ event on February 3, 1996 (No. 52), and a historical event of $M7.8$ in 1515 (No. 9), respectively. Most recently, on October 27, 2001, a moderate size earthquake with $M_S 6.0$ (No. 92) occurred in the Yongsheng-Ninglang-Muli-Jiulong seismic belt (R4). The Tonghai-Shiping seismic belt (R5) at the southeastern corner of the diamond-shaped block features a series of strong earthquakes (Nos. 12, 21, 22, 28, 32 and 42) both historically and recently, including the 1970 Tonghai earthquake of $M_S 7.8$ (No. 42) on the Honghe Fault (F1) described above. It can be said that the western part of this study area is one with complex tectonic structures and a very high level of seismicity.

In contrast to the western part of the study area, the eastern part shows a much lower level of seismic activity. The large earthquakes to the north of the Northern Qinling Fault (F6) are mainly historical events (Nos. 1, 8 and 11), which occurred along the Weihe Fault (F7). In the southern part, no large events have been documented, but a relatively large number of small to moderate size earthquakes have occurred over the past years (Figure 2). These events are located around the Baise-Hepu Fault (F8), but do not show us very clear association to the major fault system.

3 Earthquake relocation

In the study area a total of 15,092 earthquakes were routinely located for the period from 1992 to 1999 in central-western China (Figure 2) using data from four provincial seismic networks in the Sichuan, Yunnan, Shaanxi, and Guangxi provinces, plus the CNSN. Among these earthquakes, 10,057 events were recorded by at least 4 stations. We used P- and S-phase picks from the 10,057 earthquakes recorded at 193 stations and reported in bulletins of the above-mentioned five seismic networks. A total of 79,706 P-phase picks and 72,169 S-phase picks were used from the 10,057 earthquakes. The individual seismic networks only reported focal depths for 4,876 events (49% of the 10,057 events). If we exclude the events whose depths are not reported, then the mean depth is 14.3 km, with depths ranging from 1 km to 53 km. These network locations are determined on a routine basis, adopting oversimplified propagation models for different provincial seismic networks.

Before relocation the epicenter locations of the resulting catalog located by the above-mentioned five seismic networks indicate a diffuse distribution across the study area (Figure 2).

The relative event location procedure is used to increase the precision of hypocenter locations. Relative earthquake location methods depend on the seismic wave velocity structure in the source region rather than on the structure along the entire ray path from a particular hypocenter to a station as in absolute location procedures. In the master event technique (Fukao, 1972; Fitch, 1975; Chung, 1976; Yang et al., 1999, 2002), travel time differences are taken between an assigned (master) event and its neighboring events, improving locations for clusters of events that have dimensions smaller than the scale-length of the velocity heterogeneities between sources and receivers. Got et al. (1994) extended this idea to include travel time differences not only relative to one event, but between all neighboring events. They used differential times measured by cross-correlation to solve for interevent distances relative to the centroid of a cluster of events. This approach limits the spatial scale of possible applications. Based on these earlier studies Waldhauser and Ellsworth (2000) have designed a double-difference algorithm to optimally relocate seismic events across large areas, using routinely picked arrival times of standard phases, high precision cross-correlation measurements (if available), or a combination of phase picks and cross-correlation measurements.

The fundamental equation of the double-difference algorithm relates the differences between the observed and predicted phase travel times for pairs of earthquakes observed at common stations to changes in the vector connecting their hypocenters. In this approach, hundreds or thousands of earthquakes can be 'linked' together through a chain of near neighbors. By choosing only relative phase travel times for events that are close together (i.e., closer than the scale-length of the surrounding velocity heterogeneity), wave paths outside the source region are similar enough that common model travel time errors are canceled for each pair of events. It is then possible to obtain high-resolution hypocenter locations over large areas without the use of station corrections.

We used the program hypoDD (Waldhauser, 2001) that implements the DD algorithm to refine the available event locations. We searched the phase pick data for an appropriate network of phase travel time differences that efficiently links together as many events as possible through a chain of neighboring events. In this process event pairs with interevent distances up to 20km were considered, to account for possible large mislocations in the initial locations. The P-phase travel time differences are initially weighted 1, S-phases 0.5. During the iterative least-squares procedure, the data is reweighted after each iteration depending on the distances between the events, and depending on the residuals between the observed and the calculated travel time differences. The conjugate gradient method is used to solve the large system of normal equations. Travel time differences are predicted using a 1-D layered velocity model used for relocations in this study (Table 1). The 1-D layered seismic wave velocity model used for predicting the travel time differences is based on previous studies of seismic wave velocity structure for the study region (Zhao et al., 1987, 1997; Yang et al., 2004). This model is appropriate for the purpose for predicting the travel time differences because the DD algorithm is a relative earthquake location method which is much less dependent on the models, compared with "absolute" methods (Waldhauser and Ellsworth, 2000). The search for continuous areas of seismicity with well linked events resulted in more than 200 clusters, including a total of 8,208 events (82% of the total amount of 10,057 events in the catalogue). 18% of the total number of events were not recorded by enough stations that are common to at least one neighboring event within 20km distance. Most of the clustered events occur in isolated areas and include only a few events. In total, the hypocentral parameters of 6,496 relocated events were obtained. The six largest clusters have between 85 and 4,041 events, including a total of 5,827 events (Table 2). After relocation, the root-mean-square (RMS) residuals were reduced significantly. As an example, Figure 3 shows the reduction of RMS residuals after iteration for each of the six largest clusters. For the largest cluster, for example, the RMS residual is reduced from 1.83 s to 0.47 s. Similar results are obtained for other clusters (Table 2). The average RMS residual of the six largest clusters is reduced from the original 1.76 s to the final 0.57 s (Table 1, Figure 3). The relative uncertainties of relocation for the six largest clusters, on average, are 1.2 km, 1.2 km and 1.4 km for E-W, N-S and vertical directions, respectively (Table 2).

Table 1 Velocity model used for relocations in this study

Layer	Depth to top of layer/km	P-wave velocity/(km/s)
1	0.0	5.00
2	7.5	5.48
3	16.0	5.93
4	20.0	6.43
5	30.0	6.60
6	50.0	8.30

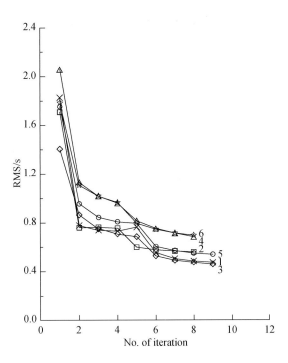

Figure 3 RMS residuals as a function of iteration steps, displayed for each of the six largest clusters

RMS residuals for cluster 1 to cluster 6 are shown with cross (line 1), open square (line 2),
open diamond (line 3), open triangle (line 4), open circle (line 5) and open star (line 6)

Table 2 Reduced residuals and uncertainties of relocation for the six largest clusters

Cluster	Event number		Centroid			RMS residuals/s		Uncertainty of relocation/m		
	Initial	Relocated	Lat. /°N	Long. /°E	Depth/km	Initial	Relocated	E-W	N-S	Vertical
1	4,041	3,130	30.737	103.315	15.5	1.83	0.47	856.8	869.4	1,210.2
2	1,010	763	29.346	105.231	8.7	1.71	0.56	1,164.4	1,527.7	1,600.6
3	241	209	27.080	102.794	10.5	1.41	0.46	933.4	809.0	1,094.0
4	228	157	27.205	100.915	6.0	2.06	0.68	1,678.0	1,627.7	1,795.4
5	222	179	25.885	102.214	10.0	1.75	0.54	1,321.0	1,168.1	1,432.4
6	85	76	28.495	101.085	8.2	1.79	0.70	1,220.7	1,437.5	1,480.9
Average						1.76	0.57	1,195.7	1,239.9	1,435.6

Further information on the uncertainties in relative location are given in Table 3, and the mean of the relative uncertainties of all the relocated events is 1.4 km, 1.5km and 1.7 km for E-W, N-S and vertical directions, respectively. These errors are derived from bootstrapping the unweighted residual vector. They are compatible with the fact that the distribution of seismic stations in the study area is sparse, with spacing of about 50km to 100km.

Table 3 Uncertainties of the 6,496 relocated events

	0⩽Uncertainty⩽2km		0⩽Uncertainty⩽3km		0⩽Uncertainty⩽5km		Uncertainty>5km	
	Event number	%	Event number	%	Event number	%	Event number	%
E-W	5,397	83	5,850	91	6,294	97	202	3
N-S	5,309	82	5,851	90	6,277	97	219	3
Vertical	4,904	76	5,732	88	6,259	96	237	4

4 Relocation results

Figure 4 shows the epicentral distribution of all the 6,496 relocated events (open circles). A much sharper image of the seismicity is obtained, compared to the initial locations of the all 6,496 events before relocation (gray circles in Figure 4) as well as that of the 15,092 routinely located events (gray circles in Figure 2). A comparison between the epicentral distributions of earthquakes before and after relocation indicates that when analysing seismicity, based on the data before relocation, one must use caution. In general, seismicity pattern represented by the epicentral distribution before relocation is diffuse and very likely biased, and hinders understanding the links between microseismicity and active tectonics in the study area, especially when using data

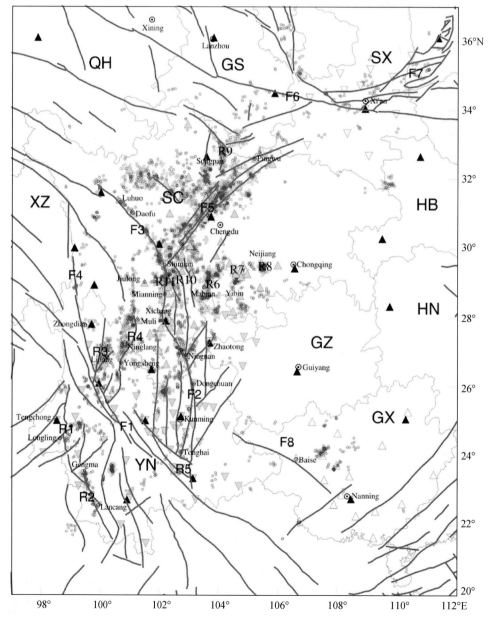

Figure 4　Epicentral distribution of all the 6,496 relocated earthquakes in central-western China using the double-difference algorithm
Open circles represent relocated events. Gray circles represent the same 6,496 events before relocation. Note that in the figure events before relocation are rather diffuse and that many of the relocated events are overlapped. Other symbols as in Figure 2

recorded by the analog-recording seismic networks. We note that in the final results of this study, almost all the aftershock sequences are excluded in the relocation. The only exception is the aftershock sequence of the 1996 Lijiang, Yunnan, earthquake of $M_S7.0$, in which about 120 aftershocks are included in the relocation. The relocated seismicity correlates strongly with tectonic activity expressed and mapped at the surface, and with the location and focal mechanism of large historic and recent earthquakes. Only about 630 events (10% of the total) occurs off-recognized faults. In particular, the correlation of the relocated seismicity correlates with the major seismic belts. Below, we discuss in some detail the following seismic belts: Lancang-Gengma (R2), Lijiang (R3) and Yongsheng-Ninglang-Muli-Jiulong (R4); Longmenshan (F5), Mabian (R6), Yibin (R7), and Neijiang (R8); and Songpan (R9), Shimian (R10) and Mianning (R11).

The focal depth of all events is relocated referred to the hypocentroid of the 274 routinely well-located events. These routinely well-located events are chosen by the criterion that at least one seismic station is located within the epicentral distance of travel time difference of P- and S-waves equal to or less than 2.0 s. Thus the resultant focal depth is independent of the zero depth initial value of the events without reported focal depths. In the same sense as the 274 well-located events, the relocated focal depth is regarded as "absolute" focal depth. Figure 5a shows relocated focal depths and uncertainties of all the 6,496 relocated earthquakes projected along a N-S profile. Most earthquakes (91%) in central-western China are situated at a depth interval between 0 and 20km. This can also be seen in a histogram of the focal depths of all the relocated earthquakes (Figure 5b). The mean depth for all relocated events is about 11.7 km, significantly shallower than the mean depth reported in previous studies (Ma and Xue, 1983; Zhang et al., 2002).

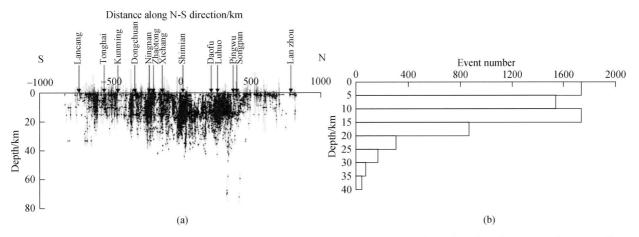

Figure 5 (a) Vertical cross-section of focal depths and uncertainties of all the 6,496 relocated earthquakes in central-western China along the S-N direction; (b) Histogram of the focal depth distribution of the relocated earthquakes in central-western China using the double-difference algorithm. Solid circles and error-bars represent the relocated focal depths and uncertainties, respectively. The reference point is 29.00°N and 105.00°E

Previous studies indicated that focal depths of the earthquakes in central-western China are predominantly distributed between 10km and 25km (Ma and Xue, 1983). Zhang et al. (2002) made a statistical analysis of routinely determined focal depths of 31,282 earthquakes with $M_L \geq 2.0$ which occurred in western China. In their study, the reported depth uncertainties of 60% of these earthquakes are less than 4 km and are grade I earthquake location (a state-specified standard used in seismic networks in China for specifying the quality of earthquake location in which grade I refers to the epicenter and depth uncertainties less than 5km and 10km, respectively). Zhang et al. (2002) concluded that the focal depths of about 90% of the earthquakes in western China are located at 5km to 34 km. Zhang et al. (2002), however, find an average focal depth of 18km, with 68% of the earthquakes located between 10km to 26km. The results obtained in this study indicate an average focal depth of 11km, with about 91% events located in the depth range from 0 to 20km, and about 77% events in the depth range from 0 to 15km. This indicates that earthquakes in central-western China are of shallower depth and fall in a narrower depth range than that reported in previous studies.

5 Discussion

The improved locations obtained in this study provide important constraints for studies of the lithosphere in general and the seismogenic layer in particular. The lithosphere consists of two layers: the seismogenic upper layer, or schizosphere, characterized by elastobrittle deformation, and the aseismic lower layer, or plastosphere (Scholz, 1982, 1990), in which ductile deformation is predominant. The seismogenic layer along the San Andreas Fault Zone in California, for example, features an average thickness of about 15km (Scholz, 1990; Pacheco et al., 1992). The results obtained in this study indicate that the seismogenic layer in central-western China reaches, on average, a depth of about 20km. With an average crustal thickness of about 45km to 50km and an average upper crustal thickness of about 20km in this area of China (Kan and Lin, 1986; Liu et al, 1989; Sun and Liu, 1991; Fan and Chen, 1992; Wang et al., 2002, 2003), most of the stress appears to be released by brittle failure in the upper crust. It is worthy pointing out that the estimates of seismogenic depths depend on the initial focal depth estimates and the completeness of the catalog. The estimates of seismogenic depths in central-western China in this study are superior to that reported in previous studies; however, further study is needed using a more complete catalog and more accurate initial depth estimates. In the following we analyze the hypocentral distribution along the major seismic belts in order to understand the dimensions and depth extensions of the active faults, and we explore their correlation with past large earthquakes and the relocated seismicity.

5.1 Lancang-Gengma seismic belt (R2)

On November 6, 1988, two earthquakes with $M_S 7.4$ and $M_S 7.2$ (Nos. 50 and 51 in Figure 1b and Table A1) occurred along this fault system within 12 min (Figure 1b). The Lancang earthquake of $M_S 7.4$ (No. 50 in Figure 6a) ruptured the southern part of this fault; while the Gengma earthquake of $M_S 7.2$ (No. 51 in Figure 6a) ruptured the northern part about 60km to the NW of the Lancang earthquake. Focal mechanisms indicate that both earthquakes are mainly right-lateral strike-slip events (Chen and Wu, 1989; Mozaffari et al., 1998). The fault plane of the Lancang earthquake strikes at 144° and dips 79° towards SW, while the Gengma earthquake indicates a strike of 158° and a dip of 77° towards SW (Jiang, 1993). Aftershocks located by the Yunnan Province Seismic Network (YNSN) show a 175° (NNW-SSE) trend, 120km long and 50km wide band, strongly correlating with the rupture areas of the Lancang and Gengma events and with the background seismicity (Jiang, 1993). The relocations obtained in this study define a 150km long and 40km wide band (Figure 6a) that coincides with the aftershock area (130km×50km) of the 1988 Lancang-Gengma earthquakes and with the surface traces of mapped faults. The relocated focal depths fall in the range from 0 to 30km (Figures 6 (b & c)), with the deepest events occurring near the transition zone between the northern and southern segments. These features indicate that the relocated seismicity in this belt is clearly related to the same seismogenic structure as that of the 1988 Lancang-Gengma earthquakes.

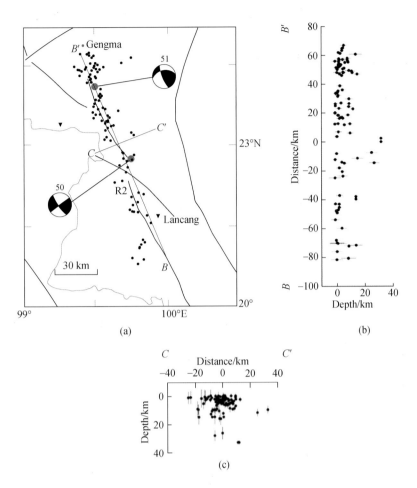

Figure 6 Relocated hypocentral distribution of seismicity (1992-1999) along the Lancang-Gengma seismic belt (R2)
(a) Epicentral distribution of relocated earthquakes. Solid circles represent relocated epicenters of earthquakes, upside-down gray triangles indicate seismic stations, thick solid lines are surface traces of active faults. Focal mechanisms are shown for the two 1988 Lancang-Gengma earthquakes (gray circles) of $M_S 7.4$ (No. 50) and $M_S 7.2$ (No. 51). (b) Vertical cross-section along the profile B-B' showing focal depths and uncertainties of the relocated earthquakes along N25°W direction. (c) Vertical cross-section along the profile C-C' showing focal depths and uncertainties of the relocated earthquakes along N65°E direction. Solid circles and error-bars in (b) and (c) represent the relocated focal depths and uncertainties, respectively. The reference point is 23.00°N and 99.60°E

5.2 Lijiang seismic belt (R3)

The N-S trending Lijiang seismic belt (R3) is bounded by the Lijiang Basin. About 450 aftershocks with $M_L \geq 2.5$ followed the $M_S 7.0$ event (No. 52 in Figure 1b and Table 1) from February 3, 1996 until December 1996. According to an investigation by Western Yunnan Earthquake Prediction Experiment Site, Seismological Bureau of Yunnan Province (1998, hereafter cited as WYEPES), the epicentral location of the mainshock did not correlate with any previously known major fault in this area, nor did the epicentral distribution of the mainshock and larger aftershocks appear to co-locate with the observed surface rupture produced by the mainshock. Field investigations revealed that the surface rupture was about 30km long and the maximum horizontal and vertical offsets were 30-50cm and 25-40cm, respectively (WYEPES, 1998). The meizoseismal area shows a maximum intensity area of grade IX on the Chinese Earthquake Intensity Scale (roughly equivalent to Modified Mercalli Earthquake Intensity Scale), with major and minor axes of 75km and 25km length in N-S and E-W directions (Figure 7a), respectively. The epicenters (open circles denoted by CNSN and YNSN in Figure 7a) of the Lijiang earthquake determined by both the CNSN and the YNSN are located at the northern portion of the meizoseismal area, near the isoseismal contour marking the transition from intensity grade IX to grade VIII. The double-difference location of the mainshock (gray solid circles denoted by DD in Figure 7a) locates near the center of the area with intensity grade IX, which also coincides with the area (open circles denoted by XC in Figure 7a) where maximum dislocation was obtained from waveform inversion (Xu and Chen, 1998). As mentioned above, the 1996 Lijiang earthquake and its aftershock sequence are included in the relocation. The relocated epicenters are distributed in an area of 70km length and 40km width; most of the relocated earthquakes are of focal depths less than 30km, and roughly show two conjugate planes, one dipping east and the another dipping west, both of which are a fit to the N-S striking, normal dip-slip focal mechanism of the mainshock (Figures 7 (a-c)).

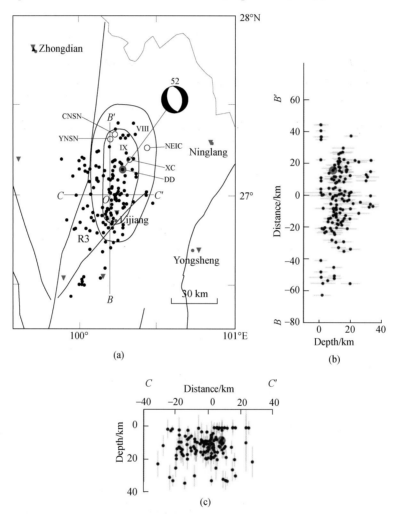

Figure 7 The 1996 Lijiang earthquake of $M_S7.0$ and relocated hypocentral distribution of seismicity within the Lijiang seismic belt (R3)

(a) Epicentral location, isoseismals, focal mechanism of the 1996 Lijiang earthquake and epicentral distribution of relocated earthquakes in the Lijiang seismic belt; (b) Vertical cross-section along the profile B-B' showing focal depths and uncertainties of the relocated earthquakes along S-N direction; (c) Vertical cross-section along the profile C-C' showing focal depths and uncertainties of the relocated earthquakes along W-E direction. Solid circles and error-bars in (b) and (c) represent the relocated focal depths and uncertainties, respectively. Open circles denoted by NEIC, CNSN, YNSN and XC show epicentral locations given by NEIC, YNSN, CNSN, and Xu and Chen (1998), respectively. Gray solid circle shows relocated epicenter of the Lijiang earthquake of $M_S7.0$ obtained in this study. The reference point is 27.00°N and 100.20°E

5.3 Yongsheng-Ninglang-Muli-Jiulong seismic belt (R4)

As shown in Figure 8a, the nearly N-S trending Yongsheng-Ninglang-Muli-Jiulong seismic belt (R4) consists of three segments: a southern segment (from 26.2°N to 26.9°N), a central segment (from 26.9°N to 28.1°N), and a northern segment (from 28.1°N to 29.5°N). The central and southern segments of the Yongsheng-Ninglang-Muli-Jiulong seismic belt (R4) as exposed at the surface has a length of 210km (from 26.2°N to 28.1°N) and a width of about 50km (from 100.6°E to 101.1°E) and is known as the Yongsheng-Ninglang seismic belt. The relocated hypocenters indicate that this seismic belt probably extends at least 150km further to the north and terminates at about 29.5°N (Figure 8a). The northward extending segment is known as the Muli-Jiulong seismic belt. It has a width of about 20km (roughly from 101.0°E to 101.2°E), compared to a width of about 50km along the central and southern segments. On the central part, the active faults mapped previously by field geological study (Li and Wang, 1975; Guo, 1984) define an arc of about 120km long (roughly from 26.9°N to 28.1°N) which in part is convex to the east (see fault traces in Figures 1b and 8a). While the epicentral distribution of the relocated earthquakes in the southern segment trend approximately north and

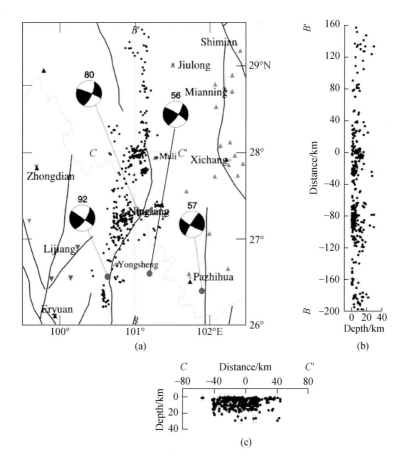

Figure 8 Relocated hypocentral distribution of Yongsheng-Ninglang-Muli-Jiulong seismic belt (R4)

(a) Epicentral distribution of relocated earthquakes (solid circles) and focal mechanisms of earthquakes Nos. 56, 57, 90 (gray solid circles); (b) Vertical cross-section along the profile B-B' showing focal depths and uncertainties of the relocated earthquakes along S-N direction; (c) Vertical cross-section along the profile C-C' showing focal depths and uncertainties of the relocated earthquakes along W-E direction. Solid circles and error-bars in (b) and (c) represent the relocated focal depths and uncertainties, respectively. The reference point is 28°N and 101°E

correlate directly to the surface trace of the fault, the earthquakes in the central and northern segments trend also approximately north but do not correlate directly to the surface trace of the fault. At the surface, no fault is currently mapped in the central segment of the seismic belt. Relative to the central segment (from 26.9°N to 28.1°N) of the seismic belt, the previously mapped fault is located about 30km to the east. Relative to the northern segment (from 28.1°N to 29.5°N) of the seismic belt, the previously mapped fault is located about 50km to the west. The striking difference between the epicentral distribution and location of the previously mapped faults in both the northern and the central segments of the Yongsheng-Ninglang-Muli-Jiulong seismic belt was not due to a systematic error in the relocation process. Thus the relocated epicenters indicate that as a whole the Yongsheng-Ninglang-Muli-Jiulong seismic belt (R4) is nearly N-S trending, is broader to the south of Muli and narrower to the north of Muli, and the structure imaged by the relocated earthquakes in the central and southern segments of the seismic belt from 26.2°N to 28.1°N is likely to image a blind active fault.

The Yongsheng-Ninglang-Muli-Jiulong seismic belt (R4) and the above-mentioned Lijiang seismic belt (R3) are within the Sichuan-Yunnan diamond-shaped tectonic block (Figure 9). Recent field geological study (Xu et al., 2003) suggests that an NE trending active fault, the Xiaojinhe-Lijiang fault, is likely cutting through the Sichuan-Yunnan diamond-shaped tectonic block and dividing the block into two sub-blocks, the southern and northern sub-blocks (Figure 9). Along the Xiaojinhe-Lijiang fault, various kinds of offset landforms have

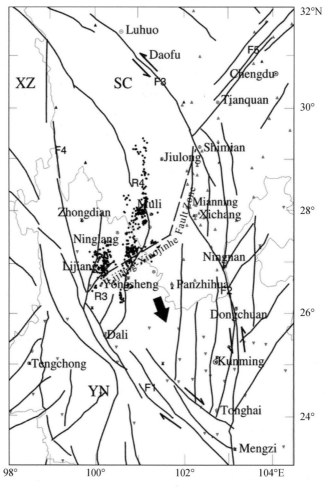

Figure 9 Yongsheng-Ninglang-Muli-Jiulong seismic belt (R4), Lijiang seismic belt (R3), and NE trending Xiaojinhe-Lijiang Fault (thick gray dash lines) as suggested by Xu et al. (2003)

Other symbols as in Figures 2 and 4

developed, showing that the fault is dominated by left-lateral strike-slip with minor reverse dip-slip components in the late Quaternary. As Figures 4 and 9 shown, the epicenters of relocated earthquakes within the Sichuan-Yunnan diamond-shaped tectonic block have not indicated a clear lineation associated with the NE trending Xiaojinhe-Lijiang fault. Instead, both the Lijiang (R3) and the Yongsheng-Ninglang-Muli-Jiulong (R4) seismic belts are two nearly parallel, N-S trending seismic belts. The focal mechanisms of earthquakes Nos. 56, 80, 92 show a NW to NNW compression (Figure 8a), and have one of the nodal planes nearly NNE striking, in good agreement with the overall trend of the Yongsheng-Ninglang-Muli-Jiulong seismic belt (R4) and mainly left-lateral strike-slip faulting with minor reverse dip-slip components. The earthquakes Nos. 56, 57 and 80, as well as the more recent M_S 6.0 earthquake (No. 92) of October 27, 2001, were not included in this relocation study. Thus the relationship between the seismicity of the Yongsheng-Ninglang-Muli-Jiulong seismic belt (R4), the Lijiang seismic belt (R3) and the Xiaojinhe-Lijiang Fault as suggested by Xu et al. (2003), and the occurrences of major earthquakes within the interior of the Sichuan-Yunnan diamond-shaped tectonic block remains unsolved and need further study.

5.4 Longmenshan (F5), Mabian (R6), Yibin (R7), Neijiang (R8) seismic belts

As shown in Figure 2, the Sichuan (SC) region has the highest levels of seismicity in the study area. The epicentral distribution of the routine earthquake locations in this region is diffuse across an extensive area (see F5, R6, R7, R8, R9, R10 and R11 in Figure 2), and no clear correlation between the microseismicity and the mapped active faults in this region emerges. After relocation, however, distinct lineation is revealed that can be associated with the Longmenshan seismic belt (F5), the Mabian seismic belts (R6), the Yibin seismic belts (R7), and the Neijiang structure (R8) (Figures 4 and 10). All the 1,398 relocated events along the Longmenshan Fault are distributed within a band of 470km length and 100km width. The band clearly coincides with the N40°E strike of the faults in this belt. Vertical cross-sections of focal depths indicate that the relocated hypocenters are mainly distributed in a layer not deeper than 30km (Figures 10 (b & c)).

Seismicity in the Mabian area (R6 in Figure 2) is characterized by a temporal clustering of earthquakes. From December 1935 to May 1936, a cluster of 11 earthquakes with M 6 to M 6.8 occurred in this area (Wang et al., 1999). Another cluster, consisting of 12 events with M_S 4.6 to M_S 5.9, occurred in this area from August to November of 1971 (Wang et al., 1999). Double-difference relocations of the recent seismicity (Figures 4 and 10 (a, d & e)) indicate a clear N-S trending lineation feature that coincides with the assumed locations of the two clusters of moderate size earthquakes. The relocated focal depths are mainly in the range of 0 to 20km (Figure 10 (d & e)). East of the Mabian seismic belt (R6) are the Yibin seismic belt (R7 in Figures 4 and 10) and the Neijiang seismic belt (R8 in Figures 4 and 10). The Mabian seismic belt (R6) is about 70km long and 20km wide and the Neijiang seismic belt (R8) is about 30km long and 20km wide. Relocated earthquakes of both belts (Figures 4 and 10 (f-i)) indicate strong spatial clustering, outlining the seismically active parts of the faults. In particular, the relocated focal depths of the Yibin seismic belt exhibits two clusters, a larger and shallower one in the south and a smaller and deeper one in the north (Figures 10 (f and g)), and the relocated focal depths of the Neijiang seismic belt also exhibits two earthquake clusters in the depth direction, one in the shallower range from 0 to 20km and another in the deeper range from 20km to 40km.

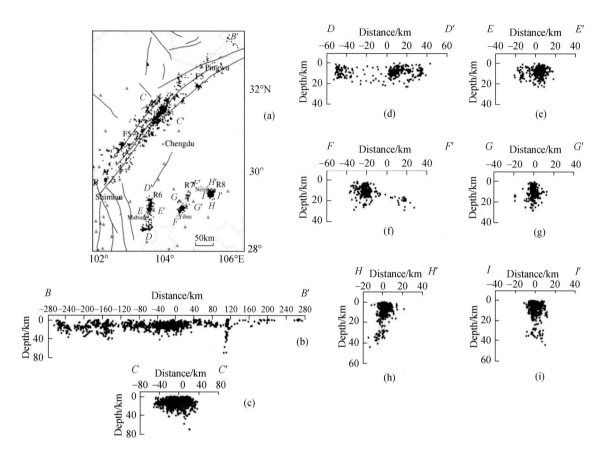

Figure 10 Focal depth distribution of the relocated earthquakes in several seismic belts

(a) Epicentral distribution of relocated earthquakes (solid circles); (b) Vertical cross-section along the profile B-B' showing focal depths and uncertainties of the relocated earthquakes along SW-NE direction of the Longmenshan seismic belt (F5); (c) Vertical cross-section along the profile C-C' showing focal depths and uncertainties of the relocated earthquakes along NW-SE direction of the Longmenshan seismic belt (F5); (d) Vertical cross-section along the profile D-D' showing focal depths and uncertainties of the relocated earthquakes along S-N direction of the Mabian seismic belt (R6); (e) Vertical cross-section along the profile E-E' showing focal depths and uncertainties of the relocated earthquakes along W-E direction of the Mabian seismic belt (R6); (f) Vertical cross-section along the profile F-F' showing focal depths and uncertainties of the relocated earthquakes along N30°E direction of the Yibin seismic belt (R7); (g) Vertical cross-section along the profile G-G' showing focal depths and uncertainties of the relocated earthquakes along N120°E direction of the Yibin seismic belt (R7); (h) Vertical cross-section along the profile H-H' showing focal depths and uncertainties of the relocated earthquakes along S-N direction of the Neijiang seismic belt (R8); (i) Vertical cross-section along the profile I-I' showing focal depths and uncertainties of the relocated earthquakes along W-E direction of the Neijiang seismic belt (R8). Solid circles and error-bars in (b) through (i) represent the relocated focal depths and uncertainties, respectively. The reference points for the Longmenshan, Mabian, Yibin and Neijiang seismic belts are (31.53°N, 104.10°E), (29.00°N, 103.62°E), (29.25°N, 104.76°E) and (29.44°N, 105.49°E), respectively

5.5 Songpan seismic belt (R9)

Songpan seismic belt (R9 in Figures 2, 4 and 11a) is to the north of the Longmenshan seismic belt (F5 in Figures 2, 4 and 10a). Relocated events along the Songpan seismic belt (R9) indicate a north trending seismic structure that appears to continue along a more NW trending fault system (Figure 11). Several strong earthquakes occurred such as the A. D. 1630 earthquake with M 6.5 (epicentral location: 32.6°N, 104.1°E) (Min, 1995) and the August 16, 22 and 23, 1976, earthquakes with M_S 7.2 (No. 47), M_S 6.7 (No. 78) and M_S 7.2 (No. 48) (Seismological Bureau of Sichuan Province, 1979), respectively. An investigation of the 1976 Songpan earthquake sequence by the Seismological Bureau of Sichuan Province (1979) indicates that the aftershocks occurred along an elongated area trending NNW, about 70km in length and 30km in width (rectangle enclosed by thin dashed line in Figure 11a), with the seismogenic fault striking N-S. The epicenters of the relocated earthquakes in this study coincide with the epicentral distribution of these events and their aftershocks. The relocated hypocenters indicate that the Songpan seismic belt extends about 50km further to the N40°W (Figure 11a), and that most earthquakes in the Songpan seismic belt are located in depth range of 0 to 20km. A few of them are at greater focal depth, and there is a tendency for deeper events toward the south (Figures 11 (b-e)). The focal mechanisms show nearly vertical fault planes, striking 0°, 334° and 325° for the preferred fault plane of these three events (Figure 11a), respectively. The relocated hypocenters in this study coincide not only with the epicentral distribution of these three events and their aftershocks (Figure 11), but also with the strikes of the preferred fault planes given by fault plane solutions (Figure 11 (a-c)).

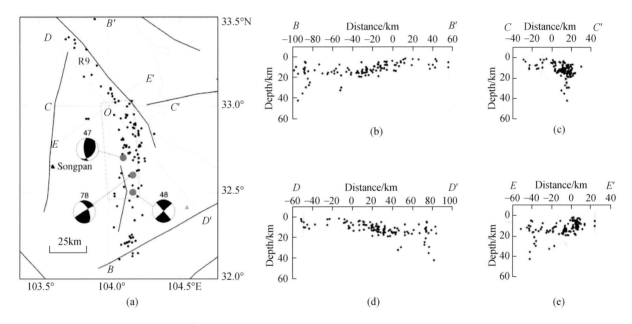

Figure 11 Relocated hypocentral distribution of Songpan seismic belt (R9) and focal mechanisms of the 1976 Songpan earthquakes (a) Epicentral distribution of relocated earthquakes (solid circles) and focal mechanisms of the 1976 Songpan earthquakes (gray circles); (b) Vertical cross-section along the profile B-B' showing focal depths and uncertainties of the relocated earthquakes along S-N direction of the Songpan seismic belt (R9); (c) Vertical cross-section along the profile C-C' showing focal depths and uncertainties of the relocated earthquakes along E-W direction of the Songpan seismic belt (R9); (d) Vertical cross-section along the profile D-D' showing focal depths and uncertainties of the relocated earthquakes along N40°W direction of the Songpan seismic belt (R9); (e) Vertical cross-section of the profile E-E' showing focal depths and uncertainties of the relocated earthquakes along N50°E direction of the Songpan seismic belt (R9). Solid circles and error bars represent focal depths and uncertainties, respectively. The reference point for the profiles B-B', C-C', D-D' and E-E' is 33.00°N and 104.00°E

5.6 Shimian and Mianning seismic belts (R10, R11)

We note that the Shimian and Mianning seismic belts, shown in Figure 12, have focal depths somewhat greater than typical for the broad region. Both belts are located in an area where the Xianshuihe Fault (F3 in Figures 1 and 4) and Xiaojiang Fault (F2 in Figures 1 and 4) meet. After relocation, two seismic belts, Shimian seismic belt (R10 in Figures 1 and 4) and Mianning seismic belt (R11 in Figures 1 and 4), are resolved. The relocation results unambiguously show that essentially the Shimian seismic belt (R10) in the east is the northern segment of the Xiaojiang Fault and trends 330° (N30°W), and the Mianning seismic belt in the west is the northern segment of Anninghe Fault and trends N-S (R11 in Figures 1 and 4). In the north, the Mianning seismic belt merges into the Shimian seismic belt. The focal depths of these two seismic belts are relatively deep (Figures 12 (b-e)). The earthquakes are distributed predominately in the depth range from 15km to 35km for the Shimian seismic belt (Figures 12 (b & c)) and from 10km to 25km for the Mianning seismic belt, respectively (Figures 12 (d & e)).

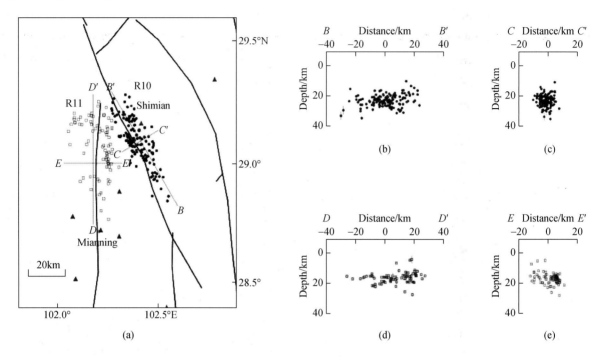

Figure 12 Hypocentral distribution of relocated earthquakes of Shimian (R10) and Mianning (R11) seismic belts
(a) Epicentral distribution of relocated earthquakes. Solid circles represent earthquakes of Shimian seismic belts (R10) and open squares represent earthquakes of Mianning seismic belts (R11). (b) Vertical cross-section along the profile B-B' showing focal depths (solid circles) and uncertainties (error bars) of the relocated earthquakes along N30°W direction in Shimian seismic belt. (c) Vertical cross-section along the profile C-C' showing focal depths (solid circles) and uncertainties (error bars) of the relocated earthquakes along N60°E direction in Shimian seismic belt. (d) Vertical cross-section along the profile D-D' showing focal depths (open squares) and uncertainties (error bars) of the relocated earthquakes along S-N direction in Mianning seismic belt. (e) Vertical cross-section along the profile E-E' showing focal depths (open squares) and uncertainties (error bars) of the relocated earthquakes along W-E direction in Mianning seismic belt. The reference points for Shimian and Mianning seismic belts are (29.09°N, 102.41°E) and (29.00°N, 102.18°E), respectively

6 Conclusions

In this study the double-difference technique (Waldhauser and Ellsworth, 2000) has been applied to earthquakes in central-western China from 21°N to 36°N and 98°E to 111°E, to obtain a clearer view of seismicity distribution. It is demonstrated that the seismicity pattern based on the routine earthquake location is diffuse and shows ambiguous features, and is not well correlated to the expression of active faults at the surface. For most seismic belts of central-western China, such as Lancang-Gengma, Lijiang, Yongsheng-Ninglang-Muli-Jiulong, Longmenshan, Mabian, Yibin, Neijiang, Songpan, Shimain and Mianning seismic belts, we find a close relationship between the relocated seismicity and the expression of active faults at the surface in general. But for some seismic belts of the study area, such as the central and northern segments of the Yongsheng-Ninglang-Muli-Jiulong seismic belt and the Xiaojinhe-Lijiang Fault, we find a significant difference between the epicentral distribution of relocated earthquakes and locations of the mapped faults. In such cases, the relocated seismicity is clustered and can be interpreted the presence of active faults. Moreover, our relocated events allow to indicate the Shimian seismic belt and Mianning seismic belt as distinct features.

We find most earthquakes are distributed between depths of 0 to 20km, indicating that the thickness of the seismogenic layer in this region is less than 20km. However, earthquakes with deeper focal depth, such as earthquakes of the Neijiang seismic belt with focal depths from 0 to 40km, are also found in some locations. These results provide important constraints for studies of the seismogenic layer, and for the mechanism of earthquake generation in central-western China.

Acknowledgments

The authors would like to thank two anonymous reviewers for their constructive comments and the authors would like to thank the Monitoring Center of the Seismological Bureau of Yunnan Province, Monitoring Center of the Seismological Bureau of Sichuan Province, Seismological Bureau of Shaanxi Province, Seismological Bureau of Guangxi Province and the Sub-Center of China National Digital Seismic Network, Institute of Geophysics, China Seismological Bureau, for providing the observational data of seismic networks. We would also like to acknowledge Dr. Yu Xiangwei and Ms. Zheng Yuejun for their help in data pre-processing for this study, and Ms. Shen Siwei, Mr. Chen Shulin and Ms. Long Xiaofan, Seismological Bureau of Yunnan Province, and Ms. Feng Dongying, Institute of Crustal Dynamics, China Seismological Bureau, for their help in data collection for this study. This study is sponsored by the key project "Process, Mechanism and Prediction of Geological Hazard" (2001CB711005-1-3) and the project "Mechanism and Prediction of Continental Earthquakes" of Ministry of Science and Technology, People's Republic of China (G19980407/95-13-02-04). Contribution No. 04FE1011, Institute of Geophysics, China Seismological Bureau.

Appendix

See Table A1.

Table A1　Catalog of major earthquake of M≥7 from 780 B.C. to April 2003 in central-western China

No.	yy	mm	dd	Lat. /(°N)	Long. /(°E)	Depth /km	M	NP1 Strike /(°)	NP1 Dip /(°)	NP1 Rake /(°)	NP2 Strike /(°)	NP2 Dip /(°)	NP2 Rake /(°)	P-axis Az. /(°)	P-axis Pl. /(°)	B-axis Az. /(°)	B-axis Pl. /(°)	T-axis Az. /(°)	T-axis Pl. /(°)
1	−780			34.5	107.8		7												
2	143	10		35.0	104.0		7												
3	734	3	23	34.6	105.6		7												
4	814	4	6	27.9	102.2		7												
5	1216	3	24	28.4	103.8		7												
6	1352	4	26	35.6	105.3		7												
7	1500	1	3	24.9	103.1		7												
8	1501	1	29	34.8	110.1		7												
9	1515	6	27	26.7	100.7		$7\frac{3}{4}$												
10	1536	3	29	28.1	102.2		$7\frac{1}{2}$												
11	1556	2	2	34.5	109.7		$8\frac{1}{4}$												
12	1588	8	9	24.0	102.8		7												
13	1652	7	13	25.2	100.6		7												
14	1654	7	21	34.3	105.5		8												
15	1713	9	4	32.0	103.7		7												
16	1718	6	19	35.0	105.2		$7\frac{1}{2}$												
17	1725	8	1	30.0	101.9		7												
18	1733	8	2	26.3	103.1		$8\frac{3}{4}$												
19	1786	6	1	29.9	102.0		$7\frac{3}{4}$												
20	1786	6	10	29.4	100.6		7												
21	1789	6	7	24.2	102.9		7												
22	1789	8	27	23.8	102.4		$7\frac{1}{2}$												
23	1816	12	8	31.4	100.7		7												
24	1833	9	6	25.0	103.0		8												
25	1850	9	12	27.7	102.4		$7\frac{1}{2}$												
26	1870	4	11	30.0	99.1		$7\frac{1}{4}$												
27	1879	7	1	33.2	104.7		8												
28	1887	12	16	23.7	102.5		7												
29	1893	8	29	30.6	101.5		7												
30	1896	3		32.5	98.0		7												
31	1904	8	30	31.00	101.10		7												

Continued

No.	yy	mm	dd	Lat. /(°N)	Long. /(°E)	Depth /km	M	NP1 Strike /(°)	NP1 Dip /(°)	NP1 Rake /(°)	NP2 Strike /(°)	NP2 Dip /(°)	NP2 Rake /(°)	P-axis Az. /(°)	P-axis Pl. /(°)	B-axis Az. /(°)	B-axis Pl. /(°)	T-axis Az. /(°)	T-axis Pl. /(°)
32	1913	12	21	24.15	102.45		7												
33	1923	3	24	31.30	100.80		7.3												
34	1925	3	16	25.70	100.20		7												
35	1933	8	25	32.00	103.70		7½	14.00	60.00	-167.00	277.00	79.00	-30.00	52.00	29.00	259.00	57.00	149.00	13.00
36	1941	5	16	23.70	99.40		7												
37	1941	12	26	22.10	100.10		7	52.00	85.00	179.00	141.00	89.00	5.00	96.00	3.00	334.00	85.00	186.00	4.00
38	1947	3	17	33.30	99.50		7.7												
39	1948	5	25	29.50	100.50		7.3	54.00	89.00	-179.00	144.00	89.00	1.00	99.00	1.00			189.00	1.00
40	1950	2	3	21.70	100.10		7	149.00	85.00	180.00	59.00	90.00	-5.00						
41	1955	4	14	30.00	101.90		7½	263.00	72.00	164.00	358.00	75.00	20.00	310.00	2.00	215.00	66.00	41.00	24.00
42	1970	1	5	24.00	102.70	13	7.8	197.00	67.00	-10.00	291.00	80.00	-157.00	337.00	24.00	136.00	65.00	242.00	10.00
43	1973	2	6	31.48	100.53	11	7.6	29.00	80.00	179.00	119.00	89.00	-10.00	74.00	4.00	307.00	80.00	164.00	8.00
44	1974	5	11	28.20	103.90	14	7.1	237.00	81.00	176.00	328.00	86.00	10.00	83.00	1.00	170.00	85.00	3.00	5.00
45	1976	5	29	24.50	99.00	24	7.3	139.00	69.00	159.00	229.00	69.00	20.00	183.00	10.00	270.00	60.00	90.00	30.00
46	1976	5	29	24.45	98.86	21	7.4	31.00	77.00	9.00	119.00	83.00	166.00	346.00	4.00	90.00	75.00	255.00	15.00
47	1976	8	16	32.61	104.13	15	7.2	38	35	128	173	65	67	101	16	225	63	5	21
48	1976	8	23	32.50	104.30	23	7.2	60	85	240	333	58	6	112	18	53	57	12	27
49	1981	1	24	30.98	101.10	12	7.0												
50	1988	11	6	22.92	99.75	13	7.4	144	79	179	54	89	-11	10	9	224	79	100	8
51	1988	11	6	23.37	99.50	16	7.2	158	77	136	259	49	15	215	18	324	46	110	39
52	1996	2	3	27.34	100.25	10	7.0	353	44	-78	157	48	-102	346	80	165	10	255	0
53	1936	4	1	22.50	109.40	7	6.8	110	65	-178	19	86	-25	331	19	192	24	67	25
54	1951	12	21	26.00	100.00		6.3	311	60	-24	209	71	-32	167	36	262	7	1	53
55	1952	9	30	28.40	102.20		6.8	245	75	130	354	40	30	128	21	235	36	13	45
56	1955	6	7	26.60	101.20		6.0	34	75	10	301	80	166	348	4	89	72	257	18
57	1955	9	23	26.40	101.90		6.8	120	80	173	210	87	20	165	5	49	80	256	9
58	1958	2	8	31.70	104.30		6.2	6	45	90	6	45	90	346	59	250	14	141	27
59	1960	11	9	32.78	103.66	20	6.8	60	50	-130	293	55	-52	84	59	269	31	177	2
60	1961	6	27	27.76	99.68	10	6.0	118	75	150	36	61	18	349	10	92	57	253	52
61	1962	6	24	25.30	101.10		6.2	120	74	-162	205	73	-17	342	24	161	65	252	1
62	1963	4	23	25.80	99.50	20	6.0	184	88	-172	93	83	-3	48	6	202	83	318	3

Continued

No.	yy	mm	dd	Lat. /(°N)	Long. /(°E)	Depth /km	M	NP1 Strike /(°)	NP1 Dip /(°)	NP1 Rake /(°)	NP2 Strike /(°)	NP2 Dip /(°)	NP2 Rake /(°)	P-axis Az. /(°)	P-axis Pl. /(°)	B-axis Az. /(°)	B-axis Pl. /(°)	T-axis Az. /(°)	T-axis Pl. /(°)
63	1965	7	3	22.50	101.40		6.1	302	73	−165	28	77	−18	164	21	352	68	255	3
64	1966	2	5	26.20	103.20		6.5	159	85	−11	249	80	−175	294	11	130	79	24	3
65	1966	2	13	26.10	103.10		6.2	330	82	43	232	49	168	94	21	337	48	198	34
66	1966	9	28	27.50	100.10		6.4	320	49	118	178	48	−64	161	70	158	20	69	1
67	1967	8	30	31.61	100.33		6.8	24	82	−179	294	88	−10	249	7	101	82	339	5
68	1967	8	30	31.70	100.33		6.0	11	80	178	282	87	12	236	5	119	80	327	9
69	1970	2	7	22.90	100.80	15	6.2	309	78	160	43	70	16	357	5	92	70	265	23
70	1970	2	24	30.60	103.20	15	6.2	260	75	169	352	80	17	125	4	25	72	217	18
71	1971	3	24	35.45	98.00	13	6.3	164	30	90	344	60	90	74	15	164	0	254	75
72	1971	4	28	22.90	101.00	15	6.7	31	81	23	117	70	170	343	7	234	68	75	21
73	1971	9	14	23.00	100.80		6.2	327	50	70	172	43	108	249	3	158	13	353	77
74	1973	2	8	31.60	100.50		6.0	212	65	−90	32	25	−90	122	70	32	0	302	25
75	1973	8	11	32.93	103.90	19	6.5	252	77	173	343	85	15	297	6	180	76	28	13
76	1973	8	16	23.00	101.08	7	6.3	328	50	−118	188	48	−61	175	69	347	21	78	2
77	1975	1	15	29.40	101.90	25	6.2	118	70	−158	21	69	−21	340	29	158	60	250	1
78	1976	8	22	32.60	104.40	21	6.7	240	85	147	333	58	6	112	18	23	57	12	27
79	1976	11	7	27.45	101.08	21	6.7	22	50	−3	115	88	−138	167	29	298	50	62	25
80	1976	12	13	27.31	101.05	21	6.4	292	84	160	24	70	3	340	10	95	69	246	18
81	1979	3	15	23.20	101.10	10	6.8	42	73	1	312	89	163	167	29	298	50	62	25
82	1981	1	24	30.98	101.10	12	6.9	142	85	−13	233	77	−175	178	11	311	73	8	12
83	1981	9	19	22.98	101.36	33	6.0	319	73	146	58	55	20	277	13	121	76	58	6
84	1982	6	16	31.86	99.75	15	6.0	20	79	−170	287	79	−10	184	16	62	75	334	1
85	1983	6	24	21.72	103.38	18	6.8	114	80	−172	23	82	−10	338	13	163	77	69	1
86	1985	4	18	25.91	102.80	5	6.2	104	47	125	236	52	56	350	1	269	26	82	64
87	1989	4	16	29.92	99.20	12	6.6	273	29	−73	74	62	−99	323	72	78	8	171	16
88	1989	4	25	30.00	99.37	7	6.6	245	40	−114	96	54	−71	59	73	264	16	172	7
89	1989	5	3	30.11	99.54	14	6.3	240	44	−117	95	52	−66	67	71	260	18	169	4
90	1995	10	24	25.95	102.30	15	6.5	13	75	−9	105	81	−165	320	40	112	85	56	5
91	2000	1	14	25.67	101.18	32	6.4	27	28	−6	118	84	−168	338	35	100	82	252	2
92	2001	10	27	26.36	101.01	15	6.0	35	69	1	305	89	159	341	60	111	72	258	15

References

Chen, Y. T. and Wu, F. T., 1989, Langcang-Gengma earthquake. A preliminary report on the November 6, 1988 event and its aftershocks, *Eos. Trans. Amer. Geophys. Union* (12), 1527–1540.

Chen, Y. T., Chen, Z. L. and Wang, B. Q., 1992, Seismological studies of earthquake prediction in China: A review. In: Dragoni, M. and Boschi, E. (eds.), *Earthquake Prediction*. IlCigno Galileo Galilei, Roma, 71–109.

Chung, W. Y. and Kanamori, H., 1976, Source process and tectonic implications of the Spanish deep-focus earthquake of March 29, 1954, *Phys. Earth Planet. Interi.* **13**, 85–96.

Deng, Q., Xu, X. and Yu, G., 1994, Characteristics of regionalization of active faults in China and their genesis. In: Department of Science and Technology, State Seismological Bureau of China (ed.), *Active Fault Research in China* 1–14. (in Chinese)

Ding, G. Y. (ed.), 1991, *Lithospheric Dynamics of China. Explanatory Notes for the Atlas of Lithospheric Dynamics of China*, Seismological Press, Beijing, 600 pp. (in Chinese)

Ding, G. Y., 1996, The research on active fault in China, *J. Earthq. Pred. Res.* **5**, 317–325. (in Chinese)

Fan, C. J. and Chen, Y. K., 1992, Crustal structure of western Yunnan. In: Kan, R. J. (ed.), *Geophysical Studies of Yunnan Province*, Yunnan University Press, Kunming, 102–109. (in Chinese)

Fitch, T. J., 1975, Compressional velocity in source regions of deep earthquakes: An application of the master event technique, *Earth Planet. Sci. Lett.* **26**, 156–166.

Fukao, Y., 1972, Source process for a large earthquake and its tectonic implications-the western Brazil earthquake of 1963, *Phys. Earth Planet. Interi.* **5**, 61–76.

Got, J.-L., Frechet, J. and Klein, F. W., 1994, Deep fault plane geometry inferred from multiplet relocation beneath the south flank of Kilauea, *J. Geophys. Res.* **99**, 15375–15386.

Gu, G. X. (ed.), 1983, *Catalogue of Chinese Earthquakes*, **1** and **2**, Science Press, Beijing, 1, 228 pp. (in Chinese)

Guo, S. M., Zhang, J., Li, X. G., Xiang, H. F., Chen, T. N. and Zhang, G. W., 1984, Fault displacement and recurrence intervals of earthquakes at the northern segment of Honghe fault zone, *Seismology and Geology* **6**, 1–12. (in Chinese with English abstract)

Jiang, K. (ed.), 1993, *The 1988 Lancang-Gengma Earthquakes (M=7.6, 7.2) in Yunnan, China*, Yunnan University Press, Kunming, 387 pp. (in Chinese)

Kan, R. J., Zhang, S. C., Yan, F. T. and Yu, L. S., 1977, Present tectonic stress and its relation to the characteristics of recent tectonic activity in southwestern China, *Acta Geophysica Sinica* **20**(2), 96–109. (in Chinese with English abstract)

Kan, R. J. and Lin, Z. Y., 1986, A preliminary study on crustal and upper mantle structures in Yunnan, *Earthq. Res. China* **2**(4), 50–61. (in Chinese)

Li, P. and Wang, L. M., 1975, Discussion on the seismo-geological characteristics of the Yunnan-west Sichuan region, *Scientia Geologica Sinica* (4), 308–325. (in Chinese with English abstract)

Li, P. (ed.), 1993, *The Xianshuihe-Xiaojiang Fault Zone*, Seismological Press, Beijing, 267 pp. (in Chinese)

Liu, J. H., Liu, F. T., Wu, H., Li, Q. and Hu, G., 1989, Three-dimension velocity image of the crust and upper mantle beneath North-South Zone in China, *Acta Geophysica Sinica* **32**(2), 152–161.

Ma, X. Y. (ed.), 1987, *Explanatory Notes of the Lithosphireic Dynamics Map of China and Adjacent Seas*, Scale 1: 4000000, Geological Publishing House, Beijing. (in Chinese)

Ma, X. Y. (ed.), 1989, *Lithospheric Dynamics Atlas of China*, China Cartographic Publishing House, Beijing, 68pp. (in Chinese)

Ma, Z. J. and Xue, F., 1983, Depth distribution and a preliminary discussion on the "seismic layer" in the Chinese continent, *Research in Seismological Science* **3**(3), 43–46. (in Chinese)

Min, Z. Q. (ed.), 1995, *Catalogue of Chinese Historical Strong Earthquakes*, Seismological Press, Beijing, 514pp. (in Chinese)

Molnar, P. and Tapponnier, P., 1975, Cenozoic tectonics of Asia: Effects of a continental collision, *Science* **159**(4201), 419–426.

Mozaffari, P., Wu, Z. L. and Chen, Y. T., 1998, Rupture process of November 6, 1988, Lancang-Gengma, Yunnan, China, earthquake of $M_S = 7.6$ using empirical Green's function deconvolution method, *Acta Seismologica Sinica* **11**(1), 1–12.

Pacheco, J. F., Scholz, C. H. and Sykes, L. R., 1992, Changes in frequency-size relationship from small to large earthquakes, *Nature* **355**, 71–73.

Qin, C., Papazachos, C. and Papadimitriou, E., 2002, Velocity field for crustal deformation in China derived from seismic moment tensor summation of earthquakes, *Tectonophysics* **359**(1/2), 29–46.

Scholz, C. H., 1982, Scaling laws for large earthquakes: Consequences for physical models, *Bull. Seism. Soc. Amer.* **72**(6), 1–14.

Scholz, C. H., 1990, *The Mechanics of Earthquakes and Faulting*, Cambridge University Press, Cambridge, 439pp.

Seismological Bureau of Sichuan Province (ed.), 1979, *The 1976 Songpan Earthquake*, Seimological Press, Beijing, 112pp. (in Chinese)

Su, Y. J. and Qin, J. Z., 2001, Strong earthquake activity and relation to regional neo-tectonic movement in Sichuan-Yunnan Region, *Earthq. Res. China* **17**(1), 24–34. (in Chinese with English abstract)

Sun, R. M. and Liu, F. T., 1991, Seismic tomography of Sichuan, *Acta Geophysica Sinica* **34**(6), 708–716.

Teng, C. T., Chang, Y. M. and Hsu, K. L., 1979, Tectonic stress field in China and its relation to plate movement, *Phys. Earth Planet. Interi.* **18**, 257–273.

Waldhauser, F. and Ellsworth, W. L., 2000, A double difference earthquake location algorithm: Method and application to the Northern Hayward Fault, California, *Bull. Seism. Soc. Amer.* **90**(6), 1353–1368.

Waldhauser, F., 2001, *HypoDD: A Computer Program to Compute Double-difference Earthquake Location*, U. S. Geol. Surv. Openfile report, 01–113, Menlo Park, California.

Wang, C. Y., Mooney, W. D., Wang, X. L., Wu, J. P., Lou, H. and Wang, F., 2002, A study on 3-D velocity structure of upper mantle in Sichuan and Yunnan region, *Acta Seismologica Sinica* **15**(1), 1–17.

Wang, C. Y., Wu, J. P., Lou, H., Zhou, M. D. and Bai, Z. M., 2003, P-wave crustal velocity structure in western Sichuan and eastern Tibetan region, *Science in China* (Ser. D) **46**(Suppl.), 254–265.

Wang, Q., Zhang, P. Z., Freymueller, J. T., Bilham, R., Larson, K. M., Lai, X. A., You, X. Z., Niu, Z. J., Wu, J. C., Li, Y. X., Liu, J. N., Yang, Z. Q. and Chen, Q. Z., 2001, Present-day crustal deformation in continental China constrained by Global Positioning System measurements, *Science* **249**, 574–577.

Wang, S. Y., Wu, G. and Shi, Z. L., 1999, *Catalogue of Chinese Current Earthquakes (1912-1990, $M_S \geq 4.7$)*, Chinese Science and Technology Press, Beijing, 637pp. (in Chinese)

Wen, X. Z., 1998, Assessment of time-dependent seismic hazard on segment of active fault, and its problem. *Chinese Science Bulletin* **43**(23), 41–50.

Western Yunnan Earthquake Prediction Experiment Site, Seismological Bureau of Yunnan Province, China (ed.), 1998, *The 1996 Lijiang Earthquake in Yunnan, China*, Seismological Press, Beijing, 188 pp. (in Chinese)

Xie, Y. S. and Cai, M. B., 1983–1987, *Compilation of Historical Materials of Chinese Earthquakes*, **1** to **5**, Science Press, Beijing, 4, 471 pp. (in Chinese)

Xu, L. S., Chen, Y. T. and Fasthoff, S., 1998, Inversion for rupture process of the 1996 Lijiang, Yunnan, China M_S 7.0 earthquake by empirical Green's function technique, In: Western Yunnan Earthquake Prediction Experiment Site, Seismological Bureau of Yunnan Province, China (ed.), *The 1996 Lijiang Earthquake in Yunnan, China*, Seismological Press, Beijing, pp. 79–81. (in Chinese)

Xu, X. W., Wen, X. Z., Zheng, R. Z., Ma, W. T., Song, F. M. and Yu, G. H., 2003, Pattern of latest tectonic motion and its dynamics for active blocks in Sichuan-Yunnan region, China, *Science in China* (Ser. D) **46**(Suppl.), 210–226.

Xu, Z. H., Wang, S. Y., Huang, Y. R. and Gao, A. J., 1989, The tectonic stress field of Chinese continent deduced from a great number of earthquakes, *Acta Geophysica Sinica* **32**(6), 636–647. (in Chinese with English abstract)

Yang, Z. X., Chen, Y. T. and Zhang, H. Z., 1999, Relocation of the Zhangbei-Shangyi earthquake sequence, *Seismological and Geomagnetic Observation and Research* **20**, 6-9. (in Chinese with English abstract)

Yang, Z. X., Chen, Y. T. and Zhang, H. Z., 2002, Relocation and seismogenic structure of the 1998 Zhangbei-Shangyi earthquake sequence, *Acta Seismologica Sinica* **15**(4), 383-394.

Yang, Z. X., Yu, X. W., Zheng, Y. J., Chen, Y. T., Ni, X. X. and Chen, W., 2004, Earthquake relocation and 3-dimensional crustal structure of P-wave velocity in central-western China, *Acta Seismologica Sinica* **26**(1), 19-29. (in Chinese with English abstract)

Zhang, G. M., Wang, S. Y., Li, L., Zhang, X. D. and Ma, H. S., 2002, Focal depth and its tectonic implications of the continental earthquakes in China, *Chinese Science Bulletin* **47**(9), 663-668. (in Chinese with English abstract)

Zhao, Z., Fan, J., Zeng, S. H., Hasegawa, A. and Horiuchi, S., 1997, Crustal structure and accurate hypocenter determination along the Longmenshan fault zone, *Acta Seismologica Sinica* **19**(6), 615-622. (in Chinese with English abstract)

Zhao, Z. and Zhang, R. S., 1987, Primary study of crustal and upper mantle velocity structure of Sichuan Province, *Acta Seismologica Sinica* **9**(2), 154-166. (in Chinese with English abstract)

Zhou, S. Y., Zhang, Y. G., Ding, G. Y., Wu, Y., Qin, X. J., Shi, S. Y., Wang, Q., Yiu, X. Z., Qiao, X. J., Shuai, P. and Deng, G. J., 1998, A preliminary research establishing the present-time intraplate blocks movement model on the Chinese mainland based on GPS data, *Acta Seismologica Sinica* **20**(4), 347-355. (in Chinese with English abstract)

2007年云南宁洱M_S6.4地震震源过程*

张 勇[1,2]　许力生[2]　陈运泰[2,1]　冯万鹏[2]　杜海林[2]

1. 北京大学地球物理学系, 北京　100871; 2. 中国地震局地球物理研究所, 北京　100081

摘要　通过反演全球范围内20个地震台的宽频带波形资料, 获得了2007年6月3日在云南宁洱发生的M_S6.4地震的矩张量解、震源时间函数和断层面上滑动随时间和空间的变化过程. 根据反演结果, 这次地震的标量地震矩为$5.51×10^{18}$Nm, 相当于矩震级M_W6.4. 震源机制解中, 最佳双力偶对应的节面Ⅰ的走向、倾角和滑动角分别为152°, 54°和166°, 节面Ⅱ的走向、倾角和滑动角分别为250°, 79°和37°. 结合震后考察得到的烈度等震线分布特征以及当地的地质构造, 可以判定这次地震的发震断层的走向为152°, 倾角为54°, 滑动角为166°, 是一次以右旋走滑为主的地震. 从震源时间函数的形态来看, 震源破裂持续时间为14s, 地震矩的释放主要集中在前11s, 在11~14s之间释放的地震矩很少. 震源的时空破裂过程图像表明, 破裂过程分为3个阶段, 在前4s的时间段内, 破裂主要沿着走向方向和朝深处发展; 在4~7s间, 破裂呈扇形向着深处扩展; 在7s之后的时间段, 破裂点比较零散. 地震破裂总体上表现为双侧破裂方式, 但在走向方向和深度方向上的滑动略占优势. 破裂较强的区域呈菱形, 长约为19km. 地震断层面上最大滑动量为1.2m, 平均滑动量为0.1m, 最大滑动速率为0.4m/s, 平均滑动速率为0.1m/s. 由反演得到的静态位错模型计算的震中区地表位移场的特征与地震的烈度分布特征具有很好的一致性.

北京时间2007年6月3日凌晨5点33分48秒, 中国云南宁洱地区发生了M_S6.4级地震. 根据美国地质调查局国家地震信息中心(NEIC/USGS)的测定结果, 这次地震的发震时刻为6月2日21时33分48秒(UTC), 震中位置为23.03°N, 101.05°E, 震源深度约5km. 许多地震监测和研究单位, 如云南省地震局、中国国家地震台网中心和哈佛大学也给出了相类似的定位结果(见图1, 其中哈佛大学结果为矩心矩张量解(CMT)的位置). 这次地震造成大量民房倒损, 导致3人死亡, 300多人受伤, 震区受灾人口达53.6万(http://news.sohu.com/20070604/n250391529.shtml).

云南宁洱地区为地震频发区, 历史上发生过多次强震. 自1884年至今, 在距此次地震震中50km范围内, 先后发生过9次5级以上地震, 其中6次地震的震级超过6级. 震级最大的地震是发生于1979年3月1日的6.8级地震. 这说明该区域的应力构造释放主要以6级左右的强震为主. 已有的研究结果表明[1,2], 云南宁洱地区位于无量山活动构造带上, 此断裂带以北北西向的右旋走滑运动为主(图1); 此外, 这一地区还分布着一系列以左旋走滑为主的北东东向的小断层群[1,2]. 此次M_S6.4地震发生后的现场考察结果表明[2],①, 极震区呈带状分布, 为北北西-南南东向. 虽然地震断层没有明显地出露到地表, 但从震中附近, 即22.98°N, 101.05°E周围的一些地表裂痕的分布仍可观察到明显的右旋象①.

---

* 本文发表于《中国科学: D辑 地球科学》, 2008年, 第38卷, 第6期, 683-692.
① 冯万鹏, 杜海林, 许力生, 等. 2007年云南宁洱M_S6.4级地震现场考察与震源模式. 2008 (待发表)

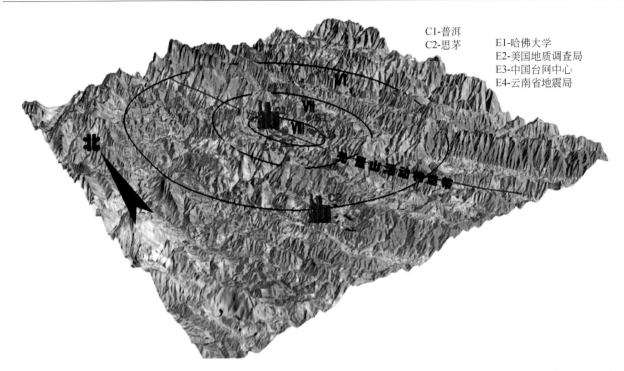

图1 2007年6月3日在云南宁洱M_S6.4级地震震中位置（红点）、烈度线（黑线）以及震中区内的主要活动断层（红线）
其中红色线条表示无量山活动构造带. 图中给出了普洱和思茅2个城市位置. E1 (23.05°N, 101.13°E), E2 (23.03°N, 101.05°E), E3 (23.00°N, 101.1°E) 和 E4 (23.03°N, 101.01°E) 分别为哈佛大学、美国地质调查局、中国台网中心和云南省地震局的定位结果

为了深入认识这次地震的运动学和几何学特性，我们收集了全球范围内宽频带数字波形资料，借助于波形反演方法，获得了这次地震的震源机制、震源时间函数和断层面上滑动量随时间和空间的演化过程，计算了地震断层的静态位错在震中区引起的地表位移分布，并与等震线进行了比较，对比结果表明，本文的结论具有相当的可靠性.

1 数据及其预处理

本项研究采用的数据取自地震学联合研究院（Incorporated Research Institution for Seismology）数据中心. 在全球范围内宽频带数字地震台中，只有如图2所示的20个台的震中距在30°~90°之间，且资料具有较高的信噪比. 这些记录的采样率均为20sps. 由于地震矩张量反演和震源时空破裂过程反演所使用的频段不同，所以，我们对原始记录只作了去除直流分量和倾斜分量的处理. 在扣除数据中含有的仪器响应之后，将其重新采样到5sps.

2 矩张量反演

在频率域里，用矩张量描述的位于坐标原点的地震点源在某一观测点 r 引起的地动位移可以表示为[3,4]

$$U_i(\boldsymbol{r},\omega) = G_{ij,k}(\boldsymbol{r},\omega) \cdot M_{jk}(\omega), \tag{1}$$

式中，ω 为角频率，$U_i(\boldsymbol{r},\omega)$ 为地震的观测位移谱，$M_{jk}(\omega)$ 为地震矩张量谱，$G_{ij,k}(\boldsymbol{r},\omega)$ 为格林函数谱. 即，地震位移谱等于格林函数的谱和矩张量的谱的乘积；反过来，矩张量谱等于地震位移谱和格林函数谱的商.

图2 2007年云南宁洱M_S6.4地震震中（六角星）和用于本项研究的宽频带记录的台站（三角形）分布

对于中小地震而言，由于震源持续时间较短，我们总可以找到一个合适的频段，使得在这个频段内可以把实际地震的震源时间函数当作一个简单的脉冲，从而可以用狄拉克（Dirac）δ-函数来代表震源时间函数。在这种情况下，(1) 式可以写成：

$$U_i(\boldsymbol{r},\omega) = G_{ij,k}(\boldsymbol{r},\omega) \cdot M_{jk}, \qquad (2)$$

即描述震源的地震矩张量与时间或频率无关。

由于一个地震的地震矩主要由相对低频的信号决定，因此，我们对用于矩张量反演的波形资料使用了 0.05～0.1Hz 的 3 阶 Butterworth 滤波器。反演得到的地震矩张量的 6 个分量以及由此计算的各向同性（EP）分量、双力偶（DC）分量和补偿线性矢量偶极（CLVD）分量如表1所示，双力偶成分为 5.51×10^{18} Nm，占绝对优势，由此计算的这次地震的矩震级 $M_W=6.4$。根据矩张量解计算得到的最佳双力偶解参数如表2所示。矩张量解和最佳双力偶解的几何表示见图3。图4比较了观测波形与合成波形的形状，可以看出，绝大多数台站的观测波形和合成波形都非常相似。

表1 2007年云南宁洱M_S6.4地震矩张量解 （单位：10^{18}Nm）

M_{11}	M_{12}	M_{13}	M_{22}	M_{23}	M_{33}	M_{ep}	M_{dc}	M_{clvd}	M_W	来源
−5.16	−2.86	−2.68	1.92	1.08	0.92	−0.77	5.51	0.52	6.4	本研究
−1.36	0.54	−0.21	0.48	−0.74	0.88	0	1.64	0.34	6.1	哈佛大学
−1.70	−0.85	−0.33	0.87	0.19	0.83	0	1.52	−0.02	6.1	美国地质调查局

表 2 2007 年云南宁洱 M_S6.4 地震的最佳双力偶解 （单位:°）

节面 I			节面 II			P 轴		T 轴		B 轴		资料来源
走向	倾角	滑动角	走向	倾角	滑动角	方位	倾角	方位	倾角	方位	倾角	
152	54	166	250	79	37	16	16	117	34	264	52	本研究
148	64	160	247	72	27	16	5	110	32	278	58	哈佛大学
138	57	140	252	58	40	15	0	105	50	285	40	NEIC/USGS

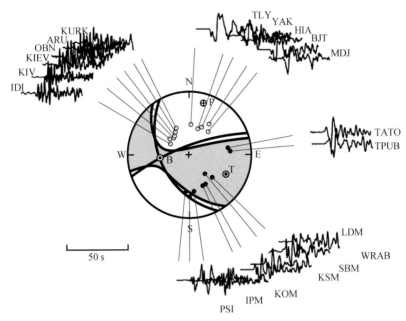

图 3 2007 年云南宁洱 M_S6.4 地震的矩张量解、最佳双力偶解、台站位置在震源机制球上的投影以及各台的垂直分向的 P 波波形

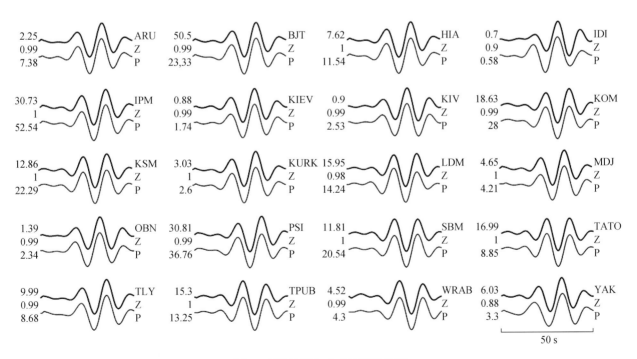

图 4 观测波形与合成波形的比较

在进行矩张量反演时，我们对观测波形和格林函数都做了频率为 0.05~0.1Hz 的零相移带通滤波．图中上面的线表示观测波形，下面的线为合成波形．每组波形的左边的数字自上而下分别为观测波形的最大振幅、相关系数和合成波形的最大振幅，振幅单位为 10^{-7}m/s．每组波形的右边的符号分别是台站名、分向和震相名

利用地震矩张量解能够确定最佳双力偶解所对应的 2 个节面. 为了从 2 个节面中确定发生地震的断层面, 需要其他补充信息, 如当地的地质构造、震害的空间分布等. 根据虢顺民等[1]的研究结果, 宁洱地区的断裂主要分布在北北西向, 且以右旋走滑为主. 根据苗崇刚等[2]和冯万鹏等①的现场考察结果, 地震断层的走向应为北北西向. 因此, 可以判定走向、倾角和滑动角分别为 152°, 54°和 166°的节面为 2007 年云南宁洱 $M_S6.4$ 地震发震断层的断层面.

3 破裂过程反演

对于一个有限断层, 某个台站的观测记录等于有限断层上各个子断层在观测点处贡献的和[5~7]. 设在第 i 个台的观测位移为 $u_i(t)$, 第 j 个子断层的地震矩率函数为 $s_j(t)$, 第 j 个子断层与第 i 个台之间的介质响应为 $g_{ij}(t)$, 它可以近似地表示为 $g_{ij}(t) \cong g_{ij_0}(t-\tau_{ij})$, $g_{ij_0}(t)$ 为参考点到台站的格林函数, 它们之间的关系可以写成如下形式:

$$u_i(t) = \sum_{i=1}^{J} g_{ij_0}(t - \tau_{ij}) \cdot s_j(t), \tag{3}$$

式中, J 为子断层的数目, τ_{ij} 表示第 j 个子断层相对于参考点到第 i 个台的走时差. 由式 (3) 可知, 每个子断层的地震矩率函数可以通过观测资料和表示两点之间介质响应的格林函数加以确定. 如果确定了有限断层上的每个子断层的地震矩率函数, 那么便可得出地震的时空破裂过程. 不过, 这样的反演往往是欠定问题, 是不稳定的. 为了使反演保持稳定, 还需要引入约束条件. 在 (3) 式表述的资料方程的基础上, 我们增加了空间光滑约束方程组[8,9]、时间光滑约束方程组[8~10]以及地震矩最小约束方程组[11]. 由 (3) 式表述的资料方程组和其他约束方程组可以组成如下的矩阵方程组:

$$\begin{bmatrix} \lambda_0 U \\ 0 \\ 0 \\ 0 \end{bmatrix} = \begin{bmatrix} \lambda_0 G \\ \lambda_1 D \\ \lambda_2 T \\ \lambda_3 Z \end{bmatrix} [s], \tag{4}$$

式中, G 为格林函数矩阵, D 为空间光滑约束矩阵, T 为时间光滑约束矩阵, Z 为地震矩最小约束矩阵. λ_0, λ_1, λ_2 和 λ_3 代表不同方程组的权重, 其中 λ_0 是一个稀疏矩阵, 而 λ_1, λ_2 和 λ_3 为常数. 在反演过程中, 通过设计和调整 λ_0, 使矩阵 $\lambda_0 G$ 的绝对值的平均数值在 1 左右. 这样一来, λ_1, λ_2 和 λ_3 的数值便可以以 1 为参考. 由于 (4) 式的系数矩阵往往很大, 所以, 我们选择反演计算效率较高的非负约束的共轭梯度法[12]来求解 (4) 式表示的线性方程组.

在反演震源破裂过程时, 与反演地震矩张量时采用的频率范围不同, 我们使用了 0.025~0.5Hz 的频率范围. 这是由于地震矩张量主要由观测资料的低频成分决定, 而震源的时空破裂过程的许多细节则与观测资料的较高频率范围相对应.

我们沿走向方向取长 45km、倾向方向取长 30km 作为反演区域, 将这个 45km×30km 的区域均匀划分为 15×10=150 个子断层. 每个子断层为 3km×3km. 鉴于震源深度为 5km, 所以我们将初始破裂点设定在走向方向上的第 8 个子断层、倾角方向上的第 3 个子断层. 我们对每个子断层的破裂时间和破裂速度作了约束, 限定每个子断层破裂时间不超过 8s, 破裂速度不超过 2.5km/s. 反演过程中, 我们不对子断层地震矩率函数的形状作任何限制, 也不预先人为地给定子断层的破裂时刻, 而是通过反演自动确定. 在考虑各台站资料的权重时, 我们调整 λ_0, 给予各台站资料相同的权重, 以期同时拟合所有的观测数据. 经过多

---

① 冯万鹏, 杜海林, 许力生, 等. 2007 年云南宁洱 $M_S6.4$ 级地震现场考察与震源模式. 2008 (待发表)

次调整参数和反演试解,我们最终选取 $\lambda_1=15$, $\lambda_2=10$, $\lambda_3=1$.

反演得到的断层面上的静态位移分布如图5所示,最大滑动量约为1.2m,平均滑动量约为0.1m;断层面上发生滑动的区域不大规则,但可以看出,主要发生破裂的区域为一个向深度方向倾斜的菱形,长约为19km. 总体上看,破裂起始点的右边滑动量、即震中的南南东方向上的滑动量较大(图5所示的震源右边的红色区域). 因此,破裂区域的矩心位置应该位于震源的南南东方向. 较大的滑动或破裂主要向深度方向发展. 从图5可以看出,滑动量超过0.8m的区域基本集中在震源右侧(图5中的红色区域),但

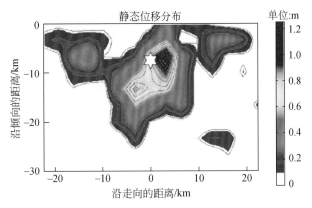

图 5 2007 年云南宁洱 $M_S 6.4$ 地震断层面上的静态位移分布

白色六角星表示初始破裂点的位置

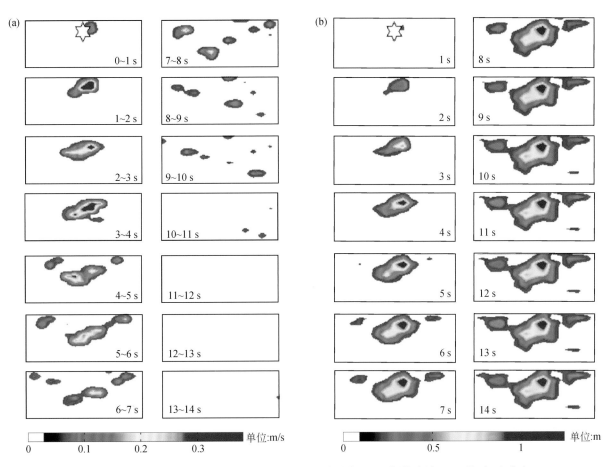

图 6 2007 年云南宁洱 $M_S 6.4$ 地震断层面上的滑动速率(a)与滑动量(b)的时–空分布

六角星表示初始破裂点的位置. (a)中每一个子图表示对应时间段内的滑动速率在断层面上的分布;
(b)中每一个子图表示迄止该时刻断层面上的滑动位移累计量的分布

在其左下方也有一点较小滑动量的区域（图5中的黄色区域）. 滑动量大于0.6m的区域主要分布在震源的右方和左下方（图5中的红色和黄色区域）. 除了主要的破裂区域外，断层面上还散布着3个滑动量较小的区域，其最大滑动量都不超过0.5m, 释放的地震矩也比较小.

根据图5所示的断层面上的静态位移分布，我们采用布龙（Brune）模型[13,14]近似地估算了应力降. 得到断层面上的最大应力降为31MPa, 平均应力降为3MPa.

从图6可以清楚地看到地震断层面上的滑动率和滑动量随时间和空间的变化图像. 根据滑动率和滑动量的时空变化特征，可以将破裂过程分为3个阶段，在前4s的时间段内，破裂主要沿着走向方向朝深处发展. 在第4s至第7s间，破裂呈扇形向着深度方向扩展；在第7s之后的时间段，破裂点比较零散. 在整个破裂中，最大滑动率约为0.4m/s, 平均滑动速率约0.1m/s. 根据破裂前锋随时间和空间变化的图像估计破裂速度约为2.3~2.5km/s.

反演结果能否很好地解释观测资料，是检验反演结果是否正确和可靠的重要判据之一. 为此，我们按如图6所示的动态破裂模型计算了合成地震图，并与观测地震图进行了比较（图7）. 从图7可以看出，绝大多数台站的观测地震图与相应的合成地震图都非常符合.

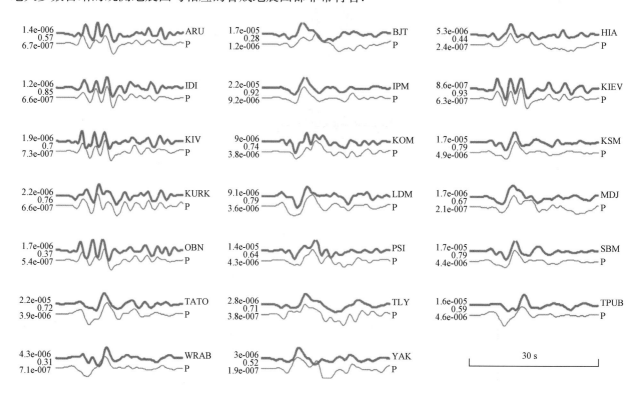

图7　全球范围内20个宽频带地震台的观测波形与合成波形的对比

每个子图中，上方的粗实线表示观测波形，下方的细实线表示合成波形. 子图左方的数字从上到下依次为观测波形最大幅度、观测波形与合成波形之间的相关系数、合成波形最大幅度，振幅的单位为m/s. 子图右边的符号从上到下分别表示对应的台站名和震相名

4　震源时间函数

如果一个地震的震源可以被当作一个点源时，这个点的震源时间函数即可代表这个地震的震源时间函数. 如果一个地震断层不能被当作一个点源时，这个地震的震源时间函数为有限断层面上所有点的震源时间函数的叠加：

$$S(t) = \sum_{j=1}^{J} s_j(t - \tau_j), \tag{5}$$

式中，$s_j(t-\tau_j)$ 为有限断层面上第 j 个点的震源时间函数，τ_j 表示第 j 个点相对于起始破裂点的时间延迟，J 表示子断层个数.

对于 2007 年云南宁洱 M_S 6.4 地震，我们把有限的断层面划分成 150 个子断层，每个子断层可以当作一个点源. 我们以断层面上起始破裂点的起始时间为参考时间，把所有子断层的地震矩率函数叠加后可以得到如图 8 所示的震源时间函数. 从图 8 可以看出，从时间上看，地震矩的释放主要集中在前 11s，第 11s 至第 14s 之间地震矩释放很少. 自破裂开始并经历 2 个峰值后，于第 5s 左右地震矩释放率达到峰值，从第 5s 至第 11s 时间段内地震矩释放率变小，但也出现了 2 个幅度较大的峰值，第 11s 后，地震矩的释放率处于一个很低的水平上，直到第 14s 破裂过程（地震矩释放过程）完全结束.

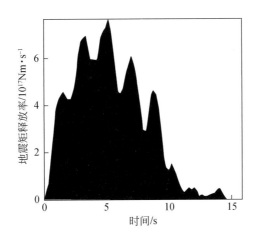

图 8 2007 年云南宁洱 M_S 6.4 地震地震震源时间函数

5 震中区合成地表位移场

一个矩形位错源在均匀各向同性完全弹性半空间中任意一点的位移可以用解析式表示[15~17]，据此，可以计算震源的有限断层模型引起的位移场. 在我们给定的 150 个 3km×3km 的正方形子断层中，只有部分子断层发生了错动（图 5）. 我们把这些发生错动的子断层在震中区地表引起的位移迭加起来，便可得到图 9 所示的震源模型在震中区引起的地表位移场. 图 9 展示了这次地震的等震线、震中区位移场的不同分量以及位移量的空间分布特征. 位移场东-西分量的最大值约 13cm，在烈度Ⅶ和Ⅷ度区范围内的位移量均大于 6cm. 位移场南-北分量的最大值约 18cm，在烈度Ⅶ和Ⅷ度区范围内的位移值量均大于 7cm. 位移场垂直分量的最大值约 12cm，在烈度Ⅶ和Ⅷ度区范围内的位移量均大于 6cm. 相比之下，南北向的位移量最大. 位移场综合反映了震中区位移量在不同空间位置处的相对大小. 从图 9 可以看出，总位移量的最大值达 21cm，位于断层的西南端，在烈度Ⅵ度区之外的位移量大都小于 1cm. 从位移场的空间分布特征可以看出，断层上盘（西盘）总位移量较大，总位移量随距离的衰减较快；断层的下盘（东盘）总位移量较小，且总位移量随距离的衰减较慢. 这一特征与等震线的分布特征是一致的.

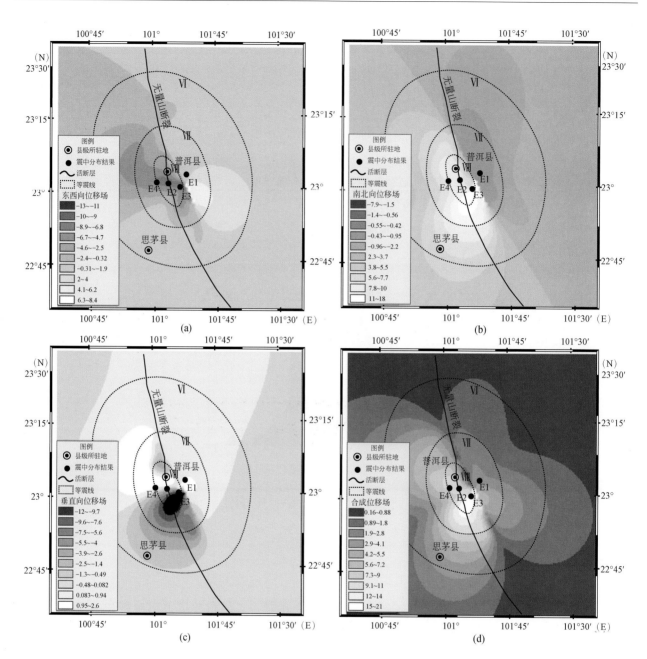

图9 2007年云南宁洱 M_S6.4 地震断层引起的震中区地表位移场

(a) 东-西向分量；(b) 南-北向分量；(c) 垂直分量；(d) 总位移量. 说明见图1

6 结论与讨论

本文通过反演全球范围内的宽频带波形资料获得了 2007 年 6 月 3 日在云南宁洱发生的 M_S6.4 地震的震源机制解、震源时间函数和断层面上滑动随时间和空间的变化过程，并利用反演得到的断层面上的静态位错模型计算了震中区地表位移场. 反演结果表明，这次地震的标量地震矩为 5.51×10^{18} Nm，相当于矩震级 M_W6.4. 震源机制解表明，节面Ⅰ的走向、倾角和滑动角分别为 152°，54°和 166°，节面Ⅱ的走向、倾角和滑动角分别为 250°，79°和 37°. 结合震后考察得到的等震线特征以及当地的地质构造，可以判定这次地震的发震断层的走向为 152°，倾角为 54°滑动角为 166°，是一次以右旋走滑为主的地震. 震源时间函数表明,震源破裂时间为14s,地震的矩释放主要集中在前11s,在第11s至第14s之间释放的地震矩很少.震源的时空破裂过程图像表明,破裂过程分为3个阶段,在前4s的时间段内,破裂主要沿着走向方向和朝深处发展,尤其是沿着走向;在第4s至第7s间,破裂呈扇形状向着深处扩展;在第7s之后的时间段,破裂点比较零散.地震破裂总体上表现为双侧破裂方式,但南南东方向上和深度方向上的滑动略占优势.破裂较强的区域区呈菱形,长约为19km.地震断层面上最大滑动量约为1.2m,平均滑动量约0.1m,最大滑动速率为0.4m/s,平均滑动速率为0.1m/s.根据反演得到的静态位错模型计算的震中区地表位移场的特征与地震的烈度分布特征具有很好的一致性.

2007年云南宁洱M_S6.4地震在全球范围内只有20个台站的记录具有较高信噪比,所有台站的方位角张角为250°,约有110°的方位内无资料可用(图2),因此,断层面上某些破裂细节可能无法完全通过反演过程重现;同时,反演得到的断层面上的某些破裂细节可能是误差所导致的假象.然而,已有的数值试验表明,即使在台站数目不足的情况下,依然能够通过反演获得破裂行为的主要特征[18],而且,分布于破裂前方的台站通常能更好地分辨出断层面上的破裂细节[18].在本研究中,分布于断层走向方向上的台站数目较多,因此,反演得到的沿着走向的破裂分布结果具有更高的可靠性.

本文得到的双力偶节面之一(152°/54°/166°)与云南宁洱地区的地质构造特性相当符合.说明当地断层主要为NNW-SSE向,滑动方式多为右旋走滑的地质构造特征是可靠的,这与已有的研究结果能够很好地相互印证[1,2].前文已经说过,宁洱地区的一个主要的发震特征为中强震频度较高(最大地震震级6.8级),地质构造积累的能量主要以多次的中强震形式释放,本次宁洱M_S6.4地震正好是其中的一次典型事件.

参 考 资 料

[1] 虢顺民,汪洋,计凤桔.云南思茅-普洱地区中强震群发生的构造机制.地震研究,2001,**22**(2):105-115

[2] 苗崇刚,胡永龙,周光全,等.云南宁洱6.4级地震应急性动及灾害特征.国际地震动态,2007,**342**:5-11

[3] 陈运泰,吴忠良,王培德,等.数字地震学.北京:地震出版社,1999

[4] Aki K, Richards P. *Quantitative Seismology: Theory and Method*, Vols Ⅰ and Ⅱ. San Francisco: W H Freeman, 1980. 1-932

[5] Chen Y T, Xu L S. A time-domain inversion technique for the tempo-spatial distribution of slip on a finite fault plane with applications to recent large earthquakes in Tibetan Plateau. *Geophys J Int*, 2000, **143**(2): 407-416

[6] Xu L S, Chen Y T, Teng D L, et al. Tempo-spatial rupture process of the 1999, M_S7.6, Chi-chi, earthquake from IRIS and GEOSCOPE long period waveform data using aftershocks as empirical Green's functions. *Bull Seism Soc Amer*, 2002, **92**(8): 3210-3228

[7] 许力生,陈运泰.震源时间函数与震源破裂过程.地震地磁观测与研究,2002,**23**(6):1-8

[8] Horikawa H. Earthquake doublet in Kagoshima, Japan: rupture of asperities in a stress shadow. *Bull Seism Soc Amer*, 2001, **91**: 112-127

[9] Yagi Y, Mikumo T, Pacheco J, et al. Source rupture process of the Tecomán, Colima, Mexico earthquake of 22 January 2003, determined by joint inversion of teleseismic body-wave and near-source data. *Bull Seism Soc Amer*, 2004, **94**(5): 1795-1807

[10] Dreger D S. Empirical green's function study of the January 17, 1994 Northridge, California earthquake. *Geophys Res Lett*, 1994, **21**(24): 2633-2636

[11] HartzellS H, Heaton T H. Inversion of strong ground motion and teleseismic waveform data for the fault rupture history of the 1979 Imperial Valley. California, earthquake. *Bull Seism Soc Amer*, 1983, **73**: 1553-1583

[12] Ward S N, Barrientos S E. An inversion for slip distribution and fault shape from geodetic observations of the 1983, Borah Park, Idaho, earthquake. *J Geophys Res*, 1986, **91**(B5): 4909-4919

[13] Brune J N. Tectonic stress and the spectra of seismic shear waves from earthquakes. *J Geophys Res*, 1970, **75**: 4997-5009

[14] Brune J N. Correction. Tectonic stress and the spectra of seismic shear waves from earthquake. *J Geophys Res*, 1971, **76**: 5002

[15] 陈运泰, 林邦慧, 林中洋, 等. 根据地面形变的观测研究1966年邢台地震的震源过程. 地球物理学报, 1975, **18**(3): 164-182

[16] 陈运泰, 黄立人, 林邦慧, 等. 用大地测量资料反演的1976年唐山地震的位错模式. 地球物理学报, 1979, **22**(3): 201-216

[17] Okada Y. Surface deformation due to shear and tensile faults in a half-space. *Bull Seism Soc Amer*, 1985, **75**(4): 1135-1154

[18] Saraò S D, Suhadolc P. Effect of non-uniform station coverage on the inversion for earthquake rupture history for a Haskell-type source model. *J Seism*, 1998, **2**: 1-25

2008年汶川大地震的时空破裂过程*

张勇[1,2]　冯万鹏[2]　许力生[2]　周成虎[3]　陈运泰[1,2]

1. 北京大学地球与空间科学学院，北京　100871；2. 中国地震局地球物理研究所，北京　100081；
3. 中国科学院地理科学与资源研究所，北京　100101

摘要　利用全球地震台网（GSN）记录的长周期数字地震资料反演了2008年5月12日四川汶川M_S8.0地震的震源机制和动态破裂过程，并在反演所得结果的基础上定量分析了汶川大地震同震位移场的特征，探讨了汶川大地震近断层地震灾害的致灾机理. 反演中采用了单一机制的有限断层模型，使用了从全球范围内挑选的、方位覆盖较均匀的21个长周期地震台垂直向记录的P波波形资料. 通过反演得出：汶川大地震的发震断层走向为225°、倾角为39°、滑动角为120°，是一次以逆冲为主、兼具小量右旋走滑分量的断层；这次地震所释放的标量地震矩为$9.4×10^{20} \sim 2.0×10^{21}$ Nm，相当于矩震级M_W7.9～8.1. 汶川大地震是在破裂长度超过300km的发震断层上发生的、破裂持续时间长达90s的一次复杂的震源破裂过程. 整个断层面上的平均滑动量约2.4m，但断层面上滑动量（位错）的分布很不均匀. 有4个滑动量集中且破裂贯穿到地表的区域，其中最大的两个，一个在汶川-映秀一带下方，最大滑动量（也是本次地震的最大滑动量）所在处在震源（初始破裂点）附近，达7.3m；另一个位于北川一带下方，一直延伸到平武境内下方，其最大滑动量所在处在北川地面上，达5.6m. 其余2个滑动量集中的区域规模较小，一个在康定以北下方，最大滑动量达1.8m；另一个位于青川东北下方，最大滑动量达0.7m. 汶川地震整个断层面上的平均应力降约18MPa，最大应力降约53MPa. 由反演得到的断层面上滑动量分布计算得出的汶川大地震震中区地表同震位移场表明，汶川大地震地表同震位移场的分布特征与该地震烈度分布的特征非常一致，表明了汶川大地震的大面积、大幅度、贯穿到地表的、以逆冲为主的断层错动是致使近断层地带严重地震灾害在震源方面的主要原因.

根据中国国家地震台网测定，2008年5月12日14时28分4秒（北京时间），在我国四川省汶川县境内的映秀镇附近（31.0°N，103.4°E，震源深度15km）发生了面波震级M_S8.0地震. 地震引发大规模的山体滑坡和泥石流，造成了多处河流淤塞，形成了3000个以上的堰塞湖（卫星影像图1（b）和（c））；汶川大地震使位于龙门山断裂带附近的上百座城镇遭受严重破坏，大量房屋损毁，公路桥梁坍塌（卫星影像图1（d）和（e）），造成了近9万人死亡或失踪.

汶川大地震震中位于青藏高原东缘的龙门山断裂带上. 龙门山断裂带是一条长约500km、宽约30～50km沿NE-SW方向展布的巨大断裂带，其断层滑动以逆冲为主、兼具右旋走滑分量[1]. 按照由西向东的顺序，龙门山断裂带主要包含龙门山后山断裂（茂县-汶川断裂）、中央断裂（映秀-北川断裂）和山前断裂（彭县-灌县断裂）（图2）. 这些断裂都以逆冲滑动为主、兼具一定的右旋走滑分量；在龙门山断裂带的东北段，右旋走滑分量更大[1]. 在龙门山断裂带上，近期中、小地震（震级$M<7$的地震）活动频繁[2]，但历史上未有发生过7级以上大地震的记载. 与上述龙门山断裂带上的地震活动特征形成强烈反差，在我国西南地区、包括龙门山断裂带附近区域的断裂带上，不但历史上而且近期均发生过多次强烈地

---

* 本文发表于《中国科学：D辑 地球科学》，2008年，第38卷，第10期，1186–1194.

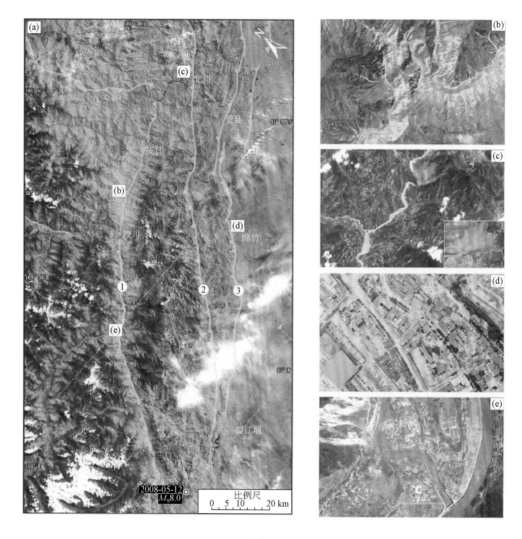

图 1

由 MODIS 卫星影像（(a)～(e)）显示出的汶川大地震（红色圆圈）造成沿龙门山断裂带山体滑坡、泥石流及堰塞湖（影像(b)，(c)）以及附近城镇遭受严重破坏、大量房屋损毁倒塌、公路桥梁坍塌（影像(d)，(e)）的情景. 图 1(a) 中带锯齿的浅黄色线表示逆冲断裂，锯齿所指的方向表示断层面的倾向；1 表示茂县-汶川断裂，2 表示映秀-北川断裂，3 表示彭县-灌县断裂

震[3,4]（图 2），但震级都不超过 8 级，其中震级最大的一次为 1933 年 8 月 25 日发生在茂县叠溪的 M_S7.5 级地震. 汶川大地震的发生是平静多年的龙门山断裂带的一次集中的能量释放.

地震发生后，作者利用全球地震台网（Global Seismographic Network，简写为 GSN）的长周期数字地震资料，反演了汶川大地震的震源机制和动态破裂过程，在震后数小时内测定完毕并随即于翌日公布了相关的震源参数（http://www.cea-igp.ac.cn/汶川地震专题/地震情况/初步研究及考察结果（一）.pdf），及时地为抗震救灾工作提供了重要参考. 分析结果表明，这次地震的断层长度超过 300km，破裂开始于汶川县的映秀镇地面下方约 15km 处，终止于震中东北方向的青川县，地震破裂持续时间长达 90s，最大滑动量发生于汶川和北川附近.

本文将叙述作者以单一机制有限断层模型反演方法[5]反演全球地震台网记录的长周期数字地震资料得到的汶川大地震的震源机制和动态破裂过程，并在反演所得结果的基础上定量分析汶川大地震同震位移场的特征.

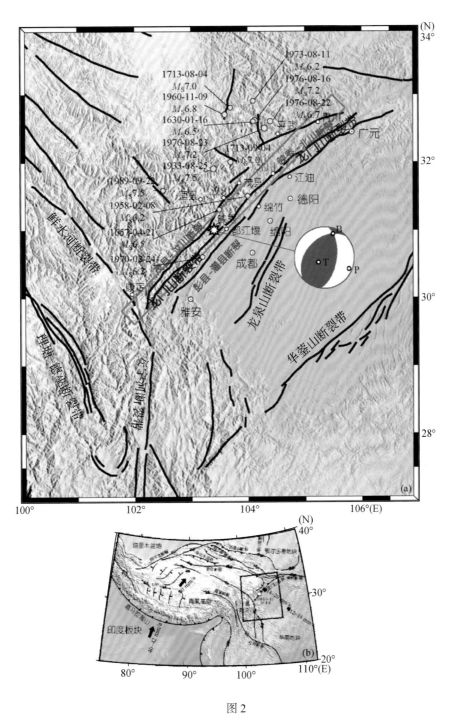

图 2

(a) 2008 年汶川大地震震中（白色八角星）位置和震中区的主要断裂（深紫色线）、历史地震（黄色圆点）和沿龙门山断裂带及其附近的主要城市（白色圆点）. 浅紫色矩形框表示本研究所采用的平面断层模型在地面上的投影，"海滩球"为本文得到的汶川大地震震源机制解（走向 225°/倾角 39°/滑动角 120°）在震源球下半球的等面积投影. (b) 汶川大地震的构造背景

1 数据

本文选取震中距在 55°~90°（1°约为 111.1 km）范围内的台站的直达 P 波的长周期波形记录反演汶川大地震的震源机制和动态破裂过程. 为使台站相对于震中的方位分布均匀，我们按照大约 5°的方位角

间隔选取了 21 个台站（图 3）的资料用于反演.

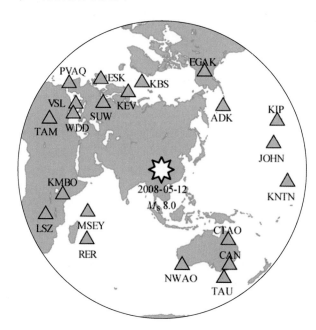

图 3 2008 年汶川大地震震中（八角星）和本研究中使用的长周期地震台（三角形）的空间分布
三角形旁的字母表示台站名称

鉴于水平向 P 波记录受干扰较大，反演中只使用信噪比高的垂直向记录，并只采用 0.002～0.2Hz 的 3 阶 Butterworth 带通滤波器对波形记录进行滤波以确保波形记录包含尽可能完整的震源信息. 采用全球标准的速度结构模型[6]，利用反射率法[7]计算相应的格林函数.

2 断层面参数的确定

地震发生后，美国哈佛（Harvard）大学（http://www.globalcmt.org/CMTsearch.html）、美国地质调查局（USGS）（http://earthquake.usgs.gov/eqcenter/eqinthenews/2008/us2008ryan/#scitech）以及陈运泰等（http://www.cea-igp.ac.cn/汶川地震专题/地震情况/初步研究及考察结果（一）.pdf）很快测报了这次地震的矩张量解的反演结果（表 1）. 3 种结果一致表明，汶川地震是一次以逆冲为主、具有一定右旋走滑分量的断层错动. 不过，上述结果都是在假设震源时间函数为三角形函数的前提下得到的，不涉及震源破裂时间过程的复杂性. 为研究震源破裂时间过程的复杂性，我们运用新发展的方法[8]，通过对震源时间函数的适当约束，在假定矩张量的各分量都具有相同时间历史的前提下进行波形反演，直接得到震源时间函数和矩张量的 6 个独立元素，进而确定断层面参数[9].

表 1 汶川大地震的地震矩 M_0，矩震级 M_W 和断层面解

来源	$M_0/10^{21}$ Nm	M_W	节面 I			节面 II			T 轴		B 轴		P 轴	
			走向/(°)	倾角/(°)	滑动角/(°)	走向/(°)	倾角/(°)	滑动角/(°)	方位/(°)	倾角/(°)	方位/(°)	倾角/(°)	方位/(°)	倾角/(°)
哈佛大学	0.94	7.9	229	33	141	352	70	63	227	57	2	25	114	9
美国地质调查局	0.75	7.9	238	59	128	2	47	45	202	57	36	31	110	16
刘超等[9]	2.0	8.1	220	32	118	8	63	74	245	69	16	14	302	6
本文	0.94	7.9	225	39	120	8	57	68	230	69	21	18	103	20

反演结果表明（表1），这次地震释放的标量地震矩 $M_0 = 2.0\times10^{21}$ Nm，相当于矩震级 $M_W = 8.1$，最佳双力偶解的两个节面的参数分别为：节面Ⅰ，走向220°/倾角32°/滑动角118°；节面Ⅱ，走向8°/倾角63°/滑动角74°. 这一结果与全球矩心矩张量（GCMT）结果相近，但略有差异[9].

反演得到的节面Ⅰ的走向（220°）与倾向（倾向西北）与龙门山断裂带的走向（NE-SW）与倾向（倾向西北）一致，也与 NE-SW 走向的余震震中分布一致. 据此可以确定节面Ⅰ（走向220°/倾角32°/滑动角118°）为汶川地震的断层面. 在以下叙述的工作中，将以这个结果作为震源破裂过程反演的初始模型.

3 矩阵方程与反演参数

震源破裂过程反演的研究工作开始于20世纪80年代初期，经过20多年的发展，逐渐形成了多种不同的反演方法，其中包括 Kikuchi 和 Kanamori[10] 与 Hartzell 和 Heaton[11] 发展的波形反演方法. 这些方法都对子断层的震源时间函数（地震矩率的时间历史）的"形状"作了假定，在一定程度上限制了破裂传播模式与破裂传播速度反演的客观性. 本文采用作者在研究2007年云南宁洱地震时所发展的方法[5]，在未对子断层的震源时间函数作任何先验假设的前提下，以资料方程、空间光滑方程、时间光滑方程和地震矩最小约束方程构成如下的矩阵方程：

$$\begin{bmatrix} \lambda_0 U \\ 0 \\ 0 \\ 0 \end{bmatrix} = \begin{bmatrix} \lambda_0 G \\ \lambda_1 D \\ \lambda_2 T \\ \lambda_3 Z \end{bmatrix}[s], \tag{1}$$

式中，U 为经过处理后的地震台站的记录资料，G 为格林函数矩阵，s 为所有子断层震源时间函数，是破裂过程反演待求解的参数，D 为空间光滑约束矩阵，T 为时间光滑约束矩阵，Z 为地震矩最小约束矩阵. λ_0，λ_1，λ_2 和 λ_3 代表不同方程组的权重，其中 λ_0 是一个稀疏矩阵，而 λ_1，λ_2 和 λ_3 为常数. 采用非负约束的共轭梯度法[12] 求解式（1）所示的矩阵方程.

取沿走向方向长510km（在震中东北和西南方向分别为305km和205km）、沿倾向长50km的矩形区域作为断层面，将这个510km×50km的面积均匀划分为 51×5 = 255 个子断层. 每个子断层长10km，宽10km. 鉴于震源（初始破裂点）深度为15km，在此断层模型中，将初始破裂点置于沿走向第31个、沿倾向第3个的子断层. 对每个子断层的破裂时间和破裂速度做了限定每个子断层破裂时间不超过25s、破裂速度不超过4.5km/s的约束. 反演中，对子断层地震矩率函数的时间历史不作任何限制，也没有预先给定子断层的破裂起始时间[13]. 在考虑各台站资料的权重时，通过调整 λ_0，给予各台站资料相同的权重以拟合所有的观测数据，并调节各台站资料的最大幅值在1左右. 在此基础上，约束方程的权重可以以 λ_0 作为参考. 经过多次调整参数和反演试解，最终选取 $\lambda_1 = 30$，$\lambda_2 = 80$，$\lambda_3 = 0.4$.

按本节叙述的方法，我们以上节提及的、反演得到的节面Ⅰ（走向220°/倾角32°/滑动角118°）作为时空破裂过程反演的初始模型，运用试错法对不同的震源机制时空破裂过程进行反演，最后得出残差最小的断层面解为：走向225°/倾角39°/滑动角120°（表1第4行与图2中的"海滩球"）.

4 静态滑动量分布

由图4可以看出，汶川大地震的断层面上的滑动量的分布很不均匀. 有4个滑动量集中的区域. 最大的一个滑动量集中的区域在汶川-映秀一带下方，沿断层走向长达180km，沿断层倾向宽达50km. 最大滑动量达 7.3m，位于震源（初始破裂点）附近．第二大滑动量集中的区域位于北川一带下方，一直延伸到平武境内下方，沿断层走向方向长达 60km，沿断层倾斜方向宽达 35km，最大滑动量达 5.6m．第三大滑动量集中的区域在康定以北下方，位于震中西南 120～170km 之间，最大滑动量达 1.8m．除此之外，在青川东北也存在一个较小的滑动量集中区域，其最大滑动量为 0.7m．整个断层面上的平均滑动量约为 2.4m．

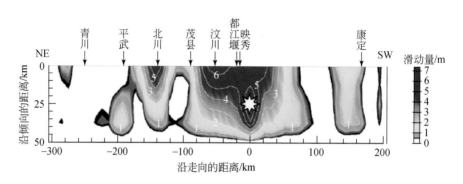

图 4 2008 年汶川大地震断层面上的静态（最终）滑动量分布

白色八角星表示震源（初始破裂点）的位置，白色线条和线条上的白色数字分别为滑动量等值线和滑动量幅值（单位：m）．图上方的箭头给出了各重灾县、市在断层线（断层面与地面的交线）上投影的位置．图中纵坐标与横坐标采用不同的比例

根据图 4 所示的断层面上的静态（最终）滑动量分布，采用布龙（Brune）震源模型[14,15]计算了断层面上的应力降，得出断层面上的最大应力降为 53MPa，平均应力降为 18MPa．这个结果与板内地震的典型应力降（约 10MPa）在数量级上是一致的，但其数值大约是板内地震的典型应力降的 2 倍[16]．从应力降的数量级来看，汶川大地震与典型的板内地震没有明显的不同．

5　破裂的时空变化

图 5 表示地震断层面上的滑动量随时间和空间的变化图像．从破裂开始（发震时间）到发震后 12s，破裂主要表现为双侧破裂形式，即同时向东北和西南两个方向扩展，其中 5s 时错动最快．随后停顿了大约 4s．在发震后 16～30s 期间，在震中东北方向约 80km 处开始新的破裂，并快速向着西南方向传播．在这个阶段破裂涉及的范围大，是汶川大地震的一个主要过程．在发震后 30～42s，在震中东北方向和西南方向都有一些零星的破裂，但规模较小、幅度较弱．在接下来的 6s 内，没有明显的破裂发生．在发震后的 48～58s 内，在震中东北 140km 的北川附近和震中西南 150km 的康定附近下方相继发生破裂．在发震后 60～66s 内，震中东北 200km 处下方的断层面上有一次较小的破裂事件．此后，在震中西南方向的破裂基本结束，而在震中东北 280km 处则零星地发生了一些破裂．

如前所述，汶川大地震的震源过程错综复杂，断层面上位错的分布也很不均匀．图 5 的红色细线表示了地震破裂过程中破裂前锋随时间的变化的进程（图 5）．从图 5 可以看出，在整个地震破裂过程中破裂前锋的扩展速度（破裂传播速度）是随时间和空间变化的．我们对几个典型的时段估算了相应的破裂传播速度的数值（如图 5 中表示破裂前锋的红色细线旁的黑色数字所示）．根据破裂前锋沿断层走向的扩展情况，可以估算出朝东北方向和朝西南方向的平均破裂速度分别约为 3.4km/s 和 2.2km/s．

6　震源时间函数

根据图 5 的破裂时空分布图像，可以计算出如图 6 所示的地震矩的释放率随时间变化曲线（震源时间

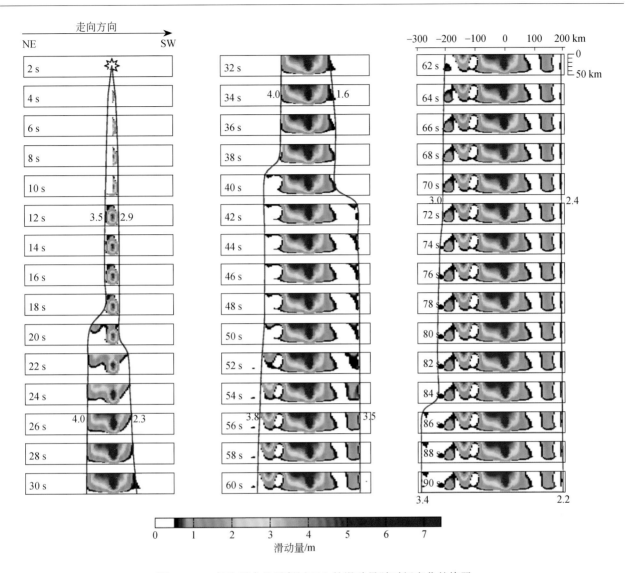

图 5 2008年汶川大地震断层面上的滑动量随时间变化的快照

八角星表示震源（初始破裂点）的位置. 图中每个矩形子图表示在长510km、宽50km的断层面上、在矩形内左下角所示的发震后的时刻的累积滑动量的分布. 发震后的第90s的累积滑动量即静态滑动量分布（参见图4）. 红色细线表示地震破裂过程中破裂前锋随时间变化的进程，旁边的黑色数字表示相应的破裂传播速度的数值（单位：km/s）

函数)[4]. 由震源时间函数的时间积分可以得到整个地震过程中释放的标量地震矩为 9.4×10^{20} Nm, 相当于矩震级为 $M_W 7.9$. 汶川大地震的整个时间过程有5个主要的能量释放阶段，即由5次子事件组成. 第一次子事件发生在发震后的最初14s, 在这个时间段内释放了汶川大地震释放的全部地震矩的约9%的地震矩; 第二次事件介于发震后14~34s之间，是最主要的一次事件，释放了全部地震矩的约60%的地震矩; 第三次事件开始于发震后34s, 结束于43s, 释放了全部地震矩的约8%的地震矩; 第四次事件为发震后43~58s, 释放了全部地震矩的约17%的地震矩; 第五次事件开始于发震后58s至全部地震破裂过程结束（发震后90s), 仅释放了全部地震矩的约6%的地震矩.

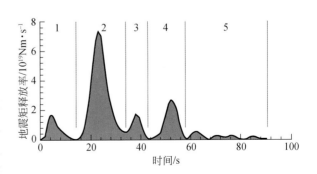

图 6 2008年汶川大地震的震源时间函数

7 地表位移场与近断层地区震灾的致灾机理

在均匀各向同性完全弹性半空间中任一矩形断层引起的位移可以用解析式表示[17~19]. 据此可以计算有限断层震源模型引起的同震位移场. 将反演地震破裂过程得到的 255 个 10km×10km 的正方形子断层的位错（滑动量）在震中区地表面引起的同震位移叠加，便可得到汶川大地震在震中区引起的同震位移场（图7）. 图 7 表示汶川大地震在震中区引起的同震位移场的空间分布. 作为比较，图中还给出了汶川大地震的等震线[20]. 从震中区同震位移场的水平向位移（图 7（a））可以看到非常清楚的右旋运动，即断层的西北盘向东北方向运动以及断层的东南盘向西南方向运动. 从震中区同震位移场的垂直向位移（图 7（b））可以清楚地看到，在 NE-SW 走向的汶川大地震发震断层的上盘（西北盘），地面隆升，而在下盘（东南盘），地面下沉. 需要特别指出的是，由图 7 可见，我们计算得出的地表水平向和垂直向同震位移的空间分布的特征都与震中区等震线的特征非常接近. 计算得出的地表垂直与水平位移最大的两个地区即汶川和北川地区正好对应于本次地震中受灾最严重的、烈度都同为Ⅺ度的两个极震区. 在汶川地区，地面最大水平向位移为 3.2m，最大垂直向位移为 2.8m；在北川地区，地面最大水平相位移为 2.9m，最大垂直向位移为 2.6m. 地面最大位移均发生于出露至地表的断层面上. 在最大相对位移即位错所在地点（震中东北约 50km 处），计算得出的滑动量为 6.1m，与地震现场调查的结果非常接近. 这些情况表明，汶川大地震贯穿到地表面的逆冲断层错动是近断层地区震灾的主要原因.

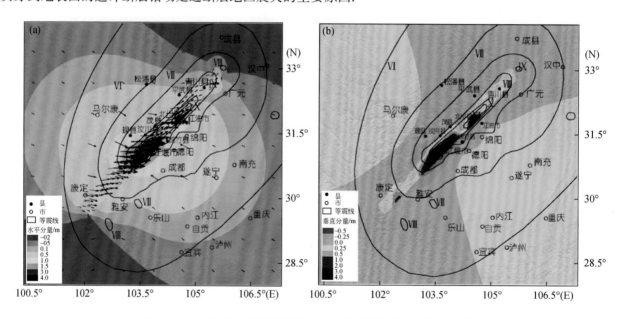

图 7　2008 年汶川大地震断层错动引起的同震位移场与等震线[20]
（a）水平向位移；（b）垂直向位移

8 讨论与结论

我们在全球范围内从全球地震台网的数字地震资料中挑选了 21 个方位分布较均匀的长周期台站的垂直向 P 波记录，通过波形反演得到了 2008 年 5 月 12 日 M_S8.0 地震的时空破裂过程. 反演结果的质量在一定程度上可以从观测波形与理论（合成）波形的拟合程度得到反映. 为此，我们利用反演得到的动态破裂过程模型计算了所用的 21 个台站所在观测点的理论（合成）地震图，并与相应的观测地震图进行了对比.从图8可以看出,理论(合成)地震波形与观测地震波形的拟合很好,大多数(多达13个台站)的理论(合成)地震波形与观测地震波形的相关系数在0.8以上,有3个台站的理论(合成)地震波形与观测地震波形的相关系数在0.7~0.8之间.可以认为,反演得到的汶川大地震动态破裂过程模型较好地解释了观测地震图.

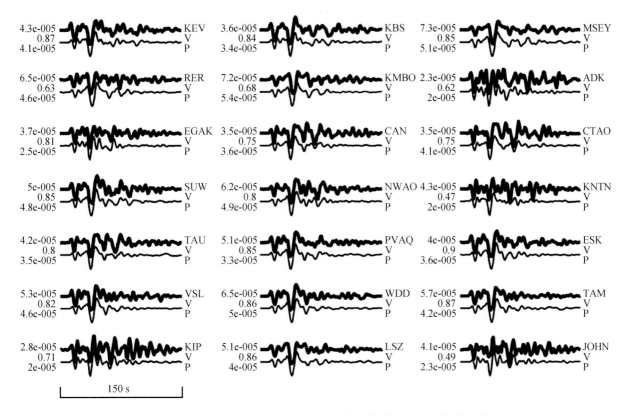

图8 全球地震台网(GSN)记录的观测地震图与理论地震图的比较

每个子图中,上方的粗实线表示观测地震图,下方的细实线表示理论地震图.子图左方的数字从上到下依次为:观测地震波形的最大振幅、观测地震图与理论合成地震图之间的相关系数、理论地震波形的最大振幅.地震波形振幅的单位为m.子图右边的字符从上到下依次为对应的台站名、分量名和震相名

由反演得到的静态(最终)滑动量分布模型计算得出的地表同震位移场与野外地震灾害调查的结果对比表明,计算得到的地表位移场的分布特征与沿断层的地震灾害分布特征非常接近,地表位移值较大的两个地区正好对应于两个极震区,显示了极震区与贯穿到地面的逆冲断层错动的密切联系.

若干细节尚待进一步深入研究.例如,本文的反演结果表明,在震中西南的康定东北方向也有明显的、滑动量最大达1.8m的断层错动,这点尚有待野外地震调查的印证.又如,尽管本文反演得到的汶川大地震动态破裂过程模型比较好地解释了观测地震图,但在有的台站,理论(合成)地震图与观测地震图的拟合程度并不高,其原因有待查明.

本文通过反演得到了汶川$M_S8.0$大地震的震源机制解、震源时间函数和断层面上滑动量随时间和空间的变化过程,并利用反演得到的断层面上的静态位错模型计算了震中区地表位移场.反演得到的震源机制和破裂过程表明,汶川大地震是一次以逆冲断层错动为主的地震事件,地震破裂过程以朝北东向破裂为主的不对称双侧破裂方式进行,最大错动量达到7.3m,且大幅度、大面积的破裂在多个区域贯穿到地表.根据反演得到的静态(最终)滑动量分布模型计算的震中区地表位移场特征与地震的烈度分布特征具有很好的一致性,表明了汶川大地震的大面积、大幅度、贯穿到地表的以逆冲为主的断层错动是致使近断层地区严重地震灾害在震源方面的主要原因.

致谢 作者对中国科学院遥感应用研究所王世新研究员在图 1 准备中给予的帮助以及两位审稿专家提出的建设性意见表示衷心的感谢.

参 考 资 料

1 陈国光，计凤桔，周荣军，等. 龙门山断裂带晚第四纪活动性分段的初步研究. 地震地质，2007，**29**(3)：657-673

2 Yang Z X, Waldhauser F, Chen Y T, et al. Double-difference relocation of earthquakes in central-western China, 1992—1999. *J Seismol*, 2005, **9**: 241-264

3 闵子群，主编. 中国历史强震目录（公元前 23 世纪—公元 1911 年）. 地震出版社，1995. 1-514

4 闵子群，主编. 中国历史强震目录（公元前 1912 年—公元 1990 年，$M_S \geq 4.7$）. 地震出版社，1995. 1-636

5 张勇，许力生，陈运泰，等. 2007 年云南宁洱 M_S6.4 地震震源过程. 中国科学 D 辑：地球科学，2008，**38**(6)：683-692

6 Kennett B L N, Engdahl E R. Travel times for global earthquake location and phase identification. *Geophys J Int*, 1991, **105**: 429-465

7 Kennett B L N. *Seismic Wave Propagation in Stratified Media*. Cambridge: Cambridge University Press, 1983. 1-339

8 张勇. 震源破裂过程反演方法研究. 北京：北京大学博士学位论文，2008. 1-158

9 刘超，张勇，许力生，等. 一种矩张量反演新方法及其对 2008 年汶川 M_S8.0 地震序列的应用. 地震学报，2008，**30**(4)：329-339

10 Kikuchi M, Kanamori H. Inversion of complex body waves. *Bull Seism Soc Am*, 1982, **72**: 491-506

11 Hartzell S H, Heaton T H. Inversion of strong ground motion and teleseismic waveform data for the fault rupture history of the 1979 Imperial Valley, California, earthquake. *Bull Seism Soc Am*, 1983, **73**: 1553-1583

12 Ward S N, Barrientos S E. An inversion for slip distribution and fault shape from geodetic observations of the 1983, Borah Park, Idaho, earthquake. *J Geophys Res*, 1986, **91**(B5): 4909-4919

13 Chen Y T, Xu L S. A time domain inversion technique for the tempo-spatial distribution of slip on a finite fault plane with applications to recent large earthquakes in Tibetan Plateau. *Geophys J Int*, 2000, **143**(2): 407-416

14 Brune J N. Tectonic stress and the spectra of seismic shear waves from earthquakes. *J Geophys Res*, 1970, **75**: 4997-5009

15 Brune J N. Correction. Tectonic stress and the spectra of seismic shear waves from earthquakes. *J Geophys Res*, 1971, **76**: 5002

16 Kanamori H. Mechanics of earthquakes. *Ann Rev Earth Planet Sci*, 1994, **22**: 207-237

17 陈运泰，林邦慧，林中洋，等. 根据地面形变的观测研究 1966 年邢台地震的震源过程. 地球物理学报，1975，**18**(3)：164-182

18 陈运泰，黄立人，林邦慧，等. 用大地测量资料反演的 1976 年唐山地震的位错模式. 地球物理学报，1979，**22**(3)：201-216

19 Okada Y. Surface Deformation due to shear and tensile faults in a half-space. *Bull Seism Soc Amer*, 1985, **75**(4): 1135-1154

20 国家汶川地震专家委员会. 汶川地震灾区地震地质灾害图集. 中国地图出版社，2008. 1-105

地震预测: 回顾与展望[*]

陈运泰[1,2]

1. 北京大学地球与空间科学学院, 北京大学–中国地震局现代地震科学技术研究中心, 北京 100871;
2. 中国地震局地球物理研究所, 北京 100081

摘要 本文概要回顾自 20 世纪 60 年代以来国际地震预测研究与地震预报实践的进展情况, 指出地震预测这一既紧迫要求予以回答、又需要通过长期探索方能解决的地球科学难题目前尚处于初期的科学探索阶段, 虽然总体水平仍然不高, 特别是短期与临震预测的水平与社会需求相距甚远, 但是近半个世纪以来并非毫无进展. 文中以板块边界大"地震空区"的确认、"应力影区"、地震活动性图像、图像识别等方法以及美国帕克菲尔德 (Parkfield) 的地震预报实践为例, 说明在中期与长期地震预测方面, 地震预测研究均取得了一些有意义的进展. 文中分析了地震预测在科学上面临的困难, 阐述了为解决这些困难所应当采取的科学途径, 展望了地震预测的前景, 指出地震预测的进展主要受到地球内部的"不可入性"、大地震的"非频发性"以及地震物理过程的复杂性等困难的制约; 地震预测虽然困难, 但并不是不可能的; 依靠科技进步, 强化对地震及其前兆的观测, 选准地点、开展并坚持以地震预测试验场为重要方式的地震预测科学试验, 坚持不懈地、系统地进行基础性的对地球内部及对地震震源区的观测、探测与研究, 对实现地震预测的前景是可以审慎地乐观的.

地震是一种会给人类社会带来巨大灾难的自然现象. 在众多的自然灾害中, 特别是在造成人员伤亡方面, 全球地震灾害造成的死亡人数占全球各类自然灾害造成的死亡人数总数的 54%, 堪称群灾之首. 在 20 世纪 (1900~1999 年), 全球有高达 180 多万人被地震夺去了生命, 平均一年约 1.8 万余人死于地震, 经济损失达数千亿美元[1-5]. 进入新世纪以来, 地震灾害不断, 似乎还有愈演愈烈之势. 2001 年印度古杰拉特 (Gujarat) 地震 (矩震级 M_W7.6) 造成了 3.5 万人死亡、6.7 万人受伤、60 万人无家可归和约 100 多亿美元的经济损失. 2003 年 12 月 26 日伊朗巴姆 (Bam) 地震只有 M_W6.6 级 (面波震级 M_S6.8), 却造成了 3.1 万人死亡、3.0 万人受伤, 使具有千年历史的巴姆古城毁于一旦. 2005 年 10 月 8 日巴基斯坦地震 (M_W7.6), 造成了 8.6 万人死亡、1 万余人受伤、9 千余人失踪, 数百万人无家可归. 在 2004 年 12 月 26 日发生的印尼苏门答腊–安达曼 (Sumatra-Andaman) 特大地震 (M_W9.1) 及其引发的印度洋特大海啸更使约 28.3 万人死亡与失踪, 令全世界为之震惊! 迄今仍余震不断, 继续危及生灵. 2008 年 5 月 12 日我国汶川 M_W7.8 (M_S8.0) 地震, 造成了 8.7 万人死亡与失踪, 迄今不但余震不断, 而且滑坡和泥石流等次生灾害亦时有发生.

作为一种自然现象, 地震最引人注目的特点是它的猝不及防的突发性与巨大的破坏力. 关于这一点, 中外古人根据经验均已有深刻的认识. 早在 2000 多年前, 在《诗经·小雅·七月之交》中就有关于地震的突发性及其巨大的破坏力的生动描述[6]:

> 烨烨震电, 不宁不令.
> 百川沸腾, 山冢崒崩.
> 高岸为谷, 深谷为陵.
> 哀今之人, 胡憯莫惩?!

---

[*] 本文发表于《中国科学: D 辑 地球科学》, 2009 年, 第 **39** 卷, 第 12 期, 1633–1658.

"不宁"指地不宁,即地动;"不令"是不预先通告给人们周知,突如其来[7]. 诗中惊叹地震突如其来,势如闪电,声如雷鸣,其力足以令山川变易. 译成白话文,就是(参阅周锡镎[8]):

  耀眼的雷霆闪电,

  地震突如其来.

  无数江河在沸腾,

  山峰碎裂崩塌.

  高耸的崖岸陷落为山谷,

  深邃的山谷隆升为丘陵.

  可怜今天的人啊,

  为何竟不知自省?!

  1835 年 2 月 20 日 15 时 30 分 UTC(协调世界时),在智利康塞普西翁(Concepción)–瓦尔帕莱索(Valparaiso)发生了一次 $M_S 8.1$ 地震,震中位置 36.0°S,73.0°W. 地震毁灭了康塞普西翁城. 1835 年 3 月 5 日,伟大的博物学家、进化论的创始人达尔文(Darwin C.,1809~1882)在他著名的贝格尔(H. M. S. Beagle)号环球旅行途中到达了康塞普西翁,经历了这次大地震的多次余震. 达尔文以进化论的创始人闻名于世,但可能鲜为人知的是他也是现在称为地震地质学(earthquake geology)的一位创始人和先驱者. 康塞普西翁–瓦尔帕莱索大地震破坏的惨烈景象给予达尔文强烈的震撼,他写道[9]:

  "通常在几百年才能完成的变迁,在这里只用了一分钟. 如此巨大场面所引起的惊愕情绪,似乎还超过了对于受灾居民的同情心."

  通过地质调查已经知道,地球在整个地质时期都发生过地震. 相传在帝舜时期(约公元前 23 世纪)"三苗欲灭时,地震泉涌";夏帝发七年(约公元前 1831 年)"泰山震"[10]. 在《史记·周本纪》中,就有关于地震的历史记载[10]:"周幽王二年(公元前 780 年),西周三川(泾水、渭水、洛水)皆震,……,是岁也,三川竭,岐山崩".

  正如日食、月食和彗星等天象一样,地震曾被中外古人归于超自然的原因,被当作是上天的惩戒. 甚至到了公元 1750 年,还有人在英国《伦敦皇家学会哲学丛刊》(*Phil. Trans. Roy. Soc. London*)上发表文章,认为把地震归于自然成因的人应当向那些因此被冒犯的人道歉[11]!也和古人从来没有放弃过对日食、月食、彗星等天象的"天意"的"窥测"的努力一样,数千年来,古人对地震的成因及其预测的探索从来没有停止过. 只是由于地质学严重缺乏物理学原理的解释,对地质构造运动与地震关系的认识长期裹足不前. 直到 18 世纪牛顿《自然哲学的数学原理》出版,牛顿力学问世,才为包括地震在内的地球上的所有运动的统一解释提供了物理基础. 在牛顿力学的影响下,地震学逐渐发展成为一门现代的科学. 到了 19 世纪 70 年代后期,现代地震仪研制成功,地震学步入了一个新的时代[12,13].

  无情的大地震激发了人们对地震的成因及其预测的探索. 自 19 世纪 70 年代后期现代地震学创立以来的 130 余年里,地震预测一直是地震学研究的主要问题之一,多少地震学家莫不苦思预测地震、预防与减轻地震灾害的方法(例如,Milne[14]). 特别是自 20 世纪 50 年代中期以来,作为一个非常具有现实意义的科学问题,地震预测一直是世界各国政府和地震学家深切关注的焦点之一[3,15-22].

  地震预测是公认的世界性的科学难题,是地球科学的一个宏伟的科学研究目标. 如能同时准确地预测出未来大地震的地点、时间和强度,无疑可以拯救数以万计乃至数十万计生活在地震危险区人民的生命;并且,如果能预先采取恰当的防范措施,就有可能最大限度地减轻地震对建筑物等设施的破坏,减少地震造成的经济损失,保障社会的稳定和促进社会的和谐发展[23-27].

  通过世界各国地震学家长期不懈的努力,地震预测、特别是中长期地震预测取得了一些有意义的进展. 但是地震预测是极具挑战性尚待解决的世界性的科学难题之一,目前尚处于初期的科学探索阶段,总体水平仍然不高, 特别是短期与临震预测的水平与社会需求相距甚远.

本文是在作者有关地震预测的几篇文章[23-27]的基础上汇编增删写成的, 意在向关注这一问题的广大读者简明地评介国际地震预测研究进展情况, 分析地震预测在科学上遇到的困难, 阐述解决这些困难应采取的科学途径, 展望地震预测的前景.

1 地震预测研究进展

1.1 预测与预报

地震预测 (prediction) 或预报 (forecast) 不是指像 "在某地最近要发生大地震" 这类含糊的 "预测"、"预报" 或说法. 不同时指明地震发生的地点、时间和大小 (简称为地震 "三要素") 并对其区间加以明确界定的 "预测"、"预报", 几乎没有什么意义. 此外, 还须要用发震概率来表示预测的可信程度. 所以, 地震学家把地震预测定义为 "同时给出未来地震的位置、大小、时间和概率4种参数, 每种参数的误差 (不确定的范围) 小于、等于下列数值[28]:

位置: ±破裂长度;

大小: ±0.5 破裂长度或震级±0.5 级;

时间: ±20% 地震复发时间;

概率: 预测正确次数/(预测正确次数+预测失误次数).

地震预测通常分为长期 (10 年以上)、中期 (1～10 年)、短期 (1 日至数百日及1 日以下)[29]. 有时还将短期预测细分为短期 (10 日至数 100 日) 和临震 (1～10 日及1 日以下) 预测. 长、中、短、临地震预测的划分主要是根据 (客观) 需要、但却是人为 (主观) 地划分的, 并不具有物理基础, 界线既不是很明确, 也并不完全统一. 在我国, 以数年至 10 年、20 年为长期; 1 年至数年为中期; 数月为短期; 数日至十几日为临震[30]. 在国外, 也有以数年至数 10 年为长期、数周至数年为中期、数周以下为短期的[31~33]. 实际上, 许多地震预测方法所用的地震前兆涉及的时间尺度并不正好落在上述划分法规定的范围内, 而是跨越了上述划分法规定的界线. 在公众的语言中, 甚而在专业人士中, 对 "地震预测" 和 "地震预报" 通常不加区分, 并且通常指的就是这里所说的 "地震短、临预测". 在国际上 (例如 Wyss[28]), 一些地震学家把不符合上述定义的 "预测"、"预报" 等等称作 "预报", 亦称概率性 (地震) 预报, 而把符合上述定义的 "预测" 称作 "确定性的 (地震) 预测". 例如对在一段长时期内的某一不确定的时间发生某一震级范围地震的概率做出估计就属于这种类型的 "预测"——按这种说法便应当叫做 "预报". 预测美国加州中部帕克菲尔德 (Parkfield) 在 (1988±4.3) 年间会有一次 6 级地震[34~36], 按这种说法也是一种 "预报". 若照这种说法, "长期预测" 和 "中期预测" 便应当称作 "长期预报" 和 "中期预报". 在我国, 习惯于把科学家和研究单位对未来地震发生的地点、时间和大小所做的相关研究的结果称作 "地震预测", 而把由政府主管部门依法发布的有关未来地震的警报称作 "地震预报". 地震长期预测 (长期预报) 通常只涉及在正常情况下地震发生的概率. 这种 "预测" 并非是广大公众最为关注的、能有足够的时间采取紧急防灾措施 (如让居民有足够时间撤离到安全地带等等) 的 "地震短、临预报". 即使如此, 这种 "预测" 对于地震危险性评估、地震灾害预测预防、抗震规范制定、地震保险、等等, 也是十分有用的. 为避免混淆起见, 除非特别说明, 本文采用我国的习惯说法. 在评估地震预测 (地震是真报对了还是碰运气 "撞上" 的?) 时, "目标震级" 的大小是很重要的. 理由很简单: 因为小地震要比大地震多得多 (一般地说, 在某一地区某一时间段内, 某一震级地震的数目是震级比它大 1 级地震的数目的 8～10 倍)、因而更容易碰巧报对! 在给定的地区和给定的时间段内要靠碰运气报对

一个 $M_W6.0$ 的地震并非易事，而靠碰运气"对应上"（"撞上"）一个 $M_W5.0$ 的地震的"预测"还是很有可能的.

从更广泛的意义讲，从预防和减轻地震灾害的目标考虑，地震预测还应包括对地震发生时指定地点的地面运动强烈程度的预测. 强（烈）地面运动（地震工程学家亦称之为强地震动）的预测是地震学与工程学交叉的重要学科领域，近30多年来发展很快；限于篇幅，本文暂不涉及这一重要问题.

1.2 地震长期预测

1.2.1 地震空区

在地震长期预测方面，最突出的进展是板块边界大地震空区的确认. 在环太平洋地震带，几乎所有的大地震都发生在利用"地震空区"方法预先确定的空区内[37-41]. 在我国，板内（板块内部）地震空区的识别也有一些成功的震例[30,42,43].

地震是地下岩石中的"应变缓慢积累-快速释放"的过程[44,45]. 对地震过程的这一认识是"地震空区"方法的物理基础. 基于这一认识可以推知：在指定的一段断层上，将会准周期性地发生具有特征大小与平均复发时间的地震. 这种地震称作"特征地震"（日本地震学家称之为"固有地震"）. 特征地震的大小（震级）既可以由在该段断层上已发生过的特征地震的震级予以估计，也可以根据该段断层的长度或面积予以估计. 特征地震的平均复发时间既可以由相继发生的两次特征地震的时间间隔予以估计，也可以由地震的平均滑动量除以断层的长期滑动速率予以估计. "地震空区"指的是在时间上已超过了平均复发时间、但仍未以特征地震的方式破裂过的一段断层. 1906年，地震预测的先驱者、国际著名的日本地震学家今村明恒（Imamura A.）在他所写的一篇论文中曾确认东京近海的相模湾（Sagami Bay）为地震空区，成功地预报了1923年 $M_S8.2$ 日本关东（Kanto）大地震（亦称东京大地震）. 今村明恒还曾经成功地预报了1944~1946年日本南海道（Nankaido）大地震[40,46]. 苏联的费道托夫（Федотов С. А.）是第一位用现代地震科学原理阐明地震空区概念的地震学家[47]. 他研究了1904~1963年间沿日本-千岛群岛-堪察加岛弧一带 $M_S \geq 7\frac{3}{4}$ 浅源地震震源区的空间分布，发现这些大地震的震源区基本上是连续分布的. 他认为大地震震源区之间的空隙区便是未来最可能发生大地震的地区即"地震空区". 费道托夫在1965年发表的论文[47]的一幅地图中指出了未来可能发生大地震的地区；他的预测很快就在3个地方得到验证，即1968年5月16日日本十胜-隐歧（Tokachi-Oki）$M_W8.3$ 地震，1969年8月11日南千岛群岛 $M_W8.2$ 地震以及1971年12月15日堪察加中部 $M_W7.8$ 地震.

20世纪60年代板块大地构造学说的确立为根据板块边界的地形变与历史地震活动性"收支"平衡情况估算在地质年代里板块边界的地形变速率提供了精确的运动学参考框架. Sykes等[37-41]将1957，1964和1965年发生于阿留申海沟的3次地震的滑动量除以北美板块与太平洋板块之间的相对运动速率，得出在发生这3次地震的3段断层上地震的平均复发时间都大约为100年. 他们运用海底磁异常条带资料以及经过准确定年的地磁场反向时间表等全球性的地球物理观测资料，在1973和1979年得出了近期可能会发生大地震的有关板块的边界段的预报结果[48,49]. 后来，他们又出版了经改进后的预报结果[41].

每条断层或断层的每段的表现都是不同的[50]. 按照特征地震的概念，对于特定的一段断层，断层上的滑动量主要是通过具有类似的震级、破裂面积和平均滑动量的特征地震释放出来的. 这样一来，相对于比它大的和比它小的地震，特征地震必定比按古登堡（Gutenberg B., 1889~1960）-里克特（Richter C. F., 1900~1985）定律（关系式）预期的多得多[51,52]，可是这与迄今在所有的地区几乎都观测到地震服从古登堡-里克特定律所表示的分布相矛盾. 对此，Wesnousky等[53-55]解释说，由于断层段服从幂律分布，所以在一个地区的地震还是按古登堡-里克特定律分布的.

特征地震的概念对于地震物理学与地震灾害评估有着重要的意义. 在地震灾害的评估中, 特征地震的平均复发时间是一个很重要的物理量. 因为上一个特征地震的发震时间好比是一只"地震钟"的"零时". 从这个"零时"开始, 与这个特征地震类似的下一个特征地震的发生概率即可予以估计. 但是, 对于按古登堡-里克特定律分布的地震来说, 就不能用"地震钟"这样一种简单方法来估算下一个地震发生的概率, 因为对于任何一个震级的地震来说, 便应当有许多个震级比它略小、但其特征并无不同的地震. 不过, 以目前实际震例的观测资料的状况, 特征地震的频度应当比按古登堡-里克特定律分布的地震的频度高, 以及特征地震的震级这两个特征也很难用实际震例的资料予以检验.

作为地震长期预测的一种方法, 特征地震方法取得了一定程度的成功. 用这个方法预测大地震原理很直观, 看上去很简单, 做起来似乎也很容易. 但是要把它推广应用仍有一定的困难, 因为不易确定特征地震的震级并且缺少估计复发时间所需的完整的地震记录资料. 此外, 由于地震过程内禀的不规则性以及地震的发生具有"空间-时间群聚"的趋势, 所以在实际应用地震空区假说同时预测特征地震的震级与发震时间时仍有困难. 地下岩石中的"应变缓慢积累-快速释放"的概念意味着在指定的一段断层上错动将周期性地发生, 这个结果是基于依次发生的地震的应力降和两次地震间应力积累的速率两者都是常量的假定 (图1(a)). 但是, 在实验室内做的岩石粘-滑实验表明两次地震事件之间的时间间隔是变化的, 应力降是不完全、不规则的, "初始应力"(震前应力) 和"最终应力"(震后应力) 都是不均匀的. 如果初始应力 σ_2 均匀但最终应力 σ_1 不均匀, 从而应力降 ($\sigma_2-\sigma_1$) 也不均匀, 那么只有地震发生的时间是可以预测的 (这种情形称作"时间可预测模式") (图1(b))[56]; 如果初始应力 σ_2 不均匀但震后应力 σ_1 均匀, 从而应力降 ($\sigma_2-\sigma_1$) 也不均匀, 那么只有地震的震级是可以预测的 (这种情形称作"震级可预测模式") (图1(c))[57].

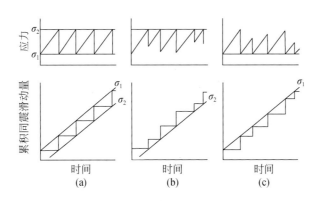

图1 在构造应力积累的速率保持恒定的情况下的地震预测模式

(a) 时间与震级均可预测的模式; 构造应力积累的速率保持恒定、"初始应力"(震前应力) σ_2 和"最终应力"(震后应力) σ_1 都均匀, 从而应力降 ($\sigma_2-\sigma_1$) 也均匀、地震按严格的周期性重复地发生. (b) 时间可预测模式; 初始应力均匀但最终应力不均匀, 从而应力降不均匀、只有地震发生的时间是可以预测的. (c) 震级可预测模式; 初始应力不均匀但最终应力均匀, 从而应力降不均匀、只有地震的震级是可以预测的 (据 Shimazaki 和 Nakata[57])

1.2.2 "东海大地震"

沿着日本西南海岸的南海海槽-相模海槽, 在过去的 500 多年间重复发生过多次震级 $M \sim 8$ 的大地震, 包括 1498, 1605, 1707, 1854 和 1944~1946 年地震, 平均复发时间约为 117 年 (图2). 在 20 世纪 70 年代初期, 一些日本地震学家指出, 1944~1946 年间发生的几次大地震比 1854 和 1707 年的地震小. 他们认为, 1944~1946 年地震的破裂并没有到达南海海槽的东北部叫做骏河 (Suruga) 海槽的地方. 于是他们推断在板块边界的这一地段、现在称为"东海大地震空区"的地方, 不久的将来将有可能发生一次震级 $M \sim 8$ 的地震. 这就是日本地震学家预报中的"东海大地震"[22,58-65]. 1978 年 6 月, 日本政府通过了一个以地震预报为前提的、预防和减轻地震灾害为目的的大型地震对策法案,称作"大规模地震对策特别措置法",从 1978 年 12 月 14 日开始实行. 该法案制定了很详细的应急反应计划以及发布短期预报的步骤,其中最重要的一点是:当监测前兆的网络观测到异常后,由专家组成的专门委员会(原先称作"东海地震判定会",现在改称作"地震防灾对策强化地区判定委员会")最迟不超过 1 个小时就得举行会议,并且会议在至多 30 分钟内就得做出判定,判定该异常是不是所预测的"东海大地震"的前兆. 如果判定是"东海大地震"的前兆,就应整理成"地震预报情况"材料经由气象厅长报告内阁总理大臣. 内阁总理大臣收到报告后,要立即在内阁会议上发布"警戒宣言",并启动应急反应计划.

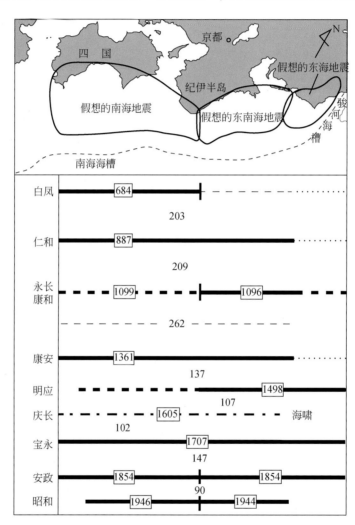

图 2　沿着日本西南海岸的南海海槽–相模海槽(在骏河海槽东北面,图中未绘出)在过去的 500 多年间重复发生过的震级 $M\sim 8$ 的浅源大地震的破裂区

沿着骏河海槽的东海地区即是日本地震学家预报中的"东海大地震"空区(据日本地震学会地震予知检讨会[22])

自 1978 年到现在已过了 31 年,迄今仍未检测到需要启动应急反应计划的异常,一次也没有开过"判定委员会"紧急会议(不过,"判定委员会"还是每月召开一次例行的碰头会). 国际著名的专家、"判定委员会"主席、东京大学地震研究所原所长茂木清夫(Mogi K.)教授对该委员会能否履行其判定东海大地震短临前兆的功能表示怀疑,并于 1997 年在地震预测饱受诟病的批评声中辞去该委员会主席职务,黯然下台(令人欣慰的是,与此形成强烈反差,2003 年在日本札幌举行的第 23 届国际地球物理与大地测量联合会(IUGG)大会上,茂木清夫教授做了题为"地震预测"的大会报告后,与会世界各国科学家报以热烈的长时间的掌声向这位地震预测的先驱者致以崇高的敬意). 然而,继任的新主席溝上惠(Mizoue

M.)教授也持有类似观点. 日本文部省国土地理院于1997年公布了一个报告[66]；报告说，在日本目前还做不到像"大规模地震对策特别措置法"所要求的短期预报，并且什么时间能做到也不得而知.

东海大地震的预报实践表明，即使对于像这样一种发生于板块边界的、看上去很有规律的历史地震序列，准确的预报也是很困难的.

1.2.3 帕克菲尔德地震

帕克菲尔德地震的预测也是基于"地震空区"理论. 在美国西海岸圣安德列斯断层靠近帕克菲尔德（在1980年代时是一个居民仅37人的小镇）的一段断层上，有仪器记录以来发生过3次震级$M \sim 6$的地震，即：1922，1934，1966年帕克菲尔德地震（图3中序数为4，5，6的地震事件）；而在有仪器记录以前，也发生过3次$M \sim 6$的地震，即：1857，1881和1901年帕克菲尔德地震（图3中序数为1，2，3的地震事件）. 平均每22年便规则地发生一次帕克菲尔德地震. 帕克菲尔德平均22年便发生一次$M \sim 6$的地震的规则性以及1934年与1966年的帕克菲尔德地震的前震活动性图像之间的相似性使得地震学家相信这些帕克菲尔德地震是以大约相同的滑动量、相隔大约22年在同一段断层的破裂. 由"同震位移"与断层滑动速率的比值求出的地震复发时间也是大约22年[34-36,67]. 根据这些资料以及其他有关资料，美国地质调查局（United States Geological Survey，缩写为USGS）在1984年发出正式的地震预报（据Shearer[68]），明确指出在圣安德列斯（San Andreas）断层靠近帕克菲尔德的一段断层上，在（1988±4.3）年（即最晚在1993年1月之前）将发生一次$M \sim 6$的地震，发震概率约为95%.

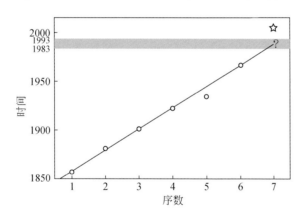

图3 帕克菲尔德地震的预测

预测帕克菲尔德在（1988±4.3）年（即最晚在1993年1月之前）将发生一次$M \sim 6$地震. 图中序数1~6的圆圈表示历史上发生的6次帕克菲尔德地震，问号表示所预测的帕克菲尔德地震，灰色条带表示预报帕克菲尔德发生的时间窗（1983~1993）. 帕克菲尔德$M_W 6.0$地震（五角星）迟至2004年9月28日才发生，比预报的时间晚了至少11年（据日本地震学会地震予知检讨会[22]）

到了1993年年底，预报中的帕克菲尔德地震还没有发生. 美国地质调查局于是宣布"关闭"帕克菲尔德地震预报的"窗口". 年复一年，"盼望"中的帕克菲尔德地震一直不来，为此，有的地震学家认为这次预报本身就是一种错误[69]，认为帕克菲尔德地震序列可能根本就不是特征地震，而是一种随机发生的事件[55,70,71]. 一些地震学家则对帕克菲尔德地震迟迟未发生提出了许多解释. 例如，一种解释是：1983年5月2日发生于加州科林佳（Coalinga）的$M_W 6.4$地震可能缓解了帕克菲尔德地区的应力[72]. 另一种解释是：1906年旧金山大地震后应力的松弛效应推迟了帕克菲尔德地震的发生[73].

1.2.4 1989年洛马普列塔地震

1989年10月18日美国加州洛马普列塔（Loma Prieta）$M_W 6.9$（$M_S 7.1$）地震被认为是一次成功预报了的地震. 这次地震发生于1906年旧金山大地震破裂带南端的一段断层上. 在洛马普列塔地震发生前，许多地震学家和研究集体曾经对这段断层做过详细的研究，他们注意到：在这段断层上，1906年旧金山大地震的地表破裂的滑动量比北段的小,表明在这一段断层积累的应变在1906年旧金山大地震时没有完全释放完. 并且在这一段小地震又明显地少,这种现象通常是大地震之前的地震活动图像. 据此他们发出了中长期地震预报,认为在未来20年内在这段断层上会发生一次 $M6.5$ 地震,发震概率是30%[74,39,75]. 在有关这次地震预报的报告发表2年后便发生了洛马普列塔地震,因此这次地震被认为是成功预报了的. 不过也有人认为,因为洛马普列塔地震与所预报的地震并不准确地相符,所以仍然不能排除是碰运气碰上的. 的确,这次地震不是发生在圣安德列斯断层系的主断层上,而是发生在一条叫做萨尔根特(Sargent)的次要断层上,该断层以约70°的倾角向西南倾斜,与圣安德列斯断层既不相交,并且断层错动的倾滑分量也相当大,与以右旋走滑为主的圣安德列斯断层不一致. 此外,历史应变资料表明,这一段断层的地表破裂的错动量虽然比北段小,但从地壳深部的情况看,在1906年旧金山大地震时,在地壳深部已发生了相当大的滑动,所以可能并没有"节余"下多少滑动量给所预报的这次地震[76].

1.3 地震中期预测

1.3.1 应力影区

由地震空区模式可以推知,地震的发生受到先前发生的地震所引起的应力变化的影响而加速或减速. 如果大地震的发生降低了破裂带附近某区域的应力,从而降低了该区域发生地震(既包括比该地震大的地震、也包括比该地震小的地震)的可能性,直至该区域内的应力得以恢复为止. 这便是"应力影区"模式[77~79]. 应力影区模式不同于地震空区模式,它不仅涉及断层段,而且也涉及其周围区域. 此外,由于应力是张量,所以地震的发生既可能使某些断层段上应力增加,也可能使某些断层段上应力减小. 在靠近已破裂的断层段的某些区域,应力实际上是增加的,从而应力影区模式对"地震群聚"现象提供了一种物理上说得通的解释[80]. 目前,运用应力影区模式对许多地震序列做了很有意义的回溯性的研究[81],不过尚未被用于地震预报试验. 这是因为,地壳中的应力分布的图像与先前发生过的地震破裂的详细情况、断层的几何情况、地壳中的应力-应变关系、地下流体的流动对地震引起的应力变化的响应以及其他诸多目前尚难以测定的因素有关.

1.3.2 地震活动性图像

地震活动性图像是用得最多的一种地震预测方法. 之所以用得多,原因是这个方法直观明了,并且比较可靠的地震活动性资料几乎随处可得. 茂木清夫(Mogi[61])提出,一次大地震之后接着是频度随时间逐渐减少的余震,然后是长期平静期(第一次平静期),这个平静期后依次是:未破裂带地震活动性增加,中期平静期(第二次平静期),前震活动期,短期平静期(第三次平静期),最后是大地震. 这就是地震活动性图像的"茂木模式". 茂木清夫(Mogi[61])描述过一系列地震活动性图像的实例以至许多人以为可以简单地用他所描述的地震活动性图像来确认大地震轮回演化的阶段,从而预报地震. 日本的 Ohtake 等[82,83]曾利用茂木模式成功地预报了1978年墨西哥南部瓦哈卡(Oaxaca) $M7.7$ 地震. 不过,也有人认为这只是一次表面上的成功,因为在1967年全球有一些大的地震台网停止运作致使全球地震记录的总体情况发生了重大的变化,由此得出的地震活动性图像所反映的前兆的真实性被复杂化了[84]. 特别需要指出的是,在实际发生的地震震例中,茂木模式所描述的任何一个阶段都有可能缺失;并且,尚未有一致公认的、可被客观地运用的鉴别各个阶段的定义;此外,迄今也还没有对茂木模式进行过全面的检验.

在地震活动性图像研究方面,除了"茂木模式"即茂木清夫的方法外,其他研究者还提出了一些不同的分析方法,例如,同时考虑地震的时-空-强三要素具有不同权重效应的"区域-时间-长度方法"(Region-Time-Length 方法,简称 RTL 方法)[85-93];基于复杂系统统计力学的地震物理预测模型的"图像信息学方法"(Pattern Informatics 方法,简称 PI 方法)[94-96];用于比较在某一地区、某一时间段内地震活动性的平均变化率与该地区地震活动性的总平均变化率以检测大地震前近震中地区的地震活动性的可能的异常平静期,并通过与其他所有在随机的时间与随机的地点可能发生的地震活动性变化率减少加以比较,估计这种"平静期"的统计意义的"Z-值法"[97];通过测量在一给定的3维空间与给定的时间"点"上微震活动的重叠程度以检测微震活动性的空间群聚、地震活动高发期以及地震平静期的"微震重叠法"(SEISMic OverLAPing 法,简称 SEISMOLAP 法)[98];等等. 上述研究者运用这些方法在一些震例研究中均检测到了地震活动性的异常变化.

1.3.3 图像识别

曾任 IUGG 主席的国际著名地球物理学家克依利斯-博罗克(Кейлис-Борок В. И.)及其俄国同事基于对"地震流"特性的分析,提出了一种称作强震发生"概率增加的时间"(亦称"增加概率的时间",Time of Increased Probability,缩写为 TIP)的中期预测方法,运用计算机进行图像识别,以识别出大地震即将来临前的信息[99-105]. 他们提出了为预测全球 8 级以上大地震而设计的"M8 算法"以及为预测美国加州(California)和内华达州(Nevada)地震而设计的"CN 算法". 他们在地震活动区中预先给定的范围(圆圈)内对地震目录进行扫描,寻找地震发生率的变化、大小地震比例的变化、余震序列的活动度与持续时间以及可用作诊断的各种标志. 他们报告说,在他们预测的未来比较可能发生大地震的那些范围(圆圈)内取得了意义重大的成功. 自 1999 年起,他们对阈值为 $M7.5$ 和 $M8.0$ 的地震做提前 6 个月的实时预测. 运用这个方法,克依利斯-博罗克及其同事对 2003 年 9 月 25 日发生在日本北海道的 $M8.1$ 大地震以及加利福尼亚中部 2003 年 12 月 22 日圣西蒙(San Simeon) $M6.5$ 地震在震前做出了预报,并取得了成功[106,107]. 特别是,预报圣西蒙地震的、题为"关于加州岩石层的现状"的报告是在该地震前 6 个月,即 2003 年 6 月 21 日提交给一个由著名科学家组成的专家组的. CN 算法也被用在如意大利等其他地区的地震预测并取得了成效[108].

尽管克依利斯-博罗克在国际上具有崇高的学术地位,声名鼎盛,他与他的俄国同事提出的算法也可谓相当地成功,不但如此,该算法还被别人应用于股票行情预测乃至总统大选的预测并且获得更为巨大的成功,但毋庸讳言,他们所预测的未来可能发生大地震的范围实在太大,其线性尺度是所预报的未来可能发生的大地震的破裂长度的大约 5~10 倍;更重要的是,该算法主要是根据对地震目录做回归统计分析的经验算法,对所采用的诊断函数的物理意义及其与地震孕育过程的关系缺乏深入的分析与探讨.

1.4 地震短、临预测

1.4.1 地震前兆

在地震发生前,常常可以观测到一些异常,如地应变加速或地面隆升、重力场变化、磁场变化、电场变化、地下电阻率变化、地下水位变化、地下流体流动、地下水化学成分变化、大气化学成分变化以及其他一些可能对应力、对岩石中的裂纹或岩石的摩擦特性的变化敏感的参数的变化. 这些异常称作地震前兆,或者说,可能的地震前兆[109-122]. 通常认为,地震前兆反映的可能是地下岩石临近破裂时的应力状态. 在地震预测中用于检测地震前兆的主要方法是地球物理方法,此外还有大地形变测量和地球化学等方法. 这些地震前兆统称"微观"地震前兆,相应的方法称为"微观"地震前兆方法. 除了上述"微观"地震前兆外,还有不依靠精密仪器、能为人们在地震前所感知的"宏观"地震前兆("宏观"异常),如动物行为异常、地下水和温度变化等性质[123-127]. 在地震预测实践中,多年来,地震学家一直在致力于探索"确定性的地震前兆",即任何一种在地震之前必被无一例外地观测到、并且一旦出现必无一例外地发生大地震的异常.

美国在 1964 年 3 月 27 日阿拉斯加 $M_W 8.5$ 大地震之前并不重视地震预测工作. 阿拉斯加大地震后,美国开始重视并逐渐加强地震预测研究. 1965 年 Press 等[18]提出了地震预测和震灾预防研究十年计划——《地震预测:十年研究计划建议书》. 1977 年美国国会通过了《减轻地震灾害法案》,把地震预测工作列为美国政府地震研究的正式目标[128]. 特别是在 20 世纪 70 年代,紧接着苏联报道了地震波波速比(纵波速度 V_P 与横波速度 V_S 的比值 V_P/V_S) 在地震之前降低之后[129,130],美国纽约兰山湖地区观测到了震前波速比异常[131,132],随之而来的大量有关震前波速异常、波速比异常等前兆现象的报道和膨胀-扩散模式、膨胀-失稳模式等有关地震前兆的物理机制的提出[133-135],以及 1975 年中国海城地震的成功预报,在美国乃至全世界范围内掀起了地震预测研究的热潮,甚而乐观地认为"即使对地震发生的物理机制了解得不是很透彻(如同天气、潮汐和火山喷发预测那样),也可能对地震做出某种程度的预报"[136,137,20]. 当时,连许多著名的地球物理学家都深信:系统地进行短、临地震预测是可行的,不久就可望对地震进行常规的预测,关键是布设足够的仪器以发现并测量地震前兆. 但是很快就发现地震预测的观测基础和理论基础都有问题:对先前报道的波速比异常(Whitcomb 等[138])重新做测量时发现结果重复不了(Allen 等[139]);对震后报道的大地测量、地球化学和电磁异常到底是不是与地震有关的前兆产生了疑问;由理论模式以及实验室做的岩石力学膨胀、微破裂和流体流动实验的结果得不出早些时候提出的前兆异常随时间变化的进程[140,141]. 到了 20 世纪 70 年代末,大多数早先提出的可能的"微观"地震前兆都被确认为对地震短、临预测价值不大. 至于"宏观"地震前兆,通常认为只要做出适当的处理,至少在一定的程度上可以用它们来做出地震震级、震中区和发震时间的实际预报[125]. 然而,在对宏观地震前兆做了系统研究后,力武常次(Rikitake T.[125])提出,宏观地震前兆的特征尚待阐明,因为宏观地震前兆常常很可能被许多"噪声"所干扰,需要对其可靠性做认真的评价. 对动物行为异常等宏观地震前兆异常的成因,对动物行为异常等宏观地震前兆为什么会、以及如何对数量级为 $10^{-7} \sim 10^{-6}$ 的地壳应变的变化产生反应的,迄今仍未得出结论性的意见[125,126,142].

从 1989 年开始,IUGG 所属的 7 个协会之一的国际地震学与地球内部物理学协会(IASPEI)下属的地震预测分委员会,组织了由 13 名专家参加的工作小组,对各国专家自己提名的有意义的地震前兆进行了严格的评审[28,143-145]. 这个专家小组把地震前兆明确地定义为"地震之前发生的、被认为是与该主震的孕震过程有关联的一种环境参数的、定量的、可测量的变化". 第一轮(1989~1990)对各国专家本人自由提名的认为是有意义的 28 项地震前兆作了评审,第二轮(1991~1996)10 项,两轮共 37 项(第二轮中有一项在第一轮中已评审过). 按照这个专家小组评定的结果,只有 5 项被通过认定. 这 5 项可分为 3 类. 第一类是地震活动性图像,包括:①震前数小时至数月的前震(foreshocks),例如 1975 年 2 月 4 日中国辽宁海城 M_S7.3 地震的前震[146];②震前数月至数年的"预震"(preshocks),例如 1988 年 1 月 22 日 M_S6.7 澳大利亚 Tennant Creek 地震[147];③强余震之前的地震"平静"[148]. 第二类是地下水的特性,只有一项,即:④1978 年 1 月 14 日日本伊豆-大岛近海 M_S7.0 地震前地下水中氡气含量减少、水温下降[149,150]. 第三类是地壳形变,也只有一项:⑤地壳形变,例如 1985 年 8 月 4 日美国加州 Kettleman 山地震前地下水上升反映的地壳形变[151]. 对于地应变、地倾斜和地壳运动等则未能做出决定,而对于尾波、Q 值、S 波分裂、潮汐应变振幅、震群、自然电位、地电阻率和地磁场、电磁辐射、应变对降水量的响应、高程变化、地面垂直运动、断层蠕动、地壳形变(海平面变化-地震)和干旱-地震等则未予以认定. 评审未予以通过并非断然否定所提名的这些前兆方法,只表明根据评审专家和专家小组的意见,该方法目前尚未成熟、或者说尚不能完全确信所提名的前兆是否真是地震前兆. 即使被确认为"有意义的地震前兆"的 5 项,并不意味着即可用以预报地震. 例如,前震无疑是地震的前兆,但是如何识别前震、特别是在震前实时地识别前震,仍然是一个待解决的问题.

1.4.2 帕克菲尔德地震预测试验场

20 世纪 80 年代以后,国际上对地震前兆的研究重点转移到寻求大地震前的暂态滑移前兆. 基于详细的实验室滑移实验和模拟计算以及对 1966 年帕克菲尔德地震现场的定性的野外考察,一些地震学家认为大地震前会有暂态滑移前兆[152,140,141]. 在实验室条件下观测到的震前暂态滑移量是很微小的, 但是理论计算表明, 在有利的条件下, 如果在实验室里观测到的临界暂态滑移可以随着岩石样品中的裂纹放大到天然地震断层那么大而成比例地放大, 那么临界暂态滑移在野外是可能被观测到的. 为研究这些问题, 美国地质调查局 (USGS) 在帕克菲尔德建立了地震预测试验场, 在靠近所预测的帕克菲尔德地震未来震中的地方用大地测量方法、应变仪和倾斜仪等前兆仪器作长期、连续、精确的地壳应变测量, 希望能记录下任何可能的前兆性滑移的应变资料以验证理论.

如前所述, 到了 1993 年年底, 预报中的帕克菲尔德地震一直没有发生, 美国地质调查局于是宣布"关闭"了这个地震预报的"窗口". 虽然如此, 幸运的是, 对预报中的帕克菲尔德地震的监测工作并没有"关闭", 设置在帕克菲尔德地震预测试验场的台网继续坚持地震前兆的监测工作[153]. 2004 年 9 月 28 日 17 时 15 分 24 秒 UTC, 地震学家在加州中部帕克菲尔德地震试验场守候多年的 $M_W 6.0$ 地震 (震中位置 35.815°N, 120.374°W, 震源深度 7.9km) 终于发生了 (见图 3). 帕克菲尔德地震姗姗来迟, 比预测的时间晚了整整 11 年, 但是无论如何还是来了. 虽然在震前未检测到、至今也仍未见分析出有地震前兆, 但是由多种仪器设备构成的复杂的前兆台网记录下了有史以来记录最为翔实的一次地震从发震前至发震时乃至发震后的全过程, 取得了地震活动性、地应力、地磁场、地电场、地下水和地震引起的强烈地面运动等等的完整的记录. 这些记录对于了解地震破裂是如何开始的、如何传播的、又是如何停止的, 对于增进对断层、地形变、震源物理过程、地震预测、预防和减轻地震灾害的认识, 提供了很有价值的资料[154,155].

1.4.3 帕克菲尔德地震预测试验的启示

帕克菲尔德地震的预测试验经历了从预测研究 (1979~1984)、发布预报 (1985)、全面展开地震监测 (1985)、地震迟迟不发生于是关闭预报"窗口" (1993) 直至预报中的地震发生 (2004), 长达四分之一世纪的漫长历程. 从帕克菲尔德地震试验场的地震预报实践, 我们可以看到:

① 与以前发生的 6 次帕克菲尔德地震比较, 2004 年 9 月 28 日帕克菲尔德地震的震级 ($M_W 6.0$) 与这些帕克菲尔德地震的震级相近; 地点一致, 破裂也发生在同一段断层上; 此外, 这次地震的余震与 1934 年、1966 年帕克菲尔德地震的余震很相似. 所以可以说, 帕克菲尔德地震的震级和破裂范围是预报对了. 但是发震时间很明显没有报对, 晚了整整 11 年! 这表明, 地震学家迄今对于特征地震在一段断层上重复发生的时间 (地震复发时间) 为什么会有这么大的起伏变化仍缺乏认识, 特征地震的中长期预测模型有待改进.

② 2004 年帕克菲尔德地震破裂段的两个端点与以前发生的帕克菲尔德地震一样; 破裂方式也一样, 都是从一端起始, 然后往另一端扩展的"单侧破裂方式". 但是, 与以前发生的帕克菲尔德地震不同, 这次地震的破裂不是从西北端起始、然后往东南端扩展的; 正相反, 它是从东南端开始, 然后往西北端扩展的, 地地道道的"南辕北辙"! 这说明, 地震学家对地震破裂起始与扩展的规律尚缺乏了解, 单凭经验是无法正确预测未来地震的破裂起始点、终止点以及破裂扩展方向的. 这是从帕克菲尔德地震试验得到的新的认识. 这一新的认识对于地震灾害预测、对防震减灾至关重要! 它告诉我们: 今后在地震灾害预测中, 再不能只根据以往的震例轻易假定未来地震的破裂扩展方向; 要加强对地震破裂起始、终止与扩展规律的研究.

③ 2004 年帕克菲尔德地震发生在预先精心设计的密集的地震观测台网与前兆观测台网内. 布设这些台网的目的本来就是为了检测前震及其他各种可能的地震前兆的. 但是从震前直至今天仍未检测到地震前兆. 诚然, 一方面, 仍需进一步仔细分析记录资料; 但是, 另一方面, 这种情况至少表明帕克菲尔德地震没有明显的地震前兆, 或者是现有的地震前兆观测手段和方法尚不足以发现、检测出地震前兆. 联想到在世界各地, 在像美国、日本这样的经济实力雄厚、科学技术先进的发达国家的地震危险区内, 地震观测台

网与前兆观测台网密布,地震区内的地质构造情况一般都认为研究得相当透彻,这些国家的地震学家一直在努力寻找、检测在中等(M_W5~7)与中等以上地震之前可能的地震前兆,特别是前兆性的应变异常变化. 既然迄今未能检测到这种变化,这说明可靠的地震前兆按现有的地震前兆观测手段和方法的确是很不容易检测出来的. 沿着这一方向继续寻找前兆的努力固然不能轻言放弃;但是,另辟蹊径、提出新的思路、探索新的方法,无疑应当予以提倡和鼓励.

④帕克菲尔德地震预测试验表明"特征地震"的概念对于地震预测可能很有意义. 不过,上面已经提到,关于"特征地震"仍有不少争议(例如,Wesnousky 等[53];Simpson 等[72];Ben-Zion 等[73];Savage 等[69];Wesnousky[54];Kagan 和 Wesnousky[55];Kagan[70];Rong 等[156];Jackson 和 Kagan[71]. 同时,我们也不要忘记,在一个地区成功的经验不一定适用于其他地区,就像即使是在我国,1975 年海城地震的经验性预报成功的经验不适用于相距约 400km 的 1976 年唐山地震一样. 所以,在我国,乃至在像日本、土耳其等地震活跃的国家或地区,选准试验场所,开展并长期坚持像帕克菲尔德地震预测试验场那样的地震预测试验研究,是非常有必要的. 这样做,可望获得在不同构造环境下断层活动、地形变、地震前兆和地震活动性等等的十分有价值的资料,从而有助于增进对地震的了解、攻克地震预测难关.

2 地震预测的困难与地震的可预测性

2.1 地震预测的困难

地震预测是公认的科学难题. 那么,它究竟难在哪里?它为什么那么难?归纳起来,地震预测的困难主要有如下三点(陈运泰[24~27]):地球内部的"不可入性"、大地震的"非频发性"和地震物理过程的复杂性.

2.1.1 地球内部的"不可入性"

地球内部的"不可入性"是古希腊人的一种说法. 我们在这里指的是人类目前还不能深入到处在高温高压状态的地球内部设置台站、安装观测仪器对震源直接进行观测. "地质火箭"、"地心探测器"已不再是法国著名科幻小说作家儒勒·凡尔纳小说中的科学幻想,科学家已经从技术层面提出了虽然大胆、然而比较务实的具体构想[157],只不过是目前尚未提到实施的议事日程上罢了. 迄今最深的钻井是苏联科拉半岛的超深钻井,达 10km,德国-捷克边境附近进行的"德国大陆深钻计划"预定钻探 15km. 和地球(平均)半径(6370km)相比,超深钻所达到的深度还是"皮毛",况且这类深钻并不在地震活动区内进行,虽然其自身有重大的科学意义,但还是解决不了直接对震源进行观测的问题. 国际著名的地震学家、俄国的一位王子伽利津(Галицын Б. Б.)曾经说过(据 Галицын[158];Саваренский 和 Кирнос[159]):

"可以把每个地震比作一盏灯,它燃着的时间很短,但照亮着地球的内部,从而使我们能观察到那里发生了些什么. 这盏灯的光虽然目前还很暗淡,但毋庸置疑,随着时间的流逝,它将越来越明亮,并将使我们能明了这些自然界的复杂现象……"

这句话非常动人!这个比喻十分贴切!不过,话虽然可以这么说,真要做起事情来却没有这么简单. 地震的地理分布并不是均匀的,全球的地震主要发生在环太平洋地震带、欧亚地震带以及大洋中脊地震带这三条地震带,并不是到处都有"灯";地震这盏"灯"也没有能够把地球内部的每个角落全照亮!何况地球表面的 70% 为海洋所覆盖,地震学家只能在地球表面(在许多情况下是在占地球表面面积仅约 30% 的陆地上)和距离地球表面很浅的地球内部(至多是几千米深的井下)、用相当稀疏、很不均匀的观测台网进行观测,利用由此获取的、很不完整、很不充足、有时甚至还是很不精确的资料来反推("反演")地球内部的情况. 地球内部是很不均匀的,也不怎么"透明",地震学家在地球表面上"看"地球内部连

"雾里看花"都不及,他们好比是透过浓雾去看被哈哈镜扭曲了的地球内部的影像.凡此种种都极大地限制了人类对震源所在环境及对震源本身的了解(尽管如此困难,近半个世纪以来,地震学家在地球内部的层析成像方面还是取得了与其他学科相比毫不逊色的巨大的成功).

2.1.2 大地震的"非频发性"

大地震是一种稀少的"非频发"事件,大地震的复发时间比人的寿命、比有现代仪器观测以来的时间长得多,限制了作为一门观测科学的地震学在对现象的观测和对经验规律的认知上的进展.迄今对大地震之前的前兆现象的研究仍然处于对各个震例进行总结研究阶段,缺乏建立地震发生的理论所必需的切实可靠的经验规律,而经验规律的总结概括以及理论的建立验证都由于大地震是一种稀少的"非频发"事件而受到限制.作为一种自然灾害,人们痛感震灾频仍,可是等到要去研究它的规律性时,又深受"样本"稀少之限(当然,这句话的意思不是说希望多来大地震)!

大地震是一种稀少的"非频发"事件,不等于说地震是一种稀少的"非频发"事件.在地震学中,表示地震数目多少(频度)与地震大小(震级)之间关系的古登堡-里克特定律表明地震频度与震级遵从幂律关系[51].简单地说,地震越小,频度越高(或者说数量越大);通俗地说,在某一地区某一时间段内,某一震级地震的数目是震级比它大1级地震的数目的8~10倍;具体地说,全球平均每3年大约发生2个$M_S \geq 8.0$地震,平均每年大约发生17个$M_S 7.0$~7.9地震,134个$M_S 6.0$~6.9地震,1319个$M_S 5.0$~5.9地震,13000个$M_S 4.0$~4.9地震,130000个以上$M_S 3.0$~3.9地震,1300000个以上$M_S 2.0$~2.9地震,等等[4,5].显然,倘若不论其大小,地震的确不但不是"非频发"的,而且是"频发"的!在地震学中,大地震通常指的是$M_S \geq 7.0$的地震[52,160].就全球而言,与平均一年发生上千个中等大小的地震($5.0 \leq M_S < 7.0$的地震)相比,平均一年发生17~18次大地震($M_S \geq 7.0$地震)尚属"非频发"事件.大地震的"非频发性"与不论其大小的地震的"频发性"不能混为一谈[161].

也不要把大地震的"非频发性"与灾害性地震的频繁发生("频发性")相混淆.近年来,不仅是大地震,连中等大小的地震($5.0 \leq M_S < 7.0$的地震)也频频袭击人口稠密的地区,造成相当严重的地震灾害.影响地震造成人员伤亡和财产损失的因素很多,除了地震大小外,还有震源深度、地理位置、发震时间、结构物与建筑物的质量、地质土层条件等因素.一个最新的典型例子当推前面已提及的2003年12月26日伊朗巴姆地震.巴姆地震的震级只是$M_W 6.6$($M_S 6.8$),地震不大,却因当地绝大多数的民居和古老的建筑物均系土砖结构、抗震性能极差、震源又浅(震源深度只有10km)、地质土层条件很差、地震震中正好就在巴姆城的正下方、地震又发生在当地时间凌晨(发震时间为1:56:52UTC,当地时间为上午5:26:52,当地时间与协调世界时的时差为3.5h),诸多不利因素叠加在一起,以至造成约3.1万人死亡,约3万人受伤,约7.56万人无家可归,85%以上的建筑物与基础设施毁坏,具有千年历史的巴姆古城毁于一旦.

2.1.3 地震物理过程的复杂性

从常识上笼统地说,不言而喻,地震是发生于极为复杂的地质环境中的一种自然现象,地震过程是高度非线性的、极为复杂的物理过程.地震前兆出现的复杂性和多变性可能与地震震源区地质环境的复杂性以及地震过程的高度非线性、复杂性密切相关.

从专业技术的层面具体地说,地震物理过程的复杂性指的是地震物理过程在从宏观至微观的所有层次上都是很复杂的.例如,宏观上,地震的复杂性表现在:在同一断层段上两次地震破裂之间的时间间隔长短不一,变化很大,地震的发生是非周期性的[162,163];地震在很宽的震级范围内遵从古登堡-里克特定律;在同一断层段上不同时间发生的地震其断层面上滑动量的分布图像很不相同;大地震通常跟随着大量的余震,而且大的余震还有自己的余震;等等.就单个地震而言,地震也是很复杂的,如:发生地震破裂时,破裂面的前沿的不规则性;地震发生后断层面上的剩余应力(亦称最终应力、震后应力)分布的不均匀性,等等. 在微观上,地震的复杂性表现在:地震的起始也是很复杂的,先是在"成核区"内缓慢地演化,然后突然快速地动态破裂、"级联"式地骤然演化成一个大地震. 这些复杂性是否彼此有关联? 如果有,是什么样的一种关系? 都是非常值得深究的问题. 从基础科学的观点来看,研究地震的复杂性有助于深入理解地震现象以及类似于地震的其他现象的普适性;反过来,对于地震现象以及类似于地震的其他现象的普适性的认识必将有助于深化对地震现象的认识,从而有助于对预防和减轻地震灾害.

2.2 地震的可预测性

在物理学中,把物理系统的演化对初始条件高度敏感的、非线性的依赖性称为"混沌". 混沌对于许多物理现象的可预测性是一种内禀的限制. 在混沌这个物理概念广为应用之前,地震学家凭借直觉早已熟知这一概念[51,52]. 有一些专家认为(例如,Bak 等[164,165], Bak 和 Tang[166], Bak[167]),地震系统与其他许多系统一样,都属于具有"自组织临界性"(self-organized criticality,常缩写为 SOC)的系统,即在无临界长度标度的临界状态边缘涨落的系统. 在具有"自组织临界性"的系统中,任何一个小事件都有可能以一定的概率"级联"式地演变成大事件. "级联"是否发生与整个系统内的所有细节有关,而不仅仅是与大事件及其邻近区域的细节有关. 从理论上说,虽然整个系统内的所有细节是可以测量的,但是因为需要测量的细节的数量是如此之多以至于实际上是不可能一一准确地测量的;况且迄今人们仍然不了解其中的物理定律. 因此,从本质上说,具有自组织临界性的现象是不可预测的. 值得注意的是,具有自组织临界性的系统中的临界现象普遍都遵从像地震学中的古登堡–里克特定律(古登堡–里克特关系式)那样的幂律分布.

地震是地下岩石的快速破裂过程. 在地震学中,表示地震数目多少(频度)与地震大小(震级)之间关系的古登堡–里克特定律表明地震频度与破裂的尺度遵从幂律关系,这意味着地震在空间域上的分布是分形的. 在地震学中,还有另外一条定律、即表示余震的频度随时间作指数衰减的、经宇津德治(Utsu T., 1928~2004)改进的大森房吉(Omori F., 1868~1923)定律(参见 Bullen[11]). "改进的大森定律"现在亦称"大森–宇津定律"或"大森–宇津定标律". 大森–宇津定律意味着地震在时间域上的分布也是分形的. 据此,一些专家认为,无论是在空间域上还是在时间域上,地震都具有典型的分形结构,并不存在一个特征的长度标度,所以地震是一种自组织临界(SOC)现象;他们还认为,地震系统与其他许多系统一样,都属于具有"自组织临界性"的系统. 这些专家进一步推论说:既然自组织临界现象具有内禀的不可预测性,所以地震是不可预测的;既然地震预测很困难,甚至是不可预测的,那么就应当放弃它,不再去研究它[33]. 言之凿凿,应和者甚众. 一时间,"地震不可预测"论甚嚣尘上.

可是,地震是不是一种自组织临界现象,这不是一个靠"民主表决"、"少数服从多数"可以解决的问题! 多数人认为地震是一种自组织临界现象,并不能说明地震就是一种自组织临界现象[168]! 这是因为地震的自组织临界性的最重要的观测依据是由古登堡–里克特定律推导出的幂律,而这个幂律实际上只是一种表观现象. 从这个实际上是表观现象的幂律出发,便得出地震除了受到一个地区所能支持的最大震级的限制以外,不存在特征尺度、具有"标度不变性"的结论,这是错误的. 产生这一错误的关键是没有考虑到余震的效应. Knopoff[168]指出,在我们通常看到的地震活动性图像中,很多地震实际上是过去发生的大地震的余震,必须把这些余震的"账"算清楚,算到大地震——它们的主震的头上,才能给出符合真实情况的地震分布的图像. 在细致地研究了余震的效应之后,Knopoff[168]发现地震现象并非不存在特征尺度,而是至少存在 4 个特征尺度:①相应于"大"地震与"小"地震分界即发震层(亦称易震层、孕震层)的厚度(约 15km,相当于 6.5 级地震)的特征尺度(Scholz[169]);②相应于"大"余震与"中"、"小"余震分界(约 5 级地震)的特征尺度;③相应于余震区的空间范围(1~3km)的特征尺度;④相应于断层带宽度(100~200m)的特征尺度.

耐人寻味的是，在研究地震的自组织临界性时，许多研究者运用的理论模型恰恰是 Knopoff 和他的学生 Burridge 在 40 多年前提出的 Burridge-Knopoff 弹簧–滑块模型（简称 B-K 模型）[170]. 这些研究者以 B-K 模型或其他与 B-K 模型大同小异的、非常简单的、类似于地震的模型做的数值模拟理论研究得出了"地震不可预测"的结论，如：一个小地震事件是否生长、是否发展为大地震事件不可预测地依赖于整个系统内的弹性性质、断层长度以及所贮存的弹性能的微小变化[166,171-173]；如果任何一个小地震都有可能演变为大地震，那么地震预测将是不可能的[174]；对单个地震的发震时间和震级做确定性的地震预测是不可能的[175]；等等. 对地震预测持否定意见的 Geller[33] 概括说，这些数值模拟采用的都是非常简单的、类似于地震的模型，唯其简单，更表明对于一个确定性的模式来说是何等容易成为不可预测的；因此没有理由认为这些理论研究得到的结论不适用于地震.

Knopoff[168] 则认为这些研究者滥用了他的模型（B-K 模型），他认为，这些研究者由于没有恰当地考虑地震的物理问题，所以"他们虽然模拟了某些现象，但他们模拟的不是地震现象."他指出，地震表观上遵从的幂律对应的只是一种过渡现象，而不是系统最终演化到的自组织临界状态；地震现象是自组织（SO）的，但并不临界（C）. 地质构造复杂的几何性质使主震和余震遵从大致相同的、类似于分形的分布，这使得人们很容易将它们混为一谈，而不考虑幂律的可靠性问题，从而简单地从幂律出发推出地震具有自组织临界性、进而推出"地震不能预测"的结论. Knopoff[168] 尖锐地指出主张"地震不可预测"的研究者在逻辑推理上的谬误. 他指出，主张"地震不可预测"的研究者的逻辑推理好比说是："哺乳动物（自组织临界现象）有 4 条腿（遵从幂律分布），桌子（地震现象）也有 4 条腿（遵从幂律分布），所以桌子（地震现象）也是一种哺乳动物（自组织临界现象）或哺乳动物（自组织临界现象）也是桌子（地震现象）".

对地震的可预测性这一与地震预测实践以及自然界的普适性定律密切相关的理论性问题的探讨或论争还在继续进行中[171-184]. 既然地震的可预测性的困难是源于人们不可能以高精度测量断层及其邻区的状态以及对于其中的物理定律仍然几乎一无所知. 那么如果这两方面的情况能有所改善，将来做到提前几年的地震预测还是有可能的. 提前几年的地震预测的难度与气象学家目前做提前几小时的天气预报的难度差不多，只不过做地震预测所需要的地球内部的信息远比做天气预报所需要的大气方面的信息复杂得多，而且也不易获取，因为这些信息都源自地下（地球内部的"不可入性"）. 不过，这样一来，对地震的可预测性（预测地震的能力）的限制可能与确定性的混沌理论没有什么关系，而是因为得不到极其大量的信息.

3 实现地震预测的科学途径

3.1 依靠科技进步、依靠科学家群体

地震预测是一个多世纪以来世界各国地震学家最为关注的目标之一. 如前已述，在 20 世纪 70 年代中期以前，由于膨胀–扩散模式[133,134]、膨胀–失稳模式[135]的发展以及 1973 年美国纽约兰山湖地震和 1975 年中国海城地震的成功预报使得国际地震学界对地震预测一度弥漫了极其乐观的情绪[136,137,20]. 然而，运用经验性的地震预报方法未能对 1976 年中国唐山大地震做出短、临预报以及到了 20 世纪 80~90 年代，美国地震学家预报的圣安德列斯断层上的帕克菲尔德地震、日本地震学家预报的日本东海大地震都不发生（前者推迟了 11 年于 2004 年 9 月 28 日才发生，后者迄今尚未发生），又使许多人感到沮丧悲观. 一个多世纪以来，对地震预测从十分乐观到极度悲观什么观点都有，不同的观点一直在辩论，从未有止息（例如，Geller[33,176,177]；Geller 等[178,179]；Hamada[180]；Turcotte[181]；Knopoff[29,168,182]；Knopoff 等[184]）. 相应

地，地震预测预报的"行情"亦大起大落. 从 20 世纪 80 年代中期开始，一直到不久以前，正如曾对帕克菲尔德地震预测做出重大贡献的科学家之一的林德（Lindh A. G.）博士[153]感叹道："当前被视为在做地震预测工作是极其不时尚的，以至于人们调侃说如果你想得到资助那么在你的基金申请书中切勿有任何涉及地震预测的字眼."

尽管如此，和坊间流传的说法大相径庭，国际地震学界对地震预测预报以及预防与减轻地震灾害的关注与研究从来没有停止和放弃过. 近一二十年来，特别是近年来，地震预测预报问题在世界范围内重新引起各界的关注[185-187]，不幸的是，这至少部分地是以频繁发生的灾害性地震（如前面已提到的 2001 年印度古杰拉特地震、2003 年伊朗巴姆地震、2005 年巴基斯坦地震以及迄今余震不断的、引发了印度洋特大海啸的 2004 年印尼苏门答腊–安达曼特大地震等）造成的灾难为代价！各国地震学家正在加紧努力，以更广阔、更新颖的视野审视地震预测预报（例如，Linde 和 Sacks[188]；Thanassoulas[189]；Crampin 等[190]；Beroza 和 Ide[191]；Bromirski[192]；Liu 等[193]；Borghi 等[194]）.

地震预测面临的困难（或者说性质、特点）是客观存在的困难，既不是今天才冒出来的，也不是最新的"发现"；地震预测研究的这些性质或特点本质上也是包括地震学在内的固体地球科学的性质或特点. 困难既是挑战，也是机遇. 事实上，一部现代地震学的历史也就是地震学家不断迎接挑战、不断克复困难、不断前进的历史. 地震预测的确困难，但并非是不可能的. 讨论地震预测的困难是为了找准问题，以便对症下药，战胜困难. 我们在下文（3.4 节等）将要述及的许多地震预测的研究方法便是地震学家针对上述困难提出来的方法. 地震预测预报面临的困难既不能作为放松或放弃地震预测研究的借口；也不能作为放弃地震预测研究、片面强调只要搞抗震设防的理由. 面对地震灾害，地震学家要勇于迎接挑战，知难而进；增强防御与减轻地震灾害的能力，包括做好抗震设防工作、增强工程抗震设防的能力（如提高地震动预测水平）等等，离不开地震预测预报水平的提高，也离不开对地震发生规律及其致灾机理的认识. 解决地震预测面临的困难的出路既不能单纯依靠经验性方法，也不能置迫切的社会需求予不顾、翘首企盼几十年后的某一天基础研究的飞跃进展和重大突破. 在这方面，地震预测、特别是短临预测与医学以及军事科学或许有类似之处，而与纯基础研究不完全一样[24-27]. 这就是：①时间上的"紧迫性"，即必须在第一时间回答问题，不容犹豫，无可推诿；②对"敌情"、"病情"、"震情"所掌握的信息的"不完全性"；③决策的"高风险性"——一个决策动辄涉及成千上万、甚至是几十万人的生命，几十亿、上百亿元或以上的经济损失以及难于用金钱计算的社会安定和谐问题. 医学和军事科学都十分重视理论和基础研究，但病人的家属不会把病人送给只懂（即使是学问高超的）医学理论知识但缺乏临床经验的医生动手术，军队也不会喜欢赵括那样的指挥员. 地震预测的这些特点既不意味着对地震预测可以降低严格的科学标准，也不意味着可以因为对地震认识不够充分、对震情所掌握的信息不够完全（极而言之，永远没有"充分"、"完全"的时候）而置地震预测于不顾. 在这方面，值得回顾一下地震学界的先辈们给我们留下的十分宝贵的然而也是极其惨痛的经验教训[195-197]：

寺田寅彦（Terada, T.）是 20 世纪初一位国际著名的日本地球物理学家，也是一位优秀的散文作家. 他是 1923 年 9 月 1 日关东大地震发生后于 1925 年 11 月建立的东京帝国大学（今东京大学）地震研究所的创建者之一（创建者还有首任代所长末廣恭二、長岡半太郎、石原純等国际地震学界熟知的地震学家），现在仍然矗立在东京大学地震研究所门前的铜版上的碑文便是出自他的手笔. 他的一句据说在日本是家喻户晓的警句是：

"天灾总是在人们将其淡忘时来临."

这句极富哲理的警句虽然并不具体涉及天灾的科学内涵，但无论是对广大公众还是对负责公共安全的政府官员，时至今日都是一个极有教益的警示. 可是，由尚有争议的科学原理得出的对公众的地震警告可能起不到什么实际作用，反而会给科学家本人带来不幸. 在日本地震界广为人知的大森房吉和今村明恒的故事就是一个例子. 今村明恒深信地震空区理论, 并曾预测东京附近的相模湾将发生大地震. 该地区处于一个大地震活动带上, 但历史上还没有发生过大地震. 1906 年, 时为东京帝国大学地震学教研室助手(助教授) 的今村明恒发表了一篇文章, 预测在 50 年内相模湾将发生大地震; 并对东京缺乏防火设施提出警告, 指出如果相模湾发生的大地震袭击东京, 东京将有 10 万人会死于火灾. 可是, 这篇文章受到了时为东京帝国大学地震学教研室主任的大森房吉教授的猛烈抨击. 大森房吉认为今村明恒的文章缺乏可靠的科学依据并会引起社会的恐慌. 在大森房吉抨击今村明恒的文章发表后的 17 年间, 今村明恒的处境十分悲惨. 直到 1923 年 9 月 1 日, 他的预测不幸言中: 地震真的发生了! 关东大地震 (日本气象厅(JMA)震级 M_{JMA} 7.9, M_S 8.2, 震中位置 35.2°N, 139.5°E) 夺走了 14.3 万人的生命, 受伤人数达 10 余万人, 房屋全部倒塌 12.8 万余间, 部分倒塌 12.6 万余间, 烧毁 44.7 万余间, 受灾人口达 340 余万人, 经济损失 55 亿日圆, 成为了日本有史以来最严重的一次自然灾害. 关东大地震发生时, 大森房吉正在澳大利亚参加第二届泛太平洋科学大会, 教研室主任的工作暂由今村明恒代理. 得知大地震的消息后, 大森房吉提前结束澳洲之行回国. "雪上加霜", 在乘船回国途中, 他的健康状况因脑瘤急剧恶化. 大森房吉于 10 月 4 日抵达横滨, 前去迎接他的正是今村明恒! 大森房吉除向前去迎接他的今村明恒表示感谢外, 还为关东大地震震灾深为自责. 大森房吉在抵达日本不久后 (11 月 8 日) 去世, 终年仅 55 岁. 逝世之前, 大森房吉将东京帝国大学地震学教研室主任的工作托付给了曾经被他猛烈抨击过的今村明恒.

我们从睿智的寺田寅彦、勇敢的今村明恒和谨慎的大森房吉的故事可以得到什么样的启示呢? 寺田寅彦的警示虽然没有从科学上对天灾做出具体的预报, 但直至今日仍然正确地反映了当今社会的状况, 他的警示是长鸣的警钟, 告诫人们时刻不要忘记防灾减灾. 今村明恒基于地震专业知识而做出的具体的警告, 虽然事后被证明是正确的但在当时却被与他同时代的更有名望的科学家以科学依据不可靠 (前面提到的信息的 "不完全性") 并会引起了社会的恐慌 (决策的 "高风险性") 为由所 "打压", 他的正确的预报竟以社会遭受严重损失、今村明恒个人长期承受巨大的压力为惨重代价才得以证明! 需要指出的是, 大森房吉的 "打压" 并非出自个人的恩怨, 而是认为当时今村明恒的科学依据尚不可靠、不足于得出相模湾将发生大地震的结论, 并且, 按照现在的流行说法, 预报的 "时间窗" 长达 50 年; 大森房吉本人也并非没有注意到东京近邻地区可能发生地震, 例如, 他于关东大地震发生前一年 (1922 年) 曾经写道 (Bolt[12]):

"现在东京近邻地区保持地震平静, 但距东京平均 60km 距离的周围山区地震频频发生, 虽然在城里常常明显感到这些地震, 但因为该区不属于严重破坏的地震带, 并不构成危险. 然而随着时间的流逝, 目前地震活动区的地震活动将逐渐平静下来, 而作为补偿, 东京湾可能再度发生地震活动, 并可能发生一次地震. 这样, 一个震源距东京一定距离的地震将产生局部变动和部分破坏."

大森房吉卓越地预报了未来即将发生地震的地点 ("东京湾可能再度发生地震活动, 并可能发生一次地震"), 但低估了地震的强度 ("一个震源距东京一定距离的地震将产生局部变动和部分破坏"), 而且没有料到不久 (翌年) 就发生!

一个世纪以来, 经过几代地震学家的努力, 对地震的认识的确大有进步, 然而不了解之处仍甚多. 目前地震预测尚处于初期的科学探索阶段, 地震预测的能力、特别是短、临地震预测的能力还是很低的, 与迫切的社会需求相距甚远. 解决这一既紧迫要求予以回答、又需要通过长期探索方能解决的地球科学难题唯有依靠科学与技术的进步、依靠科学家群体. 一方面, 科学家应当倾其所能把代表当前科技最高水平的知识用于地震预测预报; 另一方面, 科学家 (作为一个群体, 而不仅是某个人) 还应勇负责任, 把代表当前科技界认识水平的有关地震的信息 (包括正、反两方面的信息) 如实地传递给公众, 应当说实话, 永远说实话! 决不能重演像当年大森房吉压制今村明恒观点那样的悲剧.

3.2 强化对地震及其前兆的观测与研究

为了克服地震预测预报面临的观测上的困难,在地震观测与研究方面,多少年来,地震学家不但在陆地上,在海岛上,而且向海洋进军,在海底大量布设地震观测台网,形成从全球性至区域性直至地方性的多层次的地震观测系统. 例如,截至2003年,属于全球性("台距"即台站之间的距离约2000km)的全球地震台网(Global Seismic Network,缩写为 GSN)的地震台已达126个;截至2005年,中国国家数字地震台网(National Digital Seismograph Network)的地震台已达152个,区域数字地震台网(Regional Digital Seismograph Network)的地震台已达678个[198]. 即使如此,地震观测台网仍然是很稀疏的. 在大多数地区,限于财力和自然条件,台网密度仍较低,台距较大. 这种情况造成了一方面是"信息过剩",即:现有的数字地震台网产出的大量数据使用得不够,不能充分发挥其作用,积压浪费;而另一方面,则是"信息饥渴",即:由于台网密度低、台距较大,资料不便于分析研究,以至于在监测地震或开展地震研究时,捉襟见肘,感到资料不足. 有鉴于此,地震学家应努力变"被动观测"为"主动观测",在规则地加密现有固定式台网的基础上,在重点监测与研究地区布设流动地震台网(台阵),进一步加密观测,改善由于台距过大、不利于分析解释地震记录的状况;并且不但利用天然地震震源,而且也运用包括爆破在内的人工震源对地球内部进行探测以获得有关震源特征和地震波传播路径效应的更多的、更精细的信息.

在地震前兆的观测与研究方面,应继续强化对地震前兆现象的监测、拓宽对地震前兆的探索范围,以期在可靠的和丰富的前兆现象基础上,构制自由度较小的定量的物理模式进行模拟、反复验证,逐渐地、然而实效上可能会是较快地阐明地震前兆与地震发生的内在联系,实现地震预测. 地震是发生在地球内部的自然现象. 一个7级大地震释放的应变能的数量级达10^{15}J,很难置信在如此巨大的应变能释放之前不出现任何"讯号". 已知的地震前兆包括直接与地震过程相联系和不十分直接与地震过程相联系的两大类. 前者如地震活动性、地震空区、b值、Q值、波速与波速比等地震学前兆[112,199,123-126,200-203]以及如通过地倾斜、地应变、地应力和重力变化等形式表现出来的地形变、地应力与重力前兆,后者如地磁、地电[204-206]、地下水位、地下水化学[110,111,149,150,207-209]和动物异常[142,126]等. 它们涉及地球物理、大地测量、地质和地球化学等众多的学科和广阔的领域. 一方面,可沿着已有的方向继续努力寻找地震前兆;但是,另一方面,应当努力探索新的前兆. 例如,现在已经发现的洋中脊转换断层上的"慢地震",在俯冲带上以及圣安德列斯断层上的"寂静地震",还有在大型的逆冲断层下方周期性的缓慢滑动事件[210-212,188,191]及与其相关的间歇性的简谐颤动[192],台风对地震的触发作用[193],以及利用地震剪切波分裂监测地壳应力变化以预报地震的时间和震级的"应力预报地震"(stress-forecast earthquake)方法[190,213-221],地震活动性的"临界加速模型"[162,222]、"加速矩释放模型"(acceleratory moment release 模型,简称 AMR 模型)[223]、"短期丛集模型"[224]、"应力触发模型"[80,81]等等,均应引起重视.

为了推进地震预报(earthquake forecasting)稳步向前发展,有必要对地震预报的质量进行科学的、客观的评估. 为此,在美国南加州大学的南加州地震中心(Southern California Earthquake Center,缩写为 SCEC)、美国地质调查局(USGS)以及美国自然科学基金会(National Science Foundation,缩写为 NSF)的共同支持下,南加州地震中心的"区域性地震似然模型工作组"(Working Group on Regional Earthquake Likelihood Models,缩写为 RELM)倡议开展名为"地震可预测性合作研究"(Collaboratory Studies for Earthquake Predictability,缩写为 CSEP)的国际合作,计划先从美国加州开始试验,然后推广至全球,对预测模型和预报模型进行严格的评估[225]. 迄今,除 SCEC 外,已有瑞士苏黎世高等工业学校(Eidgenössische Technische Hochschule,Zurich)、新西兰地质与核科学研究所(Institute of Geological and Nuclear Sciences,New Zealand)以及日本东京大学地震研究所(Earthquake Research Institute,University of Tokyo)参加这一国际合作计划,成为其试验中心. 这一国际合作计划已从2009年8月1日开始实施,将对新西兰、日本、西太平洋北部及南部以及全球的地震预测模型和预报模型进行严格的评估. 这种全球性的预测预报模型严格的客观的评估必将有助于改进预测预报模型, 提高预测预报地震的能力.

20 世纪 90 年代以来, 空间对地观测技术和数字地震观测技进步, 使得观测 (现代地壳运动、地球内部结构、地震震源过程以及地震前兆的) 技术, 在分辨率、覆盖面和动态性等方面都有了飞跃式的发展, 高新技术如全球定位系统 (GPS)、卫星孔径雷达干涉测量术 (InSAR) 等空间大地测量技术、用于探测地震前兆的 "地震卫星" 等在地球科学中的应用为地震预测研究带来了新的机遇 (例如, Borghi 等[194]), 多学科协同配合和相互渗透是寻找发现与可靠地确定地震前兆的有力的手段.

3.3 坚持地震预测科学试验——地震预测试验场

地震既发生在板块边界 (板间地震)、也发生在板块内部 (板内地震), 地震前兆出现的复杂性和多变性可能与地震发生场所的地质环境的复杂性密切相关. 因地而异, 即在不同地震危险区采取不同的 "战略", 各有侧重地检验与发展不同的预测方法, 不但在科学上是合理的, 而且在经济效益上也是比较高的. 应重视充分利用我国的地域优势, 总结包括我国的地震预测试验场在内的世界各国的地震预测试验场经验教训, 通过地震预测试验场这样一种行之有效的方式, 开展在严格的、可控制的条件下进行的、可用事先明确的可接受的准则予以检验的地震预测科学试验研究; 选准地区, 多学科互相配合, 加密观测, 监测、研究、预测预报三者密切结合, 坚持不懈, 可望获得在不同构造环境下断层活动、形变、地震前兆和地震活动性等的十分有价值的资料, 从而有助于增进对地震的了解、攻克地震预测难关.

3.4 系统地开展基础性、综合性的对地球内部及对地震的观测、探测与研究计划

为了克服地震预测面临的观测上的困难, 除了前面已经提及的强化对地震及其前兆的观测外, 还应考虑: ①在地震活动地区进行以探测震源区为目的的科学钻探, 钻探到发震层所在深度对震源区作直接观测. 科学钻探常因代价昂贵因而只能是 "一孔"、顶多是若干个 "孔" 而被讥为 "一孔之见", 而且还只是 "皮毛". 但是能够到达发震层 (孕震层) 所在的深度对震源区作直接的观测, 尽管是 "皮毛" 的 "一孔之见", 还是十分宝贵的, 所得的结果对于克服 "不可入性" 带来的困难、对于验证理论是很有意义的. ②在断层带开挖探槽研究古地震, 延伸对地震 "观测" 的时间窗的长度以克服大地震的复发时间比人的寿命、比有现代仪器观测以来的时间长得多 (大地震的 "非频发性") 带来的困难. ③在实验室中进行岩石样品在高温高压下的破裂实验, 模拟单个地震的孕育、发生、扩展、停止, 地震序列的形成与发生以及地震的轮回过程, 等等. 通过对岩石样品中的微小破裂事件的实验研究藉以了解作为大的破裂事件的天然地震的发生发展规律, 以克服被动地等待观测复发时间漫长的天然地震的发生 (大地震的 "非频发性") 带来的困难. ④利用计算机做地震数值模拟, 模拟地震的孕育、发生、扩展、停止, 地震序列的形成与发生, 地震的轮回过程以及地震的致灾过程, 等等. 通过在计算机中进行数值模拟, 既能再现复发时间以数十年、数百年计、甚至更长的天然地震的孕育与发生过程 (大地震的 "非频发性")、地震序列的形成与发生以及地震的轮回过程 (大地震的 "非频发性"), 又能模拟破裂时间以仅仅数秒、数十秒计的天然地震的快速破裂过程 (地震的 "突发性").

为此, 应当实施旨在对地球内部及地震系统地进行基础性的、综合性的观测、探测与研究的大型的科学计划. 目前美国正在实施的 "地球透镜计划" (EarthScope) 是一个很有创新意义的例子 (EarthScope Working Group, EarthScope Project Plan, 2001)[226], 值得借鉴. "地球透镜计划" 旨在通过观测、研究北美大陆的活动构造和岩石层结构, 以发展地震科学, 促进地震科学在减轻地震灾害中的应用. 该计划由 4 个部分组成:

(1) 美国台阵 (USArray) 计划. "美国台阵计划" 拟对美国大陆、阿拉斯加及其邻区的岩石层 (地

壳和上地幔）结构做高分辨度的地震成像．该计划拟动用 400 套宽频带地震仪组成大型的遥测地震台阵，有规则地进行实时观测；再用 2400 套便携式地震仪组成移动式台阵，借助于天然与人工两种震源，对在上述大型遥测台阵"脚印"内的关键目标做高密度的短期观测；并以固定式的地震台网——美国地质调查局国家地震台网（USGS/National Seismic Network）进行长期、连续的地震观测．

（2）圣安德列斯断层深部观测计划（San Andreas Fault Observatory at Depth，缩写为 SAFOD）计划．该计划是圣安德列斯断层深部的钻探计划，拟在圣安德列斯断层带 1966 年帕克菲尔德地震震源的上方打一个 4km 深的钻．钻探将直达发震层所在深度，获取断层带的岩石及流体的样品供实验室分析地球物理参数，包括地震活动性、孔隙压、温度和应变等；该计划拟对井下和邻区的流体活动性、地震活动性和形变等进行长达 20 年的长期监测．

（3）板块边界观测计划（Plate Boundary Observatory，缩写为 PBO）计划．该计划拟用应变仪和超高精准的 GPS 仪对美国西部进行地形变测量，包括：用"骨干台网"、即台距 100～200km 的、连续记录的 GPS 遥测台网获取从阿拉斯加直至墨西哥的整条板块边界的、在空间上是长波长、在时间上则是长周期的地形变信息的概况；并在构造活动地区（如主要断层带和活动岩浆系统）集中进行 GPS、井下应变及地震观测．

（4）合成孔径雷达干涉测量（Interferometric Synthetic Aperture Radar，缩写为 InSAR）计划．用星基 InSAR 对地面形变成像、特别是对与活动断层和火山有关的地面形变场成像．美国国家航空航天局（NASA），美国国家科学基金会（NSF）和美国地质调查局（USGS）三家合作，以卫星对广阔的地域做空间上是连续性的、时间上则是间歇性的应变测量．InSAR 测量是面上的测量，它将与 PBO 的 GPS 点上的测量形成互补，"点面结合"．对于所有类型的地形地貌，空间测量将达到密集的空间复盖（100m）与时间复盖（8 天），矢量解将准确到 2mm．预计用于整个"地球透镜计划"仪器设备的经费分别是：USArray，$64.0 百万；SAFOD，$17.4 百万；PBO，$91.3 百万；InSAR，$245.0 百万；用于数据分析及运行管理的费用是每年 $15～20 百万．

3.5　加强国内合作与国际合作

地震预测研究深受缺乏作为建立地震理论的基础的经验规律所需的"样本"太少所造成的困难（大地震的"非频发性"）之限制．目前在刊登有关地震预测实践的论文的绝大多数学术刊物中几乎都不提供相关的原始资料，语焉不详，以致其他研究人员读了之后也无从作独立的检验与评估；此外，资料又不能共享．这些因素加剧了上述困难．应当正视并改变地震预测研究的实际上的封闭状况，广泛深入地开展国内、国际学术交流与合作；加强地震信息基础设施的建设，促成资料共享；充分利用信息时代的便利条件，建立没有围墙的、虚拟的、分布式的联合研究中心，使得从事地震预测的研究人员，地不分南北东西，人不分专业机构内外，都能使用仪器设备、获取观测资料、使用计算设施和资源、方便地与同行交流切磋，等等．

4　讨论与结论

以上从正反两个方面概要评述了国际地震预测预报研究的情况，分析了地震预测预报在科学上的遇到的困难，阐述了为解决这些困难应当采取的科学途径．我们指出，自 20 世纪 60 年代以来，中期和长期地震预测取得了一些有意义的进展，如：板块边界大地震空区的确认、"应力影区"、地震活动性图像、图像识别以及美国帕克菲尔德地震在预报期过了 11 年后终于发生，等等．目前地震预测的总体水平、特别是短期与临震预测的水平仍然不高，与社会需求相距仍甚遥远．我们还指出，地震预测作为一个既紧迫要求予以回答、又需要通过长期探索方能解决的地球科学难题尽管非常困难, 但并非不可能; 困难既不能作为放松或放弃地震预测研究的借口; 也不能作为放弃地震预测研究、片面强调只要搞抗震设防的理由. 地震作为一种自然现象, 是人类所居住的地球这颗太阳系中独一无二的行星生机勃勃的表现, 它的发生是不可避免的; 但是, 地震灾害, 不但应当而且也是可以通过努力予以避免或减轻的. 面对地震灾害, 地震学家要勇于迎接挑战, 知难而进; 要加强对地震发生规律及其致灾机理的研究, 提高地震预测预报水平, 增强防御与减轻地震灾害的能力. 解决地震预测面临的困难的出路既不能单纯依靠经验性方法, 也不能置迫切的社会需求予不顾、坐等几十年后的某一天基础研究的飞跃进展和重大突破. 特别需要乐观地指出的是, 与 40 多年前的情况相比, 地震学家今天面临的科学难题依旧, 并未增加; 然而这个难题却比先前暴露得更加清楚, 并且 20 世纪 60 年代以来地震观测技术的进步、高新技术的发展与应用为地震预测预报研究带来了历史性的机遇. 依靠科技进步、强化对地震及其前兆的观测, 选准地点、开展并坚持以地震预测试验场为重要方式的地震预测预报科学试验, 坚持不懈地、系统地开展基础性的对地球内部及对地震的观测、探测与研究, 对实现地震预测的前景是可以审慎地乐观的. 正如著名科学家、液态燃料火箭发明人戈达德（Goddard R. H., 1882~1945）所言:

"慎言不可能. 昨日之梦想, 今日有希望, 明日变现实."

致谢 两位审稿专家提出了宝贵的修改意见和建议, 谨表谢忱.

参 考 文 献

[1] Kanamori H. Global seismicity. In: Kanamori H, Boschi E, eds. *Earthquakes: Observation, Theory and Interpretation*. New York: Elsevier North-Holland Inc., 1983. 596–608

[2] Utsu T. A list of deadly earthquakes in the world: 1500—2000. In: Lee W H K, Kanamori H, Jennings P, et al, eds. *International Handbook of Earthquake and Engineering Seismology*, Part *A*. San Diego: Academic Press, 2002. 691–717

[3] National Research Council of the National Academies. Living on an Active Earth. Washington DC: The National Academies Press, 2003. 418

[4] Engdahl E R, Villasenor A. Global seismicity: 1900—1999. In: Lee W H K, Kanamori H, Jennings P, et al, eds. *International Handbook of Earthquake and Engineering Seismology*, Part *A*. San Diego: Academic Press, 2002. 665–690

[5] http://earthquake.usgs.gov/eqcenter/

[6] 孔丘, 编订. 诗经. 北京: 北京出版社, 2006. 383

[7] 李善邦. 中国地震. 北京: 地震出版社, 1981. 612

[8] 周锡䪖, 选注. 诗经选. 广州: 广东人民出版社, 1984. 307

[9] Darwin C. Journal of Researches into the National History and Geology of the Countries during the Voyage of H. M. S. Beagle Round the World under the Command of Capt. Fitz Roy, R. N. D. New York: Appleton and Co., 1898

[10] 谢毓寿, 蔡美彪, 主编. 中国地震历史资料汇编（第一卷）. 北京: 科学出版社, 1983. 227

[11] Bullen K E. *An Introduction to the Theory of Seismology*. 3rd ed. Cambridge: Cambridge University Press, 1963. 381

[12] Bolt B A. *Earthquakes and Geological Discovery*. New York: W. H. Freeman, 1993. 229

[13] Agnew D C. History of seismology. In: Lee W H K, Kanamori H, Jennings P, et al, eds. *International Handbook of Earthquake and Engineering Seismology Part A*. San Diego: Academic Press, 2002. 3–11

[14] Milne J. Seismic science in Japan. *Trans Seism Soc Jpn*, 1880, **1**: 3–33

[15] 傅承义, 刘恢先. 天然地震的灾害及其防御. 1956-1967 年科学技术发展远景规划, 第 33 项. 1956

[16] Tsuboi C, Wadati K, Hagiwara T. Prediction of Earthquakes—Progress to Date and Plans for Future Development. Rep. Earthquake Prediction Research Group Japan. Tokyo: Earthquake Research Institute, University of Tokyo, 1962. 21

[17] 傅承义. 有关地震预告的几个问题. 科学通报, 1963, (3): 30–36

[18] Press F, Benioff H, Frosch R A, et al. *Earthquake Prediction: A Proposal for A Ten Year Program of Research*. Washington

DC: White House Office of Science and Technology, 1965. 134

[19] Press F, Brace W F. Earthquake prediction. *Science*, 1966, **152**: 1575–1584

[20] National Research Council of the National Academies. Predicting Earthquakes: A Scientific and Technical Evaluation—With Implication for Society. Washington DC: National Academy Press, 1976. 62

[21] Садовский М А. *Прогноз Землетрясений*. Том 1-6, Изд. "Дониш". 1982–1986

[22] 日本地震学会地震予知検討委員会, 編. 地震予知の科学 (日文). 東京: 東京大学出版会, 2007. 218

[23] Chen Y T, Chen Z L, Wang B Q. Seismological studies of earthquake prediction in China: a review. In: Dragoni M, Boschi E, eds. *Earthquake Prediction*. Roma: Il Cigno Galileo Galilei, 1992. 71–109

[24] 陈运泰. 地震预测研究概况. 地震学刊, 1993, (1): 17–23

[25] 陈运泰. 地震预测现状与前景. 见: 中国科学院, 编著. 2007 科学发展报告. 北京: 科学出版社, 2007. 173–182

[26] 陈运泰. 地震预测——进展、困难与前景. 地震地磁观测与研究, 2007, **28**(2): 1–24

[27] 陈运泰. 地震预测要知难而进. 求是, 2008, (15): 58–60

[28] Wyss M. *Evaluation of Proposed Earthquake Precursors*. Washington DC: American Geophysical Union, 1991. 94

[29] Knopoff L. Earthquake prediction: the scientific challenge. In: Knopoff L, Aki K, Allen C R, et al, eds. *Earthquake Prediction: The Scientific Challenge*, Colloquium Proceedings. Proc Nat Acad Sci USA, 1996, **93**: 3719–3720

[30] 梅世蓉, 冯德益, 张国民, 等. 中国地震预报概论. 北京: 地震出版社, 1993. 498

[31] Wallace R E, Davis J F, McNally K C. Terms for expressing earthquake potential, prediction, and probability. *Bull Seism Soc Amer*, 1984, **74**: 1819–1825

[32] Kisslinger C. Potents and predictions. *Nature*, 1989, **339**: 337–338

[33] Geller R J. Earthquake prediction: a critical review. *Geophys J Int*, 1997, **131**: 425–450

[34] Bakun W H, McEvilly T V. Earthquakes near Parkfield, California: comparing the 1934 and 1966 sequences. *Science*, 1979, **205**: 1375–1377

[35] Bakun W H, McEvilly T V. Recurrence models and Parkfield, California earthquakes. *J Geophys Res*, 1984, **89**: 3051–3058

[36] Bakun W H, Lindh A G. The Parkfield, California earthquake prediction experiment. *Science*, 1985, **229**: 619–624

[37] Sykes L R. Aftershock zones of great earthquakes, seismicity gaps and prediction. *J Geophys Res*, 1971, **76**: 8021–8041

[38] Sykes L R. Intraplate seismicity reactivation of preexisting zones of weakness, alkaline magmatism, and other tectonism postdating continental fragmentation. *Rev Geophys Space Phys*, 1978, **16**: 621–688

[39] Sykes L R, Nishenko S P. Probabilities of occurrence of large plate rupturing earthquakes for the San Andreas, San Jacinto, and Imperial faults, California. *J Geophys Res*, 1984, **89**: 5905–5927

[40] Nishenko S P. Earthquakes, hazards and predictions. In: James D E, ed. *The Encyclopedia of Solid-Earth Geophysics*. New York: Van Nostrand Reinhold, 1989. 260–268

[41] Nishenko S P. Circum-Pacific seismic potential: 1989—1999. *Pure Appl Geophys*, 1991, **135**: 169–259

[42] 陈章立, 刘蒲雄, 黄德瑜, 等. 大震前区域地震活动性特征. 见: 丁国瑜, 马宗晋, 主编. 国际地震预报讨论会论文选. 北京: 地震出版社, 1981. 97–205

[43] 陆远忠, 陈章立, 王碧泉, 等. 地震预报的地震学方法. 北京: 地震出版社, 1985. 268

[44] Reid H F. *The California Earthquake of April 18, 1906*. Publication 87, **2**. Reprinted 1969. Washington: Carnegie Institution of Washington, 1910. 192

[45] Reid H F. The elastic-rebound theory of earthquakes. Univ Calif Pub *Bull Dept Geol Sci*, 1911, **6**: 413–444

[46] Imamura A. On the seismic activity of central Japan. Jpn J Astron Geophys, 1928, **6**: 119–137

[47] Федотов С А. Закономерности распределения сильных землетрясений Камчатки, Курильский островов и северо-восточной Японии. Тр. Ин-та Физики Земли АН СССР, Изд. Наука, 1965, **36**: 66–93

[48] Kelleher A, Sykes L R, Oliver J. Possible criteria for predicting earthquake locations and their application to major plate boundaries of the Pacific and the Caribbean. *J Geophys Res*, 1973, **78**: 2547–2585

[49] McCann W R, Nishenko S P, Sykes L R, et al. Seismic gaps and plate tectonics: seismic potential for major boundaries. *Pure*

[50] Schwartz D P, Coppersmith K J. Fault behavior and characteristic earthquakes: examples from the Wasatch and San Andreas faults. *J Geophys Res*, 1984, **89**: 5681–5698

[51] Gutenberg B, Richter C F. Magnitude and energy of earthquakes. *Ann Geofis*, 1956, **9**: 1–15

[52] Richter C F. *Elementary Seismology*. San Francisco: W. H. Freeman, 1958. 768

[53] Wesnousky S, Scholz C, Shimazaki K. Earthquake frequency distribution and the mechanics of faulting. *J Geophys Res*, 1983, **88**: 9331–9340

[54] Wesnousky S. The Gutenberg-Richter or characteristic earthquake distribution: Which is it? *Bull Seism Soc Amer*, 1994, **84**: 1940–1959

[55] Kagan Y, Wesnousky S. The Gutenberg-Richter or characteristic earthquake distribution. Which is it? Discussion and Reply. *Bull Seism Soc Amer*, 1996, **86**: 274–291

[56] Bufe C G, Harsh P W, Burford R O. Steady-state seismic slip—A precise recurrence model. *Geophys Res Lett*, 1977, **4**: 91–94

[57] Shimazaki K, Nakata T. Time-predictable recurrence model for large earthquakes. Geophys Res Lett, 1980, **7**: 279–282

[58] Mogi K. Recent horizontal deformation of the Earth's crust and tectonic activity in Japan (1). *Bull Earthq Res Inst Univ Tokyo*, 1970, **48**: 413–430

[59] Mogi K. Seismicity in western Japan and long-term earthquake forecasting. In: Simpson D W, Richards P, eds. *Earthquake Prediction—An International Review*. Maurice Ewing Monograph Series **4**. Washington DC: Amer Geophys Union, 1981. 43–51

[60] Mogi K. Earthquake prediction program in Japan. In: Simpson D W, Richards P, eds. *Earthquake Prediction—An International Review*. Maurice Ewing Monograph Series **4**. Washington DC: Amer Geophys Union, 1981. 635–666

[61] Mogi K. *Earthquake Prediction*. Tokyo: Academic Press, 1985. 355

[62] Ando M. Possibility of a major earthquake in the Tokai district, Japan, and its pre-estimated seismotectonic effects. *Tectonophysics*, 1975, **25**: 69–85

[63] Utsu T. Possibility of a great earthquake in the Tokai district, Japan. *J Phys Earth*, 1977, **25**: S219–S230

[64] Ishibashi K. Specification of a soon-to-occur seismic faulting in the Tokai district, central Japan, based upon seismotectonics. In: Simpson D W, Richards P, eds. *Earthquake Prediction—An International Review*. Maurice Ewing Monograph Series 4. Washington DC: Amer Geophys Union, 1981. 297–332

[65] Matsumura S. Focal zone of a future Tokai earthquake inferred from the seismicity pattern around the plate interface. *Tectonophysics*, 1997, **273**: 271–291

[66] Subcommittee for Review Drafting, Special Committee for Earthquake Prediction, Geodesy Council of the Ministry of Education, Science, and Culture. *State-of-the-Art Review of the National Programs for Earthquake Prediction*. Tokyo, 1997. 137

[67] Roeloffs E, Langbein J. The earthquake prediction experiment at Parkfield, California. Rev Geophys, 1994, **32**(3): 315–336

[68] Shearer C F. Southern *San Andreas Fault Geometry and Fault Zone Deformation: Implications for Earthquake Prediction* (National Earthquake Prediction Council Meeting, March, 1985). U. S. Geol. Surv. Open-file Rep. 85-507, Reston, Virginia, 1985. 173–174

[69] Savage J C. The Parkfield prediction fallacy. *Bull Seism Soc Am*, 1993, **83**: 1–6

[70] Kagan Y. Statistical aspects of Parkfield earthquake sequence and Parkfield prediction. *Tectonophysics*, 1997, **270**: 207–219

[71] Jackson D D, Kagan Y Y. The 2004 Parkfield earthquake, the 1985 prediction, and characteristic earthquakes: lessons for the future. *Bull Seism Soc Amer*, 2006, **96**: S397–S409

[72] Simpson R W, Schulz S S, Dietz L D, et al. The response of creeping parts of the San Andreas fault to earthquakes on nearby faults: two examples. *Pure Appl Geophys*, 1988, **126**: 665–685

[73] Ben-Zion Y, Rice J R, Dmowska R. Interaction of the San Andreas fault creeping segment with adjacent great rupture zones, and earthquake recurrence at Parkfield. *J Geophys Res*, 1993, **98**: 2135–2144

[74] Lindh A G. *Preliminary Assessment of Long-Term Probabilities for Large Earthquakes along Selected Fault Segments of the San Andreas Fault System in California*. U. S. Geol. Surv. Open-File Report 83-63, Menlo Park, California, 1983. 15

[75] Working Group on California Earthquake Probabilities. *Probabilities of Large Earthquakes Occurring in California on the San Andreas Fault*. U. S. Geological Survey Open-File Report 88-398, Reston, Virginia, 1988. 62

[76] Harris R A. Forecasts of the 1989 Loma Prieta, California earthquake. *Bull Seism Soc Amer*, 1998, **88**: 898–916

[77] Harris R A, Simpson R W. Stress relaxation shadows and the suppression of earthquakes: some examples from California and their possible uses for earthquake hazard estimates. *Seism Res Lett*, 1996, **67**: 40

[78] Harris R A, Simpson R W. Suppression of large earthquakes by stress shadows: a comparison of Coulomb and rate-and-state failure. *J Geophys Res*, 1998, **103**: 24439–24451

[79] Deng J, Sykes L. Evolution of the stress field in southern California and triggering of moderate-size earthquakes: a 200-year perspective. *J Geophys Res*, 1997, **102**: 9859–9886

[80] Felzer K R, Brodsky E E. Decay of aftershock density with distance indicates triggering by dynamic stress. *Nature*, 2006, **441**: 735–738

[81] Mallman E P, Parson T. A global search for stress shadows. *J Geophys Res*, 2008, **113**: B12304

[82] Ohtake M, Matumoto T, Latham G. Seismicity gap near Oaxaca, southern Mexico, as a probable precursor to a large earthquake. *Pure Appl Geophys*, 1977, **113**: 375–385

[83] Ohtake M, Matumoto T, Latham G. Evaluation of the forecast of the 1978 Oaxaca, southern Mexico earthquake based on a precursory seismic quiescence. In: Simpson D W, Richards P, eds. *Earthquake Prediction—An International Review*. Maurice Ewing Monograph Series **4**. Washington DC: American Geophysical Union, 1981. 53–62

[84] Habermann R E. Precursory seismic quiescence: past, present, and future. *Pure Appl Geophys*, 1988, **126**: 277–318

[85] Sobolev G A. The examples of earthquake preparation in Kamchatka and Japan. *Tectonophysics*, 2001, **338**: 269–279

[86] Sobolev G A, Chelidze T L, Zavyalov A D, et al. The maps of expected earthquakes based on a combination of parameters. *Tectonophysics*, 1991, **193**: 255–265

[87] Sobolev G A, Huang Q, Nagao T. Phases of earthquake's preparation and by chance test of seismic quiescence anomaly. *J Geodyn*, 2002, **33**: 413–424

[88] Sobolev G A, Tyupkin Y S. Low-seismicity precursors of large earthquakes in Kamchatka. *Volcanol Seismol*, 1997, **18**: 433–446

[89] Sobolev G A, Tyupkin Y S. Precursory phases, seismicity precursors, and earthquake prediction in Kamchatka. *Volcanol Seismol*, 1999, **20**: 615–627

[90] Huang Q. Search for reliable precursors: a case study of the seismic quiescence of the 2000 western Tottori prefecture earthquake. *J Geophys Res*, 2006, **111**: B04301, doi: 10.1029/2005JB003982

[91] Huang Q. Seismicity changes prior to the $M_S 8.0$ Wenchuan earthquake in Sichuan, China. *Geophys Res Letts*, 2008, **35**: L23308, doi: 10.1029/2008GL036270

[92] Huang Q, Sobolev G A, Nagao T. Characteristics of the seismic quiescence and activation patterns before the $M = 7.2$ Kobe earthquake, January 17, 1995. *Tectonophysics*, 2001, **337**: 99–116

[93] Huang Q, Öncel A O, Sobolev G A. Precursory seismicity changes associated with the $M_W = 7.4$ 1999 August 17 Izmit (Turkey) earthquake. *Geophys J Int*, 2002, **151**: 235–242

[94] Rundle J B, Klein W, Tiampo K F, et al. Linear pattern dynamics in nonlinear threshold systems. *Phys Rev E*, 2000, **61**: 2418–2431

[95] Rundle J B, Turcotte D L, Shcherbakov R, et al. Statistical physics approach to understanding the multiscale dynamics of earthquake fault systems. *Rev Geophys*, 2003, **41**: 1019, doi: 10.1029/2003RG000135

[96] Chen C C, Rundle J B, Holliday J R, et al. The 1999 Chi-Chi, Taiwan, earthquake as a typical example of seismic activation and quiescence. *Geophys Res Letts*, 2005, **32**: L22315, doi: 10.1029/2005GL023991

[97] Wyss M, Habermann R E. Precursory seismic quiescence. *Pure Appl Geophys*, 1998, **126**: 319–332

[98] Zshau J. SEISMOLAP: a quantification of seismic quiescence and clustering. IASPEI, XXI General Assembly, Boulder, Colorado, July 2–14, 1995. A389

[99] Keilis-Borok V I, Knopoff L, Rotwain I M, et al. Intermediate-term prediction of occurrence times of strong earthquakes. *Nature*, 1988, **335**: 690–694

[100] Keilis-Borok V I, Kossobokov V G. Premonitory activation of seismic flow: algorithm M8. *Phys Earth Planet Inter*, 1990, **61**: 73–83

[101] Keilis-Borok V I, Rotwain I M. Diagnosis of time of increased probability of strong earthquake in different regions of the world: algorithm CN. *Phys Earth Planet Inter*, 1990, **61**: 57–72

[102] Healy T H, Kossobokov V G, Dewey I W. *A Test to Evaluate the Earthquake Prediction Algorithm M 8*. U. S. Geological Survey Open-File Report 92–401, Denver, Colo., 1992. 23

[103] Kossobokov V G, Keilis-Borok V I, Smith S W. Localization of intermediate-term earthquake prediction. *J Geophys Res*, 1990, **95**(B12): 19763–19772

[104] Kossobokov V G, Romashkova L L, Keilis-Borok V I, et al. Testing earthquake prediction algorithms: statistically significant advance prediction of the largest earthquakes in the circum-Pacific, 1992—1997. *Phys Earth Planet Inter*, 1999, **111**: 187–196

[105] Keilis-Borok V I, Shebalin P, Gabrielov A, et al. Reverse tracing of short-term earthquake precursors. *Phys Earth Planet Inter*, 2004, **145**: 75–85

[106] Keilis-Borok V I, Shebalin P. Short-term advance prediction of the San Simeon earthquake, California, December 22, 2003, magnitude 6. 5. Personal communication, 2003

[107] Shebalin P. Increased correlation range of seismicity before large events manifested by earthquakes chains. *Tectonophysics*, 2006, **424**(3): 335–349

[108] Peresan A, Costa G, Panza G F. Seismotectonic model and CN earthquake prediction in Italy. *Pure Appl Geophys*, 1999, **154**: 281–306

[109] Kisslinger C, Suzuki Z. *Earthquake Precursors*. Tokyo: Japan Scietific Societies Press, 1978. 296

[110] Wakita H. Precursory changes in groundwater prior to the 1978 Izu-Oshima-Kinkai earthquake. In: Simpson D W, Richards P, eds. *Earthquake Prediction—An International Review*. Maurice Ewing Monograph Series 4. Washington DC: American Geophysical Union, 1981. 527–532

[111] Wakita H. Changes in groundwater level and chemical composition. In: Asada A, ed. *Earthquake Prediction Techniques: Their Application in Japan*. Tokyo: University of Tokyo Press, 1982. 175–216

[112] Asada T. *Earthquake Prediction Techniques: Their Application in Japan*. Tokyo: University of Tokyo Press, 1982. 317

[113] Varotsos P, Alexopoupos K, Nomicos K. Seismic electric currents. *Proc Greek Acad Sci*, 1981, **56**: 277–286

[114] Varotsos P, Alexopoupos K, Nomicos K. Electric telluric precursors to earthquakes. *Proc Greek Acad Sci*, 1982, **57**: 341–362

[115] Varotsos P, Alexopoupos K, Nomicos K, et al. Official earthquake prediction procedure in Greece. *Tectonophysics*, 1988, **152**: 193–196

[116] Varotsos P, Alexopoupos K. Physical properties of the variations of the electric field of the Earth preceding earthquakes. I. *Tectonophysics*, 1984, **110**: 73–98

[117] Varotsos P, Alexopoupos K. Physical properties of the variations of the electric field of the Earth preceding earthquakes. II. Determination of epicenter and magnitude. *Tectonophysics*, 1984, **110**: 99–125

[118] Evison F F. Earthquake prediction. *Proceedings of the International Symposium in Earthquake Prediction*. Paris: Terra Scientific Publishing Co., 1984. 995

[119] Varotsos P, Lazaridou M. Latest aspects of earthquake prediction in Greece based on seismic electric signals. *Tectonophysics*, 1991, **188**: 321–347

[120] Hayakawa M, Fujinawa Y. *Electromagnetic Phenomena Related to Earthquake Prediction*. Tokyo: Terra Scientific Publishing Co., 1994. 677

[121] Hayakawa M. *Atmospheric and Ionospheric Electromagnetic Phenomena Associated with Earthquakes*. Tokyo: Terra Scientific

Publishing Co., 1999. 996

[122] Gokhberg M B, Morgounov V A, Pokjotelov O A. *Earthquake Prediction: Seismo-Elecromagnetic Phenomena*. New Jersey: Gordon and Breach Publishers, 2000. 193

[123] Rikitake T. Practical approach to earthquake prediction and warning. In: Rikitake T, ed. *Current Research in Earthquake Prediction* I. Dordrecht: Reidel, 1981. 1−56

[124] Rikitake T. Anomalous animal behavior preceding the 1978 earthquake of magnitude 7.0 that occurred near Izu-Oshima, Japan. In: Rikitake T, ed. *Current Research in Earthquake Prediction* I. Dordrecht: Reidel, 1981. 67−80

[125] Rikitake T. *Earthquake Forecasting and Warning*. Tokyo: Center for Academic Publications Japan, 1982. 402

[126] Rikitake T. Nature of macro-anomaly precursory to an earthquake. *J Phys Earth*, 1996, **42**: 149−164

[127] 力武常次. 動物は地震を予知するか異常行動が教えるもの（日文）. 東京: 東京株式会社講談社, 1982. 215

[128] Allen C R. Earthquake prediction—1982 overview. *Bull Seism Soc Amer*, 1982, **72**: S331−S335

[129] Семенов А Н. Изменение отношения времен пробега поперечных и продольных волн перед сильными землетрясениями. *Изв АН СССР Физика Земли*, 1969, (4): 72−77

[130] Нерсесов И Л, Семенов А Н, Симбирева И Г. Пространственно-врменное распределение отоношений времен пробега поперечных и прдольных волн в Гармском районе. Сб. *Физические Основания Предвестников Землетрясений*. Изд Наука, 1969. 88−89

[131] Aggarwal Y P, Sykes L R, Armsbruster J, et al. Premonitory changes in seismic velocities and prediction of earthquakes. *Nature*, 1973, **241**: 101−104

[132] Aggarwal Y P, Sykes L R, Simpson D W, et al. Spatial and temporal variations in t_S/t_P and in P-wave residuals at Blue Mountain Lake, New York: Application to earthquake prediction. *J Geophys Res*, 1975, **80**: 718−732

[133] Nur A. Dilatancy, pore fluids, and premonitory variations of t_s/t_p travel times. *Bull Seism Soc Amer*, 1972, **62**: 1217−1232

[134] Scholz C H, Sykes L R, Aggarwal Y P. Earthquake prediction: a physical basis. *Science*, 1973, **181**: 803−810

[135] Mjachkin V, Brace W, Sobolev G, et al. Two models of earthquake forerunners. *Pure Appl Geophys*, 1975, **113**: 169−181

[136] Press F. Earthquake prediction. *Sci Am*, 1975, **232**(5): 134−137

[137] Press F. Heicheng and Los Angeles: a tale of two cities. *Eos Trans Amer Geophys Union*, 1976, **57**: 435−436

[138] Whitcomb J H, Garmany J D, Anderson D L. Earthquake prediction: variation of seismic velocities before the San [Fernando] earthquake. *Science*, 1973, **180**: 632−635

[139] Allen C R, Helmberger D V. Search for temporal changes in seismic velocities using large explosions in southern California. In: Kovach R L, Nur A, eds. *Proceedings of the Conference on Tectonic Problems of the San Andreas Fault System*. California: Stanford University Publications in Geological Science 13, 1973. 436−452

[140] Rice J R. Theory of precursory processes in the inception of earthquake rupture. *Gerlands Beitr Geophys*, 1979, **88**: 91−127

[141] Rice J R, Rudnicki J W. Earthquake precursory effects due to pore fluid stabilization of a weakening fault zone. *J Geophys Res*, 1979, **84**: 2177−2193

[142] 蒋锦昌, 陈德玉. 地震生物学概论. 北京: 地震出版社, 1993. 346

[143] Wyss M. Second round of earthquake of proposed earthquake precursors. *Pure Appl Geophys*, 1997, **149**: 3−16

[144] Wyss M, Booth D C. The IASPEI procedure for the evaluation of earthquake precursors. *Geophys J Int*, 1997, **131**: 423−424

[145] Wyss M, Dmowska R. *Earthquake Prediction—State of the Art*. Basel: Birkhäuser Verlag, 1997. 284

[146] 吴开统, 岳明生, 武宦英, 等. 海城地震序列的特征. 地球物理学报, 1976, (2): 95−109

[147] Bowman J R. Case 22: a seismicity precursor to a sequence of M_S 6.3—6.7 midplate earthquakes in Australia. *Pure Appl Geophys*, 1997, **149**: 61−78

[148] Matsu'ura R S. Precursory quiescence and recovery of aftershock activity before some large aftershocks. *Bull Earthq Res Inst, Univ Tokyo*, 1986, **61**: 1−65

[149] Wakita H, Nakamura Y, Sano Y. Short-term and intermediate-term geochemical precursors. *Pure Appl Geophys*, 1988, **126**: 267−278

[150] Wakita H, Nakamura Y, Sano Y. Short-term and intermediate-term geochemical precursors. In: Wyss M, ed. *Evaluation of Proposed Earthquake Precursors*. Washington DC: American Geophysical Union, 1991. 15-20

[151] Roeloffs E, Quilty E. Case 21: water level and strain changes preceding and following the August 4, 1985 Kettleman Hills, California. *Pure Appl Geophys*, 1997, **149**: 21-60

[152] Dieterich J H. Preseismic fault slip and earthquake prediction. *J Geophys Res*, 1978, **83**: 3940-3948

[153] Lindh A G. The nature of earthquake prediction. *Seism Res Lett*, 2003, **74**: 723-735

[154] Bakun W H, Aagaard B, Dost B, et al. Implications for prediction and hazard assessment from the 2004 Parkfield earthquake. *Nature*, 2005, **437**: 969-974, doi: 10.1038/nature04067

[155] Lindh A G. Success and failure at Parkfield. *Seism Res Lett*, 2005, **76**: 3-6

[156] Rong Y, Jackson D D, Kagan Y Y. Seismic gaps and earthquakes. *J Geophys Res*, 2003, doi: 10.1029/2002JB002334

[157] Stevenson D J. Mission to Earth's core—A modest proposal. *Nature*, 2003, **423**: 423-424

[158] Галицын Б Б. Работы по сейсмология. Б. Б. Галицын Избранные Труды, Том **2**, 1912. 231-465, Издательство Академии Наук СССР, Москва, 1960

[159] Саваренский Е Ф, Кирнос Д П. *Елементы Сейсмологии и Сейсмометриии*. Государственое Изд., 1955. 543

[160] 陈运泰, 刘瑞丰. 地震的震级. 地震地磁观测与研究, 2004, **25**(6): 1-12

[161] 赵纪东, 张志强. 地震能否预测. 见: 中国科学院, 主编. 2009 科学发展报告. 北京: 科学出版社, 2009. 195-201

[162] Sornette D, Sammis G G. Complex critical exponents from renormalization field theory of earthquakes: implications for earthquake prediction. *J Phys Int*, 1995, **5**: 607-619

[163] Sornette D, Knopoff L. The paradox of the expected time until the next earthquake. *Bull Seism Soc Amer*, 1997, **87**: 789-798

[164] Bak P, Tang C, Wiesenfeld K. Self-organized criticality: an explanation. *Phys Rev Lett*, 1987, **59**: 381-384

[165] Bak P, Tang C, Wiesenfeld K. Self-organized criticality. *Phys Rev*, 1988, **A38**: 364-374

[166] Bak P, Tang C. Earthquakes as a self-organized critical phenomenon. *J Geophys Res*, 1989, **94**: 15635-15637

[167] Bak P. *How Nature Works: The Science of Self-Organized Criticality*. New York: Springer-Verlag, 1996. 226

[168] Knopoff L. Earthquake prediction is difficult but not impossible. *Nature*, March 11, 1999. http://www.nature.com/nature/debates/earthquake/equake_6.html (Debates)

[169] Scholz C H. *The Mechanics of Earthquakes and Faulting*. Cambridge: Cambridge University Press, 1990. 439

[170] Burridge B, Knopoff L. Model and theoretical seismicity. *Bull Seism Soc Amer*, 1967, **57**: 341-371

[171] Otsuka M. A chain-reaction-type source model as a tool to interpret the magnitude-frequency relation of earthquakes. *J Phys Earth*, 1972, **20**: 35-45

[172] Otsuka M. A simulation of earthquake recurrence. *Phys Earth Planet Inter*, 1972, **6**: 311-315

[173] Ito K, Matsuzaki M. Earthquakes as self-organized critical phenomena. *J Geophys Res*, 1990, **95**: 6853-6860

[174] Brune J N. Implications of earthquake triggering and rupture propagation for earthquake prediction based on premonitory phenomena. *J Geophys Res*, 1979, **84**: 2195-2198

[175] Kittl P, Diaz G, Martinez V. Principles and the uncertainty principles of the probabilistic strength of materials and their applications to seismology. *ASME Appl Mech Rev*, 1993, **46**: S327-S333

[176] Geller R J. Shake-up for earthquake prediction. *Nature*, 1991, **352**: 275-276

[177] Geller R J. Unpredictable earthquakes. *Nature*, 1991, **353**: 612

[178] Geller R J, Jackson D D, Kagan Y Y, et al. Earthquake cannot be predicted. *Science*, 1997, **275**: 1616-1617

[179] Geller R J, Jackson D D, Kagan Y Y, et al. Cannot earthquake be predicted. *Science*, 1997, **278**: 488-490

[180] Hamada K. Unpredictable earthquakes? *Nature*, 1991, **353**: 611-612

[181] Turcotte D L. Earthquake prediction. *Ann Rev Earth Planet Sci*, 1991, **19**: 263-281

[182] Knopoff L. A selective phenomenology of the seismicity of Southern California. In: Knopoff L, Aki K, Allen C R, eds. *Earthquake Prediction: The Scientific Challenge, Colloquium Proceedings*. Proc Nat Acad Sci USA, 1996, **93**: 3756-3763

[183] Knopoff L. The organization of seismicity on fault networks. In: Knopoff L, Aki K, Allen C R, et al, eds. *Earthquake Prediction*: *The Scientific Challenge*, *Colloquium Proceedings. Proc Nat Acad Sci USA*, 1996, **93**: 3830-3837

[184] Knopoff L, Aki K, Allen C R, et al, eds. *Earthquake Prediction*: *The Scientific Challenge*, *Colloquium Proceedings. Proc Nat Acad Sci USA*, 1996, **93**: 3719-3837

[185] Kerr R A. Seismology: continuing Indonesian quakes putting seismologists on edge. *Science*, 2007, **317**: 1660-1661

[186] Kerr R A. After the quake, in search of the science—or even a good prediction. *Science*, 2009, **324**: 322

[187] Johnson B F. Earthquake prediction: gone and back again. *Earth*, 2009, **4**: 30-33

[188] Linde A T, Sacks I S. Slow earthquakes and great earthquakes along the Nankai trough. *Earth Planet Sci Lett*, 2002, **203**: 265-275

[189] Thanassoulas C. *Short-term Earthquake Prediction*. Greece: H. Dounias & Co., 2007. 374

[190] Crampin S, Gao Y, Peacock S. Stress-forecasting (not predicting) earthquakes: a paradigm shift. 2008, **36**: 427-430

[191] Beroza G C, Ide S. Deep tremors and slow quakes. *Science*, 2009, **324**: 1025-1026, doi: 10.1126/science1171231

[192] Bromirski P D. Earth vibrations. *Science*, 2009, **324**: 1026-1028

[193] Liu C C, Linde A T, Sacks I S. Slow earthquakes triggered by typhoons. *Nature*, 2009, **459**: 833-836

[194] Borghi A, Aoudia A, Riva R E M, et al. GPS monitoring and earthquake prediction: a success story towards a useful integration. *Tectonophysics*, 2009, **465**: 177-189

[195] Aki K. Possibilities of seismology in the 1980s. *Bull Seism Soc Amer*, 1980, **70**: 1969-1976

[196] Aki K. Synthesis of earthquake science information and its public transfer: a history of the Southern California Earthquake Center. In: Lee W H K, Kanamori H, Jennings P, et al, eds. *International Handbook of Earthquake and Engineering Seismology*, Part **A**. San Diego: Academic Press, 2002. 39-49

[197] 石橋克彦. 大地動乱の時代——地震学者は警告する（日文）. 岩波新書, 1994. 235

[198] 刘瑞丰, 吴忠良, 阴朝民, 等. 中国地震台网数字化改造的进展. 地震学报, 2003, **25**(5): 535-540

[199] Rikitake T. Earthquake prediction. *Bull Seism Soc Amer*, 1975, **65**: 1133-1162

[200] Rikitake T. Earthquake precursors in Japan: precursor time and detectability. *Tectonophysics*, **1987**, 136: 265-282

[201] Rikitake T. Earthquake prediction: an empirical approach. *Tectonophysics*, **1988**, 148: 195-210

[202] Simpson D, Richards P. *Earthquake Prediction—An International Review*, Maurice Ewing Series **4**. Washington DC: American Geophysical Union, 1981. 680

[203] Kisslinger C, Rikitake T. *Practical Approaches to Earthquake Prediction and Warning*. D Reidel Pub. Co. /Dordrecht/Boston, 1985. 685

[204] Mazzella A, Morrison H F. Electrical resistivity variations associated with earthquakes on the San Andreas fault. *Science*, 1974, **185**: 855-857

[205] Mizutane H. Earthquakes and electromagnetic phenomena. In: Asada A, ed. *Earthquake Prediction Techniques*. Tokyo: University of Tokyo Press, 1982. 217-246

[206] Zhao Y L, Qian F Y, Xu T C. The relationship between resistivity variation and strain in a load-bearing rock-soil layer. *Acta Seismol Sin-Engl Ed*, 1991, **4**: 127-137

[207] King C Y. Radon monitoring for earthquake prediction in China. *Earthq Predict Res*, 1985, **3**: 47-68

[208] King C Y. Gas geochemistry applied to earthquake prediction: an overview. *J Geophys Res*, 1986, **91**: 12269-12281

[209] 车用太, 鱼金子. 地震地下流体学. 北京: 气象出版社, 2005. 498

[210] Sacks I S, Suyehiro S, Linde A T, et al. Slow earthquakes and stress redistribution. *Nature*, 1978, **275**: 599-602

[211] Sacks I S, Suyehiro S, Linde A T, et al. Stress redistribution and slow earthquakes. Tectonophysics, 1982, **81**: 311-318

[212] Linde A T, Gladwin M T, Johnston M J S, et al. A slow earthquake sequence on the San Andreas fault. *Nature*, 1996, **383**: 65-68

[213] Crampin S. The basis for earthquake prediction. *Geophys J R Astr Soc*, 1987, **91**: 331-347

[214] Crampin S. Developing stress-monitoring sites using cross-hole seismology to stress-forecast the times and magnitudes of future

earthquakes. *Tectonophysics*, 2001, **338**: 233-245
[215] Crampin S, Evans R, Atkinson B K. Earthquake prediction: a new basis. *Geophys J R Astr Soc*, 1984, **76**: 147-156
[216] Crampin S, Volti T, Stefánsson R. A successfully stress-forecast earthquake. *Geophys J Int*, 1999, **138**: F1-F5
[217] Crampin S, Zatsepin S V, Browitt C W A, et al. GEMS: the opportunity for forecasting all damaging earthquakes worldwide. Proc Evison Symp, *Pure Appl Geophys*, 2009, in press
[218] Gao Y, Crampin S. Shear-wave splitting and earthquake forecasting. Terra Nova, 2008, **20**(6): 440-448, doi: 10.1111/j.1365-3121.2008.00836.x
[219] Gao Y, Wang P, Zheng S, et al. Temporal changes in shear-wave splitting at an isolated swarm of small earthquakes in 1992 near Dongfang, Hainan Island Southern China. *Geophys J Int*, 1998, **135**: 102-112
[220] Wu J, Crampin S, Gao Y, et al. Smaller source earthquakes and improved measuring techniques allow the largest earthquakes in Iceland to be stress-forecast (with hindsight). *Geophys J Int*, 2006, **166**: 1293-1298
[221] Wu J, Gao Y, Chen Y T. Shear-wave splitting in the crust beneath the southeast Capital area of North China. *J Seismol*, 2009, **13**(2): 277-286
[222] Varnes D J. Predicting earthquakes by analyzing accelerating precursory seismic activity. *Pure Appl Geophys*, 1989, **130**(4): 661-686
[223] Migna A, Bowman D D, King G C P. A mathematical formulation of accelerating moment release based on the stress accumulation model. *J Geophys Res*, 2007, **111**: B11304
[224] Ogata Y. Space-time point models for earthquake occurrences. *Ann Ins Statist Math*, 1998, **50**: 379-402
[225] Field E H. Overview of the working group for the development of regional earthquake likelihood models (RELM). *Seism Res Lett*, 2007, **78**: 7-16
[226] EarthScope Working Group. EarthScope Project Plan. *EarthScope: A New View into the Earth*, 2001.36

提取视震源时间函数的 PLD 方法及其对 2005 年克什米尔 $M_W 7.6$ 地震的应用[*]

张 勇[1,2] 许力生[1] 陈运泰[1,2]

1. 中国地震局地球物理研究所，北京 100081；2. 北京大学地球物理学系，北京 100871

摘要 通过重构用于确定视震源时间函数有效持续时间的判别函数，对提取视震源时间函数的 PLD 方法进行了改进；利用合成资料和实际资料，验证了改进后 PLD 方法的可行性和稳定性. 将 PLD 方法应用于 2005 年克什米尔 $M_W 7.6$ 地震及其 11 个余震的 1887 条记录，在 84 个台站处获得了这次地震的视震源时间函数. 分别平均从不同台站的 P 波、S 波、Rayleigh 波和 Love 波中得到的视震源时间函数，获取了主震的平均视震源时间函数. 对视震源时间函数的分析表明，2005 年克什米尔 $M_W 7.6$ 地震的持续时间大约为 25s，这是一次"急始型"地震，总体上表现为圆盘形破裂. 但有迹象表明，破裂在初期有向西北方向发展的单侧传播趋势.

1 引言

视震源时间函数（ASTF Apparent Source Time Function）是描述地震震源的重要参量之一，它代表地震的能量或地震矩在地震过程中随时间的变化过程[1,2]. 不同的地震具有不同的视震源时间函数，不同类型的地震具有不同特征的视震源时间函数. 因此，根据视震源时间函数可以估计地震的大小和震源的尺度[3,4]，也可以根据视震源时间函数的特征鉴别天然地震与其他类型的地震[5].

利用观测资料提取视震源时间函数的过程就是从观测资料中消除震源与台站之间的介质响应的过程. 震源与台站之间的介质响应通常被称为格林函数. 格林函数有两种，一种是基于介质模型计算得到的理论格林函数，另一种是基于实际观测获得的小震记录——经验格林函数[5~10]. 由于利用经验格林函数获得的视震源时间函数是相对于作为"格林函数"的小震而言的，所以，通常又把利用经验格林函数获得的视震源时间函数称作视相对震源时间函数（ARSTF Apparent Relative Source Time Function）.

提取视震源时间函数的具体方法很多. 最基本也是最常见的方法是频率域的水平线方法[6,11~12]，其优点是操作简单、效率高，但缺点是解的稳定性依赖于给定的水平线大小，而且边瓣效应无法避免. 与之相近的还有光滑频谱方法[13~15]和正则化方法[7]. 提取视震源时间函数的另外一类方法是时间域的矩阵线性反演方法[16,17]和非线性反演方法[18]. 这类方法的优点是容易添加约束，其缺点是计算效率较低，需要的计算机内存较大. Bertero[19]提出的 PLD（Projected Landweber Deconvolution）方法，吸纳了上述两类方法的优点，很大程度上消除了它们的缺点. 有研究者将这种方法应用于视震源时间函数的提取[20~23]. 在使用 PLD 方法估计视震源时间函数的有效长度时，需要一个判别函数. 在这方面 Lanza[20]的判别函数考虑了观测资料与合成资料之间的相对误差，Fischer[4]考虑了二者之间的绝对误差和相关系数. 前者没有考虑误差的相位信息，后者绝对误差不便于判别结果的优劣. 在本文中，我们设计了新的判别函数，既考虑了相位信息又考虑了幅度差别、同时将判别函数取值定义在 0 与 1 之间. 新的判别函数更有利于形象地判别观测资料与合成资料之间的差别.

---

[*] 本文发表于《地球物理学报》，2009 年，第 **52** 卷，第 3 期，672–680.

2005年10月8日在巴基斯坦克什米尔地区发生的M_W7.6地震的震源性质研究广受关注,但利用不同的方法和从不同的资料得到的结果却不尽相同. 基于波形反演方法的远场体波资料分析表明,最大滑动区域位于震中西北部,其最大滑动位移约6.5m[24];基于滑动块模型方法的远场体波分析表明,最大滑动区域位于震中东南,其最大滑动量为12m[25];同时还有结果表明,在震中的东南和西北同时存在着滑动较大的区域,且最大滑动位移均超过10m[26];对ASTER图像和远震资料的综合分析表明,最大滑动区域位于震中西北,整个破裂持续了25s,断层上最大滑动量和平均滑动量分别为7m和4m[27];对ENVISAT SAR图像的分析表明,断层长度约为80km,最大滑动区域集中在浅于10km的地壳内[28];通过地震现场考察研究,Kaneda等[29]发现此次地震的断层长度为70km,垂直方向的最大滑动量和平均滑动量分别为7m和3m.

为正确认识这次地震的震源复杂性,我们利用改进的PLD方法,以11次强余震作为经验格林函数,提取这次地震的视震源时间函数,并通过反演视震源时间函数得到这次地震的时空破裂过程. 关于这次地震的破裂过程详见文献[30]. 本文将重点介绍PLD方法及其对提取视震源时间函数的应用.

2 提取震源时间函数的PLD方法

在远场情况下,当小震与大震具有相同的震中位置和相类似的震源机制时,作为经验格林函数的小震记录$U'(t)$与大震记录$U(t)$可分别表示为[10]

$$\begin{cases} U(t) = Ms(t) * G(t) * I(t) \\ U'(t) = M's'(t) * G(t) * I(t) \end{cases}, \quad (1)$$

其中M和M'分别表示大震和小震的最大地震矩;$s(t)$和$s'(t)$分别为大震和小震的幅度归一化的视震源时间函数;$G(t)$为表示介质响应的格林函数;$I(t)$为台站仪器响应;"$*$"表示时间域的褶积.

与大地震相比,小地震震源持续时间较短,因此,式(1)中的$s'(t)$可以用δ函数近似. 此时,作为经验格林函数的小震记录$U'(t)$可以改写成$G_E(t)$,则式(1)变为

$$\begin{cases} \left[\dfrac{M}{M'}s(t)\right] * G_E(t) = U(t) \\ \left[\dfrac{M}{M'}s(\omega)\right] G_E(\omega) = U(\omega) \end{cases}. \quad (2)$$

(2)式分别为时间域和频率域的方程. 由此可见,通过大、小震记录在时间域里的反褶积或在频率域里的谱除法,不但可以得到大震的视震源时间函数或视震源时间函数谱,还可以得到大、小震地震矩的相对大小.

求解式(2)的最直接的方法就是频率域的水平线方法[6,11,12]和时间域线性[16,17]或非线性[18]的反演方法. 但是,这两类方法都有各自的缺点. 为了克服这些缺点,Bertero[20]首先引入PLD方法,这是一种可以在频率域进行的迭代方法,它在约束情况下的收敛性已为Eicke[31]所证明. 因此,PLD方法似乎是目前提取视震源时间函数最理想的方法.

式(2)的Landweber表示式可简写如下:

$$f_{n+1}(t) = f_n(t) + \tau G_E^T(t) * [U(t) - G_E(t) * f_n(t)], \quad (3)$$

式中$f_n(t) = [(M/M')s(t)]_n$;τ是松弛因子,为了保证迭代的收敛性,其取值范围为$\tau \in (0, 2/|G_{max}(\omega)|^2)$(常用的取值为$1/|G_{max}(\omega)|^2$);$G_E^T(t) = G_E(-t)$.

式(3)在频率域内的形式如下:

$$f_{n+1}(\omega) = f_n(\omega) + \tau G_E^*(\omega)[U(\omega) - G_E(\omega) f_n(\omega)], \quad (4)$$

其中$G_E^*(\omega)$是$G_E(\omega)$的复共轭.

由式(3)和式(4)可以看出,Landweber迭代可以在时间域里进行,也可以在频率域里进行. 但是,在时间域内的迭代需要构建矩阵,占用较多的内存,一般不被采用.

针对提取视震源时间函数的实际问题,我们还需要加上若干约束条件:视震源时间函数的因果性约束、单边性或非负约束以及持续时间有限性约束[19]. Vallée[21,22]曾经在研究中加入了各个台站标量地震矩相同的约束,但我们发现由于不同台站的波形记录的噪声水平的差异,可能导致每个台站处得到的标量地震矩大小并不十分一致,因此这种约束会导致有效震源信息的损失或扭曲. 综合式(4),我们得到如下迭代方程:

$$\begin{cases} g_{n+1}(\omega) = f_n(\omega) + \tau G_E^*(\omega)[U(\omega) - G_E(\omega)f_n(\omega)] \\ f_{n+1}(t) = P_C[g_{n+1}(t)] \\ \qquad = \begin{cases} g_{n+1}(t) & \text{当 } 0<t<T \text{ 和 } g_{n+1}(t)>0 \\ 0 & \text{其他} \end{cases} \end{cases} \quad (5)$$

式中"P_C"表示映射(Project),实为约束之意. 尽管频率域的迭代要比时间域的迭代省时,但是在待处理的资料比较多的情况下,所用的计算时间仍会较长. 所以,我们把上述方法进行进一步的拓展,通过构建频率域矩阵的技术途径,达到同时迭代处理多道记录的目的.

从理论上讲,在没有约束和误差的情况下,PLD 经过迭代总能得到精确解 $U(\omega)/G_E(\omega)$. 但是由于误差的存在,这个精确解并不是符合实际情况的最佳解. 理论数值试验表明[19],在添加约束的情况下,在经过有限次迭代之后,可以得到接近真实且能解释观测数据的最佳解. 虽然,我们无法从理论上回答这种真实性与迭代次数的关系,但是,根据对资料的解释情况,可以做一个大致的估计. Bertero[19]在对实际地震数据进行试验之后,认为 100 次迭代就可以得到很好的结果. 不过,也有研究者采用 200 次迭代[19]. 为保证不丢失必要的震源信息,在我们的研究中也采用 200 次迭代.

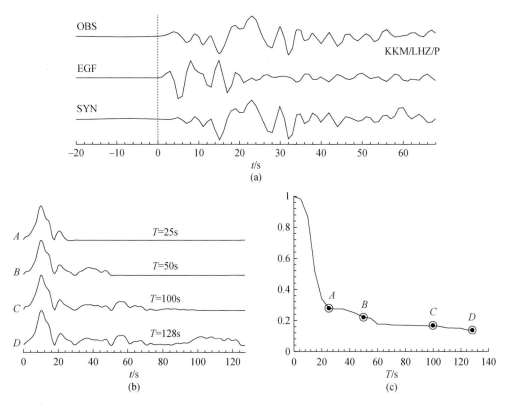

图 1 (a) 主震观测波形、作为经验格林函数的余震波形和用视震源时间函数 A 计算的合成波形;
(b) 具有不同有效持续时间的视震源时间函数; (c) 不同的有效持续时间对应的误差曲线

Fig. 1 (a) Observed waveform of the main shock, the observed waveform of the aftershock as empirical-Green's-function event and the synthetic waveforms calculated using the ASTF A;
(b) ASTFs with different efficient duration times; (c) Error curve for different efficient duration times

需要特别指出的是，由于在迭代过程中不易添加持续时间 T 的约束，所以，关于视震源时间函数有限持续时间的约束是在迭代过程完成后实现的. 此时，需要构造一个判别函数. 考虑到已有的判别函数的不足[4,20]，我们认为采用如下判别函数更好：

$$E = 1 - \frac{1}{2}\left(\frac{\sum_{t=1}^{N} U(t)U^{\text{syn}}(t)}{\sqrt{\sum_{t=1}^{N} U(t)^2 \sum_{t=1}^{N} U^{\text{syn}}(t)^2}} + 1\right) \times \left(1 - \frac{\sqrt{\sum_{t=1}^{N}(U(t)-U^{\text{syn}}(t))^2}}{\sqrt{\sum_{t=1}^{N}(|U(t)|+|U^{\text{syn}}(t)|)^2}}\right). \quad (6)$$

式中右端最后一项第一个括号内表示观测波形与合成波形的相关系数加1，这反映了两列波形的相似程度；第二个括号内的因子表示两列波形的幅度相对误差. 即式（6）综合考虑了波形之间差别的相位信息和幅度信息，是一个更为综合的量度，且其值介于 0 和 1 之间，便于判别结果的优劣. 为方便起见，本文中我们也称 E 为相对误差，但需要注意它已经包含了相位信息，比一般的相对误差要求更严格一些.

在具体确定视震源时间函数的有效持续时间时，按照时间逆序计算判别函数值. 我们首先给定最大的震源持续时间，然后逐步减少震源持续时间长度，并计算相应的判别函数值. 当发现函数值突然增大时，即确定当前给定的长度为最佳持续时间长度. 如图 1 所示，利用图 1a 中的观测资料和格林函数得到图 1b 的视震源时间函数 D，其持续时间为 128s. 用这个视震源时间函数得到的判别函数值为图 1c 中的 D. 依此类推，用图 1b 中的视震源时间函数 C，B 和 A 得到的判别函数值分别为图 1c 中的 C，B 和 A. 我们可以看到，视震源时间函数 A 所对应的判别函数值突然增高，由此断定，视震源时间函数 A 的持续时间长度是最佳长度. 当然，震源函数 A 便是最佳视震源时间函数.

3 PLD 方法的有效性及稳定性检验

3.1 基于合成资料的检验

为不失一般性，我们考虑一个包含噪声的褶积关系：

$$\begin{cases} U(t) = G_E(t) * s(t) + n(t) \\ n(t) = \alpha \max[G_E(t) * s(t)]N(0,1) \end{cases} \quad (7)$$

式中，α 为噪声水平；$n(t)$ 为表示噪声的一定大小的高斯误差函数；$N(0,1)$ 表示一个均值为 0、方差为 1 的高斯随机序列. 我们预先设定视震源时间函数 $s(t)$，与格林函数 $G_E(t)$ 相褶积得到理想的观测记录 $U(t)$. 然后，我们分别对观测记录加上不同大小的误差 $n(t)$，在不同的噪声水平下用水平线方法与 PLD 方法求解视震源时间函数，从而考察两种方法的特点.

如图 2 所示，我们预先设定理论的视震源时间函数 $s(t)$ 为一个长度 30s 的半周期正弦函数，同样采用 2005 年克什米尔 $M_W 7.6$ 地震的一次余震事件（发震时刻：2005-10-08 10:46:30.4，$M_W 6.4$）在 KKM 台站垂直向 P 波波形作为经验格林函数 $G_E(t)$，二者相褶积得到理想的观测记录. 然后，我们分别对观测记录加上不同水平的噪声，在不同的噪声水平下用两种方法求解视震源时间函数.

由图 3 可以看出，在相同噪声水平下，用 PLD 方法得到的视震源时间函数总是比水平线方法得到的结果更接近实际，且通过 PLD 方法搜索得到的视震源时间函数的长度也非常接近真实长度. 当然，随着噪声水平的提高，从资料中得到的结果逐渐偏离实际的结果，噪声水平越高，偏离就越大. 而且，可以明显地看出，随着噪声的增大，用水平线方法得到的结果中，主瓣的高度逐渐降低，主瓣附近的负值边瓣也逐渐加大；用 PLD 方法得到的结果中则没有这种虚假的边瓣. 由上可见，在 PLD 方法中加入适当的约束条件，可以部分抑制由观测资料的误差所导致的虚假信息，保留震源破裂的有效信息，进而实现对视震源时间函数的"提纯".

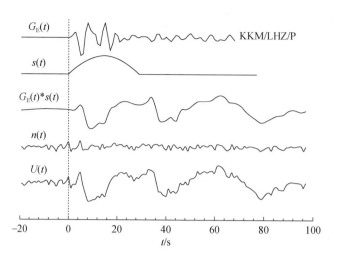

图 2　理论数值试验示意图

由上到下分别为作为经验格林函数 $G_E(t)$ 的 KKM 台站处的长周期高增益垂直向（LHZ）P 波记录、由半周期持续时间为 30s 的正弦窗定义的理论视震源时间函数 $s(t)$、经验格林函数与视震源时间函数的褶积结果，高斯误差序列 $n(t)$ 以及加上高斯误差序列的合成记录 $U(t)$

Fig. 2　Schematic diagram of theoretical numerical test

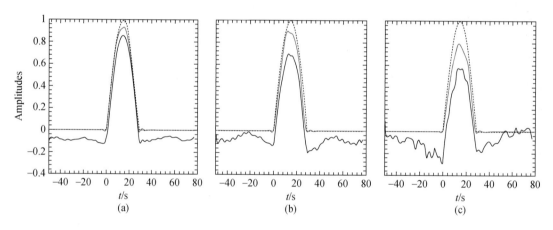

图 3　资料包含噪声时用水平线方法和 PLD 方法得到的结果的比较

细虚线表示给定的真实结果，灰色粗线表示 PLD 方法结果，黑色粗线表示水平线方法结果. (a), (b), (c) 分别对应的误差水平为：$\alpha=0.05$, $\alpha=0.1$ 和 $\alpha=0.3$

Fig. 3　Comparison of the results by the water-level method and the PLD method in case the data contains noise
Thin lines denote the defined result. Gray thick lines denote the results obtained by the PLD method. Black thick lines are the results obtained by the water-level method. (a), (b) and (c) correspond the error $\alpha=0.05$, $\alpha=0.1$ and $\alpha=0.3$, respectively

3.2　基于观测资料的检验

在试验中，我们仍然采用图 1 中的主震和余震波形作为观测波形和经验格林函数. 图 4 给出了由水平线方法和 PLD 方法分别得到的视震源时间函数. 我们可以看出，在主瓣处二者具有一定的相似性；但在具体细节上，用水平线方法得到的结果抖动幅度较大，尤其在主瓣附近，有着较大幅度的负值存在.

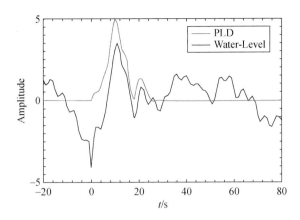

图 4 水平线方法得到的视震源时间函数（黑线）和 PLD 方法得到的视震源时间函数（灰线）之间的比较

Fig. 4 Comparison of the ASTFs obtained by the water-level method (black line) and the PLD method (gray line)

4 M_W7.6 主震与余震资料及其处理

我们从 IRIS 数据中心下载了主震和 11 个可以作为经验格林函数余震（见表 1）的远震波形数据，得到了主震和余震的 1887 条（三分向累计）全波形记录.

由于体波和面波的特征周期明显不同，所以，我们对 P 波和 S 波资料使用了频段为 0.01~0.2Hz 的带通滤波，而对于面波资料，则使用了频段为 0.005~0.1Hz 的带通滤波.

在提取视震源时间函数时，首先将主震记录和作为经验格林函数的余震记录用 PLD 方法，经过 200 次迭代，得到视震源时间函数的原始结果. 此时的原始结果往往包含有超出使用频带宽度的高频成分. 然后利用低通滤波器剔出那些超出给定频带的高频成分. 最后通过利用视震源时间函数和（余震）格林函数的褶积得到的合成地震图与（主震）观测地震图的比较，剔除品质差（相对误差大于 0.45）的结果，得到比较可靠的视震源时间函数.

表 1 克什米尔 M_W7.6 地震的主震（1）和余震序列（2~12）

Table 1 The main shock (1) and after shocks (2~12) of the Kashmir M_w7.6 earthquake sequence

序号	发震日期 年-月-日	发震时刻 时:分:秒	震中 (°N, °E)	震级 (M_S/M_W)	深度 /km	可用 台站数	节面 I 走向 /(°)	节面 I 倾向 /(°)	节面 I 滑动角 /(°)	节面 II 走向 /(°)	节面 II 倾向 /(°)	节面 II 滑动角 /(°)
1	2005-10-08	03:50:38.6	(34.43, 73.54)	7.6/7.6	10.00	84	334	40	123	114	57	65
2	2005-10-08	10:46:30.4	(34.69, 73.07)	6.3/6.4	16.20	79	328	39	107	127	53	77
3	2005-10-08	12:08:28.1	(34.57, 73.18)	5.5/5.6	10.00	40	321	31	108	121	60	80
4	2005-10-08	12:25:22.1	(34.79, 3.14)	5.9/5.7	20.30	54	321	53	122	96	47	55
5	2005-10-08	21:45:10.0	(34.68, 73.22)	5.5/5.4	10.00	30	338	33	121	122	62	71
6	2005-10-09	07:09:19.0	(34.56, 73.20)	5.5/5.3	10.00	37	315	29	104	119	62	82
7	2005-10-09	08:30:02.1	(34.64, 73.15)	5.7/5.7	10.00	69	344	37	121	127	59	69
8	2005-10-12	20:23:38.7	(34.87, 73.13)	5.6/5.3	10.00	40	315	42	97	126	49	84
9	2005-10-19	02:33:29.8	(34.82, 72.97)	5.8/5.6	10.00	63	303	31	105	106	60	81
10	2005-10-19	03:16:22.0	(34.81, 73.04)	5.6/5.4	10.00	44	308	44	103	110	47	78
11	2005-10-23	15:04:21.3	(34.88, 73.02)	6.0/5.4	10.00	61	307	32	110	105	60	78
12	2005-10-28	21:34:15.2	(34.73, 73.14)	5.5/5.2	10.00	28	357	38	132	129	63	63

注：发震时刻、震中位置、速报震级和震源深度皆来自于 IRIS 数据中心；矩震级、断层面参数来自于哈佛大学 CMT 解.

我们知道,同一个台站不但记录到了主震,也记录到了余震.所以,在同一个台站,利用主震和多个余震记录,可以得到多个视震源时间函数.理想情况下,利用不同余震记录在同一个台站得到的视震源时间函数是相同的.但是,实际上,如图 5 所示,得到的结果会由于路径和震源机制的微小差异以及噪声等原因而有所不同.为了在每个台站得到可信度较高的惟一的视震源时间函数,我们对利用不同余震记录得到的结果按照它们对应的相对误差进行加权平均,从而得到在该台站的视震源时间函数.图 5 显示了利用 GNI 台站的不同震相得到的视震源时间函数及其加权平均的结果.可以看到,尽管所使用的经验格林函数不同,但在同一台站处得到视震源时间函数结果都比较相似,而加权平均得到的结果则更加凸显了视震源时间函数的共性,抑制了某些结果中的"个性"变化.其他台站的视震源时间函数都是通过上述处理过程得到的.

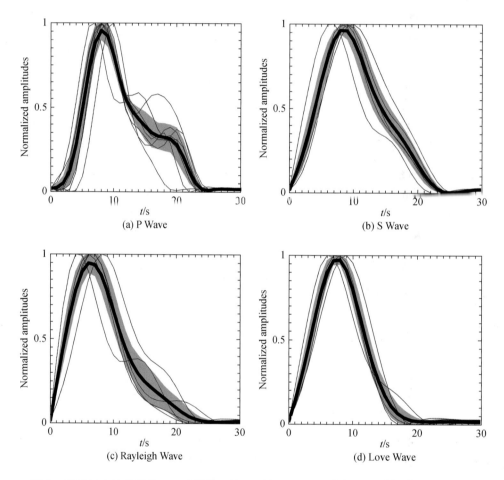

图 5 用 PLD 方法得到的 GNI 台站的 P 波、S 波、Rayleigh 波和 Love 波视震源时间函数
(深灰色细线)及其加权平均结果(黑色粗线)
浅灰色区域表示加权平均结果的可信范围

Fig. 5 The ASTFs from the P waves, S waves, Rayleigh waves and Love waves (dark gray lines) and the weighted-average results (black thick lines)
Light gray area denotes the confidence range of weighted-average results

5 2005 年克什米尔 M_W7.6 地震的视震源时间函数及其解释

按照上节描述的资料处理过程,我们分别处理了 P 波、S 波、Rayleigh 波和 Love 波震相资料,得到了

相应震相在不同台站的视震源时间函数（图6）. 从所得到的视震源时间函数的形态来看, 从同一个震相中得到的视震源时间函数在不同台站处的形态不同, 而且呈现出规律性的变化, 在东南方向观测到的视震源时间函数的变化要缓慢一些, 而西北方向观测到的视震源时间函数的峰值出现得早一些, 这说明破裂在初期有着向西北方向发展的趋势. 这种视震源时间函数随方位变化的现象是天然地震断层运动学和几何学的典型特征. 这一特征与人工爆破源的视震源时间函数结果截然不同.

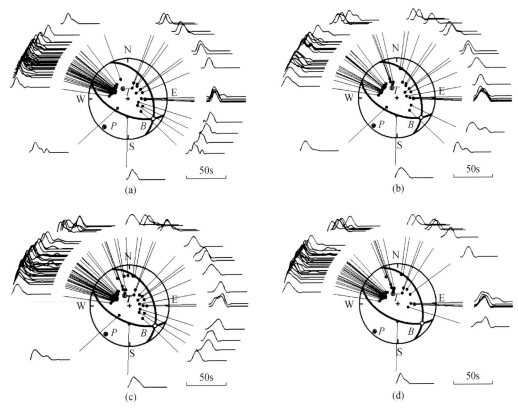

图6 用PLD方法从不同震相中得到的视震源时间函数

(a) P波; (b) S波; (c) Rayleigh波; (d) Love波

Fig. 6 The ASTFs of different phase obtained by the PLD method

对比从不同震相中得到的视震源时间函数（图6）, 我们可以认定, 此次地震不同震相的视震源时间函数的持续时间长度均在30s以内, 平均长度约为25s. 如图7所示, 从不同震相中得到的平均视震源时

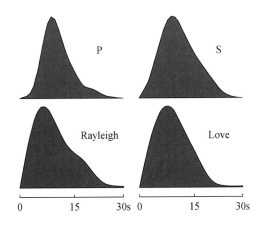

图7 P波、S波、Rayleigh波和Love波视震源时间函数平均结果

Fig. 7 The average ASTFs from the P wave, S wave, Rayleigh wave and Love wave

间函数形态大致相似,只不过细节略有不同.从 P 波视震源时间函数还可以看出,这次地震属于"急始型"地震,地震矩的释放速率在第 10s 左右达到极值,然后逐渐下降.这种形态,尤其是 S 波和 Rayleigh 波视震源时间函数的形态,与圆盘形断层的视震源时间函数非常相似,这意味着此次地震在总体上具有圆盘形破裂的特点.

6 结论与讨论

视震源时间函数的提取可以在频率域实现,也可以在时间域实现,不同的技术途径具有不同优点和缺点.用 PLD 方法提取视震源时间函数是一种新的技术途径,这种技术继承了上述两类技术的优点,克服了它们的缺点,是一种具有良好应用前景的方法.为了使这种方法在实际应用中具有更高的效率,也为了使结果更客观,我们改进了计算过程,并提出了新的持续时间有限性判别函数.数值试验和对实际资料的应用均表明,我们所使用的 PLD 方法及其计算过程是目前最好的.

2005 年 10 月 8 日在巴基斯坦克什米尔地区发生的 M_W7.6 地震伴有很多 5 级以上的强余震,其中,能够用作经验格林函数事件的余震就有 11 次.这为检验我们的 PLD 方法及其计算过程提供了良好的机遇.我们用 11 次余震在全球范围内 84 个台站记录的波形资料,分别从三个分向的 P 波、S 波、Rayleigh 波和 Love 波中提取视震源时间函数.对如此大量的资料的处理表明,这种方法在计算效率以及处理结果的稳定性方面都有良好的表现.

根据我们提取的视震源时间函数,2005 年 10 月 8 日在巴基斯坦发生的 M_W7.6 地震的持续时间大约为 25s,与 Avouac[27]得到的结果比较一致.这次地震属于"急始型"地震,总体特征上表现为圆盘形破裂,但有迹象表明,破裂在初期有向西北方向发展的单侧传播趋势,这一认识与已有的研究结果相印证[24,26,27].

需要说明的是,在构建和讨论主震的视震源时间函数时,我们只使用了垂直分向的 P 波、S 波和 Rayleigh 波以及 T 分向的 Love 波,尽管我们提取了所有分向上所有震相的视震源时间函数.这是因为,垂直分向的记录具有最高信噪比,其结果无疑是最可靠的.当然,也有人将不同分向上同一震相中提取的视震源时间函数叠加平均[32],但我们认为,这种处理会损伤好的结果.

参考文献(References)

[1] 许力生,陈运泰. 视震源时间函数与震源破裂过程. 地震地磁观测与研究,2002,**23**(6):1-8
Xu L S, Chen Y T. Apparent source time function and source rupture process. *Seismological and Geomagnetic Observation and Research* (in Chinese),2002,**23**(6):1-8

[2] 陈运泰,林邦慧,李兴才等. 巧家,石棉的小震震源参数的测定及其地震危险性的估计. 地球物理学报,1976,**19**(3):206-233
Chen Y T, Lin B H, Li X C, et al. The determination of source parameters for small earthquakes in Qiaojia and Shimian and the estimation of potential earthquake danger (in Chinese). *Chinese J. Geophys.* (*Acta Geophysica Sinca*)(in Chinese),1976,**19**(3):206-233

[3] Kikuchi M, Kanamori H. Inversion of complex body waves. *Bull. Seism. Soc. Amer.*,1982,**72**(2):491-506

[4] Fischer T. Modeling of multiple events using empirical Green's functions: method, application to swarm earthquakes and implications for their rupture propagation. *Geophy. J. Int.* 2005,**163**:991-1005

[5] 何永峰,陈晓非. 利用经验格林函数识别地下核爆炸与天然地震. 中国科学(D 辑),2006,**36**(2):177-181
He Y F, Chen X F. Identification of underground nuclear explosion and natural earthquake using empirical Green's function (in Chinese). *Science in China* (Ser. D),2006,**36**(2):177-181

[6] Mori J, Hartzell S H. Source inversion of the 1988 upland, California, earthquake: determination of a fault plane for a small event. *Bull. Seism. Soc. Amer.*, 1990, **80**: 507−517

[7] Chen Y T, Zhou J Y, Ni J C. Inversion of near-source-broadband accelerograms for the earthquake source-time function. *Tectonophysics*, 1991, **197**: 89−98

[8] Hartzell S H. Earthquake aftershocks as Green's functions. *Geophys. Res. Lett.*, 1978, **5**(1): 1−4

[9] Ammon C J, Lay T, Velasco A A, et al. Routine estimation of earthquake source complexity: the 18 October 1992 Colombian Earthquake. *Bull. Seism. Soc. Amer.*, 1994, **84**(4): 1266−1271

[10] Chen Y T, Xu L S. A time-domain inversion technique for the tempo-spatial distribution of slip on a finite fault plane with applications to recent large earthquakes in the Tibetan Plateau. *Geophys. J. Int.*, 2000, **143**(2): 407−416

[11] 许力生, 陈运泰, 高孟潭. 1997年中国西藏玛尼 M_S7.9 地震的时空破裂过程. 地震学报, 1999, **21**(5): 449−459
    Xu L S, Chen Y T, Gao M T. Earthquake rupture process of the 1997 Mani, in Tibet, China. *Chinese J. Geophys.* (in Chinese), 1999, **21**(5): 449−459

[12] 许力生, 陈运泰, 高孟潭. 2001年1月26日印度古杰拉特 M_S=7.8 地震时空破裂过程. 地震学报, 2002, **24**(5): 447−461
    Xu L S, Chen Y T, Gao M T. Spatial and temporal rupture process of the January 26, 2001, Gujarat, India, M_S=7.8 earthquake. *Acta Seismologica Sinica* (in Chinese), 2002, **24**(5): 447−461

[13] Radulian M, Popa M. Scaling of source parameters for Vrancea (Romania) intermediate depth earthquake. *Tectonophysics*, 1996, **261**: 67−81

[14] Popescu E, Radulian M. Source Characteristics of the seismic sequences in the Eastern Carpathians for deep region (Romania). *Tectonophysics*, 2001, **338**: 325−337

[15] Venkataraman A, Rivera L, Kanamori H. Radiated Energy from the 16 October 1999 Hector Mine Earthquake: Regional and Teleseismic Estimates. *Bull. Seism. Soc. Amer.*, 2002, **92**(4): 1256−1265

[16] Gurrola H, Baker G E, Minster J B. Simultaneous time-domain deconvolution with application to the computation of receiver functions. *Geophys. J. Int.*, 1995, **120**: 537−543

[17] Kraeva N. Tikhonov's regularization for deconvolution in the empirical Green function method and vertical directivity effect. *Tectonophysics*, 2004, **383**: 29−44

[18] Courboulex F, Deichmann N, Gariel J C. Rupture complexity of a moderate intraplate earthquake in the Alps: the 1996 M5 Epagny-Annecy earthquake. *Geophys. J. Int.*, 1999, **139**: 152−160

[19] Bertero M, Bindi D, Boccacci P, et al. Application of the Projected Landweber Method to the estimation of the source time function in seismology. *Inverse Problems*, 1997, **13**: 465−486

[20] Lanza V, Spallarossa, D, Cattaneo M, et al. Source parameters of small events using constrained deconvolution with empirical Green's functions. *Geophys. J. Int.*, 1999, **137**: 651−662

[21] Vallée M, Bouchon M. Imaging coseismic rupture in far field by slip patches. *Geophys. J. Int.*, 2004, **156**: 615−630

[22] Vallée M. Stabilizing the empirical Green's function analysis: development of the Projected Landweber Method. *Bull. Seism. Soc. Amer.*, 2004, **94**(2): 394−409

[23] Vallée M. Rupture process of the giant Sumatra earthquake imaged by empirical Green's functions analysis. *Bull. Seism. Soc. Amer.*, 2007, **97**(1A): 103−114

[24] Ji C. Finite-fault slip model preliminary result for the 05/10/08 PAKISTAN earthquake. 2006, http://earthquake.usgs.gov/eqcenter/eqinthenews/2005/usdyae/finitefault/

[25] Vallée M. M_W=7.7 05/10/08 Pakistan earthquake, 2005. http://www-geoazur.unice.fr/SEISME/PAKISTAN081005/note1.html

[26] Parsons T, Yeats R S, Yagi Y, et al. Static stress change from the 8 October, 2005 M=7.6 Kashmir earthquake. *Geophysical Research Letters*, 2006, **33**: L06304, doi: 10.1029/2005GL025429

[27] Avouac J P, Ayoub F, Leprince S, et al. The 2005 M_W7.6 Kashmir earthquake: Sub-pixel correlation of ASTER images and seismic waveforms analysis. *Earth and Planetary Science Letters*, 2006, **249**: 514−528

[28] Pathier E, Fielding E J, Wright T J, et al. Displacement field and slip distribution of the 2005 Kashmir earthquake from SAR imagery. *Geophysical Research Letters*, 2006, **33**: L20310, doi: 10.1029/2006GL027193

[29] Kaneda H, Nakata T, Tsutsumi H, et al. Surface rupture of the 2005 Kashmir, Pakistan, earthquake and its active tectonic implication. *Bull. Seism. Soc. Amer.*, 2008, **98**(2): 521-527

[30] Zhang Y, Chen Y T, Xu L S. Rupture process of the 2005 southern Asian (Pakistan) M_W7.6 earthquake from long-period waveform data. *AOGS*, 2006, **PS** Volume: 13-22

[31] Eicke B. Iteration methods for convexly constrained ill-posed problems in Hilbert space. *Num. Funct. Anal. Opt.*, 1992, **12**: 423-429

[32] Dreger D S. Empirical Green's function study of the January 17, 1994 Northridge, California earthquake. *Geophys. Res. Lett.*, 1994, **21**: 2633-2636

PLD method for retrieving apparent source time function and its application to the 2005 Kashmir M_W7.6 earthquake

Zhang Yong[1,2], Xu Li-Sheng[1] and Chen Yun-Tai[1,2]

1. Institute of Geophysics, China Earthquake Administration, Beijing 100081, China;
2. Geophysics Department of Peking University, Beijing 100871, China

Abstract The PLD method for retrieval of apparent source time function (ASTF) is improved by reconstructing the criteria function to be used for determination of the effective duration time of the ASTF. Using the synthetic and observed data, the feasibility and stability of the improved PLD method are verified. By applying the PLD method to 1887 waveform recordings of the 2005 Kashmir M_W7.6 earthquake and its 11 aftershocks, the ASTFs of the main shock are obtained at 84 stations. By averaging the ASTFs from the P waves, S waves, Rayleigh waves and Love waves from different stations, the average ASTFs of the main shock are constructed. The analysis of the ASTFs indicates that the duration time of the 2005 Kashmir M_W7.6 earthquake is about 25 s, and it is a quickly-starting earthquake. In general feature, the STF appears to be like a STF of a circular source model, however, some details of the ASTFs suggest that the rupture at the beginning was unilateral extending toward NW direction.

利用阿拉斯加台阵资料分析 2008 年汶川大地震的破裂过程*

杜海林　许力生　陈运泰

中国地震局地球物理研究所，北京　100081

摘要　2008 年 5 月 12 日，四川省汶川县发生 $M_S 8.0$ 地震．我们利用美国阿拉斯加区域台网的部分宽频带地震台构成广义台阵，应用非平面波台阵技术——迁移叠加方法，获得了这次地震的高频（>0.1Hz）能量辐射源随时间和空间变化的图像．图像表明，这次地震破裂从震中开始向北东方向扩展约 300km，震源过程至少长达 90s，平均破裂速度为 3.4km/s；整个过程可分为两大阶段，前段持续时间 50s，破裂长度约 110km，破裂传播的平均速度为 2.2km/s，后段持续时间约 40s，长度约 190km，平均破裂速度为 4.8km/s．这意味着地震过程的后期似乎发生了超 S 波破裂，而且后段很可能为前段动态触发所致．

1 引言

2008 年 5 月 12 日 14：28（北京时间），在我国四川省汶川县发生了 $M_S 8.0$ 级大地震[1]．地震造成了巨大的人员伤亡和财产损失，成为自 1976 年唐山大地震以来发生于我国大陆的、人员伤亡和财产损失最严重的一次地震．

汶川大地震的震中（31.0°N，103.4°E）位于青藏高原东缘的龙门山断裂带．龙门山断裂带是一条 NE-SW 方向展布的巨大断裂带，由后山断裂（茂县—汶川断裂）、中央断裂（映秀—北川断裂）和前山断裂（彭县—灌县断裂）组成[2]．震后现场考察结果表明，汶川大地震很可能发生于中央断裂和前山断裂上，其地表破裂带长度分别为 240km 和 72km，宽度多为 21～45m，最大可达 70～100m[3-5]．

汶川大地震发生后，陈运泰等（http：//www.csi.ac.cn/sichuan/chenyuntai.pdf）通过多种分析手段，对这次地震的震源机制、破裂过程、同震位移场等进行了初步分析，并于地震翌日公布了相关分析报告．随后，张勇等[6]给出了利用全球地震台网（GSN）记录的长周期波形数据反演得出的这次地震的发震断层的机制、破裂长度、破裂持续时间以及破裂速度等更详细结果．王卫民等[7]利用远场宽频带 P 波和 SH 波波形记录及 37 个 GPS 观测值反演分析了这次地震的机制、破裂长度、破裂持续时间以及破裂速度等．这些结果在总体特征上具有很好的一致性，例如，汶川地震的破裂长度至少为 300km，其震源机制是以逆冲为主兼具右旋走滑分量等等，但在有些方面还存在不可忽视的差异，例如，破裂扩展方式、破裂持续时间和破裂速度等等．

我们在本研究中将利用另外一种技术手段——广义台阵技术，运用另外一套资料——美国阿拉斯加区域地震台网的宽频带记录，分析这次地震的高频（>0.1Hz）能量辐射源随时间和空间的变化特征，从另外一个视角，为全面深入认识此次地震的震源破裂过程提供更多的观测依据．

2 方法

设一组台站的分布相对于信号源满足台阵的条件，即各台站记录到的信号具有良好的相关性，那么台

---

* 本文发表于《地球物理学报》，2009 年，第 **52** 卷，第 2 期，372-378.

阵中的任一台站记录的信号 $x_n(t)$ 可以用参考台的信号 $s(t)$ 描述为[8]:

$$x_n(t) = s(t - \Delta t_n), \tag{1}$$

式中, Δt_n 表示任意一个台的记录信号相对于参考台信号的时间延迟. 在通常情况下, Δt_n 是未知的.

假设任意一个台站相对于参考台的到时差为 $\Delta t'_n$, 那么, 由 N 个台站构成的台阵的聚束为:

$$y(t) = \frac{1}{N} \sum_{n=1}^{N} x_n(t + \Delta t'_n), \tag{2}$$

将式 (1) 带入上式, 得

$$y(t) = \frac{1}{N} \sum_{n=1}^{N} s(t + \Delta t'_n - \Delta t_n), \tag{3}$$

那么, 台阵的输出能量为上式的平方对时间的积分:

$$E(t) = \int_{-\infty}^{\infty} y^2(t) \mathrm{d}t. \tag{4}$$

从式 (3) 和式 (4) 可以看出, 当 $\Delta t'_n = \Delta t_n$ 时, 台阵的输出能量最大. 因此, 我们可以通过寻求台阵的最大能量来确定各台相对于参考台的时间延迟.

在各台相对于参考台的时间延迟确定的情况下, 不难确定这些台接收的信号的走时, 也不难确定这些台相对于信号源的距离, 从而也就不难确定信号源的位置. 在实际工作中, 为了确定台阵最大能量输出, 可将目标区域划分成密集的网格[9], 采用 IASPEI 91 地球速度标准模型[10], 计算网格点到各个台站的走时差, 根据走时差将各个台站的记录延迟叠加, 然后计算相应的台阵聚束能量. 能量最大值所对应的点即为信号源位置.

对于一个具有有限尺度的大地震而言, 它的震源可以看成是多个点源的集合; 相应地, 有限尺度的震源产生的信号在台站上记录的地震波形可以看作是点源地震记录的延迟叠加. 反过来, 我们可以用滑动时间窗将台站的地震记录分成一些可视为点源记录的片段, 通过确定不同时间窗的信号得到的各个点源的位置来构建整个大地震的能量辐射源随时间和空间变化的图像.

Ishii 等[11]、Krüger 和 Ohrnberger[12] 应用类似的台阵技术成功地分析研究了 2004 年苏门答腊大地震的破裂过程. 本文作者应用上述方法成功地分析研究了 2004 年苏门答腊大地震和其他一些地震的破裂过程[13,14].

3 资料

美国阿拉斯加区域台网 (Alaska Regional Network, ARN) 由 29 个配备宽频带地震仪的台站构成, 其中有 24 个台记录到了 2008 年汶川 $M_S 8.0$ 大地震. 我们从 IRIS 数据中心下载了 ARN 的宽频带记录, 并根据台站的分布挑选了其中 17 个台站构成图 1a 所示的"广义台阵". ARN 台阵距离汶川大地震震中约 71°, 在这个距离上, PcP-P 大约 10s, PP-P 大约 160s, 即在前 160s 的时间窗内除 PcP 外没有其他震相的干扰. 而 PcP 震相频率较低、幅度较小. 换言之, 前 160s 的高频波形几乎是纯直达 P 波, 这对于分析地震的时空破裂过程非常有利. 另外, ARN 台阵位于汶川大地震震中的北东方向, 与龙门山断层走向大体一致. "地震多普勒效应"使得这一方向的台站记录的高频信号得到加强, 也有利于震源破裂过程的分析.

在原始记录中, 有一部分台站的记录漂移较大. 在使用资料之前, 我们首先对原始数据进行了去漂移处理, 得到如图 1b 所示的波形记录. 可以看出, 数据质量相当好.

由于 PP-P 大约为 160s, 所以, 我们只截取了前 160s 的记录. 从图 2a 可以看出, 0.03～1Hz 的信号是这些记录的主要成分. 根据 ARN 台阵的增益特性[15,9] (图 2b), 台阵大大抑制了大于 1Hz 的信号, 但

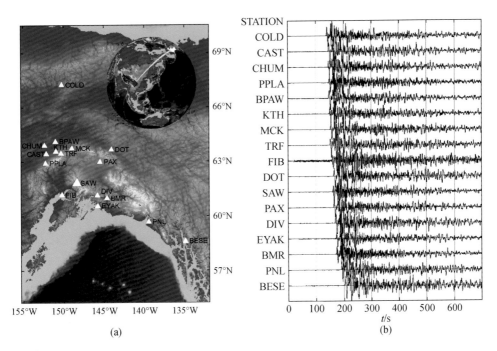

图 1 (a) 美国阿拉斯加宽频带区域数字地震台构成的广义台阵（ARN）（白色三角形）和 2008 年汶川 $M_S 8.0$ 地震的震中（红色菱形）；其中，较大的三角形为本研究使用的参考台；(b) 图 a 中台站记录的宽频带垂直分量

Fig. 1 (a) The generalized array configured with the broadband seismic stations (white triangle) of Alaska Regional Network (ARN), USA, and the epicenter (red diamond) of the 2008 Wenchuan $M_S 8.0$ earthquake. The larger triangle is specially for the reference station. (b) The vertical component waveform recordings from the stations shown in the Fig. 1a

对于小于 0.1Hz 的信号却没有很好的分辨能力. 能量辐射源的空间分辨率依赖于信号的频率, 信号的频率越高, 空间分辨率越高. 考虑到实际信号的幅频特性、台阵的增益特性以及信号的空间分辨能力, 我们对台阵各台的记录都使用了 0.1Hz 的高通滤波.

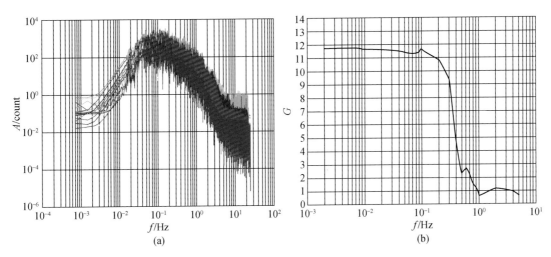

图 2 (a) ARN 台阵各台的垂直分向 160s 直达 P 波的幅频（A）特性；(b) ARN 台阵的增益（G）特性

Fig. 2 (a) Amplitude spectrum (A) of the 160s-long direct P-waves of all the ARN stations. (b) Frequency-gain property (G) of the ARN

4 高频能量辐射源的分析与结果

4.1 高频辐射源的确定

在进行滑动窗分析时,我们将滑动窗的宽度设置为15s.这个时间长度大于我们所用信号的最长周期,大体相当于一次6.0级地震的持续时间,相应的破裂尺度大约10km.为了保证辐射源在时间上的连续性,时间窗的移动步长设置为1s.为了保证辐射源在空间上的连续性,将目标区域(28°N~34°N,100°E~107°E)划分成0.05°×0.05°的网格,即空间搜索步长为0.05°.利用上述参数设置对台阵记录进行分析,得到了如图3a所示的145个能量辐射源的时空分布图像.

需要说明的是,由于台阵位置仅在震中位置的某一个方向,由台阵资料确定的能量辐射源的绝对位置总是存在系统偏差.此外,在这145个能量辐射源中,有些可能由于背景噪声或其他干扰而发生畸变,还有待进一步甄别.所以,图3a中能量辐射源的位置还有待校正,有些畸变结果还有待删除.

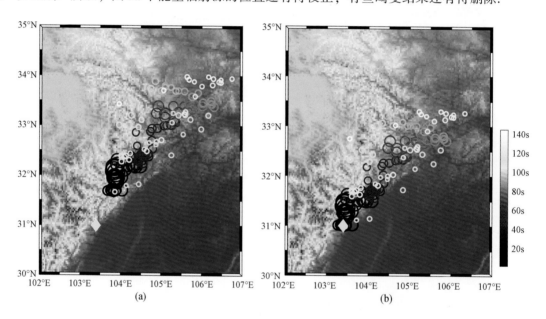

图3 从160s的记录中得到的高频能量辐射源的时空分布

(a)校正前的结果;(b)利用主震震中位置校正后的结果.圆圈的大小表示辐射能量的相对大小(圆圈的面积正比于聚束的能量),圆圈的颜色表示时间进程,色标表示时间

Fig. 3 Tempo-spatial distribution of the high-frequency energy radiation sources obtained from the 160s recordings (a) The results before calibration; (b) The results calibrated with the main shock epicenter. Sizes of the circles represent the relative sizes of the radiation energy (the circle area is proportional to the beam energy), and color shading denotes time progression

4.2 能量辐射源的位置校正

考虑到ARN台阵的孔径约1800km,远大于汶川大地震约300km的震源尺度,而且台阵位于断层的走向方向,所以,我们只采用了主震校正法[9]对图3a所示的能量辐射源的位置进行校正.简单说来,就是将直接利用台阵资料确定的所有能量辐射源的原始位置进行整体移动,使第一个能量辐射源的位置或者初始破裂点的位置与地方台网资料确定的主震的震中,即初始破裂点位置重合.图3b是利用主震校正法校正后的能量辐射源的位置分布.

4.3 畸变结果的删除与破裂持续时间

聚束信号与参考台信号的相关系数和振幅比常常作为信号源的评价指标,二者的乘积也被作为信号源是否有效或是否发生畸变的判别因子[9].但是,在有些情况下,这个评价指标似乎是不够的,因为,即使振幅较小的信号,其相关系数也可能较大,振幅比也可能较大.所以,作为一个重要改进,在本研究中又考虑了能量的衰减.当地震的能量辐射衰减到一个较低的水平时,我们可以认为地震已结束或结果发生了畸变,尽管聚束信号和参考台信号具有较高的相关性和振幅比.因此,在原来基础上本研究对能量辐射源有效性的判别因子作了进一步改进.我们定义判别因子

$$\gamma = \alpha\beta\kappa, \tag{5}$$

式中,α 为台阵聚束记录的子波与参考台站的记录对应子波之间的相关系数的归一化函数,β 为台阵聚束记录的子波与参考台站记录的对应子波之间的振幅比归一化函数,κ 为台阵聚束能量的归一化函数.

图 4a 是相关系数归一化函数、振幅比归一化函数和台阵聚束的能量归一化函数,图 4b 是能量辐射源有效性判别因子随时间的变化.从图中可以看出,判别因子值大约在 90s 时下降到 0.05 并保持在这个较低的水平.我们删去判别因子小于这个值的能量辐射源,得到如图 5 所示能量辐射源随时间和空间变化的图像.我们认为利用上述条件挑选出的能量辐射源是可以信赖的,且可以认定它们中最后一个源的辐射时间 90s 为地震停止的时间.换言之,2008 年汶川大地震的震源过程至少持续了大约 90s,这个结果与张勇等[6]根据全球台网的波形反演得到的结果 (90s) 一致,但小于王卫民等[7]得到的 110s 持续时间.

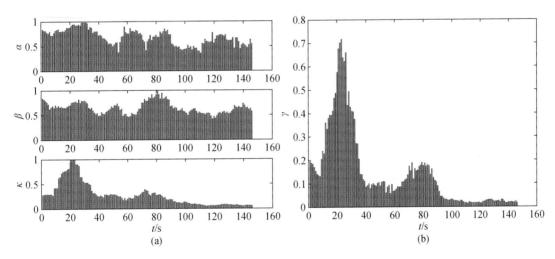

图 4 (a) 台阵聚束与参考台记录的相关系数 α,振幅比 β,聚束能量 κ 随时间的变化;(b) 判别因子 γ 随时间的变化

Fig. 4 (a) The normalized correlation coefficients between the beamed recordings and the reference-station recordings, ratio of amplitude of the beamed recording to the reference-station recording and beam energy of the array as a function of time; (b) Judgment factor as a function of time

将得以确认的高频能量辐射源叠加在具有背景构造的地图上 (图 5),我们可以清楚地看出,高频能量辐射源沿龙门山断裂带展布;地震伊始能量辐射来源于断层的西南端,随着时间的推移,能量辐射源逐渐向东北方向移动,最终停止于龙门山断裂带的东北端.

4.4 破裂尺度

为了估计汶川大地震的破裂尺度,我们计算了震源过程中任意时刻能量辐射源的位置离初始破裂点的距离.如图 6 所示,在 90s 的破裂时间内最远的能量辐射源距初始破裂点 304km.换句话说,根据高频能

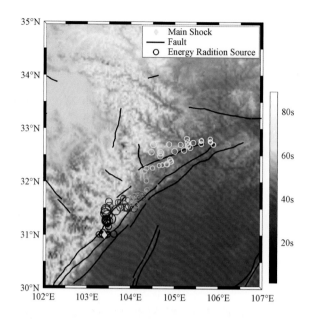

图 5 2008 年汶川 $M_S8.0$ 地震有效能量辐射源的时空分布与震中区的构造背景（色标为时间）

Fig. 5 Tempo-spatial distribution of the confirmed high-frequency energy radiation sources of the 2008 Wenchuan $M_S8.0$ earthquake and the tectonic setting in the epicenter area

量辐射源随时间和空间的变化特征估计，汶川大地震的破裂长度约 300km. 这个长度与通过震源破裂过程的反演得到的断层长度非常一致[6].

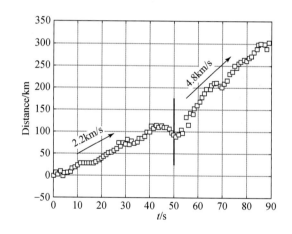

图 6 2008 年汶川 $M_S8.0$ 地震过程中高频能量辐射源相对于初始破裂点的距离随时间的变化以及根据能量辐射源位置的时空变化估计的分段平均破裂速度

Fig. 6 Varying distances of the high-frequency energy radiation sources from the initial point versus time during the 2008 Wenchuan $M_S8.0$ earthquake and estimation of the average rupture speeds over the two time intervals based on the rupture distance versus time

从图 5 还可以看出，高频能量辐射源的位置并不是严格分布在一条直线或曲线上，而是分布在具有一定宽度的条带内，并且在断层的西南端分布较宽，在东北端分布较窄，与野外调查的结果相符[3,4]. 这一特征可能是断层的宽度也即破裂面向断层倾斜方向（NW 方向）延展的反映. 不过，由于远场台阵数据有限的空间分辨度，无法据此准确估计断层的宽度和深度.

4.5 破裂速度

汶川大地震的能量辐射源的位置距初始破裂点的距离随时间大体是单调增加的（图6），呈现出明显的单侧破裂特征. 这一特点非常有利于分析地震过程中破裂速度随时间和空间的变化. 从图6可以看出，整个地震过程至少可以分为前50s和后40s两大阶段. 根据破裂长度和破裂时间计算，整个地震过程中的平均破裂速度为304km/90s=3.4km/s，其中前50s时间段内的平均破裂速度为110km/50s=2.2km/s，后40s内的平均破裂速度为190km/40s=4.8km/s. 根据这个结果，汶川大地震开始时破裂速度较小；但是，大约50s之后的破裂速度增大，超过了当地的S波速度（3.55km/s），并接近于当地沉积层的P波速度（4.88km/s）[16]. 这意味着汶川地震过程的后期似乎出现了超S波破裂，而且后段破裂很可能为前段破裂的动态触发所致.

5 结论与讨论

根据以上分析，2008年汶川$M_S8.0$地震的破裂持续时间至少90s，总破裂长度约300km，整个地震过程中的平均破裂速度为3.4km/s. 这个速度值与张勇等[6]利用波形反演估计的破裂速度（朝东北方向的平均破裂速度为3.4km/s）十分吻合. 整个过程至少可以分为两大阶段，前段持续时间约50s，破裂长度约110km，破裂传播的平均速度约2.2km/s，后段持续时间约40s，长度约190km，平均破裂速度约4.8km/s.

需要强调的是，在确定每一个能量辐射源的位置时，我们允许时间窗在2.5s的时间尺度内发生10次随机移动，10次结果的平均被确定为该能量辐射源的最终结果，这就大大避免了由于设定时间窗起始时间的人为因素造成的结果不稳定性.

本研究中所使用的美国阿拉斯加区域台网位于震中的北东方向，与破裂扩展的方向基本一致. 地震的多普勒效应使得在这个方向上记录的信号得到加强，这一点有益于求解的稳健性；同时，这种一致性使得沿断层任意两个破裂点的辐射信号由于震中距的变化在台站记录上引起的相位差最大，这一点有利于空间分辨. 不过，由于所用台阵近乎位于破裂的正前方，本研究得到的破裂持续时间应该是这次地震持续时间的下限，因此，关于震源过程后期出现超剪切破裂的认识也有待更全面和更深入的论证.

利用台阵技术追踪大地震高频能量辐射源的时空过程的技术是利用地震记录研究震源过程的重要途径之一. 与波形反演方法相比，这种技术无需计算格林函数，因此可以在较短的时间内完成资料分析、得到地震破裂过程的图像；也无需预先假定断层的长度，而破裂的长度和破裂持续时间可以通过资料的分析获得；再有，在利用波形资料反演确定震源过程时，常常需要预先假设破裂速度，而利用台阵技术确定破裂过程时，地震断层的破裂速度可以通过随时间变化的断层尺度和破裂时间加以确定，可以独立地给出地震的破裂速度. 不过，利用单一台阵技术确定的能量辐射源的绝对位置总是存在一定的偏差，同时，其空间分辨能力也不适于分辨断层宽度方向上信号位置的时空变化. 这些都是有待进一步研究解决的问题.

致谢 非常感谢两位评审专家为改善本稿质量提出的宝贵意见和建议！本研究所用数据来源于IRIS数据管理中心.

参考文献（References）

[1] 陈运泰. 汶川大地震的震级和断层长度. 科技导报, 2008, **26**(10): 26-27

Chen Y T. On the magnitude and the fault length of the great Wenchuan earthquake. *Science & Technology Review* (in Chinese), 2008, **26**(10): 26-27

[2] 陈国光, 计凤吉, 周荣军等. 龙门山断裂带第四纪活动性分段的初步研究. 地震地质, 2007, **29**(3): 657-673

Chen G G, Ji F J, Zhou R J, et al. Primary research of activity segmentation of Longmenshan fault zone since late-quaternary. Seismology and Geology (in Chinese), 2007, **29**(3): 657-673

[3] 张培震, 徐锡伟, 闻学泽等. 2008年汶川8.0级地震发震断裂的滑动速率、复发周期和构造成因. 地球物理学报, 2008, **51**(4): 1066-1073

Zhang P Z, Xu X W, Wen X Z, et al. Slip rates and recurrence intervals of the Longmen Shan active fault zone, and tectonic implications for the mechanism of the May 12 Wenchuan earthquake, 2008, Sichuan, China. *Chinese J. Geophys.* (in Chinese), 2008, **51**(4): 1066-1073

[4] 徐锡伟, 闻学泽, 叶建青等. 汶川M_S8.0地震地表破裂带及其发震构造. 地震地质, 2008, **30**(3): 597-629

Xu X W, Wen X Z, Ye J Q, et al. The M_S8.0 Wenchuan earthquake surface ruptures and its seismogenic structure. *Seismology and Geology* (in Chinese), 2008, **30**(3): 597-629

[5] 马保起, 张世民, 田勤俭. 汶川8.0级地震地表破裂带. 第四纪研究, 2008, **28**(4): 513-517

Ma B Q, Zhang S M, Tian Q J. The surface rupture of Wenchuan earthquake (M8.0). *Quaternary Sciences* (in Chinese), 2008, **28**(4): 513-517

[6] 张勇, 冯万鹏, 许力生等. 2008年汶川大地震的时空破裂过程. 中国科学: D辑, 2009, **52**(2): 145-154

Zhang Y, Feng W P, Xu L S, et al. Spatio-temporal rupture process of the 2008 great Wenchuan earthquake. *Science in China* (Series D), 2009, **52**: 145-154

[7] 王卫民, 赵连锋, 李娟等. 四川汶川8.0级地震震源过程. 地球物理学报, 2008, **51**(5): 1403-1410

Wang W M, Zhao L F, Li J, et al. Rupture process of the M_S8.0 Wenchuan earthquake of Sichuan, China. *Chinese J. Geophys.* (in Chinese), 2008, **51**(5): 1403-1410

[8] Rost S, Thomas C. Array seismology: methods and applications. Rev. Geophys., 2002, **40**(3): 1-27

[9] 杜海林. 2004年苏门答腊——安达曼大地震能量辐射源的时间域台阵技术分析[硕士论文]. 北京: 中国地震局地球物理研究所, 2007

Du H L. Analysis of the energy radiation sources of the 2004 Sumatra-Andaman Earthquake using time-domain array techniques [M. S. thesis] (in Chinese). Beijing: Institute of Geophysics, China Earthquake Administration, 2007

[10] Kennett B L N, Engdahl E R. Travel times for global earthquake location and phase identification. *Geophys. J. Int.*, 1991, **105**: 429-465

[11] Ishii M, Shearer P M, Houston H. Extend, duration and speed of the 2004 Sumatra-Andaman earthquake imaged by the Hi-Net array. *Nature*, 2005, **435**: 933-936

[12] Krüger F, Ohrnberger M. Tracking the rupture of the $M_W = 9.3$ Sumatra earthquake over 1150km at teleseismic distance. *Nature*, 2005, **435**: 937-939

[13] Du H L, Xu L S, Chen Y T, et al. Tracking the high-frequency energy radiation source of the 2004 Sumatra-Andaman M_W9.0 earthquake using the short-period seismic data: preliminary result. In: Chen Y T ed. *Advances in Geosciences*, **9**: Solid Earth, (SE), Singapore: World Scientific Publishing Co., 2007. 3-11

[14] 许力生, 杜海林, 张红霞等. 2007年7月苏门答腊岛近海三次大地震能量辐射源时空特征. 科学通报, 2008, **53**(17): 2085-2090

Xu L S, Du H L, Zhang H X, et al. Spatio-temporal characteristics of the energy radiation sources of the three earthquake near Sumatra Island in September 2007. *Chinese Science Bulletin*, 2008, **53**(15): 2364-2370

[15] Schweitazer J, Fyen J, Mykkeltveit S, et al. Seismic Arrays. In: Bormann P ed. *New Manual of Seismological Observatory Practice*. Germany: GeoForschungsZentrum Potsdam, 2002. Chapter 9

[16] 赵珠, 范军, 郑斯华等. 龙门山断裂带地壳速度结构和震源位置的精确修订. 地震学报, 1997, **19**(6): 615-622

Zhao Z, Fan J, Zheng S H, et al. Crustal structure and accurate hypocenter determination along Longmenshan fault zone. *Acta Seismological Sinica*, 1997, **10**(6): 761-768

Rupture process of the 2008 great Wenchuan earthquake from the analysis of the Alaska-array data

Du Hai-Lin, Xu Li-Sheng, Chen Yun-Tai

Institute of Geophysics, China Earthquake Administration, Beijing 100081, China

Abstract On 12th of May, 2008, an $M_S 8.0$ earthquake occurred in Wenchuan county, Sichuan province of China. We configured a generalized array with the broadband seismic stations of the Alaska Regional Network, USA, and imaged the temporal and spatial variation of high-frequency (>0.1Hz) energy radiation sources by means of one of the non-plane wave array techniques—migration stacking technique. The image shows that the earthquake faulting initiated at the instrumental epicenter and extended north-eastwards about 300km, the whole rupture process lasted for at least 90s, and the average rupture speed was about 3.4 km/s. The whole process may be divided into two major periods. The first period is 50s and in this period the rupture propagated about 110km, with an average rupture speed of 2.2km/s; the other period is 40s and in this period the rupture propagated about 190km, with an average speed of 4.8km/s. It seems that supershear rupture appeared in the later process of the earthquake, and the second segment might be dynamically triggered by the first one.

2008年汶川大地震震源机制的时空变化

张勇[1,2]　许力生[1]　陈运泰[1,2]

1. 中国地震局地球物理研究所，北京　100081；2. 北京大学地球与空间科学学院，北京　100871

摘要　本文提出了一种基于恒定破裂速度和固定子事件震源时间函数的假定、利用远场地震波形资料获取大地震震源机制的时空变化图像的线性反演方法，并利用这种方法及全球范围内48个台站的长周期波形资料反演建立了2008年汶川$M_S 8.0$地震的震源机制随时间和空间变化的图像．根据这个图像可知，汶川大地震断层的西南端震源机制接近于逆冲，随着破裂向东北方向延伸，震源机制的走滑分量逐渐增大，走滑分量超过逆冲分量的转折点在震中东北大约190km的位置．为了检验反演方法的有效性和反演结果的可靠性，我们特别设计了一个数值试验对反演结果进行了检验．检验结果表明，我们在本文中提出的反演方法是有效的，关于汶川大地震的反演结果也是可靠的（除长周期信号较弱的一段外）．通过比较发现，反演结果与震后野外考察的结果也相当吻合．

1　引言

　　特大地震震源破裂涉及的范围很广．由于大范围内介质性质存在非均匀性和地质构造的复杂多变性，发震断层面的几何形态往往十分复杂，并非是一个简单的平面；特别是，如果一次特大地震的破裂过程涉及多个断层的活动，则发震断层并非是单一的断层平面，而是由多个断层面组合而成，这就导致在破裂传播过程中震源机制随时间和空间的变化．近年来，非平面复杂断层的地震破裂过程的研究成为震源研究中广受关注的内容[1-3]．2008年5月12日发生的汶川$M_S 8.0$地震的发震断层沿龙门山断裂带向北东方向延伸，破裂区域尺度超过300km[4,5]．在这个破裂区域内，龙门山断裂的地质构造十分复杂[6]，多条断层（山前断裂、中央断裂和后山断裂等）都可能发生活动，导致在地震破裂过程中震源机制发生改变．震后野外调查结果也已表明，在汶川大地震过程中至少中央断裂和山前断裂都发生了错动[7]．

　　到目前为止，关于汶川地震震源破裂过程研究关注的重点主要是预先假定平面断层面上的滑动随时间和空间的变化[4,5]．作为震后的快速响应工作的重要内容，陈运泰等在地震发生后翌日发布了当时可资利用（因而数量较少）的资料确定的震源机制随时间和空间变化的结果[4]．随后，作者搜集整理了更多的资料，运用不久前新提出的反演方法[8]重新处理了资料，获得了这次地震的震源机制随时间和空间变化图像的更为详细的信息．本文将介绍这种侧重于研究震源机制在震源过程中随时间和空间变化的反演方法，并报告从更多的资料中获得的新结果．

　　目前反演大地震的震源机制随时间和空间变化的方法主要有两种，一种是波形"剥除"的方法[9-11]；另一种是多点矩心矩张量反演方法[12]．这些方法的求解过程都是逐个搜索子事件震源机制的非线性过程．除了计算效率的因素外，逐个搜索的方法往往导致后续子事件震源机制的准确性强烈依赖于非线性反演收敛的精度和可靠性．本文将从位移表示定理出发，引入一种新的线性反演方法，可以同时求解所有子事件的震源机制，避免了迭代求解可能造成的不稳定问题，并将其应用于汶川大地震的震源过程研究．由于采

---

* 本文发表于《地球物理学报》，2009年，第**52**卷，第2期，379–389.

用的是线性反演,此方法在计算效率和收敛精度上具有一定的优势.

2 方法

根据位移表示定理,地震震源激发的位移在时间域可以用组成地震震源的子事件的矩张量和相应的格林函数的褶积表示[13]:

$$U_n(\boldsymbol{x},t) = \sum_{k=1}^{K} G_{np,q}(\boldsymbol{x},t;\xi_k) * M_{pq}(\xi_k,t), \tag{1}$$

式中,$U_n(\boldsymbol{x},t)$ 为观测点 $\boldsymbol{x}$ 处 n 分向的地动位移;$G_{np,q}(\boldsymbol{x},t;\xi_k)$ 为格林函数,是在 ξ_k 处的第 k 个子事件的单位强度的力偶或偶极 (p,q) 在 x 处激发的 n 方向的地动位移;$M_{pq}(\xi_k,t)$ 为第 k 个子事件依赖于时间的地震矩张量分量;K 为构成大地震震源的子事件数目;星号"$*$"表示时间域的褶积.

由于子事件有较小的空间尺度和较短的破裂持续时间,我们不妨假定各子事件的震源机制不随时间发生变化,即所有子事件都是同步震源,从而表示子事件震源的矩张量的 6 个元素都具有相同的随时间变化的历史(震源时间函数):$s_{pq}(\xi_k,t)=s(\xi_k,t)$,可以将时间因子从各子事件的地震矩张量分量中提取出来,于是式(1)可以写成

$$U_n(\boldsymbol{x},t) = \sum_{k=1}^{K} [G_{np,q}(\boldsymbol{x},t;\xi_k) \cdot M_{pq}(\xi_k)] * s(\xi_k,t), \tag{2}$$

从式(2)可以看出,如果整个地震的震源机制固定不变且可以事先予以确定,那么通过线性反演可以确定子事件的震源时间函数 $s(\xi_k,t)$ 及其幅度.式(2)即是震源机制不变情况下利用观测资料反演震源破裂过程的理论依据[14,15].

如果事先知道所有子事件的震源时间函数 $s(\xi_k,t)$,式(1)可以写成

$$U_n(x,t) = \sum_{k=1}^{K} [G_{np,q}(\boldsymbol{x},t;\xi_k) * s(\xi_k,t)] \cdot M_{pq}(\xi_k), \tag{3}$$

根据上式,构成大地震震源的子事件的矩张量 $M_{pq}(\xi_k)$ 就可以通过反演地震记录得以确定.式(3)是我们在本研究中利用地震记录反演震源机制时空变化过程的理论依据.

在远场情况下,当格林函数为单位阶跃函数集中力所造成的位移、且涉及的地震波长远大于子事件震源尺度时,子事件的时间历史可以近似为一个狄拉克(Dirac)δ-函数;而且,当破裂速度给定时,每个子事件的时间函数的延迟时间可以惟一地确定.此时,待解参数仅为各子事件的地震矩张量 $M_{pq}(\xi_k)$ 的 6 个分量,并且与观测记录呈线性关系.因此,我们可以将上述问题写成一般的矩阵方程形式

$$\mathbf{U} = \mathbf{G}_S \mathbf{M}, \tag{4}$$

式中 $\mathbf{U}$ 为所有台站的观测资料组成的资料矢量;$\mathbf{G}_S$ 为子事件的矩张量元素对应的格林函数与子事件震源时间函数共同构筑的矩阵;$\mathbf{M}$ 为所有子事件的地震矩张量元素组成的未知数矢量.三者的具体形式为:

$$\mathbf{U} = [U(1),U(2),\cdots,U(l),\cdots,U(L)]^T, \tag{5a}$$

$$\mathbf{G}_S = \begin{bmatrix} G_S^{(1,1)} & G_S^{(1,2)} & & & G_S^{(1,K)} \\ G_S^{(2,1)} & G_S^{(2,2)} & & & G_S^{(2,K)} \\ & & \cdots & & \\ & & & G_S^{(l,k)} & \\ & & & & \cdots \\ G_S^{(L,1)} & G_S^{(L,2)} & & & G_S^{(L,K)} \end{bmatrix}, \tag{5b}$$

$$\mathbf{M} = [M(1),M(2),\cdots,M(k),\cdots,M(K)]^T, \tag{5c}$$

L 为台站数目；K 为子事件数目；$U(l)$ 为第 l 个台站的地震记录组成的资料矢量；$G_S^{(l,k)}$ 为第 k 个子事件到第 l 个台站的格林函数与第 k 个子事件的震源时间函数所构成的块矩阵；$M(k)$ 为第 k 个子事件的地震矩张量的 6 个独立分量组成的子矢量.

原则上，根据方程 (4) 即可通过反演确定各子事件的地震矩张量，但是，由于计算格林函数时采用的地球模型与实际地球介质之间的差别以及观测资料中不可避免的噪声，直接求解式 (4) 往往不稳定，因此有必要在方程 (4) 的基础上添加一些约束条件. 采用拉普拉斯方程对解进行光滑约束是此类反演问题中常用的有效手段之一[16]. 在本研究涉及的问题中，我们假定相邻子事件的震源机制不发生突变. 因此，我们将拉普拉斯方程分别作用于地震矩张量的各元素，以保证相邻子事件的震源机制是逐渐变化的. 此外，为了防止格林函数矩阵可能存在奇异性导致的大范数解，我们还引入标量地震矩最小约束. 考虑到以上两个因素，可将方程 (4) 扩展为如下矩阵方程

$$\begin{bmatrix} \lambda_0 U \\ 0 \\ 0 \end{bmatrix} = \begin{bmatrix} \lambda_0 G_S \\ \lambda_1 D \\ \lambda_2 I \end{bmatrix} M, \qquad (6)$$

式中，λ_0 用于调节各台站观测数据之间的相对权重和资料方程的权重大小；D 和 λ_1 分别为拉普拉斯方程对应的光滑矩阵和光滑权重因子；I 为单位矩阵，λ_2 为相应的权重因子，二者对应为标量地震矩最小约束矩阵及其相应权重. 方程 (6) 的系数矩阵往往很大，并且 G_S 为稠密矩阵，而 D 和 I 皆为稀疏矩阵，因此，本文采用共轭梯度方法[17]求解此方程. 在迭代过程中计算梯度方向和残差时，通过分别计算稠密矩阵 G_S 与稀疏矩阵 D 和 I 的方法[8]，可大幅提高计算效率.

通过求解方程 (6)，我们可以直接得到各子事件矩张量的 6 个元素，由此可进一步得到各子事件的标量地震矩和最佳双力偶解或震源机制，根据子事件震源时间函数的时间延迟便可以构建地震矩释放随时间和空间的变化过程以及震源机制随时间和空间的变化过程.

需要说明的是，在本研究中，为了进一步确保解的可靠性，我们还把断层两端的子事件的标量地震矩约束为 0，把各子事件的矩张量解中的各向同性成分和补偿线性矢量偶极成分也约束为 0，以达到增强约束、减少待解参数数目的目的.

3 2008 年汶川大地震的震源机制反演

3.1 观测资料

为了反演 2008 年 5 月 12 日汶川大地震震源机制的时空变化过程，本文选取了全球范围内的长周期数据. 考虑到这次地震的破裂时间大约为 90s[5]，为了保证有足够时间长度的纯直达 P 波波形记录，我们仅选择震中距范围为 40°~90°的台站；同时为了保证台站相对于震源具有均匀的方位分布，我们按照台站间最小方位角间隔为 2°的限制条件来挑选台站，最终得到了可资利用的 48 个台站的资料 (图 1).

由于在震源机制可变的震源破裂过程反演中的解空间通常存在着多个局部极小，而这些局部极小解恰巧对应于经过半周期"相移"之后的观测波形，从而表现为真实结果刚好相反的"伪解"[10]. 地震波形的频率越高，半周期"相移"长度越小，越容易出现这种现象. 采用较低频波形资料有利于避免这种"伪解"的出现，有利于反演过程稳定地收敛至全局极小，同时也可以更好地满足资料的波长远大于子事件震源尺度的条件. 因此本研究中我们使用了频率范围为 0.002~0.025Hz 的波形资料.

图 1 2008 年汶川大地震震中（白色八角星）和 48 个长周期台站（黑色三角形）分布

Fig. 1 Distribution of the epicenter (white aniseed star) and 48 teleseismic stations (black triangles) of the 2008 great Wenchuan earthquake

3.2 断层模型与格林函数的计算

关于汶川大地震断层的走向已经有不少研究结果[5,18]. 因此，在本研究中采用如图 2 所示的一条延伸方向为 N45°E 由 A 至 B 的线源断层作为断层初始模型. 此线源断层长为 320km，由 32 个间距为 10km 的子事件组成. 换句话说，每个子事件的断层尺度为 10km. 考虑到整个地震断层的几何特征，我们把地震破裂的起始点（震中）置于第 2 个子事件的位置.

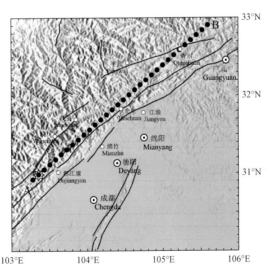

图 2 2008 年汶川大地震震中（白色八角星）与本研究中采用的断层 AB 及其子事件（黑点）分布

图中红色线表示震中区及其周围主要断层

Fig. 2 Distribution of the epicenter (white aniseed star) of the 2008 great Wenchuan earthquake, line-fault model and its sub-events (black dots) in this study

Red curves denote the major faults in epicenter and its surrounding area

本研究利用反射率方法[19]和全球标准速度模型 IASPEI91[20] 计算格林函数.

3.3 破裂速度

为了确定每个子事件震源时间函数的时间延迟,需要给定一个合理的破裂速度.地震的破裂速度在震源过程中可能是变化的,但是,根据前人的研究[16,21],预先给定一个合适的平均破裂速度不会影响震源破裂过程的一般特征.为了确定地震过程中最佳平均破裂速度,我们选择了一个较大的速度范围,以一定的步长进行格点搜索,并根据搜索结果所对应的残差来确定最佳平均破裂速度.对于汶川地震,我们选定的破裂速度范围为 2~5km/s,搜索步长为 0.1km/s,反演得到的残差-破裂速度曲线如图 3 所示.从残差曲线可以看出,残差全局极小值对应的破裂速度为 3.2km/s.

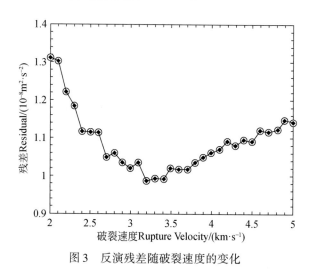

图 3 反演残差随破裂速度的变化

Fig. 3 Variation of the inversion residuals with rupture velocity

3.4 反演结果

图 4 显示了线断层上的震源机制随空间位置的变化分布.从图中可以看出,从断层的西南端至东北端震源机制在逐渐变化,逆冲分量越来越小而走滑分量越来越大.这个变化特征表明震源机制的变化大体上可以分为 4 段. A 段(震中至汶川)的震源机制接近纯逆冲,相当于一次 M_W7.5 地震; B 段(汶川至茂县)具有一定的走滑分量,相当于一次 M_W8.0 地震; C 段(北川至平武)的走滑分量较第二地段更大,相当于一次 M_W7.5 地震;而 D 段(平武至青川)的走滑分量最大,明显超过了逆冲分量,相当于一次 M_W7.7 地震. 4 段的震源机制如表 1 所示,这 4 段(A,B,C,D)恰好与标量地震矩释放的 4 个阶段(1,2,3,4)相对应(图 5).

图 5 显示了标量地震矩释放随时间和空间的变化,由于假定了恒定的破裂传播速度,图 5a 和图 5b 具有很强的相似性.从时间上看,地震的持续时间约为 95s,包括 4 个主要的地震矩释放阶段:第一阶段为从发震开始至震后 10s,释放了总地震矩的 7%;第二阶段为震后 10s 至震后 42s,释放了总地震矩的 61%;第三阶段为震后 42s 至震后 60s,释放了总地震矩的 9%;最后一个阶段为震后 60s 至震后 95s,释放了总地震矩的 23%.从空间上看,标量地震矩的释放分布也集中在 4 个主要区域:震中西南 10km 至震中东北 30km;震中东北 30km 至震中东北 120km;震中东北 120km 至震中东北 190km;震中东北 190km 至震中东北 300km.地震矩释放峰值出现在震后第 23s(图 5a),位于震中东北 70km 处(图 5b).

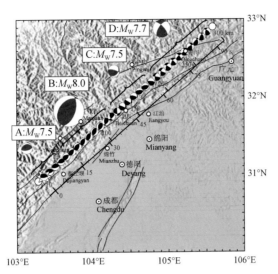

图 4 反演得到的线断层各子事件的震源机制随时间和空间的变化

黑白沙滩球为子事件震源机制,红白沙滩球为根据震源机制的时空变化划分的 4 阶段（A,B,C,D）的震源机制,黑色与红色区域表示初动为压缩的象限,白色区域表示初动为膨胀的象限. 白框内数字为各阶段相当的矩震级. 红色线表示主要断层

Fig. 4 Spatio-temporal variation of the inverted focal mechanisms of the sub-events of the line-fault model

Black-white beach balls denote source mechanisms of sub-events. Red-white beach balls denote source mechanisms of the 4 typical stages divided into in term of the spatio-temporal variation of source mechanism. And the numbers in white rectangular are moment magnitudes calculated for each stage. Red curves denote the major faults

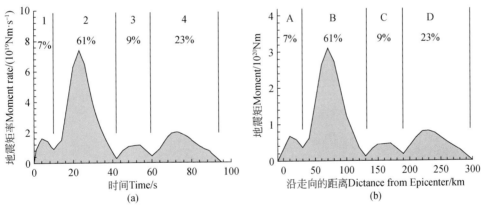

图 5 (a) 2008 年汶川大地震的地震矩释放率随时间的变化；(b) 2008 年汶川大地震的地震矩沿断层走向方向距离的变化

注意 (a) 与 (b) 图形的相似性, 以及横坐标与纵坐标所表示的物理量的不同

Fig. 5 (a) Variation of the seismic moment rate of the 2008 great Wenchuan earthquake as a function of time; (b) Variation of the seismic moment of the 2008 great Wenchuan earthquake as a function of the along-strike distance from the epicenter

Pay attention to similarity of the two sub-plots as well as to difference in physical meaning of ordinate and abscissa

需要指出的是,尽管北川在地震中受损最严重,但从反演结果看,北川附近（距震中 130～190km）释放的地震矩并不是最大. 同时我们还注意到,尽管采用了光滑约束,但北川附近的震源机制变化仍然较大. 我们将在下文分析其原因.

图 6 比较了利用反演得到的结果计算的合成波形与观测波形. 比较表明, 48 个台站最小相关系数 0.74, 其中, 96%（46 个）台站的波形相关系数大于 0.8, 80%（38 个）台站的波形相关系数超过 0.9, 说明反演结果相当好地解释了几乎所有的观测波形.

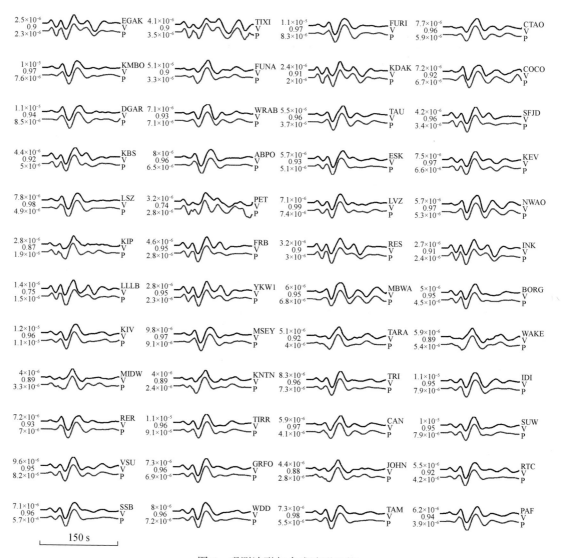

图 6 观测波形与合成波形比较

每个子图中，上方的黑线表示观测波形，下方的灰线表示合成波形．子图左方的数字从上到下依次为观测波形最大幅度、观测波形与合成波形之间的相关系数、合成波形最大幅度，振幅的单位为 m/s．子图右边的符号从上到下分别为台站名、分量名和震相名

Fig. 6 Comparison of the observed with synthetic seismograms

In each panel, the upper trace is the observed waveform and the lower trace is the synthetic waveform, on the left are the maximum amplitude of the observed waveform, the correlation coefficient and the maximum amplitude of the synthetic waveform, respectively, and on the right, from top to bottom, are the station code, component name and phase name, respectively

4 可靠性检验

为了检验反演结果的可靠性，我们采用了和上面相同的台站分布和断层模型，进行了数值试验．

4.1 测试模型

采用和上面相同的线断层模型，设破裂起始点位于第 2 个子断层，破裂速度为 3km/s；在破裂过程中，所有子断层的标量地震矩、走向、倾角和滑动角都发生缓慢而稳定的变化．

表1 汶川大地震的4个主要子事件的震源机制参数

Table 1 Source mechanism parameters of the 4 major sub-events of the great Wenchuan earthquake

地震事件	M_0/Nm	M_W	节面Ⅰ			节面Ⅱ			T-轴		B-轴		P-轴	
			走向/(°)	倾角/(°)	滑动角/(°)	走向/(°)	倾角/(°)	滑动角/(°)	方位角/(°)	俯角/(°)	方位角/(°)	俯角/(°)	方位角/(°)	俯角/(°)
1	1.9×10^{20}	7.5	256	27	88	79	63	91	351	72	258	1	168	18
2	1.4×10^{21}	8.0	222	37	107	21	55	77	251	76	28	10	119	9
3	2.0×10^{20}	7.5	284	74	115	45	29	35	224	55	96	24	355	25
4	4.6×10^{20}	7.7	215	63	157	316	69	29	177	35	349	55	85	4
总体	1.8×10^{21}	8.1	221	40	115	10	54	71	228	73	21	15	113	7

4.2 观测波形资料

为了保证与汶川大地震的资料情况相同,基于上述测试模型,利用反射率方法[19]和全球标准速度模型 IASPEI91[20]计算了图1所示的48个台站处频带范围为 0.002~0.025Hz 的合成地震图,然后,在合成地震图上添加20%的随机噪声.添加噪声后的合成地震图用作"观测"波形资料.

4.3 最佳平均破裂速度

已经知道给定的线断层震源模型的破裂速度为3km/s,所以,我们以0.2km/s的步长在2~4km/s的范围内进行反演搜索.通过反演搜索得到的破裂速度—残差曲线如图7所示.从图中可以清楚地看到,最小残差对应的破裂速度为3km/s,和预先假定的破裂速度完全一致.

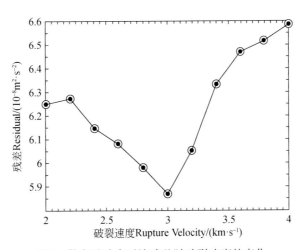

图7 数字试验中反演残差随破裂速度的变化

Fig. 7 Residual variation with rupture velocity in the numerical experiment

4.4 震源机制时空变化

基于3km/s的破裂速度,反演"观测"波形资料得到各子事件标量地震矩、断层走向、倾角和滑动角如图8所示.与预先设定的值相比,标量地震矩大小在线断层上的分布依然保留了原来的主要特征,各子事件的断层面参数与预先的设定值基本接近.但相对而言,具有较大标量地震矩的子断层的断层面参数得到了更好地分辨;而反演结果中具有较小标量地震矩的子断层的断层面参数的可靠程度不是很高,主要表现在：反演得到的走向、倾角和滑动角与预先设定数值差别较大，且反演结果中相邻子断层间的机制变化相对比较明显，即光滑约束条件对标量地震矩较小的子断层的机制约束并不好．

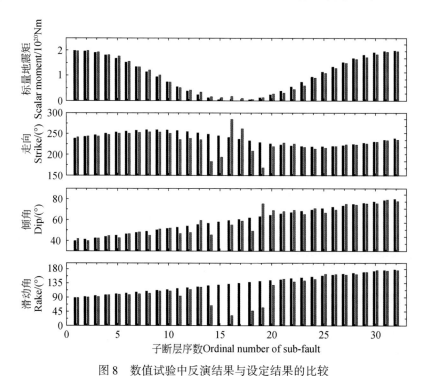

图 8　数值试验中反演结果与设定结果的比较

黑色柱表示设定值，灰色柱表示反演结果

Fig. 8　Comparison of the inverted results with the model values in the numerical experiment

Black bars are the model values and the gray bars are the inverted values

图 9 给出了"观测"波形与合成波形的比较．可以看到，跟上文中实际的观测波形和合成波形的拟合情况相比，由于高频噪声的影响，这里的拟合情况略差一些．但主要的波形特征保持了很好的一致性．

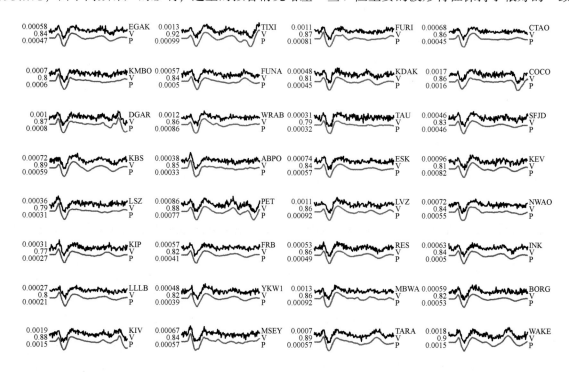

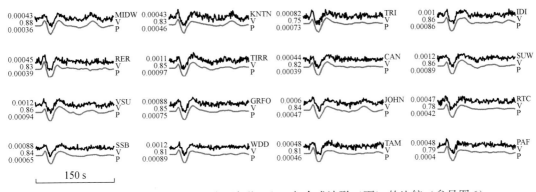

图 9　20% 噪声水平下的"观测"波形（上）与合成波形（下）的比较（参见图 6）

Fig. 9　Comparison of the observed (top) with synthetic (bottom) seismograms at the 20% noise level (Refer to Fig. 6)

上面的检验表明，在当前的震源模型和台站分布情况下，利用本文提出的方法获得的汶川大地震的震源机制的时空变化图像中，由于 C 段即第三次事件相对较小的标量地震矩和变化较大的震源机制，其结果的可靠性相对较弱，而其他阶段的参数是可信赖的.

5　讨论与结论

从位移表示定理出发，借助于地震矩张量的概念，本文提出了一种利用远场地震波形资料确定大地震可变机制的震源破裂过程的线性反演方法. 这个方法可用于三维有限震源、二维有限震源或者一维有限震源. 但是，在一般情况下，这类问题的未知数较多、系数矩阵较大，所以实际上更适合于一维有限震源的情况，即正如本研究中使用的线断层的情况.

为了使反演问题变成一个线性问题，我们使用了两个前提条件：一是震源过程平均破裂速度已知；二是组成地震的所有子事件的震源时间函数为狄拉克 δ-函数. 从上面的数值试验或对实际地震的应用可以看出，获取震源过程中的平均破裂速度并不困难. 事实上，已经有不少关于震源破裂过程的研究中也使用了类似的技术途径[16,21]. 实际地震的震源时间函数可能是各种各样的，断层面上任意两点的滑动历史也可能截然不同，但是，如果所使用的观测波形的波长远大于子事件的震源尺度，那么，所有子事件震源时间函数都用狄拉克 δ-函数代替是合理的. 因此，本文提出的方法是一种计算效率高、有良好应用前景的方法.

利用本文提出的反演方法，基于 32 个子事件组成的线断层模型，作者反演了 2008 年汶川大地震在全球范围内 48 个长周期台的直达 P 波记录，获得了震源机制在地震过程中随时间和空间的变化图像，并通过针对这次地震的观测数据和断层模型的数值试验检验了结果的可靠性.

对长周期波形资料的反演结果表明，汶川大地震的震源过程持续了约 95s，释放总地震矩 1.8×10^{21} Nm，相当于矩震级 $M_W 8.1$. 根据震源机制的变化，整个地震过程可以分为 4 个阶段. 在第一个阶段，在时间上对应于地震开始后 10s，在空间上大体对应于震中至汶川之间的 30km 长的断层段（A 段），释放了总地震矩的约 7%，相当于一次 $M_W 7.5$ 地震，震源机制接近于纯逆冲. 第二阶段，在时间上对应于地震开始后 10s 至 42s，在空间上大体对应于汶川至茂县之间的 100km 长的断层段（B 段），释放了总地震矩的约 61%，相当于一次 $M_W 8.0$ 地震，震源机制以逆冲为主兼小量走滑分量. 第三阶段，在时间上对应于地震开始后 42s 至 60s，在空间上大体对应于北川至平武段附近的 50km 的断层段（C 段），释放了总地震矩的约 9%，相当于一次 $M_W 7.5$ 地震，震源机制为逆冲兼走滑. 但由于地震矩相对较小，震源机制参数的可信度相对较差. 第四阶段，在时间上对应于地震开始后 60s 至 95s，在空间上大体对应于平武至青川附近

110km 长的断层段（D 段），释放了总地震矩的约 23%，相当于一次 M_W7.7 地震，震源机制以走滑为主兼具小量逆冲分量. 总体看来，地震断层的南端为近乎逆冲，随着破裂向东北方向延伸，断层的走滑分量逐渐加大. 从以逆冲为主到以走滑为主的转折点在震中东北大约 190km 的位置，与震后野外考察结果相当一致[7].

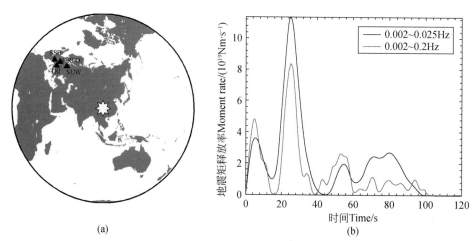

图 10 （a）4 个方位角接近垂直于断层走向方向的台站位置分布（震中距范围为 6673~8269km，方位角范围为 311°~318°）；（b）基于反演得到的总体机制对应的合成格林函数，采用 PLD 方法得到的 4 个台站的两个频带对应的平均视震源时间函数结果

Fig. 10 (a) Distribution of locations of the 4 seismic stations in the direction perpendicular to the fault strike (with distances of 6673~8269km and azimuths of 311°~318°); (b) The average apparent source time functions retrieved by means of the PLD method from the 4 stations, which have 2 different frequency bands, using the synthetic Green's functions calculated with the inverted overall mechanism

需要指出的是，根据本研究的结果，震源过程可分为 4 个阶段或 4 次子事件. 表面上，这个结果与基于有限断层震源破裂过程反演确定的 5 个阶段或 5 次子事件有所不同[5]. 根据本研究的结果，震源过程的最后一个阶段（60~95s）释放的地震矩要多于第三个阶段（42~60s）释放的地震矩，这也与基于有限断层震源破裂过程反演确定的结果有所不同[5]. 实际上这种差别是表观的. 为了说明这一点，我们基于前文反演得到的整个地震整体机制（表 1）合成的格林函数，选取了如图 10a 所示的近垂直于断层走向、震中距在 6673~8269km 的 4 个台站的观测资料，分别使用 0.002~0.025Hz 和 0.002~0.2Hz 的带通滤波，然后采用 PLD 方法[22] 分别得到了如图 10b 所示的平均视震源时间函数. 由于与走向方向接近垂直，且震中距远大于震源尺度，因此获取的平均视震源时间函数将非常接近真实的震源时间函数. 从图 10 可以看出，从 0.002~0.2Hz 频带的资料中提取的震源时间函数略显复杂并接近于基于有限断层震源破裂过程反演确定的结果[5]；从 0.002~0.025Hz 频带的资料中提取的震源时间函数中第三个阶段幅度小于第四个阶段幅度，接近本研究所得结果. 这些比较清楚地表明，造成震源时间函数差别的原因主要是所选频带的不同. 同时也表明，平武—青川段激发的低频成分较强而高频成分较弱，北川—平武段激发的低频成分较弱而高频成分较强. 有限断层震源破裂过程反演结果[5]还表明，北川—平武段存在大幅度的错动出露到地表. 因此，结合本文结果和有限断层震源破裂过程反演结果，我们推测出露地表的大幅度断层错动及其辐射的大幅度的高频地震波是北川一带遭受相对更为严重破坏的原因之一.

参考文献（References）

[1] Lin A M, Kikuchi M, Fu B H. Rupture segmentation and process of the 2001 M_W7.8 central Kunlun, China, earthquake. *Bull. Seis. Soc. Am.*, 2003, 93: 2477-2492

[2] Antolik M, Aberrcrombie R E, Ekström G. The 14 November 2001 Kokoxili (Kunlunshan), Tibet, earthquake: rupture transfer through a large extensional step-Over. *Bull. Seis. Soc. Am.*, 2004, **94**: 1173–1194

[3] Tocheport A, Rivera L, Woerd J V. A study of the 14 November 2001 Kokoxili earthquake: history and geometry of the rupture from teleseismic data and field observations. *Bull. Seis. Soc. Am.*, 2006, **96**: 1729–1741

[4] 陈运泰, 许力生, 张勇等. 汶川地震专题/地震情况/初步研究及考察结果（一）. http://www.cea-igp.ac.cn/pdf (2008-05-02)
Chen Y T, Xu L S, Zhang Y, et al. Reports on the source characters of Wenchuan great earthquake occurred on 12th May 2008. China seismic information network, 2008, http://www.cea-igp.ac.cn/special_issue/earthquakesituation/preliminary_results(1).pdf (2008-05-02)

[5] 张勇, 冯万鹏, 许力生等. 2008年汶川大地震的时空破裂过程. 中国科学 (D辑: 地球科学), 2008, 38(10): 1186–1194
Zhang Y, Feng W P, Xu L S, et al. Spatio-temporal rupture process of the 2008 great Wenchuan earthquake. *Science China (Ser D-Earth Sci)*, 2009, **52**(2): 145–154

[6] 陈国光, 计凤桔, 周荣军等. 龙门山断裂带晚第四纪活动性分段的初步研究. 地震地质, 2007, 29(3): 657–673
Chen G G, Ji F J, Zhou R J, et al. Primary research of activity segmentation of Longmenshan fault zone since Late-Quaternary. *Seismology and Geology* (in Chinese), 2007, **29**(3): 657–673

[7] 徐锡伟, 闻学泽, 叶建青等. 汶川M_S8.0地震地表破裂带及其发震构造. 地震地质, 2008, 30(3): 597–629
Xu X W, Wen X Z, Ye J Q, et al. The M_S8.0 Wenchuan earthquake surface ruptures and its seismogenic structure. *Seismology and Geology* (in Chinese), 2008, **30**(3): 597–629

[8] 张勇. 地震震源破裂过程反演方法研究 [博士论文]. 北京: 北京大学地球与空间科学学院, 2008. 1–158
Zhang Y. *Study on the Inversion Methods of Source Rupture Process* [Ph. D. dissertation] (in Chinese). Beijing: School of Earth and Space Science, Peking University, 2008. 1–158

[9] Kikuchi M, Kanamori H. Inversion of complex body waves—II. *Phys. Earth Planet. Interiors*, 1986, **43**: 205–222

[10] Kikuchi M, Kanamori H. Inversion of complex body waves—III. *Bull. Seism. Soc. Am.*, 1991, **81**: 2335–2350

[11] Zahradní K J, Serpetsidaki A, Sokos E, et al. Iterative deconvolution of regional waveforms and a double-event interpretation of the 2003 Lefkada earthquake, Greece. *Bull. Seism. Soc. Am.*, 2005, **95**: 159–172

[12] Ekström G. A very broadband inversion method for the recovery of earthquake source parameters. *Tectonophysics*, 1989, **166**: 73–100

[13] Aki K, Richards P. *Quantitative Seismology: Theory and Methods*. Vols I and II. San Francisco: W H Freeman, 1980. 1–932

[14] Chen Y T, Xu L S. A time-domain inversion technique for the tempo-spatial distribution of slip on a finite fault plane with applications to recent large earthquakes in the Tibetan Plateau. *Geophys. J. Int.*, 2000, **143**: 407–416

[15] Xu L S, Chen Y T, Teng T L, Patau G. Temporal and spatial rupture process of the 1999 Chi-Chi earthquake from IRIS and GEOSCOPE long period waveform data using aftershocks as empirical Greens functions. *Bull. Seism. Soc. Am.*, 2002, **92**: 3210–3228

[16] Yagi Y, Kikuchi M. Source rupture process of the Kocaeli, Turkey, earthquake of August 17, 1999, obtained by joint inversion of near-field data and teleseismic data. *Geophys. Res. Lett.*, 2000, **27**: 1969–1972

[17] Ward S N, Barrientos S E. An inversion for slip distribution and fault shape from geodetic observations of the 1983, Borah Park, Idaho, earthquake. *J. Geophys. Res.*, 1986, **91**(B5): 4909–4919

[18] Liu C, Zhang Y, Xu L S, et al. A new technique for moment tensor inversion with applications to the 2008 Wenchuan M_S8.0 earthquake sequence. *Acta Seismologica Sinica*, 2008, **21**(4): 333–343

[19] Kennett B L N. *Seismic Wave Propagation in Stratified Media*. Cambridge: Cambridge University Press, 1983. 1–339

[20] Kennett B L N, Engdahl E R. Travel times for global earthquake location and phase identification. *Geophys. J. Int.*, 1991, **105**: 429–465

[21] Hartzell S H, Heaton T H. Inversion of strong ground motion and teleseismic waveform data for the fault rupture history of the

1979 Imperial Valley, California, earthquake. *Bull. Seism. Soc. Am.*, 1983, **73**: 1553-1583

[22] 张勇, 许力生, 陈运泰. 提取视震源时间函数的 PLD 方法及其对 2005 年克什米尔 M_W7.6 地震的应用. 地球物理学报, 2009, **52**(3): 672-680

Zhang Y, Xu L S, Chen Y T. PLD method for retrieving apparent source time function and its application to 2005 Kashmir M_W 7.6 earthquake. *Chinese J. Geophys.* (in Chinese), 2009, **52**(3): 672-680

Spatio-temporal variation of the source mechanism of the 2008 great Wenchuan earthquake

Zhang Yong[1,2], Xu Lisheng[1] and Chen Yuntai[1,2]

1. Institute of Geophysics, China Earthquake Administration, Beijing 100081, China;
2. School of Earth and Space Science, Peking University, Beijing 100871, China

Abstract A linear inversion technique based on assumption of constant rupture velocity and fixed source time functions of sub-events is proposed in this paper for inverting teleseismic waveform data to image spatio-temporal variation of source mechanism of large earthquakes, and applied to the 2008 Wenchuan M_S8.0 earthquake, obtaining its image of the spatio-temporal variation of source mechanism by inverting 48 worldwide stations of long period waveform data. The image shows that the mainly thrust mechanism appeared in the southwest segment of the fault, and the strike-slip component was getting more and more with the fault extending northeastwards, with the turning point at which the strike-slip component became larger than the thrust-slip component appearing around 190km northeast to the epicenter. In order to inspect the validity of the used method and reliability of the inverted results a specially designed numerical test was performed. The inspection indicated that the inversion method proposed in this paper was valid and the inverted results regarding the great Wenchuan earthquake were trustable (except those on one segment where the long-period component is weak). A comparison shows that the inverted results were coincident with the field-investigation results.

Single-couple component of far-field radiation from dynamical fractures

Leon Knopoff[1] and Yun-Tai Chen[2]

1. Institute of Geophysics and Planetary Physics, University of California, Los Angeles;
2. Institute of Geophysics, China Earthquake Administration

Abstract We reexamine two canons of the seismological literature, that elastic displacements in the far field are proportional to slip velocities on the dynamical fault surface, and that dynamical in-plane slip on an earthquake fault has a double-couple body force equivalent. We show that if faulting takes place on a fault of finite thickness, and there is a strength-weakening zone near the advancing crack tip, there is an additional single-couple term in the body force equivalence and additional terms in the far-field displacement, which are proportional to the time rate of increase of stress drop in the advancing weakening zone. We also show that the single-couple equivalent does not violate principles of Newtonian mechanics because the torque imbalance in the single couple is counterbalanced by rotations within the fault zone; the crack therefore radiates torque waves and a rotational deformation field.

1 Introduction

The proportionality between elastic displacements in the far field and slip velocities on a dynamical fault surface (Knopoff and Gilbert, 1960; Haskell, 1964, 1966) and the statement that dynamical in-plane slip on an earthquake fault has a double-couple body force equivalent (Knopoff and Gilbert, 1960; Maruyama, 1963; Burridge and Knopoff, 1964; Stauder and Bollinger, 1964, 1966) are two canons of the seismological literature. The first is a basis for calculations of seismic energy radiated from earthquake events (Knopoff and Gilbert, 1960; Haskell, 1964, 1966; Kostrov, 1970, 1974), and the second is a basis for seismic moment calculations (Aki, 1966). These two statements are directly related. We show that both statements must be modified if a fracture takes place on a fault of finite thickness and if there is a strength-weakening zone near the advancing crack tip.

The statement of proportionality between the far-field displacement and the slip velocity would appear to be contradicted by classical integrations of the scalar wave equation by Kirchhoff in 1882 (Born and Wolf, 1959, pp. 374-377) and the elastic wave equation by Knopoff (1956) and Knopoff and Gilbert (1960), in which the field at great distance would seem to be dependent not only on the value of the time derivative of the field on a surface Σ, which we take to be the fracture surface of an elastic rupture but also on the spatial gradient of the slip. The Kirchhoff integration of the scalar wave equation is the more transparent and illustrative of the two. The solution to the scalar wave equation for radiation from sources on a closed surface Σ enclosing a volume V is (Stratton, 1941, p. 427, equation 22)

$$\psi(\boldsymbol{x},t) = \frac{1}{4\pi}\int_{\Sigma}\left\{\frac{1}{R}[\nabla\psi] - \frac{1}{R^2}[\psi]\boldsymbol{\gamma} - \frac{1}{cR}[\dot{\psi}]\boldsymbol{\gamma}\right\}\cdot\boldsymbol{n}(\boldsymbol{\xi})\,\mathrm{d}\Sigma(\boldsymbol{\xi})\,,$$

where the brackets $[f(t)]$ denote time retardation by R/c, $[f(t)]=f(t-R/c)$; c is the wave velocity; the vector $\boldsymbol{R}$ extends from a point $\boldsymbol{\xi}$ on the surface to the point of observation at $\boldsymbol{x}$, $\boldsymbol{R}=\boldsymbol{x}-\boldsymbol{\xi}$, $R=|\boldsymbol{R}|$, and $\boldsymbol{\gamma}$ is the unit vector along $\boldsymbol{R}$, $\boldsymbol{\gamma}=\boldsymbol{R}/R$; $\boldsymbol{n}(\boldsymbol{\xi})$ is the outward drawn unit normal to the surface. The surface integral includes the contribution from sources outside V. If ψ and its spatial gradient and time derivative are known everywhere on Σ, ψ is completely determined at all interior points of the volume enclosed by Σ (provided that it is not possible to assign these values arbitrarily). In the far field the solution for the wave function includes the first and third of the terms just listed,

$$\psi(\boldsymbol{x},t) = \frac{1}{4\pi}\int_{\Sigma}\left\{\frac{1}{R}[\nabla\psi] - \frac{1}{cR}[\dot{\psi}]\boldsymbol{\gamma}\right\}\cdot\boldsymbol{n}(\boldsymbol{\xi})\,\mathrm{d}\Sigma(\boldsymbol{\xi})\,.$$

The second of these latter terms is the expected, conventional component, which is proportional to its time derivative at the surface. The first term, which depends on the normal component of the spatial gradient on the surface, has been neglected in the treatment of radiation in the seismological literature to date.

In the elastic wave case Knopoff's solution (Knopoff, 1956, p. 223, equation 43) is quite complicated but consists of terms similar to those just discussed, which we do not reproduce here for simplicity. The solution has two classes of terms with retardations having P- and S-wave velocities; each class includes terms with the two types of spatial (vector) differentiation (divergence and curl) and terms with the time derivative of the displacement on the surfaces. We refrain from being more explicit regarding the operators, which are relatively complex. Kirchhoff's solution for the scalar wave equation and Knopoff's solution for the elastic (vector) wave equation are for a one-sided surface. Though they are not given explicitly for the two-sided fault surface, they have the same essential properties, which is that their solutions consist not only of the wave function and its time derivative on the surface but also the vector spatial derivatives of the wave function (displacement). Our concern in this article is with the significance of the spatial derivative terms on a two-sided fault surface, which have been neglected heretofore. We give the solution to the elastic wave equation in a more useful form in the next section.

The familiar double-couple force equivalent (de Hoop, 1958, pp. 14-19; Knopoff and Gilbert, 1960; Maruyama, 1963; Burridge and Knopoff, 1964) is derivable directly from the properties of the terms in the time derivatives of the displacements. In this article we will show that the contribution of the spatial derivative of slip is indeed zero in the far field if the fault zone has zero thickness. However, it is nonzero if the fault zone has finite thickness, and there is a strength-weakening zone near an advancing crack tip. These hitherto neglected terms radiate as a single couple. The single-couple equivalent does not violate principles of Newtonian mechanics because the torque imbalance in the single couple is counterbalanced by rotations within the fault zone; the crack therefore radiates torque waves, which are shear waves manifested as a radiated rotational field at great distance.

2　Green's function

Stokes' retarded solution (Stokes, 1849) for the Green's function G_{ij} for homogeneous elasticity in the far field is (Love, 1927, p. 305; Aki and Richards, 1980, p. 73)

$$4\pi\rho G_{ij} = \frac{\gamma_i\gamma_j}{\alpha^2 R}\delta\left(t-\tau-\frac{R}{\alpha}\right) + \frac{(\delta_{ij}-\gamma_i\gamma_j)}{\beta^2 R}\delta\left(t-\tau-\frac{R}{\beta}\right), \tag{1}$$

where the Green's function $G_{ij}=G_{ij}(\boldsymbol{x},t;\boldsymbol{\xi},\tau)$ is the displacement in the j direction at a distant point $P(\boldsymbol{x})$ at *time* t due to a unit impulse in the i direction at $Q(\xi)$ on a fault at time τ. ρ is the density of the medium, and α and β are the P- and S- wave velocities, respectively. We adopt the convention that the first index is the component of the force at the source, and the second is the component of the motion at (great) distance $R=|\mathbf{R}|$. The γ_i s are the direction cosines of the vector $\mathbf{R}=\mathbf{x}-\boldsymbol{\xi}$ from an element of the source $Q(\boldsymbol{\xi})$ to the point of observation $P(\mathbf{x})$ (Fig. 1). We restrict the problem to the determination of the elastic wave radiation at P due to slip on the internal surface Σ.

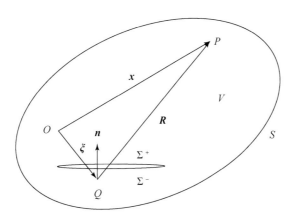

Figure 1 An elastic solid, with volume V and closed external and internal surfaces S and Σ, respectively O is the origin of coordinates, $P(\mathbf{x})$ is a point of observation within V, $Q(\boldsymbol{\xi})$ is a slipping element on Σ, $n_k(\boldsymbol{\xi})$ is the outward drawn normal to surface Σ. The inner surface Σ has two nearby sheets Σ^+ and Σ^-

The stress Green's function σ_{ijk} derived from equation (1) is

$$\sigma_{ijk} = \lambda \delta_{jk} G_{im,m} + \mu(G_{ij,k} + G_{ik,j}) ,$$

where λ and μ are the Lamé elastic parameters; we have dropped the arguments on σ_{ijk} and G_{ij}. From equation (1)

$$4\pi\rho\sigma_{ijk}(\mathbf{x},t;\boldsymbol{\xi},\tau) = -\frac{\gamma_i(\lambda\delta_{jk}+2\mu\gamma_j\gamma_k)}{\alpha^3 R}\dot{\delta}\left(t-\tau-\frac{R}{\alpha}\right) - \frac{\mu(\psi_{ij}\gamma_k+\psi_{ik}\gamma_j)}{\beta^3 R}\dot{\delta}\left(t-\tau-\frac{R}{\beta}\right), \qquad (2)$$

where $\psi_{ij} = \delta_{ij} - \gamma_i\gamma_j$ is a rotation operator, we use $\gamma_m\gamma_m = 1$, $\psi_{im}\gamma_m = 0$, and where we have differentiated only with respect to the second index in equation (1).

3 Integration of the wave equation

We consider sources only on the internal surface Σ. The wave equation for elasticity in the absence of body forces is

$$\tau_{jk,k} - \rho \ddot{u}_j = 0. \qquad (3)$$

We solve (3) in a region V bounded by an extremely large outer surface S and the two surfaces of the zone of faulting, which is the internal surface Σ. The remoteness of S means that motions generated at Σ never reach S in time to influence the fields at P (Fig. 1). To solve (3) for slip on a fault, we let Σ have two sheets, Σ^+ and Σ^-, which are very close to one another. In what follows, we allow Σ^+ and Σ^- to have a small but finite separation.

The Green's function $G_{ij}(\mathbf{x}, t; \boldsymbol{\xi}, \tau)$ in equation (1) satisfies

$$\sigma_{ijk,k} - \rho \ddot{G}_{ij} = -\delta_{ij}\delta(\mathbf{x}-\boldsymbol{\xi})\delta(t-\tau), \qquad (4)$$

where we have dropped the space and time arguments on σ_{ijk} and G_{ij}; $\sigma_{ijk} = \sigma_{ijk}(\mathbf{x}, t; \boldsymbol{\xi}, \tau)$ is the stress tensor Green's function in equation (2). In the usual way, multiply equation (3) by $G_{ij}(\mathbf{x}, t; \boldsymbol{\xi}, \tau)$ from equation

(1), multiply equation (4) by $u_j(\boldsymbol{\xi}, \tau)$, and subtract. It follows that

$$(\tau_{jk,k}G_{ij} - u_j\sigma_{ijk,k}) - \rho(\ddot{u}_j G_{ij} - u_j \ddot{G}_{ij}) = u_i\delta(\boldsymbol{x}-\boldsymbol{\xi})\delta(t-\tau). \tag{5}$$

We integrate equation (5) over $V(\boldsymbol{\xi})$, $-\infty < \tau < \infty$, and apply Gauss' theorem. We obtain

$$u_i(\boldsymbol{x},t) = \int_{-\infty}^{\infty} d\tau \int_{\Sigma} (u_j\sigma_{ijk} + \tau_{jk}G_{ij})n_k d\Sigma(\boldsymbol{\xi}), \tag{6}$$

where $u_j = u_j(\boldsymbol{\xi}, \tau)$, $\tau_{jk} = \tau_{jk}(\boldsymbol{\xi}, \tau)$ and $n_k = n_k(\boldsymbol{\xi})$ in the integrand. Equation (6) is the representation theorem for surface sources and is the point of departure for the remainder of this article.

4 The usual problem

We consider the first integral of equation (6),

$$u_i(\boldsymbol{x},t) = \int_{-\infty}^{\infty} d\tau \int_{\Sigma} u_j\sigma_{ijk}n_k d\Sigma(\boldsymbol{\xi}). \tag{7}$$

We substitute the stress tensor Green's function σ_{ijk} from equation (2) into equation (7) and get the motion at P,

$$u_i(\boldsymbol{x},t) = -\int_{\Sigma} \frac{\gamma_i(\lambda\delta_{jk} + 2\mu\gamma_j\gamma_k)n_k}{4\pi\rho\alpha^3 R} \frac{\partial u_j\left(\boldsymbol{\xi}, t - \frac{R}{\alpha}\right)}{\partial t} d\Sigma(\boldsymbol{\xi}) - \int_{\Sigma} \frac{\mu(\psi_{ij}\gamma_k + \psi_{ik}\gamma_j)n_k}{4\pi\rho\beta^3 R} \frac{\partial u_j\left(\boldsymbol{\xi}, t - \frac{R}{\beta}\right)}{\partial t} d\Sigma(\boldsymbol{\xi}). \tag{8}$$

We have used the identity

$$\int_{-\infty}^{\infty} u(\tau)\dot{\delta}\left(t - \tau - \frac{R}{\alpha}\right) d\tau = \dot{u}\left(t - \frac{R}{\alpha}\right).$$

The two terms in equation (8) have easily identifiable P- and S-wave retardations; we consider them separately.

Let the two sheets, Σ^+ and Σ^-, be separated by a small distance ΔW. It is easy to demonstrate that

$$\gamma_i^+ \approx \gamma_i^- \left\{1 + O\left(\frac{\Delta W}{R}\right)\right\}$$

$$\frac{1}{R^+} \approx \frac{1}{R^-} \left\{1 + O\left(\frac{\Delta W}{R}\right)\right\},$$

where R^+ is the distance R between a point $\boldsymbol{\xi}^+$ in the (upper) surface Σ^+ and the point of observation $\boldsymbol{x}$, which is therefore a function of $\boldsymbol{\xi}^+$. R^- is the same quantity for a point $\boldsymbol{\xi}^-$ in the (lower) surface Σ^-. If we neglect all terms of order higher than the zeroth in $\Delta W/R$ and notice that the two normals point in opposite directions $n_k^+ = -n_k^-$, we have

$$u_i^P(\boldsymbol{x},t) = \int_{\Sigma} \frac{\gamma_i(\lambda\delta_{jk} + 2\mu\gamma_j\gamma_k)n_k}{4\pi\rho\alpha^3 R} \frac{\partial \left\langle u_j\left(\boldsymbol{\xi}, t - \frac{R}{\alpha}\right)\right\rangle}{\partial t} d\Sigma(\boldsymbol{\xi})$$

$$u_i^S(\boldsymbol{x},t) = \int_{\Sigma} \frac{\mu(\psi_{ij}\gamma_k + \psi_{ik}\gamma_j)n_k}{4\pi\rho\beta^3 R} \frac{\partial \left\langle u_j\left(\boldsymbol{\xi}, t - \frac{R}{\beta}\right)\right\rangle}{\partial t} d\Sigma(\boldsymbol{\xi}),$$

where we now integrate only over one surface Σ^- of the two and drop the superscript, and $\langle u_j \rangle$ denotes the difference in the quantity in brackets across the fault zone. Let

$$\left\langle u_j\left(\boldsymbol{\xi}, t - \frac{R}{c}\right)\right\rangle = e_j \left\langle u\left(\boldsymbol{\xi}, t - \frac{R}{c}\right)\right\rangle,$$

where e_j is the unit vector in the direction of slip, and $\langle u \rangle$ is the jump in $u(\boldsymbol{\xi}, t-R/c)$ across the fault zone,

$$\left\langle u\left(\boldsymbol{\xi}, t - \frac{R}{c}\right)\right\rangle = u\left(\boldsymbol{\xi}^+, t - \frac{R^+}{c}\right) - u\left(\boldsymbol{\xi}^-, t - \frac{R^-}{c}\right).$$

Thus,

$$u_i^P(\mathbf{x},t) = \int_\Sigma \frac{\gamma_i(\lambda\delta_{jk} + 2\mu\gamma_j\gamma_k)n_k e_j}{4\pi\rho\alpha^3 R} \frac{\partial \left\langle u\left(\xi, t-\frac{R}{\alpha}\right)\right\rangle}{\partial t} d\Sigma(\xi) \quad (9.1)$$

$$u_i^S(\mathbf{x},t) = \int_\Sigma \frac{\mu(\psi_{ij}\gamma_k + \psi_{ik}\gamma_j)n_k e_j}{4\pi\rho\beta^3 R} \frac{\partial \left\langle u\left(\xi, t-\frac{R}{\beta}\right)\right\rangle}{\partial t} d\Sigma(\xi). \quad (9.2)$$

It is easy to show that

$$\left\langle u\left(\xi, t-\frac{R}{c}\right)\right\rangle \approx \Delta u\left(\xi^-, t-\frac{R^-}{c}\right) + \frac{\partial u\left(\xi^-, t-\frac{R^-}{c}\right)}{\partial t}\frac{\gamma_k^- n_k^-}{c}\Delta W\left\{1+O\left(\frac{\Delta W}{cT_S}\right)\right\}, \quad (10)$$

Where

$$\Delta u\left(\xi^-, t-\frac{R^-}{c}\right) = u\left(\xi^+, t-\frac{R^-}{c}\right) - u\left(\xi^-, t-\frac{R^-}{c}\right)$$

is the usual jump in the displacement (dislocation) in $u(\xi, t-R/c)$ across the fault zone at the same instant of time $t-R^-/c$, and T_s is the rise time, $c=\alpha$ or β.

It is worthy to note that the dislocation

$$\Delta u\left(\xi^-, t-\frac{R^-}{c}\right) \approx D$$

and the slip velocity $\partial u(\xi^-, t-R^-/c)/\partial t$ is related to the final slip D by

$$\frac{\partial u\left(\xi^-, t-\frac{R^-}{c}\right)}{\partial t} \approx \frac{D/2}{T_S}.$$

Thus,

$$\left[u\left(\xi, t-\frac{R}{c}\right)\right] \approx \Delta u\left(\xi^-, t-\frac{R^-}{c}\right)\left[1+O\left(\frac{\Delta W}{cT_S}\right)\right]. \quad (11)$$

Because the direction of slip e_j is perpendicular to the normal fault surface n_k for in-plane slip, $n_k e_k = 0$. Hence, equation (9) becomes

$$u_i^P(\mathbf{x},t) = \int_\Sigma \frac{\beta^2\gamma_i\gamma_j e_j\gamma_k n_k}{2\pi\alpha^3 R} \frac{\partial \Delta u\left(\xi, t-\frac{R}{\alpha}\right)}{\partial t} d\Sigma(\xi), \quad (12.1)$$

$$u_i^S(\mathbf{x},t) = \int_\Sigma \frac{(\psi_{ij}\gamma_k + \psi_{ik}\gamma_j)n_k e_j}{4\pi\beta R} \frac{\partial \Delta u\left(\xi, t-\frac{R}{\beta}\right)}{\partial t} d\Sigma(\xi) \quad (12.2)$$

for in-plane slip. Equation (12) holds for the case of finite thickness of fault zone ΔW if $\Delta W/R \ll 1$ and $\Delta W/cT_s \ll 1$, where $c=\alpha$ or β. Equation (12) is formally in agreement with equation (14.6) of Aki and Richards (1980, p. 802) for the case of zero fault width. From the condition $\gamma \times \gamma = 0$ and the orthogonality condition $\gamma_i\psi_{ij} = 0$, we have that u_i^P and u_i^S are radially and transversely polarized with P-wave and S-wave retardations respectively. The radiation pattern $\gamma_j e_j\gamma_k n_k$ in equation (12.1) is the expected double-couple result for the P-wave radiation pattern.

For slip in the x_1 direction and the normal to Σ in the x_3 direction, in spherical polar coordinates (R, θ, ϕ) centered on the source,

$$\mathbf{e} = (1,0,0), \quad \mathbf{n} = (0,0,1),$$
$$\boldsymbol{\gamma} = (\sin\theta\cos\phi, \sin\theta\sin\phi, \cos\theta),$$

measuring θ from the x_3 direction and taking the (x_1, x_3) plane as $\phi=0$. Thus,

$$u_R^P(\bm{x},t) = \int_\Sigma \frac{\beta^2 \sin 2\theta \cos\phi}{4\pi\alpha^3 R} \frac{\partial \Delta u\left(\bm{\xi}, t-\frac{R}{\alpha}\right)}{\partial t} \mathrm{d}\Sigma(\bm{\xi}). \qquad (13.1)$$

We also calculate the amplitude of the S-wave term. The S-wave terms in the θ and ϕ directions are

$$u_\theta^S(\bm{x},t) = \int_\Sigma \frac{\cos 2\theta \cos\phi}{4\pi\beta R} \frac{\partial \Delta u\left(\bm{\xi}, t-\frac{R}{\beta}\right)}{\partial t} \mathrm{d}\Sigma(\bm{\xi}), \qquad (13.2)$$

$$u_\phi^S(\bm{x},t) = -\int_\Sigma \frac{\cos\theta \sin\phi}{4\pi\beta R} \frac{\partial \Delta u\left(\bm{\xi}, t-\frac{R}{\beta}\right)}{\partial t} \mathrm{d}\Sigma(\bm{\xi}). \qquad (13.3)$$

In the (x_1, x_3) plane ($u_\phi^S = 0$), u_θ^S is the expected quadrifoliate radiation pattern rotated by 45° from the P-wave pattern in the double-couple case.

This concludes our treatment of the usual term. This part of the result, the double-couple force equivalent, holds whether we have a strength-weakening zone or not because the first term of equation (6) does not depend on the stresses. In this case the far-field radiation is proportional to the slip velocity on the fault and is the canonical result. The ratio of the amplitudes of the P- and S- wave terms is $(\beta/\alpha)^3$ except for the angular polarization coefficients.

5 The strength-weakening zone

We return to the problem of the second term in the integral of equation (6). As remarked, this term is zero if $\langle \tau_{jk}(\bm{\xi}, \tau)\rangle = \tau_{jk}(\bm{\xi}^+, \tau) - \tau_{jk}(\bm{\xi}^-, \tau) = 0$. In this case, the only term in the solution is the conventional double couple of the preceding section. However, in a strength-weakening zone (such as near the edge of an advancing crack), the stress is time-dependent and nonzero because the stress drop must vary between the critical shear stress and its final value at the edge of the strength-weakening zone (Fig. 2). The terms $\tau_{jk} n_k$ are the tractions on the two surfaces; they are oppositely directed and each nonzero. Thus, the radiation from the nucleation site and from the evolution of the crack at its edge must have a component that depends on the stress in these regions. The mathematical problem is simpler than that of the preceding section.

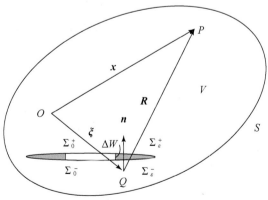

Figure 2 An elastic solid with volume V, closed external surface S, and an internal surface $\Sigma = \Sigma_0 + \Sigma_\varepsilon$

The internal surface consists of the dislocation surface $\Sigma_0 = \Sigma_0^+ + \Sigma_0^-$ and the surface of the strength-weakening zone (area in light gray) $\Sigma_\varepsilon = \Sigma_\varepsilon^+ + \Sigma_\varepsilon^-$.

ΔW is the thickness of the strength-weakening zone, O is the origin of coordinates, $P(\bm{x})$ is a point of observation within V, $Q(\bm{\xi})$ is a slipping element on Σ_ε, $n_k(\bm{\xi})$ is the outward drawn normal to surface Σ

In this problem we start with

$$u_i(\boldsymbol{x},t) = \int_{-\infty}^{\infty} d\tau \int_{\Sigma_\varepsilon} G_{ij}\tau_{jk}n_k d\Sigma(\boldsymbol{\xi}) \tag{14}$$

from equation (6), where Σ_ε is the surface of the strength-weakening zone. From equation (1)

$$u_i(\boldsymbol{x},t) = \int_{-\infty}^{\infty} d\tau \int_{\Sigma_\varepsilon} \left\{ \frac{\gamma_i\gamma_j}{4\pi\rho\alpha^2 R}\delta\left(t-\tau-\frac{R}{\alpha}\right) + \frac{(\delta_{ij}-\gamma_i\gamma_j)}{4\pi\rho\beta^2 R}\delta\left(t-\tau-\frac{R}{\beta}\right) \right\} \tau_{jk}(\boldsymbol{\xi},\tau)n_k d\Sigma(\boldsymbol{\xi}).$$

As before the P-wave term is radially polarized, and the S-wave terms are orthogonally polarized to the P waves. The two terms are

$$u_i^P(\boldsymbol{x},t) = \int_{\Sigma_\varepsilon} \frac{\gamma_i\gamma_j}{4\pi\rho\alpha^2 R}\tau_{jk}\left(\boldsymbol{\xi},t-\frac{R}{\alpha}\right) n_k d\Sigma(\boldsymbol{\xi}) \tag{15.1}$$

$$u_i^S(\boldsymbol{x},t) = \int_{\Sigma_\varepsilon} \frac{(\delta_{ij}-\gamma_i\gamma_j)}{4\pi\rho\beta^2 R}\tau_{jk}\left(\boldsymbol{\xi},t-\frac{R}{\beta}\right) n_k d\Sigma(\boldsymbol{\xi}). \tag{15.2}$$

Let $\tau_{jk}(\boldsymbol{\xi},t) n_k = e_j T(\boldsymbol{\xi},t)$, where $e_j T(\boldsymbol{\xi},t)$ is the traction acting on the surface element $d\Sigma$ having the outward drawn unit normal n_k, e_j is the unit vector in the direction of the traction. Thus,

$$u_i^P(\boldsymbol{x},t) = \int_{\Sigma_\varepsilon} \frac{\gamma_i\gamma_j e_j}{4\pi\rho\alpha^2 R} T\left(\boldsymbol{\xi},t-\frac{R}{\alpha}\right) d\Sigma(\boldsymbol{\xi}) \tag{16.1}$$

$$u_i^S(\boldsymbol{x},t) = \int_{\Sigma_\varepsilon} \frac{(\delta_{ij}-\gamma_i\gamma_j)e_j}{4\pi\rho\beta^2 R} T\left(\boldsymbol{\xi},t-\frac{R}{\beta}\right) d\Sigma(\boldsymbol{\xi}). \tag{16.2}$$

We take the conditions just presented, that is, $\boldsymbol{e}=(1, 0, 0)$, $\boldsymbol{n}=(0, 0, 1)$, $\boldsymbol{\gamma}=(\sin\theta\cos\phi, \sin\theta\sin\phi, \cos\theta)$. The P-wave term is

$$u_R^P(\boldsymbol{x},t) = \int_{\Sigma_\varepsilon} \frac{\sin\theta\cos\phi}{4\pi\rho\alpha^2 R} T\left(\boldsymbol{\xi},t-\frac{R}{\alpha}\right) d\Sigma(\boldsymbol{\xi}). \tag{17.1}$$

The S-wave terms are

$$u_\theta^S(\boldsymbol{x},t) = \int_{\Sigma_\varepsilon} \frac{\cos\theta\sin\phi}{4\pi\rho\beta^2 R} T\left(\boldsymbol{\xi},t-\frac{R}{\beta}\right) d\Sigma(\boldsymbol{\xi}), \tag{17.2}$$

$$u_\phi^S(\boldsymbol{x},t) = -\int_{\Sigma_\varepsilon} \frac{\sin\phi}{4\pi\rho\beta^2 R} T\left(\boldsymbol{\xi},t-\frac{R}{\beta}\right) d\Sigma(\boldsymbol{\xi}). \tag{17.3}$$

The integrands are the radiation from a point force in the x_1 direction.

In equation (15) if we take the contribution from the internal surfaces $\Sigma_\varepsilon = \Sigma_\varepsilon^+ + \Sigma_\varepsilon^-$ into account, the point sources point in opposite directions, and we have a vector point pointing in the e_j direction plus a couple of oppositely directed forces separated by a distance equal to the thickness of the fault. This is a torque whose axis of rotation points is in the $\boldsymbol{e}\times\boldsymbol{n}$ direction (i.e., a single couple).

Let $\tau_{jk}(\boldsymbol{\xi},t)n_k = e_j T(\boldsymbol{\xi},t)$ and $n_k^+ = -n_k^-$ show as before. From equation (15) we have

$$u_i^P(\boldsymbol{x},t) = -\int_{\Sigma_\varepsilon} \frac{\gamma_i\gamma_j e_j}{4\pi\rho\alpha^2 R} \left\langle T\left(\boldsymbol{\xi},t-\frac{R}{\alpha}\right)\right\rangle d\Sigma(\boldsymbol{\xi}) \tag{18.1}$$

$$u_i^S(\boldsymbol{x},t) = -\int_{\Sigma_\varepsilon} \frac{(\delta_{ij}-\gamma_i\gamma_j)e_j}{4\pi\rho\beta^2 R} \left\langle T\left(\boldsymbol{\xi},t-\frac{R}{\beta}\right)\right\rangle d\Sigma(\boldsymbol{\xi}) \tag{18.2}$$

as $\Delta W/R \ll 1$, where we now integrate over only the lower surface and drop the superscript. In equation (18) $\langle T(\boldsymbol{\xi}, t-R/c)\rangle$ is the jump in $T(\boldsymbol{\xi}, t-R/c)$ across the fault zone of thickness ΔW,

$$\left\langle T\left(\boldsymbol{\xi},t-\frac{R}{c}\right)\right\rangle = T\left(\boldsymbol{\xi}^+,t-\frac{R^+}{c}\right) - T\left(\boldsymbol{\xi}^-,t-\frac{R^-}{c}\right).$$

It is easy to show that

$$\left\langle T\left(\xi, t-\frac{R}{c}\right)\right\rangle \approx \Delta T\left(\xi^-, t-\frac{R^-}{c}\right) + \frac{\partial T\left(\xi^-, t-\frac{R^-}{c}\right)}{\partial t}\frac{\gamma_k^- n_k^-}{c}\Delta W\left\{1 + O\left(\frac{\Delta W}{cT_s'}\right)\right\}, \tag{19}$$

where

$$\Delta T\left(\xi^-, t-\frac{R^-}{c}\right) = T\left(\xi^+, t-\frac{R^-}{c}\right) - T\left(\xi^-, t-\frac{R^-}{c}\right)$$

is the stress dislocation, and T_s' is the characteristic time for stress change, $c = \alpha$ or β.

Although equations (10) and (19) are similar in form, they are essentially quite different. Unlike equation (10), where $\Delta u(\xi^-, t-R^-/c)$ and $\partial u(\xi^-, t-R^-/c)/\partial t$ are related due to the fact that $\Delta u(\xi^-, t-R^-/c) \approx D$ and $\partial u(\xi^-, t-R^-/c)/\partial t \approx D/2T_s$ in equation (19) $\Delta T(\xi^-, t-R^-/c)$ is the stress dislocation while $\partial T(\xi^-, t-R^-/c)/\partial t$ is the time rate of stress; the latter quantities are not related as they are in the displacement dislocation case. The stress dislocation $\Delta T(\xi^-, t-R^-/c)$ may be zero or a certain finite quantity, regardless of whether the thickness of the fault zone is finite or zero. There is no similar relationship relating the stress dislocation $\Delta T(\xi^-, t-R^-/c)$ with the time rate of stress $\partial T(\xi^-, t-R^-/c)/\partial t$ such as $\Delta u(\xi^-, t-R^-/c) \approx D$ and $\partial u(\xi^-, t-R^-/c)/\partial t \approx D/2T_s$.

In the case $\Delta W/R \ll 1$, $\Delta W/cT_s \ll 1$ and $\Delta W/cT_s' \ll 1$, where $c = \alpha$ or β, we have

$$\left\langle u\left(\xi, t-\frac{R}{c}\right)\right\rangle \approx \Delta u\left(\xi^-, t-\frac{R^-}{c}\right) \tag{20}$$

and

$$\left\langle T\left(\xi, t-\frac{R}{c}\right)\right\rangle \approx \Delta T\left(\xi^-, t-\frac{R^-}{c}\right) + \frac{\partial T\left(\xi^-, t-\frac{R^-}{c}\right)}{\partial t}\frac{\gamma_k^- n_k^-}{c}\Delta W. \tag{21}$$

The second term of the right-hand side of equation (21) cannot be neglected although the second term of the righthand side of equation (10) can be in the case of $\Delta W/R \ll 1$, $\Delta W/cT_s \ll 1$, and $\Delta W/cT_s' \ll 1$. The difference in the two cases arises because the first term in equation (10) is nonzero and persists as $\Delta W \to 0$, while in the case of equation (21), the first term may be zero if the stress drop is continuous across the fault zone.

The first term on the right-hand side of equation (21) contributes the usual solution for a stress dislocation,

$$u_i^P(\mathbf{x}, t) = -\int_{\Sigma_\varepsilon} \frac{\gamma_i \gamma_j e_j}{4\pi\rho\alpha^2 R}\Delta T\left(\xi, t-\frac{R}{\alpha}\right)d\Sigma(\xi), \tag{22.1}$$

$$u_i^S(\mathbf{x}, t) = -\int_{\Sigma_\varepsilon} \frac{(\delta_{ij} - \gamma_i \gamma_j)e_j}{4\pi\rho\beta^2 R}\Delta T\left(\xi, t-\frac{R}{\beta}\right)d\Sigma(\xi). \tag{22.2}$$

As in equation (16) this is the radiation from a point source of strength of $-\Delta T$ pointing in the e_j direction. The solution for the stress dislocation is zero if the traction on Σ_ε is continuous.

The second term of the right-hand side of equation (21) is the solution for a single couple of oppositely directed, time-dependent forces separated by a distance equal to the thickness of the fault and is nonzero even if the traction is continuous across the fault,

$$u_i^P(\mathbf{x}, t) = -\int_{\Sigma_\varepsilon} \frac{\gamma_i \gamma_j e_j \gamma_k n_k}{4\pi\rho\alpha^3 R}\Delta W \frac{\partial T\left(\xi, t-\frac{R}{\alpha}\right)}{\partial t}d\Sigma(\xi), \tag{23.1}$$

$$u_i^S(\mathbf{x}, t) = -\int_{\Sigma_\varepsilon} \frac{(\delta_{ij} - \gamma_i \gamma_j)e_j \gamma_k n_k}{4\pi\rho\beta^3 R}\Delta W \frac{\partial T\left(\xi, t-\frac{R}{\beta}\right)}{\partial t}d\Sigma(\xi), \tag{23.2}$$

where ΔW is the thickness of the strength-weakening zone, $-\partial T/\partial t$ is the time rate of stress-drop increase, and the integration takes place only over the lower surface of the strength-weakening zone. Equation (23) holds for similar conditions as equation (12) for the case of finite thickness of fault zone ΔW if $\Delta W/R \ll 1$ and $\Delta W/cT_s' \ll 1$, where $c = \alpha$ or β.

For the same conditions listed in the previous paragraph, that is, $\mathbf{e} = (1, 0, 0)$, $\mathbf{n} = (0, 0, 1)$, $\boldsymbol{\gamma} = (\sin\theta\cos\phi, \sin\theta\sin\phi, \cos\theta)$, the P-wave term is

$$u_R^P(\mathbf{x},t) = -\int_{\Sigma_\varepsilon} \frac{\sin 2\theta \cos\phi}{8\pi\rho\alpha^3 R} \Delta W \frac{\partial T\left(\boldsymbol{\xi}, t - \frac{R}{\alpha}\right)}{\partial t} d\Sigma(\boldsymbol{\xi}), \tag{24.1}$$

the S-wave terms in the θ and ϕ directions are

$$u_\theta^S(\mathbf{x},t) = -\int_{\Sigma_\varepsilon} \frac{\cos^2\theta \cos\phi}{4\pi\rho\beta^3 R} \Delta W \frac{\partial T\left(\boldsymbol{\xi}, t - \frac{R}{\beta}\right)}{\partial t} d\Sigma(\boldsymbol{\xi}), \tag{24.2}$$

$$u_\phi^S(\mathbf{x},t) = \int_{\Sigma_\varepsilon} \frac{\cos\theta \sin\phi}{4\pi\rho\beta^3 R} \Delta W \frac{\partial T\left(\boldsymbol{\xi}, t - \frac{R}{\beta}\right)}{\partial t} d\Sigma(\boldsymbol{\xi}). \tag{24.3}$$

This is the radiation from a single couple with time-dependent forces in the $\pm x_1$ direction and torque axis in the x_3 direction. The solution, which gives a time-dependent single couple, is consistent with that obtained by Knopoff and Gilbert (1960) for the problem of radiation from a fault of finite thickness with stress drop across the fault.

The solution for the radiation from a strength-weakening zone depends on the rate of stress drop. A comparison between equations (12) and (23) or (13) and (24) shows that the expressions for the radiation from an element of the strength-weakening zone are equivalent to the radiation from a single couple with torque, while the radiation from an element of slip is equivalent to the radiation from a double couple without torque. They have opposite signs, expressing the fact that the radiation from the strength-weakening zone arises from the varying (decreasing) stress within the strength-weakening zone, which is opposite to the direction of the increasing stress drop before breakdown, while the radiation from the completely fractured crack is a consequence of the breakdown of the weakening zone and the increasing dislocation.

6 Torque waves

The radiation of a torque wave is part of the deformation field, which results from the relative displacement (Bullen, 1953, pp. 13–14). Consider an element of solid material within which displacements $\mathbf{u}(\mathbf{x})$ have occurred. Let the particle initially at position $\mathbf{x}$ be moved to position $\mathbf{x} + \mathbf{u}(\mathbf{x})$. The displacement of a point at position $\mathbf{x} + \delta\mathbf{x}$ is

$$u_i(\mathbf{x}+\delta\mathbf{x}) \approx u_i(\mathbf{x}) + \frac{\partial u_i(\mathbf{x})}{\partial x_j} \delta x_j, \tag{25}$$

where $\delta\mathbf{x}$ is infinitesimal and where the partial derivatives are evaluated at $\mathbf{x}$. By adding and subtracting $(1/2)(\partial u_j(\mathbf{x})/\partial x_i) \delta x_j$ to equation (25), the displacement of the point at $\mathbf{x}+\delta\mathbf{x}$ can be separated into three parts

$$u_i(\mathbf{x}+\delta\mathbf{x}) \approx u_i(\mathbf{x}) + e_{ij}\delta x_j - \omega_{ij}\delta x_j, \tag{26}$$

where e_{ij} is the symmetric strain tensor, and ω_{ij} is the anti-symmetric rotation tensor

$$e_{ij} = \frac{1}{2}\left(\frac{\partial u_j}{\partial x_i} + \frac{\partial u_i}{\partial x_j}\right), \quad \omega_{ij} = \frac{1}{2}\left(\frac{\partial u_j}{\partial x_i} - \frac{\partial u_i}{\partial x_j}\right).$$

The rotation tensor can be written as

$$-\omega_{ij}\delta x_j = (\mathbf{\Omega} \times \delta \mathbf{x})_i, \tag{27.1}$$

$$\mathbf{\Omega} = \frac{1}{2}\nabla \times \mathbf{u}, \tag{27.2}$$

where $\mathbf{\Omega}$ is the rotation vector.

The elastic wave equation (3) for homogeneous elasticity is

$$(\lambda + 2\mu)\nabla(\nabla \cdot \mathbf{u}) - \mu \nabla \times (\nabla \times \mathbf{u}) = \rho \ddot{\mathbf{u}} - \mathbf{f} \tag{28}$$

for which equation (6) is the solution. We take the curl of equation (28) and get

$$\mu \nabla^2 \mathbf{\Omega} = \rho \ddot{\mathbf{\Omega}} - \frac{1}{2}\nabla \times \mathbf{f}.$$

Hence, the rotations are S waves and can be expected to be orthogonal to both the P- and, via equation (27.2), the S-wave components of the motion.

Equation (26) shows that in an elastic solid, the deformation in the vicinity of $\mathbf{x}$ consists of three parts. The first part, which is given by the first term of equation (26) ($u_i(\mathbf{x})$), is equal to the displacement of $\mathbf{x}$ and thus, corresponds to a pure translation of matter near $\mathbf{x}$, which produces no deformation or rotation. The second part, represented by the e_{ij} term, is the true elastic distortion resulting from the differential motion within the body. The third part, represented by the $\mathbf{\Omega}$ term, corresponds to the pure rotation of a small volume element containing the point $\mathbf{x}$ about an axis parallel to $\mathbf{\Omega}$. This is a local rotation and should not be confused with the rigid rotation of the whole body, which has been excluded from $\mathbf{u}$ *ab initio* or with the microscopic rotational motion in which a typical particle is considered not as a material point but as an infinitesimal rigid body (Nowacki, 1986, p.9).

In the case of in-plane slip the $u_i^P(\mathbf{x}, t)$ term in the far field of equation (12.1) is irrotational and makes no contribution to rotational motion, while the $u_i^S(\mathbf{x}, t)$ term of equation (12.2) is rotational and contributes rotational waves or torque-waves

$$\Omega_i(\mathbf{x},t) = -\int_\Sigma \frac{\varepsilon_{ikl}\gamma_j\gamma_k(n_j e_l + e_j n_l)}{8\pi\beta^2 R} \frac{\partial^2 \Delta u\left(\boldsymbol{\xi}, t - \frac{R}{\beta}\right)}{\partial t^2} d\Sigma(\boldsymbol{\xi}), \tag{29}$$

where ε_{ikl} is the usual permutation symbol.

From equation (29) the far-field rotational waves or torque waves for in-plane slip depend on the slip acceleration on the fault. From the orthogonality condition that $\gamma_i \Omega_i = 0$ and equation (29), we have that the rotation is orthogonal to both the P-wave motions $u_i^P(\mathbf{x}, t)$ and the S-wave motions $u_i^S(\mathbf{x}, t)$ and travels with S-wave velocity.

For the coordinate system just presented, that is, $\mathbf{e} = (1, 0, 0)$, $\mathbf{n} = (0, 0, 1)$, $\boldsymbol{\gamma} = (\sin\theta\cos\phi, \sin\theta\sin\phi, \cos\theta)$ the rotational or torque waves from a double-couple source are

$$\Omega_R(\mathbf{x},t) = 0, \tag{30.1}$$

$$\Omega_\theta(\mathbf{x},t) = -\int_\Sigma \frac{\cos\theta\sin\phi}{8\pi\beta^2 R} \frac{\partial^2 \Delta u\left(\boldsymbol{\xi}, t - \frac{R}{\beta}\right)}{\partial t^2} d\Sigma(\boldsymbol{\xi}), \tag{30.2}$$

$$\Omega_\phi(\mathbf{x},t) = -\int_\Sigma \frac{\cos 2\theta\cos\phi}{8\pi\beta^2 R} \frac{\partial^2 \Delta u\left(\boldsymbol{\xi}, t - \frac{R}{\beta}\right)}{\partial t^2} d\Sigma(\boldsymbol{\xi}). \tag{30.3}$$

Equation (30) is in agreement with the far-field term of equation (30) obtained by Cochard et al. (2006).

As in the case of in-plane slip (i.e., the case of a double-couple source), in the case of time-dependent stress drop in the strength-weakening zone, which is the case of a single couple, the $u_i^P(\mathbf{x}, t)$ term expressed by equation (23.1) makes no contribution in the far field to rotational motions. The $u_i^S(\boldsymbol{x}, t)$ term expressed in (23.2) contributes rotational waves

$$\Omega_i(\boldsymbol{x},t) = \int_{\Sigma_\varepsilon} \frac{\varepsilon_{ikl}\gamma_k e_l \gamma_j n_j}{8\pi\beta^2 R} \Delta W \frac{\partial^2 T\left(\boldsymbol{\xi}, t - \frac{R}{\beta}\right)}{\partial t^2} d\Sigma(\boldsymbol{\xi}). \tag{31}$$

The far-field rotational waves or torque waves from the stress drop in the strength-weakening zone depend on the stress acceleration $\partial^2 T/\partial t^2$ in the fault plane. In this case of strength weakening from the orthogonality condition $\gamma_i \Omega_i = 0$ and equation (31), we have that Ω_i is orthogonal to both the P-wave motions $u_i^P(\boldsymbol{x}, t)$ and S-wave motions $u_i^S(\boldsymbol{x}, t)$ and with S-wave retardation in the far field.

Under the same conditions presented in the last paragraph $\boldsymbol{e} = (1, 0, 0)$, $\boldsymbol{n} = (0, 0, 1)$, $\boldsymbol{\gamma} = (\sin\theta\cos\phi, \sin\theta\sin\phi, \cos\theta)$, the rotation from the single-couple source is

$$\Omega_R(\boldsymbol{x},t) = 0, \tag{32.1}$$

$$\Omega_\theta(\boldsymbol{x},t) = -\int_{\Sigma_\varepsilon} \frac{\cos\theta\sin\phi}{8\pi\beta^2 R} \Delta W \frac{\partial^2 T\left(\boldsymbol{\xi}, t - \frac{R}{\beta}\right)}{\partial t^2} d\Sigma(\boldsymbol{\xi}), \tag{32.2}$$

$$\Omega_\phi(\boldsymbol{x},t) = \int_{\Sigma_\varepsilon} \frac{\cos^2\theta\cos\phi}{8\pi\beta^2 R} \Delta W \frac{\partial^2 T\left(\boldsymbol{\xi}, t - \frac{R}{\beta}\right)}{\partial t^2} d\Sigma(\boldsymbol{\xi}). \tag{32.3}$$

7 Seismic Effects of Finite Fault Thickness

The presence of a single-couple source is not in conflict with the usual interpretation of double-couple force equivalents: the double couple is the lowest order combination of forces, which has no net force and no torque. How then can we have a single-couple solution in this case? The answer comes from an appreciation of the fact that our fault has a finite thickness and that the relaxation of static torques within the fault zone is exactly compensated by the radiation of torque waves from the fault. The double-couple solution is the appropriate solution if the fault has zero thickness (Fig. 3). If the fault zone has zero thickness, the torque, which is equal to the product of the traction and fault thickness, is zero. During an earthquake event, a transition zone, also known as a breakdown or cohesive zone, is found at the edge of the crack. Within this zone the medium undergoes a gradual transition from the continuum state of static elasticity to the nonlinear ruptured state. The stress across the fault plane is related to the slip by the constitutive relation, as shown schematically in Figure 4 (Ohnaka and Yamashita, 1989; Ohnaka et al. (1997); Venkataraman and Kanamori, 2004). In this figure, σ_0 is the initial stress or prestress (i.e., the stress before the earthquake or the stress in the surrounding medium). σ_p is the yield stress or peak strength (i.e., the upper limit of static frictional stress). At this level instantaneous instability begins, and strength weakening occurs; σ_d is the dynamic frictional stress, and σ_1 is the final or static frictional stress.

With this constitutive relation we can analyze the moment release in dynamic earthquake ruptures. Let us assume that a dynamically ruptured fault propagates with variable rupture velocity v_f in the x_1 direction (see Fig. 5c). In the elastic solid, dynamical rupture will initiate at a point where the strength excess is least; after initiation a stress concentration is located at the crack tip. The shear stress on the fault plane ($x_3 = 0$) far ahead of the crack tip is the initial stress σ_0. The shear stress σ rises from the initial stress σ_0 to the yield stress or peak strength σ_p and then decreases to the dynamic frictional stress σ_d (Fig. 5a). During the slip-weakening part of the

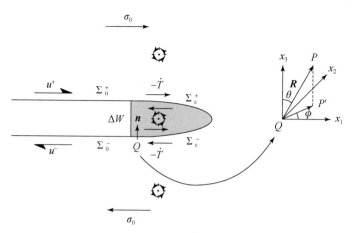

Figure 3 Radiation from a strength-weakening zone of a finite thickness of ΔW is equivalent to a single couple $-T$ is the time rate of increase of stress drop. The axis of rotation for the torque exerted by $-T$ points in the $-e \times n$ direction (x_2 direction in the present example). This radiation of torque wave (small circles rotated counterclockwise outside the strength-weakening zone) from the fault exactly compensates the relaxation of torques within the fault and points in the $e \times n$ direction (negative x_2 direction in the present example) as schematically shown by the single couple and the small circle rotated counterclockwise within the strength-weakening zone. The figure to the right represents the spherical polar coordinate system centered at a slipping element $Q(\xi)$ on Σ_ε (not to scale). P is a point of observation, P' is the projection of P onto the (x_1, x_2) plane. For explanation of the other symbols, see text. The net force, which is the sum of the tractions T, may be zero, but nevertheless they exert a torque because of the finite separation

process, the shear stress drops from σ_p to the dynamical frictional stress σ_d with continuing, increasing slip. The critical slip D_c marks the transition from the decreasing stress to the steady, dynamical friction. Slip continues to the final slip D resisted by the dynamic sliding friction σ_d where healing begins (Fig. 5b). When the slip stops the stress on the newly locked portion of the fault continues to vary (increase) with time, ultimately approaching the final (static) frictional stress σ_1. σ_1 may be greater than, equal to, or smaller than σ_d, but in Figures 4 and 5 only the case $\sigma_1 > \sigma_d$ is shown for simplicity.

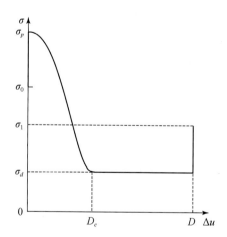

Figure 4 Constitutive relation of shear stress across the fault plane versus slip (after Ohnaka and Yamashita, 1989; Ohnaka et al., 1997; Venkataraman and Kanamori, 2004)

In contrast to Figure 5, which presented a picture in space of the variations in stress, Figure 6 illustrates schematically the stress change with time at a representative point on the fault plane (Yamashita, 1976) as the stress-change pattern associated with crack growth moves across the point. At time t_0 rupture initiates at the

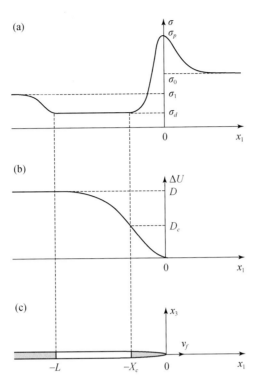

Figure 5 Distribution of stress (a) and slip (b) within the breakdown zone for a dynamically ruptured fault (c) Area in light gray ($-X_c \leq x_1 \leq 0$) represents the breakdown zone, and area in dark gray ($x_1 \leq -L$) represents the healing portion of the fault (after Heaton, 1990; Rice et al., 2005). For explanations of other symbols, see text. The patterns in this figure travel rightward

hypocenter on the fault plane. As rupture progresses from the point of initiation to this point, the stress begins to increase from σ_0 to peak stress σ_p at time t_p. As slip increases from zero to D_c, strength-weakening occurs, and stress drops from σ_p to the dynamic frictional stress σ_d at time t'_s. For slip greater than D_c, the stress on the fault plane is σ_d until the slip comes to a stop at time t_s. D_c is the breakdown slip or critical slip-weakening distance. After the slip stops, the stress increases to the final stress σ_1 at time t_1. Our concern in this article has been with the time-dependent stress during the slip-weakening process. In addition to the increase in slip, which accompanies the stress decrease in the early stages of fracturing, there is an increase of stress (as noted in Fig. 6), which takes place in the example of the figure after the slip stops. If the increase in stress takes place before slip stops completely, there will be a contribution to the single-component term from the stopping phase. In this article we consider only the first of the two episodes of single-couple radiation and assume that strength hardening takes place only after all slip has been completed and is nonradiative.

We do not enter into present day arguments concerning the cause of the stoppage of slip in self-healing pulses, whether it is due to an increase of sliding friction as slip decelerates (Heaton, 1990; Cochard and Madariaga, 1996; Zheng and Rice, 1998) or due to an encounter of the crack with an extended strong region in the fault surface, which is a region of high strength excess (Mikumo and Miyatake, 1978; Day, 1982; Wald and Heaton, 1994), and which causes the increase in the friction in Figure 6. As in Figures 4 and 5, only the case of $\sigma_1 > \sigma_d$ is shown in Figure 6. $\Delta\sigma_b = \sigma_p - \sigma_d$ is the effective shear stress. $\Delta\sigma_d = \sigma_0 - \sigma_d$ is the dynamic stress drop, and $\Delta\sigma = \sigma_0 - \sigma_1$ is the static stress drop. Thus, the breakdown zone is a zone within which the stress changes dramatically.

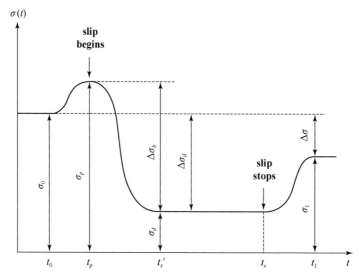

Figure 6　Time dependence of stress at a representative point on the fault plane

For explanation of the symbols, see text (after Yamashita, 1976)

On a ruptured portion of the fault of area ΔA, the far-field radiation due to the increase of the stress drop is approximately proportional to the time rate of torque moment (equation 23)

$$\Delta \dot{M}_t = \Delta W \Delta \dot{\sigma} \Delta A, \tag{33}$$

where ΔW is the thickness of the breakdown zone, $\Delta \dot{\sigma}$ is the time rate of increase of effective shear stress $\Delta \sigma(t)$ (i.e., the rate of increase of the stress drop),

$$\Delta \sigma(t) = \sigma_p - \sigma(t), \tag{34}$$

$\sigma(t)$ is the shear stress as a function of time. From the integration of $\Delta \dot{M}_t$ over the complete time of fracture and the entire fractured area, we get the total torque moment release

$$M_t = \Delta W \Delta \sigma_b A, \tag{35}$$

where

$$\Delta \sigma_b = \sigma_p - \sigma_d \tag{36}$$

is the effective shear stress, and A is the area of the total ruptured plane of the fault.

The far-field radiation from the beginning of slip to final slip D for the ruptured portion of the fault of area ΔA is proportional to the time rate of change of seismic moment

$$\Delta \dot{M}_0 = \mu \Delta \dot{u} \Delta A. \tag{37}$$

If we also integrate $\Delta \dot{M}_0$ over the entire time interval of rupture and the entire rupture surface, we obtain the usual representation for scalar seismic moment

$$M_0 = \mu D A, \tag{38}$$

where M_0 is the scalar seismic moment due to a dislocation source of final slip D.

The dimensionless ratio k of the torque moment rate in the radiation from the decrease of stress in the breakdown process with an effective shear stress change of $\Delta \sigma_b$, thickness of breakdown zone of ΔW to the usual seismic moment rate from the same dislocation source with rigidity μ, is

$$k = \frac{\Delta \dot{M}_t}{\Delta \dot{M}_0}. \tag{39}$$

To estimate the torque moment rate and seismic moment rate, we find from equations (33) and (37)

$$\Delta \dot{M}_t \approx \frac{\Delta W \sigma_b \Delta A}{T'_s}, \quad \Delta \dot{M}_0 \approx \frac{\mu D \Delta A}{T_s},$$

where $T'_s = t'_s - t_p$ is the characteristic time for stress change (i.e., time for completing the breakdown), and $T_s = t_s - t_p$ is the usual rise time (i.e., the time for the completion of slip at a point on the fault). We estimate T_s and T'_s roughly by $T_s \approx D/v_f$, $T'_s \approx D_c/v_f$, where v_f is the advancing velocity of the crack tip (i.e., the rupture velocity). Thus,

$$k = \frac{\Delta W \Delta \sigma_b}{\mu D_c}, \tag{40}$$

where we have used $T'_s/T_s \approx D/D_c$ by virtue of $T_s \approx D/v_f$ and $T'_s \approx D_c/v_f$.

The effective shear stress $\Delta \sigma_b$ has several estimates. According to an early estimate by Kanamori (1994), $\Delta \sigma_b \approx 2$ to 20MPa. The estimate of Ohnaka (2003) is $\Delta \sigma_b \approx 1$ to 100MPa. Recently, Rice et al. (2005) gave $\Delta \sigma_b \approx 100$MPa. In our numerical estimate, we adopt

- $D_c \approx 0.5$m (Mikumo and Yagi, 2003; Fukuyama and Mikumo, 2007).
- $\Delta W \approx 200$m (Li and Leary, 1990; Li et al., 1990; Li and Vidale, 1996; Li et al., 1997).
- $\Delta \sigma_b \approx 60$MPa (Kanamori, 1994; Ohnaka, 2003; Rice et al., 2005).
- $\mu \approx 3 \times 10^4$MPa.

Substitution of these values into equation (40) yields $k \approx 4/5$. Field investigations from some great earthquakes show that the width of the fault zone or the thickness of the strength-weakening zone ΔW ranges from several hundred meters to several kilometers. In view of these comments the k value, at least for some larger earthquakes, may be even larger than the present estimate.

The total moment from the torque moment M_t and seismic moment M_0 for the entire rupture process and the entire ruptured area is equivalent to a dislocation source with seismic moment

$$M'_0 = \frac{1}{2} M_t + M_0, \tag{41}$$

where the factor of 1/2 is introduced because the radiation from a single couple with unit moment is half the radiation of a double couple with the same amount of seismic moment for each pair of couples.

If the contribution to the total moment from the torque moment is not taken into account, the seismic moment due to a dislocation source is estimated by equation (38). Thus, the ratio of seismic moment estimated from far-field radiation, taking into account the contribution from the torque moment to that simply from the dislocation source, is

$$\frac{M'_0}{M_0} = 1 + \frac{M_t}{2M_0} \approx 1 + \frac{k}{2} \cdot \frac{D_c}{D}. \tag{42}$$

The ratios M'_0/M_0 are listed in Table 1 for some representative values of the final slip D for the case $k = 0.8$. The overestimate of the total moment magnitude is only about $(2/3) \lg(M'_0/M_0) = (2/3) \lg(1.2) \approx 0.06$ in the most optimistic case of importance of the torque moment. The influence is not pronounced on a logarithmic scale.

The results obtained here have two implications. The first is that in usual waveform analysis, the use of the dislocation model alone would overestimate seismic moment and consequently the final slip D if the contribution to the far-field radiation from the torque moment is not taken into account. The overestimate may be as much as a factor of 1.2 to 1.04 as D ranges from 1 to 5m. These results may explain the discrepancies between seismic moments or final slips estimated from usual waveform analysis using the dislocation model alone and field observations or geodetic measurements. The second is that in usual waveform analysis, the use of the dislocation model alone would introduce an extra seismic moment rate and consequently an extra slip rate of as much as a factor of $k/2 \approx 0.4$ of the usual seismic moment rate and the usual slip rate during the time interval for completion of the breakdown process if the contribution to the far-field radiation from the torque moment is not taken into account. This will in turn influence the calculation of seismic energy radiated from earthquake events.

Table 1 Ratios of M_0'/M_0

D/m	D/D_c	M_0'/M_0
1	2	1.20
2	4	1.10
3	6	1.07
4	8	1.05
5	10	1.04

8 Discussion and Conclusions

We have reexamined the two canons of the seismological literature that elastic displacements in the far-field are proportional to slip velocities on the dynamical fault surface, and that dynamical in-plane slip on an earthquake fault has a double-couple body force equivalent. Taking into account the fact that if faulting occurs on a fault of finite thickness and if a strength-weakening zone exists near the advancing crack tip, we have shown that in addition to the usual double-couple term, there is a single-couple term in the body force equivalence in the far-field displacement, which is proportional to the time rate of increase of stress drop. We have shown that the single-couple equivalent does not violate principles of Newtonian mechanics because the torque imbalance in the single couple is counterbalanced by rotations within the fault zone. The crack therefore radiates torque waves.

We have estimated the ratio of the torque moment rate radiated during the breakdown process to the usual seismic moment rate from the same rupturing fault and the ratio of seismic moment released with contribution from the torque moment released to that simply from the dislocation source. These results imply that if the contribution to the far-field radiation from the torque moment is not taken into account in usual waveform analysis, the use of the dislocation model alone would overestimate seismic moment and consequentially the final slip. It would introduce an extra seismic moment rate and consequentially an extra slip rate. This will in turn affect calculations of seismic energy radiated from earthquake events. We have also shown that frictional torques accumulated in a fault zone of finite width during an inter-earthquake interval are relaxed through the development of torque or rotation waves radiated as shear waves during the time-dependent, frictional, or stress-weakening (relaxation) part of the fracture process near the tips of advancing cracks. Torque waves are small during the more familiar frictional sliding interval, where the dynamical friction remains relatively constant.

The relaxation of torques within the fault may play an important role in driving the rotation of material in the fault zone and dramatically changing the dynamic friction in the fault zone, and the radiation of torque waves from an advancing strength-weakening zone before the occurrence of a large earthquake may give a clue to account for rotational phenomena, which have been reported in some historical documents (Galitzin, 1912, p. 75; Bullen, 1953, pp. 135, 251; Richter, 1958, p. 213; Bouchon and Aki, 1982; Takeo and Ito, 1997; Takeo, 1998; Teisseyre et al., 2003; Igel et al., 2007).

Data and Resources

All data used in this article came from published sources listed in the references.

Acknowledgments

Yun-Tai Chen was supported by the National Natural Science Foundation of China (No. 40774021).

References

Aki, K. (1966). Generation and propagation of G waves from Niigata earthquake of June 16, 1964. 2. Estimation of earthquake movement, released energy, and stress-strain drop from G wave spectrum, Tokyo University, *Bull. Earthq. Res. Inst.* **44**, 23–88.

Aki, K., and P. Richards (1980). *Quantitative Seismology Theory Methods*, Vols. **1** and **2**, W. H. Freeman, San Francisco, 932 pp.

Born, M., and E. Wolf (1959). *Principles of Optics*, Pergamon Press, London, 803 pp.

Bouchon, M., and K. Aki (1982). Strain, tilt, and rotation associated with strong ground motion in the vicinity of earthquake faults, *Bull. Seismol. Soc. Am.* **72**, 1717–1738.

Bullen, K. E. (1953). *An Introduction to the Theory of Seismology*, Second Ed., Cambridge University Press, New York, 296 pp.

Burridge, R., and L. Knopoff (1964). Body force equivalents for seismic dislocations, *Bull. Seismol. Soc. Am.* **51**, 69–84.

Cochard, A., and R. Madariaga (1996). Complexity of seismicity due to highly rate-dependent friction, *J. Geophys. Res.* **101**, no. B11, 25, 321-25, 366.

Cochard, A., H. Igel, B. Schuberth, W. Suryanto, A. Velikoseltsev, U. Schreiber, I. Wassermann, F. Scherbaum, and D. Vollmer (2006). Rotational motions in seismology: Theory, observation, simulation, in *Earthquake Source Asymmetry, Structural Media and Rotation Effects*, R. Teisseyre, M. Takeo, and E. Majewski (Editors), Springer-Verlag, Berlin, Heidelberg, New York, The Netherlands, 582 pp.

Day, S. M. (1982). Three-dimensional simulation of spontaneous rupture: the effect of nonuniform prestress, *Bull. Seismol. Soc. Am.* **72**, 1881–1902.

de Hoop, A. T. (1958). *Representation Theorems for the Displacement in An Elastic Solid and Their Application to Elastodynamic Diffraction Theory*, Doctoral Thesis, Technische Hogeschool, Delft, The Netherlands, 84 pp.

Fukuyama, E., and T. Mikumo (2007). Slip-weakening distance estimated at near-fault stations, *Geophys. Res. Lett.* **34**, L09302, doi 10.1029/2006 GL029203.

Galitzin, B. B. (1912). Lecture on seismometry, in *Selected Works of B. B. Galitzin*, Vol. **2**, 1–228, Academy of Sciences, USSR, Moscow, 487 pp. (in Russian).

Haskell, N. A. (1964). Total energy and energy spectral density of elastic wave radiation from propagating faults, Part I, *Bull. Seismol. Soc. Am.* **54**, 1811–1841.

Haskell, N. A. (1966). Total energy and energy spectral density of elastic wave radiation from propagating faults, Part II, A statistical source model, *Bull. Seismol. Soc. Am.* **56**, 125–140.

Heaton, T. (1990). Evidence for and implication of self-healing pulses of slip in earthquake rupture, *Phys. Earth Planet. Interi.* **64**, 1–20.

Igel, H., A. Cochard, J. Wassermann, A. Flaws, U. Schreiber, A. Velikoseltsev, and N. Pham Dinh (2007). Broad-band observations of earthquake-induced rotational ground motions, *Geophys. J. Int.* **168**, 182–196.

Kanamori, H. (1994). Mechanics of earthquakes, *Ann. Rev. Earth Planet Sci.* **22**, 207–237.

Knopoff, L. (1956). Diffraction of elastic waves, *J. Acous. Soc. Am.* **28**, 217–229.

Knopoff, L., and F. Gilbert (1960). First motions from seismic sources, *Bull. Seismol. Soc. Am.* **50**, 117–134.

Kostrov, B. V. (1970). The theory of the focus for tectonic earthquakes, *Izv. Phys. Solid Earth* 258–267, (in Russian).

Kostrov, B. V. (1974). Seismic moment and energy of earthquake and seismic flow of rock, *Izv. Phys. Solid Earth*, 13–21, (in Russian).

Li, Y.-G., and P. C. Leary (1990). Fault zone trapped seismic waves, *Bull. Seismol. Soc. Am.* **80**, 1245–1271.

Li, Y.-G., and J. E. Vidale (1996). Low-velocity fault-zone guided waves: numerical investigations of trapping efficiency, *Bull. Seismol. Soc. Am.* **86**, 371–378.

Li, Y.-G., W. L. Ellsworth, C. H. Thurber, P. E. Malin, and K. Aki (1997). Fault-zone guided waves from explosions in the San Andreas fault at Parkfield and Cienega valley, California, *Bull. Seismol. Soc. Am.* **87**, 210–222.

Li, Y.-G., P. C. Leary, K. Aki, and P. E. Malin (1990). Seismic trapped modes in the Oroville and San Andreas fault zones, *Science* **249**, 763–766.

Love, A. E. H. (1927). *A Treatise on the Mathematical Theory of Elasticity*, 4th ed., Cambridge University Press, New York, 643 pp.

Maruyama, T. (1963). On the force equivalents of dynamical elastic dislocations with reference to the earthquake mechanism, Tokyo University *Bull. Earthq. Res. Inst.* **41**, 46–86.

Mikumo, T., and T. Miyatake (1978). Dynamic rupture process on a three-dimensional fault with non-uniform frictions and near-field seismic waves, *Geophys. J. R. Astr. Soc.* **54**, 417–458.

Mikumo, T., and Y. Yagi (2003). Slip-weakening distance in dynamic rupture of in-slab normal-faulting earthquakes, *Geophys. J. Int.* **155**, 443–455.

Nowacki, W. (1986). *Theory of Asymmetric Elasticity*, Pergamon Press, Oxford, 384 pp.

Ohnaka, M. (2003). A constitutive scaling law and a unified comprehension for frictional slip failure, shear fracture of intact rock, and earthquake rupture, *J. Geophys. Res.* **108**, no. B2, 2080, doi 10.1029/2000JB000123.

Ohnaka, M., and T. Yamashita (1989). A cohesive zone model for dynamic shear faulting based on experimentally inferred constitutive relation and strong motion source parameters, *J. Geophys. Res.* **94**, 4089–4104.

Ohnaka, M., H. Akatsu, A. Mochizuki, F. Tagashira, and Y. Yamamoto (1997). A constitutive law for the shear failure of rock under lithospheric condition, *Tectonophysics* **277**, 1–27.

Rice, J. R., C. G. Sammis, and R. Parsons (2005). Off-fault secondary failure induced by a dynamic slip-pulse, *Bull. Seismol. Soc. Am.* **95**, 109–134, doi 10.1785/0120030166.

Richter, C. F. (1958). *Elementary Seismology*, W. H. Freeman, San Francisco, 768 pp.

Stauder, W., and G. A. Bollinger (1964). The S-wave project for focal mechanism studies, earthquakes of 1962, *Bull. Seismol. Soc. Am.* **54**, 2198–2208.

Stauder, W., and G. A. Bollinger (1966). The S-wave project for focal mechanism studies, earthquakes of 1963, *Bull. Seismol. Soc. Am.* **56**, 1363–1371.

Stokes, G. G. (1849). On the dynamical theory of diffraction, *Trans. Camb. Phil. Soc.* **9**, 1–62, Reprinted in *Stokes' Mathematical and Physical Papers*, **2** (1883). Cambridge, 243–328.

Stratton, J. A. (1941). *Electromagnetic Theory*, McGraw-Hill, New York, 615 pp.

Takeo, M. (1998). Ground rotational motions recorded in near-source region of earthquake, *Geophys. Res. Lett.* **25**, 789–792.

Takeo, M., and H. M. Ito (1997). What can be learned from rotational motions excited by earthquakes? *Geophys. J. Int.* **129**, 319–329.

Teisseyre, R., J. Suchcicki, K. P. Teisseyre, J. Wiszniowski, and P. Palangio (2003). Seismic rotation waves: Basic elements of theory and recording, *Ann. Geophys.* **46**, 671–685.

Wald, D. J., and T. H. Heaton (1994). Spatial and temporal distribution of slip for the 1992 Landers, California earthquake, *Bull. Seismol. Soc. Am.* **84**, 668–691.

Venkataraman, A., and H. Kanamori (2004). Observational constraints on the fracture energy of subduction zone earthquakes, *J. Geophys. Res.* **109**, B05302, doi 10.1029/2003JB002549.

Yamashita, T. (1976). On the dynamic process of fault motion in the presence of friction and inhomogeneous initial stress. Part I. Rupture propagation, *J. Phys. Earth* **24**, 417–444.

Zheng, G., and J. R. Rice (1998). Conditions under which velocity weakening friction allows a self-healing versus a cracklike mode of rupture, *Bull. Seismol. Soc. Am.* **88**, no. 6, 1466–1483.

Teleseismic receiver function and surface-wave study of velocity structure beneath the Yanqing-Huailai Basin Northwest of Beijing[*]

Rong-Mao Zhou[1], Brian W. Stump[2], Robert B. Herrmann[3], Zhi-Xian Yang[4] and Yun-Tai Chen[4]

1. Geophysics Group, MS D443, Los Alamos National Laboratory, Los Alamos, New Mexico 87545;
2. Roy M. Huffington Department of Earth Sciences, Southern Methodist University, Dallas, Texas 75275;
3. Department of Earth and Atmospheric Sciences, Saint Louis University, St. Louis, Missouri 63108;
4. Institute of Geophysics, China Earthquake Administration, Beijing, P. R. China, 100081

Abstract Shear-wave velocities beneath the Yanqing-Huailai Basin, 90-140km northwest of Beijing, are estimated from the joint inversion of surface-wave phase velocities and teleseismic receiver functions. The data set is from a temporary broadband seismic network supported by the Program for Array Seismic Studies of the Continental Lithosphere (PASSCAL) in the basin and includes 34 teleseismic events from 2003 to 2005.

Receiver functions from the teleseismic events are similar for the stations around the Yanqing-Huailai Basin and exhibit little variation with azimuth. The velocity models constrained by receiver functions and surface-wave dispersion curves are also similar. The resulting models reflect the low-velocity basin sediments to 2km followed by a positive velocity gradient to 15km with shear-wave velocity increasing from 2.0 to 3.55km/sec. Evidence of a midcrust low-velocity layer starts at 15km with a shear velocity decrease to 3.3km/sec that extends to approximately 25km. The total crustal thickness is 38–42km with a smooth Moho transition to an upper-mantle shear velocity of 4.3km/sec. The low-velocity zone is consistent with recent extension, geothermal activity, and earthquake locations above this depth.

The average shear velocity model for the basin has similarities to other regional and global models but provides more detailed structure in the uppermost and lower portions of the crust. The new model includes the effect of the sediments in the basin, the low-velocity layer, and the gradual Moho transition. Predicted P- and S- travel times are 1–3.5 sec slower than the previous models at regional distances.

1 Introduction

The Yanqing-Huailai Basin is the geographic focus of this study and is one of the extensional features in the North China Block (NCB). It is located between 40°00′N–40°38′N and 115°04′E–116°14′E, 90–140km northwest of Beijing, China (Fig. 1b, c), and is part of the active Zhangjiakou-Bohai seismic zone that passes through the northern part of China (Zhao et al., 2005). Eight historic earthquakes with magnitude greater than 5 have occurred in the Yanqing-Huailai region since 294 A. D. (Zhang et al., 1996); and thus earthquake risk

---

[*] 本文发表于 *Bull. Seism. Soc. Amer.*, 2009 年, 第 **99** 卷, 第 3 期, 1937–1952.

assessment is important in light of the population of near-by Beijing and other large cities. Numerous underground mines in the region regularly experience rock bursts and collapses resulting in the disruption of mining operations. Development of crust and upper-mantle models for this region are designed to improve the understanding of crustal development, the occurrence of earthquakes and mine related events, and the locations of seismic events in the region. The political and economic importance of this region has motivated a number of geophysical studies in the area, including refraction/wide-angle reflection surveys (Zhang et al., 1996; Zhu et al., 1997; Zhao et al., 2005), a large earthquake risk study (Liu et al., 1997), a neotectonics and seismicity study (Wu et al., 1979), and a tectonic stress study (Xu et al., 1997), that provide a foundation for our study.

2 Regional tectonic setting

The Yanqing-Huailai Basin is in the NCB, part of the Sino-Korean Block, the largest and oldest cratonic region in China, with a total area greater than 1,500,000km² (Fig. 1). It is bordered to the north by the Yinshan-Yanshan orogenic belt, to the south by the Qingling-Dabie belt, and extends into the Gulf of Bohai and the northern Yellow Sea to the east (Rowley et al., 1997; Zhao et al., 1999, 2002; Lin et al., 2004). As discussed and modeled by Lin et al. (2004), the NCB has undergone two distinct phases of tectonics and resulting crustal deformation that

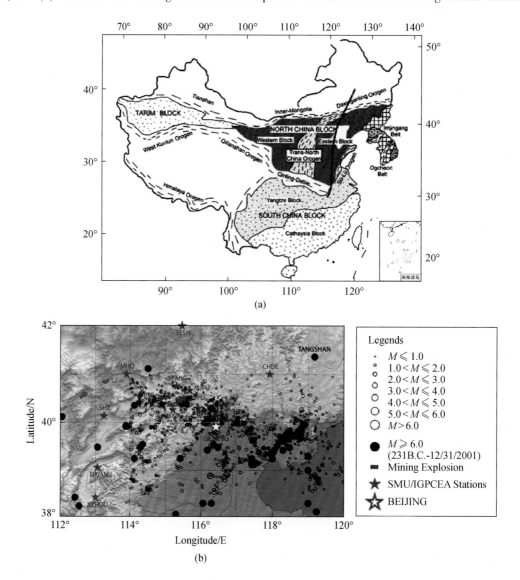

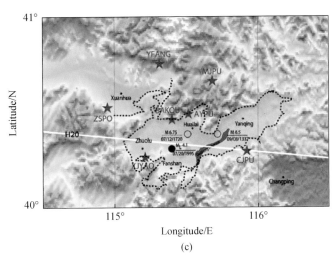

Figure 1 (a) Schematic tectonic map of China showing the major Precambrian blocks and Late Neoproterozoic and Paleozoic fold belts (courtesy of Guochun Zhao at the University of Hong Kong) and the north-northeast-south-southwest extended thick line indicating the Tan-Lu fault zone (modified from Xu et al., 2005). (b) Map of SMU-IGPCEA Huailai Seismic Network and seismicity (open circles) for the time period of 01 January 2002 through 31 December 2006 from the Capital Circle Seismic Network (see the Data and Resources section). Black solid circles are locations of earthquakes with magnitude greater than 6 from 231 B.C. to 31 December 2001. White dashed ectangle is the Yanqing-Huailai Basin and adjacent area (Fig. 1c), and red solid rectangle denotes the source location of the 1.3 kt mining explosion in 2002. (c) Topographic map of Yanqing-Huailai Basin (black dashed line) and adjacent area. Broadband seismic stations of the SMU-IGPCEA Huailai Seismic Network are designated as stars. The white line is the Beijing-Huailai-Fengzhen (H-20) refraction/wide-angle reflection profile (Zhu et al., 1997). The open circles are the locations of two historical earthquakes in 1337 and 1720. The solid circle is the epicenter of an M_L 4.1 earthquake on 20 July 1995. The solid dots are towns in the area

affects the velocity structure and seismicity in the region. Their modeling shows that the first phase involved north-south compression and explains compressional structures including the two bounding east-west orogenic belts and crustal thickening before the late Jurassic. The second phase of deformation is responsible for north-northeast-trending extensional basins in the region and suggests east-west-directional crustal stretching and thinning during the late Mesozoic moderated by the Tan-Lu fault to the east.

3 Seismicity—NCB

The NCB is currently a region of intraplate seismicity and active crustal deformation. The China Historical Earthquake Catalogue Database (see the Data and Resources section) and the China Seismograph Network (CSN) Catalog (see the Data and Resources section) list a total of 323 earthquakes with magnitude greater than 4.0 in the region occurring between 231 B.C. and 2002, with 36 events larger than magnitude 6.0 including the destructive Tangshan earthquake of 1976. Recent seismicity (01 January 2002 through 31 December 2006) in the Capital Circle Region (longitude: 114°E–120°E; latitude: 38.5°N–41°N) is well constrained by the Capital Circle Seismic Network (see the Data and Resources section) (Fig. 1b).

4 Yanqing-Huailai Basin

The Yanqing-Huailai Basin consists of four intermountain basins: Yanqing, Huailai, Fanshan, and Zhuolu, extending in a north-northeast-south-southwest direction. These represent subparallel elongated half-grabens bounded by normal faults to the north-northwest (Pavlides et al., 1999) and reflect the crustal stretching introduced earlier. The basin basement is composed of Archaean metamorphic rocks, Proterozoic and Palaeozoic dolomites, gneisses and clastic rocks, interbedded with thin layers of coal, as well as Mesozoic volcanics (andesite, rhyolite, and tuff), and pyroclastics. Granitic and granodioritic bodies intruded into the Proterozoic and Palaeozoic rocks of the area from the late Jurassic to the Cretaceous. Mafic dykes are aligned along north-northeast-south-southwest tectonic structures. Sedimentation in the basins, initiated in the Pliocene, produced deposits of cemented gravels, overlain by clays, marls, sands, and unconsolidated conglomerates in the middle and upper parts (Pavlides et al., 1999). An unconformity exists between Pliocene and Pleistocene sediments with the younger deposits consisting of fluvial and lacustrine sands, gravels, silts, and clays interbedded with palaeosoil layers. Total estimated Pliocene sediment thickness reaches 1500 m, with Pleistocene thicknesses of 800 m (Wu et al., 1979). These layer thicknesses decrease as the southern edge of the basin is approached in the half-graben structure (Pavlides et al., 1999).

Two, large historic earthquakes have occurred in the Yanqing-Huailai Basin: the Huailai earthquake (magnitude 6.5) on 8 September 1337 and the Shacheng earthquake (magnitude 6.75) on 12 July 1720 (Fig. 1c). The Beijing Telemetered Seismograph Network operated by the China Seismological Bureau (now China Earthquake Administration) provides enhanced earthquake monitoring capability in this area and records 15–20 earthquakes with magnitude equal or greater than 2.0 every year in the Yanqing-Huailai Basin and its adjoining area (Chen et al., 1998). An M_L4.1 earthquake occurred in the Yanqing-Huailai Basin on 20 July 1995, followed by approximately 450 aftershocks (Chen et al., 1998). Xu et al. (1999, 2001) and Chen et al. (2005) have studied the rupture process and focal mechanism of this earthquake and the following earthquake sequence.

5 Analysis approach

We follow previous approaches and develop velocity models for the study area using the joint inversion of surface-wave dispersion and receiver functions. Data were recorded by seven temporary broadband seismic stations deployed in and around the Basin (Fig. 1, Table 1) that were part of the temporary broadband seismic network operated by the Southern Methodist University (SMU) and the Institute of Geophysics, China Earthquake Administration (IGPCEA) with support from the Incorporated Research Institutions for Seismology (IRIS) Program for Array Seismic Studies of the Continental Lithosphere (PASSCAL) Instrument Center (Zhou, Stump, Chen, Hayward, et al., 2003).

Teleseismic receiver functions have been used to provide constraints on crust and upper-mantle shear-wave structure in a number of regions (Langston, 1977, 1979; Owens et al., 1984; Zheng et al., 2005), but the resulting models using typical problem formulation are nonunique as they contain little absolute velocity information (Ammon et al., 1990). The inability to constrain absolute velocity leads to trade-offs between velocity and depth and results in a range of final velocity models with different average velocities that fit the observed receiver functions equally well. Özalaybey et al. (1997) addressed this nonuniqueness by proposing the joint inversion of receiver functions and surface-wave phase velocities. Several authors (Du and Foulger, 1999; Herrmann et al., 2000; Julia et al., 2000; Hermann et al., 2001; Julia et al., 2003; Yoo et al., 2007) have determined shear-wave structures using the joint inversion of receiver functions and surface waves, and their results have illustrated the value of including even a limited band of dispersion data to reduce the trade-offs between crustal thickness and velocity. Our analysis combines local phase velocity estimates from teleseismic events with the teleseismic receiver functions in order to develop improved crust and upper-mantle velocity models.

Table 1 Geographical distribution of the broadband seismic stations

Station Name	Station Code	Latitude/(°N)	Longitude/(°E)	Elevation/m
An Ying Pu	AYPU	40.4936	115.5168	815
Ba Kou	BAKOU	40.4650	115.4025	741
Chen Jia Pu	CJPU	40.3019	115.9233	675
Ma Jia Pu	MJPU	40.6667	115.6837	1075
Xu Jia Yao	XJYAO	40.2679	115.2133	789
You Fang	YFANG	40.7544	115.3188	1109
Zhai Shan Po	ZSPO	40.5278	114.9500	865

6 Data set and analysis tools

The seven temporary stations in the study consist of STS-2 seismometers, Quanterra Q330 data acquisition systems, and Quanterra PB14 Packet Balers. Two separate vaults were constructed at each site, one for the seismometer and the other for the rest of the equipment in order to reduce noise. Seismometers were deployed in hard rock after excavating the $1 \times 1 \times 1$m vaults. They were covered with a 10cm thick foam box for additional temperature stability (Zhou, Stump, Chen, Hayward, et al., 2003). Data were archived at the IRIS-Data Management Center (DMC).

6.1 Seismic events

Thirty-four high signal-to-noise ratio teleseismic events with great circle epicentral distances of $30° - 85°$ occurring between 2003 and 2005 were chosen for analysis. Source parameters were obtained from the Preliminary Determination of Epicenters (PDE) bulletins provided by the U.S. Geological Survey (USGS) National Earthquake Information Center (NEIC) (Table 2). The distribution of the 34 teleseismic events is displayed in an azimuthal equidistant projection centered on station AYPU (Fig. 2).

6.2 Receiver functions

Receiver functions were extracted from the three-component broadband recordings of the teleseismic P waveforms with epicentral distances between 30° and 85° (Özalaybey et al., 1997) (Fig. 2). The nearly vertical incident P waves dominate the vertical component, while P-to-S converted waves are recorded primarily on the horizontal component. The receiver functions are sensitive to the shear-wave velocity structure beneath the station because P-to-S conversions make significant contributions to the horizontal components corrected for source effects (Owens et al., 1984). Ammon (1991) argues that the absolute amplitude of a receiver function provides additional constraint on the near-surface shear-wave velocity and can be used to assess the presence of dipping layers (Cassidy, 1992).

Table 2 List of event parameters

Event Code	Date yyyy-mm-dd	Time hhmmss.mm	Latitude /(°N)	Longitude /(°E)	Magnitude	Depth/km	Distance to AYPU/(°)	AYPU Back Azimuth/(°)
2003076	2003-03-17	163617.31	51.27	177.98	7.1	33	43.4	53.8
2003089	2003-03-30	181334.09	−3.17	127.54	6.2	33	45.0	162.9
2003141	2003-05-21	184420.10	36.96	3.63	6.9	12	80.6	311.3
2003166	2003-06-15	192433.15	51.55	176.92	6.5	20	42.7	53.6
2003174	2003-06-23	121234.47	51.44	176.78	7.0	20	42.6	53.8
2003339	2003-12-05	212609.48	55.54	165.78	6.7	10	35.9	48.0
2003360	2003-12-26	015652.44	29.00	58.31	6.8	10	47.5	274.7
2004039	2004-02-08	085851.80	−3.66	135.34	6.9	25	47.7	152.8
2004107	2004-04-16	215705.41	−5.21	102.72	6.0	44	47.2	197.5
2004114	2004-04-23	015030.22	−9.36	122.84	6.7	65	50.3	170.6
2004162	2004-06-10	151957.75	55.68	160.00	6.9	188	32.6	47.1
2004180	2004-06-28	094947.00	54.80	−134.25	6.8	20	67.7	35.8
2004316	2004-11-11	212641.15	−8.15	124.87	7.5	10	49.4	167.8
2004353	2004-12-18	064619.87	48.84	156.31	6.2	11	29.8	59.9
2005022	2005-01-22	203017.35	−7.73	159.48	6.5	29	62.9	129.4
2005023	2005-01-23	201012.15	−1.20	119.93	6.3	11	41.9	173.4
2005036	2005-02-05	033425.73	16.01	145.87	6.6	142	35.9	124.1
2005036-2	2005-02-05	122318.94	5.29	123.34	7.1	525	35.9	166.6
2005046	2005-02-15	144225.85	4.76	126.42	6.6	39	37.1	161.8
2005050	2005-02-19	000443.59	−5.56	122.13	6.5	10	46.5	170.9
2005061	2005-03-02	104212.23	−6.53	129.93	7.1	201	48.9	160.8
2005089	2005-03-30	161941.10	2.99	95.41	6.4	22	41.7	211.1
2005093	2005-04-03	005921.42	0.37	98.32	6.0	30	43.1	205.7
2005093-2	2005-04-03	031056.47	2.02	97.94	6.3	36	41.6	207.0
2005099	2005-04-09	151627.89	56.17	−154.52	6.0	14	57.3	41.4
2005100	2005-04-10	102911.28	−1.64	99.61	6.7	19	44.6	203.0
2005100-2	2005-04-10	172439.40	−1.59	99.72	6.4	30	44.5	202.9
2005106	2005-04-16	163803.90	1.81	97.66	6.4	31	41.9	207.3
2005118	2005-04-28	140733.70	2.13	96.80	6.3	22	41.9	208.7
2005130	2005-05-10	010905.10	−6.23	103.14	6.4	17	48.1	196.6
2005134	2005-05-14	050518.48	0.59	98.46	6.8	34	42.8	205.6
2005139	2005-05-19	015452.85	1.99	97.04	6.9	30	42.0	208.3
2005165	2005-06-14	171016.64	51.23	179.41	6.8	51	44.3	53.6
2005186	2005-07-05	015202.95	1.82	97.08	6.8	21	42.1	208.1

Teleseismic receiver function and surface-wave study of velocity structure beneath the Yanqing-Huailai Basin Northwest of Beijing

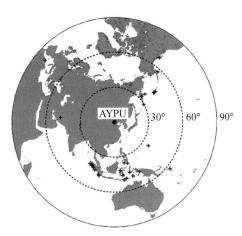

Figure 2 Map (azimuth equidistant projection centered at AYPU) of events (plus)
Circles are 30, 60, and 90 great circle distances range from AYPU (dot)

Receiver functions were computed using the iterative, time-domain deconvolution technique of Ligorría and Ammon (1999), an implementation of the Kikuchi and Kanamori (1982) procedure. Three-component seismograms at six stations from a magnitude 6.2 event (event code: 2003089; location: −3.17°N and 127.54° E on 30 March 2003) are presented in Figure 3. The radial receiver functions for the event with Gaussian filter parameter, $\alpha = 1$, are included, illustrating the strong similarity of receiver functions across the region. Low-amplitude transverse components are an indication of limited anisotropic effects or nonplane layering near the stations.

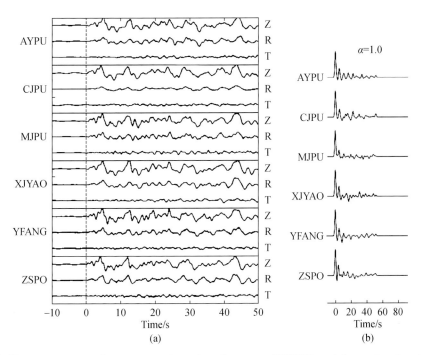

Figure 3 (a) Three-component seismograms at six stations from event 2003089 used for receiver function calculation.
(b) Receiver functions from event 2003089 with a Gaussian window parameter, α, of 1.0

The radial component receiver functions ($\alpha = 1.0$) for all the events in this study at AYPU and XJYAO are plotted as a function of source azimuth in Figure 4. These receiver functions exhibit very modest azimuthal variation. Similar comparisons at other stations motivated the development of different plane-layer velocity models for each station as a first step in the analysis. For the 34 events, AYPU, BAKOU, CJPU, MJPU, XJYAO, YFANG, and ZSPO recorded 27, 25, 6, 15, 27, 23, and 10 events, respectively.

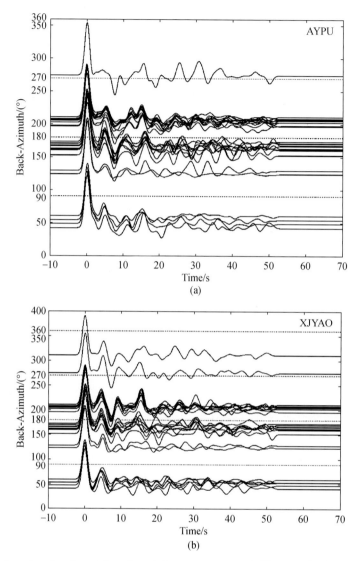

Figure 4   Plot of receiver functions (Gaussian window parameter α=1.0) versus back azimuth for all events recorded at (a) station AYPU (27 events) and (b) station XJYAO (27 events)

6.3 Surface waves

Phase velocities under the network are estimated by applying the technique of McMechan and Yedlin (1981) to teleseismic fundamental mode Rayleigh waves. Observations at each station are projected onto a pseudolinear array using a great circle path assumption. A p-τ stack is applied and followed by transformation into the p-ω domain in order to estimate the phase velocity dispersion curves. By simultaneously analyzing the data from the seven stations in our deployment, we assume a uniformity of structure beneath the network at depths greater than about one-third of the shortest wavelength observed.

Fundamental mode Rayleigh waves for the 34 events are extracted using multiple filter analysis (Dziewonski et al., 1969) followed by phase matched filtering (Herrin and Goforth, 1977). An example set of fundamental mode Rayleigh waves extracted using this procedure for event 2003089 in the 20-100 sec period range is displayed in Figure 5. Phase velocity dispersion is estimated by applying the p-τ stacking technique. The resulting mean phase velocities and standard deviations at a total of 288 frequencies for all of the events are plotted in Figure 6. The estimated phase velocities range from 3.06 to 4.06 km/s over the period range of 15–95 s with standard deviations between 0.01 and 0.28 km/s. The depth sensitivity of the Rayleigh-wave phase velocity dispersion is between one-third and one-half of a wavelength (Beaty et al., 2002; Li and Detrick, 2006). The long-period phase velocities help constrain the upper-mantle structure. The estimated phase velocity dispersion curves display limited scatter attributable to the small array aperture, local site effects, and local departure from the assumption of a plane wave across the array. We use the estimated mean phase velocity at each frequency with the standard deviations as weights in the joint inversions.

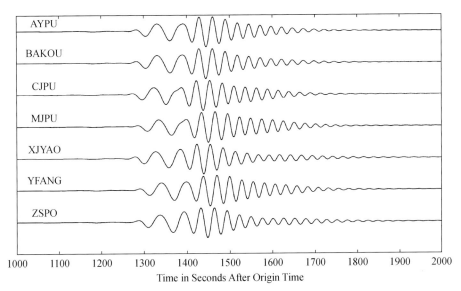

Figure 5 Fundamental Rayleigh waves (20–100 s) extracted from vertical component seismograms using multiple filter analysis and phase matched filter for event 2003089

7 Joint inversion

7.1 Starting model

The receiver function inversion utilizes a linearized and iterative, least-squares waveform fitting procedure (Herrmann and Ammon, 2002). Surface-wave dispersion partial derivatives are computed analytically while receiver function partial derivatives are computed numerically. The shear-wave velocities in each layer are the only parameters in the inversion as a result of fixing the V_p/V_s ratio in each layer and computing layer density from the P-wave velocity after each iteration. A differential smoothing constraint is applied so that the simplest model that fits the observations is identified. The smoothing constraint is implemented by solving for the change in the velocity contrast at boundaries. The starting model has 42 plane layers extending from surface to a depth of 200 km. Layer thicknesses are 1 km for the top 3 km (3 layers), 2 km to a depth of 5 km (1 layer), 2.5 km for 45 km (18 layers), 5 km for 50 km (10 layers), and 10 km below 100 km (10 layers). The upper 50 km of the starting model has the single velocity value of AK135-F at 50 km depth and below 50 km it follows the AK135-F continental model (Kennett et al., 1995). The high-initial crustal velocities provide a uniform unbiased starting model for the *inversion*. No *a priori* assumptions were made concerning the Moho depth or the location of any internal layering in the crust. Starting with a continental upper mantle and only permitting significant departure from the starting model in the upper 80km ensures that the lower part of the model does not depart from the experience of global seismology. We tested the model's sensitivity to layer thicknesses in the top 50km using (1) 3 layers with 1km thickness, 1 layer with 2.5km thickness, and 18 layers with 2km thickness; (2) 25 layers with 2km thickness; and (3) 2km for the top layer and 4.0km for the rest to 50km. These three test cases produced similar final velocity models below 2km. Results from the first case are used for illustration, taking into account the possible effects of the basin where the stations are located.

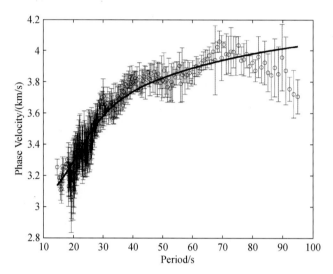

Figure 6 Comparison of measured teleseismic phase velocities (thin gray circles) with the prediction (thick black curve) based on the model determined by the joint inversion

The measured data are plot with ±1 standard deviation

7.2 Weighting the receiver functions and surface-wave dispersion curves in the inversions

The goal of the joint inversion is to find a model that fits the two different types of data simultaneously. The joint inversion approach of Herrmann and Ammon (2002) includes a parameter p to control the relative weights applied to the receiver functions and surface-wave dispersion curves. This parameter varies from 0.0 to 1.0, with 0.0 corresponding to an inversion with only the receiver functions and 1.0 to an inversion with only the surface-wave dispersion data. Figure 7 compares the inverted models using different p values and provides some idea of the importance of each type of data in constraining the velocity model. The partial derivatives of the surface-wave dispersion curves with respect to the model parameters have values about five times larger than the partial derivatives of the receiver functions with respect to the model parameters. Based on these characteristics, we argue that inversions using p values close to 0.15 may provide the best weighting between the two types of data used in this study. This p value has been used by one of the authors, Herrmann, in a separate study of the Korean Peninsula (Yoo et al., 2007). They found that synthetic seismograms computed from the model with a p value of 0.15 gave the best fit to observed waveforms from an earthquake further motivating our preference for this p value.

7.3 Inversion results

A joint inversion of the receiver functions and surface-wave dispersion curves at each station across the

Yanqing-Huailai Basin was conducted. Figure 8 illustrates the starting model in gray and the final velocity model in black with AK135F included for comparison. Observed (black line) and predicted receiver functions (gray line) at AYPU are also reproduced. The predicted receiver functions match the observed with 90% of the filtered radial signal power explained by the predictions for events at all azimuth and distance ranges. Observed phase velocities from teleseismic events are compared to predicted phase velocities in Figure 6.

An average velocity model for the region was determined by inverting all receiver functions from the seven stations with the phase velocities from the teleseismic events.

Figure 9 plots the station specific velocity models (dark line) against the averaged model for the region (gray line). The individual velocity models at depth exhibit little or no difference across the network as expected based on the common surface-wave dispersion data. The models have a positive velocity gradient from the surface to approximately 15km with shear-wave velocity increasing from 2.0 to 3.55km/sec. A slight negative gradient in velocity starts at about 15km resulting in an extended low-velocity layer to approximately 25km with velocity near 3.3km/sec. Subtle differences in the velocity models from ZSPO may reflect the smaller number of the events recorded at this station and its location outside of the Huailai Basin. There is evidence of a low-velocity layer in the midcrust at all stations. There is no sharp Moho interface with the Moho represented as a transition in velocity ranging from 38 to 42km, consistent with other studies in the region (Zhang et al., 1996; Zhao et al., 2005). The transitional Moho, imaged with many thin layers, may be real or may be a consequence of the particular data sets available, the bandlimited receiver functions, and dispersion data, as well as the assumption of a plane-layered underlying structure. The transitional Moho provides a good match to observed first-arrival times used for earthquake location. Other authors (e.g., Fnais, 2004) have found a transitional, rather than a sharp, Moho to be the norm.

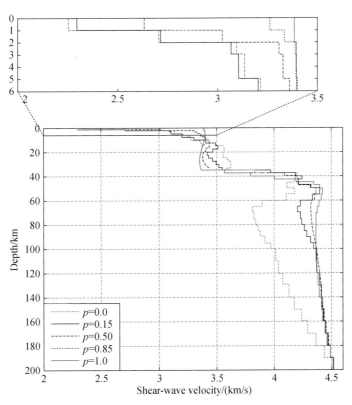

Figure 7 Comparison of models resulting from inversions with different p values that control the weighting of the receiver functions and surface-wave phase velocities at AYPU

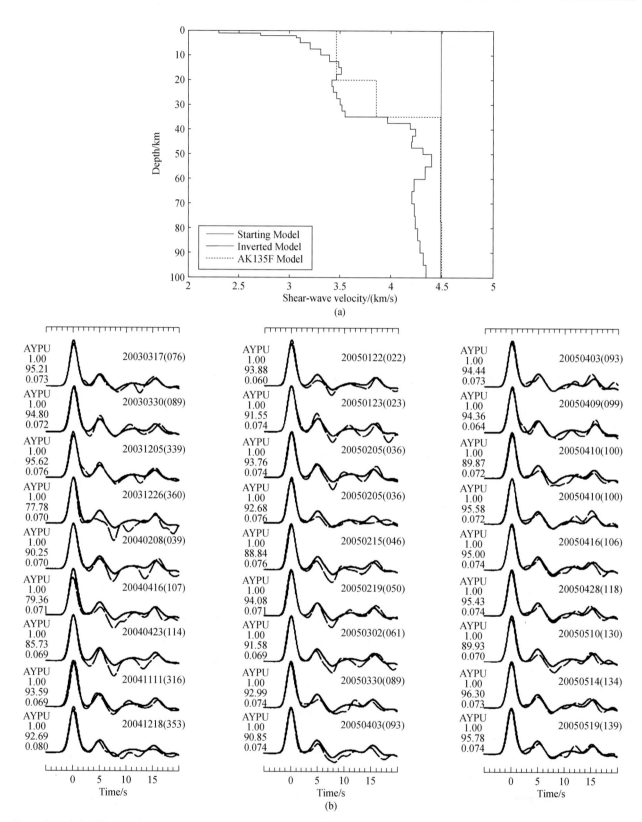

Figure 8 (a) Plot of the shear-wave velocity starting model (gray line) and the final model (black line) at AYPU. The AK135F model is plotted as a dashed line. (b) Comparison between observed (black solid curve) and predicted receiver functions (gray dashed curve) after inversion. The receiver functions are labeled with the year/month/day (day of year) to the upper right of each trace and with the station name, Gaussian filter parameter, the percentage of fit, and the ray parameter (s/km) to the left of each trace

8 Model comparison and discussion

During the instrument deployment, a 1.3 kiloton mining explosion occurred 240-280km southeast of the network and generated intermediate period surface waves (2-16 sec) (Zhou, Stump, and Chen, 2003; Zhou et al., 2006). The vertical component seismograms from this shallow explosion superimposed with the extracted fundamental Rayleigh waves after using multiple filter analysis (Dziewonski et al., 1969) and phase matched filtering (Herrin and Goforth, 1977) are reproduced in Figure 10. This figure also includes the mean group velocities and their standard deviations for the intermediate period surface waves estimated from the four stations that recorded this unique event. To explore the use of these high-frequency surface waves to further constrain the shallow structures, we conducted an inversion that included these regional group velocities in addition to the teleseismic receiver functions and phase velocities. Figure 11 compares the resulting models with and without these high-frequency group velocities for both AYPU and the average model for the region. The inverted models including the regional group velocities produce higher shear velocities from the surface to depth of 15km and slightly lower velocities to the Moho. This comparison illustrates the importance of careful data set selection for the meaningful joint inversion. Julia et al. (2000) point out that, for joint inversion of surface-wave and receiver functions, consistency requires that both signals sample the same portion of the propagating medium, so that the information contained in the waveforms references the same part of the Earth. The high-frequency group velocities obtained from the mining explosion reflect the entire path effects across the Yanshan Uplift between the mine and the stations, which are different from the part of the crust sampled by the receiver functions and phase velocities estimates using the stations in and around the Huailai Basin. The fit of the dispersion curves by the resulting model that is included in Figure 10 also illustrates that the predicted group velocities are still underestimated and require even higher velocities models to fit the path effects of the Yanshan Uplift, thus suggesting regional variations in the crust.

Three refraction/wide-angle reflection profiles were conducted in 1993 in order to constrain the crust-mantle structures northwest of Beijing. One of these, the Beijing-Huailai-Fengzhen profile (H-20), passes in a west-to-east direction across the Yanqing-Huailai area (Fig.1; Liu et al., 1997; Zhao et al., 2005). The publication of the velocity models resulting from this data interpretation (Zhang et al., 1996; Liu et al., 1997; Zhu et al., 1997; Zhao et al., 2005) provides the basis for the comparison of these velocity models with our results in Figure 12. These earlier models are somewhat simpler than those produced in this study and do not include the effect of shallow basin structure nor the mid to lower crust low-velocity zone.

Crustal thickness in the region identified in the refraction/reflection study was estimated between 37 and 41km, which is consistent with our depth estimates from the joint inversion. Liu et al. (1997) inverted both the P-wave velocity (V_P) and the V_P/V_S ratio using the Seis83 program package from the Beijing-Huailai-Fengzhen (H-20) profile. Their results indicated that the crust in the Yanqing-Huailai Basin could be divided into an upper and lower part. Their model gives S-wave velocities in the depth range 10-14km that are similar to our models and are consistent with AK135F to 20km depth. Predicted first-arrival times of P and S waves from their models are compared to those from our work and the difference reflects the slower velocities in the very shallow crust in our models. The shallow details of our model are consistent with the known basin geology with total sediment thicknesses exceeding 2km (Wu et al., 1979). Although this study was not focused on earthquake location, our detailed velocity model could impact studies of regional earthquake location. Focal depths of historical earthquakes in the region range from 5 to 15km and are assumed to be located within the brittle upper crust, above the low-velocity region identified in this study and others (Zhu et al., 1997). The lack of a low-velocity region in the refraction interpretation illustrates the advantage of adding the receiver function and surface-wave analysis to the interpretation. Che et al. (2001) suggest that the strength of the upper crust has been decreased by the action of hot fluids consistent with the small and moderate earthquake locations and the existence of hot springs in the region. Detailed modeling and moment tensor inversion for the Huailai M_L event in 1995 (Xu et al., 1999, 2001) led to a source depth estimate of 5.5km again in the higher velocity layers of our model. Sun et al. (1987) propose that the geothermal anomaly and tectonic extension are possibly related to the existence of the low-velocity layers. Alternatively, Zhou and He (2002) hypothesize that the crustal low-velocity layers in north China are controlled by Cenozoic rifting and mantle uplifting, which caused the rise of temperature and, hence, plastic deformation of the lower-middle and lower crust. Our models are consistent with the existence of this midcrustal region as a reflection of the tectonic evolution of the NCB, possibly the extensional processes discussed earlier.

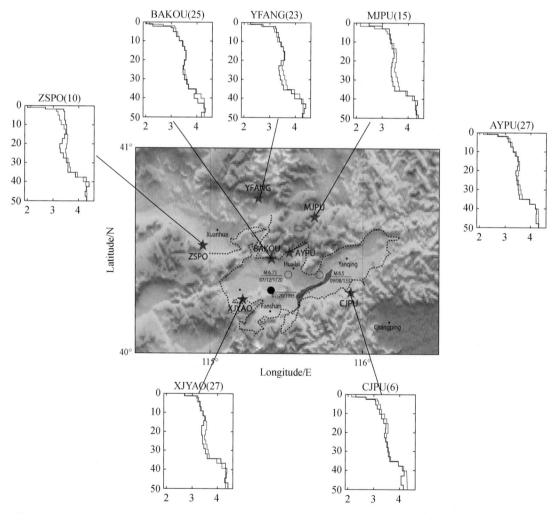

Figure 9　Map of stations and shear-wave velocity models from joint inversions at each of the seven stations (black lines) compared to the averaged model inverted using all seven stations (gray lines) simultaneously

The number after each station is the number of events used for station specific inversion

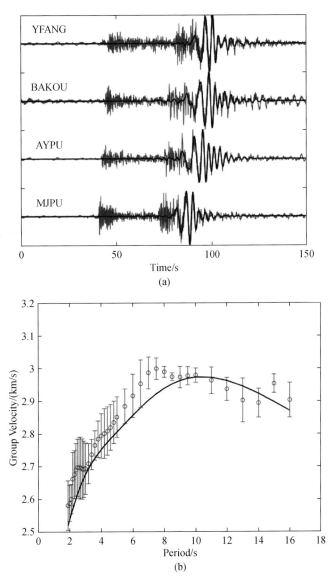

Figure 10 (a) Raw vertical component seismograms (thin gray curve) and fundamental mode Rayleigh waves (thick black curve) from the 1.3 kt mining explosion. (b) Comparison of measured regional group velocities (thin gray circles) with a prediction (thick black curve) based on the velocity model from the joint inversion. The measured data are plot with±1 standard deviation

9 Conclusion

Thirty-four teleseismic events recorded on seven broadband seismic stations in and around the Yanqing-Huailai Basin have been acquired to constrain the shear velocity structure beneath and around the basin using the inversion of receiver functions and surface-wave dispersion curves. The similarity of the receiver functions between stations supports the plane-layered models used to represent the structure of this region. The inverted models indicate that the top 2km beneath the Yanqing-Huailai area has a low-shear velocity between 2.0 and 2.5km/sec. Low-velocity layers are identified in upper and midcrustal layers that do not appear in region specific models based on wide-angle reflection/refraction analysis. Inversions with a variety of weighting between the receiver functions and phase velocity estimates illustrate the stability of this low-velocity layer. Our estimates of a Moho depth of approximately

40km are consistent with the results from refraction/wide-angle reflection profiles. Predicted receiver functions fit well with the observed for events at all azimuth and distance range with inverted models predicting on average 90% of filtered radial signal power. The calculated phase velocities match the observations. An average velocity model beneath the Basin is obtained from the joint inversion using all receiver functions and dispersion curves at the seven stations. The predicted first-arrival times of the P and S wave from this average model when compared with results from the refraction/wide-angle reflection and global AK135F model are delayed by 1 – 3.5 sec at regional distances. The shear-wave velocities of the average model are consistent with the two other models from 10 to 20km and give more details for the shallow basin and the mid to lower crust (low-velocity zone). The low-velocity zone is consistent with recent extension, geothermal activity, and earthquake locations above this depth.

10 Data and resources

The China Historical Earthquake Catalogue was obtained from the database of the World Data Center for Seismology, Beijing (http://210.72.96.21:8080/wdc/cezhen/ historycatalog_query.jsp, last accessed October 2007), and the CSN Catalog was acquired from Web site http://210.72.96.165/wdcd/csn_catalog_p001.jsp (last accessed October 2007). Recent seismicity (01 January 2002 through 31 December 2006) in the Capital Circle Region was obtained from the Capital Circle Seismic Network (http://www.csndmc.ac.cn/newweb/data.htm#, last accessed October 2007). Source parameters of 34 teleseismic events (Table 2) were obtained from the PDE bulletins provided by the USGS NEIC. Figures 1b, c, 2, and 9 were created using the Generic Mapping Tools (GMT) software (Wessel and Smith, 1998). All seismograms were recorded by the temporary broadband seismic network operated by SMU and the IGPCEA with support from the IRIS PASSCAL Instrument Center, and they are archived at the IRIS Data Management Center.

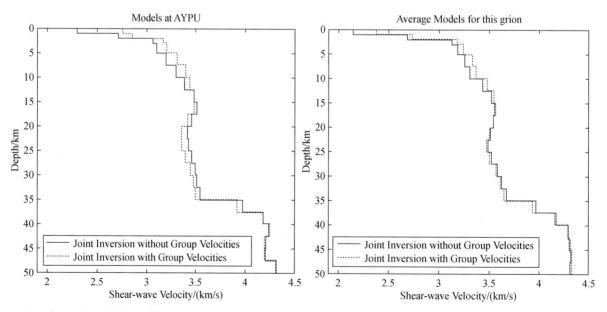

Figure 11 Comparison of the model from the joint inversion using receiver function, phase velocities and with (dashed line) and without (solid line) group velocities from the mining events observed at AYPU (left-hand panel)

Comparison of average models from the joint inversions for this region using receiver functions, phase velocities and with (dashed line) and without (solid line) group velocities from the mining events (right-hand panel)

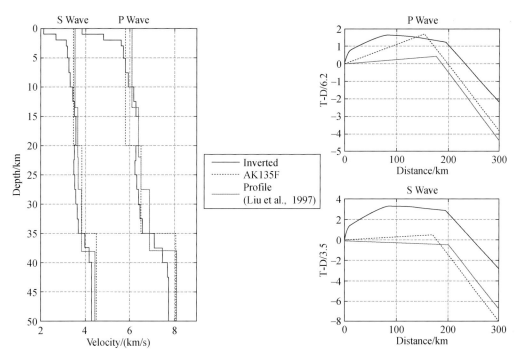

Figure 12 Comparison between the model resulting from the joint inversion (receiver functions and phase velocities) using all station data (solid black line) and the model from a refraction/wide-angle reflection survey (solid gray line) and AK135F model (dotted line) (left-hand panel)

Predicted first-arrival times of P (upper right-hand panel) and S waves (lower right-hand panel) from the three different models

Acknowledgments

The authors would like to thank Chris Hayward, Mary Templeton, Xiang-Wei Yu, Xiang-Tong Xu, and Shi-Yu Bai for their help with network installation and data collection. A large portion of the success of this experiment is due to the outstanding help and support of the seismological bureaus and seismic station operators in this region. We thank Arthur Rodgers and one anonymous reviewer and the associate editor Anton Dainty for numerous comments and suggestions that have improved the manuscript. George Randall's comments on the manuscript are also acknowledged. R. Zhou is grateful to Guochun Zhao for providing schematic tectonic map as Figure 1a. All topographic maps and Figure 2 were produced using the GMT software of Wessel and Smith (1998). This work was supported by the Air Force Research Laboratory Contract Numbers DTRA01-02-C-0003 and FA8717005-C-0020. Z.-X. Y. and Y.-T. C. acknowledge the supports of NSFC (National Natural Science Foundation of China, Grant Number 40574025) and MOST (Ministry of Science and Technology, Grant Number 2001CB711005) of China. IRIS PASSCAL Instrument Center provided the STS-2 seismometers, Quanterra Q-330, and Baler systems.

References

Ammon, C. J. (1991). The isolation of receiver effects from teleseismic P waveforms, *Bull. Seismol. Soc. Am.* **81**, 2504–2510.

Ammon, C. J., G. E. Randall, and G. Zandt (1990). On the nonuniqueness of receiver functions, *J. Geophys. Res.* **95**, 15, 303–15, 318.

Beaty, K. S., D. R. Schmitt, and M. Sacchi (2002). Simulated annealing inversion of multimode Rayleigh-wave dispersion curves for geological structure, *Geophys. J. Int.* **151**, 622–631.

Cassidy, J. F. (1992). Numerical experiments in broadband receiver function analysis, Bull. Seismol. Soc. Am. 82, 1453–1474.

Che, Y.-T., J.-H. Wang, J.-Z. Yu, and W.-Z. Liu (2001). Character of thermal fluids in upper crust and relationship with seismicity in Yanqing-Huailai, *Seismol. Geol.* **23**, 49–54.

Chen, Y.-T., X.-T. Xu, X.-W. Yu, and P.-D. Wang (1998). Observations and interpretation of seismic ground motion and earthquake hazard mitigation in Beijing area, *South China J. Seismol.* **18**, 2–8 (in Chinese).

Chen, X.-Z., X.-T. Xu, and W.-J. Zhai (2005). Variation of stress during the rupture process of the 1995 M_L = 4.1 Shacheng, Hebei, China, earthquake sequence, *Acta Seismol. Sin.* **18**, 297–302.

Du, Z.-J., and G. R. Foulger (1999). The crustal structure beneath the northwest fjords, Iceland, from receiver functions and surface waves, *Geophys. J. Int.* **139**, 419–432.

Dziewonski, A., S. Bloch, and M. Landisman (1969). A technique for the analysis of transient seismic signals, *Bull. Seismol. Soc. Am.* **59**, 427–444.

Fnais, M. S. (2004). *The Crustal and Upper Mantle Shear Velocity Structure of Eastern North America from the Joint Inversion of Receiver Function and Surface-Wave Dispersion*, Ph. D. Thesis, Saint Louis University.

Herrin, E., and T. Goforth (1977). Phase-matched filters: Application to the study of Rayleigh waves, *Bull. Seismol. Soc. Am.* **67**, 1259–1275.

Herrmann, R. B., and C. J. Ammon (2002). *Computer Programs in Seismology: Surface Waves, Receiver Functions and Crustal Structure*, Version 3.15: available at http://www.eas.slu.edu/People/RBHerrmann/CPS330.html (last accessed March 2007).

Herrmann, R. B., C. J. Ammon, and J. Julia (2000). Joint inversion of receiver functions and surface-wave dispersion for crustal structure, in Proc. of the 22nd Annual DoD/DOE Seismic Research Symposium: *Planning for Verification of and Compliance with the Comprehensive Nuclear-Test-Ban Treaty (CTBT)*, 13–15 September, New Orleans, Louisiana, 43–53.

Herrmann, R. B., C. J. Ammon, and J. Julia (2001). Application of joint receiver-function surface-wave dispersion for local structure in Eurasia, in Proc. of the 23rd Seismic Research Review: *Worldwide Monitoring of Nuclear Explosions*, October 2001, LA-UR-01-4454, Los Alamos National Laboratory, Los Alamos, New Mexico, 46–54.

Julia, J., C. J. Ammon, R. B. Herrmann, and A. M. Correig (2000). Joint inversion of receiver function and surface-wave dispersion observations, *Geophys. J. Int.* **143**, 99–112.

Julia, J., C. J. Ammon, and R. B. Herrmann (2003). Lithospheric structure of the Arabian Shield from the joint inversion of receiver functions and surface-wave group velocities, *Tectonophysics* **37**, 1–21.

Kennett, B. L. N., E. R. Engdahl, and R. Buland (1995). Constraints on seismic velocities in the earth from travel times, *Geophys. J. Int.* **122**, 108–124.

Kikuchi, M., and H. Kanamori (1982). Inversion of complex body waves, Bull. Seismol. Soc. Am. 72, 491–506.

Langston, A. C. (1977). The effect of planar dipping structure on source and receiver responses for constant ray parameter, *Bull. Seismol. Soc. Am.* **67**, 1029–1050.

Langston, A. C. (1979). Structure under Mount Rainier, Washington, inferred from teleseismic body waves, *J. Geophys. Res.* **84**, 4749–4762.

Li, A., and R. S. Detrick (2006). Seismic structure of Iceland from Rayleigh-wave inversions and geodynamic implications, *Earth Planet. Sci. Lett.* **241**, 901–912.

Ligorría, J. P., and C. J. Ammon (1999). Iterative deconvolution and receiver function estimation, *Bull. Seismol. Soc. Am.* **89**, 1395–1400.

Lin, G., Y.-H. Wang, F. Guo, Y.-J. Wang, and W.-M. Fan (2004). Geodynamic modeling of crustal deformation of the North China block: A preliminary study, *J. Geophys. Eng.* **1**, 63–69, doi 10.1088/1742-2132/1/1/008.

Liu, C.-Q., S.-X. Jia, M.-J. Liu, and C.-F. Li (1997). Analysis and study of the large earthquake risk in Yanqing-Huailai Basin, *Acta Seismol. Sin.* **10**, 639–647.

McMechan, G. A., and M. J. Yedlin (1981). Analysis of dispersive waves by wave field transformation, *Geophysics* **46**, 869–874.

Owens, J. T., G. Zandt, and S. R. Taylor (1984). Seismic evidence for an ancient rift beneath the Cumberland Plateau, Tennessee: A detailed analysis of broadband teleseismic P waveforms, *J. Geophys. Res.* **89**, 7783–7795.

Özalaybey, S., M. K. Savage, A. F. Sheehan, J. N. Louie, and J. N. Brune (1997). Shear-wave velocity structure in the Northern Basin and Range Province from the combined analysis of receiver functions and surface waves, *Bull. Seismol. Soc. Am.* **87**, 183–199.

Pavlides, B. S., N. C. Zouros, Z.-J. Fang, S.-P. Cheng, M. D. Tranos, and A. A. Chatzipetros (1999). Geometry, kinematics, and morphotectonics of the Yanqing-Huailai active faults (northern China), *Tectonophysics* **308**, 99–118.

Rowley, D. B., F. Xue, R. D. Tucker, Z. X. Peng, J. Baker, and A. Davis (1997). Ages of ultrahigh pressure metamorphism and protolith orthogneisses from the Eastern Dabie Shan: U/Pb zircon geochronology, *Earth Planet. Sci. Lett.* **151**, 191–203.

Sun, W.-C., S.-L. Li, L.-L. Luo, and H.-F. Yue (1987). A preliminary study on low-velocity layer in the crust in North China, *Seismol. Geol.* **9**, no. 1, 17–26.

Wessel, P., and W. H. F. Smith (1998). New, improved version of the Generic Mapping Tools released, *EOS Trans. AGU* **79**, 579.

Wu, Z.-R., B.-Y. Yuan, J.-Z. Sun, and Z.-S. Liu (1979). Neotectonics and seismicity of Yan-Huai basin in Hebei Province, *Seismol. Geol.* **1**, no. 2, 46–56.

Xu, X.-T., Y.-T. Chen, and P.-D. Wang (1997). The tectonic stresses in the Huailai basin, *Seismol. Geomagn. Obs. Res.* **18**, 1–8 (in Chinese with English abstract).

Xu, X.-T., Y.-T. Chen, and P.-D. Wang (1999). Rupture process of the M_L=4.1 earthquake in Huailai basin on July 20, *Acta Seismol. Sin.* **12**, 618–631.

Xu, X.-T., Y.-T. Chen, and P.-D. Wang (2001). Precise determination of focal parameters for July 20, 1995 M_L=4.1 earthquake sequence in the Huailai basin, *Acta Seismol. Sin.* **14**, 237–250.

Xu, Y.-G., J.-L. Ma, F. A. Frey, M. D. Feigenson, and J.-F. Liu (2005). Role of lithosphere-asthenosphere interaction in the genesis of Quaternary alkali and tholeiitic basalts from Datong, western North China Craton, *Chem. Geol.* **224**, 247–271.

Yoo, H. J., R. B. Herrmann, K. H. Cho, and K. Lee (2007). Imaging the three-dimensional crust of the Korean Peninsula by joint inversion of surface-wave dispersion and teleseismic receiver functions, *Bull. Seismol. Soc. Am.* **97**, 1002–1011, doi 10.1785/0120060134.

Zhang, X.-K., C.-Y. Wang, G.-D. Liu, J.-L. Song, L.-L. Luo, T. Wu, and J.-C. Wu (1996). Fine crustal structure in Yanqing-Huailai region by deep seismic reflection profiling, *Chin. J. Geophys.* **39**, 356–364 (in Chinese with English abstract).

Zhao, G.-C., S. A. Wilde, P. A. Cawood, and L.-Z. Lu (1999). Thermal evolution of two textural types of mafic granulites in the North China craton: Evidence for both mantle plume and collisional tectonics, *Geol. Mag.* **136**, 223–240.

Zhao, G.-C., S. A. Wilder, P. A. Cawood, and M. Sun (2002). Shrimp U-Pb zircon ages of the Fuping Complex: Implications for late archean to paleoprotezoic accretion and assembly of the North China Craton, *Am. J. Sci.* **302**, 191–226.

Zhao, J.-R., X.-K. Zhang, C.-K. Zhang, J.-S. Zhang, B.-F. Liu, Q.-F. Ren, S.-Z. Pan, and Y. Hai (2005). The heterogeneous characteristics of crust-mantle structures and the seismic activities in the northwest Beijing region, *Acta Seismol. Sin.* **18**, 125–134.

Zheng, T.-Y., L. Zhao, and L. Chen (2005). A detailed receiver function image of the sedimentary structure in the Bohai Bay Basin, *Phys. Earth Planet. Interiors* **152**, 129–143.

Zhou, Y.-S., and C.-R. He (2002). The relationship between low-velocity layers and rheology of the crust in North China and its effect on strong earthquake, *Seismol. Geol.* **24**, no. 1, 125–131.

Zhou, R.-M., B. W. Stump, Y.-T. Chen, C. T. Hayward, Z.-X. Yang, M. Templeton, X.-W. Yu, S.-Y. Bai, and X.-T. Xu (2003). Network installation in the Yanqing-Huailai Basin and preliminary study of natural and man-induced events, *Proc. of the 25th Seismic Research Review Nuclear Explosion Monitoring: Building the Knowledge Base*, September 2003, LA-UR-03-6029, Los Alamos National Laboratory, Los Alamos, New Mexico, 504–513.

Zhou, R.-M., B. W. Stump, and Y.-T. Chen (2003). A comparative study of intermediate-period surface waves from kiloton-size mining explosions in Northeast China and Western United States (Abstract S32B-0848), *EOS Trans. AGU*, **84**, no. 46 (Fall

Meet. Suppl.), S32B-0848.

Zhou, R.-M., B. W. Stump, and C. T. Hayward (2006). M_S: m_b discrimination study of mining explosions in Wyoming, USA, and in Qianan, China, *Bull. Seismol. Soc. Am.* **96**, no. 5, 1742–1752, doi 10.1785/0120050178.

Zhu, Z.-P., X.-K. Zhang, J.-S. Zhang, C.-K. Zhang, J.-R. Zhao, and Y.-J. Gai (1997). Study on crust-mantle tectonics and its velocity structure along the Beijing-Huailai-Fengzhen profile, *Acta Seismol. Sin.* **10**, 615–623.

Shear-wave splitting in the crust beneath the southeast Capital area of North China

Jing Wu[1], Yuan Gao[1] and Yun-Tai Chen[2]

1. Institute of Earthquake Science, China Earthquake Administration, Beijing 100036, China;
2. Institute of Geophysics, China Earthquake Administration, Beijing 100081, China

Abstract This study focuses on the southeast Capital area of North China (38.5–39.85°N, 115.5–118.5°E). Shear-wave splitting parameters at 20 seismic stations are obtained by a systematic analysis method applied to data recorded by the Capital Area Seismograph Network (CASN) between the years 2002 and 2005. Although some differences in the results are observed, the average fast-wave polarization is N88.2°W±40.7° and the average normalized slow wave time delay is 3.55±2.93ms/km. The average polarization is consistent with the regional maximum horizontal compressive stress and also with the maximum principal strain derived from global positioning system measurements in North China. In spite of the uneven distribution of faults around the array stations that likely introduce some amount of scatter in the shear-wave splitting measurements, site-dependent polarizations of fast shear wave are clearly observed: in the northern half of the study area, the polarizations at CASN stations show E-W direction, whereas in the southern half the polarizations exhibit a variety of possible azimuths, thus suggesting dissimilar stress field and tectonic frame in both areas. Comparing the splitting results with those previously obtained in the northwest part of the region, we find a difference in polarization of about 20° between the southeast and northwest parts of the Capital area; also, in the southeast Capital area the average time delay is smaller than in the northwest Capital area, thus making clear that the magnitude of crustal seismic anisotropy is not the same in the two zones. Being the shear-wave splitting polarizations in the southeast Capital area, which lies on the basin, clearly different from the observed polarizations in the northwest Capital area, where uplifts and basin converge, it is quite evident that the shear-wave splitting results are consequence of the tectonics and stress field affecting the two regions.

1 Introduction

Most researches show that seismic anisotropy exists widely in both the crust and the upper mantle (Crampin 1978; Gao et al. 1995, 1999; Zhang et al. 2000; Müller 2001; Liu et al. 2001). Extensive-dilatancy anisotropy microcracks are typical sources of seismic anisotropy in the crust, whose dynamic characteristics can be modeled by anisotropy poroelasticity (Zatsepin and Crampin 1997). The spatial distribution of the predominant polarization of fast shear-waves is consistent with the maximum horizontal compressive stress in the region. However, the polarizations of fast shear-wave at stations near to active faults are commonly parallel to the strikes of faults (Peng and Ben-Zion 2004; Shi et al. 2006; Wu et al. 2007; Gao et al. 2008). Furthermore, the polarizations of fast shear-wave are much scattered in complex tectonic areas, such as crossing fault zones. These anisotropy features can provide detailed information about the regional tectonics and lead to a better understanding of the real stress field in situ (Gao and Crampin 2006; Gao et al. 2008).

The main tectonic units around the Capital area of North China are Yanshan Uplift, Taihang Uplift, and North China Basin (Fig. 1). It is a region with a relatively complex tectonic and strong seismic activity. It is found from global positioning system observations that the stress in the east part is quite different from the stress in the west part (Jiang et al. 2000). The crustal seismic velocity structure in Taihang Uplift and Yanshan Uplift is that of a relatively simple crust of stable paleoland, very different from the Neozoic crust beneath the North China Basin (Huang and Zhao 2005; Jia et al. 2005). 3D tomography of the upper crust in the Beijing area (Wang et al. 2005) based on deep seismic sounding data reveals that seismic velocity structure and fault movements are quite related to each other. Due to the requirements of the shear-wave splitting analysis and the limitations of the seismic network, the research about shear-wave splitting in the Capital area is still at present an interesting study matter. Lai et al. (2006) have studied by shear-wave splitting analysis the seismic anisotropy and stress field of the crust in the Capital area from events occurring between May 2002 and March 2003 recorded by permanent stations and events occurring between March 2002 and November 2002 recorded by temporary stations. More recently, Wu et al. (2007), from data recorded during 2 years, have obtained preliminary results about the seismic anisotropy of the crust in the northwest Capital area. In this paper, we analyze the seismic anisotropy of the crust beneath the southeast Capital area based on data recorded during a period of more than 3 years and a half and waveforms selected with an extended shear-wave window of 55°. We further discuss the relation between shear-wave splitting and regional faulting.

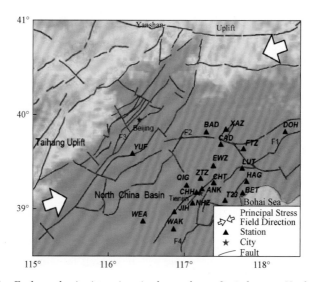

Fig. 1 Faults and seismic stations in the southeast Capital area, North China

Short lines represent faults; *black triangles* are stations; *star* marks the city, and *arrows* indicate the direction of horizontal principal compressive stress in the area. Key to symbols: F1, Tangshan-Dacheng fault; F2, Baodi fault; F3, Tongxian-Nanyuan fault; F4, Cangdong fault

2 Data

The Capital Area Seismograph Network (CASN) array, which spans about 500km from east to west and some 400km from north to south, was installed in 1999, but it is in operation since the first of October 2001. The array is formed by 107 stations which keep an average interdistance between stations of about 40km: 53 stations equipped with short-period seismometers are placed in North China Basin, while other 54 short-period stations are mostly distributed in Taihang Uplift and Yanshan Uplift. CASN is a very large regional seismic network and one of the denser seismic networks in China (Zhuang 1999). Figure 1 shows a map with the regional faults, the principal stress field direction, and the locations of 20 CASN stations in the southeast Capital area (except for DOH, the other 19 stations are put in boreholes).

The data used in this study are waveforms generated by local events of M_L-magnitude from 0.5 to 4.3 and depth from 5 to 30km (average depth of 15km), which occurred during the period January 2002-August 2005 in the southeast Capital area of North China (38.5–39.85°N, 115.5–118.5°E) and recorded by the 20 control stations belonging to the CASN array.

Two main requirements for data acquisition were taken into account: high signal-to-noise ratio and small incidence angle within the shear-wave window. This last means that the incidence angle is less than critical angle, i.e., less than $\theta = \sin^{-1}(V_s/V_p)$, where V_p and V_s are the P- and S-wave velocities, respectively. Because of the low-velocity sediment layer, this effective shear-wave window is typically as much as 45° or 50° (Crampin and Peacock 2005). Thus, those records of quality and with incidence angle less than 45° were selected for shear-wave splitting analysis. As an example, Fig. 2 shows the filtered seismic waveforms generated by the 13 October 2004 earthquake of M_L-magnitude 2.0 and depth 22km recorded at station LUT at an epicentral distance of 13.32km from the source. The signal can be clearly picked from the seismograms. Figure 3 (left column) shows the east-west and north-south horizontal components of the ground motion including shear waves and the nonlinear particle motion.

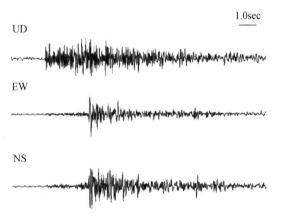

Fig. 2 Filtered waveforms generated by the 13 October 2004 earthquake of M_L-magnitude 2.0 and depth 22km recorded at station LUT

From top to bottom, vertical (UD), east-west (EW), and north-south (NS) ground motion components. A 2-20-Hz Butterworth band pass filter was used

3 Method

In order to obtain the seismic anisotropy parameters in the crust, a systematic analysis method (Gao et al. 2004) of split waveforms was used in this study. The method mainly includes time-delay correction, computation of the particle motion, and polarization analysis of split shear waves. As is well known, when a shear wave travels through an anisotropic medium, it will split into a fast shear wave and a slow shear wave, so that the polarization

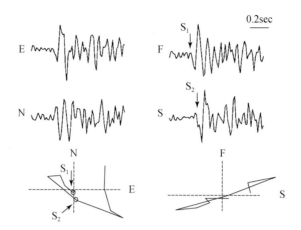

Fig. 3 Shear-wave splitting analysis

Left and from *top* to *bottom*: east-west (E) and north-south (N) shear-wave components and particle motion as extracted from the original seismic signal (Fig. 2). *Right* and from *top* to *bottom*: fast shear wave (F, S_1) and slow shear wave (S, S_2) and particle motion as derived from polarization analysis. The fast-wave polarization and the slow-wave time delay are 135° and 0.08s, respectively. The normalized time delay is 3.11ms/km

of fast shear wave is parallel to the vertically aligned microcracks, while the polarization of slow shear wave is nearly perpendicular to the aligned microcracks. The key parameters in shear-wave splitting analysis are the polarization of fast shear wave and the time delay of slow shear wave that gives the magnitude of seismic anisotropy of the medium. Theoretically, both the fast shear wave and the slow shear wave originate from the same source and therefore they should be similar as to the form. Based on this consideration, the correlation analysis is used in order to estimate the splitting parameters. The north-south and east-west shear-wave components are rotated and shifted in time. Then, the correlation coefficients are calculated for possible values of time shift of the split shear waves. Lastly, the splitting parameters are determined from the maximum of the correlation coefficients (Gao et al. 1995, 1999).

Many factors may however influence the results, such as crustal structure, surface topography, geological-tectonic conditions around the station, waveform data, applied method, etc. (Crampin and Peacock 2005; Gao and Crampin 2006). In some cases, the shear-wave splitting parameters cannot be obtained properly by cross-correlation only, and polarization analysis is then necessary (Gao et al. 1995, 1999). Let us suppose that the polarization angle of fast shear wave is α; then, after rotating the north-south and east-west components by angle α, these horizontal components become the fast and slow shear-wave components (Fig. 3, right column). If the time delay of slow shear wave is Δt, the slow shear wave can be moved forward just this time Δt to eliminate the time delay, and the particle motion becomes more linear as can be seen in the polarization diagram (Fig. 3, bottom). Therefore, by rotation of waveforms, time-delay correction, computation of the particle motion, and polarization analysis of split shear waves, it is possible to control the process and to estimate reliable shear-wave splitting parameters.

Here, these splitting parameters are always systematically adjusted by these operations (that we identify by its acronym SAM in a previous work by Gao et al. 2004) since the direct calculation often results in some mistakes (Crampin and Gao 2006). A very efficient semiautomatic method based on Expert System has been developed to analyze shear-wave splitting from small earthquakes; by now, it is ready for only specific seismic data from Iceland (Gao et al. 2006). The shear-wave splitting parameters are so obtained at 20 CASN stations (Fig. 1). Table 1 contains the station parameters and the shear-wave splitting results from measurements made at these 20 control stations.

Table 1 Station parameters and shear-wave splitting parameters in the southeast Capital area

Station name	Station code	East longitude	North latitude	Number of records	Polarization (in degrees East of North)	Standard error /(±degrees)	Time delay /(ms/km)	Standard error of time delay /(±ms/km)
Ankang Hospital	ANK	117.2	39.2	2	60.00	10.00	2.05	0.46
Baodi	BAD	117.3	39.7	16	94.69	31.45	4.77	6.41
Beitang	BET	117.7	39.1	3	161.67	4.71	0.98	0.34
Caodian	CAD	117.5	39.6	19	95.11	28.18	3.57	2.26
Changhong Park	CHH	117.2	39.1	3	96.67	9.43	2.43	0.50
Chitu	CHT	117.4	39.2	5	93.00	31.24	2.20	0.86
Douhe	DOH	118.3	39.7	121	90.12	33.83	3.88	2.93
Erwangzhuang	EWZ	117.4	39.4	2	110.00	0.00	3.53	1.14
Fengtai Town	FTZ	117.8	39.6	11	80.91	37.10	1.99	1.11
Hangu	HAG	117.8	39.2	18	83.89	29.89	3.93	2.05
Jinghai	JIH	116.9	38.9	6	179.67	30.26	2.78	0.31
Lutai	LUT	117.7	39.4	22	124.77	27.41	3.91	2.31
Nanhe Town	NHZ	117.1	39.0	4	47.50	43.95	3.98	2.96
Qingguang	QIC	117.0	39.2	6	154.17	39.20	2.12	1.63
Tang23	T23	117.5	39.1	2	5.00	35.00	1.90	0.65
Wangkuang	WAK	116.9	38.8	3	45.00	21.21	1.76	0.19
Wenan	WEA	116.5	38.8	3	123.33	6.24	4.53	1.45
Xinan Town	XAZ	117.5	39.8	13	90.62	30.67	1.96	0.91
Yufa	YUF	116.3	39.5	2	90.00	30.00	5.94	0.06
Zhutangzhuang	ZTZ	117.2	39.3	5	87.00	20.64	3.03	0.98

4 Results

From the statistical analysis of the shear-wave splitting results obtained for the southeast Capital area, the average polarization of fast shear waves is N88.2°W±40.7° (Fig. 4), whereas the average normalized time delay of slow shear waves is 3.55±2.93ms/km. The average slow wave time delay in the northwest Capital area is 4.44±2.93ms/km (Wu et al. 2007), which is comparatively higher than the average time delay in the southeast Capital area.

The average polarization estimated for the southeast Capital area is close to the maximum horizontal principal compressive stress in North China (Xu 2001) and consistent with the maximum compressive strain in North China (Zhang et al. 2004). However, the average polarization of fast shear waves in the northwest Capital area is N69.9°W±44.5° (Wu et al. 2007) and consequently there is a difference in average polarization between the southeast and northwest parts of the Capital area of about 20°. In addition, the standard error affecting the fast-shear-wave polarization in the southeast part is much smaller than the scatter observed in the northwest Capital area. Figure 5 shows the equal-area rose diagram (lower hemisphere projection) which describes the polarizations of the faster shear waves recorded at the 20 CASN stations. In this diagram, the center of any circle represents the position of the respective station and within each circle the midpoint and direction of a short segment mark the position of an event and the fast-shear-wave polarization, respectively. It is obvious that the polarizations estimated here show different patterns at CASN stations.

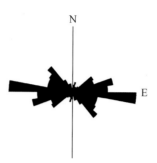

Fig. 4 Rose diagram of fast-shear-wave polarizations observed in the southeast Capital area

The results correspond to the waveforms recorded at 20 CASN stations during the period January 2002 to August 2005

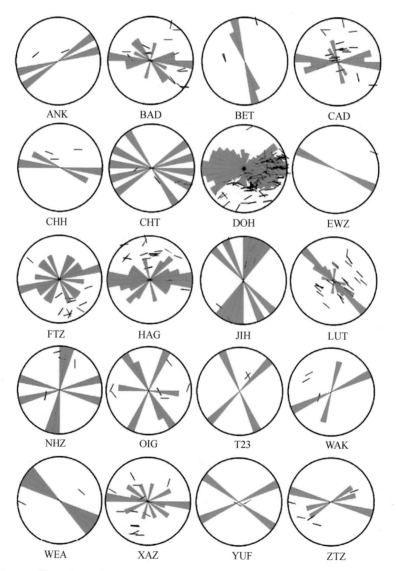

Fig. 5 Equal-area rose diagram (lower hemisphere projection) of fast-shear-wave polarizations at the 20 CASN stations installed in the southeast Capital area

In this diagram, the center of a circle represents the position of the respective station and within each circle the midpoint and direction of a short segment mark the position of an event and the fast-shear-wave polarization, respectively

5 Spatial distribution of polarizations

The spatial distribution of fast-shear-wave polarizations at CASN stations is displayed in Fig. 6, wherein the directions of the short segments indicate the average fast-shear-wave polarization at the respective station, while the lengths of the segments are proportional to the average time delay of the slow shear wave at each station. Site-dependent polarizations of fast shear wave are clearly observed. North of the study area, the average polarizations at stations such as BAD, XAZ, CAD, FTZ, and DOH are nearly in E-W direction (Fig. 6). Station YUF presents only two records (Table 1) which provide different polarizations (Fig. 5) and despite of it the average polarization direction is too E-W (Fig. 6). However, some stations, EWZ and LUT, in the central part of the explored area show polarization W-NW, although other stations (ZTZ, CHT, HAG) show clearly polarizations in E-W direction in spite of the observed scatter, for example, at station CHT (Fig. 5). In the northern half of the study area, the average polarizations differ ~20° from the horizontal principal compressive stress in North China (N71.6°E), thus being roughly consistent with the horizontal principal compressive stress in the region (Fig. 6).

In contrast, a variety of polarization patterns can be observed in the southern half of the test area (Fig. 6). With the exception of station CHH which also shows clear fast-shear-wave polarization in E-W direction, other stations such as JIH, T23, and BET exhibit average polarization nearly in N-S direction, and another stations such as ANK, NHZ, and WAK in N-E direction. Lastly, station WEA exhibits fast-shear-wave polarization in nearly N-W direction. Station QIG shows very scattered polarizations (Fig. 5) and the reliability of the polarization data is not good enough.

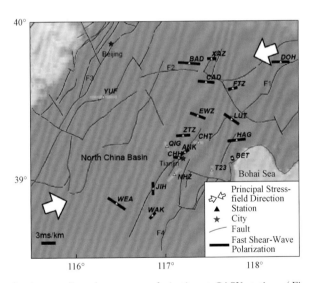

Fig. 6 Average fast-shear-wave polarization at CASN stations (Fig. 1)

The directions of the short segments show the average fast-shear-wave polarization at the respective station, while the lengths of the segments are proportional to the average slow-wave time delay at each station. *Black segments* correspond to stations with two or more records. Four stations (YUF, QIG, CHT, NHZ, T23) show polarizations with high standard errors. A time-delay scale is inserted on the *bottom left* corner. Local faults are the same as in Fig. 1

Most of the 20 control stations used in this study provide consistent fast-shear-wave polarizations and predominant directions of polarization (Fig. 5). However, some stations present scattered polarizations, large standard errors, and low reliability, such as YUF, CHT, QIG, NHZ, and T23 (Figs. 5 and 6). There are many possible reasons by which scattered polarization may happen independently from the data quality. At stations such as BAD, CAD, FTZ, DOH, and HAG, some analyzed waveforms taken in the limit of the shear-wave window can be the cause of biased results and therefore of less accuracy (Fig. 5). Anyhow, because of the large number of useful records, the final results are not influenced by this reason. Other scattering factor may be the complex stress field owing to crossing faults, which can result in an unclear pattern with more than one predominant polarization. For instance, the admissible reason for such a situation at stations NHZ and T23 (Fig. 5) is the existence of nearby crossing faults in N-S and E-W directions, respectively (Fig. 6). Moreover, unknown deep faults may be the other possible reason too. Station QIG shows, for example, various polarizations (Fig. 5) which are not easy to understand. Complex scattered polarizations patterns at stations such as QIG and NHZ (Fig. 5) need to be studied in detail in the future and likely with more data.

6 Preliminary observations on the splitting results

The results support that the fast-shear-wave polarizations at stations in the northern half of the Capital area are rather consistent and exhibit nearly E-W direction (Fig. 6). In particular, stations BAD, XAZ, CAD, FTZ, and DOH are all very close to the Baodi fault (F2 in Fig. 1) and their respective polarizations are very similar and parallel to the strike of this fault. Even a station with so low reliability as YUF, which is far from these five stations, presents a similar polarization in E-W direction. But in all these cases the polarizations are different from the maximum horizontal principal compressive stress in the zone. In consequence of the splitting results, the wave polarizations at stations on or nearly-around active faults are usually parallel to the strike of the faults (Crampin et al. 2002; Peng and Ben-Zion 2004; Gao et al. 2008; Shi et al. 2006; Wu et al. 2007).

In change, the predominant fast-shear-wave polarizations at stations in the southern half of the Capital area are quite different ones from others (Figs. 5 and 6), likely in accordance with local variations regarding the tectonic of this area. Stations WAK, JIH, NHZ, ANK, CHH, and QIG show certainly different polarizations, albeit the first four of them depict a predominant polarization in N-NE or nearly N-S direction, following more or less the strike of a nearby long fault which crosses the region from south to north. As before, the polarizations are all different from the maximum horizontal principal compressive stress in the zone but nearly parallel to the strike of the long fault, so that the influence of this fault on the shear-wave splitting results seem to be a real possibility.

With respect to station WEA, the average fast-wave polarization is N123.3°E±6.2° (Fig. 5), obviously very far from the dominant direction of the regional stress field. It is interesting to observe that the predominant polarization at WEA (Fig. 6) is also different from the wave polarizations at other nearby stations (WAK, JIH), which possibly is related with recent seismic activity: a M_L 5.1 earthquake occurred in Wen-an in 2006. At present, it is not possible to say whether the splitting result at WEA is due to a nearby active fault in W-NW direction or not.

Furthermore, there is a clear wave polarization in E-W direction at station CHH which is the nearest one to the city. But it is indeed premature to attribute this result to a hidden active fault with strike in the same direction. A conclusion of this nature needs more research.

7 Conclusions

By applying systematically a method based on time-delay correction, computation of the particle motion, and polarization analysis of split shear waves to seismograms recorded during the period 2002 – 2005, we have determined the splitting parameters at 20 stations belonging to the CASN array installed in the southeast Capital area of North China. The average polarization of fast shear wave is N88.2°W±40.7° and the normalized average time delay of slow shear wave is 3.55±2.93ms/km.

Since the horizontal principal compressive stress in North China is N71.6°E, it is evident that the observed predominant polarization in the northern half of the test area is not quite different from the direction of the regional stress field and is roughly consistent with the regional maximum compressive strain, which indicates that shear-wave splitting depicts reasonably well the principal compressive stress and strain affecting the region.

An overall view of the fast-shear-wave polarizations allows us to appreciate that they are very distinct depending on the location of the CASN stations either in the northern half or in the southern half of the study area. Unlike this last area where the polarizations exhibit a variety of possible azimuths, in the north, the polarizations are rather consistent and show E-W direction, thus suggesting dissimilar stressfield and tectonic frame in both areas.

As the polarization of fast shear wave is consequence of the principal compressive stress in situ and the time delay of slow shear wave express the degree of seismic anisotropy in situ, the faults near the measurement sites can influence the shear-wave splitting results; thus, the uneven distribution of faults in the explored region may be the cause of the scatter affecting some estimates of polarization, as expected.

Furthermore, different tectonic structures can result in different shear-wave splitting measurements. So, since the average fast-wave polarization in the northwest Capital area is N69.9°W±44.5° (Wu et al. 2007), there is a difference in polarization of about 20° between the southeast and northwest parts of the Capital area. Furthermore, the average time delay of slow shear wave in the southeast Capital area, that is 3.55ms/km, is obviously less than the normalized time delay of 4.44ms/km in the northwest Capital area (Wu et al. 2007), which implies that the magnitude of crustal seismic anisotropy is not the same in the two zones. In addition, the standard polarization errors in the southeast part are smaller than the standard errors in the northwest part. As the southeast Capital area lies on the basin and the northwest Capital area lies where uplifts and basin converge, the shear-wave splitting results permit to do a distinction regarding the tectonics and stress field affecting the two regions.

Acknowledgements

We thank Dr. Zhan-wu Gao, Prof. Jin-li Huang and Prof. Rui-feng Liu for their help in geological setting and data acquisition. Helpful comments and suggestions from two anonymous referees that led to significant improvement of the early manuscript are gratefully acknowledged. The IES-CEA Project 2007 – 13, the Seismic Professional Science Foundation (Grant 200708008), and the NSFC Project 40674021 supported this research.

References

Crampin S (1978) Seismic wave propagation through a cracked solid: polarization as a possible dilatancy diagnostic. *Geophys J R Astron Soc* **53**: 467–496

Crampin S, Peacock S (2005) A review of shear-wave splitting in the compliant crack-critical anisotropic Earth. *Wave Motion* **41**: 59–77

Crampin S, Gao Y (2006) A review of techniques for measuring seismic shear-wave splitting above small earthquakes. *Phys Earth Planet Inter* **159**: 1–14

Crampin S, Volti T, Chastin S, Gudmundsson A, Stefánsson R (2002) Indication of high pore-fluid pressures in a seismically-active fault zone. *Geophys J Int* **151**: F1-F2

Gao Y, Crampin S (2006) A further stress-forecast earthquake (with hindsight), where migration of source of earthquakes causes anomalies in shear-wave polarizations. *Tectonophysics* **426**: 253-262

Gao Y, Hao P, Crampin S (2006) SWAS: a shear-wave analysis system for semi-automatic measurement of seismic shear-wave splitting above small earthquakes. *Phys Earth Planet Inter* **159**: 71-89

Gao Y, Zheng S, Sun Y (1995) Crack-induced anisotropy in the crust from shear wave splitting observed in Tangshan region, North China. *Acta Seismol Sin* **8**(3): 351-363

Gao Y, Zheng S, Zhou H (1999) Polarization patterns of fast shear wave in Tangshan region and their variations. *Chinese J Geophys* **42**(2): 228-232 (in Chinese)

Gao Y, Liu X, Liang W, Hao P (2004) Systematic analysis method of shear-wave splitting: SAM software system. *Earthq Res China* **18**(4): 365-372

Gao Y, Wu J, Cai J, Shi Y, Lin S, Bao T, Li Z (2008) Shear-wave splitting in the Southeast of Cathaysia Block, North China. *J Seismol* **13** (2): 267-275

Huang J, Zhao D (2005) Three dimensional P wave velocity tomography and deep structure related to strong earthquake in Capital area. *Chin Sci Bull* **50**(4): 348-355 (in Chinese)

Jia S, Qi C, Wang F, Chen Q, Zhang X, Chen Y (2005) Three-dimensional crustal gridded structure of the Capital area. *Chin J Geophys* **48**(6): 1316-1324 (in Chinese)

Jiang Z, Zhang X, Chen B, Xue F (2000) Characteristics of recent horizontal movement and strain-stress field in the Crust of North China. *Chin J Geophys* **43**(5): 657-665 (in Chinese)

Lai Y, Liu Q, Chen J, Liu J, Li S, Guo B, Huang Z (2006) Shear wave splitting and the features of the Crustal stress field in the Capital circle. *Chin J Geophys* **49**(1): 189-196 (in Chinese)

Liu K, Zhang Z, Hu J, Teng J (2001) Frequency band-dependence of S-wave splitting in China mainland and its implications. *Sci China*, Ser D **44**(7): 659-665

Müller C (2001) Upper mantle seismic anisotropy beneath Antarctica and the Scotia Sea region. *Geophys J Int* **147**: 105-122

Peng Z, Ben-Zion Y (2004) Systematic analysis of crustal anisotropy along the Karadere-Duzce branch of the North Anatolian fault. *Geophys J Int* **159**: 253-274

Shi Y, Gao Y, Wu J, Luo Y, Su Y (2006) Seismic anisotropy of the crust in Yunnan, China: polarizations of fast shear-waves. *Acta Seismol Sin* **19**(6): 620-632

Wang F, Zhang X, Chen Q, Chen Y, Zhao J, Yang Z, Pan S (2005) Fine tomographic inversion of the upper crust 3-D structure around Beijing. *Chin J Geophys* **48**(2): 359-366 (in Chinese)

Wu J, Gao Y, Chen Y, Huang J (2007) Seismic anisotropy in crust in Northwest Capital area. *Chin J Geophys* **50**(1): 209-220 (in Chinese)

Xu Z (2001) A present-day tectonic stress map for eastern Asia region. *Acta Seismol Sin* **14**(5): 524-533

Zatsepin SV, Crampin S (1997) Modelling the compliance of crustal rock. I. Response of shear-wave splitting to differential stress. *Geophys J Int* **129**: 477-494

Zhang Z, Li Y, Lu D, Teng J, Wang G (2000) Velocity and anisotropy structure of the crust in the Dabieshan orogenic belt from wide-angle seismic data. *Phys Earth Planet Inter* **122**: 115-131

Zhang G, Ma H, Wang H, Li L (2004) The relation between active blocks and strong earthquakes in China. *Sci China*, Ser. D, **34**(7): 591-599 (in Chinese)

Zhuang C (1999) Technical composition of the digital telemetry seismic network in Capital Zone. *Seismological and Geomagnetic Observation and Research* **20**(5): 23-28 (in Chinese)

2010年青海玉树地震震源过程

张 勇 许力生 陈运泰

中国地震局地球物理研究所，北京 100081

2010年4月14日7时49分（北京时间），在我国青海省玉树藏族自治州发生了M_S7.1地震（简称玉树地震）. 根据中国地震台网中心最新测定结果，玉树地震震中位置为33.2°N，96.6°E，震源深度约14km，位于玉树城区西北约44km处. 截至2010年5月30日，玉树地震已造成三千多人死亡或失踪、一万多人受伤、大量房屋倒塌.

玉树地震发生于巴颜喀拉地块（松潘-甘孜地块）南缘的甘孜-玉树断裂带. 在这条沿北西-南东方向延伸、以左旋走滑为主的断裂带上，历史上地震频发[1,2]，地震活动性较强[3,4]. 玉树地震是该断裂带北西段最近一百年来发生的最大的一次地震事件. 作为震后应急响应的重要工作内容之一，作者在震后约2.5小时得出并发布了这次地震破裂过程的初步结果；随着波形数据的逐渐增加，在震后约5小时和2天后先后两次对上述结果进行了修订（http://www.csi.ac.cn）.

为深入理解和认识玉树地震的震源过程，在应急响应工作结束后，作者利用余震记录，采用经验格林函数方法，从勒夫波（Love wave）中提取了主震的视震源时间函数，并据此估计了破裂速度；在此基础上，重新对P波波形进行反演，得到了比震后应急响应工作更为准确的玉树地震破裂过程结果. 以上工作概述如下.

1 勒夫波视震源时间函数分析

玉树地震发生后大约1.5小时，在主震震中附近发生了一次M_S6.3余震，其震中位置为33.23°N，96.58°E，震源深度约10km（据青海地震局测定结果）. 这次余震具有与主震相近的震源位置和几乎完全相同的震源机制[6]（图1），因此可以用它的波形记录作为经验格林函数来提取主震的视震源时间函数，

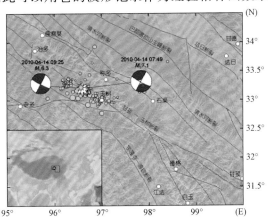

图1 玉树地震位置及其构造背景

白色大八角星和青色小八角星分别表示玉树地震及M_S6.3余震的震中位置，红色实线为断层[5]，两个"海滩球"分别表示玉树地震主震及其最大余震震源机制解的下半球投影[6]，青色圆圈为玉树地震5天后发生的余震震中位置（据青海省地震局张晓青研究员）

---

* 本文发表于《中国科学：D辑：地球科学》，2010年，第40卷，第7期，819-821.

分析主震的破裂特征. 鉴于在很多台站的记录中该余震的P波和S波的信噪比较低, 作者选用了勒夫波的波形记录.

基于图2（a）所示的24个台站的勒夫波波形记录, 采用PLD方法[7,8]提取了如图2（b）所示的勒夫波视震源时间函数. 由图2（b）可见, 方位角相近的台站的勒夫波视震源时间函数的形状非常相似, 具有一定的共同性; 随着方位角的变化, 持续时间和两个峰值的位置均呈规律性的变化, 表现出明显的方向性. 方位角位于100°～150°范围内的台站的视震源时间函数的持续时间最短, 约为10s; 而方位角位于300°左右的台站的视震源时间函数持续时间最长, 接近30s. 视震源时间函数持续时间随方位的变化显示出强烈的"地震多普勒效应", 表明这次地震总体上是由震中朝东南方向（100°～150°）扩展的单侧破裂事件.

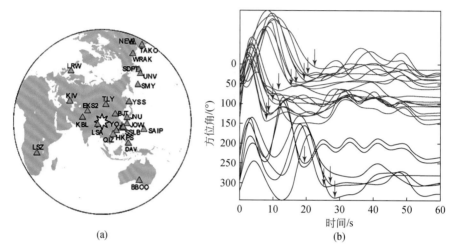

图2 震中与台站分布（a）和勒夫波视震源时间函数随台站方位角的变化（b）

（a）中的白色八角星表示玉树地震的震中位置, 青色三角形为提取勒夫波视震源时间函数所选用的台站; （b）中红色箭头标示了位于不同方位角的台站处的视震源时间函数持续时间, 蓝色线条表示用于确定两次子事件峰值所对应的时空参数的视震源时间函数

由方位角230°～330°范围的台站的视震源时间函数（图2（b））可以清楚地分辨出两次子事件; 但在其他方位角范围的台站上, 两次子事件相距很近而聚合在一起. 为确定两次子事件的峰值所对应的时空参数, 根据方位角位于230°～330°范围的台站的视震源时间函数（图2（b））所显示的两次子事件峰值所对应的时间, 建立如下目标函数:

$$\Delta(R,T) = \sum_i \left| T - \frac{R\cos(\phi_i - \phi_s)}{V} - t_i \right| = \min, \quad (1)$$

式中, i 表示台站序数, R 和 T 分别为与子事件峰值相对应的空间和时间位置, V 为震中区的勒夫波群速度, ϕ_i 为第 i 个台站相对于震中的方位角, ϕ_s 为断层走向, t_i 为第 i 个台站的视震源时间函数中子事件峰值出现的时间. 根据刘超等[6]确定的玉树地震的震源机制解, 取 $\phi_s=119°$; 根据彭艳菊等[9]关于勒夫波层析成像的研究结果, 取 $V=3.4$km/s. 利用遗传算法求解式（1）, 得出: 与第一次子事件峰值对应的空间位置位于震中东南3.3km, 出现在破裂开始后2.1s, 因此, 在0～2.1s内的平均破裂速度约为1.6km/s; 与第二次子事件峰值对应的空间位置位于震中东南32.9km, 出现在破裂开始后8.3s, 因此, 在0～8.3s内的平均破裂速度约为4.0km/s. 可见, 玉树地震破裂的扩展是一个由缓到急的过程, 其中破裂开始后0～8.3s内的平均破裂速度甚至超过了当地的S波速度[10].

2 P波资料分析

根据刘超等[6]确定的玉树地震的断层面参数,作者选用走向119°/倾角83°的节面作为玉树地震发震断层所在的平面,在这个长96km、宽30km的平面上,沿走向方向和倾向方向上划分为32×10 = 320 个3km×3km的子断层,由玉树地震的震源参数可确定初始破裂点位于走向方向上第5块、倾向方向上第5块子断层处。选用如图3(a)所示的16个台站的垂直向P波资料,基于滑动角可变的线性反演技术[11],利用全球标准速度模型[12]和格林函数的反射率计算方法[13],得到了断层面上的时空破裂过程。

根据P波反演结果(图3),玉树地震释放的标量地震矩约为2.7×10^{19} Nm,相当于矩震级$M_W 6.9$,断层面上平均滑动量约0.6m,平均应力降约15MPa,与板内地震典型的应力降(10MPa)水平大体相当;整个破裂过程的持续时间约16s,主要由两次子事件组成(图3(b))。若以初始破裂发生的时刻为时间原点,则第一次子事件发生于0~5s,对应的破裂分布在震中西北10km至震中东南10km处,最大静态滑动量约为0.8m。第二次子事件发生于5~16s,对应的破裂主要发生在走向方向(震中东南方向)上距震中17~54km处(图3(c)和(e)中的红-橙-黄色区域),最大静态滑动量达1.8m,破裂明显贯穿到地表。

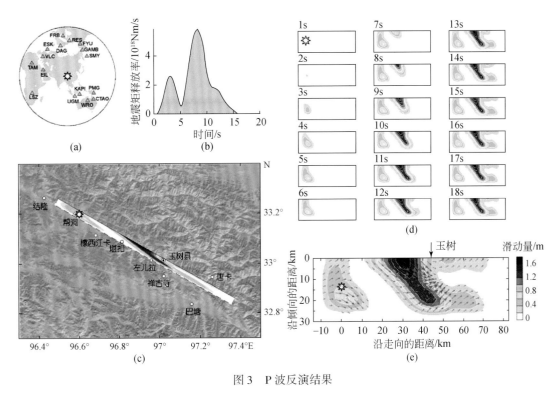

图3 P波反演结果

(a)震中与台站分布;(b)震源时间函数;(c)断层面上静态滑动量分布在地表面的投影;(d)断层面上滑动量的时空分布;(e)断层面上静态滑动量分布。(a)和(c)中的白色八角星表示震中位置;(d)和(e)中的八角星表示初始破裂点在断层面上的位置;(e)中的灰色箭头表示滑动矢量

3 讨论与结论

本文通过勒夫波和P波综合分析得出2010年4月14日青海玉树$M_S 7.1$($M_W 6.9$)地震的破裂过程主要由两次子事件组成。从规模上讲,第一次子事件小于第二次子事件;从时间上讲,第一次子事件发生于地震破裂起始后0~5s,第二次子事件发生于地震破裂起始后5~16s;从空间位置讲,两次子事件分别发

生在震中附近与玉树城区附近，二者相距约 30km，且第二次子事件的破裂在玉树城区西北附近贯穿至地表；从破裂速度上讲，与两次子事件峰值对应的平均破裂速度分别约为 1.6km/s 和 4.0km/s，其中 4.0km/s 的速度超过了震中附近的剪切波速度.

玉树地震属于大地震（震级 $M \geqslant 7$ 的地震）下限边缘的地震，其震源破裂过程中出现的超剪切破裂现象表明，超剪切破裂并非仅可能发生于特大地震[14,15]. 以超剪切波速度传播的破裂导致的地震能量聚焦效应以及规模较大、破裂贯穿到地表的第二次子事件是玉树城区遭受严重破坏在震源方面的主要原因.

致谢 IRIS 数据中心提供全球宽频带地震数据，审稿专家提出宝贵意见，在此一并致谢.

参 考 文 献

[1] 谢毓寿，蔡美彪. 中国地震历史资料汇编第三卷（下）. 北京：科学出版社，1987. 235-236
[2] 周荣军，闻学泽，蔡长星，等. 甘孜-玉树断裂带的近代地震与未来地震趋势估计. 地震地质，1997，**19**：115-124
[3] 李闽锋，邢成起，蔡长星，等. 玉树断裂活动性研究. 地震地质，1995，**17**：218-224
[4] 闻学泽，徐锡伟，郑荣章，等. 甘孜-玉树断裂的平均滑动速率与近代大地震破裂. 中国科学 D 辑：地球科学，2003，**33**：199-208
[5] 邓起东，张培震，冉永康，等. 中国活动构造基本特征. 中国科学 D 辑：地球科学，2002，**32**：1020-1030
[6] 刘超，许力生，陈运泰. 2010 年 4 月 14 日青海玉树地震快速矩张量解. 地震学报，2010，**32**：366-368
[7] Bertero M, Bindi D, Boccacci P, et al. Application of the projected Landweber method to the estimation of the source time function in seismology. *Inv Problems*, 1997, **13**: 465-486
[8] 张勇，许力生，陈运泰. 提取视震源时间函数的 PLD 方法及其对 2005 年克什米尔 M_W7.6 地震的应用. 地球物理学报，2009，**52**：572-580
[9] 彭艳菊，苏伟，郑月军，等. 中国大陆及海域 Love 波层析成像. 地球物理学报，2002，**45**：792-803
[10] Bassin C, Laske G, Masters G. The current limits of resolution for surface wave tomography in North America. *EOS Trans AGU*, 2000, **81**: F897
[11] 张勇，冯万鹏，许力生，等. 2008 年汶川大地震的时空破裂过程. 中国科学 D 辑：地球科学，2008，**38**：1186-1194
[12] Kennett B L N, Engdahl E R. Travel times for global earthquake location and phase identification. *Geophys J Int*, 1991, **105**: 429-465
[13] Kennett B L N. *Seismic Wave Propagation in Stratified Media*. Cambridge：Cambridge University Press, 1983. 1-339
[14] 杜海林，许力生，陈运泰. 利用阿拉斯加台阵资料分析 2008 年汶川大地震的破裂过程. 地球物理学报，2010，**52**：372-378
[15] Das S. The need to study speed. *Science*, 2007, **317**: 905-906

京津唐地区地壳三维 P 波速度结构与地震活动性分析*

于湘伟[1]　陈运泰[2]　张 怀[1]

1. 中国科学院研究生院计算地球动力学实验室，北京　100049；2. 北京大学地球与空间科学学院，北京　100871

摘要　本文利用华北遥测地震台网和首都圈数字地震台网 112 个台站记录到的 1993～2004 年发生在首都圈地区 3983 次地震的 P 波绝对到时资料和相对到时资料，采用双差地震层析成像方法联合反演了京津唐地区地壳三维 P 波速度结构和震源参数。京津唐地区的三维 P 波速度结构图像在浅层上很好地反映了地表地质、地形的特征。在平原和凹陷的盆地处呈现 P 波低速速度异常，而在隆起的山区或基岩出露区显示为 P 波高速速度异常。在研究区域内震级 $M \geq 6.0$ 历史地震和经过重新定位后的震级 $M_L \geq 3.0$ 的地震的震源位置在 10km 深度和 15km 深度处的 P 波相对速度扰动图上的投影都显示出相似的特点，即：绝大部分的地震的震源位置在 P 波相对速度扰动图上的投影分布在低、高速异常的交界地带，且偏高速体一侧，只有极少数的地震分布在 P 波速度异常体内部。

1 引言

在过去的几十年里，北京、天津和唐山地区（以下简称京津唐地区），由于其地理位置的重要性以及地质构造的复杂性，一直是许多地震研究关注的地区。该地区的地壳和上地幔的三维速度结构迄今已有许多具有参考价值和科学意义的研究结果[1-10]。近年来，国际上地震定位（相对定位和绝对定位）方法有显著的进展，例如双差地震定位方法的提出及其广泛应用与成功验证[11,12]。在我国，运用一维速度模型，双差地震定位方法在地震定位方面亦得到了一些有参考价值和科学意义的研究结果[13-16]。

事实上，双差地震定位方法可以与常规地震层析成像方法相结合，既联合各自的优点，又克服各自的缺点和局限性。基于上述思想，Zhang 和 Thurber[17] 提出了基于双差地震定位的地震层析成像方法，简称双差地震层析成像方法。与常规地震层析成像方法相比，用双差地震层析成像方法得到的三维速度结构模型更精确可信。在常规地震层析成像方法中，由于绝对到时数据的观测误差的影响，使得联合反演的震源位置会有某种程度上的散开现象，而在双差地震层析成像方法中由于使用了相对到时数据则去除了这些误差，从而避免了地震定位中出现的弥散现象，继而在反演速度模型时避免了某些失真部分。

本文利用双差地震层析成像方法研究京津唐地区更加精细的三维速度结构和地震的精确定位，借以探索该地区的三维速度结构和地震活动性之间的关系。

2 方法与数据

2.1 双差地震层析成像方法

根据射线理论，对于微扰问题，震源 i 到地震观测台站 k 的地震波观测到时和理论到时之间的残差 r_k^i

---

* 本文发表于《地球物理学报》，2010 年，第 **53** 卷，第 8 期，1817-1828。

和地震波走时对震源和速度结构的微扰有如下关系：

$$r_k^i = \sum_{l=1}^{3} \frac{\partial T_k^i}{\partial x_l^i} \Delta x_l^i + \Delta \tau^i + \int_i^k \delta u \, ds ,\tag{1}$$

式中，T_k^i 是震源 i 到地震观测台站 k 的体波观测到时，τ^i 是震源 i 的初始发震时刻，u 是波的慢度（速度的倒数），ds 表示沿射线路径的线元，震源坐标为 (x_1, x_2, x_3)，积分路径从震源 i 到观测台站 k. 在式（1）中，初始发震时刻、射线路径和波的慢度（模型参数）均为未知.

同样，对于震源 j 到地震观测台站 k 地震波观测到时和理论到时之间的残差 r_k^j 和方程（1）类似：

$$r_k^j = \sum_{l=1}^{3} \frac{\partial T_k^j}{\partial x_l^j} \Delta x_l^j + \Delta \tau^j + \int_j^k \delta u \, ds ,\tag{2}$$

因此，对于同一个地震观测台站 k，震源 i 和震源 j 之间有：

$$r_k^i - r_k^j = \sum_{l=1}^{3} \frac{\partial T_k^i}{\partial x_l^i} \Delta x_l^i + \Delta \tau^i + \int_i^k \delta u \, ds - \sum_{l=1}^{3} \frac{\partial T_k^j}{\partial x_l^j} \Delta x_l^j - \Delta \tau^j - \int_j^k \delta u \, ds .\tag{3}$$

如果震源 i 和震源 j 彼此相距足够接近，则震源到同一个观测地震台站 k 的射线路径可以认为是近似相同的，此时若射线经过区域的速度结构为已知，则方程（3）可进一步写为

$$\mathrm{d}r_k^{ij} = r_k^i - r_k^j = \sum_{l=1}^{3} \frac{\partial T_k^i}{\partial x_l^i} \Delta x_l^i + \Delta \tau^i - \sum_{l=1}^{3} \frac{\partial T_k^j}{\partial x_l^j} \Delta x_l^j - \Delta \tau^j ,\tag{4}$$

式中，$\mathrm{d}r_k^{ij}$ 称为双差（double-difference）[11]，即为震源 i 和震源 j 的观测到时之差和理论到时之差的残差，即

$$\mathrm{d}r_k^{ij} = r_k^i - r_k^j = (T_k^i - T_k^j)^{\mathrm{obs}} - (T_k^i - T_k^j)^{\mathrm{cal}} .\tag{5}$$

式（5）即为双差地震定位方法（简称 DD 法）[11]. 式中，观测到时差 $(T_k^i - T_k^j)^{\mathrm{obs}}$ 可以通过地震目录中绝对到时计算得到. 方程（3）能简化成方程（4）是有其隐含的路径假定限制的，即由地震对所在的震源区附近的速度不均匀性所引起的射线路径异常与震源对 (i, j) 的相对位置近似无关. 这一假定在震源彼此之间相距很近时才成立，而对于彼此相距较远的震源对则不成立，在此情况下会造成相距较远的地震对的相对定位有很大的系统性偏差.

Zhang 和 Thurber[17]提出了一种基于双差地震定位方法的地震层析成像方法，简称双差地震层析成像方法. 利用到时数据直接求解方程（3），考虑地震对之间的路径异常，即该方程右端第三项和第六项. 若两个震源相距很近，它们到同一个台站的射线路径基本上是重叠的，此时方程（3）中震源区以外的模型偏导数部分可以互相抵消，震源区以外的速度结构在反演中通过地震波的绝对到时数据来确定. 这样，既使用了相对到时数据，又使用了绝对到时数据，经过联合反演同时得到了研究区的三维速度结构模型、地震的绝对定位和地震的相对定位，这是以往的常规地震层析成像方法所不能做到的.

2.2 数据来源

鉴于现有资料和地震台站的分布情况，本文的研究区域确定为 114°E ~ 120°E，37.5°N ~ 41.5°N（简称京津唐地区）. 该地区主要位于华北遥测地震台网和首都圈数字地震台网所覆盖的区域内. 研究中共收集了 1993 年 1 月至 2001 年 12 月间华北遥测地震台网分布在京津唐地区的 46 个台站记录的、发生在研究区域内的近震直达 P 波初动到时数据，和 2002 年 1 月至 2004 年 8 月间首都圈数字地震台网分布在京津唐地区 105 个地震台站（部分台站与华北遥测地震台网的台站重叠）记录到的发生在研究区域内的近震直达 P 波初动到时数据. 将华北遥测地震台网和首都圈数字地震台网相互重叠的台站合并后，两个地震台网分布在研究区域内的共有 112 个地震台站（图 1）. 华北遥测地震台网和首都圈数字地震台网提供的这些高质量、大批量的地震观测数据为京津唐地区的双差地震层析成像研究创造了很好的条件.

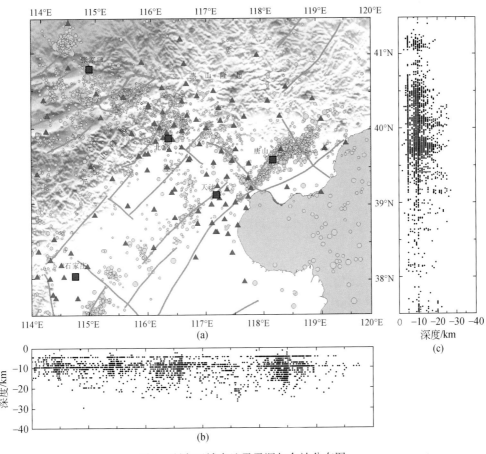

图 1 研究区域内地震震源与台站分布图

(a) 研究区域内地震震中与台站分布，图中三角形表示地震台站，正方形表示主要城市，粗线表示活动断层，圆圈表示本研究所用的地震的初始震中；(b) 研究区域内地震震源沿纬度剖面的深度分布；(c) 研究区域内地震震源沿经度剖面的深度分布

Fig. 1 Distribution of seismic stations and earthquake hypocenters used in this study

(a) Distribution of earthquake epicenters (circles) and seismic stations (triangles). Squares show the major cities. Thick lines denote active faults; (b) Cross sectional view of focal depth along latitude profile; (c) Cross sectional view of focal depth along longitude profile

经过筛选，共筛选出 $M1.0 \sim 6.2$ 范围内 3983 次地震（每次地震的观测记录数至少 6 个，图 1），43408 条直达 P 波初动绝对到时数据. 在挑选出来的数据中，每个地震对之间最大的空间距离为 10km，每次地震最多可以和 20 次地震组成地震对，每个震相对的到时残差小于 0.5s，地震对的平均空间距离为 5.3km，若地震对有 6 个以上震相对则认为是"强匹配地震对"，其平均空间距离为 5.0km. 在筛选出的地震中，震源深度的分布在地表到地表以下约 30km 深，从这些地震对的 P 波绝对到时数据中共挑选出 200665 条 P 波相对到时数据，与 P 波绝对到时数据联合用于本文中的双差地震层析成像计算. P 波初动到时数据的读数误差平均为 $\pm 0.2 \sim \pm 0.3$s. 由于唐山地

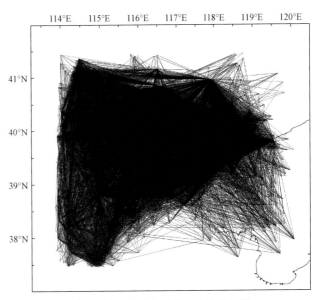

图 2 研究区域内射线二维分布情况

Fig. 2 Distribution of P-wave ray paths used in this study

区地震分布密集,为了保持整个研究区域内的地震分布的均衡,对该地区的地震做了更严格的筛选,即每次地震至少有8个观测记录,使得筛选后的地震尽可能均匀地分布在整个研究区域内,又保证在地震活动性较强的区域有足够多且记录足够好的地震.图2是3983次地震在研究区内的二维射线分布图,在5km到15km深处,穿过北京地区的射线数约为4000~6000条,而在天津和唐山地区约为300~5500条;在20~30km深处,穿过北京地区的射线数最低约为2000条,最高约为3500条;天津和唐山地区约为3000条.可以看出,京津唐地区是射线分布最为密集的地区.

我们根据所选用的地震资料的数量和质量、地震震中和台站分布情况确定节点间距,以便使尽可能多的节点处有较好的分辨率.通过反复实验,比较反演结果,将研究区域中节点间距水平向按0.5°×0.5°划分,垂直向节点分别放置在0~40km(步长5km)处,这样92%以上的节点周围有几百条以上射线穿过.初始一维速度模型(图3)采用于湘伟等[6]所得到的京津唐地区的一维初始速度模型,35km以下部分速度模型参考人工爆破方面的结果以及已有的地震层析成像方面的研究结果[18].

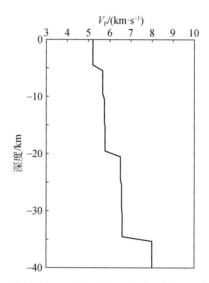

图3 用于双差地震层析成像反演的一维初始速度模型

Fig. 3 The starting 1-D velocity models used for double difference tomographic inversion in this study

我们在双差地震层析成像中使用了两种类型的到时数据:地震波绝对到时数据和相对到时数据.为了把这两种数据联合在一个反演体系中,在双差地震层析成像方法中使用了分级加权方法.在联合反演的前2次迭代过程中,赋予绝对到时数据较大的加权,以便建立一个"大"尺度的计算结果.具体的权重比为:绝对到时数据的权重为1,相对到时数据的权重为0.1.在反演的后2次迭代过程中,赋予相对到时数据更大的权重,进一步改进地震定位和震源区附近的速度结构,所选用的权重比为:相对到时数据的权重为1,绝对到时数据的权重为0.1.

3 层析成像结果与讨论

3.1 模型验证

我们用3983次地震的43408条直达P波初动绝对到时数据做了检测板实验[19],对反演结果的分辨率进行了检测,结果如图4所示.图中蓝色代表正的速度异常,红色代表负的速度异常,扰动值取为正常值的±3%.由图可见,研究区内各深度层上均能正确恢复正负速度的相对变化,特别是研究区内的重点地区北京、天津和唐山，由于有较多台站和事件的分布，使得每个深度层上这些地区的分辨率更高．密集数字地震台网记录的大量高精度数据为反演结果的可靠性提供了良好的保证．

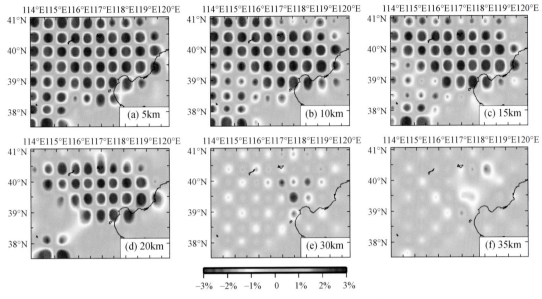

图 4 研究区域内不同深度处的检测板结果分布

Fig. 4 Map view of results of the checkerboard resolution test for the P-wave structure at different depth in the study area

除检测板的检验外，每个节点附近通过的射线数也可以作为衡量解的可靠性的一个估计．穿过全部节点的射线数分布如图 5 所示，有 92% 的节点有足够的射线数（射线数≥10）参与反演．穿过每个网格节点的射线数介于 0～7000 之间，其中，穿过北京、天津、唐山地区的节点的射线数大于 3000．

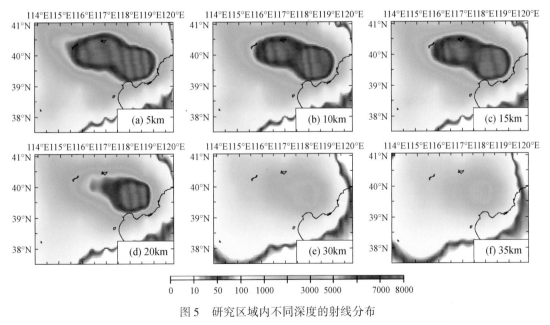

图 5 研究区域内不同深度的射线分布

层的深度示于每个子图的右下角；通过每个节点的射线数如图底部的色标所示

Fig. 5 Distribution of the number of the P-wave rays passing through each grid node (hit counts)

The depth of the layer is shown at the lower right corner of each map. The hit count scale is shown at the bottom

3.2 成像结果与讨论

经过四次迭代,用双差地震层析成像方法得到了以百分比表示的京津唐地区各个深度的 P 波速度扰动(图6). 图 6 显示了在不同深度处速度结构的横向变化,可以看出在研究区域内的中上地壳中, P 波速度结构存在明显的横向不均匀性.

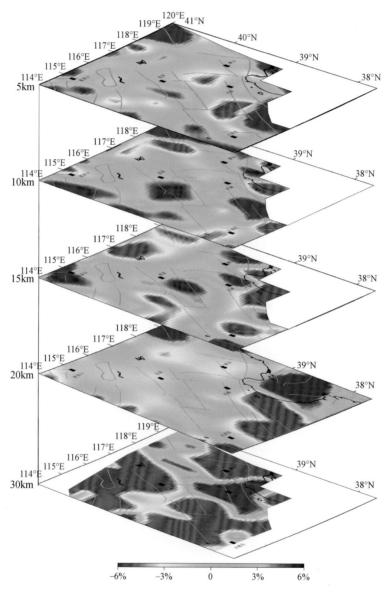

图 6 研究区域内各个深度层上的 P 波速度扰动

Fig. 6 P-wave velocity image at each depth slice (in percentage from the average velocity)

将上述结果与京津唐地区的地质构造图(图1)进行对比后可以看出,在 5km 到 10km 深度处, P 波相对速度扰动异常带的走向与各构造单元内地表地质构造和活动断裂的方向具有很好的一致性,总的来说浅层速度图像很好地反映了地表地质、地形的变化,即山脉和隆起区的下部表现为高速异常,凹陷和山间盆地等处的下部则表现为低速异常. 这与以往的地震层析成像结果[1~10]基本是一致的,但是在 P 波相对速度的高速异常和低速异常的交界带地区,本文的高、低速速度异常对比更为剧烈. 唐山地区虽位于华北平原,但却是张(家口)渤(海)地震带上一个隆起的块体[20],在上地壳层位上(图 6 中 5km 和 10km 深度层)都显示为高速异常. 在深部,在 35km 深度层上,太行山区则表现为低速异常,而在燕山和华北平原的大部地区呈现高速. 这些结果与 Pn 波层析成像结果[7~10]是类似的.

图 7 为研究区域内沿 115.5°E, 116.5°E, 117.5°E, 118.5°E (图 7a) 与沿 40.5°N, 39.8°N, 39.0°N (图 7b) 垂直剖面上的 P 波相对速度扰动分布 (以百分比表示). 可以看出, 在浅部, P 波相对速度高

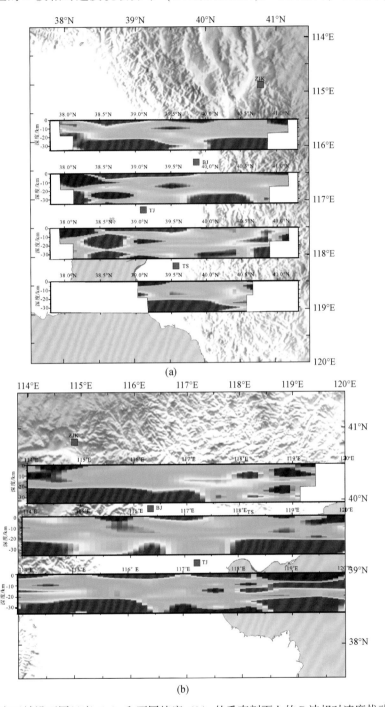

图 7 研究区域沿不同经度 (a) 和不同纬度 (b) 的垂直剖面上的 P 波相对速度扰动分布

图中 BJ 表示北京, TJ 表示天津, TS 表示唐山, ZJK 表示张家口. (a) 中每个剖面的上边缘经度位置即为沿经度方向的 P 波相对速度扰动的垂直剖面, 从上到下依次为沿 115.5°E, 116.5°E, 117.5°E 和 118.5°E 的垂直剖面上的 P 波相对速度扰动分布 (以百分比表示); (b) 中每个剖面的上边缘纬度位置即为沿纬度方向的 P 波相对速度扰动的垂直剖面. 从上到下依次为沿 40.5°N、39.8°N 和 39.0°N 的垂直剖面上的 P 波相对速度扰动分布 (以百分比表示)

Fig. 7 Vertical cross-sections of P-wave velocity perturbations at each longitude (115.5°E, 116.5°E, 117.5°E and 118.5°E) (a) and each latitude (40.5°N, 39.8°N and 39.0°N) (b)

BJ, Beijing; TJ, Tianjin; TS, Tangshan; ZJK, Zhangjiakou

速异常区和低速异常区的交界地带很好地反映了隆起区与平原、盆地的交界，如在 115.5°E 剖面上 39.5°N 处，117.5°E 剖面上 40.0°N 处（图 7a），40.5°N 剖面上 118.0°E，39.8°N 剖面上 118.0°E 处（图 7b），均为平原或凹陷区（低速异常区）与隆起区（高速异常区）的交界地区，两者非常一致．此外，从 116.5°E，117.5°E 和 118.5°E 垂直剖面上（图 7a）可以看出，北京、天津和唐山地区的下部在 15～20km 深度有一个尺度约 50～90km 的高速异常带，与周围速度对比剧烈．在研究区域的西部，在 115.5°E 剖面上（图 7a），燕山隆起和平原地区在 20～30km 深度都呈现出大范围的低速异常，而唐山地区（118.5°E 垂直剖面，图 7a）在 20～30km 深度范围内也呈现出低速异常，并且与上层的速度结构对比剧烈．39.8°N 剖面中（图 7b），唐山以东地区作为张渤地震带上一个隆起的块体表现为高速异常，太行山隆起区下部的高速异常区从地表向东延伸直到北京下部的约 20km 处，与周围的速度结构形成强烈对比．

3.3 地震活动性与 P 波速度结构之间的关系

本文用双差地震层析成像方法联合反演了京津唐地区的三维 P 波速度和震源参数（绝对位置和相对位置），在得到京津唐地区 P 波三维速度结构的同时，也得到了京津唐地区所用的 3983 次地震中 2809 次地震的重新定位结果[21]．

用所得到的 P 波三维速度结构对地震做重新定位后，地震的均方根残差由初始的 1.2s 降为 0.3s，有了显著的改善．在整个研究区域内，地震重新定位后得到的 2809 次地震的震源位置测定误差在 E-W 方向为 1.0km，在 N-S 方向为 1.1km，深度方向为 1.5km．

为进一步探讨地震活动性与 P 波速度结构之间的关系，我们将用双差地震层析成像方法联合反演得到的震级 $M_L \geq 3.0$，震源深度在 5～15km 范围内的地震的震源投影在 10km 深度处的 P 波相对速度扰动分布图上（图 8a），将震级 $M_L \geq 3.0$，震源深度在 10～20km 范围内的地震投影在 15km 深度处的 P 波相对速度扰动图上（图 8b）．根据臧绍先等[22]对华北地区地震活动的统计分析，该区大多数较大地震都发生在研究区的中上地壳，主要在 10～15km 深度范围内．我们将发生在研究区域内，公元前 780～2002 年间震级 $M \geq 6.0$ 的历史地震的震源位置分别投影在 10km 深度和 15km 深度处的 P 波相对速度扰动分布图上（图 8a，8b）．

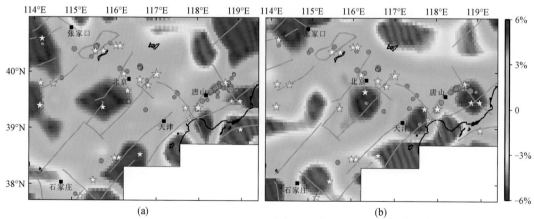

图 8 （a）研究区内地震重新定位后的震源（$M_L \geq 3.0$，5km≤深度≤15km，灰色圆圈）和历史地震（$M \geq 6.0$，白色五角星）在 10km 深度处 P 波相对速度扰动（以百分比表示）图上的分布；（b）研究区内地震重新定位后的震源（$M_L \geq 3.0$，10km≤深度≤20km，灰色圆圈）和历史地震（$M \geq 6.0$，白色五角星）在 15km 深度处 P 波相对速度扰动（以百分比表示）图上的分布

Fig. 8 (a) Map view of P-wave velocity perturbations (in percentage) and distribution of both relocated earthquake hypocenters ($M_L \geq 3.0$, 5km≤ focal depth≤15km, gray circles) and historic earthquake hypocenters ($M \geq 6.0$, white stars) at depth 10km; (b) Map view of P-wave velocity perturbations (in percentage) and distribution of both relocated earthquake hypocenters ($M_L \geq 3.0$, 10km≤ focal depth≤20km, gray circles) and historic earthquake hypocenters ($M \geq 6.0$, white stars)

由震源在 10km 和 15km 深度 P 波相对速度扰动图上的投影（图 8a，8b）可以看出，重新定位后，大部分地震的震源投影位于 P 波低速异常区与高速异常区的对比强烈的交界地带，较少地震的震源投影位于偏 P 波低速异常区的过渡带（黄色区域）上，较多地震的震源投影位于偏 P 波高速异常区的过渡带（浅蓝色区域）上. 历史地震的震源投影（白色五角星）也有类似的分布，其中，唐山地震、滦县地震、三河-平谷地震的震源投影均位于高速异常区的过渡带内. 值得注意的是，在 10km 深度上（图 8a），北京地区、唐山地区的 $M_L \geqslant 3.0$ 的地震的震源位置投影全部沿着 P 波高速异常区的边缘分布，震源的展布方向与高速异常带的走向非常一致.

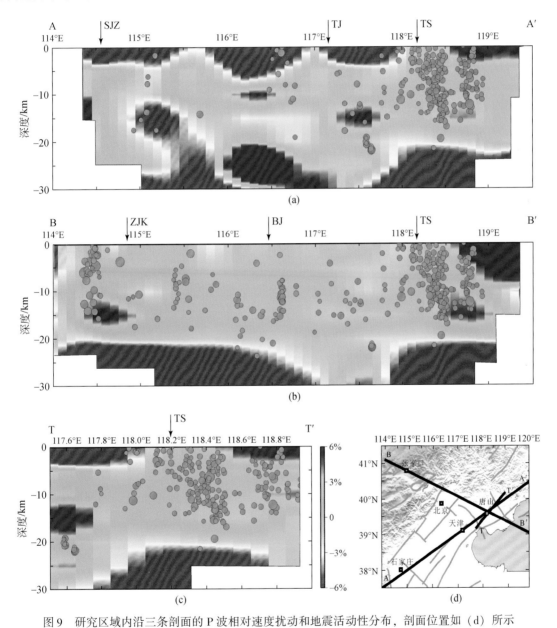

图 9　研究区域内沿三条剖面的 P 波相对速度扰动和地震活动性分布，剖面位置如（d）所示

Fig. 9　Distribution of P-wave velocity perturbations (in percentage) and seismicity along three vertical cross-sections, (d) the locations of cross-sections

根据研究区域内地震活动性的分布，我们在研究区内做了三个剖面来探讨地震活动性和 P 波速度异常之间的关系（图 9）. 在各个剖面的 P 波相对速度扰动分布图中，我们选取了震中位置距该剖面在纬度上相差 ±0.5° 范围内 $M_L \geqslant 2.5$ 地震，将其震源投影在该剖面上（图中灰色圆圈）.

在剖面 AA′ 的 P 波速度扰动分布和地震震源分布图（图 9a）中，0~10km 深度范围内 P 波速度异常和浅层的地表、地形特征很吻合. 在 10~20km 深度处沿唐山到天津一线存在一个沿水平方向尺度约 100km 的高速异常区，而在 20~30km 范围内唐山及其以北地区则存在大范围的低速异常区，这与 Huang 等[8,9]的研究结果相一致. 在 Huang 等[8,9]的工作中，使用了非天然地震和定点爆破资料，在本文的工作中使用的全部是天然地震，但是取得了相同的地震层析成像结果，这说明：使用双差地震层析成像方法提高了地震的定位精度，更加精确的地震定位使得我们仅用天然地震数据也可以得到更好的地震层析成像结果. 从剖面 AA′ 的 P 波相对速度变化中可以看出，唐山-河间-磁县地震活动断裂带在上地壳均处于不连续的低速异常区内，而在 10~20km 深度范围内则表现为一条断续的南浅北深的高速异常带，在浅层的低速异常体中和深部的高速异常体边缘分布有地震. 唐山地区地震的最大深度与该地区中下地壳的低速异常的上缘相一致. 此外石家庄以北地区的震源深度在 10~20km 的地震均分布在该深度范围内的低速异常的边缘地带.

在沿着张家口—渤海地震带所做的北西西向剖面 BB′（图 9b）中，唐山地区仍然处于低、高速异常的交界地带，其东南侧为高速异常区，而西北侧为低速异常区，但低速异常的幅度没有东南侧高速异常的幅度大. 在唐山地区下部 20~30km 范围内存在低速异常区，低速异常的幅度达到 6%，该低速异常的顶部深度西北深东南浅. 唐山地区的地震的震源位置处于低、高速异常带的交界部位，震源深度的变化也表现为西北深东南浅，两者相一致. 北京地区下部在该剖面上表现为扰动约 3% 的高速异常. 在该剖面上，重新定位后张北地区震级 $M_L \geq 4.0$ 的地震震源最大深度在 15km 左右，位于张北地区下部的高速异常体边缘.

为了更清楚地研究唐山地区地震活动性与 P 波速度异常之间的关系，我们将唐山地区也做了剖面（TT′剖面，图 9c）. 于湘伟等[21]通过将地震台网、两种常规地震层析成像、双差法和双差地震层析成像不同方法确定的唐山地区地震震源参数做了详细对比后得出，双差地震层析成像得到的震源参数（地震绝对位置和地震相对位置）的精度高于常规地震层析成像方法和双差法得到的震源参数的精度，指出唐山地区的地震沿北东方向剖面在深度上呈现明显的三个小震群的特征. 将双差地震层析成像联合反演得到的唐山地区地震的震源投影在 P 波速度剖面 TT′（图 9c）上，这三个小震群分布在 P 波低、高速异常区的交界地带和高速体内. 靠近唐山地区下部低速异常体的地震的震源深度变化与展布与低速异常体上边缘的深度变化和走向很一致. 唐山地区东北侧迁安地区的地震的震源深度分布与该地区 10~20km 深度范围内的高速异常带的走向很一致，唐山地区的地震最大深度为 25.4km，位于唐山地区下部 20~30km 范围内低速异常区的上边缘，其中震级 $M_L \geq 4.0$ 的地震的最大震源深度为 19.4km（±1.1km），该地震（$M_L = 4.0$）震源位置位于唐山地区下部的低速异常区的上边缘（浅黄色区域）（该地震在地震目录中给出的震源深度为 10km）. 在以往的地震层析成像结果中[8,9]，唐山地区下部的低速异常带上边缘的深度为近 20km，而在本文得到的唐山地区下部的低速异常带的上边缘要深于 20km，约在 25km，与该地区地震震源的最大深度相一致. 由于双差地震层析成像反演中采用了相对到时数据，改进了震源区的精细速度结构，因而更精细地刻画了震源区的速度结构特征.

4 结论

本文联合使用华北遥测地震台网和首都圈数字地震台网所记录的大量高精度区域地震的 P 波绝对到时数据和相对到时数据，用双差地震层析成像方法联合反演了京津唐地区精细的地壳三维 P 波速度结构模型和地震的绝对定位和相对定位. 由反演结果得到以下主要结论：

（1）联合反演得到的京津唐地区的地震绝对定位结果显示了研究区域内地震震源分布的更加精细的图像，沿主要断裂带的地震的震中呈现出更清晰的条带状分布，同时给出了部分地震台网未确定震源深度的地震的震源参数.

(2) 京津唐地区的三维速度结构图像在浅层上很好地反映了地表地质、地形和岩性的特征. 在平原和凹陷的盆地处呈现 P 波低速速度异常, 而在隆起的山区或基岩出露区显示为 P 波高速速度异常.

(3) 研究区域内震级 $M \geq 6.0$ 历史地震的震源位置和重新定位后震级 $M_L \geq 3.0$ 的地震在 10km 深度和 15km 深度处的 P 波相对速度扰动图上的投影都显示出相似的特点. 在所统计的地震中, 绝大部分的地震的震源位置投影分布在低、高速异常的交界地带, 且偏高速体一侧, 极少数的地震分布在 P 波速度异常体内部.

(4) 唐山地区三个小震群均分布在 P 波低、高速异常区的交界地带和高速体内, 地震最大深度为 25.4km, 位于唐山地区下部 20~30km 范围内低速异常区的上边缘, 两者非常吻合. 说明双差地震层析成像反演中采用了相对到时数据, 改进了震源区的精细速度结构, 更精细地刻画了震源区的速度结构特征.

致谢 感谢郑秀芬研究员在数据分析上给予的帮助和周元泽副教授及两位审稿专家对本文提出的宝贵意见.

参考文献（References）

[1] 金安蜀, 刘福田, 孙永智. 北京地区地壳和上地幔的三维 P 波速度结构. 地球物理学报, 1980, **23**(2): 172-182
    Jin A S, Liu F T, Sun Y Z. Three-dimensional P wave velocity structure of the crust and upper mantle under Beijing region. *Chinese J. Geophys. (Acta Geophysica Sinica)* (in Chinese), 1980, **23**(2): 172-182

[2] 刘福田, 曲克信, 吴华等. 华北地区的地震层面成像. 地球物理学报, 1986, **29**(5): 442-449
    Liu F T, Qu K X, Wu H, et al. Seismic tomography of North China region. *Chinese J. Geophys. (Acta Geophysica Sinica)* (in Chinese), 1986, **29**(5): 442-449

[3] Sheldlock K M, Roecker S W. Elastic wave velocity structure under the crust and upper mantle beneath the North China. *J. Geophys. Res.*, 1987, **92**: 9327-9350

[4] 朱露培, 曾融生, 刘福田. 京津唐张地区地壳上地幔三维 P 波速度结构. 地球物理学报, 1990, **33**(3): 267-277
    Zhu L P, Zeng R S, Liu F T. Three-dimensional P wave velocity structure under the Beijing network area. *Chinese J. Geophys. (Acta Geophysica Sinica)* (in Chinese), 1990, **33**(3): 267-277

[5] 丁志峰. 近震层析成像的理论及应用[博士论文]. 北京: 中国地震局地球物理研究所, 1999. 154
    Ding Z F. Regional seismic tomography: theory and applications [Ph. D. thesis] (in Chinese). Beijing: Institute of Geophysics, China Earthquake Administration, 1999. 154

[6] 于湘伟, 陈运泰, 王培德. 京津唐地区中上地壳三维 P 波速度结构. 地震学报, 2003, **25**(1): 1-14
    Yu X W, Chen Y T, Wang P D. Three-dimensional P wave velocity structure in Beijing-Tianjin-Tangshan area. *Acta Seismologica Sinica* (in Chinese), 2003, **25**(1): 1-14

[7] 汪素云, 许忠淮, 裴顺平. 华北地区上地幔顶部 P_n 波速度结构及其构造含义. 中国科学, 2003, **33**(增刊): 91-98
    Wang S Y, Xu Z H, Pei S P. Velocity structure of uppermost mantle beneath North China from Pn tomography and geological structure. *Science in China (Series D)* (in Chinese), 2003, **33**(Suppl.): 91-98

[8] Huang J L, Zhao D P. Crustal heterogeneity and seismotectonics of the region around Beijing, China. *Tectonophysics*, 2005, **385**: 159-180

[9] Huang J L, Zhao D P. Seismic imaging of the crust and upper mantle under Beijing and surrounding regions. *Phys. Earth Planet. Interi.*, 2009, **173**: 330-348

[10] Pei S P, Zhao J M, Sun Y S, et al. Upper mantle seismic velocities and anisotropy in China determined through Pn and Sn tomography. *J. Geophys. Res.*, 2007, **112**, B05312, doi: 10.1029/2006/JB004409

[11] Waldhauser F, Ellsworth W L. A double-difference earthquake location algorithm: method and application to the Northern Hayward Fault, California. *Bull. Seism. Soc. Am.*, 2000, **90**: 1353-1368

[12] Waldhauser F, Ellsworth W. Fault structure and mechanics of the California, from double-difference earthquake

locations. *J. Geophys. Res.*, 2002, **107**(B3): ESE3-11-3-14

[13] 杨智娴, 陈运泰, 郑月军等. 双差地震定位法在我国中西部地区地震精确定位中的应用. 中国科学, 2003, **33**(增刊): 129-134

Yang Z X, Chen Y T, Zheng Y J, et al. Accurate relocation of earthquakes in central-western China using the double difference earthquake location algorithm. *Science in China (Series D)* (in Chinese), 2003, **33**(Suppl.): 129-134

[14] Yang Z X, Waldhauser F, Chen Y T, et al. Double-difference relocation of earthquakes in central-western China, 1992~1999. *Journal of Seismology*, 2005, **9**(2): 241-264

[15] 朱艾斓, 徐锡伟, 胡平等. 首都圈地区地震重新定位及其在地震构造研究中的应用. 地质论评, 2005, **51**(3): 268-274

Zhu A L, Xu X W, Hu P, et al. Relocation of small earthquakes in Beijing area and its implication to seismotectonics. *Geological Review* (in Chinese), 2005, **51**(3): 268~274

[16] 张宏志, 刁桂苓, 陈祺福等. 1976年唐山7.8级地震震区现今地震震源机制分析. 地震研究, 2008, **31**(1): 2-6

Zhang H Z, Diao G L, Chen Q F, et al. Focal mechanism analysis of the recent earthquake in Tangshan seismic region of M7.6 in 1976. *Journal of Seismological Research* (in Chinese), 2008, **31**(1): 2-6

[17] Zhang H J, Thurber C H. Double-difference tomography: the method and its application to the Hayward fault, California. *Bull. Seism. Soc. Am.*, 2003, **93**(5): 1875-1889

[18] 滕吉文, 姚虹, 周海南. 北京、天津、唐山和张家口地区的地壳结构. 地球物理学报, 1979, **22**(3): 218-236

Teng J W, Yao H, Zhou H N. Crustal structure in the Beijing-Tianjin-Tangshan-Zhangjiakou region. *Chinese J. Geophys.* (*Acta Geophysica Sinica*) (in Chinese), 1979, **22**(3): 218-236

[19] Leveque J, Rivera L, Wittlinger G. On the use of the checkerboard test to assess the resolution of tomographic inversions. *Geophys. J. Int.*, 1993, **115**: 313-318

[20] 高文学, 马瑾. 首都圈地震地质环境与地震灾害. 北京: 地震出版社, 1993. 452

Gao W X, Ma J. *Seismo-Geological Background and Earthquake Hazard in Beijing Area* (in Chinese). Beijing: Seismological Press, 1993. 452

[21] 于湘伟, 陈运泰, 张怀. 京津唐地区中小地震重定位结果及对比. 地震学报, 2010, **32**(3): 257-269

Yu X W, Chen Y T, Zhang H. Relocation of earthquakes in Beijing-Tianjin-Tangshan region by double-difference tomography technique. *Acta Seismologica Sinica* (in Chinese), 2010, **32**(3): 257-269

[22] 臧绍先, 杨军亮. 我国华北等地区板内地震的深度分布及其物理背景. 地震地质, 1984, **6**(5): 67-75

Zang S X, Yang J L. The depth distribution of North China intraplate earthquakes and their physical background. *Seismol. Geol.*, 1984, **6**(5): 67-75

Three-dimensional crustal P-wave velocity structure and seismicity analysis in Beijing-Tianjin-Tangshan Region

Yu Xiang-Wei[1], Chen Yun-Tai[2] and Zhang Huai[1]

1. Laboratory of Computational Geodynamics, Graduate University of Chinese Academy of Sciences, Beijing 100049, China;
2. School of Earth and Space Sciences, Peking University, Beijing 100871, China

Abstract In this paper, we present a joint inversion investigation of both three-dimensional crustal P-wave structures and hypocenter parameters using double-difference seismic tomography method in Beijing-Tianjin-Tangshan (BTT) region. The double-difference seismic tomography method was adopted to refine the absolute and relative event locations and P-wave velocity structures simultaneously with the direct use of both absolute arrival times and the more accurate differential travel times. These data came from the earthquake catalog records of totally 3983 events recorded by 112 stations from the Northern China Telemetry Seismic Network (NCTSN) and Capital Digital Seismic Network (CDSN) in BTT region during the years of 1993 to 2004. The inversion results indicated that this method produced a more pronounced and detailed velocity model of the BTT region, which was more consistent with the features of the local geological setting. It revealed the low P-wave velocity anomaly in the plains and basins. In the uplifts of mountain and exposed rock areas, it appeared as high P-wave velocity anomalies. In the BTT region, after being projected to the horizontal P-wave relative velocity perturbation maps in 10km and 15km depths, the relocated hypocenters of magnitude $M_L \geqslant 3.0$ earthquakes showed identical features with those of the historic earthquakes with magnitude $M \geqslant 6.0$. The majority of the hypocenters were located in the conjunctional areas of low and high P-wave velocity anomalies. And they were slightly closer to the high P-wave velocity abnormal areas. Only few earthquakes have ever occurred within the high P-wave velocity anomalies.

The 2009 L'Aquila M_W 6.3 earthquake: a new technique to locate the hypocentre in the joint inversion of earthquake rupture process[*]

Yong Zhang[1], Wanpeng Feng[1,2], Yuntai Chen[1], Lisheng Xu[1], Zhenhong Li[2] and David Forrest[2]

1. Institute of Geophysics, China Earthquake Administration, Beijing, China;
2. School of Geographical and Earth Sciences, University of Glasgow, Glasgow G12 8QQ, UK

Summary A new technique is presented to jointly invert the teleseismic and Interferometric Synthetic Aperture Radar (InSAR) data by simultaneously searching for the hypocentre and the relative weight of InSAR data. In this technique, the parameters of causative fault is determined by using InSAR data first, and then the hypocentre is searched for by jointly inverting teleseismic and InSAR data with the assumption that each subfault on the fault is a potential hypocentre. With this technique, we investigated the source rupture process of the 2009 April 6 L'Aquila M_W 6.3 earthquake without the use of the existing hypocentre locations. Our estimated hypocentre is 42.366° N, 13.385° E, depth 6.9km, with an uncertainty of 1~2km, similar to the hypocentre (42.348°N, 13.380°E, depth 9.5km) determined by National Institute of Geophysics and Volcanology (INGV) using arrival time data within the epicentral distance of 50km. Our joint inversion suggests a scalar moment of 3.5×10^{18} Nm, equivalent to a moment magnitude of M_W 6.3. The source process consists of two subevents with a total duration of 7.7s. The first event (0~3s) corresponds to the slip patch near the hypocentre and the second (in the next 4.7s) ruptured the other slip patch at 4~16km along the strike direction. These unilateral rupture characteristics of the L'Aquila earthquake are confirmed by the apparent source time functions (ASTFs) analysis. In addition, two resolution tests are performed to check the reliability of this work, clarifying the differences between the inversion results with teleseismic data only, InSAR data only and joint inversion, indicating that a higher resolution can be achieved through the joint inversion.

1 Introduction

Joint inversion of seismic data and near field geodetic data (e.g. GPS, InSAR and levelling data) is an effective way to study and understand earthquake source process. Seismic data provides information on the rupture history since it contains the Doppler effects of the spatiotemporal rupture behaviours on the fault, while the near field geodetic data well constrains the fault parameters and fault static slip distribution. Thus, the combined use of the two data sets greatly improves the resolution of the complex rupture propagation and slip history.

In recent years, some studies have jointly inverted these two kind of data sets to investigate the rupture process of significant earthquakes (e.g. Delouis et al. 2002; Kaverina et al. 2002; Salichon et al. 2003; Konca et

---

[*] 本文发表于 *Geophys. J. Int.*, 2012 年, 第 **191** 卷, 1417–1426.

al. 2007; Kim & Dreger 2008). However, there remain some important issues in the work of joint inversion. For example, to improve the use of both the seismic and geodetic data sets, proper weights should be allocated for each data sets. In this aspect, Kim&Dreger (2008) have calculated and compared the variance reduction (VR) for each type of data in detail for the weight allocations. In addition, to carry out a joint inversion, it is usually assumed that several parameters are needed to build a finite fault model: (1) fault geometric parameters (e. g. strike, dip), (2) fault position and (3) position of the hypocentre on the fault plane. When using seismic data only, the fault model is built with the hypocentre and the causative fault obtained based on the best double couple of moment tensor solution. But the hypocentre and moment tensor result usually have uncertainties (Billings et al. 1994; Helffrich 1997), often making the fault model too rough to be used in the joint inversion with near field geodetic data. On the other hand, among the near field geodetic data, the Interferometric Synthetic Aperture Radar (InSAR) data can better constrain the fault position and fault parameters for shallow earthquakes because of its dense coverage over the fault, but it cannot distinguish the position of the hypocentre. Therefore, it is optimal to make use of the hypocentre location determined with seismic data, and the fault position and fault parameters determined with InSAR data. However, the hypocentre determined with seismic data often does not lie on the 2-dimension fault plane which is determined with geodetic data, and thus it is hardly to be used as a reasonable hypocentre, especially for the moderate or small earthquakes whose fault dimension is not much larger than the seismic location uncertainties. The 2003 Bam M_W 6.6 earthquake is an example: with the arrival times of strong motion data, focal depth determined with teleseismic data and assumption that the hypocentre should be on the fault plane, Jackson et al. (2006) got a new hypocentre for this earthquake, because the existing hypocentres were not located on the fault whose position and parameters were obtained using InSAR data.

It is well-known that the teleseismic data and InSAR data have different resolution power on the rupture process. The teleseismic data provides the information of the slip distribution and the relative position of the slip distribution compared to the hypocentre. InSAR data also contains the information of slip distribution, and can better constrain the position of the slip distribution, but it has no resolution on the position of the hypocentre. Thus, if proper weights are allocated to the teleseismic and InSAR data sets, it is possible for us to obtain a new position of the hypocentre by jointly inverting these two data sets. In the next of this paper, we will demonstrate how to determine a new hypocentre by jointly inverting the teleseismic data and InSAR data for rupture process.

The 2009 April 6 L'Aquila M_W 6.3 earthquake is the largest event that occurred near L'Aquila in the last 300 years since the 1703 earthquake (Boschi et al. 2000; Fig. 1). This earthquake killed about 300 people, injured ~ 1000 people and damaged or destroyed thousands of buildings in the epicentral region, leaving 66000 people homeless. Institutions around the world have released a series of hypocentre locations (Fig. 1) for this earthquake. These locations differ from each other by about 4 ~ 10km, which is nearly half of the fault dimension of 10 ~ 20km (Anzidei et al. 2009; Cirella et al. 2009; Walters et al. 2009; Cheloni et al. 2010; Scognamiglio et al. 2010). It is obvious that different hypocentre locations will lead to different results in the joint inversion for the rupture process, although it is believed that the INGV location result is the most reliable and has been adopted in some follow-up work (Anzidei et al. 2009; Cirella et al. 2009) since it is determined by using of the arrival time within the epicentral distance of 50km picked from waveforms recorded by high sensitivity network (http://portale.ingv.it).

The L'Aquila earthquake has drawn a great deal of interests on its fault mechanism and source characteristics because of the terrible resulting damage. The causative fault has been found to strike about 133° ~ 144° to the SE

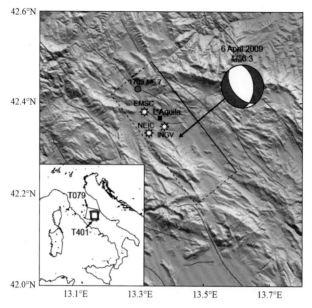

Figure 1 Map of the epicentral region of the 2009 L'Aquila earthquake

The red 'beach ball' denotes the lower hemisphere projection of focal mechanism (strike 141.3°/dip 50°/rake −114°) used in this earthquake. The rectangle is the surface projection of the causative fault plane, while the full black line is the intersection of the fault and the ground surface, and dashed black lines are the other three edges. Three stars denote the epicentre released by National Earthquake Information Center (NEIC, 42.334°N, 13.334°E, depth 8.8km), European Mediteranean Seismological Centre (EMSC, 42.38°N, 13.32°E, depth 2km) and National Institute of Geophysics and Volcanology (INGV, 42.348°N, 13.380°E, depth9.5km). Red lines are the active normal faults in the epicentral region (Roberts & Michetti 2004).

In the subgraph, the two black rectangles are the two tracks of InSAR data, and red rectangle outlines the area shown in the main plot

and dip about 45° ~ 55° to the SW by studying the location of aftershocks (Chiarabba et al. 2009), focal mechanism determination with regional seismic data (Pondrelli et al. 2010), and geodetic data analysis (Anzidei et al. 2009; Atzori et al. 2009; Cirella et al. 2009; Walters et al. 2009). The slip distribution on the fault has been investigated and the maximum slip on the fault plane is estimated to be 0.6 ~ 1.1 m using GPS data (Anzidei et al. 2009; Cheloni et al. 2010), InSAR data (Walters et al. 2009) and/or joint inversion (Cirella et al. 2009; Scognamiglio et al. 2010). Among these studies, it seems that there is only one slip patch in most inversion of geodetic data (Atzori et al. 2009; Walters et al. 2009) except Cheloni et al. (2010) who got three slip patches using nearfield GPS data, but two major slip patches or subevents found in seismic data analysis (Pino & Di Luccio 2009) and joint inversion (Cirella et al. 2009; Scognamiglio et al. 2010). This is an interesting phenomenon requiring further analysis.

In this paper we introduce a new technique to perform the joint inversion of teleseismic and InSAR data, and determine the hypocentre of the 2009 L'Aquila earthquake during its joint rupture process inversion with this technique. It is believed that this technique is suitable for any moderate to strong inland shallow earthquakes ($M_W > 6$) when both teleseismic and InSAR data are available.

2 Teleseismic and InSAR data

In this study, the teleseismic records with the epicentral distance of 30° ~ 90° were selected, and the broadband data was downloaded from the Incorporated Research Institutions for Seismology (IRIS) website. Only P waves on the vertical components were considered to ensure good signal to noise ratios of the seismic waves. We selected the stations with an azimuth interval of 5° to provide a good homogeneous azimuthal coverage of the stations on the epicentre. Finally seismograms of 19 stations were selected (Figs 2 and 6). Following a correction of direct components and instrument response, we applied a bandpass filter of 0.01~0.4Hz to the waveforms, and then resampled them to 3 sps.

We measured the coseismic ground displacement signals of the 2009 L'Aquila earthquake using one pair of Envisat ASAR ascending track 079 data and one pair of ASAR descending track 401 data (Table 1). These SAR images were processed to two independent interferograms (Figs 1 and 7) using JPL/Caltech ROI_PAC (Rosen et al. 2004) combined with the 3-arc-second (~90m) Shuttle Radar Topographic Mission (SRTM) Digital Elevation Model (DEM; Farr et al. 2007). The track 079 interferogram shows two clear lobes of deformation near the epicentre, while track 401 gives only one lobe on the hangingwall of the seismic fault. To reduce computational expense, the resolution-based sampling algorithm (Lohman & Simons 2005) was employed to resample the two interferograms into 701 points (T079) and 765 points (T401), respectively.

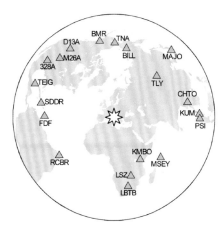

Figure 2 Epicentre (white star) and seismic stations (cyan triangles)

Table 1 Envisat InSAR data used in this study

Model	Track	Date1	Date2	Interval/days	Baseline[a]	Residual[b]
Descending	T401	20090223	20090504	70	108.4	1.03/1.09
Ascending	T079	20080427	20090412	350	26.2	1.08/1.13

[a] Perpendicular baseline in the Image Centre (unit: m).

[b] Standard deviation of the residual interferograms (unit: cm): the first one is for the model from InSAR only, the second is for the joint model.

3 Method

For a finite fault, the seismic waves on the nth component observed at the kth station $u_n^k(t)$ can be expressed as (Aki & Richards 1980):

$$u_n^k(t) = \sum_{l=1}^{L} [g_{n,pq}^{kl}(t) * M_{pq}^l(t)], \qquad (1)$$

where $g_{n,pq}^{kl}(t)$ is the Green's function, denoting the seismic wave on the nth component at the kth station caused by a unit couple (p, q) at the lth subfault with a source time function of Dirac-delta function; $M_{pq}^l(t)$ denotes the time history of the seismic moment tensor of the lth subfault; '*' is the convolution operator. For the Green's function $g_{n,pq}^{kl}(t)$, supposing $g_{n,pq}^{(H)kl}(t)$ as its response of Heaviside function instead of Dirac-delta function, we have

$$u_n^k(t) = \sum_{l=1}^{L} [g_{n,pq}^{(H)kl}(t) * \dot{M}_{pq}^l(t)], \tag{2}$$

where $\dot{M}_{pq}^l(t)$ is the moment rate history.

If only the pure shear is concerned

$$\dot{M}_{pq}^l(t) = M_0^l(t) \cdot [\dot{e}_p(t)v_q + \dot{e}_q(t)v_p], \tag{3}$$

where $M_0^l(t)$ is the scalar moment history of the lth subfault; e is the slip vector and v is the normal direction of the fault. We can convert the dummy index p and q into any Cartesian coordinates systems (x, y, z). As an example, we define z as the normal direction of the fault, and x and y as two orthogonal slip directions on the fault plane. We thus have $v_x=0$, $v_y=0$, $v_z=1$ and $e_z=0$. By supposing $m_x^l(t) = M_0^l(t) \cdot \dot{e}_x(t)$ and $m_y^l(t) = M_0^l(t) \cdot \dot{e}_y(t)$, eq. (2) can be rewritten as

$$u_n^k(t) = \sum_{l=1}^{L} [g_{n,xz}^{(H)kl}(t) * m_x^l(t) + g_{n,yz}^{(H)kl}(t) * m_y^l(t)], \tag{4}$$

where $m_x^l(t)$ and $m_y^l(t)$ are the scalar moment rate history (i.e. far field source time function) of the lth subfault in x and y directions, respectively. Eq. (4) can be rewritten in a matrix form

$$[U] = [G_x \quad G_y] \begin{bmatrix} m_x \\ m_y \end{bmatrix}, \tag{5}$$

where m_x and m_y are the unknown vectors to be solved, they can describe the time histories of slip rate amplitudes and slip angles of all subfaults on the fault plane.

Because of the inherited errors in the seismograms and the inaccuracy of the Green's functions, directly solving eq. (5) will lead to unphysical results. Therefore, smoothness and scalar moment minimization constraints are introduced to stabilize the solution of eq. (5) as follows

$$\begin{bmatrix} \lambda_0 U \\ 0 \\ 0 \\ 0 \end{bmatrix} = \begin{bmatrix} \lambda_0 [G_x \quad G_y] \\ \lambda_1 \begin{bmatrix} D & 0 \\ 0 & D \end{bmatrix} \\ \lambda_2 \begin{bmatrix} T & 0 \\ 0 & T \end{bmatrix} \\ \lambda_3 \begin{bmatrix} Z & 0 \\ 0 & Z \end{bmatrix} \end{bmatrix} \begin{bmatrix} m_x \\ m_y \end{bmatrix}, \tag{6}$$

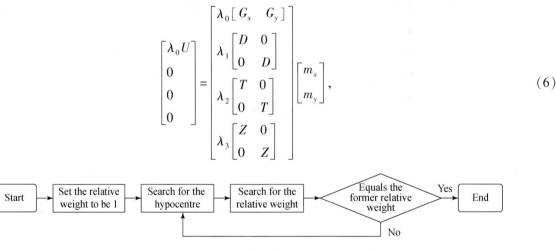

Figure 3 Operation flow used to search for the hypocentre and relative weight

where D and T are spatial and temporal smoothness matrixes, respectively, which are constructed with Laplace's equations (Horikawa 2001; Yagi et al. 2004); Z is the matrix for minimizing scalar moment (Hartzell 1990). λ_0 is usually a sparse matrix to allocate the weights of seismographs at each station, while λ_1, λ_2 and λ_3 are scalar constants and denote the corresponding weights of constraints.

In the Cartesian coordinates systems (x, y, z) defined before, the equation of InSAR data inversion can be described as,

$$[Q] = [B_x \quad B_y] \begin{bmatrix} f_x \\ f_y \end{bmatrix}, \tag{7}$$

where Q is the observed InSAR data; B is matrix of the surface deformation caused by a unit slip on subfaults; f is the unknown parameter which contains the slip amplitudes of subfaults. To perform the joint inversion, the inversion of InSAR data and seismic data must have the same unknown vector, thus we make a transform for eq. (7). Because $f = \frac{1}{\mu A} \int_0^T m(t) \mathrm{d}t$ (μ is the shear modulus around the fault and is set to be 3×10^{10} Nm in this paper; A is the area of each subfault; T is the source duration), the transformed eq. (7) together with the spatial smoothing constraints is,

$$\begin{bmatrix} Q \\ 0 \end{bmatrix} = \begin{bmatrix} [K_x \quad K_y] \\ \lambda_4 \begin{bmatrix} D & 0 \\ 0 & D \end{bmatrix} \end{bmatrix} \begin{bmatrix} m_x \\ m_y \end{bmatrix}, \tag{8}$$

where K is the expansion of B in rows direction, $[K] = \frac{1}{\mu A} [B, B, \cdots, B]$; λ_4 is the weight of spatial smoothing constraint for InSAR data inversion.

Combining eqs (6) and (8), we obtain the equation for the joint inversion

$$\begin{bmatrix} \begin{bmatrix} \lambda_0 U \\ 0 \\ 0 \\ 0 \end{bmatrix} \\ \lambda_I \begin{bmatrix} Q \\ 0 \end{bmatrix} \end{bmatrix} = \begin{bmatrix} \begin{bmatrix} \lambda_0 [G_x \quad G_y] \\ \lambda_1 \begin{bmatrix} D & 0 \\ 0 & D \end{bmatrix} \\ \lambda_2 \begin{bmatrix} T & 0 \\ 0 & T \end{bmatrix} \\ \lambda_3 \begin{bmatrix} Z & 0 \\ 0 & Z \end{bmatrix} \end{bmatrix} \\ \lambda_I \begin{bmatrix} [K_x \quad K_y] \\ \lambda_4 \begin{bmatrix} D & 0 \\ 0 & D \end{bmatrix} \end{bmatrix} \end{bmatrix} \begin{bmatrix} m_x \\ m_y \end{bmatrix}, \tag{9}$$

where λ_I is the relative weight of InSAR data compared to seismic data. This is the final equation of our joint inversion, and is solved by a gradient method (Ward & Barrientos 1986), which can invoke a positivity constraint on the solution.

We first performed the inversion of seismic data only and InSAR data only, respectively, and empirically determined proper weights of constraints (λ_1, λ_2, λ_3 and λ_4) to ensure a result smooth enough making the observed data well fit. Then in the next step of joint inversion, the weights of constraints were fixed, making the relative weight of InSAR data λ_I the only unknown parameter of weight to be determined.

Fig. 3 shows the iteration procedure used to search for the hypocentre and relative weight λ_I, as follows

(1) Set the relative weight λ_I to be 1.

(2) Make a grid search under the assumption that the centre of each subfault is a potential hypocentre, solve eq. (9) for each subfault, find the one with the minimum residual, and take it as an initial hypocentre.

(3) Based on the initial hypocentre, make a grid search for different values of λ_I by solving eq. (9), calculate the corresponding VR (variation reduction) curve, and find the initial value of relative weight λ_I with the

maximum VR.

(4) Based on the initial value of λ_I, go to step (2) and repeat until the hypocentre obtained in step (2) and the relative weight obtained in step (3) stabilize.

By following the iteration scheme described above, we aim to retrieve the rupture process as well as the relative weight of InSAR data and a reasonable position of hypocentre on the fault plane.

In the inversion, the theoretical seismograms were calculated with the reflectivity method (Kennett 1983) based on the standard based on the global velocity model (IASPEI91) (Kennett & Engdahl 1991). The surface deformation is calculated using the method of Okada (1985).

4　Results

We use the fault position and fault parameters obtained from InSAR data by Feng et al. (2010). In their study, the fault, with the central point of the upper edge at (42.370°N, 13.470°E), strikes 141.3° and dips 50° to the SW (Feng et al. 2010). Based on this fault, we chose a rectangular fault area with a size large enough (34km × 20km) to make sure that the rupture can propagate more freely on this fault, and divided it into 17 and 10 subfaults (total number is 17 × 10 = 170), each having the size of 2km × 2km in strike and dip direction, respectively. To stabilize the linear inversion with the consideration of the trade-off between rupture velocity and slip distribution, we limit the rupture velocity and rupture duration for the subfault source time functions. A maximum rupture velocity of 2.8km s^{-1} was specified, which is approximately 0.8 times the basement shear wave velocity in the upper crust of the epicentral area and has been confirmed in near fault strong motion data analysis (Cirella et al. 2009); and a maximum rupture duration of 3s for each subfault is preferred. Within the rupture duration of each subfault, the source time function can have any shape, and is determined by observed data so that errors induced by improper assumptions on the shapes of subfaults can be reduced as much as possible. With these two constrains, the rupture velocity of subfaults can change with a range of $[d/(d/2.8 + 3), 2.8]$, where d is the distance of subfault to the hypocentre.

Following the iteration scheme mentioned before, we search for the hypocentre (step 2) three times and for the relative weight (step 3) two times, respectively, and get the optimal solution. We found that the 7th and 5th subfault in strike and dip directions, whose centre corresponds to a location of 42.366°N, 13.385°E, depth 6.9km, has the minimum residual (Fig. 4c); and the relative weight of InSAR data was 1.4 (Fig. 5). Fig. 4 (c) shows the surface projection of the residual distribution on the fault based on the searched relative weight 1.4. Compared with the three released hypocentre locations, the searched hypocentre (42.366°N, 13.385°E, depth 6.9km) is most consistent with the result of INGV (42.348°N, 13.380°E, depth 9.5km). On the basis of the searched hypocentre, VR curves of the seismic data, InSAR data and their averaged values are shown in Fig. 5, indicating that the relative weight 1.4 gives a good compromise of VR for both of the seismic data and InSAR data (Figs 6 and 7).

From the joint inversion results, the released scalar seismic moment is about 3.5×10^{18} Nm, equivalent to a moment magnitude of M_W 6.3. The average rake on the fault is about −114°, which coincides with the result of Atzori et al. (2009) and Cirella et al. (2009). Based on the Brune's model (1970), the calculated average stress drop is about 13MPa. It is consistent with the typical stress drop (~10MPz) of intraplate earthquakes (Kanamori & Anderson 1975).

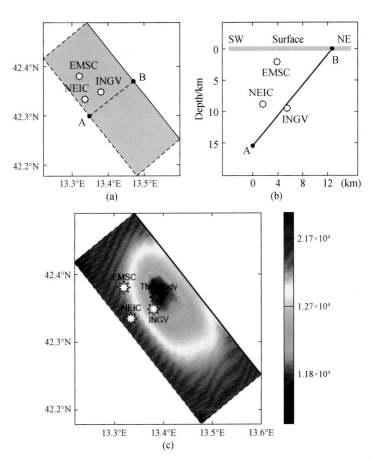

Figure 4 (a) Surface projection of fault plane and three epicentral locations; (b) The cross-section of (a); (c) Surface projection of the residual distribution on the fault obtained by using the iteration scheme. The centre of the subfault with the minimum residual is marked with a red star and is regarded as the hypocentre obtained in the joint inversion. There is no uniform unit for the residual in (c) since the seismic data is in 'm s^{-1}' while InSAR data is in 'm'

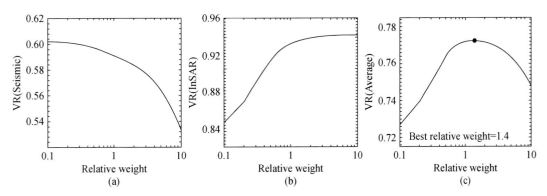

Figure 5 (a) VRs of seismic data; (b) VRs of InSAR data; (c) The average VRs

The inverted static slip distribution on the fault by using seismic data only, InSAR data only, and joint inversion, are shown in Fig. 7. There are two distinguishable slip patches in the seismic result, but only one in InSAR result. The slip distribution of the joint inversion lies between the seismic and InSAR results. The InSAR result has a smaller slipped area, but a larger peak slip value of 0.9m than the seismic and joint results whose peak slips are both about 0.7m. From the joint inversion, there are two slip patches located near the hypocentre and 4–16km away in strike direction, respectively, indicating a unilateral rupture of the L'Aquila earthquake, which coincides with the study of Pino & Di Luccio (2009), and Cirella et al. (2009).

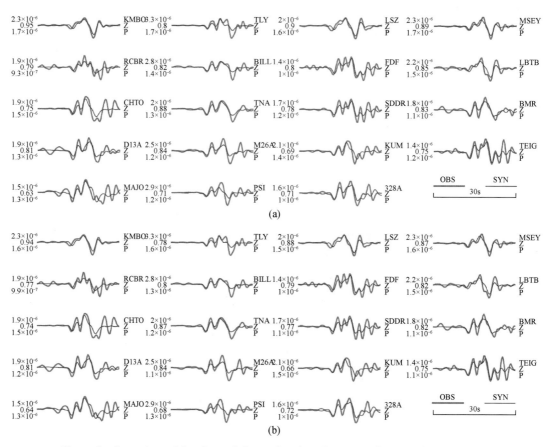

Figure 6 Comparison of the observed (green lines) and synthetic (red lines) seismograms
(a) Inversion of seismic data only; (b) Joint inversion. Note: In each subgraph, from top to bottom on the left are the maximal amplitudes of the observed waves, correlation coefficients of the observed with synthetic waves, and the maximal amplitudes of the synthetic waves, respectively. The amplitude unit is m s^{-1}. And from top to bottom on the right are the stations names, component names and phase names, respectively

The source time function (released moment rate history) obtained in joint inversion, with a length of about 7.7s, consists of two subevents (Fig. 9a), one occurred in 0~2.7s and the other occurred in the next 5s. The two subevents correspond well to two rupture stages (Fig. 9b). In the first small stage, rupture mainly occurred at the slip patches near the hypocentre in 0~3s; and in the second larger stage, it occurred at the slip patches about 4–16km away in strike direction in 3~7.7s (Fig. 9b).

5 Discussion

In this paper, we developed a technique by designing an iteration scheme to search for the hypocentre on the fault plane. During this technique, we limit the rupture velocity and rupture duration by giving their maximum values. These are two common constraints in rupture process inversion since the instability of the solution (Yagi et al. 2004; Honda et al. 2011). By fixing the rupture duration (which is usually not very sensitive for the inversion), one can make a grid search for the maximum rupture velocity (Yagi et al. 2004). In this study, we directly use the existing value of 2.8km s^{-1} since it has been confirmed by analysing the near fault strong motion data (Cirella et al. 2009). There remain some uncertainties of the position of the hypocentre. If only concern the size (2km×2km) of the subfault and dip angle (50°), they are about ±1km in strike direction (half of the size of subfault), ±0.6km in horizontal dip direction [half of the size of subfault multi-cosine (dip)] and ±0.8km in depth [half of the size of subfault multi-sine (dip)]. Furthermore, the uncertainties also depend closely on the resolution power of the teleseismic data on the rupture details, and that of the InSAR data on fault position and geometric parameters. To check this, numerical experiments are needed. As we know, the InSAR data should be the best choice to constrain the position of shallow ruptured faults. Therefore, by fixing the position of the central point on the upper edge of the fault plane which has been determined with InSAR data, we change the fault strike and dip ±5° for the numerical tests. With the strike range of 136.3° ~ 146.3° and dip range of 45° ~ 55°, which cover most of the values having been obtained (Anzidei et al. 2009; Atzori et al. 2009; Chiarabba et al. 2009; Cirella et al. 2009; Walters et al. 2009) for the fault parameters of the L'Aquila earthquake, we search for the hypocentre and the relative weight four times. The results are shown in Table 2. It is found that latitude and relative weight are usually stable, but longitude and depth mainly vary with strike and dip, respectively. The maximum difference of the searched positions (Table 2) from each other is about 1.6km in horizontal surface and 2.5km in depth, which can be regarded as an estimation of the uncertainties of the hypocentre obtained before (42.366°N, 13.385°E, depth 6.9km), suggesting a high precision and stability of the technique used in this study.

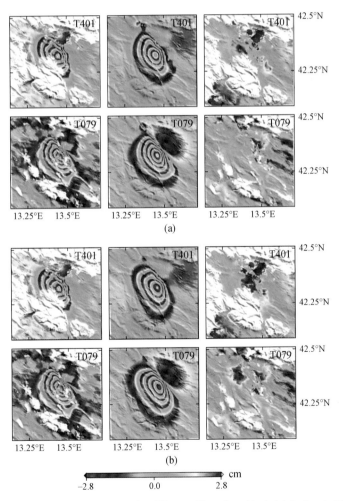

Figure 7 Observed (left-hand side), synthetic (middle panel) and residual (right-hand side) interferograms for ENVISAT Track 401 (top panel) and Track 079 (bottom panel)
(a) Inversion of InSAR data only; (b) Joint inversion

Table 2 The searched latitude, longitude and depth of the hypocentre, and the relative weight, based on different fault parameters

	Latitude/(°N)	Longitude/(°E)	Depth/km	Relative weight
Strike136.3°/Dip45°	42.378	13.376	4.9	1.3
Strike136.3°/Dip55°	42.377	13.374	7.4	1.2
Strike146.3°/Dip45°	42.375	13.393	4.9	1.3
Strike146.3°/Dip55°	42.374	13.391	7.4	1.2

For the 2009 L'Aquila earthquake, it is interesting that only one slip patch was found in most studies with geodetic data (GPS and InSAR data) (Atzori et al. 2009; Walters et al. 2009) except for the study of Cheloni et al. (2010), but usually two slip patches corresponding to two subevents were retrieved in the inversion of seismic data and joint inversion (Cirella et al. 2009; Pino & Di Luccio 2009). One reason may exist in that the first slip patch near the hypocentre is located more deeply than the other. Consequently, it causes strong seismic waves, but relatively little deformation on the ground surface.

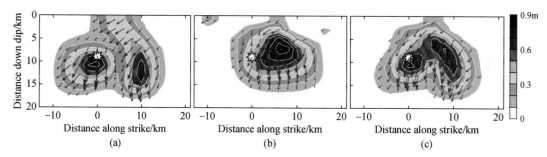

Figure 8 (a) Static slip distribution obtained by inversion of seismic data only; (b) static slip distribution obtained by inversion of InSAR data only; (c) static slip distribution obtained by joint inversion. White star denotes the hypocentre on the fault plane

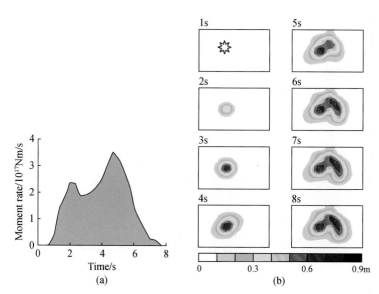

Figure 9 (a) Source time function; (b) Temporal variation of the slip distribution on the fault. The white star in (b) denotes the hypocentre

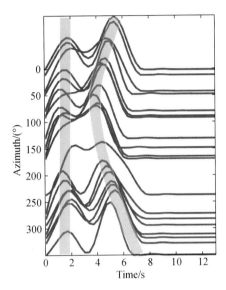

Figure 10 Azimuth-dependent ASTFs of the 2009 L'Aquila earthquake
The two broad grey curves denote the ivariation of the two peaks of the ASTFs

To further analyse these special source characteristics, with the fault mechanism of strike 141.3°/dip 50°/rake −114°, we retrieved the apparent source time functions (ASTFs) by using of the projected Landweber deconvolution (PLD) method (Bertero et al. 1997) with an iteration number of 200. Fig. 10 shows the azimuth-dependent ASTFs obtained. From the Doppler Effect contained in the ASTFs, we can get a rough estimation of the rupture characteristics of the major subevents. Two subevents are distinguished and confirmed from the ASTFs. The first subevent, whose peak is at almost the same time for all azimuth-different stations, should be a bilateral rupture event. And the second, whose peaks are earliest at the azimuth between 100° ~ 150° (around the strike direction), should be somewhere in strike direction. From the fact that the second subevent is larger, we can infer that the L'Aquila earthquake is mainly a unilateral rupture event, starting from the hypocentre and propagating southeastwards along strike direction. The ASTF analysis coincides with the results of regional relative source time functions (RSTFs) by means of empirical Green's function (EGF) method (Pino & Di Luccio 2009).

Additionally, we perform two resolution tests to examine the resolution power of the teleseismic data, InSAR data and joint inversion, as well as the effectiveness of this technique. In the two tests, a uniform rupture velocity of 2.8km s^{-1} and a source time function of half-cycle sine function with a length of 3s are given for each subfault. In the first test, to discuss the inversion results of L'Aquila earthquake, we give two slip patches on the fault with the hypocentre at the 7th and 5th subfault (42.366°N, 13.385°E, depth 6.9km) in strike and dip direction, respectively. Then the synthetic seismograms and synthetic surface deformation were calculated and added with 10 per cent Gaussian noise. Also, the iteration scheme was carried out again. After executing step (2) twice and step (3) once, we get the final solution. The hypocentre is found to be that assumed in this test (42.366°N, 13.385°E, depth 6.9km), and the relative weight is found to be 0.8. Fig. 11 shows the static slip distribution obtained in the resolution test. From the test of seismic data inversion, two slip patches are distinguished but the depth and maximum slip of the second slip patch were slightly poorly imaged (Fig. 11b). For the InSAR data inversion, the first slip patch cannot be distinguished (Fig. 11c). It should be because of its greater depth, which causes weaker surface deformation than the second. Thus, the InSAR data inversion, which always likes to find a solution with a small norm, prefers to attribute the surface deformation to the second slip patch at the shallow depth instead of the first. The second test shows a more general situation of bilateral slip distribution, two slip patches with different depths and slip values are fixed on the left and right of the hypocentre, respectively (Fig. 12a). After executions of step (2) and step (3) once, respectively, the solution stabilized at the real hypocentre and a relative weight of 0.6 (also with 10 per cent Gaussian noise). It is clear that the shallower and deeper patches are mainly distinguished by InSAR data and seismic data, respectively (Figs 12b and c), and joint inversion is the best since it synthesizes the advantages of seismic data and InSAR data inversion (Fig. 12d).

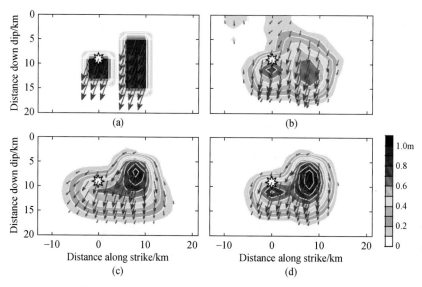

Figure 11　The first resolution test with 10 per cent noise on the observed data

(a) Real slip distribution; (b) Slip distribution obtained by inversion of seismic data only; (c) Slip distribution obtained by inversion of InSAR data; (d) Slip distribution obtained by joint inversion of seismic data and InSAR data

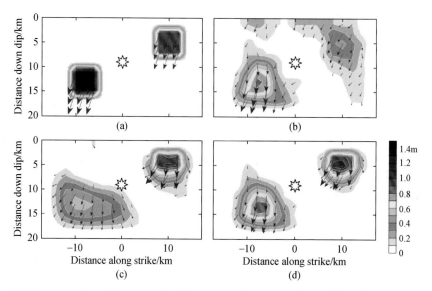

Figure 12　The second resolution test with 10 per cent noise on the observed data (Refer to Fig. 11.)

From comparison with the results of the seismic data only and InSAR data only, joint inversion has a better resolution power on the position, scale and maximum slip values of the two slip patches, suggesting that it is more effective to image the rupture process than the inversion of seismic data or InSAR data only.

We note that there are some differences of the best-fit relative weights among the real data inversion and resolution tests. In our opinion, the best relative weight in joint inversion depends on three factors: the data quality (signal-to-noise ratio), the accuracy of the path description, and the resolution ability of the data on the ruptured slip model. The three factors influent the VR curves and lead to different relative weights. In the real data inversion of the L'Aquila earthquake, the signal-to-noise ratios of the teleseismic data (of a M_W 6.3 earthquake) is low, and its Green's functions also are not very accurate; but for InSAR data, the surface deformation calculation corresponding to the zero-frequency response of earth structure is of higher accuracy. This may be why the relative weight of InSAR data is slightly larger (1.4). In the resolution tests, however, the Green's functions are accurate since they are used to synthesize the seismograms, but the signal-to-noise ratios of teleseismic data and InSAR data are almost the same (both are of 10 per cent noise), thus the poor resolution power of the InSAR data on the slip in depth may account for the slightly smaller relative weight (0.6 and 0.8).

6 Conclusions

We demonstrate a new technique to investigate the source rupture process by jointly inverting teleseismic data and InSAR data. This technique does not need any existing hypocentre, but offers another efficient way to locate a new hypocentre. The specialty is that the hypocentre is confined to be on the 2-D fault plane and is searched for by a waveform match instead of arrival times. Because of the use of the teleseismic data and InSAR data, this technique can be applied widely to the moderate to strong inland shallow earthquakes ($M_W > 6$) that occur in the area with sparse seismic stations (e.g. Tibetan Plateau) where earthquake hypocentres are hardly to be well located.

Acknowledgments

This work is supported by China NSFC Projects (ID: 41090291; 90915012). Part of this work is supported by the Natural Environmental Research Council (NERC) through the GAS project (Ref: NE/H001085/1) as well as by a China NSFC Project (ID: 41074005). The ENVISAT images were supplied through the ESA-MOST Dragon 2 Cooperation Program (ID: 5343). We thank JPL/Caltech for the use of ROI_PAC in our interferometric processing, and A. Singleton for useful discussions. We especially thank two anonymous reviewers who have offered very good suggestions, which are of great help to improve this manuscript.

References

Aki, K. & Richards, P. G., 1980. *Quantitative Seismology*, Freeman and Co., New York.

Anzidei, M. et al., 2009. Coseismic deformation of the destructive April 6, 2009 L'Aquila earthquake (central Italy) from GPS data, *Geophys. Res. Lett.*, **36**, L17307, doi: 10.1029/2009GL039145.

Atzori, S. et al., 2009. Finite fault inversion of DInSAR coseismic displacement of the 2009 L'Aquila earthquake (central Italy), *Geophys. Res. Lett.*, **36**, L15305, doi: 10.1029/2009GL039293.

Bertero, M., Bindi, D., Boccacci, P., Cattaneo, M., Eva, C. & Lanza, V., 1997. Application of the projected Landweber method to the estimation of the source time function in seismology, *Inv. Probl.*, **13**, 465-486.

Billings, S. D., Sambridge, M. S. & Kennett, B. L. N., 1994. Errors in hypocentre location: picking, model, and magnitude dependence, *Bull. Seism. Soc. Am.*, **84**(6), 1978-1990.

Boschi, E., Guidoboni, E., Ferrrari, G., Mariotti, D., Valensise, G. & Gasperini, P., 2000. Catalogue of strong Italian earthquakes from 461 B. C. to 1997, *Ann. Geofis.*, **43**, 609-868.

Brune, J. N., 1970. Tectonic stress and the spectra of seismic shear waves from earthquakes, *J. geophys. Res.*, **75**(26), 4997-5009, doi: 10. 1029/JB075i026p04997.

Cheloni, D. et al., 2010. Coseismic and initial post-seismic slip of the 2009 M_W 6.3 L'Aquila earthquake, Italy, from GPS measurements, *Geophys. J. Int.*, **181**, 1539-1546.

Chiarabba, C. et al., 2009. The 2009 L'Aquila (central Italy) M_W6.3 earthquake: main shock and aftershocks, *Geophys. Res. Lett.*, **36**, L18308, doi: 10. 1029/2009GL039627.

Cirella, A., Piatanesi, A., Cocco, M., Tinti, E., Scognamiglio, L., Michelini, A., Lomax, A. & Boschi, E., 2009. Rupture history of the 2009 L'Aquila (Italy) earthquake from non-linear joint inversion of strong motion and GPS data, *Geophys. Res. Lett.*, **36**, L19304, doi: 10. 1029/2009GL039795.

Delouis, B., Giardini, D., Lundgren, P. & Salichon, J., 2002. Joint inversion of InSAR, GPS, teleseismic, and strong-motion data for the spatial and temporal distribution of earthquake slip: application to the 1999 I. zmit mainshock, *Bull. Seism. Soc. Am.*, **92**(1), 278-299.

Farr, T. G. et al., 2007. The shuttle radar topography mission, *Rev. Geophys.*, **45**, RG2004, doi: 10. 1029/2005RG000183.

Feng, W. P., Li, Z. H. & Li, C. L., 2010. Optimal source parameters of the 6 April 2009 M_W6.3 L'quila, Italy earthquake from InSAR observations (in Chinese with English abstract), *Prog. Geophys.*, **25**(5), doi: 10. 3969/j. issn. 1004-2903. 2010. 05. 001.

Hartzell, S., 1990. Source complexity of the 1987 Whittier Narrows, California, earthquake from the inversion of strong motion records, *J. geophys. Res.*, **95**(B8), 12475-12 485.

Helffrich, G. R., 1997. How good are routinely determined focal mechanisms? Empirical statistics based on a comparison of Harvard, USGS and ERI moment tensors, *Geophys. J. Int.*, **131**, 741-750.

Honda, R. et al., 2011. A complex rupture image of the 2011 off the Pacific coast of Tohoku Earthquake revealed by the MeSO-net, *Earth Planets Space*, **63**, 583-588.

Horikawa, H., 2001. Earthquake doublet in Kagoshima, Japan: rupture of asperities in a stress shadow, *Bull. Seism. Soc. Am.*, **91**, 112-127.

Jackson, J. et al., 2006. Seismotectonic, rupture process, and earthquake hazard aspects of the 2003 December 26 Bam, Iran, earthquake, *Geophys. J. Int.*, **166**, 1270-1292.

Kanamori, H. & Anderson, D. L., 1975. Theoretical basis of some empirical relations in seismology, *Bull. Seism. Soc. Am.*, **65**(5), 1073-1095.

Kaverina, A., Dreger, D. & Price, E., 2002. The combined inversion of seismic and geodetic data for the source process of the 16 October 1999 M_W7.1 Hector Mine, California, Earthquake, *Bull. Seism. Soc. Am.*, **92**(4), 1266-1280.

Kennett, B. L. N., 1983. *Seismic Wave Propagation in Stratified Media*, Cambridge University Press, Cambridge, pp. 1-339.

Kennett, B. L. N. & Engdahl, E. R., 1991. Travel times for global earthquake location and phase identification, *Geophys. J. Int.*, **105**, 429-465.

Kim, A. & Dreger, D. S., 2008. Rupture process of the 2004 Parkfield earthquake from near-fault seismic waveform and geodetic records, *J. Geophys. Res.*, **113**, B07308, doi: 10. 1029/2007JB005115.

Konca, A. O. et al., 2007. Rupture Kinematics of the 2005 M_W 8.6 Nias-Simelue Earthquake from the Joint Inversion of Seismic and Geodetic Data, *Bull. Seism. Soc. Am.*, **97**(1A), S307-S322, doi: 10. 1785/0120050632.

Lohman, R. B. & Simons, M., 2005. Some thoughts on the use of InSAR data to constrain models of surface deformation: noise structure and data downsampling, *Geochem. Geophys. Geosyst.*, **6**, Q01007, doi: 10. 1029/2004GC000841.

Okada, Y., 1985. Surface deformation due to shear and tensile faults in a half-space, *Bull. Seism. Soc. Am.*, **75**(4), 1135-1154.

Pino, N. A. & Di Luccio, F., 2009. Source complexity of the 6 April 2009 L'Aquila (central Italy) earthquake and its strongest aftershock revealed by elementary seismological analysis, *Geophys. Res. Lett.*, **36**, L23305, doi: 10. 1029/2009GL041331.

Pondrelli, S., Salimbeni, S., Morelli, A., Ekstrom, G., Olivieri, M. & Boschi, E., 2010. Seismic moment tensors of the April 2009, L'Aquila (Central Italy), earthquake sequence, *Geophys. J. Int.*, **180**(1), 238-242.

Roberts, G. P. & Michetti, A. M., 2004. Spatial and temporal variations in growth rates along active normal fault systems: an example from the Lazio-Abruzzo Apennines, central Italy, *J. Struct. Geol.*, **26**, 339–376.

Rosen, P. A., Henley, S., Peltzer, G. & Simons, M., 2004. Updated repeat Orbit Interferometry Package released, *EOS, Trans. Am. Geophys. Un.*, **85**(5), doi: 10.1029/2004EO050004.

Salichon, J., Delouis, B., Lundgren, P., Giardini, D., Costantini, M. & Rosen, P., 2003. Joint inversion of broadband teleseismic and interferometric synthetic aperture radar (InSAR) data for the slip history of the $M_W = 7.7$, Nazca ridge (Peru) earthquake of 12 November 1996, *J. Geophys. Res.*, **108**(B2), 2085, doi: 10.1029/2001JB000913.

Scognamiglio, L., Tinti, E., Michelini, A., Dreger, D. S., Cirella, A., Cocco, M., Mazza, S. & Piatanesi, A., 2010. Fast determination of moment tensors and rupture history: what has been learned from the 6 April 2009 L'Aquila Earthquake Sequence, *Seism. Res. Lett.*, **81**(6), 892–906, doi: 10.1785/gssrl.81.6.892.

Walters, R. J. et al., 2009. The 2009 L'Aquila earthquake (central Italy): a source mechanism and implications for seismic hazard, *Geophys. Res. Lett.*, **36**, L17312, doi: 10.1029/2009GL039337.

Ward, S. N. & Barrientos, S. E., 1986. An inversion for slip distribution and fault shape from geodetic observations of the 1983, Borah Park, Idaho, earthquake, *J. geophys. Res.*, **91**, 4909–4919.

Yagi, Y., Mikumo, T., Pacheco, J. & Reyes, G., 2004. Source rupture process of the Tecomán, Colima, Mexico earthquake of 22 January 2003, determined by joint inversion of teleseismic body-wave and near-source data, *Bull. Seism. Soc. Am.*, **94**, 1795–1807.

An inversion of Lg-wave attenuation and site response in the North China region*

Xin-Yun Zhu[1] and Yun-Tai Chen[2]

1. Zhejiang Province Earthquake Administration; 2. Institute of Geophysics, China Earthquake Administration

Abstract A joint inversion method for determining the Lg-wave attenuation along the propagation path and site response based on seismic spectral ratios was developed and applied to invert the Lg-wave attenuation γ and site response in the North China region: γ is related to the quality factor Q by $\gamma = \pi f / UQ$, where U is the velocity. Waveform data of 200 earthquakes during 2006–2008 with M_L 1.7–5.5 in the area of 38°–41°N, 114°–120°E were selected for this study. More than 1500 three-component records chosen from 81 stations on the basis of the signal-to-noise ratio were used in the inversion. The waveform data were processed in the frequency range of 1–6Hz with an interval of 0.2Hz. The results indicate that there is remarkable heterogeneity in Lg-wave attenuation in the study area. We divided the area into four regions according to the Lg-wave propagation paths. Lower γ values are found in the tectonic mountain region, higher values occur in the tectonic basin region, and moderate γ values are found for the propagation paths passing through both the mountain and basin regions.

On the vertical component, γ was found to be frequency dependent and in the form $\gamma(f) = 0.0029 f^{0.4}$ (1/km) and $\gamma(f) = 0.0070 f^{0.4}$ (1/km) for propagation paths in the mountain and basin regions, respectively, and $\gamma(f) = 0.0043 f^{0.22}$ (1/km) and $\gamma(f) = 0.0046 f^{0.24}$ (1/km) for paths passing through both mountain and basin regions with seismic ray direction from the mountain to the basin and vise versa, respectively. The absolute quality factor Q_0 ranges from 125 to 306.

The response of rock sites is smooth and flat and less than five at low frequency, while that for soil sites is higher than for rock sites. For soil sites, the amplification of low-frequency waves is higher than that for high-frequency waves. For both site response and attenuation, no significant differences were observed between different components.

1 Introduction

A seismogram couples multiple response factors, such as path effect, site response, instrument response, and directivity of seismic radiation (Chen et al., 2000). The geometric spreading, instrument response, and seismic radiation directivity can be modeled directly, calculated easily, or even neglected if appropriate (Chun et al., 1987). But to separate the effects of site response and wave attenuation is very difficult. For simplicity, the site response was often completely neglected in previous studies of seismic-wave attenuation. However, station site response can differ greatly even among closely spaced stations and is strongly frequency dependent (King and Tucker, 1984; Tucker et al., 1984; Campillo et al., 1985). Error arising from neglecting the site response was

---

* 本文发表于 *Bull. Seism. Soc. Amer.*, 2012 年, 第 **102** 卷, 第6期, 2600–2610.

introduced into the attenuation parameters. In order to acquire more accurate attenuation parameters, Chun et al. (1987) proposed the reversed two-station spectral ratio (RTSR) method. This method is a reliable method for calculating Lg-wave attenuation parameters, but it is hard to apply widely because it requires collinear alignment of event and receiver pairs (Chun et al., 1987; Chung and Lee, 2002; Chung et al., 2007).

The site response is related to the density and seismic velocity of near-surface formations, but it is independent of the distance of seismic-wave propagation (Shearer, 1999). Typically, there are several algorithms applied for evaluating site response (Bonilla et al., 1997; Riepl et al., 1998). Among them, the standard spectral ratio (SSR) method (Borcherdt, 1970; Parolai et al., 2000, 2010; Nath et al., 2001) and the horizontal-to-vertical spectral ratio (HVSR) method (Borcherdt, 1970; Lachet et al., 1996; Parolai et al., 2010) are the major and most widely-used algorithms. The SSR method assumes that the records from the reference station contain the same source and path effects as the target stations; therefore, the differences observed between the target and the reference stations lie mainly in local site response. No site response is presumed for the reference station; that is, there is no effect on seismic waves. But selecting a reference station is very subjective and difficult (Steidl et al., 1996). The HVSR method assumes that the vertical component of ground motion is free of near-surface influence, and the site response is calculated by dividing the horizontal spectrum by the vertical spectrum of the same earthquake. An alternative method for calculating site response is the generalized inversion method, which solves the records of a number of earthquakes for the source, path, and site responses simultaneously (Hartzell, 1992).

A number of studies on the inversion of source, path, and site responses have been published (Field and Jacob, 1995; Raoof et al., 1999; Bay et al., 2003). In this paper, based on the characteristic that path attenuation is related to seismic propagation distance and independent of site response, a joint inversion method that allows for the simultaneous determination of attenuation coefficient and site response from seismic spectral ratios was developed and applied to invert the parameters of Lg-wave attenuation and site response of the North China region.

Observational and theoretical studies of Lg-wave propagation have been greatly motivated by the potential value of the Lg wave in seismic nuclear monitoring research (Pomeroy et al., 1982; Nuttli, 1986). Lg waves are short-period surface waves propagating in the continental crust at a group velocity close to 3.5km/s (Knopoff et al., 1973; Panza and Calcagnile, 1975), and they are the result of multiple reflection and superposition of surface waves in the crust (Campillo et al., 1985). The Lg wave is the main component of seismic waves from local and near-field earthquakes (Hasegawa, 1985; Kennett, 1986). According to Rapine et al. (1997), the Chinese continent and its continental shelf are known for efficient Lg-wave propagation. In a study of Lg-wave attenuation in South Korea, Chung et al. (2005) found efficient Lg-wave propagation all over the region and its peripheries except for ocean basins of depths over 1000 m and volcanic regions. Robert et al. (1999) clearly identified Lg-wave phases from seismograms of the Red Sea, Persian Gulf, and Mediterranean recorded by seismic stations in the Arabian shield. In an analysis of seismic records from southern California and the Nevada Test Site, Zhang et al. (2002) showed that the Lg wave is insensitive to seismic source radiation pattern. Mitra et al. (2006) investigated the Lg-wave attenuation of India platform earthquakes based on the insensitivity of Lg waves to the source radiation pattern. Numerous studies have shown that seismic attenuation coefficients reflect the regional tectonics and seismic activities (Sato and Fehler, 1998). The parameter Q_{Lg}, or equivalently γ, is a basic parameter describing Lg-wave attenuation; a high Q_{Lg} value is associated with low seismic attenuation and weak seismicity, and vice versa

(Aki, 1980; Jin and Aki, 1988; Xie and Mitchell, 1990).

The objectives of this paper are as follows: (1) to introduce a simple but reliable method for calculating the frequency-dependent attenuation coefficients, (2) to obtain an improved understanding of the characteristics of seismic attenuation parameters in the North China region, and (3) to measure the site response of the stations in this region.

2 Theoretical basis

The displacement spectral amplitude $A(f,d)$ of a seismic wave recorded by a seismic station can be expressed as (Hasegawa, 1985)

$$A(f,d) = So(f)R(f,\theta)G(d)\psi(f,d)I(f)Si(f), \tag{1}$$

where f is the frequency, d is the epicentral distance, θ is the station azimuth, $So(f)$ is the source factor term, $R(f, \theta)$ is the source radiation pattern factor, $I(f)$ is the instrument response, $Si(f)$ is the site response, and $G(d)$ is the geometric spreading factor, expressed as

$$G(d) = d^{\mu}, \tag{2}$$

where μ is the parameter of geometric spreading. For the Lg wave, it is difficult to estimate geometric spreading and attenuation simultaneously. Simulations in a simple plane-layered velocity structure (Herrmann and Kijko, 1983) demonstrate the fact that the spectral amplitudes of Lg waves behave as cylindrical waves with a geometrical spreading of $d^{-0.5}$. In this regard, we assume a frequency-independent geometric spreading factor of $\mu = -0.5$ in this study. $\psi(f,d)$ represents the media attenuation term and can be expressed as

$$\psi(f,d) = e^{\gamma d}, \tag{3}$$

where γ is the attenuation coefficient of the Lg wave and can be expressed as (Chun et al., 1987; Shih et al., 1994; Chung et al., 2007)

$$\gamma = \frac{\pi f}{QU}, \tag{4}$$

where U is group velocity of Lg waves and Q is the dimensionless quality factor.

$I(f)$ is known, so it can be calibrated. $Si(f)$ is frequency dependent. Assuming that the number of earthquakes is m and the number of stations is n then the ratio of the seismic spectra recorded by the ith and jth stations ($i, j = 1, 2, 3, \cdots, n$) for the kth earthquake ($k = 1, 2, 3, \cdots, m$) can be expressed as

$$\frac{A_{k,i}(f,d_{k,i})}{A_{k,j}(f,d_{k,j})} = \frac{So_k(f)R_{k,i}(f,\theta_{k,i})G(d_{k,i})\psi(f,d_{k,i})I_i(f)Si_i(f)}{So_k(f)R_{k,j}(f,\theta_{k,j})G(d_{k,j})\psi(f,d_{k,j})I_j(f)Si_j(f)}, \tag{5}$$

where i, j denote the station number and k denotes the earthquake number; for the same earthquake, the ratio between the source factor terms can be removed. Considering the Lg wave's insensitivity to the source radiation pattern (Mitra et al., 2006) and following Sereno (1990), we assume that both the source radiation pattern and site amplification of the station are independent of azimuth. This assumption is empirically supported by a rapidly accumulating body of Lg-wave amplitude observations (Sereno, 1990; Shih et al., 1994; Mitra et al., 2006). Then,

$$R_{k,i}(f,\theta_{k,i}) \approx R_{k,j}(f,\theta_{k,j}). \tag{6}$$

The term for the ratio between the source radiation pattern factors can be removed. Substituting equation (3) into equation (5) and reorganizing the formula yields

$$\frac{A_{k,i}(f,d_{k,i})/[I_i(f)G(d_{k,i})]}{A_{k,j}(f,d_{k,j})/[I_j(f)G(d_{k,j})]} = \frac{Si_i(f)}{Si_j(f)} e^{-\gamma(d_{k,i}-d_{k,j})}. \quad (7)$$

Taking the natural logarithm of both sides of equation (7) yields

$$\ln Y_{k,i,j}(f,\Delta_{k,i,j}) = \ln Si_{i,j}(f) - \gamma \Delta_{k,i,j}, \quad (8)$$

where

$$Y_{k,i,j}(f,\Delta_{k,i,j}) = \frac{A_{k,i}(f,d_{k,i})/[I_i(f)G(d_{k,i})]}{A_{k,j}(f,d_{k,j})/[I_j(f)G(d_{k,j})]}, \quad (9)$$

$$Si_{i,j}(f) = \frac{Si_i(f)}{Si_j(f)}, \quad (10)$$

and

$$\Delta_{k,i,j} = d_{k,i} - d_{k,j}. \quad (11)$$

Let

$$a_{k,i,j} = \ln Y_{k,i,j}(f,\Delta_{k,i,j}), \quad (12)$$

and

$$b_{i,j}(f) = \ln Si_{i,j}(f). \quad (13)$$

Thus

$$a_{k,i,j}(f) = b_{i,j}(f) - \gamma(f) \cdot \Delta_{k,i,j}, \quad (14)$$

where $\Delta_{k,i,j}$ represents the difference in the epicentral distance between two stations for the same earthquake. Because the velocity spectrum computed for a particular seismogram is $2\pi f$ times the displacement spectrum, the displacement spectra on the left side of equation (5) can be replaced directly by the corresponding velocity spectra. Unless otherwise specified, velocity spectra are used in subsequent equations, because we shall deal only with spectral ratios rather than with spectral amplitudes.

We make the following hypotheses: (1) In the period of the observation time, there was no large tectonic movement in the study region, that is, the seismic ray transmission and the attenuation parameters were stable. (2) Site response of stations is only related to frequency; thus, for any earthquake, the ratio of the earthquake response amplitudes recorded at two stations is only affected by frequency. (3) Among the stations examined, there is always a station whose site response is ideal for a given frequency; that is, the site response of the station corresponding to this frequency should be the ideal rock site response, and the response amplitude is one unit.

Based on these three hypotheses, we initially fixed the first station as the reference station, that is, $j=1$; then,

$$a_{k,i,1}(f) = b_{i,1}(f) - \gamma(f) \cdot \Delta_{k,i,1}, \quad (15)$$

where the meanings of the subscripts are the same as the equations shown previously in this paper. Matrix equations were constructed for each frequency to calculate the unknown (Zhu and Chen, 2007). $\gamma(f)$ is the parameter corresponding to the attenuation, and $b_{i,1}(f)$ is the natural logarithm ln of the site response ratio between the reference station and the other stations at corresponding frequency f. The site response ratio of the reference station itself is set at one for all frequencies.

The aforementioned reference station is the initial reference station set in the calculation process and does not necessarily have the characteristic of site response for a true reference station. In the relevant studies on site response using the SSR method (Borcherdt, 1970; Parolai et al., 2000, 2010; Nath et al., 2001), a reference station does not have any amplification effect, or its amplification effect is minimal and consistent at all frequencies. However, many studies have shown that site response shows strong frequency dependence (King and

Tucker, 1984; Tucker et al., 1984; Campillo et al., 1985). Our study takes into account both the minimum site response of a reference station and the frequency dependence of site response. $b_{i,1}(f)$ is the natural logarithm ln of the ratio of all stations involved in the study to the initial reference station at frequency f. Based on equation (13), $b_{i,1}(f)$ is converted to $S_{i,1}(f)$. We seek the minimum value for $S_{i,1}(f)$ and take the station corresponding to this minimum as the optimal reference station at this frequency. Then, for each station, we convert the site response ratios based on the initial reference station to the site response ratio based on the optimal reference station to obtain the final site response of this station at this frequency. Repeating this for all frequencies, the site response of the stations at each frequency can be obtained.

3 Survey of the study area and data sets

3.1 A survey of the study area

The study area (38°–41°N, 114°–120°E), North China region, is located in the North China active block (Fig. 1). The area has complex geological structures with a number of crisscrossing active faults and is one of the tectonically active regions in the Chinese continent. Northeast-trending faults on both sides of the Taihang uplift and on the eastern side of the Cangxian uplift intersect near vertically with northwest-trending faults. The nearly east-west trending Yanshan uplift is in the north, and the rifted basin of the North China Plain is in the southeast. In the rifted basin, different tectonic units have been developed, including the central Hebei depression, the Cangxian uplift, and the Huanghua depression. From historical earthquake records, the earliest earthquake in this area is the M 6.5 southeast Pingyin, Hebei, earthquake of 231 BC, the largest is the M 8.0 Sanhe-Pinggu earthquake of 2 September 1679 AD, and the largest in the last 40 years is the M_S7.8 Tangshan earthquake of 28 July 1976. In the study area, high-density earthquake observation stations have been established to support earthquake research using high-quality data (Fig. 1). Studies of the regional tectonics reveal that the tectonics of the region show lateral partitioning (Luo et al., 2008); the crust thickens gradually from southeast to northwest, with an average thickness of 34km.

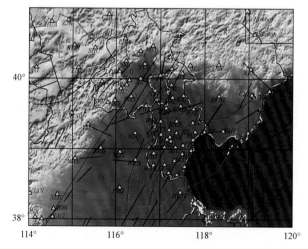

Figure 1   Tectonics and seismic stations in the study area

The tectonic regions are indicated by red solid lines: I, Shanxi graben fault zone; II, Beijing depression; III, Taihang uplift; IV, Jizhong uplift; V, Daxing uplift; VI, Cangxian uplift; VII, Chengning uplift; VIII, Huanghua depression; and IX, Yanshan uplift

We divide the data used in evaluating the attenuation parameters into four mutually independent sets based on seismic ray paths: the first set includes the records of earthquakes in the northwestern mountainous region recorded by stations in the same region; the second set includes the records of earthquakes in the southeastern basin region recorded by stations in the same region (Fig. 2a); the third set includes the records of earthquakes in the southeastern basin recorded by stations in the northwestern mountainous region; and the fourth set includes the records of earthquakes in the northwestern mountainous region recorded by stations in the southeastern basin region (Fig. 2b). The third and fourth sets of data both pass through the abrupt change zones in the crust thickness along opposite paths. These regions are called the northwest mountain zone, southeast basin, southeast-northwest transition zone, and northwest-southeast transition zone, respectively.

3.2 Digital seismic waveform data

The waveform data used in this study are from the local and regional earthquakes recorded by the three-component seismic stations of the China Digital Seismograph Network (CDSN; Zheng et al., 2010). The CDSN stations, managed by the China Earthquake Networks Center (CENC) and equipped with three-component velocity-type seismometers, have a sampling frequency of 50Hz (equivalent to a Nyquist frequency of 25Hz) and a dynamic range from 90 to 120 dB. The stations can be classified into three groups based on the bandwidth of instruments. For 44 of the stations, equipped with a broadband FBS-3A seismometer, frequency response of the instrument system is roughly flat in the range from 0.5 to 20Hz; for 58 stations, equipped with a short-period FSS-3 seismometer, frequency response of the instrument system is flat in the range from 1 to 20Hz; and for the other 5 stations, equipped with a ultrabroadband CTS-1 seismometer, frequency response of the instrument system is roughly flat in the range from 0.01 to 20Hz. Figure 3 shows the amplitude-frequency and phase-frequency characteristics of the velocity response for the vertical component of the instrument systems used in this study.

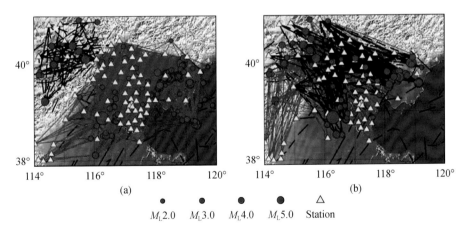

Figure 2 Distribution of seismic stations, seismic ray paths, and earthquakes used in this study

(a) The rays of earthquakes in the mountain region recorded by stations in the same region (black lines) and the rays of earthquakes in the southeastern basin region recorded by stations in the same region (red lines); (b) the rays of earthquakes in the southeastern basin recorded by stations in the northwestern mountain region (black lines) and the rays of earthquakes in the northwestern mountain region recorded by stations in the southeastern basin (red lines). Ray coverage regions are called the northwest mountain zone, southeast basin, southeast-northwest transition zone, and northwest-southeast transition zone, respectively

Waveform records for a total of 200 earthquakes with M_L 1.7–5.3, focal depth of less than 30km, and epicentral distance of 90–400km occurring between 2006 and 2008 were collected for this paper; for a complete

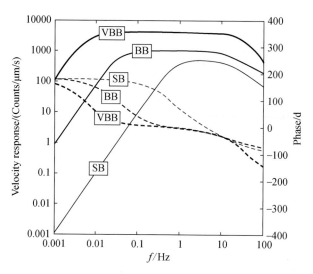

Figure 3 Amplitude-frequency response (solid lines) and phase-frequency response (dashed lines) of the seismographs equipped at CDSN stations in the North China region

list of earthquakes ⓔ see Table S1 in the electronic supplement to this paper. Each of the earthquakes was recorded by at least two stations, and the signal-to-noise ratio is higher than two. More than 1500 three-component records chosen from 81 stations were used in this study.

4 Data processing

The fixed minimum group velocity method was used when windowing the Lg wave. The starting points were determined according to a clearly picked Lg-wave onset time for each record. But for records whose Lg-wave onset time could not be picked clearly, the starting points were selected by the average velocity. The velocity inferred from the clear Lg-wave onset times is 3.6km/s, the lower limit of the Lg-wave velocity is 2.6km/s, and Lg-waves were intercepted in this velocity window (Fig. 4a). Visual inspection of all Lg-wave phases used in this study shows that the peak amplitudes all lie within this window. The velocity window used in this study is compatible with that considered by others (Murphy and Bennett, 1982; Campillo et al., 1985). In a study of Lg-wave attenuation in central France, Campillo et al. (1985) found that the Lg-wave attenuation parameters estimated using different velocity windows (2.3–2.6, 2.6–3.1, and 3.1–3.6km/s) were essentially identical, indicating that varying the length of the Lg-wave velocity window used to define the Lg-wave phase has a negligible effect on the results. After removal of mean and trend, the windowed Lg waves were padded with zeros to the nearest power of integer 2 to allow for the application of the fast Fourier transform algorithm with cosine windowing of 2% on both ends. Then, the instrument response and geometric attenuation were corrected, and the data were smoothed at three units of the frequency step (Fig. 4b) and fitted using equation (14); see Figure 5.

Linear regressions according to equation (15) of $\ln Y_{k,i,1}(f, \Delta_{k,i,1})$ versus $\Delta_{k,i,1}$ were made separately for each region, each component, and each frequency from 1 to 10Hz at increments of 0.2Hz in the following manner: first, regressions were made for each station i over all earthquakes k recorded at that station. The slope of these regressions is the estimate of the attenuation coefficient $\gamma(f)$ for that station, and the intercept is $Si_{i,1}(f)$ for that station. Second, the regional final attenuation coefficient was calculated as the weighted average of the estimates of $\gamma(f)$ over all of the stations, with the weighting proportional to the number of records at each

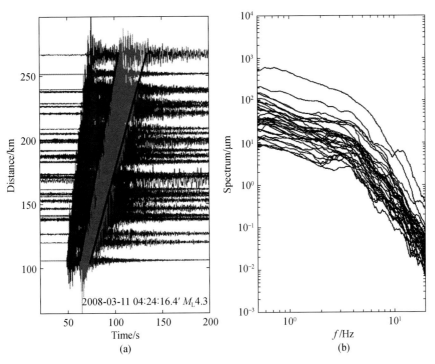

Figure 4 Lg-wave velocity window and example

(a) Vertical component seismograms of the 11 March 2008 $M_L 4.3$ earthquake, arranged according to the epicentral distance from top to bottom; the two diagonals correspond to the upper and lower limits of the Lg-wave velocity, the segment between the diagonals is the Lg-wave data window, and the amplitude of the waves is normalized for better illustration. (b) Fourier spectra corresponding to the Lg wave

station. An illustration of the procedure is given in Figure 5.

Conducting the same process for the four data sets, we get the attenuation coefficients and the site responses of the corresponding stations for each area. In data grouping, most stations are involved in two sets of data

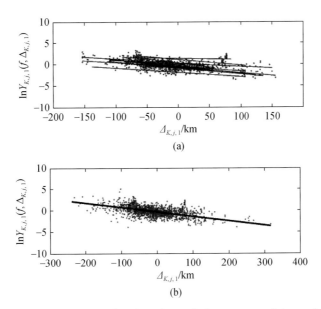

Figure 5 Example of data fitting using data from the vertical component of the southeast basin at 1Hz

The dots are observations. (a) Each line shows the fit for a single station i over all earthquakes k; the slope is the estimate of attenuation coefficient $\gamma(f)$, and the intercept is $\ln Si_{i,1}(f)$. (b) The slope of the solid line shows the final estimate of the attenuation coefficient $\gamma(f)$ as the weighted average of the estimates from each station. See Discussion and Conclusions

simultaneously, so the same station will have two site responses after data processing. By comparison, we find that the difference between them is fractional. Therefore, we simply take the average value as the final site effect of the station.

The residual is within ±10% of the principal value of the corresponding attenuation coefficient.

5 Results

5.1 Attenuation Coefficient

The curve of $\gamma(f)$ versus f was plotted with a frequency range 1–10Hz and a frequency sampling interval of 0.2Hz (Fig. 6). There is a distinct change in the general trend of the results for all three components at a frequency

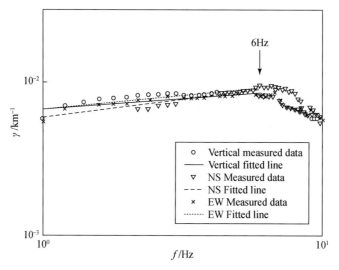

Figure 6 Seismic attenuation relationship corresponding to the second data set

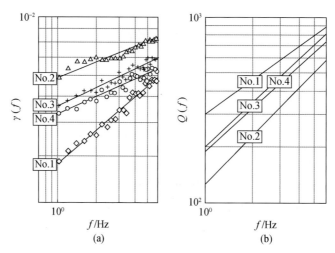

Figure 7 North-south component attenuation coefficient and linear regression fit at frequencies of 1–6Hz for all data sets (for better illustration, only one component result is plotted for each data set)

(a) Relationship between $\gamma(f)$ and f. The squares, triangles, crosses, and circles represent the results determined by the first, second, third, and fourth data sets, respectively. (b) Relationship between $Q(f)$ and f based on equation (4)

of 6Hz. The same phenomenon was reported by Shin and Herrmann (1987), Domínguez et al. (1997), and Raúl et al. (2008). Shin and Herrmann (1987) suggested that the Lg wave was contaminated by the Sn and Pn coda waves at higher frequencies, and, therefore, an Lg wave of <7Hz or a seismogram of an earthquake with an epicentral distance of less than 600km is recommended for the study of Lg-wave attenuation. In this paper, we will focus on the calculation of the attenuations for 1–6Hz and their relationships with frequency.

Table 1 Seismic attenuation relationship

Data Set (Data Quantity)	Component	$\gamma(f) = \gamma_0 f^v$		$Q(f) = Q_0 f^\eta$	
		γ_0	v	Q_0	η
No. 1 (84) Northwest Mountain	Vertical	0.0029	0.40	303	0.60
	East-West	0.0039	0.41	227	0.59
	North-South	0.0029	0.69	306	0.31
No. 2 (1036) Southeast Basin	Vertical	0.0070	0.14	125	0.86
	East-West	0.0066	0.15	132	0.85
	North-South	0.0059	0.25	149	0.75
No. 3 (182) Southeast-Northwest Transition Zone	Vertical	0.0046	0.24	190	0.76
	East-West	0.0047	0.24	187	0.76
	North-South	0.0048	0.26	188	0.74
No. 4 (196) Northwest-Southeast Transition Zone	Vertical	0.0043	0.22	201	0.78
	East-West	0.0044	0.22	201	0.78
	North-South	0.0042	0.22	209	0.78

It is observed that the attenuation coefficient is positively correlated with frequency in the frequency range of 1–6Hz. We fit an exponential relationship between attenuation coefficient and frequency, $\gamma(f) = \gamma_0 f^v (1/km)$, where γ_0 is the value of γ when the frequency is 1Hz and v is the frequency-dependent index of the attenuation coefficient (Aki, 1980; Herrmann and Kijko, 1983; Chun et al., 1987).

There were only 86 ray paths available for the northwestern mountain region due to the small number of earthquakes and sparse density of stations, and the attenuation coefficients obtained are $\gamma(f) = 0.0029 f^{0.4}$, $\gamma(f) = 0.0038 f^{0.41}$, and $\gamma(f) = 0.0029 f^{0.69}$ corresponding to the vertical, east-west, and north-south components, respectively (Fig. 7a).

The regional seismicity is high in the southeastern basin region. There were 1036 ray paths for this region, and the attenuation coefficients obtained are $\gamma(f) = 0.0070 f^{0.14}$, $\gamma(f) = 0.0066 f^{0.15}$, and $\gamma(f) = 0.0059 f^{0.25}$ corresponding to the vertical, east-west, and north-south components, respectively.

The third data set includes 182 ray paths. The attenuation coefficients obtained are $\gamma(f) = 0.0046 f^{0.24}$, $\gamma(f) = 0.0047 f^{0.24}$, and $\gamma(f) = 0.0048 f^{0.26}$ corresponding to the vertical, east-west, and north-south components, respectively. The fourth data set includes 196 ray paths, and the attenuation coefficients obtained corresponding to the vertical, east-west, and north-south components are $\gamma(f) = 0.0043 f^{0.22}$, $\gamma(f) = 0.0044 f^{0.22}$, and $\gamma(f) = 0.0042 f^{0.22}$, respectively.

Based on equation (4), assuming the velocity is 3.5km/s, the attenuation coefficient is converted from $\gamma(f) = \gamma_0 f^v$ to $Q(f) = Q_0 f^\eta$, where Q_0 is the absolute quality factor, and the corresponding results are shown in Table 1 and Figure 7b.

As shown in Table 1 and Figure 7, there are higher Q_0 values for seismic ray paths located in the northwestern

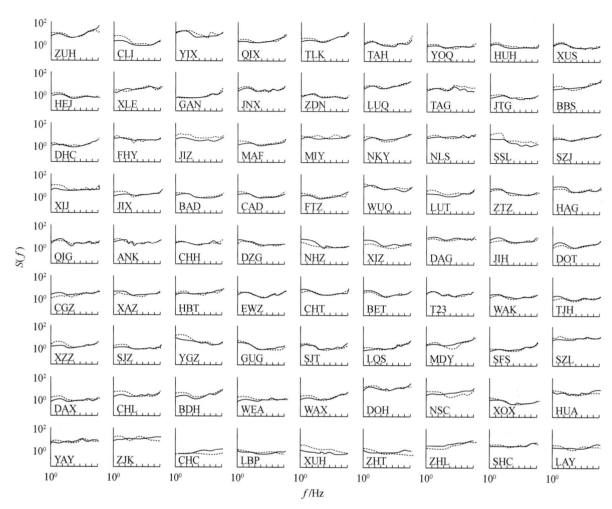

Figure 8 Site response

Solid lines represent the result from vertical components, dotted lines represent the north-south components, and dashed lines represent the east-west components

mountain region, lower Q_0 values in the southeastern basin region, and moderate Q_0 values for ray paths traveling through both the mountain and basin regions. This indicates that seismic attenuation is low when propagating in the northwestern mountain region, high in the southeastern basin region, and moderate when propagating through both the mountain and basin regions. There are certain differences in Q_0 values inferred from the third and fourth sets of ray paths that traveled through both the mountain and basin regions along opposite directions. As shown in Figure 2b, the areas covered by these two sets of ray paths are different. The magnitude of the earthquakes in the northwestern mountain region was small, only those stations adjacent to the northwestern part in the southeastern basin region recorded good waveforms, and most of the ray paths were still located in the mountain region; while earthquakes in the southeastern basin region are larger in magnitude, most of the ray paths of the good waveforms recorded by stations in the northwestern mountain region still propagated in the basin region. Similarly, the difference between the Q_0 values inferred from the third and the fourth sets of ray paths resulted from the difference between the paths located in the high-attenuation southeastern basin and the low-attenuation northwestern mountains. The results obtained by this study are consistent with Liu et al. (2004), who obtained the Lg-wave coda attenuation parameters in the same region with the stacked spectral ratio method.

5.2 Site response

Stable site responses were obtained for a total of 81 stations in this study. Eleven stations located in the northwestern mountain region are surface stations on a hard rock base; the other 70 stations are located in the southeastern basin region, in which 8 are surface stations on a hard bedrock base, 10 are borehole stations in mudstones, and the rest are borehole stations located in loess sediments. The results show that all stations have site amplifications lower than 5 at low frequency. For rock site stations, the site response is mainly flat and stable in the frequency range of 1–6Hz; only a few stations, such as ZHU and TLK, show obvious amplification at high frequencies (5–6Hz), where the value reached 25. At low frequencies (1–2Hz), the site response of stations on a soil base is generally higher than that on a rock base (Fig. 8). For stations based on soil, the site amplification of low-frequency waves is higher than that of high-frequency waves. Site response describes the amplification effect of near-surface media on seismic ground motion, according to Shearer (1999); at low frequencies (< 2Hz), soft sediment sites have a greater amplification effect on seismic ground motion when compared with hard bedrock sites. Our results are consistent with the previously mentioned results.

6 Discussion and conclusions

Based on seismic-wave spectral ratios, we present a joint inversion method to determine the parameters of site response and seismic-wave attenuation. Unlike the Moya method (Moya et al., 2000), the present method removes the source effects and is independent of the correctness of the source model. The method uses less assumptions than other methods and thus decreases errors caused by such assumptions. In selecting the data, the method does not have the limitations of the RTSR method (Chun et al., 1987) and thus increases greatly the number of data. The study indicates that the present method is reasonable in theory and operational in practice (Zhu and Chen, 2007).

The Q_{Lg} values of Lg-wave attenuation in the study area have significant lateral heterogeneity, which is clearly related to regional geological structure, in particular, a higher Q_{Lg} value on the propagation paths located in the northwestern mountain region and a lower Q_{Lg} value in the southeastern basin region; the Q_{Lg} value on the paths propagating through both mountain and basin regions is between the Q_{Lg} values inferred for the paths restricted to the respective regions. The results are consistent with the quality factor of the Lg-wave coda Q_c in the same area (Liu et al., 2004).

No significant difference was observed in both attenuation coefficient and site response for the three components. In order to verify the reliability of the results, the results obtained in this study were compared with other Lg-wave attenuation results in 13 regions where the frequency range is roughly similar to ours (Fig. 9, Table 2). The comparison shows that Q_{Lg} is low in a seismically active region (Paul et al., 1996; Benz et al., 1997; Domínguez and Rebollar, 1997; Domínguez et al., 1997) and high in a seismically inactive region (Chun et al., 1987; Sereno et al., 1988; Shi et al., 1996); for areas with moderate seismicity, Q_{Lg} values are between the two (Campillo et al., 1985; Chávez and Priestley, 1986; Sereno, 1990; McNamara et al., 1996). The attenuation relationship that we obtained matches the attenuation relationship obtained by other authors for the seismically active regions. Our values are reasonable for the North China region, which is a seismically active region.

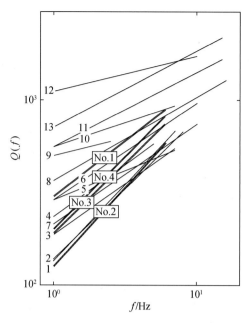

Figure 9 Comparison of Q_{Lg} values between regions

The four thick solid lines are the results obtained by this paper; Nos. 1-4 are the same as in Figure 7. The thin lines are the results obtained in other regions, and the numbers correspond to the region code listed in Table 2

Table 2 Comparison of Lg-wave attenuation relationship between regions

Region Code		Regions	Hypocentral Distance/km	Q_0	η	Frequency Range/Hz	Seismicity*
1		Subduction Zone, Mexico (Domínguez et al., 1997)	285-640	134	0.83	2-7	A
2		Southern Sierra Nevada, California (Paul et al., 1996)	150-400	138	0.76	1-8	A
3		Southern California (Benz et al., 1997)	150-1000	187	0.55	1-7	A
4		Basin and Range Province (Benz et al., 1997)	150-1000	235	0.56	1-5	A
5		Northern Baja California (Domínguez and Rebollar, 1997)	135-420	288	0.32	1-7	A
6		Central France (Campillo et al., 1985)	289-576	290	0.52	0.5-10	L
7		Great Basin (Chávez and Priestley, 1986)	200-500	214.08	0.54	0.3-10	L
8		Tibetian Platean (McNamara et al., 1996)	150-700	366	0.45	0.5-16	L
9		Eastern Kazakhstan (Sereno, 1990)	200-1300	500	0.19	0.5-2.5	L
10		Norway (Sereno et al., 1988)	200-1400	560	0.26	1-7	I
11		Central Appalachian (Shi et al., 1996)	41-1394	561	0.40	1-15	I
12		Eastern Canada (Chun et al., 1987)	90-867	1111.1	0.19	0.6-10	I
13		New England Appalachians (Shi et al., 1996)	41<1394	714.29	0.41	1-15	I
No. 1 (84)	Vertical	Northwest Mountain Zone (this study)	90-350	303	0.60	1-6	A
	East-West			227	0.59		

Continued

Region Code		Regions	Hypocentral Distance/km	Q_0	η	Frequency Range/Hz	Seismicity*
	North-South			306	0.31		
No. 2 (1036)	Vertical	Southeast Basin Zone (this study)	90–400	126	0.86		A
	East-West			132	0.85		
	North-South			149	0.75		
No. 3 (182)	Vertical	Southeast-Northwest Transition Zone (this study)	90–400	190	0.76		A
	East-West			187	0.76		
	North-South			188	0.74		
No. 4 (196)	Vertical	Northwest-Southeast Transition Zone (this study)	90–400	201	0.78		A
	East-West			201	0.78		
	North-South			209	0.78		

* Active region, A; less active region, L; inactive region, I.

7 Data and resources

Seismic data used in this study were collected by the China Digital Seismograph Network (CDSN) and managed by the China Earthquake Networks Center (CENC). They are available online at http://www.csndmc.ac.cn/newweb/ (last accessed September 2009). The figures in this paper were made using Generic Mapping Tools software written by Wessel and Smith (1991, 1998, last accessed September 2011).

Acknowledgments

The authors would like to express their sincere thanks to Associate Editor Anton M. Dainty and the two anonymous reviewers for their thoughtful comments and suggestions. Sincere thanks also go to Xiaoping Wang of the China Earthquake Networks Center (CENC) for her kind help in preparing the manuscript. This study was supported by the National Natural Science Foundation of China (Grant Number 41090291) and China Earthquake Administration (Grant Number 200808068, Grant Number XH12023).

References

Aki, K. (1980). Scattering and attenuation of shear waves in the lithosphere, *J. Geophys. Res.* **85**, 6496–6504.

Bay, F., D. Fäh, L. Malagnini, and D. Giardini (2003). Spectral shear-wave ground-motion scaling in Switzerland, *Bull. Seismol. Soc. Am.* **93**, 414–429.

Benz, H. M., A. Frankel, and D. M. Boore (1997). Regional *Lg* attenuation for the continental United States, *Bull. Seismol. Soc. Am.* **87**, 606–619.

Bonilla, L. F., H. J. Steidl, G. T. Lindley, A. G. Tumarkin, and R. Archuleta (1997). Site amplification in the San Fernando Valley, California: Variability of site-effect estimation using the S-wave, coda, and H/V methods, *Bull. Seismol. Soc. Am.* **87**, 710–730.

Borcherdt, R. D. (1970). Effects of local geology on ground motion near San Francisco Bay, *Bull. Seismol. Soc. Am.* **60**, 29–61.

Campillo, M., J.-L. Plantet, and M. Bouchon (1985). Frequency-dependent attenuation in the crust beneath central France from Lg waves: Data analysis and numerical modeling, *Bull. Seismol. Soc. Am.* **75**, 1395–1412.

Chávez, D. E., and K. F. Priestley (1986). Measurement of frequency dependent Lg attenuation in the Great Basin, *Geophys. Res. Lett.* **16**, 551–554.

Chen, Y. T., Z. L. Wu, P. D. Wang, L. S. Xu, H. J. Li, and Q. D. Mu (2000). *Digital Seismology*, Seismological Press, Beijing, 171 pp (in Chinese).

Chun, K.-Y., G. F. West, R. J. Kokoski, and C. Samson (1987). A novel technique for measuring Lg attenuation—Results from eastern Canada between 1 to 10Hz, *Bull. Seismol. Soc. Am.* **77**, 398–419.

Chung, T. W., and K. Lee (2002). Attenuation of high-frequency Lg waves around the Yangsan fault area, the southeast Korea, *J. Korean Geophys. Soc.* **5**, 1–8.

Chung, T. W., M.-H. Noh, J.-K. Kim, Y.-K. Park, H.-J. Yoo, and J. M. Lees (2007). A study of the regional variation of low-frequency Q_{Lg}^{-1} around the Korean Peninsula, *Bull. Seismol. Soc. Am.* **97**, 2190–2197.

Chung, T. W., Y.-K. Park, I. B. Kang, and K. Lee (2005). Crustal Q_{Lg}^{-1} in South Korea using the source pair/receiver pair method, *Bull. Seismol. Soc. Am.* **95**, 512–520.

Domínguez, T., and C. J. Rebollar (1997). Regional variations of seismic attenuation from coda and Lg waves in northern Baja California, *J. Geophys. Res.* **102**, 15, 259–15, 268.

Domínguez, T., C. J. Rebollar, and R. R. Castro (1997). Regional variations of seismic attenuation of Lg waves in southern Mexico, *J. Geophys. Res.* **102**, 27, 501–27, 509.

Field, E. H., and K. H. Jacob (1995). A comparison and test of various site response estimation techniques, including three that are not reference site dependent, *Bull. Seismol. Soc. Am.* **85**, 1127–1143.

Hartzell, S. H. (1992). Site response estimation from earthquake data, *Bull. Seismol. Soc. Am.* **82**, 2308–2327.

Hasegawa, H. S. (1985). Attenuation of Lg waves in the Canadian Shield, *Bull. Seismol. Soc. Am.* **75**, 1569–1582.

Herrmann, R. B., and A. Kijko (1983). Modeling some empirical vertical component Lg relations, *Bull. Seismol. Soc. Am.* **73**, 157–171.

Jin, A., and K. Aki (1988). Spatial and temporal correlation between coda Q and seismicity in China, *Bull. Seismol. Soc. Am.* **78**, 741–769.

Kennett, B. (1986). Lg waves and structural boundaries, *Bull. Seismol. Soc. Am.* **76**, 1133–1141.

King, J. L., and B. E. Tucker (1984). Observed variations of earthquake motions across a sediment-filled valley, *Bull. Seismol. Soc. Am.* **74**, 137–152.

Knopoff, L., F. Schwab, and E. Kauselt (1973). Interpretation of Lg, *Geophys. J. Roy. Astron. Soc.* **33**, 389–404.

Lachet, C., D. Hatzfeld, P. Bard, N. Theodulidis, C. Papaioannou, and A. Savvaidis (1996). Site effects and microzonation in the city of Thessaloniki (Greece) comparison of different approaches, *Bull. Seismol. Soc. Am.* **86**, 1692–1703.

Liu, J. H., F. T. Liu, and X. W. Yan (2004). A study of Lg coda attenuation beneath North China: Seismic imaging Lg coda Q_0. *Chin. J. Geophys.* **47**, 1044–1052 (in Chinese).

Luo, Y., J. J. Chong, and S. D. Ni (2008). Moho depth and sedimentary thickness in Capital region, *Chin. J. Geophys.* **51**, 1135–1145 (in Chinese).

McNamara, D. E., T. J. Owens, and W. R. Walter (1996). Propagation characteristics of Lg across the Tibetian Plateau, *Bull. Seismol. Soc. Am.* **86**, 457–469.

Mitra, S., K. Priestley, V. K. Gaur, and S. Rai (2006). Frequency-dependent Lg attenuation in the Indian platform, *Bull. Seismol. Soc. Am.* **96**, 2449–2456.

Moya, C. A., J. Aguirre, and K. Irikura (2000). Inversion of source parameters and site response from strong ground motion records using genetic algorithms, *Bull. Seismol. Soc. Am.* **90**, 977–992.

Murphy, J. R., and T. J. Bennett (1982). A discrimination analysis of short-period regional seismic data recorded at Tonto Forest Observatory, *Bull. Seismol. Soc. Am.* **72**, 1351–1366.

Nath, S. K., N. N. Biswas, and M. Dravinski (2001). Determination of S-wave site response in Anchorage, Alaska in the 1–9Hz frequency band, *Pure Appl. Geophys.* **159**, 2673–2698.

Nuttli, O. W. (1986). Yield estimation of Nevada Test Site explosions obtained from seismic Lg waves, *J. Geophys. Res.* **91**, 2137–2151.

Panza, G. F., and G. Calcagnile (1975). Lg, Li, and Rg from Rayleigh models, *Geophys. J. Roy. Astron. Soc.* **40**, 475–487.

Parolai, S., D. Bindi, and P. Augliera (2000). Application of the generalized inversion technique (GIT) to a microzonation study: Numerical simulations and comparison with different site-estimation techniques, *Bull. Seismol. Soc. Am.* **90**, 286–297.

Parolai, S., S. Orunbaev, and D. Bindi (2010). Site effects assessment in Bishkek (Kyrgyzstan) using earthquake and noise recording data, *Bull. Seismol. Soc. Am.* **100**, 3068–3082.

Paul, A., D. Jongmans, M. Campillo, P. Malin, and D. Baumont (1996). Amplitude of regional seismic phases in relation to crustal structure of the Sierra Nevada, California, *J. Geophys. Res.* **101**, 25,243–25,254.

Pomeroy, P. W., W. J. Best, and T. V. McEvilly (1982). Test ban treaty verification with regional data—A review, *Bull. Seismol. Soc. Am.* **72**, S89–S129.

Raoof, M., R. B. Herrmann, and L. Malagnini (1999). Attenuation and excitation of three-component ground motion in southern California, *Bull. Seismol. Soc. Am.* **89**, 888–902.

Rapine, R. R., J. F. Ni, and T. M. Hearn (1997). Regional wave propagation in China and its surrounding regions, *Bull. Seismol. Soc. Am.* **87**, 1622–1636.

Raúl, R. C., C. Cristina, and R. Oscar (2008). Seismic attenuation in northeastern Sonora, Mexico, *Bull. Seismol. Soc. Am.* **98**, 722–732.

Riepl, J., P.-Y. Bard, and D. Hatzfeld (1998). Detailed evaluation of site-response estimation methods across and along the sedimentary valley of Volvi (EURO-SEISTEST), *Bull. Seismol. Soc. Am.* **88**, 488–502.

Robert, J. M., V. E. Camp, F. L. Vermon, and M. S. Abdullah (1999). Regional waveform propagation in the Arabian Peninsula, *J. Geophys. Res.* **104**, 20,221–20,235.

Sato, H., and M. C. Fehler (1998). *Seismic Wave Propagation and Scattering in the Heterogeneous Earth*, Springer-Verlag, New York, 308 pp.

Sereno, T. J. (1990). Frequency-dependent attenuation in eastern Kazakhstan and implications for seismic detection thresholds in the Soviet Union, *Bull. Seismol. Soc. Am.* **80**, 2089–2105.

Sereno, T. J., S. R. Bratt, and T. C. Bache (1988). Simultaneous inversion of regional wave spectra for attenuation and seismic moment in Scandinavia, *J. Geophys. Res.* **93**, 2019–2035.

Shearer, P. M. (1999). *Introduction to Seismology*, Cambridge University Press, New York, 260 pp.

Shi, J., W. Y. Kim, and P. G. Richards (1996). Variability of crustal attenuation in the northeastern United States from Lg waves, *J. Geophys. Res.* **101**, 25,231–25,242.

Shih, X. R., K.-Y. Chun, and T. Zhu (1994). Attenuation of 1–6 sec. Lg waves in Eurasia, *J. Geophys. Res.* **99**, 23,859–23,874.

Shin, T. C., and R. B. Herrmann (1987). Lg attenuation and source studies using 1982 Miramichi data, *Bull. Seismol. Soc. Am.* **77**, 384–397.

Steidl, J. H., A. G. Tumarkin, and R. J. Archuleta (1996). What is a reference site? *Bull. Seismol. Soc. Am.* **86**, 1733–1748.

Tucker, B. E., J. L. King, and D. Hatzfeld (1984). Observations of hard-rock effects, *Bull. Seismol. Soc. Am.* **74**, 121–136.

Wessel, P., and W. H. F. Smith (1991). Free software helps map and display data, *Eos Trans. AGU* **72**, no. 441, 445–446.

Wessel, P., and W. H. F. Smith (1998). New, improved version of the Generic Mapping Tools released, *Eos Trans. AGU* **79**, 579.

Xie, J., and B. J. Mitchell (1990). Attenuation of multiphase surface waves in the Basin and Range province, part I: Lg and Lg coda, *Geophys. J. Int.* **102**, 121–137.

Zhang, J., T. Lay, J. Zaslow, and W. R. Walter (2002). Source effects on regional seismic discriminate measurements, *Bull. Seismol. Soc. Am.* **92**, 2926–2945.

Zheng, X. F., Z. X. Yao, J. H. Liang, and J. Zheng (2010). The role played and opportunities provided by IGP DMC of China National Seismic Network in Wenchuan earthquake disaster relief and researches, *Bull. Seismol. Soc. Am.* **100**, 2866–2872.

Zhu, X. Y., and Y. T. Chen (2007). An inversion of site response and Lg attenuation using Lg waveform, *Acta Seismol. Sinica* **20**, 605–616 (*in Chinese with English abstract*).

2008年5月12日汶川M_W7.9地震的震源位置与发震时刻[*]

杨智娴[1,2]　陈运泰[2,1]　苏金蓉[3]　陈天长[3]　吴　朋[3]

1. 中国地震局地球物理研究所　中国北京 100081；2. 北京大学-中国地震局现代地震科学技术研究中心　中国北京 100871；
3. 四川省地震局　中国成都　610041

摘要　综合运用四川省地震台网与紫坪铺水库地震台网的观测资料，精确地测定了2008年5月12日汶川M_W7.9地震的震源位置与发震时刻．指出，对汶川地震这样的大地震精准定位，必须克服或尽量减少远台观测对地震精确定位的局限性、地壳介质模型的不完善性以及识别与检测初至波震相的不一致性等因素的影响．通过分析对比、反复试验，从上述台网中精心选取了方位分布均匀、具有近震源台站约束、直达P波震相确系由初始破裂辐射出的15个地震台的直达P波到时数据，反演得出精确度比区域性地震台网常规测定的精确度高一个数量级的汶川大地震的定位结果，即：发震时刻（北京时间）：2008年5月12日14：27：57.59±0.03s；震中位置：31.018°N±0.3km，103.365°E±0.3km；震源深度：15.5km±0.3km.

引言

2008年5月12日汶川M_W7.9地震发生后，中国地震台网中心（China Earthquake Network Center，CENC）于14时41分（北京时间，以下若非特别说明，均为北京时间）发出"地震初报"信息：2008年5月12日14时28分在四川省汶川县（31.0°N，103.4°E）发生M_S7.6地震（表1第1行）. 14时47分该中心发出"地震速报"信息：2008年5月12日14时28分在四川省汶川县（31.0°N，103.4°E）发生M_S8.0地震（表1第2行）. 这个"地震速报"结果与中国地震局（2009）[①]所列的结果（发震时刻14：28：04.1，震中位置30.95°N，103.40°E）（表1第3行），在发震时刻精确到分、震中位置精确到0.1°时是一致的，因此，"地震速报"可以看作是表1第3行所列结果面向公众的"简略版". 汶川地震发生一个月以后，中国地震台网中心运用中国地震台网93个地震台的记录资料，将汶川地震的震源参数修订为：发震时刻14：27：58.9，震中位置30.95°N，103.57°E，震源深度14km，面波震级M_S8.2，但未给出定位误差（表1第4行）. 不同作者用同一数据源、同一定位方法给出的结果与表1的第4行给出的结果略有不同：发震时刻14：27：58.7，震中位置30.97°N，103.57°E，震源深度14km，面波震级M_S8.3，也未给出定位误差（陈培善，2008；表1第5行）. 概括地说，与表1第3行相比，第4，第5行所列的结果，发震时刻分别提前了5.2s和5.4s，震中位置分别北移了0.00°和0.02°，东移了0.17°和0.17°.

刘启元等（2008）根据川西流动地震台阵中距汶川地震震中最近的KCD05台的直达P波到时为14：28：03.33（早于中国地震台网中心测定的发震时刻）判断，中国地震台网中心给出的14：28：04.1发震时刻是不可能的，该文认为，汶川地震的发震时刻应当接近于美国地质调查局的"地震初定报告"PDE/USGS给出的14：28：01.5；进而又认为PDE/USGS给出的汶川地震的19km震源深度可能更接近实际情况. 不过，该文只是对中国地震台网中心给出的结果与PDE/USGS给出的结果进行分析判断，既没

---

[*] 本文发表于《地震学报》，2012年，第**34**卷，第2期，127—136.
[①] 中国地震局. 2009. 四川5·12汶川8.0级地震科学总结与反思报告，第三章表3.1.1.

有对汶川地震进行定位，也没有给出定位的结果（所以在本文表1中没有列入）.

陈九辉等（2009）对汶川地震余震序列重新定位，其中也包含对汶川 $M_S 8.0$ 地震重新定位，结果如表1第6行所列（有效数字照抄）：发震时刻 14：28：0.39，震中位置 30.9607°N，103.3523°E，震源深度 18.8 km，水平方向平均标准差 0.85 km，垂直方向平均标准差 0.75 km.

黄媛等（2008）也对汶川 $M_S 8.0$ 地震及其余震序列重新定位. 认为其重定后的精度大致为：东西向 0.6 km，南北向 0.7 km，垂直向 2.5 km（黄媛等，2008，p.1246）. 虽然该文题为《汶川 8.0 级大地震及其余震序列重新定位研究》，但在文章的正文中并未给出主震震源参数（发震时刻、震中位置、震源深度），只在该文摘要中提到"主震震源深度重定后为 16.0 km"（所以在本文表1第7行中未能列出震中位置的数值）.

表1 一些机构或作者确定的汶川大地震的参数

Table 1 Source parameters of the great Wenchuan earthquake determined by various institutions or authors

序号	发震时刻（北京时间）		震中位置		震源深度 /km	震级			测定机构或作者	说明
	年-月-日	时：分：秒	纬度/°N	经度/°E		m_b	M_S	M_W		
1	2008-05-12	14：28	31.0	103.4			7.6		中国地震台网中心	14：41 发出的"地震初报"信息
2	2008-05-12	14：28	31.0	103.4			8.0		中国地震台网中心	14：47 发出的"地震速报"信息
3	2008-05-12	14：28：04.1	30.95	103.40	14		8.0		中国地震台网中心	中国地震局（2009）[①]
4	2008-05-12	14：27：58.9	30.95	103.57	14	6.4	8.2		中国地震台网中心	由93个台测定，水平与深度定位误差均≤5km；m_b 由63个台测定；M_S 由57个台测定
5	2008-05-12	14：27：58：7	30.97	103.57	14	6.4	8.3		陈培善（2008）	由93个台测定，水平与深度定位误差均≤5km
6	2008-05-12	14：28：0.39	30.9607	103.3523	18.8				陈九辉等（2009）	水平标准差 0.85km，垂直标准差 0.75km
7	2008-05-12	14：27：58.7			16				黄媛等（2008）	定位精度：东西向 0.6km，南北向 0.7km，垂直向 2.5km
8	2008-05-12	14：27：57.7	31.01	103.38	15		8.0	7.9	杨智娴等（2008）	定位精度：发震时刻 0.2s，东西向 1km，南北向 1km，垂直向 2km
9	2008-05-12	14：27：57.59 ±0.03	31.018 ±0.003 (±0.3km)	103.365 ±0.003 (±0.3km)	15.5±0.3		8.0	7.9	本文	

[①] 中国地震局. 2009. 四川 5·12 汶川 8.0 级地震科学总结与反思报告，第三章表 3.1.1.

杨智娴等（2008）在汶川地震发生后不久，综合运用全国地震台网与四川省地震台网记录的观测资料对汶川大地震的震源参数进行了修订，得出：发震时刻 14：27：57.7，震中位置 31.01°N, 103.38°E，震源深度15km，面波震级 M_S8.0，矩震级 M_W7.9（陈运泰，2008；表1 第8行）.

在地震定位中，当台距远大于地震的震源深度时，由于走时数据对深度的变化不敏感，使得震源深度得不到有效的约束．这种震源深度与发震时刻的折中（trade-off）效应致使反演得到震源深度与发震时刻具有较大的不确定性（Bormann，2006）．为了能够有效地约束震源深度，最近台站的震中距应约为震源深度的分数．考虑到表1第1行至第8行列出的测定结果表明汶川地震的震源深度介于14—19km之间，应当运用震中距大约5km或者5km以下的近台记录，才可望最有效地约束震源深度与发震时刻．有鉴于此，本文将综合运用四川省地震台网的记录资料以及最靠近震中、其震中距小至约5km的紫坪铺水库地震台网的记录资料，反演汶川地震的精准震源位置与发震时刻．

1 汶川地震震源参数的精准测定

1.1 影响大地震精准定位的主要因素

大地震的精准定位受到诸多因素的影响与制约，其中最主要的因素有如下三点：

（1）远台观测的局限性．单用远台直达P波到时资料对地震定位时，在震中距远大于震源深度的情况下，远台到时数据所含震源深度的信息极为微弱．这种内在的困难导致了反演的不稳定性以及震源深度与发震时刻的不确定性.

（2）地壳介质模型的不完善性．运用大范围的远台资料进行地震定位，采用简单的一维地壳介质模型不能很好地表示地震波的传播效应．在远震定位以及区域性地震定位中，地壳介质模型的不完善性有时可导致地震波观测走时（O）与理论走时（C）的差（O—C）即残差（Res）大至1秒至数秒，从而使定位精度大大降低．

（3）所确认的震相的不一致性．震源位置与发震时刻就其定义指的是地震破裂起始点的位置与破裂起始的时刻．由近台观测检测出的初至波是与地震破裂起始点及其附近的较小范围的破裂相联系的，而由远台观测检测出的初至波则是与包含破裂起始点在内的较大范围的破裂相联系的．能为近台观测到并且检测出的地震破裂起始点及其附近的小范围的破裂所辐射出的初至波在到达远台时常常已经衰减殆尽、十分微弱，从而被淹没在背景噪声之中以致不能被辨认识别，使得由远台记录中读出的、与较大范围的破裂相联系的"初至波"震相与由近台记录中读出的、与较小范围的破裂相联系的"初至波"震相在严格的意义上并非来自同一地点.

在这种情况下，采用台距较小、震中距较小并且有近震中的地震台的观测资料对汶川大地震的震源重新定位，可望获得高精确度的结果，为对地震定位的精确度与准确度有较高要求的研究工作提供重要的基础性数据（杨智娴等，2008；陈运泰，2008）.

1.2 资料

为了准确测定汶川大地震的震源参数，我们选取了四川省地震台网中仪器运行正常、数据可靠、记录清晰的16个地方性地震台（图1a）以及紫坪铺水库地震台网中4个靠近震中的地震台（图1b）在地震仪器运行正常时段，即强烈地面运动尚未使记录"出格"甚而毁坏地震仪器之前的记录数据进行定位.

由图1a可见，地震台的方位与震中距分布在1, 3, 4象限比较均匀，在 $\Delta<250$km 范围内，每个象限有方位与震中距分布两者都比较均匀的台站4—6个. 相比之下，在第2象限即震中西北台站较少.

图2是上述四川省地震台网的震中距 $\Delta \leq 160$km 的11个地震台与紫坪铺水库地震台网的震中距约5—20km 的4个近震中的地震台记录的2008年汶川大地震的走时（图2a）与波形（图2b）图. 在图2a中，横坐标是震中距 Δ，纵坐标是直达P波的走时 t_p. 作为参考，过原点的斜率为 $1/v$ 的直线是在P波速度为 v 的均匀介质中表面震源的直达P波的走时曲线，图中 $v=5.89$km/s. 图2b显示了初至波到达后的最初几秒（约3秒）垂直向的记录波形，横坐标是震中距 Δ，纵坐标是直达P波的折合走时 $t_p-\Delta/v$.

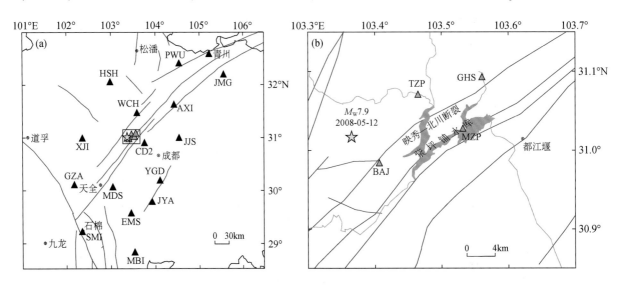

图1 重新定位后的汶川大地震震中与地震台分布图

(a) 用于汶川大地震精确定位的四川省地震台网的16个地方性地震台（黑三角形）与紫坪铺水库地震台网的4个靠近震中的地震台（灰色三角形），以及重新定位后的汶川大地震的震中（黄色五角星）图；(b) 图(a) 中的方框所示区域的放大图

Fig. 1 Distribution of the relocated epicenter of the Wenchuan earthquake and seismic stations

(a) The 16 local seismic stations of Sichuan Province Seismic Network (black triangles) and the 4 near-by stations of Zhipinpu Reservoir Seismic Network (gray triangles) used for source parameter determination of the great Wenchuan earthquake and the relocated epicenter of the Wenchuan earthquake (yellow star); (b) Enlarged map of the rectangular area in Fig. 1a

1.3 地壳速度结构模型

四川位于青藏高原的东部边缘，以北东—南西走向的龙门山断裂带为界，东、西两部分无论是地形、地貌还是结构、地质条件差异都很大. 东部是以平原、丘陵为主的盆地，海拔高程低，是稳定的大陆地块；西部是高原，海拔高程高，东西两部分海拔高程相差3700m. 东部与西部的地震波走时存在明显的系统偏差，与地壳厚度由西向东突然减薄、波速陡增密切相关（赵珠，张润生，1987a，b；赵珠等，1997；Mooney et al，2002）. 考虑到汶川地震发生于将东、西两部分分开的龙门山断裂带上，其震源深度在20km以内，参考上述作者的结果，经简化调整，采用如表2所示的地壳速度结构模型对汶川大地震进行定位.

表2 用于汶川大地震重新定位的四川地壳速度结构模型

Table 2 Velocity structure model of the Sichuan, China used in the relocation of the great Wenchuan earthquake in this study

层序	深度/km	P波速度/km·s^{-1}	S波速度/km·s^{-1}
1	0.0—7.5	5.00	2.89
2	7.5—16.0	5.48	3.17
3	16.0—20.0	5.93	3.43
4	20.0—30.0	6.43	3.72
5	30.0—50.0	6.60	3.82
6	≥50.0	8.30	4.80

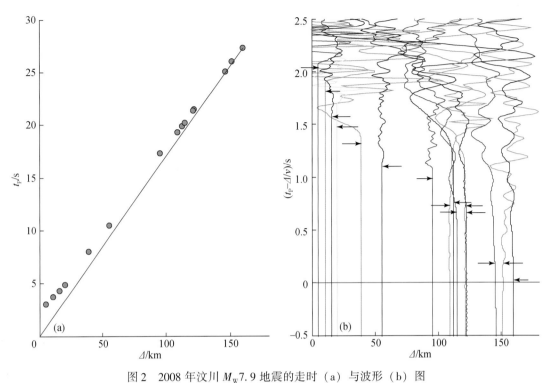

图 2 2008 年汶川 M_W7.9 地震的走时（a）与波形（b）图

图中，Δ 表示震中距，t_P 表示直达 P 波的走时，$t_P-\Delta/v$ 表示直达 P 波的折合走时．图（a）中过原点的斜率为 $1/v$ 的直线是在 P 波速度为 v 的均匀介质中表面震源的直达 P 波的走时曲线，$v=5.89$km/s．图（b）中显示了初至波到达后的最初几秒（约 3s）垂直向的记录波形．小箭头表示经过追踪对比确认的 P 波到时

Fig. 2 Travel times (a) and waveforms (b) of the 2008 M_W7.9 Wenchuan earthquake

Δ, t_P and $t_P-\Delta/v$ represent epicentral distance, travel time and reduced travel time of the direct P waves, respectively. In Fig. 2a, the straight line with slope $1/v$, pasing through the coordinate origin is the travel time curve of direct P waves for a surface source in a homogeneous medium of P wave velocity $v=5.89$km/s. In Fig. 2b, waveforms of about 3s of the first P arivals recorded in the vertical components are displayed. Small arows show the onsets of the first P arivals identified through tracing and comparison

2 汶川大地震的精确震源参数测定结果与分析

以下用震中距 39—250km 的 16 个地方性地震台和 4 个震中距 5—20km 的近台的 P 波到时数据、运用经典的盖革（Geiger）法（Geiger，1910；杨智娴，2003；Hermann，Ammon，2004；Bormann，2006）反演汶川地震主震的震源参数．分析试验过程可以概括为下述 4 种情况．

情形 A（图 3a）采用震中距 39—250km 的 16 个地方性地震台的资料定位．情形 B（图 3b）在情形 A 的基础上加入 4 个震中距 5—20km 的近台资料．情形 C（图 3c）不用远台，只用震中距 39—160km 的 11 个地方台的资料．情形 D（图 3d）在情形 C 的基础上加入 4 个震中距 5—20km 的近台．这 4 种情形的结果如表 3 与图 3 所示．

由上述 4 种情形反演得出的结果可以看出，情形 A（表 3 的第 1 行与图 3a）反演结果的均方根误差最大，达 0.026s，并且残差（Res）最大的 3 个台站（MBI，JMG 和 QCH 台）正是震中距最大的台站．在情形 B（表 3 的第 2 行与图 3b）中，4 个震中距 5—20km 的近台资料的加入，减小了均方根误差，由情形 A 的 0.026s 减为 0.020s；残差最大的仍是震中距最大的 3 个台站（MBI，JMG 和 QCH 台）；近台资料的加入使震源深度与发震时刻得到了更好的约束，其误差分别由 0.8km 与 0.07s 减为 0.3km 与 0.04s. 由此可看出，近台观测资料的应用可使震源深度与发震时刻得到更好约束，显示出误差减小.

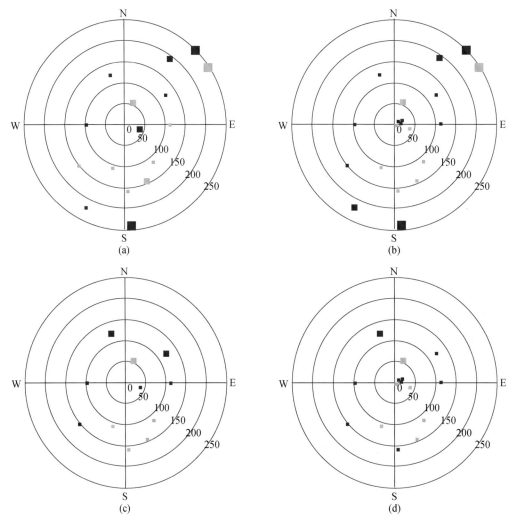

图 3 残差（Res）的地理分布

（a）情形 A. 采用震中距 39—250km 的 16 个地方性地震台的资料定位；（b）情形 B. 在 A 的基础上加入 4 个震中距 5—20km 的近台资料定位；（c）情形 C. 只用震中距 39—160km 的 11 个地方台的资料定位；（d）情形 D. 在 C 的基础上加入 4 个震中距 5—20km 的近台定位.（a）—（d）各图以震中为坐标原点，台站用相对于震中的方位角和震中距表示，圆圈上的数字表示震中距（km）. 蓝色方块表示正残差，浅蓝色方块表示负残差，方块大小表示残差大小，小符号表示 Res≤0.15s，中符号表示 0.15s<Res<0.40s，大符号表示 Res≥0.40s

Fig. 3 Geographical distribution of residuals（Res）

（a）Case A. Source parameter determination by solely using data recorded at 16 local seismic stations with epicentral distance $\Delta \approx$ 39—250km; （b）Case B. Source parameter determination by using both data as case A and that recorded at 4 near-by seismic stations with epicentral distance $\Delta \approx$ 5—20km; （c）Case C. Source parameter determination by solely using data recorded at 11 local seismic stations with epicentral distance $\Delta \leqslant$ 160km; （d）Case D. Source parameter determination by using both data as case C and that recorded at 4 near-by seismic stations. In Fig. 3a through Fig. 3d, the inverted epicentral location for each case is taken as coordinate origin, location of seismic station is represented by its epicentral azimuth and distance. Numerical figures at circles stand for epicentral distances in km. Blue and light blue squares represent positive and negative residuals, respectively. Small symbols stand for Res≤0.15s, moderate ones for 0.15s<Res<0.40s, and large ones for Res≥0.40s

表3 由地方性地震台与近震中的地震台资料反演汶川大地震的震源参数结果

Table 3 Source parameters of the great Wenchuan earthquake inverted by using both data recorded at local and near-by seismic stations

情形	震中位置 纬度/°N	震中位置 经度/°E	震源深度 /km	发震时刻（北京时间） 时：分：秒	均方根误差/s	说明
A	31.020±0.003 (±0.4km)	103.362±0.005 (±0.4km)	13.5±0.8	14：27：57.46±0.07	0.026	16个地方性地震台（Δ≈39—250km）
B	31.018±0.003 (±0.4km)	103.365±0.004 (±0.4km)	15.3±0.3	14：27：57.61±0.04	0.020	情形A加4个近台（Δ≈5—20km）
C	31.021±0.003 (±0.3km)	103.366±0.004 (±0.4km)	15.4±0.7	14：27：57.56±0.06	0.011	11个地方性地震台（Δ≈39—160km）
D	31.018±0.003 (±0.3km)	103.365±0.003 (±0.3km)	15.5±0.3	14：27：57.59±0.03	0.010	情形C加4个近台（Δ≈5—20km）

在情形C（表3的第3行与图3c）中，只用震中距39—160km、方位分布较均匀的11个地方性地震台的资料进行反演．结果表明，其均方根误差大大减小，与同是无近台约束的情形A相比，均方根误差从0.026s减为0.011s，约为情形A的2/5，说明了远台观测资料是导致反演结果的均方根误差大的重要原因．但是震源深度与发震时刻的误差与情形A相比减小量都不大，分别由0.8km与0.07s减为0.7km与0.06s，反映了如果缺乏震中距与震源深度相当、甚至比震源深度小得多的近台控制，震源深度与发震时刻不能得到有效的约束．

情形D与前3种情形不同．在情形D（表3的第4行与图3d）中，在震中距39—160km、方位分布较均匀的11个地方性地震台资料的基础上加入了4个震中距小至5—20km的近台资料．反演结果表明，均方根误差可以小到0.010s，约为情形A的2/5；相应地，震中位置与震源深度的误差也都减至0.3km，发震时刻的误差减至0.03s，均约为情形A与情形C的1/2．

图4综合表示了所计算的4种情形的误差椭圆（严格地说，这里所说的"误差椭圆"是误差椭球在过震源的水平面上的截面）（图4a）与沿垂直方向的误差（图4b）．图4a在以情形D得到的震中为中心的2km×2km范围内表示计算所得的4种情形的震中位置及其误差椭圆，图4b是在沿纬度方向（即由西向东）的剖面上表示计算所得的4种情形的震源深度及其沿垂直方向的误差棒．由图4可见，在所有4种情形中，情形D是67%概率的置信区域（误差椭圆内的面积）最小、最圆以及沿垂直方向的误差最小的情形．对于情形A至情形D的所有情形，误差椭圆都比较圆，反映了台站的分布总体上较均匀，特别是情形D．我们还可以看到，尽管所有情形的误差椭圆都比较圆，但还不是圆形，它们都是短轴沿近北南方向至北北东方向（方位角分别为4.3°，31.2°，13.7°，25.3°）的椭圆，反映了台站的分布对于上述方位控制较好而对与上述方位垂直的近东西方向至北北西方向的方位控制稍差一些．而沿垂直方向的误差以及发震时刻的误差的大小的系统变化（对于情形A，B，C，D，垂直方向的误差分别为0.8km，0.3km，0.7km和0.3km，发震时刻的误差分别为0.07s，0.04s，0.06s和0.03s）说明了近台资料的应用可在相当大的程度上改善对震源深度、从而对发震时刻的约束．

由以上分析可见，在情形D（表3的第4行与图3d）中，由于近台与地方性地震台观测资料的综合运用，并且在震相识别时，通过追踪对比来自初始破裂位置的震相，使震相同一性的确认得到切实的保证，大大减小了反演时震源深度与发震时刻之间的折中效应，使反演更趋稳定，获得了水平向和垂直向误差都减少到约为0.3km的震源位置，而发震时刻的误差也大大减小，减至0.03s.因此情形D的计算结果确认为经重新进行精确定位后得到的最佳结果，即：2008年5月12日汶川大地震的发震时刻为14：27：

57.59±0.03s,震中位置为31.018°N±0.3km,103.365°E±0.3km,震源深度15.5km±0.3km. 精确定位得到的汶川地震的震中位于八角地震台（BAJ）北西方向约5.6km处,距离紫坪铺水库最西端约7.5km,比早先其他结果要远离紫坪铺水库的西端（由约4.0km变至约7.5km）（图1b）.

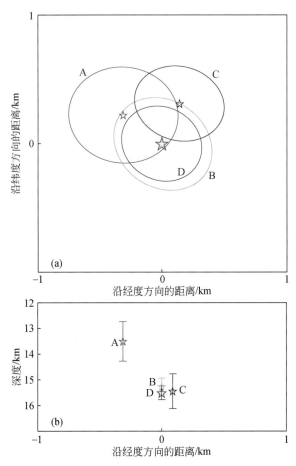

图4 所计算的4种情形的误差椭圆（a）与沿垂直方向的误差（b）分布

（a）是在以表3的情形D得到的震中为中心的2km×2km范围内表示所计算得到的4种情形的震中位置及其误差椭圆. 图中, 绿色、黄色、紫色与红色符号依次表示表3所列的情形A, B, C, D等4种情形的相应结果, 五角星表示震中（a）或震源（b）, 情形D得到的震中（a）或震源（b）用较大的红色五角星表示

Fig. 4 Error ellipses (a) and error bars in vertical direction (b) for the four cases as described in the text

(a) is an area of 2km×2km with epicentral location from case D as coordinate origin. Green, yellow, purple and red symbols stand for the results of cases A, B, C and D, respectively. Star stands for epicenter (a) or hypocenter (b), and large red star, for the results of case D

3 讨论与结论

以上对于影响和制约测定像汶川地震这样一种破裂长达300km以上的大地震的震源位置与发震时刻的因素做了详细的分析,特别指出：除了在通常地震定位中已熟知的台距、台站方位分布和地壳模型不完善性等因素的影响之外,单用远台直达P波的观测资料的局限性以及读取的震相数据的不一致性也是影响大地震精准定位的重要因素. 在此分析研究的基础上,作者从四川省地震台网的观测资料与紫坪铺水库地震台网的记录资料中,选用台站方位分布较均匀、震中距较小（$\Delta \leqslant 160$km）的地方性地震台以及震中距近至5km的近台观测资料,并对震相进行跟踪对比,确保所提取的震相数据是与主震初始破裂点的辐射相关联的. 这样的处理方法最大限度地避免了远台观测数据的局限性、地壳模型的不完善性以及震相数

据读取的不一致性对汶川大地震定位带来的影响,得到了比已有的定位精度高一个数量级的2008年5月12日汶川M_W7.9地震的震源参数的测定结果:震中位置31.018°N±0.3km,103.365°E±0.3km,震源深度15.5km±0.3km,发震时刻(北京时间):14:27:57.59±0.03s.

致谢 博士研究生朱新运、杨志高参加了部分图的绘制工作,谨表谢忱.

参 考 文 献

陈九辉,刘启元,李顺成,郭凤灸,李昱,王峻,齐少华. 2009. 汶川M_S8.0地震余震序列重新定位及其地震构造研究[J]. 地球物理学报,**52**(2):390-397.

陈培善. 2008. 全球大震和中国及邻区中强震地震活动(2008年4—5月)[J]. 地震学报,**30**(5):545-549.

陈运泰. 2008. 汶川地震的震源破裂过程[C]//海峡两岸防震减灾学术研讨会,2008年6月27日—28日. 北京:中国科学院:10-11.

黄媛,吴建平,张天中,张东宁. 2008. 汶川8.0级大地震及其余震序列重定位研究[J]. 中国科学:D辑,**38**(10):1242-1249.

刘启元,陈九辉,李顺成,李昱,郭凤灸,王峻,齐少华. 2008. 汶川M_S8.0地震:川西流动地震台阵观测数据的初步分析[J]. 地震地质,**30**(3):584-596.

杨智娴. 2003. 地震定位[G]//中国地震局监测预报司(编). 地震参数:数字地震学在地震预测中的应用. 北京:地震出版社:120-140.

杨智娴,陈运泰,苏金蓉,陈天长. 2008. 汶川大地震主震震源参数的初定情况报告[R]. 北京:北京大学-中国地震局现代地震科学技术研究中心.

赵珠,范军,郑斯华,长谷川昭,崛内茂木. 1997. 龙门山断裂带地壳速度结构和震源位置的精确修定[J]. 地震学报,**19**(6):615-622.

赵珠,张润生. 1987a. 四川地区地壳上地幔速度结构的初步研究[J]. 地震学报,**9**(2):154-166.

赵珠,张润生. 1987b. 四川地区分区走时表的编制[J]. 四川地震,(2):29-35.

Bormann P (ed). 2006. *New Manual of Seismological Observatory Practice* [M]. Vol. **1** and **2**:1-1003.

Geiger L. 1910. Herdbestimming bei erdbeben aus ankunftszeiten[J]. K Ges Wiss Got,**4**:331-349.

Herrmann R B, Ammon C J. 2004. *Computer Programs in Seismology*:GSAC(Generic Seismic Application Coding)[CP]. Version 3.30.

Mooney W D, Prodehl C, Pavlenkova N I. 2002. Seismic velocity structure of the continental lithosphere from controlled source data [M]//Lee W H K, Kanamori H, Jennings P C, Kisslinger C eds. *International Handbook of Earthquake and Engineering Seismology*. Part **A**:887-910.

The hypocenter and origin time of the M_W 7.9 Wenchuan earthquake of May 12, 2008

Yang Zhixian[1,2], Chen Yun-Tai[2,1], Su Jinrong[3], Chen Tianchang[3] and Wu Peng[3]

1. Institute of Geophysics, China Earthquake Administration, Beijing 100081, China;
2. PKU-CEA Joint Research Center for Modern Seismology, Peking University, Beijing 100871, China;
3. Earthquake Administration of Sichuan Province, Chengdu 610041, China

Abstract The hypocenter and origin time of the M_W7.9 Wenchuan earthquake of May 12, 2008, were relocated precisely using the observation data from both of the Sichuan Province Seismic Network and the Zhipinpu Reservoir Seismic Network. It is pointed out that the factors affected significantly to the precise and accurate determination of the hypocenter and origin time of the great earthquakes such as the Wenchuan earthquake are the limitation of far stations, incompleteness of the crust model used for earthquake location, and the inconsistency existed in the direct P phases detected from the near-by seismic stations and the far stations being generated from different rupture areas and at different times. To diminish the limitation caused by these factors, data from the local seismic stations were analyzed, and from these stations, 15 near-by stations of epicentral distance of 5—160km, with well azimuthal distribution, and with accurately correlated direct P phases were selected for inversion. The preferred relocated source parameters, obtained from inversion, have coordinates latitude 31.018°N±0.3km, longitude 103.365°E±0.3km, focal depth 15.5km±0.3km, origin time (Beijing Time) May 12, 2008, 14:27:57.59±0.03s.

芦山 4.20 地震破裂过程及其致灾特征初步分析*

张 勇　许力生　陈运泰

中国地震局地球物理研究所，北京 100081

摘要　采用全球地震台网的远震地震波数据快速反演了 2013 年 4 月 20 日雅安芦山地震（芦山 4.20 地震）的破裂过程，在震后 3 小时得到并发布了相关测定结果。结果显示这次地震的震级约 M_W6.8，包括两次破裂子事件，但都发生在断层面上震源（破裂起始点）附近，整个地震没有表现出明显的破裂方向性。雅安芦山地震的破裂没有大规模出露地表，主要地震灾害集中在断层上盘，位于芦山县和宝兴县一带。

1　引言

据中国地震台网中心测定，2013 年 4 月 20 日上午 08 点 02 分（北京时间），四川雅安市芦山县境内发生了 M_S7.0 地震。这次地震是继 2008 年 5 月 12 日汶川地震之后在龙门山断裂带上发生的又一次灾害性地震事件。地震之后，作者立即开展了破裂过程的快速反演研究工作，在震后 3 小时左右成功确定了结果，并通过电子邮件方式和中国地震局相关渠道进行了发布和上报。在这一工作中，我们力求通过确定震源破裂特征来分析地震灾害分布，并希望可以为地震救援和灾害评估等工作的开展提供力所能及的支持。

2　数据和方法

我们采用由 IRIS 数据中心提供的全球地震台网（GSN）远震宽频带数据，筛选了全球范围内震中距介于 30°～90°的方位角分布均匀的 39 个台站（图 1a）垂直向 P 波波形。基于 AK135 大陆速度结构模型[1]，采用 Wang[2] 的方法计算得到格林函数。在反演中，我们对数据和格林函数同时进行了 0.01～0.4Hz 的带通滤波。

此次快速测定工作作为汶川地震之后国内外大地震破裂过程快速反演系列工作的延续[3]，作者沿用了本次地震之前发展的破裂过程线性反演方法进行反演[4-6]。该方法在近年来国内外的一系列大地震研究，如 2008 年汶川地震和 2010 年玉树地震的快速反演和后续研究中[6-7]，都得到了成功的应用。由于线性反演在计算效率方面的优势，这一方法可以有效保证我们在震后以最快的速度确定和发布破裂过程结果，为地震应急救援提供相关信息。

3　破裂过程快速反演

快速反演采用了 USGS 最早发布的 WPhase 矩张量反演结果，在走向 219°、倾角 33°的断层平面上，选取了一个足够大的长 63km、宽 48km 的平面作为潜在的破裂断层面，将其划分为 21×16＝336 个子断层，每个子断层的尺度为 3km×3km。采用 USGS 发布的震中位置（30.284°N，102.956°E）和震源深度（12.3km），将其置于走向方向第 11 个、倾向方向第 8 个子断层处。

---

* 本文发表于《地球物理学报》，2013 年，第 **56** 卷，第 4 期，1408-1411.

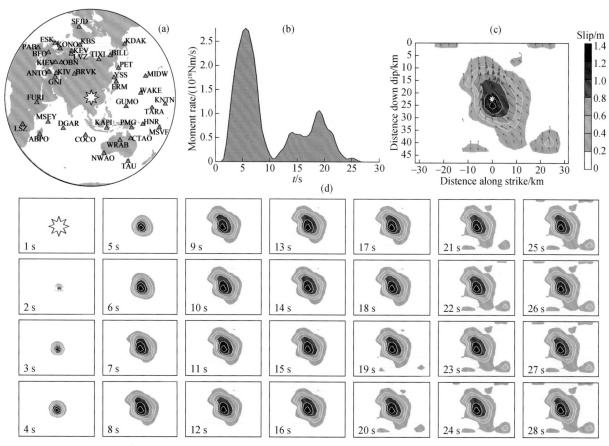

图 1 雅安芦山地震破裂过程快速反演结果

(a) 震中-台站位置分布；(b) 震源时间函数；(c) 断层面上静态滑动量分布；(d) 断层面上滑动量分布随时间的变化

Fig. 1 Fast inversion results of rupture process of the Lushan 4.20 earthquake

(a) The epicenter and used stations; (b) Source time function; (c) Static fault slip distribution; (d) Temporal variation of fault slip distribution

快速反演得到的结果见图1，我们得到的这次地震的地震矩约 $1.69 \times 10^{19} \mathrm{N \cdot m}$，对应的矩震级约为 $M_\mathrm{W}6.8$。震源时间函数结果（图1b）表明此次地震包括两次主要的子事件，其中第一次子事件发生在 0~10.5s，是这次地震最大也最主要的一次子事件，矩震级规模 $M_\mathrm{W}6.6$，释放的地震矩 $1.12 \times 10^{19} \mathrm{N \cdot m}$，约占总地震距的 2/3；第二次子事件发生在 10.5~27s，其规模约为第一次子事件的一半。整体而言，两次子事件都没有表现出明显的破裂方向性，因此断层面上的静态滑动量分布比较简单（图1c），只存在一个长宽皆约为25km的滑动集中区域，且主要分布在震源附近，其最大滑动量约1.3m。从断层面上滑动分布随时间的变化过程看（图1d），主要滑动在第一次子事件期间（0~10.5s）就已经形成；相比之下，第二次子事件所对应的滑动分布比较零散。

反演得到的合成地震波形与观测波形的比较见图2，平均相关系数约0.92，反演结果很好地解释了绝大多数台站的地震波形。

4 致灾特征初步分析

图3显示了断层面上静态滑动量分布在地表的投影。如果我们采用的震源位置基本准确，那么滑动分布主要集中在震源附近，基本没有大规模的地表出露。因此，断层下盘临近断层地表出露线的区域，如雅安、名山和邛崃等地区，估计地震灾害不会特别严重；而在断层上盘位于主要滑动区域上方的芦山、宝兴

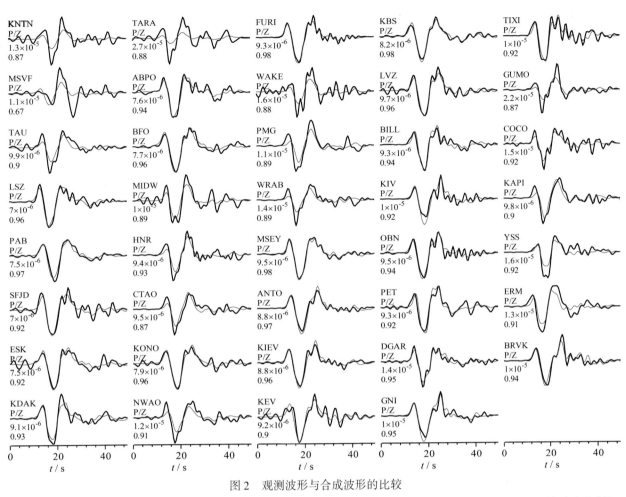

图 2 观测波形与合成波形的比较

黑线表示观测波形,红线表示合成波形. 子图左方的字符和数字从上到下依次为:台站名、震相名与分量名、观测地震与合成波形的最大振幅(单位为 m/s)、以及观测波形与合成波形之间的相关系数

Fig. 2 Comparison of observed and synthetic waves

Black lines are the observed waves and red lines are the synthetic waves. In each sub-graph, on the left from top to bottom are stations code, phase and component name, maximum amplitude of observed and synthetic waves (in m/s), and the correlation coefficient of observed and synthetic waves, respectively

等地,由于破裂的上盘效应,人员伤亡和财产损失都可能会比较严重. 这一灾害分布特征与 2008 年汶川地震有明显区别. 汶川地震中,由于主要破裂大规模出露地表[6],地震灾害最严重的区域都位于在断层线附近,如北川、映秀、都江堰等地;而在断层上盘处的汶川县城和茂县县城,破坏程度相对较弱.

因此,破裂是否出露地表是影响和判定地震灾害分布特征非常重要的一个因素,也是我们进行破裂过程快速反演所重点关注的一大震源特征. 所幸的是,此次雅安芦山地震没有发现破裂大规模出露地表,否则造成的人员伤亡和财产损失都将比实际情况严重得多.

5 讨论和结论

作者快速反演得到的矩震级约为 $M_W6.8$,略大于 USGS 通过地震矩张量反演所得到的 $M_W6.6$,这可能是因为矩张量反演主要反映的是我们得到的第一次子事件的信息. 如果我们只考虑第一次子事件,那么得到的矩震级约为 $M_W6.6$,与 USGS 的结果一致. 第二次子事件的规模较小,所释放的地震矩约为第一次子

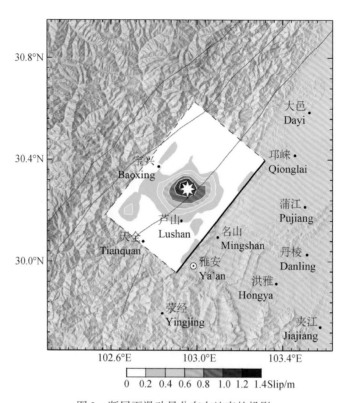

图 3 断层面滑动量分布在地表的投影

红线为震中区域的活断层[8]，白色八角形星表示震中位置

Fig. 3 Surface projection of fault slip distribution

Red lines are the active faults[8] in the epicentral area. White aniseed star denotes the epicenter

事件的一半．由于其对应的破裂空间分布也比较零散，其真实性尚需进一步研究确证．

本文介绍了作者在 2013 年 4 月 20 日雅安芦山地震之后 3 小时得到和发布的破裂过程快速反演结果．这一结果没有发现地震破裂大规模出露地表的迹象，因此推测主要灾害区域集中在发震断层上盘的芦山、宝兴一带，而在断层下盘处的雅安、名山和邛崃等地，灾害损失将相对较轻．

参考文献（References）

[1] Kennett B L N, Engdahl E R, Buland R. Constraints on seismic velocities in the earth from traveltimes. *Gepphys. J. Int.*, **1995**, 122: 108-124.

[2] Wang R J. A simple orthonormalization method for stable and efficient computation of Green's functions. *Bull. Seism. Soc. Amer.*, 1999, **89**(3): 733-7410.

[3] Zhang Y, Chen Y T, Xu L S. Fast and robust inversion of earthquake source rupture process and its application to earthquake emergency response. *Earthq. Sci.*, 2012, **25**(2): 121-128.

[4] Chen Y T, Xu L S. A time-domain inversion technique for the tempo-spatial distribution of slip on a finite fault plane with applications to recent large earthquakes in the Tibetan Plateau. *Geophys. J. Int.*, 2000, **143**: 407-416.

[5] Xu L S, Chen Y T, Teng T L, Patau G. Temporal and spatial rupture process of the 1999 Chi-Chi earthquake from IRIS and GEOSCOPE long period waveform data using aftershocks as empirical Green's functions. *Bull. Seism. Soc. Am.*, 2002, **92**: 3210-3228.

[6] 张勇，冯万鹏，许力生等．2008 年汶川大地震的时空破裂过程．中国科学 D 辑：地球科学，2008，**38**(10)：1186-1194．

Zhang Y, Feng W P, Xu L S, et al. Spatio-temporal rupture process of the 2008 great Wenchuan earthquake. *Sci. China Ser.*

D. ,2009, **52**(2): 145-154.

[7] 张勇, 许力生, 陈运泰. 2010 年青海玉树地震震源过程. 中国科学 D 辑: 地球科学, 2010, **40**(7): 819-821.
Zhang Y, Xu L S, Chen Y T. 2010. Source process of the 2010 Yushu, Qinghai, earthquake. *Sci. China* Ser. D., **53**(9): 1249-1251.

[8] 邓起东, 张培震, 冉勇康, 等. 中国活动构造基本特征. 中国科学 D 辑: 地球科学, 2002, 32 (12): 1020-1030.
Deng Q D, Zhang P Z, Ran Y K, et al. Basic characteristics of active tectonics of China. *Sci. China Ser. D.*, 2003, **46**(4): 356-372.

Rupture process of the Lushan 4.20 earthquake and preliminary analysis on the disaster-causing mechanism

Zhang Yong, Xu Li-Sheng and Chen Yun-Tai

Institute of Geophysics, China Earthquake Administration, Beijing 100081, China

Abstract Fast inversion for the rupturing process of the 20 April 2013 Lushan, Ya'an earthquake was conducted about 3 hours after the earthquake occurrence by using the GSN teleseismic waveform data. The obtained moment magnitude is $M_W 6.8$. It included two sub-events during the rupturing process, which occurred near the hypocenter or rupture initiation point and did not show significant rupture direction. The rupture of the earthquake did not break through the ground surface, and seismic disaster is concentrated on the hanging wall of the fault, around Lushan county and Baoxing county of Sichuan Province.

从汶川地震到芦山地震

陈运泰 杨智娴 张 勇 刘 超

中国地震局地球物理研究所，北京 100081

摘要 本文概述作者在龙门山断裂带中、小地震精确定位、地震活动性以及2008年汶川$M_W7.9$($M_S8.0$)地震和2013年芦山$M_W6.7$($M_S7.0$)地震破裂过程等方面所做的研究工作. 这些工作表明，青藏高原东缘的龙门山断裂带不但是一条规模宏大的断裂带，也是一条非常活跃的地震带. 通过对地震构造、地震活动性、地震矩释放"亏空"区以及余震活动规律的分析，作者在汶川地震后提出了龙门山断裂带西南段宝兴–小金一带存在发生$M_W6.7\sim7.3$地震的潜在危险性的地震趋势估计. 芦山地震的发生初步验证了这一估计. 芦山地震发生后作者进一步做的分析结果表明，芦山地震的发生并没有显著地缓解龙门山断裂带西南段的地震危险性，该地段整体上仍存在发生$M_W7.2\sim7.3$地震的潜在危险性；特别是，其北段（即邛崃大邑西–宝兴北–汶川南一带）存在发生$M_W6.8$地震的潜在危险性；其南段（即天全–荥经–泸定–康定一带）存在发生$M_W7.2$地震的潜在危险性. 作者认为，应当强化对上述具有潜在地震危险性区域的监测与多学科综合研究.

据中国地震台网中心测定，2013年4月20日北京时间上午08时02分46秒，四川省雅安市芦山县境内发生了面波震级$M_S7.0$地震，震中位置：30.3°N，103.0°E，震源深度：13.0km. 另据美国地质调查局国家地震信息中心（USGS/NEIC）报道，发震时刻：00时02分47.5秒UTC（协调世界时）；震中位置：30.308°N，102.888°E；震源深度：14.0km；矩震级$M_W6.6$. 截止至05月17日16时00分，共记录到余震9294次，其中$M≥3.0$余震132次，包括$3.0≤M<3.9$地震106次，$4.0≤M<4.9$地震22次，$5.0≤M<5.9$地震4次.

芦山地震是继2008年5月12日汶川$M_W7.9$地震以来在青藏高原东缘的龙门山断裂带上发生的又一次灾害性地震事件. 截止至24日14时30分，芦山强烈地震已造成196人死亡，失踪21人，11470人受伤.

芦山地震发生于龙门山断裂带上的西南段，该段是汶川地震时龙门山断裂带没有发生破裂的特殊地段. 为什么会发生芦山地震？芦山地震是如何发生的？它的发生对于龙门山断裂带及其周边地区的地震活动性究竟有何影响？等等，都是亟待研究解决的问题. 为此，本文通过概述作者自2003年以来开展的与龙门山断裂带地震精确定位、地震活动性以及汶川地震和芦山地震的破裂过程有关的研究工作，对芦山地震的发生及其可能的影响做一初步分析与探讨.

1 芦山地震的地震构造背景与龙门山断裂带的地震活动性

芦山地震发生于青藏高原东缘的龙门山断裂带西南段. 龙门山断裂带是一条长约500km、宽约30~50km沿NE-SW方向展布的巨大断裂带，其断层滑动以逆冲为主，兼具右旋走滑分量. 按照由西向东的顺序，龙门山断裂带主要包含龙门山后山断裂（茂县–汶川断裂）、中央断裂（映秀–北川断裂）和山前断

---

* 本文发表于《中国科学：D辑地球科学》，2013年，第**43**卷，第6期，1064–1072.

裂（安县–灌县断裂，亦称彭县–灌县断裂、江油–灌县断裂）（参见文献［1］中的图1）．这些断裂都以逆冲滑动为主，兼具一定的右旋走滑分量；在龙门山断裂带的东北段，右旋走滑分量更大[1-3]．

杨智娴等[4-6]在汶川地震之前，曾经运用国际上刚开始兴起的最先进的"双差定位法"[7]对包括松潘–平武、龙门山断裂带等在内的我国中西部地区1992～1999年发生的10057个中、小地震重新定位．在10057个地震中，经过重新精确定位的6496个地震显示了该地区的清晰的中、小地震活动性图像，揭示出以往的定位结果未能揭示出的该地区的中、小地震活动性图像与断裂带的关系（见文献［6］中的图4），表明在龙门山断裂带上，虽然未有发生过7级以上大地震的历史记载，但近期中、小地震（震级$M<7$的地震）活动非常活跃，分布在一条长约470km、宽约50km的地带上，使得龙门山断裂带成为非常具有地震危险性的一条断裂带（见文献［6］中的图4和图10）．

汶川地震的发生是沉寂多年的龙门山断裂带的一次集中的能量释放．汶川地震发生后，Yang等[8,9]运用同一方法对汶川地震的余震重新精确定位（图1（b））．余震重新精确定位结果表明，汶川地震的余震（图1（b）主图中的绿色实心圆）主要分布在NE-SW向的龙门山断裂带从映秀到青川约350km的地段，清楚地表明汶川地震便是发生在这条长达470km的龙门山断裂带上约350km的地段上的大规模的破裂．龙门山断裂带西南段即从汶川地震的余震区南端至康定大约120km的地段（以下简称"西南段"），是汶川地震的余震在主震西南方向未延伸到"历史上"曾经发生过地震的范围，是在汶川地震破裂中未破裂的、值得警惕的特殊地段．

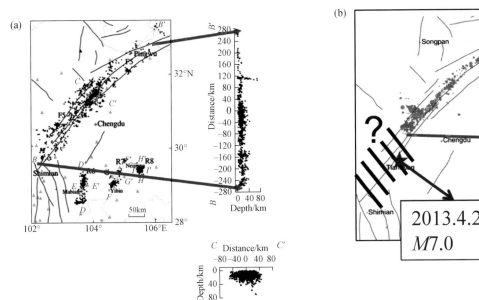

图1　龙门山断裂带在汶川地震前（1992～1999）的地震活动性[6]（a）和汶川地震的余震分布[8]（b）

图1（a）摘引自文献［6］的图10，表示龙门山断裂带及其附近地区在1992～1999年发生的地震（黑色实心圆）的分布，右侧是沿主图所示的BB'剖面的地震分布图，右下角是沿主图所示的CC'剖面的地震分布图．在图1（a）中，文献［6］原图10中的其他地震带如冕宁地震带、宜宾地震带、内江地震带的剖面图均从略．汶川地震是发生在长达470km的龙门山断裂带（图1（a））上长约350km的地段（图1（b））上的大规模破裂．图1（b）是汶川地震发生后用同一方法对汶川地震的余震重新精确定位的结果，引自文献［8］中的原图．在图1（b）主图中，绿色实心圆表示汶川地震余震震中，左侧是沿SW-NE剖面即图1（a）的BB'剖面的余震（黑色实心圆）分布图，右下角是沿NW-SE剖面即图1（a）的CC'剖面的余震（黑色实心圆）分布图．在图1（b）主图中，在汶川地震破裂中未破裂的龙门山断裂带西南段（即从汶川地震的余震区南端至康定大约120km的地段），以蓝色斜线及问号予以突出

（图中表示芦山地震震中的红色星号及红色框中的文字系芦山地震后新加）

2 汶川地震与芦山地震的震源特性

2.1 汶川地震

从 2003 年开始,作者致力于研究发展地震破裂过程快速反演的实用方法,目标是在重大地震发生后能以尽可能快的速度、准实时地反演地震观测资料得出表征地震震源特性的参数,包括:表示地震机制的矩张量,表示地震大小的地震矩或与其相当的矩震级,地震破裂面的大小、地震断层滑动量即错距的方向与大小的分布、地震破裂的动态进程及其致灾效应,等等,为震后快速反应、应急救援决策提供参考.该方法完成于 2008 年 5 月汶川地震前数日,在汶川地震后首次得到成功应用(张勇[10]和刘超等[11]).汶川地震发生后 4 个多小时,张勇等[10]和刘超等[11]根据全球地震台网(GSN)的数字地震记录用各自独立的方法得到了相当一致的结果(据陈运泰等[12]),这就是:汶川地震的发震断层走向为 225°、倾角为 39°、滑动角为 120°(据张勇等[13]),或:走向为 220°、倾角为 32°、滑动角为 118°(据刘超等[11]).结果表明,汶川地震是一次以逆冲为主、兼具小量右旋走滑分量的规模宏大的地震事件,地震破裂以既有朝 NE 方向破裂、也有朝 SW 方向破裂的方式进行,但是是以朝 NE 方向破裂为主的不对称双侧破裂,地震破裂面即断层面长达 300 多千米,以 32°~39°的倾角向西北倾斜,从地面斜向地下延伸,宽度将近 50km,相当于深度达约 30km;汶川地震是在破裂长度超过 300km 的发震断层上发生的、破裂持续时间长达 90s 的一次复杂的震源破裂过程,整个断层面上的平均滑动量即平均错距约 2.4m,但断层面上滑动量的分布很不均匀,有 4 个滑动量集中且破裂贯穿到地表的区域,其中大的两个:一个在汶川-映秀一带下方,最大滑动量(也是本次地震的最大滑动量)所在处在震源(破裂起始点)附近,达 7.3m;另一个位于北川一带下方,一直延伸到平武境内下方,其最大滑动量所在处在北川地面上,达 5.6m(据张勇[10]、张勇等[13]、刘超等[11]和陈运泰等[12]).

由反演得到的结果进一步计算得到的地表永久位移场的分布特征与沿断层的地震灾害分布特征非常接近,地表永久位移值较大的两个地区正好对应于地震后数月发表的烈度分布图显示的两个极震区,表明了汶川地震的大面积、大幅度、贯穿到地表的、以逆冲为主的断层错动是致使近断层地带严重地震灾害在震源方面的主要原因,显示了极震区与贯穿到地面的逆冲断层错动的密切联系(据陈运泰等[12]),也表明了由地震破裂过程快速反演得到的有关地震震源的许多信息不仅具有科学意义,而且在为震后快速反应、应急救援服务方面也具有明显的应用价值.

通过对最终(静态)滑动量分布的分析(图2),作者在汶川地震震后数日公布于互联网上的"2008 年 5 月 12 日汶川特大地震震源特性分析报告"[12]及学术研讨会上[14~16]提出:"在汶川地震时发生滑动的区域之间存在着地震矩释放的亏空区,迄今已经发生的较大余震无论是数量还是强度也都未能基本上填充这些亏空区.通常认为这种矩释放的亏空区是阻碍破裂的'障碍体','障碍体'的破裂即为余震.因此,这些矩释放的亏空区很可能是未来强余震发生的地点."并指出"在汶川地震的破裂带上有两个比较大的破裂亏空区,即迄今基本上没有发生过破裂的地方.这两个破裂亏空区分别位于汶川东北约 100km(北川-平武间)和西南约 100km(宝兴-小金间),是很有可能发生强余震的地区."根据这两个破裂亏空区面积的大小,作者估计了这两个亏空区如果一旦破裂可能的震级的大小分别为 $M_W7.1$ 与 $M_W7.3$.位于汶川东北约 100km 的亏空区在北川-平武间,距离 1933 年 8 月 25 日叠溪 $M7.5$ 地震与 1976 年 8 月 16 日和 8 月 23 日松潘-平武两次震级均为 $M_S7.2$ 的地震震中只有 80~100km.对于这些地震的发生与汶川地震及该亏空区的关系,以及该亏空区的地震危险性作者未做进一步的分析.但是,毫无疑问,这是一个亟待研究的问题.

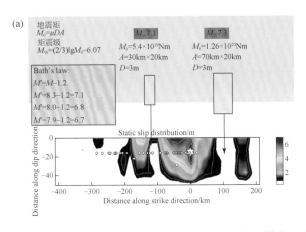 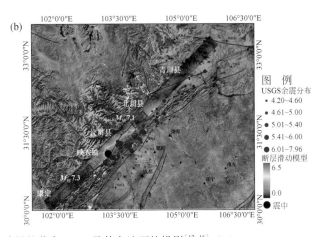

图 2 汶川地震时断层面上最终（静态）滑动量的分布（a）及其在地面的投影[12-16]（b）

滑动量集中且破裂贯穿到地表的两个区域（色标为红色与黄色的区域）一个位于汶川–映秀一带下方，另一个位于北川一带下方，一直延伸到平武境内下方. 方框表示在汶川地震时发生滑动的区域之间存在着地震矩释放的"亏空"区. 图 2（a）与 2（b）两张图片引自 2008 年陈运泰[14-16]所做学术报告的 PPt 原始图片

对于位于汶川西南约 100km 在宝兴–小金间（小金县城在宝兴县城西面约 60km 处，但其辖区最东处与宝兴县城仅相距 25km，固当时有此表述）的破裂亏空区，按照其规模，如果一旦发生一次性破裂，应当相当于一个 $M_W 7.3$ 地震（图 2（a），（b））.

汶川地震发生后余震活动最强烈的一周间，一直没有发生特别大的余震，都是 $M_S 5.0 \sim 6.0$ 左右（现在已经可以确认，直至芦山地震发生前，最大的一次余震是 2008 年 5 月 25 日发生的面波震级 $M_S 6.4$ 地震）. 汶川地震的面波震级是 $M_S 8.0$，矩震级是 $M_W 7.9$. 按照巴特定律[17]，最大余震的震级比主震的震级平均小约 1.2 级. 如果汶川地震是主震型地震的话，其最大的强余震应该在 $M_W 6.7$ 左右. 但是注意到巴特定律离散相当大；同时考虑到当时国际上不同的权威机构对主震震级的测定结果也有些分散，在 $M_W 7.9 \sim 8.3$ 间，因此作者当时提出："如果汶川地震是主震型地震，发生强余震的震级将很高，约 $M_W 6.7 \sim 7.1$（图 2（a）左边的浅红色方框表示按主震震级 $M_W = 7.9$，8.0，8.3 估算，最大余震的震级相应地应为 $M_W = 6.7$，6.8，7.1）；如果是双震型或强震群型地震，未来发生的地震可能比 $M_W 6.7$ 还要大. 这两种情况都警示一定要特别加强对强余震的监测，并高度警惕'晚期强余震'".

综合以上分析，作者把汶川地震发生后龙门山断裂带西南段发生汶川地震最大余震的震级估计为 $M_W 6.7 \sim 7.3$. 2013 年芦山 $M_W 6.7$（$M_S 7.0$）地震的发生验证了上述结论.

上述结果是汶川地震发生后作者对龙门山断裂带西南段的地震危险性的综合分析与趋势估计，并非对该地段的"地震预测预报". 如作者在过去发表的一些论文已述[18,19]，按照现今国际地震科学界普遍采用的、严格意义上的地震预测指的是"同时给出未来地震的位置、大小、时间和概率四种参数，每种参数的误差（不确定的范围）小于、等于可以接受的合理数值"（据 Wyss[20] 和 Jordan 等[21]）. 虽然这个分析结果不是对于龙门山断裂带的西南段的严格意义上的"地震预测预报"，但对于汶川地震发生后该地段存在发生 $M_W 6.7 \sim 7.3$ 地震的危险性的分析判断无疑可为强化该地段的地震监测与研究提供有益的参考.

2.2 芦山地震

五年来，地震破裂过程快速反演的方法以及地震矩张量快速反演方法已经列入作者对国际国内重要地震应急反应的工作范围，在诸如国内云南宁洱地震、青海玉树地震、国际意大利拉奎拉（L'Aquila）地震、海地（Haiti）地震等有重要影响的地震发生后第一时间内提供地震的机制、规模（矩震级）、破裂过程、可能的重灾区等信息，为救灾减灾服务（据刘超等[11,22-24]、张勇等[13,25,26]、Liu 等[27] 和 Zhang

等[28]）. 从2008年10月5日至2013年5月15日已经提供了国内外132个有重要影响的地震的快速矩张量解，其中震级大于等于5.5的国内地震58次，震级大于等于7.0的国外地震74次. 目前，地震矩张量快速反演从地震发生到发布的平均时间约为2.5h；地震破裂过程快速反演从地震发生到发布的时间为3～5h，平均约为4h（据Zhang等[28]）；这些时间均包括记录资料下载时间在内，并且主要是用于资料下载.

2.2.1 芦山地震的震源机制

芦山地震发生（2013年4月20日上午8:02北京时间）后，刘超等[29]利用全球地震台网的数字地震观测资料，迅速开展了地震矩张量快速反演工作，在约一个小时后（9:04北京时间）发布和上报了芦山地震矩张量反演结果（图3）. 反演结果表明（表1第2行），芦山地震释放的标量地震矩 $M_0 = 1.6 \times 10^{19}$ Nm，相应的矩震级为 $M_W = 6.7$，最佳双力偶解的两个节面的参数分别为：节面Ⅰ：走向34°/倾角55°/滑动角87°；节面Ⅱ：走向220°/倾角35°/滑动角95°.

反演得到的芦山地震的断层面参数和主应力轴参数（图3，表1第2行）与汶川地震的断层面参数和主应力轴参数（表1第1行）基本上一致，它们的节面Ⅱ与龙门山断裂带的走向（NE-SW）与倾向（倾向西北）一致，并且与NE-SW走向的余震震中分布一致（图5主图）. 据此可以确定节面Ⅱ（走向220°/倾角35°/滑动角95°）为芦山地震的断层面，这表明，芦山地震是一个与汶川地震震源机制一致的以逆冲为主、但兼具比汶川地震还要小的右旋走滑分量的地震.

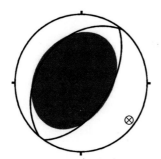 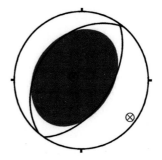

图3 由矩张量快速反演得到的芦山地震矩张量解

本图展示2013年4月20日芦山地震后约1h在互联网上发布的由快速反演得到芦山地震的矩张量解的原始图件（见文献[12]）. 图中按国际通用的规范格式分别以中文（右图）与英文（左图）给出地震的矩张量解. 芦山地震的矩张量解表明其震源机制为：走向220°/倾角35°/滑动角95°，与汶川地震的震源机制（走向225°/倾角39°/滑动角120°）相当一致，是一个以逆冲为主、但兼具比汶川地震还要小的右旋走滑分量的地震

2.2.2 芦山地震的破裂过程及其特征

2013年4月20日芦山地震发生后约3h（北京时间11时01分08秒），张勇等[30]按照大震应急的要求，利用全球地震台网的数字地震记录反演了该地震震源破裂过程，并对该地震震源的特性做了初步的分析（图4）. 根据这一初步结果，这次地震的标量地震矩 $M_0 = 1.7 \times 10^{19}$ Nm，相应的矩震级约为 $M_W 6.8$，与刘超等[29]由矩张量反演得到的结果（$M_W 6.7$）相近，与美国地质调查局国家地震信息中心（USGS/NEIC）的结果（$M_W 6.6$）亦相近；断层面上大滑动量约为1.3m. 进一步分析表明，芦山地震的震源及其破裂过程具有如下与汶川地震不同的特征：

（i）芦山地震滑动量较大的区域没有达到地面但较靠近地面. 这意味着总体上地震破裂面没有扩展到地表（图4（b），（d）），因此在极震区可能不出现破裂带；或者即使有，规模不会像汶川地震那么大. 预期可能替代破裂带出现的是在极震区有可能出现与较强烈变形相联系的"形变带"，例如观察到喷沙冒水鼓包等宏观现象.

（ii）芦山地震具有长度方向与宽度方向等尺度的中、小地震的震源特征. 由于地震震级为 $M_W 6.7 \sim 6.8$，没有达到但接近大震的下限 $M_W 7.0$，属于中震里较大者（按照国际上通行的称谓，$5 \leq M_W < 7$ 的地震称为中震，$7 \leq M_W < 8$ 的地震称为大震，$M_W \geq 8$ 的地震称为特大地震，等等）. 不但总体上芦山地震的破裂没有扩展到地表面，而且地震破裂面呈现长度方向与宽度方向等尺度（即大约相当）的中、小地震的震源特征. 具体地说，总体上地震破裂面上滑动量集中的面积其长与宽均接近于30km. 芦山地震的破裂面

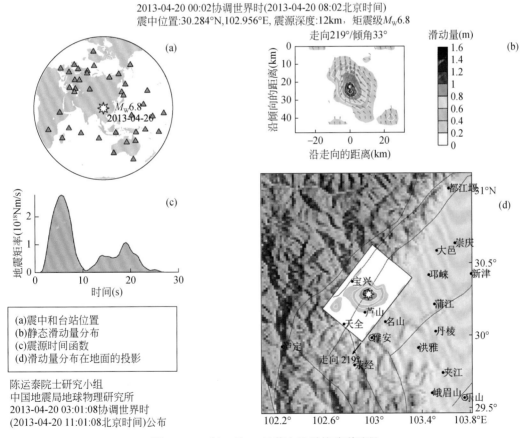

图4 2013年4月20日芦山地震的破裂过程

图片展示在震后约3h于互联网上发布的由快速反演得到芦山地震的破裂过程. (a) 芦山地震震中（白色星号）与台站（绿色三角形）分布图；(b) 芦山地震断层面上静态（最终）滑动量分布图；(c) 芦山地震的地震矩释放率-时间关系图（震源时间函数图）；(d) 芦山地震断层面上滑动量分布在地面的投影图

向下延伸到深度约 20km 的地方. 这个深度也就是 Yang 等[6] 通过对我国中西部地震震源的重新精确定位得到的我国中西部脆裂层 (schizosphere) 即孕震层 (seismogenic layer) 的下界面的深度, 说明芦山地震的破裂向下扩展到了脆裂层的下界面, 是一个处于地震的自相似性分段处的地震. 与美国圣安德烈斯断层不同, 那里脆裂层厚度为约 15km, 自相似性在 $M_W 6.0 \sim 6.5$ 分段; 与脆裂层较厚的俯冲带不同, 那里自相似性可在 $M_W \approx 7.5$ 分段 (据 Shimazaki[31] 和 Scholz[32]).

(iii) 芦山地震的破裂过程没有呈现明显的方向性. 破裂起始点接近于地震滑动量集中区的中心 (图 4 (b), (d)), 破裂过程没有呈现明显的方向性. 这一点实际上亦是中、小地震的震源的一个特征. 在 27s 的破裂持续时间中, 前 10.5s 完成了大部分的矩释放 (图 4 (c)), 在这段时间里, 破裂扩展速度约为 3.0km/s.

表 1 由矩张量反演得到的汶川地震与芦山地震的地震矩 M_0, 矩震级 M_W, 断层面参数和主应力轴参数

地震	M_0/Nm	M_W	节面 I			节面 II			T 轴		B 轴		P 轴	
			走向/(°)	倾角/(°)	滑动角/(°)	走向/(°)	倾角/(°)	滑动角/(°)	方位/(°)	倾角/(°)	方位/(°)	倾角/(°)	方位/(°)	倾角/(°)
汶川	2.0×10^{21}	8.1	8	63	74	220	32	118	246	68	16	14	110	16
芦山	1.6×10^{19}	6.7	34	55	87	220	35	95	292	79	36	3	126	10

3 芦山地震与龙门山地震带西南段未来地震危险性

芦山地震发生于青藏高原东缘的龙门山断裂带上的西南段. 如前已述, 龙门山断裂带不但是一条规模宏大的断裂带, 也是一条非常活跃的地震带. 无论是从地质构造看, 还是从地震活动性看, 龙门山断裂带 (地震带) 都是一个整体, 是一条统一的、系统的构造带, 地质学家、地震学家对此均有共识.

2008 年 5 月 12 日汶川 $M_W 7.9$ 地震虽然很大, 但是, 根据 Yang 等[6] 在汶川地震前对龙门山地震带大量地震精确定位结果、对汶川地震余震重新精确定位结果 (据 Yang 等[8], Yang 和 Chen[9])、汶川地震后留下的"地震矩释放"的"亏空区"(据陈运泰等[12]和张勇等[13]), 以及汶川地震较大余震的"缺失"(据陈运泰等[12]) 所做的分析都清楚地指向同一结论: 汶川地震只是发生在长达 470km 的统一的、复杂的龙门山断裂带上从映秀到青川、沿 NE-SW 方向约 350km 地段上的断裂, 留下了龙门山断裂带大约 120km 的西南段 (以下简称"西南段") 没有在汶川地震时破裂. 根据地震震级与地震断层长度的统计关系 (据 Wells 和 Coppersmith[33]), 120km 的西南段如果一次性地破裂, 不管人们愿意称其为"主震"还是"余震", 将相当于一次 $M_W 7.5$ 的地震.

然而, 在芦山地震中, 西南段并没有发生一次性的破裂, 而是在其中段"拦腰"发生了芦山地震. 由矩张量反演与地震破裂过程反演得到的芦山地震矩震级为 $M_W 6.7 \sim 6.8$, 震源尺度仅约为长 30km×宽 30km. 长度为 30km 的断层如果一次性地破裂将相当于大约 $M_W 6.8$ 的地震, 因此可以认为, 芦山地震基本上完成了龙门山断裂带西南段中段 (图 5 中的 C 段) 的破裂. 在汶川地震破裂区最西南端至芦山地震破裂区的最北端, 尚有约 30km 长的地段"北段"(图 5 中的 N 段) 没有破裂; 在芦山地震破裂区的最南端至龙门山断裂带的最南端, 尚有约 70km 长的地段"南段"(图 5 中的 S 段) 也没有破裂. 如果 30km 长的北段一次性地发生破裂, 将相当于一次 $M_W 6.8$ 地震; 如果 70km 长的南段一次性地发生破裂, 将相当于一次 $M_W 7.2$ 地震. 在历史上以及近代, 龙门山断裂带西南段发生过 3 次 $M6.0$ 以上地震, 即 1327 年 9 月天全 $M6.0$ 地震 (震中位置: 30.1°N, 102.7°E) (据闵子群[34])、1941 年 6 月 12 日 07:13:31 泸定-天全一带 $M6.0$ 地震 (震中位置: 30.1°N, 102.5°E)、1970 年 2 月 24 日大邑 $M_S 6.2$ 地震 (震中位置: 30.61°N,

103.2°E，震源深度15km）（据谢毓寿和蔡美彪[35]、顾功叙[36,37]）．就释放的地震波能量而言，按照古登堡（Gutenberg）-里克特（Richter）震级-能量关系式（Richter[38]）估计，$M_S 6.0 \sim 6.2$ 地震所释放的地震波能量仅为 $M_W 6.8$ 地震所释放的地震波能量的大约 6%～13%，计及上述3次地震的发生，从所释放的地震波能量看，还是太小，不足以改变上述估计值．此外，这一估计与作为对地震现象的"回顾性"的描述的地震序列类型的判别无涉（据Jordan等[21]）．

如果换一个角度，还可以从汶川地震后留下的"地震矩释放"的"亏空"区以及汶川地震较大余震的"缺失"来分析．按照陈运泰等[12] 2008年5月所做的估计，汶川地震最大余震的震级估计应为 $M_W 6.7 \sim 7.3$．芦山地震的矩震级为 $M_W 6.7 \sim 6.8$，只达到这一估计值的下限．若按上述估计值的上限 $M_W 7.3$ 估计，芦山地震的发生并没有显著地缓解龙门山断裂带西南段的地震危险性：一次 $M_W 6.7 \sim 6.8$ 地震只释放了约 $7.1 \times 10^{14} \sim 1.0 \times 10^{15}$ J 的能量，一次 $M_W 7.3$ 地震将释放约 5.6×10^{15} J 的能量，即 $M_W 6.7 \sim 6.8$ 芦山地震的发生只释放了 $M_W 7.3$ 地震的大约 13%～18% 的能量，可谓杯水车薪，剩下的还未释放的能量估计约为 $4.6 \times 10^{15} \sim 4.9 \times 10^{15}$ J，即相当于 $M_W 7.2 \sim 7.3$ 的"余震"．

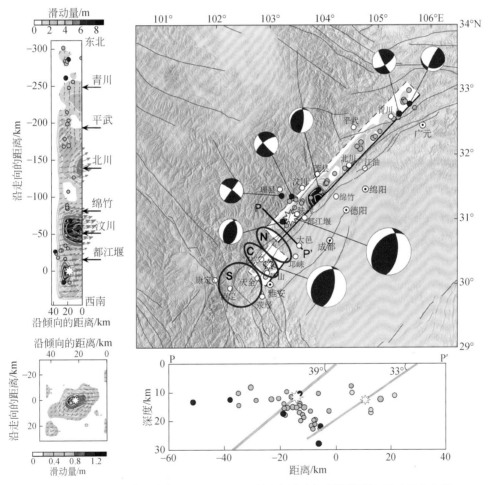

图5 汶川地震及其较大余震（$M_S \geq 5.0$）的空间分布、震源机制与滑动量分布图

大、小白色星号分别表示汶川地震、芦山地震，圆圈表示汶川地震余震，小圆圈表示 $5.0 \leq M_S < 6.0$ 地震，大圆圈表示 $M_S \geq 6.0$ 地震，红色圆圈表示在图中给出表示其震源机制解的红白色"海滩球"（震源球下半球投影）的汶川地震的较大余震，青色圆圈表示其他较大余震，黄色圆圈表示芦山地震的余震．汶川地震与芦山地震在其断层面上的滑动量分布表示于图的左侧，其色标分别表示于左侧最上方与最下方．滑动量分布在地面的投影图表示于主图上．龙门山断裂带西南段的北段（N段）、中段（C段）与南段（S段）以红色圈圈出．详细解释参见正文

4 讨论与结论

以上简要回顾了作者自 2003 年以来在龙门山断裂带中、小地震精确定位、地震活动性,以及对 2008 年汶川 $M_W7.9$ 地震和 2013 年芦山 $M_W6.7$ 地震破裂过程快速反演等方面所做的研究工作. 指出:青藏高原东缘的龙门山断裂带不但是一条巨大断裂带,也是一条非常活跃的地震带. 通过对地震构造、地震活动性、地震矩释放"亏空"区以及余震活动规律的分析,在汶川地震发生后随即提出了龙门山断裂带西南段宝兴-小金一带存在发生 $M_W6.7 \sim 7.3$ 地震的潜在危险性的地震趋势估计. 2013 年芦山 $M_W6.7$ ($M_S7.0$) 地震的发生初步验证了上述估计. 芦山地震发生后作者进一步做的分析结果表明,芦山地震的发生并没有显著地缓解龙门山断裂带西南段的地震危险性,该地段整体上仍存在发生 $M_W7.2 \sim 7.3$ 地震的潜在危险性;特别是西南段的北段(即邛崃大邑西-宝兴北-汶川南一带)存在发生 $M_W6.8$ 地震的潜在危险性;南段(即天全-荥经-泸定-康定一带)存在发生 $M_W7.2$ 地震的潜在危险性. 作者认为,应当强化对上述具有潜在地震危险性区域的监测与多学科综合研究.

上述研究结果是芦山地震发生后作者在前一阶段工作基础上对龙门山断裂带(地震带)西南段未来地震危险性的估计,不包含时间因素,并不是对该地段的严格意义上的"地震预测预报"[18,19]. 虽然如此,本文叙述的结果表明,虽然实现地震预测预报的科学目标可能还须假以时日,但如能恰当地运用迄今已有的地震科学知识,朝着地震预测预报这个科学目标一步一个脚印地努力[19,21,39,40],还是可以有所作为,可为防震减灾贡献一些有意义的信息.

参 考 文 献

[1] Xu Z, Ji S, Li H, et al. Uplift of the Longmen Shan range and the Wenchuan earthquake. *Episodes*, 2008, **31**: 291–301

[2] 邓起东, 陈社发, 赵小麟. 龙门山及其邻区的构造和地震活动及动力学. 地震地质, 1994, **16**: 389–403

[3] 陈国光, 计凤桔, 周荣军, 等. 龙门山断裂带第四纪活动性分段的初步研究. 地震地质, 2007, **29**: 657–673

[4] 杨智娴, 陈运泰, 郑月军, 等. 双差地震定位法在我国中西部地区地震精确定位中的应用. 中国科学 D 辑:地球科学, 2003, **33**(增刊): 129–134

[5] 杨智娴, 于湘伟, 郑月军, 等. 中国中西部地区地震的重新定位和三维地震波速度结构. 地震学报, 2004, **26**: 19–29

[6] Yang Z X, Waldhauser F, Chen Y T, et al. Double-difference relocation of earthquakes in central-western China, 1992-1999. *J Seismol*, 2005, **9**: 241–264

[7] Waldhauser F, Ellsworth W L. A double difference earthquake location algorithm: Method and application to the Northern Hayward Fault, California. *Bull Seism Soc Amer*, 2000, **90**: 1353–1368

[8] Yang Z X, Chen Y T, Stump B, et al. *International Geological Congress in Oslo*, August 6-14, 2008. Abstract, EID-03

[9] Yang Z X, Chen Y T. Relocations of the $M_W7.9$ Wenchuan mainshock and its aftershock sequences. *IASPEI 2009 General Assembly*, Cape Town, SA. 2009

[10] 张勇. 地震震源破裂过程反演方法研究. 北京:北京大学博士学位论文. 2008. 1–158

[11] 刘超, 张勇, 许力生, 等. 一种矩张量反演新方法及其对 2008 年汶川 $M_S8.0$ 地震序列的应用. 地震学报, 2008, **30**: 329–339

[12] 陈运泰, 许力生, 张勇, 等. 2008 年 5 月 12 日汶川特大地震震源特性分析报告. http://www.cea-igp.ac.cn/汶川地震专题/地震情况/初步研究及考察结果(一). 2008

[13] 张勇, 冯万鹏, 许力生, 等. 2008 年汶川地震的时空破裂过程. 中国科学 D 辑:地球科学, 2008, **38**: 1186–1194

[14] 陈运泰. 汶川地震的成因断层、破裂过程与成灾机理. 见:中国科学院, 编. 中国科学院第十四次院士大会学部学术报告汇编. 2008 年 6 月, 北京. 38–39

[15] 陈运泰. 汶川大地震的震源破裂过程. 海峡两岸防震减灾学术研讨会. 2008 年 6 月 27–28 日, 北京. 10–11

[16] 陈运泰. 汶川地震的成因断层、破裂过程与成灾机理."科学技术与抗震救灾"科学技术论坛. 2008 年 7 月 24–26 日, 成都

[17] Båth M. Lateral inhomogeneities in the upper mantle. *Tectonophysics*, 1965, **2**: 483–514

[18] 陈运泰. 地震预测现状与前景. 见: 中国科学院, 编. 2007 科学发展报告. 北京: 科学出版社, 2007. 173–182

[19] 陈运泰. 地震预测: 回顾与展望. 中国科学 D 辑: 地球科学, 2009, **39**: 1633–1658

[20] Wyss M. *Evaluation of Proposed Earthquake Precursors*. Washington DC: American Geophysical Union, 1991. 1–94

[21] Jordan T H, Chen Y T, Gasparini P, et al. Operational earthquake forecasting: State of knowledge and guidelines for utilization. *Annals Geophys*, 2011, **54**: 315–391

[22] 刘超, 许力生, 陈运泰. 2010 年 4 月 14 日青海玉树地震快速矩张量解. 地震学报, 2010, **32**: 366–368

[23] 刘超, 许力生, 陈运泰. 2008 年 10 月至 2009 年 11 月 32 次中强地震的快速矩张量解. 地震学报, 2010, **32**: 619–624

[24] 刘超, 许力生, 陈运泰. 2009 年 11 月至 2011 年 11 月 27 次中强地震的快速矩张量解. 地震学报, 2011, **33**: 550–552

[25] 张勇, 许力生, 陈运泰, 等. 2007 年云南宁洱 M_S6.4 地震震源过程. 中国科学 D 辑: 地球科学, 2008, **38**: 683–692

[26] 张勇, 许力生, 陈运泰. 2008 年汶川地震震源机制的时-空变化. 地球物理学报, 2009, **52**: 379–389

[27] Liu C, Xu L, Chen Y T. Quick moment tensor solution for 6 April 2009, L'Aquila, Italy, earthquake. *Earthq Sci*, 2009, **22**: 449–450

[28] Zhang Y, Chen Y T, Xu L S. Fast and robust inversion of earthquake source rupture process and its application to earthquake emergency response. *Earthq Sci*, 2012, **25**: 121–128

[29] 刘超, 许力生, 陈运泰. 地震矩张量反演结果. 中国地震局地球物理研究所. 2013 年 4 月 20 日四川芦山 7.0 级地震 (1.0 版). 2013 年 4 月 20 日. http://www.cea-igp.ac.cn/tpxw/266824.shtml

[30] 张勇, 许力生, 陈运泰. 芦山 4·20 地震破裂过程及其致灾特征初步分析. 地球物理学报, 2013, **56**: 1408–1411

[31] Shimazaki K. Small and large earthquakes: The effects of the thickness of the seismogenic layer and free surface. In: Das S, Boatwright J and Scholz C H (eds), *Earthquake Source Mechanics*. AGU Geophys Mono 37. Washington DC: *Am Geophys Union*. 1986. 209–216

[32] Scholz C H. *The Mechanics of Earthquakes and Faulting*. 2nd ed. Cambridge: Cambridge University Press, 2002. 1–471

[33] Wells D L, Coppersmith K J. New empirical relations among magnitude, rupture length, rupture width, rupture area, and surface displacement. *Bull Seismol Soc Am*, 1994, **84**: 974–1002

[34] 闵子群. 中国历史强震目录(公元前 23 世纪–公元 1911 年). 北京: 地震出版社, 1995. 1–514

[35] 谢毓寿, 蔡美彪. 中国历史地震资料汇编. 第一卷. 北京: 科学出版社, 1983. 1–227

[36] 顾功叙. 中国地震目录(公元前 1831 年–公元 1969 年). 北京: 科学出版社, 1983. 1–894

[37] 顾功叙. 中国地震目录(公元 1970–1979). 北京: 地震出版社, 1983. 1–334

[38] Richter C F. *Elementary Seismology*. San Francisco: W. H. Freeman, 1958. 1–769

[39] Jordan T H. Earthquake predictability, brick by brick. *Seismol Res Lett*, 2006, **77**: 3–6

[40] Peresan A, Kossobokov V G, Panza G F. Operational earthquake forecast/prediction. *Rend Fis Acc Lincei*, 2012, doi: 10.1007/s12210-012-0171-7

Kinematic rupture model and hypocenter relocation of the 2013 M_W 6.6 Lushan earthquake constrained by strong-motion and teleseismic data[*]

Yong Zhang[1,2], Rongjiang Wang[1,3], Yun-tai Chen[2], Lisheng Xu[2], Fang Du[4], Mingpei Jin[5], Hongwei Tu[1] and Torsten Dahm[1]

1. GFZ German Research Centre for Geosciences, Telegrafenberg, Potsdam 14473, Germany;
2. Institute of Geophysics China Earthquake Administration Beijing 100081, China;
3. Earthquke Administration of Qinghai Province, Xining 810001, China;
4. Earthquake Administration of Sichuan Province Chengdu 610041, China;
5. Earthquake Administration of Yunnan Province Kunming 650041, China

Online Material Tables of 1D velocity structure, hypocentral locations, and kinematic rupture model.

1 Introduction

On 20 April 2013, an M_W 6.6 earthquake struck Lushan, Sichuan province, China (hereafter as Lushan earthquake), which caused 196 deaths and 21 missing. As another disastrous earthquake occurred on the Longmenshan fault after the 2008 M_W 7.9 Wenchuan earthquake, the Lushan earthquake stimulated an extensive discussion about its seismogenesis. In particular, whether it was a large aftershock of the 2008 M_W 7.9 Wenchuan earthquake has been heatedly debated (Chen et al., 2013; Du et al., 2013; Liu, Yi, et al., 2013; Wang et al., 2013; Jia et al., 2014). Actually, after the Wenchuan earthquake, an increased seismic hazard for the southwest of the Longmenshan fault has been commonly recognized (Parsons et al., 2008; Toda et al., 2008). The occurrence of the Lushan earthquake seems to have confirmed this common understanding, and further attracts attention on the reevaluation of future seismic activities on the southwest Longmenshan fault (Chen et al., 2013). As the basis of these discussions, the source characteristics of the Lushan earthquake need to be well understood.

As a routine work for fast source earthquake information, Zhang et al. (2013) released the preliminary teleseismic rupture model about three hours after the Lushan earthquake. Their results show that the earthquake has no dominant rupture direction. The major slip area is located around the hypocenter and does not reach the surface.

It is known, however, that teleseismic data can only constrain the ruptured area relatively to the hypocenter (rupture initiation point). In contrast, near-field strong-motion data can provide information useful in precisely and accurately locating the fault rupture. Conversely, selection of the hypocenter location may significantly impact the results of the strong-motion data inversion. If the assumed hypocenter location is incorrect, the inversion will result in an unreliable rupture direction and distortion of Doppler effects, leading to a large misfit between synthetic and observed waveforms. This problem has been encountered already by Hartzell and Mendoza (1991) in their

[*] 本文发表于 *Seismol. Res. Lett.*, 2014年, 第85卷, 第1期, 15–22.

study of the 1978 Tabas, Iran, earthquake. Actually, the inaccurate hypocenter location is always a problem in the analysis of near-field data (both seismic and geodetic data; Hartzell and Mendoza, 1991; Jackson et al., 2006; Zhang et al., 2012), especially when the hypocenter uncertainty is not negligible compared with the source dimension (i.e., fault size). In the source inversion of the 2009 L'Aquila earthquake, Zhang et al. (2012) have shown that it is feasible to relocate the hypocenter by a joint inversion of the rupture process with teleseismic and near-field Interferometric Synthetic Aperture Radar data.

In this work, we relocate the hypocenter of the Lushan earthquake by inverting the near-field strong-motion data. Based on the relocated hypocenter, we then update the kinematic rupture model of the earthquake by jointly inverting the strong-motion and teleseismic data. The approach helps to constrain the nonuniqueness of sole teleseismic or sole near-field strong-motion kinematic inversions. Finally, we discuss the increased seismic hazard on the southwest Longmenshan fault after the Lushan earthquake.

2 Hypocenter relocation with near-field strong-motion data

The southwest part of the Longmenshan fault zone consists of at least three major parallel faults, all of which are dominated by similar thrust mechanisms (Fig. 1a) (Deng et al., 2003), resulting from the continental collision of the Tibetan plateau from northwest to southeast against the Sichuan basin. The focal mechanism of the Lushan earthquake is given by strike = 218°, dip = 39°, and rake = 103° according to the W-phase moment tensor solution of the U.S. Geological Survey (USGS; http://earthquake.usgs.gov/earthquakes/eqarchives/fm/neic_b000gcdd_wmt.php, last accessed October 2013). The hypocenter was located at 30.3°N, 103.0°E, depth 13km by China Earthquake Networks Center (CENC), at 30.308°N, 102.888°E, depth 14km by USGS, and at 30.27°N, 102.96°E, depth 19km by German Research Center for Geosciences (GFZ), all indicating that the earthquake activated a southeast branch to the Sichuan basin side of the Longmenshan fault in the Lushan region.

In our fault-slip inversions, a sufficiently large potential fault plane of 63km long and 45km wide (Fig. 1a) is chosen and divided into 21×15 subfaults. Three-component accelerograms recorded by 14 strong-motion stations (Fig. 1a) are used for the inversions, except for the north-south component of station TQLL which shows large baseline shifts during the strong shaking period. Thus, totally we have data including 41 accelerograms for the inversions. These accelerograms are integrated to velocity seismograms and then filtered by a bandpass of 0.02-0.40Hz. The lower-frequency cutoff is necessary to remove the effects caused by baseline shifts, which exist in all strong-motion records, whereas the upper frequency cutoff of 0.40Hz is consistent with the subfault resolution of 3km×3km.

Synthetic Green's functions are calculated using the code of Wang (1999). Because the earthquake occurred on the edge between Tibetan plateau and Sichuan basin, the velocity structure and topography is quite complicated within the spatial coverage of the strong-motion stations. Moreover, the strong-motion seismograms in the frequency range we used are dominated by the S waves for which waveform and arrival are more sensitive to the velocity structure than the P waves. To take into account for the complex velocity structure, we used two different crustal velocity structures (Ⓔ Table S1 available as an electronic supplement to this article) for the stations located in the Tibetan plateau (cyan triangles in Fig. 1a) and those in the Sichuan basin (blue triangles in Fig. 1a), respectively, when calculating Green's functions. In addition, the topography effect on the wave arrivals is minimized using a source-depth correction depending on the station elevation.

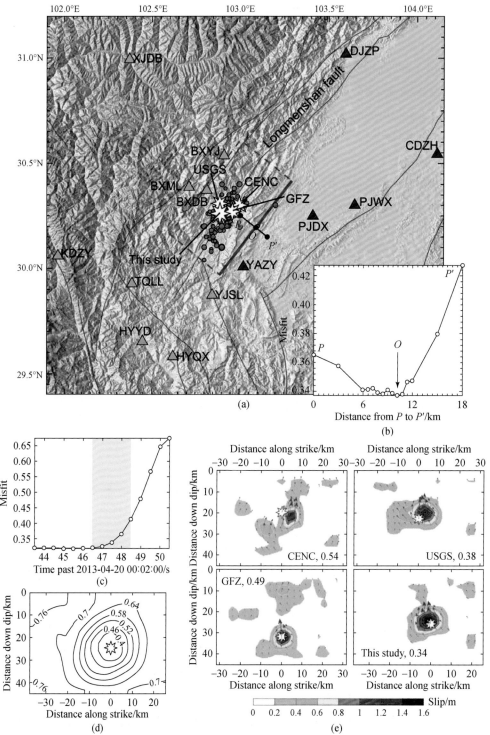

Figure 1 (a) Strong-motion stations (cyan and blue triangles are those located in Tibetan plateau and Sichuan basin, respectively), surface projection of final determined fault position (pink rectangular), aftershocks (orange circles), and hypocenters (white stars) of the mainshock determined by CENC, USGS, GFZ, and this study. The red lines are the active faults (Deng et al., 2003) in the epicentral area. (b) Misfit curve of the fault position. (c) Misfit curve of the origin time. (d) Misfit map of the hypocenter location by fixing the origin time and fault position as shown in (a). (e) Fault slip distributions obtained with the hypocenters of CENC, USGS, GFZ, and this study, respectively. The misfits based on these four hypocenters are marked by the lower right number in each subplot

To invert the strong-motion data for the kinematic rupture process of the earthquake, we use the linear inversion approach of Zhang et al. (2012) based on the principle of least-squares optimization. During the linear inversion, several earthquake parameters need to be fixed, which are (1) the hypocenter location on the fault plane; (2) the origin time of the earthquake; and (3) the fault geometry parameters including the fault position, the strike, and dip angles. Results from the inversion for each subfault are time histories of the moment rate, also called the source time function (STF) and slip direction (rake angle). For numerical stability, the linear inversion approach needs generally to restrict the degree of freedom. This is realized using certain *a priori* constraints on the rupture process, which include in the present case that (1) the rupture starts at the hypocenter and propagates outward with a velocity not faster than 3km/s (i.e., no super-shear rupture is considered); (2) the maximum rupture duration of each subfault is 6s, in which there are 12 equidistant triangles (with varying but non-negative amplitudes) used to describe the slip time process; and (3) the rake angle is allowed only to vary within ±45° around the USGS estimate of 103°. Differently to most nonlinear inversion approaches, we do not need any constraints on the form of the STFs, but determine them automatically by the data.

For each given fault geometry (strike, dip, and location) and origin time, we perform a number of inversions by varying the hypocenter location on the fault plane. The optimal hypocenter location is then determined by the one leading to the best fit between synthetic and observed strong-motion waveforms. Actually, this is another approach for the hypocenter location; it uses the complete near-field waveform data instead of only the arrival times of seismic waves.

In principle, we next need to vary the fault geometry and origin time to find the best-fit hypocenter location. However, this would mean a grid search in a 6D parameter space (strike, dip, origin time, 1D fault position perpendicular to the strike, and 2D hypocenter position on the fault plane), which requires a huge computational effort, besides a number of trade-off problems among the six parameters. Considering that the teleseismic focal solution is almost independent of the hypocenter location, we only optimize the fault position and origin time, but discuss uncertainties in the strike and dip angles as well as their impacts later. Fixing the fault strike and dip, the location problem is reduced to a grid search in a 4D parameter space. Our results show that there is no strong trade-off between the origin time and the fault position. The misfit curves for these two parameters are nearly independent from each other and are shown in Figure 1b and c, respectively. Because of the relatively conservative *a priori* constraints on the rupture area, the rupture velocity, and the rupture duration, it appears that the earlier the origin time is selected, the better are the waveform fits. The phenomenon can be easily explained. In the extreme case, that is, without the *a priori* constraints, the model space with an earlier origin time has always a larger degree of freedom than that with a subsequent origin time, and thus can fit the data more easily. Based on the synthetic tests, which will be discussed later, we found that the optimal origin time should appear where the misfit begins to significantly increase. Because the cutoff frequency used is 0.40Hz, the resolution of the origin time cannot be higher than 1-2s. Actually, a range of about 1-2s around 20 April 2013, 00:02:47.5 can be identified, after when the misfit starts to increase rapidly. Because this range of potential origin time is not sensitive to the fault position, and most origin times lead to similar hypocenter locations (Ⓔ Table S2 available as an electronic supplement to this article), we therefore fix the origin time at 47.5 s, which is equal to that released by the USGS. With this origin time, we get the fault position with the minimum waveform misfit as shown in Figure 1b. The corresponding best-fit hypocenter is located at 30.261°N, 102.889°E, depth 16.0km (Fig. 1d).

Our relocated hypocenter is about 11.9, 5.6, and 7.5km apart from those released by the institutions of CENC, USGS, and GFZ, and 8.6 and 9.0km from those by Lv et al. (2013) and Zhang and Lei (2013), respectively. The latter two groups relocated the hypocenter using the double-difference method. To demonstrate the influence of the hypocenter on the inversion results, we compare the slip distributions obtained based on three different hypocenters of CENC, USGS, and GFZ (Fig. 1e). They show non-negligible differences in rupture direction, slip area, maximum slip, and misfit. The fault slip distribution based on our relocated hypocenter (Fig. 1e) shows a circular rupture event without predominant rupture direction, which is consistent with the manual teleseismic inversion results by Liu, Zheng, et al. (2013) and Zhang et al. (2013) and the automatic teleseismic inversion results by S. Heimann (personal comm., 2013).

3 Joint inversion with strong-motion and teleseismic data

The teleseismic data used in this study are provided by Incorporated Research Institutions for Seismology (IRIS). We selected 31 stations at epicentral distances between 30° and 90° with a roughly homogeneous coverage on the azimuth and take-off angles (Fig. 2). Only vertical components of the P waves are used, differently to the strong-motion data, from which all three components of the complete waveform are used. Green's functions for the teleseismic stations are generated based on the AK135 continental earth model (Kennett et al., 1995).

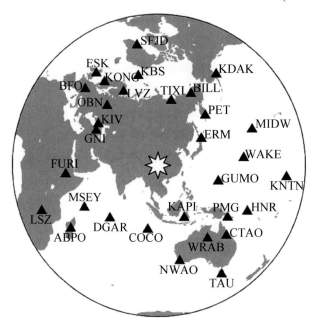

Figure 2 Epicenter (white star) and teleseismic stations (black triangles)

We totally have 41 and 31 waveform channels from the strong-motion and teleseismic datasets, respectively. In the joint inversion, each channel of the data is normalized by its signal energy (integral of squared velocity seismograms) so that they are equally weighted in the sense of waveform. The results of the joint inversion are shown in Figure 3. The scalar seismic moment is estimated to be about 1.26×10^{18} N·m, equivalent to a moment magnitude of $M_W 6.67$. The source process is characterized by a single subevent (Fig. 3a), with most energy released within the first 10s, consistent with the existing studies (Liu, Zheng, et al., 2013; Zhang et al., 2013). The fault slip is mainly concentrated around the hypocenter with a dimension of about 20km (Fig. 3b and c). The maximum slip is about 1.2m, slightly smaller than that inverted from the strong-motion data only. The strong-motion and teleseismic waveforms are both well fitted by the synthetics (Fig. 3d-e). Ⓔ The digital rupture model obtained by the joint inversion is available in the electronic supplement to this article.

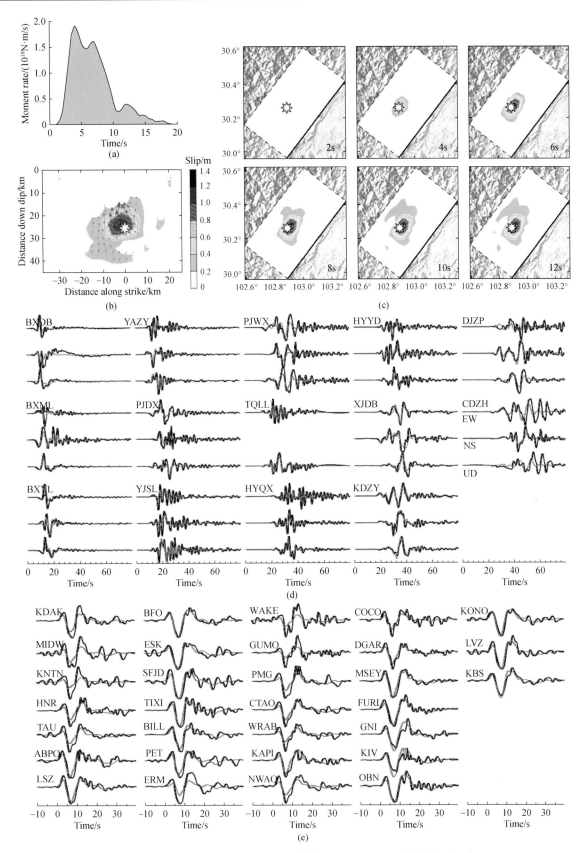

Figure 3　Results from joint inversion of the strong-motion and teleseismic data
(a) Source time function, (b) static fault slip distribution, (c) temporal variations of the surface projection of fault slip distributions, (d) comparisons between observed (black) and synthetic (red) strong-motion waveforms, (e) comparisons between observed (black) and synthetic (red) teleseismic waveforms. The time zeros in (d) and (e) are the earthquake origin time and P arrivals, respectively

Unlike the nonlinear inversion, the rupture front cannot be uniquely determined from the linear inversion results. Following the work of Lee et al. (2006), we assume that the rupture front reaches the subfault when the accumulated slip exceeds 0.05 m. The slip-weighted average rupture velocity found in this way is about 2.3km/s, or 0.6-0.7 of the S-wave velocity in the upper crust.

The joint inversion shows a smoother slip distribution and smaller maximum slip than those derived from the strong-motion data only. It could be attributed to the relative poor spatial resolution ability of the teleseismic data. In addition, it suggests that the second subevent obtained by inverting only the teleseismic data (Zhang et al., 2013) is not evident.

4 Discussion and conclusions

Recently many near-field strong-motion data are available shortly after an earthquake. Such data, combined with teleseismic records and a proper relocation of the hypocenter, have the potential to significantly improve kinematic rupture and slip models, which may otherwise be inherently nonunique. In the case of the Lushan earthquake, we have demonstrated significant influences of hypocenter uncertainties on the inversion results using near-field strong-motion data. The problem is less serious for large earthquakes if the fault extension is much larger than the hypocenter uncertainty. However, once the fault size is comparable to, or even smaller than, the hypocenter uncertainty, different hypocenters certainly result in different apparent rupture processes.

In this paper, we use the linear inversion method to optimize the hypocenter of the Lushan earthquake by trying different fault positions and origin times with fixed strike and dip angles. In principle, this work can also be done by a nonlinear network inversion method, which, however, would cost much more in computational efforts. In fact, if using the nonlinear inversion approach, at least two more parameters for each subfault, that is, the rupture time and the rupture duration, need to be optimized in a nonlinear way. In addition, it should be noticed that the nonlinear inversion approach uses generally a preselected shape for the subfault STF, which may lead to certain artifact. To check the reliability of the result, a resolution test is performed. A rectangular slip area is assumed on the fault, and then synthetic strong-motion and teleseismic waveform data are generated and added with 10% Gaussian noise. It is shown in Figure 4 that the fault position and hypocenter are both well determined, but the origin time is relatively less resolved (Fig. 4a-c). The hypocenter determined is stable when varying the origin time by ±1s, suggesting that uncertainties in the origin time may not be critical for our rupture location.

In the resolution test, the fault slip can be generally well retrieved by either jointly inverting the synthetic strong-motion and teleseismic data or inverting only the strong-motion data (Fig. 4d-f). It is noticed that the slip distribution of the joint inversion is smoother than that from the strong-motion data inversion. As mentioned above, it may be attributed to the relative poor spatial resolution ability of the teleseismic data. Though the synthetic test does not show any obvious advantage of the joint inversion, it is always useful in practice because results by only using the near-field strong-motion data are more sensitive to the fault location and fault irregularities. The latter are usually not well known. Therefore, although the strong-motion data can resolve more details of source ruptures, it usually leads to large uncertainties caused by the local 3D structure effect. In contrast, teleseismic data result in a smoother slip model, but it is less sensitive to the hypocenter location and local crust structure, and thus can help to estimate the major rupture pattern (relative to the hypocenter) more reliably. The complementary advantages of the two datasets can be combined by the joint inversion. In addition, the joint use of the near-field strong-motion and teleseismic data can increase the coverage of take-off angle on the ruptured fault and therefore improve the resolution of the slip distribution particularly with depth.

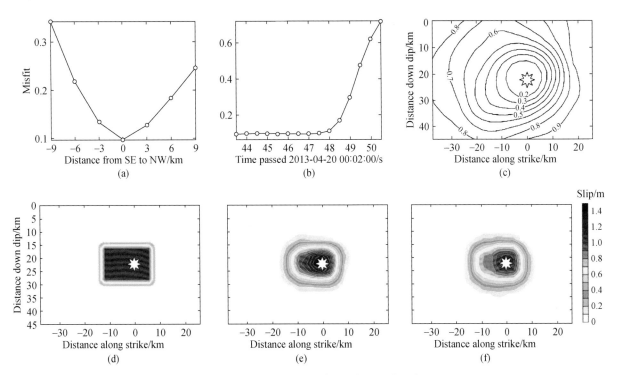

Figure 4 Resolution test for the Lushan earthquake.
(a) Grid search for the fault position through the strong-motion inversion. The assumed fault position corresponds to 0. (b) Grid search for the origin time with strong-motion data. The assumed origin time is 47.5 s. (c) Misfit of all subfaults as hypocenters in the strong-motion inversion based on the optimized origin time and fault position. (d) The input fault slip model. (e) Slip distribution from the strong-motion inversion. (f) Slip distribution from the joint inversion of the strong-motion and teleseismic data

In the above analysis, we do not consider the impacts of the fault strike and dip angles. Nowadays, uncertainties in teleseismic focal solutions (i.e., strike and dip) are about 10° (Helffrich, 1997). To check their influences on the hypocenter location through the strong-motion inversion, we change the strike and dip of the Lushan earthquake by ±10° and compare the corresponding location results. Ⓔ As shown in Table S3 available as an electronic supplement to this article, these changes do not significantly disturb the hypocenter location. The maximum difference among the five hypocenters including the preferred one is about 6km in the horizontal distance and 4km in depth, which could be considered as the upper limit of uncertainties in our hypocenter location.

The joint inversion based on the relocated hypocenter shows a simple rupture process of the Lushan earthquake with a comparable spatial extension along the strike and dip directions, which is typical of most moderate earthquakes. The ruptures occurred in the upper crust, but did not breach the surface. This could explain why the earthquake damages mainly concentrated on the hanging wall, instead of the areas around the extrapolated surface trace of the fault (Zhang et al., 2013).

To investigate how the local bilateral velocity structure influences the hypocenter location, we also carried out the inversions using an average crust model (Ⓔ Table S1 available as an electronic supplement to this article) based on CRUST2.0 (Bassin et al., 2000). No substantial changes on the rupture pattern are found, but the source becomes systematically shallower by about 2km and the root mean square misfit to the strong-motion data increases by 20%. Additionally, when using the average velocity model, the hypocenter coincides with the slip centroid and no dominant rupture directivity is visible. In comparison, the bilateral velocity model shifts the hypocenter 2-3km apart from the slip centroid, so that slight up-dip rupture directivity appears (Fig. 3b). However, we notice that this directivity effect is not significant because the uncertainty of our hypocenter location is twice larger than 2-3km.

It is widely accepted that the Lushan earthquake is closely related to the 2008 Wenchuan earthquake (Chen et al., 2013; Du et al., 2013; Liu, Yi, et al., 2013; Wang et al., 2013). In this work, it is found that the major slip area of the Lushan earthquake is about 40-50km apart from the southwest end of the 2008 Wenchuan earthquake fault, suggesting that there is a wide area unruptured southwest of Longmenshan fault. Because the Lushan earthquake only ruptured a small fault segment, increased seismic hazards in its adjacent regions, particularly in the gap to the Wenchuan earthquake, should be noticeable.

Acknowledgments

This work is supported by the Towards Real-Time Earthquake Risk Reduction (REAKT) project of the European Seventh Framework Programme (Grant Agreement Number 282862), the National Natural Science Foundation of China (NSFC) projects ID 41090291 and ID 90915012, and the China Earthquake Administration (CEA) expedition project "Scientific Investigation of 20 April 2013 M7.0 Lushan, Sichuan Earthquake." The aftershock and strong-motion data were provided by China Earthquake Networks Center (CENC) and Sichuan Earthquake Administration (SCEA), respectively. We thank the guest editor and the anonymous reviewers for their valuable comments.

References

Bassin, C., G. Laske, and G. Masters (2000). The current limits of resolution for surface wave tomography in North America, *Eos Trans. AGU* **81**, F897.

Chen, Y. T., Z. X. Yang, Y. Zhang, and C. Liu (2013). From 2008 Wenchuan earthquake to 2013 Lushan earthquake (in Chinese), *Sci. China Ser. D.* **43**, 1064–1072.

Deng, Q., P. Zhang, Y. Ran, X. Yang, W. Min, and Q. Chu (2003). Basic characteristics of active tectonics of China, *Sci. China Ser. D.* **46**, no. 4, 356–372.

Du, F., F. Long, X. Yuan, G. Yi, Y. Gong, M. Zhao, Z. Zhang, H. Qiao, Z. Wang, and J. Wu (2013). The M7.0 Lushan earthquake and the relationship with the M8.0 Wenchuan earthquake in Sichuan, China (in Chinese), *Chin. J. Geophys.* **56**, no. 5, 1772–1783.

Hartzell, S. H., and C. Mendoza (1991). Application of an iterative least-squares wave-form inversion of strong-motion and teleseismic records to the 1978 Tabas, Iran, earthquake, *Bull. Seismol. Soc. Am.* **81**, no. 2, 305–331.

Helffrich, G. R. (1997). How good are routinely determined focal mechanisms? Empirical statistics based on a comparison of Harvard, USGS and ERI moment tensors, *Geophys. J. Int.* **131**, 741–750.

Jackson, J., M. Bouchon, E. Fielding, G. Funning, M. Ghorashi, D. Hatzfeld, H. Nazari, B. Parsons, K. Priestley, M. Talebian, M. Tatar, R. Walker, and T. Wright (2006). Seismotectonic, rupture process, and earthquake-hazard aspects of the 2003 December 26 Bam, Iran, earthquake, *Geophys. J. Int.* **166**, 1270–1292.

Jia, K., S. Zhou, J. Zhuang, and C. Jiang (2014). Analysis on the possibility of the independence between the 2013 Lushan earthquake and the 2008 Wenchuan earthquake on Longmen Shan fault, Sichuan, China, *Seismol. Res. Lett.* **85**, no. 1, 60–67, doi: 10.1785/0220130115.

Kennett, B. L. N., E. R. Engdahl, and R. Buland (1995). Constraints on seismic velocities in the earth from traveltimes,

Geophys. J. Int. **122**, 108–124.

Lee, S., K. Ma, and H. Chen (2006). Three-dimensional dense strong motion waveform inversion for the rupture process of the 1999 Chi-Chi, Taiwan, earthquake, *J. Geophys. Res.* **111**, no. B11308, doi: 10.1029/2005JB004097.

Liu, C., Y. Zheng, C. Ge, X. Xiong, and H. Xu (2013). Rupture process of the *M*7.0 Lushan earthquake, *Sci. China Earth Sci.* **43**, no. 6, 1020-1026, doi: 10.1007/s11430-013-4639-9.

Liu, J., G. Yi, Z. Zhang, Z. Guan, X. Ruan, F. Long, and F. Du (2013). Introduction to the Lushan, Sichuan *M* 7.0 earthquake on 20 April 2013 (in Chinese), *Chin. J. Geophys.* **56**, no. 4, 1404–1407.

Lv, J., X. Wang, J. Su, L. Pan, Z. Li, L. Yin, X. Zeng, and H. Deng (2013). Hypocentral location and source mechanism of the M_S 7.0 Lushan earthquake sequence (in Chinese), *Chin. J. Geophys.* **56**, no. 5, 1753–1763.

Parsons, T., J. Chen, and E. Kirby (2008). Stress changes from the 2008 Wenchuan earthquake and increase hazard in the Sichuan basin, *Nature* **454**, 509–510.

Toda, S., J. Lin, M. Meghraoui, and R. S. Stein (2008). 12 May 2008 $M = 7.9$ Wenchuan, China, earthquake calculated to increase failure stress and seismicity rate on three major fault systems, *Geophys. Res. Lett.* **35**, L17305, doi: 10.1029/2008GL034903.

Wang, R. (1999). A simple orthonormalization method for stable and efficient computation of Green's functions, *Bull. Seismol. Soc. Am.* **89**, no. 3, 733–7410.

Wang, W., J. Hao, and Z. Yao (2013). Preliminary result for rupture process of Apr. 20, 2013, Lushan earthquake, Sichuan, China (in Chinese), *Chin. J. Geophys.* **56**, no. 4, 1412–1417.

Zhang, G., and J. Lei (2013). Relocation of the Lushan, Sichuan strong earthquake (M_S 7.0) and its aftershocks (in Chinese), *Chin. J. Geophys.* **56**, no. 5, 1764–1771.

Zhang, Y., Y. T. Chen, and L. Xu (2013). Rupture process of the Lushan 4.20 earthquake and preliminary analysis on the disaster-causing mechanism (in Chinese), *Chin. J. Geophys.* **56**, no. 4, 1408–1411, doi: 10.6038/cjg20130435.

Zhang, Y., W. Feng, Y. T. Chen, L. Xu, Z. Li, and D. Forrest (2012). The 2009 L'Aquila M_W 6.3 earthquake: A new technique to locate the hypocenter in the joint inversion of earthquake rupture process, *Geophys. J. Int.* **191**, 1417–1426, doi: 10.1111/j.1365246X.2012.05694.x.

Comment on the paper "Normal and shear stress acting on arbitrarily oriented faults, earthquake energy, crustal GPE change, and the coefficient of friction" by P. P. Zhu[*]

Yun-Tai Chen

Institute of Geophysics, China Earthquake Administration, Beijing 100081, China

In his paper "Normal and shear stress acting on arbitrarily oriented faults, earthquake energy, crustal GPE change, and the coefficient of friction", Zhu (2013) tried to theoretically solve the calculation of the fundamental parameters in seismology. These parameters include the normal and shear stresses acting on arbitrary oriented faults, the earthquake energy, crustal gravitational potential energy (GPE) change (hereon abbreviation or symbol in parenthesis is that used in Zhu's paper), and the coefficient of friction. He especially emphasized the importance of gravitational potential energy (GPE) change, and stated that "many geophysicists accepted the idea of partitioning earthquake energy into radiated seismic energy (E_R), friction energy (E_F), and rupture energy (E_{RP}) without consideration of crustal gravitational potential energy (GPE) change (e.g. Kanamori 2001; Abercrombie et al. 2006; Kanamori and Rivera 2006)," and that "the research on the earthquake energy budget stayed in qualitative analysis for a long time (e.g. Abercrombie et al. 2006)." In this short note, I will comment on the fundamental idea presented in Zhu's paper (Zhu 2013) and will point out that his statement is incorrect.

The energy balance in faulting was discussed very early and completely by Kostrov et al. (1969), Kostrov (1974, 1975), Dahlen (1977), Kostrov and Das (1988), and Dahlen and Tromp (1998), and a simplified and easily understood discussion of the energy changes involved in faulting within a self-gravitational body was given by Savage and Walsh (1978).

The model earth Savage and Walsh (1978) used is a simple one. In their model, the fault is represented by a displacement dislocation buried in a self-gravitating earth subject to initial strain but free from external forces. They found for the total change in elastic strain energy (ESE) ΔE_{el} is

$$\Delta E_{el} = \oint_{\Sigma^1} \left(\sigma_{ij}^0 + \frac{\tau_{ij}}{2} \right) u_i^1 n_j dS + \iiint_V f_i^0 u_i^1 dV \qquad (1)$$

where dS is the surface element of the fault surface Σ^1 consisting of two closely adjacent sheets, $\Sigma^1 = \Sigma^{1+} + \Sigma^{1-}$, dV the volume element, $\tau_{ij} = \sigma_{ij}^1 - \sigma_{ij}^0$ the incremental stress, i.e., the difference between final stress σ_{ij}^1 and initial stress σ_{ij}^0, u_i^1 the displacement on Σ^1, n_j the outward drawn normal to surface Σ^1, respectively. Throughout this paper, the superscript "0" and "1" refer to the quantity before and after rupture process, respectively. The total change in gravitational potential energy (GPE) ΔE_g is simply the work done against the gravitational body force f_i^0:

---

[*] 本文发表于 J. Seismol., 2014 年, 第 18 卷, 第 4 期, 795–797.

$$\Delta E_g = -\iiint_V f_i^0 u_i^1 \mathrm{d}V. \tag{2}$$

During an earthquake, the total change in potential energy ΔE is the sum of total change in elastic strain energy ΔE_{el} and the total change in gravitational potential energy ΔE_g:

$$\Delta E = \Delta E_{el} + \Delta E_g. \tag{3}$$

Substitution of Eqs. (1) and (2) into Eq. (3) yields

$$\Delta E = \oiint_{\Sigma^1} \left(\sigma_{ij}^0 + \frac{\tau_{ij}}{2}\right) u_i^1 n_j \mathrm{d}S. \tag{4}$$

Equations (1), (2), and (4) are essentially the same as Eqs. (15), (16), and (17) given by Savage and Walsh (1978), respectively. As Savage and Walsh (1978) noted this result is the same as that given by Dahlen (1977). In the expression for the total change in elastic strain energy (ESE) ΔE_{el} (Eq. (1)) the explicit term $\iiint_V f_i^0 u_i^1 \mathrm{d}V$ is the work done by the gravitational forces, while in the expression for the total change in gravitational potential energy ΔE_g (Eq. (2)) the explicit term $-\iiint_V f_i^0 u_i^1 \mathrm{d}V$ is the work against the gravitational forces. These two explicit terms cancel out of the expression for the total potential energy change (Eq. (4)). Equation (4) is the expression for the total potential energy change including both the total change in elastic strain energy (ESE) ΔE_{el} and the total change in gravitational potential energy (GPE) ΔE_g. The gravitational effects are included but still implicit in Eq. (4), however, through σ_{ij}^0 in the surface integral.

Equation (4) can be re-written as

$$\Delta E = -\iint_{\Sigma^1} \left(\sigma_{ij}^0 + \frac{\tau_{ij}}{2}\right) \Delta u_i^1 n_j \mathrm{d}S \tag{5}$$

or the total potential energy released during an earthquake, E, can be written as

$$E = -\Delta E \tag{6}$$

$$E = \frac{1}{2} \iint_{\Sigma^1} (\sigma_{ij}^0 + \sigma_{ij}^1) \Delta u_i^1 n_j \mathrm{d}S \tag{7}$$

where Δu_i^1 is the displacement dislocation on Σ^1, and in Eqs. (5) and (7), we now integrate only over one sheet Σ^{1-} of the two sheets of the fault surface and drop the superscript "$-$". Equation (5) is the same as that obtained by Kostrov (1974, Eq. (2.13), 1975, Eq. (IV.4.4)) and also Kostrov and Das (1988, Eq. (4.4.4)).

Using the energy conservation law, Kostrov (1974, 1975) and Kostrov and Das (1988, Eq. (4.4.24)) as well as Rudnicki and Freund (1981) derived the following expression for the radiated seismic energy in a homogeneous space:

$$E_S = -\frac{1}{2} \iint_{\Sigma^1} (\sigma_{ij}^0 - \sigma_{ij}^1) \Delta u_i^1 n_j \mathrm{d}S + \int_0^{t_m} \mathrm{d}t \iint_{\Sigma(t)} \dot{\sigma}_{ij} \Delta u_i n_j \mathrm{d}S - 2\gamma_{\mathrm{eff}} A \tag{8}$$

where integrations are taken only over one sheet Σ^{1-} of the two sheets of the fault surface $\Sigma^1 = \Sigma^{1+} + \Sigma^{1-}$ and the superscript "$-$" are dropped, γ_{eff} the effective surface energy, and A the fracture area, respectively. This equation can be re-written as

$$E_S = \frac{1}{2} \iint_{\Sigma^1} (\sigma_{ij}^0 + \sigma_{ij}^1) \Delta u_i^1 n_j \mathrm{d}S - \iint_{\Sigma^1} \mathrm{d}S \int_0^{\Delta u_i^1} \sigma_{ij} n_j \mathrm{d}(\Delta u_i) - 2\gamma_{\mathrm{eff}} A. \tag{9}$$

The first term on the right-hand side of Eq. (9) is the total potential energy released (sum of the total elastic strain energy released and the total gravitational energy released) given by Eq. (7). The second term without negative sign on the right-hand side of the Eq. (9) is the work done against the dynamic frictional stress σ_{ij} on the fault plane during earthquake faulting E_f:

$$E_f = \iint_{\Sigma^1} dS \int_0^{\Delta u_i^1} \sigma_{ij} n_j d(\Delta u_i). \tag{10}$$

The third term without negative sign on the right-hand side of Eq. (9) is the energy used to create the new fracture surface on the edge of the expanding fault, i.e., the fracture energy

$$E_\gamma = 2\gamma_{eff} A. \tag{11}$$

Hence, the radiated seismic energy can be expressed as follows:

$$E_S = E - E_f - E_\gamma \tag{12}$$

In the other words, the equation of energy balance can be written as (Kostrov et al. 1969; Kostrov 1974, 1975, Eq. (IV.4.19); Kostrov and Das 1988, Eq. (4.4.19); Kanamori and Heaton 2000, Eq. (5))

$$E = E_S + W_f + E_\gamma \tag{13}$$

where E is the total potential energy released during an earthquake given by Eq. (7).

Equation (13) is similar only in form to but essentially different from Eq. (4) of Zhu's paper (Zhu 2013, p. 988). The E in Eq. (13) stands for the total potential energy released, which is the sum of $-\Delta E_{el}$ and $-\Delta E_g$, while in Zhu's paper, E is defined as the total elastic strain energy (ESE) released from an earthquake (Zhu 2013, p. 987, Eq. (2); p. 988, Eq. (4)) or $-\Delta E_{el}$ in the symbol of this paper, without taking into account the contribution of gravitational potential energy (GPE) released or $-\Delta E_g$ in the symbol of this paper. Zhu (2013) misinterpreted the physical meaning of E in Eq. (13) of this paper or Eq. (4) in his paper (Zhu 2013, p. 988). By introducing crustal gravitational potential energy (GPE) change (ΔP_G) to the equation of energy balance, Zhu (2013) duplicates the contribution of GPE change. Hence, Eqs. (44a) and (44b) in Zhu's paper (Zhu 2013, p. 993) are incorrect.

In summary, Zhu (2013) correctly emphasized the importance of the change in gravitational potential energy during earthquake faulting which has been correctly taken into account by the pioneer classical authors (Kostrov et al. 1969; Kostrov 1974; Dahlen 1977; Husseini 1977; Savage and Walsh 1978; Kostrov and Das 1988; Kanamori 1994; Kanamori and Heaton 2000; Rivera and Kanamori 2005; Abercrombie et al. 2006; Cocco et al. 2006; Kanamori and Rivera 2006; among others). But due to the fact that he misinterpreted the physical meaning of the total potential energy released E expressed by Eq. (13) of this paper or Eq. (4) of his paper which already has taken into account the contribution of gravitational potential energy (GPE) change ΔE_g. The introduction of the GPE change term into Eqs. (44a) and (44b) in Zhu's paper (Zhu 2013, p. 993) is redundant and leads to erroneous conclusion.

Acknowledgments

Very helpful discussions with Professors Shamita Das and Raul Madagiaga were greatly appreciated. The research was supported by the National Natural Science Foundation of China (No. 41090291).

References

Abercrombie R, McGarr A, Toro G D, Kanamori H (2006) *Earthquakes: Radiated Energy and the Physics of Faulting*. AGU Geophysical Monograph **170**. AGU, Washington, DC, 327pp

Cocco M, Spudich P, Tinti E (2006) On the mechanical work absorbed on faults during earthquake ruptures. In: Abercrombie R, McGarr A, Toro G D, Kanamori H (eds) *Earthquakes: Radiated Energy and the Physics of Faulting*. AGU Geophysical Monograph **170**. AGU, Washington, D C, pp 237–254

Dahlen F A (1977) The balance of energy in earthquake faulting. *Geophys J R Astron Soc* **48**: 239–260

Dahlen F A, Tromp J (1998) *Theoretical Global Seismology*. Princeton University Press, Princeton, 1025pp

Husseini M I (1977) Energy balance for formation along a fault. *Geophys J R Astron Soc* **49**: 699–714

Kanamori H (1994) Mechanics of earthquakes. *Ann Rev Earth Planet Sci* **22**: 207–237

Kanamori H (2001) Energy budget of earthquakes and seismic efficiency. In: Teisseyre R, Majewski E (eds) *Thermodynamics and Phase Transformations in the Earth's Interior*. Chapter 11. Academic, New York, pp 293–305

Kanamori H, Heaton T H (2000) Microscopic and macroscopic physics of earthquakes. In: Rundle J B, Turcotte D L, Klein W (eds) *GeoComplexity and the Physics of Earthquakes*. AGU Monograph **120**. AGU, Washington, D C, pp 147–163

Kanamori H, Rivera L (2006) Energy partitioning during an earthquake. In: Abercrombie R, McGarr A, Toro G D, Kanamori H (eds) *Earthquakes: Radiated Energy and the Physics of Faulting*. AGU Geophysical Monograph **170**. AGU, Washington, D C, pp 3–13

Kostrov B V (1974) Seismic moment and energy of earthquake and seismic flow of rock (in Russian). *Izv Phys Solid Earth* **1**: 23–40

Kostrov B V (1975) *The Mechanics of the Focus of Tectonic Earthquake* (in Russian). Nauka Publisher, Moscow, 176pp

Kostrov B V, Das S (1988) *Principles of Earthquake Source Mechanics*. Cambridge University Press, Cambridge, 286pp

Kostrov B V, Nikitin L V, Flitman L M (1969) The mechanics of brittle fracture. *MTT* **3**: 112–125 (English transl)

Rivera L, Kanamori H (2005) Representations of the radiated energy in earthquakes. *Geophys J Int* **162**: 148–155

Rudnicki J W, Freund L B (1981) On energy radiation from seismic sources. *Bull Seismol Soc Am* **71**: 583–595

Savage J C, Walsh J B (1978) Gravitational energy and faulting. *Bull Seismol Soc Am* **68**: 1613–1622

Zhu P P (2013) Normal and shear stress acting on arbitrarily oriented faults, earthquake energy, crustal GPE change, and the coefficient of friction. *J Seismol* **17**(3): 985–1000

断层厚度的地震效应和非对称地震矩张量

刘超[1]　陈运泰[1,2]

1. 中国地震局地球物理研究所，北京 100081；2. 北京大学地球与空间科学学院，北京 100871

摘要　本文导出了具有厚度和滑动弱化区域的断层的非对称地震矩张量表示式，指出要求地震矩张量具有对称性不是一个绝对必要的限制．在非对称地震矩张量中，位错项对应于对称地震矩张量，拉力项对应于非对称地震矩张量．由于拉力项等效于单力偶，所以在非对称地震矩张量解的两个节面上，沿滑动矢量方向的力偶强度不再相同，与较大力偶相联系的节面为断层面，与较小力偶相联系的节面为辅助面．这一性质可用以从两个正交的节面中判断哪一个节面是断层面．如果忽略拉力项，会高估与位错对应的标量地震矩．只有满足相应的约束条件的非对称地震矩张量才能表示具有厚度和滑动弱化区域的断层模型，并从中分离出与位错和拉力对应的地震矩张量．

1　引言

1970 年代以来，震源的地震矩张量表示及其反演获得了相当大的成功（Gilbert and Dziewonski，1975；McCowan，1976；Fitch et al.，1980；Dziewonski et al.，1981；Ekström，1989）．常规地震矩张量反演结果表明，大多数的地震是以剪切位错为主的地震，其矩张量解以最佳双力偶成分为主，只有少量的补偿线性矢量偶极（Compensated Linear Vector Dipole，缩写为 CLVD）成分和各向同性成分．已有一些研究工作试图揭示非最佳双力偶成分的起因和特点（Backus and Mulcahy，1976a；Kanamori et al.，1984a；Eissler and Kanamori，1987；Ukawa and Ohtake，1987；Kuge and Kawakatsu，1990；Kawakatsu，1991；Takei and Kumazawa，1995；Julian et al.，1998；Miller et al.，1998；Chouet，2003；Knopoff and Chen，2009）．有一些研究工作表明，火山地震以单力或 CLVD 成分为主（Kanamori et al.，1984b；Eissler and Kanamori，1987；Ukawa and Ohtake，1987；Chouet，2003），而大规模低倾角滑坡引发的地震可能以单力为主（Takei and Kumazawa，1995）．此外，发震断层构造的复杂性也会导致较大的 CLVD 分量（Kuge and Kawakatsu，1990；Kawakatsu，1991）．Knopoff 和 Chen（2009）指出，如果考虑到地震是发生在具有厚度、而不是厚度等于零的断层上，由于存在滑动弱化区域，在远场，地震波所对应的震源项要比不计及断层厚度的剪切位错源多出一个单力偶项．

在震源物理的研究中，作为二阶对称张量的地震矩张量已被广泛接受并且得到了成功的应用，但对地震矩张量为非对称矩张量的情况则鲜有涉及．理论和实践两方面因素造成非对称地震矩张量在过去的研究中被忽视甚而被遗忘．在震源理论方面，通常基于天然地震是发生于地球内部的震源（内源）的前提，从角动量守恒得出地震矩张量必定对称的结论（Backus and Mulcahy，1976a，1976b；Aki and Richards，1980；Kennett，1983）．然而，正如 Takei 和 Kumazawa（1994）通过严格的论证所指出的，非对称地震矩张量是可以合理存在的，它的存在并不违背角动量守恒原理．Knopoff 和 Chen（2009）指出，地震矩张量的对称性实际上是在一定条件下引入的简化和限制的结果（Burridge and Knopoff，1964），并不是绝对必要的．一方面，剪切位错与无矩双力偶的等效性在理论上得到严格的证明后（Burridge and Knopoff，

---

* 本文发表于《地球物理学报》，2014 年，第 57 卷，第 2 期，509–517.

1964),大部分的研究实践便以对称矩张量为基础展开,并取得了相当大的成功,表明对称矩张量是地震震源的很好近似.但是,另一方面,如果研究更为精细的震源模型,例如考虑断层具有厚度和滑动弱化区域的存在(Knopoff and Chen,2009)或考虑震源区的质量的流动(Takei and Kumazawa,1994),则超出了对称矩张量的范畴,便需要引入非对称矩张量以表示相应的地震效应.

本文将在 Knopoff 和 Chen(2009)工作的基础上进一步研究与上述单力偶项对应的地震矩张量问题,引入非对称地震矩张量表示具有厚度和滑动弱化区域的断层模型,分析讨论与之相联系的震源时间函数、断层面解的不确定性、标量地震矩和震源模型对非对称地震矩张量的约束等问题.

2 厚度为零的断层的地震矩张量

大部分天然地震是由断层错动引起.在震源模型中,假设震源区是由内曲面 $\Sigma(\boldsymbol{\xi})$ 所包围的区域,$\boldsymbol{\xi}$ 是内曲面 $\Sigma(\boldsymbol{\xi})$ 上的一点的空间坐标,断层的错动可由内曲面 $\Sigma(\boldsymbol{\xi})$ 上的位移间断即位错表示.

根据表示定理,如果略去体力项,则由内曲面 $\Sigma(\boldsymbol{\xi})$ 上的 $\boldsymbol{\xi}$ 点、τ 时刻的拉力 $T_j(\boldsymbol{\xi},\tau)$ 和位移 $u_j(\boldsymbol{\xi},\tau)$ 在 $\boldsymbol{x}$ 点、t 时刻引起的位移 $u_i(\boldsymbol{x},t)$ 可表示为(Aki and Richards,1980):

$$u_i(\boldsymbol{x},t)=\int_{-\infty}^{\infty}\mathrm{d}\tau\iint_{\Sigma(\boldsymbol{\xi})}[T_j(\boldsymbol{\xi},\tau)G_{ij}(\boldsymbol{x},t;\boldsymbol{\xi},\tau)-u_j(\boldsymbol{\xi},\tau)\times c_{jkpq}(\boldsymbol{\xi})n_k(\boldsymbol{\xi})G_{ip,q}(\boldsymbol{x},t;\boldsymbol{\xi},\tau)]\mathrm{d}\Sigma(\boldsymbol{\xi}),\quad(1)$$

式中,$G_{ij}(\boldsymbol{x},t;\boldsymbol{\xi},\tau)$ 表示格林函数,$c_{jkpq}(\boldsymbol{\xi})$ 表示弹性系数张量,$n_k(\boldsymbol{\xi})$ 表示曲面 $\Sigma(\boldsymbol{\xi})$ 上的外法向单位矢量,下标逗号","表示对空间坐标 $\boldsymbol{\xi}$ 求偏导数.

设内曲面 $\Sigma(\boldsymbol{\xi})$ 系由分开一定间隔的两叶单曲面 $\Sigma^+(\boldsymbol{\xi})$ 和 $\Sigma^-(\boldsymbol{\xi})$ 组成.如果假定断层的厚度为零,即内曲面 $\Sigma(\boldsymbol{\xi})$ 的厚度为零、也即 $\Sigma^+(\boldsymbol{\xi})$ 和 $\Sigma^-(\boldsymbol{\xi})$ 的间距为零.若设拉力 T_j 在 $\Sigma(\boldsymbol{\xi})$ 面上连续,格林函数 $G_{ij}(\boldsymbol{x},t;\boldsymbol{\xi},\tau)$ 及其一阶导数在 $\Sigma^+(\boldsymbol{\xi})$ 和 $\Sigma^-(\boldsymbol{\xi})$ 上相等,那么利用在 $\Sigma^+(\boldsymbol{\xi})$ 与 $\Sigma^-(\boldsymbol{\xi})$ 上的外法向单位矢量 $n_k^+(\boldsymbol{\xi})$ 与 $-n_k^-(\boldsymbol{\xi})$ 方向相反的性质 $n_k^+(\boldsymbol{\xi})=-n_k^-(\boldsymbol{\xi})$ 可将 $\Sigma^+(\boldsymbol{\xi})$ 面上的积分变换为 $\Sigma^-(\boldsymbol{\xi})$ 面上的积分.对于含有位移 $u_j(\boldsymbol{x},t)$ 的项(式(1)右边被积函数的第 2 项),由式(1)可得:

$$u_i(\boldsymbol{x},t)=\int_{-\infty}^{\infty}\mathrm{d}\tau\iint_{\Sigma}[u_j]c_{jkpq}n_kG_{ip,q}\mathrm{d}\Sigma=\int_{-\infty}^{\infty}\mathrm{d}\tau\iint_{\Sigma}m_{pq}^uG_{ip,q}\mathrm{d}\Sigma,\quad(2)$$

式中,略去 Σ^- 的上角标,以 Σ 表示 Σ^-,并且,为简明起见,将自变量略去不写;$[u_j]$ 表示位错,m_{pq}^u 为与位错 $[u_j]$ 对应的矩密度张量,用上标 u 表示.

如果用 Δu 表示位错的幅值,用 e_j 表示位错的方向即滑动矢量,则可将位错 $[u_j]$ 表示为:

$$[u_j]=u_j^+-u_j^-=\Delta u e_j.$$

在均匀、各向同性和完全弹性的介质中,弹性系数张量为:

$$c_{jkpq}=\lambda\delta_{jk}\delta_{qp}+\mu(\delta_{jp}\delta_{kq}+\delta_{jq}\delta_{kp}),$$

式中,λ 和 μ 为拉梅(Lamé)弹性系数,δ_{jk} 为克朗内克(Kronecker)δ,

$$\delta_{jk}=1,\quad 若 j=k,$$

$$\delta_{jk}=0,\quad 若 j\neq k.$$

利用剪切位错的滑动矢量 e_j 与位错面的法向矢量 n_k 垂直的条件,$e_m n_m=0$,可得到矩密度张量 m_{pq}^u 的表示式:

$$m_{pq}^u=[u_j]c_{jkpq}n_k=\mu(e_pn_q+e_qn_p)\Delta u.$$

由该式不难看出,与剪切位错相联系的地震矩密度张量具有对称性:

$$m_{jk}^u=m_{kj}^u.$$

对于式(1)中含有 $T_j(\boldsymbol{\xi},\tau)$ 的项(式(1)右边被积函数的第 1 项),因 $n_k^+=-n_k^-$,$T_j^-=\tau_{jk}^-n_k^-=T^-e_j$,$T_j^+=$

$\tau_{jk}^+ n_k^+ = -\tau_{jk}^- n_k^- = -T^+ e_j$，$\tau_{jk}^+$ 与 τ_{jk}^- 分别是作用于 Σ^+ 与 Σ^- 面上的剪切应力，所以，$(T_j^+ + T_j^-) = -(T^+ - T^-) e_j = -[T] e_j$，相应的位移场 $u_i(\boldsymbol{x}, t)$ 可表示为：

$$u_i(\boldsymbol{x},t) = -\int_{-\infty}^{\infty} d\tau \iint_{\Sigma} [T] e_j G_{ij}(\boldsymbol{x},t;\boldsymbol{\xi},\tau) d\Sigma, \quad (3)$$

式中，与式（2）类似，以 Σ 表示 Σ^-，$[T] = T_j^+ - T_j^-$ 即应力错. 若 T 在 Σ 面上连续，则 $[T] = 0$. 所以在此情况下，（3）式中应力错 $[T]$ 的积分、即应力错对位移场的贡献为零.

以上分析表明，厚度为零的断层的矩密度张量 m_{jk}^u 只与位错 $[u_j]$ 相关，与拉力项 T_j 无关，m_{jk}^u 为二阶对称矩张量. 与这一震源模型对应的矩张量即迄今被广泛运用的地震矩张量.

3 具有厚度和滑动弱化区域的断层的非对称地震矩张量表示

3.1 非对称地震矩张量表示式

Knopoff 和 Chen（2009）考虑到断层具有厚度和滑动弱化区域，在图 1 所示的震源模型中以破裂面（断层面）的两侧 Σ^+ 和 Σ^- 分开一段距离 ΔW 表示. 他们指出，位移项引起的远场位移与断层厚度为零时相同，拉力项引起的远场位移不再为零. 在这种震源模型中，应力的变化会辐射出地震波，因此震源的矩张量表达式中出现与应力变化相联系的项.

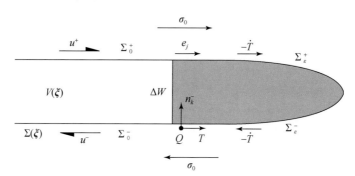

图 1　具有厚度和滑动弱化区的断层模型

σ_0 为背景构造应力，u 为断层的滑动量，e_j 为滑动矢量. 震源区 $V(\boldsymbol{\xi})$ 由内曲面 $\Sigma = \Sigma_0 + \Sigma_\varepsilon$ 包围. 内曲面 $\Sigma(\boldsymbol{\xi})$ 由位错面 $\Sigma_0 = \Sigma_0^+ + \Sigma_0^-$ 和包围强度弱化区（浅灰色区）的曲面 $\Sigma_\varepsilon = \Sigma_\varepsilon^+ + \Sigma_\varepsilon^-$ 组成. ΔW 表示强度弱化区的厚度，$n_k(\boldsymbol{\xi})$ 表示内曲面 Σ 的外法向. T 为滑动单元 $Q(\boldsymbol{\xi})$ 上所受拉力（据 Knopoff and Chen, 2009）

Fig. 1　Fault model of finite thickness and slip-weakening zone

σ_0 is the initial stress or the ambient stress in the surrounding media, u is the slip, e_j is slip vector. The source region with volume $V(\boldsymbol{\xi})$ is enclosed by internal surface $\Sigma = \Sigma_0 + \Sigma_\varepsilon$. The internal surface $\Sigma(\boldsymbol{\xi})$ consists of the dislocation surface $\Sigma_0 = \Sigma_0^+ + \Sigma_0^-$ and the surface of strength-weakening zone (area in light gray) $\Sigma_\varepsilon = \Sigma_\varepsilon^+ + \Sigma_\varepsilon^-$. ΔW is the thickness of the strength-weakening zone, $n_k(\boldsymbol{\xi})$ is the outward drawn normal to surface Σ. T is the traction on the slip element $Q(\boldsymbol{\xi})$

(After Knopoff and Chen, 2009)

与拉力项相关的远场位移可表示为（据 Knopoff and Chen, 2009, p1096, 公式 (23.1), (23.2)）：

$$u_i(\boldsymbol{x},t) = -\iint_{\Sigma_\varepsilon} \frac{\gamma_i \gamma_j e_j}{4\pi\rho\alpha^2 R} \frac{\partial T\left(\boldsymbol{\xi}, t - \frac{R}{\alpha}\right)}{\partial t} \frac{\gamma_k n_k}{\alpha} \Delta W d\Sigma(\boldsymbol{\xi}) - \iint_{\Sigma_\varepsilon} \frac{\psi_{ij} e_j}{4\pi\rho\beta^2 R} \frac{\partial T\left(\boldsymbol{\xi}, t - \frac{R}{\beta}\right)}{\partial t} \frac{\gamma_k n_k}{\beta} \Delta W d\Sigma(\boldsymbol{\xi}), \quad (4)$$

式中，积分区域为包围强度弱化区的曲面 Σ_ε，与式（2）类似，略去 Σ_ε^- 的上角标以 Σ_ε 表示 Σ_ε^-. ρ 表示介质密度，α 和 β 分别表示 P 波和 S 波的速度. R 表示由源点 $Q(\boldsymbol{\xi})$ 指向观测点 $P(\boldsymbol{x})$ 的距离矢量 $\boldsymbol{R} = \boldsymbol{x} - \boldsymbol{\xi}$，

γ_i 为 $\mathbf{R}$ 的方向余弦 $\gamma_i = R_i/R$. $\psi_{ij} = \delta_{ij} - \gamma_i\gamma_j$ 是旋转算子.

与拉力相联系的远场位移表达式, 需要满足断层厚度 ΔW 远小于源点到观测点的距离 $R (\Delta W/R \ll 1)$ 的近似条件和断层厚度 ΔW 远小于波速 c 与应力变化特征时间 T'_s 之积的近似条件 $(\Delta W/cT'_s \ll 1)$ $(T'_s = t'_s - t_p,$ 参见图 2). 远场位移 $u_i(\mathbf{x}, t)$ 的幅值与拉力减小的变化率 $-\partial T/\partial t$ 成正比, 与强度弱化区的厚度 ΔW 成正比.

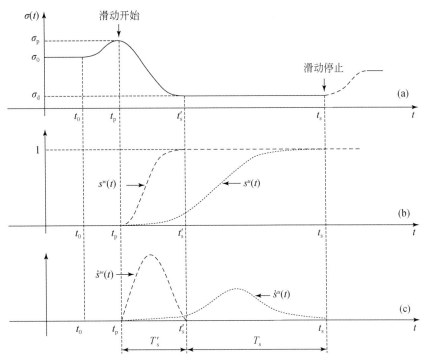

图 2 (a) 断层面上某一具有代表性的点 $\boldsymbol{\xi}$ 上的应力随时间的变化 (Yamashita, 1976). 破裂开始时, 应力从初始值 σ_0 增加到峰值 σ_p. 随着滑动量从零增加到临界滑动弱化距离, 应力由 σ_p 降至动态摩擦应力 σ_d, 并一直保持到滑动停止. (b) 与矩密度张量 $m^u_{jk}(\boldsymbol{\xi}, t)$ 和 $m^w_{jk}(\boldsymbol{\xi}, t)$ 对应的归一化震源时间函数 $s^u_{jk}(\boldsymbol{\xi}, t)$ 和 $s^w_{jk}(\boldsymbol{\xi}, t)$. t_p 表示破裂的起始时刻, t'_s 表示应力急剧变化的停止时刻、也即滑动急剧变化的起始时刻, t_s 表示滑动的终止时刻. 震源时间函数 $s^w(\boldsymbol{\xi}, t)$ 用虚线表示, $s^u(\boldsymbol{\xi}, t)$ 用点线表示. (c) 归一化震源时间函数 $s^u(\boldsymbol{\xi}, t)$ 和 $s^w(\boldsymbol{\xi}, t)$ 的变化率 $\dot{s}^u(\boldsymbol{\xi}, t)$ 和 $\dot{s}^w(\boldsymbol{\xi}, t)$

Fig. 2 (a) Time dependence of stress at a representative point $\boldsymbol{\xi}$ on the fault plane (Yamashita, 1976). When rupture initiates, stress increase from σ_0 to peak stress σ_p. As slip increase from zero to critical slip-weakening distance, stress drops from σ_p to dynamic frictional stress σ_d until slip comes to a stop. (b) Normalized source time functions $s^u_{jk}(\boldsymbol{\xi}, t)$ and $s^w_{jk}(\boldsymbol{\xi}, t)$ corresponding to moment density tensor $m^u_{jk}(\boldsymbol{\xi}, t)$ and $m^w_{jk}(\boldsymbol{\xi}, t)$, respectively. t_p represents rupture initial time, t'_s and t_s represent stopping times for stress and slip changes, respectively. The source time functions $s^w(\boldsymbol{\xi}, t)$ and $s^u(\boldsymbol{\xi}, t)$ are schematically showed by dashed lines and dotted lines respectively. (c) $\dot{s}^u(\boldsymbol{\xi}, t)$ and $\dot{s}^w(\boldsymbol{\xi}, t)$ are the time derivatives of normalized source time functions $s^u(\boldsymbol{\xi}, t)$ and $s^w(\boldsymbol{\xi}, t)$

断层面上某一具有代表性的点上的应力 σ 在地震过程中的变化如图 2a 所示. 如果只考虑应力 σ 从峰值应力 σ_p 下降到动态摩擦应力 σ_d 的滑动弱化过程 $(\sigma_p \geq \sigma \geq \sigma_d)$, 用 $\sigma(\boldsymbol{\xi}, t)$ 表示滑动弱化过程中曲面 Σ_ε 上的拉力值, $\sigma(\boldsymbol{\xi}, t) = T(\boldsymbol{\xi}, t)$, 则有关系:

$$\Delta\sigma(\boldsymbol{\xi}, t) = \sigma_p - \sigma(\boldsymbol{\xi}, t) = \sigma_p - T(\boldsymbol{\xi}, t),$$
$$\frac{\partial \Delta\sigma(\boldsymbol{\xi}, t)}{\partial t} = \frac{\partial(\sigma_p - \sigma(\boldsymbol{\xi}, t))}{\partial t} = -\frac{\partial \sigma(\boldsymbol{\xi}, t)}{\partial t} = -\frac{\partial T(\boldsymbol{\xi}, t)}{\partial t}. \tag{5}$$

式中 $\Delta\sigma(\boldsymbol{\xi}, t)$ 为滑动弱化过程中的有效剪切应力. 由 $\sigma_p \geq \sigma(\boldsymbol{\xi}, t)$ 可知 $\Delta\sigma(\boldsymbol{\xi}, t) > 0$. 滑动弱化过程中, 拉力急剧减小, 拉力减小的变化率即有效剪应力的变化率大于零, $-\partial T/\partial t = \partial\Delta\sigma/\partial t > 0$. 有效剪切应力 $\Delta\sigma$ 具有与位错幅度 Δu 相似的性质. 在破裂发生前, $\Delta\sigma = 0$, $\Delta u = 0$; 一旦发生破裂, 应力降 $\Delta\sigma$ 与位错幅度

Δu 及其变化率 $\Delta\dot{\sigma}$ 与 $\Delta\dot{u}$ 均大于零：$\Delta\sigma>0$，$\Delta u>0$ 且 $\Delta\dot{\sigma}>0$，$\Delta\dot{u}>0$. 将式 (5) 代入式 (4)，得：

$$u_i(\boldsymbol{x},t) = -\iint_{\Sigma_\varepsilon} \frac{\gamma_i\gamma_j e_j}{4\pi\rho\alpha^2 R} \frac{\partial \Delta\sigma\left(\boldsymbol{\xi}, t-\frac{R}{\alpha}\right)}{\partial t} \frac{\gamma_k n_k}{\alpha} \Delta W \mathrm{d}\Sigma(\boldsymbol{\xi}) + \iint_{\Sigma_\varepsilon} \frac{\psi_{ij} e_j}{4\pi\rho\beta^2 R} \frac{\partial \Delta\sigma\left(\boldsymbol{\xi}, t-\frac{R}{\beta}\right)}{\partial t} \frac{\gamma_k n_k}{\beta} \Delta W \mathrm{d}\Sigma(\boldsymbol{\xi}).$$

利用狄拉克 (Dirac) δ-函数的性质，可将该式化为：

$$u_i(\boldsymbol{x},t) = \int_{-\infty}^{\infty} \mathrm{d}\tau \iint_{\Sigma_\varepsilon} \Delta\sigma(\boldsymbol{\xi},\tau) \Delta W e_j n_k \frac{1}{4\pi\rho} \times \left\{ \frac{\gamma_i\gamma_j\gamma_k}{\alpha^3 R}\dot\delta\left(t-\tau-\frac{R}{\alpha}\right) + \frac{\psi_{ij}\gamma_k}{\beta^3 R}\dot\delta\left(t-\tau-\frac{R}{\beta}\right) \right\} \mathrm{d}\Sigma(\boldsymbol{\xi}). \quad (6)$$

在均匀、各向同性和完全弹性的无限介质中，格林函数 G_{ij} 的远场表示式为 (Love, 1927; Aki and Richards, 1980)：

$$4\pi\rho G_{ij} = \frac{\gamma_i\gamma_j}{\alpha^2 R}\delta\left(t-\tau-\frac{R}{\alpha}\right) + \frac{\psi_{ij}}{\beta^2 R}\delta\left(t-\tau-\frac{R}{\beta}\right),$$

式中，格林函数 $G_{ij}=G_{ij}(\boldsymbol{x}, t; \boldsymbol{\xi}, \tau)$ 表示 τ 时刻作用在 $Q(\boldsymbol{\xi})$ 点 j 方向的单位脉冲集中力在 t 时刻 $P(\boldsymbol{x})$ 点 i 方向引起的位移. 利用该式计算 G_{ij} 的空间偏导数 $G_{ij,k}$ 可得：

$$G_{ij,k} = \frac{1}{4\pi\rho} \frac{\partial}{\partial \xi_k} \left\{ \frac{\gamma_i\gamma_j}{\alpha^2 R}\delta\left(t-\tau-\frac{R}{\alpha}\right) + \frac{\psi_{ij}}{\beta^2 R}\delta\left(t-\tau-\frac{R}{\beta}\right) \right\}.$$

利用等式 $\partial R/\partial \xi_k = -\gamma_k$ 和 $\partial \gamma_j/\partial \xi_k = -\psi_{jk}/R$，只保留与 $1/R$ 同阶的远场项，可得 $G_{ij,k}$ 的远场近似表示式：

$$4\pi\rho G_{ij,k} \approx \frac{\gamma_i\gamma_j\gamma_k}{\alpha^3 R}\dot\delta\left(t-\tau-\frac{R}{\alpha}\right) + \frac{\psi_{ij}\gamma_k}{\beta^3 R}\dot\delta\left(t-\tau-\frac{R}{\beta}\right).$$

将其代入式 (6) 即得：

$$u_i(\boldsymbol{x},t) = \int_{-\infty}^{\infty} \mathrm{d}\tau \iint_{\Sigma_\varepsilon} \Delta\sigma(\boldsymbol{\xi},\tau) \Delta W e_j n_k G_{ij,k} \mathrm{d}\Sigma(\boldsymbol{\xi}) = \int_{-\infty}^{\infty} \mathrm{d}\tau \iint_{\Sigma_\varepsilon} m_{jk}^w G_{ij,k} \mathrm{d}\Sigma(\boldsymbol{\xi}), \quad (7)$$

式中，

$$m_{jk}^w = \Delta\sigma \Delta W e_j n_k, \quad (8)$$

即是与有效剪应力相联系的矩密度张量 m_{jk}^w 的表达式，用上标 w 表示. m_{jk}^w 为二阶非对称矩张量，$m_{jk}^w \neq m_{kj}^w$，m_{jk}^w 不具有对称性.

在具有厚度和滑动弱化区域的断层模型中，如果近似条件 $\Delta W/R \ll 1$ 和 $\Delta W/cT_s \ll 1$ 得到满足 ($T_s = t_s - t_p$ 为参考点滑动的特征时间，参见图 2a)，与位移项 u_j 对应的矩密度张量与 $\Delta W = 0$ 时的表示式相同，用带有上标 u 的量 m_{jk}^u 表示. m_{jk}^u 的推导过程与 m_{jk}^w 类似 (刘超，2011)，在此不再赘述. 利用 m_{jk}^u 的表示式：

$$m_{jk}^u = \mu\Delta u(e_j n_k + e_k n_j),$$

可得与位移项 u 相联系的远场位移表示式：

$$u_i(\boldsymbol{x},t) = \int_{-\infty}^{\infty} \mathrm{d}\tau \iint_{\Sigma} m_{jk}^u G_{ij,k} \mathrm{d}\Sigma. \quad (9)$$

需要特别指出的是，虽然式 (9) 和式 (2) 具有相同的数学形式，但式 (9) 的得出利用了 $G_{ij,k}$ 在远场的近似表达式，只在远场条件下成立，而式 (2) 中并未涉及格林函数的具体表达式和远场近似，不受远场近似条件的限制.

对于一次地震事件，考虑整个断层的错动，将矩密度张量 $m_{jk}^u(\boldsymbol{\xi},t)$ 与 $m_{jk}^w(\boldsymbol{\xi},t)$ 在断层面上积分，可得到随时间变化的地震矩张量表示式：

$$M_{jk}^u(t) = \iint_\Sigma m_{jk}^u(\boldsymbol{\xi},t)\mathrm{d}\Sigma(\boldsymbol{\xi}) = \mu\Delta u(t) A(e_j n_k + e_k n_j),$$

$$M_{jk}^w(t) = \iint_{\Sigma_\varepsilon} m_{jk}^w(\boldsymbol{\xi},t)\mathrm{d}\Sigma(\boldsymbol{\xi}) = \Delta\sigma(t) \Delta W A_\varepsilon e_j n_k,$$

$$M_{jk}(t) = M_{jk}^u(t) + M_{jk}^w(t) = \mu\Delta u(t) A(e_j n_k + e_k n_j) + \Delta\sigma(t) \Delta W A_\varepsilon e_j n_k. \quad (10)$$

式中，A 为断层的面积，A_ε 为滑动弱化区域的面积. M_{jk} 为与震源对应的地震矩张量，简称矩张量. 由于

M_{jk}^u 为二阶对称张量，M_{jk}^w 为二阶非对称张量，所以 M_{jk} 为二阶非对称张量，不具有对称性 $M_{jk} \neq M_{kj}$.

考虑位移项和拉力项的作用，由式（7），式（9）和式（10）可将远场位移 $u_j(\boldsymbol{x}, t)$ 表示为：

$$\begin{aligned}u_i(\boldsymbol{x},t) &= \int_{-\infty}^{\infty} \mathrm{d}\tau \iint_{\Sigma} m_{jk}^u G_{ij,k} \mathrm{d}\Sigma + \int_{-\infty}^{\infty} \mathrm{d}\tau \iint_{\Sigma} m_{jk}^w G_{ij,k} \mathrm{d}\Sigma \\ &= (M_{jk}^u(t) + M_{jk}^w(t)) * G_{ij,k}(\boldsymbol{x},t;\boldsymbol{0},0) \\ &= M_{jk}(t) * G_{ij,k}(\boldsymbol{x},t;\boldsymbol{0},0),\end{aligned} \quad (11)$$

式中，"$*$"表示褶积.

矩张量 M_{jk}^u 具有对称性是因为位错与无矩双力偶等效，净力和净力矩为零. M_{jk}^u 代表的震源项不与震源区域外的介质发生动量和角动量交换，从而震源区的动量和角动量守恒. 矩张量 M_{jk}^w 不具有对称性是因为考虑断层厚度后，应力项等效于单力偶，其净力为零，但净力矩不为零. M_{jk}^w 代表的震源项与震源区域外的介质发生了角动量交换，震源区的角动量不守恒. 通常认为地震震源作为地球介质的内源，角动量应该守恒的结论是震源区的厚度（从而体积）为零的必然结果. 在地震过程中，震源区向外辐射地震波，与周围介质发生相互作用，若只考虑震源区"外"的介质所在区域（Σ 与 Σ_ε 之间的区域）或只考虑震源区本身（Σ_ε 内的区域），即在这两个区域内动量和角动量都可以不守恒. 若将震源区和整个地震波传播介质视为一个系统，所受的外力和力矩为零，则动量和角动量守恒.

3.2 震源时间函数

对于一次地震事件，矩张量 $M_{jk}^u(t)$ 与 $M_{jk}^w(t)$ 分别是矩密度张量 $m_{jk}^u(\boldsymbol{\xi}, t)$ 与 $m_{jk}^w(\boldsymbol{\xi}, t)$ 按位错面面积和滑动弱化区域面积的加权叠加，地震矩张量的震源时间函数既取决于矩密度张量的时间历史，又取决于整个断层面的破裂过程，其中后者起主要作用.

首先考虑断层面上发生剪切滑动的某一具有代表性的点，矩密度张量 $m_{jk}^u(\boldsymbol{\xi}, t)$ 与断层的位错相关，$m_{jk}^w(\boldsymbol{\xi}, t)$ 与断层面上的等效剪应力相关，二者起源不同，时间历史也不相同. 用 $s^u(\boldsymbol{\xi}, t)$ 和 $s^w(\boldsymbol{\xi}, t)$ 分别表示与矩密度张量 $m_{jk}^u(\boldsymbol{\xi}, t)$ 和 $m_{jk}^w(\boldsymbol{\xi}, t)$ 对应的归一化震源时间函数. 如图 2 所示，理论上，$s^u(\boldsymbol{\xi}, t)$ 与 $s^w(\boldsymbol{\xi}, t)$ 均起始于滑动起始时刻 t_p，$s^u(\boldsymbol{\xi}, t)$ 终止于 t_s'，持续时间为 $T_s' = t_s' - t_p$；$s^w(\boldsymbol{\xi}, t)$ 起始于 t_p，终止于 t_s，但是，实际上主要集中地发生在 t_s' 至 t_s，持续时间约为 $T_s \approx t_s - t_s'$，所以 $T_s' + T_s \approx t_s - t_p$.

再考虑整个地震事件，用 $S(t)$ 表示断层面破裂过程的时间函数. 若以 $\Omega^u(t)$ 与 $\Omega^w(t)$ 分别表示与矩张量 $M_{jk}^u(t)$ 和 $M_{jk}^w(t)$ 对应的震源时间函数. 则 $\Omega^u(t) = S(t) * s^u(t)$ 与 $\Omega^w(t) = S(t) * s^w(t)$. 今以 T^{Total} 表示整个断层面破裂过程 $S(t)$ 的持续时间，则由于断层面上一点的滑动持续时间 $T_s \approx t_s - t_s'$，$T_s' \approx t_s' - t_p$ 均远小于整个断层面的破裂时间（前者约等于后者的 10%）（Thomas, 1990; Nielsen and Madariaga, 2003），$T_s \approx t_s - t_s' \ll T^{\text{Total}}$，$T_s' \approx t_s' - t_p \ll T^{\text{Total}}$，$T_s' + T_s \approx t_s - t_p \ll T^{\text{Total}}$，因此 $\Omega^u(t) \approx S(t)$，$\Omega^w(t) \approx S(t)$.

利用断层面破裂过程的时间函数 $S(t)$ 将地震矩张量 $M_{jk}(t)$ 中含有时间的部分分离：

$$\begin{aligned}M_{jk}(t) &= M_{jk}^u \Omega^u(t) + M_{jk}^w \Omega^w(t) \\ &\approx (M_{jk}^u + M_{jk}^w) S(t) \\ &= M_{jk} S(t).\end{aligned}$$

由该式可见，对于非对称地震矩张量 $M_{jk}(t)$，仍可假设其各个分量具有相同的时间历史（Lay and Wallace, 1995），从而将 $S(t)$ 与不含时的 M_{jk} 分离. 在地震矩张量反演中，其时间历史 $S(t)$ 和各个分量的大小 M_{jk} 可分离出来单独反演.

3.3 断层面解的不确定性

对称地震矩张量 M_{jk}^u 求得的断层面解，含两个节面，无法区分哪一个是断层面，哪一个是辅助面. 利用 M_{jk} 中非对称地震矩张量 M_{jk}^w 含有的额外信息，可望通过比较两个节面上沿位错矢量方向单力偶的大小，以区分断层面和辅助面.

由式（8）可知，m_{jk}^w 等效于强度为 $\Delta\sigma\Delta W$，力的作用方向沿 $\pm e_j$ 方向，力臂沿 n_k 方向，力矩沿 $-e\times n$ 的单力偶（图1）. 与 $m_{jk}^u=\mu\Delta u(e_j n_k+n_j e_k)$ 对比可发现，m_{jk}^w 与 m_{jk}^u 中的单力偶 $\mu\Delta u e_j n_k$ 具有相同的力的作用方向 $\pm e_j$，力臂方向 n_k 和力矩方向 $-e\times n$. 在 Σ_ε^+ 面上，拉力 T 沿 $-e_j$ 方向，当 T 减小时，等效于一个沿 e_j 方向作用的力；在 Σ_ε^- 面上，拉力 T 沿 e_j 方向，当 T 减小时，等效于一个沿 $-e_j$ 方向作用的力. 由于 Σ_ε^+ 和 Σ_ε^- 分开一段距离 ΔW，与 Σ_ε 面上拉力 T 减小等效于一个力臂为 ΔW 的单力偶，在 Σ_ε^+ 面上的力沿 e_j 方向，在 Σ_ε^- 面上的力沿 $-e_j$ 方向. 这个单力偶，与位错等效的无矩双力偶中沿位错矢量方向作用的那个单力偶具有相同的力矩方向 $-e\times n$. 这使得沿断层面方向（即滑动矢量 e_j 的方向）的单力偶强度较大，与其相应的矩密度为 $\mu\Delta u+\Delta\sigma\Delta W$；沿辅助面方向的单力偶强度较小，与其相应的矩密度为 $\mu\Delta u$. 利用断层面和辅助面方向单力偶强度不同的性质，如果从地震矩张量中求出两个节面方向的单力偶，则可根据断层的厚度效应判定单力偶较大的节面为断层面，单力偶较小的节面为辅助面.

对于任意非对称矩张量 M_{jk}，总可以将它分解为一个对称张量和一个反对称张量：

$$M_{jk}=M_{(jk)}+M_{[jk]},$$

式中，

$$M_{(jk)}=M_{(kj)}=\frac{1}{2}(M_{jk}+M_{kj}),$$

$$M_{[jk]}=-M_{[kj]}=\frac{1}{2}(M_{jk}-M_{kj}).$$

由对称张量 $M_{(jk)}$ 可得到主轴参数和两个互相正交的节面解. 在图3所示的平面直角坐标系中，用 t 和 p 分别表示与张应力轴（T 轴）和压应力轴（P 轴）对应的归一化本征矢量. 两个节面为剪切破裂面及与其正交的辅助面，其交线沿零轴（B 轴）方向，以 b 表示，并且上述两个节面与 T 轴和 P 轴的夹角均为 $45°$. 可将上述两节面的外法向 $\boldsymbol{n}^1$，$\boldsymbol{n}^2$ 以及矢量 $\boldsymbol{b}$ 表示为：

$$\boldsymbol{n}^1=\frac{\sqrt{2}}{2}(\boldsymbol{t}-\boldsymbol{p}),$$

$$\boldsymbol{n}^2=\frac{\sqrt{2}}{2}(\boldsymbol{t}+\boldsymbol{p}),$$

$$\boldsymbol{b}=\boldsymbol{p}\times\boldsymbol{t}.$$

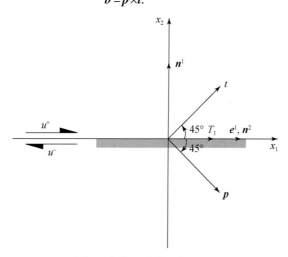

图3　灰色区域表示断层面

其外法向如图中的 $\boldsymbol{n}^1$ 所示. 半箭头分别示意在 Σ^+ 和 Σ^- 上的断层错动方向 u^+ 和 u^- 与幅度. 其他参数的解释参见正文

Fig. 3　Area in light gray denotes fault plane.

The outward draw normal is $\boldsymbol{n}^1$. u^+ and u^- are the slips on Σ^+ and Σ^-, respectively. Other parameters are referred to the text

在外法向为 $\boldsymbol{n}^1$ 的节面（节面1）上，滑动矢量的方向 $\boldsymbol{e}^1$ 与 $\boldsymbol{n}^2$ 相同，即 $\boldsymbol{e}^1 = \boldsymbol{n}^2$，沿 $\boldsymbol{n}^2$ 方向的单力偶强度（偶极矩）可表示为：

$$T_1 = n_j^1 M_{jk} e_k^1 = n_j^1 M_{jk} n_k^2 = \mu \Delta u A + \Delta\sigma \Delta W A_\varepsilon. \tag{12}$$

在外法向为 $\boldsymbol{n}^2$（节面2）的节面上，滑动矢量的方向 $\boldsymbol{e}^2$ 与 $\boldsymbol{n}^1$ 相同，即 $\boldsymbol{e}^2 = \boldsymbol{n}^1$，沿 $\boldsymbol{n}^1$ 方向的单力偶强度（偶极矩）可表示为：

$$T_2 = n_j^2 M_{jk} e_k^2 = n_j^2 M_{jk} n_k^1 = \mu \Delta u A. \tag{13}$$

$T_i (i=1, 2)$ 的单位为 N·m。与 T_i 对应的单力偶中，一个单力为作用在 $\boldsymbol{n}^i$ 面上的拉力，方向与 $\boldsymbol{e}^i$ 相同；另一个单力为作用在 $-\boldsymbol{n}^i$ 面上的拉力，方向与 $\boldsymbol{e}^i$ 相反.

若仅考虑对称张量 $M_{(jk)}$ 时，$T_1 = T_2 = \mu \Delta u A$；考虑到非对称地震矩张量 M_{jk} 时，T_1 和 T_2 的大小一般不再相等. 根据具有厚度和滑动弱化区域的地震断层模型，与较大力偶相联系的节面即为断层面，其力偶的矩为 T_1；与较小力偶相联系的节面为辅助面，相应力偶的矩为 T_2. 比较 T_1 和 T_2 的大小，即可区分断层面和辅助面.

为表示 T_1 和 T_2 大小的差异，定义标量 λ：

$$\lambda = \frac{T_1 - T_2}{T_2} = \frac{T_1}{T_2} - 1, \quad 0 \leq \lambda,$$

量 λ 反映了两个节面上，沿位错矢量方向力偶强度差异的大小. 当 $\lambda = 0$ 时，$T_1 = T_2$，矩张量 M_{jk} 退化为对称矩张量 $M_{(jk)}$，在此情况下，无法分辨断层面和辅助面. 随着反对称部分 $M_{[jk]}$ 比重的增大，T_1 逐渐增大，T_2 逐渐减小，λ 也随之增大.

3.4 标量地震矩

在具有厚度和滑动弱化区域的断层模型中，非对称地震矩张量 M_{jk} 可分解为一个对称矩张量 $M_{(jk)}$ 和一个反对称矩张量 $M_{[jk]}$. 因为 M_{jk}^u 对称，M_{jk}^w 非对称，所以对称分量 $M_{(jk)}$ 中有 M_{jk}^u 和 M_{jk}^w（通过 M_{jk}^w 的对称分量 $M_{(jk)}^w$）两者的贡献. 在此情况下，若假设震源对应的矩张量为对称矩张量，即假定 $M_{jk} = M_{(jk)}$，此矩张量仍然受到断层厚度的地震效应的影响（即受 M_{jk}^w 项的影响）. 将由 $M_{(jk)}$ 求得的标量地震矩记为 M_0'，这便是忽略断层厚度的地震效应得到的标量地震矩.

考虑断层厚度的地震效应后，将位错对应的标量地震矩记为 M_0，拉力项对应的矩记为 M_0^w，由式 (13) 和式 (12) 可得：

$$T_2 = \mu \Delta u A = M_0, \tag{14}$$

$$T_1 = \mu \Delta u A + \Delta\sigma \Delta W A_\varepsilon = M_0 + M_0^w. \tag{15}$$

由于

$$T_1 = n_j^1 (M_{(jk)} + M_{[jk]}) e_k^1 = M_0' + \frac{1}{2} M_0^w,$$

$$T_2 = n_j^2 (M_{(jk)} + M_{[jk]}) e_k^2 = M_0' - \frac{1}{2} M_0^w,$$

代入式 (14) 和式 (15) 后，两式相加得：

$$M_0' = M_0 + \frac{1}{2} M_0^w. \tag{16}$$

(16) 式表明考虑断层厚度的地震效应后，与对称矩张量相应的标量地震矩由 M_0 增大为 M_0'，二者之差为 $M_0^w/2$（Knopoff and Chen, 2009, p1100, 公式 (41)）. 这一结果可用于标量地震矩的对比和修正.

3.5 具有厚度和滑动弱化区域的断层模型对非对称地震矩张量的约束

具有厚度和滑动弱化区域的断层模型与非对称地震矩张量对应，但只有特定的非对称矩张量才能用于表示该震源模型. 或者说, 断层模型对非对称地震矩张量有约束. 相关的约束可作为地震矩张量反演的约束条件.

对于与位错项对应的矩张量 M_{jk}^u, 需要将其约束为对称矩张量 $M_{jk}^u = M_{(jk)}^u$, 还可进一步约束其各向同性分量为零 $M_{kk}^u = 0$.

对于与拉力项对应的矩张量 M_{jk}^w, 由式（8）易得约束条件:

$$M_{jk}^w = M_0^w e_j n_k.$$

该式将单力偶 M_{jk}^w 的力的方向约束为 $\pm e_j$, 力臂的方向约束为 n_k, 力矩的方向约束为 $-e \times n$. 若反演 M_{jk}^w, M_{jk}^w 的力臂方向, 作用力方向和力矩方向应由约束条件控制, 与震源模型保持一致.

对于反演得到矩张量 M_{jk}, 可按 3.3 节中的方法判断出断层面, 再利用约束条件 $M_{jk}^w = M_0^w e_j n_k$ 求出 M_0, M_0^w 和 M_{jk}^w. 这样矩张量 M_{jk} 在约束后变为 M_{jk}': $M_{jk}' = M_{jk}^u + M_{jk}^w = M_0(e_j n_k + e_k n_j) + M_0^w e_j n_k$.

利用约束条件, 我们才可以将 M_{jk}^u 和 M_{jk}^w 从非对称矩张量 M_{jk} 中分离并求解, 得到满足约束条件的地震矩张量 M_{jk}', 通常 $M_{jk}' \neq M_{jk}$. 在从 M_{jk} 求解 M_{jk}' 的过程中, M_{jk} 中与震源模型不一致的部分被舍弃, 使约束后的地震矩张量 M_{jk}' 与模型保持一致.

由厚度的断层模型与非对称地震矩张量的约束可以看出, 只有特定的非对称矩张量 M_{jk}' 才能表示具有厚度和滑动弱化区域的断层模型, 并从中分离出与位错和拉力对应的矩张量 M_{jk}^u 和 M_{jk}^w.

4 讨论与结论

考虑具有厚度和滑动弱化区域的断层模型, 位移项 u 引起的远场位移与断层厚度为零时相同, 拉力项 T 引起的远场位移不再为零. 利用格林函数的远场近似, 可得到震源对应的非对称地震矩张量表示式:

$$M_{jk}(t) = M_{jk}^u(t) + M_{jk}^w(t)$$
$$= \mu \Delta u A(e_j n_k + e_k n_j) + \Delta \sigma \Delta W A_\varepsilon e_j n_k.$$

$M_{jk}^u(t)$ 与断层的位错相关, $M_{jk}^w(t)$ 与断层面上的应力变化相关, 二者起源不同, 时间历史也不相同. 但对于一次地震事件, 考虑整个断层的破裂, $M_{jk}^u(t)$ 和 $M_{jk}^w(t)$ 的时间历史主要是由断层的破裂过程决定的, 因此近似地具有相同的归一化震源时间函数.

非对称地震矩张量与对称地震矩张量不同, 它消除了断层面的不确定性. 由于 M_{jk}^w 的存在, 两个节面上, 沿位错矢量方向, 力偶强度不再相同. 根据具有厚度和滑动弱化区域的断层模型, 可判断与较大力偶相联系的节面为断层面, 与较小力偶相联系的节面为辅助面.

考虑断层厚度的地震效应后, 拉力项对标量地震矩也有贡献, 标量地震矩不再是 $M_0 = \mu \Delta u A$, 而是 M_0':

$$M_0' = M_0 + \frac{1}{2} \Delta \sigma \Delta W A_\varepsilon,$$

式中, 因子 1/2 的引进是因为单位矩的单力偶的辐射是单位矩的双力偶的辐射的 1/2. 因此, 若忽略拉力项, 只按通常的标量地震矩计算, 会高估位错项对应的标量地震矩: $(1/2)\Delta \sigma \Delta W A_\varepsilon$ (Knopoff and Chen, 2009).

并非所有的非对称地震矩张量都能用于表示有厚度的断层. 在对对称矩张量的常见约束基础上, 由 $M_{jk}^w = M_0^w e_j n_k$. 可知, 震源模型对非对称矩张量的约束为: 与 M_{jk}^w 对应的单力偶, 力的方向为 $\pm e_j$, 力臂的方向为 n_k, 力矩的方向为 $-e \times n$. 利用约束条件, 人们才可以将 M_{jk}^u 和 M_{jk}^w 从非对称矩张量 M_{jk} 中分离出来并求解, 从而得到与具有厚度和滑动弱化区的断层模型协调的地震矩张量 M_{jk}'.

本文在得到非对称地震矩张量表示式后, 分析讨论了震源时间函数、断层面解的不确定性、标量地震矩和震源模型对非对称地震矩张量的约束. 这些结果为非对称地震矩张量反演工作提供了理论基础.

References

Aki K, Richards P G. 1980. *Quantitative Seismology: Theory and Methods*. San Francisco: W H Freeman.

Backus G, Mulcahy M. 1976a. Moment tensors and other phenomenological descriptions of seismic sources—I. Discontinuous displacements. *Geophys. J. R. A. S.*, **46**(2): 341-361.

Backus G, Mulcahy M. 1976b. Moment tensors and other phenomenological descriptions of seismic sources—II. Discontinuous displacements. *Geophys. J. R. A. S.*, **47**(2): 301-329, doi: 10.1111/j.1365-246X.1976.tb01275.x.

Burridge R, Knopoff L. 1964. Body force equivalents for seismic dislocations. *Bull. Seism. Soc. Amer.*, **54**(6A): 1875-1888.

Chouet B, Dawson P, Ohminato T, et al. 2003. Source mechanisms of explosions at Stromboli volcano, Italy, determined from moment-tensor inversions of very-long-period data. *J. Geophys. Res.*, **108**(B1): ESE 7-1-ESE 7-25, doi: 10.1029/2002jb001919.

Dziewonski A M, Chou T A, Woodhouse J H. 1981. Determination of earthquake source parameters from waveform data for studies of global and regional seismicity. *J. Geophys. Res.*, **86**(B4): 2825-2852, doi: 10.1029/JB086iB04p02825.

Eissler H K, Kanamori H. 1987. A single-force model for the 1975 Kalapana, Hawaii, earthquake. *J. Geophys. Res.*, **92**(B6): 4827-4836, doi: 10.1029/JB092iB06p04827.

Ekström G. 1989. A very broad band inversion method for the recovery of earthquake source parameters. *Tectonophysics*, **166**(1-3): 73-100, doi: 10.1016/0040-1951(89)90206-0.

Fitch T J, McCowan D W, Shields M W. 1980. Estimation of the seismic moment tensor from teleseismic body wave data with applications to intraplate and mantle earthquakes. *J. Geophys. Res.*, **85**(B7): 3817-3828, doi: 10.1029/JB085iB07p03817.

Gilbert F, Dziewonski A M. 1975. An application of normal mode theory to the retrieval of structural parameters and source mechanisms from seismic spectra. *Phil. Trans. Roy. Soc. London*, **278**(1280): 187-269.

Julian B R, Miller A D, Foulger G R. 1998. Non-double-couple earthquakes 1. Theory. *Rev. Geophys.*, **36**(4): 525-549, doi: 10.1029/98rg00716.

Kanamori H, Given J W, Lay T. 1984a. Analysis of seismic body waves excited by the mount st. Helens eruption of May 18, 1980. *J. Geophys. Res.: Solid Earth*, **89**(B3): 1856-1866.

Kawakatsu H. 1991. Enigma of earthquakes at ridge-transform-fault plate boundaries-distribution of non-double couple parameter of Harvard CMT solutions. *J. Geophys. Res.*, **18**(6): 1103-1106.

Kennett B L N. 1983. *Seismic Wave Propagation in Stratified Media*. Cambridge: Cambridge University Press.

Knopoff L, Chen Y T. 2009. Single-couple component of far-field radiation from dynamical fractures. *Bull. Seism. Soc. Amer.*, **99**(2B): 1091-1102, doi: 10.1785/0120080288.

Kuge K, Kawakatsu H. 1990. Analysis of a deep "non double couple" earthquake using very broadband data. *Geophys. Res. Lett.*, **17**(3): 227-230, doi: 10.1029/GL017i003p00227.

Lay T, Wallace T C. 1995. *Modern Global Seismology*. New York: Academic Press.

Liu C. 2011. *Seismic Effect of Finite Fault Thickness and Asymmetric Moment Tensor* [Ph. D. thesis] (in Chinese). Beijing: Institute of Geophysics, CEA.

Love A E H. 1927. *A Treatise on the Mathematical Theory of Elasticity*. New York: Cambridge University Press.

McCowan D W. 1976. Moment tensor representation of surface wave sources. *Geophys. J. Int.*, **44**(3): 595-599, doi: 10.1111/j.1365-246X.1976.tb00295.x.

Miller A D, Foulger G R, Julian B R. 1998. Non-double-couple earthquakes 2. Observations. *Rev. Geophys.*, **36**(4): 551-568, doi: 10.1029/98rg00717.

Nielsen S, Madariaga R. 2003. On the self-healing fracture mode. *Bull. Seism. Soc. Amer.*, **93**(6): 2375-2388, doi: 10.1785/0120020090.

Takei Y, Kumazawa M. 1994. Why have the single force and torque been excluded from seismic source models? *Geophys. J. Int.*, **118**(1): 20-30.

Takei Y, Kumazawa M. 1995. Phenomenological representation and kinematics of general seismic sources including the seismic vector *modes. Geophys. J. Int.*, **121**(3): 641–662, doi: 10.1111/j.1365-246X.1995.tb06428.x.

Thomas H H. 1990. Evidence for and implications of self-healing pulses of slip in earthquake rupture. *Phys. Earth Planet. Int.*, **64**(1): 1–20, doi: 10.1016/0031-9201(90)90002-f.

Ukawa M, Ohtake M. 1987. A monochromatic earthquake suggesting deep-seated magmatic activity beneath the Izu-Ooshima volcano, Japan. *J. Geophys. Res.*, **92**(B12): 12649–12663, doi: 10.1029/JB092iB12p12649.

Yamashita T. 1976. On the dynamical process of fault motion in the presence of friction and inhomogeneous initial stress. I. Rupture propagation. *Journal of Physics of the Earth*, **24**(4): 417–444.

附中文参考文献

刘超. 2011. 断层厚度的地震效应和非对称矩张量 [博士论文]. 北京: 中国地震局地球物理研究所.

Seismic effect of fault thickness and asymmetric seismic moment tensor

Liu Chao[1] and Chen Yun-Tai[1,2]

1. Institute of Geophysics, China Earthquake Administration, Beijing 100081, China;
2. School of Earth and Space Sciences, Peking University, Beijing 100871, China

Abstract Asymmetric seismic moment tensor representation of fault with finite thickness and slip-weakening zone was introduced. It is pointed out that the symmetry constraint on seismic moment tensor is not necessary. In asymmetric seismic moment tensor, the displacement dislocation term corresponds to a symmetric moment tensor, while the traction term corresponds to an asymmetric moment tensor. As their duration times are approximately equal, we can assume that they have the same normalized source time function. Due to the traction term is equivalent to a single-couple, the strengths of two single-couples are no longer the same (along the slip direction on two nodal planes). Fault ambiguity can be resolved by the fact that the real fault plane is with a larger couple, while the auxiliary plane is with a smaller couple. The scalar seismic moment corresponding to dislocation will be overestimated if the traction term is not taken into account. Appropriate constraints should be imposed when one use asymmetric moment tensors to represent the earthquake source of finite thickness and slip-weakening zone, and derive the moment tensor corresponding to the traction separately from that corresponding to the displacement.

Automatic imaging of earthquake rupture processes by iterative deconvolution and stacking of high-rate GPS and strong motion seismograms[*]

Yong Zhang[1,2], Rongjiang Wang[2], Jochen Zschau[2], Yun-tai Chen[1,3], Stefano Parolai[2] and Torsten Dahm[2]

1. School of Earth and Space Sciences, Peking University, Beijing, China;
2. Helmholtz Centre Potsdam, GFZ German Research Centre for Geosciences, Potsdam, Germany;
3. Institute of Geophysics, China Earthquake Administration, Beijing, China

Abstract By combining the complementary advantages of conventional network inversion and backprojection methods, we have developed an iterative deconvolution and stacking (IDS) approach for imaging earthquake rupture processes with near-field complete waveform data. This new approach does not need any manual adjustment of the physical (empirical) constraints, such as restricting the rupture time and duration, and smoothing the spatiotemporal slip distribution. Therefore, it has the ability to image complex multiple ruptures automatically. The advantages of the IDS method over traditional linear or nonlinear optimization algorithms are demonstrated by the case studies of the 2008 Wenchuan and 2011 Tohoku earthquakes. For such large earthquakes, the IDS method is considerably more stable and efficient than previous inversion methods. Additionally, the robustness of this method is demonstrated by comprehensive synthetic tests, indicating its potential contribution to tsunami and earthquake early warning and rapid response systems. It is also shown that the IDS method can be used for teleseismic waveform inversions. For the two major earthquakes discussed here, the IDS method can provide, without tuning any physical or empirical constraints, teleseismic rupture models consistent with those derived from the near-field GPS and strong motion data.

1 Introduction

In recent decades, seismologists have expended great efforts in developing methods for the rapid determination of earthquake source parameters, which include the hypocenter, the origin time, the magnitude, the moment tensor and/or focal mechanism, and the kinematic rupture process. Nowadays, the estimation of the hypocenter and magnitude has become an automated routine of seismological agencies around the world. Some institutions, such as the United States Geological Survey (USGS), the Global Centroid Moment Tensor (GCMT) group [Ekström et al., 2012], and the German Research Centre for Geosciences, can release their automatic (sometimes manually revised) moment tensor solutions within 1h to several hours. For the rapid response of earthquake emergency and/or earthquake early warning (EEW), however, hypocentral location, magnitude,

---

[*] 本文发表于 *J. Geophys. Res., Solid Earth*, 2014 年, 第 **119** 卷, 第 7 期, 1–18.

and moment tensor solution are still not enough to describe all source characteristics, since they cannot offer the slip distribution and the rupture kinematics, which are important for simulating strong ground motion and/or tsunami waves. Therefore, the timely identification of all of these rupture characteristics is one of the most essential requirements for EEW [Allen et al., 2009]. If information about the spatiotemporal evolution of the rupture is available soon after an event's occurrence, intensities and/or tsunami heights can then be predicted for rapid response.

Currently, there are two most commonly used methods for imaging the rupture process: the conventional network inversion and the backprojection. In the past, these methods have significantly improved our knowledge of earthquake ruptures and their complexities. The conventional network inversion method can use both near-field and teleseismic waveform data [e.g., Kikuchi and Kanamori, 1982; Olson and Apsel, 1982; Hartzell and Heaton, 1983; Ji et al., 2002; Sekiguchi and Iwata, 2002; Yagi et al., 2004; Vallée, 2004]. The various algorithms of the conventional network inversion can be classified into linear and nonlinear types. The only difference lies in the parameterization of the source time function (STF) of the subfaults. In linear methods, the subfault source time function is also discretized. Normally it is expressed by multiple triangles, equidistantly distributed within a presumed time window [Sekiguchi and Iwata, 2002; Yagi et al., 2004]. The unknowns are the scaling factors of the triangles, which are related linearly to the data. In contrast, nonlinear methods use a preselected elementary source time function (ESTF) for each subfault. Usually, the ESTF is simplified by a single triangular function. In recent years, there have been studies that consider dynamically consistent ESTFs, but they are more commonly used for forward modeling strong ground motion [Pitarka et al., 2000; Guatteri et al., 2003; Tinti et al., 2005; Liu et al., 2006; Graves and Pitarka, 2010], with a few applications for finite-fault inversion [Ji et al., 2002; Wei et al., 2013]. In general, all ESTFs are defined with three free parameters: amplitude, rupture time, and risetime, all of which can be subfault dependent. In particular, the latter two need to be searched for through a nonlinear approach, such as by the use of the simulated annealing [Hartzell and Liu, 1996; Ji et al., 2002]. To achieve a stable and physically reasonable solution, a priori knowledge and additional empirical constraints at the expense of a degree of data fit are necessary in both linear and nonlinear algorithms, for example, restricting the variation range of the rupture velocity and smoothing the spatial-temporal rupture distribution. The weights of these constraints, which determine the compromise among the data fit, the source resolution, and the solution stability, often need to be optimized manually. The strong influence of these constraints on the inversion results makes it difficult to fully automatize the conventional inversion methods and therefore reduces their application potential in early warning and rapid response systems.

The backprojection method is based on the phase interference principle [e.g., Ishii et al., 2005; Krüger and Ohrnberger, 2005]. Generally, it uses single-wave phase data, mostly the direct P waves recorded by a teleseismic seismometer array of limited aperture. Recently, there have been a few studies successfully applying backprojection to high-frequency S waves recorded by strong motion networks [Allmann and Shearer, 2007; Honda and Aoi, 2009; Honda et al., 2008, 2011]. In contrast to the conventional network inversion, backprojection works with fewer free parameters and, in particular, does not need Green's functions. On the other hand, however, the coherency stacking technique used in the backprojection approach does not allow one to retrieve actual slip or slip rate values, but just a relative measure of radiation strength in the frequency band of interest.

In this study, we present a new approach which benefits from the complementary advantages of the conventional network inversion and backprojection methods. To serve in earthquake rapid response and tsunami early warning, the new approach should be robust and efficient so that it can be implemented for the automatic analysis of strong motion and high-rate GPS data. After an introduction to the basic theory and the technical details of the new approach, we will demonstrate the value of the new method using synthetic tests and case studies, namely, for the 2008 Wenchuan and 2011 Tohoku earthquakes. Finally, we will discuss the potential contribution of the new approach to tsunami and earthquake early warning and rapid response systems.

2 Method

In the backprojection approach, seismic records from different stations are stacked after applying appropriate time shifts that depend on their distances to a target subfault. The time shifting is in principle comparable with tuning a satellite antenna. There are two limitations for the backprojection approach. One is that the spatial extent of the seismic network (the antenna aperture) should be much smaller than the distance to the earthquake source so that the seismic signals propagate through about the same path from the source to the network and therefore have coherent waveforms. The second limitation is that the approach can only use a single phase of seismic waves (mostly the direct P waves), because the overlapping of different seismic phases may make the received seismic records incoherent, even within a small network. Because of these two limitations, the backprojection approach is mostly applied to a local or regional seismic array at teleseismic distances ($30°$-$90°$ for direct P waves).

To extend the backprojection technique so that it can work for complete waveform data with arbitrary spatial coverage and also in the low-frequency band, we propose correcting the path and radiation effects by using synthetic (or empirical, if available) Green's functions. Instead of stacking the seismic records directly, we stack apparent source time functions (ASTFs) observed at different stations. The ASTFs of a selected subfault can be obtained by deconvolving its Green's functions from the observed seismograms. For each observation site, the ASTF of the selected subfault consists in principle of two pieces of information. One is the true source signal from this subfault and the other is the aliasing signal induced from the remaining subfaults. The former is coherent and in phase for all stations, but the latter is generally incoherent and in different phases for different stations. Through the stacking, the true source signal will be amplified, and the aliasing signals will be mostly canceled by each other. At this step, the new method works in a similar way to the backprojection approach. Note that the presented scheme differs from how the ASTF deconvolution has been carried out, in many previous studies, e.g., Mori and Hartzell [1990], Dreger [1994], Velasco et al. [1994], and Chen and Xu [2000], where the earthquake is approximated by a point source. In those previous studies, by stacking the ASTFs, one can obtain a rough estimate of the STF for the whole earthquake without considering spatial rupture variability [Xu et al., 2002]. In this study, however, we stack the ASTFs for each subfault aiming to retrieve the subfault STF.

Figure 1 shows a simple synthetic test which describes how the new method works. The input source model includes two ruptured subfaults (A and B), whose STFs are described by a triangle and a trapezoid, respectively (Figure 1, middle). Synthetic data (velocity seismograms) are generated for a network of observation sites around the assumed fault. The network coverage relative to the fault is designed to be comparable with the case of the 2011 Tohoku earthquake. For any location on the fault plane, the local ASTFs are calculated by deconvolving the data with the synthetic Green's function based on the focal mechanisms of strike = $193°$, dip = $14°$, and rake = $81°$, released by USGS (http://comcat.cr.usgs.gov/earthquakes/eventpage/pde20110311054624120_29#

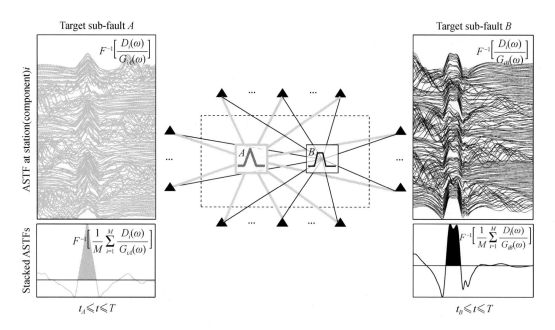

Figure 1 Description of the principle behind the deconvolution and stacking approach

In the input model, two subfaults (left) A and (right) B within a potential fault are ruptured, whose (middle) STFs are described by a triangle and a trapezoid, respectively. Synthetic waveform data (velocity seismograms) are generated for a network of observation sites (black triangles) around the potential fault (dashed rectangle). The network coverage relative to the fault is designed to be comparable with the case of the 2011 Tohoku earthquake (see Figure 5). (left and right) The comparisons of the ASTFs and their stacked results for the target subfaults A and B, respectively. Parameters t_A and t_B are the earliest possible rupture times of subfaults A and B, respectively, and T is the preestimated time window. For each subfault, the maximum positive wavelet (shadowed part) of the stacked ASTF is used as the inferred STF in the IDS method presented in this study

scientific). The results show that almost all the ASTFs for each subfault exhibit a coherent signature of their own input STF and some incoherent noise induced by the other subfault STFs. After stacking the ASTFs, the incoherent noise becomes vanishingly small, while the coherent part is amplified, leading to an inferred STF close to the true one. Through this deconvolution and stacking approach, the whole potential rupture area can be scanned patch by patch. When the subfault STFs are detected, their contributions to the observed data are estimated through forward modeling and then removed from the data. Usually, the procedure is repeated until the cumulative seismic moment converges or the misfit between the data and the synthetics cannot be further reduced. In this sense, the new method may be called the iterative deconvolution and stacking (IDS) method. In the following, we will outline the theory and the developed algorithm of this IDS method.

We define $d_i(t)$ as the data from an observation site (station) i ($i=1, 2, \cdots, M$), $s_j(t)$ as the STF (slip rate time history) of subfault ($j=1, 2, \cdots, N$), and $g_{ij}(t)$ as the Green's function, i. e., the synthetic data at station i caused by a delta slip rate impulse at subfault j, where t ($0 \leq t \leq T$) is the time since the earthquake's occurrence. The observation equations then read as

$$d_i(t) = \sum_{j}^{N} \int_{0}^{t} g_{ij}(t-\tau) \dot{s}_j(\tau) \mathrm{d}\tau, \quad (i=1,2,\cdots,M). \tag{1}$$

If the Fourier spectra of $d_i(t)$, $\dot{s}_j(t)$, and $g_{ij}(t)$ are denoted by $D_i(\omega)$, $S_j(\omega)$, and $G_{ij}(\omega)$, respectively, the frequency domain observation equations then read as

$$D_i(\omega) = \sum_{j=1}^{N} [G_{ij}(\omega) S_j(\omega)], \quad (i=1,2,\cdots,M), \tag{2}$$

where ω is the angular frequency.

For any selected subfault J, we may stack the ASTFs observed at all the stations and obtain

$$\frac{1}{M}\sum_{i=1}^{M}\frac{D_i(\omega)}{G_{iJ}(\omega)} = S_J(\omega) + \frac{1}{M}\sum_{i=1}^{M}\sum_{j\neq J}^{N}\frac{G_{ij}(\omega)S_j(\omega)}{G_{iJ}(\omega)}. \tag{3}$$

The results obtained after stacking include two pieces of information. The first is the STF of subfault J, which is what we want to extract. The second is the disturbance caused by other subfault STFs, which can be rewritten in the form

$$\Delta_J(\omega) = \frac{1}{M}\sum_{j\neq J}^{N}S_j(\omega)\sum_{i=1}^{M}\frac{G_{ij}(\omega)}{G_{iJ}(\omega)}. \tag{4}$$

In the case of an ideal network coverage around the fault, $G_{ij,j\neq J}$ is on average incoherent with G_{iJ}. Thus, we suppose that Δ_J can be negligible compared to S_J and suggest the approximation

$$S_J(\omega) \approx \frac{1}{M}\sum_{i=1}^{M}\frac{D_i(\omega)}{G_{iJ}(\omega)}. \tag{5}$$

Equation (5) means that we can obtain an approximation of each subfault's STF by stacking the deconvolved seismograms. In practice, the deconvolution given in equation (5) can be stabilized by using the so-called water level approach [Helmberger and Wiggins, 1971],

$$S_J(\omega) \approx \frac{1}{M}\sum_{i=1}^{M}\frac{D_i(\omega)G_{iJ}^*(\omega)}{\max(|G_{iJ}(\omega)|^2, |G_{iJ}(\omega)|^2_{\max}\cdot\varepsilon^2)}, \tag{6}$$

where G_{iJ}^* is the complex conjugate of G_{iJ} and ε is the dimensionless water level parameter. The time domain STFs (slip rate time history) are obtained by the inverse Fourier transform,

$$\dot{s}_J(t) \approx F^{-1}[S_J(\omega)]. \tag{7}$$

To ensure stability, we choose ε in equation (6) to be as large as 0.1. The large water level is helpful for minimizing disturbances in imaging the rupture propagation, because it leads to only the most significant rupture signals being retrieved. On the other hand, a complex rupture process that involves temporally discrete episodes of rupture on the same fault patch cannot be retrieved through a single deconvolution and stacking process. Therefore, the deconvolution and stacking procedures normally need to be performed iteratively. Additionally, to minimize artifacts from large oscillatory ASTFs for stations (components) close to the theoretical nodal planes, the STF estimated using equation (7) needs to be examined based on sensitivity tests. For these reasons outlined above, we suggest the following pseudo algorithm:

1. *Discretization.* Discretize the potential rupture area into a number of subfaults identified by $j=1, 2, \cdots, N$, each of which is represented by a point source (so far with uniform source mechanisms).

2. *Deconvolution.* Calculate the ASTFs for each subfault by deconvolving the current residual data with synthetic Green's functions (with known focal mechanism). The current residual data are denoted by $\Delta d_i(t)$ [$=d_i(t)$ in the first iteration] for all observation sites $i=1, 2, \cdots, M$.

3. *Stacking.* Stack the ASTFs of all the stations for each subfault and select the maximum positive wavelet (see Figure 1) as an incremental STF $\Delta\dot{s}_j(t)$ for the concerned subfault in the current iteration.

4. *Scaling.* Scale all the incremental STFs separately, $\Delta\dot{s}_j(t) \Rightarrow A_j\Delta\dot{s}_j(t)$, so that the synthetic data associated with each other best fit the current residual data. In the least squares sense, A_j is calculated by

$$A_j = \frac{\sum_{i=1}^{M}\int_0^T \Delta d_i(t)\Delta y_{ij}(t)\,dt}{\sum_{i=1}^{M}\int_0^T [\Delta y_{ij}(t)]^2\,dt}, \tag{8}$$

where $\Delta y_{ij}(t)$ is the incremental synthetic data at station i produced by $\Delta \dot{s}_j(t)$,

$$\Delta y_{ij}(t) = \int_0^t g_{ij}(t-\tau) \Delta \dot{s}_j(\tau) \, d\tau. \tag{9}$$

Negative A_j values need to be set to zero (positivity constraint). After $\Delta \dot{s}_j(t)$ has been scaled, the updated $\Delta y_{ij}(t)$ has the property that

$$\sum_{i=1}^M \int_0^T [\Delta y_{ij}(t)]^2 \, dt = \sum_{i=1}^M \int_0^T \Delta d_i(t) \Delta y_{ij}(t) \, dt. \tag{10}$$

5. *Regularization*. Modify $\Delta \dot{s}_j(t)$ with a sensitivity factor $f_j \propto R_j^2 \gamma_j$, where R_j^2 is the relative data fit of $\Delta \dot{s}_j(t)$ to $\Delta d_i(t)$,

$$R_j^2 = 1 - \frac{\sum_{i=1}^M \int_0^T [\Delta d_i(t) - \Delta y_{ij}(t)]^2 \, dt}{\sum_{i=1}^M \int_0^T [\Delta d_i(t)]^2 \, dt}, \tag{11}$$

and γ_j is the correlation coefficient between $\Delta y_{ij}(t)$ and the original data $d_i(t)$,

$$\gamma_j = \frac{\sum_{i=1}^M \int_0^T d_i(t) \Delta y_{ij}(t) \, dt}{\sqrt{\sum_{i=1}^M \int_0^T [d_i(t)]^2 \, dt} \sqrt{\sum_{i=1}^M \int_0^T [\Delta y_{ij}(t)]^2 \, dt}}. \tag{12}$$

Negative f_j values also need to be set to zero (again the positivity constraint). Note that by using equation (10), it can be easily proved that R_j^2 is equal to the squared correlation coefficient between $\Delta y_{ij}(t)$ and $\Delta d_i(t)$. Thus, R_j^2 is also called the coefficient of determination.

6. *Smoothing*. In the first iteration, calculate the cumulative slip $\Delta s_j = \int_0^T \Delta \dot{s}_j(t) \, dt$ ($j = 1, 2, \cdots, N$), find the largest asperity (the one including the peak slip value, max (Δs_j)), and determine the azimuthally averaged and normalized height-distance curve $h_o(r)$ for the slip within this asperity, where r is the distance to the peak slip. Parameter $h_o(r)$ will be used as a reference for the characteristic slip roughness. In all the following iterations, the obtained Δs_j are examined for their characteristic roughness, $h(r)$. If $h(r)$ is below $h_o(r)$, i.e., indicates a larger roughness, a space domain low-pass filter (e.g., via a 2-D moving average window) is applied to $\Delta \dot{s}_j(t)$ repeatedly until the resulted $h(r)$ best fits $h_o(r)$.

7. *Re-scaling*. Re-scale $\Delta \dot{s}_j(t)$ globally with a uniform scaling factor A, which is determined so that the synthetic data associated with all $\Delta \dot{s}_j(t)$ collectively best fit the current residual data. Add them to the cumulative STFs $\dot{s}_j(t)$ and then update the residual data for the next iteration. The global scaling factor A is calculated by

$$A = \frac{\sum_{i=1}^M \int_0^T \Delta d_i(t) \sum_{j=1}^N \Delta y_{ij}(t) \, dt}{\sum_{i=1}^M \int_0^T [\sum_{j=1}^N \Delta y_{ij}(t)]^2 \, dt}. \tag{13}$$

8. *Iteration*. Repeat steps 2-7 until the misfit between the cumulative synthetic data and the original data cannot be further reduced, where the misfit function is defined by the ratio of unfitted observed waves compared to the complete observed waves in the least squares sense, i.e.,

$$\sigma^2 = \frac{\sum_{i=1}^{M} \int_0^T \left[d_i(t) - \sum_{j=1}^{N} y_{ij}(t)\right]^2 dt}{\sum_{i=1}^{M} \int_0^T [d_i(t)]^2 dt}.$$

Practically, it is thought to be excellent for the relative misfit values between 0 and 0.2, good for 0.2-0.4, acceptable for 0.4-0.6, and unsatisfactory above 0.6.

Note that steps 5 and 6 are both necessary. Step 5 is the sensitivity-based regularization. By using this step, possible overestimation can be prevented for subfaults located too far from the network or at the sites, which are poorly resolved by the network. For these subfaults, the amplitudes of the Green's functions are relatively small so that the deconvolution results (i.e., the ASTFs) may include large artifacts. We recall that in the classic singular-value decomposition method, the regularization is solved by excluding the eigenvectors associated with zero or very small eigenvalues, while in the conventional inversion methods, which are most used currently, it is solved through the preestimated fault size and/or the moment minimization. Step 6 is the smoothing constraint, which is also commonly used in most geophysical inversions. Using the smoothing rule, the update slip distributions during the iterations have about the same characteristic roughness as obtained after the first iteration without smoothing. The idea behind this rule is that a single iteration of the IDS approach is comparable with backprojection, leading to a result which roughly reflects the resolution ability of the network. It is noted that the frequency bands employed in the IDS and backprojection are different. If the network coverage around the fault is adequate and the rupture is simple enough, a single iteration might be sufficient to obtain the major characteristics of slip distribution.

Additionally it should be noted that in the above algorithm, the time/frequency domain is not expressed in a discrete form. In fact, the IDS method can work with arbitrarily small time intervals without any negative effect on the solution stability. This feature is different to the conventional linear algorithm. On the other hand, it differs from the conventional nonlinear algorithm; in that, the form of the ESTFs is not preselected, but derived directly from the data.

A priori source information used in the IDS method includes the hypocenter location, the focal mechanism, and the fault geometry. In addition, we assume that the rupture propagation can never exceed the P wave velocity and that each subfault can rupture until the time when there is no significant waveform energy left in the observations. In this study, if not otherwise specified, the Green's functions are calculated using the code "Qseis" of Wang [1999] and are based on the seismic reference model AK135 [Kennett et al., 1995], modified with the local crustal structure adopted from CRUST2.0 [Bassin et al., 2000].

In the following, we apply the new method to the 2008 Wenchuan and 2011 Tohoku earthquakes, since their ground motion was well recorded by near-field strong motion and/or high-rate GPS networks.

3 The 2008 M_W7.9 Wenchuan earthquake

The hypocenter of the 12 May 2008 Wenchuan earthquake, which occurred at 06:27:57.59 UTC, is located at (30.018°N, 103.365°E) at a depth of 15.5km [Yang et al., 2012]. We adopt the focal mechanism of strike=225°, dip=39°, and rake=120° from Zhang et al. [2009] and select a large enough potential rupture area of about 810km long and 50km wide, consisting of 81×5=405 subfaults, each 10km×10km in size (Figure 2a). The Wenchuan earthquake was well recorded by the local strong motion network distributed along the ruptured fault. We choose 36 stations with an average station distance of about 50km (Figure 2a). The selected accelerograms are integrated to velocity seismograms and then filtered by a band-pass filter of 0.02-0.10Hz.

Using the IDS method, the moment magnitude of the earthquake convergesat $M_W 7.97$ after only six iterations, which is consistent with the GCMT solution (about 0.05 more). Although the potential fault length is chosen to be as large as 810km, most of the detected rupture is located along the northeast segment of the fault, confirming that the earthquake fault ruptured unilaterally. The fault slip extends about 300km, and the slip maxima appear mostly in the shallowest patches, indicating that the fault slip may have broken the surface. The peak slip value is about

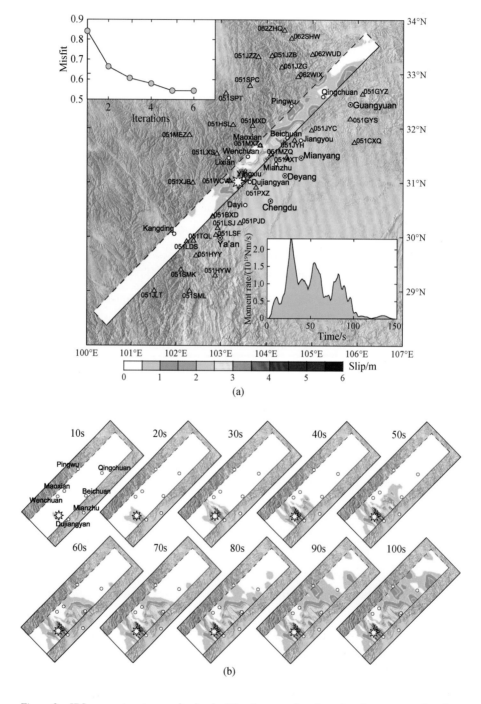

Figure 2　IDS source imaging results for the Wenchuan earthquake using the strong motion data
(a) The misfit curve of (top left) the iterations and (bottom right) the source time function, strong motion stations (cyan triangles), and the surface projection of fault slip distribution. (b) Snapshots of the temporal variations of the fault slip distribution. The stars in Figures 2a and 2b denote the epicenter

5.8m, 30km northeast of the epicenter, where the largest surface rupture was observed [Xu et al., 2009]. Three major slip patches were found to be located near Beichuan, Mianzhu, and Yingxiu, respectively, consistent with reports that the most serious damage occurred in these regions. There is also a slip gap on the fault about 100km northeast of the epicenter (near station 051MZQ; Figure 2a), consistent with other studies [Zhang et al., 2009; Shen et al., 2009; Xu et al., 2010; Wang et al., 2011; G. Zhang et al., 2012]. In the first 20s, the slip initiated around the hypocenter and did not show any dominant rupture directions (Figure 2b). From 30s, the ruptures began to propagate to the NE unilaterally at a shallow depth, reaching Mianzhu and Beichuan at 40-50s and 50-60s, respectively. After 70s, the rupture became relatively weak and scattered, but still propagated to the NE. However, our results show that the rupture did not continuously propagate to the northeast. As shown in Figure 3, some subfault STFs around the hypocenter demonstrate two peaks, indicating that the rupture first propagated from the hypocenter to the northeast and then began to rupture to the southwest. The repeated northeast-southwest rupture around the hypocenter is significant and resulted in the largest subevent occurring around 25s after the rupture initiation, which is consistent with previous studies [Zhang et al., 2009]. Because there are at least two parallel faults near the epicenter that ruptured during the earthquake [Xu et al., 2009], the northeast and southwest rupture propagations in the first 40s may not have occurred on the same fault [Hartzell et al., 2013]. The peak slip rate reached about 0.6m/s and occurred at locations about 30km and 130km to the northeast of the hypocenter. The maximum slip is (5.8m) less than the value of 8-10m determined from the near-field geodetic data [Shen et al., 2009; Xu et al., 2010] but is more consistent with the observed surface ruptures [Xu et al., 2009]. The source time function obtained in this study (lower right inset, Figure 2a) shows that the rupture process lasted about 100s.

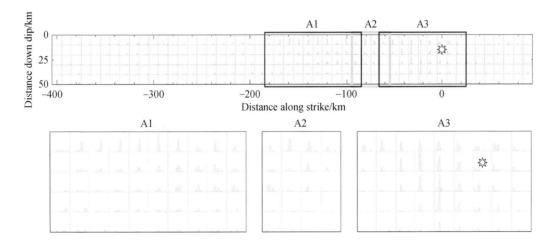

Figure 3 (top) Subfault STFs (slip rate) obtained from the IDS inversion of the strong motion data for the Wenchuan earthquake. In each patch, the x axis is the time, and the y axis is the STF value. The time window showing the STFs is 100s, and the maximum slip rate is 0.6m/s. (bottom) The three zoomed slip asperities

The misfit (σ^2) is 0.54 (i.e., acceptable according to our criteria above; Figure 4). Large discrepancies appear for a few stations, such as 051WCW and 051JYC, whose baseline shifts on the NS component could not be sufficiently removed by the band-pass filter, and 051MZQ, which is very close to the fault trace such that any uncertainties in the fault location and geometry can cause large errors in the synthetic seismograms.

Although only strong motion data were used for the 2008 Wenchuan earthquake, both the slip distribution and

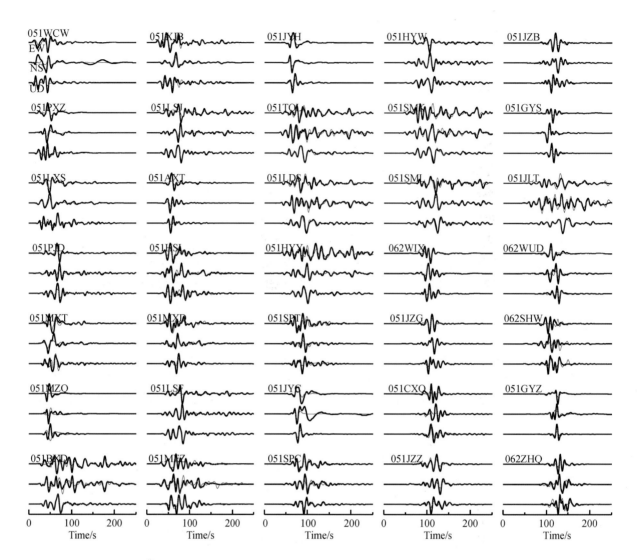

Figure 4　Comparisons of the observed (black) and synthetic (red) seismograms of the strong motion data for the Wenchuan earthquake

the source time function are consistent with the results obtained by the joint inversion of the GPS and interferometric synthetic aperture radar data [Shen et al., 2009; Wang et al., 2011] or all available seismic and geodetic data [Hartzell et al., 2013]. However, there is a significant difference in the peak slip, which is about 5.8 m in our results, less than most of the previous studies but is closer to the results from the field investigations [Xu et al., 2009]. It should be pointed out that we have chosen a long enough potential fault; i.e., no constraint on the fault size was assigned. This would be helpful for the rapid response of earthquake emergency and EEW, in which a very little information about the rupture length and direction is available in the earliest stages. Additionally, it should be noted that for the purpose of rapid response, the rupture process is usually estimated on a single-fault plane. In the case of the Wenchuan earthquake, many previous studies indicated that its rupture process involved multiple complex faults. This feature is reflected by some apparent multiple or repeated ruptures in our single-fault results.

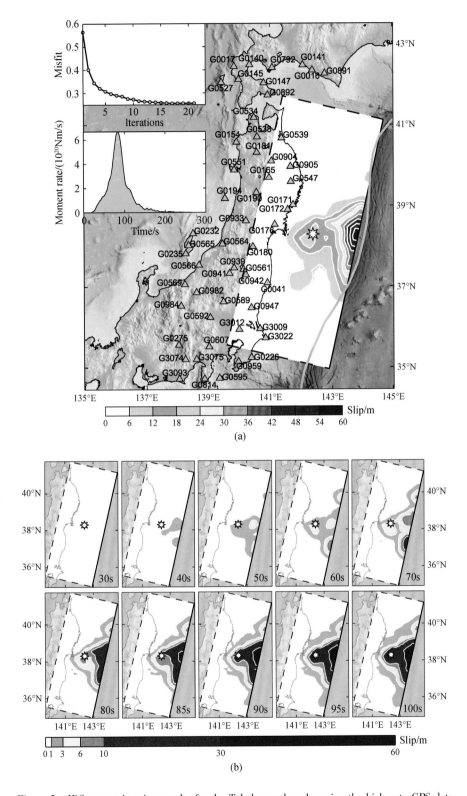

Figure 5 IDS source imaging results for the Tohoku earthquake using the high-rate GPS data
(a) The misfit curve of (upper left) the iterations and (lower right) the source time function, GPS stations (cyan triangles), and the surface projection of fault slip distribution. (b) Snapshots of the temporal variations of the fault slip distribution projected to the surface. The stars in Figures 5a and 5b denote the epicenter

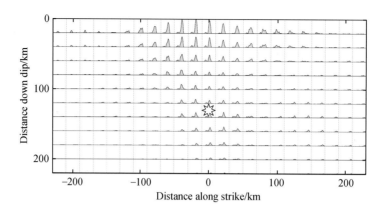

Figure 6 Subfault STFs (slip rate) of the Tohoku earthquake

In each patch, the x axis is the time, and the y axis is the STF value. The time window showing the STFs is 150s, and the maximum slip rate is 2.4m/s. The star denotes the hypocenter

4 The 2011 M_W9.0 Tohoku earthquake

The Tohoku earthquake nucleated at (38.297°N, 142.372°E) at a depth of 30km on 11 March 2011 at 05:46:24 UTC, with a fault plane of strike=193°, dip=14°, and rake=81° (W phase solution of USGS). The latter is consistent with the tectonics of the Japan Trench as a well-known tsunamigenic subduction thrust fault. We select a large potential rupture area on this fault with a length of 700km and width of 300km, consisting of 35×15 = 525 subfaults, each 20km×20km in size (Figure 5a).

We choose 55 high-rate GPS stations, each of which contains three-component displacement seismograms, randomly from 414 stations with an average interstation distance of about 70km (Figure 5a). The displacement records are differentiated to velocity seismograms and then filtered by a low-pass filter of 0-0.05Hz.

Figure 5 shows the inversion results for the Tohoku earthquake. After 21 iterations (inset in Figure 5a), the seismic moment converges at about 4.65×10^{22} Nm, equivalent to a moment magnitude of M_W 9.05. The total rupture process lasted as long as 200s, with 80% of the energy released between 50 and 120s (inset in Figure 5a). The major rupture area (Figure 5a) is located at a very shallow depth, consistent with other studies [Lee et al., 2011; Shao et al., 2011; Y. Zhang et al., 2012a]. The maximum slip determined is close to 60 m, which is larger than some results derived from jointly inverting teleseismic waveform and inland coseismic GPS data [Y. Zhang et al., 2012a], but close to the result of Wang et al. [2013] based on the inversion of inland and seafloor coseismic GPS data and that of Lee et al. [2011], who jointly inverted teleseismic, strong motion, and high-rate GPS waveform data.

Many subfault STFs of the Tohoku earthquake show multirupture features (Figure 6), supporting the findings of Lee et al. [2011]. The whole rupture process can be roughly characterized by two events (Figure 5b). The first one initiated around the hypocenter, propagated along the updip direction, and broke the seafloor. The cumulative slip of this event is estimated to be less than 6 m. The second is the major event of the earthquake, which occurred at a shallow depth and lead to the maximum slip (~57m) near the trench. Because of the coverage of the GPS network on only one side of the rupture, earlier signals near the trench are not easily distinguished from later rupture signals in the deeper area toward the coast. This may explain why the peak moment rate at 85 s appears later here than is resolved from teleseismic inversions [Y. Zhang et al., 2012a; Lee et al., 2011].

The misfit of the inversion is about 0.26 (i.e., good, according to our criteria discussed above). From Figure 7, the data fits with the high-rate GPS seismograms are fairly good, except for the three northeastmost GPS stations (G0141, G0016, and G0891; Figure 5b), whose records show very long coda waves, probably caused by local basin effects (Figure 7).

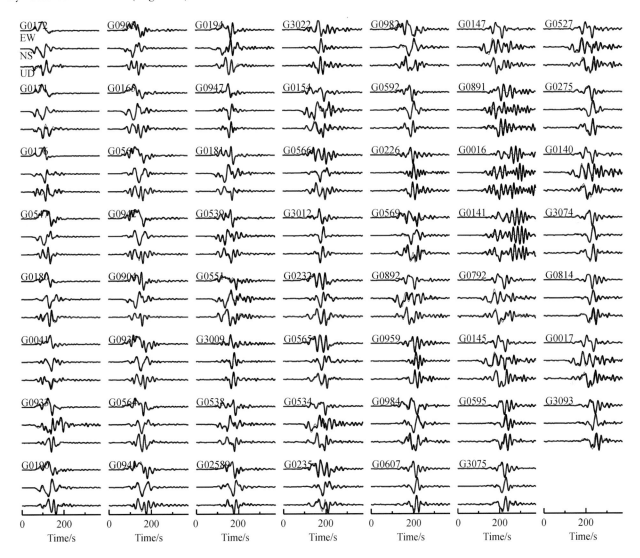

Figure 7　Comparisons of the observed (black) and synthetic (red) high-rate GPS seismograms from the Tohoku earthquake

The Tohoku earthquake was also well recorded by the strong motion network KiK-net. For the rupture process inversion, the only difference in the data processing between the high-rate GPS and the strong motion data is that the latter needs to be additionally high-pass filtered to remove the bias caused by their baseline errors [Wang et al., 2013]. We also applied the IDS method to the strong-motion-based velocity seismograms in the frequency band of 0.02-0.05Hz and obtain results similar to those from the high-rate GPS network (see Figures 12-14 in the supporting information), noting that the earth structure used for the high-rate GPS and strong motion data inversions are the same. The obtained scalar moment is about 4.30×10^{22} Nm, equivalent to a moment magnitude of M_W 9.03. The peak slip is estimated to be 51 m. Probably because of the lack of very low frequency content (<0.02Hz) in the strong motion data, both the peak slip and the moment magnitude are slightly smaller than those from the high-rate GPS data.

An interesting question for the Tohoku earthquake is whether the slip reaches the trench. This source characteristic could not be estimated well using the static coseismic displacement data from the GPS measurements, since all of the GPS Earth Observation Network stations are located on the westside of the trench. In comparison, the waveform inversion has a better resolution because of the Doppler effects contained in the high-rate GPS and strong motion seismograms. Based on our results, the slip maxima occurred at a very shallow depth and reached at least 50 m. This indicates that the ruptures might have broken the seabed, consistent with the observation of results of repeated multibeam bathymetric surveys [Fujiwara et al., 2011].

5 Resolution tests for the 2008 Wenchuan and 2011 Tohoku earthquakes

We conducted resolution tests on the strong motion network for the Wenchuan earthquake and the high-rate GPS network for the Tohoku earthquake, respectively. In these tests, synthetic data are generated based on the assumed rupture models, to which we added 10% Gaussian noise. A constant rupture velocity (3km/s for the Wenchuan earthquake and 2km/s for the Tohoku earthquake) is used in all the input models. When inverting the synthetic data with the IDS method, the maximum rupture velocity is allowed to be as large as the P wave velocity and the rupture duration as long as necessary, implying that practically no assumption on the rupture process is needed.

In the first resolution test (Figure 8) on the strong motion network for the Wenchuan earthquake, four rectangular slip asperities, each with a single triangular subfault STF, are defined on the fault (Figures 8a and 8c). The input and estimated slip and subfault STFs are compared in Figures 8a-8d. The comparison shows that the moment magnitude, fault slip, and the shapes of the subfault STFs are all well retrieved using the IDS method, indicating the reliability of the rupture process that we have reconstructed for this event.

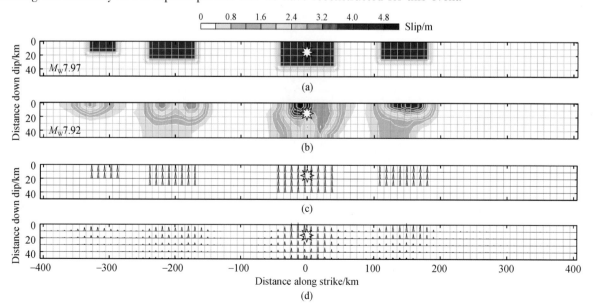

Figure 8 Resolution test for the Wenchuan earthquake

(a) The input slip distribution. (b) The resolved slip distribution by the IDS method. (c) The input STFs of all the ruptured subfaults. (d) The resolved STFs. The stars denote the hypocenter. In each patch in Figures 8c and 8d, the x axis is the time, and the y axis is the STF value. The time window showing STFs is 160s

For the other resolution test performed on the high-rate GPS network of the Tohoku earthquake (Figure 9), the earthquake is represented by two rectangular slip asperities at different depths (Figure 9a). To simulate the multiruptures, the subfault STF, in contrast to the Wenchuan earthquake test, is represented by twin triangles (Figure 9c). Additionally, we recognize that the hypocenter location can be biased by up to several tens of kilometers for subduction events because of the one-sided network coverage. Therefore, the "true" hypocenter location in this test is offset 45km (grey star in Figures 9b and 9d), corresponding to a conservative estimate of the rapid location error for such subduction zone earthquakes. Comparing the input and estimated rupture models shows that even with the strongly biased hypocenter estimate, the two slip asperities and their twin triangular STFs are well resolved.

6 Comparison with the conventional inversion method

We compare the IDS method with the conventional linear network inversion (LNI) method for the rupture imaging of the Wenchuan and Tohoku earthquakes. To stabilize the solution, the LNI method generally needs to specify the weights of different physical constraints. In the code of Y. Zhang et al. [2012b] used for the following tests, three physical constraints, which contain the spatial smoothing, temporal smoothing, and moment minimization, have to be specified with three weighting factors. After certain scaling, the two smoothing factors usually vary between 0 and 50, and the moment-minimizing factor takes a value between 0 and 1. Generally these weighting factors depend on the resolution capability of the network used. A poorly distributed network requires relatively stronger physical constraints to ensure the reasonability and stability of the inversion results. In practice, however, the weighting factors can only be optimized empirically using the trial-and-error approach so that a certain subjective influence on the results is often not avoidable.

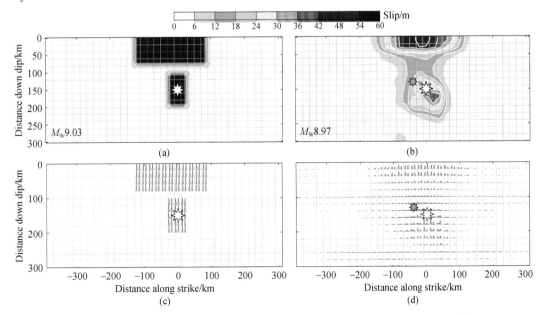

Figure 9 Resolution test for the Tohoku earthquake with a large bias in the hypocenter location
(grey star, 45km away from the true location)

(a) The input slip distribution. (b) The resolved slip distribution by the IDS method. (c) The input STFs of all the ruptured subfaults. (d) The resolved STFs. The stars denote the hypocenter. In each patch in Figures 9c and 9d, the x axis is the time, and the y axis is the STF value. The time window showing STFs is 200s

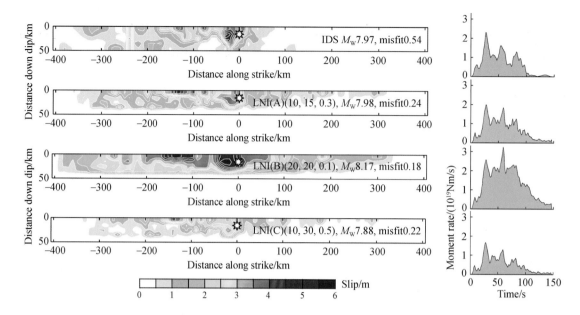

Figure 10 Comparisons of (left) the slip models and (right) the source time functions obtained with the IDS and LNI methods applied to the strong motion data for the Wenchuan earthquake

The first row shows the results of the IDS method, and the second row is the preferred results of the LNI method with proper constraint weights. The third and fourth rows show the other two results of the LNI method obtained during the trials carried out to optimize the constraint weights. The three weighting factors for spatial and temporal smoothing and moment minimizing are given in the parentheses. The stars denote the hypocenter

We use the same parameterizations for the LNI method as used for the IDS method. The maximum rupture velocity is set uniformly to 6km/s (the P wave velocity in the upper crust), and the rupture duration is not limited for each subfault. In the case of the Tohoku earthquake, we downsampled the high-rate GPS data from 1 to 0.25 samples per second in order to reduce the required computational effort for the LNI method. Since the cutoff frequency is 0.05Hz, this downsampling would not miss any effective waveform information.

For each of the two earthquakes, the LNI method is tested for three different combinations of the three constraining weighting factors (denoted as LNI (A), LNI (B), and LNI (C)), and the results obtained are compared with those of the IDS method (Figures 10 and 11). For the Wenchuan earthquake, the combination of 20, 20, and 0.1 for the three constraining weighting factors leads to a smoothed fault slip distribution and overestimated moment magnitude, while the combination of 10, 30, and 0.5 results in an underestimated moment magnitude compared to the model of IDS model. After several trials by empirically adjusting the three weights, we arrive at the combination of 10, 15, and 0.3, which best estimates the fault slip distribution and moment magnitude (Figure 10) and appears to be the most reasonable based on recent studies [Shen et al., 2009; Wang et al., 2011] and the IDS results. For the Tohoku earthquake, the finally preferred weighting factor combination is 0.1, 15, and 0.0, yielding a rupture area of 300-400km along the strike and 150-200km along the dip, with the largest asperity at a shallow depth close to the trench and a moment magnitude of M_W 9.0, all of which are consistent with the results of other independent studies (Figure 11).

Since the LNI method is based on least squares optimization, its data fits are generally better than those of the IDS method. In the case of the Tohoku earthquake, for example, the relative misfit we obtained with the IDS method is 0.26. When using the LNI method, the relative misfit is slightly improved to 0.10-0.19, depending on the choice of empirical parameters for the physical constraints. In the case of the Wenchuan earthquake, the data fit achieved by the LNI method is about 0.3 less compared with that derived by the IDS method. The reason is the narrow and relatively higher frequency range (0.02-0.10Hz) of the strong motion data, which can be better explained by the least squares optimization, although not necessarily also more reasonably than the direct imaging. In fact, the slip models obtained by the LNI method depend strongly on the choice of the three physical constraints, and most of them appear less consistent with the results of recent studies if the three weighting factors were not properly specified. The trade-off problem related to the physical constraints is well known for conventional inversions with either linear or nonlinear algorithms. So far, the weighting of the physical constraints is often optimized manually, leading to difficulties when attempting the automatic use of this method. We note that there have been several studies suggesting the so-called BIC (Bayesian information criterion) optimization of the physical constraints [e.g., Akaike, 1980; Yagi et al., 2004; Fukuda and Johnson, 2008], which can, in principle, be automatized. In the present case, however, the BIC approach would mean a grid search process for a multiparameter space that would generally require a considerably larger computational effort.

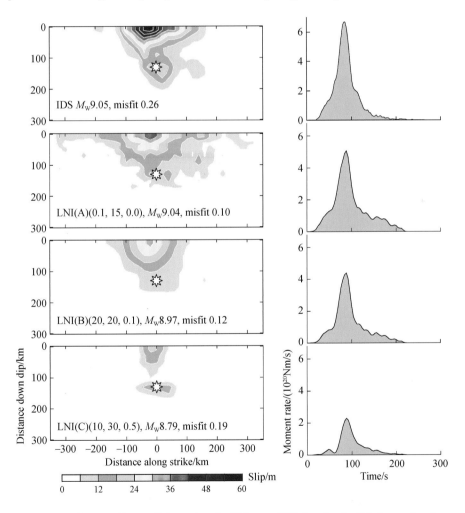

Figure 11 Same as Figure 10 but using the high-rate GPS data for the Tohoku earthquake

7 Discussion and conclusions

One particular advantage of the IDS method over traditional methods is that it can easily retrieve complex rupture processes in a straightforward way, which cannot be easily done when using the nonlinear inversion method. Normally, the linear inversion method can be adopted to resolve multiple ruptures, although such capability is generally limited because it can affect the numerical stability. Using the IDS method, no empirical constraint needs to be changed dependent on different earthquakes. The two case studies and the resolution tests presented above have demonstrated the robustness of this new approach for the inversion of complete near-field seismograms.

In addition, the IDS method is considerably more efficient than the LNI method. The IDS and LNI codes we used for the comparison are written in MATLAB. Neither is specially parallelized but can make use of all CPU cores by using the distributed computing toolbox of MATLAB. Without any optimization of the empirical constraints, the CPU time required by the LNI method on an Intel i7-3770 PC is 1140s for the Wenchuan earthquake and 1060s for the Tohoku earthquake, in comparison to 5 s and 35 s for the IDS method, respectively. It should be noted that the deconvolution and stacking procedures are computed independently from subfault to subfault. Since such calculations are easily parallelized, the computational effort required by the IDS algorithm should not be a problem when integrated into an EEW system.

In general, we can make a theoretical comparison of the computational efforts required by the IDS and linear network inversion (LNI) methods. There are three factors impacting upon the computation time, namely, the number of waveform data channels N_w, the number of subfaults N_s, and the number of time samples N_1 of both waveform data and subfault STFs. For the IDS method, the computation effort is estimated to be $O(N_w \cdot N_s \cdot N_1 \log N_1)$, where the factor $N_1 \log N_1$ is known to be related to the fast Fourier transform needed for the convolution and deconvolution. For the LNI method, since the inversion of a $M \times N$ matrix needs the time $O[\min(M^2 \cdot N, M \cdot N^2)]$, the total computation effort becomes $O[N_1^3 \cdot \min(N_s^2 \cdot N_w, N_s \cdot N_w^2)]$. Note that to achieve as high temporal resolution as possible, the subfault STFs should have the same sampling rate as the waveform data. Thus the number of time samples impacts upon the number of both rows and columns of the inversion matrix, leading to the factor N_1^3. Additionally, in contrast to the LNI method, the major computation part (i.e., convolution and deconvolution) of the IDS method is independent for different stations and subfaults, implying that it can be easily parallelized. If the nonlinear network inversion method is used, the computation effort is even considerably larger than the LNI inversion method.

An issue of the IDS method to be considered is the physical constraint on the roughness of the inferred slip distribution. We solved this problem by an empirical rule (see step 6 of the pseudo algorithm), which has been tested in many synthetic tests and the present two case studies. For both the Wenchuan and Tohoku earthquakes, it yields peak slip values that are acceptable when considering other independent studies. Without this smoothing step, the peak slip would be overestimated by a factor of 2-3. The peak slip value can also be affected by the choice of the sensitivity factor (see step 5 of the pseudo algorithm). However, other relevant source characteristics, such as magnitude, fault size, location of the major slip asperity, rupture propagation, and duration, which are all useful for early warning and hazard assessments, are not significantly dependent on the empirical constraints used here.

The water level is the only parameter that could be changed but is fixed in the IDS method. For numerical stability, we suggest a relatively large water level parameter for deconvolving the data with synthetic Green's functions, because this leads to only the most significant rupture signals being retrieved during the iterations. If the water level is too small, overestimated slip may occasionally appear in earlier iterations, which cannot be corrected in the subsequent iterations because of the imposed positivity constraint (see step 3 of the pseudo algorithm). By using a large water level parameter, this effect can be sufficiently minimized. In the present two case studies (and many other tests not shown here), we found that no substantial change in the inversion results if the water level parameter ε used in equation (6) is larger than 0.05 (see Figures 15 and 16 in the supporting information). Therefore, we can fix the water level at 0.1 to ensure the automaticity of IDS method.

We have also tested the IDS method for teleseismic inversions. In this case, when using conventional inversion methods, a general problem is that the synthetic Green's functions based on a standard Earth model have large uncertainties in travel times. Consequently, it is difficult to accurately resolve the rupture distribution absolutely. The same problem also exists when using the IDS method. Hence, similarly to conventional inversion methods, we shift the P wave arrival time of the data so that it is consistent with the Earth model used and attempt to resolve the rupture distribution relative to the given hypocenter. For the Wenchuan and Tohoku earthquakes, the teleseismic rupture models produced by the IDS method are consistent with those from the near-field data inversions, except for slightly underestimated magnitudes, lower spatial resolutions, and underestimates of the peak slips (see Figures 17-22 in the supporting information). These underestimations may be associated with the low resolution of the teleseismic networks compared to the near-field networks. For very large earthquakes such as the Tohoku event, because of the lack of low and zero frequencies, the moment magnitude as well as the peak slip were underestimated using teleseismic waveform data. We note that in order to test the robustness of the new method for teleseismic inversions, all of the parameters for the physical (empirical) constraints are fully adopted from those used for the near-field data inversions. The same is also true for the cutoff frequencies applied to the data (0.01-0.1Hz for the Wenchuan earthquake and 0.01-0.05Hz for the Tohoku earthquake) and the model parameterizations. In particular, we note the significant differences between the teleseismic results of the IDS method and the backprojection method. The latter generally only recovers the high-frequency source signals along the deep edge of the ruptured area [Honda et al., 2011; Meng et al., 2011; Yao et al., 2011]. Such information is interesting for understanding the earthquake nucleation process but less relevant for tsunami early warning.

We recognize that there are still a few practical issues and limitations for the application of the IDS method to EEW. First, while existing EEW systems can determine the hypocenter in real time, they are generally not able to provide an equally rapid focal mechanism. Fortunately, many large earthquakes occur on existing faults with well-known mechanisms. For example, the focal mechanisms of the Wenchuan earthquake and the Tohoku earthquake are both consistent with the tectonic background of their epicentral regions. Especially, for most tsunamigenic subduction earthquakes, the fault geometries have been well determined [Hayes et al., 2012]. There are also some special earthquakes with an unexpected mechanism controlled by the regional tectonics that were not very clear previously, such as the 2013 M_W8.6 Sumatra earthquake [Yue et al., 2012]. Therefore, for EEW in a given area, the mechanisms of the seismogenic faults should be carefully investigated (e.g., by geological investigations into inland earthquakes). Once an earthquake over a certain threshold magnitude (for example, M_W6.5) is detected and well located, we can therefore immediately determine which fault has ruptured and thus assume a corresponding mechanism. For this case, uncertainties of the hypocenter location may impact upon the selections of the ruptured fault and its mechanism. Thus, a densely distributed network would be required to reduce the location uncertainties. In addition, it would be useful to extend the IDS method with a module for rapid mechanism estimation using, e.g., the approach suggested by Melgar et al. [2013]. Second, it should be noted that the fault size, subfault size, and the cutoff frequency used in the present case studies are fixed. In EEW, however, since we do not have the knowledge of the final magnitude of the earthquake, these three parameters need to be adjusted during the real-time inversions. In most cases, a large enough potential fault size can be fixed based on the tectonic setting of interest. To achieve an appropriate resolution, the subfault size and frequency band can be changed dynamically, depending on the increased earthquake magnitude detected during the real-time analysis. For example, supposing that the lowest moment magnitude of disastrous earthquakes is about $M_W 6.5$, we can first use the subfault size of 5km and cutoff frequency of 0.2Hz ($f_{cut} \sim \frac{1}{4}\frac{V_s}{\Delta L}$, where V_s is the S wave velocity and ΔL is the subfault length [see, e.g., Heimann, 2010], respectively. Once the moment magnitude exceeds $M_W 7.5$, for example, the subfault size can be increased to 10km, and the cutoff frequency can be decreased correspondingly to 0.1Hz, and the inversion performed once more. If the moment magnitude further exceeds $M_W 8.5$, the subfault size and cutoff frequency can be updated again until the moment magnitude converges (Table 1).

Table 1 Subfault sizes and cutoff frequencies suggested for the real-time source imaging

M_W	Subfault Size/km	Cutoff Frequency/Hz
6.5-7.5	5	0.2
7.5-8.5	10	0.1
8.5-9.5	20	0.05

In summary, the IDS method presented in this paper shows its robustness both in synthetic tests and in its application to the case studies of the 2008 Wenchuan and 2011 Tohoku earthquakes. In particular, this approach does not need any manual adjustment of the physical (empirical) constraints and has the ability to automatically image earthquake sources using either near-field or teleseismic waveform data. This may have an important impact on the development of future earthquake and tsunami early warning and rapid response systems.

Acknowledgments

This paper is funded by the REAKT project (Towards Real-Time Earthquake Risk Reduction) of the European Seventh Framework Programme (grant agreement 282862). The strong motion and GPS data for the Tohoku earthquake were provided by the National Research Institute for Earth Science and Disaster Prevention and the Geospatial Information Authority of Japan, respectively. The strong motion data for the Wenchuan earthquake were provided by the China Strong Motion Networks Center. We thank Martin Mai, Gavin Hayes, and another anonymous reviewer for their constructive comments. Kevin Fleming proofread the manuscript.

References

Akaike, H. (1980), Likelihood and Bayes procedure, in *Bayesian Statistics*, edited by J. M. Bernardo et al., pp. 143–166, Univ. Press, Valencia, Spain.

Allen, M. R., P. Gasparini, O. Kamigaichi, and M. Bose (2009), The status of earthquake early warning around the world: An introductory overview, *Seismol. Res. Lett.*, **80**(5), 682–693, doi: 10.1785/gssrl.80.5.682.

Allmann, B. P., and P. M. Shearer (2007), A high-frequency secondary event during the 2004 Parkfield earthquake, *Science*, **318**, 1279–1283, doi: 10.1126/science.1146537.

Bassin, C., G. Laske, and G. Masters (2000), The current limits of resolution for surface wave tomography in North America, *Eos*

Trans. AGU, **81**, F897.

Chen, Y. T., and L. S. Xu (2000), A time-domain inversion technique for the tempo-spatial distribution of slip on a finite fault plane with applications to recent large earthquakes in the Tibetan Plateau, *Geophys. J. Int.*, **143**, 407–416.

Dreger, D. S. (1994), Empirical Green's function study of the January 17, 1994 Northridge, California earthquake, *Geophys. Res. Lett.*, **21**, 2633–2636, doi: 10.1029/94GL02661.

Ekström, G., M. Nettles, and A. M. Dziewoński (2012), The global CMT project 2004-2010: Centroid-moment tensors for 13,017 earthquakes, *Phys. Earth Planet. Inter.*, 200–201, 1–9, doi: 10.1016/j.pepi.2012.04.002.

Fujiwara, T., S. Kodaira, T. No, Y. Kaiho, N. Takahashi, and Y. Kaneda (2011), The 2011 Tohoku-Oki earthquake: Displacement reaching the trench axis, *Science*, **334**, 1240, doi: 10.1126/science.1211554.

Fukuda, J., and K. M. Johnson (2008), A fully Bayesian inversion for spatial distribution of fault slip with objective smoothing, *Bull. Seismol. Soc. Am.*, **98**, 1128–1146.

Graves, R. W., and A. Pitarka (2010), Broadband ground-motion simulation using a hybrid approach, *Bull. Seismol. Soc. Am.*, **100**, 2095–2123.

Guatteri, M., P. M. Mai, G. C. Beroza, and J. Boatwright (2003), Strong ground-motion prediction from stochastic-dynamic source models, *Bull. Seismol. Soc. Am.*, **93**(1), 301–313.

Hartzell, S. H., and T. H. Heaton (1983), Inversion of strong ground motion and teleseismic waveform data for the fault rupture history of the 1979 Imperial Valley, California, earthquake, *Bull. Seismol. Soc. Am.*, **73**, 1553–1583.

Hartzell, S. H., and P. Liu (1996), Calculation of earthquake rupture histories using a hybrid global search algorithm: Application to the 1992 Landers, California, earthquake, *Phys. Earth Planet. Inter.*, **95**, 79–99.

Hartzell, S. H., C. Mendoza, L. Ramirez-Guzman, Y. Zeng, and W. Mooney (2013), Rupture history of the 2008 M_W7.9 Wenchuan, China, earthquake: Evaluation of separate and joint inversions of geodetic, teleseismic, and strong-motion data, *Bull. Seismol. Soc. Am.*, **103**, 353–370, doi: 10.1785/0120120108.

Hayes, G. P., D. J. Wald, and R. L. Johnson (2012), Slab1.0: A three-dimensional model of global subduction zone geometries, *J. Geophys. Res.*, **117**, B01302, doi: 10.1029/2011JB008524.

Heimann, S. (2010), *A Robust Method to Estimate Kinematic Earthquake Source Parameters*, PhD Thesis of University Hamburg, Germany. [Also see http://ediss.sub.uni-hamburg.de/volltexte/2011/5357/.]

Helmberger, D. V., and R. A. Wiggins (1971), Upper mantle structure of midwestern United States, *J. Geophys. Res.*, **76**, 3229–3245, doi: 10.1029/JB076i014p03229.

Honda, R., and S. Aoi (2009), Array back-projection imaging of the 2007 Niigataken Chuetsu-oki earthquake striking the world's largest nuclear power plant, *Bull. Seismol. Soc. Am.*, **99**, 141–147, doi: 10.1785/0120080062.

Honda, R., S. Aoi, H. Sekiguchi, and H. Fujiwara (2008), Imaging an asperity of the 2003 Tokachi-oki earthquake using a dense strong-motion seismograph network, *Geophys. J. Int.*, **172**, 1104–1116, doi: 10.1111/j.1365-246X.2007.03702.x.

Honda, R., et al. (2011), A complex rupture image of the 2011 off the Pacific coast of Tohoku earthquake revealed by the MeSO-net, *Earth Planets Space*, **63**, 583–588.

Ishii, M., P. Shearer, H. Houston, and J. E. Vidale (2005), Extent, duration and speed of the 2004 Sumatra-Andaman earthquake imaged by the Hi-Net array, *Nature*, **435**, 933–936.

Ji, C., D. J. Wald, and D. V. Helmberger (2002), Source description of the 1999 Hector Mine, California, earthquake, Part I: Wavelet domain inversion theory and resolution analysis, *Bull. Seismol. Soc. Am.*, **92**, 1192–1207.

Kennett, B. L. N., E. R. Engdahl, and R. Buland (1995), Constraints on seismic velocities in the earth from traveltimes, *Geophys. J. Int.*, **122**, 108–124.

Kikuchi, M., and H. Kanamori (1982), Inversion of complex body waves, *Bull. Seismol. Soc. Am.*, **72**, 491–506.

Krüger, F., and M. Ohrnberger (2005), Tracking the rupture of the $M_W = 9.3$ Sumatra earthquake over 1,150 km at teleseismic distance, *Nature*, **435**, 937–939.

Lee, S. J., B. S. Huang, M. Ando, H. C. Chiu, and J. H. Wang (2011), Evidence of large scale repeating slip during the 2011

Tohoku-Oki earthquake, *Geophys. Res. Lett.*, **38**, L19306, doi: 10.1029/2011GL049580.

Liu, P., R. Archuleta, and S. H. Hartzell (2006), Prediction of broadband ground-motion time histories: Frequency method with correlated random source parameters, *Bull. Seismol. Soc. Am.*, **96**, 2118–2130.

Melgar, D., W. Crowell, Y. Bock, and J. S. Haase (2013), Rapid modeling of the 2011 M_W 9.0 Tohoku-oki earthquake with seismogeodesy, *Geophys. Res. Lett.*, **40**, 2963–2968, doi: 10.1002/grl.50590.

Meng, L., A. Inbal, and J.-P. Ampuero (2011), A window into the complexity of the dynamic rupture of the 2011 M_W 9 Tohoku-Oki earthquake, *Geophys. Res. Lett.*, **38**, L00G07, doi: 10.1029/2011GL048118.

Mori, J., and S. H. Hartzell (1990), Source inversion of the 1988 Upland, California, earthquake: Determination of a fault plane for a small event, *Bull. Seismol. Soc. Am.*, **80**, 507–517.

Olson, A. H., and R. J. Apsel (1982), Finite faults and inverse theory with applications to the 1979 Imperial Valley earthquake, *Bull. Seismol. Soc. Am.*, **72**, 1969–2001.

Pitarka, A., P. Somerville, Y. Fukushima, T. Uetake, and K. Irikura (2000), Simulation of near-fault strong-ground motion using hybrid Green's functions, *Bull. Seismol. Soc. Am.*, **90**(3), 566–586.

Sekiguchi, H., and T. Iwata (2002), Rupture process of the 1999 Kocaeli, Turkey, earthquake estimated from strong-motion waveforms, *Bull. Seismol. Soc. Am.*, **92**, 300–311.

Shao, G., X. Li, C. Ji, and T. Maeda (2011), Focal mechanism and slip history of the 2011 M_W 9.1 off the Pacific coast of Tohoku earthquake, constrained with teleseismic body and surface waves, *Earth Planets Space*, **63**, 559–564.

Shen, Z. K., J. Sun, P. Zhang, Y. Wan, M. Wang, R. Bürgmann, Y. Zeng, W. Gan, H. Liao, and Q. Wang (2009), Slip maxima at fault junctions and rupturing of barriers during the 2008 Wenchuan earthquake, *Nat. Geosci.*, **2**, 718–724.

Tinti, E., E. Fukuyama, A. Piatanesi, and M. Cocco (2005), A kinematic source-time function compatible with earthquake dynamics, *Bull. Seismol. Soc. Am.*, **95**(4), 1211–1223.

Vallée, M. (2004), Stabilizing the empirical Green's function analysis: Development of the projected Landweber method, *Bull. Seismol. Soc. Am.*, **94**, 394–409.

Velasco, A., C. Ammon, and T. Lay (1994), Empirical Green's function deconvolution of broadband surface waves: Rupture directivity of the 1992 Landers, California (M_W 7.3) earthquake, *Bull. Seismol. Soc. Am.*, **84**, 735–750.

Wang, Q., X. J. Qiao, Q. G. Lan, F. Jeffrey, S. Yang, C. Xu, Y. L. Yang, X. Z. You, K. Tan, and G. Chen (2011), Rupture of deep faults in the 2008 Wenchuan earthquake and uplift of the Longmen Shan, *Nat. Geosci.*, **4**, 634–640, doi: 10.1038/ngeo1210.

Wang, R. (1999), A simple orthonormalization method for stable and efficient computation of Green's functions, *Bull. Seismol. Soc. Am.*, **89**(3), 733–7410.

Wang, R., S. Parolai, M. Ge, M. Jin, T. R. Walter, and J. Zschau (2013), The 2011 M_W 9.0 Tohoku earthquake: Comparison of GPS and strong-motion data, *Bull. Seismol. Soc. Am.*, **103**(2B), 1336–1347, doi: 10.1785/0120110264.

Wei, S., D. Helmberger, Z. Zhan, and R. Graves (2013), Rupture complexity of the M_W 8.3 Sea of Okhotsk earthquake: Rapid triggering of complementary earthquakes? *Geophys. Res. Lett.*, **40**, 5034–5039, doi: 10.1002/grl.50977.

Xu, C., Y. Liu, Y. Wen, and R. Wang (2010), Coseismic slip distribution of the 2008 M_W 7.9 Wenchuan earthquake from joint inversion of GPS and InSAR data, *Bull. Seismol. Soc. Am.*, **100**(5B), 2736–2749, doi: 10.1785/0120090253.

Xu, L. S., G. Patau, and Y. T. Chen (2002), Source time functions of the 1999, Taiwan, M_S 7.6 earthquake retrieved from IRIS and GEOSCOPE long period waveform data using aftershocks as empirical Green's functions, *Acta Seismol. Sin.*, **15**(2), 121–133.

Xu, X. W., X. Z. Wen, G. H. Yu, G. H. Chen, Y. Klinger, J. Hubbard, and J. Shaw (2009), Coseismic reverse- and oblique-slip surface faulting generated by the 2008 M_W 7.9 Wenchuan earthquake, China, *Geology*, **37**, 515–518.

Yagi, Y., T. Mikumo, J. Pacheco, and G. Reyes (2004), Source rupture process of the Tecomán, Colima, Mexico earthquake of 22 January 2003, determined by joint inversion of teleseismic body-wave and near-source data, *Bull. Seismol. Soc. Am.*, **94**, 1795–1807.

Yang, Z. X., Y. T. Chen, J. R. Su, T. C. Chen, and P. Wu (2012), The hypocenter and origin time of the M_W 7.9 Wenchuan earthquake of May 12, 2008, *Acta Seismol. Sin.*, **34**(2), 127–136.

Yao, H., P. Gerstoft, P. M. Shearer, and C. Mecklenbräuker (2011), Compressive sensing of the Tohoku-Oki M_W9.0 earthquake: Frequency dependent rupture modes, *Geophys. Res. Lett.*, **38**, L20310, doi: 10.1029/2011GL049223.

Yue, H., T. Lay, and K. D. Koper (2012), En échelon and orthogonal fault ruptures of the 11 April 2012 great intraplate earthquakes, *Nature*, **490**, 245–249, doi: 10.1038/nature11492.

Zhang, G., M. Vallée, X. Shan, and B. Delouis (2012), Evidence of sudden rupture of a large asperity during the 2008 M_W7.9 Wenchuan earthquake based on strong motion analysis, *Geophys. Res. Lett.*, **39**, L17303, doi: 10.1029/2012GL052516.

Zhang, Y., W. P. Feng, L. S. Xu, C. H. Zhou, and Y. T. Chen (2009), Spatio-temporal rupture process of the 2008 great Wenchuan earthquake, *Sci. China Ser.* D, **52**(2), 145–154.

Zhang, Y., L. S. Xu, and Y. T. Chen (2012a), Rupture process of the 2011 Tohoku earthquake from the joint inversion of teleseismic and GPS data, *Earthquake Sci.*, **25**, 129–135, doi: 10.1007/s11589-012-0839-1.

Zhang, Y., W. P. Feng, Y. T. Chen, L. S. Xu, Z. H. Li, and D. Forrest (2012b), The 2009 L'Aquila M_W6.3 earthquake: A new technique to locate the hypocentre in the joint inversion of earthquake rupture process, *Geophys. J. Int.*, **191**, 1417–1426, doi: 10.1111/j.1365-246X.2012.05694.x.

2014年云南鲁甸 $M_W 6.1$ 地震：一次共轭破裂地震

张 勇[1]　陈运泰[1,2]　许力生[2]　魏 星[3]　金明培[4]　张 森[4]

1. 北京大学地球与空间科学学院，北京　100871；2. 中国地震局地球物理研究所，北京　100081；
3. 中国地震台网中心，北京　100045；4. 云南省地震局，昆明　650041

摘要　烈度与余震分布显示2014年云南鲁甸 $M_W 6.1$($M_S 6.5$) 地震的发震构造较复杂. 为深入了解鲁甸地震的发震断层与破裂特征，本文考虑了单一断层破裂和共轭断层破裂的情况，对震中距250km范围内的近震资料（宽频带资料和强震资料）和远震体波资料进行了反演，得到了鲁甸地震的破裂过程，探讨了滑动分布与余震分布之间的关系. 根据反演得到的滑动分布、震源时间函数和波形拟合，认为鲁甸地震是一次在北西向主压应力与北东向主张应力的统一应力场下发生的两条共轭断层先后破裂的一次复杂地震事件. 在破裂开始后0~2s，破裂主要发生在 ENE—WSW 向（近东西向）的断层上，随后 NNW—SSE 向（近南北向）断层开始破裂，释放了大部分的地震矩. 由于近南北向断层南段（即震中以南）的破裂规模较大，且以左旋走滑为主，对近东西向断层的西段起到了一定程度的解锁作用，可能是震中以西无明显主震破裂但存在密集余震分布的主要原因.

1 引言

2014年8月3日发生在云南鲁甸的 $M_W 6.1$($M_S 6.5$) 地震是一次中等强度的地震事件，但余震分布却相当复杂. 根据最新的余震精定位结果，鲁甸地震的余震分布在 NNW—SSE（以下简称近南北向）和 ENE—WSW（以下简称近东西向）两个方向上都有一定尺度的延伸（王未来等，2014；张广伟等，2014），表明此次地震的发震构造和破裂过程可能具有一定的复杂性. 这一现象引起了地震学界的广泛关注（徐锡伟等，2014；张勇等，2014；刘成利等，2014；张振国等，2014；王未来等，2014；张广伟等，2014）. 如何判别这次地震的发震断层，成为了相关研究工作的一大热点. 目前，烈度分布和破裂过程反演等研究都更倾向于近南北向的小河—包谷垴断裂是此次地震的发震构造（徐锡伟等，2014；张勇等，2014；刘成利等，2014）. 而视震源时间函数分析表明，鲁甸地震可能是共轭断层先后破裂的结果（许力生等，2014）. 另外，张广伟等（2014）根据余震分布和震源机制解结果也提出可能存在共轭破裂的情况. 为更定量地研究两条断层的破裂属性，有必要在破裂过程反演中，进一步考虑共轭断层破裂的可能性.

实际上，主震破裂和余震呈现出复杂分布的地震并不鲜见. 比如2011年盈江地震（Lei et al., 2012）和2012年苏门答腊北部海域 $M_W 8.6$ 地震. 后者的破裂过程中，由于多达三条共轭断层先后发生了破裂（Meng et al., 2012；H. Zhang et al., 2012；Yue et al., 2012），余震分布呈现出比较散乱的特征. 然而，主震破裂与余震分布之间也不总是严格地相互对应. 发生破裂的区域一般伴随着余震分布，但有余震分布的区域不一定有破裂发生. 一个典型的例子是2010年青海玉树地震. 该地震是一次以单侧破裂为主的地震事件，破裂主要朝东南方向传播（Zhang et al., 2010），但在主要破裂区域的另一端，即震中西北约40km处，也存在着比较集中的余震分布（Wang et al., 2013）. 对于2014年鲁甸地震，余震在近东西向和

---
* 本文发表于《地球物理学报》，2015年，第 **58** 卷，第1期，1804–1811.

近南北向都呈现出明显的条带状分布,是两条断层都发生了破裂,还是只有一条断层发生破裂,是震源研究需要关注和解决的问题.

本文搜集了距震中250km范围内强震与近震宽频带的全波形数据,以及远震宽频带体波数据,分别基于单一断层和两条交叉的共轭断层模型,对鲁甸地震的破裂过程进行了反演.通过比较与分析反演结果与反演得到的波形残差,确定鲁甸地震是在北西向主压应力方向与北东向主张应力方向的统一应力场下发生的两条共轭的断层先后破裂的一次复杂地震事件.最后,我们根据得到的震源破裂模型,探讨了共轭断层破裂与形成复杂余震分布的原因.

2 数据

本文选用了13个强震台站(图1青色三角形)和较近距离上11个未限幅的宽频带台站(图1黄色三角形)的三分量全波形记录,作为参与反演的近震数据.在分别对强震数据的加速度记录积分两次与对宽频带数据的速度记录积分一次后,得到相应的位移记录.采用张勇等(2014)使用的远震数据,经过进一步筛选,去掉了信噪比较低的2个P波台站和1个SH波台站,得到了26条垂直向P波波形和31条SH波波形(图2),作为参与反演的远震数据.与近震数据一样,将远震数据积分得到位移记录.为去掉近震数据存在的低频基线漂移,并且考虑到宽频带仪器的有效频段响应范围,我们对近震和远震数据

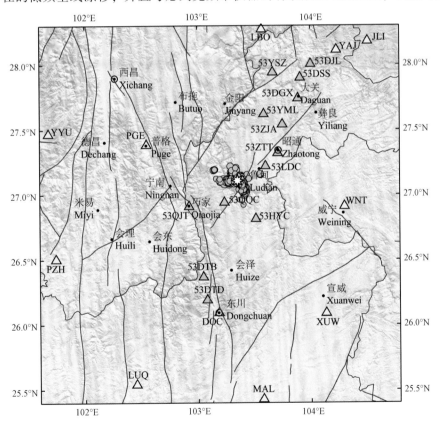

图1 鲁甸地震震中附近区域与近震台站分布

八角形为震中位置,青色圆圈为余震分布(王未来等,2014),青色三角形为强震台站分布,黄色三角形为近震宽频带台站分布,红线为活动断层(Deng et al.,2003)

Fig. 1 Map of the epicentral area and local seismic stations

Aniseed star is the epicenter. Cyan circles are the aftershocks (Wang et al., 2014). Cyan and yellow triangles are the strong-motion and local broadband stations, respectively. Red lines denote the active faults (Deng et al., 2003)

都进行了0.02~0.5Hz的带通滤波，保证在去除近震数据基线漂移的同时，得到的数据信息位于宽频带仪器有效频段响应范围内．

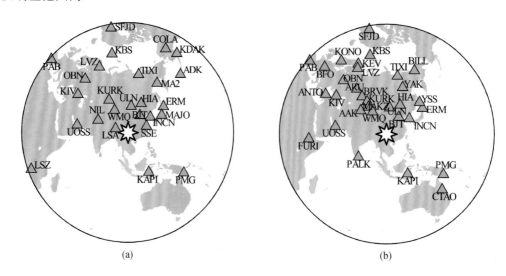

图2　震中与远震资料台站位置
(a) 震中与P波台站位置分布；(b) 震中与SH波台站位置分布
Fig. 2　The epicenter and teleseismic stations
(a) Epicenter and seismic stations of P waves. (b) Epicenter and seismic stations of SH waves

本文采用Wang（1999）的方法计算格林函数．鲁甸地震震中所在的川滇地区是地震多发区域，其地下速度结构得到了广泛的关注和比较系统的研究（熊绍柏等，1993；王椿镛等，2002；Wang et al.，2003）．在反演中，我们选用王未来等（2014）给出的区域速度结构，作为近震数据反演的介质模型；远震数据反演采用了AK135全球速度结构模型（Kennett et al.，1995）．与观测记录一样，对格林函数位移波形记录也进行了相同的带通滤波处理．

3　反演

3.1　反演方法与参数

根据王未来等（2014）对鲁甸地震序列的定位结果，以及张广伟等（2014）反演区域地震波资料得到的震源机制解，我们给定了两条共轭断层的位置、走向、倾角等参数．其中，近南北向断层为：走向162°/倾角70°/滑动角-14°，近东西向断层为：走向257°/倾角77°/滑动角-159°；主压应力轴（P轴）方位角121°/俯角24°；主张应力轴（T轴）方位角28°/俯角5°；中间主应力轴（B轴）方位角288°/俯角66°．主压应力轴方向与鲁甸周边区域的水平运动方向一致，但断层面走向与昭通断裂的走向有所差别（闻学泽等，2013），表明鲁甸地震的这一特殊的震源机制可能是在区域应力背景下，因昭通断裂的走向在鲁甸附近发生局部改变而形成的．另一方面，由这两个断层面参数确定的断层位置与余震分布（王未来等，2014；张广伟等，2014）相当一致，表明震源机制结果是可靠的．与作者之前的工作一样（张勇等，2014），两个断层面的尺度都给定得足够大，即长42km、宽20km，并将其划分为21×10=210个2km×2km大小的子断层．初始破裂点均位于沿走向方向第11个、沿倾向方向向下第6个子断层处，在两个断层面上的深度分别约为10km和11km．这与中国地震台网中心得到的结果（12km）相仿，但略浅于精定位得到的15km和13km（王未来等，2014；张广伟等，2014）．

本文采用Y. Zhang等（2012）的线性反演方法进行反演. 在这一方法中，无须预先假定子断层震源时间函数的形式. 相反地，子断层震源时间函数是作为待定参数由波形反演自动给出. 因此，可根据反演结果探讨断层两盘相对错动过程的时间复杂性. 同时，作为约束条件，这一方法需要给定破裂速度和破裂持续时间上限. 鉴于此次地震规模较小，震源尺度不大，我们给定破裂速度上限为$3km·s^{-1}$，即不考虑超剪切破裂的情况，并限定最大破裂持续时间不超过10s. 在这两个参数相对宽松的约束下，断层面上距震源（初始破裂点）d km处破裂速度大于$d/(d/3+10)$ $km·s^{-1}$小于$3km·s^{-1}$的滑动信息都可以被顺利提取. 同时，这一相对比较宽松的破裂速度范围可在一定程度上减弱震源位置（包括震源深度）的不确定性对破裂分布求解的干扰.

在破裂过程反演中，本文考虑了三种可能的断层破裂模式：①破裂只发生在近南北向断层上；②破裂只发生在近东西向断层上；③破裂可同时在两条共轭断层上自由传播. 根据这三种断层破裂模式进行反演，得到不同的破裂模型，结合反演得到的波形拟合情况，可以推测真实的发震构造和破裂过程是否具备多断层特征.

3.2 近震数据反演

首先采用近震数据，分别针对以上三种情形进行反演. 反演结果如图3所示. 首先，从反演得到

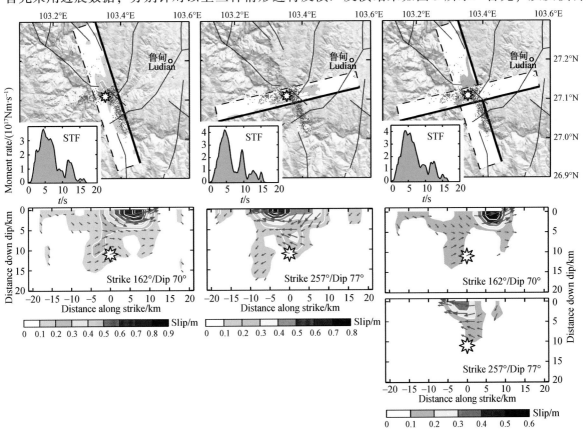

图3 近震数据反演得到的滑动量分布与震源时间函数

从左至右分别为：近南北向断层反演结果，近东西向断层反演结果，以及两条共轭断层反演结果. 每个反演结果中，从上到下依次为断层面滑动量分布在地表的投影，震源时间函数（插图），以及断层面上的静态滑动量分布. 地图中的红线表示断层（闻学泽等，2013）

Fig. 3 Inversion results of local seismic waves

From left to right are the inversion results of NNW-SSE fault, ENE-WSW fault, and two conjugated fault. From top to bottom for each panel of inversion results are the surface projection of fault slip distribution, source time function (STF, see subset), and fault slip distribution. Red lines on the upper maps denote the faults (Wen et al., 2013)

的滑动分布和震源时间函数看，共轭断层破裂模型反演结果都更接近于近南北向断层结果．表明在完全自由的条件下，破裂更倾向于发生在近南北向断层上，意味着近南北向断层应当是此次地震的主要发震断层．其次，采用近南北向和近东西向断层进行反演得到的波形残差分别为0.60和0.64，表明如果只考虑单一断层破裂的情形，则近南北向断层更有可能是真实的发震断层，这也与现有的研究结果相符（徐锡伟等，2014；许力生等，2014；张勇等，2014；刘成利等，2014）．相比之下，共轭断层模型波形残差为0.57，意味着除近南北向断层外，近东西向断层的破裂对波形拟合也存在一定贡献．在共轭断层模型反演结果中，近南北向断层释放了62%的地震矩，其对波形拟合的贡献约60%；近东西向断层释放了38%的地震矩，其对波形拟合的贡献约为40%．可见，尽管近东西向断层上发生的破裂规模不占优势，但单位地震矩对资料拟合的贡献程度还略大于近南北向断层，表明它在地震破裂过程中可能也发生了破裂．

需要说明的是，以上结果（图3）与作者之前采用远震资料反演得到的结果（张勇等，2014）在主要特征上一致：近南北向断层反演结果中，破裂主要朝震中以南传播，而近东西向断层面的反演结果则显示，震中以东的滑动分布更占优势．不同的是，在远震资料反演结果中，由于地震资料的信噪比较差等因素，只能根据得到的滑动分布与余震分布的一致性来推断真实发震断层；但在本文开展的近震资料反演中，由于数据相对更高的空间分辨能力，我们可以通过资料残差来确定发震断层．

3.3 联合反演

在近震数据反演基础上，我们增加了远震地震数据，分别对单一断层和共轭断层模型进行了联合反演．由于近震与远震数据的联合使用提高了对断层的离源角覆盖范围，理论上可以增强反演在深度方向上的空间分辨能力（Zhang et al.，2014）．另外，一些强震数据位于土层上，波形记录受台站下方局部效应的影响，其幅度有可能被放大．在格林函数计算中，这种非常局部的台站效应很难予以考虑．因此，近震数据和远震数据的联合，还可更好地约束矩震级．在联合反演中，两种数据被赋予了相同的权重．反演结果如图4所示．在主要破裂特征上，联合反演结果与近震数据单独反演结果基本相似．只是在矩震级和最大滑动量上，联合反演结果稍小于近震数据反演结果．上文的近震资料反演与分析表明，近东西向断层也可能发生了破裂，因此我们倾向于基于共轭断层模型的联合反演结果，为本文的最终破裂模型．根据该模型，地震的矩震级约为$M_W6.1$．近南北向断层上最大滑动量约为0.5m，发生在震中以南约8km处；近东西向断层上最大滑动量约为0.3m，位于震中附近．如果只考虑滑动量大于0.2m的区域（橙色—红色区域），则近南北向断层上的破裂尺度约为12km，近东西向断层仅约为3km，前者远大于后者．

图5显示了采用共轭断层模型进行联合反演得到的滑动速率时空分布图像．由图可见，近东西向断层上的破裂开始较早，表现为从深部到浅部的扩展过程，其破裂在6s左右结束．近南北向断层上的破裂开始稍晚一些，但规模更大，主要破裂存在从深部向浅部和南南东方向传播的过程．综合两条断层上的破裂特征，在破裂开始后0~2s，滑动主要发生在近东西向断层上；2s之后，近南北向断层开始破裂．由此推测，在破裂过程的早期，近南北向断层的破裂可能是受近东西向断层上的破裂触发所致．从破裂开始后2s直至破裂结束，两条断层上的滑动速率变化分布比较独立，没有表现出明显的相干性．其中在破裂开始后4~7s，破裂抵达浅地表处，并在两条断层上形成了各自的最大滑动区域．破裂开始8s之后，主要滑动区域已经形成，虽然从震源时间函数看，地震矩率的大小还比较可观，但断层面上已经没有比较集中的滑动分布，意味着破裂已经进入愈合阶段．

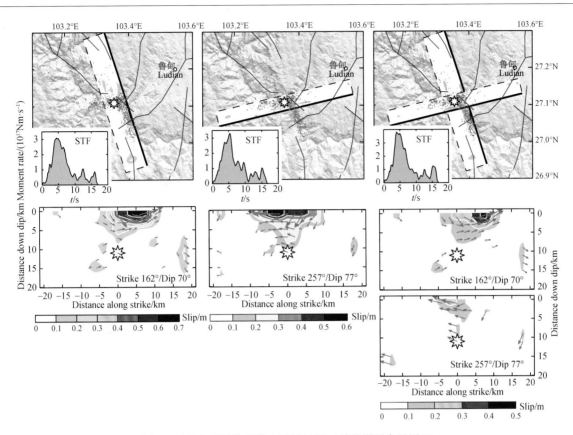

图 4 近震和远震数据联合反演结果（其他说明参见图 3）

Fig. 4 Same as Fig. 3, but for joint inversion of local and teleseismic data

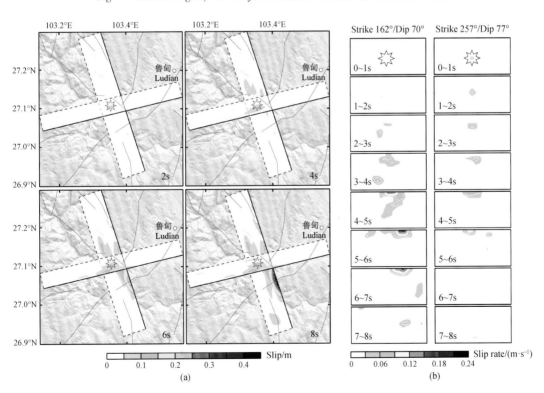

图 5 基于共轭断层模型的联合反演结果

(a) 断层面上滑动量累积分布在地面投影的时空变化；(b) 断层面上滑动速率的时空分布

Fig. 5 Results of joint inversion based on the two conjugated faults

(a) Spatial and temporal variations of surface projection of cumulative fault slips; (b) Spatial and temporal distribution of slip-rate on the faults

图 6 比较了以上破裂模型的观测与合成波形. 总体上, 近震和远震记录的主要波形特征都得到了较好地解释. 其中, 一些台站处的近震记录受台站效应的影响, 表现出一定的高频振荡特征. 由于不是有效的震源破裂信号, 且局部场地效应很难在地球模型中予以考虑, 这部分波形的拟合相对较差. 在远震波形拟合方面, 联合反演与远震记录单独反演 (张勇等, 2014) 基本相当. 这意味着在需要兼顾近震资料拟合的情况下, 远震记录的资料解释并未明显变差, 表明我们从近震和远震记录中有效提取了二者所共有的震源破裂信息.

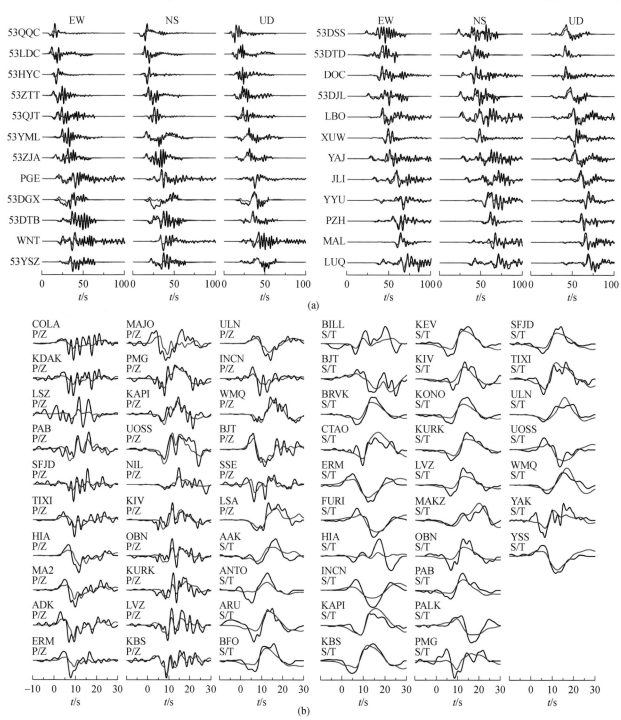

图 6 基于共轭断层模型的联合反演得到的观测地震波形 (黑线) 与合成地震波形 (红线) 的比较
(a) 近震数据拟合; (b) 远震数据拟合

Fig. 6 Comparisons between observed (black lines) and synthetic (red lines) waves of joint inversion based on the conjugate-fault model
(a) Local seismic data; (b) Teleseismic data

4 讨论与结论

本文考虑了单一断层破裂和共轭断层同时破裂的几种可能的情形，分别采用近震资料，以及近震与远震资料对 2014 年鲁甸地震的破裂过程进行了反演。近震资料单独反演表明，鲁甸地震的两条共轭断层——近南北向断层和近东西向断层——都发生了破裂，但主要地震矩释放过程发生在近南北向的断层上，且破裂主要朝南南东方向扩展。近、远震数据反演显示，破裂最初可能发生在近东西向断层上，并在破裂开始后 2~3s 触发了与其共轭的近南北向断层的破裂。此后，两条断层上的破裂相对独立地扩展。近东西向断层的破裂主要向浅地表处延伸，而近南北向断层上的破裂则朝地表和南南东方向传播，最终形成了复杂的滑动分布。本文得到的这一破裂特征与区域距离上独立得到的视震源时间函数分析结果可以很好地相互印证（许力生等，2014）。

需要注意的是，在共轭断层模型的近震资料反演和联合反演中，破裂在近南北向断层上的传播距离与余震分布比较吻合；但在近东西向断层上的西段，破裂延伸范围明显小于余震分布的尺度，不能解释该方向上余震密集分布的现象。根据本文反演结果，我们认为近南北向断层上位于震中以南的破裂可能是形成这一异常余震活动的主要原因。由于该段破裂以左旋走滑为主，且是整个地震中占主导的破裂，导致在破裂区域以东形成向北的推挤作用（图 7），增大了近东西向断层东段上下盘间的法向压力与摩擦阻力，造成其进一步闭锁，因而该方向没有密集余震发生；而在近南北向断层以西，情况正好相反，主震破裂形成了向南的拉伸作用，导致近东西向断层西段在一定程度上解锁。由于地震前该断层上闭锁程度（如断层面粗糙度）的非均匀性，以及积累的应变能的有限性，这种解锁作用没有导致大规模破裂的发生，而是形成了密集的小规模破裂事件（余震）。这是近东西向断层西段没有破裂发生但却存在密集的余震分布的主要原因。

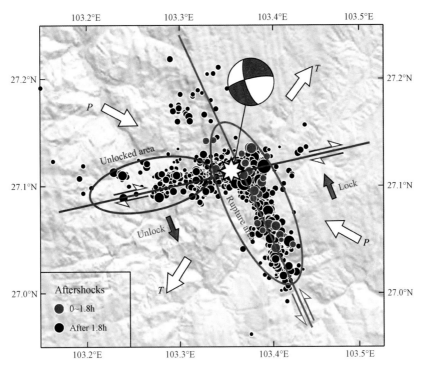

图 7 共轭断层破裂与余震关系示意图

Fig. 7 Sketch of unlocking process, which was caused by the left lateral strike-slips on the south part of the NNW-SSE fault, and impacts the west part of the ENE-WSW fault

另外，精定位结果也显示，余震分布尽管呈"倒 L"型，但在不同方向上的发生时间存在差别（王未来等，2014；张广伟等，2014）. 早期余震主要沿近南北向断层分布，表明是在主震破裂区域附近介质应力集中情况下发生的后续小规模破裂事件；震中以西（尤其是最西端）余震活动开始于震后 1.8h，且在震后 1.8h 至震后 1 天的活动性都较弱，可能是主震破裂在近东西向断层上的解锁作用所致（图 7）. 可见这种共轭断层的余震触发模式可以与两个方向上余震发生的先后次序很好地对应. 实际上，位于主震破裂区域之外的余震发生较晚的现象也可以从其他地震中找到例证，比如 2010 年青海玉树 M_W6.9 地震，震中东南且位于主震破裂区域周边的余震发生较早，而位于主震破裂区域之外（震中西北约 40km 处）的余震则大都发生较晚（王未来等，2012）.

综上所述，结合本文确定的地震共轭断层破裂模型，以及鲁甸地震余震精定位结果（王未来等，2014；张广伟等，2014），我们确定鲁甸地震是在北西—南东向主压应力方向与北东—南西向主张应力方向的统一应力场下发生的两条共轭的断层先后破裂的一次复杂地震事件，该地震破裂过程和余震活动可以概括为以下 5 个阶段：

① 破裂开始后 1~2s，开始阶段，近东西向断层上发生破裂，且其破裂一直集中在震中附近；

② 破裂开始后 2~8s，主要破裂阶段，近南北向断层和近东西向断层同时破裂，其中近南北向断层上的破裂朝南传播，并释放了大部分地震矩；

③ 破裂开始后 8~17s，破裂愈合阶段，断层面上破裂分布比较零散；

④ 破裂开始 17s 到震后 1.8h，受主震破裂的影响，余震活动主要沿近南北向断层分布；

⑤ 地震发生 1.8h 后，由于主震破裂导致的解锁作用，余震逐渐开始在震中以西的近东西向断层上分布.

致谢 本文所用远震宽频带地震数据取自 IRIS 数据中心；强震数据来自西南强震动台网中心；近震宽频带数据来自云南省地震局. 审稿专家对本文提供了有益的修改意见；中国科学院测量与地球物理研究所刁法启博士对本文的数据处理工作提供了有益的意见. 作者谨向以上单位和个人表示由衷的感谢.

References

Deng Q D, Zhang P Z, Ran Y K, et al. 2003. Basic characteristics of active tectonics of China. *Sci. China Ser. D.*, **46**(4): 356-372.

Kennett B L N, Engdahl E R, Buland R. 1995. Constraints on seismic velocities in the earth from traveltimes. *Geophys. J. Intl.*, **122**(1): 108-124.

Lei J S, Zhang G W, Xie F R, et al. 2012. Relocation of the 10 March 2011 Yingjiang, China, earthquake sequence and its tectonic implications. *Earthq. Sci.*, **25**(1): 103-110.

Liu C L, Zheng Y, Xiong X, et al. 2014. Rupture process of M_S6.5 Ludian earthquake constrained by regional broadband seismograms. *Chinese J. Geophys.* (in Chinese), **57**(9): 3028-3037, doi: 10.6038/cjg20140927.

Meng L, Ampuero J P, Stock J, et al. 2012. Earthquake in a Maze: Compressional rupture branching during the 2012 M_W8.6 Sumatra earthquake. *Science*, **337**(6095): 724-726.

Wang C Y, Mooney W D, Wang X L, et al. 2002. Study on 3-D velocity structure of crust and upper mantle in Sichuan-Yunnan region, China. *Acta Seismol. Sinica* (in Chinese), **24**(1): 1-16.

Wang C Y, Chan W W, Mooney W D. 2003. Three-dimensional velocity structure of crust and upper mantle in southwestern China and its tectonic implications. *J. Geophys. Res.*, **108**(B9): 2442.

Wang R J. 1999. A simple orthonormalization method for stable and efficient computation of Green's functions. *Bull. Seismol. Soc. Am.*, **89**(3): 733-741.

Wang W L, Wu J P, Fang L H, et al. 2013. Relocation of the Yushu M_S7.1 earthquake and its aftershocks in 2010 from HypoDD. *Science China: Earth Sciences*, **56**(2): 182-191, doi: 10.1007/s11430-012-4450-z.

Wang W L, Wu J P, Fang L H, et al. 2014. Double difference location of the Ludian M_S6.5 earthquake sequences in Yunnan province in 2014. *Chinese J. Geophys.* (in Chinese), **57**(9): 3042-3051, doi: 10.6038/cjg20140929.

Wen X Z, Du F, Yi G X, et al. 2013. Earthquake potential of the Zhaotong and Lianfeng fault zones of the eastern Sichuan-Yunnan border region. *Chinese J. Geophys.* (in Chinese), **56**(10): 3361-3372, doi: 10.6038/cjg20131012.

Xiong S B, Zheng Y, Yin Z X, et al. 1993. The 2-D structure and it's tectonic implications of the crust in the Lijiang-Panzhihua-Zhehai region. *Chinese J. Geophys.* (in Chinese), **36**(4): 434-444.

Xu L S, Zhang X, Yan C, et al. 2014. Analysis of the Love waves for the source complexity of the Ludian M_S6.5 earthquake. *Chinese J. Geophys.* (in Chinese), **57**(9): 3006-3017, doi: 10.6038/cjg20140925.

Xu X W, Jiang G Y, Yu G H, et al. 2014. Discussion on seismogenic fault of the Ludian M_S6.5 earthquake and its tectonic attribution. *Chinese J. Geophys.* (in Chinese), **57**(9): 3060-3068, doi: 10.6038/cjg20140931.

Yue H, Lay T, Koper K D. 2012. En échelon and orthogonal fault ruptures of the 11 April 2012 great intraplate earthquakes. *Nature*, **490**(7419): 245-250, doi: 10.1038/nature11492.

Zhang G W, Lei J S, Liang S S, et al. 2014. Relocations and focal mechanism solutions of the 3 August 2014 Ludian, Yunnan M_S6.5 earthquake sequence. *Chinese J. Geophys.* (in Chinese), **57**(9): 3018-3027, doi: 10.6038/cjg20140926.

Zhang H, Chen J, Ge Z. 2012. Multi-fault rupture and successive triggering during the 2012 M_W8.6 Sumatra offshore earthquake. *Geophysical Research Letters*, **39**, L22305, doi: 10.1029/2012GL053805.

Zhang Y, Xu L S, Chen Y T. 2010. Source process of the 2010 Yushu, Qinghai, earthquake. *Science China: Earth Sciences*, **53**(9): 1249-1251.

Zhang Y, Feng W P, Chen Y T, et al. 2012. The 2009 L'Aquila M_W6.3 earthquake: a new technique to locate the hypocentre in the joint inversion of earthquake rupture process. *Geophys. J. Int.*, **191**(3): 1417-1426, doi: 10.1111/j.1365-246X.2012.05694.x.

Zhang Y, Wang R J, Chen Y T, et al. 2014. Kinematic rupture model and hypocenter relocation of the 2013 M_W6.6 Lushan earthquake constrained by strong-motion and teleseismic data. *Seismol. Res. Lett.*, **85**(1): 15-22.

Zhang Y, Xu L S, Chen Y T, et al. 2014. Rupture process of the 3 August 2014 Ludian, Yunnan, M_W6.1 (M_S6.5) earthquake. *Chinese J. Geophys.* (in Chinese), **57**(9): 3052-3059, doi: 10.6038/cjg20140930.

Zhang Z G, Sun Y C, Xu J K, et al. 2014. Preliminary simulation of strong ground motion for Ludian, Yunnan earthquake of 3 August 2014, and hazard implication. *Chinese J. Geophys.* (in Chinese), **57**(9): 3038-3041, doi: 10.6038/cjg20140928.

附中文参考文献

刘成利, 郑勇, 熊熊等. 2014. 利用区域宽频带数据反演鲁甸M_S6.5级地震震源破裂过程. 地球物理学报, **57**(9): 3028-3037, doi: 10.6038/cjg20140927.

王椿镛, Mooney W D, 王溪莉等. 2002. 川滇地区地壳上地幔三维速度结构研究. 地震学报, **24**(1): 1-16.

王未来, 吴建平, 房立华等. 2014. 2014年云南鲁甸M_S6.5地震序列的双差定位. 地球物理学报, **57**(9): 3042-3051, doi: 10.6038/cjg20140929.

闻学泽, 杜方, 易桂喜等. 2013. 川滇交界东段昭通、莲峰断裂带的地震危险背景. 地球物理学报, **56**(10): 3361-3372, doi: 10.6038/cjg20131012.

熊绍柏, 郑晔, 尹周勋等. 1993. 丽江—攀枝花—者海地带二维地壳结构及其构造意义. 地球物理学报, **36**(4): 434-444.

许力生, 张旭, 严川等. 2014. 基于勒夫波的鲁甸M_S6.5地震震源复杂性分析. 地球物理学报, **57**(9): 3006-3017, doi: 10.6038/cjg20140925.

徐锡伟, 江国焰, 于贵华等. 2014. 鲁甸6.5级地震发震断层判定及其构造属性讨论. 地球物理学报, **57**(9): 3060-3068, doi: 10.6038/cjg20140931.

张广伟, 雷建设, 梁姗姗等. 2014. 2014年8月3日云南鲁甸M_S6.5级地震序列重定位与震源机制研究. 地球物理学报, **57**(9): 3018-3027, doi: 10.6038/cjg20140926.

张勇, 许力生, 陈运泰等. 2014. 2014年8月3日云南鲁甸M_W6.1 (M_S6.5) 地震破裂过程. 地球物理学报, **57**(9): 3052-3059, doi: 10.6038/cjg20140930.

张振国, 孙耀充, 徐建宽等. 2014. 2014 年 8 月 3 日云南鲁甸地震强地面运动初步模拟及烈度预测. 地球物理学报, 57(9): 3038-3041, doi: 10.6038/cjg20140928.

The 2014 M_W 6.1 Ludian, Yunnan, earthquake: a complex conjugated ruptured earthquake

Zhang Yong[1], Chen Yun-Tai[1,2], Xu Li-Sheng[2], Wei Xing[3], Jin Ming-Pei[4], Zhang Sen[4]

1. School of Earth and Space Sciences, Peking University, Beijing 100871, China;
2. Institute of Geophysics, China Earthquake Administration, Beijing 100081, China;
3. China Earthquake Networks Center, Beijing 100045, China;
4. Earthquake Administration of Yunnan Province, Kunming 650041, China

Abstract Both intensity and aftershock distributions have shown that the seismogenic structure of the 2014 M_W6.1 (M_S6.5) Ludian, Yunnan, earthquake was quite complex. Especially, aftershocks distribute in a reverse L-shape on conjugated faults (i.e., the NNW-SSE fault and the ENE-WSW fault), making it difficult to determine the causative fault and rupture process. For better understanding the seismogenic structure of the Ludian earthquake, we invert local seismic (strong-motion and broadband data) as well as teleseismic data to investigate the rupture kinematics.

Three cases of fault ruptures were tested. (1) Ruptures occur only on the NNW-SSE fault. (2) Ruptures occur only on the ENE-WSW fault. (3) Ruptures can propagate freely on the conjugated faults. A linear technique was used in these inversions. We have two steps in this work. First, local seismic data were inverted for rupture process to identify the rupture faults. Second, with the determined rupture faults, the rupture model was renewed by jointly inverting local seismic and teleseismic waves.

Local seismic data inversions show that data fitting of the NNW-SSE fault is better than that of the ENE-WSW fault, suggesting the NNW-SSE fault is the major rupture fault of the Ludian earthquake. Meanwhile, the best data fitting was found in inversion of conjugated faults, in which data fitting per seismic moment of the ENE-WSW fault was even slightly better than that of the NNW-SSE fault, indicating that both conjugated faults may have ruptured during the earthquake. From the joint inversion of local and teleseismic data, rupture initiated on the ENE-WSW fault in the first two seconds, and after 2s, the NNW-SSE fault began to rupture. The source duration is about 17s, but most seismic moments were released in the first 10s after the rupture initiation. On the conjugated faults, ruptures propagated toward up-dip directions and mainly occurred in very shallow depth, which may account for the terrible earthquake damages. However, rupture extent of the NNW-SSE fault is much larger than that of the ENE-WSW fault. As a result, rupture area on the NNW-SSE fault is comparable with aftershocks along that direction, but on the ENE-WSW fault, the dimension of rupture area is much smaller than that of aftershock distribution.

The above inversion results show that the Ludian earthquake is a complex conjugated ruptured earthquake. Ruptures firstly initiated on the ENE-WSW fault, and then triggered the NNW-SSE fault. Consecutive ruptures on conjugated faults constitute the complex rupture process of the earthquake. The left-lateral strike-slips in the south of the NNW-SSE fault could unlock the west part of the ENE-WSW fault, leading to aftershocks in that area where no significant ruptures were observed.

2015年尼泊尔M_W7.9地震破裂过程：
快速反演与初步联合反演[*]

张 勇[1]　许力生[2]　陈运泰[1,2]

1. 北京大学地球与空间科学学院，北京　100871；2. 中国地震局地球物理研究所，北京　100081

摘要　本文介绍了2015年4月25日尼泊尔M_W7.9（M_S8.1）地震发生后的破裂过程快速反演工作，以及后续开展的地震波与少量GPS资料的初步联合反演工作．两项工作得到的反演结果尽管在最大滑动量估计方面存在一些差别，但都一致地显示此次地震是发生在低倾角俯冲断裂上的一次单侧破裂事件，破裂主要朝东南方向传播；断层滑动主要发生在震中至加德满都一带．在加德满都附近区域，其下方破裂与朝东南传播的地震波的多普勒聚焦效应可能造成较强的震感和较大的破坏．对比历史大地震发现，2015年尼泊尔M_W7.9地震的浅部破裂紧邻1934年M_W8.2地震的地表破裂，余震分布与1833年M7.6地震的宏观震中基本重合，其破裂填补了前两次地震破裂以西100km左右的空区，表明此次地震是1934年M_W8.2地震与1833年M7.6地震向西继续延伸的结果．

1　引言

北京时间2015年4月25日14时09分，在邻近我国西藏自治区的尼泊尔境内发生了M_W7.9（M_S8.1）地震．截至2015年5月4日，地震共造成逾7000余人遇难，1.4万人受伤．此次尼泊尔地震发生在青藏高原南部与印度板块的边界处，位于全球陆地碰撞和垂直变形最剧烈的区域．在此区域，历史上曾经发生过多次大地震（Molnar and Pandey，1989；Bilham et al.，2001），其中距此次地震最近的是1833年的M7.6地震和1934年的比哈尔邦M_W8.2地震（Bilham et al.，2001；Sapkota et al.，2013）．21世纪以来，青藏高原周边也先后发生了多起灾害性大地震事件，其中2005年克什米尔M_W7.6地震与2008年汶川M_W7.9地震分别位于青藏高原西部和东部边缘，导致的死亡或失踪人数都接近9万人，是两次影响非常巨大的地震灾难．2015年尼泊尔地震是继这两次地震之后，21世纪以来青藏高原周边发生的第三次灾害性大地震事件（M_W>7.5）．

尼泊尔地震发生后，作为地震应急快速响应的工作之一，作者获取了远震地震波数据，在震后2小时09分快速确定和发布了此次地震的破裂过程模型．此后，在快速反演工作基础上，采用了全球地震台网（GSN，Global Seismographic Network）震中距90°以内的宽频带地震数据，以及同震位移明显的两个台站的GPS位移数据，对破裂模型进行了更新，并根据反演结果与历史大地震的破裂分布情况，简单定性讨论了此次地震对加德满都附近区域地震活动性和危险性的影响．

2　破裂过程快速反演

快速反演工作采用IRIS（Incorporated Research Institutions for Seismology，美国地震学研究联合会）数

---

[*] 本文发表于《地球物理学报》，2015年，第**58**卷，第5期，1804-1811.

据中心提供的 GSN 远震宽频带地震波形数据，选用了全球范围内震中距介于 40°~90° 且方位角分布均匀的 32 个台站的垂直向 P 波波形（图 1a）. 基于 AK135 全球大陆速度结构模型（Kennett et al.，1995），采用 Wang（1999）的方法计算得到格林函数. 在反演中，我们对数据和格林函数同时进行了 0.01~0.1Hz 的带通滤波. 采用作者发展的计算效率较高的线性反演方法，快速确定此次地震的破裂过程（Chen and Xu，2000；Xu et al.，2002；Zhang et al.，2012）. 在快速反演开始阶段（约震后 1 小时），只有 USGS（United States Geological Survey，美国地质调查局）提供了可供参考震源机制. 根据震中区域的地质构造特征，我们采用了 USGS 发布的震源机制中断层面参数为走向 301°/倾角 11°/滑动角 115° 的节面，作为此次地震的发震断层.

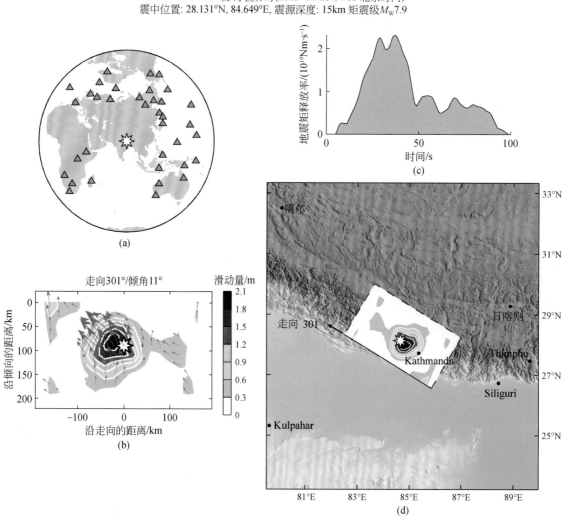

图 1 2015 年尼泊尔地震快速反演结果

(a) 震中与台站位置分布；(b) 断层面上静态滑动量分布；(c) 震源时间函数；(d) 滑动量分布在地面的投影

Fig. 1 Fast inversion results of rupture process of the 2015 Nepal earthquake

(a) Epicenter and seismic stations; (b) Static fault slip distribution; (c) Source time function; (d) Surface projection of fault slip distribution

破裂过程快速反演得到的矩震级为 $M_W 7.9$，断层面上滑动分布相对集中，存在一个主要滑动区域，大部分滑动分布于震中的东南方向（图 1）. 破裂过程持续近 100s，其中最主要的一次子事件发生在破裂开始后前 50s，50s 以后，后续破裂规模减弱，地震矩释放率的幅度相对较小，断层滑动与破裂进入缓慢愈合阶段. 从断层面上滑动分布在地面的投影看，破裂主要朝震中东南方向传播，断层滑动主要发生在震中至加德满都一带. 在加德满都附近区域, 由于下方断层的错动, 以及朝东南方向传播的破裂形成的地震波多普勒聚焦效应, 可能导致地震波能量在该地区集中释放, 形成较强的震感或破坏.

尽管快速反演工作确定了尼泊尔地震的初步震源特征, 但也存在一些问题. 比如, 由于破裂面积和滑动量大小之间存在权衡 (trade-off) 关系, 远震地震波单独反演得到的滑动量大小存在较强的不确定性. 尤其地, 2015 年尼泊尔地震是发生在板块边界处一次低倾角的逆冲型地震, 由于倾角小、倾向方向延伸范围较大, 绝对滑动量与破裂面积之间的权衡相对一般地震更为严重, 确定反演滑动量绝对大小面临的不确定性也更强. 比如, 在对 2011 年日本东北 (Tohoku) $M_W 9.0$ 地震的研究中, 反演远震地震波和陆地 GPS 同震位移数据得到的最大滑动量约为 30m (Ide et al., 2011; Wang et al., 2013), 明显小于反演海底 GPS 位移和海啸波数据得到的 50~60m 的结果 (Lay et al., 2011a; Wang et al., 2013). 再如, 在 2010 年 Mentawai $M_W 7.8$ 地震的反演工作中, 不同作者得到的最大滑动量可介于 3.5~20m 之间 (Lay et al., 2011b; Yue et al., 2014), 显示出非常强的不确定性. 因此, 有必要选用覆盖更好和类型更多的数据, 尤其是测量资料的同震位移数据, 对破裂模型进行进一步修订.

3 破裂过程后续反演

为提高台网的空间分辨能力, 我们增加了 GSN 全球台网中震中距更近的资料, 并按照 10° 的方位角和离源角间隔筛选台站, 最终得到了 37 个台站的宽频带 P 波波形 (图 2a). 此外, 从 UNAVCO 网站 (www.unavco.org) 下载获取了 2 个 GPS 台站的同震位移数据 (图 2b), 以更好地约束静态滑动量的绝对大小.

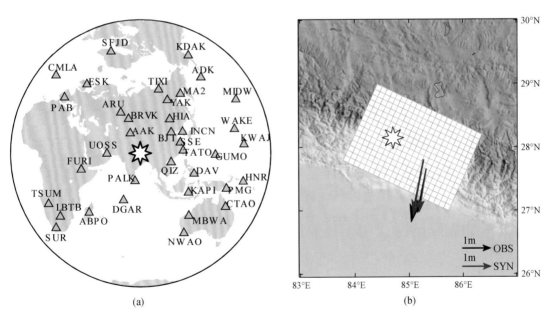

图 2 本研究所用的资料: (a) 地震台站; (b) GPS 台站以及断层面在地面的投影
(b) 中黑色箭头为观测 GPS 位移, 红色箭头为根据本文联合反演模型计算得到的合成位移
Fig. 2 Data used in this work: (a) Seismic waveform stations; (b) Two GPS stations
Rectangular area in (b) is the surface projection of fault plane. Black arrows are observed displacements, and red arrows are displacements synthesized with the joint rupture model of this work

后续反演采用了 CRUST2.0 (Bassin et al., 2000) 给出的震中附近区域的地壳速度结构模型. 在较近震中距距离上, 存在多种震相混杂的情况. 为更好地模拟所有震相波形, 我们考虑和计算了每个子断层到每个台站之间的格林函数. 对于 GPS 资料反演, 采用 Okada (1985) 方法计算静态位错的格林函数. 地震

资料和 GPS 资料联合反演的技术细节可参考 Zhang 等 (2012). 联合反演中, 基于资料拟合程度与权重平衡的原则, GPS 资料整体权重固定为地震波资料的 1/2. 震源机制采用 USGS 更新后的震源机制, 选用的断层面参数为: 走向 295°/倾角 11°/滑动角 108°.

图 3 显示了地震波资料单独反演与联合反演分别确定的震源时间函数与断层面上的静态滑动量分布. 二者得到的矩震级皆约为 $M_W7.9$, 与快速反演结果相当; 震源时间函数形态也与快速反演结果一致, 但最大地震矩释放率略高 (图 1c). 联合反演得到的静态滑动量分布比地震波单独反演结果更紧凑, 滑动分布范围有所缩小, 但滑动量幅度增加, 最大滑动量由地震波单独反演的 3.2m 提高到联合反演的 5.2m. 考虑到同震 GPS 位移对滑动量大小的约束更好, 5.2m 的最大滑动量可能更接近实际情况. 此外, 地震波反演模型显示了较强的右旋走滑分量, 且距震中越远, 走滑分量越强; 而在联合反演中, 没有观察到这一现象. 但由于本文采用的 GPS 资料偏少, 这一现象还需引入更多数据予以核实.

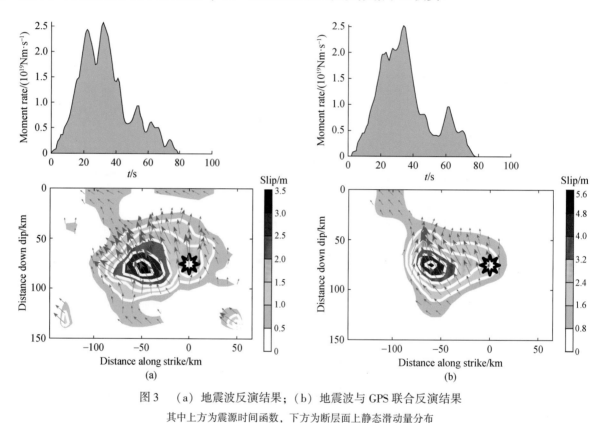

图 3 (a) 地震波反演结果; (b) 地震波与 GPS 联合反演结果

其中上方为震源时间函数, 下方为断层面上静态滑动量分布

Fig. 3 (a) Results of seismic data inversion; (b) Results of joint inversion with seismic and GPS data

In both (a) and (b), on the top and bottom are the source time functions (STFs) and static fault slip distributions, respectively

联合反演得到的时空破裂图像见图 4. 滑动速率和滑动量的时空累积图像都显示, 破裂从震源处以单侧破裂的形式朝东南方向传播. 规模最大的一次子事件发生在破裂开始后 0~45s, 位于断层走向方向上的深部区域; 45s 以后, 滑动速率的强度及其空间分布的整体性都有所减弱, 但仍显示了朝东南传播的破裂方向性. 60s 左右, 在震中东南 90~120km 处疑有较小规模的破裂抵达浅部区域.

联合反演模型很好地同时拟合了地震数据 (图 5) 和 GPS 数据 (图 2b). 地震波观测波形与合成波形的平均相关系数约为 0.85, 其中接近一半的台站 (17 个台) 的相关系数超过了 0.9. GPS 数据的拟合程度高达 99%. 由于只有两个台站存在明显的同震位移, 数据数量和覆盖范围非常有限, 这一拟合程度不能说明更多问题. 但这两个台站的引入显著提高了最大滑动量, 在一定程度上增强了对破裂强度和分布的有效约束.

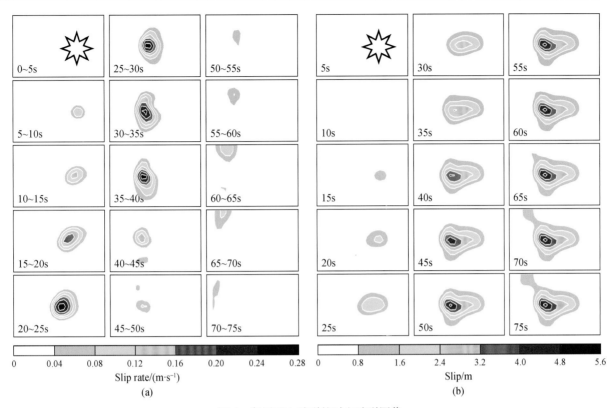

图 4 断层面上破裂的时空破裂图像

(a) 滑动速率时空分布；(b) 滑动量时空分布

Fig. 4 Images of spatial-temporal rupture processes on the fault

(a) Spatial-temporal distribution of slip-rate on the fault；(b) Spatial-temporal distribution of slip on the fault

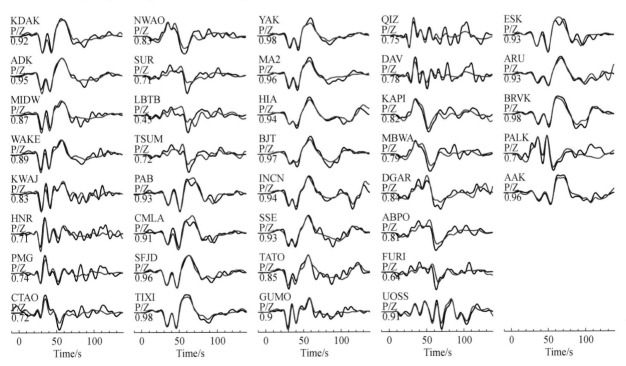

图 5 联合反演得到的观测与合成地震图的比较

每个子图的左边从上到下依次为台站名、震相名/分量名、以及观测波形与合成波形之间的相关系数

Fig. 5 Comparison between observed and synthetic seismograms of joint inversion

On the left of each sub-graph from top to bottom are the station code, the phase name/component name, and the correlation coefficient of observed and synthetic waves, respectively

4 讨论与结论

本文报道了 2015 年尼泊尔地震后的破裂过程快速反演工作,以及采用 GSN 全球台网的宽频带地震波形与目前可用的两个台站的 GPS 同震位移,联合反演确定的破裂模型. 结果显示,采用地震波资料快速反演得到的破裂模型尽管在滑动量绝对大小方面存在一定的低估,但仍然能够较好地确定地震破裂方向和破裂延伸范围等与地震灾害密切联系的震源特征. 通过地震波资料和 GPS 资料的联合反演,滑动分布更加紧凑,滑动量绝对大小也得到了更好地约束.

根据反演结果,地震破裂朝东南方向传播,形成的静态滑动主要分布在震中至加德满都(Kathmandu)一带. 从余震分布看(图 6),几乎所有余震都分布在震中东南方向,与本文确定的主震破裂方向与图像一致. 具体地说,余震主要分布在主震破裂区域周围,余震区与主震破裂区形成了互补的关系. 这再一次说明余震的发生是主震破裂导致的周围应力增强静态触发的结果. 同时我们也注意到,尼泊尔地震之后,在我国西藏自治区日喀则地区的定日县境内也发生了多次地震,其中最大震级为 $M_S 5.9$,由于距主震破裂相对较远,存在破裂远程动态触发的可能. 图 6 同时显示了此次地震主震破裂与余震分布,以及 1833 年

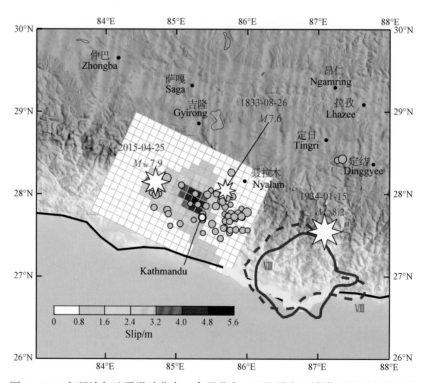

图 6 2015 年尼泊尔地震滑动分布、余震分布,以及震中区域附近的历史大地震

青色圆圈为 USGS 2015 年 5 月 1 日发布的 2015 年尼泊尔地震的余震震中($M>3.8$),圆圈直径正比于地震震级,红色实线和虚线为不同工作得到的 1934 年 $M_W 8.2$ 地震的烈度Ⅷ区等震线(Pandey and Molnar, 1988; Delcailleau, 1992),黄色粗线标示了 1934 年地震的地表破裂位置(Sapkota et al., 2013),黑线为印度板块与欧亚板块的边界. 除 2015 年尼泊尔地震外,另外两个白色八角星分别表示 1833 年和 1934 年地震的宏观震中(Delcailleau, 1992)与微观震中(Hough and Bilham, 2008)

Fig. 6 Fault slip distribution and aftershocks of the 2015 Nepal earthquake, and historical large earthquakes nearby

Cyan circles are aftershocks ($M>3.8$) up to 1 May 2015 (from USGS), whose radius are proportional to magnitudes. Red solid and dashed lines are isoseismals of intensity Ⅷ determined from different studies (Pandey and Molnar, 1988; Delcailleau, 1992). Yellow bold line marks the surface ruptures of the 1934 earthquake (Sapkota et al., 2013), and black line is the boundary between the India Plate and the Euro-Asia Plate. In addition to the epicenter of the 2015 Nepal earthquake, other two white stars are macroseismic and instrumental epicenters of the 1833 and 1934 earthquakes, respectively

M7.6 地震与 1934 年 M_W8.2 地震的大致位置. 从地震活动的空间位置看, 2015 年尼泊尔地震的破裂区域位于 1833 年 M7.6 地震与 1934 年 M_W8.2 地震以西, 其深部破裂与余震活动紧邻 1833 年 M7.6 地震的宏观震中, 且接近 1934 年 M_W8.2 地震的Ⅷ度区; 而浅部破裂基本衔接了 1934 年 M_W8.2 地震的地表破裂, 在很大程度上表明这三次地震的破裂区域可能存在一定的空间连续性. 表明此次地震之后, 发生破裂的加德满都地区原来存在的破裂空区在一定程度上得到了"填充". 然而另一方面, 大震之后主震破裂区与余震分布区的应力释放与调整过程往往会延续一段时间, 短期内仍然存在发生强余震的可能.

需要说明的是, 本文反演都采用了既有的震源位置和震源机制. 由于目前近断层强震资料和大地测量资料的匮乏, 这些基本震源参数不能得到很好地约束, 可能存在不确定性, 并在一定程度上影响破裂过程结果. 其中, 宽频带地震波反演只能获取破裂相对于震中的相对位置, 一旦震中位置出现偏差, 滑动分布也会发生相应的偏移. 另外, 2015 年尼泊尔 M_W7.9 地震是一次比较典型的板间低倾角逆冲型地震事件, 对于此类地震, 断层面倾角有可能随深度发生改变, 即断层几何模型存在相当的复杂性, 可能降低采用简单断层模型对破裂过程成像的精细和可靠程度. 在下一步工作中, 迫切需要搜集更多的近断层地震波形资料和大地测量资料, 对此次地震的震源位置和断层几何模型进行深入全面的分析, 提高破裂过程模型的准确度与可靠性.

致谢 本文地震波数据来自 IRIS 数据中心, GPS 数据下载自网页 www.unavco.org. 中国科学院测量与地球物理研究所的刘成利博士在 GPS 数据方面给予了帮助, 两位审稿专家提出了非常中肯的修改意见, 作者对他们致以诚挚谢意!

References

Bassin C, Laske G, Masters G. 2000. The current limits of resolution for surface wave tomography in North America., *EOS Trans. AGU* **81**, F897.

Bilham R, Gaur V K, Molnar P. 2001. Himalayan seismic hazard. *Science*, **293**(5534): 1442–1444.

Chen Y T, Xu L S. 2000. A time-domain inversion technique for the tempo-spatial distribution of slip on a finite fault plane with applications to recent large earthquakes in the Tibetan Plateau. *Geophys. J. Int.*, **143**: 407–416.

Delcailleau B. 1992. Les Siwaliks du Népal oriental. Presses du CNRS (Editions du Centre National de la Recherche Scientifique).

Hough S E, Bilham R. 2008. Site response of the Ganges basin inferred from re-evaluated macroseismic observations from the 1897 Shillong, 1905 Kangra, and 1934 Nepal earthquakes. *J. Earth Syst. Sci.*, **117**: 773–782.

Ide S, Baltay A, Beroza G C. 2011. Shallow dynamic overshoot and energetic deep rupture in the 2011 M_W9.0 Tohoku-Oki earthquake. *Science*, **332**(6036): 1426–1429.

Kennett B L N, Engdahl E R, Buland R. 1995. Constraints on seismic velocities in the earth from traveltimes. *Geophys. J. Int.*, **122**: 108–124.

Lay T, Ammon C J, Kanamori H, et al. 2011a. Possible large near-trench slip during the 2011 M_W9.0 off the Pacific coast of Tohoku earthquake. *Earth Planet and Space*, **63**(7): 687–692.

Lay T, Ammon C J, Kanamori H, et al. 2011b. The 25 October 2010 Mentawai tsunami earthquake (M_W7.8) and the tsunami hazard presented by shallow megathrust ruptures. *Geophys. Res. Lett.*, **38**(6): L06302, doi: 10.1029/2010GL046552.

Molnar P, Pandey M R. 1989. Rupture zones of great earthquakes in the Himalayan region. *Proceedings of the Indian Academy of Sciences-Earth and Planetary Sciences*, **98**(1): 61–70.

Okada Y. 1985. Surface deformation due to shear and tensile faults in a half-space. *Bull. Seism. Soc. Am.*, **75**(4): 1135–1154.

Pandey M R, Molnar P. 1988. The distribution of intensity of the Bihar-Nepal earthquake of 15 January 1934 and bounds on the extent of the rupture zone. *J. Nepal Geol. Soc.* **5**: 22–44.

Sapkota S N, Bollinger L, Klinger Y, et al. 2013. Primary surface ruptures of the great Himalayan earthquakes in 1934 and 1255. *Nature Geoscience*, **6**(1): 71–76.

Wang R J. 1999. A Simple orthonormalization method for stable and efficient computation of Greene's functions. *Bull. Seism. Soc. Am.*, **89**(3): 733-741.

Wang R, Parolai S, Ge M, et al. 2013. The 2011 M_W 9.0 Tohoku earthquake: comparison of GPS and strong-motion data. *Bull. Seism. Soc. Am.*, **103**(2B): 1336-1347.

Xu L S, Chen Y T, Teng T L, et al. 2002. Temporal and spatial rupture process of the 1999 Chi-Chi earthquake from IRIS and GEOSCOPE long period waveform data using aftershocks as empirical Green's functions. *Bull. Seism. Soc. Am.* **92**: 3210-3228.

Yue H, Lay T, Rivera L, et al. 2014. Rupture process of the 2010 M_W7.8 Mentawai tsunami earthquake from joint inversion of near-field hr-GPS and teleseismic body wave recordings constrained by tsunami observations. *J. Geophys. Res. Solid Earth*, **119**: 5574-5593.

Zhang Y, Feng W P, Chen Y T, et al. 2012. The 2009 L'Aquila M_W6.3 earthquake: a new technique to locate the hypocentre in the joint inversion of earthquake rupture process. *Geophys. J. Int.*, **191**: 1417-1426, doi: 10.1111/j.1365-246X.2012.05694.x.

Rupture process of the 2015 Nepal M_W 7.9 earthquake: Fast inversion and preliminary joint inversion

Zhang Yong[1], Xu Li-Sheng[2] and Chen Yun-Tai[1,2]

1. School of Earth and Space Sciences, Peking University, Beijing 100871, China;
2. Institute of Geophysics, China Earthquake Administration, Beijing 100081, China

Abstract As a routine work for earthquake emergency response, we have performed fast inversions of rupture processes for large earthquakes since the 2008 Wenchuan M_W7.9 earthquake. About 2.15 hours after the 2015 Nepal M_W7.9 earthquake, we determined the rupture model by inverting teleseismic P waves. From the obtained rupture images, the earthquake was a unilateral rupture event and ruptures mainly propagated to the southeast. So we concluded that the earthquake may cause relatively terrible damages around Kathmandu area due to nearby ruptures and Doppler effects of the southeastward rupture propagations.

After the fast inversion, we collected more seismic data and a few GPS data (two stations), and carried out a joint inversion of rupture process. The joint model well confirms the major source characteristics, e.g. southeastward rupture propagation, and displays a larger peak slip (about 5.2m) than that of the fast estimated model (about 2.0m). The source duration is about 80s, with the major sub-event occurred at 0~45s. At 60~70s, a small-scale sub-event occurs in very shallow depth about 100km along the strike direction, and may cause a few surface breakages.

From comparisons with historical large earthquakes, the 2015 Nepal M_W7.9 earthquake ruptured an area, with a dimension of about 100km, to the west of the 1833 *M*7.6 earthquake and the 1934 M_W8.2 earthquake. The shallow ruptures are very close to the surface ruptures of the 1934 M_W8.2 earthquake, and the aftershocks are around the macroseismic epicenter of the 1833 *M*7.6 earthquake. It is highly probable that ruptures of the three earthquakes have connected. From this perspective, the 2015 Nepal earthquake is the result of the westward rupture extensions of the 1833 and 1934 events.

Further studies on the focal mechanism and source rupture process of the 2012 Haida Gwaii, Canada, 7.8 moment magnitude earthquake

Dariush Motazedian[1], Yong Zhang[2], Shutian Ma[1] and Yun-tai Chen[3]

1. Department of Earth Sciences, Carleton University, 1125 Colonel By Drive, Ottawa, ON K1S 5B6, Canada;
2. Department of Geophysics, Peking University, Beijing, China;
3. Institute of Geophysics, China Earthquake Administration, Beijing, China

Abstract The 28 October 2012 Haida Gwaii, British Columbia, Canada, earthquake with a moment magnitude (M_W) of 7.8 occurred along an east-dipping poorly known thrust fault beneath the Queen Charlotte Terrace. It was the largest thrust event ever recorded in this dominated by strike-slip motion region. We studied the focal mechanism and the source rupture process for the event. The retrieved geometric parameters of the fault plane were a strike of 329°, dip of 24°, and slip of 114°. The isotropic moment was negative, and its value was about one-fifth of the total seismic moment released. The earthquake ruptured an area of about 160km×60km, and major slip occurred in an area of about 100km× 60km. The maximum slip was about 5.8m. The slip distribution on the fault plane was highly heterogeneous, with four slip patches. The main slip lay on a large zone above the hypocentre to the sea floor. The maximum and average stress drops calculated using the Brune model were 16.5 and 4.6MPa, respectively. The major rupture occurred about 10s after the rupture initiation, and lasted about 25s. During a subducting earthquake, the leading edge of the overriding plate is assumed to spring seaward and upward, while the landward portion is assumed to extend and drop down, and the generated rapid motions set off a tsunami. The falling-down process seems to be consistent with a negative isotropic moment.

Résumé Le tremblement de terre de Haida Gwaii du 28 octobre 2012, en Colombie-Britannique, avait un moment de magnitude (M_W) de 7,8 et il s'est produit le long d'une faille de chevauchement, peu connue, à pendage est, située sous la terrasse de la Reine Charlotte. C'était le plus grand événement de chevauchement jamais enregistré dans cette région dominée par des mouvements de failles de décrochement. Nous avons étudié le mécanisme au foyer et le processus de rupture à la source de cet événement. Les paramètres géométriques obtenus pour le plan de faille étaient de : une direction de 329°, un pendage de 24° et un rejet de 114°. Le moment isotrope était négatif et sa valeur était d'environ un cinquième du moment sismique total relâché. Le tremblement de terre a rompu un secteur d'environ 160km sur 60km et le rejet principal a eu lieu sur un secteur d'environ 100km sur 60km. Le rejet maximal a été d'environ 5,8m. La distribution du rejet sur le plan de faille était très hétérogène, avec quatre parcelles de concentration du glissement. Le glissement principal se trouvait sur une grande zone au-dessus de l'hypocentre par rapport au plancher océanique. Les chutes de contraintes maximales et moyennes, calculées en utilisant le modèle de Brune étaient respectivement de 16, 5 and 4, 6MPa. La rupture principale s'est produite environ 10s après l'initiation de la rupture et elle a duré environ 25 s. Durant un tremblement de terre avec subduction, il est présumé que la bordure avant de la plaque supérieure rebondit vers le large et vers le haut, alors que la portion vers le continent est présumée s'étendre et aller vers le bas; les mouvements rapides générés déclenchent un tsunami. Le processus de chute semble concorder avec un moment isotrope négatif. [Traduit par la Rédaction]

1 Introduction

On 28 October 2012, an earthquake with moment magnitude M_W 7.8 occurred off the western Moresby Island, Haida Gwaii, British Columbia. The Haida Gwaii region is one of the most seismically active areas along the west coast of North America. Four large earthquakes (magnitude $M>7$) have occurred during the past 100 years, and hundreds of moderate and small earthquakes occur in the region each year. The strike-slip type was the focal mechanism for most earthquakes.

The tectonic setting in this dominated by strike-slip faulting region is complex. However, oblique convergence is a significant element. The near-vertical Queen Charlotte Fault accommodates about 52mm/year of right-lateral motion between the Pacific and North American plates (e.g., Cassidy et al. 2014). The high and steep topography along the west coast of Haida Gwaii is probably the result of the initiation of oblique convergence. In the Haida Gwaii region, the component of convergence has been interpreted to be mainly accommodated by the oblique underthrusting of the seafloor beneath Moresby and Graham Island.

The 2012 Haida Gwaii, Canada, M_W7.8 earthquake has received great attention, with a special issue of Bulletin of the Seismological Society of America (James et al. 2015) addressing the research results of this earthquake (and the 5 January 2013 M_W7.5 Craig strike-slip earthquake). James et al. (2015) provided the introduction to this special issue.

As an earthquake is a representation of the tectonics and tectonic stress in the epicentral region, a detailed study of the behaviour of the Haida Gwaii earthquake may help us understand tectonic movements and complexities and provide information for seismic hazard assessment in the Haida Gwaii region. The earthquake was also a tsunami-generating event. An understanding of this earthquake's behaviour may also help us to better understand the tsunami threat to the west coast of Canada.

To obtain stable solutions in retrieving a focal mechanism for moderate, strong, and large earthquakes in most regions through out the world, it is a popular practice to confine the trace of the moment tensor to zero. However, a negative or positive trace indicates a different stress status of the earth materials in the source region. For a large thrust event such as the Haida Gwaii earthquake, the material in the source region experiences great compressive and tensile forces. A negative trace value may suggest that the source region is still in a state of considerable compression.

A moment tensor solution for an earthquake shows the average stress state imposed on the source region, while the source rupture process inversion can provide more details of the source. Rupture details may provide information for seismic-hazard modeling and tsunami simulation.

The moment tensor solutions for the Haida Gwaii earthquake have been calculated (e.g., Global Centroid

Fig. 1　Distribution of the 34 seismic stations from which the recorded mantle wave seismograms were selected for inversion
The solid circle at the centre shows the epicentre

Moment Tensor, G-CMT), and the source rupture process models were also derived (e.g., Lay et al. 2013). These issues are complex, and cannot be completely understood; therefore, we performed further studies. In this article, we introduce our full moment tensor solution and the source rupture process inversion results for the Haida Gwaii earthquake, which generated the strongest tsunami in Canada's written history (Leonard and Bednardski 2014).

2　Introduction to methods

In this section, we briefly introduce the methods used for the inversions of the full moment tensor and the source rupture process.

2.1　Method used to retrieve a full moment tensor

An earthquake source can be described using a moment tensor, which is a 3×3 matrix. From linear algebra, we know that a complex matrix can be expressed by the summation of several simple, independent matrixes. Following Kikuchi and Kanamori (1991), an arbitrary moment tensor can be expressed by summing up six different constant moment tensors:

$$M = \sum_{k=1}^{6} a_k M_k \quad (1)$$

in which $M_1 - M_5$ indicate different constant shear dislocation sources, M_6 indicates an explosion source, and a_k ($k=1, 2, \cdots, 6$) are summation coefficients.

The relationship between the moment tensor and the coefficients can be written as

$$M = \begin{bmatrix} a_2 - a_5 + a_6 & a_1 & a_4 \\ a_1 & -a_2 + a_6 & a_3 \\ a_4 & a_3 & a_5 + a_6 \end{bmatrix} \quad (2)$$

Given an earthquake hypocentre and earth model, each of the aforementioned constant sources can be used to generate synthetic seismograms, called Green's functions, at a specific seismic station. We assume that the three component Green's functions generated at time t by M_k are $U_k(t)$, $V_k(t)$, and $W_k(t)$. The synthetic three component seismograms at the station generated by the moment tensor can be calculated by the following formulas:

$$U(t) = \sum_{k=1}^{6} a_k U_k(t) = a_0 \sum_{k=1}^{6} a'_k U_k(t)$$
$$V(t) = \sum_{k=1}^{6} a_k V_k(t) = a_0 \sum_{k=1}^{6} a'_k V_k(t) \qquad (3)$$
$$W(t) = \sum_{k=1}^{6} a_k W_k(t) = a_0 \sum_{k=1}^{6} a'_k W_k(t)$$

in which $a'_k(=a_k/a_0)$ is confined in the range of -1 and $+1$ for the inversion, and the factor a_0 is a constant, determined using the following formula:

$$a_0 = \frac{1}{N} \sum_{j=1}^{N} \frac{O_j(t)_{\max}}{X_j(t)_{\max}} \times 10^{20} \qquad (4)$$

(in dyn·cm; 1 dyn·cm = 10μN·cm) in which subscript j is an ordinal number for N digital records used in the inversion, $O_j(t)$ is a segment of a digital waveform record, and $X_j(t)$ is the segment of a synthetic seismogram corresponding to $O_j(t)$.

Once the six coefficients are obtained, the moment tensor can be calculated using eq. (2). The simultaneous waveform shape and amplitude inversion procedure was used (Ma and Adams 2002).

Table 1 Source parameters, isotropic moments, maximum and minimum double-couple moments, and total moments in the inversion tests

Depth/km	Isotropic/10^{20} Nm	Maximum/10^{20} Nm	Minimum/10^{20} Nm	Total moment/10^{20} Nm
15	−0.1918	7.2245	0.2474	7.6637
16	−1.0241	5.8749	0.1976	7.0966
17	−1.2214	5.2620	0.3466	6.8299
18	−1.5122	4.8073	0.5219	6.8414
19	−1.7807	4.4913	0.6803	6.9524
20	−2.0580	4.2321	0.8338	7.1239
21	−2.2692	4.0460	0.9510	7.2661
22	−2.4695	3.9032	1.0526	7.4253
23	−2.5234	3.8326	1.0946	7.4506
24	−2.7425	3.7116	1.1699	7.6240
25	−2.8829	3.6543	1.2328	7.7700

Note: Date, 28 October 2012; time, 03:04:37; latitude 52.61°; longitude −132.06°. Boldface data correspond to the dashed line in Fig. 2.

There are several ways to generate Green's functions, depending on the wave types used. As the Haida Gwaii earthquake was very large (the length of its source was more than 100km), we used the mantle Rayleigh wave fitting to retrieve its moment tensor. The Green's functions were generated by the normal modes summation method (Gilbert and Masters 1989).

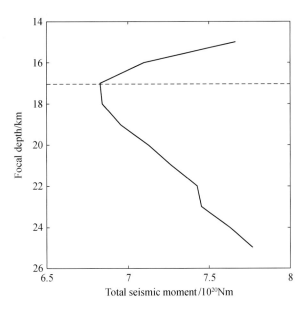

Fig. 2 Total seismic moment change with focal depth

2.2 Method used for the source rupture process inversion

To better understand the source rupture process, we used a finite-fault model inversion procedure to estimate the source rupture process. The procedure was developed by Zhang et al. (2012).

For moment tensor or focal mechanism inversions, an earthquake source is treated as a point. For a finite source model, the source rupture area may be divided into small areas, usually small rectangles. Each small area can be treated as a point source.

The synthetic seismic waves at a seismic station can be calculated using the following formula introduced by Aki and Richards (1980):

$$u_n^k(t) = \sum_{l=1}^{L} [g_{n,pq}^{kl}(t) * M_{pq}^l(t)] \tag{5}$$

Table 2 Moment tensor solutions for the Haida Gwaii earthquake

			Strike/(°)	Dip/(°)	Rake/(°)	Axis	Azimuth/(°)	Plunge/(°)	Strength/10^{20} Nm
This article		NP1	329	24	114	T	1	62	4.915
		NP2	112	70	76	N	117	13	0.347
						P	213	24	-5.262
	Element		M_{RR}	$M_{\theta\theta}$	$M_{\varphi\varphi}$	$M_{R\theta}$	$M_{R\varphi}$	$M_{\theta\varphi}$	$M_0 = 5.262$
	/10^{20} Nm		1.77	-3.16	-2.27	3.64	-1.17	2.12	ISO = -1.221
G-CMT		NP1	318	25	104	T	19	69	5.505
		NP2	122	66	83	N	125	6	0.331
						P	217	20	-5.846
	Element		M_{RR}	$M_{\theta\theta}$	$M_{\varphi\varphi}$	$M_{R\theta}$	$M_{R\varphi}$	$M_{\theta\varphi}$	$M_0 = 5.676$
	/10^{20} Nm		4.08	-2.53	-1.56	3.26	-1.78	2.40	—

Note: ISO, isotropic portion of the moment.

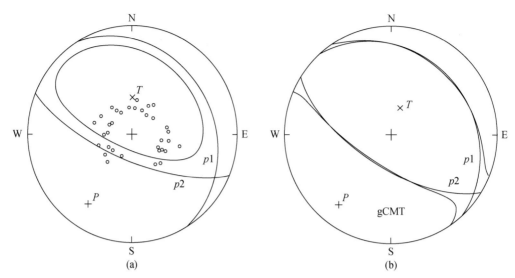

Fig. 3 Comparison between the two sets of moment tensor solutions

(a) The lower hemispherical projection of the moment tensor solution obtained by the research outlined in this article; (b) the lower hemispherical projection of the G-CMT. The nodal planes 1 and 2 are indicated with p1 and p2. The area in which the T axis is located is much smaller in projection (a) than that in (b)

where $n(=1, 2, \text{or } 3)$ is a three-component index; k is the station index; l and L are the small rectangular indices and their total number; p and $q(=1, 2, \text{or } 3)$ are orthogonal coordinates; $g_{n,pq}^{kl}$ is Green's function, denoting the seismic wave on the nth component at the kth station caused by a unit force couple (p, q) at the lth sub-fault (a rectangular area) with a Dirac-delta type source time function; M_{pq}^{l} denotes the time history of the seismic moment tensor of the lth sub-fault; and $*$ is the convolution operator.

To better describe the source rupture process, the moment rate history $\dot{M}_{pq}^{l}$ is often used, so eq. (1) can also be expressed as

$$u_n^k(t) = \sum_{l=1}^{L} [g_{n,pq}^{(H)kl}(t) * \dot{M}_{pq}^{l}(t)] \qquad (6)$$

where $g_{n,pq}^{(H)kl}$ is the response of the Heaviside (step) function. If only the pure shear rupture is considered, we have

$$\dot{M}_{pq}^{l}(t) = M_0^l(t)[\dot{e}_p(t)v_q + \dot{e}_q(t)v_p] \qquad (7)$$

where $M_0^l(t)$ is the scalar moment history of the lth sub-fault, e is the slip vector, and v is the normal direction vector of the fault. The dummy (p, q) index can be converted into any Cartesian coordinate system (i.e., x, y, z). If, for example, z is defined as the normal direction of the fault plane and the x and y axes are on the fault plane, $v_x = 0$, $v_y = 0$, $v_z = 1$, and $e_z = 0$.

The x and y components of the scalar moment rate history of the lth sub-fault are

$$\begin{aligned} m_x^l(t) &= M_0^l(t)\dot{e}_x(t) \\ m_y^l(t) &= M_0^l(t)\dot{e}_y(t) \end{aligned} \qquad (8)$$

From eqs. (6) and (8), we have

$$u_n^k = \sum_{l=1}^{L} [g_{n,xz}^{(H)kl}(t) * m_x^l(t) + g_{n,yz}^{(H)kl}(t) * m_y^l(t)] \qquad (9)$$

To use eq. (9) for rupture process inversion, it can be expressed in matrix form as

$$[\mathbf{G}_x \quad \mathbf{G}_y] \times \begin{bmatrix} \mathbf{m}_x \\ \mathbf{m}_y \end{bmatrix} = [\mathbf{U}] \qquad (10)$$

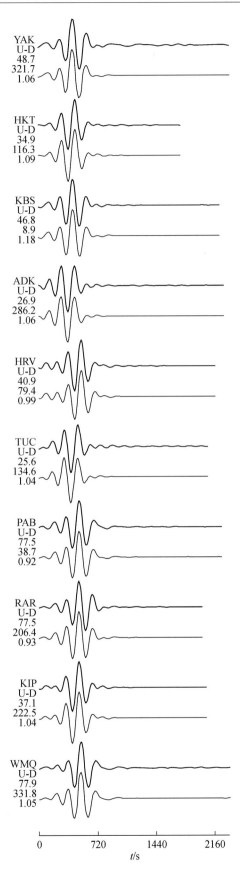

Fig. 4 Comparison between the first 10 observed and synthetic seismograms for the Haida Gwaii earthquake

For each pair of waveforms, the upper trace is the recorded waveform; the lower trace is the synthetic waveform generated with the moment tensor solution in Fig. 3 (a). Both the observed and the synthetic waveforms were filtered with a pass-band of 135~500s. At the left side of each pair, the symbols and numbers from top to bottom indicate station name, vertical component, distance in degrees, station azimuth in degrees, and the ratio between the observed and synthetic maximum amplitudes

where $\mathbf{G}_x$ and $\mathbf{G}_y$ are Green's function matrixes, $\mathbf{U}$ is a vector formed by the observed seismograms, and $\mathbf{m}_x$ and $\mathbf{m}_y$ are the unknown vector components to be solved for and which will describe the time histories of slip rate amplitudes and slip angles of all sub-faults on the fault plane.

Given the inherent errors in the observed seismograms and the inaccuracy of Green's functions, directly solving eq. (10) leads to nonphysical solutions. Therefore, smoothness and scalar moment minimization constraints are imposed to stabilize the solutions. The equation format after the constraints are added can be expressed as

$$\begin{bmatrix} \mathbf{C}_d^{-1}[\mathbf{G}_x & \mathbf{G}_y] \\ \lambda_1 \begin{bmatrix} \mathbf{D} & 0 \\ 0 & \mathbf{D} \end{bmatrix} \\ \lambda_2 \begin{bmatrix} \mathbf{T} & 0 \\ 0 & \mathbf{T} \end{bmatrix} \\ \lambda_3 \begin{bmatrix} \mathbf{I} & 0 \\ 0 & \mathbf{I} \end{bmatrix} \end{bmatrix} \begin{bmatrix} \mathbf{m}_x \\ \mathbf{m}_y \end{bmatrix} = \begin{bmatrix} \mathbf{C}_d^{-1}\mathbf{U} \\ 0 \\ 0 \\ 0 \end{bmatrix} \qquad (11)$$

where $\mathbf{D}$ and $\mathbf{T}$ are spatial and temporal smoothness matrixes (Horikawa 2001; Yagi et al. 2004); $\mathbf{I}$ is an identity matrix for minimizing scalar moment (Hartzell and Iida 1990; Mendoza and Hartzell 2013); $\mathbf{C}_d$ is an a priori data covariance matrix that is used to normalize each data record to a peak amplitude of 1.0; and, λ_1, λ_2, and λ_3, are scalar constants that denote the corresponding weights of the constraints.

Equation (11) can be solved with a gradient method (Ward and Barrientos 1986), which can invoke a nonnegative constraint on the solution.

3 Full moment tensor inversion for the Haida Gwaii earthquake

The full moment tensor inversion for the Haida Gwaii earthquake was performed using the method outlined earlier.

3.1 Mantle Rayleigh wave records

As the Haida Gwaii earthquake was very large, it excited very strong mantle waves, which were recorded at seismic stations throughout the world with hundreds of wave form records on LHZ channel (long-period high-gain vertical component). We selected 34 vertical records, with approximately one record for every 10° of azimuth around the epicentre. We filtered these records with a band-pass filter (the period range is 135 ~ 500s), and decimated the sampling rate from 1 to 0.1 sps (samples per second). When we assume the velocity of the mantle waves was in the order of 3km/s, the shortest wavelength was in the order of 400km, which was approximately three times of the rupture length of the earthquake. For such long-period mantle waves, the source can be treated as a "point". In the following section, we introduce the moment tensor results obtained using the procedure presented earlier. Figure 1 displays the distribution of the seismic stations that recorded the mantle waves we used.

3.2 Full moment tensor inversion tests

Moment tensor inversions require source parameter values. Errors in those values can generate errors in the moment tensor solution. Compared with the error in the epicentre of an earthquake, an error in focal depth can generate a larger error in a moment tensor solution.

Fig. 5 Distribution of the 25 seismic stations from which P wave vertical segments were selected for rupture process inversion

The solid circle at centre shows the epicentre

We conducted the following tests using focal depths from 15 to 25km with an increment. For each focal depth, we calculated Green's functions; aligned the observed mantle Rayleigh waves with the synthetic ones that were calculated using an initial moment tensor; and performed a full moment tensor inversion. The obtained full moment tensor was then decomposed into the isotropic seismic moment, and maximum, minimum double-couple moment tensors, which are listed in Table 1, as well as the coordinates and origin of the Haida Gwaii earthquake, used by the Global CMT (G-CMT) Project group. The focal depth used for the G-CMT solution was 12km. We thought that this value may be too shallow for our inversion using mantle waves.

Table 3 Crustal model CN06 used for the rupture process study

Depth/km	V_p/(km/s)	V_s/(km/s)
0	5.5	3.18
1	6.0	3.47
10	6.5	3.76
20	6.8	3.93
29	8.1	4.68

Note: V_p, P wave velocity; V_s, S wave velocity.

The source time function we used was two overlapped triangles. The time length of the bottom side was 20s.

The far right column in Table 1 provides the total seismic moment, which is the summation of the isotropic and maximum and minimum double-couple moments. The total moment obtained using a focal depth of 17km was the smallest, although the total moment obtained using a focal depth of 18km was also small. Figure 2 shows the total moment change with focal depth. The minimum of the total moment at around 17 and 18km indicated that the solutions at these depths were stable, therefore, we took the solution obtained using focal depth 17km as the preferred solution.

Table 2 presents our full moment tensor inversion solution, together with the solution obtained by the G-CMT

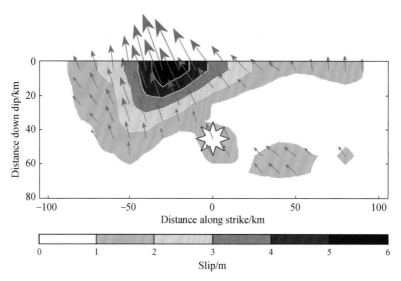

Fig. 6 Static final slip distribution on the fault plane of the Haida Gwaii earthquake

The eight-pointed star denotes the location of the hypocentre (rupture initiation point). The direction of the arrows shows the slip direction; the lengths of the arrows indicate the magnitude of the slip. The largest slips occurred at the shallow portion of the fault in the depth range of about 0~15km

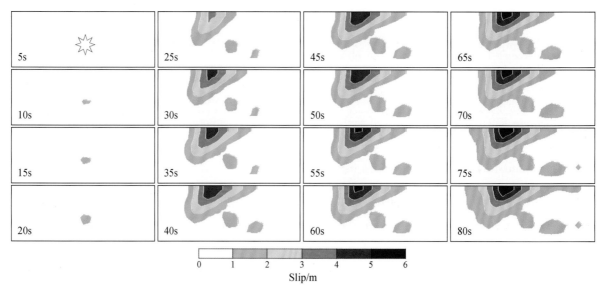

Fig. 7 Snapshots of slip variation with time on the fault plane of the Haida Gwaii earthquake

The star denotes the location of the hypocentre. Each rectangle denotes the slip distribution at the time indicated in the bottom-left corner of each snapshot. The last snapshot in the bottom-right corner is the static final slip distribution at 80s

Project group, retrieved from the website of the Inc. Research Institutions for Seismology (IRIS; see data sources section).

In the moment tensor solutions provided by the G-CMT group, and other agencies, the trace of the moment tensor was confined to 0 ($M_{RR}+M_{\theta\theta}+M_{\varphi\varphi}=0$). The G-CMT trace in Table 2 was calculated as 4.08+(-2.53)+(-1.56)=0.01~0. However, in our inversion process, the trace $M_{RR}+M_{\theta\theta}+M_{\varphi\varphi}$ was not confined. After several inversion tests, we found that the trace or in other words, the isotropic part moment was negative (-1.22). This implies that the earth material's volume in the source region became smaller after this large earthquake.

Figure 3 shows that the best double-couple solutions from the two sets of moment tensors were similar (the curves indicated with $p1$ and $p2$). The non-double-couple solutions varied somewhat (i. e., the areas in which P or T axis was located were different). In our solution, the area in which the P axis was located was larger than that in the G-CMT solution. This shows that the isotropic portion of the moment was negative. In other words, the Haida Gwaii earthquake caused a negative volume change during its source process.

A necessary and important criterion for assessing the quality of the solution is a comparison between the synthetic and observed seismograms. Figure 4 shows the first 10 pairs of seismograms for the comparison. Other pairs from the remaining 24 stations were of similar quality. The similarity between the synthetic and the observed seismograms in both the waveform shapes and the maximum amplitudes were excellent. This implies our moment tensor solution was likely reasonable.

4 Source rupture process inversion for the Haida Gwaii earthquake

Following the inversion procedure using the inversion code developed by Zhang et al. (2012), we performed the source rupture process inversion for the Haida earthquake.

4.1 Teleseismic waveform records

In this study, teleseismic records with an epicentral distance of $30° \sim 90°$ were selected, and the waveform records on the BHZ channel (broadband high-gain vertical component) were downloaded from the IRIS website. To ensure good signal-to-noise ratios, only the P wave segments were considered. Stations with an azimuth interval of $15°$ were chosen to provide an even azimuthal coverage of the stations around the epicentre. Figure 5 shows the distribution of the selected 25 stations. Following an instrument response correction, a band-pass filter of 0.01-0.1Hz was applied to the observed waveforms.

4.2 Crustal model used in the epicentral region of the 2012 Haida Gwaii earthquake

When Green's functions are generated for teleseismic stations, an earth model is needed, since the P wave segments include the P phase, and the converted and reflected phases, sP and pP. The influence of these two phases on the P wave segments is related to the crustal structure in the epicentral region. A crustal model close to the epicentral region (CN06) was provided by the Geological Survey of Canada (GSC, J. Cassidy, personal communication, 29 April 2014). The model is used for locating earthquakes in the Hecate Strait region, which is close to the epicentre of the 2012 Haida Gwaii earthquake. As the crustal model in the epicentral region is currently not available, the CN06 model (listed in Table 3) is the best one available and was used in the rupture process inversions.

4.3 Source rupture process inversion results

In the inversion procedure used, the element values in matrix C_d were dependent on the data set prepared for each specific earthquake. The matrix was calculated and imposed on our data set. The maximum amplitude of each seismogram was normalized to around 1. Other weights were also determined by the matrix. The three constraint weights (i. e., spatial smooth, temporal smooth, and moment minimization) were optimized through a trial-and-error procedure.

Following the procedure of Zhang et al. (2012) as previously described, we performed the inversion for the spatiotemporal rupture process using the focal mechanism with a strike of 329°, a dip of 24°, and a rake of 114°, as obtained in the section "Full moment tensor inversion for the Haida Gwaii earthquake". The epicentral location (latitude of 52.742° and longitude of −132.131°) used in the inversion was obtained from the U.S. Geological Survey (USGS). The fault model dimensions were 210km×80km, while the sub-fault size was 10km×10km. The dimensions of the fault rupture model had been tested for several times and were finally determined to be reasonably large enough for the inversion. Thus, slip along the fault edges of the fault's rupture area except for the uppermost edge were fixed at zero.

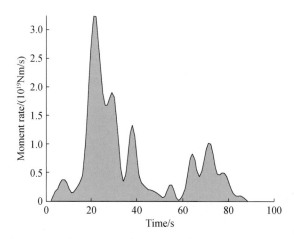

Fig. 8 Source time (moment rate) function of the Haida Gwaii earthquake

In addition, we assumed that no supershear ruptures occurred during the earthquake, and we did not allow the rupture velocity to exceed 3.5km/s. For each sub-fault, the maximum rupture duration was set to 60s. It suggests that, for a sub-fault with a distance $d$km from the hypocentre, its rupture velocity can vary between $3.5d/(d+210)$ and 3.5km/s. The rupture velocity and the waveforms used in the inversion determine the sub-fault source time functions (i.e., the shape and amplitude).

4.4 Static slip distribution

Figure 6 shows that the 2012 Haida Gwaii earthquake produced a highly heterogeneous slip distribution on the fault plane. There was only one major slip patch located upper left of the hypocentre in Fig. 6 (i.e., south-southwest of the epicentre), which released more than 80% of the total moment of the earthquake. There were also several scattered small slip patches distributed in depth, but their credibility was relatively weak due to their small amplitudes and scales. The average slip on the fault plane was about 2.1m, and the maximum slip on the fault plane, close to the seafloor, was about 5.8m.

Based on the static final slip distribution in Fig. 6, the stress drop on the fault was calculated using the Brune model (Brune 1970, 1971). The results indicate that the maximum and average values over the fault plane were 16.5 and 4.6MPa, respectively.

4.5 Spatiotemporal variation of rupture process

Figure 7 shows the slip variation with space and time on the fault. In the first 15s after the rupture initiation, the rupture propagated around the hypocentre. The major rupture then began, reaching the seafloor about 10s later. The major rupture lasted for about 25s, with the maximum slip of about 5.8m near the seafloor, located south-southwest of the epicentre. At about 60s after the rupture initiation, a rupture at the right side of the major rupture occurred in the shallow part. By supposing that a sub-fault began to rupture when its cumulated slip exceeded 0.1m, the slip-weighted average rupture velocity was estimated to be 2.3km/s.

4.6 Source time function

Figure 8 shows the moment rate function, also called the source time function, which can be obtained based on the spatiotemporal rupture images shown in Fig. 7 (Zhang et al. 2009). The source time function indicates that the total seismic moment released in the whole earthquake rupture process was about 5.3×10^{20} Nm, corresponding to a moment magnitude $M_W 7.8$.

The whole time history of the Haida Gwaii earthquake can be divided into four stages or events. The first event occurred in the first 10s after the rupture initiation. The second event, the main event, was from 11 to 35s. The third one started at 35s and ended at 45s. The fourth occurred in the time period of 60–80s. The whole rupture process lasted about 90s.

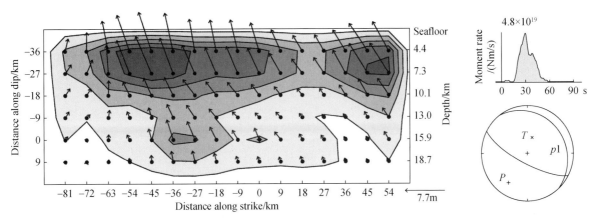

Fig. 9 Finite-fault slip model from teleseismic body wave inversions for the Haida Gwaii earthquake

The left panel shows the slip distribution on the fault plane. The arrows' directions indicate the average rake of each sub-fault, and the contours show the magnitude of the slip. The lengths of the arrows show the peak slip. The upper-right panel shows the moment rate (source time) function, and the bottom-right panel shows the lower hemisphere equal area projection of the focal mechanism. These results are all from Lay et al. (2013), with permission from Elsevier

4.7 Comparison with results obtained by other authors

Lay et al. (2013) performed a focal mechanism inversion for the 2012 Haida Gwaii event using long-period W-phase signals (Kanamori and Rivera 2008) in the frequency band of 1.67–5.0mHz for a range of source depths. The preferred solution was determined to have a centroid depth of 9.5km, a seismic moment of 6.9×10^{20} Nm ($M_W 7.8$), and a best double couple with a strike of 317.1°, a dip of 18.5°, and a rake of 103.3°.

They also performed finite-fault model inversion for the 2012 event using teleseismic broadband P waves and SH waves with constraints from W-phase solutions, information from shortperiod back-projections, and tsunami observations. Figure 9 summarizes the resulting model of slip distribution from linear kinematic inversions with a constant rupture velocity of 2.3km/s for model grids with a strike of 320° and a dip of 18.5°.

The sub-fault source time functions were parameterized by Lay et al. (2013) using eight 2s rise-time triangles with 2s time lags, giving 18s sub-fault rupture durations. The velocity structure in the source region was adapted from Smith et al. (2003), with the addition of a 3km water layer. The rupture expanded from the hypocentre at about 2.3km/s over a total fault length of about 150km, and most of the slip was up-dip from a 15.9km deep hypocentre below western Haida Gwaii.

With an 18.5° dip, the model extended to the toe of the offshore sedimentary terrace about 40km to the southwest, with relatively uniformly distributed shallow slip in the upper 10km. The estimated seismic moment was 6.7×10^{20} Nm (M_W 7.8); and the average slip for the sub-faults, with well-resolved seismic moments of at least 20% of the peak sub-fault moment, was 3.3m, corresponding to a seismic moment of 6.4×10^{20} Nm.

A comparison of Figs. 6 and 9 indicates that there were four rupture patches resolved by the two inversions, at similar locations. Along deeper sections of the fault rupture, there were two small patches, and another small patch in the shallow portion; the major rupture occurred within the shallow section of the fault.

Figures 6 and 10 show that the maximum slips broke to the seafloor. From the contours in Fig. 9, the maximum slips occurred immediately beneath the seafloor. As the 2012 Haida Gwaii earthquake was a large and very shallow thrust event, it is more reasonable that the maximum slips intersected with the seafloor.

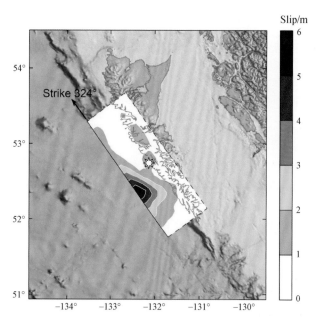

Fig. 10 Surface projections of the fault plane and the rupture distribution

The star denotes the epicentre

Good data fitting of the observed waveforms is an necessary condition for the rupture model to be considered reliable. Figure 11 shows the comparison between the observed and synthetic waveforms in our rupture process inversion. The match between the shapes of the observed and synthetic P wave segments was very good. We calculated correlation coefficients between the observed and synthetic waveforms. As Fig. 11 shows, correlation coefficients at nearly all stations were more than 0.90, except for GUMO whose value is about 0.89. As a whole, the average correlation coefficient is 0.95, suggesting an excellent data fit of the seismograms, implying that our rupture model may be reasonable.

5 Discussion

The Haida Gwaii earthquake occurred in the Moresby Island region, on a fault between the Pacific and North

America plates. As a thrust event in an oblique subduction zone, this earthquake generated a tsunami, and can be considered a subduction event. During the earthquake, the leading edge of the overriding plate sprang seaward and upward, while the landward portion of the overriding plate stretched and dropped down. These movements then set off a tsunami.

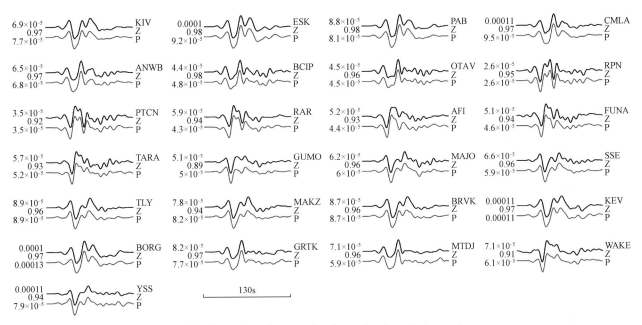

Fig. 11　Comparisons between the observed and synthetic waveforms

For each pair, the upper trace is the observed P wave segment, and the bottom trace is the synthetic P wave segment. On the left of each pair from the top to bottom are the maximum amplitude of the observed waves (m), the correlation coefficient of the observed and synthetic waveforms, and the maximum amplitude of synthetic waves. On the right are the station name, component, and phase name

Based on the tsunami-generating model discussed earlier, the process of the plate dropping downward is similar to the volume of the materials in the earthquake source region becoming smaller. In our full moment tensor solution, the isotropic moment was negative. This also shows that the material volume in the source region became smaller. The phenomenon implies that a percentage of the negative isotropic moment in the total moment may be used to study tsunami-generating earthquakes.

The seismograms used in our source rupture process inversion were P wave segments recorded at teleseismic stations. Major components of these segments include direct P waves, reflected P waves (pP depth phase), and converted P waves (sP depth phase). At each station, the differential times between sP and P and between pP and P depend on the focal depth for a point source (or the rupture patch depth on a finite fault) and the crustal structure at the epicentre region. The amplitudes of the P wave segments at each station also depend on the event's focal mechanism, as well as the hypocentral depth for a point source and the local crustal structure at the epicentral region.

For these reasons, if a better crustal model in the epicentral region becomes available, a better constrained source rupture model may be obtained.

Different inversion methods can be used to retrieve the source rupture process for a single earthquake. The actual rupture process for a specific earthquake is unique, but the rupture models retrieved using different methods can be different. However, the major rupture features obtained by different methods should be similar.

It can be seen in Figs. 6 and 9 that both rupture models had one major rupture and several smaller ruptures.

The rupture process of this earthquake was also studied using an empirical Green's function method (Hobbs et al. 2015). Three rupture stages were retrieved. The rupture azimuths and velocities for these stages were 202°, 323°, and 295° and 0.9, 1.6, and 1.6km/s, respectively. The average azimuth and velocity were 308.5° and 1.1km/s. Both we and Lay et al. (2013) used a rupture velocity of 2.3km/s. The factors related to this rupture velocity discrepancy are complex and need further studies.

The peak point of synthetic Rayleigh waves generated at each station usually has a time shift from the observed one, as the earth model used to generate the synthetics has a bias from the Earth. To reduce the error in the moment tensor solution caused by this bias, we applied a time correction by first generating synthetics using a reasonable initial moment tensor, and then aligning the synthetic and the observed peaks manually, prior to performing the inversion.

Any observed seismogram has noise signal components that are not generated by the earthquake. The noise signal components tend to make a contribution to the isotropic moment in the inversion. To decrease this influence, we carefully selected high-quality seismograms with the assistance of a graphical program.

As there are many high-quality waveform records available, it is possible to retrieve reliable isotropic moment. This information is useful for earthquake source, geodynamic, tsunami, and other studies.

6 Conclusion

The 2012 Haida Gwaii M_W 7.8 earthquake generated a large amount of high-quality seismic waveform records. We carefully selected 34 mantle Rayleigh wave vertical records (LHZ channel) and 25 P wave vertical segment records (BHZ channel) at teleseismic stations. Using waveform modelling methods, we retrieved a full moment tensor solution and established a source rupture process model for the earthquake. The focal mechanism calculated from the moment tensor solution was a thrust type. The isotropic moment was negative, and its absolute value was more than one-fifth of the total seismic moment. In the source rupture process model, there were four rupture patches. Two small rupture patches occurred at the focal depth level, and another smaller patch was close to the seafloor. The major rupture patch began at depth about 20km and ran to the seafloor; and it was initiated about 10s following the initiation of the small rupture at the hypocentral area, and lasted about 25s. The major rupture length along the seafloor was about 100km, in the northwest direction with the maximum slip occurred beneath the seafloor. The total rupture lasted about 90s.

Data sources

Seismograms used in this study were collected from the Inc. Research Institutions for Seismology (IRIS) Consortium at http://www.iris.edu/hq/ (accessed 31 May 2014). G-CMT solutions were retrieved from http://www.iris.edu/spud/momenttensor/1059989 (accessed 31 May 2014).

Acknowledgements

This research was supported by the Natural Sciences and Engineering Research Council of Canada (NSERC) under the Discovery Grants program and National Natural Science Foundation of China (41574035 and

41541035). We gratefully acknowledge the constructive comments, suggestions, and text revisions from Alison Bird and Maurice Lamontagne at the Geological Survey of Canada. The waveform records were processed using SAC2000, *redseed*, and *geotool* programs. A special thanks to John Cassidy at Geological Survey of Canada, who provided a useful crustal model.

References

Aki, K., and Richards, P. G. 1980. *Quantitative Seismology*. Freeman and Co., New York.

Brune, J. N. 1970. Tectonic stress and the spectra of seismic shear waves from earthquakes. *Journal of Geophysical Research*, **75**: 4997–5009. doi: 10.1029/JB075i026p04997.

Brune, J. N. 1971. Correction. Tectonic stress and the spectra of seismic shear waves from earthquake. *Journal of Geophysical Research*, **76**: 5002–5002.

Cassidy, J. F., Rogers, G. C., and Hyndman, R. D. 2014. An Overview of the 28 October 2012 M_W7.7 Earthquake in Haida Gwaii, Canada: a tsunamigenic thrust event along a predominantly strike-slip margin. *Pure and Applied Geophysics*, **171**: 3457–3465. doi: 10.1007/s00024-014-0775-1.

Gilbert, F., and Masters, G. 1989. Low-frequency seismology. University of California, San Diego, USA. [Lecture notes, Chap. 4.]

Hartzell, S, and Iida, M 1990. Source complexity of the 1987 Whittier Narrows, California, earthquake from the inversion of strong motion records. *Journal of Geophysical Research*, **95**(B8): 12475–12485. doi: 10.1029/JB095iB08p12475.

Hobbs, T., Cassidy, J., and Dosso, S. 2015. Rupture Process of the 2012 M_W7.8 Haida Gwaii Earthquake from an Empirical Green's Function Method. *Bulletin of the Seismological Society of America*, **105**: 1219–1230. doi: 10.1785/0120140175.

Horikawa, H. 2001. Earthquake doublet in Kagoshima, Japan: rupture of asperities in a stress shadow. *Bulletin of the Seismological Society of America*, **91**: 112–127. doi: 10.1785/0119990131.

James, T., Cassidy, J., Rogers, G., and Haeussler, P. 2015. Introduction to the Special Issue on the 2012 Haida Gwaii and 2013 Craig Earthquakes at the Pacific-North America Plate Boundary (British Columbia and Alaska). *Bulletin of the Seismological Society of America*, **105**(2B): 1053–1057. doi: 10.1785/0120150044.

Kanamori, H., and Rivera, L. 2008. Source inversion of W phase: speeding up seismic tsunami warning, *Geophysics Journal International*, **175**: 222–238. doi: 10.1111/j.1365-246X.2008.03887.x.

Kikuchi, M., and Kanamori, H. 1991. Inversion of complex body waves-III. *Bulletin of the Seismological Society of America*, **81**: 2335–2350.

Lay, T., Ye, L., Kanamori, H., Yamazaki, Y., Cheung, K. F., Kwong, K., and Koper, K. 2013. The October 28, 2012 M_W 7.8 Haida Gwaii underthrusting earthquake and tsunami: Slip partitioning along the Queen Charlotte Fault transpressional plate boundary. *Earth and Planetary Science Letters*, **375**: 57–70. doi: 10.1016/j.epsl.2013.05.005.

Leonard, L. J., and Bednardski, J. M. 2014. Field survey following the 28 October 2012 Haida Gwaii tsunami. *Pure and Applied Geophysics*, **171**(12): 3467–3482. doi: 10.1007/s00024-014-0792-0.

Ma, S. and Adams, J. 2002. Estimation of moment tensor for moderate earthquakes in eastern Canada and its vicinity by modeling surface waveforms. Open file draft to Natural Resources Canada, 110 pp.

Mendoza, C., and Hartzell, S. 2013. Finite-fault source inversion using teleseismic P waves: Simple parameterization and rapid analysis. *Bulletin of the Seismolological Society of America*, **103**(2A): 834–844. doi: 10.1785.0120120069.

Smith, A. J., Hyndman, R. D., Cassidy, J. F., and Wang, K. 2003. Structure, seismicity, and thermal regime of the Queen Charlotte Transform Margin. *Journal of Geophysical Research*, **108**(B11). doi: 10.1029/2002JB002247.

Ward, S. N., and Barrientos, S. E. 1986. An inversion for slip distribution and fault shape from geodetic observations of the 1983, Borah Peak, Idaho, earthquake. *Journal of Geophysical Research*, **91**: 4909–4919. doi: 10.1029/JB091iB05p04909.

Yagi, Y., Mikumo, T., Pacheco, J., and Reyes, G. 2004. Source rupture process of the Tecomán, Colima, Mexico earthquake of 22 January 2003, determined by joint inversion of teleseismic body-wave and near-source data. *Bulletin of the Seismological Society* of America, **94**: 1795–1807. doi: 10. 1785/012003095.

Zhang, Y., Feng, W. P., Xu, L. S., Zhou, C. H., and Chen, Y. -T. 2009. Spatio-temporal rupture process of the 2008 great Wenchuan earthquake. *Science in China Series D: Earth Science*, **52**: 145–154. doi: 10. 1007/s11430-008-0148-7.

Zhang, Y., Feng, W. P., Chen, Y. -T., Xu, L. S., Li, Z. H., and Forrest, D. 2012. The 2009 L'Aquila M_W 6. 3 earthquake: a new technique to locate the hypocentre in the joint inversion of earthquake rupture process. *Geophysics Journal International*, **191**: 1417–1426. doi: 10. 1111/j. 1365-246X. 2012. 05694. x.

非对称地震矩张量时间域反演：理论与方法

刘 超[1] 陈运泰[1,2]

1. 中国地震局地球物理研究所 中国北京 100081；2. 北京大学地球与空间科学学院 中国北京 100871

摘要 本文在对称地震矩张量反演的基础上，进一步研究了非对称地震矩张量时间域反演的理论与方法。结果表明：非对称地震矩张量反演与对称地震矩张量反演类似，只需将对称地震矩张量反演方法略加改动，即增加3个待解参数，便可实现非对称地震矩张量反演。为了判断非对称地震矩张量反演相对于对称地震矩张量反演是否存在过度拟合，运用了 AIC 准则（赤池信息准则）。为了定量地描述地震矩张量之间的差异，引入了地震矩张量的矢量表示法。通过分析格林函数与地震矩张量各分量之间的关系，得出：在非对称地震矩张量反演时，若仅用垂直向地动位移数据，将无法区分 M_{xy} 与 M_{yx} 这两个分量，需要同时运用垂直向与水平向地动位移数据进行联合反演才能区分 M_{xy} 与 M_{yx}；若采用不同的速度结构模型或不同的格林函数计算方法，则需重新评估地震矩张量各分量的分辨度问题。为检验非对称地震矩张量反演方法的可行性，利用合成地震图进行了一系列数值试验。数值试验结果表明，在非对称地震矩张量反演中，有必要引入 S 波进行 P 波与 S 波联合反演以提高反演的准确性和判定断层面的能力。

0 引言

在震源物理研究中，作为二阶对称张量的地震矩张量已被广泛接受并且得到了成功的应用（Gilbert, Dziewonski, 1975；McCowan, 1976；Fitch et al, 1980；Dziewonski et al, 1981；Ekström, 1989）。基于天然地震是发生于地球内部的震源（内源）这一前提，从角动量守恒可得出地震矩张量具有对称性的结论（Backus, Mulcahy, 1976a, b；Aki, Richards, 1980；Kennett, 1983）。然而，地震矩张量的对称性不过是在一定条件下引入的简化和限制，并不是绝对必要的（Takei, Kumazawa, 1995）。在平面断层的假设下，对称矩张量是对地震震源的很好近似。但是，若研究更为接近实际的震源模型，例如考虑震源区的质量流动（Takei, Kumazawa, 1994），或断层具有厚度和强度弱化区（Knopoff, Chen, 2009），则需要引入非对称矩张量以表示相应的地震效应。

若考虑到地震是发生在具有厚度的强度弱化区的剪切断层上，那么与其远场地震波对应的震源项，要比不考虑具有厚度的强度弱化区的剪切断层（Burridge, Knopoff, 1964）多出一个单力偶项（Knopoff, Chen, 2009）。由于与该单力偶项对应的地震矩张量是非对称张量，因而与计及断层厚度和强度弱化区域的剪切位错震源所对应的地震矩张量也是非对称张量。与对称地震矩张量不同的是，非对称地震矩张量中的反对称部分使得断层面解中方向相反的两对正交单力偶的矩不再相等。因此，根据具有厚度的断层模型，可判定与力偶矩较大的单力偶相联系的节面为断层面，另一节面则为辅助面（刘超，陈运泰，2014）。

本文将在上述工作的基础上，进一步研究时间域内的非对称地震矩张量反演的理论与方法，特别是研究从对称地震矩张量反演到非对称地震矩张量反演，是否存在过度拟合，以及影响非对称地震矩张量反演的因素等问题；并将通过数值试验，检验非对称地震矩张量反演方法的可行性.

1 理论与方法

1.1 理论

非对称地震矩张量反演的理论与方法和对称地震矩张量反演类似，只需在其基础上略加改动即可实现. 参考对称地震矩张量反演的思路，可将作为观测点位置 x 和时间 t 函数的地面运动位移 $u_i(\boldsymbol{x},t)$ 表示为表征震源特性的、含时非对称地震矩张量 $M_{jk}(t)$ 与其相对应的格林函数 $G_{ij,k}(\boldsymbol{x},t;\boldsymbol{0},0)$ 卷积的线性组合（刘超等，2008；刘超，2011；刘超，陈运泰，2014）：

$$u_i(\boldsymbol{x},t) = G_{ij,k}(\boldsymbol{x},t;\boldsymbol{0},0) * M_{jk}(t), \tag{1}$$

式中"*"表示卷积. 在具有厚度的地震断层模型中（Knopoff，Chen，2009；刘超，2011；刘超，陈运泰，2014）

$$M_{jk}(t) = M_{jk}^u(t) + M_{jk}^T(t), \tag{2}$$

式中，$M_{jk}^u(t)$ 和 $M_{jk}^T(t)$ 分别为与断层面上的位错 u 和应力错 T 相联系的、含时地震矩张量，

$$\begin{cases} M_{jk}^u(t) = \iint_\Sigma m_{jk}^u(\boldsymbol{\xi},t) \mathrm{d}\Sigma(\boldsymbol{\xi}), \\ M_{jk}^T(t) = \iint_\Sigma m_{jk}^T(\boldsymbol{\xi},t) \mathrm{d}\Sigma(\boldsymbol{\xi}), \end{cases} \tag{3}$$

式中，$m_{jk}^u(\boldsymbol{\xi},t)$ 和 $m_{jk}^T(\boldsymbol{\xi},t)$ 分别为与断层面上的位错 u 和应力错 T 相联系的、作为位置 $\boldsymbol{\xi}$ 与时间 t 函数的地震矩张量密度（又称"地震矩密度张量"）. 由式（2）可知，由于 $M_{jk}^u(t)$ 为二阶对称张量，$M_{jk}^T(t)$ 为二阶非对称张量，所以 $M_{jk}(t)$ 为二阶非对称张量，即 $M_{jk}(t) \neq M_{kj}(t)$.

由式（2），可将式（1）改写为

$$u_i(\boldsymbol{x},t) = G_{ij,k}(\boldsymbol{x},t;\boldsymbol{0},0) * [M_{jk}^u(t) + M_{jk}^T(t)]. \tag{4}$$

若以 $S^u(t)$ 和 $S^T(t)$ 分别表示与 $M_{jk}^u(t)$ 和 $M_{jk}^T(t)$ 对应的归一化震源时间函数（以下简称震源时间函数），M_{jk}^u 和 M_{jk}^T 分别表示与 $M_{jk}^u(t)$ 和 $M_{jk}^T(t)$ 对应的不含时非对称地震矩张量：

$$\begin{cases} M_{jk}^u(t) = M_{jk}^u S^u(t), \\ M_{jk}^T(t) = M_{jk}^T S^T(t), \end{cases} \tag{5}$$

则

$$u_i(\boldsymbol{x},t) = G_{ij,k}(\boldsymbol{x},t;\boldsymbol{0},0) * [M_{jk}^u S^u(t) + M_{jk}^T S^T(t)]. \tag{6}$$

对于断层面上的某一点 $\boldsymbol{\xi}$，地震矩张量密度 $m_{jk}^u(\boldsymbol{\xi},t)$ 与 $m_{jk}^T(\boldsymbol{\xi},t)$ 的时间历史不同，即 $m_{jk}^u(\boldsymbol{\xi},t)$ 对应滑动开始至滑动停止的过程，持续时间为 T_s；$m_{jk}^T(\boldsymbol{\xi},t)$ 则对应强度弱化过程，持续时间为 T_s'，且 $T_s > T_s'$. 对于某一次地震事件，$M_{jk}^u(t)$ 的持续时间为整个断层面的破裂时间 T_s^t，$M_{jk}^T(t)$ 的持续时间为 $T_s^t - (T_s - T_s')$. 由于断层面上某一点的滑动持续时间 T_s 远小于整个断层面的破裂时间 T_{st}（Heaton，1990；Nielsen，Madariaga，2003），则有 $T_s \ll T_s^t$，所以 $T_s - T_s' \ll T_s^t$，也即 $M_{jk}^T(t)$ 的持续时间近似等于 T_s^t. 因此，$M_{jk}^u(t)$ 与 $M_{jk}^T(t)$ 具有近似的、由断层面的破裂过程决定的震源时间函数：$S^u(t) \approx S^T(t) \approx S(t)$，故由式（6）可得

$$u_i(\boldsymbol{x},t) = M_{jk} G_{ij,k}(\boldsymbol{x},t;\boldsymbol{0},0) * S(t). \tag{7}$$

这样，利用上式即可将对含时非对称地震矩张量 $M_{jk}(t)$ 的反演转换为分别对不含时非对称地震矩张量 M_{jk} 和震源时间函数 $S(t)$ 进行的线性反演. 若反演 M_{jk}，则有 9 个待解参数；若反演 $S(t)$，则当用 N 个离散

点表示 $S(t)$ 时，便有 N 个待解参数．理论上，震源时间函数 $S(t)$ 是由断层面的破裂过程决定的，与波传播速度无关．实际上，由于破裂传播效应，在不同台站（因而不同方位）、由不同波型的波（如 P 波、S 波）反演得到的 $S(t)$ 是视震源时间函数，与波型（波传播速度）有关，因此，我们分别称由 P 波和 S 波反演得到的震源时间函数为 P 波震源时间函数 $S^P(t)$ 和 S 波震源时间函数 $S^S(t)$．

1.2 方法

与对称地震矩张量反演对比可见，从对称地震矩张量反演到非对称地震矩张量反演，只是增加了 3 个待解参数．因此，对称地震矩张量反演的方法和步骤可以应用于非对称地震矩张量反演，只是当地震矩张量对称时，$M_{jk} = M_{kj}$，地震矩张量的待解参数减为 6 个；当地震矩张量非对称时，地震矩张量的待解参数为 9 个．因此，只要将对称地震矩张量反演方法和程序作适当改动，便可用于非对称地震矩张量反演．

在反演非对称地震矩张量时，首先利用对称地震矩张量反演，分别得到垂直向 P 波和水平向 S 波的震源时间函数，然后再联合反演非对称地震矩张量，具体步骤如下：

1) 选择 9 个地震矩张量作为基本地震矩张量 M'_{jk} 进行一次正演，得到与 M'_{jk} 相应的格林函数 $G_{ij,k}$. 这 9 个基本地震矩张量 M'_{jk} 为

$$\left\{ \begin{aligned} M'_{11} &= \begin{bmatrix} 1 & 0 & 0 \\ 0 & 0 & 0 \\ 0 & 0 & 0 \end{bmatrix}, M'_{12} = \begin{bmatrix} 0 & 1 & 0 \\ 0 & 0 & 0 \\ 0 & 0 & 0 \end{bmatrix}, M'_{13} = \begin{bmatrix} 0 & 0 & 1 \\ 0 & 0 & 0 \\ 0 & 0 & 0 \end{bmatrix}, \\ M'_{21} &= \begin{bmatrix} 0 & 0 & 0 \\ 1 & 0 & 0 \\ 0 & 0 & 0 \end{bmatrix}, M'_{22} = \begin{bmatrix} 0 & 0 & 0 \\ 0 & 1 & 0 \\ 0 & 0 & 0 \end{bmatrix}, M'_{23} = \begin{bmatrix} 0 & 0 & 0 \\ 0 & 0 & 1 \\ 0 & 0 & 0 \end{bmatrix}, \\ M'_{31} &= \begin{bmatrix} 0 & 0 & 0 \\ 0 & 0 & 0 \\ 1 & 0 & 0 \end{bmatrix}, M'_{32} = \begin{bmatrix} 0 & 0 & 0 \\ 0 & 0 & 0 \\ 0 & 1 & 0 \end{bmatrix}, M'_{33} = \begin{bmatrix} 0 & 0 & 0 \\ 0 & 0 & 0 \\ 0 & 0 & 1 \end{bmatrix}; \end{aligned} \right. \tag{8}$$

2) 合成对称地震矩张量反演所需的格林函数 $G_{ij,k}(\boldsymbol{x},t;\boldsymbol{0},0)$；

3) 利用 P 波进行对称地震矩张量反演，得到 P 波震源时间函数 $S^P(t)$；

4) 利用 S 波进行对称地震矩张量反演，得到 S 波震源时间函数 $S^S(t)$；

5) 用 u_i^P 和 $G_{ij,k}^P$ 分别表示 P 波观测数据和相应的格林函数，用 u_i^S 和 $G_{ij,k}^S$ 分别表示 S 波观测数据和相应的格林函数，令

$$u_i = \begin{bmatrix} u_i^P \\ u_i^S \end{bmatrix}, G_{ij,k} = \begin{bmatrix} G_{ij,k}^P \\ G_{ij,k}^S \end{bmatrix}, S = \begin{bmatrix} S^P \\ S^S \end{bmatrix}, \tag{9}$$

将其代入式（7），可以看出，此时在式（7）中，仅有地震矩张量 M_{jk} 未知；

6) 将式（7）中的 $S(t)$ 与 $G_{ij,k}(\boldsymbol{x},t;\boldsymbol{0},0)$ 卷积得到

$$G'_{ij,k}(\boldsymbol{x},t) = G_{ij,k}(\boldsymbol{x},t;\boldsymbol{0},0) * S(t), \tag{10}$$

即可将式（7）进一步简化为

$$u_i(\boldsymbol{x},t) = G'_{ij,k}(\boldsymbol{x},t) M_{jk}, \tag{11}$$

据此即可通过线性反演得出地震矩张量 M_{jk}．

由式（7）可知，在反演非对称地震矩张量 M_{jk} 时，需要分别计算出 M_{jk} 中 9 个分量对应的格林函数 $G_{ij,k}$ 用于构建反演问题的系数矩阵．非对称地震矩张量的各个分量能否被精确地反演，与位移数据是否包含有相应的震源信息密切相关．在对称地震矩张量反演时，利用垂直向 P 波位移数据即可反演其 6 个分量．然而，在非对称地震矩张量反演时，若仅用垂直向 P 波位移数据将会无法区分 M_{xy} 与 M_{yx} 这两个分量，

需加入水平向信号进行联合反演（刘超，2014）．这是非对称地震矩张量反演与对称地震矩张量反演的最大区别．

考虑到S波的水平向地震信号有较高的信噪比，因此可运用S波水平向数据与P波垂直向数据联合反演非对称地震矩张量．对于S波，可将计算的格林函数旋转至N—S分量和E—W分量后，再用于反演．不用SH波而分别选用SH波的N—S分量和E—W分量，是因为在数据处理实践中，直接用N—S分量和E—W分量可获得更多可靠的S波信号．SH波需要通过旋转N—S向和E—W向的水平记录得出，这要求两个水平分向的仪器响应参数、标定参数、极性和授时等均准确无误．然而在处理S波波形数据时，两个水平分量记录的不一致性很难识别（如极性反转，标定不准确，仪器响应误差或授时不一致等），使得部分错误的SH波会被用于反演，增大了反演的不确定性．另一方面，即使识别出不一致的水平向地震数据，也只能将两条数据都舍弃，从而减少了可用于反演的SH波数据量．但是，如果我们单独选择可用的N—S分量或E—W分量数据用于反演，则只需旋转理论计算的水平向格林函数，从而避免对N—S分量或E—W分量数据一致性的检查，并且使得可用的S波波形数据增多．在S波与P波的联合反演中，数据的反演权重会自动调整，使得每一条单独的P波或S波波形数据，均具有相同的反演权重．

1.3 非对称地震矩张量反演是否过度拟合

即使不涉及震源的理论模型，单从数据反演的角度，需要回答的一个重要问题是：非对称地震矩张量反演是否只是因为增加了3个模型参数而提高了对数据的拟合程度，因此增加的3个参数并不是必需的？或者说，从对称地震矩张量反演到非对称地震矩张量反演，是否存在过度拟合问题？

为回答上述问题，本文引入赤池信息准则（Akaike Information Criterion，简称AIC准则）．按照AIC准则，衡量是否过度拟合的AIC值可表示为（Akaike，1974）

$$\text{AIC} = N_c N \ln E + 2 N_{\text{MT}}, \tag{12}$$

式中已略去常数项，N_c为波形数，N为每条波形的抽样点数，N_{MT}为反演地震矩张量的自由度（对于对称地震矩张量$N_{\text{MT}}=6$，非对称地震矩张量$N_{\text{MT}}=9$），E为反演残差，其定义为

$$E = \sum_{i=1}^{N_c} \sum^{N} (u_i - s_i)^2, \tag{13}$$

式中，u_i为观测波形，s_i为拟合波形．利用AIC值可挑选同一反演条件下的最佳模型．所谓最佳模型是指该模型既能很好地拟合数据，又具有最少的模型参数．AIC值的大小不具有绝对意义，仅具有相对意义．在比较不同模型的AIC值时，较小的AIC值对应较优模型．

在非对称地震矩张量反演时，可利用同一组数据进行对称地震矩张量反演．比较两次反演结果的AIC值即可判定对称地震矩张量与非对称地震矩张量孰为更优的模型，从而判定非对称地震矩张量反演是否存在过度拟合的问题．

2 地震矩张量的矢量表示法

在反演结果的误差分析中，需要定量地描述地震矩张量之间的差异，这涉及地震矩张量的分解和分类方法．地震矩张量的分解是不唯一的，地震矩张量的分解与分类应视问题的物理内涵而定．在设定地震矩张量的分解分类标准时，往往希望这种分类方法既能直接服务于特定的研究内容，又能具有清晰直观的物理图像．

常见的地震矩张量分类表示方法通常具有以下一些不足：对地震矩张量间的差异无法完整统一地定量描述；用于描述地震矩张量的参数的权重难以定量确定；方法难以推广至非对称地震矩张量的分析中．在

地理坐标系中，可通过定义走向、倾角和滑动角来表示地震矩张量中的直流分量（"零频"分量）所对应的断层面参数，但这3个参数间的权重难以定量地确定：倾角较小时，震源类型的相似性较大；倾角较大时，震源类型的相似性较小. Kagan（1991）提出利用两个地震矩张量的直流分量对齐所需的最小空间旋转角（通常称为 Kagan 角）来描述其间的差异；但是，考虑到两个完全相反的直流分量（T 轴与 P 轴互换），绕 B 轴转动 90°即可对齐，而另一些直流分量之间则需转动 120°才能对齐，可见 Kagan 角并不是一个能够特别合理地描述地震矩张量之间差异的参数. 利用地震矩张量的 T-k 分类法（Hudson et al, 1989）可以定量地分析各向同性分量（isotropic component，简写为 ISO），双力偶分量（double couple component，简写为 DC）和补偿线性矢量偶极分量（compensated linear vector dipole component，简写为 CLVD），其中 T 表示震源中体积为常量（即剪切）分量的类型，k 表示体积变化（即膨胀）分量的大小. 虽然如此，但仍然难以确定 T-k 参数与本征轴参数之间的权重. 尤为重要的是，上述方法均很难运用于非对称地震矩张量分析.

为克服上述困难，这里引入地震矩张量的矢量表示法（Willemann, 1993），利用地震矩张量内积的性质，将地震矩张量之间的距离定义为矢量之间的夹角. 地震矩张量的矢量表示法还可以自然地推广至非对称地震矩张量的分析中.

在以 e_i 为基矢量的直角坐标系中，张量 $\mathbf{M}=M_{ij}e_ie_j$ 和 $\mathbf{N}=N_{ij}e_ie_j$ 的内积，即

$$\mathbf{M}:\mathbf{N}=M_{ij}N_{ij}. \tag{14}$$

对于迹为零（ISO = 0）的地震矩张量，其自身的内积作为第三不变量，与坐标系选择无关（Kagan, Knopoff, 1985）. 利用标量地震矩 M_0 可以定义归一化地震矩张量 m_{ij}，使得对于任意地震矩张量 M_{ij}，均满足

$$M_{ij}=M_0 m_{ij}, \tag{15}$$
$$m_{ij}m_{ij}=2, \tag{16}$$

即归一化地震矩张量 m_{ij} 自身的内积为 2. 很自然地，可利用归一化地震矩张量的内积 J_3 描述 m_{ij} 与 n_{ij} 之间的差异，即

$$J_3 = m_{ij}n_{ij}, \tag{17}$$

J_3 在 [-2, 2] 上变化：当 $m_{ij}=n_{ij}$ 时，$J_3=2$；当 $m_{ij}=-n_{ij}$ 时，$J_3=-2$.

定义 m_{ij} 与 n_{ij} 之间的距离 D 为

$$D=\arccos\left(\frac{J_3}{2}\right)=\arccos\left(\frac{m_{ij}n_{ij}}{2}\right), \tag{18}$$

这样定义的 D 具有角度的量纲，与地震矩张量在矢量空间中的旋转相联系. D 在 [0, π] 上变化：当地震矩张量相同（$m_{ij}=n_{ij}$）时，距离最小，$D=0$；当地震矩张量相反（$m_{ij}=-n_{ij}$）时，距离最大，$D=\pi$.

选择这种定义距离的方法，使得 D 在数学上有以下优点：
1）距离 D 与坐标系选择无关，是地震矩张量之间差异的客观量度；
2）D 是非负的；
3）D 满足三角不等式 $D_{ij}+D_{jk} \geq D_{ik}$；
4）相同的地震矩张量之间的距离最小，为 0；相反的地震矩张量之间的距离最大，为 π；
5）可自然地推广至非对称地震矩张量.

从距离 D 的定义还可以看出，与 m_{ij} 距离为 D 的地震矩张量 n_{ij} 是不唯一的，这一特点在地震矩张量的矢量表示中需要注意.

3 格林函数对地震矩张量反演的影响

本文运用广义反射、透射系数矩阵法（Kennett, Kerry, 1979; Kennett, 1980, 1983）和相关程序

(李旭，1993；李旭等，1994；李旭，陈运泰，1996a，b）计算格林函数．将原程序用Matlab重新编写并优化，以提高计算速度．利用该方法可较快地求出分层均匀介质中的全波理论地震图，因受限制因素较少，适于反演中格林函数的计算．

本节将通过分析格林函数与矩张量各分量之间的关系说明：在非对称矩张量反演时，若仅用垂直向地动位移数据，将无法区分M_{xy}与M_{yx}这两个分量；需要同时运用垂直向与水平向地动位移数据进行联合反演才能区分M_{xy}与M_{yx}．

利用地球展平变换，将地球的球形分层速度结构模型变换为水平分层速度结构模型．采用柱坐标系(r, θ, z)，坐标原点取在震中，z轴取垂直向下为正，相应的基矢量为$(\hat{r}, \hat{\theta}, \hat{z})$，坐标系遵循右手法则．

引入一组面谐矢量$(\boldsymbol{R}_k^m, \boldsymbol{S}_k^m, \boldsymbol{T}_k^m)$

$$\begin{cases} \boldsymbol{R}_k^m = \hat{z} Y_k^m(r,\theta), \\ \boldsymbol{S}_k^m = \frac{1}{k} \nabla_1 Y_k^m(r,\theta), \\ \boldsymbol{T}_k^m = -\frac{1}{k} \hat{z} \nabla_1 Y_k^m(r,\theta), \end{cases} \quad (19)$$

式中，$Y_k^m(r, \theta) = J_m(kr) \exp(im\theta)$，$\nabla_1 = \hat{r}\partial_r + \hat{\theta}(1/r)\partial_\theta$，$J_m(kr)$是第一类$m$阶贝塞尔函数，$m$是整数．位移$\boldsymbol{u}$可表示为

$$\boldsymbol{u}(r,\theta,z) = u_r\hat{r} + u_\theta\hat{\theta} + u_z\hat{z} = \frac{1}{2\pi}\int_{-\infty}^{\infty} d\omega \exp(-i\omega t) \int_0^{\infty} dk k \sum_m (U\boldsymbol{R}_k^m + V\boldsymbol{S}_k^m + W\boldsymbol{T}_k^m), \quad (20)$$

式中，位移$\boldsymbol{u}$各分量的表示式为

$$\begin{cases} u_z(r,\theta,z,t) = \frac{1}{2\pi}\int_{-\infty}^{\infty} d\omega \exp(-i\omega t) \int_0^{\infty} dk k \sum_{m=-2}^{2} U(k,m,z,\omega) J_m(kr) \exp(im\theta), \\ u_r(r,\theta,z,t) = \frac{1}{2\pi}\int_{-\infty}^{\infty} d\omega \exp(-i\omega t) \int_0^{\infty} dk k \sum_{m=-2}^{2} \left[V(k,m,z,\omega) \frac{\partial J_m(kr)}{\partial(kr)} + W(k,m,z,\omega) im \frac{J_m(kr)}{kr} \right] \exp(im\theta), \\ u_\theta(r,\theta,z,t) = \frac{1}{2\pi}\int_{-\infty}^{\infty} d\omega \exp(-i\omega t) \int_0^{\infty} dk k \sum_{m=-2}^{2} \left[V(k,m,z,\omega) im \frac{J_m(kr)}{kr} - W(k,m,z,\omega) \frac{\partial J_m(kr)}{\partial(kr)} \right] \exp(im\theta), \end{cases} \quad (21)$$

由上式可以看出，经过面谐矢量展开，在位移$\boldsymbol{u}$的表示式中，u_z只与U有关，与V，W无关；u_r和u_θ则均与V，W有关，与U无关：

$$\begin{cases} u_z = f_z(U), \\ u_r = f_r(V, W), \\ u_\theta = f_\theta(V, W). \end{cases} \quad (22)$$

假设在水平分层均匀介质内部$z = z_S$处有一点源，它在自由表面$z = z_0$所产生的位移为（Kennett，Kerry，1979）

$$W(z_0) = \boldsymbol{R}_{EV}(\boldsymbol{I} - \boldsymbol{R}_D^{RS}\tilde{\boldsymbol{R}})^{-1}\boldsymbol{T}_U^{RS}(\boldsymbol{I} - \boldsymbol{R}_D^{SL}\boldsymbol{R}_U^{FS})^{-1}[\boldsymbol{R}_D^{SL}\boldsymbol{\Sigma}_D(z_S) - \boldsymbol{\Sigma}_U(z_S)], \quad (23)$$

式中，$\boldsymbol{R}_{EV}$为自由表面的接收系数矩阵，$\tilde{\boldsymbol{R}}$为自由表面的反射系数矩阵，$\boldsymbol{R}_D^{RS}$为自由表面与震源之间下行波的广义反射系数矩阵，$\boldsymbol{T}_U^{RS}$为自由表面与震源之间上行波的广义透射系数矩阵，$\boldsymbol{R}_D^{SL}$为震源与底界面之间下行波的广义反射系数矩阵，$\boldsymbol{R}_U^{FS}$为震源与自由表面之间上行波的广义反射系数矩阵（包括自由表面的反射作用），$\boldsymbol{\Sigma}_D$和$\boldsymbol{\Sigma}_U$为表示震源的波矢量间断矩阵．

由式（23）可看出，面谐矢量位移解 $\mathbf{W}(z_0)$ 包含了自由表面的反射、层间的多次反射、透射以及 PS-V 波的相互转换等所有作用，是完全响应的位移解．为突出震源项与位移解的关系，可将其改写为

$$\mathbf{W}(z_0) = f_W(\mathbf{\Sigma}_D, \mathbf{\Sigma}_U). \tag{24}$$

对于 P 波和 SV 波，

$$\mathbf{W} = \begin{bmatrix} U \\ V \end{bmatrix}, \quad \mathbf{\Sigma}_D = \begin{bmatrix} \varphi^D \\ \psi^D \end{bmatrix}, \quad \mathbf{\Sigma}_U = \begin{bmatrix} \varphi^U \\ \psi^U \end{bmatrix}, \tag{25}$$

将其代入式（23）得

$$\begin{bmatrix} U \\ V \end{bmatrix} = f_W\left(\begin{bmatrix} \varphi^D \\ \psi^D \end{bmatrix}, \begin{bmatrix} \varphi^U \\ \psi^U \end{bmatrix} \right), \tag{26}$$

将上式改写为

$$\begin{cases} U = f_U(\varphi^D, \psi^D, \varphi^U, \psi^U), \\ V = f_V(\varphi^D, \psi^D, \varphi^U, \psi^U). \end{cases} \tag{27}$$

对于 SH 波，

$$\mathbf{W} = W, \quad \mathbf{\Sigma}_D = \chi^D, \quad \mathbf{\Sigma}_U = \chi^U, \tag{28}$$

将其代入式（23）得

$$W = f_W(\chi^D, \chi^U). \tag{29}$$

由式（27）和式（29）可以看出，面谐矢量位移解 $\mathbf{W}$ 中，U 和 V 与震源项 $\varphi^D, \psi^D, \varphi^U, \psi^U$ 有关，W 与震源项 χ^D 和 χ^U 有关．利用式（27）和式（29），可将式（22）改写为

$$\begin{cases} u_z = f'_z(\varphi^D, \psi^D, \varphi^U, \psi^U), \\ u_r = f'_r(\varphi^D, \psi^D, \varphi^U, \psi^U, \chi^D, \chi^U), \\ u_\theta = f'_\theta(\varphi^D, \psi^D, \varphi^U, \psi^U, \chi^D, \chi^U). \end{cases} \tag{30}$$

在直角坐标系中（李旭，1993；李旭等，1994；李旭，陈运泰，1996a）：
当 $m=0$ 时，

$$\begin{cases} \varphi^U = (2\rho q_\alpha)^{-1}\left\{ -q_\alpha \omega^{-1}\varepsilon_z + \mathrm{i}\left[\dfrac{1}{2}p^2 M_1 + \dfrac{1}{\alpha^2}M_{zz}\right]\right\}, \\ \varphi^D = (2\rho q_\alpha)^{-1}\left\{ -q_\alpha \omega^{-1}\varepsilon_z - \mathrm{i}\left[\dfrac{1}{2}p^2 M_1 + \dfrac{1}{\alpha^2}M_{zz}\right]\right\}, \\ \psi^U = (2\rho q_\beta)^{-1}\left\{ \mathrm{i}p\omega^{-1}\varepsilon_z - \dfrac{1}{2}p q_\beta M_1 \right\}, \\ \psi^D = (2\rho q_\beta)^{-1}\left\{ -\mathrm{i}p\omega^{-1}\varepsilon_z - \dfrac{1}{2}p q_\beta M_1 \right\}, \\ \chi^U = (2\rho\beta q_\beta)^{-1}\left\{ \dfrac{1}{2}\mathrm{i}p\omega^{-1}N_1 \right\}, \\ \chi^D = (2\rho\beta q_\beta)^{-1}\left\{ -\dfrac{1}{2}\mathrm{i}p\omega^{-1}N_1 \right\}; \end{cases} \tag{31}$$

当 $m=\pm 1$ 时，

$$\begin{cases}
\varphi^{U} = (2\rho q_{\alpha})^{-1}\left\{\dfrac{1}{2}ip\omega^{-1}(\mp\varepsilon_{x}+i\varepsilon_{y})+pq_{\alpha}\left[p_{\pm}-\dfrac{1}{2}(iN_{3}\pm N_{2})\right]\right\}, \\
\varphi^{D} = (2\rho q_{\alpha})^{-1}\left\{-\dfrac{1}{2}ip\omega^{-1}(\mp\varepsilon_{x}+i\varepsilon_{y})+pq_{\alpha}\left[p_{\pm}-\dfrac{1}{2}(iN_{3}\pm N_{2})\right]\right\}, \\
\psi^{U} = (2\rho q_{\beta})^{-1}\left\{-\dfrac{1}{2}q_{\beta}\omega^{-1}(\mp\varepsilon_{x}+i\varepsilon_{y})+\dfrac{1}{2}i(\beta^{-2}-2p^{2})P_{\pm}+\dfrac{1}{2}ip^{2}(iN_{1}\pm N_{2})\right\}, \\
\psi^{D} = (2\rho q_{\beta})^{-1}\left\{-\dfrac{1}{2}q_{\beta}\omega^{-1}(\mp\varepsilon_{x}+i\varepsilon_{y})-\dfrac{1}{2}i(\beta^{-2}-2p^{2})P_{\pm}-\dfrac{1}{2}ip^{2}(iN_{1}\pm N_{2})\right\}, \\
\chi^{U} = (2\rho\beta q_{\beta})^{-1}\left\{\dfrac{1}{2}\omega^{-1}(-\varepsilon_{x}\pm i\varepsilon_{y})+\dfrac{1}{2}q_{\beta}Q_{\mp}\right\}, \\
\chi^{D} = (2\rho\beta q_{\beta})^{-1}\left\{-\dfrac{1}{2}\omega^{-1}(-\varepsilon_{x}\pm i\varepsilon_{y})+\dfrac{1}{2}q_{\beta}Q_{\mp}\right\};
\end{cases} \quad (32)$$

当 $m=\pm 2$ 时,

$$\begin{cases}
\varphi^{U} = (2\rho q_{\alpha})^{-1}\left\{\dfrac{1}{4}ip^{2}(N_{4}\pm iM_{2})\right\}, \\
\varphi^{D} = (2\rho q_{\alpha})^{-1}\left\{-\dfrac{1}{4}ip^{2}(N_{4}\pm iM_{2})\right\}, \\
\psi^{U} = (2\rho q_{\beta})^{-1}\left\{-\dfrac{1}{4}pq_{\beta}(N_{4}\pm iM_{2})\right\}, \\
\psi^{D} = (2\rho q_{\beta})^{-1}\left\{-\dfrac{1}{4}pq_{\beta}(N_{4}\pm iM_{2})\right\}, \\
\chi^{U} = (2\rho\beta q_{\beta})^{-1}\left\{\dfrac{1}{4}p(\pm N_{4}+iM_{2})\right\}, \\
\chi^{D} = (2\rho\beta q_{\beta})^{-1}\left\{-\dfrac{1}{4}p(\pm N_{4}+iM_{2})\right\}.
\end{cases} \quad (33)$$

式中,

$$\begin{cases}
M_{1}=M_{xx}+M_{yy}-M_{zz}, & M_{2}=M_{xy}+M_{yx}, \\
N_{1}=M_{xy}-M_{yx}, & N_{2}=M_{xz}-M_{zx}, \\
N_{3}=M_{zy}-M_{yz}, & N_{4}=M_{yy}-M_{xx}, \\
P_{\pm}=\pm M_{xz}-iM_{yz}, & Q_{\pm}=\pm M_{yz}-iM_{xz}.
\end{cases} \quad (34)$$

将含有地震矩张量的表示式代入震源波矢量间断的表示式,并略去与单力 ε_{x}, ε_{y} 和 ε_{z} 相关的项,得到:
当 $m=0$ 时,

$$\begin{cases}
\varphi^{U} = (2\rho q_{\alpha})^{-1}\left\{\dfrac{ip^{2}}{2}(M_{xx}+M_{yy})+i\left(\dfrac{1}{\alpha^{2}}-\dfrac{p^{2}}{2}\right)M_{zz}\right\}, \\
\varphi^{D} = (2\rho q_{\alpha})^{-1}\left\{-\dfrac{ip^{2}}{2}(M_{xx}+M_{yy})-i\left(\dfrac{1}{\alpha^{2}}-\dfrac{p^{2}}{2}\right)M_{zz}\right\}, \\
\psi^{U} = (2\rho q_{\beta})^{-1}\left(-\dfrac{1}{2}pq_{\beta}\right)\{M_{xx}+M_{yy}-M_{zz}\}, \\
\psi^{D} = (2\rho q_{\beta})^{-1}\left(-\dfrac{1}{2}pq_{\beta}\right)\{M_{xx}+M_{yy}-M_{zz}\}, \\
\chi^{U} = (2\rho\beta q_{\beta})^{-1}\left(\dfrac{1}{2}ip\right)\{M_{xy}-M_{yx}\}, \\
\chi^{D} = (2\rho\beta q_{\beta})^{-1}\left(-\dfrac{1}{2}ip\right)\{M_{xy}-M_{yx}\};
\end{cases} \quad (35)$$

当 $m=\pm 1$ 时，

$$\begin{cases} \varphi^{\mathrm{U}} = (2\rho q_\alpha)^{-1}\left(\frac{1}{2}pq_\alpha\right)\{\pm(M_{xz}+M_{zx})-\mathrm{i}(M_{yz}+M_{zy})\}, \\ \varphi^{\mathrm{D}} = (2\rho q_\alpha)^{-1}\left(\frac{1}{2}pq_\alpha\right)\{\pm(M_{xz}+M_{zx})-\mathrm{i}(M_{yz}+M_{zy})\}, \\ \psi^{\mathrm{U}} = (2\rho q_\beta)^{-1}\frac{1}{2}\{\pm\mathrm{i}[(\beta^{-2}-p^2)M_{xz}-p^2M_{zx}]+[(\beta^{-2}-p^2)M_{yz}-p^2M_{zy}]\}, \\ \psi^{\mathrm{D}} = (2\rho q_\beta)^{-1}\left(-\frac{1}{2}\right)\{\pm\mathrm{i}[(\beta^{-2}-p^2)M_{xz}-p^2M_{zx}]+[(\beta^{-2}-p^2)M_{yz}-p^2M_{zy}]\}, \\ \chi^{\mathrm{U}} = (2\rho\beta q_\beta)^{-1}\left(\frac{1}{2}\mathrm{i}q_\beta\right)\{\mp M_{yz}-\mathrm{i}M_{xz}\}, \\ \chi^{\mathrm{D}} = (2\rho\beta q_\beta)^{-1}\left(\frac{1}{2}\mathrm{i}q_\beta\right)\{\mp M_{yz}-\mathrm{i}M_{xz}\}; \end{cases} \quad (36)$$

当 $m=\pm 2$ 时，

$$\begin{cases} \varphi^{\mathrm{U}} = (2\rho q_\alpha)^{-1}\left(\frac{1}{4}\mathrm{i}p^2\right)\{(M_{yy}-M_{xx})\pm\mathrm{i}(M_{xy}+M_{yx})\}, \\ \varphi^{\mathrm{D}} = (2\rho q_\alpha)^{-1}\left(-\frac{1}{4}\mathrm{i}p^2\right)\{(M_{yy}-M_{xx})\pm\mathrm{i}(M_{xy}+M_{yx})\}, \\ \psi^{\mathrm{U}} = (2\rho q_\beta)^{-1}\left(-\frac{1}{4}pq_\beta^2\right)\{M_{yy}-M_{xx}\pm\mathrm{i}(M_{xy}+M_{yx})\}, \\ \psi^{\mathrm{D}} = (2\rho q_\beta)^{-1}\left(-\frac{1}{4}pq_\beta^2\right)\{M_{yy}-M_{xx}\pm\mathrm{i}(M_{xy}+M_{yx})\}, \\ \chi^{\mathrm{U}} = (2\rho\beta q_\beta)^{-1}\left(\frac{1}{4}p\right)\{\pm(M_{yy}-M_{xx})+\mathrm{i}(M_{xy}+M_{yx})\}, \\ \chi^{\mathrm{D}} = (2\rho\beta q_\beta)^{-1}\left(-\frac{1}{4}p\right)\{\pm(M_{yy}-M_{xx})+\mathrm{i}(M_{xy}+M_{yx})\}. \end{cases} \quad (37)$$

式（35）—（37）即为仅含矩张量项的震源波矢量间断的表示式.

现在来分析垂直向位移的表示式. 由式（30）的第一式

$$u_z = f_z'(\varphi^{\mathrm{D}}, \psi^{\mathrm{D}}, \varphi^{\mathrm{U}}, \psi^{\mathrm{U}})$$

可以看出：在 φ^{D}，ψ^{D}，φ^{U}，ψ^{U} 的表示式（35）—（37）中，式（35）和式（36）中不含有与 M_{xy} 和 M_{yx} 有关的项，式（37）中的 M_{xy} 和 M_{yx} 只以 $M_{xy}+M_{yx}$ 的形式出现，所以垂直向位移 u_z 仅与 $M_{xy}+M_{yx}$ 有关，不携带任何单独与 M_{xy} 或 M_{yx} 有关的信息；若单用垂直向位移 u_z 进行反演，则只能获得 $M_{xy}+M_{yx}$ 的大小，而不能单独分辨出 M_{xy} 和 M_{yx} 的大小. 在对称地震矩张量反演中，由于 $M_{xy}=M_{yx}$，利用 $M_{xy}+M_{yx}$ 可得 $M_{xy}=M_{yx}=(M_{xy}+M_{yx})/2$；但是，在非对称地震矩张量反演中，由于 $M_{xy}\neq M_{yx}$，所以仅利用垂直向的位移 u_z 无法单独反演出 M_{xy} 和 M_{yx}.

若引入水平向位移 u_r 或 u_θ 与垂直向位移 u_z 进行联合反演，由式（30）的第二和第三式

$$\begin{cases} u_r = f_r'(\varphi^{\mathrm{D}}, \psi^{\mathrm{D}}, \varphi^{\mathrm{U}}, \psi^{\mathrm{U}}, \chi^{\mathrm{D}}, \chi^{\mathrm{U}}), \\ u_\theta = f_\theta'(\varphi^{\mathrm{D}}, \psi^{\mathrm{D}}, \varphi^{\mathrm{U}}, \psi^{\mathrm{U}}, \chi^{\mathrm{D}}, \chi^{\mathrm{U}}). \end{cases}$$

可以看出：u_r 和 u_θ 不仅与震源项 φ^{D}，ψ^{D}，φ^{U}，ψ^{U} 有关，还与震源项 χ^{D} 和 χ^{U} 有关. 在 χ^{D} 和 χ^{U} 的表示式式（35）中，M_{xy} 和 M_{yx} 以 $M_{xy}-M_{yx}$ 的形式出现. 这表明 u_r 和 u_θ 所携带的与 M_{xy} 和 M_{yx} 有关的信息不仅以 $M_{xy}+M_{yx}$ 的形式出现，还以 $M_{xy}-M_{yx}$ 的形式出现，这使得利用 u_r 和（或）u_θ 与 u_z 进行联合反演，有望单独反演出 M_{xy} 和 M_{yx}.

利用 u_z 即可单独反演地震矩张量的其他分量，不存在上述分辨问题，分析方法与对 M_{xy} 分量和 M_{yx} 分

量的分析类似. 在下一节数值试验中将通过数值计算验证本节的结论, 即仅利用 u_z 无法单独反演出 M_{xy} 分量和 M_{yx} 分量, 需要利用 u_r 和 (或) u_θ 与 u_z 进行联合反演, 才有可能单独反演出 M_{xy} 和 M_{yx}.

4 数值试验

4.1 用于数值试验的数据

为了检测上述非对称地震矩张量反演方法的可行性, 本文采用合成数据进行了数值试验. 数值试验中设地震震源深度为 10km, 标量地震矩 $M_0 = 1 \times 10^{19}$ N·m, 相当于矩震级 $M_W 6.6$ 的地震. 选择两组具有代表性的非对称地震矩张量用于数值试验: 第一组中的非对称地震矩张量的对称部分, 对应于左旋走滑为主的震源机制, 有少量 CLVD 分量, 用 $\mathbf{M}_1$ 表示; 第二组中的非对称地震矩张量的对称部分, 对应于逆冲为主的震源机制, 含有较大的 CLVD 分量, 用 $\mathbf{M}_2$ 表示. 同一组矩张量, 均具有相等的对称部分和幅值不同的反对称部分.

第一组 $\mathbf{M}_1$ 中的非对称地震矩张量的对称部分 $\mathbf{M}_{1\text{sym}}$ 为

$$\mathbf{M}_{1\text{sym}} = \begin{bmatrix} 0.03 & -1 & -0.01 \\ -1 & -0.1 & 0.25 \\ -0.01 & 0.25 & 0.7 \end{bmatrix}, \tag{38}$$

利用与 $\mathbf{M}_{1\text{sym}}$ 有关的反对称地震矩张量 $\mathbf{M}_{1\text{anti\_sym}}$

$$\mathbf{M}_{1\text{anti\_sym}} = \begin{bmatrix} 0 & 1 & 0.01 \\ -1 & 0 & -0.25 \\ -0.01 & 0.25 & 0 \end{bmatrix} \tag{39}$$

可以合成得到非对称地震矩张量 $\mathbf{M}_{1\text{asym}}$

$$\mathbf{M}_{1\text{asym}} = \mathbf{M}_{1\text{sym}} + \mu \mathbf{M}_{1\text{anti\_sym}}, \tag{40}$$

式中, μ 为标量, 分别取 0, 0.05, 0.1 和 0.21. 由上式计算出 $\mathbf{M}_{1\text{asym}}$ 并归一化 (刘超, 2011), 即可得到 M_1 中的 4 个非对称地震矩张量, 分别用 $\mathbf{M}_1^{(1)}$, $\mathbf{M}_1^{(2)}$, $\mathbf{M}_1^{(3)}$ 和 $\mathbf{M}_1^{(4)}$ 表示:

$$\mathbf{M}_1^{(1)} = \begin{bmatrix} 0.029 & -0.967 & -0.010 \\ -0.967 & -0.097 & 0.242 \\ -0.010 & 0.242 & 0.068 \end{bmatrix}, \mathbf{M}_1^{(2)} = \begin{bmatrix} 0.029 & -0.917 & -0.009 \\ -1.014 & -0.097 & 0.229 \\ -0.010 & 0.253 & 0.068 \end{bmatrix},$$
$$\mathbf{M}_1^{(3)} = \begin{bmatrix} 0.029 & -0.866 & -0.009 \\ -1.058 & -0.096 & 0.216 \\ -0.011 & 0.265 & 0.067 \end{bmatrix}, \mathbf{M}_1^{(4)} = \begin{bmatrix} 0.028 & -0.758 & -0.008 \\ -1.138 & -0.095 & 0.190 \\ -0.011 & 0.284 & 0.066 \end{bmatrix}, \tag{41}$$

式中, $\mathbf{M}_1^{(1)}$ 为对称地震矩张量. 若考虑具有厚度的滑动弱化区地震断层模式 (刘超, 2011; 刘超, 陈运泰, 2014), 那么与较大力偶相联系的节面即为断层面, 其矩为 T_1; 与较小力偶相联系的节面即为辅助面, 其矩为 T_2. 为了表示力偶矩大小的差异, 定义标量 λ 为

$$\lambda = \frac{T_1 - T_2}{T_2} = \frac{T_1}{T_2} - 1, \quad \lambda \geq 0, \tag{42}$$

λ 反映的是在两个节面上, 沿位错矢量方向力偶矩大小的差异. 当 $\lambda = 0$ 时, $T_1 = T_2$, 矩张量退化为对称矩张量. 在此情形下, 无法分辨出断层面和辅助面. 随着反对称部分比重的增加, T_1 逐渐增大, T_2 逐渐减小, λ 也随之增大. 由式 (42) 可计算出 $\mathbf{M}_1^{(1)}$, $\mathbf{M}_1^{(2)}$, $\mathbf{M}_1^{(3)}$ 和 $\mathbf{M}_1^{(4)}$ 对应的 λ 值分别为 0, 0.09, 0.19 和 0.43.

类似地, 在生成第二组 ($\mathbf{M}_2$) 非对称地震矩张量时, 其对称部分 $\mathbf{M}_{2\text{sym}}$ 和反对称部分 $\mathbf{M}_{2\text{anti\_sym}}$ 分别为

$$\mathbf{M}_{2\text{sym}} = \begin{bmatrix} 0.25 & -1 & -0.1 \\ -1 & -1 & 0.25 \\ -0.1 & 0.25 & 0.75 \end{bmatrix}, \mathbf{M}_{2\text{anti\_sym}} = \begin{bmatrix} 0 & 1 & 0.1 \\ -1 & 0 & -0.25 \\ -0.1 & 0.25 & 0 \end{bmatrix}, \qquad (43)$$

μ 分别取 0,0.05,0.1 和 0.21. 由式（40）可计算出 $\mathbf{M}_{2\text{asym}}$ 并归一化，即得到 $\mathbf{M}_2$ 中的 4 个非对称地震矩张量，分别用 $\mathbf{M}_2^{(1)}$，$\mathbf{M}_2^{(2)}$，$\mathbf{M}_2^{(3)}$ 和 $\mathbf{M}_2^{(4)}$ 表示：

$$\mathbf{M}_2^{(1)} = \begin{bmatrix} 0.182 & 0.728 & -0.073 \\ 0.728 & -0.728 & 0.182 \\ -0.073 & 0.182 & 0.546 \end{bmatrix}, \mathbf{M}_2^{(2)} = \begin{bmatrix} 0.182 & 0.691 & -0.069 \\ 0.764 & -0.728 & 0.173 \\ -0.076 & 0.191 & 0.546 \end{bmatrix},$$
$$\mathbf{M}_2^{(3)} = \begin{bmatrix} 0.182 & 0.654 & -0.065 \\ 0.799 & -0.726 & 0.163 \\ -0.080 & 0.200 & 0.545 \end{bmatrix}, \mathbf{M}_2^{(4)} = \begin{bmatrix} 0.180 & 0.576 & -0.057 \\ 0.864 & -0.720 & 0.144 \\ -0.086 & 0.216 & 0.540 \end{bmatrix}, \qquad (44)$$

式中，$\mathbf{M}_2^{(1)}$ 为对称地震矩张量. 由式（42）可计算出 $\mathbf{M}_2^{(1)}$，$\mathbf{M}_2^{(2)}$，$\mathbf{M}_2^{(3)}$ 和 $\mathbf{M}_2^{(4)}$ 对应的 λ 值分别为 0，0.06,0.11 和 0.24.

数值试验选用了 37 个震中距处于 30° 与 90° 之间的虚拟台站（图 1）. 对每一个台站，均采用归一化的、半周期为 15s 的正弦函数作震源时间函数，与格林函数卷积合成地震图. 对地震图分别加上 0,20%，40% 和 80% 的高斯噪声，再用 4 阶巴特沃斯（Butterworth）滤波器滤波，滤波频率上限为 0.03Hz，下限为 0.002Hz. 截取滤波后合成地震图 P 波或 S 波到时前 10s 至到时后 90s 的数据，组成非对称地震矩张量反演数值试验的合成地震图.

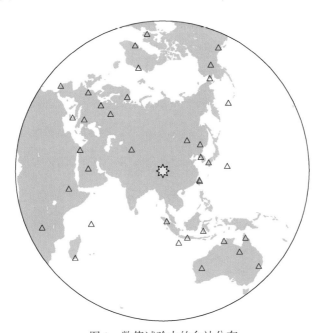

图 1 数值试验中的台站分布

八角星表示震中，三角形表示地震台

Fig. 1 Station distribution in numerical tests

The star denotes earthquake epicenter, and triangles denote seismic stations

4.2 结果

4.2.1 $\mathbf{M}_1$ 组中地震矩张量反演结果

以矩张量 $\mathbf{M}_1^{(1)}$ 为震源，反演结果如图 2 左上图和图 3 所示. 由图 2 左上图可知，在不同噪声强度水平下，反演结果的误差 D 随噪声增大而增加，其中以单用 P 波反演（以下简称 P 波反演）的误差最大，P 波和 S 波联合反演（以下简称联合反演）的误差较小。由图 3 可知：在 40% 噪声下，P 波反演结果的误差均超过 0.03，最高可达 0.07；在 80% 噪声下，单用 S 波反演（以下简称 S 波反演）结果的误差和联合反演结果的误差才超过 0.03，分别达到 0.04 和 0.05。$\mathbf{M}_1^{(1)}$ 为对称地震矩张量，其 $\lambda=0$。反演得到的 λ 最大为 0.02，相对较小，反演方法对 λ 的分辨较高，没有产生较大的误差。P 波反演和 S 波反演分别得出的震源时间函数 $S^p(t)$ 和 $S^s(t)$ 均与 $S(t)$ 基本上一致，准确稳定。

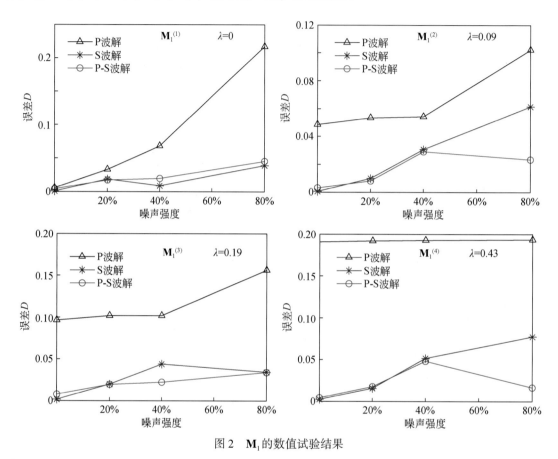

图 2 $\mathbf{M}_1$ 的数值试验结果

Fig. 2 Results of numerical test for $\mathbf{M}_1$

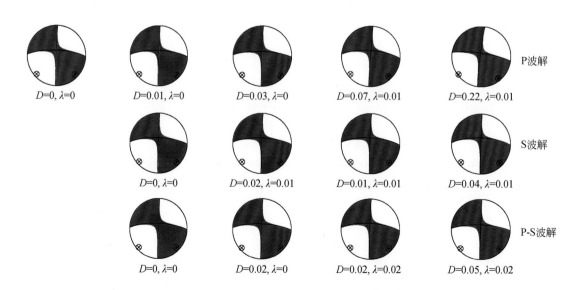

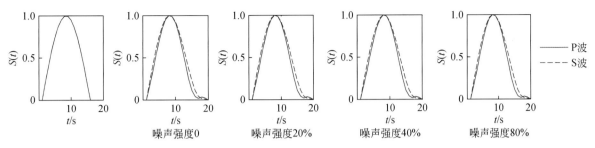

图 3　$\mathbf{M}_1^{(1)}$ 的数值试验结果

第 1 列为"真实"的地震矩张量和震源时间函数 $S(t)$，第 2 列到第 5 列为不同噪声强度水平下的反演结果；
第 1 行为 P 波解，第 2 行为 S 波解，第 3 行为 P 波和 S 波联合反演解，第 4 行为震源时间函数 $S(t)$

Fig. 3　Results of numerical test for $\mathbf{M}_1^{(1)}$

Column 1 shows the 'real' seismic moment tensor and source time function $S(t)$. Columns 2 to 5 show inversion results with difference noise level. Rows 1 to 4 show the inverted seismic moment tensor solutions solely from P wave, S wave, and both P and S waves, and the source time function $S(t)$, respectively

以矩张量 $\mathbf{M}_1^{(2)}$ 为震源，反演结果如图 2 右上图和图 4 所示. 由图 2 右上图可知：在不同噪声强度水平下，反演结果的误差 D 随噪声增大而增加，其中以 P 波反演的误差最大，联合反演的误差较小. 由图 4

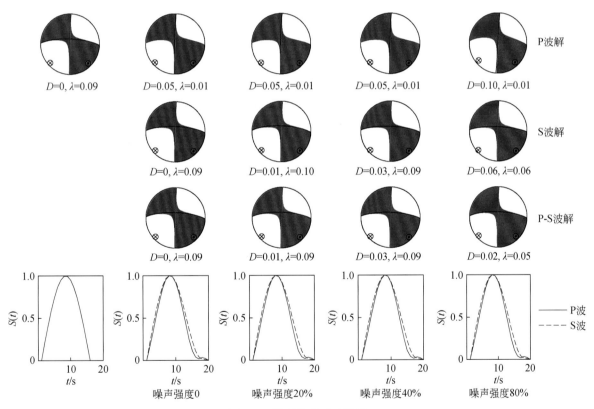

图 4　$\mathbf{M}_1^{(2)}$ 的数值试验结果

第 1 列为"真实"的地震矩张量和震源时间函数 $S(t)$，第 2 列至第 5 列为不同噪声强度水平下的反演结果；
第 1 行为 P 波解，第 2 行为 S 波解，第 3 行为 P 波和 S 波联合反演解，第 4 行为震源时间函数 $S(t)$

Fig. 4　Results of numerical test for $\mathbf{M}_1^{(2)}$

Column 1 shows the 'real' seismic moment tensor and source time function $S(t)$. Columns 2 to 5 show inversion results with different noise level. Rows 1 to 4 show the inverted seismic moment tensor solutions solely from P wave, S wave, and both P and S waves, and the source time function $S(t)$, respectively

可知：无论噪声大小，P 波反演的误差均超过 0.03，最小为 0.05，最大可达 0.10；S 波和联合反演结果的误差在 40% 噪声下，达到 0.03. $\mathbf{M}_1^{(2)}$ 为非对称地震矩张量，其 $\lambda = 0.09$. P 波反演结果均未能正确地分辨出断层面，但 S 波反演和联合反演结果均能正确地分辨出断层面，λ 值的误差随噪声增大而增加，在 0，20% 和 40% 噪声下 λ 值误差较小，在 80% 噪声下 λ 值有一定的误差. P 波反演和 S 波反演分别得出的震源时间函数 $S^p(t)$ 和 $S^s(t)$ 均与 $S(t)$ 基本上一致，准确稳定.

以矩张量 $\mathbf{M}_1^{(3)}$ 为震源，反演结果如图 2 左下图和图 5 所示. 由图 2 左下图可知，在不同噪声强度水平下，反演结果的误差 D 大致随噪声增大而增加，其中以 P 波反演的误差最大，联合反演的误差较小. 由图 5 可知：无论噪声大小，P 波反演结果的误差均超过 0.03，最大可达 0.16；S 波反演结果的误差在 40% 噪声下，可达 0.04，在 80% 噪声下，又降至 0.03；联合反演结果的误差在 80% 噪声下达到 0.03. $\mathbf{M}_1^{(3)}$ 为非对称地震矩张量，其 $\lambda = 0.19$. P 波反演结果均未能正确地分辨出断层面；但 S 波和联合反演结果均能正确分辨出断层面，λ 值误差随噪声增大而增加，在 0，20% 和 40% 噪声下 λ 值误差较小，在 80% 噪声下 λ 值的误差较大. P 波反演和 S 波反演分别得出的震源时间函数 $S^p(t)$ 和 $S^s(t)$ 均与 $S(t)$ 基本上一致，准确稳定.

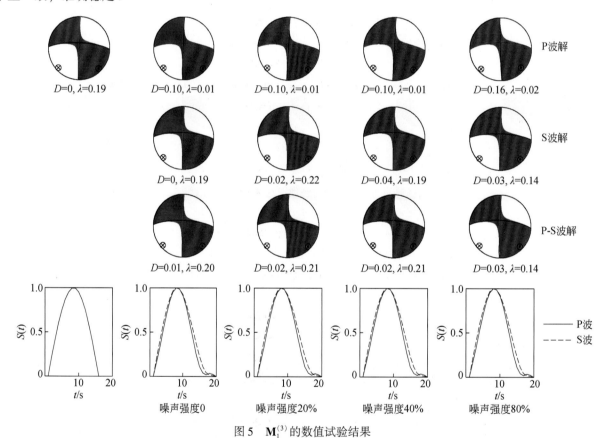

图 5 $\mathbf{M}_1^{(3)}$ 的数值试验结果

第 1 列为"真实"的地震矩张量和震源时间函数 $S(t)$，第 2 列至第 5 列为不同噪声强度水平下的反演结果；第 1 行为 P 波解，第 2 行为 S 波解，第 3 行为 P 波和 S 波联合反演解，第 4 行为震源时间函数 $S(t)$

Fig. 5 Results of numerical test for $\mathbf{M}_1^{(3)}$

Column 1 shows the 'real' seismic moment tensor and source time function $S(t)$. Columns 2 to 5 show inversion results with different noise level. Rows 1 to 4 show the inverted seismic moment tensor solutions solely from P wave, S wave, and both P and S waves, and the source time function $S(t)$, respectively

以矩张量 $\mathbf{M}_1^{(4)}$ 为震源，反演结果如图 2 右下图和图 6 所示. 由图 2 右下图可知，在不同噪声强度水平下，反演结果的误差 D 基本上随噪声增大而增加，其中以 P 波反演的误差最大，联合反演的误差较小. 由图 6 可见：无论噪声大小，P 波反演结果的误差均超过 0.03，为 0.19；S 波和联合反演结果的误差在

40%噪声下均达到0.05. $\mathbf{M}_1^{(4)}$为非对称矩张量,其$\lambda=0.43$. P波反演结果均未能正确分辨出断层面,但S波反演结果和联合反演结果均能正确分辨出断层面,λ值误差大致随噪声增大而增加. P波反演和S波反演分别得出的震源时间函数$S^p(t)$和$S^s(t)$均与$S(t)$基本上一致,准确稳定.

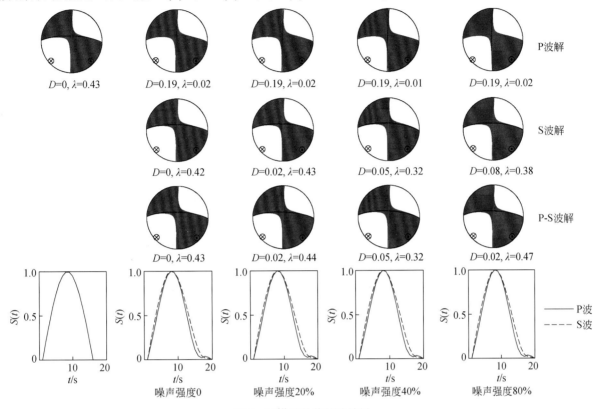

图6 $\mathbf{M}_1^{(4)}$的数值试验结果

第1列为"真实"的地震矩张量和震源时间函数$S(t)$,第2列至第5列为不同噪声强度水平下的反演结果;
第1行为P波解,第2行为S波解,第3行为P波和S波联合反演解,第4行为震源时间函数$S(t)$

Fig. 6 Results of numerical test for $\mathbf{M}_1^{(4)}$

Column 1 shows the 'real' seismic moment tensor and source time function $S(t)$. Columns 2 to 5 show inversion results with different noise level. Rows 1 to 4 show the inverted seismic moment tensor solutions solely from P wave, S wave, and both P and S waves, and the source time function $S(t)$, respectively

4.2.2 $\mathbf{M}_2$组中地震矩张量的反演结果

以矩张量$\mathbf{M}_2^{(1)}$为震源,反演结果如图7左上图和图8所示. 由图7左上图可知,在不同噪声强度水平下,反演结果的误差D大致随噪声增大而增加,其中以P波反演的误差最大,联合反演的误差较小. 由

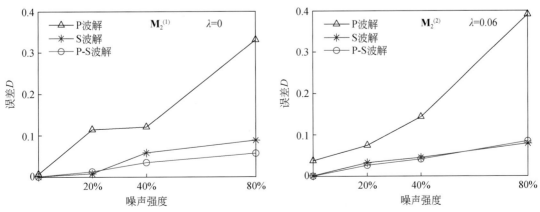

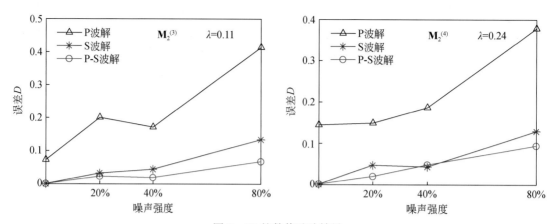

图 7 $\mathbf{M}_2$ 的数值试验结果

Fig. 7 Results of numerical test for $\mathbf{M}_2$

图 8 可知,在 20% 噪声下,P 波反演结果的误差超过 0.03,可达 0.11;在 40% 噪声下,S 波反演的结果误差超过 0.03,可达 0.06;在 40% 噪声下,联合反演结果的误差可达 0.03. $\mathbf{M}_2^{(1)}$ 为对称矩张量,其 $\lambda=0$. 反演得到的 λ 值随噪声的增大而增加,在 40% 和 80% 噪声下,λ 值误差相对较大. P 波反演和 S 波反演分别得出的震源时间函数 $S^P(t)$ 和 $S^S(t)$ 均与 $S(t)$ 基本上一致,准确稳定.

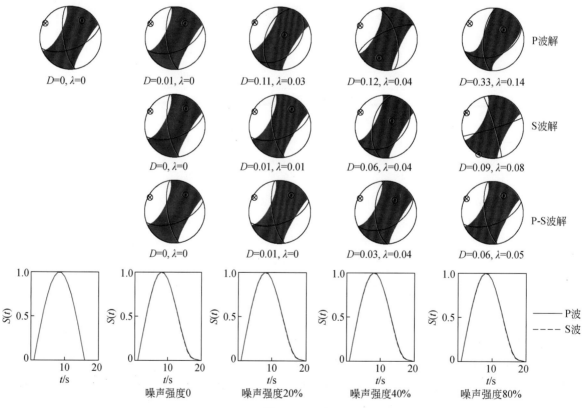

图 8 $\mathbf{M}_2^{(1)}$ 的数值试验结果

第 1 列为"真实"的地震矩张量和震源时间函数 $S(t)$,第 2 列至第 5 列为不同噪声强度水平下的反演结果;
第 1 行为 P 波解,第 2 行为 S 波解,第 3 行为 P 波和 S 波联合反演解,第 4 行为震源时间函数 $S(t)$

Fig. 8 Results of numerical test for $\mathbf{M}_2^{(1)}$

Column 1 shows the 'real' seismic moment tensor and source time function $S(t)$. Columns 2 to 5 show inversion results with different noise level. Rows 1 to 4 show the inverted seismic moment tensor solutions solely from P wave, S wave, and both P and S waves, and the source time function $S(t)$, respectively

以矩张量 $\mathbf{M}_2^{(2)}$ 为震源,反演结果如图7右上图和图9所示. 由图7右上图可知,在不同噪声强度水平下,反演结果的误差 D 随噪声增大而增加,其中以 P 波反演的误差最大,联合反演的误差较小. 由图9可知:无论噪声大小,P 波反演结果的误差均超过 0.03,最小为 0.04,最大达 0.39;S 波反演和联合反演结果的误差在 40% 噪声下,均超过 0.03. $\mathbf{M}_2^{(2)}$ 为非对称矩张量,其 $\lambda=0.06$. P 波反演结果均未能正确分辨出断层面. 在无噪声情况下,S 波反演和联合反演结果均能正确分辨出断层面;在 20% 噪声下,只有联合反演的结果能正确分辨出断层面,λ 值误差随噪声增大而增加. P 波反演和 S 波反演分别得出的震源时间函数 $S^p(t)$ 和 $S^s(t)$ 均与 $S(t)$ 基本上一致,准确稳定.

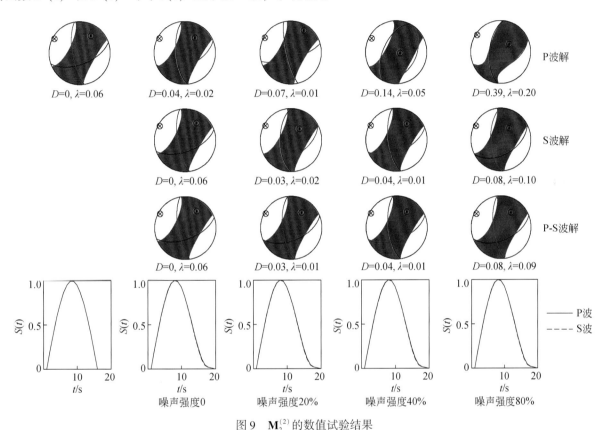

图 9 $\mathbf{M}_2^{(2)}$ 的数值试验结果

第 1 列为"真实"的地震矩张量和震源时间函数 $S(t)$,第 2 列至第 5 列为不同噪声强度水平下的反演结果;
第 1 行为 P 波解,第 2 行为 S 波解,第 3 行为 P 波和 S 波联合反演解,第 4 行为震源时间函数 $S(t)$

Fig. 9 Results of numerical test for $\mathbf{M}_2^{(2)}$

Column 1 shows the 'real' seismic moment tensor and source time function $S(t)$. Columns 2 to 5 show inversion results with different noise level. Rows 1 to 4 show the inverted seismic moment tensor solutions solely from P wave, S wave, and both P and S waves, and the source time function $S(t)$, respectively

以矩张量 $\mathbf{M}_2^{(3)}$ 为震源,反演结果见图 7 左下图和图 10. 由图 7 左下图可知,在不同噪声强度水平下,反演结果的误差 D 随噪声增大而增加,其中以 P 波反演的误差最大,联合反演的误差较小. 由图 10 可知:无论噪声大小,P 波反演结果的误差均超过 0.03,最小为 0.07,最大达 0.41;S 波反演结果误差在 20% 噪声下达到 0.03;联合反演结果误差在 80% 噪声下超过 0.03,可达 0.07. $\mathbf{M}_2^{(3)}$ 为非对称矩张量,其 $\lambda=0.11$. P 波反演结果均未能正确分辨出断层面,但 S 波反演和联合反演结果均能正确分辨出断层面. P 波反演和 S 波反演分别得出的震源时间函数 $S^p(t)$ 和 $S^s(t)$ 均与 $S(t)$ 基本上一致,准确稳定.

以矩张量 $\mathbf{M}_2^{(4)}$ 为震源,反演结果如图 7 右下图和图 11 所示. 由图 7 右下图可知,在不同噪声强度水平下,误差随噪声增大而增加,P 波反演的误差最大,联合反演的误差较小. 由图 11 可知:无论噪声大小,P 波反演结果的误差超过 0.03,最小为 0.14,最大为 0.38;S 波反演结果误差在 20% 噪声下超过

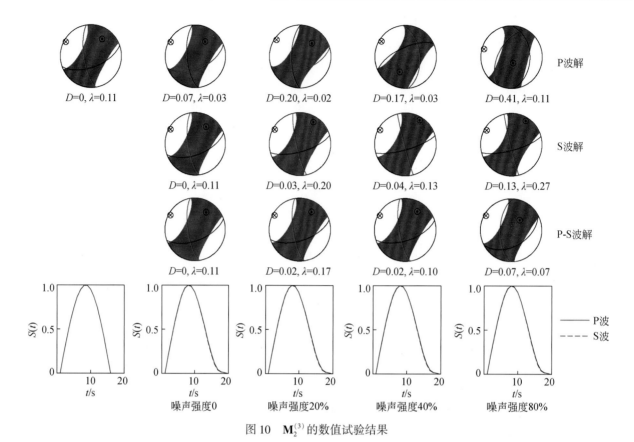

图 10 $\mathbf{M}_2^{(3)}$ 的数值试验结果

第 1 列为"真实"的地震矩张量和震源时间函数 $S(t)$，第 2 列至第 4 列为不同噪声下的反演结果；
第 1 行为 P 波解，第 2 行为 S 波解，第 3 行为 P 波和 S 波联合反演解，第 4 行为震源时间函数 $S(t)$

Fig. 10 Results of numerical test for $\mathbf{M}_2^{(3)}$

Column 1 shows the 'real' seismic moment tensor and source time function $S(t)$. Columns 2 to 5 show inversion results with different noise level. Rows 1 to 4 show the inverted seismic moment tensor solutions solely from P wave, S wave, and both P and S waves, and the source time function $S(t)$, respectively

0.03，达到 0.05；联合反演结果误差在 40% 噪声下超过 0.03，达到 0.05. $\mathbf{M}_2^{(4)}$ 为非对称矩张量，其 λ = 0.24. P 波反演结果均未能正确分辨出断层面，但 S 波反演和联合反演结果在噪声小于等于 40% 时，均能正确地分辨出断层面. P 波反演和 S 波反演分别得出的震源时间函数 $S^P(t)$ 和 $S^S(t)$ 均与 $S(t)$ 基本上一致，准确稳定.

4.3 非对称地震矩张量反演方法的可行性

以上通过数值试验验证了非对称地震矩张量反演方法的可行性. 数值试验涉及具有代表性的非对称地震矩张量，包括以走滑为主、含有少量 CLVD 分量，以及以倾滑为主、含有较大 CLVD 分量两种情形，并且这两种情形下的矩张量均具有相等的对称部分和幅值不同的反对称部分. 数值试验涉及范围广泛，试验结果具有相当大的适用性. 由 P 波反演和 S 波反演分别得出的震源时间函数 $S^P(t)$ 和 $S^S(t)$ 均与 $S(t)$ 基本上一致，准确稳定；同时，$S^P(t)$ 与 $S^S(t)$ 之间又有主要由破裂传播效应产生的微小差别，有力地说明了反演结果的可靠性与反演方法的可行性.

观察 P 波反演结果误差与 λ 值的关系即可知，随着 λ 值的增大，P 波反演结果的误差也增大. 这是由于 P 波无法分辨非对称矩张量中 M_{xy} 和 M_{yx} 这两个分量所造成（刘超，2011）. 由 P 波反演得到的矩张量，其 M_{xy} 分量与 M_{yx} 分量始终相等. 随着 λ 增大，矩张量的非对称性越大，M_{xy} 与 M_{yx} 的差异也越大. 但因 P 波无法分辨 M_{xy} 与 M_{yx}，若在反演中始终令 M_{xy} 与 M_{yx} 相等，会造成结果误差也随 λ 的增大而增大. 在 S 波反演以及联合反演中,便不存在类似的问题.

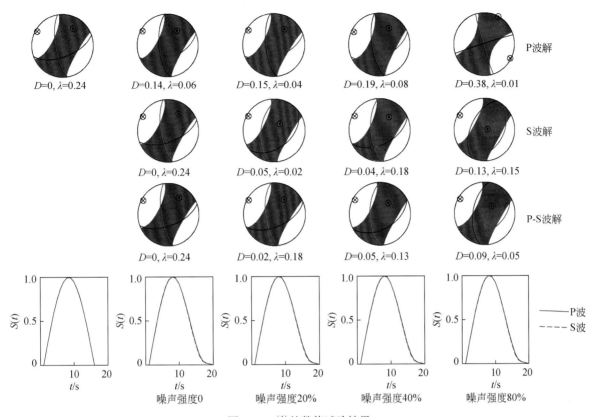

图 11 $\mathbf{M}_2^{(4)}$ 的数值试验结果

第 1 列为"真实"的地震矩张量和震源时间函数 $S(t)$,第 2 列到第 5 列为不同噪声下的反演结果;
第 1 行为 P 波解,第 2 行为 S 波解,第 3 行为 P 波和 S 波联合反演解,第 4 行为震源时间函数 $S(t)$

Fig. 11 Results of numerical test for $\mathbf{M}_2^{(4)}$

Column 1 shows the 'real' seismic moment tensor and source time function $S(t)$. Columns 2 to 5 show inversion results with different noise level. Rows 1 to 4 show the inverted seismic moment tensor solutions solely from P wave, S wave, and both P and S waves, and the source time function $S(t)$, respectively

除了对称地震矩张量 $\mathbf{M}_1^{(1)}$ 和 $\mathbf{M}_2^{(1)}$ 以外,$\mathbf{M}_2^{(2)}$ 对应的 λ 值最小,为 0.06. 当 $\lambda = 0.06$ 时,联合反演能在噪声小于等于 20% 的条件下正确地分辨出断层面;当 $\lambda \geq 0.09$ 时,由 $\mathbf{M}_1^{(2)}$ 的结果可知,联合反演能在噪声小于等于 40% 的条件下正确分辨出断层面.

这些数值试验结果表明:P 波反演的误差最大,S 波反演和联合反演的误差较小;S 波携带的震源信息与 P 波不同,P 波与 S 波联合反演可提高结果的准确性;与 P 波反演和 S 波反演的结果比较,联合反演结果有较高的断层面判定正确率,充分说明了在非对称地震矩张量反演中,引入 S 波进行 P 波与 S 波联合反演的必要性.

5 讨论与结论

本文研究了非对称地震矩张量时间域反演的理论与方法. 研究表明,虽然非对称地震矩张量反演与对称矩张量反演类似,但存在诸如是否过度拟合、如何定量表示地震矩张量之间的差异以及地震矩张量各分量的分辨等问题. 为此,本文引入了赤池信息准则(AIC 准则)以判断非对称地震矩张量反演相对于对称地震矩张量反演,是否存在过度拟合的问题;引入了地震矩张量的矢量表示法以定量地描述地震矩张量之间的差异;通过分析格林函数与地震矩张量各个分量之间的关系,得出需要同时运用垂直向与水平向地动位移数据进行联合反演才能区分 M_{xy} 与 M_{yx}. 为了检验非对称地震矩张量反演方法的可行性，本文还利用合成地震图进行了数值试验，验证了引入 S 波进行 P 波与 S 波联合反演对于提高非对称地震矩张量反演的准确性以及提高对断层面判定能力的必要性. 这些研究结果对于运用实际观测资料反演非对称地震矩张量的工作均是很有益的参考.

参 考 文 献

李旭. 1993. 用地震波波形资料反演 1990 年青海共和地震的震源过程 [D]. 北京：中国地震局地球物理研究所：1−59.

Li X. 1993. Inversion of Seismic Body Wave Data for the Source Process of the Gonghe, Qinghai, China Earthquake [D]. Beijing: Institute of Geophysics, China Earthquake Administration: 1−59 (in Chinese).

李旭，陈运泰，王培德. 1994. 水平层状介质中理论地震图计算程序的使用说明 [R]. 北京：中国地震局地球物理研究所：1−20.

Li X, Chen Y T, Wang P D. 1994. User's Manual of Computer Program of Theoretical Seismogram in Horizontally Layered Media [R]. Beijing: Institute of Geophysics, China Earthquake Administration: 1−20 (in Chinese).

李旭，陈运泰. 1996a. 合成地震图的广义反射透射系数矩阵方法 [J]. 地震地磁观测与研究，**17**(3)：1−20.

Li X, Chen Y T. 1996a. The generalized reflection-transmission coefficient matrix method for synthetic seismograms [J]. Seismological and Geomagnetic Observation and Research, **17**(3): 1−20 (in Chinese).

李旭，陈运泰. 1996b. 用长周期地震体波波形资料反演 1990 年青海共和地震的震源过程 [J]. 地震学报，**18**(3)：279−286.

Li X, Chen Y T. 1996b. Inversion of long period body wave data for the source process of the Gonghe, Qinghai, China earthquake [J]. Acta Seismologica Sinica, **9**(3): 4−12.

刘超，张勇，许力生，陈运泰. 2008. 一种矩张量反演新方法及其对 2008 年汶川 M_S8.0 地震序列的应用 [J]. 地震学报，**30**(4)：329−339.

Liu C, Zhang Y, Xu L S, Chen Y T. 2008. A new technique for moment tensor inversion with applications to the 2008 Wenchuan M_S 8.0 earthquake sequence [J]. Acta Seismologica Sinica, **21**(4): 333−343.

刘超. 2011. 断层厚度的地震效应和非对称矩张量 [D]. 北京：中国地震局地球物理研究所：1−110.

Liu C. 2011. Seismic Effect of Fault Thickness and Asymmetric Seismic Moment Tensor [D]. Beijing: Institute of Geophysics, China Earthquake Administration: 1−110 (in Chinese).

刘超，陈运泰. 2014. 断层厚度的地震效应和非对称地震矩张量 [J]. 地球物理学报，**57**(2)：509−517.

Liu C, Chen Y T. 2014. Seismic effect of fault thickness and asymmetric seismic moment tensor [J]. Chinese Journal of Geophysics, **57**(2): 509−517 (in Chinese).

Akaike H. 1974. A new look at the statistical model identification [J]. IEEE Trans Automat Contr, **19**(6): 716−723.

Aki K, Richards P G. 1980. Quantitative Seismology: Theory and Methods [M]. San Francisco: W H Freeman: 1−932.

Backus G, Mulcahy M. 1976a. Moment tensors and other phenomenological descriptions of seismic sources: Ⅰ. Continuous displacements [J]. Geophys J R A S, **46**(2): 341−361.

Backus G, Mulcahy M. 1976b. Moment tensors and other phenomenological descriptions of seismic sources: Ⅱ. Discontinuous displacements [J]. Geophys J R A S, **47**(2): 301−329.

Burridge R, Knopoff L. 1964. Body force equivalents for seismic dislocations [J]. Bull Seismol Soc Am, **54**(6A): 1875−1888.

Dziewonski A M, Chou T A, Woodhouse J H. 1981. Determination of earthquake source parameters from waveform data for studies of global and regional seismicity [J]. J Geophys Res, **86**(B4): 2825−2852.

Ekström G. 1989. A very broad band inversion method for the recovery of earthquake source parameters [J]. Tectonophysics, **166**(1/2/3): 73−100.

Fitch T J, McCowan D W, Shields M W. 1980. Estimation of the seismic moment tensor from teleseismic body wave data with applications to intraplate and mantle earthquakes [J]. J Geophys Res, **85**(B7): 3817−3828.

Gilbert F, Dziewonski A M. 1975. An application of normal mode theory to the retrieval of structural parameters and source mechanisms from seismic spectra [J]. *Phil Trans Roy Soc London*, **278**(1280): 187–269.

Heaton T H. 1990. Evidence for and implications of self-healing pulses of slip in earthquake rupture [J]. *Phys Earth Planet Interi*, **64**(1): 1–20.

Hudson J A, Pearce R G, Rogers R M. 1989. Source type plot for inversion of the moment tensor [J]. *J Geophys Res*, **94**(B1): 765–774.

Kagan Y Y, Knopoff L. 1985. The two-point correlation function of the seismic moment tensor [J]. *Geophys J R A S*, **83**(3): 637–656.

Kagan Y Y. 1991. 3-D rotation of double-couple earthquake sources [J]. *Geophys J Int*, **106**(3): 709–716.

Kennett B L N, Kerry N J. 1979. Seismic waves in a stratified half space [J]. *Geophys J R A S*, **57**(3): 557–583.

Kennett B L N. 1980. Seismic waves in a stratified half space: II. Theoretical seismograms [J]. *Geophys J R A S*, **61**(1): 1–10.

Kennett B L N. 1983. *Seismic Wave Propagation in Stratified Media* [M]. Cambridge: Cambridge University Press: 1–339.

Knopoff L, Chen Y T. 2009. Single-couple component of far-field radiation from dynamical fractures [J]. *Bull Seismol Soc Am*, **99**(2B): 1091–1102.

McCowan D W. 1976. Moment tensor representation of surface wave sources [J]. *Geophys J R A S*, **44**(3): 595–599.

Nielsen S, Madariaga R. 2003. On the self-healing fracture mode [J]. *Bull Seismol Soc Am*, **93**(6): 2375–2388.

Takei Y, Kumazawa M. 1994. Why have the single force and torque been excluded from seismic source models? [J]. *Geophys J Int*, **118**(1): 20–30.

Takei Y, Kumazawa M. 1995. Phenomenological representation and kinematics of general seismic sources including the seismic vector modes [J]. *Geophys J Int*, **121**(3): 641–662.

Willemann R J. 1993. Cluster analysis of seismic moment tensor orientations [J]. *Geophys J Int*, **115**(3): 617–634.

A time-domain inversion for the asymmetric seismic moment tensor: Theory and method

Liu Chao[1] and Chen Yun-tai[1,2]

1. Institute of Geophysics, China Earthquake Administration, Beijing 100081, China;
2. School of Earth and Space Sciences, Peking University, Beijing 100871, China

Abstract In this paper the theory and method of time-domain inversion for the asymmetric seismic moment tensor are studied based on the symmetric seismic moment tensor inversion. It is shown that the asymmetric seismic moment tensor inversion is analogous to the symmetric one, and that only a minor modification on the method of symmetric seismic moment tensor inversion, i.e., increasing three unknown parameters is required to realize the asymmetric seismic moment tensor inversion. The Akaike Information Criterion (AIC) is introduced to judge whether the asymmetric seismic moment tensor inversion is over fitting as compared with the symmetric seismic moment tensor inversion. In analyzing the asymmetric seismic moment tensor the vector representation of the moment tensor is introduced to describe the difference between seismic moment tensors quantitatively. By analyzing the relation between Green's function and each of moment tensor's components, it is found that, in asymmetric seismic moment tensor inversion, the components M_{xy} and M_{yx} could not be distinguished if only vertical displacement data were used in inversion. The vertical and horizontal displacement data should be used jointly in inversion to distinguish M_{xy} from M_{yx}. It is also found that if a different velocity structure model or a different method for calculating the Green's function was adopted for the inversion, the resolutions between any seismic moment tensor components need to be re-analyzed. A series of numerical tests were performed to test the feasibility of the present inversion method by using synthetic data. It is shown that the present inversion method is feasible and that it is necessary to use jointly both the P and S wave data to invert for the asymmetric seismic moment tensor.

Inversion of earthquake rupture process: theory and applications*

Yun-tai Chen[1,2], Yong Zhang[2,3] and Li-sheng Xu[3]

1. College of Earth and Planetary Sciences, University of Chinese Academy of Sciences-Beijing 100049, China;
2. School of Earth and Space Sciences, Peking University-Beijing 100871, China;
3. Institute of Geophysics, China Earthquake Administration-Beijing 100081, China

Summary We present the theory and methods of earthquake rupture process inversion by using seismic and geodetic data, and their applications to scientific researches and earthquake emergency responses. It is shown that the knowledge obtained from these studies has much improved our understanding of the complexities of the earthquake source and causative mechanism of the earthquake disaster, and is of important reference value in earthquake disaster mitigation such as rapid earthquake emergency response. Especially since the 2008 $M_W 7.9$ ($M_S 8.0$) Wenchuan, Sichuan, earthquake, fast and routine determination of the earthquake rupture process has been performed for significant earthquakes ($M_S \geqslant 6.5$ in China and $M_S \geqslant 7.5$ worldwide), and the results obtained are timely reported to the authorities and released to the public on the web site. The time consumed by the inversion has been reduced from more than 5 hours in 2009 to approximately 1-3 hours at present. The timely released rupture model was routinely used by the China Earthquake Administration and other authorities during the earthquake emergency responses period for destructive earthquakes, such as the 2010 $M_W 6.9$ Yushu earthquake, the 2013 $M_W 6.6$ Lushan earthquake, the 2014 $M_W 6.1$ Ludian earthquake, and the 2015 $M_W 7.8$ Gorkha, Nepal, earthquake, among the others.

1 Introduction

As the continuous developments in digital seismic network and earthquake source theory[1-3], it has been feasible to investigate the rupture process on a finite fault since the 1980s. Many significant earthquakes were studied and their rupture models have been determined. A detailed rupture model of an earthquake provides important source information, *i.e.*, the fault dimension, the source duration, the rupture directivity, the slip distribution, and the source time function, etc. The source information renews and improves our understanding of the earthquake source physics, and is necessary to constrain the source dynamic process. In addition, the rupture model is valuable in the simulation of strong ground motion and thus is useful in the estimation of intensities which are critical in earthquake emergency response (EER)[4] and even in earthquake early warning (EEW)[5].

The pioneering works of rupture process inversion can be referred to Kikuchi and Kanamori[6], Hartzell and Heaton[7], and Das and Kostrov[8], who estimated the rupture models by inverting local strong motion and/or

---

* 本文发表于 *Rivista Del Nuovo Cimento*,2019 年,第 **42** 卷,第 8 期,367-406.

teleseismic seismograms. The techniques were further developed by the subsequent researchers (e.g., [9]). An alternative method is to invert the apparent source time functions (ASTFs), which should have been obtained by deconvolving the Green's function (sometimes a small shock served as the empirical Green's function) from the mainshock seismograms[10-13]. In principle, the seismogram inversion and the ASTF inversion are equally effective to estimate the rupture model[14-17].

The fault slip model can be also determined by inverting geodetic deformation data. Compared with the seismic data inversion, geodetic inversion of the co-seismic deformation data has fewer unknown parameters and thus has a higher stability. Meanwhile, the disadvantage is that it cannot constrain the temporal evolution of slip accumulation. Combination of the seismic and geodetic data can synthesize their advantages and results in a better resolution power at different depths. In the past two and more decades, joint inversion of seismic and geodetic data has been a powerful tool to study the earthquake source process. All seismic and geodetic datasets, e.g., strong-motion data, broad-band data at local, regional and teleseismic distances, high-rate/campaign GNSS (Global Navigation Satellite System) data, InSAR (Interferometric Synthetic Aperture Radar) data, and leveling data, have been used and inverted for that purpose (e.g., [18-19]). The joint inversion much expands the frequency band of source inversions, and provides better estimates of source parameters, such as magnitude, source dimension and duration, rupture velocity and rupture directivity, etc.

On the other hand, as large and disastrous earthquakes frequently occurred in the 21st century, more and more needs have been raised for the source inversion by disaster mitigations. Fast estimation of earthquake sources can effectively provide valuable information for intensity estimation especially at near-fault distances. An example is the 12 May 2008 Wenchuan earthquake. The two meizoseismal areas of intensity XI, which are located around the epicenter and the Beichuan County, are closely associated with the two slip-concentrated patches[20,21]. However, it is regrettable that should the meizoseismal area in the Beichuan County as identified by the inverted fault slip distribution be noticed and corresponding measurements adopted at that time, the casualties in that area would be able to be greatly reduced. Another example is the 2011 M_W9.1 Tohoku earthquake. Although the earthquake early warning (EEW) system of Japan was well operated, the fast measured magnitude stabilized at M_S8.1 rather than M_W9.1 two minutes after the earthquake initiation (http://www.eqh.dpri.kyoto-u.ac.jp/~masumi/ecastweb/110311/index.htm), from which the height of the tsunami wave approaching the coast was empirically estimated to be 3m to 6m[22]. Actually, the M_W9.1 shock caused a height of the tsunami wave as large as 37.9m in Miyako. The unexpected big wave caused the majority of loss of life during the earthquake. The problem, as has been widely discussed by the international seismological community, is that the EEW system of JMA treated the earthquake as a point source, rather than a finite-sized fault plane (https://www.nature.com/news/2011/110329/full/471556a.html). Therefore, to well estimate the intensity at near-source distance and the height of the tsunami wave, the finite-fault rupture model should be considered.

In this paper, we first summarize the theory and methods of finite-fault source inversions. We will derive the equations of seismogram inversion and ASTF inversion. In application to the 2009 M_W6.3 L'Aquila, Italy, earthquake, we will compare and discuss the advantages and disadvantages of the two kinds of the inversion methods. The method of joint inversion with seismic and geodetic data is also presented. By also taking the 2009 M_W6.3 L'Aquila earthquake as an example, we will show the improved resolution power of the joint inversion. Then in the following, we present the source studies of three significant earthquake occurred around the Tibet Plateau, i.e., the 2011 M_W7.8 Kunlun earthquake, the 2008 M_W7.9 Wenchuan earthquake, and the 2010 M_W6.9 Yushu earthquake. In addition, we will introduce the progress we have made in fast-source inversion for earthquake emergency responses. The success in prediction of the rupture direction and even the meizoseismal area has provided valuable information for intensity estimation. This suggests that at the present time the earthquake source inversion has been able to play an important role in earthquake disaster mitigations.

2 Theory and methods

2.1 Seismic inversion

The seismic waves radiated from a finite-fault can be represented as (fig. 1)

$$U_n(P,t) = \iint_\Sigma G_{np,q}(P,t;Q,0) * m_{pq}(Q,t) d\Sigma', \tag{1}$$

where $U_n(P, t)$ is the seismogram on the n-th component observed at P, $G_{np,q}(P, t; Q, 0)$ is the Green's function at P caused by a force couple (p, q) at Q, $m_{pq}(Q, t)$ is the time history of the force couple (moment tensor element). Equation (1) is the basis of the source inversion.

2.1.1 Inversion with fixed rake

If a purely shear dislocation is assumed and the double-couple mechanism is known, we have

$$m_{pq}(Q,t) = M_0(Q,t)(e_p v_q + e_q v_p), \tag{2}$$

where M_0 is the scalar seismic moment, e and v are the slip vector and the normal direction of the fault plane, respectively. By putting eq. (2) into eq. (1), we can get

$$U_n(P,t) = \iint_\Sigma G_{np,q}(P,t;Q,0) * [M_0(Q,t)(e_p v_q + e_q v_p)] d\Sigma'. \tag{3}$$

If assuming

$$g_n(P,t;Q,0) = G_{np,q}(P,t;Q,0) \cdot (e_p v_q + e_q v_p), \tag{4}$$

where $g_n(P, t; Q, 0)$ is the Green's functions caused by a point source with a given focal mechanism. Then eq. (3) becomes

$$U_n(P,t) = \iint_\Sigma g_n(P,t;Q,0) * M_0(Q,t) d\Sigma'. \tag{5}$$

In eq. (5), the unknown is $M_0(Q, t)$, i.e., spatial and temporal variations of the scalar seismic moment.

For a case of spatial point source ($R \gg L$, L is the source dimension), a path approximation can be assumed

$$g_n(P,t;Q,0) \cong g_n(P,t-\tau;O,0), \tag{6}$$

where O is a reference point on the fault (e.g., the hypocenter), τ is the difference of travel time between Q-P and O-P, and equals $(R - R')/c$, where c is the phase velocity. Sometimes for convenience, $R - R'$ is approximated as $r' \cdot e_R$ when the minimum wavelength λ, the hypocentral distance R, and the source dimension L satisfy $\lambda R/2 \gg L^2$ (notice $R \gg L$ is also demanded here).

By putting eq. (6) into eq. (5), we get

$$U_n(P,t) = \iint_\Sigma [g_n(P,t-\tau;O,0) * M_0(Q,t)] d\Sigma'. \tag{7}$$

It can be rewritten as

$$U_n(P,t) = g_n(P,t;O,0) * \iint_\Sigma [\delta(t-\tau) * M_0(Q,t)] d\Sigma', \tag{8}$$

the integration on the right-hand side actually the apparent source time function (ASTF)

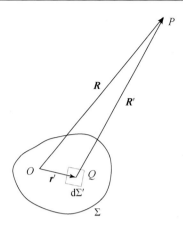

Fig. 1 The finite fault model. The origin of coordinates is taken on the finite fault plane O is a reference point (e. g., the hypocenter) on the fault Σ. R and R' are the position vectors $\overline{OP}$ and $\overline{QP}$, of the observation point P, respectively. r' is the position vector $\overline{OQ}$

$$S_A(P,t) = \iint_\Sigma [\delta(t-\tau) * M_0(Q,t)] d\Sigma'. \quad (9)$$

From eq. (7), we can estimate the rupture process by calculating only one Green's function for each station, which much reduce the calculation cost. Equations (8) and (9) suggest another way to determine the rupture model: First determine the ASTF at each station by deconvolving the Green's function from the observation, and then invert the ASTFs for rupture process based on eq. (9). Two important issues should be noted for the path approximation described in eq. (6) and used in eq. (7) to eq. (9). First, it works at only far-field distances ($R \gg L$). And second, single phase should be considered since the lengths of ray path (R and R') and the phase velocity c are needed in calculation of the travel time difference $\tau = (R-R')/c$.

To determine spatio-temporal rupture process of an earthquake, it is commonly to discretize a finite fault into sub-faults (fig. 2). The unknowns to be solved are sub-fault source time functions or moment rate functions. Because the Green's function is calculated based on a point source, the sub-faults should be reasonably small enough. This point source approximation is only tenable when the both the hypocentral distance (R) and the minimum wavelength (λ) are very much larger than the sub-fault size (L_S), where $R \gg L_S$ means Green's functions from different places on the sub-fault to the station are similar in waveform (i.e., the spatial point source), and $\lambda \gg L_S$ means the phase differences of the Green's functions are negligible (i.e., the temporal point source). With the point source approximation, the integrations in eqs. (7) to (9) can be replaced as summations

$$U_n^m(t) = \sum_k [g_n^m(t-\tau^{mk}) * M_0^k(t)], \quad (10)$$

$$U_n^m(t) = g_n^m(t) * S_A^m(t), \quad (11)$$

$$S_A^m(t) = \sum_k [\delta(t-\tau^{mk}) * M_0^k(t)], \quad (12)$$

where the superscripts m and k are indexes of the station and sub-fault, respectively. They suggest that the spatio-temporal rupture process can be imaged by inverting either the station waveforms $U_n^m(t)$ or the ASTFs $S_A^m(t)$. In the following, these two kinds of inversions will be introduced, respectively.

1) Apparent Source Time Function Inversion

Equation (12) can be also rewritten as the matrix equation

$$[\mathbf{S}_A] = [\boldsymbol{\delta}][\mathbf{M}_0], \quad (13)$$

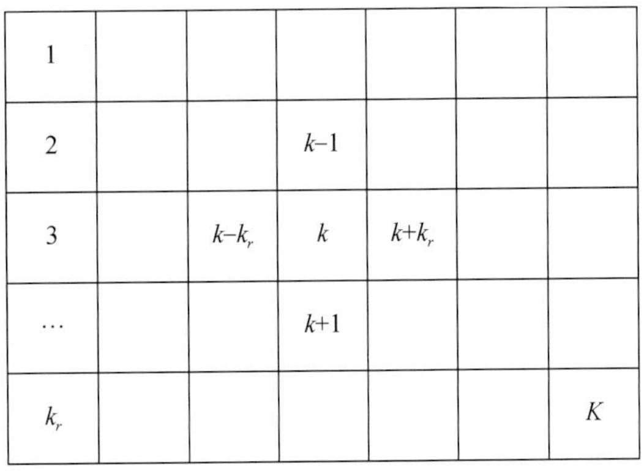

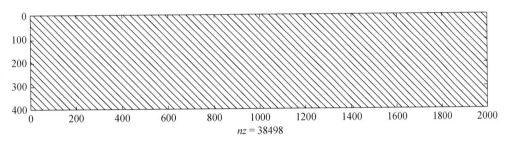

Fig. 2 Discretization of a fault plane

Here k_r is the number of sub-fault along the down-dip direction of the fault

Fig. 3 An example of the matrix $[\boldsymbol{\delta}]$

Blue dots denote the non-zero elements

where $[\mathbf{S}_A]$ is the vector of all ASTFs, and $[\boldsymbol{\delta}]$ is a sparse matrix, which consists of block diagonally dominant matrices. An example of the sparse matrix $[\boldsymbol{\delta}]$ is shown in fig. 3. The matrix $[\boldsymbol{\delta}]$ consists of equal-sized block matrices $[\boldsymbol{\delta}_B]$. For each block matrix $[\boldsymbol{\delta}_B]$, there would be an identity matrix if $\tau^{mk}=0$, that is

$$[\boldsymbol{\delta}_B(t)] = \begin{bmatrix} 1 & & & \\ & 1 & & \\ & & \ldots & \\ & & & 1 \end{bmatrix}.$$

When τ^{mk} is not equal to 0, the non-zero elements will be moved upward or downward, e.g.,

$$[\boldsymbol{\delta}_B(t-1)] = \begin{bmatrix} 0 & & & \\ 1 & 0 & & \\ & \ldots & \ldots & \\ & & 1 & 0 \end{bmatrix},$$

and

$$[\boldsymbol{\delta}_B(t+1)] = \begin{bmatrix} 0 & 1 & & \\ & 0 & \ldots & \\ & & \ldots & 1 \\ & & & 0 \end{bmatrix}.$$

2) Waveform Inversion

Equation (10) can be rewritten as the matrix equation

$$[\mathbf{U}] = [\mathbf{g}][\mathbf{M}_0], \tag{14}$$

where $[\mathbf{U}]$ is the vector of all waveform seismograms, $[\mathbf{M}_0]$ is the sub-fault source time functions, and $[\mathbf{g}]$

consists of block matrices as

$$[\mathbf{g}] = \begin{bmatrix} g^{11} & g^{12} & \cdots & g^{1K} \\ g^{21} & g^{22} & \cdots & g^{2K} \\ \cdots & \cdots & g^{mk} & \cdots \\ g^{M1} & g^{M2} & \cdots & g^{MK} \end{bmatrix}, \quad (15)$$

where $[\mathbf{g}^{mk}]$ is the convolution matrix for $g_n^{mk}(t-\tau^{mk})$, i.e., a lower triangular matrix,

$$[\mathbf{g}^{mk}(t)] = \begin{bmatrix} g^{mk}(1) & & & \\ g^{mk}(2) & g^{mk}(1) & & \\ \cdots & & \cdots & \\ g^{mk}(L_g) & \cdots & g^{mk}(2) & g^{mk}(1) \end{bmatrix}. \quad (16)$$

Equation (16) is for the case that the maximum source duration is equal to the length of Green's functions (L_g). For more general cases, a shorter source duration (T_S) is assumed for the inversion, then only the first T_S columns of $[g^{mk}(t)]$ are preserved. Because over half of the elements are occupied, $[\mathbf{g}]$ is generally a dense matrix.

2.1.2 Inversion with Rake Variation

In some case, the rake variation is considered in the inversion. The rake variation case represents the case that the slip vector on the fault Plane e is allowed to change with space and time:

$$m_{pq}(Q,t) = M_0(Q,t)[e_p(t)v_q + e_q(t)v_p]. \quad (17)$$

By putting eq. (17) into eq. (1), we get

$$U_n(P,t) = \iint_\Sigma G_{np,q}(P,t;Q,0) * \{M_0(Q,t) \cdot [e_p(t)v_q + e_q(t)v_p]\} d\Sigma'. \quad (18)$$

The projections of the slip vector and normal direction of the fault can be made in any coordinate system. Without loss of generality, we define a coordinate system on the faultplane (fig. 4) as: x_1-axis is the strike-slip direction, x_2-axis is the reverse-slip direction, x_3-axis is the normal direction of the fault plane. For a purely shear dislocation, $e_3 = 0$, $v_1 = v_2 = 0$. This makes eq. (18) become

$$U_n(P,t) = \iint_\Sigma [g_{n1}(P,t;Q,0) * M_{01}(Q,t) + g_{n2}(P,t;Q,0) * M_{02}(Q,t)] d\Sigma', \quad (19)$$

where

$$\begin{aligned} g_{n1}(P,t;Q,0) &= G_{n1,3}(P,t;Q,0) \cdot v_3, & M_{01}(Q,t) &= M_0(Q,t) \cdot e_1(t), \\ g_{n2}(P,t;Q,0) &= G_{n2,3}(P,t;Q,0) \cdot v_3, & M_{02}(Q,t) &= M_0(Q,t) \cdot e_2(t), \end{aligned} \quad (20)$$

$M_{01}(Q, t)$ and $M_{02}(Q, t)$ are moment histories along the strike slip and reverse-slip directions, respectively. Similar to eq. (14), eq. (18) can be also written as

$$[\mathbf{U}] = [\mathbf{g}_1 \quad \mathbf{g}_2] \begin{bmatrix} \mathbf{M}_{01} \\ \mathbf{M}_{02} \end{bmatrix}. \quad (21)$$

If $M_{01}(Q, t)$ and $M_{02}(Q, t)$ are solved, the rake angle and its variations will be obtained. Compared with the inversion with fixed rake, the rake variation inversion has double unknown parameters.

Equations (13), (14) and (21), correspond to the ASTF inversion, seismogram inversion with fixed rake, and seismogram inversion with rake variation, respectively. It is worth noticing that the unknown parameters in the three equations are the history of scalar seismic moment. When the Green's functions are calculated by considering the step response, the unknowns become the moment rate function or the far-field source time function. The unknowns of moment rate can be transferred into fault slip-rate by multiplying the Green's functions with the shear modulus and the sub-fault area.

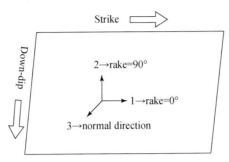

Fig. 4 The coordinate system on the fault plane used in the inversion of earthquake rupture process

2.1.3 Limitations and constraints

Since the ill-conditioning of matrix of Green's functions, and the usually poorly distributed seismic stations, directly solving eqs. (13), (14) or (21) is likely to result in a misleading rupture model. Hence it is necessary to introduce some effective limitations or constraints during the inversion. Commonly, several constraints are considered.

1) Limitation of maximum rupture velocity

In the inversions with far-field data, especially teleseismic data, it is difficult to well constrain the rupture velocity. A limitation would be helpful to make the rupture velocity more reasonable. In principle, the rupture velocity does not exceed P-wave velocity, and usually less than 0.85 times of shear wave velocity. For a sub-fault d km away from the hypocenter, the maximum rupture velocity v_r means it does not rupture until $\tau_r = d/v_r$ (see fig. 5).

2) Limitation of maximum rupture duration of sub-fault

The number of unknowns is largely decided by the sub-fault rupture duration, that is, the time differences between the rupture initiation and stop of a sub-fault. In principle, a larger duration leads to better data fit, but also causes more instabilities. A reasonable limitation of maximum rupture duration (D, see fig. 5) is usually needed for an inversion. It needs to be reasonably large when there exist multiple ruptures.

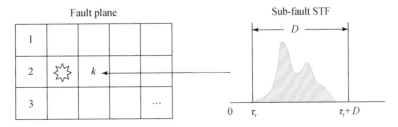

Fig. 5 A sketch showing the rupture initiation time (τ_r) and rupture duration (D) of a sub-fault STF

3) Non-negative constraint

Earthquake fault slips occur at depth under a high level of confining pressure. The friction is likely to prevent the stress from being fully released. It makes the fault hard to slip reversely. Therefore, non-negative constraint is commonly imposed to the fault slips.

4) Spatial smoothing

Based on the assumption that the residual stress after an earthquake is homogeneous to some extent, the fault slips should be reasonably smooth. A Laplace equation is often preferred for this purpose (e.g., [23]). For the k-th sub-fault, the equation is

$$4s^k(t) - [s^{k-1}(t) + s^{k+1}(t) + s^{k-k_r}(t) + s^{k+k_r}(t)] = 0, \quad (22)$$

where $s(t)$ is the slip velocity.

5) Temporal smoothing

In the inversion, the point source approximation demands that the minimum wavelength is much more than the sub-fault size, and the high-frequency waves are hard to be well modeled in the Green's functions due to our relatively poor knowledge on the 3D Earth structure. Only low-frequency waveform data are inverted, and thus the resulted source time functions of sub-faults (moment rate history or slip-rate function) should be smooth enough. A Laplace equation is also applied to the k-th sub-fault source time functions

$$2s^k(t) - [s^k(t-1) + s^k(t+1)] = 0. \quad (23)$$

6) Scalar moment minimization

The ideal case for the source inversion is that the seismic stations evenly distribute on both the azimuthal and take-off angles. Sometimes in real data inversions, however, the station distribution is usually uneven. It causes the inversion problem ill-conditioned. A feasible way to solve this problem is to suppress the solution norm by using a scalar moment minimization[24].

2.1.4 Equations for the three kinds of inversions

The constrain conditions of the spatial and temporal smoothing, and the scalar moment minimization can be formulated. Together with eqs. (13), (14), and (21), the equations of the three kinds of inversions are

$$\begin{bmatrix} \mathbf{S}_A \\ 0 \\ 0 \end{bmatrix} = \begin{bmatrix} \boldsymbol{\delta} \\ \kappa_1 \mathbf{D} \\ \kappa_2 \mathbf{T} \end{bmatrix} [\mathbf{s}], \quad (24)$$

$$\begin{bmatrix} \mathbf{U} \\ 0 \\ 0 \\ 0 \end{bmatrix} = \begin{bmatrix} g \\ \kappa_1 \mathbf{D} \\ \kappa_2 \mathbf{T} \\ \kappa_3 \mathbf{Z} \end{bmatrix} [\mathbf{s}], \quad (25)$$

and

$$\begin{bmatrix} \mathbf{U} \\ 0 \\ 0 \\ 0 \end{bmatrix} = \begin{bmatrix} \mathbf{g}_1 & \mathbf{g}_2 \\ \kappa_1 \begin{bmatrix} \mathbf{D} & \\ & \mathbf{D} \end{bmatrix} \\ \kappa_2 \begin{bmatrix} \mathbf{T} & \\ & \mathbf{T} \end{bmatrix} \\ \kappa_3 \begin{bmatrix} \mathbf{Z} & \\ & \mathbf{Z} \end{bmatrix} \end{bmatrix} \begin{bmatrix} \mathbf{s}_1 \\ \mathbf{s}_2 \end{bmatrix}, \quad (26)$$

where the unknowns [s] are sub-fault slip-rate functions, [D], [T] and [Z] are matrices of spatial smoothing, temporal smoothing, and scalar moment minimization, respectively, κ_1, κ_2 and κ_3 are the corresponding weights, which can be optimized by the Bayesian criterion or with experience. No scalar moment minimization appears in the

ASTF inversion since $[\delta]$ is a sparse matrix which has a small condition number.

2.1.5 An Example: The 2009 M_W6.3 L'Aquila, Italy, earthquake

We take the 6 April 2009 M_W6.3 L'Aquila, Italy, earthquake as an example to compare the three kinds of inversions of eqs. (24) to (26). We used P-waves on vertical components of 24 teleseismic stations (fig. 6a) for these analysis. The ASTFs were retrieved by using the Projected Landweber Deconvolution (PLD)[13,25] method and based on the fault parameters strike $\phi = 132°$/dip $\delta = 53°$/rake $\lambda = -103°$[26]. The azimuth-dependent ASTFs are shown in fig. 6b. Two sub-events can be clearly identified. The first one appears at almost the same time (~2s) at all stations, suggesting that it is around the hypocenter. The peak value of the second sub-event appears earliest at azimuth of approximately 150°, indicating that it is located to the southeast of the epicenter. The rupture directions of the two sub-event imply that the earthquake was a unilateral rupture event.

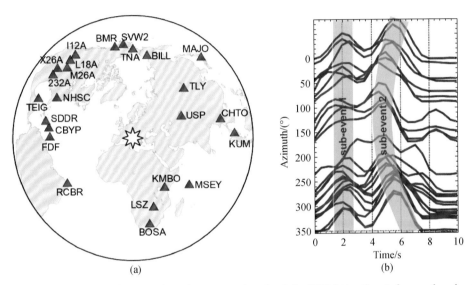

Fig. 6　Teleseismic stations (triangles) and epicenter (star) of the 2009 L'Aquila, Italy, earthquake (a).
Azimuth-dependent ASTFs retrieved (b)

With the ASTFs and the P-wave seismograms, three rupture models were estimated by solving eqs. (24) to (26), respectively (figs. 7 to 10). They are basically consistent with each other, suggesting that both the ASTF inversion and seismogram inversion are effective to estimate the source rupture processes. However, because of the difference in inversion workflow, the two inversion ways are somewhat different in the following aspects (see table I).

1) Stability

For ASTF inversion, the instability mainly exists in the deconvolution for ASTFs. When inverting ASTFs for rupture process, since the condition number of the matrix $[\delta]$ is not very large, the inversion is basically stable. However, in the inversion of the seismograms, the matrix $[G]$ is usually ill-conditioned, leading to instabilities in the inversion.

2) Efficiency

In case of the ASTF inversion, the coefficient matrix is a sparse matrix. We can use the multiplication of sparse matrix to solve the equation very efficiently. In the application to the L'Aquila earthquake, the computation time for the three kinds of the inversion are 0.8s (including 0.05s for retrieving ASTFs), 3.6s and 7.3s, respectively (performed on a laptop with Intel CPU T9300). The computation time of the ASTF inversion is almost one order less than the seismogram inversions.

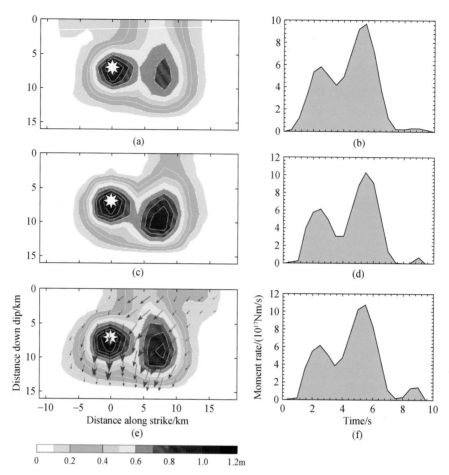

Fig. 7 Static-slip distributions and STFs obtained by solving eqs. (23) to (25), respectively, *i.e.*, the results by the ASTF inversion (a, b) and by the seismogram inversion with fixed rake (c, d), and with variable rake (e, f)

3) Scope of application

In the ASTF inversion, an approximation of spatial point source (the hypocentral distance is much larger than the source dimension) is needed, and waveform data of single phase are demanded. These limit the scope of its application. In the seismogram inversion these limitations are not needed, thus the seismogram inversion technique is suitable for more general cases.

4) Since the ASTF deconvolution is performed separately for each station, any misestimate on the ASTF duration may cause the rupture behaviors loss or some errors introduced. In contrast, the rupture behaviors were determined by inverting all waves simultaneously in the seismogram inversion. It would help to minimize the problems existing in the ASTF inversion (table I).

Table I Comparison of the ASTF and seismogram inversions

	ASTF inversion	Seismogram inversion
Stability	√	
Efficiency	√	
Scope of application		√
Misleading of rupture behaviors		√

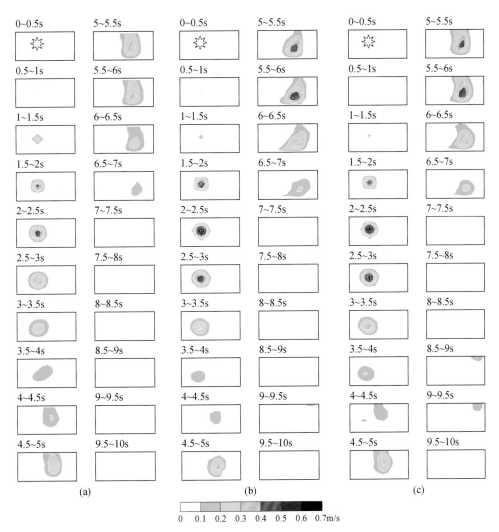

Fig. 8 Snapshots showing spatio-temporal variations of fault slip-rate obtained by solving eqs. (23) to (25), respectively, i.e., the results by the ASTF inversion (a), and by the seismogram inversion with fixed rake (b), and with variable rake (c)

By synthesizing the advantages of ASTF and seismogram inversions, we suggest a technical flow to effectively and robustly determine the rupture model. The first is to retrieve the ASTFs through deconvolution, which helps to get the overall estimates on the source duration and rupture direction. Then based on the overall estimates, we can invert the seismograms directly for the detailed rupture model.

2.2 Joint inversion of seismic and geodetic data

Source inversion with seismic data Solely often shows instabilities. A major reason is the trade-off between the rupture velocity and fault slips. The geodetic co-seismic deformation data have no resolution on the temporal rupture process, but can strongly constrain the fault slip distribution. Therefore, the combination of seismic and geodetic data can help to better estimate the spatio-temporal rupture process.

By considering the variable-rake case, the equation of co-seismic deformation data inversion can be represented as

$$[\mathbf{E}] = [\mathbf{B}_1 \quad \mathbf{B}_2]\begin{bmatrix}\mathbf{f}_1\\\mathbf{f}_2\end{bmatrix}, \tag{27}$$

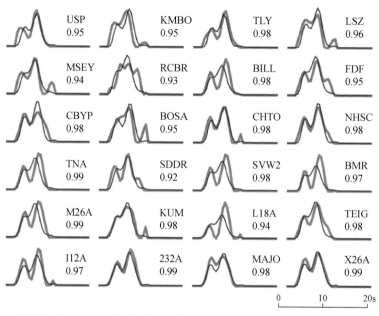

Fig. 9 Comparison of the observed (green) and synthetic (red) ASTFs

In each sub-graph, the station codes (top) and correlation coefficients (bottom) between the observed and synthetic ASTFs are labeled, respectively

where $[\mathbf{E}]$ is the co-seismic deformation data, $[\mathbf{B}]$ is the Green's function, i.e., surface deformation cause by an unit slip on the rectangular sub-fault, $[\mathbf{f}]$ is the fault slips. The fault slip $[\mathbf{f}]$ is equal to the integration of the slip-rate $[\mathbf{s}]$, which can be approximately written as summations:

$$\begin{bmatrix} \mathbf{f}_1 \\ \mathbf{f}_2 \end{bmatrix} = \begin{bmatrix} \mathbf{J} & 0 \\ 0 & \mathbf{J} \end{bmatrix} \begin{bmatrix} \mathbf{s}_1 \\ \mathbf{s}_2 \end{bmatrix}, \tag{28}$$

where $[\mathbf{J}]$ consists of row vectors with elements equal to 1, i.e., $[1, 1, \cdots, 1]$.

Putting eq. (28) into eq. (27), we can get

$$[\mathbf{E}] = [\mathbf{H}_1 \quad \mathbf{H}_2] \begin{bmatrix} \mathbf{s}_1 \\ \mathbf{s}_2 \end{bmatrix}, \tag{29}$$

where $[\mathbf{H}_1] = [\mathbf{B}_1] * [\mathbf{J}], [\mathbf{H}_2] = [\mathbf{B}_2] * [\mathbf{J}]$.

The equation of joint inversion can be obtained by putting eqs. (26) and (29) together

$$\begin{bmatrix} \mathbf{U} \\ 0 \\ 0 \\ 0 \\ \kappa\mathbf{E} \end{bmatrix} = \begin{bmatrix} \mathbf{g}_1 & \mathbf{g}_2 \\ \kappa_1 \begin{bmatrix} \mathbf{D} & \\ & \mathbf{D} \end{bmatrix} \\ \kappa_2 \begin{bmatrix} \mathbf{T} & \\ & \mathbf{T} \end{bmatrix} \\ \kappa_3 \begin{bmatrix} \mathbf{Z} & \\ & \mathbf{Z} \end{bmatrix} \\ \kappa[\mathbf{H}_1 \quad \mathbf{H}_2] \end{bmatrix} \begin{bmatrix} \mathbf{s}_1 \\ \mathbf{s}_2 \end{bmatrix}, \tag{30}$$

where κ is the relative weight of the co-seismic deformation data compared with the seismic data. Since the difference in values of the seismic data $[\mathbf{U}]$ and the deformation data may reach several orders of magnitude, a normalization is necessary to ensure that they are comparable in the joint inversion. Usually we normalize them by their square root of energy, i.e.

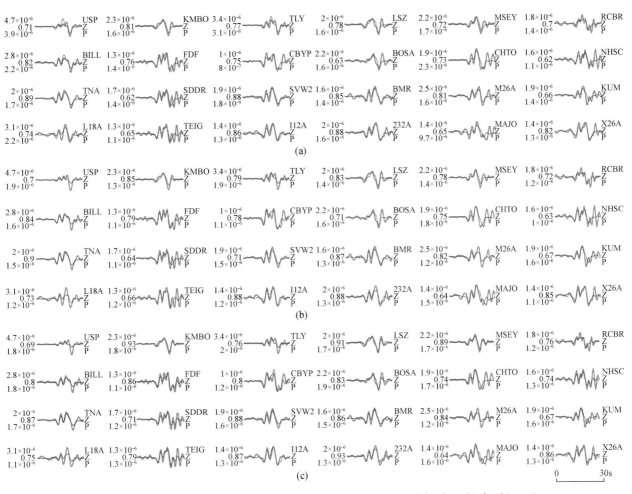

Fig. 10 Comparison between the observed and synthetic seismograms of the three kinds of inversions
(a) The results by the ASTF inversion; (b) the results by the seismogram inversion with fixed rake; (c) the results by the seismogram inversion with variable rake

$$U = \frac{U}{\sqrt{\int U^2 \, dt}}, \quad E = \frac{E}{\sqrt{\int E^2 \, dt}}. \tag{31}$$

This makes the two datasets be equally weighted in the least-square optimizations if $\kappa = 1$.

There may be not an objective criterion to choose the best relative weight κ. For our experiences, equally weighting the datasets is usually enough to recover the major rupture behaviors for most cases.

In addition to the constraint on the static fault slip distribution, the co-seismic deformation data can better constrain the shallow slip. In contrast, seismic data may have better resolution in depth since some seismic waves are radiated downward away from the source. Here we demonstrate two numerical tests for the 2009 L'Aquila earthquake by use of teleseismic data and Interferometric Synthetic Aperture Radar (InSAR) deformation data. In the first test, two slip patches with the identical area but different slips and depths are assumed: the shallow patch slips less than the deep one (fig. 11a). Seismic data inversion mainly recovers the deep slip patch which has larger seismic moment (fig. 11b). In contrast, the InSAR slip model has better resolution at shallow depth (fig. 11c). The joint inversion synthesizes the advantages of both the two datasets (fig. 11d).

In the second test (fig. 12), we assumed two adjacent slip patches which have different extensions along the down dip direction (fig. 12a). Seismic data inversion well distinguishes the two slip patches, but losses some details of shapes of the patches (fig. 12b). InSAR data inversion well recovers the shallow slips but cannot distinguish the slip gap (fig. 12c). The joint inversion provides the best estimates compared with the single data inversion (fig. 12d).

The fault slips of the 2009 L'Aquila earthquake may be similar to the slip model shown in fig. 12a. Seismic data inversion, as well as the ASTF analysis, clearly finds 2 sub-events and 2 slip patches (fig. 13a). However, InSAR data inversion only results in 1 slip patch (fig. 13b). The reason may be that the smaller patch causes little surface deformation and thus can hardly be distinguished. Through joint inversion, the two slip patches are well imaged (fig. 13c).

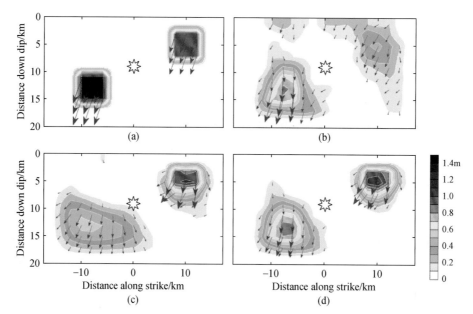

Fig. 11 Numerical test for the case of two slip patches with identical area but different slips and depths

The shallower patch has smaller slip than the deeper one. Fault slip distributions (a) obtained by teleseismic data inversion (b); InSAR data inversion (c); and joint inversion (d)

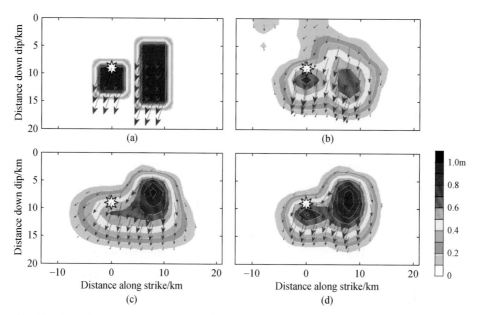

Fig. 12 Two adjacent slip patches which have different extensions along the down dip direction

Fault slip distributions (a) obtained by teleseismic data inversion (b); InSAR data inversion (c); and joint inversion (d)

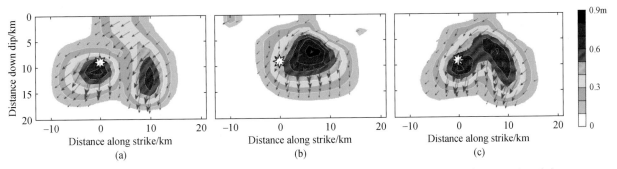

Fig. 13 Fault slip distributions of the 2009 L'Aquila earthquake obtained by teleseismic data inversion (a), by InSAR data inversion (b), and by joint inversion (c), respectively

3 Applications

During the last three decades, several significant earthquakes occurred around the boundaries of the Bayan Har Block (Table II, fig. 14). By significant earthquake we mean the earthquake of magnitude 7.5 or greater or the one that caused fatalities, injuries or substantial damage. These include the M_W 7.8 Kunlun Mountain Pass earthquake (KMPE) of 14 November 2001, the M_W 7.9 Wenchuan, Sichuan, earthquake of 12 May 2008, and the M_W 6.9 Yushu, Qinghai, earthquake of 14 April 2010[27-30]. In the following these earthquakes will be taken as examples to illustrate briefly how the knowledge on the earthquake source rupture process retrieved from seismic recordings be used in the studies of earthquake rupture complexities and the earthquake disaster emergency response[31].

Table II Catalog of the focal mechanisms of recent significant earthquakes in the Tibetan Plateau. Symbols ϕ, δ and λ denote strike, dip and rake, respectively

No.	Date a:m:d	Time h:min:s	Lat. /(°)	Long. /(°)	h /km	M_W	M_S	M_0 /Nm	NP1 ϕ/(°)	NP1 δ/(°)	NP1 λ/(°)	NP2 ϕ/(°)	NP2 δ/(°)	NP2 λ/(°)	Place	Source
1	1990-04-26	17:37:15	35.986	100.245	8.1	6.5	6.9	9.4×10^{18}	113	68	89	294	22	91	Gonghe, Qinghai	[28]
2	1997-11-08	18:02:55	35.260	87.33	40	7.5	7.4	3.4×10^{20}	250	88	19	159	71	178	Mani, Tibet	[31]
3	2001-11-14	17:26:12	35.880	90.580	15	7.8	8.1	3.2×10^{20}	290	85	−10	21	80	−175	Kunlun Mt. Pass	[32]
4	2008-03-21	06:32:58	35.490	81.467	10	7.2	7.3	8.3×10^{19}	353	29	−131	219	69	−68	Yutian, Xinjiang	USGS
5	2008-05-12	14:28:01	31.002	103.322	19	7.9	8.0	2.0×10^{21}	220	32	118	8	63	74	Wenchuan, Sichuan	[33]
6	2010-04-14	07:49:37	33.271	96.625	14	6.9	7.1	3.2×10^{20}	119	83	−2	209	88	−173	Yushu, Qinghai	[30]
7	2013-04-20	08:02:48	30.314	102.934	13	6.7	7.0	1.6×10^{19}	34	55	87	220	35	95	Lushan, Sichuan	[34]
8	2014-02-12	17:19:48	35.922	82.558	12.5	6.9	7.3	1.5×10^{19}	160	80	167	252	77	11	Yutian, Xinjiang	[35]

3.1 The M_W 7.8 Kunlun Mountain Pass earthquake of 14 November 2001

3.1.1 Tectonic settings

The first example is the M_W 7.8 Kunlun Mountain Pass earthquake (KMPE) of 14 November 2001. This earthquake occurred on the northern margin of the Bayan Har Block of the eastern Tibetan Plateau. The epicenter of this earthquake was located at the west-to-east striking, left-lateral strike-slip Kunlun Mountain Fault (KMF) (fig. 15).

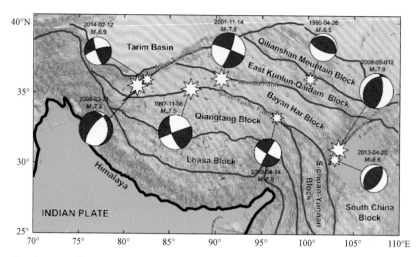

Fig. 14 Significant earthquakes around the boundaries of the Bayan Har block in the past three decades

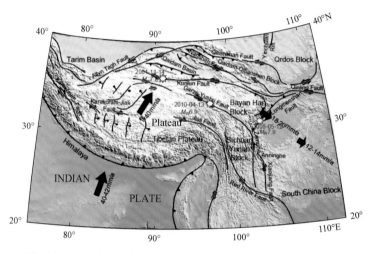

Fig. 15 Tectonics of the Tibetan plateau and its surrounding regions

3.1.2 Aftershocks

Despite its large magnitude of $M_W 7.8$, the 2001 KMPE caused no casualty or major damage, because it occurred in a nearly uninhabited mountainous area with elevation ranging from 4500 m to 6860 m. This earthquake was followed by a large number of aftershocks. Till 27 November 2001 there had been more than 2000 aftershocks with magnitude $M_L \geqslant 1.0$ recorded and located by the local seismograph network of Qinghai Province Earthquake Administration. These aftershocks mainly distributed about 100km away to the east of the mainshock epicenter along the KMF (fig. 16).

3.1.3 Focal mechanism

The focal mechanism of the KMPE was determined by moment tensor inversion and with first motions of P-waves. The inversion results show that this earthquake has a focal mechanism with a strike of 290°, a dip of 85° and a rake of −10° (fig. 16), and a moment released of 3.5×10^{20} Nm ($M_W 7.6$). These results indicate that the KMPE ruptured a left-lateral strike-slip and nearly vertical fault. The mechanism is consistent with the southeastward movement of the Bayan Har Block relative to the East Kunlun-Qaidam Block (fig. 14).

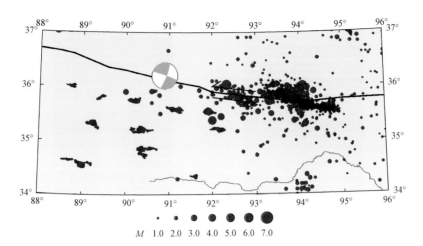

Fig. 16 Focal mechanism of the M_W7.8 Kunlun Mountain Pass earthquake of 14 November 2001, and the distribution of more than 2000 aftershocks till 27 November 2001 recorded and located by the local seismograph network of Qinghai Province Earthquake Administration

The black line denotes the Kunlun Mountain Fault

3.1.4 Distribution of static slip

The method of ASTF inversion in time domain as described in[12,36] was used for imaging the rupture process of the KMPE. The distribution of static slip on the fault plane obtained by the inversion is shown in fig. 17. In this figure, the white star denotes the hypocenter or rupture initiation point. As shown in the fig. 17, the rupture area with slip amplitude greater than 1.0 m was 420km in length, which extended from 120km to the west of the epicenter to 300km to the east of the epicenter. There are two slip-concentrated patches, i.e., areas with slip amplitude greater than 1.5m. One slip-concentrated patch extended from 30km to the east of the epicenter to 80km to the west of the epicenter, with a length of 110km, and another, from 150km to 280km to the east of the epicenter, with a length of 130km. The slip distribution on the fault plane was spatially heterogeneous. The peak slip amplitude was about 2.2m, and the average slip amplitude over the whole rupture area was about 1.2m. The peak stress drop was estimated to be 7.9MPa, and the average stress drop was 4.0MPa.

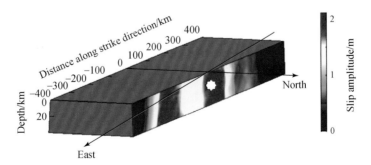

Fig. 17 Distribution of static (final) slip on the fault plane obtained by the inversion

The white star represents the hypocenter

3.1.5 Source rupture process

Figure 18 demonstrates the spatio-temporal rupture process of the 2001 KMPE mainshock. In fig. 18, left and right panels show the spatio-temporal variations of slip-rate and slip, respectively. The white star represents the hypocenter or rupture initiation point.

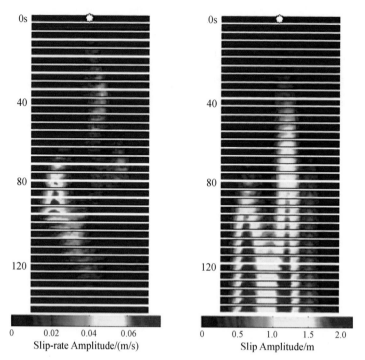

Fig. 18 The snapshots of the slip-rate (left panel) and slip (right panel) evolution on the fault plane of the 2001 KMPE, which shows the propagation of rupture on the fault plane of 40km in width and 800km in length at 4s intervals
The white star represents the hypocenter

The source process of the KMPE had spatio-temporal complexity but overall was characterized by unilateral rupture from west to east. The whole event can be divided into three sub-events in terms of the characteristics of spatio-temporal distribution of slip-rate and slip. The three sub-events started at different locations in the strike direction and at different time, and afterwards fused into a single event. During the rupture duration time of 140s and on 600km long fault, a complicated process of rupture initiation, propagation, fusion, healing and stopping occurred (fig. 18).

3.1.6 Surface ruptures

The surface ruptures of the KMPE were mostly distributed to the east of the epicenter of mainshock (35.97°N, 90.59°E) (fig. 19). The surface rupture length (SRL) to the east of the epicenter was estimated to be about 350km, the fault motion was mainly left-lateral strike-slip, the offsets on the most segments of the fault were from 2m to 4m, and the maximum offset was about 6m[35]. To the west of the epicenter there was an additional segment (Taiyang Lake segment) of surface rupture of about 90km in length with the largest horizontal offset of about 5m right near the epicenter. In total, the SRL of the M_W7.8 KMPE is about 440km[37].

This SRL of 440km well corresponds to the calculated length of 420km of the rupture area with slip amplitude greater than 1.0m. The later just corresponds to the segment to the east of the Kusai Lake Peak (Kusai Lake segment in fig. 19) where severe ground surface ruptures were found.

Based on the focal mechanism solution, the distribution of the aftershocks as well as the inverted results of the source rupture process, the KMF has been inferred to be the seismogenic fault and to have left-lateral strike-slip faulting on a nearly vertical fault. The large left-lateral strike-slip produced by the KMPE implies that the differential motion of 12-14mm/a between the north Tibetan Plateau and the south-central Tibetan Plateau is accommodated by the KLF as seismic slip in one event. Our studies confirm that large-scale strike-slip motion along the east-west-trending KLF can accommodate the continuing penetration of the Indian Plate into the Eurasian Plate by the eastward extrusion of the Tibetan Plateau.

The source process had spatio-temporal complexity but was characterized by unilateral rupture from west to east. The slip-concentrated patches which breach through the ground surface are well correspondent to the surface rupture segment or meizoseismal area found in the field investigations after the earthquake.

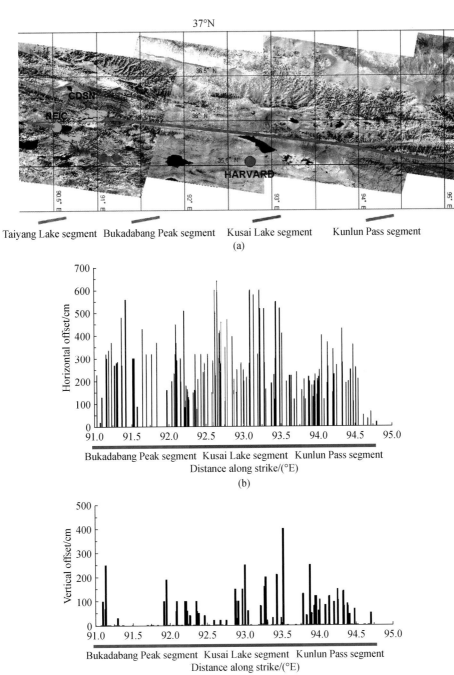

Fig. 19 Distribution of surface ruptures of the M_W7.8 Kunlun Mountain Pass earthquake of 14 November 2001
(a) Map view of surface ruptures; (b) Horizontal offset distribution along strike; (c) vertical offset distribution along strike

It is recognized that the M_W7.8 KMPE which occurred on the northern margin of the Bayan Har Block of the eastern Tibetan Plateau was caused by a left-lateral strike-slip faulting on a nearly vertical fault, being consistent with the southeastward movement of the Bayan Har Block relative to the South China Block, and that the rupture process of the KMPE is much more complicated than we had assumed, during the rupture process of the KMPE, three fault segments or three sub-events started at different locations in the strike direction and at different time, and afterwards fused into a single event of magnitude M_W7.8. The lesson we learned from the KMPE is that in the assessment of the earthquake disaster, not only the possibility of one segment of the fault, but also more than two segments rupturing at one big event should be taken into account, and that the concentrated-slip patches which breach the ground surface are well correspondent to the surface rupture segments or meizoseismal areas.

3.2 The M_W7.9 Wenchuan, Sichuan, earthquake of 12 May 2008

3.2.1 Tectonic setting

The M_W7.9 Wenchuan earthquake occurred on the eastern margin of the Bayan Har Block of the eastern Tibetan Plateau. The SW-striking Longmenshan Fault (LF) lies to the northeast of the SSE-moving Sichuan-Yunnan Block of southwest China (fig. 15). As a result of the continental collision between the Indian and Eurasian Plates at about 40-42mm/a at N20°E, and an eastward extrusion transfer of crustal material from underneath the Tibetan Plateau to the South China Block; a predominantly thrust motion with a minor component of right-lateral motion at a rate of about 18-20mm/a has been associated with NW-SE-thrusting of southwest China with respect to South China Block along the NE-SW striking LF. The southeastward motion of south China at a rate of 12-14mm/a incompletely accommodates the accumulation of stress due to the NW-SE thrusting of Bayan Har Block with respect to South China Block along the LF. Thus the LF is an active tectonic structure with the highest degree of vulnerability for earthquake ruptures.

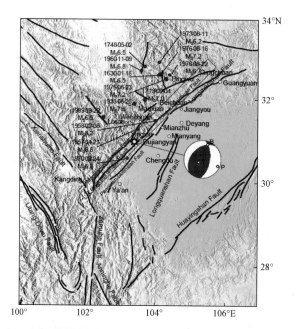

Fig. 20 Epicenter (white star) location of the 2008 M_W7.9 Wenchuan earthquake, main faults (purple lines) in the epicentral area, historical earthquakes (yellow circles), and main cities (white circles) along Longmenshan Fault

"Beach-ball" represents the lower hemisphere projection of focal mechanism (strike 225°/dip 39°/rake 120°) of the Wenchuan earthquake

Despite its active tectonic setting, historically the LF experienced only moderate seismic activity[38-40]. Frequent earthquakes of small to moderate size occur on the fault zone, but no earthquakes larger than magnitude 7 have been documented along this fault. Some earthquakes with magnitudes around 6 have been recorded (fig. 20), which include the 21 April 1657 Wenchuan earthquake ($M6.5$) (epicentral location: 31.3°N, 103.5°E), the 8 February 1958 Maowen-Beichuan earthquake ($M_S 6.2$) (epicentral location: 31.7°N, 104.3°E) occurred to the northeast of the epicenter of the 2008 Wenchuan earthquake, and the 24 February 1970 Dayi earthquake ($M_S 6.2$) (epicentral location: 30.65°N, 103.28°E, focal depth: 15km) occurred to the SSW of the epicenter of the 2008 Wenchuan earthquake. Based on the relocation of 10057 earthquakes that occurred in central-western China from 1992 to 1999, Yang et al.[41] noted that the LF is 470km in length and about 2400 earthquakes with magnitudes between 1.0 and 5.0 occurred on or near the fault zone during this period. Several magnitude-7 earthquakes have occurred nearby but outside the fault zone (fig. 20), including the 25 August 1933 $M_S 7.5$ Maowen-Diexi and the two $M_S 7.2$ Sonpan earthquakes of 16 and 23 July 1976.

Unlike the KMPE, about 89000 people were confirmed killed or missing in the Wenchuan earthquake. Soon after the occurrence of the earthquake, we started to analyze available seismic recordings to retrieve information describing the earthquake source mechanism, and released the preliminary results of the focal mechanism and the source rupture process to the authorities and public within 12 hours for earthquake disaster emergency response.

3.2.2 Focal mechanism and aftershocks

Time-domain moment-tensor inversion was used to obtain the focal mechanism of the Wenchuan earthquake. The results are that one of the nodal planes has a strike of 225° with dip 39° and rake 120°, and the other has a strike of 8° with dip 57° and rake 68° (fig. 20). The epicenters of aftershocks relocated by the double-difference earthquake location algorithm are located in a NE-SW-trending zone about 320km in length and 50km in width which clearly coincides with the strike of the first nodal plane (fig. 21, main panel).

Vertical cross-sections along the strike direction NE-SW (fig. 21, leftmost panel) and the direction perpendicular to the strike NW-SE (fig. 21, lower panel), of the projections of the relocated hypocenters also show that the aftershocks are clustered around this nodal plane. We identify the plane striking 225° as the causative fault due to its agreement with the NE-SW-trending aftershock distribution and the NE-SW strike of the LF. Overall, the 2008 Wenchuan mainshock rupture was a predominantly thrust event with a minor right-lateral strike-slip component on a plane dipping 39° with a strike of 225° and rake 117°. Our solution (strike 225°/dip 39°/rake 120°) for the Wenchuan earthquake is very close to the moment tensor solution (strike 238°/dip 59°/rake 128°) obtained by USGS/NEIC and the Global Centroid Moment Tensor (GCMT) solution (strike 229°/dip 33°/rake 141°) obtained by the GCMT Project. The scalar seismic moment M_0 we obtained is 9.4×10^{20} Nm, corresponding to a moment magnitude of $M_W 7.9$, and is in good agreement with that obtained by the USGS/NEIC ($M_0 = 7.5 \times 10^{20}$ Nm, $M_W = 7.8$) and GCMT Project ($M_0 = 9.4 \times 10^{20}$ Nm, $M_W = 7.9$).

3.2.3 Distribution of static slip

Figure 22 shows the static-slip distribution on the fault plane[17,42]. The slip vector represents the direction and amount of displacement of the hanging wall with respect to the foot wall at each sub-fault. The Wenchuan earthquake has a slip distribution with predominantly thrust faulting and a small right-lateral strike-slip component. There are four major concentrated-slip patches on the fault plane (fig. 22). From the southwest to the northeast (fig. 23), the focal mechanisms that varied from oblique thrust and right-lateral strike-slip faulting just underneath the southwestern most segment near Kangding, to predominant thrust with small right-lateral strike-slip faulting just underneath the elongated Yingxiu-Dujiangyan-Wenchuan area, to oblique thrust and right-lateral strike-slip faulting underneath the Beichuan area, and to a predominantly right-lateral strike-slip faulting with minor normal faulting

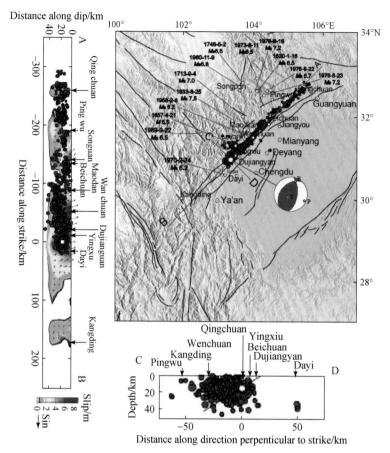

Fig. 21　Main panel: Map showing topography and major faults (solid purple lines) in northeast Sichuan, the 2008 Wenchuan M_W 7.9 earthquake epicenter (white octagonal star), the locations of relocated aftershocks (red dots), and locations of historical earthquakes (yellow dots) in and near the Longmenshan Fault (LF) which consists of three nearly parallel NE-SW-striking faults: from NW to SE, Maoxian-Wenchuan Fault (MWF), Yingxiu-Beichuan Fault (YBF) and Pengxian-Guanxian Fault (PGF). Leftmost panel: Vertical cross-section A-B along the strike direction NE-SW. Lower panel: Vertical cross-section C-D along the direction perpendicular to the strike direction NW-SE show that the aftershocks are clustered around the fault plane striking 225°, dipping 39° with a rake 117°

just underneath the Qingchuan area. The overall focal mechanism of the Wenchuan earthquake is similar to that of the second or largest concentrated-slip patch which is one of the predominan thrust faulting with a minor right-lateral strike-slip component and in good agreement with the general NE-SW strike of the LF. The overall focal mechanism of the Wenchuan earthquake is well consistent with the tectonic setting of the generally NE-SW-striking LF and is due to the NW-SE thrusting of Bayan Har Block with respect to South China Block.

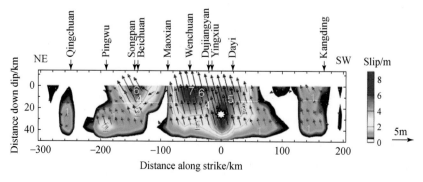

Fig. 22　The static-slip distribution on the fault plane

The inversion shows that the Wenchuan earthquake ruptured an area 350km in length by 50km in width, of the 470km-long NE-SW-striking LF. A seismic moment of about 9.4×10^{20} Nm was released in the Wenchuan earthquake, corresponding to a moment magnitude of $M_W 7.9$. The average stress drop and the maximum stress drop in this earthquake are about 18MPa and 65MPa, respectively, and are comparable with an intraplate earthquake with magnitude 8.0.

As the static-slip distribution on the fault plane of fig. 22 and the beach-balls of fig. 23 show, the $M_W 7.9$ Wenchuan earthquake is a complex event consisting of four concentrated-slip patches in which there are two substantial slip-concentrated patches on the fault plane with slips of up to 8.9m and 6.7m, just underneath the elongated area extending from Yingxiu-Dujiangyan-Wenchuan and the area near Beichuan, respectively. These patches breached the ground surface and account for the tremendous destructive damage in these two meizoseismal areas (fig. 24). In addition to the two substantial slip-concentrated patches, another two major concentrated-slip patches on the fault plane are of moderate slips of up to 2.3m and 1.6m, respectively. They are just underneath the Qingchuan and the Kangding areas, respectively, and corresponding to the areas with intensities lower than the two meizoseismal areas.

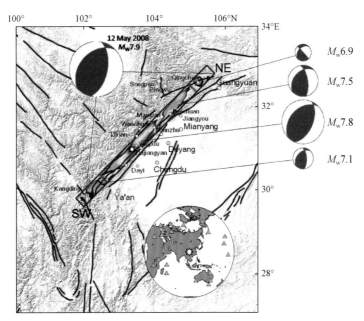

Fig. 23 Focal mechanism of the Wenchuan earthquake (white octagonal star) and horizontal projection of the static-slip distribution on the fault plane

The background shows the topography of northeast Sichuan and the major faults (purple thick lines). The inset map shows the location of the epicenter of the Wenchuan earthquake (white octagonal star) and the IRIS seismic stations (blue triangles). Differing focal mechanisms of the four concentrated-slip patches are shown by the four beach-balls to the right of the figure with corresponding moment magnitudes

During the process of rupture growth, irregular high rupture velocities (3.6km/s, 4.0km/s, 4.4km/s, 3.5km/s in the four stages) and high rupture acceleration in the northeast direction also play an important role in enhancing the devastating damage in the northeastern part of the fracture.

3.2.4 Source rupture process

The inverted spatio-temporal rupture process of the Wenchuan earthquake shows that the earthquake is a complex asymmetric bilateral rupture with overall strong northeast directivity, with high irregular rupture velocities and long rupture duration of about 90s (fig. 25). These characteristics account for the northeasternward elongated

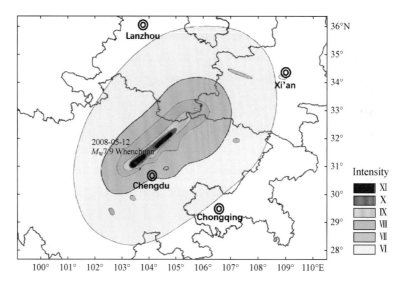

Fig. 24 Isoseismals of the 2008 M_W7.9 Wenchuan earthquake

Two substantial concentrated-slip patches of M_W7.8 and M_W7.5 on the fault plane in fig. 23 are correspondent to the two meizoseismal areas of the intensity XI

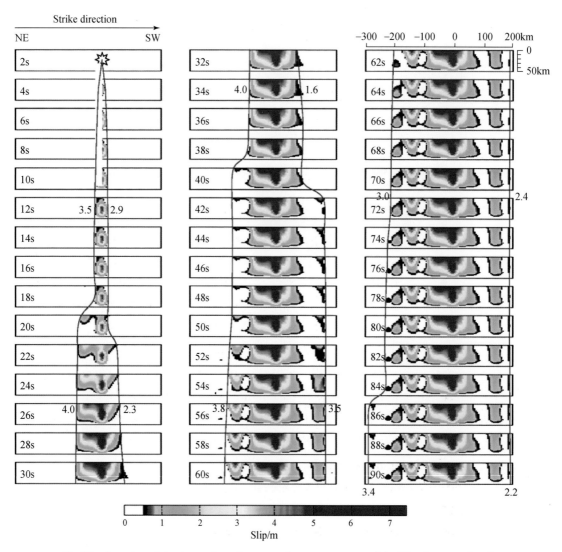

Fig. 25 Snapshot of the inverted spatio-temporal rupture process of the Wenchuan earthquake

meizoseismal areas and the asymmetric distribution of aftershocks that are significantly more numerous to the northeast of the epicenter of the Wenchuan earthquake, and much less so to its southeast. As theoretical studies have indicated, and the field observations have confirmed, in thrust earthquakes the hanging wall moves more than the foot wall with strongly peaked velocity at the fault trace. The significant difference in damage between the meizoseismal areas (Yingxiu-Dujiangyan-Wenchuan and Beichuan-Qingchuan areas) which are located in the hanging wall of the causative thrust fault, and minor damage areas such as Chengdu-Deyang-Guang'an areas which are located in the foot wall of the causative thrust fault can be attributed to the effect of the asymmetric dipping fault geometry, i.e., the hanging wall/foot wall effect.

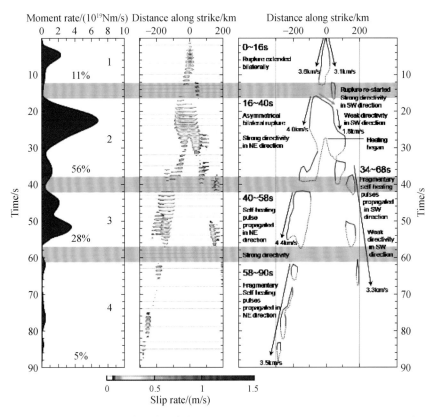

Fig. 26 (a) Source time function (STF). (b) The spatio-temporal variation of slip-rate on the fault plane. (c) Summary of the rupture process of the 2008 Wenchuan earthquake

The inverted STF suggests that the fault rupture development of the Wenchuan earthquake was divided into 4 stages (fig. 26). The first stage extends from the time of the earthquake initiation to 16s afterward; the second stage, from 16s to 40s; the third stage, from 40s to 58s, and the last stage, from 58s to 90s. The rupture propagates bilaterally in the first stage. In the second stage the rupture appears to be more complicated, starting 50km to the northeast of the epicenter, and then propagates northeastward at first and subsequently bilaterally to the southwest; the rupture is temporally discrete and spatially fragmentary. In the third stage, the rupture took place at two patches, one, 120km to the northeast of the hypocenter, and the other 160km to the southwest of the hypocenter. The rupture in the last stage appears to be unilateral to the northeast only and is temporally discrete and spatially fragmentary. The solid brown line and the dotted brown line represent leading edge and trailing edge of fracture, respectively.

It is recognized that the predominantly thrust motion with a minor component of right-lateral motion of the

M_W7.9 Wenchuan earthquake was consistent with the tectonic setting of the generally NE-SW-striking LF and was due to the NW-SE thrusting of Bayan Har Block with respect to South China Block along the NE-SW striking LF[43]. It is also recognized that the rupture process of the Wenchuan earthquake was extremely complicated[44]. During the rupture process of the Wenchuan earthquake, two substantial concentrated-slip patches on the fault plane breached the surface and were responsible for the tremendous destruction in the two meizoseismal areas. In addition, during the process of rupture growth, irregular high rupture velocities and high rupture acceleration also played an important role in enhancing the devastating damage in the meizoseismal areas. It is worth noticing that the Wenchuan earthquake is a complicated rupture event which involved rupturing of several fault segments and that in the assessment of the earthquake disaster risk, not only the possibility of one segment of the fault, but also more than two segments rupturing at one big event should be taken into account.

3.3 The M_W6.9 Yushu, Qinghai, earthquake of 14 April 2010

3.3.1 Tectonic setting

The M_W 6.9 (M_S7.1) Yushu earthquake occurred on 14 April, 2010, at 07:49 am (Beijing Time) (13 April 2010, 23:49 UTC). The epicenter of the Yushu earthquake was at (33.2°N, 96.6°E), 44km northwestern of Yushu city, and the focal depth was 14km (table II; figs. 14 and 27). By May 30th, 2010, the Yushu earthquake caused about 3000 people to be killed or missing, over 10000 people injured, and a large number of houses and buildings collapsed. The Yushu earthquake occurred on the Ganzi-Yushu Fault, a southeast-striking, left-lateral strike-slipping fault, which lies on the southern boundary of the Bayan Har Block[45-47]. Historically there has been high seismicity on the fault. The Yushu earthquake was the largest event on the northwestern segment of the Ganzi-Yushu fault during the last 100 years.

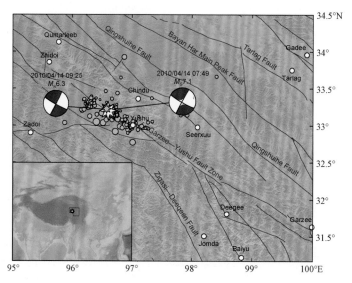

Fig. 27 Tectonic setting of the Yushu earthquake

White and cyan stars are epicenters of the M_W6.9 (M_S7.1) mainshock and the largest aftershock of M_S6.3, respectively. Red lines denote the major faults. Beach-balls represent the focal mechanisms of the mainshock and the largest aftershock of M_S6.3 (equal-area projection of the lower hemisphere of the focal sphere). Cyan circles are locations of the aftershocks in the first 5 days after the occurrence of the mainshock (from Qinghai Province Earthquake Administration). The black frame in the left-bottom inset outlines the area of the main plot

For fast emergency response to the earthquake disaster, we obtained the source rupture process by inverting the seismic recordings and had it released about 2.5 hours after its occurrence. About 5 hours and 2 days after the earthquake occurrence, as more data become available, we updated twice the results, respectively (http://www.csi.ac.cn). In order to better understand the source rupture process of the Yushu earthquake, after the fast emergency response activity, once again we inverted the carefully selected waveform data to improve the results of the source rupture process (fig. 27).

3.3.2 Focal mechanism

The inverted result (fig. 27) shows that the focal mechanism of Yushu earthquake was strike 119°/dip 83°/rake $-2°$, and that the scalar seismic moment was about 2.7×10^{19} Nm, yielding a moment magnitude of $M_W 6.9$[30].

3.3.3 Distribution of static slip

Figure 28 summarizes the inverted spatio-temporal rupture process of the 2010 Yushu earthquake[48]. As fig. 28 shows, there are two concentrated-slip patches located near the hypocenter and to the southeast of the epicenter, respectively. The first concentrated-slip patch was between 10km northwestern and 10km southeastern from the epicenter, with a maximum slip of 0.8m. The second one was in the range 17-54km southeastern from the epicenter in the strike direction (red-orange-yellow area in fig. 28c and e), with a maximum slip of 1.8m and breached the ground surface. The average slip on the whole fault plane was about 0.6m, and the average stress drop was 15MPa, which was in a good agreement with the typical stress-drop ($\sim$10MPa) for intraplate earthquake.

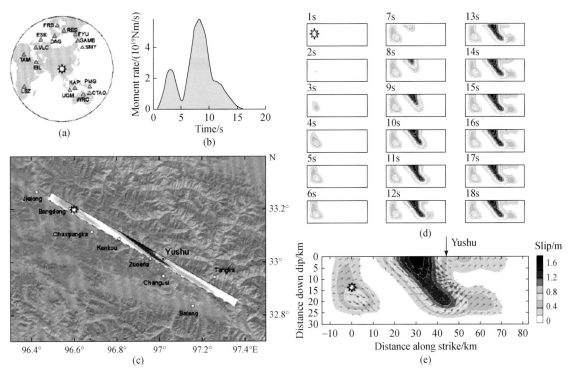

Fig. 28 Inverted spatio-temporal rupture process of the 2010 Yushu earthquake

(a) Locations of the epicenter and seismic stations; (b) source time function; (c) projection of the static (final) slip distribution on ground surface; (d) spatio-temporal distribution of slip on the fault plane; (e) Static slip distribution on the fault plane. White octagonal star represents the epicenter in (a) and (c), but the rupture initiation point in (d) and (e). Gray arrows in (e) denote the slip vectors

3.3.4 Source rupture process

The inverted results clearly show that the Yushu $M_W 6.9$ ($M_S 7.1$) earthquake consists of two distinct subevents (fig. 28b, c, e). The first one occurred in the first 5s releasing less seismic moment, and the second occurred in the later 11s releasing more seismic moment. The rupturing of the concentrated-slip patch to the southeast of the epicenter breached the ground surface. The peak-slip and peak slip-rate are about 2.1m and 1.1m/s, respectively. They indicate that the Yushu earthquake is an event with large slip velocity on the fault plane. Overall the Yushu earthquake is a unilateral rupture event with the rupture mainly propagating southeastward. It was inferred that the Yushu city, located 44km to the southeast of the epicenter, would be heavily destroyed due to the facts that the concentrated-slip patch to the southeast of the epicenter breached the ground surface, and that the strong focusing of the seismic energy to the southeast of the epicenter caused by the seismic Doppler effect. These results have been reported to the authorities and released to public about 2.5 hours after the occurrence, and confirmed later by the tremendous damage in the Yushu city and proved to be very informative in the Yushu earthquake relief work.

3.4 Applications to the earthquake emergency response

On the basis of the comparison and discussion between the ASTF inversion and seismogram inversion for rupture process (table I), we design two inversion steps to perform the fast rupture process inversion. Firstly, we retrieve the ASTFs, and get a rough estimation on the overall rupture direction by analyzing ASTFs with seismic Doppler effects. Secondly, based on the rough estimation on the source characteristics, we directly invert the seismograms for rupture model. The final determined results of seismogram inversion should satisfy two conditions: 1) They suggest similar rupture directivity. 2) The obtained source time function (STF) should be similar with the average ASTF (approximately the superposition of all ASTFs).

We construct a semiautomatic operation process to fast and stably carry out the rupture process inversion, which contains several useful program modules, such as the teleseismic data processing, fast calculation of Green's function, ASTFs analysis, inversion parameters' setting, and quick display of the inverted results. In the data processing module, the P-waves were picked automatically. The teleseismic stations were re-selected by an azimuthal filter with an interval of about 5 degrees to ensure a uniform coverage of the stations over the epicenter. The teleseismic P-waves with the epicentral distance of 30° to 90° on their vertical components are downloaded from the Incorporated Research Institutions for Seismology (IRIS) website.

The fast and robust inversion method of the earthquake source rupture process we developed in the past two and more decades proved to be effective in providing information on the source characteristics. It is found that in addition to the usual earthquake source parameters, such as the epicentral location, focal depth, focal mechanism, the fast inverted results of the spatio-temporal rupture process of the earthquake sources provided additional important information such as the possible disastrous areas and the timely release of these results were very useful to earthquake emergency response and seismic disaster relief efforts. We applied this method to some of the recent significant earthquakes worldwide. Since May 2008, the source rupture processes of more than 71 significant earthquakes worldwide were inverted by using this newly developed method and the inverted results were timely released on the website within 3 to 5 and more hours in 2009 to 1-3 hours at present after the occurrence of the earthquakes (table III; figs. 29 and 30). These earthquakes include the $M_W 7.9$ Wenchuan, Sichuan, earthquake of 12 May 2008 (no fast inversion for this earthquake), the $M_W 6.3$ L'Aquila, Italy, earthquake of 6

April 2009 (no fast inversion for this earthquake), the M_W7.0 Haiti earthquake of 12 January 2010[47], the M_W 8.8 Chile earthquake of 27 February, 2010, the M_W6.5 Jiaxian, Taiwan, earthquake of 4 March 2010, the M_W 7.2 Mexico earthquake of 4 April 2010 (poorly constrained model)[48], the M_W7.8 Sumatra earthquake of 6 April 2010, the M_W6.9 Yushu, Qinghai, earthquake of 13 April 2010 (14 April 2010, Beijing time), the 20 April 2013 M_W 6.6 Lushan, Sichuan, earthquake[49], the 21 July 2013 M_W 6.0 Mingxian-Zhangxian, Gansu, earthquake, the 3 August 2014 M_W6.1 Ludian, Yunnan, earthquake[50], the 25 April 2015 M_W7.8 Gorkha, Nepal, earthquake, and the 8 August 2017 M_W6.5 Jiuzhaigou, Sichuan, earthquake.

Table III Source parameters of the significant earthquakes for which the earthquake rupture processes were inverted and released immediately after the earthquake occurrence in the time period 2009-2018 (Part of these models can be found at http://www.cenc.ac.cn/cenc/_300651/index.html)

No.	Epicentral Region	Origin Time (UTC)	Epicentral Location (Lat., Long.)	Focal Depth /km	M_W	Time Consumed /hr
1	North of Papua	2009-01-04 04:43	(−0.5°, 132.8°)	33	7.7	6.4
2	North of Papua	2009-01-04 06:33	(−0.7°, 133.2°)	33	7.5	5.7
3	Tonga	2009-03-19 18:17	(−23.0°, −174.7°)	10	7.8	8.3
4	Offshore Honduras	2009-05-28 08:24	(16.8°, −86.2°)	15	7.2	8.6
5	Hualian, Taiwan	2009-07-13 18:05	(24.1°, 122.2°)	6	6.4	3.0
6	South Island, N.Z.	2009-07-15 09:22	(−45.7°, 166.6°)	33	7.8	3.7
7	Andaman Islands	2009-08-10 19:55	(14.1°, 92.9°)	33	7.8	4.2
8	Samoa Islands	2009-09-29 17:48	(−15.5°, −172.2°)	33	8.0	3.3
9	Southern Sumatra	2009-09-30 10:16	(−0.8°, 99.8°)	60	7.6	4.5
10	Vanuatu	2009-10-07 22:03	(−13.0°, 166.3°)	33	7.8	2.7
11	Hualian, Taiwan	2009-12-19 13:02	(23.8°, 121.7°)	30	6.6	3.5
12	Haiti	2010-01-12 21:53	(18.5°, −72.4°)	10	7.1	5.2
13	Central Chile	2010-02-27 06:34	(−35.8°, −72.7°)	33	8.6	3.2
14	Central Taiwan	2010-03-04 00:18	(23.0°, 120.7°)	5	6.5	3.5
15	Northern Mexico	2010-04-04 22:40	(32.1°, −115.5°)	10	7.2	4.8
16	Southern Sumatra	2010-04-06 22:15	(2.4°, 97.1°)	31	7.8	3.1
17	Yushu, Qinghai	2010-04-13 23:49	(33.1°, 96.7°)	10	6.9	2.5
18	Nicobar Islands	2010-06-12 19:26	(7.7°, 91.9°)	30	7.6	4.6
19	Vanuatu	2010-12-25 13:16	(−19.7°, 168.9°)	20	7.4	2.4
20	Southwestern Pakistan	2011-01-18 20:23	(28.8°, 63.9°)	10	7.1	4.1
21	Tohoku, Japan	2011-03-11 05:46	(38.3°, 142.4°)	24	9.0	2.5
22	Kermadec Islands	2011-07-06 19:03	(−29.3°, −176.2°)	10	7.7	2.9
23	Kermadec Islands	2011-10-21 17:57	(−29.0°, −176.2°)	33	7.5	2.9
24	Eastern Turkey	2011-10-23 10:41	(38.6°, 43.5°)	20	7.3	3.0
25	Mexico	2012-03-20 18:02	(16.7°, −98.2°)	20	7.5	2.7
26	Northern Sumatra	2012-04-11 08:38	(2.3°, 93.1°)	23	8.6	3.6
27	Xinyuan, Xinjiang	2012-06-29 21:07	(43.4°, 84.8°)	7	6.3	3.3
28	Costa Rica	2012-09-05 14:42	(10.1°, 85.3°)	41	7.6	2.6
29	Haida Gwaii	2012-10-28 03:04	(52.8°, −131.9°)	18	7.8	3.2
30	Southeastern Alaska	2013-01-05 08:58	(55.2°, −134.8°)	10	7.5	2.2
31	Solomon Islands	2013-02-06 01:12	(−10.8°, 165.1°)	6	7.8	2.7

Continued

No.	Epicentral Region	Origin Time (UTC)	Epicentral Location (Lat., Long.)	Focal Depth /km	M_W	Time Consumed /hr
32	Nantou, Taiwan	2013-03-27 02:03	(23.8°, 121.1°)	21	6.0	2.8
33	Khash, Iran	2013-04-16 10:44	(28.1°, 62.1°)	82	7.7	4.3
34	Lushan, Sichuan	2013-04-20 00:02	(30.3°, 103.0°)	12	6.8	3.0
35	Sea of Okhotsk	2013-05-24 05:44	(54.9°, 153.3°)	610	8.3	2.8
36	Nantou, Taiwan	2013-06-02 05:43	(23.8°, 121,1°)	20	6.2	2.6
37	Mingxian-Zhangxian, Gansu	2013-07-21 23:45	(34.5°, 104.2°)	10	6.0	2.4
38	Hualian, Taiwan	2013-10-31 12:02	(23.6°, 121.4°)	12	6.3	1.7
39	Scotia Sea	2013-11-17 09:04	(−60.3°, −46.4°)	10	7.8	1.6
40	Yutian, Xinjiang	2014-02-12 09:19	(35.9°, 82.6°)	13	6.9	3.2
41	Northern Chile	2014-04-01 23:46	(−19.6°, −70.8°)	20	8.2	2.8
42	Northern Chile	2014-04-03 02:43	(−20.4°, −70.1°)	20	7.7	1.3
43	Solomon Islands	2014-04-12 20:14	(−11.3°, 162.2°)	29	7.6	5.0
44	Solomon Islands	2014-04-13 12:36	(−11.5°, 162.1°)	35	7.5	2.4
45	Papua New Guinea	2014-04-19 13:27	(−6.7°, 154.9°)	31	7.5	1.3
46	Alaska	2014-06-23 20:53	(51.8°, 178.8°)	114	7.9	2.7
47	Ludian, Yunnan	2014-08-03 08:30	(27.1°, 103.3°)	12	6.1	2.4
48	Jinggu, Yunnan	2014-10-07 13:49	(23.4°, 100.5°)	5	6.0	1.6
49	Papua New Guinea	2015-03-29 23:48	(−4.8°, 152.6°)	18	7.5	2.6
50	Nepal	2015-04-25 06:11	(28.1°, 84.6°)	40	7.9	2.2
51	Nepal	2015-05-12 07:05	(27.8°, 86.1°)	15	7.2	2.9
52	Pishan, Xinjiang	2015-07-03 01:07	(37.5°, 78.1°)	15	6.3	1.7
53	Central Chile	2015-09-16 22:54	(−31.6°, −71.7°)	13	8.2	2.3
54	Farkhar, Afghanistan	2015-10-26 09:09	(36.4°, 70.7°)	213	7.5	2.0
55	Tarauaca, Brazil	2015-11-24 22:45	(−10.5°, −70.9°)	600	7.4	2.6
56	Sumatra	2016-03-02 12:49	(−4.9°, 94.2°)	10	7.7	1.0
57	South Georgia Island	2016-08-19 07:32	(−55.3°, −31.9°)	10	7.4	2.0
58	Zaduo, Qinghai	2016-10-17 07:14	(32.8°, 94.9°)	9	5.8	1.9
59	Norcia, Italy	2016-10-30 06:40	(42.9°, 13.1°)	10	6.3	1.7
60	South Island, N.Z.	2016-11-13 11:02	(−42.8°, 173.1°)	10	7.9	1.7
61	Aketao, Xinjiang	2016-11-25 14:24	(39.3°, 74.0°)	12	6.5	1.8
62	Hututi, Xinjiang	2016-12-08 05:15	(43.8°, 86.4°)	6	6.2	3.5
63	Chile	2016-12-25 14:22	(−43.4°, −73.8°)	40	7.5	1.6
64	Solomon Islands	2017-01-22 04:30	(−6.1°, 155.2°)	168	7.9	1.4
65	Jiuzhaigou, Sichuan	2017-08-08 13:19	(33.2°, 103.8°)	10	6.5	1.7
66	Jinghe, Xinjiang	2018-08-08 23:27	(44.3°, 82.9°)	11	6.3	1.1
67	Mexico	2017-09-08 12:49	(14.9°, −94.0°)	30	8.1	1.3
68	Milin, Tibet	2017-11-17 22:34	(29.8°, 94.9°)	10	6.4	1.2
69	Northern Honduras	2018-01-10 02:51	(17.5°, −83.5°)	10	7.7	1.9
70	Alaska	2018-01-23 09:31	(56.0°, −149.1°)	10	7.9	1.4
71	Papua New Guinea	2018-02-25 17:44	(−6.10°, 142.8°)	20	7.4	2.1

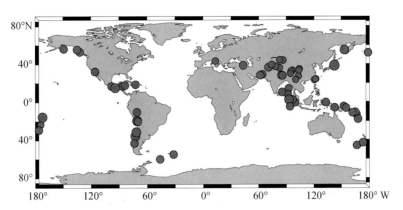

Fig. 29 Map showing the epicentral distribution of the recent significant earthquakes whose source rupture processes were timely inverted and released after the earthquake occurrence in the time period 2009-2018

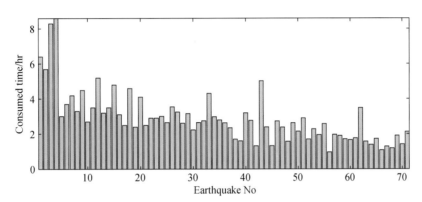

Fig. 30 The time consumed on each of the significant earthquakes in table III for which the source rupture processes were inverted and released immediately after the earthquake occurrence in the time period 2009-2018

The abscissa represents serial number (No.) of the significant earthquake. The ordinate represents the time consumed for data processing in hour

4 Summary

The earthquake rupture process inversion is not only an important part of scientific research at present, but also a necessary work in the emergency response of earthquake disasters. We have presented the theory and the methods of earthquake rupture process inversion and have shown in this paper that the earthquake rupture model can be well determined about 1-3 hours after the earthquake occurrence, which is valuable to estimate the meizoseismic areas of disastrous earthquakes. However, for early warning of near-shore earthquake-generated tsunamis, this is far from enough because the tsunami usually reaches the coast in 10-30 minutes. To further reduce the time consumed, a feasible way is to use the local or regional data, rather than the teleseismic data. As the continuous improvement of earthquake monitoring particularly in earthquake-prone area, inversion with the near-source data would be the future direction to speed up the fast and routine source inversion.

References

[1] Butler R., Lay T., Creager K., Earl P., Fischer K., Gaherty J. and Tromp J., The Global Seismographic Network surpasses its design goal, *Eos*, *Trans. Amer. Geophys. Union*, **85**(2004)225.

[2] Heidemann J., Ye W., Wills J., Syed A. and Li Y., Research challenges and applications for underwater sensor networking, in *Wireless Communications and Networking Conference*, 2006, *IEEE*, **1**(2006)228.

[3] Zheng X F., Yao Z X., Liang J H., and Zheng J., The role played and opportunities provided by IGP DMC of China National Seismic Network in Wenchuan earthquake disaster relief and researches, *Bull. Seismol. Soc. Am.* **100**(2010)2866.

[4] Zhang Y., Chen Y T., and Xu L S., Fast and robust inversion of earthquake source rupture process and its application to earthquake emergency response, *Earthquake Sci.*, **25**(2012)121.

[5] Allen R M., Gasparini P., Kamigaichi O. and Bose M., The status of earthquake early warning around the world: An introductory overview, *Seismol. Res. Lett.*, **80**(2009)682.

[6] Kikuchi M. and Kanamori H., Inversion of complex body waves, *Bull. Seismol. Soc. Am.*, **72**(1982)491.

[7] Hartzell S. H. and Heaton T. H., Inversion of strong ground motion and teleseismic waveform data for the fault rupture history of the 1979 Imperial Valley, California, earthquake, *Bull. Seismol. Soc. Am.* **73**(1983)1553.

[8] Das S., and Kostrov B. V., Inversion for seismic slip rate history and distribution with stabilizing constraints: Application to the 1986 Andreanof Islands earthquake, *J. Geophys. Res. Solid Earth*, **95**(1990)6899.

[9] Ji C., Wald D. J. and Helmberger D. V., Source description of the 1999 Hector Mine, California, earthquake, part I: wavelet domain inversion theory and resolution analysis, *Bull. Seismol. Soc. Am.*, **92**(2002)1192.

[10] Mori J. and Hartzell H. S., Source inversion of the 1988 Upland, California, earthquake: determination of a fault plane for a small event, *Bull. Seismol. Soc. Am.*, **80**(1990)507.

[11] Xu, L. S. and Chen Y. T., Tempo-spatial rupture process of the 1997 Mani, Xizang (Tibet), China earthquake of $M_S=7.9$, *Acta Seismol. Sinica*, **12**(1999)495.

[12] Chen Y. T. and Xu L. S, A time domain inversion technique for the tempo-spatial distribution of slip on a finite fault plane with applications to recent large earthquakes in the Tibetan Plateau, *Geophys. J. Int.*, **143**(2000)407.

[13] Zhang Y., Xu L. S. and Chen Y. T., PLD method for apparent source time function (ASTF) retrieving and its application to the 2005 Kashmir $M_W 7.6$ earthquake, *Chin. J. Geophys.*, **52**(2009)672 (in Chinese).

[14] Zhang Y., Chen Y. T. and Xu L. S., Rupture process of the 2005 Southern Asian (Pakistan) $M_W 7.6$ earthquake from long-period waveform data, in *Advances in Geosciences*, Vol. **9** *Solid Earth (SE)*, *Ocean Science (OS)* and *Atmospheric Science (AS)*, edited by Chen Y. T. (World Scientific Publishing Co., Singapore) 2007, pp. 13-21.

[15] Zhang Y., Xu L. S. and Chen Y. T., Rupture process of April 6 2009 L'Aquila Italy $M_W 6.3$ earthquake: a comparison between apparent source time function method and direct wave inversion method, *Chin. J. Gepphys.*, **53**(2010)1428.

[16] Zhang Y., Chen Y. T., Xu L. S., Wei X., Jin M. P. and Zhang S., The 2014 $M_W 6.1$ Ludian, Yunnan, earthquake: A complex conjugated ruptured earthquake, *Chin. J. Geophys.*, **58**(2105)153 (in Chinese with English abstract).

[17] Zhang Y., 2008 *Study on the Method of Earthquake Rupture Process Inversion*, Dissertation for Doctor Degree, Beijing, School of Earth and Space Sciences, Peking University(2008)1-158 (in Chinese).

[18] Delouis B., Giardini D., Lundgren P. and Salichon J., Joint inversion of InSAR, GPS, teleseismic, and strong-motion data for the spatial and temporal distribution of earthquake slip: application to the 1999 Izmit mainshock, *Bull. Seismol. Soc. Am.*, **92**(2002)278.

[19] Zhang Y., Feng W. P., Chen Y. T., Xu L. S., Li Z. H., and Forrest D., The 2009 L'Aquila $M_W 6.3$ earthquake: a new technique to locate the hypocentre in the joint inversion of earthquake rupture process, *Geophys. J. Int.*, **191**(2012)1417.

[20] Zhang Y., Feng W. P., Xu L. S., Zhou C. H. and Chen Y. T., Spatio-temporal rupture process of the 2008 great Wenchuan earthquake, *Sci. China*, *Ser. D: Earth Sci.*, **52**(2009)145.

[21] Chen Y. T., Yang Z. X., Zhang Y. and Liu C., From the 2008 Wenchuan earthquake to the 2013 Lushan earthquake, *Sci. China*, *Ser. D: Earth Sci.*, **43**(2013)1064 (in Chinese).

[22] Monastersky R., The next wave, *Nature*, **483**(2012)144.

[23] Yagi Y., Mikumo T., Pacheco J. and Reyes G., Source rupture process of the Tecomán, Colima, Mexico earthquake of 22 January 2003, determined by joint inversion of teleseismic body-wave and near-source data, *Bull. Seismol. Soc. Am.*, **94**(2004)

1795.

[24] Hartzell S. and Iida M., Source complexity of the 1987 Whittier Narrows, California, earthquake from the inversion of strong motion records, *J. Geophys. Res.: Solid Earth*, **95**(1990)12475.

[25] Bertero M., Bindi D., Boccacci P., Cattaneo M., Eva C. and Lanza V., Application of the projected Landweber method to the estimation of the source time function in seismology, *Inverse Problems*, **13**(1997)465.

[26] Liu C., Xu L. and Chen Y. T. Quick moment tensor solution for 6 April 2009, L'Aquila, Italy, earthquake, *Earthquake Sci.*, **22**(2009)449.

[27] Li X. and Chen Y. T., Inversion of long-period body-wave data for the source process of the Gonghe, Qinghai, China earthquake, *Acta Seismol. Sin.* (English Edition), **9**(1996)361.

[28] Xu L. S. and Chen Y. T., Source parameters of the Gonghe, Qinghai Province, China, earthquake from inversion of digital broadband waveform data, *Acta Seismol. Sin.* (English Edition), **10**(1997)143.

[29] Xu X. T., Chen Y. T. and Wang P. D., Rupture process of the $M_L = 4.1$ earthquake in Huailai basin on July 20, 195, *Acta Seismol. Sin.*, **12**(1999)618.

[30] Liu C., Xu L. S. and Chen Y. T., Quick moment solution for 14 April 2010 Yushu, Qinghai, earthquake. *Earthquake Sci.*, **32**(2010)366 (in Chinese).

[31] Chen Y. T. and Xu L. S., Source processes of recent large earthquakes in Qinghai-Xizang (Tibetan) Plateau, *J. Geol.*, **B**(1999) 216.

[32] Xu L. S. and Chen Y. T., Observed evidence for crack fusion from the 14 November 2000 Kunlun Mountain earthquake, in *Advances in Geosciences*, Vol. **1**, *Solid Earth* (SE), edited by Chen Y. T. (World Scientific Publishing Co., Singapore) 2006, pp. 51-59.

[33] Liu C., Zhang Y., Xu L. S. and Chen Y. T., A new technique for moment tensor inversion with applications to the 2008 Wenchuan $M_s8.0$ earthquake sequence, *Acta Seismol. Sin.* **21**(2008)333 (in Chinese).

[34] Liu R. F., Chen Y. T., Zou L. Y., Chen H. F., Liang J. H., Zhang L. W., Han X. J., Ren X. and Sun L., Determination of parameters for the 20 April 2013 Lushan $M_W6.7(M_S7.0)$ earthquake in Sichuan Province, *Acta Seismol. Sin.*, **35**(2013)652.

[35] Zhang Y., Xu L. S., Chen Y. T. and Wang R. -J., Fast inversion on the rupture process of the 12 February 2014 Yutian $M_W6.9$ earthquake: Discussion on the impacts of focal mechanisms on rupture process inversion, *Acta Seismol. Sin.*, **36**(2014)159.

[36] Xu L. S., Chen Y. T., Teng T. L. and Patau G., Temporal and spatial rupture process of the 1999 Chi-Chi earthquake from IRIS and GEOSCOPE long-period waveform data using aftershocks as empirical Green's functions, *Bull. Seismol. Soc. Am.*, **92**(2002) 3210.

[37] Xie Y. S. and Cai M. B., *Compilation of Historical Materials of Chinese Earthquakes*, Vol. **3** (Science Press, Beijing) 1987, pp. 235-236 (in Chinese).

[38] Min Z. Q., *Historical Strong Earthquake Catalog in China (the 23th Century BC—1911 AD)* (Seismological Press, Beijing) 1995, pp. 1-514 (in Chinese).

[39] Min Z. Q., *Historical Strong Earthquake Catalog in China (1912AD — 1990AD)* (Seismological Press, Beijing) 1995, pp. 1-636 (in Chinese).

[40] Yang Z. X., Waldhauser F., Chen Y. T. and Richards P., Double-difference relocation of earthquakes in central-western China 1992-1999, *J. Seismol.*, **9**(2005) 241.

[41] Zhang Y., Feng W. P., Xu L. S., Zhou C. H. and Chen Y. T., Spatio-temporal rupture process of the 2008 great Wenchuan earthquake, *Sci. China, Ser. D: Earth Sci.*, **52**(2009)145.

[42] Deng Q. D., Zhang P Z., Ran Y K., Yang X. P. and Zu Q. X., Basic characteristics of active tectonics of China, *Sci. China, Ser. D: Earth Sci.*, **46**(2003)356.

[43] Du H. L., Xu L. S. and Chen Y. T., Rupture process of the 2008 great Wenchuan earthquake from the analysis of the Alaska-array data. *Chin. J. Geophys.*, **52**(2009)372 (in Chinese).

[44] Li M. F., Xing C. Q., Cai C. X., Guo W. X., Wu S. X., Yuan Z. Z., Meng Y. Q., Tu D. L., Zhang R. B. and Zhou R. J.,

Research on activity of Yushu fault, *Seismol. Geol.*, **17**(1995)218 (in Chinese).

[45] Zhou R. J., Wen X. Z., Cai C. X. and Ma S. H., Recent earthquakes and assessment of seismic tendency on the Ganzi-Yushu fault zone, *Seismol. Geol.*, **19**(1997)115 (in Chinese).

[46] Wen X. Z., Xu X. W., Zheng R. Z., Xie Y. Q. and Wan C., Average slip-rate and recent large earthquake ruptures along the Garzê-Yushu fault, *Sci. China Ser. D Earth Sci.*, **46**(2003)276.

[47] Zhang Y., Xu L. S. and Chen Y. T., Fast inversion of the rupture process of 2010 January 12 Haiti earthquake, *Acta Seismol. Sin.*, **32**(2010)124 (in Chinese with English abstract).

[48] Zhang Y., Xu L. S. and Chen Y. T., Source process of the 2010 Yushu, Qinghai, earthquake, *Sci. China Ser. D: Earth Sci.*, **53**(2010)1249.

[49] Zhang Y., Wang R., Chen Y. T., Xu L., Du F., Jin M., Tu H. and Dahm T., Kinematic rupture model and hypocenter relocation of the 2013 M_W 6.6 Lushan earthquake constrained by strong-motion and teleseismic data, *Seismol. Res. Lett.*, **85**(2014)15.

[50] Zhang Y., Chen Y. T., Xu L. S., Wei X., Jin M. P. and Zhang S., The 2014 M_W 6.1 Ludian, Yunnan, earthquake: A complex conjugated ruptured earthquake, *Chin. J. Geophys.*, **58**(2105)153 (in Chinese with English abstract).

第二部分　回忆与祝贺

贺陈运泰院士八十华诞

蔡永恩

北京大学地球与空间科学学院

乙亥孟秋，博雅伟岸，未名灵秀．是时，陈运泰院士八十华诞．独步燕园，觅师尊足迹．仰塔之高雅，慕基牢础固；登临枫岛、低徊静园，感松贞柏劲；访湖轩、赏朗润，染竹气莲清．

愚识师尊于昭通地震考察，迄今已逾四十五秋．时年，师风华正茂，开拓、笔耕于地震战线，成果硕硕．“文革"劳师莅临北大讲座，欣欣然徒步来去．概念深入浅出，来龙去脉，一清二楚．改革挂帅"攀登预研""地震局重点"，躬身指导，察问建言，研讨聆教，仍历历在目，回响耳畔．每每会聚餐饮，海阔天空、溯史追今，虽越古稀，记忆堪比衡奉．

予虽未出师门，耳闻目染，日积月累，获益匪浅．幸甚至哉，书诗恭贺：

八月燕园菡萏开，
碧荷清气起蓬莱．
宏篇力作名科苑，
卓见博学育俊材．
平气平心平上下，
亦师亦友亦兄台．
身安运泰源陈酿，
华诞八十春又来．

2019 年 8 月

又《西江月》一首忆 1974 年昭通地震考察之帐篷讲学：

帐外潇潇山雨，
帐中慢道震源．
纸悬麟角解谜团，
顿启学员茅塞．

单矩横波证伪，
无旋双矩波全．
醍醐灌顶纸三篇，
山高路险觅断．

2019 年 12 月

1994 年"国际地球动力学中的力学问题学术讨论会"期间留影(一)

1994 年"国际地球动力学中的力学问题学术讨论会"期间留影(二)

蔡永恩(左)与陈运泰(右)在 2015 布拉格 IUGG 大会期间合影

回忆和陈运泰老师在一起的日子

李世愚

中国地震局地球物理研究所

今年8月10日，我们师兄弟欢聚一堂，为陈运泰老师祝贺他的八十大寿．耄耋之年，八十载风雨兼程，他与我们一路相伴相持，他就是我们真正的良师益友．

沧桑轮回，有多少往事如烟似梦．陈老师大学头两年就读北大物理系物理专业（1956～1958）．三年级（1958年）分配专业时，他原希望学习理论物理，但被分配学习地球物理．他随遇而安，愉快地投入到与生气勃勃的大自然打交道的地球物理专业的学习中．陈老师在北大物理系期间打下了深厚的数理功底，在从事地球物理研究的生涯中又具备丰富的地球物理学造诣．在陈老师的早期论文中，给我印象最深的就是1979年他在《地球物理学报》上发表的《用大地测量资料反演的1976年唐山地震的位错模式》（第22卷，第3期）．大家都知道1976年唐山大地震是没有直接前震的，而且大多数大地震都是没有直接前震的（震前平静），这给用地震活动性预报大地震带来了极大困难．陈老师的这篇论文明确回答了，大地震发生之前，地面的大规模形变是存在的，这和里德（1911年）提出的弹性回跳说是一致的．这对我们将地震前兆研究的重心从前震转移到大地形变以及相关的物理表现具有指导性意义．直到50年后的今天，它的生命力还是强劲的（见和泰名，李世愚．2017．地震的应变张量观测与应用前景．地球物理学报，**60**（11）：4327-4390）．

1980年，我在陈老师的名下开始读硕士研究生．后因陈老师去美国和洛杉矶加州大学诺波夫（Knopoff, Leon）教授合作，把我托付给尹祥础老师，但在这期间，我和陈老师始终保持着联系．1983年10月陈老师从美国回来之后，就为所里的科研人员举办"地震破裂动力学"学习班．要求每个参加者选定一篇参考文献．我们读一篇都感到吃力，而陈老师却早已对所有文献了然于胸，他甚至可以在辅导时说出哪一段文字来自哪篇文献的第几页．这给我留下了深刻的印象，也一直影响到我后来的研究生涯．

陈老师治学严谨，学识渊博，自然地成为地球物理学界名副其实的泰斗，1991年毫无悬念地当选为中国科学院院士．记得1998年美国邓大量教授来中国访问，在学术报告中提到陈老师关于地震波辐射中的旋转和旋转张量问题的最新理论成果，说他来中国就是为了得到这份理论推导的最新内容．陈老师的地球物理造诣在国际上的影响可见一斑．相信陈老师的这种学风对于大家都有影响．

陈老师的治学和育人的一大特点是胸怀开阔，开放包容．他学识渊博，厚积薄发，对我们启发甚多，我们受益匪浅．陈老师平时在和我们一起聊天时，不是干巴巴地就事论事地谈几条业务，而是天南海北、名人轶事、谈古论今．我参加过几次陈老师接待外宾或外籍华人的活动，发现陈老师也是这样侃侃而谈．这使得他在与国内外朋友的接触中有许多共同话题．我由此悟出了一个道理，就是治学如作诗，功夫在诗外．记得我在博士论文开题时，不但谈了破裂动力学的调研内容，还加上了断裂力学三维破裂的内容．这部分内容的研究一开始是不顺的，受到各方面的质疑和责难．而且和前面破裂动力学的内容似乎也不衔接．我直觉这两者一定有内在关系，但在当时又缺少充分理由．我的处境一度很孤立．陈老师一开始也不太适应，但他没有责难，而是抱着一种宽容的态度．很快，他就从肖尔茨（Scholz, Christopher H.）的专著中发现这在断层力学中是个重要的前沿课题，鼓励我抓紧研究发表，并赞同我写进了博士论文．后来的

---

* 本文写于2019年10月23日．

研究实践表明，我的博士论文的两个看似不衔接的内容，在断层自相似扩展问题中恰恰是密不可分的．我退休后继续研究，终于找到了这两个内容的衔接点，给出了进一步的推理．事实证明，陈老师的做法是有远见的．多年来，陈老师在指导研究生选题方面，对于震源物理的基础研究下了很大功夫，这使得我们这个群体在地球物理学界总是处于前沿地位．

回顾多年来的科研和从教经历，陈老师以满腔的热忱和一颗赤子之心，孜孜不倦地工作在科研和教书育人的第一线，把自己的一生奉献给地球物理科学，奉献给学生．在工作中，陈老师与同学们建立了深厚的感情．我记得有一年春节期间，陈老师的家人到老家去接孩子，他一个人在家，就特意把我和周家玉，记得还有柯兆明、倪江川和李肇仁等好几个同学请到他家里，不是我们请客送礼（陈老师从来没有让学生请过他的客），而是他亲自下厨为我们烹调．我们想帮忙都不要，他把一盘盘亲自烧制的美味菜肴端上饭桌请我们品尝．陈老师的厨艺很高，和我们亦师亦友的情谊更深．这份感情让我们深深觉得自己的人生丰沛饱满，可以说，这份浓情厚谊造就了今天健康、自信的他，也造就了自信的我们．我们每个师兄弟对陈老师的苦心指导都深有体会，毕业后各个都有出色的成就．陈老师桃李满天下，这是对他多年来呕心沥血指导每位学生的最好回报．

我本人没有什么好的东西送给陈老师，只是找到了一些和陈老师相处的照片展示于此，希望成为我们的美好回忆．

作者于1991年3月通过博士学位论文答辩时的合影（一）
前排左起：陈运泰，尹祥础，郭自强，曾融生，范天佑，朱传镇等教授．
后排左起：李世愚，许忠淮教授，张之立教授，所教育组宋昭仪老师

作者于1991年3月通过博士学位论文答辩时的合影（二）
左起：陈运泰，曾融生，李世愚，尹祥础

作者和博士生导师陈运泰院士的合影（作者获得国家地震局地球物理研究所博士学位几年之后补照的）

1995年美国洛杉矶加州大学诺波夫教授访问位于三里河的震源物理实验室时合影
诺波夫（左3），陈运泰（右3），李世愚（右2）

1994年在成都地质学院参加"国际岩石层计划地球构造层三维成像工作组"会议期间于都江堰合影
第1排：陈运泰（左1），F. Schwab（左3）；第三排：李世愚（左1）

1994年在成都地质学院参加"地球构造层三维成像"工作组会议期间与陈运泰老师于都江堰合影

1994年8月31日会见来访的日本东京大学地震研究所丸山卓男（T. Maruyama）教授（左4）和夫人（左5）

1998年8月21日会见美国洛杉矶南加州大学邓大量教授（前排右3）

1996年8月10日会见中国台湾"中央"研究院地球科学研究所徐济安教授（左5）等客人

陈运泰老师带领我们做野外观测

王培德

中国地震局地球物理研究所

地球物理学是一门观测的科学，陈运泰院士不仅在地震波传播理论和地震震源理论方面都有很深的造诣，有突出的学术成就，而且在地震学观测这最基本的科研领域也取得了众多成果．

自 20 世纪 70 年代以来，以大规模和超大规模集成电路为代表的微电子技术迅速发展，这一技术应用在地震观测中的成果之一即是产生了新一代的地震仪——数字地震仪．在 20 世纪 70 年代末和 80 年代初期，中国在数字地震观测技术方面与国际先进水平有较大的差距，而从 20 世纪 60 年代到 70 年代这一时期中国大陆的一系列强烈地震也引起了国际地震学界的关注．1981 年美国宾厄姆顿纽约州立大学地球科学系吴大铭教授与中国国家地震局地球物理研究所所长顾功叙教授议定，中美双方合作进行"近场强地面运动研究"的项目，并确定中国方面的项目负责人为陈运泰教授．这一项目得到中国国家地震局国际合作司、地震科学联合基金会和美国国家科学基金会的支持．就是从那时起，陈运泰教授就开始紧紧抓住近场数字地震的观测与研究．

1983 年年底陈运泰教授结束了在美国的合作研究回到国内，1984 年就亲自带队赴新疆马兰进行地下核试验的近场观测．我们地球物理所一行十多人，除了近场观测的课题组外，还有地磁、电磁波等项目．我们这个课题组去马兰的多数是陈老师的学生——他的博士生和硕士生，携带了 6 台 DCS-302 数字地震仪．

马兰是我国著名的核试验基地，位于我国新疆的罗布泊地区，属于戈壁滩和沙漠地区，自然条件十分恶劣．马兰是 21 基地所在地，20 世纪 60 年代之前就是一片戈壁，从 20 世纪 60 年代开始建设到我们所去的 80 年代，经过一代人的艰苦奋斗已经建设得很好了，绿树成荫、环境优美、科研和生活设施齐备．然而马兰并不是我们工作的所在地，从马兰向戈壁深处再前行 60 千米，才能到达"前进基地"，这里可就是一片荒漠了．在没有实验任务的时候前进基地是闲置的，有任务了才有人入住，因此十分简陋．前进基地有宿舍、伙房、库房等生活设施，但因为是临时性的，也就是能用就行．给我们安排的住地就是一间大房间，二十多人住在一起，陈老师那时已经升任研究室主任、研究员，但他毫不在乎生活条件的好坏，就在类似工棚的住房内和大家吃、住在一起．

我们去往马兰的目的就是记录地下核爆炸所产生的强地面运动．为此目的我们需要在试验之前把带去的数字地震仪布设在爆心周围．按课题要求，我们在爆心周围安排了 5 台地震仪，在外场安排了一台．我们借助基地提供的地图，在爆心周围适合的方位和距离寻找合适的基岩出露场地，找到场地后，再想方设法把拾震器固定在地面，调整好记录器、连接好电源并做好防护．因为是戈壁滩、半沙漠地带，气候极其干燥，加上太阳晒，嗓子像冒火，走在沙地上，脚下软绵绵的，体力消耗特别大．经过选点、探勘等一系列工作，我们终于按科研的要求把台站布设好．参加了试验场组织的一次又一次的"演练"，一切就等"零时"的到来．当然，参加场区里面的观测要比在外场危险程度更高一些，陈运泰教授身体力行，和年轻的学生们一起做场区内的观察．

"零时"就是核试验的起爆时间．临近"零时"之前的两小时，我们参加场区内任务的人员就都穿好防护服，戴上护眼的墨镜，做好一切准备在警戒线外待命．"零时"一到，一声沉闷的响声，紧跟着就是大地的上下波浪状起伏，我国又一次地下核试验成功了！

试验成功紧跟着就是紧张的"抢资料"阶段. 尽管是地下核试验,绝大部分放射性物质被封闭在地下,但仍然有极少量的放射性物质会在爆炸后 40 分钟到两小时之间沿地裂缝等通道溢出地面,对井口附近造成放射性污染. 参加核试验观测的项目就必须抢时间,在这段时间内把观测资料和设备从井口附近的观测点回收回来,这就是"抢资料".

随着三颗绿色信号弹的升起,在警戒线外待命的车辆均以最快的速度向自己的观测点快速开进. 我们的车辆也不例外,以最快的速度向观测点疾驰而去. 地下核爆破引起的强烈的地面震动,把戈壁滩上的尘土都"扬"上了天空,然后又黑压压地向我们的观测点扑来. 因为场区经历过多次核试验,这些沙尘不可避免地带有放射性,大家心里都明白,必须赶紧取回资料和仪器设备后撤离这危险地带. 参加核试验观测的有些项目只有一个观测点,而我们却有分布在不同地方的 5 个观测点,需要花费的时间更多. 车子陆续奔赴这 5 个观测点,每到一点车一停,我们就像战士冲锋一样,跑向仪器,取下记录磁带、撤下仪器并装上汽车. 陈老师也像我们年轻的战士一样,跑向观测点,和大家一起拆装设备. 等这 5 个点都完成以后,撤离观测点,经过"洗消站"冲洗、检查合格后,再返回住地. 参加完部队的庆功活动,把所有设备整理装箱、托运,然后我们返回北京,这次观测任务就算圆满完成了.

就在马兰地下核爆破近场观测完成不久,1985 年 4 月 18 日云南省禄劝地区发生 M_S 6.1 地震,地震后我们近场观测组奔赴禄劝进行近场观测. 禄劝县位于昆明北部,距昆明直线距离 100 千米,但实际路程大大超过 100 千米. 1985 年从昆明到禄劝尚没有高速公路,自己开车从昆明到震区去需用 3~4 个小时. 那时禄劝地区的经济欠发达,多数建筑质量也不好,地震造成了 22 人死亡,300 多人受伤,2 万多间房屋倒塌. 陈运泰老师在得知地震方面的消息后及时和云南省地震局的同志联系,在判断了余震的发展趋势后,决定进行这次近场观测. 我们是 4 月 24 日到达地震现场的,携带 DCS-302 数字地震仪 4 台. 我们一行 3 人先行到达,马上就架设了地震台开展了观测. 陈运泰老师随后也到达了地震现场,和我们一起进行野外工作,由于其他工作需要,陈老师在经过一段时间野外工作后先期返回北京.

地震现场的生活和工作条件都是比较艰苦的,有些当地的居民就住在临时搭建的抗震棚内. 地震灾区是农村,平时的饮用水就是井水,地震导致井水都变得浑浊了,那也没有办法,简单沉淀一下,也就只能喝这个水了. 地震后震区有各路抗震救援的队伍,比如部队、电力、道路、通信等,抗震救灾指挥部为了这些队伍更有效地工作,搭建了一些食物供应点,我们称之为"抗震饭",当然非常简单了,我们有时就在这些供应点吃点东西. 云南省地震局在震区有一支前线工作队伍,我们很多工作和生活上的事情也就和他们一起. 陈运泰老师在野外工作期间,和我们一样在抗震现场吃饭,和我们一起经历一次次余震的晃动. 十多天的工作记录到地震 400 多个,最大的余震 4.8 级. 为科研积累了资料. 当然,地震现场工作的一项必须要做的工作是地震趋势评估,协助当地政府组织抗震减灾活动,我们既然在野外,也是必须参加的,每天晚上都要做震情会商. 我们的仪器、我们的记录都发挥了作用.

禄劝地震后,陈老师领导他的科研团队利用所得到的记录开展地震矩张量反演的研究工作,并且取得了成功. 利用近场记录进行矩张量反演在世界地震学界是第一次,是地震学界中开拓性的研究成果. 它是利用波型来求解地震震源的机制解,大大扩展了对震源机制的研究. 在这之前,地震学界求取震源机制是利用 P 波初动的方向,这样求解机制解必须有足够的台站记录,这同时意味着,所研究的地震必须足够大(一般要 5 级以上),才有可能同时被几十个台站记录到. 有了近场记录和反演技术,有三、四个台的记录就可以做出来了,研究地震的震级可以下延到 2~3 级,研究范围大大扩展了. 随后,利用在新疆马兰所得到的地下核爆破的近场强地面运动的记录做的矩张量反演也获得成功. 从此,在陈运泰老师领导下地球物理研究所在地震震源方面翻开了新的一页,从地震矩张量到复杂震源过程、地震破裂过程成像,不断取得新的成绩.

此后,陈运泰老师担任了中国地震局地球物理研究所所长,工作更加繁忙、责任更加重大,再用很多时间直接到野外进行观测已不太可能.但他一直关心、支持野外工作.近场观测课题组执行过多次余震、地震危险监测等课题.震源物理研究室所执行的多项国际合作项目,例如,在怀来和欧盟合作的观测和研究,在延庆和德国合作的地震研究,都是以观测为基础的科研项目,都是在陈老师的领导和安排下进行的,并且取得了丰硕的成果.

作为一位杰出的地球物理学家,陈运泰老师身体力行、重视实际、重视观测的精神永远值得我们学习.

前排左起:姜大庆,柯兆明,钱书请,李聚臣.后排左起:王培德,陈运泰,王璋

我的导师陈运泰院士——永远年轻的科学家

陈晓非

南方科技大学

  我第一次知道我的导师陈运泰先生的名字是在大学三年级的时候，四十多年前．那时我在中国科技大学地球与空间科学系地球物理学专业读书．当时对所学的地球物理学专业不甚了解，只知道"地球物理学运用物理学原理研究地球内部的运动规律"这一抽象定义，不知道具体研究哪些科学问题？如何研究？需要哪些知识储备？为此，我经常去图书馆翻阅、浏览与地球物理学相关的期刊或书籍．虽然当时只是走马观花地浏览，但是在那些期刊的编委会中我常常看到一个与老一代著名地球物理学家并列在一起的名字：陈运泰，我想这一定是一位杰出的年轻地球物理学家，将来我就报考他的研究生．后来从我们的任课老师那里了解到，陈运泰老师是非常有影响的杰出青年科学家，我更加坚定了将来报考陈老师研究生的想法．毕业时我通过了硕士研究生考试，如愿成为陈老师的研究生．

  我第一次见到导师是在研究生入学后的第二年．陈老师结束在美国洛杉矶加州大学为期三年的进修访问后回到所里．那时我还在中科院研究生院集中学习基础课，听说导师回来了，非常高兴．没过几天，陈老师约我去所里见面并汇报学习进展．要见导师了，我很是兴奋也很紧张．由于第一次见面不知说什么好，陈老师亲切而和蔼的谈话很快就打消了先前的紧张，我向导师汇报了学习情况，得到了导师的肯定和鼓励．此后，基本上每个星期我都向导师汇报科研进展并不断得到导师的指导．在陈老师指导下，我于1985初完成了毕业论文，获得硕士学位，并同时考上了陈老师的博士研究生，开始跟随陈老师继续攻读博士学位．几乎在同时，刚刚从麻省理工学院转到南加州大学工作的世界著名地震学家 Keiiti Aki 教授有意招收中国学生，并请陈老师推荐优秀学生到他那里攻读博士学位．陈老师毫不犹豫地推荐我到南加州大学跟随 Aki 教授攻读博士学位，我非常感动．当时，我出国读书的意识还不是很强烈，觉得跟随陈老师在国内读博士很好．陈老师告诉我，到 Aki 教授那里读博士可以受到更好、更全面、更系统的科研训练，可以接触更多的科学前沿．在陈老师强烈推荐下，我被南加州大学录取为博士研究生，成为 Aki 教授的学生．1985年的最后一天，我告别了陈老师，告别了家人，告别了同学，踏上了海外留学之路．多年后，当我也成为一名研究生导师时，才深刻体会到当年陈老师对我、对学生的厚爱．陈老师总是为学生的发展、成长着想，为他们创造最有利的条件．

  在海外留学期间，我每次回国探亲都去看望陈老师，向他汇报我在国外的学习和工作进展．1994年，我随 Aki 教授一起回到北京参加由王仁先生和陈老师共同主持的地球动力学国际研讨会，再次在北京见到了陈老师．这一次，我除了向陈老师汇报我在国外科研工作进展外，还向陈老师表露了想回国工作的意愿．陈老师对我回国工作的想法非常支持，为我详细介绍了当时国内的科研环境和工作条件，并告诉我回国后短期内可能工作、生活条件不尽如意，但国家发展很快，也非常需要人才，长远来说应该大有作为．陈老师的鼓励和支持使我下定决心回国工作．回美国后，我和家人就开始着手为回国工作做准备．1996年，经过一年多的准备，我终于回到国内工作，到陈老师的母校北京大学任教．我没有辜负导师对我的培养和期望，在国外留学、工作10年后又回到祖国，亲身加入到波澜壮阔的国家建设之中，在中华民族伟大复兴的建设事业中没有缺席．

1995 年陈运泰老师参加在科罗拉多丹佛召开的 IUGG 大会后访问南加州大学，笔者与陈运泰老师合影

回国后，无论是在学校里的教学和组织管理工作，还是在学术界的科研工作，都得到了陈老师的鼎力支持. 2001 年陈老师被北京大学校长聘为新整合组建的北京大学地球与空间科学学院的院长. 北大地空学院师生对陈老师回到母校带领我们开展一流大学学科建设感到格外欣喜. 在陈老师的领导下，北京大学地球与空间科学学院发展成为国内最有影响的地球与空间科学学院.

2011 年在中国科学院大学计算地球动力学重点实验室成立大会期间，笔者与陈运泰老师合影

虽然，作为陈老师的学生，我在陈老师直接指导下学习的时间仅有 3 年. 无论与大学本科学习时间相比，还是与后来在国外读博士的时间相比都是最短的，但陈老师对我的影响却是最大的. "大音希声"，陈老师对我的影响是无形的，潜移默化的. 陈老师对待学问一丝不苟、精益求精、追求极致的严谨治学态度对我影响至深，我是从陈老师这里学会科研的. 陈老师非常关心和爱护学生、尊重学生，对待学生总是非常平易近人、和蔼可亲，学生也非常热爱和尊重陈老师. 陈老师知识渊博、虚怀若谷，品德高尚，乐观豁达，是我终身学习的榜样.

时光飞逝，光阴如梭，从我下决心报考陈老师研究生时算起，已经过去 40 年了，如今陈老师也正式入列 "80" 后了. 但在我心目中，陈老师永远都是年轻的大科学家. 衷心祝愿陈老师身体健康，永葆青春！

名师指路，终生受益——回忆陈运泰老师对我的教诲

柯兆明

北京瑞尼尔技术有限公司

回国创业十几年来，很多人问过我："柯总，你是学物理出身的，为什么能够成为有建树的金融大数据专家呢？"

从上世纪90年代初到今天，从华尔街到国内，我从事金融大数据分析和应用工作已经30年了．仔细想来，我之所以能够终生专注金融大数据工作并取得一些成就，在很大程度上是获益于我的恩师陈运泰院士对我的教诲和培养打下的扎实基础．

陈老师是国内外著名的地球物理学家．他于1962年从北京大学地球物理系毕业，1966年从中国科学院地球物理研究所研究生毕业，然后一直在中国科学院地球物理研究所（上世纪70年代改制为中国地震局地球物理研究所）工作．1981年至1983年作为访问学者，在美国洛杉矶加利福尼亚大学地球与行星物理研究所研修，1991当选为中国科学院学部委员（院士），1999年当选为第三世界科学院院士，2015年当选为国际大地测量与地球物理学联合会首批会士．

从上世纪80年代起，陈老师积极倡导和从事数字地震学研究，这一富有远见的战略思想极大地促进了高技术时代中国地震学的发展．他从1983年起，主持我国第一支用宽频带数字化地震仪装备起来的近震源强地面运动观测队伍的工作，十余年来获得大量高质量的观测资料，为改善国内外近震源强地面运动资料仍很缺乏的状况做出了重要的贡献．

我于1983年7月从中国科学技术大学本科毕业，当年考上了陈老师的硕士研究生．从1983年9月至1986年7月，我在陈老师的指导下参加了用宽频带数字化地震仪开展近震源强地面运动的数据观测、数据处理和数据分析工作，这是我首次接触数字化建设工作，是我终生从事大数据工作的开始．

1986年7月我获得硕士学位留所，继续在陈老师指导下开展近震源强地面运动的研究工作．1987年9月，在陈老师的推荐下，我赴美国纽约州立大学攻读地球物理学博士学位．1990年，我转学计量经济学，把自然科学领域的量化理念、方法和技术应用到社会科学领域．由于我具备比较好的数理基础，也掌握了一定的量化理念、方法和技术，转学计量经济学是比较容易的．

上世纪八九十年代，随着计算机技术的普及，华尔街金融机构用计算机系统记录了大量金融业务原始数据．这些数据包含着非常丰富的客户、产品、渠道等业务特点和规律信息，但必须通过开展数据分析和挖掘工作，才能把原始数据中的科学决策信息提炼出来，用量化的方法和技术不断优化金融业务的管理和经营过程，提升核心竞争力．正是从上世纪八九十年代开始，华尔街金融机构开始了持续至今的数字化转型建设，开启了全球数字化/智能化金融时代．

由于掌握量化的理念、方法和技术，从1993年起，我比较容易地在华尔街金融机构开始了金融大数据分析和应用工作，先后在摩根大通银行和美国国际集团（AIG）担任过多年的大数据定量管理建设的管理职务，将自然科学领域的量化理念、方法和技术引入到金融领域的精细化管理过程中，把金融理论、金融管理经验、大数据挖掘技术和信息技术结合起来，设计和发展了很多金融业务精细化管理策略和自动化加智能化的科学决策系统，为提高金融机构的核心竞争力做出了一定的贡献．

2004年8月我回国创建了北京瑞尼尔技术有限公司，把在华尔街掌握的数字化建设的理念、方法、技术和经验应用到国内金融机构．回国十几年来，我培养了一支非常专注和专业的金融大数据定量管理建

设团队. 这支专业团队把金融业务大数据定量管理理念、方法和技术与中国金融业务管理和经营的具体实践结合起来，帮助国内很多大中小银行、保险公司、证券公司等金融机构开展了数字化转型建设工作，以先进的大数据定量管理理念、方法和技术支持这些金融机构由第二代自动化经营阶段提升到第三代智能化管理阶段，提升了他们的核心竞争力. 由于取得了一些成绩，我被很多金融机构誉为中国金融数字化转型建设的"布道者"和"引领者".

纵观我一生的职业生涯，可以说我的价值观、思维方式和做事方式在相当大的程度上是受陈老师的影响而形成并延续至今的. 正是受到陈老师的影响，我从来不以阴谋诡计陷害人，不搞歪门邪道，对周围的人待之以诚，待之以真，老老实实做人，踏踏实实做事，以专注和专业的方式为社会做一些实事.

现在，我从四个方面回忆一些与陈老师相处的点点滴滴. 从这些具体事情中可以看到，我的成长经历确实受到了陈老师很大的影响.

1 陈老师为人非常正直，待人以诚

我刚到地球物理研究所时，很多熟悉陈老师的人就给我们这些刚刚进所的研究生们讲起很多在"文革"期间地球物理研究所发生的往事. 几乎所有讲述往事的人都说陈老师在10年"文革"动乱期间，从来不紧跟运动风，不参加任何派系斗争，没有整过任何人. 即使对所里那些受到冲击的老院士们和老科学家们，也一直非常尊敬，从不落井下石.

1983年我进所后，多次亲眼见到陈老师对傅承义院士、顾功叙院士、曾融生院士、秦馨菱院士、谢毓寿先生等老科学家的崇敬之情，对同学、同事和职工的友爱之情，多次亲耳听到陈老师对在唐山大地震中遇难的同事和朋友们的深切怀念之情.

1986年陈老师正式担任地球物理研究所所长后，我看到陈老师下很大力气整顿地球物理研究所的风气，鼓励和引导科研人员把时间和精力放在科研业务上，而不是放在钻营和经商上，使得地球物理研究所面目一新，学术气氛非常浓厚，保持了一片科研净土.

2 专注和专业，作风非常严谨，是一位真正的科学家

我刚到地球物理研究所时，很多人对我说："你的老师是书呆子，只懂念书." 我认为，恰恰是陈老师专注和专业的工作作风，才使得陈老师成为国内外著名的地球物理学家.

陈老师跟我说过，他在北京大学地球物理专业接受了6年的本科教育，数理功底训练非常严格，为他一生的科研工作打下了非常扎实的基础. 在地球物理研究所4年读研究生期间，他受到了傅承义院士、曾融生院士、顾功叙院士等直接或者间接的严格指导，科学研究功底深厚.

我听所里很多人说过，在"文革"十年动乱期间，陈老师从不卷入运动漩涡，自己置身运动事外，专心致志做研究，陈老师的很多科研成果就是在这段期间产出的. 所以，陈老师在"文革"结束后能够迅速脱颖而出，1978年被晋升为副研究员，1982年被晋升为研究员，1991年当选为中国科学院院士（当时叫做学部委员）.

我在所里时，陈老师带着第三研究室科研人员和他指导的研究生经常一起非常认真和仔细研读各种国内外专业文献，然后一起讨论，每个人给出自己的看法和想法. 陈老师亲自推导各种方程，非常认真的态度和深厚的理论功底让我非常佩服. 通过这样的实战型训练，我们的理论功底不断加深加厚，科研能力不断增强.

我记的非常清楚的事情是，当时我负责把强地面数字地震仪记录的地震原始信号和频谱分析结果画成

图.那时候,既没有激光打印机,也没有现成的绘图软件,需要我用 BASIC 语言编写出绘图程序,用绘图仪把信号打印出来,总共有 1000 多张打印出来的图.对于每一张图,陈老师带着我用尺子反复仔细测量,不断修改,尽可能把每一张图打印得非常漂亮.

陈老师有时安排我给国内专业期刊翻译国外发表的专业文献.当时,我的英文水平还不够好,翻译结果不令人满意.陈老师与我一起,非常认真和仔细地校正我的翻译结果,并给我认真讲解为什么这么翻译.

3 个人修养很高,心地非常纯净

见过陈老师的人都会对陈老师良好的个人修养印象深刻,肯定认为陈老师是一位谦谦君子.确实,我从陈老师身上看到了非常浓厚的学者作风.在陈老师身边四年,我从来没有见过陈老师对任何人耍过阴谋诡计,只是见到陈老师对大家的真诚、坦荡、友爱行为,也亲眼见到了陈老师是多么痛恨耍阴谋诡计动不动整人的行为.

陈老师内心很纯净,不喜欢把时间和精力花在搞乱七八糟的事情上.我从来没有见到过陈老师对吃喝玩乐有什么兴趣,除了本职的地球物理科研工作,陈老师的兴趣就是读各种书籍和期刊,包括文学、历史、地理、哲学等,不断丰富自己的知识库.

陈老师的精神世界一直非常丰富,他记忆力惊人,博闻强记,学识极为渊博.我仍然清晰地记得1984 年 9 月至 10 月我跟随陈老师去新疆马兰核试验基地参加地下核爆破时的点点滴滴.马兰基地位于天山南侧的戈壁滩,环境很荒凉,除了基地部队人员几乎没有人烟,周围只有骆驼草,偶尔能见到一些奔跑的黄羊.陈老师与我们一样,住在半地下的"地窑子"里.在等待 10 月初的核爆破试验的日子里,陈老师每天带领我们在"地窑子"里读书读文献,指导我们开展科学研究工作.休息时间,陈老师带领我们在戈壁滩上散步,以非常风趣幽默的方式讲解在国外的各种见闻,讲解各种中外名著,讲解历史、地理、哲学等等.

即使现在,陈老师已经是 80 岁的老人了,但我每次去探望陈老师,陈老师都会与我兴高采烈大聊几个小时,聊历史,聊地理,聊文学,聊哲学,聊各地的风土人情,比青壮年人的记忆都要清晰.

4 对学生倍加爱护,非常有人情味

陈老师的"护犊子"是非常出名的,对此,我体会很深.在陈老师身边四年,他不仅在学术上对我严格要求,让我健康成长,在日常工作和生活上也非常关心和爱护我.即使快过去四十年了,我仍然清晰地记得陈老师爱护我的点点滴滴.

1984 年 9 月至 10 月我跟随陈老师去新疆马兰核试验基地参加地下核爆破时,我们乘坐苏联制造的伊尔飞机从北京去乌鲁木齐.伊尔飞机给我的印象是质量不怎么高,噪声比较大,也很陈旧,令第一次坐飞机的我有些不安.一路上,陈老师给我很多安慰和照顾,让我消除了恐惧和不安.

"文革"期间,陈老师随着中国科学院地球物理研究所的九室迁到云南省昆明市工作了几年.所以,陈老师对云南很了解,与云南省地震局的领导非常熟悉. 1985 年我与陈老师一起去云南时,为了帮助我在云南顺利开展强地面运动观测工作,陈老师带着我拜见云南省地震局的各位领导,请求云南省地震局尽可能给予我支持.由于有了这样的安排,从 1985 年到 1986 年,我在云南将近一年的强地面观测工作非常顺利,取得了很好的观测结果.

1986 年 7 月我做硕士论文答辩.答辩前,陈老师与我一起非常仔细地准备我的答辩材料,帮助我纠正存在的各种问题．答辩时，陈老师亲自为我的答辩掌控投影仪，帮助我一张张地翻篇．

1987年8月我在地球物理研究所举办婚礼．当天陈老师在中国地震局整天开会，会后马上赶到所里主持我的婚礼，并对大家宣布："小柯不仅是我的学生，也是我的朋友．"这让我到了今天仍然非常感动．

我到云南出差时，经常需要雇用当地人员，帮助把数字地震记录仪搬到观测地点，我们只能用"白条子"付给这些人员报酬．但上世纪80年代的财务制度非常严格，这些"白条子"很难报销．为了帮助我解决困难，陈老师亲自带着我到财务科去说明情况，最后把这些问题顺利解决．

在我读研究生的时代，生活条件还是比较差的．陈老师经常把我们这些研究生叫到他家里，亲自下厨，给我们烹调各种美食，让我们这些青年小伙子们好好解馋．每次与陈老师出差，陈老师也经常自掏腰包，请我们品尝当地美食．

1984年9月我与陈老师到了乌鲁木齐，陈老师的大学同学来看望他，带来了哈密瓜和新疆西瓜．陈老师让我们与他一起吃瓜，这是我第一次吃到哈密瓜和新疆西瓜．第二天，陈老师亲自带领我们到维族人居住区去吃真正的新疆烤羊肉串、烤"皮牙子（洋葱）"肉包子等，这也是我第一次吃烤羊肉串和烤肉包子．

我们去云南时，陈老师在昆明请我们几个人到餐馆吃饭，点了汽锅鸡和过桥米线，这也是我第一次品尝这些云南美食．在云南各地考察期间，陈老师兴高采烈地给我们讲"云南十八怪"等风土人情，讲他在"文革"期间，如何参加在云南各地的地震考察工作，尤其是在昭通等原始森林地区，如何在迷路的情况下，艰难地走出来．

从以上四个方面可以看到，陈老师对我的一生影响太大了．我衷心感谢陈老师对我的教诲和培养！作为学生，我衷心祝愿陈老师健康快乐长寿！

我的老师陈运泰

周云好

我第一次见到陈运泰老师,是 1987 年 5 月,在成都.

我当时在安徽省地震局偏远的佛子岭地震台工作,这是我大学毕业 5 年后第一次出远门,到成都参加纪念松潘、平武大地震 10 周年地震预报学术讨论会. 这次地震学界的盛会令人难忘,国内专家、学者 200 多人云集,其中有丁国瑜、陈运泰、梅世蓉、许绍燮、马宗晋、马瑾、冯德益、顾浩鼎、林邦慧等.

第一天上午的大会上,作为中国地震学会理事长,陈老师作了"关于震源力学研究中的问题"的学术报告,着重讲了地震断层滑动弱化、地震断层蠕动与前兆发生的时间、地震模拟、地震破裂的动力学等四个方面的问题,反响十分强烈.

站在讲台上的他,不满 47 周岁,身材高大,目光炯烁,神采飞扬. 他一会儿面向听众,一会儿转向展示报告内容的投影大屏幕. 手里拿的指示棒时不时在大屏幕上指点. 他说话逻辑缜密,滔滔不绝. 他头发梳理得一丝不乱,白衬衣袖口扣得严严实实. 十足的绅士、学者风度.

仰慕陈老师大名已久,今天得见,我非常高兴. 我心里想,何不趁这次会议的机会和他说说话呢?我的大学同班同学和好朋友陈晓非正在跟他读硕士,他若知道我和晓非的这层关系,会乐意和我说话的. 于是,当天中午,我打听到陈老师的房间号,来到他的门前,敲开门,毕恭毕敬地向他做了自我介绍. 陈老师很高兴,说晓非在他那里很好. 他热情地让我坐下,亲切地询问我近况如何?在看什么书?做什么地震工作?等. 我一一作答. 我把将在专题会上交流的《断层泥在孕震过程中的作用》的研究工作文稿交给了陈老师,请他指正. 他一边翻看我的文稿,一边鼓励我. 他说:"您对地震震源问题感兴趣,这很好. 在地震台工作,是寂寞了点,但没有乱哄哄的人和事来打扰您,能够静下心来读书,也是好事. 趁年轻,要抓紧时间读书. 把基础打好了,将来做大事就容易了. 要看得远,来日方长." 他还说:"近年来,Aki 和 Richards 写了一本很重要的书,叫《定量地震学》,在地震学界很流行. 地震出版社刚刚出了中译本. 您到你们地震局资料室把它借来,好好读. 学习中遇到什么问题,可以给我写信,我们共同讨论."

这次见面,是我人生中的一大幸事. 陈老师的话,给我指明了人生方向,我仿佛隐约看到了远处的好境况. 他的话我牢牢记在了心里.

身为地球物理研究所所长,所里的公务太多,学术讨论会结束的第二天,陈老师就回北京了(他原本计划参加大会后期的野外地震考察). 我则跟随部分专家、学者考察了龙门山、雪山、岷江等断裂,参加了"松潘、平武地震纪念碑"落成揭碑典礼,参观了松潘、平武等地震台,到过黄龙、九寨沟、江油、绵阳等地. 一路上的见闻和感受,我终身难忘. 我第一次看见那么高的山,第一次乘车在那么险的山路上行驶.

以后的 4 年里,我按照陈老师的话去做,在佛子岭地震台默默地啃那两卷本的《定量地震学》. 这期间,陈老师没有忘记我. 大约是成都会议后的第二年,他托人给我带话,问我想不想考他的研究生?因为多种原因,我走不掉,只好把跟陈老师上学的念头推到脑后去了. 需要我做的事情一件接着一件. 1988 年底,我被选派到国家科委上海培训中心接受半年英语培训. 1992 年 5 月,被选派到新西兰地质与核科学研究所做访问学者. 半年后回国,被调到安徽省地震局分析预报中心做地震编目和地震预报工作. 一晃又四年过去了. 期间的工作经历,加上在新西兰的工作经历,使我深深感到有必要离开安徽省地震局,到研究所专心跟大地震学家学习几年. 我最终下定了决心,走,跟陈老师读博士去!

我把这个想法写信告诉了陈老师．陈老师很高兴．他从百忙中抽时间，多次从北京往合肥我家里打电话，指导我复习，准备参加博士生专业知识考试．我没有硕士研究生阶段的学习经历，1997年1月，我带着两篇能代表自己研究工作能力的已发表论文，带着中国科技大学徐果明教授和施行觉教授给我写的推荐信，以同等学力报考陈老师的博士生．3月，我顺利通过了中科院出题目的英语和政治考试及国家地震局（后改称中国地震局）地球物理研究所出题目的地震学专业知识考试．第一次见到陈老师10年后，我终于成了陈老师的脱产博士生．

到北京去读博这年，我已过40周岁了，是个年龄很大的学生了．在那个在职、带薪读博盛行的年代，很多人不理解我这个辞了工作、等毕业后再重新分配工作的大龄学生．我凭录取通知书在合肥买了一张去北京的火车学生半票，到了北京站，一个检票的工作人员把我拦下了，说我这么大了怎么可能是个学生．我向他出示博士生录取通知书争辩，他硬说我是带薪上学的在职学生，不能享受国家规定的学生半票优待．我又出示了户口迁移证等，说我不是在职的学生．也不行．这真是有理说不清．无奈之下，我补了张全票．回到地球物理所，我越想这事越生气：国家规定学生半票没有年龄限制，北京站凭什么硬要我补全票？！我找所里给我开证明，证明我是不在职的学生，我要到北京站讨回公道！这时候，陈老师知道了这件事，他很关心我，劝我不要去北京站耽误时间了，说不定又要生一肚子气．他把我的火车票要去，从他的科研经费里给我报销了．从那以后，我再也没买过学生半票．

我和地球物理所的十几个学生，先住在所办公楼大院内临时搭建的铁皮房．后来搬进了院内新建的平房．4个学生住一间房．地球物理所食堂就在办公楼大院内．我们看书，做研究工作，就到办公楼去．吃，住，学习，都在一个院子里，很方便．

开始的两年时间里，只要陈老师来地球物理所上班，他几乎每天打电话把我叫到他办公室里，不厌其烦地给我讲地震波理论，讲波谱分析，讲快速傅里叶变换理论和方法，讲地震震源的矩张量理论，讲地震矩张量反演理论和方法，讲震源破裂过程反演理论和方法，等．他每次都给我开列大量的中、英文文献目录，要我到地球物理所图书馆里去找文献，复印出来，仔细研读．他每次给我讲之前，都要求我先向他汇报我对他上一次讲的东西的理解和领悟情况．他前一天在纸上推算给我看，后一天则叫我在纸上推算给他看．陈老师的数学和物理功底我深深佩服，他在纸上推演得飞快又精准．他对我要求非常严格，他每次叫我读的多篇文献，我若有一篇没读，他就批评我：为什么不努力？我若推算错了一步，他就训问我：怎么走神了？每次进他的办公室，我都紧张得手脚冰凉．每次从他办公室出来，我都长舒一口气．到了第三年，我基本上被陈老师严格训练出来了，基本掌握了他教给我的一系列地震学理论和方法，能用计算机程序处理实际地震波资料，得出研究结果了．这时，只要他来地球物理所上班，陈老师几乎每天打电话叫我去他办公室，向他汇报最新研究结果，并和我讨论这个结果，给我提出进一步做工作的意见和建议．

2000年6月底，陈老师去美国洛杉矶加利福尼亚大学（UCLA）做了半年研究工作．临行前，陈老师把我叫到他的办公室，把办公室钥匙交给我，要我记住他的计算机密码，告诉我他计算机里有哪些重要文件，必要时，他会用email告诉我，要我用他的计算机帮他处理一些事情．

这半年里，陈老师经常给我发email，询问我的生活情况，学习情况，研究工作进展情况，等．我和他经常通过email讨论地震矩反演和震源破裂过程反演研究中遇到的一些新问题．每次讨论，他总能给我提出解决问题的好建议．

一转眼半年过去了，陈老师回来了．两个月后，我在陈老师指导下做的关于2000年6月4日印度尼西亚苏门答腊南部$M_S8.0$地震破裂过程反演的研究工作，有了较为满意的结果，可以写成文章了．陈老师给我讲了文章的大体结构，我就按照他说的去写．稿子准备投给美国的BSSA．英文稿写成后，我就兴冲冲地送给陈老师．他把稿子带回家，开夜车修改．第二天，他把修改稿还给我．我一看，立马感到很羞愧．每页纸上都被改得红红的一片又一片，删除了很多单词和句子，又添加了很多．陈老师修改得特别仔细,每个单词,每个句子,甚至连每个标点符号,都反复推敲,决不轻易放过.我对照陈老师改过的纸质稿,在计算机上一处一处地修改成清样.每改一处,我都细细琢磨,认真学习.陈老师的英文确实好!打印出清样,我又送给陈老师.几天过后,他又把修改稿还给我.纸质稿上又是大片大片的红字.经过再一次推敲,陈老师觉得前一次修改过的文字仍有修改提高的余地.就这样,稿子被他修改了5~6遍.然后,他才叫我把稿子投出去.陈老师治学就是这样严谨.不久,收到回信,这篇文章被BSSA采用了.

我跟陈老师读博读了4年零8个月,2002年5月顺利通过了题为《用远场体波反演震源破裂过程研究》的博士论文答辩.俗话说:"严师出高徒."陈老师是位非常严谨的老师,也是对学生要求非常严格的老师.但我离高徒却差得很远.令我欣慰的是,我能感觉到,总体来说,陈老师对我这个学生还是相当满意的.

博士毕业后,在陈老师的大力推荐下,我来到了南京,在江苏省地震局工作.当时,陈老师对我说:"南京比合肥好.您去吧!"

南京离北京上千千米,我不在领导岗位上,出差的机会不多,毕业近20年来,我去北京拜访陈老师的次数很少.但陈老师从没有忘记我.他紧跟时代潮流,善于接受新事物,是较早使用手机微信的人.多年前,有一天,我突然看到陈老师加我微信.我很惊喜,赶紧"接受".从那以后,陈老师经常给我发微信.

在微信里,陈老师给我讲了很多我以前不知道的事情.给我长了很多知识.

他对我说:在他成长的过程中,受到过傅承义、顾功叙、曾融生、谢毓寿等老先生的指导、帮助和提携.

他对我说:"我在1956年考大学时,距西南联大'北归'仅十年.不知为什么,我那个中学的小小图书馆居然有许多有关西南联大的书!我的志愿就是北大—清华—南开!但清华已无理科,所以我的第一志愿便是物理,学校是北大—南开.后来学校动员理科生报师范,我也积极响应号召,加填为北大—南开—北师大.我的物理老师显然有意偏袒我,听其自然,不要求我把北师大列为第一学校,否则,我的个人历史轨迹又不知会是怎样的了."

陈老师读的书太多了!知道的事情太多了!不仅是一位地球物理学家,还堪称是一位人文历史学家.他记忆力极强.有一次,我在他家里,谈话间,他脱口而出,背诵了《史记》中《项羽本纪》和《陈涉世家》的大段名句.令我羡慕至极!

我祝陈老师快快乐乐每一天!健康长寿超百年!

在中国地震局地球物理研究所2002届博士、硕士研究生毕业典礼上与陈运泰老师合影(2003年1月16日)

地震科学研究的引路人——陈运泰老师

王庆良

中国地震局第二监测中心，西安，710054

我是1997—2003年师从陈运泰老师攻读博士学位研究生的，实际上最早向陈老师学习地震科学研究可追溯到1990年的青海共和7.0级地震.30多年来，陈老师一直是我从事地震科学研究的重要引路人，他那高尚的品格、渊博的学识、宽广的视野、严谨的学风，一直是我做人、做事、为人师表和开展科学研究的楷模.今值《陈运泰院士论文选》出版之机会，特写此文回忆几段难忘的经历，感谢陈老师多年来对我科研和学习工作的热心帮助和指导.

一、走进地震科学的启蒙研究——1990年共和7.0级地震

1990年4月26日在青海省共和县塘格木农场一带发生了一次$M_S7.0$地震.该地震造成119人死亡，2049人受伤，房屋倒塌2.1万间，损坏6.68万间，经济损失达2.7亿元.共和7.0级地震发生之后，中国地震局第二监测中心（以下简称二测中心）迅速对穿过震区的两条国家一、二水准路线（倒花线、石塘线）进行了一期精密水准复测，并结合震前1978年水准观测资料计算获得了共和7.0级地震的同震垂直形变场图像（图1，巩守文等，1993）.

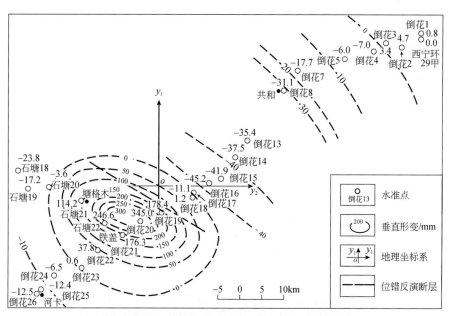

图1 1990年青海共和$M_S7.0$地震同震垂直形变图

由于共和7.0级地震发生在共和盆地内部的盲断层上，地表无明显的断层破裂形迹，因此，综合利用地震、地壳形变和余震分布等资料，研究揭示该地震的发震断层展布及其运动性质，就成为共和7.0级地震科学研究的一项重要内容，对共和盆地及龙羊峡水电站的震害防御工作也有重要的实际应用价值.

在得知共和7.0级地震同震垂直形变资料获得之后，时任中国地震局地球物理研究所所长的陈运泰老师，积极推动地球所、二测中心就共和7.0级地震测震、形变资料开展合作研究，我有幸作为二测中心成员之一,参与了该地震的合作研究工作.虽然我本人1980年9月—1984年8月本科就读于长春地质学院物探系地震勘探专业,1984年8月—1987年8月任教西安地质学院物探系,1987年9月—1990年7月攻读西安地质学院工程地质专业硕士研究生,具有一定的地质、地球物理专业基础知识和研究能力,但与陈运泰老师团队的共和7.0级地震合作研究,无疑在我个人的地震科学研究生涯中具有十分重要的启蒙引路意义,它使我第一次了解掌握了陈运泰老师1975年推导建立的弹性位错理论模型及其反演应用(陈运泰等,1975;陈运泰等,1979),第一次接触到了宽频带地震学中的矩张量解、震源时间函数等基本概念,合作研究中深为陈运泰老师渊博的学识、谦虚的人品所折服,并渐渐萌生了攻读陈老师博士研究生的最初愿望.

二、终身受益的学术报告——如何写好论文和做好学术报告

与陈运泰老师团队的共和7.0级地震合作研究没过几年,1996年5月中国地震学会青年科技工作委员会暨全国首届青年地震学术研讨会在西安召开,由二测中心承办,张培震、何昌荣、金星、朱元清、聂永安、田勤俭、赵凤新等多位青年科技工作委员会委员齐聚西安.为了提高青年地震科技工作者的学术论文写作水平和学术报告水平,时任中国地震学会理事长的陈运泰老师做了"如何写好论文 做好学术报告"的大会学术报告,在与会青年地震科技工作者中引起热烈反响.陈老师用深入浅出的语言和丰富的案例,精心指导大家如何写好论文摘要、论文引言、论文主体和论文结论,如何合理组织安排学术报告每一张投影胶片(现在都用PPt了)的主要内容、页面布局和文字大小,其报告精彩实用之处甚至具体到了如何正确使用汉语标点符号、英文标点符号等细节内容,无不体现了陈运泰老师高尚的育人品格、严谨的学术素养和扎实的主编功底.

时至今日,我一直感慨陈运泰老师当年所做的"如何写好论文 做好学术报告"之报告,是迄今为止对我个人科研工作影响和贡献最大的一次报告,真可谓一日当面聆听,终身受益匪浅!只可惜当时没有把报告的投影胶片复印下来.另外,从单位层面而言,1996年5月在西安举办的中国地震学会青年科技工作委员会以及陈运泰老师的精彩报告,对二测中心的青年科技人才成长起到了非常重要的推进作用,它首次打开了二测中心青年科技工作者的科学眼界,我本人就是此次会议之后勇敢申请并顺利获得了个人的第一项国家自然科学基金资助课题,这在之前是连想也不敢想的事情.

图2 陈运泰院士在青年地震科技工作委员会上(1996年5月,西安)
由左至右:吴荣辉研究员,陈运泰院士,陕西省地震局局长李博研究员

三、宽宏大量的科学家胸怀——自主选择博士研究方向

1996年西安青年科技工作委员会会议之后,我怀着十分忐忑的心情向陈运泰老师口头表达了想报考

其博士研究生的愿望，没想到陈老师当场就爽快地答应了，令我十分激动．由于我英语基础比较好，加之在准备专业基础课时也下了一番功夫，因此比较顺利地考取了陈运泰老师 1997 年的固体地球物理学博士研究生（在职），同年考取的还有同窗好友周云好同学、高伟同学．

由于我属于在职博士研究生，除第一年在京集中上课时间之外，有相当部分时间需要在原单位继续工作．考虑到我工作单位在西安的实际情况，加之当时我还是单位 GPS 科室的负责人，因此，在选择博士论文研究方向时，陈老师希望我能够围绕团队的震源物理大研究方向，结合几个典型震例（唐山地震、共和地震等）在震后形变方面开展一些比较深入的研究，这在当时也是一个比较热的研究方向．到了 1998 年下半年，在进行地壳形变观测资料对比分析研究时，我偶然发现了一些地壳形变观测资料与地球自转日长变化之间的十年尺度相关现象（图 3）．考虑到该十年尺度相关现象可能在地球自转动力学研究、地震前兆异常变化判别研究等方面有重要的理论和实际应用价值，经与陈老师讨论沟通，最终将博士论文研究方向确定为地壳形变—地球物理场—日长变化十年尺度相关性研究．

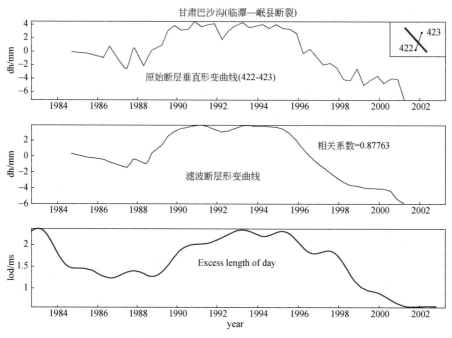

图 3 断层形变—地球自转日长十年尺度相关曲线（甘肃巴沙沟场地）

在陈老师的精心指导下，经过 6 年多的读博持久战（1997—2003 年），我的博士毕业研究论文进一步发现，地球自转日长变化的十年尺度相关性现象不但普遍存在于断层形变、地倾斜等形变观测资料中，在地磁、重力、地温等地球物理场变化资料中也较普遍存在，其物理力学机制主要来源于地球深部的核幔耦合作用和地壳的横向不均匀性效应（王庆良，2003）．

2003 年博士毕业至今，我一直十分感恩陈运泰老师对我自主选择博士研究方向、开展自由探索研究所秉持的科学宽容性，也正是这种科学宽容性，一直鼓励我持续在大地测量与地球动力学、大地测量与地震物理过程等方面开展更加深入的研究．

以上仅回忆和列举了陈运泰老师对我开展地震科学研究有重要影响的三个片段，实际上，从陈老师身上我需要学习的东西还很多，他优秀的科学组织管理能力、对地震科普宣传的积极实践热情、对地震预报摇旗呐喊的历史使命感等，时刻都在激励和鞭策我为国家防震减灾工作做更多、更扎实的贡献．

随笔——为陈老师八十岁生日作

许力生

中国地震局地球物理研究所

……"我买几个橘子去. 你就在此地, 不要走动."

……这时我看见他的背影, 我的眼泪很快地流下来了.

……他已抱了朱红的橘子往回走了.

……走到车上, 将橘子一股脑儿放在我的皮大衣上.

……等他的背影混入来来往往的人里. 我的眼泪又来了.

……

这是一个朴素男儿眼中父亲的背影.

一位年迈的父亲的背影!

我的老师总是很晚才离开办公室. 离开前, 他习惯地要把浸泡了一天的茶叶倒进开水房的废茶篓里, 把用了一天的茶杯清洗得干干净净, 整齐地摆回原来放茶杯的地方, 然后才提着沉甸甸的书包离开.

我和我老师的办公室在同一楼层, 我在西头, 他在东头, 中间有个开水房. 因此, 我经常会看到我的老师清理完茶杯回办公室时的背影. 不知从哪天开始, 我突然感觉我老师不再那样挺拔, 步履也慢了许多. 于是情不自禁地想起中学时代读过的朱自清的散文《背影》. 虽然没有流下眼泪, 但心中免不了有些惆怅. 唉! 天若有情天亦老, 谁背岁月不弯腰?!

我在我老师身边学习工作已近30年, 他一直以来对我的关爱令我终生难忘. 读书期间, 手把手教诲, 不厌其烦, 至今历历在目. 毕业后, 送我去美国、去德国、去法国, 多次创造机会养我心智、助我学有所长. 在外期间, 经常挤出时间给我写邮件, 谈专业, 谈生活, 让我感到非常亲切和温暖. 我深知, 没有过去这一次次的深造, 就不会有今日的我. 我父母的关爱造就了我体魄的健康和性格的坚毅, 我老师的关爱却为我营造了良好的职业生涯.

回想起陈老师对我无微不至的关爱, 便不由自主地想起我那些师兄弟. 他们说起陈老师, 也是津津乐道. 陈晓非院士不止一次说起陈老师和他在一起的点点滴滴以及送他出国深造的热忱, 李世愚研究员谈起陈老师给他做饭、煮面条的旧事时也是笑口常开. 还有早期送出国的张家骏博士、王璋博士, 以及后来送出国的何雄博士、赵明博士、李旭博士、周荣茂博士, 等等, 都是陈老师培养并送出国的学生. 即使在今天, 无论是周云好博士、王庆良博士, 或是张勇博士, 我们几个在国内的师兄弟, 只要和陈老师坐在一起, 或聚餐、或聊天, 总是谈笑风生, 其乐融融, 宛若一个温馨的家. 他对待学生就像自家的亲人, 却分明又是朋友, 亲切、温暖.

我的老师不但在学业和工作上影响着我, 他那与生俱来的品格和素养也无声无息滋润着我. 他始终把自己当作普普通通的社会一员, 极其自然! 孤寂的地震台, 热闹的大都市; 黄沙弥漫的西北戈壁, 富丽堂皇的国际酒会; 地震废墟中的救灾帐篷, 湖光山色中的国际讲坛, 哪里需要就出现在哪里, 从不挑剔. 陈老师出生在炮火纷飞的战争年代, 目睹或亲历过自抗日战争后期、解放战争、"反右" …… "大跃进"、"文化大革命"至改革开放的几十年, 轰轰烈烈, 可谓阅历丰富, 但违背天理、忤逆人伦的事却始终与他无缘. 他没有革命小将的"激情", 也没有乘人之危凌驾同僚的盛气, 更没有愤怒声讨老师的威风. 相反, 救人于危难之际、雪中送炭似乎是他天生的本能.

我和陈老师的第一次相遇是 1990 年研究生的课堂上．玉泉路中科院研究生院的教室里，讲台上，一位挺拔帅气的年轻人，温文尔雅．其实，那时的陈老师已经 50 岁，不过，凡见过他的人没有不承认他的年轻．我清楚地记得，我的同学曾揶揄我"你和你的老师在一起不知道谁是老师"！

如今，我已是满头灰发，一边感慨岁月的无情，一边常常回想起往日的片段．看到陈老师那不再挺拔的背影，与陈老师在一起的场景一幕接着一幕．现在已经是二十一世纪了，假如和朱自清在同一个年代，我的老师一定会买一些朱红色的橘子放在我那破旧的皮袄上．

是的！我的老师，仿佛是家人，又好像是朋友，情真意切！

2006 年 11 月 7 日至 10 日在曼谷举行的"亚洲地震委员会（ASC）第 6 届大会暨预防和减轻地震与海啸灾害学术研讨会"后考察印度洋特大海啸重灾区之一泰国普吉岛，与会中国专家合影

参加亚洲与大洋洲地球科学学会（AOGS）学术大会期间与陈运泰老师（左）合影（2005 年 6 月 23 日，新加坡）

陈运泰院士培养硕士研究生名录

序号	招生类别	姓名	性别	出生年月	入学时间	本科毕业学校	备注
1	硕士	张家骏	男	1946.06	1978.09	北京大学	协助傅承义院士指导
2	硕士	陈晓非	男	1958.02	1982.02	中国科学技术大学	
3	硕士	谢家康	男	1955.09	1982.02	北京大学	
4	硕士	王 璋	男	1956.09	1982.09	中国科学技术大学	协助谢毓寿教授指导
5	硕士	柯兆明	男	1961.05	1983.09	中国科学技术大学	
6	硕士	方 扬	男	1962	1983.09	中国科学技术大学	推荐赴德攻博
7	硕士	倪江川	男	1960.10	1984.09	江苏师范大学	
8	硕士	周家玉	男	1962.11	1984.09	北京大学	
9	硕士	李肇仁	男	1963.04	1985.09	北京大学	推荐免试生
10	硕士	倪晓希	男	1964.05	1986.09	中国科学技术大学	推荐免试生
11	硕士	翟 江	男	1965.05	1987.09	中国科学技术大学	推荐免试生
12	硕士	何 雄	男	1966.04	1989.09	中国科学技术大学	推荐免试生
13	硕士	邹 勇	男	1966.04	1988.09	北京大学	推荐免试生
14	硕士	赵 明	男	1967.12	1990.09	北京大学	推荐免试生
15	硕士	李 旭	男	1968.07	1990.09	北京大学	推荐免试生
16	硕士	许建平	男	1971.09	1994.09	北京大学	推荐免试生
17	硕士	周荣茂	男	1968.10	1995.09	北京大学	
18	硕士	于湘伟	女	1972.02	1997.09	北京大学	推荐免试生
19	硕士	白石羽	男	1978.01	2000.09	中国科学技术大学	

陈运泰院士培养博士研究生名录

序号	招生类别	姓名	性别	出生年月	入学时间	本科/硕士毕业学校	备注
1	博士	李世愚	男	1945.12	1985.12	中国科学技术大学	
2	博士	高孟潭	男	1957.02	1988.09	北京大学	与胡聿贤院士、李坪院士合作指导
3	博士	许力生	男	1965.03	1991.09	云南大学/兰州地震研究所	
4	博士	刘瑞丰	男	1962.05	1992.10	国家地震局地球物理研究所	在职，与陈培善教授合作指导
5	博士	米尔扎伊（Mirzaei, N.）	男	1962	1993.09	伊朗德黑兰大学/Tarbiat Moalem 大学	与高孟潭教授合作指导
6	博士	莫扎法里（Mozaffari, P.）	男	1953	1993.09	伊朗国立大学	
7	博士	许向彤	男	1968.04	1995.09	天津大学	在职
8	博士	涂毅敏	男	1957.11	1995.09	兰州地震研究所	
9	博士	张洪魁	男	1964.03	1996.10	国家地震局地球物理研究所	在职
10	博士	王庆良	男	1963.03	1997.09	西安工程学院	在职
11	博士	周云好	男	1957.01	1997.09	中国科学技术大学	在职
12	博士	高 伟	男	1960.02	1997.09	中国地震局地球物理研究所	在职
13	博士	成小平	男	1956.01	1990.09	重庆大学	与胡聿贤院士、李坪院士合作指导
14	博士	于湘伟	女	1972.02	2000.09	北京大学/中国地震局地球物理研究所	提前攻博
15	博士	唐传海	男	1974.08	1999.09	中国地震局地球物理研究所	
16	博士	张 勇	男	1981.08	2003.09	北京大学	
17	博士	吴 晶	女	1977.01	2004.09	中国地震局分析预报中心	在职，与高原教授合作指导
18	博士	朱新运	男	1968.06	2005.09	兰州地震研究所	在职
19	博士	郑建常	男	1978.01	2007.09	山东大学/中国地震局分析预报中心	在职
20	博士	刘 超	男	1983.10	2008.09	北京师范大学	硕博连续
21	博士	刘文兵	男	1977.07	2008.09	云南大学/天津地震局	在职
22	博士	周连庆	男	1982.09	2009.09	安庆师范大学/中国地震局地震预测研究所	在职，与陈章立教授、宋晓东教授合作指导
23	博士	杨志高	男	1981.07	2011.09	长安大学/中国地震局地震预测研究所	在职
24	博士	崔子健	男	1983.09	2011.09	济南大学/中国地震局地震预测研究所	在职，与陈章立教授合作指导
25	博士	宋 金	女	1986.03	2012.09	中国石油大学（华东）/中国地震局台网中心	在职
26	博士	王安简	女	1993.06	2013.09	中国科学技术大学（少年班）	硕博连续

陈运泰论著目录（1971–2020）

[1] 陈运泰，林邦慧，顾浩鼎，1971. 强震发生的规律性探讨. 地震战线，(8)：18-24.

[2] 陈运泰，1974. 多层弹性半空间中的地震波㈠. 地球物理学报，**17**(1)：20-43.

[3] 陈运泰，1974. 多层弹性半空间中的地震波㈡. 地球物理学报，**17**(3)：173-185.

[4] 陈颙，陈运泰，1974. 晋中南地区地壳结构的初步研究. 地球物理学报，**17**(3)：186-199.

[5] 陈运泰，林邦慧，林中洋，李志勇，1975. 根据地面形变的观测研究1966年邢台地震的震源过程. 地球物理学报，**18**(3)：164-182. 被外国刊物译载：Chen, Y. T., Lin, B. H., Lin, Z. Y. and Li, Z. Y., 1975. The focal mechanism of the 1966 Xingtai earthquake as inferred from the ground deformation observations. In：Teng, T. L. and Lee, W. H. K. (eds.), *Chinese Geophysics*, **1**(2)：263-288, AGU, Washington D C, USA.

[6] 王妙月，杨懋源，胡毓良，李自强，陈运泰，金严，冯锐，1976. 新丰江水库地震的震源机制及其成因初步探讨. 地球物理学报，**19**(1)：1-17. 又载：中国科学，**1975**(1)：85-97.

[7] 陈运泰，林邦慧，李兴才，王妙月，夏大德，王兴辉，刘万琴，李志勇，1976. 巧家、石棉的小震震源参数的测定及其地震危险性的估计. 地球物理学报，**19**(3)：206-233.

[8] 邱群（陈运泰等），1976. 1976年7月28日河北省唐山7.8级地震的发震背景及其活动性. 地球物理学报，**19**(4)：259-269. 被外国刊物译载：Qiu, Q. (Chen, Y. T. *et al.*), 1976. On the background and seismic activity of the $M = 7.8$ Tangshan earthquake, Hebei Province of July 28, 1976. In：Wu, F. T. (ed.), *Chinese Geophysics*, **1**(1), Washington D C：AGU, 67-78.

[9] 顾浩鼎，陈运泰，高祥林，赵毅，1976. 1975年2月4日辽宁省海城地震的震源机制. 地球物理学报，**19**(4)：270-285.

[10] 陈运泰，王妙月，林邦慧，刘万琴，1976. 中、小地震体波的频谱和纵、横波拐角频率比. 科学通报，(9)：414-418.

[11] 刘万琴，陈运泰，1979. 由瑞雷波方向性函数研究1974年5月11日云南省昭通地震的震源过程. 地震学报，**1**(1)：25-38.

[12] 林邦慧，陈运泰，魏富胜，李志勇，1979. 不对称双侧破裂过程的研究及其在海城地震的应用. 地震学报，**1**(2)：133-149.

[13] 陈运泰，黄立人，林邦慧，刘妙龙，王新华，1979. 用大地测量资料反演的1976年唐山地震的位错模式. 地球物理学报，**22**(3)：201-217. 被外国刊物译载：Chen, Y. T., Huang, L. R., Lin, B. H., Liu, M. L. and Wang, X. H., 1979. A dislocation model of the Tangshan earthquake of 1976 from the inversion of geodetic data. In：Wu, F. T. (ed.), *Chinese Geophysics*, **2**(1), Washington D C：AGU, 11-30.

[14] 傅承义，陈运泰，陈颙，1979. 我国的震源物理研究. 地球物理学报，**22**(4)：315-320.

[15] Chen Yun-tai, Gu Hao-ding and Lu Zao-xun, 1979. Variations of gravity before and after the Haicheng earthquake, 1975, and the Tangshan earthquake, 1976. *Phys. Earth Planet. Interior*, **18**(4)：330-338. 中文刊载：陈运泰，顾浩鼎，卢造勋，1980. 1975年海城地震与1976年唐山地震前后的重力变化. 地震学报，**2**(1)：21-31.

[16] 张之立，陈运泰，谷继成，李钦祖，靳雅敏，1981. 唐山地震震源机制. 见：国家地震局科研处（主编），唐山地震考察与研究，北京：地震出版社，81-93.

[17] 靳雅敏，陈运泰，于新昌，1982. 唐山地震余震的震源参数及地壳介质的品质因数. 地震学报，**4**(1)：62-67.

[18] 李兴才，陈运泰，1982. 唐山地震引起的剩余倾斜场的空间分布和倾斜阶跃. 地球物理学报，**25**(3)：219-226.

[19] 傅承义，陈运泰，祁贵仲，1985. 地球物理学基础. 北京：科学出版社，1-447.

[20] 陈运泰，许忠淮，1985. 地球自由振荡. 见：傅承义，秦馨菱（主编），中国大百科全书·固体地球物理学卷，北京：中国大百科全书出版社，115-116.

[21] Chen, Y. T. and Knopoff, L., 1986. Static shear crack with a zone of slip-weakening. *Geophys. J. R. astr. Soc.*, **87**(3): 1005-1024. 中文译载: 陈运泰, Knopoff, L., 1987. 具有滑动弱化区的静态剪切裂纹. 世界地震译丛, (5): 33-45.

[22] Chen, Y. T. and Knopoff, L., 1986. The quasistatic extension of a shear crack in a viscoelastic medium. *Geophys. J. R. astr. Soc.*, **87**(3): 1025-1039. 中文译载: 陈运泰, Knopoff, L., 1987. 剪切裂纹在粘弹性介质中的准静态扩展. 世界地震译丛, (6): 30-37.

[23] Chen, Y. T. and Knopoff, L., 1987. Simulation of earthquake sequences. *Geophys. J. R. astr. Soc.*, **91**(3): 693-709. 中文译载: 陈运泰, Knopoff, L., 1988. 地震序列的模拟. 世界地震译丛, (1): 9-19.

[24] Chen, Y. T., Chen, X. F. and Knopoff, L., 1987. Spontaneous growth and autonomous contraction of a two-dimensional earthquake fault. In: Wesson, R. L. (ed.), *Mechanics of Earthquake Faulting. Tectonophysics*, **144**(1/3): 5-17. 中文译载: 陈运泰, 陈晓非, Knopoff, L., 1988. 二维地震断层的自然扩展和自发收缩. 世界地震译丛, (3): 14-23.

[25] 吴大铭, 王培德, 陈运泰, 1987. 北京和云南剑川地区的 Q_P 和 Q_S. 地震学报, **9**(4): 337-346. 英文刊载: Wu, F. T., Wang, P. D. and Chen, Y. T., 1987. Q_P and Q_S in Beijing and Jianchuan, Yunnan areas. *Acta Seismologica Sinica* (English Edition), **1**(4): 1-12.

[26] Zheng, R. S., Chen, Y. T., Zhu, C. Z., Gao, L. S., Zheng, S. H. and Liu, F. T., 1987. Advances on seismology and physics of the Earth's interior in China. In: *National Report on Seismology and Physics of the Earth's Interior for the XIX-th General Assembly of IUGG*, Vancouver, Canada, 1987, 1-34. 中文刊载: 曾融生, 陈运泰, 朱传镇, 高龙生, 郑斯华, 刘福田, 1988. 地震学和地球内部物理学的进展. 地球物理学报, **31**(专辑): 1-16.

[27] Chen, Y. T., 1987. The experimental site for earthquake prediction in western Yunnan, China: A review. In: Office of the United Nations Disaster Relief Coordinator (ed.), *Earthquake Prediction and Mitigation of Earthquake Losses*, **2**: 11-33.

[28] 陈运泰, 张肇诚, 1987. 在苏联杜尚别召开的地震预报和减轻地震损失国际学术讨论会概况. 国际地震动态, (2): 8-11.

[29] 顾浩鼎, 陈运泰, 1987. 裂纹端部旋转与应力集中的相互作用. 东北地震研究, **3**(1): 1-10.

[30] 顾浩鼎, 陈运泰, 1987. 旋转矩和地震矩. 东北地震研究, **3**(3): 1-16.

[31] 顾浩鼎, 陈运泰, 1988. 旋转在地震学中的意义. 东北地震研究, **4**(2): 1-9.

[32] 陈运泰, 1988. 震源物理学研究概况和进展. 地球物理学报, **31**(专辑): 114-120.

[33] 王培德, 吴大铭, 陈运泰, 1988. 地震矩、震级、震源尺度及应力降之间相互关系的研究. 地壳形变与地震, **8**(1): 109-123.

[34] 陈运泰, 王璋, 1989. 具有滑动弱化区的二维地震断层的动态扩展. 见: 地球物理学报编辑委员会 (主编), 八十年代中国地球物理学进展——纪念傅承义教授八十寿辰, 北京: 学术书刊出版社, 17-32.

[35] 吴大铭, 王培德, 陈运泰, 1989. 用 SH 波和 P 波振幅比确定震源参数. 地震学报, **11**(3): 275-281. 英文刊载: Wu, F. T., Wang, P. D. and Chen, Y. T., 1989. Determination of focal mechanism using SH to P amplitude ratio. *Acta Seismologica Sinica* (English Edition), **3**(3): 325-334.

[36] 陈运泰, 1989. 台湾地球科学研究概况. 海峡科技交流研究, (10): 1-14.

[37] 陈运泰, 1989. 在意大利举行的"地震预测"讲习班概况. 国际地震动态, (11): 16-18.

[38] Chen, Y. T. and Wu, F. T., 1989. Lancang-Gengma earthquake. A preliminary report on the November 6, 1988, event and its aftershocks. *EOS*, **70**(49): 1527/1540.

[39] 陈运泰, 1989. 中美强地面运动研究讨论会. 国际地震动态, (4): 13-15.

[40] 陈运泰, 王鸣等, 1989. 近场强地面运动的观测和分析进展. 见:《中国地震年鉴》编辑委员会 (编), 中国地震年鉴 (1989), 北京: 地震出版社, 170-174.

[41] 李世愚, 陈运泰, 1990. 地震破裂动力学研究进展. 见:《中国地震年鉴》编辑委员会 (编), 中国地震年鉴 (1990), 北京: 地震出版社, 337-339.

[42] 陈运泰, 倪晓希, 1990. 地震空区的破裂动力学研究. 见: 地球物理研究所四十年编委会 (主编), 地球物理研究所四十年. 北京: 地震出版社, 60-67.

[43] 王鸣, 王培德, 周家玉, 翟江, 倪晓希, 倪江川, 陈运泰, 吴明熙, 吴大铭, 1990. 1988 年 11 月 6 日澜沧-耿马地震

的近场研究临时报告. 东北地震研究, **6**(1): 35-40.

[44] 华昌才, 陈运泰, 1990. 火山的重力场研究. 东北地震研究, **6**(1): 44-48.

[45] 王培德, 吴大铭, 陈运泰, 1990. 连续发生的地震破裂之空间与时间分布. 地震学报, **12**(1): 38-42. 英文刊载: Wang, P. D., Wu, F. T. and Chen, Y. T., 1990. Temporal and spatial distribution of two consecutive earthquake ruptures. *Acta Seismologica Sinica* (English Edition), **4**(1): 55-61.

[46] 吴明熙, 王鸣, 孙次昌, 柯兆明, 王培德, 陈运泰, 吴大铭, 1990. 1985年禄劝地震部分余震的精确定位. 地震学报, **12**(2): 121-129. 英文刊载: Wu, M. X., Wang, M., Sun, C. C., Ke, Z. M., Wang, P. D., Chen, Y. T. and Wu, F. T., 1990. Accurate hypocenter determination of aftershocks of the 1985 Luquan earthquake. *Acta Seismologica Sinica* (English Edition), **4**(2): 181-191.

[47] 倪江川, 陈运泰, 陈祥熊, 1991. 地震矩张量及其反演. 地震地磁观测与研究, **12**(5): 1-17.

[48] Chen, Y. T., Mu, Q. D. and Zhou, G. W., 1991. China Digital Seismograph Network: Current status and future directions. In: Boschi, E., Giardini, D. and Morelli, A. (eds.), *MedNet, The Broard-band Seismic Network for the Mediterranean*, Rome, Italy: Il Cigno Galileo Galilei, 114-120.

[49] 王培德, 陈运泰, 1991. 岩石层和震源的数字地震成像: 数字地震学发展的回顾与展望. 国际地震动态, (11): 1-8.

[50] Chen, Y. T., Zhou, J. Y. and Ni, J. C., 1991. Inversion of near-source broadband accelerograms for the earthquake source time function. *Tectonophysics*, **197**(1): 89-98.

[51] Chen, Y. T. and Niu, Z. R., 1991. Rock fracture and focal process. In: *China National Report (1987-1990) on Seismology and Physics of the Earth's Interior for the XXth General Assembly of IUGG*, Vienna, Austria, 1991, 27-41. Also In: Zeng, R. S. (ed.), *Annual Review of Geophysics* 1991, Beijing: Seismological Press, 24-35. 中文刊载: 陈运泰, 牛志仁, 1992. 岩石破裂和震源过程. 见: IUGG中国委员会(主编), 国家大地测量和地球物理联合会中国委员会国家报告, 北京: 气象出版社, 39-46.

[52] 倪江川, 陈运泰, 王鸣, 吴明熙, 周家玉, 王培德, 吴大铭, 1991. 云南禄劝地震部分余震的矩张量反演. 地震学报, **13**(4): 412-419. 英文刊载: Ni, J. C., Chen, Y. T., Wang, M., Wu, M. X., Zhou, J. Y., Wang, P. D. and Wu, F. T., 1992. Moment tensor inverison of some aftershocks of the April 18, 1985, Luquan earthquake of Yunnan Province, China. *Acta Seismologica Sinica* (English Edition), **5**(3): 459-467.

[53] 王培德, 王鸣, 周家玉, 瞿江, 倪晓希, 倪江川, 陈运泰, 吴大铭, 1991. 澜沧-耿马地震近场强地面运动特征. 见: 陈达生, 周锡元, 那向谦, 罗永康(主编), 云南澜沧耿马地震震害论文集, 北京: 科学出版社, 69-84.

[54] 王培德, 王鸣, 周家玉, 瞿江, 倪晓希, 倪江川, 陈运泰, 吴大铭, 1991. 澜沧-耿马地震强余震的反应谱. 地震学报, **13**(3): 338-343. 英文刊载: Wang, P. D., Wang, M., Zhou, J. Y., Qu, J., Ni, X. X., Ni, J. C., Chen, Y. T. and Wu, F. T., 1992. Response spectrum to strong aftershocks of the Lancang-Gengma earthquake. *Acta Seismologica Sinica* (English Edition), **5**(2): 337-342.

[55] Chen, Y. T., Chen, Z. L. and Wang, B. Q., 1992. Seismological studies of earthquake prediction in China: A review. In: Dragoni, M. and Boschi, E. (eds.), *Earthquake Prediction*, Roma, Italy: Il Cigno Galileo Galilei, 71-109.

[56] 姚陈, 王培德, 陆玉美, 陈运泰, 1992. 对大同地震横波分裂的研究. 华北地震科学, **14**(3): 12-26.

[57] 姚陈, 王培德, 陈运泰, 1992. 卢龙地区S波偏振与上地壳裂隙各向异性. 地球物理学报, **35**(3): 305-315.

[58] 赵明, 陈运泰, 巩守文, 王庆良, 1992. 用水准测量资料反演1990年青海共和地震的震源机制. 地壳形变与地震, **12**(4): 1-11.

[59] 陈运泰, 王培德, 吴忠良, 1992. 地震矩张量及其反演. 北京: 国家地震局地球物理研究所, 1-101.

[60] 陈运泰, 1992. 台湾的地震和地球物理研究. 见: 国家科委台湾事务办公室(主编), 台湾科技要览. 北京, 104-113.

[61] Yao, C., Wang, P. D., Lu, Y. M. and Chen, Y. T., 1993. Interpretation of shear-wave splitting in Datong area, northern China. *Can. J. Expl. Geophys.*, **29**(1): 341-351.

[62] 李世愚, 陈运泰, 1993. 平面内剪切断层的超S波速破裂. 地震学报, **15**(1): 9-14. 英文刊载: Li, S. Y. and Chen, Y. T., 1993. On the propagation of an in-plane shear fault with super-S-wave velocity. *Acta Seismologica Sinica* (English

Edition),**6**(2):335-341.

[63] 周家玉,陈运泰,倪江川,王鸣,王培德,孙次昌,吴大铭,1993. 用经验格林函数确定中小地震的震源时间函数. 地震学报,**15**(1):22-31. 英文刊载:Zhou, J. Y., Chen, Y. T., Ni, J. C., Wang, M., Wang, P. D., Sun, Z. C. and Wu, F. T., 1993. Determination of source-time function of intermediate and small earthquakes from empirical Green's functions. *Acta Seismologica Sinica* (English Edition),**6**(2):353-363.

[64] 王培德,陈运泰,王鸣,1993. 中国西南地区近震源强地面运动特征. 北京:地震出版社,1-120.

[65] 陈运泰,1993. 地震预测研究概况. 地震学刊,(1):17-23.

[66] Zhao, M., Chen, Y. T., Gong, S. W. and Wang, Q. L., 1993. Inversion of focal mechanism of the Gonghe, Qinghai, China, earthquake of April 26, 1990 using leveling data. In: Ding, G. Y. and Chen, Z. L. (eds.), *Continental Earthquakes*, IASPEI Publication Series for the IDNDR, **3**, Beijing, China: Seismological Press, 246-252.

[67] 陈运泰,赵明,李旭,许力生,1994. 青海共和地震震源过程的复杂性. 见:陈运泰,阚荣举,滕吉文,王椿镛(主编),中国固体地球物理学进展——庆贺曾融生教授诞辰七十周年. 北京:海洋出版社,287-304.

[68] 吴忠良,陈运泰,倪江川,王培德,王鸣,1994. 近震源宽频带记录的地震矩张量反演. 地震学报,**16**(2):141-152. 英文刊载:Wu, Z. L., Chen, Y. T., Ni, J. C., Wang, P. D. and Wang, M., 1994. Moment tensor inversion of near-source broadband data. *Acta Seismologica Sinica* (English Edition),**7**(2):187-199.

[69] 李世愚,陈运泰,1994. 由动态裂纹退化得到的静态裂纹解. 地震学报,**16**(3):304-309. 英文刊载:Li, S. Y. and Chen, Y. T., 1994. Static solution of a crack degenerated from dynamic solution of a propagating crack. *Acta Seismologica Sinica* (English Edition),**7**(3):389-395.

[70] 吴忠良,陈运泰,牟其铎,1994. 核爆炸地震学概要. 北京:地震出版社,1-140.

[71] Chen, Y. T., Mu, Q. D. and Zhou, G. W., 1994. The China Digital Seismograph Network, *Annali di Geofisica*, **37**(5):1049-1053.

[72] Wu, Z. L., Chen, Y. T. and Mu, Q. D., 1995. Digital seismology in China: Digital observation, near-source strong ground motion study, and broadband seismology. In: *1991-1994 China National Report on Seismology and Physics of the Earth's Interior for the XXIth General Assembly of IUGG, Boulder, Colorado, USA, July 1995*, Beijing: China Meteorological Press, 11-20.

[73] Chen, Y. T. and Li, S. Y., 1995. Progress in the studies of earthquake source. In: *1991-1994 China National Report on Seismology and Physics of the Earth's Interior for the XXIth General Assembly of IUGG, Boulder, Colorado, USA, July 1995*, Beijing: China Meteorological Press, 27-40.

[74] 陈运泰,1995. 全球数字地震台网的发展. 见:陈运泰(主编),地球与空间科学观测技术进展——庆贺秦馨菱院士八十寿辰,北京:地震出版社,19-23.

[75] 陈运泰,1995. 台湾的地震和地球物理研究. 地震地磁观测与研究,**16**(4):1-16.

[76] 王培德,王鸣,任道容,田玉红,张廉强,汪更滋,陈运泰,1995. 中国-欧共体地震科学合作项目"北京西北延庆怀来盆地地震学研究"进展. 地震地磁观测与研究,**16**(6):8-14.

[77] Kim, S. G., Chen, Y. T. and Wu, Z. L., 1995. Seismic moment tensor and its inversion: An overview. *J. Eng. Geol.*,**5**(2):215-231.

[78] 陈运泰,1995. 陈运泰. 见:中国科学院学部联合办公室(主编),中国科学院院士自述. 上海:上海教育出版社,607-608.

[79] Wu, Z. L., Chen, Y. T. and Kim, S. G., 1995. Measuring the fractal dimension of seismic source through the high-frequency fall-off of source spectra. International Atomic Energy Agency/United Nations Educational, Scientific and Cultural Organization, IC/95/283 Miramare-Triesters, 1-16.

[80] 陈运泰,1996. 人类能攻克地震预测的难关吗?中国科技画报,(1):34-38.

[81] Xu, L. S. and Chen, Y. T., 1996. Relative source time function of the April 26, 1990, Gonghe, China earthquake by empirical Green's function deconvolution. In: Kim, S. G. (ed.), *Modern Seismology*, Seoul: Hanyang University, 169-177.

[82] 李旭,陈运泰,1996. 合成理论地震图的广义反射透射系数矩阵法. 地震地磁观测与研究,**17**(3):1-20.

[83] 陈运泰,1996. 走向21世纪的地震学(代前言). 见:中国地震学会第六次学术大会论文摘要集. 北京:地震出版

社，1-3.

[84] Chen, Y. T., Xu, L. S., Li, X. and Zhao, M., 1996. Source process of the 1990 Gonghe, China, earthquake and tectonic stress field in the northeastern Qinghai-Xizang (Tibetan) plateau. *Pure Appl. Geophys.*, **146**(3/4): 697-715.

[85] Wu, Z. L. and Chen, Y. T., 1996. Decomposition of seismic moment tensors for underground nuclear explosions. *Pure Appl. Geophys.*, **147**(2): 357-366.

[86] Wu, Z. L., Chen, Y. T. and Kim, S. G., 1996. Physical significance of earthquake quanta. *Bull. Seism. Soc. Amer.*, **86**(5): 1623-1626.

[87] Chen, Y. T., Zhou, G. W. and Wu, Z. L., 1996. Seismological network in China celebrates first 10 years. *EOS*, **77**(47): 468.

[88] 陈运泰，1996. 北京西北延庆怀来盆地地震学研究. 见：国家科学技术委员会国际合作司（主编），中欧科技合作回顾与评价 1981-1995. 北京：清华大学出版社，117-121. 英文刊载：Chen, Y. T., 1996. Seismological study of the Yanqing-Huailai basin, NW of Beijing. In: Department of International Scientific and Technological Cooperation, SSTC, China (ed.), *Successful Sino-European Science and Technology Cooperation, 1981-1995*, Beijing: Tsinghua University Press, 134-139.

[89] 许力生，陈运泰，1996. 用经验格林函数方法从长周期数字波形资料中提取共和地震的震源时间函数. 地震学报，**18**(2): 156-169. 英文刊载：Xu, L. S. and Chen, Y. T., 1996. Source time function of the Gonghe, China earthquake retrieved from long period digital waveform data using empirical Green's function technique. *Acta Seismologica Sinica* (English Edition), **9**(2): 209-222.

[90] 李旭，陈运泰，1996. 用长周期地震波波形资料反演1990年青海共和地震的震源过程. 地震学报，**18**(3): 279-286. 英文刊载：Li, X. and Chen, Y. T., 1996. Inversion of long-period body-wave data for the source process of the Gonghe, Qinghai, China earthquake. *Acta Seismologica Sinica* (English Edition), **9**(3): 361-370.

[91] Chen, Y. T., Li, S. Y. and Wu, Z. L., 1996. Understanding the complexity of continental earthquakes: The study of earthquake sources in China in the 1990s. In: He, Y. N. (ed.), *Achievements of Seismic Hazard Prevention and Reduction in China*, Beijing: Seismological Press, 26-39.

[92] Wu, Z. L., Zhou, G. W. and Chen, Y. T., 1996. Ten years of the China Digital Seismograph Network (CDSN) from 1986 to 1996. In: He, Y. N. (ed.), *Achievements of Seismic Hazard Prevention and Reduction in China*, Beijing: Seismological Press, 40-47.

[93] 李世愚，陈运泰，1996. 震源物理研究的回顾与展望. 国际地震动态，(4): 1-6.

[94] 许力生，陈运泰，1997. 用数字化宽频带波形资料反演共和地震的震源参数. 地震学报，**19**(2): 113-128. 英文刊载：Xu, L. S. and Chen, Y. T., 1997. Source parameters of the Gonghe, Qinghai Province, China, earthquake from inversion of digital broadband waveform data. *Acta Seismologica Sinica* (English Edition), **10**(2): 143-159.

[95] 许向彤，陈运泰，王培德，1997. 怀来盆地的构造应力场. 地震地磁观测与研究，**18**(1): 1-8.

[96] 许力生，陈运泰，1997. 震源深度的误差对矩张量反演的影响. 地震学报，**19**(5), 462-470. 英文刊载：Xu, L. S. and Chen, Y. T., 1997. The effect of focal depth error on moment tensor inversion. *Acta Seismologica Sinica* (English Edition), **10**(5): 571-580.

[97] 王培德，田玉红，李春来，陈运泰，1997. 怀来盆地的地震活动与活动断裂. 地震学报，**19**(5), 551-554. 英文刊载：Wang, P. D., Tian, Y. H., Li, C. L. and Chen, Y. T., 1997. Seismic activity and active faults in Huailai basin. *Acta Seismologica Sinica* (English Edition), **10**(5): 683-687.

[98] 陈运泰，1997. 走向21世纪的地震学. 见：周光召，朱光亚（主编），共同走向科学——百名院士科技系列报告集（下），北京：新华出版社，304-326.

[99] 倪江川，陈运泰，吴忠良，王培德，王璋，柯兆明，1997. 地下爆炸的矩张量反演. 地震地磁观测与研究，**18**(6A): 5-13.

[100] Mirzaei, N., Gao, M. T. and Chen, Y. T., 1997. Evaluation of uncertainty of earthquake parameters for the purpose of seismic zoning of Iran. *Earthquake Research in China*, **11**(2): 197-212.

[101] 于湘伟, 陈运泰, 王培德, 李春来, 1997. 京西北延庆-怀来盆地的 Q_{SH} 值和小震震源参数的测定. 地震地磁观测与研究, **18**(5): 6-18.

[102] Mirzaei, N., 高孟潭, 陈运泰, 王健, 1997. 用于地震危险性分析的统一的伊朗地震目录. 地震学报, **19**(6): 574-585. 英文刊载: Mirzaei, N., Gao, M. T., Chen, Y. T. and Wang, J., 1997. A uniform catalog of earthquakes for seismic hazard assessment in Iran. *Acta Seismologica Sinica* (English Edition), **10**(6): 713-726.

[103] 刘勇卫, 陈运泰, 1997. 地球内核的转动比地壳、地幔快. 见: 中国科学院 (主编), 1997 科学发展报告, 北京: 科学出版社, 37-38.

[104] Schwab, F., Hsu, H. T. and Chen, Y. T., 1997. Three-dimensional mapping of the Earth's lithosphere and asthenosphere. *International Lithosphere Program, Scientific Committee on the Lithosphere, Annual Report*-1996, Report **21**: 22-27.

[105] 孟凡顺, 郭海燕, 陈运泰, 王椿镛, 1997. 柱坐标系下任意埋藏源层状弹性半空间问题的奇异解. 地球物理学报, **40**(4): 512-521.

[106] 雷军, 王培德, 姚陈, 陈运泰, 1997. 云南剑川近场横波特征及其与构造的关系. 地球物理学报, **40**(6): 790-801.

[107] 陈运泰, 吴忠良, 1997. 固体地球物理发展趋势讨论. 见: 叶笃正 (主编), 赵九章纪念文集, 北京: 科学出版社, 358-365.

[108] 陈运泰, 吴忠良, 李世愚, 1997. 世纪之交我国震源物理研究的进展与展望. 地球物理学报, **40**(增刊): 164-176.

[109] 顾浩鼎, 陈运泰, 1997. 旋转运动、旋转矩定律及弹性介质动力学关系. 科学, **49**(6): 37-39.

[110] 陈运泰, 1997. 地震学和地球内部物理学. 见: 国家地震局科学技术委员会 (主编), 走向 21 世纪的地震科学技术, 国际地震动态, (11): 1-21.

[111] 陈运泰, 1997. 《中国地震学研究进展——庆贺谢毓寿教授八十寿辰》前言. 见: 陈运泰 (主编), 中国地震学研究进展——庆贺谢毓寿教授八十寿辰, 北京: 地震出版社, 1-5.

[112] 顾浩鼎, 陈运泰, 1997. 地震空区的物理意义和震源系统的无标度性. 见: 陈运泰 (主编), 中国地震学研究进展——庆贺谢毓寿教授八十寿辰, 北京: 地震出版社, 37-41.

[113] 许力生, 陈运泰, Fastholf, S., 1997. 1996 年 2 月 3 日丽江地震 M_S = 7.0 地震震源过程的时空复杂性. 见: 陈运泰 (主编), 中国地震学研究进展——庆贺谢毓寿教授八十寿辰, 北京: 地震出版社, 91-105.

[114] 吴忠良, 陈运泰, 1997. 一类二维 Burridge-Knopoff 模型中的一个孤立波解: 对地震破裂过程的自愈合脉冲的一种解释. 见: 陈运泰 (主编), 中国地震学研究进展——庆贺谢毓寿教授八十寿辰, 北京: 地震出版社, 134-140.

[115] Wu, Z. L., Kim, S. G. and Chen, Y. T., 1997. High-frequency fall-off of source spectra of deep-focus earthquakes from Wigner-distribution estimation. *Phys. Earth Planet. Interior*, **99**(3/4): 221-229.

[116] Wu, Z. L., Kim, S. G. and Chen, Y. T., 1997. A theorem for the direct estimation of seismic source spectra. *Tectonophysics*, **269**(3/4): 337-341.

[117] Kim, S. G., Chen, Y. T., Wu, Z. L. and Panza, G. F., 1997. A mathematical theorem useful for the direct estimation of seismic source spectra. *Bull. Seism. Soc. Amer.*, **87**(50): 1281-1287.

[118] Chen, Y. T., Xu, X. T., Yu, X. W. and Wang, P. D., 1997. Observations and interpretation of seismic ground motion and earthquake hazard mitigation in the Beijing area. In: German IDNDR Committee for Natural Disaster Reduction (ed.), *First International Earthquakes and Megacities Workshop.* September 1-4, 1997. Seeheim, Germany. Release II, 80-90. IDNDR Series 9. The United Nations University, 80-90.

[119] Mirzaei, N., Gao, M. T. and Chen, Y. T., 1998. Seismic source regionalization for seismic zoning of Iron: Major seismotectonic provinces. *J. Earthq. Predic. Res.*, (1): 465-495.

[120] 陈运泰, 1998. 跨世纪的中国地震学 (代前言). 见: 中国地震学会第七次学术大会论文摘要集, 北京: 地震出版社, 1-6.

[121] Schwab, F., Hsu, H. T. and Chen, Y. T., 1998. Three dimensional mapping of the Earth's tectosphere. *International Lithosphere Program, Scientific Committee on the Lithosphere, Annual Report-1997*, Report **22**: 16-22.

[122] 陈运泰, 许向彤, 于湘伟, 王培德, 1998. 北京西北延庆-怀来盆地的地震观测及减轻地震灾害研究. 华南地震,

18(1): 2-8.

[123] 陈运泰, 王培德, 许向彤, 于湘伟, 1998. 地震观测中的高新技术和在减轻地震灾害中的作用. 见: 李振声（主编）, 中国减轻自然灾害研究, 北京: 中国科学技术出版社, 272-276.

[124] Mozaffari, P., 吴忠良, 陈运泰, 1998. 用经验格林函数方法研究澜沧–耿马 $M_S=7.6$ 地震的破裂过程. 地震学报, **20**(1): 1-11. 英文刊载: Mozaffari, P., Wu, Z. L. and Chen, Y. T., 1998. Rupture process of November 6, 1988, Lancang-Gengma, Yunnan, China, earthquake of $M_S=7.6$ using empirical Green's function deconvolution method. *Acta Seismologica Sinica* (English Edition), **11**(1): 1-12.

[125] 陈运泰, 1998. 地震预测: 困难但并非毫无希望（代前言）. 地震地磁观测与研究, **19**(1A): 1-3.

[126] 陈运泰, 孙枢, 吴忠良, 1998. 跨世纪的中国地球科学. 见: 中国科学院地学部"中国地球科学发展战略"研究组（主编）, 中国地球科学发展战略的若干问题——从地学大国走向地学强国, 北京: 科学出版社, 17-39.

[127] 王椿镛, 王飞, 丁志峰, 张小兵, 李强, 陈运泰, 1998. 中国东北日本海地区岩石层结构三维模拟（1）——Ⅰ数据集. 地震地磁观测与研究, **19**(1): 1-10.

[128] 涂毅敏, 陈运泰, 张德存, 1998. 中国数字地震台网数字化资料使用指南. 地震地磁观测与研究, **19**(4): 1-7.

[129] 李世愚, 滕春凯, 刘绮亮, 卢振业, 刘小红, 陈运泰, 1998. 三维破裂及其在地震和断层研究中的应用（综述）. 地震地磁观测与研究, **19**(1): 11-25.

[130] 许力生, 陈运泰, S. Fasthoff, 1998. 经验格林函数与主震破裂过程反演. 见: 云南省地震局, 滇西地震预报实验场（主编）, 一九九六年丽江地震, 北京: 地震出版社, 79-88.

[131] 吴忠良, 陈运泰, 张天中, 朱传镇, 1998. 地震预测研究的规则. 科学, **50**(5): 30-34.

[132] 陈运泰, 1998. 走向21世纪的地震学. 见: 国家地震局人事教育司（主编）, 走向世界——寄语21世纪的中国地震工作者, 北京: 地震出版社, 13-23.

[133] 陈运泰, 1998. 写好科技论文 做好科技报告. 见: 国家地震局人事教育司（主编）, 走向世界——寄语21世纪的中国地震工作者, 北京: 地震出版社, 103-110.

[134] 陈运泰, 吴忠良, 1998. 跨世纪的中国地震学. 见: 周光召（主编）, 科技进步与学科发展（上册）. 北京: 中国科学技术出版社, 115-119.

[135] 魏修成, 董敏煜, 陈运泰, 1998. 非均匀各向异性介质中弹性波的传播. 地震学报, **20**(6): 561-572. 英文刊载: Wei, X. C., Dong, M. Y. and Chen, Y. T., 1998. Elastic wave propagation in inhomogeneous anisotropic media. *Acta Seismologica Sinica* (English Edition), **11**(6): 655-667.

[136] Gao, Y., Wang, P., Zheng, S. Wang, M., Chen, Y. T. and Zhou, H. L., 1998. Temporal changes in shear-wave splitting at an isolated swarm of small earthquakes in 1992 near Dongfang, Hainan Island, southern China. *Geophys. J. Intl.*, **135**: 102-112.

[137] Wu, Z. L. and Chen, Y. T., 1998. Solitary wave in a Burridge-Knopoff model with slip-dependent friction as a clue to understanding the mechanism of the self-healing slip pulse in an earthquake rupture process. *Nonlinear Processes in Geophysics*, (5): 121-125.

[138] Mirzaei, N., Gao, M. T. and Chen, Y. T., 1998. Seismic source regionalization for seismic zoning of Iran: Major seismotectonic provinces. *J. Earthquake Prediction Research*, **7**: 465-495.

[139] 刘振兴, 钦俊德, 陶诗言, 白以龙, 陈运泰, 应崇福, 叶连俊, 蒋丽金, 李文华, 1998. 加强我院自然科学期刊工作迎接知识创新的新局面. 中国科技期刊研究, **9**(4): 201-202.

[140] Schwab, F., Hsu, H. T., Chen, Y. T., Roy, K. K. and Thakur, V. C., 1999. Three dimensional mapping of the Earth's tectosphere. *International Lithosphere Program*, *Scientific Committee on the Lithosphere*, *Annual Report-1998*, Report 23, 12-15.

[141] 陈祥雄, 陈运泰, 1999. 西太平洋地区地震台网建设及其有关工作. 地震地磁观测与研究, **20**(1): 1-8.

[142] Mirzaei, N., Gao, M. T. and Chen, Y. T., 1999. Delineation of potential seismic sources for seismic zoning of Iran. *J. Seismol.*, **3**(1): 17-30.

[143] Mozaffari, P., 吴忠良, 陈运泰, 1999. 用"无Q谱"方法测定震源谱的高频衰减. 地震学报, **21**(1): 1-9. 英文刊

载: Mozaffari, P., Wu, Z. L., Chen, Y. T. and Wang, M., 1999. High frequency fall-off of source spectra using Q-free spectra estimation. *Acta Seismologica Sinica* (English Edition), **12**(1): 1-10.

[144] 李世愚, 陈运泰, 1999. 分形断层的隧道效应和平面内剪切断层的跨S波破裂. 地震学报, **21**(1): 17-23. 英文刊载: Li, S. Y. and Chen, Y. T., 1999. Tunnel effect of fractal fault and transient S-wave velocity rupture (TSVR) of in-plane shear fault. *Acta Seismologica Sinica* (English Edition), **12**(1): 19-25.

[145] Chen, Y. T., 1999. Prevention and mitigation of earthquake hazard in moderate- to- large cities. In: Chinese Association of Science and Technology (ed.), *Proceedings of 1999 Symposium on Mitigation of Natural Hazards*, Beijing: 261-263 (in Chinese).

[146] 徐冠华, 孙枢, 陈运泰, 吴忠良, 1999. 迎接"数字地球"的挑战. 遥感学报, **3**(2): 85-89. 转载: 徐冠华, 孙枢, 陈运泰, 1999. 迎接"数字地球"的挑战. 见: 崔伟宏 (主编), 数字地球, 北京: 中国环境科学出版社, 6-12. 转载: 徐冠华, 孙枢, 陈运泰, 吴忠良, 1999. 迎接"数字地球"的挑战. 首都信息化, **12**: 31-33.

[147] 陈运泰, 1999. 祝贺《地震地磁观测与研究》创刊20周年. 地震地磁观测与研究, **20**(5): 1-2.

[148] 陈运泰, 吴忠良, 1999. 中国数字地震学的发展战略问题. 国际地震动态, (5): 1-5.

[149] 刘瑞丰, 陈运泰, 周公威, 涂毅敏, 陈培善, 1999. 地震矩张量反演在地震快速反应中的应用. 地震学报, **21**(2): 115-122. 英文刊载: Liu, R. F., Chen, Y. T., Zhou, G. W., Tu, Y. M. and Chen P. S., 1999. Applications of seismic moment tensor inversion in fast response to earthquakes. *Acta Seismologica Sinica* (English Edition), **12**(2): 129-136.

[150] 周荣茂, 陈运泰, 吴忠良, 1999. 由矩张量反演得到的海南东方震群的震源机制. 地震学报, **21**(4): 337-343. 英文刊载: Zhou, R. M., Chen, Y. T. and Wu, Z. L., 1999. Moment tensor inversion for the focal mechanism of the Dongfang (Hainan) earthquake swarm. *Acta Seismologica Sinica* (English Edition), **12**(4): 371-378.

[151] Mozaffari, P., 许力生, 吴忠良, 陈运泰, 1999. 用长周期体波数据反演1988年11月6日澜沧-耿马地震的矩张量. 地震学报, **21**(4): 344-353. 英文刊载: Mozaffari, P., Xu, L. S., Wu, Z. L. and Chen, Y. T., 1999. Moment tensor inversion of the November 6, 1988 $M_S = 7.6$, Lancang-Gengma, China, earthquake using long-period body-waves data. *Acta Seismologica Sinica* (English Edition), **12**(4): 379-389.

[152] 许力生, 陈运泰, 1999. 1997年中国西藏玛尼 $M_S = 7.9$ 地震的时空破裂过程. 地震学报, **21**(5): 449-459. 英文刊载: Xu, L. S. and Chen, Y. T., 1999. Tempo-spatial rupture process of the 1997 Mani, Xizang (Tibet), China earthquake of $M_S = 7.9$. *Acta Seismologica Sinica* (English Edition), **12**(5): 495-506.

[153] 吴忠良, 陈运泰, Mozaffari, P., 1999. 应力降的标度性质与震源谱高频衰减常数. 地震学报, **21**(5): 460-468. 英文刊载: Wu, Z. L, Chen, Y. T. and Mozaffari, P., 1999. Scaling of stress drop and high-frequency fall-off of source spectra. *Acta Seismologica Sinica* (English Edition), **12**(5): 507-515.

[154] 周荣茂, 吴忠良, 陈运泰, 1999. 由矩张量反演得到的北部湾地震的震源机制. 地震学报, **21**(6): 561-569. 英文刊载: Zhou, R. M., Chen, Y. T. and Wu, Z. L., 1999. Moment tensor inversion for the focal mechanism of the Beibuwan earthquake. *Acta Seismologica Sinica* (English Edition), **12**(6): 609-617.

[155] 许向彤, 陈运泰, 王培德. 1999. 1995年7月20日怀来盆地 $M_L = 4.1$ 地震的破裂过程. 地震学报, **21**(6): 570-582. 英文刊载: Xu, X. T., Chen, Y. T. and Wang, P. D., 1999. Rupture process of the $M_L = 4.1$ earthquake in Huailai basin on July 20, 195. *Acta Seismologica Sinica* (English Edition), **12**(6): 618-631.

[156] Chen, Y. T. and Xu, L. S., 1999. Source processes of recent large earthquakes in Qinghai-Xizang (Tibetan) Plateau. In: *Proceedings and Abstracts of the International Workshop GPA'99, Tectonics, Geodynamics and Natural Hazards in West Pacific-Asia*, Hanoi, Vietnam, November 10-11, 1999. Journal of Geology, Series **B**, 13-14(special issue): 216-217.

[157] 杨智娴, 陈运泰, 张宏志, 1999. 张北-尚义地震序列的重新定位. 地震地磁观测与研究, **20**(6): 6-9.

[158] 许力生, 杨智娴, 陈运泰, 1999. 集集地震序列和震源破裂过程初步分析. 见: 陈运泰 (主编), 中国地震学会成立20周年纪念文集, 北京: 地震出版社, 97-112.

[159] 陈运泰, 杨智娴, 1999. 台湾遭遇百年以来最大地震——"9.21"集集7.6级地震概况. 海峡科技与产业, **63**(6): 2-4.

[160] 陈运泰, 1999. 重视大中城市的防震减灾, 保障21世纪的可持续发展. 中外交流, (10): 58-59.

[161] 陈运泰, 王椿镛, 吴忠良, 1999. 面向"数字地球"的岩石层三维成像. 地震地磁观测与研究, **20**(5): 3-10.

[162] Chen, Y. T. and Wu, Z. L., 1999. Physics of seismic source. I. Presentation of seismic sources. *Earthquake Research in China*, **13**(2): 113-126.

[163] Chen, Y. T. and Wu, Z. L., 1999. Physics of seismic source. II. Mechanics of seismic source. *Earthquake Research in China*, **13**(2): 127-136.

[164] Chen, Y. T. and Wu, Z. L., 1999. Physics of seismic source. III. Statistical mechanics of earthquakes. *Earthquake Research in China*, **13**(2): 137-148.

[165] Chen, X. F. and Chen, Y. T., 1999. Review on studies of seismic wave propagation and physics of earthquake source. In: *China National Report (1995-1998) on Seismology and Physics of the Earth's Interior for the XXIInd General Assembly of IUGG, Birmingham, UK, July 1999*, 1-15.

[166] Chen, Y. T. and Xu, L. S., 1999. Source processes of recent large earthquakes in Qinghai-Xizang (Tibetan) Plateau. In: *Proceedings of the Across Strait Symposium on Exchange and Cooperation in Seismology*, Taipei: 27-28 (in Chinese).

[167] Chen, Y. T. and Wu, Z. L. 2000. Monitoring a Comprehensive Test Ban Treaty within the framework of the Digital Earth. In: Xu, G. H. and Chen Y. T. (eds.), *Towards Digital Earth*. Beijing: Science Press, 731-734.

[168] 陈运泰, 杨智娴, 2000. 从集集地震和全球的震情与灾情看大中城市的防震减灾. 见: 2000 科学发展报告. 北京: 科学出版社, 168-172.

[169] 陈运泰, 杨智娴, 2000. 大中城市与特大城市的防震减灾——从集集地震和全球的震情与灾情谈起. 见: 2000 年减轻自然灾害研讨会论文汇编, 北京: 中国科学技术协会, 101-106.

[170] 陈运泰, 杨智娴, 2000. 集集大地震及其对防震减灾的启示. 见: 2000 年减轻自然灾害白皮书, 北京: 中国科学技术协会, 144-148.

[171] 陈运泰, 刘振兴, 俞鸿儒, 等, 2000. 要重视科技期刊工作的建议. 中国科技期刊研究, **11**(1): 1.

[172] Chen, Y. T. and Xu, L. S., 2000. A time domain inversion technique for the tempo-spatial distribution of slip on a finite fault plane with applications to recent large earthquakes in Tibetan Plateau, *Geophys. J. Intl.*, **143**(2): 407-416.

[173] Kim, S. G., Kraeva, N. and Chen, Y. T., 2000. Source parameter determination of regional earthquakes in the Far East using moment tensor inversion of single-station data. *Tectonophysics*, **317**: 125-136.

[174] 曾融生, 陈运泰, 吴忠良, 2000. 探测地球内部的"雷达"——地震波. 见: 中国地震学会普及工作委员会（编）, 院士专家谈地震, 北京: 地震出版社, 1-10.

[175] 陈运泰, 曾融生, 吴忠良, 2000. 地震的分类、发生和预测. 见: 中国地震学会普及工作委员会（编）, 院士专家谈地震, 北京: 地震出版社, 27-38.

[176] Wu, Z. L. and Chen, Y. T., 2000. Potential impact of "Digital Earth" on seismological monitoring. *Seism. Res. Lett.*, **71**(5): 548-552.

[177] 陈运泰, 吴忠良, 王培德, 许力生, 李鸿吉, 牟其铎, 2000. 数字地震学, 北京: 地震出版社, 1-171.

[178] 王培德, 陈运泰, 2001. 中德合作开展流动台阵近场地震观测和北京地区防震减灾研究. 国际地震动态, (5): 8-12.

[179] 陈运泰, 吴忠良, 吕苑苑, 2001. 地震学今昔谈, 山东: 山东教育出版社, 1-163.

[180] 陈运泰, 2001. 加强防震减灾合作 造福海峡两岸人民——记第四届海峡两岸地震科技研讨会. 海峡科技, (11): 15.

[181] 陈运泰, 滕吉文, 张中杰, 2001. 地球物理学的回顾与展望. 地球科学进展, **16**(5): 634-642.

[182] 许向彤, 陈运泰, 王培德, 2001. 1995 年 7 月 20 日怀来盆地 M_L4.1 地震序列震源参数的精确测定. 地震学报, **23**(3): 225-238. 英文刊载: Xu, X. T., Chen, Y. T. and Wang, P. D., 2001. Precise determination of focal parameters for July 20, 1995 M_L = 4.1 earthquake sequence in the Huailai basin. *Acta Seismologica Sinica* (English Edition), **14**(3): 237-250.

[183] Xu, L. S., Patau, G. and Chen, Y. T., 2001. Soure tome functions of the 1999, Jiji (Chi-Chi) earthquake from GDSN long period waveform data using aftershocks as empirical Green's functions. *Acta Seismologica Sinica* (English Edition), **15**(2):

121-133.

[184] Zhou, Y. H., Xu, L. S. and Chen, Y. T., 2002. Source process of the 4 June 2000 southern Sumatra, Indonesia, earthquake. *Bull. Seism. Soc. Amer.*, **92**(5): 2027-2035.

[185] 许力生, 陈运泰, 高孟潭, 2002. 2001年1月26日印度古杰拉特（Gujarat）M_S7.8地震时空破裂过程. 地震学报, **24**(5): 447-461. 英文刊载: Xu, X. T., Chen, Y. T. and Gao, M. T., 2002. Spatial and temporal rupture process of the January 26, 2001, Gujarat, India, M_S = 7.8 earthquake. *Acta Seismologica Sinica* (English Edition), **15**(5): 469-483.

[186] 周云好, 许力生, 陈运泰, 2002. 2000年6月4日印度尼西亚苏门答拉南部M_S8.0地震的时空破裂过程. 中国地震, **18**(3): 221-229.

[187] 周云好, 许力生, 陈运泰, 2002. 2000年6月4日印度尼西亚苏门答腊南部M_S8.0地震的震源机制. 地震学报, **24**(5): 462-469. 英文刊载: Zhou, Y. H., Xu, L. S. and Chen, Y. T., 2002. Mechanism of the June 4, 2000 southern Sumatra, Indonesia, earthquake. *Acta Seismologica Sinica* (English Edition), **15**(5): 484-494.

[188] Panza, G. F., Alvarez, L., Aoudia, A., Ayadi, A., Benhallou, H., Benouar, D., Bus, Z., Chen, Y. T., Cioflan, C., Ding, Z. F., et al., 2002. Realistic modeling of seismic input for megacities and large urban areas (the UNESCO/IUGS/IGCP) project 414. *Episodes*, **25**(3): 160-184.

[189] 许力生, Patau, G., 陈运泰, 2002. 用余震作为经验格林函数从GDSN长周期波形资料中提取1999年集集地震的震源时间函数. 地震学报, **24**(2): 113-125. 英文刊载: Xu, L. S., Patau, G. and Chen, Y. T., 2001. Soure tome functions of the 1999, Jiji (Chi-Chi) earthquake from GDSN long period waveform data using aftershocks as empirical Green's functions. *Acta Seismologica Sinica* (English Edition), **15**(2): 121-133.

[190] 杨智娴, 陈运泰, 张宏志, 2002. 张北-尚义地震序列的重新定位和发震构造. 地震学报, **24**(4): 366-377. 英文刊载: Yang, Z. X., Chen, Y. T. and Zhang, H. Z., 2002. Relocation and seismogenic structure of the 1998 Zhangbei-Shangyi earthquake sequence. *Acta Seismologica Sinica* (English Edition), **15**(4): 383-394.

[191] 涂毅敏, 陈运泰, 2002. 德国大陆超深钻井注水诱发地震的精确定位. 地震学报, **24**(6): 587-598. 英文刊载: Tu, Y. M. and Chen, Y. T., 2002. *Acta Seismologica Sinica* (English Edition), **15**(6): 587-598.

[192] Panza, G. F., Alvarez, L., Aoudia, A., Ayadi, A., Benhallou, H., Benouar, D., Bus, Z., Chen, Y. T., Cioflan, C., Ding, Z. F., et al., 2002. Realistic modeling of seismic input for megacities and large urban areas (the UNESCO/IUGS/IGCP) project 414. *Episodes*, **25**(3): 160-184.

[193] 唐兆华, 陈运泰, 2002. 大型维歇特地震仪的记录特征及其在地震速报中的作用. 地震地磁观测与研究, **23**(5): 1-5.

[194] 陈运泰, 唐兆华, 2002. 维歇特地震仪在中国——纪念南京地震台建台70周年. 见: 李清河（编）, 纪念南京基准地震台建台70周年论文集, 北京: 地震出版社, 6-12.

[195] 吴忠良, 陈运泰, 2002. 地震预测与统计物理. 物理, **31**(6): 365-371.

[196] 吴忠良, 陈运泰, 2002. 从弹簧滑块到地震预测: BK模型今昔谈. 物理, **31**(11): 719-724.

[197] 陈运泰, 吴忠良, 吕苑苑, 2002. 地震与板块大地构造. 城市与减灾, （5）: 14-18.

[198] 陈运泰, 朱传镇, 吴忠良, 2003. 中国现代地震学的回顾与展望. 世界科技研究与发展, **25**(1): 12-16. 英文刊载: Chen, Y. T., Zhu, C. Z. and Wu, Z. L., 2002. Seismology in China in the 20th century. *International Symposium Science and Technology in Modern China: Retrospect and Prospect*, **1**: 331-342.

[199] 曾融生, 陈运泰, 2002. 探索地球内部的奥秘, 北京: 清华大学出版社/广州: 暨南大学出版社, 1-116.

[200] 陈运泰, 刘克人, 郑金涵, 宋胜合, 刘端法, 卢红艳, 郭凤义, 2002. 局部重力场变化与地震发生的关系合作研究课题的回顾. 见: 孙枢（编）, 理论与应用地球物理进展——庆贺郭宗汾教授八十寿辰, 北京: 气象出版社, 40-47.

[201] 许力生, 陈运泰, 2002. 震源时间函数与震源破裂过程. 地震地磁观测与研究, **23**(6): 1-8.

[202] 许力生, 陈运泰, 2002. 震源时间函数与震源破裂过程. 见: 地震参数——数字地震学在地震预测中的应用, 北京: 中国地震局监测预报司/中国地震局地球物理研究所/中国地震局分析预报中心/中国科学院研究生院, 44-50.

[203] 陈运泰, 2002. 地震能量和地震效率. 见: 地震参数——数字地震学在地震预测中的应用, 北京: 中国地震局监测

预报司/中国地震局地球物理研究所/中国地震局分析预报中心/中国科学院研究生院，51-62.

[204] Ding, Z., Chen, Y. T. and Panza, G. F., 2002. Estimation of site effects in Beijing City. *Pure Appl. Geophys.*, **200**(2): 1-10.

[205] Ding, Z., Romanelli, F., Chen, Y. T. and Panza, G. F., 2002. Realistic modeling of seismic strong ground ground motion in Beijing area. *Pure Appl. Geophys.*, **200**(2): 20-30.

[206] Xu, L. S., Chen, Y. T., Teng, T. L. and Patau, G., 2002. Temporal and spatial rupture process of the 1999 Chi-Chiearthquake from IRIS and GEOSCOPE long-period waveform data using aftershocks as empirical Green's functions. *Bull. Seism. Soc. Amer.*, **92**(8): 3210-3228.

[207] 于湘伟，陈运泰，王培德，2003. 京津唐地区中上地壳三维P波速度结构. 地震学报，**25**(1): 1-14. 英文刊载：Yu, X. W., Chen, Y. T. and Wang, P. D., 2003. Three-dimensional P velocity structure in Beijing area. *Acta Seismologica Sinica* (English Edition), **16**(1): 1-15.

[208] 陈运泰，许力生，2003. 青藏高原及其周边地区大地震震源过程成像. 地学前缘，**10**(1): 57-62.

[209] 杨智娴，陈运泰，郑月军，于湘伟. 2003. 双差地震定位法在我国中西部地区地震精确定位中的应用. 中国科学，**33**(增刊): 129-134. 英文刊载：Yang, Z. X., Chen, Y. T., Zheng, Y. J. and Yu, X. W., 2003. Accurate relocation of earthquakes in central-western China using the double-difference earthquake location algorithm, *Science in China* (Series D), **46**(Supp.): 181-188.

[210] 陈运泰，Stump, B. W., 杨智娴，Hayward, C., 2003. 天然和人为地震活动性的对比研究，地震地磁观测与研究，**24**(3): 1-9.

[211] Chen Y. T., Yang, Z. X., Stump, B., Zhou R. M. and Hayward, C., 2003. New PASSCAL instruments in the Huailai Basin, China. *IRIS 2003 Annual Report*, 6-7.

[212] 李世愚，陈运泰，2003. 地震震源的研究. 地震学报，**25**(5): 453-464. 英文刊载：Li, S. Y. and Chen, Y. T., 2003. Studies on seismic source. *Acta Seismologica Sinica* (English Edition), **16**(5): 479-491.

[213] 许力生，俞言祥，陈运泰，2003. 中国强地面运动研究进展. 地震学报，**25**(5): 475-478. 英文刊载：Xu, L. S. Yu, Y. X. and Chen, Y. T., 2003. Advances in ground motion studies in China. *Acta Seismologica Sinica* (English Edition), **16**(5): 503-507.

[214] 刘瑞丰，吴忠良，阴朝民，陈运泰，庄灿涛，2003. 中国地震台网数字化改造的进展. 地震学报，**25**(5): 535-540. 英文刊载：Liu, R. F., Wu, Z. L., Yin, C. M., Chen, Y. T. and Zhuang, C. T., 2003. Development of China Digital Seismological Observational Systems. *Acta Seismologica Sinica* (English Edition), **16**(5): 568-573.

[215] 陈运泰，2003. 板块大地构造与地震. 见：季延寿（主编），科学名家讲座（第三辑），北京：北京科海电子出版社，212-268.

[216] Wu, Z. L. and Chen, Y. T., 2003. Definition of seismic moment at a discontinuity interface. *Bull. Seismol. Soc. Am.*, **93**(4): 1832-1834.

[217] 陈运泰，吴忠良，2004. 国际地震学与地震工程学手册. 地震学报，**26**(1): 110-111. 英文刊载：Chen, Y. T. and Wu, Z. L., 2004. International Handbook of Earthquake and Engineering Seismology. *Acta Seismologica Sinica* (English Edition), **17**(1): 119-120.

[218] 杨智娴，于湘伟，郑月军，陈运泰，倪晓希，Chan, W., 2004. 中国中西部地区地震的重新定位和三维地震波速度结构. 地震学报，**26**(1): 19-29. 英文刊载：Yang, Z. X., Yu, X. W., Zheng, Y. J., Chen Y. T., Ni X. X. and Chan, W., 2004. Earthquake relocation and 3-dimensional crustal structure of P-wave velocity in central-western China. *Acta Seismologica Sinica*, **17**(1): 20-30.

[219] 杨智娴，陈运泰，2004. 用双差地震定位方法再次精确测定1998年张北-尚义地震序列的震源参数，地震学报，**26**(2): 1-6. 英文刊载：Yang, Z. X. and Chen, Y. T., 2004. Relocation of the 1998 Zhangbei-Shangyi earthquake sequence using the double difference earthquake location algorithm, *Acta Seismologica Sinica* (English Edition), **17**(2): 125-130.

[220] 陈运泰，刘瑞丰，2004. 地震的震级. 地震地磁观测与研究，**25**(6): 1-12.

[221] Wang, Q. L., Chen, Y. T., Cui, D. X. and Wang, W. P., 2004. Decade-scale correction between crustal deformation and

length of day: implication to earthquake hazard estimation. In: Chen, Y. T., Panza, G. F. and Wu, Z. L. (eds.), *Earthquake: Hazard, Risk, and Strong Ground Motion*, Beijing: Seismological Press, 151-163.

[222] Ding, Z. F., Vaccari, F., Chen, Y. T. and Panza, G. F., 2004. Deterministric seismic hazard map in North China. In: Chen, Y. T., Panza, G. F. and Wu, Z. L. (eds.), *Earthquake: Hazard, Risk, and Strong Ground Motion*, Beijing: Seismological Press, 351-359.

[223] Ding, Z. F., Romanelli, F., Chen, Y. T. and Panza, G. F., 2004. Realistic modeling of seismic wave ground motion in Beijing City. *Pure Appl. Geophys*, **161**(5/6): 1093-1106.

[224] Ding, Z. F., Chen, Y. T. and Panza, G. F., 2004. Estimation of site effects in Beijing City. *Pure Appl. Geophys*, **161**(5/6): 1107-1123.

[225] 许力生, 陈运泰, 2004. 从全球长周期波形资料反演2001年11月14日昆仑山大地震时空破裂过程. 中国科学 D 辑 地球科学, **34**(3): 256-264.

[226] 许力生, 陈运泰, 2004. 格林函数库技术与快速地震矩张量反演. 陈运泰, 滕吉文, 阚荣举, 王椿镛（主编）, 中国大陆地震学与地球内部物理学研究进展——庆贺曾融生院士八十寿辰, 北京: 地震出版社, 625-630.

[227] Chen, Y. T., Panza, G. F. and Wu, Z. L., (eds.) 2004. *Earthquake: Hazard, Risk, and Strong Ground Motion*, Beijing: Seismological Press, 1-486.

[228] Shedlock, K. M., Panza, G. F. and Chen, Y. T., 2004. Strong Ground Motion, Earthquake Hazard and Risk in Alpine-Himalayan and Pacific Regions. *Tectonophysics* (Special Issue), **390**(1-4): 1-243.

[229] 刘瑞丰, 陈运泰, 任枭, 侯建民, 邹立晔, 2005. 2001年11月14日昆仑山口西地震——一次面波震级未饱和的地震. 地震学报, **27**(5): 467-476. 英文刊载: Liu, R. F., Chen, Y. T., Ren, X., Hou, J. M. and Zou, L. Y., 2005. The November 14, 2001 west of Kunlun Mountatin Pass earthquake with unsaturated surface wave magnitude. *Acta Seismologica Sinica*, **18**(5): 499-509.

[230] 刘瑞丰, 陈运泰, Bormann, P., 任枭, 侯建民, 邹立晔, 杨辉, 2005. 中国地震台网与美国地震台网测定震级的对比（I）——体波震级. 地震学报, **27**(6): 583-587. 英文刊载: Liu, R. F., Chen, Y. T., Bormann, P., Ren, X., Hou, J. M., Zou, L. Y. and Yan, H., 2005. Comparison between earthquake magnitudes determined by China seismograph network and US seismograph networks (I): Body wave magnitude. *Acta Seismologica Sinica*, **18**(6): 627-631.

[231] Yang, Z. X., Waldhauser, F., Chen, Y. T. and Richards, P., 2005. Double-difference relocation of earthquakes in central-western China, 1992-1999. *J. Seismol.*, **9**(2): 241-264.

[232] Xu, L. S., Chen, Y. T., 2005. Temporal and spatial rupture process of the great Kunlun Mountain Pass earthquake of November 14, 2001 from the GDSN long period waveform data. *Science in China* (Ser. D: Earth Science), **48**(1): 112-122.

[233] 陈运泰, 杨智娴, 许力生, 2005. 海啸、地震海啸与海啸地震. 物理, **34**(12): 864-872.

[234] 陈运泰, 2005. 地球物理与人类社会的可持续发展. 见: "物理与人类未来"研讨会文集, 北京: 北京市科学技术协会, 北京市科学界联合会, 31-32.

[235] 陈运泰, 朱传镇, 2006. 地球: 我们的变化着的行星——IUGG 动态. 国际地震动态, （1）: 1-7.

[236] 陈颙, 陈运泰, 张国民, 等, 2005. "十一·五"期间中国重大地震灾害预测预警和防治对策. 灾害学, **20**(1): 1-14.

[237] 刘瑞丰, 陈运泰, Bormann, P., 任枭, 侯建民, 邹立晔, 杨辉, 2006. 中国地震台网与美国地震台网测定震级的对比（II）——面波震级. 地震学报, **28**(1): 1-10. 英文刊载: Liu, R. F., Chen, Y. T., Bormann, P., Ren, X., Hou, J. M., Zou, L. Y. and Yan, H., 2006. Comparison between earthquake magnitudes determined by China seismograph network and US seismograph networks (II): Surface wave magnitude. *Acta Seismologica Sinica*, **19**(1): 1-10.

[238] Knopoff, L., 陈运泰, 2006. 断层厚度的地震效应. 见: 1906梅山地震百周年纪念研讨会论文摘要集. 29.

[239] Chen, Y. T. (Ed.-in-Chief), 2006. *Advances in Geosciences*, **1**: *Solid Earth* (SE), World Scientific Publishing Co., Singapore, 255pp.

[240] Xu, L. S. and Chen, Y. T., 2006a. Source process of the 2004 Sumatra-Andaman earthquake. In: Chen, Y. T. (Ed.-in-Chief), *Advances in Geosciences*, **1**: *Solid Earth* (SE), World Scientific Publishing Co., Singapore, 27-40.

[241] Xu, L. S. and Chen, Y. T., 2006b. Observed evidence for crack fusion from the 14 November 2000 Kunlun Mountain earthquake. In: Chen, Y. T. (Ed.-in-Chief), *Advances in Geosciences*, **1**: *Solid Earth* (*SE*), World Scientific Publishing Co., Singapore, 51-59.

[242] Wu, J., Crampin, S., Gao, Y., Hao, P., Volti, T. and Chen, Y. T., 2006. Smaller source earthquakes and improved measuring techniques allow the largest earthquakes in Iceland to be stress forecast (with hindsight). *Geophys. J. Int.*, **166**: 1293-1298.

[243] Duan, Q. and Chen, Y. T., 2006. Western Pacific Geophysics Meeting in Beijing, China. *EOS*, *Trans. Amer. Geophys. Union*, **87**(51) (9 December 2006): 584.

[244] 吴晶, 高原, 陈运泰, 黄金莉, 2007. 首都圈西北部地区地壳介质地震各向异性特征初步研究. 地球物理学报, **50**(1):209-220.

[245] Chen, Y. T. (Ed.-in-Chief), 2007. *Advances in Geosciences*, **9**: *Solid Earth* (*SE*), *Ocean Science* (*OS*) *and Atmospheric Science* (*AS*), World Scientific Publishing Co., Singapore, 1-233.

[246] Du, H. L., Xu, L. S., Chen, Y. T., Li, C. L. and Stammler, K., 2007. Tracking the high-frequency energy radiation sources of the 2004 Sumatra-Andaman M_W9.0 earthquake using the short-period seismic data: Preliminary result. In: Chen, Y. T. (Ed.-in-Chief), *Advances in Geosciences*, **9**: *Solid Earth*, *Ocean Science & Atmospheric Science*, World Scientific Publishing Co., Singapore, 3-11.

[247] Zhang, Y., Chen, Y. T. and Xu, L. S., 2007. Rupture process of the 2005 Southern Asian (Pakistan) M_W7.6 earthquake from long-period waveform data. In: Chen, Y. T. (Ed.-in-Chief), *Advances in Geosciences*, **9**: *Solid Earth* (*SE*), *Ocean Science* (*OS*) *and Atmospheric Science* (*AS*), World Scientific Publishing Co., Singapore, 13-21.

[248] 朱新运, 陈运泰, 2007. 用 Lg 波资料反演场地效应与地震波衰减参数. 地震学报, **29**(6): 569-580. 英文刊载: Zhu, X. Y. and Chen, Y. T., 2007. An inversion of site response and Lg attenuation using Lg waveform. *Acta Seismologica Sinica* (English Edition), **20**(6): 605-616.

[249] Zhou, R. M., Stump, B. W., Yang, Z. X., Chen, Y. T., Herrmann, R. B., Burlacu, R., Hayward, C. T. and Pankow, K., 2007. Broadband network operation and shear velocity structure beneath the Xiuyan Area, NE China. *Proceedings of the 29th Monitoring Research Review*: *Ground-Based Nuclear Explosion Monitoring Technologies*, September 2007, LA-UR-07-5613, Los Alamos National Laboratory, Los Alamos, New Mexico, 353-362.

[250] 陈运泰, 2007. 地震预测现状与前景. 见: 中国科学院 (主编), 2007 科学发展报告, 北京: 科学出版社, 173-182.

[251] 陈运泰, 2007. 地震预测——进展、困难与前景. 地震地磁观测与研究, **28**(2): 1-24.

[252] 陈运泰, 2007. 活动的地球: 板块大地构造学说简介. 路甬祥 (主编), 科学与中国——院士专家巡讲团报告集 第四辑, 北京: 北京大学出版社, 17-60, 北京: 北京大学出版社, 171-191.

[253] 刘瑞丰, 陈运泰, 任枭, 徐志国, 孙丽, 杨辉, 梁建宏, 任克新, 2007. 中国地震台网震级的对比. 地震学报, **29**(5):467-476. 英文刊载: Liu, R. F., Chen, Y. T., Ren, X., Xu, Z. G., Sun, L., Yang, H., Liang, J. H. and Ren, K. X., 2007. Comparison between different earthguake magnitudes determined by China Seismograph Network. *Acta Seismologica Sinica*, **20**(5): 497-506.

[254] 陈运泰, 许力生, 张勇, 杜海林, 冯万鹏, 刘超, 李春来, 张红霞, 2008. 2008 年 5 月 12 日汶川特大地震震源特性分析报告. http://www.cea-igp.ac.cn/汶川地震专题/地震情况/初步研究及考察结果（一）.

[255] 张勇, 许力生, 陈运泰, 冯万鹏, 杜海林, 2008. 2007 年云南宁洱 M_S6.4 地震震源过程. 中国科学 D 辑: 地球科学, **38**(6): 683-692. 英文刊载: Zhang, Y., Xu, L. S., Chen, Y. T., Feng, W. P. and Du, H. L., 2008. Source process of M_S6.4 earthquake in Ning'er, Yunnan in 2007. *Science in China*, *Series D*: *Earth Sciences*, **52**(2): 180-188.

[256] 张勇, 冯万鹏, 许力生, 周成虎, 陈运泰, 2008. 2008 年汶川大地震的时空破裂过程. 中国科学 D 辑: 地球科学, **38**(10): 1186-1194. 英文刊载: Zhang, Y., Feng, W. P., Xu, L. S., Zhou, C. H. and Chen, Y. T., 2008. Spatio-temporal rupture process of the 2008 great Wenchuan earthquake. *Science in China*, *Series D*: *Earth Sciences*, **52**(2): 1-11.

[257] 刘超, 张勇, 许力生, 陈运泰, 2008. 一种矩张量反演新方法及其对 2008 年汶川 M_S8.0 地震序列的应用. 地震学报, **30**(4): 329-339. 英文刊载: Liu, C., Zhang, Y., Xu, L. S. and Chen, Y. T., 2008. A new technique for moment

tensor inversion with applications to the 2008 Wenchuan $M_S 8.0$ earthquake sequence. *Acta Seismologica Sinica*, **21**(4): 333-343.

[258] 陈运泰, 2008. 汶川特大地震的震级和断层长度. 科技导报, **26**(10): 26-27. 又载: 学部通讯, **2008**(5): 28-30.

[259] 陈运泰, 2008. 地震预测要知难而进. 求是, **2008**(5): 58-60.

[260] 陈运泰, 2008. 汶川地震的震源破裂过程. 海峡两岸防震减灾学术研讨会, 2008年6月27日-28日, 北京, 10-11.

[261] 陈运泰, 2008. 汶川地震的成因断层、破裂过程与成灾机理. 中国科学院第十四次院士大会学部学术报告汇编, 2008年6月, 38-39. 又载: "科学技术与抗震救灾"科学技术论坛, 2008年7月24-26日, 成都.

[262] Stump, B. W., Zhou, R. M., Kim, T.-S., Chen, Y. T., Yang, Z. X., Herrmann, R. B., Burlacu, R., Hayward, C. and Pankow, K., 2008. Shear velocity structure in NE China and characterization of infrasound wave propagation in the 1-210 kilometer range. *Proceedings of the 30th Monitoring Research Review: Ground-Based Nuclear Explosion Monitoring Technologies*, Portsmouth, VA, September 2008, LA-UR-08-05261, Los Alamos National Laboratory, Los Alamos, New Maxico, 287-296.

[263] 许力生, 蒋长胜, 陈运泰, 李春来, 张天中, 2007. 2004年首都圈地区中小地震的矩张量反演. 地震学报, **29**(3): 229-239.

[264] 杜海林, 许力生, 陈运泰, 2009. 利用阿拉斯加台阵资料分析2008年汶川大地震的破裂过程. 地球物理学报, **52**(2): 372-378. 英文刊载: Du, H. L., Xu, L. S. and Chen, Y. T., 2009. Rupture process of the 2008 Wenchuan earthquake from the analysis of the Alaska-array data. *Chinese J. Geophys.*, **52**(2): 372-378 (in Chinese).

[265] 陈运泰, 2009. 地震预测: 回顾与展望. 中国科学D辑: 地球科学, **39**(12): 1633-1658.

[266] 张勇, 许力生, 陈运泰, 2009. 提取视震源时间函数的PLD方法及其对2005年克什米尔$M_W 7.6$地震的应用. 地球物理学报, **52**(3): 672-680.

[267] 张勇, 许力生, 陈运泰, 2009. 2008年汶川大地震震源机制的时-空变化. 地球物理学报, **52**(2): 379-389.

[268] Knopoff, L. and Chen, Y. T., 2009. Single-couple component of far-field radiation from dynamical fractures. *Bull. Seism. Soc. Amer.*, **99**(2B), 1091-1103. 中文译载: Knopoff, L., 陈运泰, 2010. 动态破裂远场辐射的单力偶分量. 世界地震译丛, (1): 55-70.

[269] Zhou, R. M., Stump, B. W., Herrmann, R. B., Chen, Y. T. and Yang, Z. X., 2009. Teleseismic receiver function and surface wave study of velocity structure beneath the Yanqing-Huailai Basin, NW of Beijing. *Bull. Seism. Soc. Amer.*, **99**(3): 1937-1952.

[270] 吴晶, 高原, 陈运泰, 2008. 首都圈东南部地区地壳介质各向异性. 地震学报, **30**(1): 1-11. 英文刊载: Wu, J., Gao, Y. and Chen Y. T., 2009. Shear-wave splitting in the crust beneath the southeast Capital area of North China. *J. Seismol.*, **13**(2): 277-286.

[271] 刘超, 许力生, 陈运泰, 2009. 2009年3月19日汤加地震快速矩张量解. 地震学报, **31**(2): 230-231.

[272] 刘超, 许力生, 陈运泰, 2009. 2009年8月3日加利福利亚湾地震快速矩张量解. 地震学报, **31**(5): 583-585.

[273] 刘超, 许力生, 陈运泰, 2009. 2009年8月9日日本伊豆群岛地震矩张量解. 地震学报, **31**(5): 586-588.

[274] Liu, C., Xu, L., Chen, Y. T., 2009. Quick moment tensor solution for 6 April 2009, L'Aquila, Italy, earthquake. *Earthquake Science*, **22**(5): 449-450.

[275] 张勇, 许力生, 陈运泰, 2010. 2009年9月29日萨摩亚群岛地区地震破裂过程快速反演. 地震学报, **32**(1): 118-120.

[276] 张勇, 许力生, 陈运泰, 2010. 2009年10月7日瓦努阿图地震破裂过程快速反演. 地震学报, **32**(1): 121-123.

[277] 张勇, 许力生, 陈运泰, 2010. 2010年1月12日海地地震破裂过程快速反演. 地震学报, **32**(1): 124-126.

[278] 刘超, 许力生, 陈运泰, 2010. 2009年12月19日台湾花莲地震快速矩张量解. 地震学报, **32**(1): 127-129.

[279] 刘超, 许力生, 陈运泰, 2010. 2010年1月12日海地地震快速矩张量解. 地震学报, **32**(1): 130-132.

[280] 张勇, 刘超, 许力生, 陈运泰, 等, 2010. 地震震源机制和破裂过程快报. 中国地震局地球物理研究所(编), 2010年4月14日青海省玉树县7.1级地震分析报告.

[281] 于湘伟, 陈运泰, 张怀, 2010. 京津唐地区中小地震重新定位. 地震学报, **32**(3): 257-269.

[282] 于湘伟, 陈运泰, 张怀, 2010. 京津唐地区地壳三维 P 波速度结构与地震活动性分析. 地球物理学报, 53(8): 1817-1828.

[283] 于湘伟, 张怀, 陈运泰, 2010. 华北地区地震重新定位结果分析. 大地测量与地球动力学, 30(2): 29-33.

[284] 刘超, 许力生, 陈运泰, 2010. 2010 年 4 月 14 日青海玉树地震快速矩张量解. 地震学报, 32(3): 366-368.

[285] Zhang, Y., Xu, L., Chen, Y. T., 2010. Fast inversion of rupture process of the 14 April 2010 Yushu, Qinghai, earthquake. *Earthquake Science*, **23**(3): 201-204.

[286] 刘超, 许力生, 陈运泰, 2010. 2008 年 10 月至 2009 年 11 月 32 次中强地震的快速矩张量解. 地震学报, 32(5): 619-624.

[287] Zhang, Y., Xu, L. S. and Chen, Y. T., 2010. Fast inversion of the rupture process of 27 February 2010 Chile earthquake. *Acta Seismologica Sinica*, (2): 242-244.

[288] 陈运泰, 2010. 海啸. 见: "10000 个科学难题"地球科学编委会（编）, 10000 个科学难题·地球科学卷, 北京: 科学出版社, 529-537.

[289] 陈运泰, 2010. 地震预测. 见: "10000 个科学难题"地球科学编委会（编）, 10000 个科学难题·地球科学卷, 北京: 科学出版社, 538-545.

[290] 陈运泰, 2010. 热流佯谬. 见: "10000 个科学难题"地球科学编委会, 10000 个科学难题·地球科学卷, 北京: 科学出版社, 546-548.

[291] 张勇, 许力生, 陈运泰, 2010. 2010 年青海玉树地震震源过程. 中国科学: 地球科学, 40(7): 819-821. 英文刊载: Zhang, Y., Xu, L. S. and Chen, Y. T., 2010. Source process of the 2010 Yushu, Qinghai, earthquake. *Sci. China Earth Sci.*, **53**(9): 1249-1251.

[292] Jordan, T. H., Chen, Y.-T., Gasparini, P., Madariaga, R., Main, I., Marzocchi, W., Papadopoulos, G., Sobolev, G., Yamaoka, K., Zschau, J., 2011. Operational earthquake forecasting: State of knowledge and guidelines for utilization. *Annals of Geophysics*, **54**(4): 315-391.

[293] 杨智娴, Stump, B., 陈运泰, Herrmann, R. B., Zhou, R. M., Hayward, C., 2011. 1999 年海城-岫岩地震序列的精准定位. 地震学报, 33(3): 271-278.

[294] Yu, X. W., Zhang, W. B. and Chen, Y. T., 2011. Seismic imaging and seismicity analysis in Beijing-Tianjin-Tangshan region. *Int. J. Geophys.*, 2011, Article ID 216315, 13 pages, doi: 10.1155/2011/216315 Hindawi Publishing Corporation.

[295] 刘超, 许力生, 陈运泰, 2011. 2009 年 11 月至 2011 年 11 月 27 次中强地震的快速矩张量解. 地震学报, 33(4): 550-552.

[296] Jordan, T. H., Chen, Y.-T., Gasparini, P., Madariaga, R., Main, I., Marzocchi, W., Papadopoulos, G., Sobolev, G., Yamaoka, K., Zschau, J., 2012. Reply to "Comment on 'Operational earthquake forecasting: State of knowledge and guidelines for implementation by Jordan et al. [2011]' by Stuart Crampin". *Annals of Geophysics*, **55**(1): 13-15.

[297] Zhang, Y., Chen, Y. T. and Xu, L. S., 2012. Fast and robust inversion of earthquake source rupture process and its application to earthquake emergency response. *Earthquake Science*, **25**(2): 121-128.

[298] 杨智娴, 陈运泰, 苏金蓉, 陈天长, 吴朋, 2012. 2008 年 5 月 12 日汶川 M_W7.9 地震的震源位置与发震时刻. 地震学报, 34(2): 127-136.

[299] Zhang, Y., Xu, L. S. and Chen, Y. T., 2012. Rupture process of the 2011 Tohoku earthquake from the joint inversion of teleseismic and GPS data. *Earthquake Science*, **25**(2): 129-135.

[300] Zhang, Y., Feng, W. P., Chen, Y. T., Xu, L. S., Li, Z. H. and Forrest, D., 2012. The 2009 L'Aquila M_W6.3 earthquake: A new technique to locate the hypocenter in the joint inversion of the earthquake rupture process. *Geophys. J. Intl.*, **191**: 1417-1426.

[301] Zhu, X. Y. and Chen, Y. T., 2012. An inversion of Lg attenuation and site response in the North China region. *Bull. Seism. Soc. Amer.*, **102**(6): 2600-2610.

[302] 陳運泰, 2012. 汶川大地震解讀. 見: 聆聽大師 走近科學——澳門科技大學"大師講座"院士講演錄（第二輯）, 176-202.

[303] 陈运泰，2012. 活动的地球：板块大地构造与地震. 白春礼（主编），科学与中国·十年辉煌 光耀神州·气候与灾害科学技术集. 北京：北京大学出版社，91-128.

[304] 陈运泰，2012. 海啸与地震. 白春礼（主编），科学与中国·十年辉煌 光耀神州·气候与灾害科学技术集. 北京：北京大学出版社，151-182.

[305] 张勇，许力生，陈运泰，2013. 芦山4.20地震破裂过程及其致灾特征初步分析. 地球物理学报，**56**(4)：1408-1411.

[306] 陈运泰，杨智娴，张勇，刘超，2013. 从汶川地震到芦山地震. 中国科学 D 辑：地球科学，**43**(6)：1064-1072.

[307] 陈运泰，杨智娴，张勇，刘超，2013. 浅谈芦山地震. 地震学报，**35**(3)：285-295.

[308] 刘瑞丰，陈运泰，邹立晔，陈宏峰，梁建宏，张立文，韩雪君，任枭，孙丽，2013. 2013年4月20日四川芦山M_W6.7（M_S7.0）地震参数的测定. 地震学报，**35**(5)：652-660.

[309] 陈运泰，2013. 地震破裂过程的快速稳健反演及其在地震应急反应中的应用. 中国地震局科技委院士专家甘肃行专刊，北京，14-24.

[310] 王晓欣，刘瑞丰，邹立晔，袁菲，陈宏峰，陈运泰，张立文，梁建宏，2014. 2014年2月12日新疆于田M_S7.3地震参数测定. 地震学报，**36**(3)：522-528.

[311] 陈运泰，2014. 从苏门答腊-安达曼到日本东北：特大地震及其引发的超级海啸的启示. 地学前缘，**21**(1)：120-131.

[312] Chen, Y. T., 2014. Comment on the paper "Normal and shear stress acting on arbitrarily oriented faults, earthquake energy, crustal GPE change, and the coefficient of friction" by P. P. Zhu. *J. Seismol.*, **18**(4): 795-797.

[313] 刘超，陈运泰，2014. 断层厚度的地震效应和非对称地震矩张量. 地球物理学报，**57**(2)：509-517.

[314] Zhang, Y., Wang, R., Chen, Y. T., Xu, L, Du, F., Jin, M., Tu, H. and Dahm, T., 2014. Kinematic rupture model and hypocenter relocation of the 2013 M_W6.6 Lushan earthquake constrained by strong-motion and teleseismic data. *Seismol. Res. Lett.*, **85**(1): 15-22.

[315] 张勇，许力生，陈运泰，汪荣江，2014. 2014年2月12日于田M_W6.9地震破裂过程初步反演：兼论震源机制对地震破裂过程反演的影响. 地震学报，**36**(2)：159-164.

[316] Zhang, Y., Wang, R., Zschau, J., Chen, Y.-T. and Parolai, S., 2014. Automatic imaging of earthquake rupture processes by iterative deconvolution and stacking of high-rate GPS and strong motion seismograms. *J. Geophys. Res., Solid Earth*, **119**(7): 5633-5650.

[317] 陈运泰，2015. 地震与防震减灾. 见：白春礼（主编），科学与中国——院士专家巡讲团报告集（第九辑），北京：科学出版社，17-60.

[318] 陈运泰，2015. 昭通永善-大关地震40年：杂忆与随想. 城市与减灾，**2015**(1)：1-5.

[319] 张勇，陈运泰，许力生，魏星，金明培，张森，2015. 2014年云南鲁甸M_W6.1地震：一次共轭破裂地震. 地球物理学报，**58**(1)：153-162.

[320] 陈运泰，2015. 可操作的地震预测预报. 北京：中国科学技术出版社，1-152.

[321] 刘瑞丰，陈运泰，任枭，徐志国，王晓欣，邹立晔，张立文，2015. 震级的测定. 北京：地震出版社，1-154.

[322] 郭祥云，陈运泰，房立华，刘瑞丰，2015. 2015年4月25日尼泊尔M_W7.9地震的震源机制. 地震学报，**37**(4)：705-707.

[323] 张勇，许力生，陈运泰，2015. 2015年尼泊尔M_W7.9地震破裂过程：快速反演与初步联合反演. 地球物理学报，**58**(5)：1804-1811.

[324] Zhang, Y., Wang, R., Chen, Y.T., 2015. Stability of rapid finite-fault inversion for the 2014 M_W6.1 South Napa earthquake. *Geophys. Res. Lett.*, **42**(23): 10263-10272.

[325] 王琼，高原，钮凤林，陈运泰，2016. 利用接收函数计算地壳各向异性的可靠性分析及倾斜界面的影响. 地震，**36**(2)：14-25.

[326] Wang, Q., Niu, F., Gao, Y., Chen, Y. T., 2016. Crustal structure and deformation beneath the NE margin of the Tibetan plateau constrained by teleseismic receiver function data. *Geophys. J. Int.*, **204**(1): 167-179.

[327] Motazedian, D., Zhang, Y., Ma, S., Chen, Y. T., 2016. Further studies on the focal mechanism and source rupture process

of the 2012 Haida Gwaii, Canada, M_W 7.8 earthquake. *Can. J. Earth Sci.*, **53**(2): 129-139.

[328] 刘超, 陈运泰, 2017. 非对称地震矩张量时间域反演: 理论与方法. 地震学报, **39**(2): 155-175.

[329] 张勇, 陈运泰, 宋金, 2018. 2018年2月12日 M_W 4.3 河北永清地震的震源机制与破裂特征. 科学通报, **63**(22): 2296-2301.

[330] Zhang, Y., Chen, Y. T., Feng, W., 2019. Complex multiple-segment ruptures of the 28 September 2018, Sulawesi, Indonesia, earthquake. *Chin. Sci. Bull.*, **63**: 2296-2301.

[331] Chen, Y. T., Xu, L. S., Zhang, Y. and Zhang, X., 2019. Inversion of the earthquake rupture process: methods, case studies and applications to emergency response. In: Li, Y. G. (ed.), *Earthquake and Disaster Risk: Decade Retrospective of the Wenchuan Earthquake*, Higher Education Press/Springer, Beijing, China. 1-30.

[332] Chen, Y. T., Zhang, Y. and Xu, L. S., 2019. Inversion of earthquake rupture process: Theory and applications. In: Bizzarri, A., Das, S. and Petri, A. (eds.), Proceedings of the International School of Physics "Enrico Fermi", Course 202, *Mechanics of Earthquake Faulting*, IOS Press; Societa' Italiana di Fisica (SIF), Bologna, Italy. Also in: *La Rivista del Nuovo Cimento*, **42**(8): 367-406.

[333] 宋金, 陈运泰, 张勇, 2020. 2013年岷漳地震和2014年景谷地震的能矩比对比研究. 地球物理学报, **63**(9): 3324-3337.

[334] 陈运泰, 2019. 地震浅说. 北京: 地震出版社, 1-289.

[335] 陈运泰, 2019. 地震. 北京: 地震出版社, 1-31.

[336] 陈运泰, 2020. 地震现象与科学. 北京: 地震出版社, 1-72.